Anatomy

as a Basis for
Clinical Medicine

Anatomy
as a Basis for Clinical Medicine

by

E.C.B. Hall-Craggs

M.A., M.B., B.Chir., Ph.D.

Consultant in Anatomy, Department of Human Anatomy
University of Oxford
Formerly Professor and Head, Division of Gross Anatomy
University of Maryland School of Medicine, Baltimore

Illustrated by Diane Abeloff, M.A., A.M.I.

Second Edition

WILLIAMS & WILKINS
BALTIMORE • HONG KONG • LONDON • MUNICH
PHILADELPHIA • SYDNEY • TOKYO

Copy Editor: Mary Kidd
Production Coordinator: Kathleen C. Millet

Williams & Wilkins
428 East Preston Street
Baltimore, Maryland 21202, USA

Printed in the United States of America

First Edition 1985

Library of Congress Cataloging in Publication Data

Hall-Craggs, E. C. B.
Anatomy as a basis for clinical medicine / by E. C. B. Hall Craggs : illustrated by Diane Abeloff. — 2nd ed.
p. cm.
ISBN 0-683-03831-1
1. Human anatomy. I. Title.
[DNLM: 1. Anatomy. OS 4 H182a]
OM23.2.H34 1990
611 — dc20
DNLM/DLC 89-70495
for Library of Congress CIP

92 93 94 95 96
2 3 4 5 6 7 8 9 10

Preface to the Second Edition

This preface is particularly addressed to all the students and teachers who have expressed an interest in the first edition of this book. Their opinions and comments have allowed me to concentrate on the accuracy of both text and figures and to "fill out" a number of areas in which description and illustration have proved inadequate. However, the temptation to produce a larger volume has been resisted and the aim of the first edition, to limit the material presented, has been kept. This edition, therefore, remains an anatomical basis for clinical medicine that is compatible with independent learning and a reduction in the hours devoted to formal teaching.

The continued collaboration with Diane Abeloff has made it possible to maintain consistency where new illustrations have been needed. New images have been added; particularly where they demonstrate the capability of new techniques to display normal anatomy.

I hope that in this second edition, I have justified the encouragement I have received from many readers of the first.

E.C.B. Hall-Craggs
Somerton, England

Preface to the First Edition

This book has been written for medical students and the aim throughout has been to present an appropriate amount of material in a readable and interesting form. What is an appropriate amount is debatable and I have had to rely on my own clinical and teaching experience to determine this. No major omissions have been made intentionally, but with the time available to gross anatomy courses in mind, emphasis and space have been devoted to topics that have the greatest importance. To achieve a readable text, continuous narrative has been used wherever possible and the number of subheadings, tables, and typefaces have been reduced to a minimum (several excellent review texts already provide these features for use at the opportune moment). The primary interest, I hope, remains in the material itself but this interest is reinforced by many references to clinical relevance. These references are made only for this purpose and not to place an additional learning task on the student at this time. The opportunity to collaborate with a skilled medical artist has meant that a set of original illustrations has been created and closely integrated with the text. Every effort has been made to place these illustrations so that the reader's attention can be repeatedly drawn to them in order to add to or confirm a written description. In this way it is hoped that the student will look and understand as well as read and learn. Color has been used freely in these illustrations and the majority are representational rather than diagrammatic; the diagram being the prerogative of the lecturer and often needing his or her personal interpretation. To add to these illustrations I have had the privilege of drawing on the incomparable collection of normal radiographs published by Dr. Lothar Wicke in his *Atlas of Radiologic Anatomy*, 3rd. Ed. (© 1983, Urban & Schwarzenberg), and further radiographs, CT scans, sonograms, and nuclear scans have been generously supplied by colleagues at the University of Maryland School of Medicine.

The form of the book departs little from the traditional. An introductory chapter presents a body of general information that should enable the regional chapters to be approached in any order that suits a particular course. I do believe, however, that the sequence used here has some merit in that the fundamental segmental arrangement of the body's structure is established early, that the thorax, abdomen, and pelvis follow each other in a logical manner, and that a study of the head and neck serves as a fitting finale for the, by now, experienced student.

With the advent of new imaging techniques a need has arisen for physicians to be able to interpret normal and abnormal anatomy presented in the form of sections. Examples of these techniques have been included among the figures. Furthermore, drawings of sections are frequently used, often as a method of reviewing a region. However, I am sure that a good grasp of topographical anatomy is essential before an intelligent interpretation of the commonly used scans can be made and the emphasis on this is therefore retained. Embryology forms an important part of the teaching of human anatomy and warrants its own text. For this reason attention is drawn in only a few instances to the developmental background of the adult anatomy presented.

If this book is found of value, I look forward to making the inevitable corrections and incorporating any ideas for improvement that may be suggested.

E.C.B. Hall-Craggs
Baltimore, Maryland

Acknowledgments

In the first edition, I was most grateful to my colleagues, both in Baltimore and in Europe, for the willingness with which they met my requests for illustrations. Professor Lothar Wicke of Vienna, Austria provided the bulk of radiographs, and the textbook could only be the better for their inclusion. Professor Jacob Altaras of Giessen, West Germany also had my thanks. I remain grateful to Nancy Whitley, M.D., Lois Young, M.D., Jeremy Young, M.D., Gerald Johnston, M.D., Morgan Dunne, M.D., and William C. Gray, M.D., who, at the University of Maryland School of Medicine, had generously supplied additional radiographs, CT scans, sonographs, nuclear scans, and photographs. In this second edition my gratitude extends not only to Diane Abeloff for her artwork but also to Margaret Hall-Craggs, M.D. and H.N. Schnitzlein, Ph.D. for providing new radiographs and MR scans.

It is customary for an author to acknowledge the help that he receives from his publisher. Having seen during the preparation of the first edition the enormous labor required to turn a manuscript into the printed page, I wish to uphold this custom and thank Braxton Mitchell and Kathleen Millet who, with their associates, have brought about the second edition with continued good will and enthusiasm.

Contents

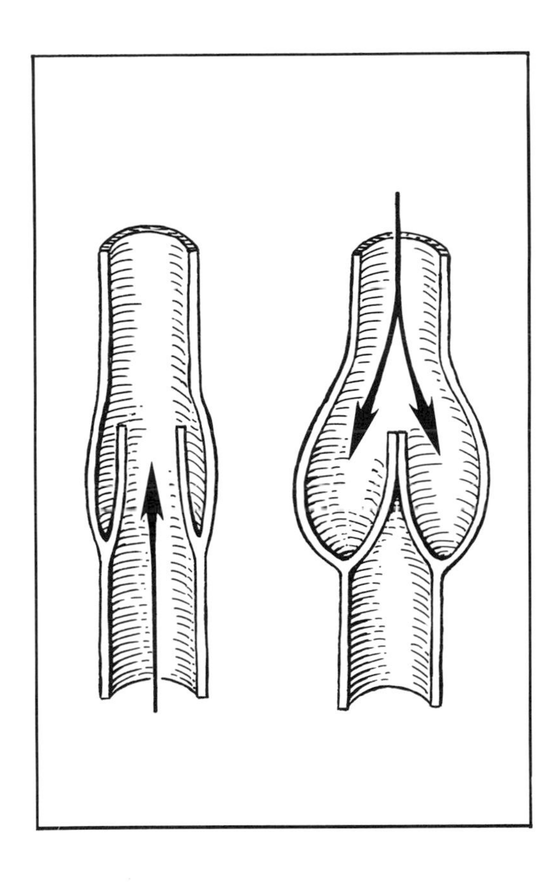

1 Introduction

For many good reasons courses in gross anatomy vary in the sequence in which each region of the body is taught and dissected. A textbook, however, can only present these regions in one order. To save repetition a number of general anatomical subjects and principles that are common to all regions are therefore presented as an introductory chapter. If time is spent reading this chapter, a vocabulary and background will be acquired that will make it possible to begin a regional study with any subsequent chapter.

The Discipline of Anatomy

Gross Anatomy has formed the foundation for studies in medicine for millennia and although later joined by other disciplines, it still retains a fundamental position among the basic medical sciences. Its importance to the surgeon has long been apparent but today such developments as new imaging techniques, biopsy procedures, and noninvasive therapeutic methods make an accurate knowledge of anatomy equally essential to the practice of all specialty areas of medicine.

Anatomy is a visual and practical subject and, for those who have the good fortune to learn it in the anatomy laboratory, each step in a dissection should be considered as an experiment and a picture should be retained of the relationships of structures exposed. While illustrations cannot replace the dissected specimen this book has been designed to make them an integral part of the text and thus of the learning process.

The Language of Anatomy

Nomina Anatomica

The ability to communicate clearly and precisely with colleagues is an important part of training in medicine; misunderstandings in giving or carrying out oral or written instructions can have unfortunate results. To assist in this communication an international vocabulary is used along with a series of conventions and terms to define positions and movements. The vocabulary is called *Nomina Anatomica,* the sixth edition of which was published in 1989*. To retain its international status the language of this publication is Latin. The close relationship of English and Latin has allowed the majority of the words to be anglicized with little change and where the Latin word is by custom retained, a little thought usually reveals an English word that provides a clue to its meaning. *Nomina Anatomica* is adhered to in this book and only in a few cases are more time-honored expressions used. Where words of Latin or Greek origin have some particular meaning a translation is given as a footnote.

Having assembled a vocabulary which identifies parts and regions of the body it is now necessary to establish the terms and conventions that allow these parts to be described in relation to each other. The most basic of these is the anatomical position.

The Anatomical Position

Patients may be examined while standing up or lying down. If, as a result of an examination, an area of pain or tenderness is described as being above the fifth left rib, a later reader can be confused if the original position of the patient was not specified. To obviate this difficulty all terms describing position are given in relation to a standard position known as the **anatomical position.** This is illustrated in Fig. 1-1. Note that the body is

* *Nomina Anatomica,* Sixth Edition. Churchill Livingstone, New York, 1989.

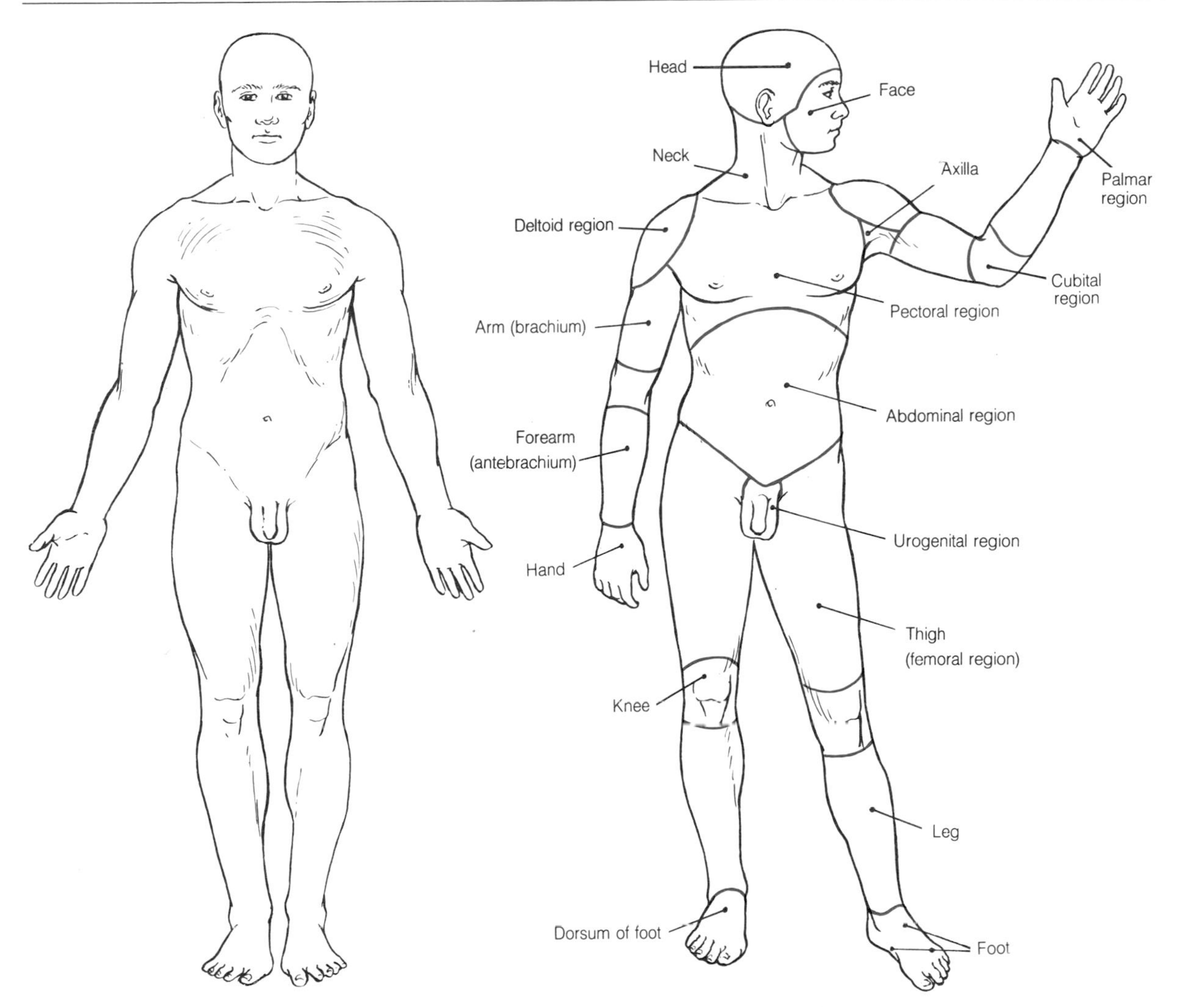

1-1 The anatomical position.

1-2 Regions of the body.

standing erect and facing toward the examiner. Look at the upper limbs and see that the palms are facing forward and the thumbs are away from the body. Although this would seem an unnatural position for the limb it is only thus that the two bones of the forearm lie parallel to each other. In this position the forearm is said to be **supine** and when rotated so that the palm faces backward it is **prone.** The movements that bring about these positions are therefore **supination** and **pronation.**

The Regions of the Body

Fig. 1-2 shows the commonly described regions of the body. These do not have precisely defined limits but are useful descriptive terms both in anatomy and in clinical medicine. Most are self explanatory but attention should be drawn to the fact that the commonly used words arm and leg do not refer to the whole upper or lower limb but designate the region above the elbow and the region below the knee.

Anatomical Planes

In Fig. 1-3 a series of planes is illustrated. These are used in anatomy to describe cuts or sections made through a cadaver for descriptive purposes. The **median plane** is a vertical plane passing from the front to the back of the body through the midline. Any plane parallel to this is a **sagittal plane.** All vertical planes which pass from side to side and are at right angles to the median plane are **coronal planes.** Planes that are at right angles to both coronal and median planes are called **horizontal.** These planes have now assumed clinical importance in that they form the basis for display in computerized tomography and magnetic resonance imaging. However, for the interpretation of horizontal sections generated by these techniques it becomes important to be able to differentiate between the left and the right side. As shown in Fig. 1-4 a section of a leg can be viewed from above or below and could thus be assigned to the left or right limb. To avoid this ambiguity the convention has been established by radiologists that all sections are viewed from below. Because the subject is normally supine *the left side of the patient therefore appears on the right side of the image.*

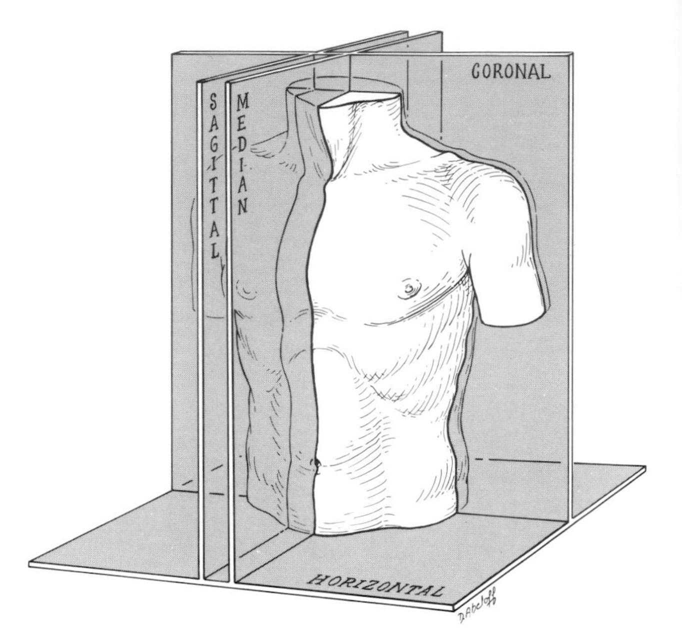

1-3 Illustrating anatomical planes.

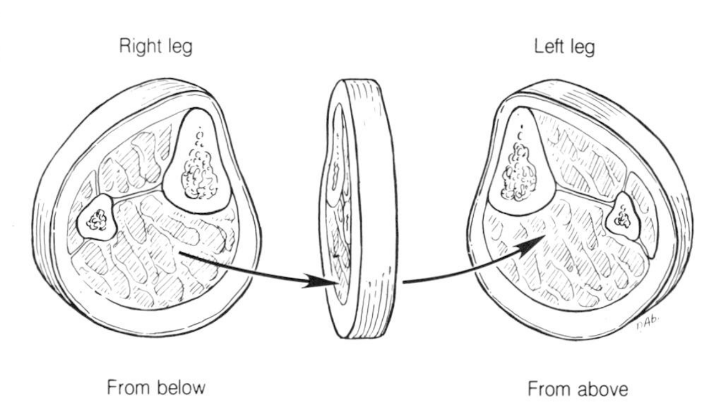

1-4 To show that a section of an asymmetrical part can be ascribed to the left or right side.

Terms of Position

A further useful group of terms is used to relate one structure to another. These also use the anatomical position as a reference. When one structure lies above another it is **superior** to it and conversely the structure below is **inferior** to the one above. A structure lying in front of another is **anterior** to it and when behind is **posterior** to it. The words **ventral** and **dorsal** have almost the same meaning as anterior and posterior but are more often applied to surfaces. When relating two structures to the surface of the body the first to be encountered is said to be **superficial** and the second **deep.** Two structures can also be related to the midline of the body; that nearest the midline is described as **medial** and that further away is **lateral.** Lastly the terms **proximal** and **distal** are useful when describing parts of structures nearer to or further from their origin or from the trunk. These terms provide a sufficient basis for most anatomical descriptions.

Terms which describe and define movements at joints are also necessary but a discussion of these is deferred until the joints themselves are considered.

Skin

There is a tendency to consider the skin as an impediment to dissection and a structure to be removed as rapidly as possible. This approach is shortsighted as the skin serves many important functions and is perhaps all that is seen of many patients. In addition, "lay" questions are asked about it which require answers that are not always easily available in textbooks.

The skin forms a limiting envelope at the external surface of the body and separates it from a very variable and hostile environment. Its superficial layer also covers the cornea of the eye and is invaginated to line, at least in part, the orifices of the ears, nose, mouth, and anal canal. In this manner it serves to protect against mechanical injury, bacterial invasion, and the effects of ultraviolet light and, when this barrier is broken, it possesses the capacity to repair itself. As a safeguard against injury it has a multitude of sensory receptors that record touch, changes in temperature, and pain. Through its superficial network of small vessels and its sweat glands it plays, with the nervous system, an active part in the regulation of body temperature, and being waterproof prevents the unregulated loss of body water. Furthermore, the irradiation of ergosterol in the skin by ultraviolet light is a source of vitamin D.

The detailed structure of the skin is the province of the histologist but reference to Fig. 1-5 will help in understanding its anatomical features. Identify the superficial and highly cellular **epidermis.** Its **basal layer** consists of dividing cells which replace the dead cells shed from the tough protective surface layer. These desquamated cells can be seen in abundance on the inner aspect of a cast that has been worn for a week or two. Beneath the epidermis lies the **dermis.** This is composed of a dense feltwork of collagen fibers which in cowhide form the basis of leather. The orientation of these collagen fibers is of practical significance to the surgeon. The simple experiment of perforating the skin of a cadaver with a rounded point shows that the resulting wound is linear not circular. This is because many of the collagen fibers are aligned parallel to each other and they split apart. Using this technique multiple perforations reveal a pattern of lines (Langer's lines) over the body and it is well known that surgical incisions made along these lines gape less and heal with less scarring. Elastic fibers which are also found in this layer give the skin its elasticity. Return to the epidermis and note the invaginations of this layer that are modified to form **sweat glands** lying in the dermis and subcutaneous tissue and similar downward projections that form the **hair follicles.** The upward projections of the dermis into the epidermis are called **dermal papillae.** Use is made of this feature in the split-skin graft, which leaves some of the germinal layer at the donor site and transfers the remainder to the skin defect that needs to be covered. Also lying in the dermis are many nerve endings and a capillary bed whose vessels

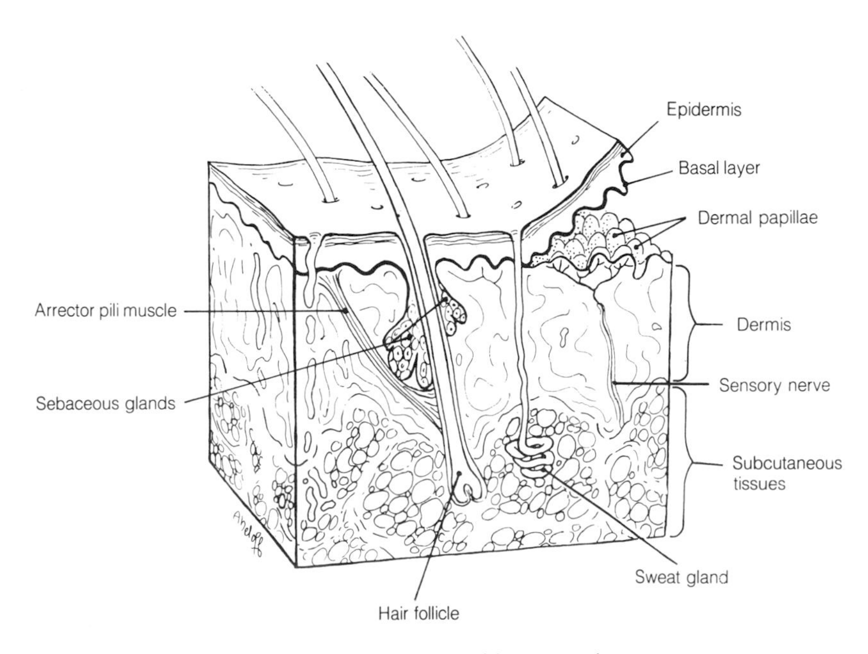

1-5 Diagram illustrating the anatomical features of the skin and its appendages.

extend in loops into the dermal papillae. The dermis blends more or less gradually with the underlying subcutaneous tissue or superficial fascia.

The skin is thicker over extensor surfaces and over the palms and soles. The palms and soles also show the well known prints which are formed by **papillary ridges.** It is on the surface of the ridges that the openings of the many sweat glands in these regions are found.

In some areas the skin is anchored more firmly than others. This is particularly so in the palms and soles where it allows a better grip. In the palms this is most apparent at the **flexure lines.**

The skin of all races contains pigment. Melanin has the most striking effect on skin color although carotene and the hemoglobin circulating in the dermal capillaries also contribute to this. In light-skinned races pigmentation is well marked in areas such as the areola around the nipple and parts of the external genitalia. Such pigmentation becomes even more pronounced during pregnancy and is never altogether lost. Dark-skinned races are usually found living in tropical climates. This would at first sight seem an adaptation of doubtful value since heat absorption is increased. However, the high incidence of cancer of the skin among the fair-skinned immigrants to Australia can be explained by the lack of a pigment layer resistant to the ill effects of ultraviolet light.

Appendages of the Skin

Hair

Although the human body may appear to be rather naked when compared with that of other animals, the greater part of the skin has a covering of fine hair of greater density than that of the anthropoid apes. The nature of individual hairs depends upon their cross-section; oval hairs curl and round hairs are straight. The hair follicles have a life span that ranges from several years for naturally long hairs to a year or less for short hairs. Normally the lost follicles are replaced but with aging there is a reduction in the total number. The loss of pigment and presence of air between the atrophic hair cells are responsible for the white appearance of hair in older people.

Associated with each hair follicle is a **sebaceous gland.** Its oily secretion called **sebum** is poured into the hair follicle to lubricate the hair shaft and provide a waterproof and bacteriocidal coating for the epidermis. Between each follicle and the epidermis lies a bundle of smooth muscle fibers. These are the **arrector pili** muscles whose contraction erects the hair and produces a "goose bump."

Nails

The nails are firm plates of dead and flattened cells derived from the most superficial layer of the epidermis. Their strength like that of hair is provided by the protein keratin. The **germinal matrix** from which the nails are formed lies tucked under the nail fold on the dorsum of the terminal phalanx. It also extends forward beyond this and can be recognized as the pale **lunula.** From the germinal matrix the nail grows forward over the vascular and pink nail bed which, unlike that of a claw, contributes nothing to the substance of the nail.

Sudoriferous Glands*

Sweat glands and their ducts are derivatives of the epidermis which invaginate through the dermis to come to lie in the subcutaneous tissue. The cooling produced by the evaporation of their watery secretion plays a major role in temperature regulation. Their activity is controlled by the nervous system. Somewhat similar glands which are not under nervous control are found in the ear, where they produce wax or **cerumen,** and in the perineum and axilla. The latter are responsible for the characteristic odor of sweat.

The breasts develop from downgrowths of the surface epithelium into the subcutaneous tissue and as such are appendages of the skin. They are described more fully in a later section.

Fascia

Beneath the skin the various structures that make up the body are held together by connective tissue. Because this tissue is in some regions found in

* Sudor, *L.* = sweat

the form of bands it is known as fascia.* This connective tissue is of a somewhat undifferentiated nature in that some of its cells retain the potential for further differentiation. However, other cell types are more specialized and, for example, may synthesize and store fat (lipocytes) or secrete collagen (fibroblasts). The proportions of cell types and the extracellular matrix secreted by the cells gives to fascias a variety of macroscopic appearances which range from the fat-laden subcutaneous tissue to the tough bands of connective tissue that retain tendons over the surfaces of joints. The fascias of the body are divided into the **superficial fascia** and the **deep fascia.**

Superficial Fascia

The superficial fascia is the fat-containing subcutaneous tissue that underlies the greater part of the skin of the body and for which it functions as an insulating layer. Within it lie the ramifications of cutaneous nerves and blood vessels that will reach the dermis of the skin. In the face and neck it contains striated musculature in the form of muscles of facial expression that gain attachment to the skin, and smooth muscle is found in the superficial fascia of the nipple and the scrotum. Over the lower anterior abdominal wall and perineum its deeper layers are condensed into a distinct sheet or membrane. In some regions such as the palms of the hands or soles of the feet the superficial fascia anchors the skin to the deep fascia. At other sites such as the flexure lines of the skin the dermis is directly adherent to the deep fascia.

Deep Fascia

Deep fascia is the wrapping material surrounding individual organs and the packing material that fills the interstices between them. As such it takes on a number of characteristics. In some regions it is in the form of thick investing sheets or bands and in others especially where it surrounds mobile or expansible structures it is more loosely arranged. Much of the evidence of its ability to allow movement between structures is lost, unfortunately, in preserved cadavers but an indication of this can usually be found between the esophagus and the prevertebral fascia on which it lies and between the muscle clothing the deep surface of the scapula and the thoracic wall on which the scapula freely moves.

Fascial planes and spaces have clinical significance for they frequently determine the direction of spread of infections or sites at which they become localized.

* Fascia, *L.* = a band

Bones

The bones of the body perform a number of functions and of these perhaps the most obvious is to serve as rigid levers in locomotion. It is through the skeleton of the limbs that the force generated by muscles is applied to the ground. Bones also support the soft tissues of the body and give it a permanent shape, and the bones of the skull, vertebral column, and thorax clearly protect underlying organs of particularly vital importance. They provide a source of ionized calcium that can be mobilized into the vascular system by the parathyroid hormone, and cavities within them contain the adult hemopoietic* tissue.

To carry out these functions bones contain calcium salts in a complex crystalline form. These give hardness to bone but its real strength lies in the matrix of collagen fibers about which the salts are deposited. Incinerated bone that has lost its collagen is highly brittle whereas a decalcified long bone retains its shape but can be easily deformed. In Fig. 1-6 a decalcified fibula has been tied into a knot.

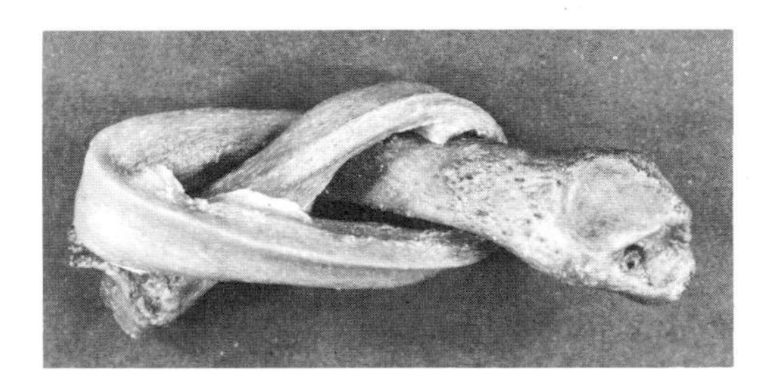

1-6 To show the flexibility of a decalcified fibula.

* Haima, *Gk.* = blood; Poiesis, *Gk.* = a making

The Classification of Bones

The bones of the body can be divided into the **axial skeleton** which consists of the skull, the vertebral column, and the ribs and sternum, and the **appendicular skeleton** which consists of the pectoral and pelvic girdles, and the bones of the upper and lower limbs.

Further classification of bones is made according to their shape in the manner set out below:
Long bones—e. g., the femur
Short bones—e. g., the bones of the wrist
Flat bones—e. g., the bones of the skull cap
Irregular bones—e. g., the vertebrae

The overall shape of a bone is genetically determined, a fact demonstrated by the growth of isolated bones in organ culture. However, the removal of muscles during normal development inhibits the appearance of prominences to which they would have been attached.

The General Structure of Bones

Bones are surrounded by a connective tissue membrane called the **periosteum** which is continuous with the capsules of joints, the attachment of muscles, and deep fascia where this passes over the surface of a bone. The outer layer of periosteum is fibrous but there is an inner more vascular layer which contains cells having the property of laying down new bone.

The bone itself has an outer layer of dense **compact bone** and an inner layer of spongy or **trabecular bone.** The latter, although appearing very fragile, is arranged along the lines of tension and compression of the normally stressed bone and gives it additional strength. In Fig. 1-7 the lattice work formed by the trabecular bone in the neck and head of the femur can be seen. Note also the outer layer of compact bone which is thick in the body ot the bone but thins out over its neck. The inner surface of the bone is lined by a layer of flattened cells called the **endosteum.** A typical long bone contains a **medullary* cavity** in which yellow fatty marrow or red hemopoietic marrow is found. Marrow also fills the spaces between trabeculae.

* Medulla, *L.* = marrow

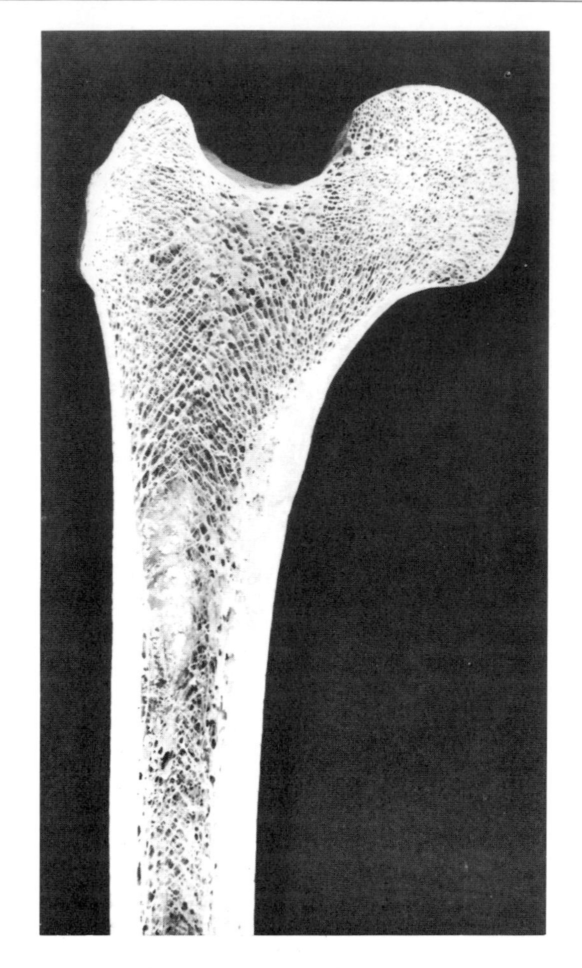

1-7 To show the arrangement of bony trabeculae in the head of the femur.

*Ossification**

During their development bones can first be recognized either as areas of embryonic connective tissue called membrane or as cartilaginous models. These preformed tissues are turned into bone by the process of **ossification.** When bone appears in membrane the ossification is called **intramembranous** and when in cartilage it is called **endochondral.** Although the process of ossification is similar in each case, bone is laid down directly in membrane, whereas the cartilage model must first be removed in endochondral ossification. The bones of the vault of the skull, the face including the mandible, and the clavicle undergo

* Os, plural ossa, *L* = a bone; hence ossification or bone formation

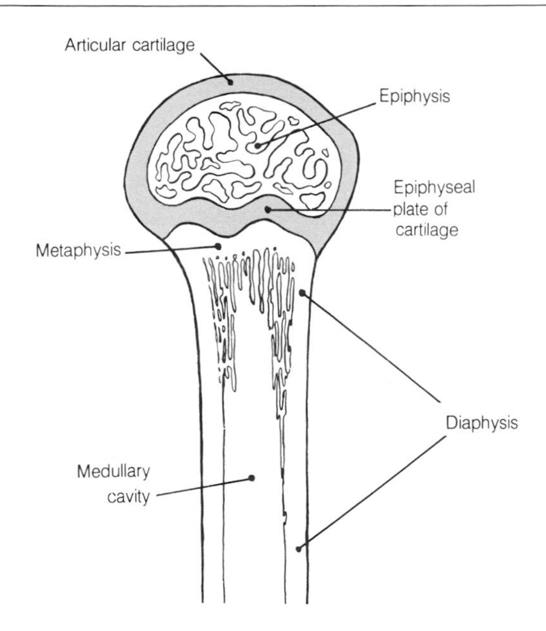

1-8 Diagram to illustrate the parts of a growing long bone.

intramembranous ossification and the remainder of the skeleton initially undergoes endochondral ossification. Endochondral ossification of a long bone begins at the end of the second month of intrauterine life near the center of the cartilage model and gradually invades this. The site at which bone first appears is known as the **primary center of ossification.** Before this invasion is complete one or more **secondary centers of ossification** typically appear at the upper and lower ends of the bone at or after birth. This region, called the **epiphysis,** is separated from the body of the bone, called the **diaphysis,** by the **epiphyseal plate of cartilage.** That part of the diaphysis adjacent to the plate is known as the **metaphysis.** These regions can be more easily identified in Fig. 1-8 and the radiographic appearance of an unfused epiphysis can be seen in Fig. 1-9. The radiolucent cartilage should not be confused with a fracture of the bone. The cartilage cells in the epiphyseal plate continue to divide and provide a substrate for the further ossification so important for the growth in length of a long bone. When this cell division ceases the epiphyseal and metaphyseal bone fuse. The epiphyseal plate that contributes most to the growth of a bone designates this end as "the growing end." The growing ends of limb bones become of particular importance in

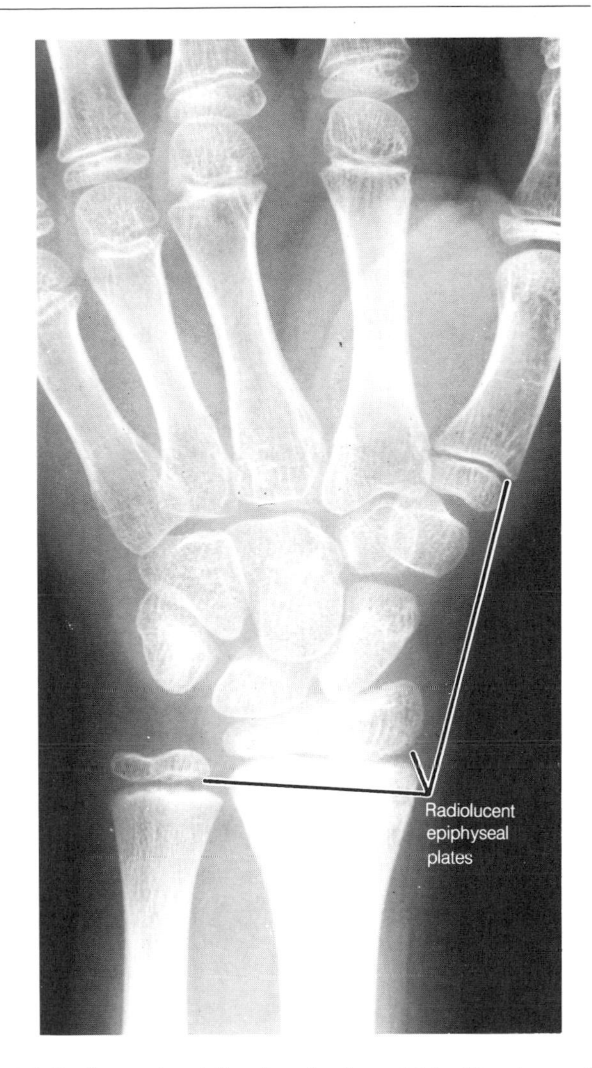

1-9 Radiograph of the hand of a child. (Courtesy of Jeremy Young, M. D., Dept. of Diagnostic Radiology, University of Maryland School of Medicine.)

the treatment of unequal leg growth in children since by surgically limiting the contribution of the more rapidly growing epiphyseal cartilages, limb length can be equalized. (If the fetal position can be visualized as in Fig. 1-10 the growing ends of the long bones are those nearest the head.) Thus the times of appearance and fusion of centers of ossification have a clinical application; they also have forensic significance. As individual bones are described in subsequent sections these times will be quoted. There are some general guidelines that help in memorizing the more important of these:

1. The primary centers of long bones appear at the end of the second month of intrauterine life.

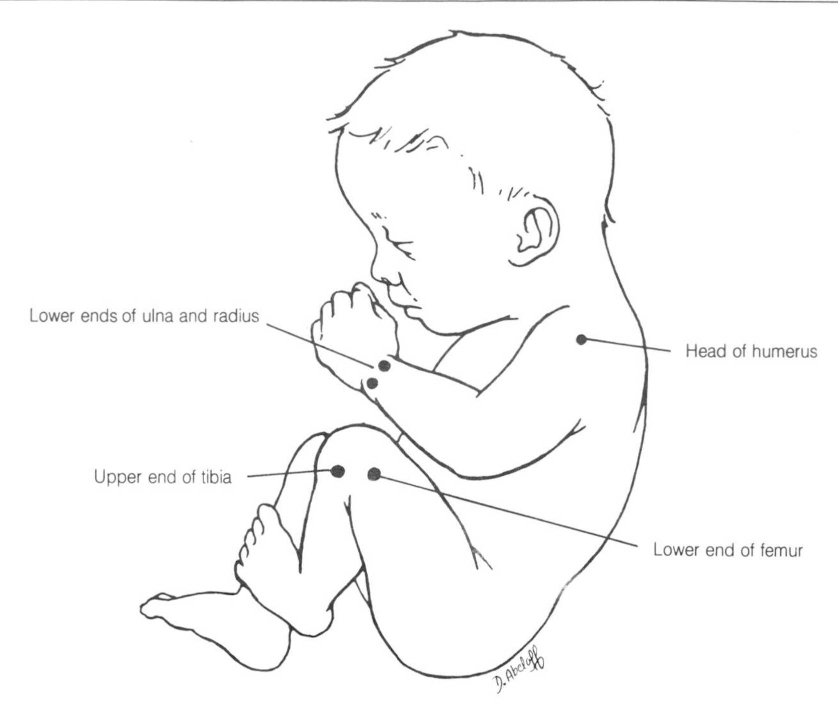

1-10 To show the growing ends of the limb bones.

2. Secondary centers appear at or after birth.
3. Secondary centers appear earlier in females.
4. Secondary centers for the growing ends of long bones appear first.
5. Secondary centers that appear early fuse late.

The Growth of Bones

Organs may grow by apposition of tissue to their surface or by interstitial growth, i. e., from within their substance. Which of these applies to bone was demonstrated 250 years ago by a series of simple observations and experiments. These explain the process of bone growth so clearly that they bear repeating here.

The story begins in 1736 at the meeting of a calico printer and a surgeon, John Belchier. The surgeon had been invited to dinner and a roast of pork was provided for the meal. Belchier noticed that the bone of the pork was stained red and on further enquiry found that the calico printer had fed madder-soaked bran from his dye vats to his pigs.

Belchier's contribution was his publication of these findings which he had also confirmed to be present in fowl. He drew no conclusions from them and was still prepared to accept the current theory that bone enlarged by interstitial growth and expansion. However, the attention of an inquiring Frenchman called Duhamel was drawn to this technique. After repeating the experiments he realized that new bone had been laid down after the diet had stopped and that this was not stained. Its distribution showed him that the new bone was laid down by superimposition on the old. By encircling the shaft of a bone with silver wire and finding that it gradually sank into the cortex he confirmed this theory and also demonstrated that the medullary cavity was continually enlarging.

Duhamel did not stop here, for his knowledge of the growth of trees suggested a further experiment. He drilled holes in bones at measured distances and found that after a further period of growth the distance between the holes was unaltered. He concluded that growth in length must occur at the extremities but although he observed the epiphyseal plates of cartilage he did not associate them with the site of growth.

A clue to the significance of the epiphysis was provided in 1769 by Stephen Hales, a country parson. He also had worked extensively on vegetable growth and, on turning to that of animal bones, "took a half-grown chick, whose leg bone was then two inches long; with a sharpened iron, at half an inch distance, I pierced two small holes through the middle of the scaly covering of the leg and shin-bone." On killing the bird later he found that this distance had not altered but in addition remarked that the growth had occurred mostly at the upper end of the bone where "a wonderful provision is made for its growth at the joining of its head to the shank, called by anatomists symphysis." He even went as far as to suggest that the bony epiphysis was there to protect the growing region from the motion of the joint.

During the 1750's and 60's the problem of bone growth and modeling was again investigated by John Hunter. He repeated the madder experiments and confirmed that growth in length occurred at the extremities by implanting metal pellets into long bones. His most significant contribution was the conclusion that not only is bone laid down on the surface but also that it must be "constantly changing its matter" and that, to account for the growth of the irregular shape of a bone's extremity, absorption is as essential as deposition.

Little need be added to this story other than to say that growth in length of a long bone occurs at the metaphysis by the endochondral ossification of the cartilage continually produced by the cells of the epiphyseal plate. When the cartilage cells cease to multiply growth ceases. Growth in girth occurs by the intramembranous ossification of a collagenous matrix laid down by the periosteum while the medullary cavity is expanded by endosteal reabsorption of bone. As Hunter pointed out the reabsorption and replacement of the expanded ends of bone (remodeling) is essential for the preservation of its shape.

The growth process is summarized in Fig. 1-11. In this the upper end of the tibia is shown to grow from the size of the thick outline to that of the thin. To achieve this, new bone has been laid down by endochondral ossification at the epiphyseal plate (*blue*) and by the periosteum (*yellow*). Original bone has been removed (*green*) to model the expanded upper end and to increase the size of the medullary cavity.

The impression should not be gained from this description that bony epiphyses are necessary for growth in length. The second to fifth metatarsal bones of the foot, for example, have epiphyses only at their distal ends. The proximal ends remain cartilaginous during the period of growth and the cells continue to multiply. However, at maturity this cartilage does not become completely ossified and a thin rim is left for articulation with the adjacent tarsal bone.

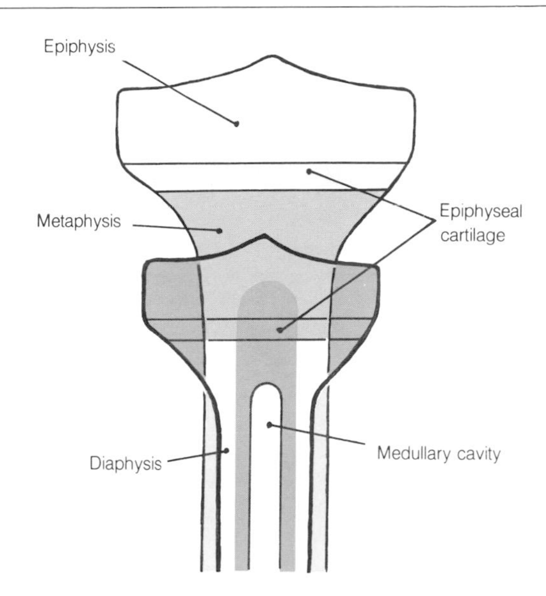

1-11 Diagram summarizing the growth and remodeling of a long bone.

The Blood Supply of Bones

Living bone is permeated by a fine vascular network. This network is supplied partly from the nutrient artery which penetrates the bone through the nutrient foramen to supply the marrow and partly from vessels in the periosteum. The importance of the latter source is demonstrated by the fact that necrosis of superficial bone may follow extensive stripping of the periosteum during surgical procedures.

In a growing long bone an epiphysis is supplied by vessels of the periarticular plexus which reach it through the attachments of the capsule and ligaments. The avascular epiphyseal cartilage forms a barrier between the vessels of the epiphysis and diaphysis. After their fusion some vascular anastomoses are formed between the two regions. It is the extensive blood supply of bone that allows it to carry out a continuing process of reconstruction and remodeling and to contribute to the metabolic activity of the body. This activity is well illustrated

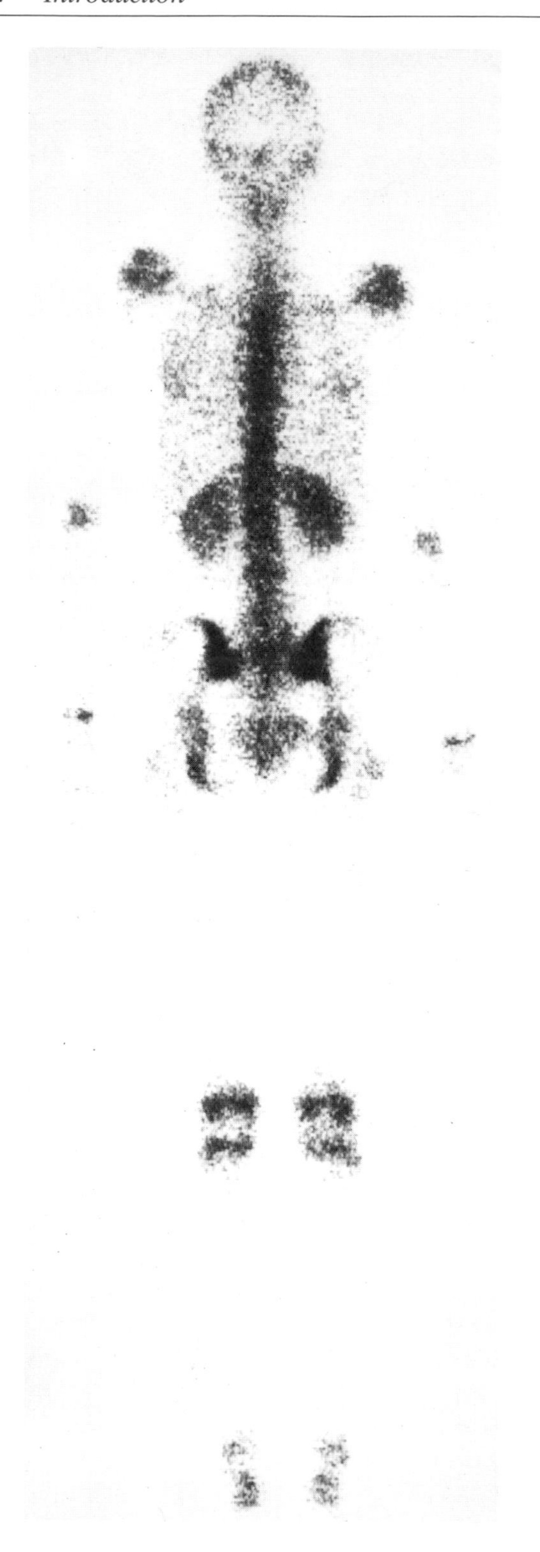

1-12 A whole body bone scan (posterior). (Courtesy of Gerald S. Johnston, M. D., Division of Nuclear Medicine, University of Maryland School of Medicine.)

in the "bone scan," an investigation used by clinicians and an example of which is shown in Fig. 1-12. The scan records the presence of technetium 99 ethylene diphosphonate that has been incorporated into the bones of the subject since its introduction two hours beforehand.

Joints or Arthroses*

When thinking of joints in biological terms there is a tendency to consider them as mobile structures associated with movement. This is of course far from the truth as many, like carpenters' joints, are immobile. If a joint is defined simply as a union between two or more bones this misunderstanding can be avoided. It is good to have a method of classifying joints and the most straightforward is presented here:

1. Fibrous joints (sutures, syndesmoses, and gomphoses)
2. Cartilaginous joints
   (a) Primary (synchondroses)
   (b) Secondary (symphyses)
3. Synovial joints

Each class of joint has its distinguishing features. **Fibrous** and **cartilaginous joints** allow minimal or limited movement, but **synovial joints** typically have a wide range. However, exceptions to these generalizations can be found. In a **fibrous joint** the bones are united by fibrous tissue only and examples would be the **sutures** joining the flat bones of the skull, the **inferior tibiofibular joint** (a **syndesmosis**) which unites the lower ends of the two bones of the leg, or the joints between both the mandible and maxilla and the teeth (**gomphoses**). In a **primary cartilaginous joint** the bones are joined by **hyaline cartilage** only (**synchondrosis**). This variety is found at the first costosternal joint (note that subsequent costosternal joints are synovial), and between the diaphysis and epiphysis of a growing bone.

A **secondary cartilaginous joint** or **symphysis** is more complex. Each bony surface is covered by a thin layer of **hyaline cartilage,** and between the layers of hyaline cartilage is sandwiched a layer of

* Arthrosis, *Gk.* = a joint; hence arthritis, an inflammation of a joint

fibrocartilage. The secondary cartilaginous joints lie in the midline of the body and examples are the joints between the bodies of the vertebrae, the pubic symphysis, and the manubriosternal joint.

Synovial joints differ from cartilaginous joints in that the articulating bony surfaces, although covered by hyaline cartilage, are separated from each other by an **articular cavity** containing **synovial fluid.** The bones are linked only at the periphery by a **fibrous capsule** and **ligaments.** These structures are illustrated diagrammatically in Fig. 1-13. It is apparent that the structure of the synovial joint would allow a great deal of movement and this, with one or two exceptions, is so. However, with mobility come the drawbacks of instability and a proneness to injury. For this reason synovial joints must be described in greater detail. Once the general features are grasped the modifications of these features found in any individual joint can be understood and an assessment of its stability made.

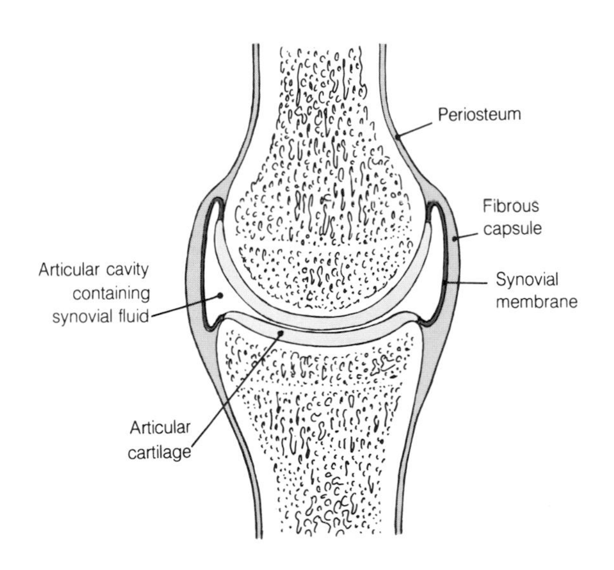

1-13 Diagram illustrating the features of a typical synovial joint.

Synovial Joints

The Articulating Bony Surfaces

It should be realized that more than two surfaces may take part in a joint, for example the femur, the tibia, and the patella (knee cap) at the knee joint. The surfaces are often congruent but sometimes congruency is aquired through means of **articular discs** or **menisci.** The shapes of the surfaces determine the types of movement possible and they are also important factors in the maintenance of stability.

The Articular Cartilage

A thin layer of **articular cartilage** covers the bony articulating surfaces. This is normally hyaline cartilage and is derived from the cartilage model from which the adjacent bone has become ossified. Where the bone has been ossified in membrane the cartilage is fibrocartilage as in the temporomandibular joint.

Articular cartilage is avascular and derives its nourishment from the synovial fluid and from the underlying vascular bone. Its metabolic requirements are small as it consists largely of matrix. After growth has ceased, cell division is infrequent and damage is repaired by fibrous tissue.

The Capsule

The **fibrous capsule** normally surrounds the joint completely although there may be deficiencies where a **synovial bursa*** communicates with the enclosed articular cavity (e. g., the suprapatellar bursa at the knee joint) or where a tendon and its **synovial sheath** pass through it (e. g., the tendon of the long head of biceps brachii at the shoulder joint). The capsule is usually attached to the bone at the margins of the articular cartilage. It may, in some joints, be reflected back over the bone for some distance before turning again to cross the joint (e. g., around the neck of the femur). The joint is frequently reinforced by **ligaments.** These may be **intracapsular** or **capsular.** Other **extracapsular** ligaments may also strengthen and stabilize the joint.

The Synovial Membrane

The **synovial membrane** is a delicate vascular layer of connective tissue covered by a discontinuous layer of irregularly shaped cells. It lines the inner aspect of the capsule and other intracapsular

* Bursa, *Gk.* = an item made of skin; hence a wineskin, i. e., a fluid-filled bag

structures but does **not** cover the articular cartilage. The synovial membrane secretes **synovial fluid** and is a source of cells responding to infection or the presence of foreign material.

The Articular Cavity

This cavity is enclosed by synovial membrane and articular cartilage. It contains only a small quantity of synovial fluid. As mentioned before it may communicate with extracapsular bursae. The clinical importance of this lies in the fact that a perforating injury may introduce infection into such a bursa which will rapidly spread to the associated joint. Air or radiopaque material may be introduced into the articular cavity for radiologic diagnostic purposes.

Synovial Fluid

This is a clear, yellowish, viscous fluid secreted by the synovial membrane into the synovial cavity. It forms a fluid medium whereby nutrients may be carried from the blood to the avascular articular cartilage and other intraarticular structures. In addition, it acts as a lubricant and protects the joint surfaces from wear and tear. Only a very small quantity of fluid is present in a normal joint.

Articular Discs and Menisci

A number of synovial joints are separated partially or completely into two separate cavities by fibrocartilaginous discs or menisci. When these are present they seem to increase the congruency of the articulating surfaces especially when two rather different types of movement occur at the joint. Good examples are found at the temporomandibular joint where the surfaces hinge and slide or at the knee joint where the movements are sliding and rotation.

The Blood Supply of Joints

Joints have a very profuse blood supply. This is derived from a large number of branches from surrounding vessels which form an elaborate network both around and within the capsule. These, in turn, anastomose with periosteal and metaphysial vessels of the adjacent bones. Articular cartilage, although avascular, receives nourishment via the synovial fluid and probably from vessels in the underlying bone and marrow cavity. However, when an **epiphyseal cartilage** lies between bony **metaphysis** and **epiphysis** in a growing bone, anastomoses between articular and metaphyseal vessels do not exist and the latter may be considered as end arteries. This fact may well account for the incidence of osteomyelitis at the metaphysis in growing children, infected emboli coming to rest at this site.

The Nerve Supply of Joints

Joints have a rich nerve supply derived from surrounding nerve trunks and their muscular branches. These nerves serve an important protective and proprioreceptive function. It is almost impossible to denervate a joint by surgical means, but a generalized disease process may do this and precipitate a gross disorganization of joint structure. A hundred years ago Hilton established that nerves supplying the muscles that operate a joint also supply the joint and the overlying skin. Hence, joint pain is often referred to overlying skin. This presents no problems until it is realized that some muscles operate two joints and pain may be referred erroneously to an unaffected joint. This applies particularly to the hip and knee joints. Pain arising in the hip joint may frequently be referred to the knee which, on examination, is found to be normal and the patient, often a child, is dismissed. The culprit here is the obturator nerve which supplies both joints.

Movements at Joints

The movements that can be performed by joints are **flexion, extension, abduction, adduction, circumduction,** and **rotation.** To define each of these in words is difficult and examples provide a much better explanation.

Flexion occurs at the elbow joint when the angle between the arm and forearm is decreased and **extension** is performed as the angle is increased again and the anatomical position is restored. These movements are illustrated in Fig. 1-14.

Abduction implies movement away from the midline of the body or limb and **adduction** movement toward the midline. In Fig. 1-15 these movements are seen at the shoulder joint as the upper limb is moved away from and toward the trunk.

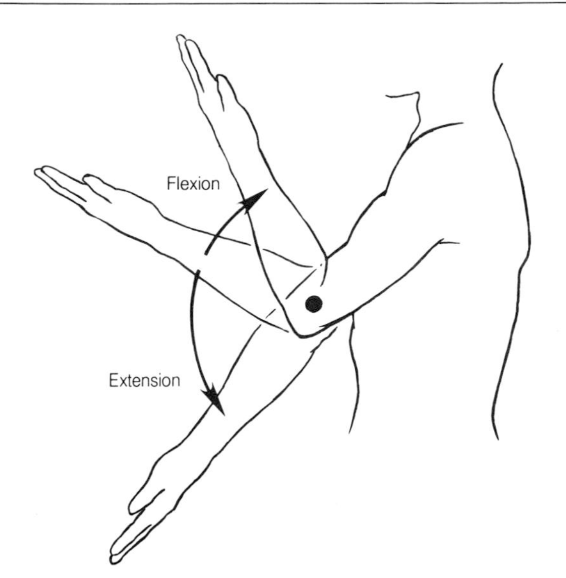

1-14 Showing the movements of flexion and extension.

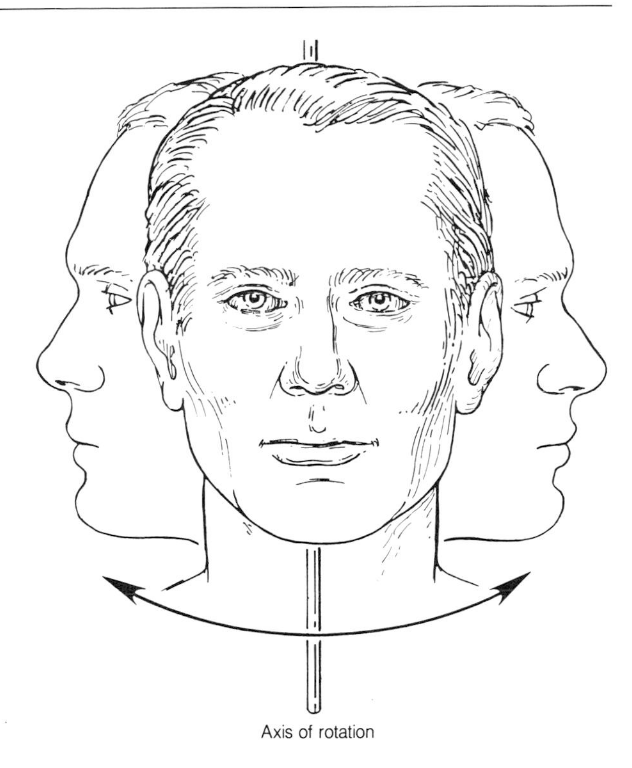

1-17 Showing the movement of rotation.

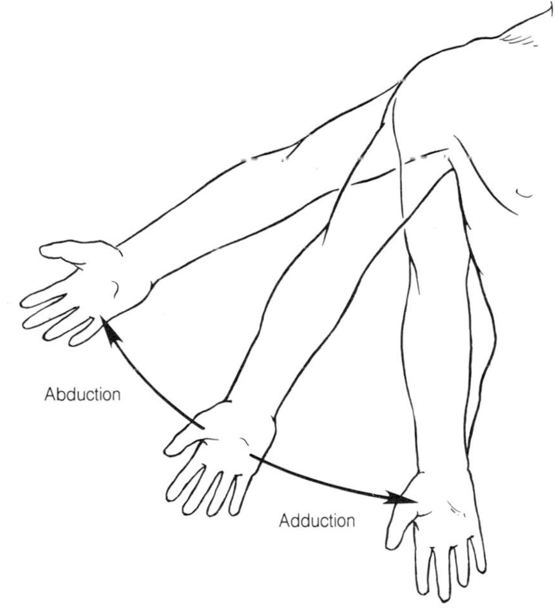

1-15 Showing the movements of abduction and adduction.

By combining these movements in continuously varying proportions **circumduction** is produced and in Fig. 1-16 this is illustrated at the wrist joint. Note in this that there is no twisting or rotation of the hand but that the long axis of the hand describes a cone and the tip of the middle finger describes a circle.

Rotation should not be confused with circumduction. In rotation movement occurs about a longitudinal axis which remains stationary and perhaps the most obvious example is the rotation of the head that signifies a negative and which is illustrated in Fig. 1-17.

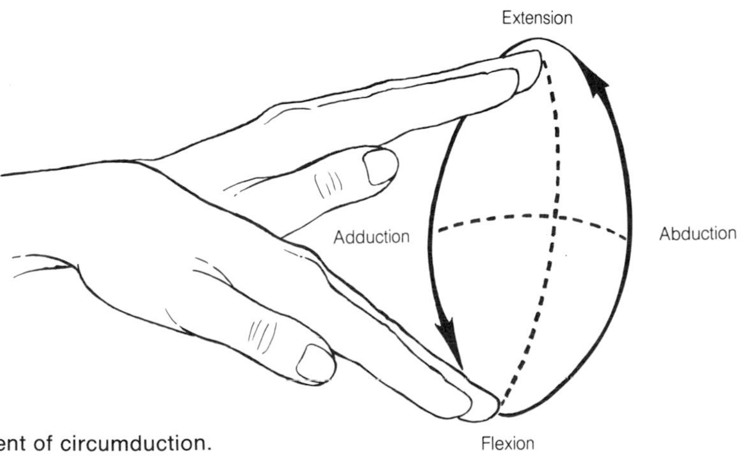

1-16 Showing the movement of circumduction.

Additional and rather more complex movements in which more than one joint is involved are described in detail with the appropriate region.

As each joint is described it will be seen that the movements it can perform are determined by the shape of the articulating surfaces and restrictions imposed by ligaments and adjacent soft parts.

The difference between **active** and **passive movements** should be clearly recognized. **Active movements** are those produced voluntarily by a subject. **Passive movements** are elicited by an outside force and may include movements not possible to perform voluntarily. If the outside force is an examining physician he or she must be aware of the normal range so as to be in a position to detect abnormal movement. This is of particular importance when examining large joints such as the knee.

Movement at articular surfaces can be reduced to two varieties, sliding or spinning (i. e., rotation about an axis at right angles to the joint surface); rolling does not occur. A combination of these gives rise to the movements of flexion, extension, rotation, circumduction, etc. However, even apparently simple movements become complex on complete analysis, hence the difficulties in designing prosthetic joints and artificial limbs. When assessing joint movements, care should be taken that movement at other joints is excluded. For example, the range of rotation of the forearm about its long axis (pronation and supination) can only be demonstrated when rotation at the shoulder is eliminated by flexion of the elbow. These features of joint movement become of importance when determining and following up the success of the treatment of a fracture and when assessing (possibly for compensation purposes) a final degree of disability.

Muscles Producing Movements at Joints

A knowledge of the muscles that move a joint is of value for a number of reasons. Inability to perform a movement may indicate a lesion of the muscles that normally carry out this task, or injury to the motor nerves that supply them. Also, the maintenance of the ability of a muscle group to function after a joint injury or surgery often forms an essential part of postoperative care and rehabilitation.

The Stability of Joints

To some extent the freedom of a joint to move is directly related to the chance of its dislocation but this is by no means always so. The stability of a joint depends upon: (1) The **shape of the articulating bony surfaces** – the shallow cup and large ball at the shoulder form a much less stable joint than the deep cup of the hip joint that almost grasps the ball-like head of femur. In fact, despite the wide range of movements possible at the hip joint, gross forces such as those encountered in an automobile accident are required before it dislocates; (2) The **ligaments surrounding a joint** – these contribute to its stability but, of course, at the same time limit its range of movement. As in the knee, they are frequently ruptured when these limits are transgressed or alternatively a bony attachment is torn off as in injuries to the ankle joint; (3) The **muscles spanning a joint** – these add to its stability and by reflex activity protect it from extremes of movement. Paralysis of the muscles of the leg leads to instability of the ankle and tarsal joints and permanent fusion (arthrodesis) of the latter may greatly increase the ability to walk.

These characteristic features are common to most synovial joints but as each joint is described those that have particular importance will be emphasized.

Skeletal Muscles and Tendons

A feature of the animal kingdom is the ability to move, gather food, fight, or flee from danger. At almost every level this movement is produced by aggregations of the proteins actin and myosin. It is this system that is embodied in the highly specialized cells that make up skeletal muscles and enable them to develop tension and, when appropriate, to shorten. These muscles cells or fibers are 50-100 μm in diameter, i. e., just visible to the naked eye, and in some instances can be up to 10 cm in length.

Although one contractile unit of a muscle cell —that is the sarcomere—is indistinguishable from another, the fibers which they form are bundled together by connective tissue in such a way as to produce a wide variety of sizes and shapes, for example the small slender muscles that move the eye or the massive quadriceps femoris muscle in the front of the thigh.

The Action of Muscles

Muscles are attached to the skeletal system by a specialized form of connective tissue called tendon. This linkage converts shortening of the muscle into movement of one part of the body relative to another or the whole body relative to the ground (air or water). However, movement is not an essential result of muscle activity. In practice more energy is probably expended by muscles to prevent movement as in maintaining the posture of the body or stabilizing the body while one part—a hand or lower limb—performs a movement. Muscles can only serve this function because they possess the property of developing and maintaining tension without shortening or doing mechanical work. This is called isometric contraction. An analogy for this would be allowing the clutch of an automobile to slip while the auto remains stationary on a hill. Force is applied to the vehicle but no work is done and heat is generated. However, continuing the metaphor, a muscle cannot go into reverse or actively lengthen and any movement produced by a muscle can only be retraced by the force of gravity or the action of an antagonistic muscle.

The Attachment of Muscles

It has been said above that muscles are linked to the skeleton by tendons and the points at which these join the bone are known as the **origin** and **insertion** of a muscle. These can be defined by saying that it is the origin that remains stationary and the insertion that is drawn toward it. Nevertheless, it is clear that this situation can easily be reversed in climbing a rope or doing "chin-ups." It is perhaps often less confusing, therefore, to talk about the **attachments**—proximal or distal—of a muscle. This is not to say that the attachments of a muscle are not important; they are, and it is a knowledge of their attachments that makes it possible to deduce the nature of the movements that a muscle will produce. It must be said at this point, however, that muscles seldom if ever function in isolation but rather in groups. Those muscles that help the prime mover or **agonist muscle** are called **synergists. Antagonists** are muscles having an action opposite to that of the agonist.

It is worthwhile looking ahead to see how the attachment of a muscle and the relation of this to a joint allows it sometimes to operate a lever and sometimes to operate around a pulley.

Fig. 1-18 shows how the biceps muscle in the arm is attached to the radius in the forearm in such a way as to operate a third-order lever when it supports a weight held in the hand. On the other hand, in Fig. 1-19, the triceps muscle can be seen to act on a first order lever when it extends the elbow to raise a weight on a pulley. Because at the knee joint there is no bony projection for its extensor muscle to pull on the quadriceps femoris muscle acts as if over a pulley to extend this joint.

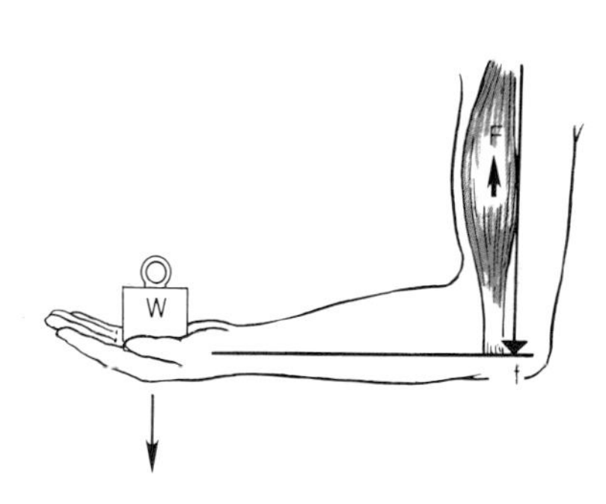

1-18 The biceps brachii muscle operating a third-order lever.

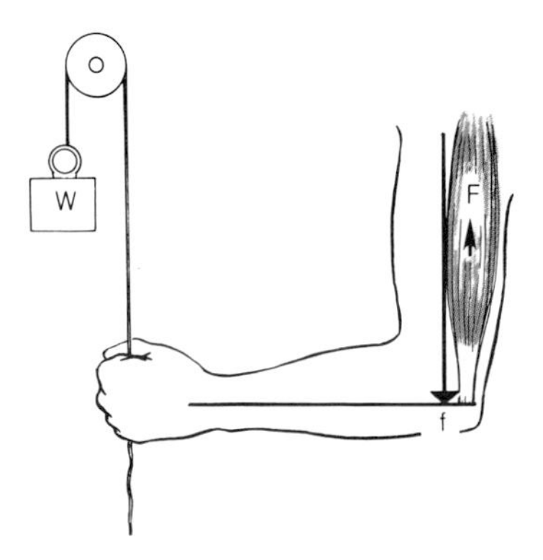

1-19 The triceps brachii muscle operating a first-order lever.

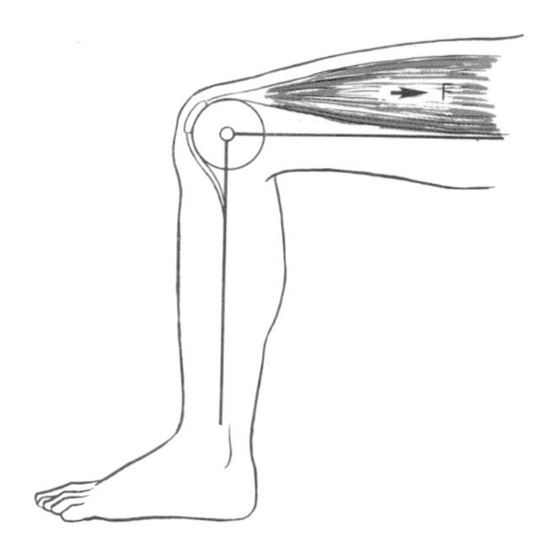

1-20 The quadriceps femoris muscle operating over a pulley at the knee joint.

This principle is illustrated in Fig. 1-20. The quadriceps femoris muscle illustrates another general principle, that when a muscle crosses a joint close to the bone it loses its fleshy fibers and these are replaced by a tendon. In some instances where perhaps wear and tear is great, a small bone is found in the tendon. This is called a **sesamoid bone** and the familiar knee cap or patella is an example.

The Design of Muscles

Varieties of muscle design depend on a number of features of their fibers. These features which are somewhat interdependent are:

1. their relationship to tendons,
2. their length, and
3. their total cross-sectional area.

In Fig. 1-21 a number of different types of muscle are illustrated whose shapes are dependent on the arrangement of their fibers. When the fibers are parallel and extend from a proxomal to a distal attachment in the "line of pull," the muscle may be said to be a strap muscle when thin (Fig. 1-21A), or fusiform when it has a distinct belly (Fig. 1-21B). When two or three fusiform muscles unite at a single attachment they are described as being bicipital or tricipital (Fig. 1-21C). Other

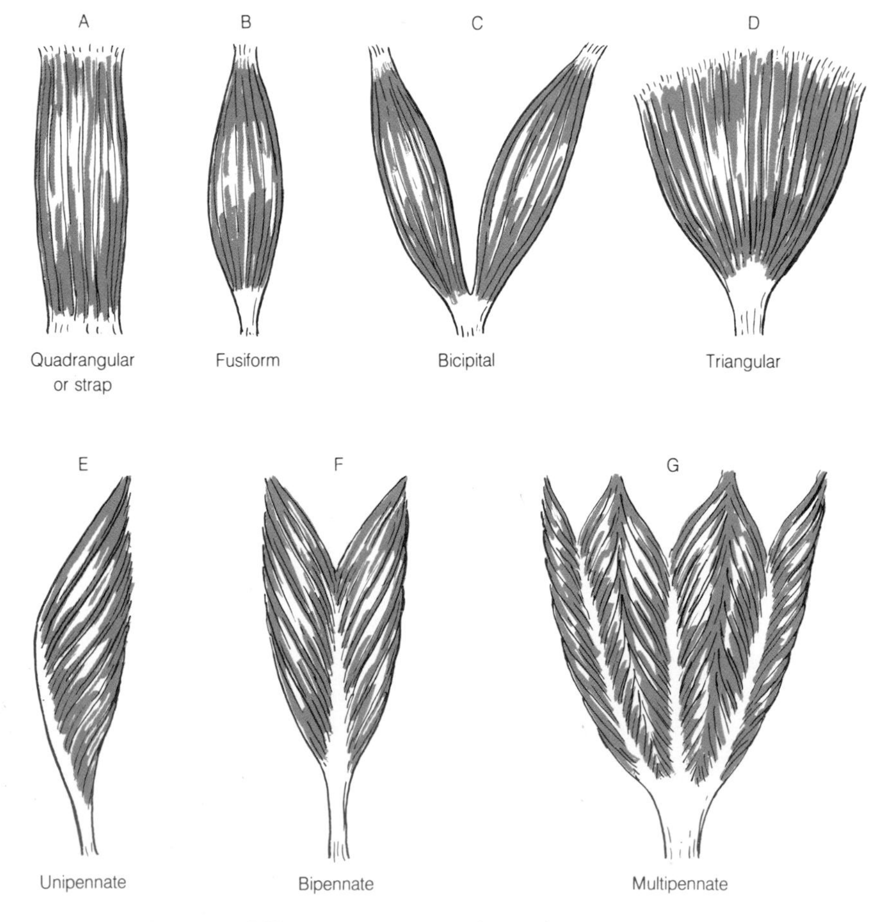

1-21 Illustrating the variety of shapes and fiber arrangements of muscles.

muscles which have one wide and one narrow attachment may be triangular or fan-shaped (Fig. 1-21D). In many muscles the fibers are not arranged along the line of pull but lie at an angle to the proximal and distal tendons. Such muscles are said to be pennate* and examples can be seen in Fig. 1-21, E-G.

It is the length of its fibers that determines how much a muscle may shorten and thus a long muscle with parallel fibers will have an advantage in this respect. The force which it can generate is, however, related to the cross-sectional area of its fibers. Thus for a unit volume of muscle more short fibers produce greater force but less ability to shorten and vice versa. This inverse relationship is illustrated in Fig. 1-22. Although the volume of muscle is identical in Fig. 1-22, A and B, that in Fig. 1-22A can develop greater force and that in Fig. 1-22B can shorten more. This principle is embodied in the design of the multipennate muscle shown in Fig. 1-21G in which many short fibers give great strength at the expense of the ability to shorten.*

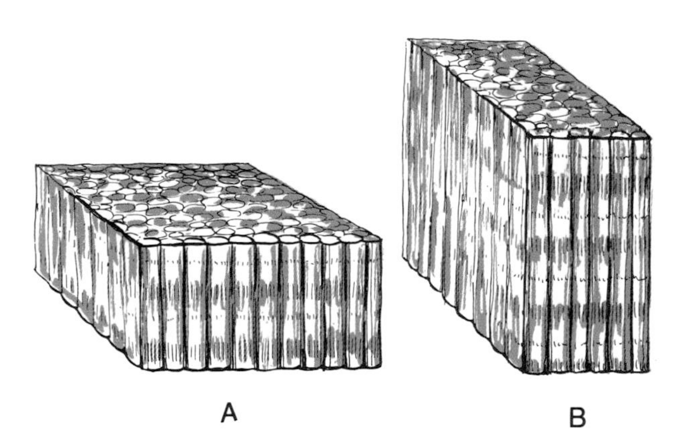

1-22 A comparison between the fiber length and cross-sectional area of two similar unit volumes of muscle.

The Blood Supply and Nerve Supply of Muscles

Typically a muscle has a neurovascular hilus where the nerve and main vessels supplying it enter its belly. However, a double nerve supply is not uncommon and a small proportion of its blood supply may reach a muscle through its tendon. The blood supply is potentially great and the flow is raised during activity by the dilatation of blood vessels. The nerve innervating a muscle carries **motor fibers** and **sensory fibers.** The motor fibers stimulate the muscle fibers through a special type of synapse called the **motor endplate.** Sensory fibers carry information about the physical state of the muscle from receptor organs back to the spinal cord. Autonomic motor fibers regulate the smooth muscle in the small arteries. Division of the nerve clearly will produce complete paralysis but it is not always realized that interruption of the sensory pathway alone renders a muscle almost equally useless.

Although on stimulation an individual muscle fiber contracts to the best of its ability (the "all or none" law), the force developed by a whole muscle can be graded. This gradation is produced by the recruitment of more or fewer **motor units.** The principle of a motor unit is an important one and will be found to have considerable clinical application to neuromuscular disease. A motor unit consists of a single motor nerve fiber, its cell body lying in the spinal cord, and the muscle fibers that it supplies. These parts of the unit are shown in Fig. 1-23. Notice that the fibers supplied are not necessarily adjacent to each other. The size of

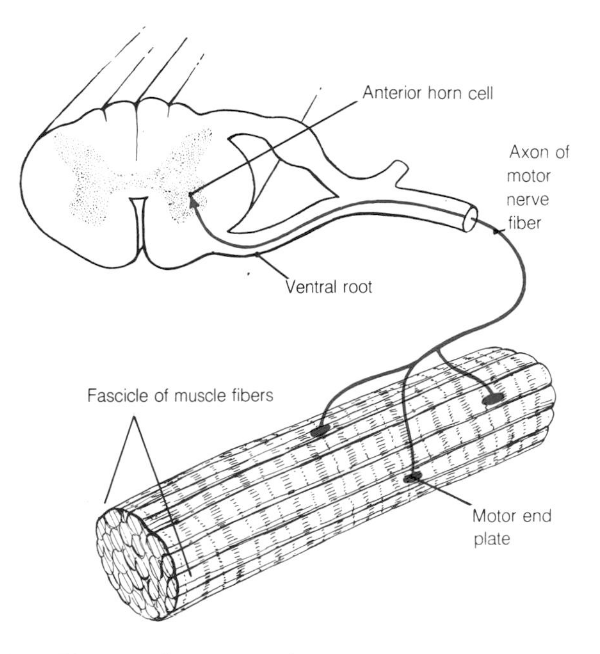

1-23 Diagram of a motor unit.

* Penna, *L.* = a feather
* The loss of force due to the angulation of the fibers is small; e.g., Cos. 30° = 0.8660

motor units varies greatly. The fibers supplied by a single nerve fiber in the large gluteus maximus muscle of the buttock may number hundreds while those in an eye muscle may be less than ten.

Tendons

The tendons found at the extremities of muscles are formed from closely packed and parallel bundles of collagen fibers. Their shape varies considerably and they may be found to be cylindrical, flattened, or as broad sheets or aponeuroses. They are sometimes very short and hardly visible in which case muscle fibers may appear to be attached directly to bone. Such attachments are given the rather misleading name of "fleshy." A further variety of muscle attachment is found in the form of a tendinous raphe.* These lie in the midline of the body where the fibers of two symmetrical muscles join each other rather than bone.

Tendons are surrounded by loose and mobile connective tissue and where they cross joints they are frequently enclosed in **synovial sheaths.** These take the form of a double layer of connective tissue lined by synovial membrane and separated by a space containing a small quantity of fluid similar to synovial fluid. The arrangement of these sheaths is more easily illustrated, as in Fig. 1-24, than described.

Bursae are in many ways similar to synovial sheaths in that they are sacs of connective tissue which have a smooth lining and are filled with synovial fluid. They normally lie between tendons and bony areas over which the tendon passes. Some bursae communicate with nearby joint cavities and thus form a pathway by which infection may be introduced into the joint.

* Raphe, *Gk.* = a seam

The Nervous System

The nervous system receives information from within the body and from the surrounding environment, integrates this and makes an appropriate response. In this it plays a major role in the homeostasis* of the body. Although the input from the five senses plays an important part in this process it is a humbling thought to realize that only a small fraction of the information received from the remainder of the sensory system reaches a conscious level.

* Homoios, *Gk.* = alike or same; Stasis, *Gk.* = a standing. Hence the sense of a state of equilibrium maintained among a variety of pressures.

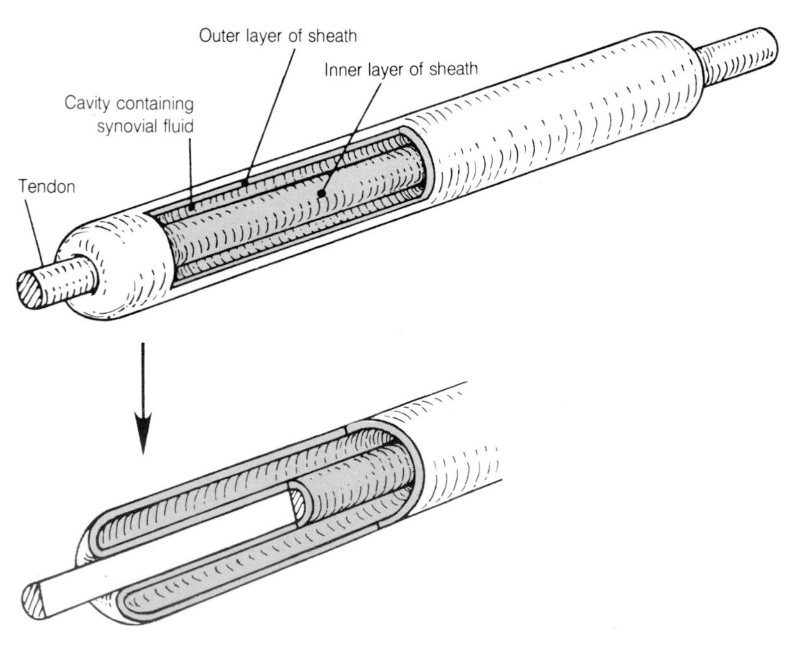

1-24 Diagrams showing the arrangement of the layers of a synovial sheath.

The **nervous system** is divided into central and peripheral parts. The **central nervous system** consists of the **brain** and the **spinal cord,** and the **peripheral nervous system** of the **cranial and spinal nerves** and the **autonomic nerves.**

The Central Nervous System

The **brain** lies in the cranial cavity where it is enclosed and protected by the bones of the skull. While a detailed study of the brain is outside the scope of this text some description of its various parts is given in Chapter 9 in relation to the cranial cavity, blood supply, and the origins of the cranial nerves.

The **spinal cord** is surrounded and protected by the vertebral canal which is formed by the neural arches and bodies of the vertebrae. It is described in Chapter 2 with the vertebral column.

The Peripheral Nervous System

The **twelve cranial nerves** arise from the brain and each has its own distinctive neural components and course. These are summarized in Chapter 9 and further details are given with the regions into which their course carries them. The anatomy and organization of the **spinal nerves** and the **autonomic nervous system** are described in Chapter 2 and discussion is limited here to the neurons which form the conducting tissue of these nerves.

Neurons and Nerve Fibers

The cell type of which the excitable tissue of the nervous system is composed is called a **neuron.** Typically this is described as having a **cell body** which contains a nucleus with a well marked nucleolus. There are several processes which convey impulses toward the cell body called **dendrites,** and a single process called an **axon** which carries impulses away from the cell body. Communication between neurons occurs at a **synapse.*** This junction is classically between the terminal process or processes of an axon and the dendrites of the subsequent neuron. However, junctions with the cell body and the axon are also found. The passage of information across the synapse is usually mediated by the release of a chemical neurotransmitter substance, for example norepinephrine or acetylcholine, by the presynaptic neuron.

Three morphological varieties of neuron which are found in the peripheral nervous system need to be described before its function can be considered. A **motor neuron** has a large cell body in the grey matter of the spinal cord. It has many short dendrites and a single long axon extending to its termination outside the spinal cord. In the case of a neuron supplying a muscle in the foot its axon may be several feet in length. A motor neuron is illustrated diagrammatically in Fig. 1-25. It has all

* Syn, *Gk.* = with or together; Haptein, *Gk.* = to clasp

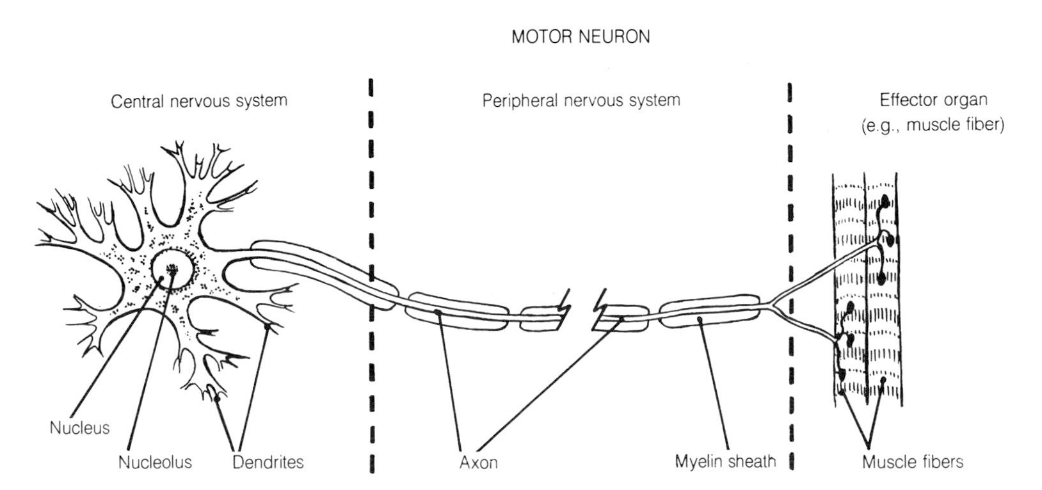

1-25 Diagram of a motor neuron.

the features of a "typical" neuron. A **sensory neuron** may be either bipolar or pseudounipolar. A **bipolar neuron** has a central and peripheral process between which lies the cell body. Unlike the motor neuron the cell body of a peripheral sensory neuron is situated outside the central nervous system, and with the cell bodies of other sensory neurons forms a **sensory ganglion.** A **pseudounipolar neuron** is similar to a bipolar neuron except that during development its cell body migrates to one side. As a result a single process leaves the cell body and divides into a central process directed toward the spinal cord and a peripheral process that is carried away from the cell body toward a receptor organ. As with a bipolar sensory neuron the cell body is located with others in a sensory ganglion. The central and peripheral processes of these sensory neurons both have the characteristics of an axon and with the axons of motor neurons are commonly referred to as nerve fibers. Diagrams of these types of sensory neurons are shown in Fig. 1-26. Peripheral nerve fibers range in diameter from less than 1.0 μm to 20 μm. Fibers of more than 1.0 μm in diameter are usually surrounded by an interrupted sheath of lipid material called myelin which is part of the **Schwann cells** wrapped around the axons. Such nerves are said to be myelinated. Motor, sensory, and, as will be seen later, autonomic nerve fibers are bundled together by connective tissue to form the **mixed peripheral nerves** visible to the gross anatomist. These mixed nerves are derived from the **spinal nerves** which are in turn formed by the union of the **dorsal and ventral roots** that leave the spinal cord. A spinal nerve and its branches typically supply the skin and muscles derived from a single embryological somite with fibers derived from a single spinal cord

SENSORY NEURONS

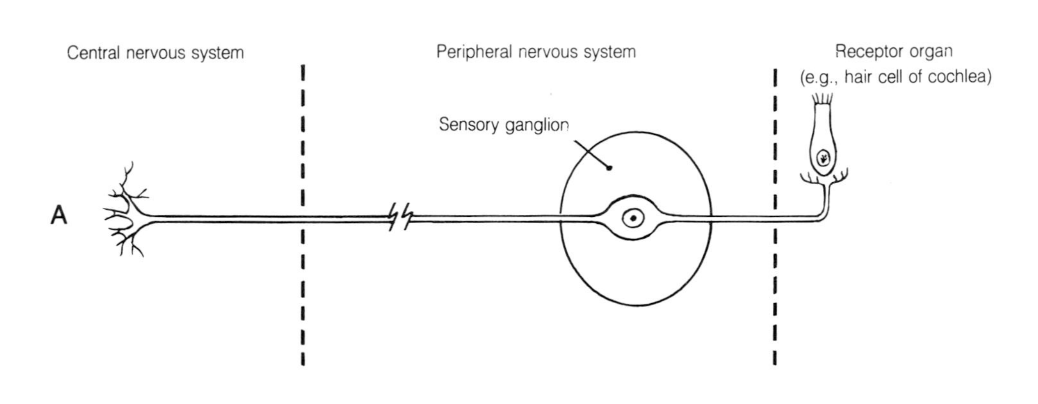

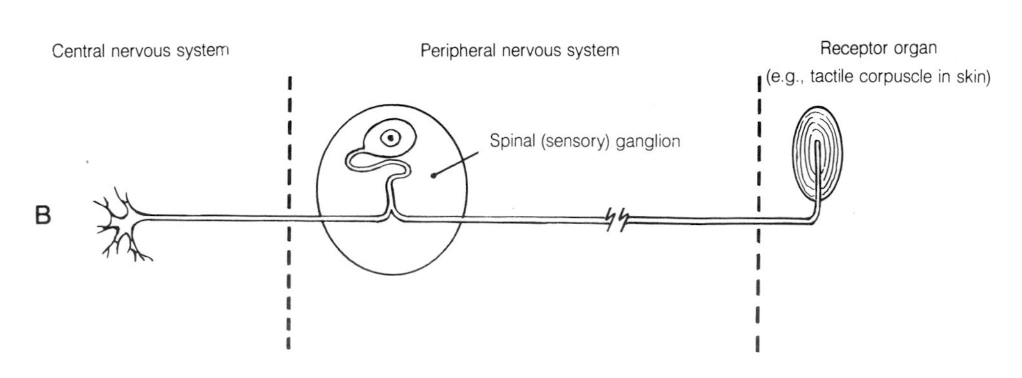

1-26 Diagrams of a bipolar sensory neuron (*A*) and a pseudounipolar sensory neuron (*B*).

segment, and this arrangement is seen most clearly in the thoracic region. However, in the cervical, lumbar, and sacral regions the spinal nerves fuse and separate from each other forming a network of nerves or **plexus.** These plexuses allow nerve fibers from several spinal cord segments to be distributed in one peripheral nerve. This principle is well illustrated in the brachial plexus of the upper limb to which four cervical and one thoracic spinal cord segments contribute. In Fig. 1-27 it can be seen that motor nerve fibers from each of these segments can be directed in the plexus to a single large peripheral nerve.

Motor or effector pathways of the **autonomic nervous system** are comprised of two neurons. The first of these has its cell body either in a nucleus* in the brainstem or in the lateral horn of the grey matter of the spinal cord. The axon leaves the central nervous system in either a cranial or spinal nerve to reach an **autonomic ganglion.** Here this **preganglionic fiber** synapses with the cell body of a **postganglionic fiber** which is distributed to an effector organ. The target organ will be either smooth muscle, cardiac muscle, or a gland.

* Here the word nucleus is used to describe a collection of nerve cell bodies in the central nervous system.

The Cardiovascular and Lymphatic Systems

The Cardiovascular System

The cardiovascular system provides the body with a means of transporting and distributing oxygen, other nutrients, and information in the form of hormones. It also plays an essential role in the excretion of waste products through the lungs, liver, and kidneys, and in the control of body temperature.

The propulsive force which drives the blood through the vessels is supplied by the muscular heart. This, although a single organ, is in fact two

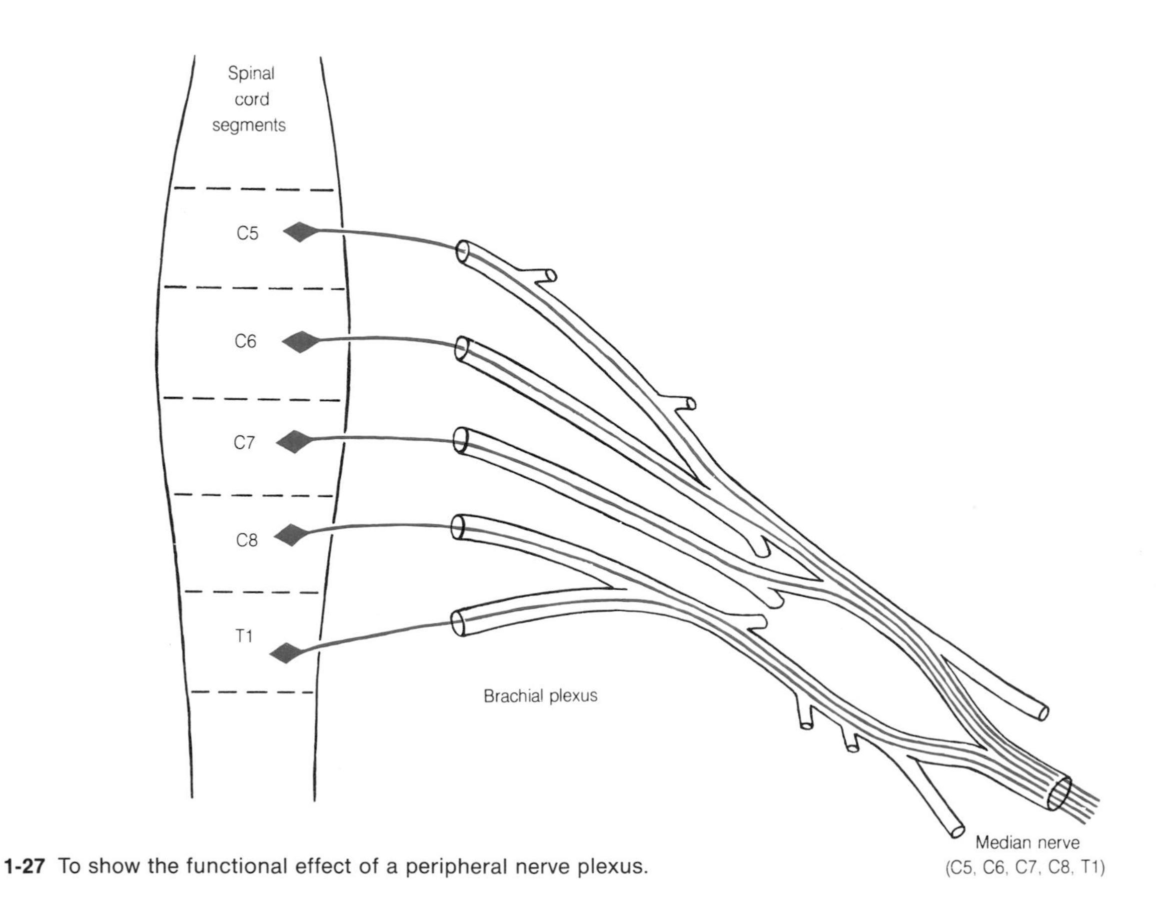

1-27 To show the functional effect of a peripheral nerve plexus.

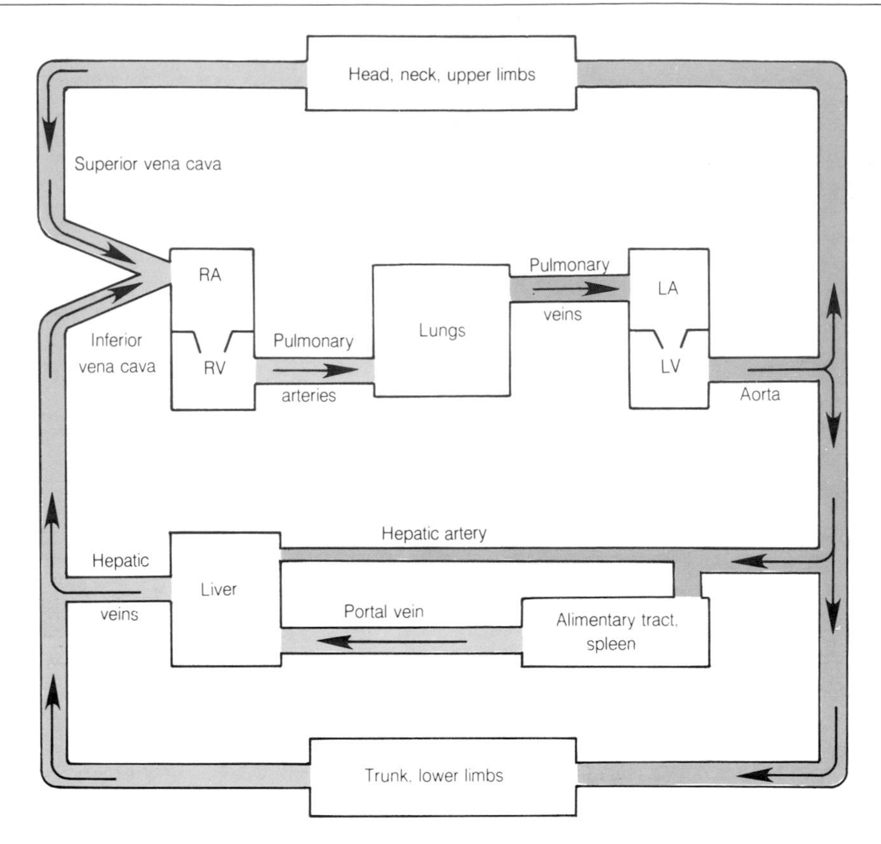

1-28 Diagrammatic representation of the cardiovascular system.

pumps and is illustrated as such in Fig. 1-28 which should be followed as the circulation is described. From the right side of the heart blood is pumped to the lungs for oxygenation through the **pulmonary circulation** and on its return to the left side is pumped to the remainder of the body through the **systemic circulation.** The orderly action of the two pumps is ensured by an intrinsic pacemaker and conducting system but superimposed upon this is a neural regulation which allows the heart to respond to varying needs of the body.

At each heart beat blood passes from the left side of the heart into the **aorta** at a maximum pressure, the **systolic pressure,** of 120 mm Hg. The elastic wall of the aorta is distended by this pressure. The potential energy it thus acquires is released at the end of the pumping stroke and thus prevents a sudden fall to its minimum or **diastolic pressure.** The blood is distributed in a branching system of **arteries** and the flow through individual organs is controlled by the neural and chemical regulation of the caliber of small vessels called **arterioles** which have a lumen of approximately 30 μm in diameter. The exchange of gases and metabolites occurs in the tissues through the walls of the finest vessels, called **capillaries,** and the blood is then returned to the right side of the heart through **veins.** A vein is a thin-walled vessel whose caliber gradually increases as it is joined by other veins or, as they are frequently called, tributaries. The majority of smaller veins follow the course of an artery and are not specifically named. It will soon be found that only the large veins and those that do not accompany an artery do receive names. The right side of the heart pumps blood to the lungs through the **pulmonary trunk** and **arteries.** These vessels are only subjected to a pressure of about 25 mm Hg and have much thinner walls than the large arteries carrying blood from the left side of the heart. The pulmonary circulation is completed by the **pulmonary veins** which return oxygenated blood to the left side of the heart.

A number of special features of vessels and their arrangement will be referred to in subsequent chapters but need further description here.

Valves

Valves are found between chambers of the heart, at the origins of the aorta and pulmonary trunk, and in many veins. The valves associated with the heart will be described with that organ. The valves of veins consist of a pair of cup-shaped flaps. In the diagram shown in Fig. 1-29A it can be seen that when the blood is flowing toward the heart the flaps collapse against the vessel wall, but if the blood attempts to flow back again as in Fig. 1-29B the cups are distended and occlude the lumen. The phlebogram* shown in Fig. 1-30 demonstrates the presence of many such valves in the **axillary vein,** the main vein returning blood from the upper limb. Valves are also present in the veins of the lower limb and their incompetence is associated with varicose veins. The blood in a segment of a vein between two valves can be emptied in a centripetal direction by the pressure of surrounding skeletal muscles and this aids in the return of blood to the heart. This **muscle pump** becomes of particular importance when standing because the effect of the force of gravity is to raise the venous pressure at the ankles. The pooling of blood in the veins of the leg that results may cause a reduction in the venous return to the heart and fainting. This can be avoided by periodic contractions of the calf muscles.

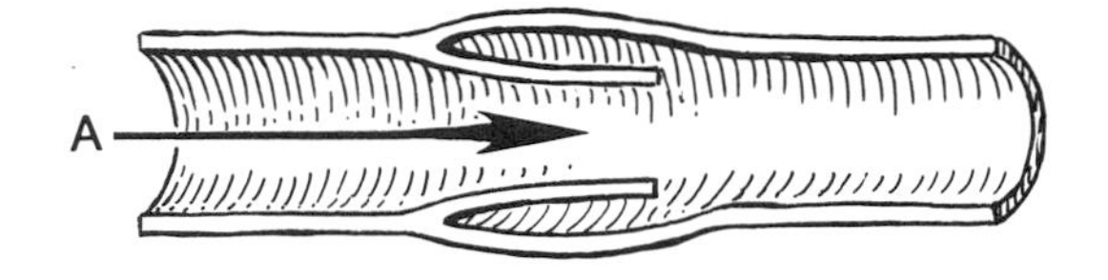

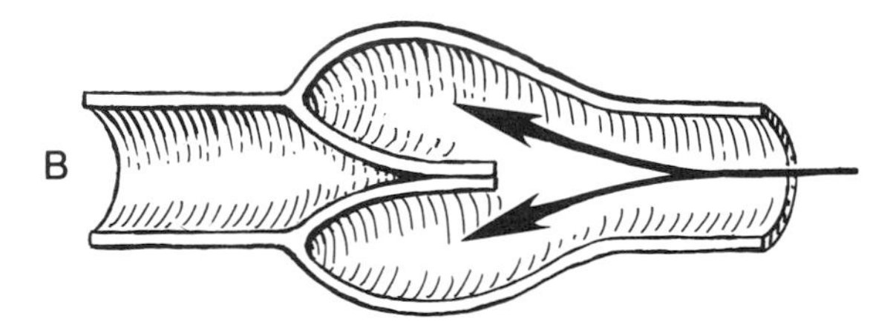

1-29 To show the function of valves in veins.

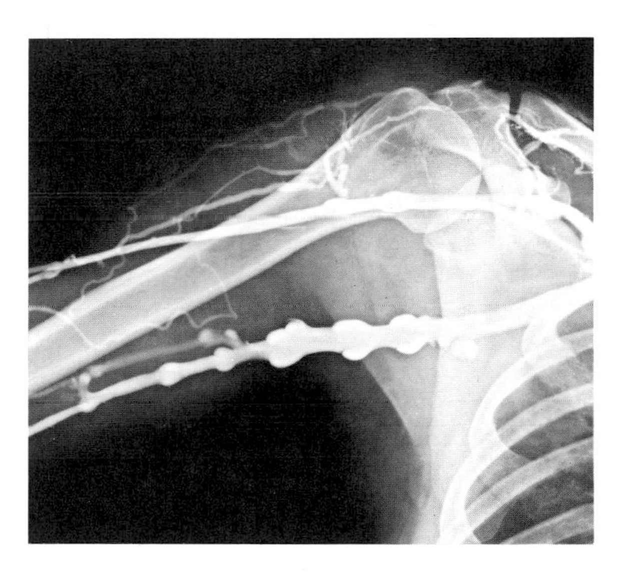

1-30 Phlebogram of the veins of the upper limb, showing numerous valves along the course of the cephalic and axillary veins. (Reproduced by permission, from Sobotta: *Atlas of Human Anatomy,* 10th Ed, Urban & Schwarzenberg, Baltimore – Munich, 1983.)

Anastomoses

An anastomosis refers to a communication between two tubes and in vascular anastomoses these tubes are arteries or veins. Anastomotic vessels can be found joining the larger arteries of the limbs or the more proximal to the more distal parts of one artery. Under normal circumstances probably little exchange of blood occurs through these anastomoses but if an artery becomes occluded they can form an alternative pathway or **collateral circulation.**

When such communications are absent or minimal an artery is called an **end artery** and occlusion of an end artery leads to oxygen lack (anoxia) and death (necrosis) of the tissue supplied by the vessel. A dramatic example of this occurs when the central artery of the retina is obstructed and total blindness of the eye follows.

Not all the blood delivered from the heart to the tissues passes through capillaries because **arteriovenous anastomoses** between small arteries and veins are found particularly in the alimentary tract and in the skin. These anastomoses whose caliber is under nervous control provide a further method for regulating the distribution of blood. Traumatic arteriovenous anastomoses sometimes follow stab and gunshot wounds. These are frequently formed between large vessels such as the

* Phleps, *Gk.* = a vein

femoral artery and vein in the groin and, as can be imagined, lead to a substantial change in the hemodynamics of the region.

Portal Systems

Blood has been described as returning to the right side of the heart from the capillaries through the systemic venous system. In this, smaller veins form **tributaries** of larger veins until the two largest veins, the **superior and inferior venae cavae,** open into the right atrium of the heart. There are, however, exceptions to this general pattern and these are the **portal systems.** Of these, one is concerned with the venous drainage of the pituitary gland and another with the drainage of the alimentary canal. Because the latter carries blood to the liver (*hepar* in Latin) it is known as the **hepatic portal system.** It is included in Fig. 1-28 and serves well to illustrate the features of a portal system. Blood from the capillaries in the alimentary tract drain in the normal manner into veins that are tributaries of the **hepatic portal vein.** On reaching the liver the portal vein then breaks up again into **branches** which lead to fine specialized venous channels called **sinusoids.** From these, nutrients absorbed by the alimentary tract are passed to the liver cells. The blood from the **sinusoids** is collected again by **hepatic veins** which return it to the systemic venous system and thence to the heart.

The Lymphatic System

At the arterial end of a capillary, where the hydrostatic pressure of the blood exceeds its osmotic pressure, fluid passes into the tissues. At the venous end, where the hydrostatic pressure of the blood has fallen below that of its osmotic pressure, the greater part of this fluid returns. The remainder, and this may be a substantial amount at times of increased tissue activity, passes as **lymph** into the lymphatic system. It is eventually returned to the systemic venous system close to the heart. In addition to this function the lymphatic system is able to engulf small particles of foreign or unwanted material from the tissues and deposit them in the lymph nodes that lie along the course of its vessels. This function will become obvious when the lymph nodes at the root of the lung are examined and found to be loaded with inhaled carbon particles. In the alimentary tract this system also performs the special function of taking up finely emulsified fat that has been absorbed.

At the periphery of the system fluid and particulate matter is taken up by a network of blindly ending lymphatic capillaries which drains into collecting lymphatic vessels that possess valves. These in turn converge on a number of major lymphatic trunks. An examination of Fig. 1-31 shows that lymph from the lower limb, the abdomen, and a large part of the thorax reaches the largest lymphatic vessel called the **thoracic duct.** This narrow but visible structure passes superiorly through the thorax to open into the junction of the large veins draining the left side of the head and neck and the upper limb. The left **jugular, subclavian,** and **bronchomediastinal trunks** also join the systemic venous system at this point. On the right side of the body lymph from the head, neck, upper limb, and the upper part of the thorax joins the venous system at the junction of the equivalent right veins. Obstruction of either the venous or lymphatic systems leads to an abnormal collection of tissue fluid and this condition is known as **edema.**

The course of lymphatic collecting vessels is periodically interrupted by groups of **lymph nodes,** one of which is illustrated in Fig. 1-32. These small bean-like structures have a connective tissue **capsule** from which partitions or **trabeculae** extend inward and divide it into smaller areas. These are filled with a mesh of fine fibers which are dense in the center and packed with lymphocytes, forming a **lymphatic nodule.** At the periphery this mesh is less dense and forms a **lymphatic sinus.** Throughout the node are also found phagocytic **reticuloendothelial cells** which take up particulate matter from the lymph. **Afferent lymphatic** vessels open into a **subcapsular sinus** and the lymph percolates through the node to leave by **efferent lymphatic** vessels at the hilus where the artery and vein of the node are also found.

One of the difficulties in understanding the lymphatic system is that, except for the thoracic duct and the lymph nodes, its component vessels cannot be seen with the naked eye. However, if a small quantity of India ink is injected into the foot of a rat the course of the vessels to the lymph

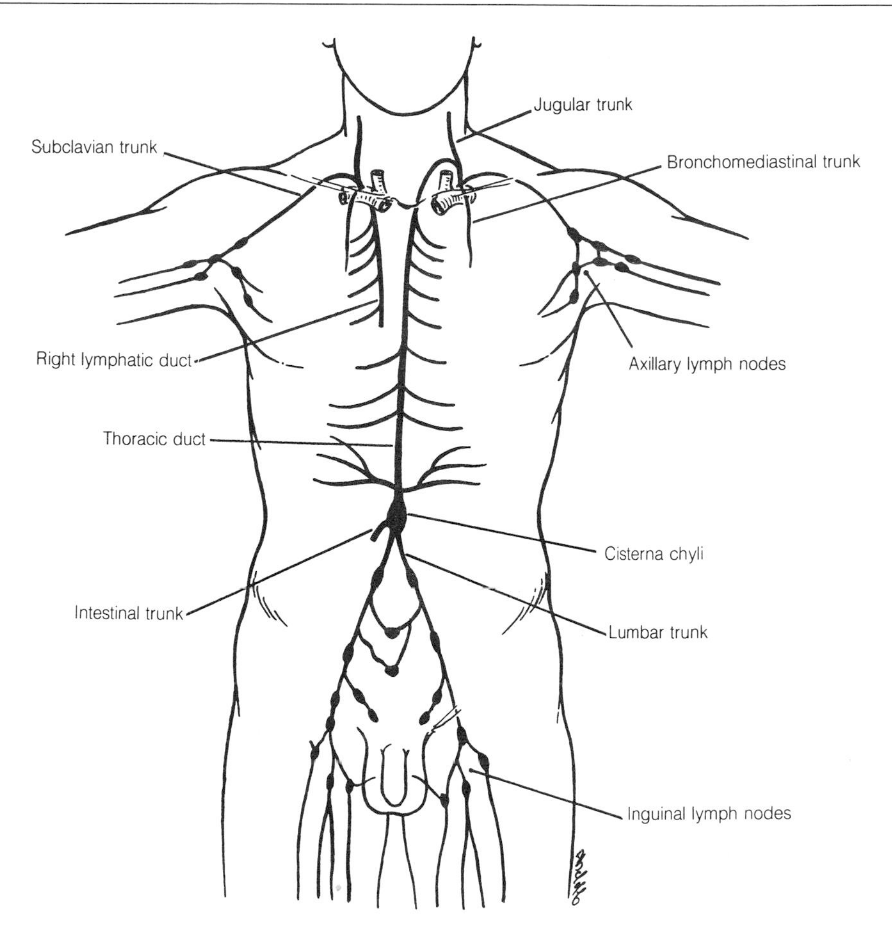

1-31 Diagram illustrating the main lymphatic vessels of the body.

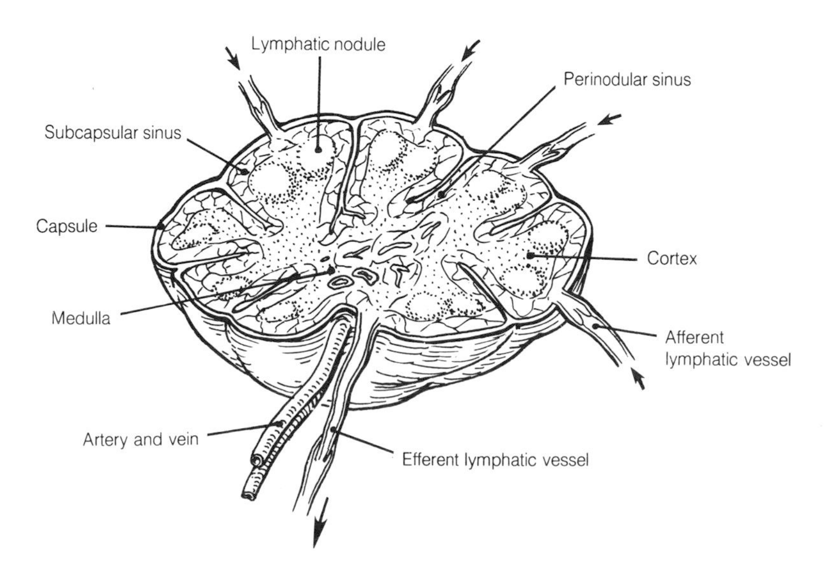

1-32 Showing the main features of a lymph node.

nodes at the groin can be traced within an hour or so. A similar technique is applied to the radiographic investigation of lymph vessels and nodes. A small quantity of dye is injected subcutaneously and after allowing it to be taken up by nearby lymphatics the area is exposed through a short incision. A suitable vessel, now demonstrated by the dye, is cannulated and a radiopaque contrast medium slowly introduced. In Fig. 1-33 a lymphangiogram obtained by this method shows a group of lymph nodes in the groin (inguinal nodes) to which fine lymphatic vessels can be traced from the lower limb. These are the afferent vessels; efferent vessels of larger caliber can be seen leav-

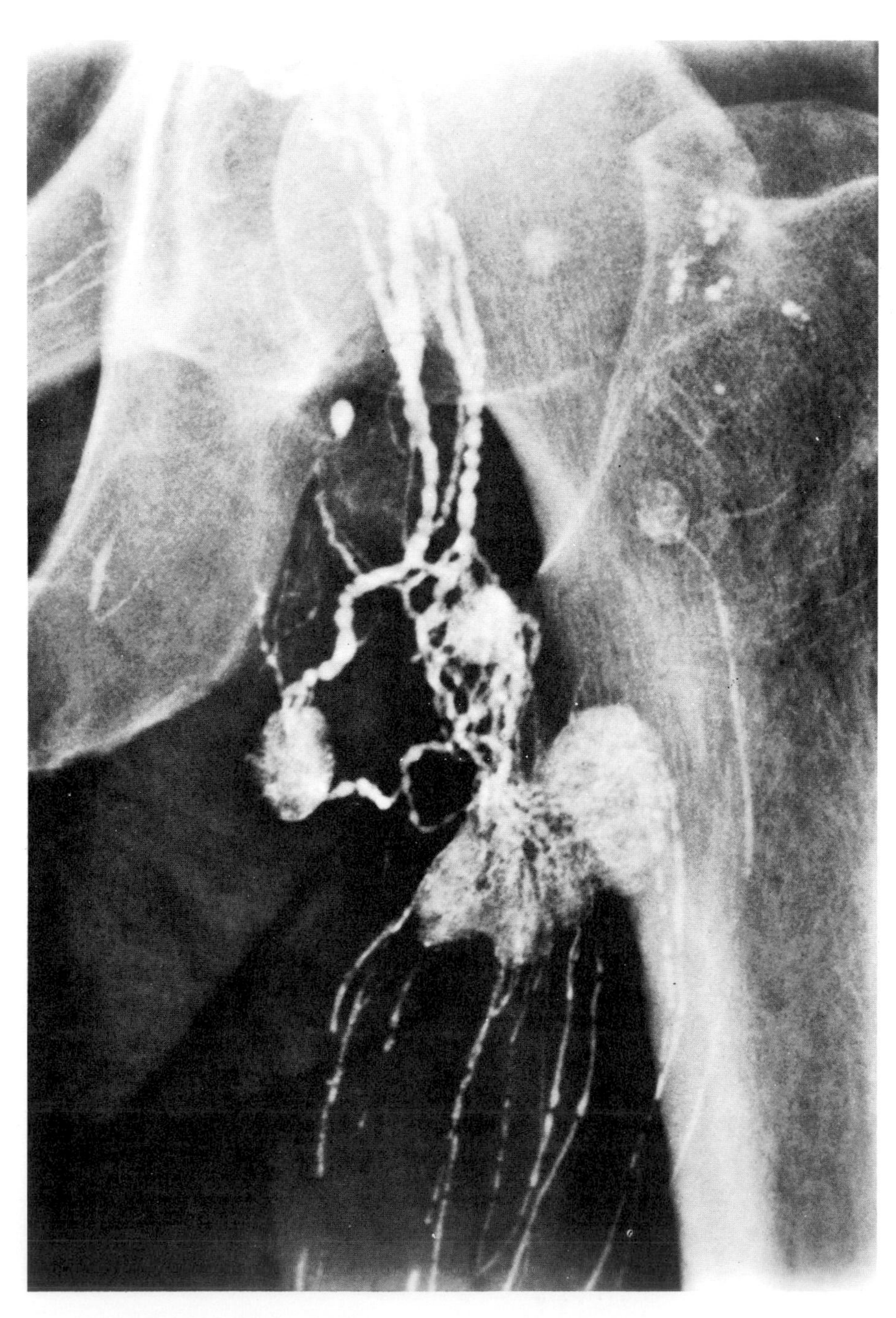

1-33 Lymphogram of the inguinal lymph nodes. (Reproduced by permission, from Wicke: *Atlas of Radiologic Anatomy,* 4th Ed, Urban & Schwarzenberg, Baltimore – Munich, 1987.)

ing the nodes to pass proximally. Note that the latter have a beaded appearance caused by many valves which are similar in structure to those of veins.

Before leaving the lymphatic system it should be noted that lymphatic nodules similar to those found in the substance of a lymph node are present in a number of other structures that can be rather loosely described as lymphoid tissues. These are the spleen, the thymus, the tonsils, and the small intestine where nodules are found alone or as aggregations (Peyer's patches). Apart from functional differences these groups of nodules are not associated with afferent lymphatics although with the exception of those of the spleen, they do, as most other organs, have a lymphatic drainage.

Radiological Anatomy

No clinical discipline illustrates the relevance of regional anatomy to the practice of medicine better than that of radiology. The wide range of cavities and vessels that can now be displayed by contrast media and the information provided by Computed Tomography (CT) scanning, sonography, and magnetic resonance imaging reinforce this relevance. To help in the interpretation of radiographs presented in this book a brief description of the manner in which they are made is now given.

The X-Ray Tube

X-rays are produced when a beam of electrons generated by a heated filament strike a tungsten target inside an evacuated glass tube. To obtain the necessary X-ray energy this system is placed between a cathode of which the filament is an integral part and an anode which is also the tungsten target. A very large potential difference is created between the cathode and anode by a high voltage circuit during the short time that the emission of X-rays is required. Much of the energy created is transformed into heat rather than X-rays and measures have to be taken to cool the tube. The tube is also surrounded by a lead shield in which there is a small window through which the X-rays to be used are allowed to pass. The main features of such an X-ray tube are shown in Fig. 1-34.

The X-Rays

It is fortunate that X-rays, when they strike a sheet of photographic emulsion, have a similar effect as a light; i. e., where the sheet of film is exposed to X-rays it turns black when developed.

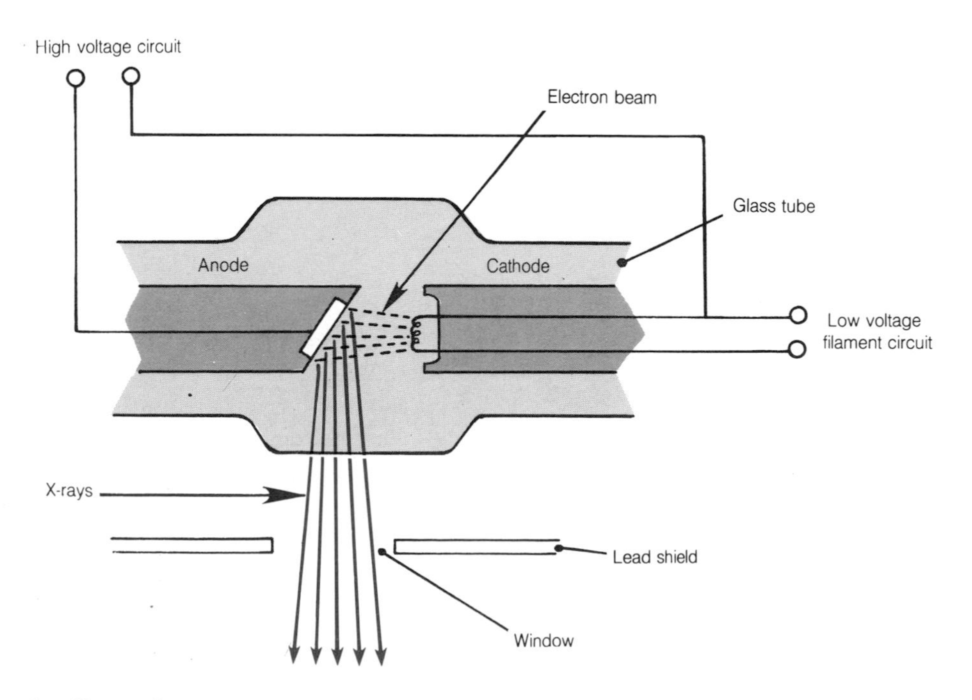

1-34 Diagram of an X-ray tube.

It is also fortunate that the tissues of the body are not all penetrated by X-rays with the same ease. For example gas and fat present little obstruction. They are said to be radiolucent and appear as dark structures on the negative. Bone on the other hand is radiopaque because of its calcium salts and appears as a light structure. Most soft tissues fall into an intermediate category. By choosing a suitable exposure time and filament and high tension voltage the gradation of a good X-ray film will range from bright white through intermediate shades to black.

The X-ray Image

To obtain an X-ray image or radiograph the subject is placed between the X-ray tube and a sheet of photographic emulsion or film. This is shown in Fig. 1-35. Note that the X-rays are emitted from a point source and diverge to form a cone. Because of this divergence the hand placed at some distance from the plate at level A will be grossly distorted in size while that close to the plate at B will show minimum distortion. Although this is an exaggeration of the effect it still exists to a lesser degree even when the subject is near the plate. When the anterior aspect of the subject is close to the plate the X-rays pass from his back to his front. For this reason the resultant image is called a posteroanterior or p. a. projection; the reverse orientation of the subject gives an a. p. projection. Lateral and oblique projections are also commonly used.

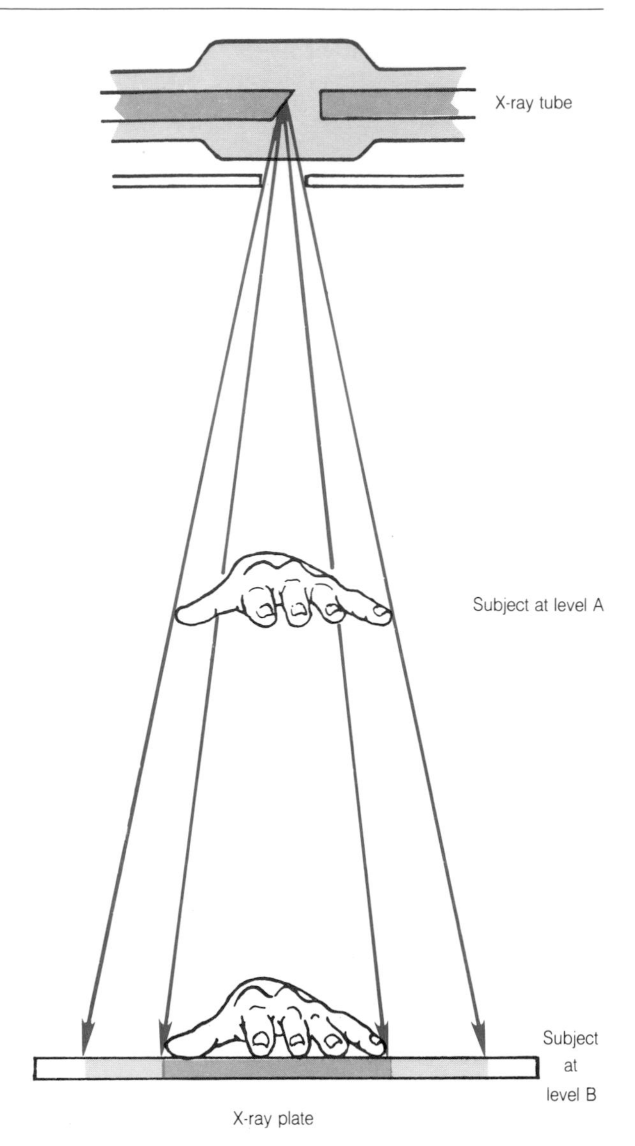

1-35 To show the effect of the position of the subject on X-ray image distortion.

The X-Ray Film

The X-rays that manage to pass through the subject "expose" the emulsion of the film which is contained in a light-tight cassette. In this, the film is sandwiched between two fluorescent intensifying screens and shielded at its back by a sheet of lead foil. After exposure the film is developed. It is this negative that is presented to the physician for examination and no positive print is normally made. As a result, as explained before, radiopaque structures such as bone appear white while radiolucent structures appear black.

Contrast Media

There are many ways in which the value of a radiograph can be enhanced or the information it provides increased. Of these perhaps the most effective is the use of contrast media. These are usually radiopaque fluids which can be introduced into hollow organs, patent vessels or ducts, and the arterial and venous systems. The resultant white image will outline the inner margins of such structures and thus, in the pathological organ, reveal surface irregularities, filling defects, and obstructions, or reveal abnormal positioning. Where adjacent structures are already radiopaque as in a joint cavity the contrast medium of choice may be radiolucent air. Examples of these techniques will be found in subsequent chapters.

Computed Tomography Images

X-rays are also used to produce the tomographic images which are now commonly referred to as CT scans. The X-ray source and its detector are rotated around the long axis of the trunk or of a limb and the information collected from this five to ten millimeter slice is stored in a computer and then assembled spatially to produce an image on a screen. As it is X-rays that are being used to construct this image, the gradations of densities produced by different tissues correspond to those found in a regular radiograph.

The interpretation of CT scans by the radiologist is based on a knowledge of regional anatomy and for the student of anatomy to rebuild them into a three-dimensional structure is a very useful exercise for review purposes.

Magnetic Resonance Images

Magnetic resonance (MR) images are obtained by placing the subject in a powerful magnetic field and then pulsing him with radiofrequency waves. These cause the nuclei of atoms to emit a radiosignal that can be detected, stored, and with the aid of a computer, reproduced as an image on a screen as in a CT scan.

One great advantage this technique has over the CT scan is that sagittal and coronal as well as cross-sectional images can be obtained. This ability is well demonstrated in Fig. 1-36 where an almost median section passes through the chambers of the heart and the thoracic and abdominal parts of the aorta. However, the nature of the image differs in that it depends upon the emission of radiofrequency waves rather than the transmission of X-rays. For example, it is found that dense cortical bone appears black, i. e. no signal is emitted, but, on certain sequences, fat of the marrow the bone may contain appears white. The absence of signals emitted from bone is also a major advantage of this technique as it allows the imaging of soft tissues such as the spinal cord which are surrounded by bone.

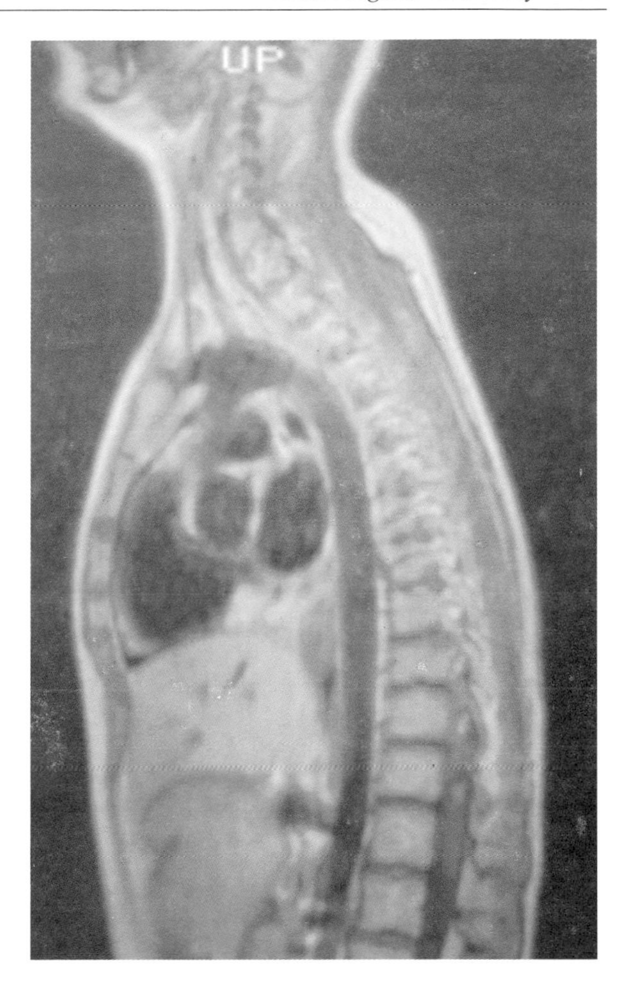

1-36 A magnetic resonance scan of the thorax and upper abdomen. (Courtesy of M. A. Hall-Craggs, B. A., M. R. C. P., F. R. C. R., and The Hospital For Sick Children, Great Ormond Street, London.)

Ultrasound Images

Ultrasound scanning uses high frequency sound waves of between 1 and 15 MHz to form an image. Energy is emitted from a transducer and is reflected, refracted, and absorbed by the transmitting region. The reflected sound is analyzed and a 2-dimensional image reconstructed. Real time ultrasound is most commonly used and involves the continuous display of sequential tomographic images, such that movement can be demonstrated.

Ultrasound is an extremely useful means of imaging soft tissues, for example those of the abdomen and pelvis. It does not produce ionization and is therefore of particular value in obstetric and pediatric practice. Gas and bone are poor transmitters of sound, hence the lungs, adult brain, and spinal cord cannot be visualized using this technique.

Anatomical Variations

Textbooks of anatomy normally describe the most common form in which structures are found in the body but after a short time spent in the anatomy laboratory it soon becomes clear that variations from the textbook description are frequent; in fact, when the complexity of structure at the cellular and macroscopic level is considered, it is surprising that they are not more so. The majority of these variations are perfectly compatible with a normal life and go unrecognized until discovered in the cadaver. Some, however, are sufficiently gross to interfere with function while others which present no symptoms may pose problems in even quite simple surgical procedures. As an example of this the ulnar artery running an abnormal and superficial course beneath the skin over the front of the elbow may be mistaken for the vein in this region from which samples of blood are commonly taken. In this book attention is drawn only to a small number of the variations that may be encountered. For a very much fuller listing and description, reference can be made to a recently published catalog.*

* *Catalog of Human Variation,* by R. A. Bergman, S. A. Thompson, A. K. Afifi. Urban & Schwarzenberg, Baltimore and Munich, 1984

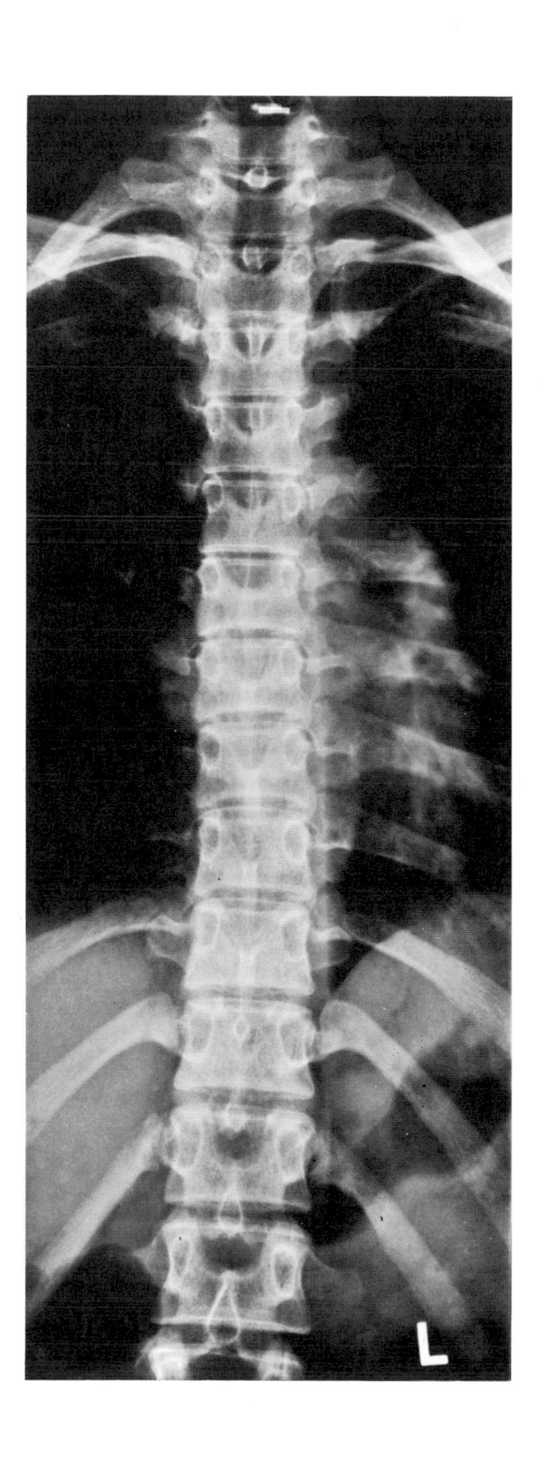

2 The Back and Spinal Cord

The anatomical region designated as the back includes the bony vertebral column and the spinal cord, and dorsal and ventral roots of spinal nerves which it encloses. Behind the vertebral column there is a large vertically oriented muscle mass that is used for its support and movement.

Dissection instructions may well refer to surface markings in planning incisions. These landmarks are not purely academic but will help in judging normal posture and thus form an important part of an examination of the back. Following on from these landmarks, the normal curvatures of the back should be appreciated in order that abnormal curvatures or deformities can be recognized. This often requires more than a superficial examination as the vertebral column tends to overcome any structural imbalance by forming a compensatory curve.

The human vertebral column has evolved in association with a habitual erect posture and a bipedal mode of locomotion and as such is unique in the animal kingdom. Instead of acting as a suspension bridge between fore and hindlimbs, it is required to support the weight of the body above the pelvic girdle and the progressively greater load born by its individual bones is reflected by their increasing size from the skull to the sacrum.

Although movements between the individual bones of the vertebral column are limited, the sum of these movements is considerable and leads to a mobile whole. It is only from a knowledge of the characteristics of the individual vertebrae, the joints by which they move, one upon another, the ligaments that stabilize and restrain them, and the muscles that act on them that these movements can be fully understood, nor can radiographs be properly interpreted without this information. Back pain, be it of muscular origin or that of a genuine prolapsed intervertebral disc, remains a perennial and largely unsolved clinical problem.

No detailed knowledge of the many anatomical parts of the deep muscles of the back is required and it should be realized that not all the muscles encountered in this region are true back muscles. A number of more superficial muscles have migrated from the upper limb, carrying their nerve supply with them. However, other muscles such as those of the anterior abdominal wall, though clearly not a part of the back, play an important role in its movements.

The surgical operation of laminectomy is performed in the anatomical laboratory to expose the spinal cord and its coverings. This vital structure is placed deep within a protective shield of muscle and bone. Despite this, injuries occur or tumors may compress the cord. The clinical signs of these conditions depend on the relationship of damaged spinal cord segments to areas of skin and groups of muscles supplied by the corresponding spinal nerves. The relationship of the cord to its coverings and the enclosed cerebrospinal fluid determines the optimum sites for the introduction of various forms of spinal anesthesia and, of course, suitable and safe levels at which to perform a spinal tap.

Surface Anatomy of the Back

Look at the drawing in Fig. 2-1. The subject's shoulders are level and the head is upright. The gluteal folds are balanced and a midline furrow over the spines of the lower thoracic and the lumbar vertebrae is vertical. By palpation of the tips of the spines, both in this position and when the back is bent forward, their alignment in a straight line can be confirmed and the posture considered normal.

Note that the most lateral palpable bony point of the shoulder is formed by the greater tubercle of the humerus. Above this and slightly nearer the midline, the lateral margin of the acromion can be felt. If the subject is not too muscular the spine and medial border of the scapula can be palpated and the latter followed down to the inferior angle.

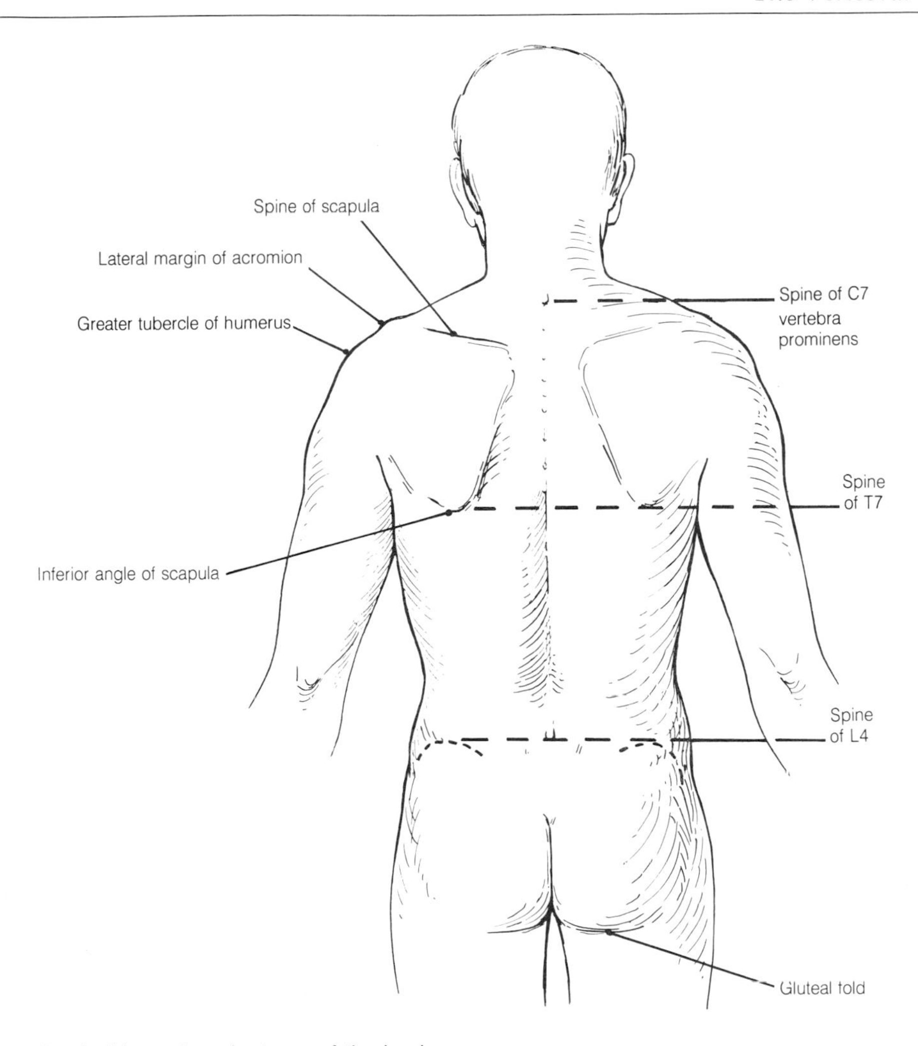

2-1 Visible and palpable surface features of the back.

This lies approximately at the level of the spine of the seventh thoracic vertebra. Vertebral spines can be identified by running the fingers down the back of the neck until they are arrested by the prominent tip of the spine of the seventh cervical vertebra. In the lower part of the trunk the crests of the iliac bones can be felt. A line joining their highest points will pass across the spine of the fourth lumbar vertebra. This is a useful landmark when a spinal tap is to be performed.

The Vertebral Column

Composition

The vertebral column forms part of the axial skeleton and supports the skull, the thoracic cage, and indirectly the pectoral girdle. The pelvic girdle is firmly anchored to the sacral portion of the column. The vertebral column is composed of 33 **vertebrae**. The upper seven are named **cervical vertebrae**, the next 12 are **thoracic vertebrae** and bear **ribs**, and below these are the **lumbar vertebrae**. The fifth lumbar vertebra is followed by five fused vertebrae which form the **sacrum**. A variable number (3-5) of fused vestigial caudal* or **coccygeal vertebrae** complete the vertebral column. The general shape of the column and the arrangement of the vertebrae can be seen in Fig. 2-2. Abnormal transitional vertebrae are occasionally encountered in the thoracolumbar region and in the lumbosacral region where, for example, a fifth lumbar vertebra may be partially fused to the sacrum. Wedge-shaped hemivertebrae may also be seen and these lead to an abnormal lateral curvature or **scoliosis.**

* Cauda, *L.* = tail

Curvatures

Fig. 2-2 B shows a lateral view of the vertebral column and it can be seen that it has a sinuous shape. Before birth the vertebral column has a single primary curvature which is concave anteriorly. By the time of birth an angle is present between the lumbar spine and sacrum and, soon after, a cervical curvature with an anterior convexity begins to appear. During the second year a similar curvature appears in the lumbar region. The cervical and lumbar curvatures are termed secondary and are superimposed on the primary curvature which is retained in the thoracic and sacral regions. The

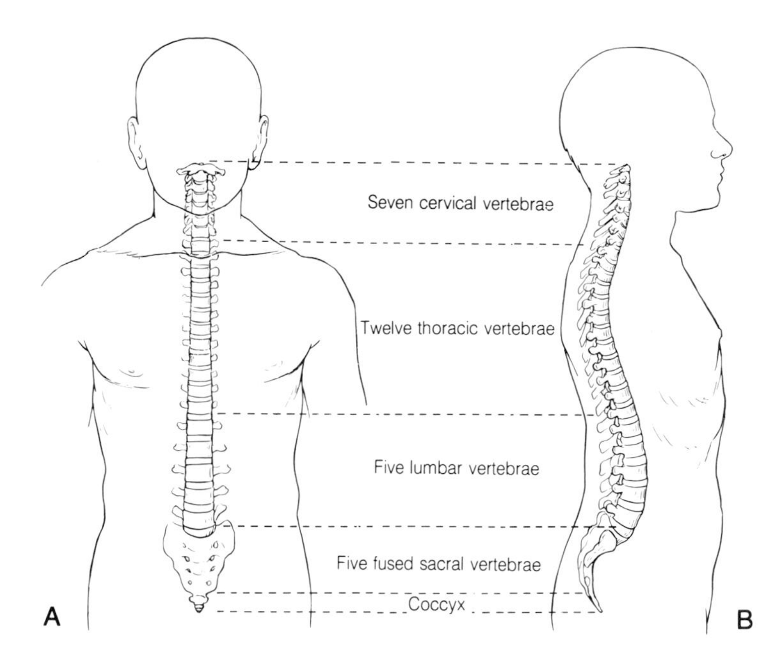

2-2 Showing anterior (*A*) and lateral (*B*) views of the vertebral column.

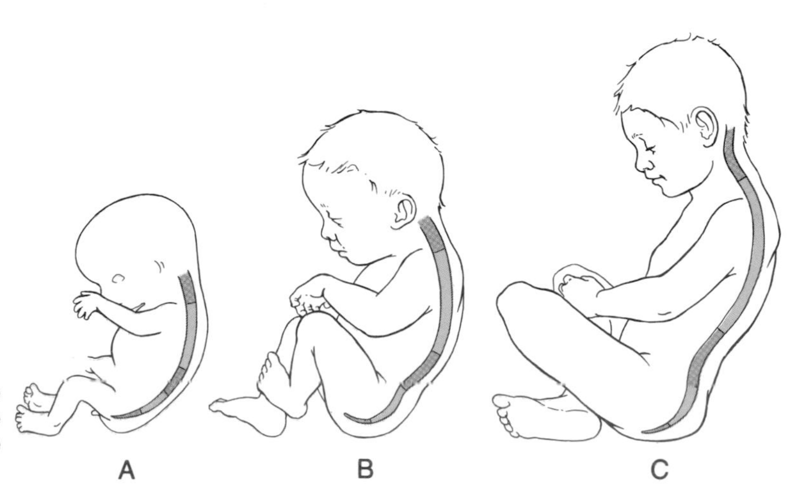

2-3 Showing the development of the normal curvatures of the spine from the fetus (*A*) through the newborn (*B*) to the child (*C*).

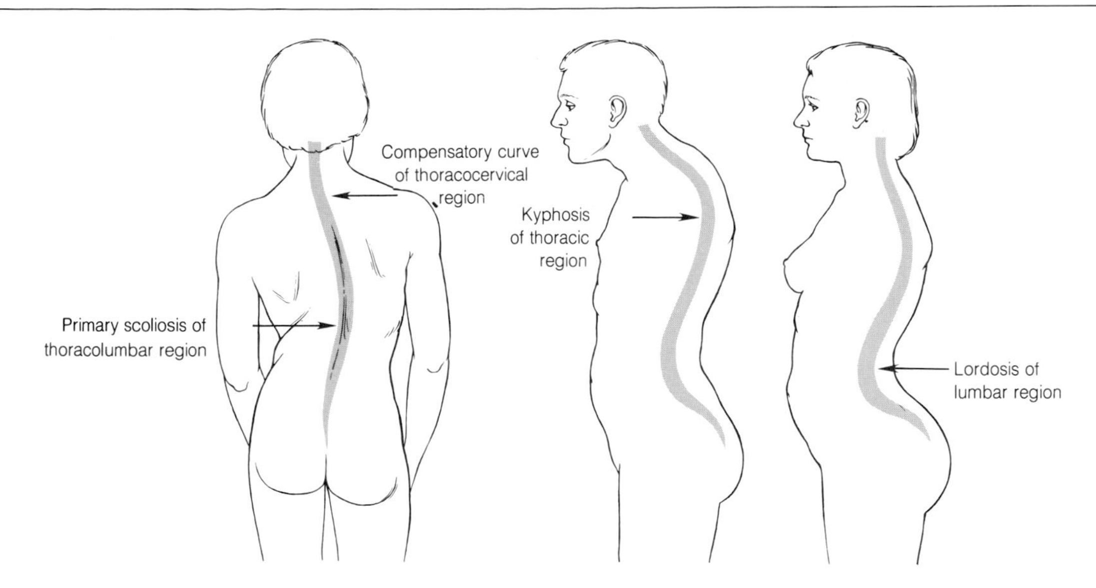

2-4 Illustrating abnormal curvatures of the vertebral column.

development of these curvatures, which are the result of differences in the anterior and posterior depths of both vertebrae and intervening discs, is illustrated in Fig. 2-3.

Disease of vertebrae, the paralysis of back muscles, or even physiological events such as pregnancy may lead to abnormal curvatures. These are illustrated in Fig. 2-4 where it can be seen that a lateral deviation is called a **scoliosis**, an increased posterior convexity a **kyphosis**,* and an increased anterior convexity a **lordosis**. Note in the illustration of a scoliosis that a compensatory curve has been formed above the primary deformity to restore the orientation of the head and shoulders. For geometrical reasons a scoliosis and its compensatory curve are always associated with some rotation of the vertebral column. This can easily be identified in the thoracic region as it is magnified by the attached ribs.

The Individual Vertebrae

The **thoracic vertebrae** show all the features of a typical vertebra and will be described first. Vertebrae from other levels will be seen to show modifications of this basic pattern. Look at Fig. 2-5 which shows a superior and lateral view of a thoracic vertebra. Note the moderately stout, heart-shaped **body** to which are applied two posteriorly projecting **pedicles**. These in turn support a pair of **laminae** which, by uniting posteriorly, form with the pedicles a **vertebral arch.** The body, pedicles, and laminae enclose the **vertebral foramen.** Superimposed on this basic structure are two laterally projecting **transverse pro-**

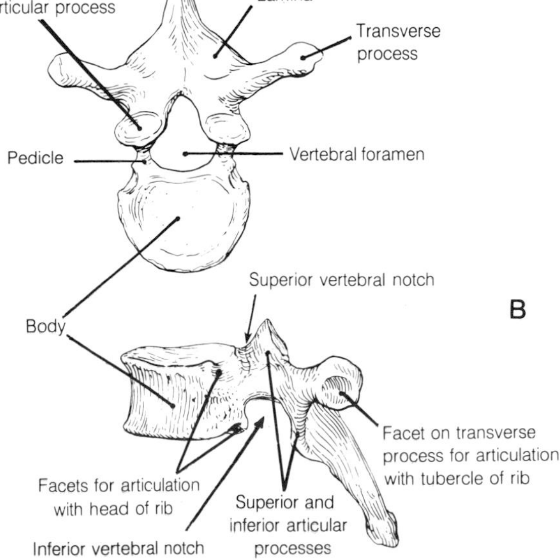

2-5 A typical thoracic vertebra viewed from above (*A*) and from the side (*B*).

* Kyphosis, *Gk.* = hunchback

cesses and a posterior midline **spinous process**. Superior and inferior **articular processes** project from the junctions of the pedicles with the laminae. They bear articular facets that take part in synovial joints which unite adjacent vertebral arches. Facets at the upper and lower borders of the body and facets on the transverse processes are for articulation with ribs. Fig. 2-6 shows how a typical rib articulates with the upper border of the body of the vertebra of corresponding number and the lower border of the body of the vertebra **above**. Note that the lower and to a lesser extent the upper borders of the pedicles are notched. When the vertebrae are articulated these **superior and inferior vertebral notches** lie opposite each other and form **intervertebral foramina** for the passage of spinal nerves. It is here that the spinal nerve may be irritated by new bone laid down in osteoarthritis.

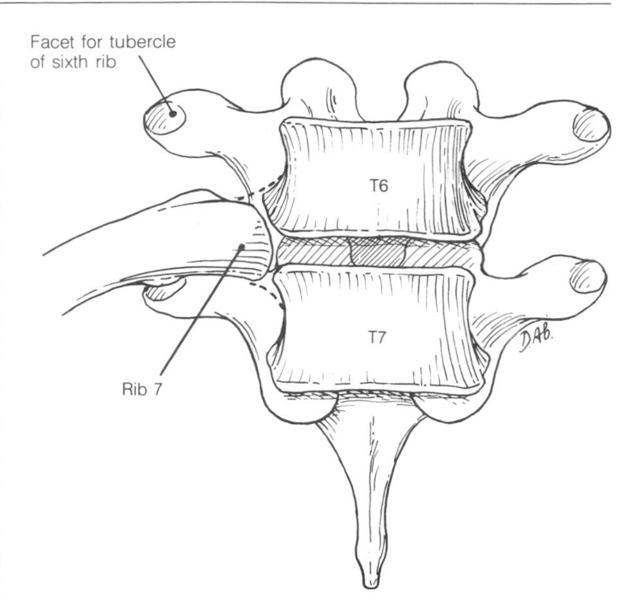

2-6 Showing the manner in which a rib articulates with two adjacent vertebrae.

This description holds good for the majority of the thoracic vertebrae but a few particular features should be noted. The first thoracic vertebra articulates with the whole of the head of the first rib and the upper part of the second and as a result its body has a complete facet at its upper border and a demifacet at its lower border. The tenth to twelfth vertebrae usually only have complete facets on their bodies for articulation with the tenth to twelfth ribs. The lower two pairs of transverse processes lack facets for ribs. The tapering spines slant downward although that of the twelfth vertebra is more horizontal and quadrangular like the spine of the first lumbar vertebra. Look carefully at the orientation of the articular processes. The superior ones face backward and outward and the inferior ones forward and inward. The joint space between each pair, therefore, lies on a circle drawn about a point within the body. This arrangement becomes of importance when considering movements of the vertebral column.

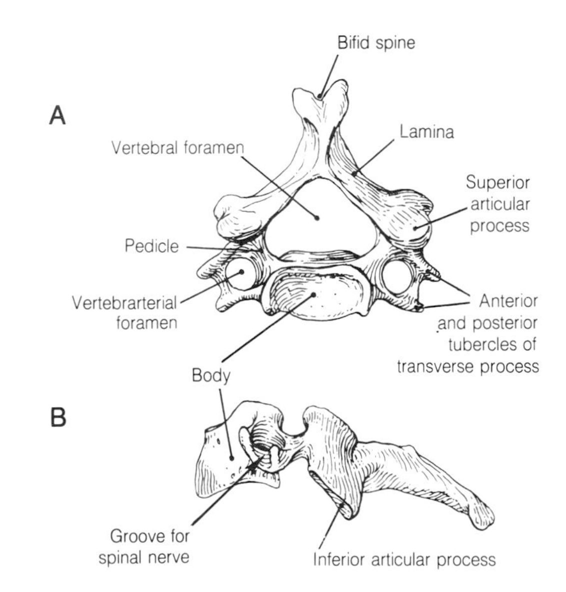

2-7 A typical cervical vertebra viewed from above (*A*) and from the side (*B*).

Of the **cervical vertebrae** the first two are highly modified and require a separate description. The vertebra illustrated in Fig. 2-7 is typical of the remainder. Compare this with the thoracic vertebra and note that the body is smaller and rather quadrangular, the spine is bifid, and the articular processes face upward and backward or downward and forward. The most striking feature is the **vertebrarterial foramen** in the transverse process which carries the vertebral vessels. In fact the anterior and lateral parts of the transverse processes are homologous with ribs and from Fig. 2-8 it can be understood how this foramen has evolved. An anterior and posterior tubercle can be seen at the extremity of the transverse process and on the upper surface of the transverse process, between the tubercles, there is a groove for

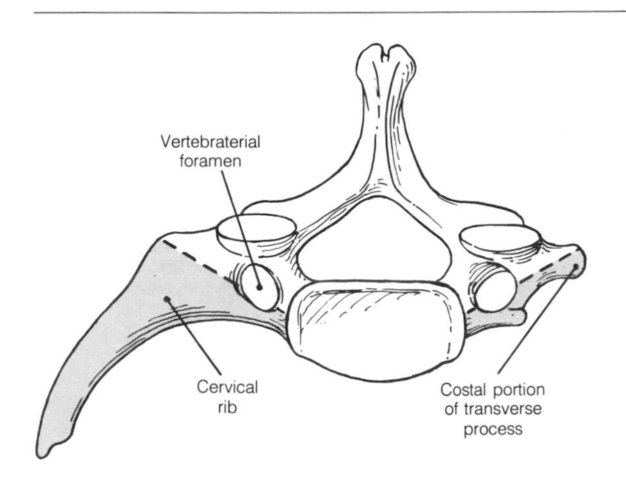

2-8 Showing how a vertebrarterial foramen is formed between costal and true portions of a cervical transverse process.

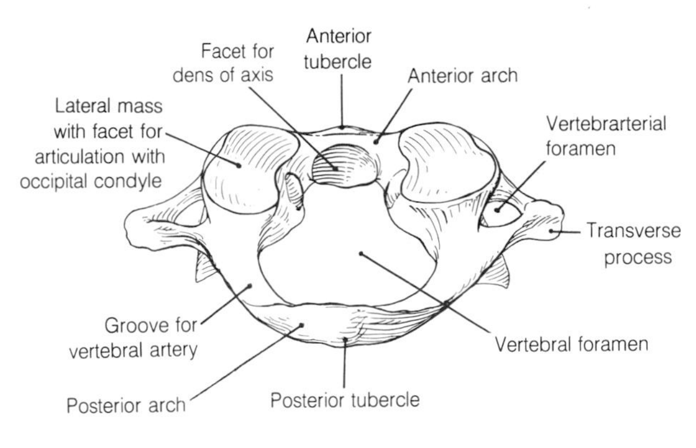

2-9 The atlas.

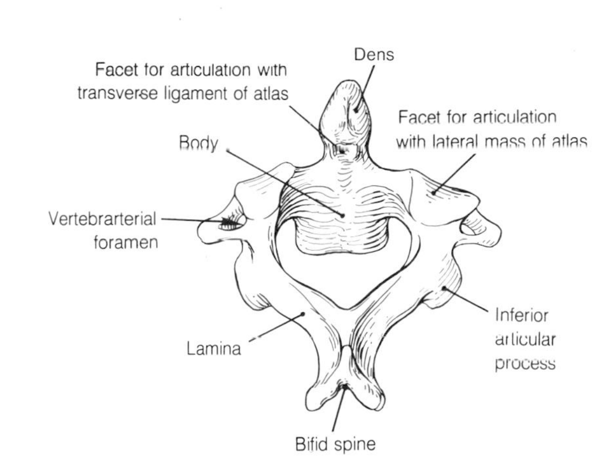

2-10 The axis.

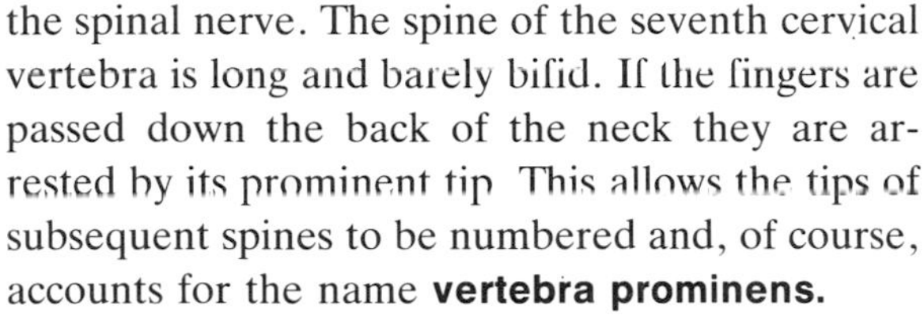

the spinal nerve. The spine of the seventh cervical vertebra is long and barely bifid. If the fingers are passed down the back of the neck they are arrested by its prominent tip. This allows the tips of subsequent spines to be numbered and, of course, accounts for the name **vertebra prominens.**

The **first cervical vertebra** or **atlas** (the mythical giant who supports the pillars of heaven) has no body. It is illustrated in Fig. 2-9: Observe the two **lateral masses** which articulate with the occipital condyles of the skull above and with a pair of complimentary facets on the **axis** below. The hourglass-shaped **superior articular surfaces** are aligned on a nearly anteroposterior axis and allow the movement of nodding. The lateral masses are linked in front of and behind the vertebral foramen by an **anterior** and a **posterior arch.** There is no spine but note the small **posterior tubercle** on the posterior arch and a larger **anterior tubercle** on the anterior arch. The upper surface of the posterior arch has a deep groove laterally which accommodates the vertebral artery. The length of the transverse processes makes the atlas the widest of the cervical vertebrae. In the living subject their tips can be felt just below and in front of the mastoid processes.

The **second cervical vertebra** or **axis** displays a stout, tooth-like process called the **dens*** which projects upward behind the anterior arch of the atlas with which it articulates. The dens is in fact the body (or more strictly the centrum) of the atlas. In Fig. 2-10 an additional articular facet can be seen on its posterior surface. This articulates through means of a bursa with a transverse ligament passing between the two lateral masses of the atlas. Thus it is about this peg that the atlas and skull rotate. The articular facets for the lateral masses of the atlas can be seen lying in front of the groove for the spinal nerve. Below, typical inferior articular processes lying behind the vertebral notches articulate with the third cervical vertebra.

* Dens, *L.* = a tooth

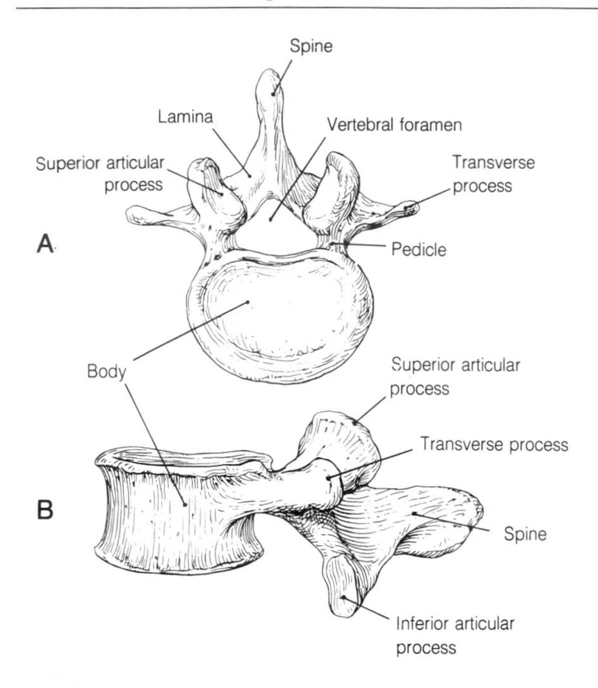

2-11 A typical lumbar vertebra seen from above (*A*) and from the side (*B*).

The **lumbar vertebrae** have large kidney-shaped bodies which can be seen in Fig. 2-11. The thickened spines project posteriorly and are quadrangular in shape. The transverse processes, especially those of the fifth vertebra, are also sturdy. Note again the orientation of the articular processes. The upper ones face medially and slightly posteriorly and the lower ones laterally and slightly anteriorly.

The five **sacral vertebrae** are fused into a solid bony mass and have no intervertebral discs between them. In Fig. 2-12 this fusion can be seen to involve not only the bodies but the transverse processes as well. As a result, **anterior and posterior sacral foramina** are present for the passage of ventral and dorsal rami of the spinal nerves. The upper four spines are fused and appear as an irregular **median sacral crest.** The spine and laminae of the fifth sacral vertebra are wanting and leave a deficiency called the **sacral hiatus.** It is at this site that epidural anesthetics may be introduced. The anterior surface of the body of the first sacral vertebra forms a well marked angle with that of the fifth lumbar vertebra and its upper lip is named the **sacral promontory.** On the upper part of the lateral surface an ear-shaped or **auricular surface** can be seen. This forms part of the **sacroiliac joint**

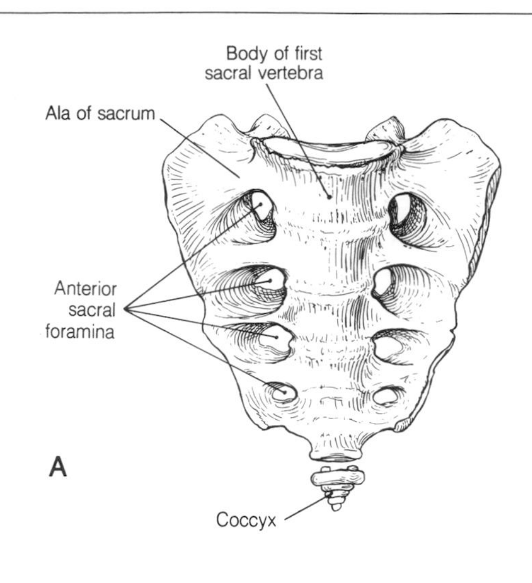

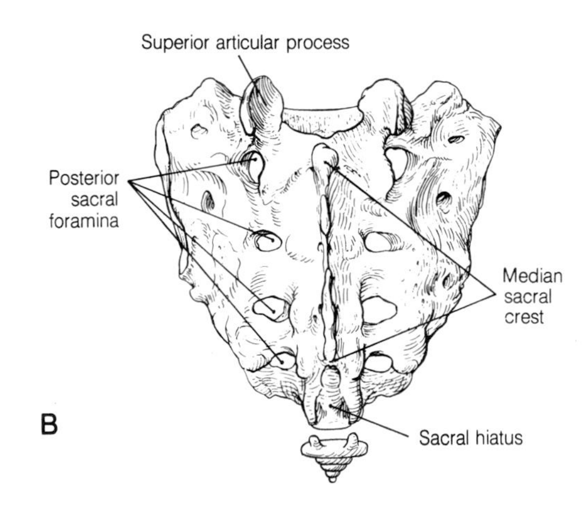

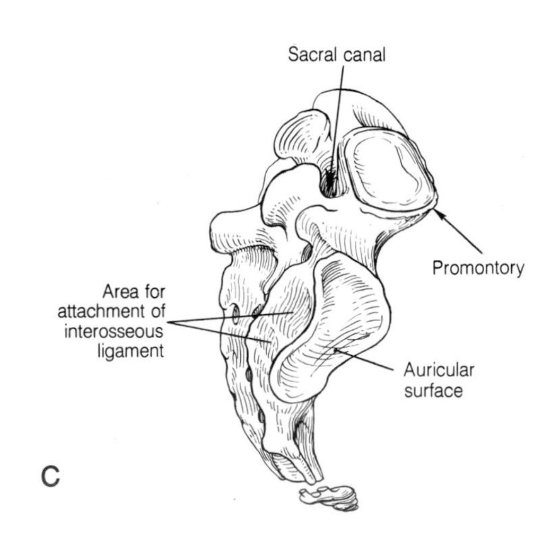

2-12 The anterior (*A*), posterior (*B*), and superolateral (*C*) aspects of the sacrum.

which, together with strong interosseous ligaments attached to the irregular surface posterior to the joint, links the sacrum to the pelvic bone.

The Ossification of Vertebrae

A typical vertebra undergoes endochondrial ossification from three primary centers. One of these is for the body and there is one for each half of the vertebral arch. Those for the arches appear first in the upper part of the vertebral column at about the 10th week of intrauterine life, and then at successively lower levels. Those for the bodies first appear in the lower thoracic vertebra also at about the 10th week, and subsequently in vertebrae above and below this level. Secondary centers for the tips of the transverse processes and spine and the upper and lower surfaces of the bodies appear at puberty. Failure of the centers for the arch to fuse is associated with the developmental abnormality known as spina bifida.

The Joints of the Vertebral Column

From what has been said so far it should be apparent that the vertebral column is a mobile structure. The loss of this mobility is a distressing disability and can be well seen in the disease ankylosing spondylitis.* In this the bodies of the vertebrae become fused together giving a radiological picture often referred to as bamboo spine. Mobility of the column is provided by the joints between the vertebrae, and these are of two types. Those between the bodies are secondary cartilaginous joints and those between the articular facets on the vertebral arches are synovial joints.

Joints Between the Vertebral Bodies

The most important structures joining the bodies of the vertebrae are the **intervertebral discs.** As can be seen in Fig. 2-13, the shape of the disc corresponds to the margins of adjacent vertebrae,

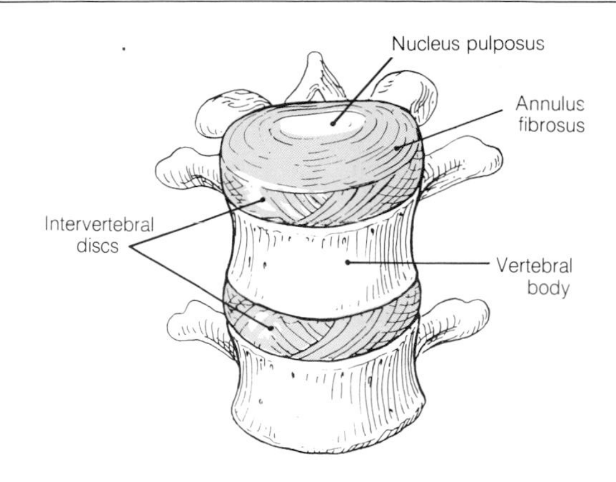

2-13 Intervertebral disc between vertebral bodies.

and differences in anterior and posterior depth contribute to the normal spinal curvatures. Between the disc and the vertebral body lies a thin layer of hyaline cartilage. When viewed in section as in Fig. 2-14, the disc can be seen to consist of an outer ring composed of concentric layers of fibrous tissue and fibrocartilage called the **annulus fibrosus** and a soft core, the **nucleus pulposus,** which lies nearer the posterior than the anterior border. The design of intervertebral discs allows them to provide a strong union between the vertebral bodies and in addition, by distortion, to absorb compression forces and allow one vertebra to rock and rotate on another. Degenerative changes in the disc may be followed by prolapse or herniation of the nucleus pulposus. The anatomy of the neurological condition to which this gives rise will be mentioned after the spinal cord and spinal nerves have been described.

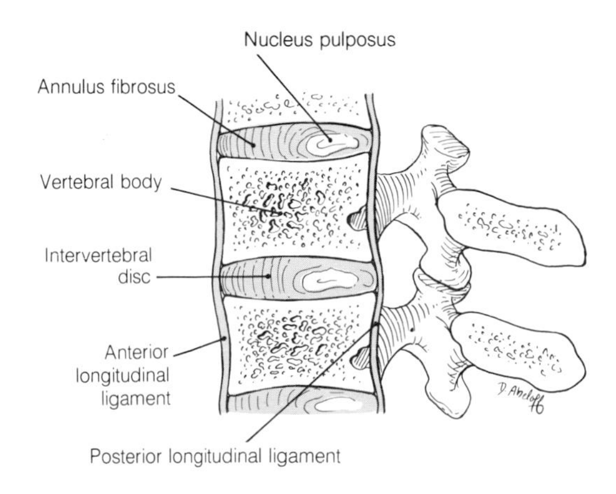

2-14 Median section of intervertebral discs and vertebrae.

* Ankylos, *Gk.* = crooked (diseased joints more often become fixed in a crooked or bent position); Spondylos, *Gk.* = vertebra

The vertebrae and discs are joined together by the **anterior** and **posterior longitudinal ligaments,** which can be seen in Fig. 2-14. The anterior ligament is broad and is attached to the anterior surfaces of the bodies and discs. The posterior ligament is narrower, and is attached to the upper and lower margins of the bodies but expands over its attachment to the discs. Note that it cannot be seen until the vertebral arches and the spinal cord are removed.

Joints Between the Vertebral Arches

The joints between the articular processes on the vertebral arches are small plane synovial joints in which one articular facet glides on the other. As well as by a thin articular capsule these joints are supported by more distant ligaments between the laminae, the spines, and the transverse processes. In Fig. 2-15 identify the **ligamenta flava**,* the **intertransverse ligaments**, and the **supraspinous ligaments**. **Interspinous ligaments** lie between the spines deep to the supraspinous ligaments. Note that the ligamenta flava between the laminae of two adjacent vertebra arise from the deep surface of the upper lamina and run to the upper margin of the lamina below. These ligaments have a yellowish tinge in life due to their content of elastic tissue.

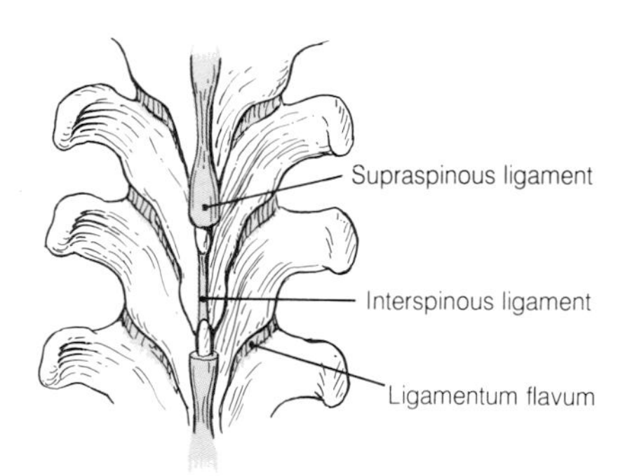

2-15 Showing the ligaments uniting the laminae and spines of vertebrae.

Having studied the individual vertebrae and the joints between them, their appearance in radiographs of the vertebral column can now be interpreted.

* Flavus, *L.* = yellow

Radiological Appearances of the Vertebral Column

The various regions of the vertebral column are frequently radiographed. Clinically these regions are known as the cervical, thoracic, and lumbar spines. In Fig. 2-16 the cervical spine is seen in a lateral projection. Note:

1. its regular but marked anterior convexity,
2. the even intervertebral spaces,
3. that no intervertebral space is seen between the first and second vertebrae; it is obscured by the dens *(D)* extending upward from the body of the axis,
4. the joint space (indicated by *arrowheads*) between the inferior articular process of

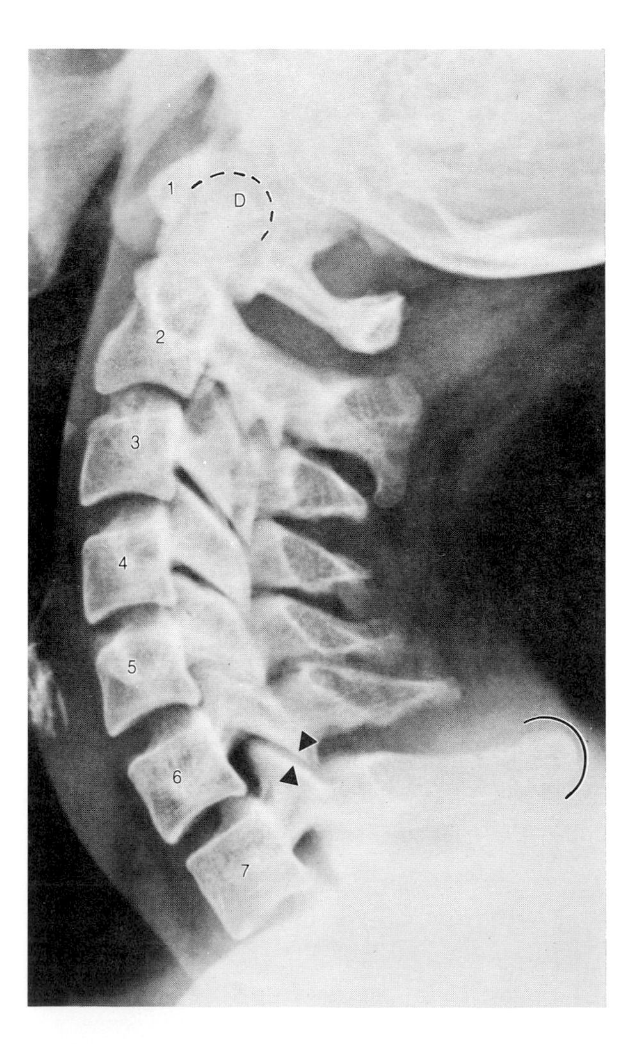

2-16 Radiograph of the cervical spine (lateral projection). (Reproduced by permission, from Wicke: *Atlas of Radiologic Anatomy,* 4th Ed, Urban & Schwarzenberg, Baltimore-Munich, 1987.)

the sixth vertebra and the superior articular process of the seventh, and
5. the long and prominent spinous process of the seventh cervical vertebra.

Fig. 2-17 is an anteroposterior projection of the thoracic spine. Note:
1. the head and tubercle of the first rib articulating with the first thoracic vertebra,
2. the midline translucency over the upper vertebrae produced by air in the trachea,
3. the midline shadow produced by a spinous process *(S)* flanked by shadows of the pedicles *(P)*,
4. the head of the eighth rib articulating with the bodies of the seventh and eighth vertebrae, and
5. a typical transverse process *(T)*.

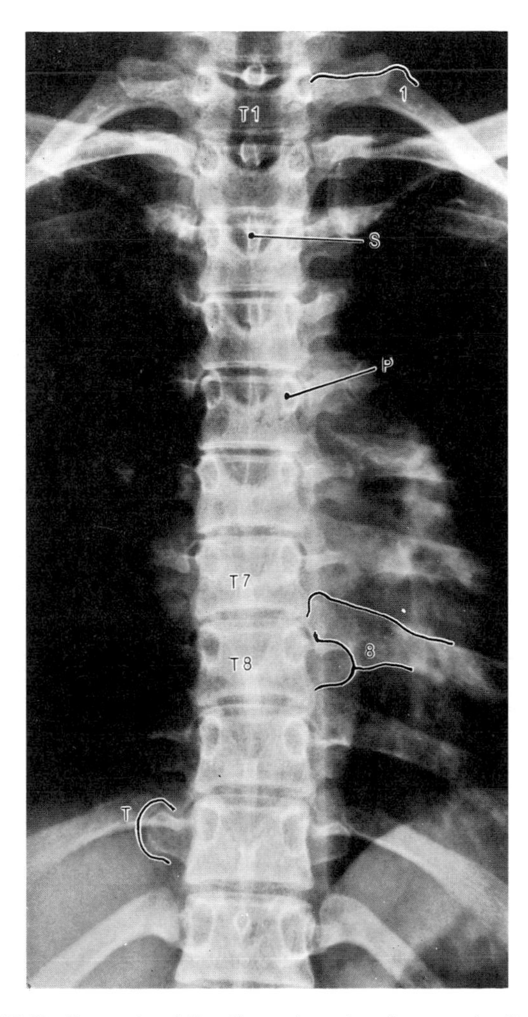

2-17 Radiograph of the thoracic spine (a.p. projection). (Reproduced by permission, from Wicke: *Atlas of Radiologic Anatomy,* 4th Ed, Urban & Schwarzenberg, Baltimore-Munich, 1987.)

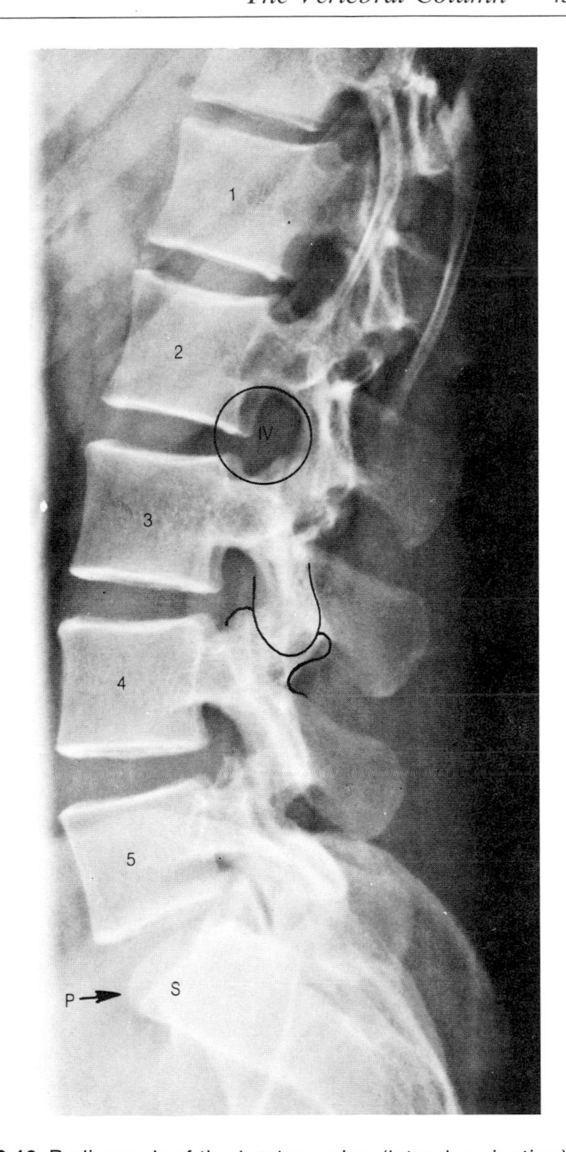

2-18 Radiograph of the lumbar spine (lateral projection). (Reproduced by permission, from Wicke: *Atlas of Radiologic Anatomy,* 4th Ed, Urban & Schwarzenberg, Baltimore-Munich, 1987.)

Fig. 2-18 shows the lumbar spine in a lateral view. Note:
1. the anterior convexity,
2. the clearly marked intervertebral spaces occupied by radiolucent intervertebral discs,
3. the almost square outlines of the bodies of the vertebrae,
4. an intervertebral foramen *(IV)*,
5. the overlapping superior and inferior articular processes (indicated between the third and fourth vertebrae),

6. the deep and rather quadrangular spinous processes, and
7. the lower border of the fifth vertebra and the upper border of the sacrum *(S)* forming the promontary *(P)*.

Fig. 2-19 shows the lumbar spine in an oblique view. Note the outline of a "Scottie dog" within the dotted line. This is produced by adjacent articular processes. In fracture dislocations of the lumbar spine this outline is lost.

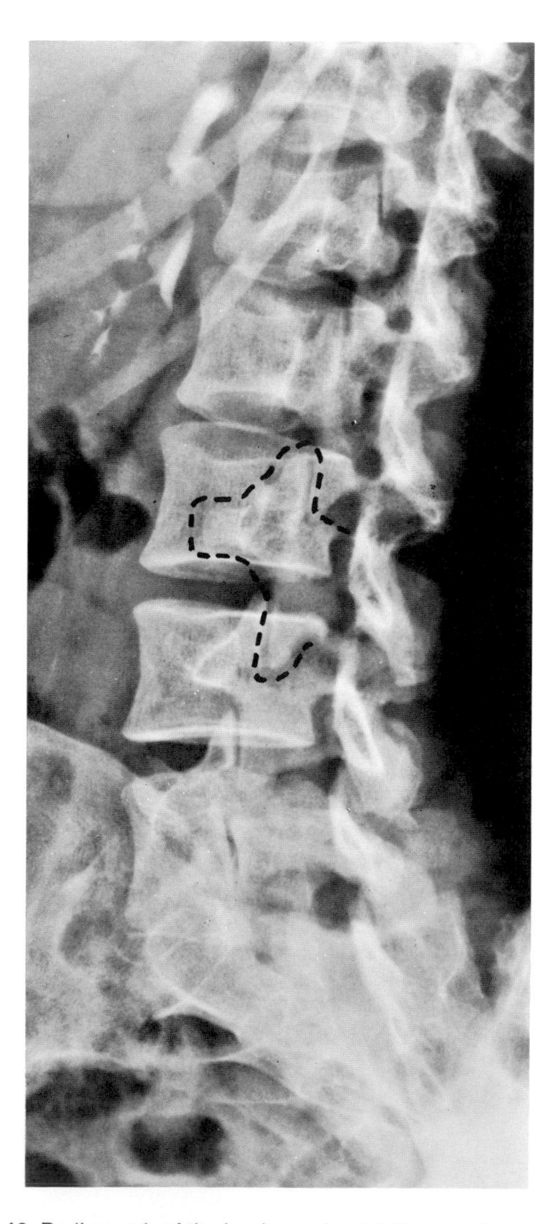

2-19 Radiograph of the lumbar spine (oblique projection). (Reproduced by permission, from Wicke: *Atlas of Radiologic Anatomy,* 4th Ed, Urban & Schwarzenberg, Baltimore-Munich, 1987.).

Movements of the Vertebral Column

The point has already been made that small movements occurring between a number of vertebrae can together produce an appreciable total range. The movements possible are flexion (bending forward), extension, lateral flexion (bending to the left or right), circumduction, and rotation about a vertical axis. Compression and distortion of the intervertebral discs allow all these movements between the bodies of the vertebrae but their extent is limited by the orientation of the joints between the vertebral arches. Flexion, extension, and lateral flexion are free in the cervical region but the articular processes which face anteriorly or posteriorly limit rotation. The exception is, of course, the specialized atlantoaxial joint at which the head and atlas rotate upon the axis. In the thoracic region all movements are possible, although to a limited degree. Rotation, suprisingly enough, is free and made possible by the alignment of articular processes on an arc about the axis of rotation. Flexion, extension, and lateral flexion all occur in the lumbar spine, but rotation is limited by the inwardly facing processes.

These movements are brought about by the deep muscles of the back and indirectly by muscles attached to the skull, thoracic cage, and the limb girdles. Extension is performed by the erector spinae muscle. Flexion and lateral flexion, when not produced by gravity against the controlled release of extensors, are performed by muscles of the neck and anterior abdominal wall. Rotation is produced by the action of short, deep back muscles running obliquely from transverse processes to spines of adjacent vertebrae, and again by muscles of the left or right anterior abdominal wall.

The Deep Muscles of the Back

The title of this section suggests that there are also superficial muscles of the back and this is so. However, these which include trapezius, latissimus dorsi, levator scapulae, and the rhomboid muscles are muscles of the upper limb whose proximal attachments have migrated to the spines of the vertebrae. When they are reflected in a dissection, they are found to be separated by a layer of

thoracolumbar fascia from the erector spinae muscle.

The Erector Spinae Muscle

Except in the region of the suboccipital triangle, which is situated between the occipital bone of the skull and the second cervical vertebra, the component parts of the **erector spinae muscle** are extremely difficult to define by dissection, and individual actions cannot be assigned to them. As a result the description of this muscle is confined to some general and readily understood features.

The left and right portions of the muscle lie in a trough on either side of the vertebral spines. The trough is limited anteriorly by transverse processes and in the thoracic region by the portions of ribs medial to their angles. The remainder of the muscle is enclosed by the **thoracolumbar fascia.** This arrangement is illustrated in Fig. 2-20.

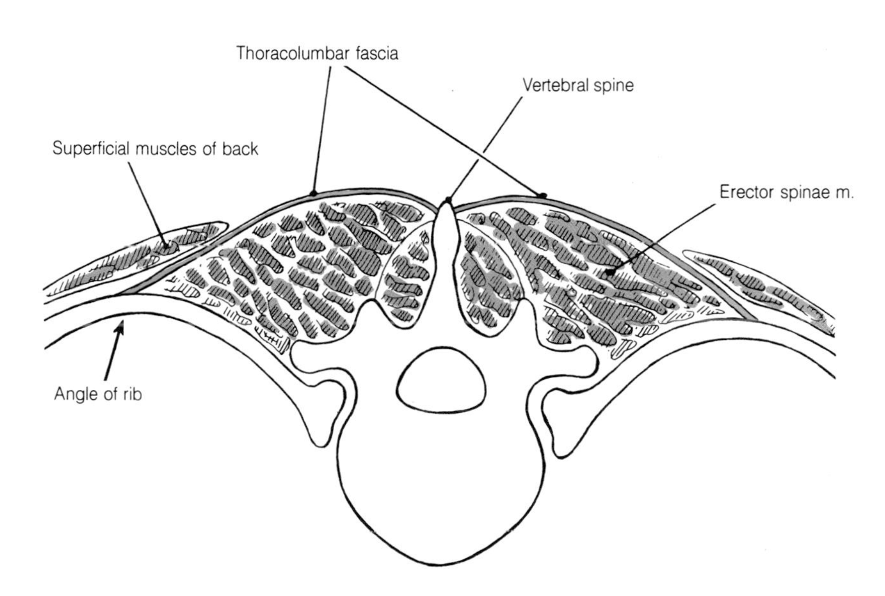

2-20 Showing the confines of the deep muscles of the back.

The muscle mass itself consists of long musculotendinous slips spanning many vertebrae. It can be divided into three portions: **iliocostalis, longissimus**, and **spinalis**. The most lateral portion, iliocostalis, arises from the sacrum, lumbar vertebral spines, and the posterior part of the iliac crest and is reinforced by further fibers arising from vertebral spines at higher levels. From this origin, laterally running bundles of fibers extend to ribs and cervical transverse processes. Intermediate bundles called the lingissimus muscle, reach ribs or transverse processes in the thoracic and cervical regions, with the highest bundles, called longissimus capitis, reaching the mastoid process of the skull. The most medial portion of the erector spinae is called spinalis. It is most evident in the thoracic region where it forms bundles of fibers spanning the gaps between varying numbers of the spinous processes. For reference purposes these muscles and their parts are tabulated below and are illustrated in Fig. 2-21.

Iliocostalis	Longissimus
-lumborum	-thoracis
-thoracis	-cervicis
-cervicis	-capitis
Spinalis	
-thoracis	
-cervicis	inconstant as
-capitis	discrete muscles

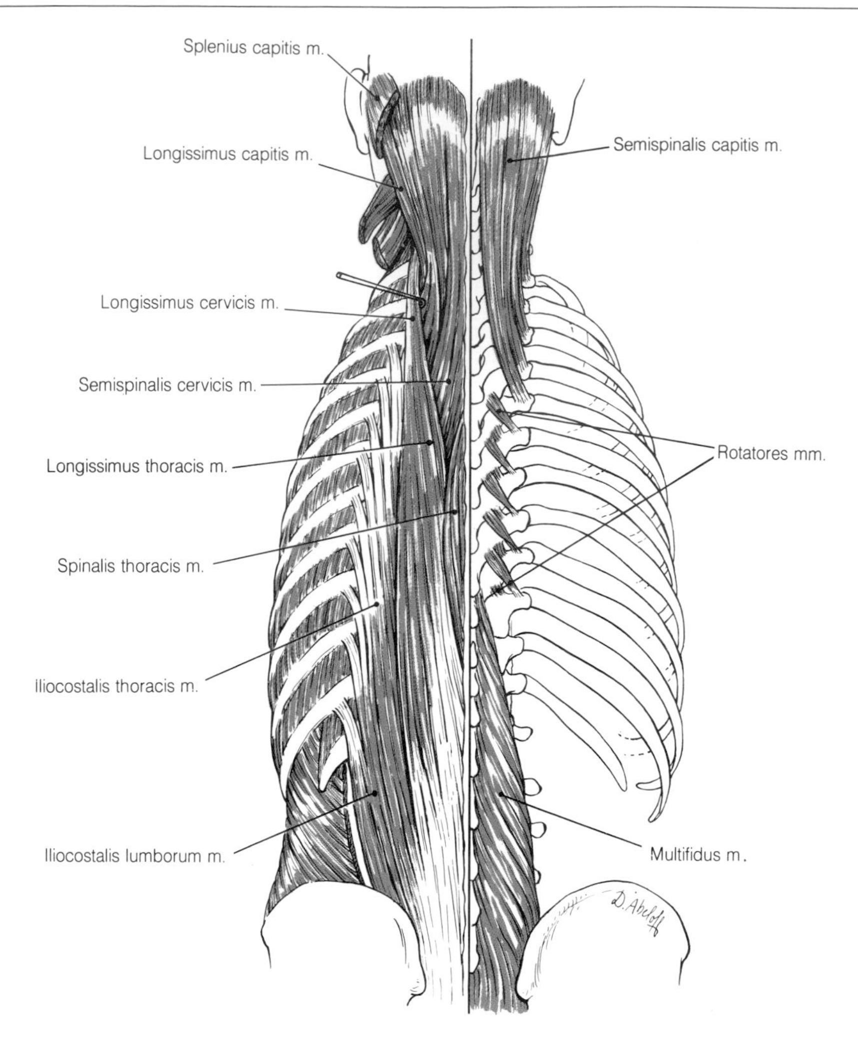

2-21 Showing the parts of the erector spinae muscle on the left side of the back and of the transversospinalis muscle on the right side.

The Transversospinalis Muscle

Included in the group of muscles called transversospinalis are **semispinalis, multifidus**, and the **rotatores**. Typically these muscles are deep to the erector spinae, are short in extent, and run from transverse processes to spines. Portions of each of these muscles can also be seen in Fig. 2-21.

Semispinalis arises as thin slips from thoracic and cervical transverse processes which are attached to spines at a higher level. The slips of the highest part, which is called **semispinalis capitis,** are more substantial and form a bulky mass which is attached to the occipital bone between the **superior** and **inferior nuchal lines.**

Deeper still are the fibers of **multifidus.** These arise inferiorly from the sacrum and iliac bone and at higher levels from transverse processes. They are attached to spines of vertebrae above.

The **rotatores** are best seen in the thoracic region. They extend from a transverse process to the lamina and spine of the vertebra above.

Additional bundles of fibers are found between spines and transverse processes of adjacent vertebrae. These are the **interspinales** and **intertransversarii.**

All the deep muscles of the back are supplied by muscular branches of the **dorsal rami of spinal nerves.** The cutaneous branches of these rami supply the skin that overlies the muscles.

The Actions of the Deep Muscles of the Back

The erector spinae muscle extends the head and vertebral column and produces lateral flexion. It is also active in flexion of the back, when the prime mover is the force of gravity. As the back is bent forward, its movement is controlled by the gradual lengthening of the erector spinae muscle. This is termed a paradoxial action.

Knowledge of the actions of the deeper short muscles is limited and theoretical. They can be considered as postural muscles which stabilize one vertebra on another and, in addition, are able to bring about extension, lateral flexion, and rotation.

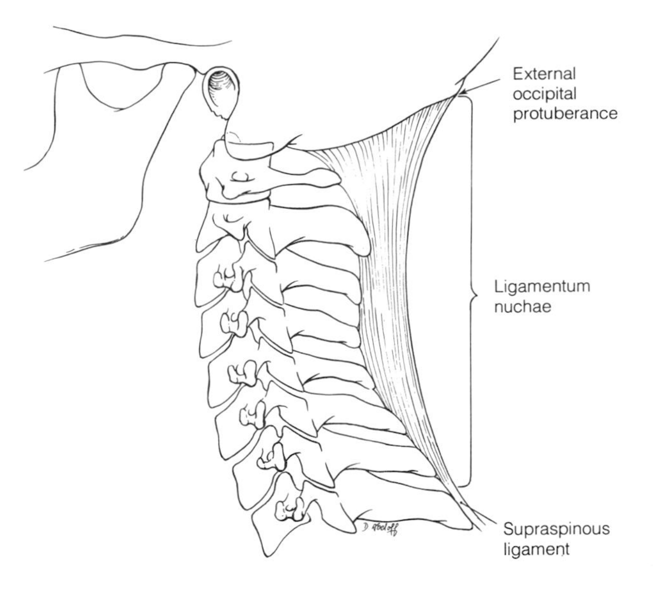

2-22 Showing the position and attachments of the ligamentum nuchae.

The Back of the Neck and the Suboccipital Triangle

The smooth contour of the neck between the skull and the back is formed by the superficial trapezius muscles and, beneath these, the prolongations to the neck and skull of the deep muscles of the back. The most bulky of the latter is semispinalis capitis. Between the deep and superficial muscles of this region lies a flat sheet of muscle called **splenius.*** This muscle arises from upper thoracic spines and the ligamentum nuchae on either side of the midline. Its fibers sweep upward and laterally to become attached to the **superior nuchal line** and **mastoid process.** As a member of the deep group of muscles it is supplied by dorsal rami of cervical spinal nerves. Its function is to help rotate the head and laterally flex the neck. Separating these muscles in the midline and in part giving attachment to them is the **ligamentum nuchae**. In Fig. 2-22 this is seen to be a shallow triangular ligament having a base extending from the external occipital protuberance to the spine of the seventh cervical vertebra, a short side attached to the occipital bone, and a long side attached to the posterior tubercle of the atlas and to the spines of the remaining cervical vertebra. It becomes continuous with the supraspinous ligament.

* Splenium, *L.* = a bandage (the muscle is wrapped around the back of the neck)

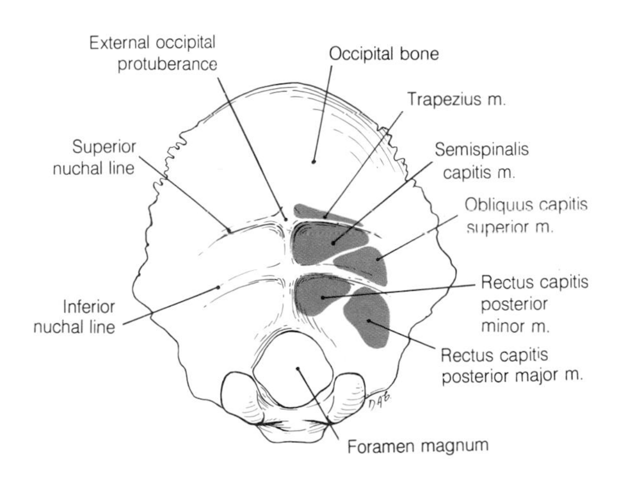

2-23 The occipital bone showing the attachments of suboccipital and nearby muscles.

When the preceding muscles are removed, a group of small discrete muscles of the **suboccipital triangle** are exposed. These muscles are closely associated with the axis, atlas, and occipital bone. Note the position and attachment of each in Figs. 2-23 and 2-24. The **rectus capitis posterior minor** (i.e., the smaller posterior upright muscle of the head) extends from the posterior tubercle of the atlas to the occipital bone on and below the inferior nuchal line. More superficially the **rectus capitis posterior major** extends

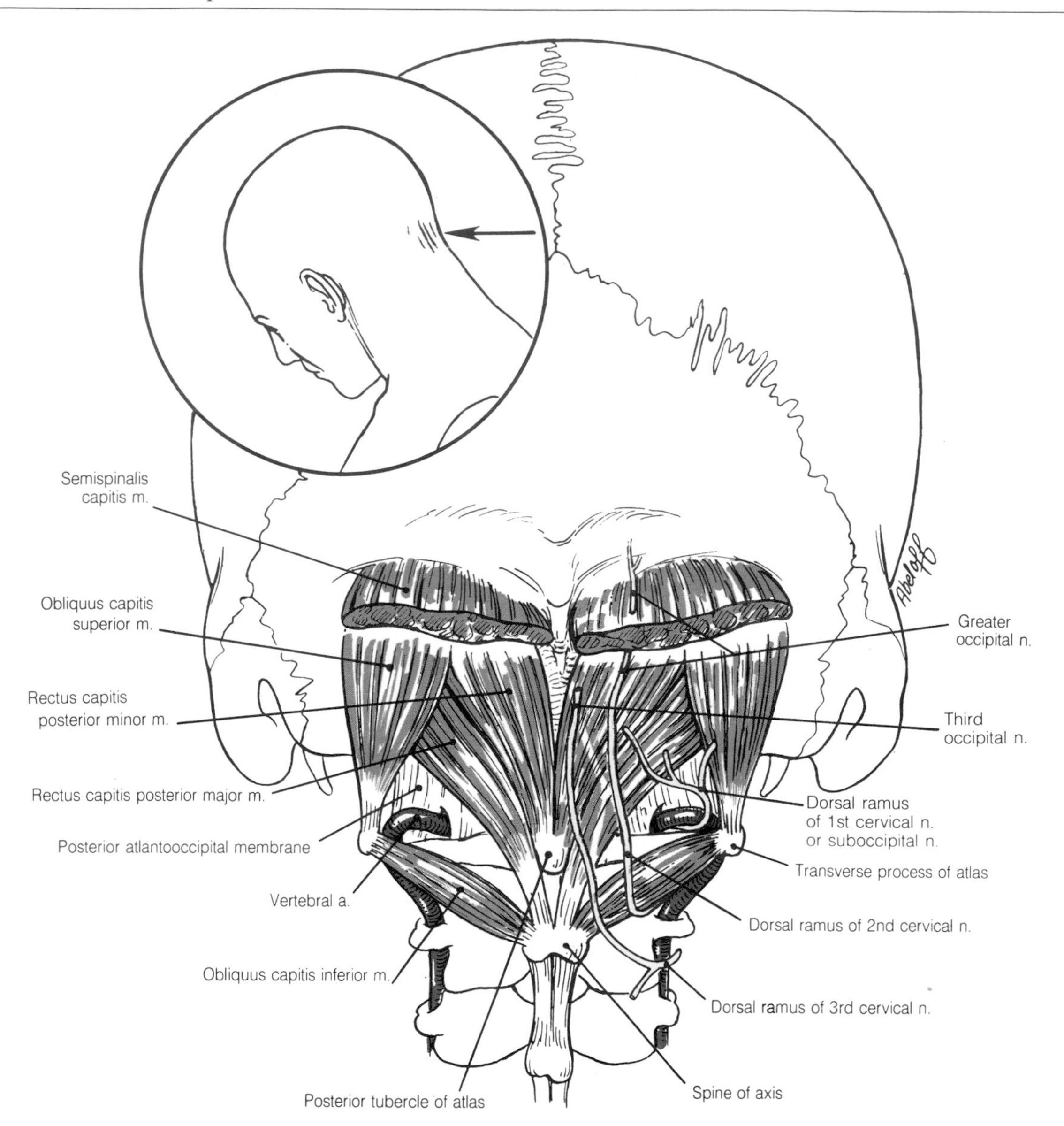

2-24 The suboccipital triangle and the occipital bone showing attachments of muscles.

from the bifid spine of the axis to the occipital bone lateral to the minor. The **obliquus capitis inferior** arises from the same spine and is attached to the transverse process of the atlas. The **obliquus capitis superior** can be followed from here to the occipital bone where it is attached between the inferior and superior nuchal lines.

The action of this group of short muscles is to extend the head on the atlas and rotate the head and atlas on the axis. They are all supplied by the **dorsal ramus of the first cervical spinal nerve.**

There are a number of other structures in the suboccipital triangle which can be seen in Fig. 2-24. Note that the dorsal ramus of the first cervical nerve, which is called the **suboccipital nerve,** appears above the posterior arch of the atlas. It has no cutaneous branches but supplies all the muscles of the triangle. The dorsal ramus of the second nerve appears below the inferior oblique muscle. Branches of this nerve supply nearby deep back muscles but the greater part of the nerve forms the **greater occipital nerve,** a cutaneous nerve. This ascends across the triangle, pierces semispinalis, splenius, and trapezius, and is then joined by the occipital artery. On becoming superficial it supplies the skin of the posterior aspect of the scalp up to the vertex. The skin of the lower part of the occipital region is supplied by a branch of the third cervical dorsal ramus which is called the **third occipital nerve.**

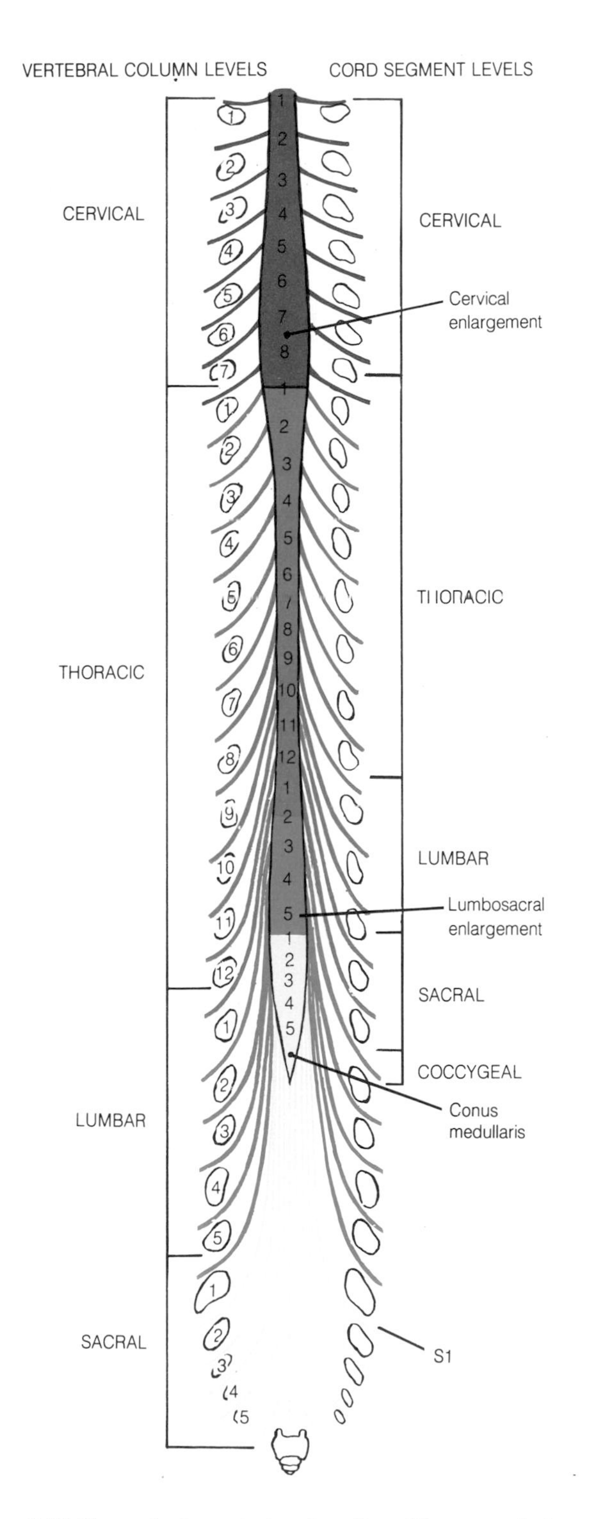

2-25 The spinal cord showing the differences between vertebral levels and segmental levels.

In the depths of the triangle a small portion of the **vertebral artery** can be seen. At this point it is passing medially from the vertebrarterial foramen of the atlas to pierce the **posterior atlantooccipital membrane**. This is described as the third part of the artery. Note that between the vertebrarterial foramina of the axis and atlas the artery must pass upward and laterally. This sinuous course produces an easily recognized feature in an angiogram.

The Spinal Cord

Within the protective covering of the vertebral column lies the spinal cord. It begins just below the foramen magnum of the skull where it is continuous above with the medulla of the hind brain. It ends opposite the **second lumbar vertebra**. In shape it is a somewhat flattened cylinder and shows two gentle swellings along its course. The upper of these, the **cervical enlargement**, is related to the nerves supplying the upper limbs; the lower is called the **lumbosacral enlargement** and is related to the nerves supplying the lower limbs. Below the lumbosacral enlargement the cord tapers to a point. This region is called the **conus medullaris**. These features are illustrated in Fig. 2-25. In a cross-section of the cord, such as that shown in Fig. 2-26, a deep **anterior median fissure** and a shallow **posterior median sulcus** can be seen. There is also a **posterolateral sulcus** along which the **dorsal roots** of spinal nerves enter the cord. The **ventral roots** of the spinal

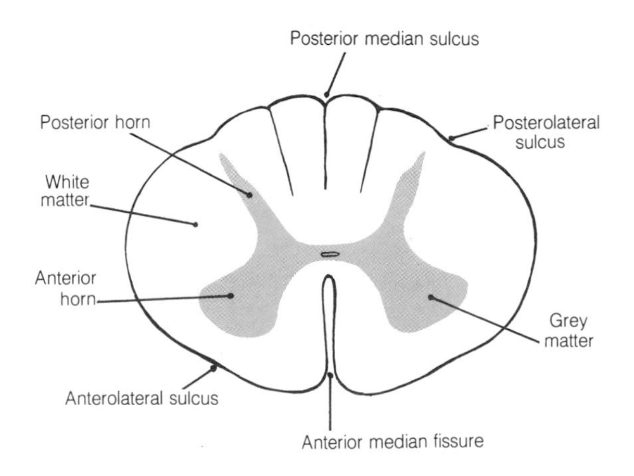

2-26 A transverse section of the spinal cord.

nerves leave the cord at the **anterolateral sulcus**. In a fresh specimen a clean cut will also show that the cord is divided into an outer white region or **white matter** surrounding a swollen H-shaped grey central region or **grey matter**. Within the crosspiece of the H is a narrow **central canal** containing cerebrospinal fluid. The vertical limbs of the H form a pair of **anterior horns** and a pair of **posterior horns**. The grey matter is largely composed of cell bodies of neurons and the white matter of longitudinally running nerve fibers. The white color is provided by the myelin sheaths of the nerve fibers.

Although there is no obvious external sign, the cord is functionally divided into segments that correspond to the spinal nerves which leave the vertebral canal between the vertebrae. In the embryo, these segments lie opposite the appropriate intervertebral foramina. But, owing to different growth rates, the cord terminates at the level of the **third lumbar vertebra** at birth and the **second** in the adult. From this it follows that the cervical and lumbar enlargements described above will lie at a higher level than would be expected and that the spinal nerve roots will have to travel downward for an increasing distance to reach their foramina of exit. Follow the first sacral nerve in Fig. 2-25 to confirm this.

Spinal nerves are formed by the fusion of **dorsal and ventral roots** at the intervertebral foramen and the way in which this occurs is shown in Fig. 2-27. Note that unlike the arrangement suggested by many textbook diagrams, the dorsal and ventral roots are formed from a continuous series of rootlets which leave the cord dorsally and ventrally. Each root is formed by the convergence of three or four rootlets in such a manner that a pair of dorsal and ventral roots is formed for each cord segment. Just before the spinal nerve is formed, each dorsal root bears a swelling called the **spinal ganglion.** Functionally, the dorsal roots carry sensory or afferent fibers to the cord from the periphery and have their cell bodies in the ganglion. The ventral roots carry efferent or motor fibers from the cord to the periphery and have their cell bodies in the anterior horns of the grey matter.

Because of the discrepancy between the level of the termination of the cord and the length of the vertebral canal, the rootlets which come together to form the lower lumbar and sacral nerves form a leash below the cord known as the **cauda equina.*** By the same token, as the cord descends, spinal segments lie increasingly further above their corresponding vertebral levels to the extent that the lower three thoracic vertebral spines lie opposite the five lumbar and first sacral segments of the cord. The differences between cord and vertebral levels assume clinical importance in the diagnosis and prognosis of cord lesions and fracture dislocations of the vertebral column.

Radiographic investigation of the spinal cord had proved difficult until the introduction of magnetic resonance imaging. The example of this technique shown in Fig. 2-28 shows the accurate anatomical visualization of this soft tissue structure that is now possible.

* Cauda equina, *L.* = a horse's tail

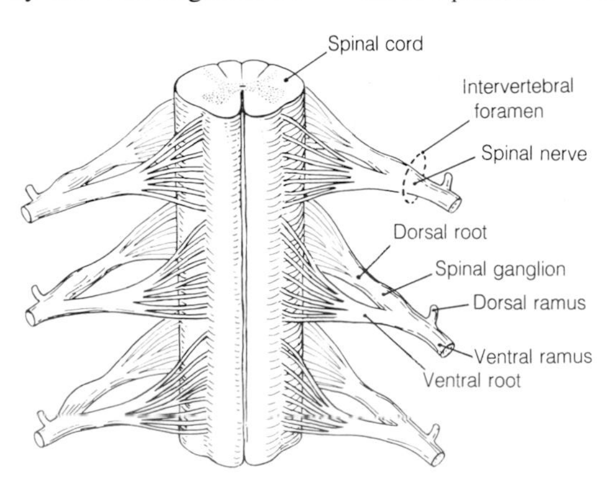

2-27 A short length of the spinal cord showing the formation of spinal nerves.

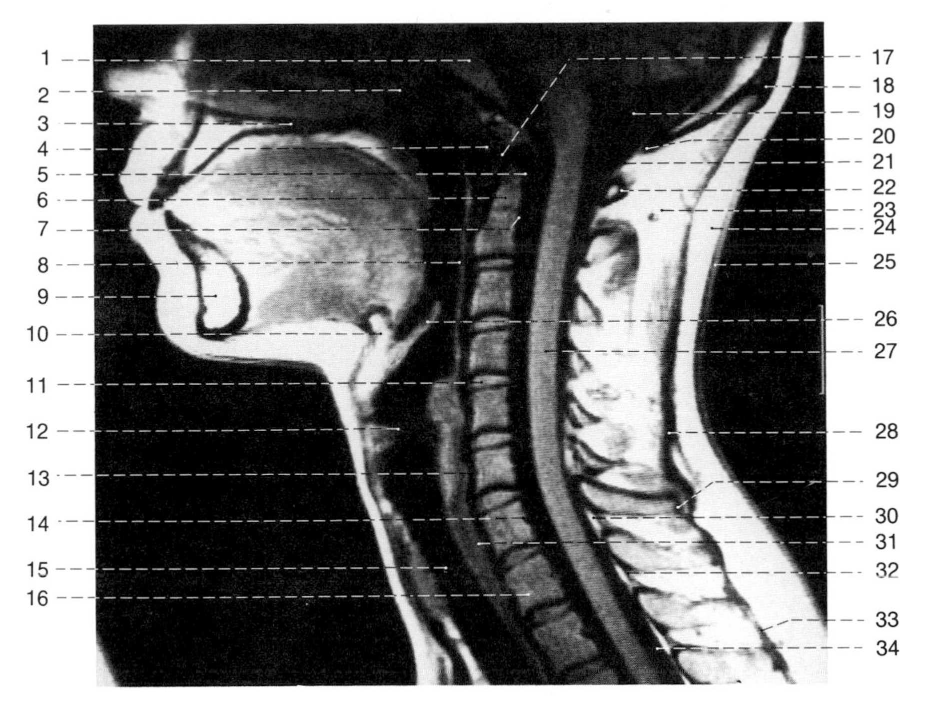

Basiocciput – 1
Nasopharynx – 2
Hard palate – 3
Anterior arch of atlas (C1) – 4
Transverse ligament – 5
Odontoid process (dens) – 6
Tectorial membrane – 7
Mucosa of oral pharynx – 8
Body of mandible – 9
Hyoid bone – 10
Nucleus pulposus of C4–C5 intervertebral disc – 11
Vocal fold – 12
Anterior longitudinal ligament and cortical bone – 13
Basivertebral vein – 14
Trachea – 15
Body (centrum) 1st thoracic vertebra – 16

17 – Anterior atlanto-dental joint
18 – External occipital protuberance (inion)
19 – Cisterna magna
20 – Posterior margin of foramen magnum
21 – Posterior atlanto-occipital membrane
22 – Posterior arch of atlas (C1)
23 – Suboccipital fat
24 – Subcutaneous fat
25 – Skin
26 – Epiglottis
27 – Spinal cord (C5)
28 – Ligamentum nuchae
29 – Spinous process C7 vertebra (vertebra prominens)
30 – Epidural fat
31 – Esophagus
32 – Ligamentum flavum
33 – Supraspinous ligament
34 – Cerebrospinal fluid

2-28 Midsagittal MR scan of the cervical region showing the spinal cord (gray) clearly separated from surrounding tissues by cerebrospinal fluid (black). (Reproduced by permission, from Schnitzlein, et al: *Imaging Anatomy of the Cervical Spine with Magnetic Resonance*, Urban and Schwarzenberg, Baltimore-Munich 1987.)

The Blood Supply of the Spinal Cord

The blood supply of the spinal cord is becoming of greater interest as research continues into the problems of spinal cord damage and regeneration. In Fig. 2-29, the position of the single **anterior spinal artery** and two pairs of **posterior spinal arteries** can be seen. These are derived in the cranial cavity from the vertebral arteries or their posterior inferior cerebellar branches. They are not large vessels but are assisted at each vertebral level by small **anterior and posterior radicular*** **branches** of **spinal branches** of the vertebral, cervical, posterior intercostal, and lumbar arteries. A variable number of radicular branches in the lower cervical and thoracolumbar regions are considerably larger than their fellows. These anastomose with the anterior spinal artery and play a major part in the supply of the cord. They are illustrated in Fig. 2-30. Venous blood leaves to join venous plexuses on the surface of the cord. These plexuses communicate with cranial veins and venous sinuses and with the **internal and external vertebral plexuses.**

As elsewhere in the central nervous system the spinal cord contains no lymphatic vessels.

* Radix, *L.* = a root

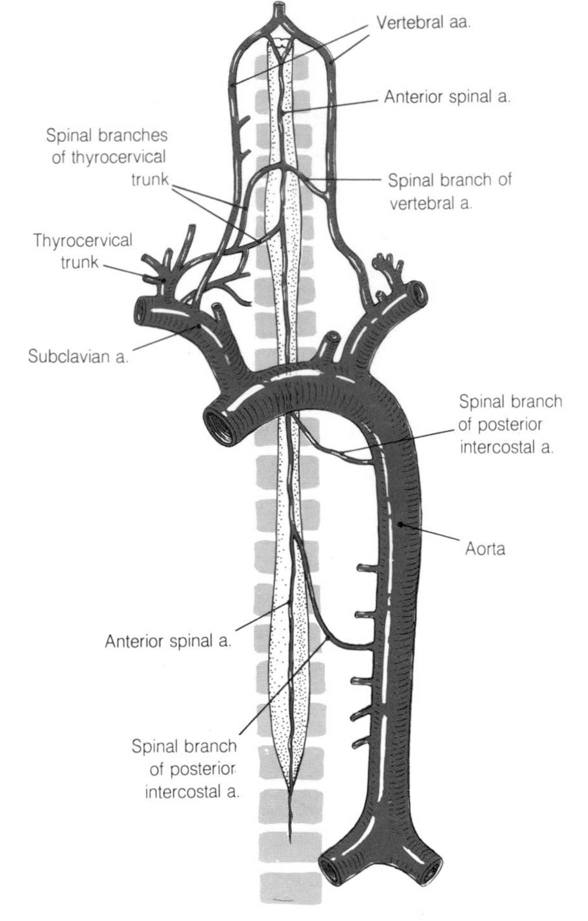

2-30 Showing the sites at which substantial anastomoses are found between spinal branches and the anterior spinal artery.

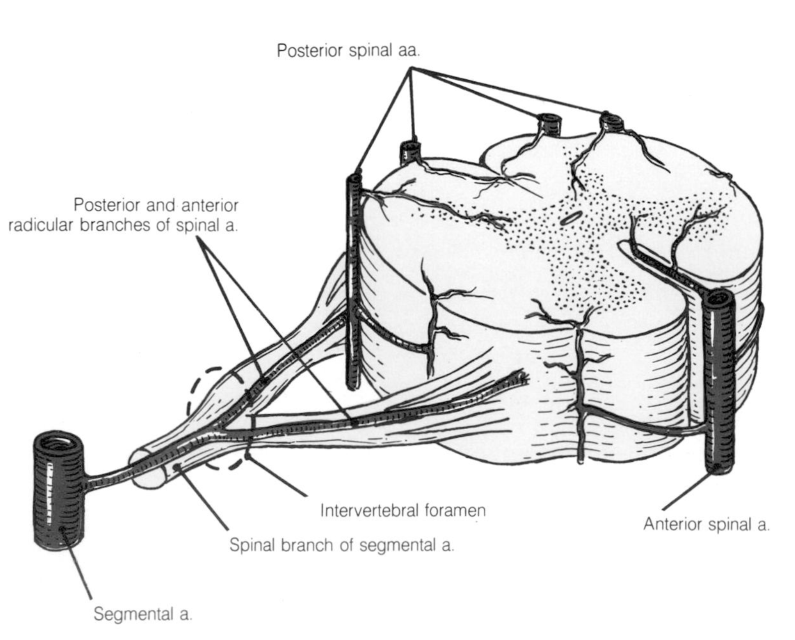

2-29 Showing the blood supply of a segment of the spinal cord.

The Meninges

The spinal cord is surrounded by three sheaths of tissue called the **meninges.*** From without inward these are known as the **dura mater**, the **arachnoid mater**, and the **pia mater**. In clinical practice the dura mater is often called the **pachymeninx*** and the latter two membranes are commonly referred to together as the **leptomeninges.***

The Dura Mater

The **dura mater** (hard mother) of the spinal cord is continuous with the dura mater surrounding the brain. However, whereas the dura surrounding the brain is tightly adherent to the overlying endosteum of the skull bones, the dura in the vertebral canal lies free and is separated from the periosteum of vertebrae by a space called the **epidural space**, which is filled by fat, and a substantial venous plexus, the **internal vertebral plexus of veins**. This plexus has considerable importance as it may, with the **basivertebral veins**, form a channel for the dissemination of malignant cells to vertebral bodies in, for example, tumors of the breast and prostate gland. The reason for this is that the plexus, the veins of which are without valves, communicates both with veins draining the hemopoietic (blood-forming) red marrow of the vertebral bodies and with the **external vertebral plexus of veins**. Direction of blood flow in these valveless vessels is determined by respiration, coughing, or changes in intraabdominal pressure, and thus would allow a few detached malignant cells traveling from the pelvis to form a secondary tumor or metastasis in a thoracic vertebra. The dura itself is a relatively tough fibrous layer which is prolonged as a sleeve around each set of dorsal and ventral rootlets as they travel to the intervertebral foramen. At the foramen this sleeve becomes attached to periosteum, thus anchoring the dura. The dural sac terminates at the level of the **second sacral vertebra**. Beyond this point the vertebral canal surrounds only epidural space and the lower dorsal and ventral rootlets.

* Meninx, *Gk.* = membrane; Pachys, *Gk.* = thick; Leptos, *Gk.* = thin or delicate

The Arachnoid Mater

Immediately beneath the dura and separated from it by no more than a potential **subdural space** is the **arachnoid** (like a spider) **mater**. This is a much more delicate membrane which sends fine extensions across a very real subarachnoid space to the underlying pia mater. The **subarachnoid space** contains **cerebrospinal fluid**, and extends downward like the dura mater to the level of the **second sacral vertebra**. This important fact is demonstrated radiographically by the myelogram shown in Fig. 2-31. In this examination an oily radiopaque contrast medium is introduced into the subarachnoid space.

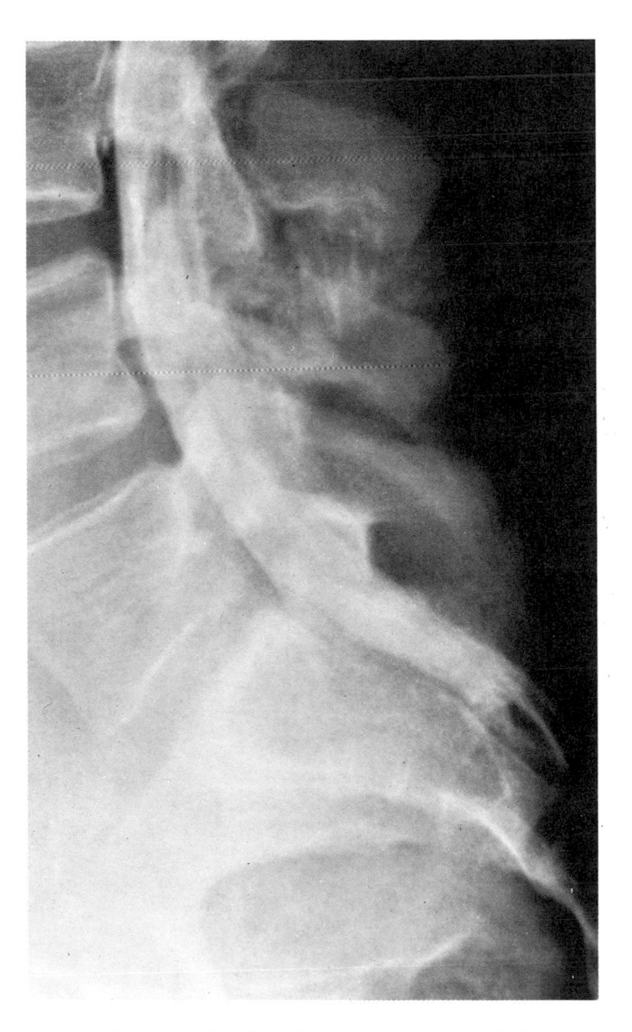

2-31 Myelogram showing the lowest extent of the subarachnoid space. (Reproduced by permission, from Wicke: *Atlas of Radiologic Anatomy,* 4th Ed, Urban & Schwarzenberg, Baltimore-Munich, 1987.)

The Pia Mater

The **pia** (dutiful) **mater** is a fine vascular layer closely investing the cord. As over the brain, the pia mater is difficult to strip off. It sends extensions around the nerve roots which blend with their outer covering. In addition, it extends laterally as interrupted tooth-like projections forming the **ligamentum denticulatum**, which anchors the cord to the dura mater. At the conus medullaris it continues downward as a fine thread, the **filum terminale**, which passes through the dura and arachnoid mater at the level of the second sacral vertebra to become attached to the coccyx.

Having read this description of the meninges, look at Fig. 2-32 and observe their relationship to each other, to surrounding spaces, to the cord, and to the spinal nerves.

The levels to which the cord and its coverings extend inferiorly have importance when deciding the optimum level at which to perform a spinal tap. The aim of the tap is to obtain a specimen of cerebrospinal fluid and record its hydrostatic pressure. While the cord terminates between the bodies of the first and second lumbar vertebrae, the arachnoid and dura mater and the subarachnoid space continue to the level of the second sacral vertebra. A spinal needle inserted between the third and fourth (most commonly) or fourth and fifth lumbar vertebrae will enter the subarachnoid space safely below the level of the cord. Remember, however, though of less importance,

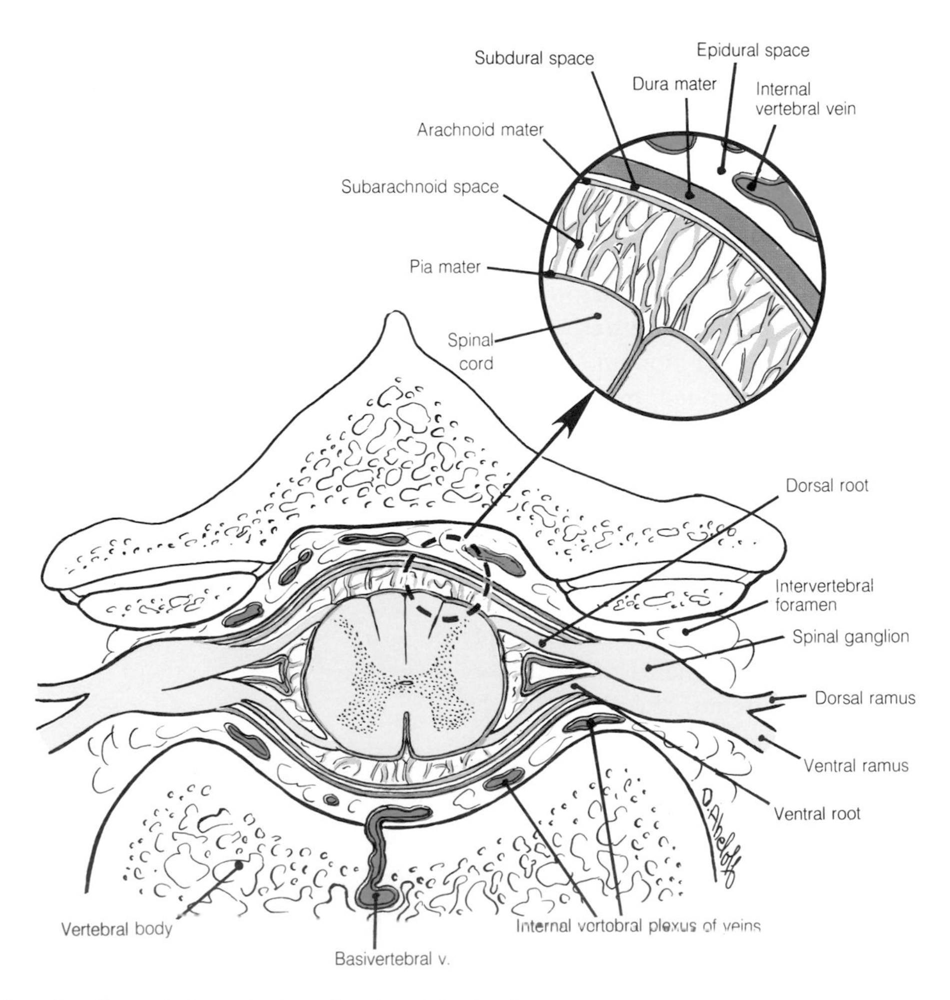

2-32 Horizontal section through the spinal cord and vertebral column.

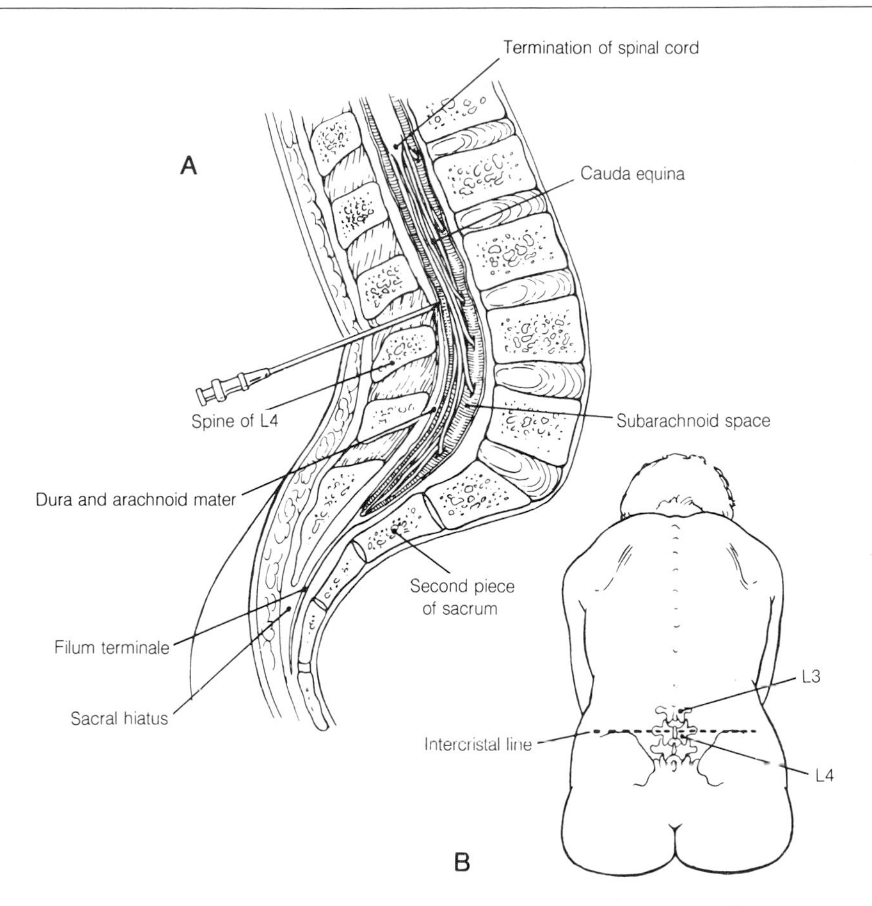

2-33 (*A*) median section through the lumbar spine and sacrum to show the extent of the spinal cord and subarachnoid space. (*B*) Surface markings of help in performing a spinal tap.

that the cauda equina is still at risk. The epidural space continues downward to the coccyx and may be injected with anesthetic solutions through the sacral hiatus as in the technique of caudal anesthesia. The anatomical features that make both a spinal tap and caudal anesthesia possible are illustrated in Fig. 2-33. It is obviously necessary to determine accurately the space between lumbar spines into which to introduce a needle. To achieve this an imaginary line is constructed between the highest points of the iliac crests as is shown in Fig. 2-33. This line crosses the spine of the fourth lumbar vertebra.

The Spinal Nerves

A spinal nerve is formed by the union of a dorsal and ventral root as has already been described. With the vertebrae between which they emerge they represent the most consistent reminder of the segmental origin of the human body. Their numbering appears confusing until it is realized that the first pair of nerves appears above the first cervical vertebra, the seventh above the seventh cervical vertebra, and that between the seventh cervical and first thoracic vertebra is numbered the eighth cervical nerve. As a result all subsequent pairs appear below the neural arch of the similarly named and numbered vertebra. There are, therefore, eight cervical, twelve thoracic, five lumbar, and five sacral pairs of spinal nerves. One or two coccygeal pairs complete the series. Look now at Fig. 2-34 and by identifying the first and second cervical nerves and the eighth cervical and first thoracic nerves confirm their relationship to the numbered arches. The spinal nerves emerge through the intervertebral foramina and after a short course divide into **dorsal and ventral rami.**

Dorsal Rami

Except for the first, the **cervical dorsal rami** divide into medial and lateral branches. All branches supply the deep muscles of the back and suboccipital triangle, but only the medial branches of the second to fifth cervical nerves supply skin. Of the **thoracic dorsal rami**, the upper medial and lower lateral branches become cutaneous. Three lateral branches of the **lumbar dorsal rami** supply skin over the upper part of the buttock or gluteal region and lateral branches of the **sacral dorsal rami** form two or three cutaneous nerves which supply the low medial part of the buttock.

VERTEBRAL COLUMN LEVELS
CORD SEGMENT LEVELS
CERVICAL
THORACIC
LUMBAR
SACRAL
COCCYGEAL
Cervical enlargement
Lumbosacral enlargement
Conus medullaris
S1

2-34 Showing the relationship of the numbered spinal nerves to adjacent numbered vertebrae.

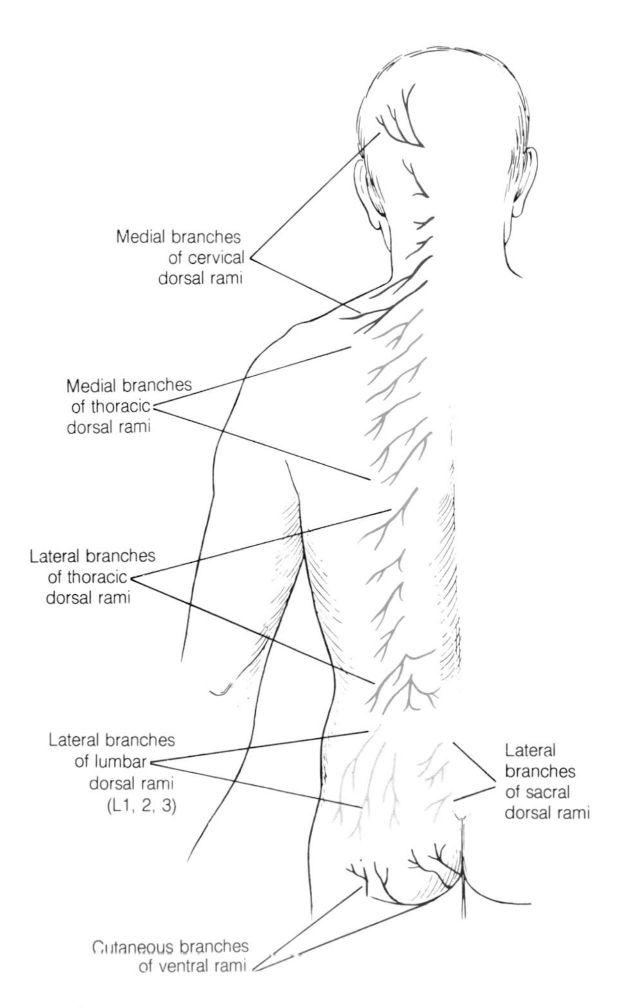

2-35 Showing the cutaneous distribution of the dorsal rami of spinal nerves.

(Cutaneous branches of ventral rami supply the remaining lower part of the buttock.) These three groups of cutaneous nerves, two from dorsal rami and one from ventral rami, are sometimes known as the superior, middle, and inferior **cluneal*** nerves. The distribution of the dorsal rami is seen in Fig. 2-35.

Ventral Rami

The ventral rami supply the remaining skin and skeletal muscle of the lateral and anterior aspect of the trunk and the skin and muscles of the limbs. Their distribution is much more complex than that of the dorsal rami because the individual rami of the cervical, lumbar, and sacral regions unite to form networks or plexuses of nerves and only in the thoracic region do they clearly retain their discrete semental distribution. A typical ventral ramus from this region sweeps round the body wall between adjacent skeletal elements (ribs) and terminates anteriorly near the midline as an anterior cutaneous branch. Along its course it will give off a lateral cutaneous branch and branches supplying the body wall musculature through which it passes. The lateral and anterior cutaneous branches supply a fairly well defined strip of overlying skin known as a **dermatome**. This fact may become very obvious in the condition herpes zoster or "shingles" where a virus infection of the nerve produces an acute inflammation of the strip of skin it supplies.

As has already been suggested, the arrangement of dermatomes is clear cut on the trunk but is modified on the limbs where they have been drawn out around the developing limb bud. This can be seen in Fig. 2-36 where the segmental distribution of dermatomes is illustrated. Note how on the trunk there is a break in the sequence between the fourth cervical and second thoracic dermatomes. This fact becomes of importance when examining sensation in the skin of a patient in order to determine the level of anesthesia following damage to the upper part of the spinal cord.

* Clunis, *L.* = buttock

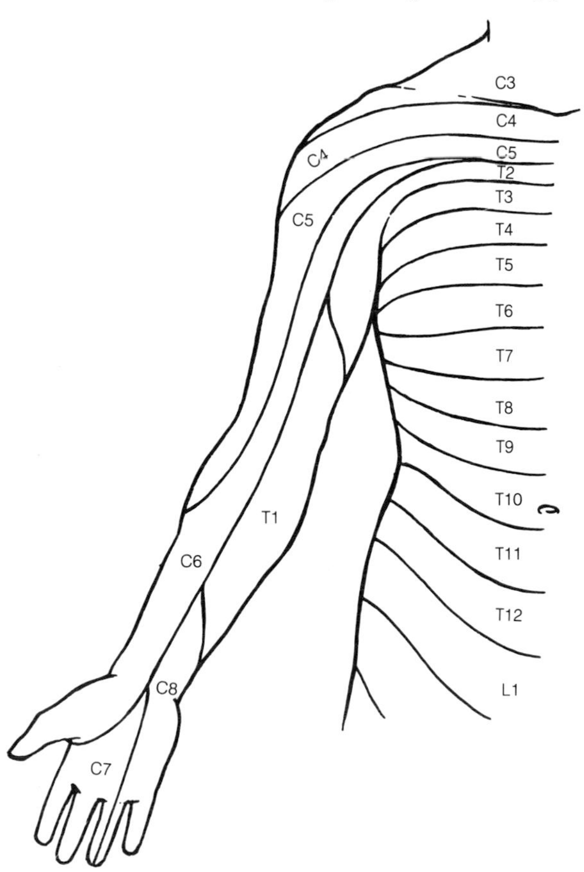

2-36 Showing the segmental distribution of dermatomes of the trunk and upper limb.

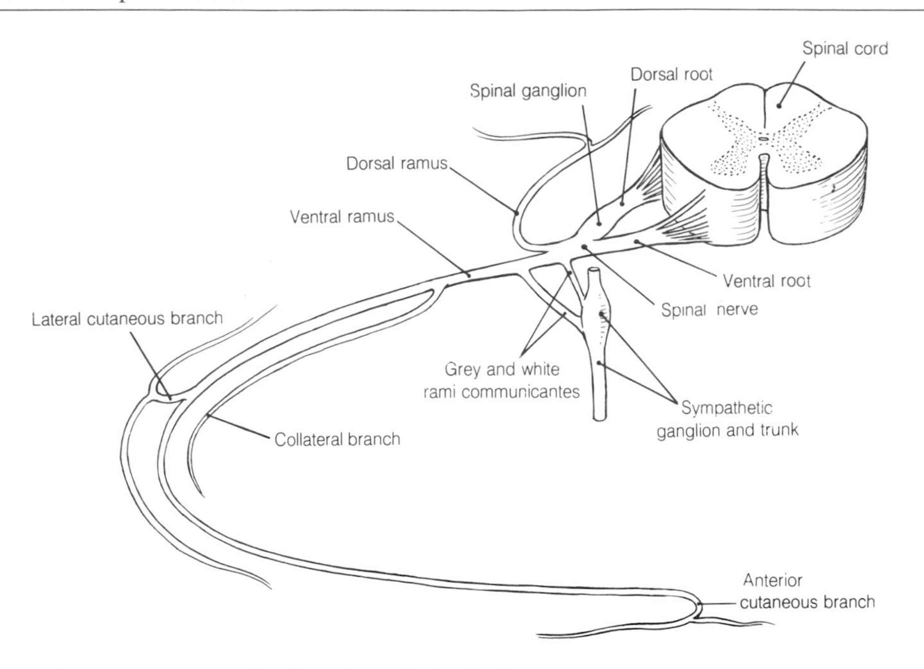

2-37 Showing the formation and distribution of a typical spinal nerve.

Using Fig. 2-37, review the main features of a spinal nerve and the distribution of a ventral ramus. Note the formation of the nerve by the union of dorsal and ventral roots of the spinal cord, its division into dorsal and ventral rami, and the muscular and cutaneous branches of these.

Nerve Fibers in the Spinal Nerve

The spinal nerves carry **motor nerve fibers** to the muscles of the body wall and limbs, and **sensory nerve fibers** returning to the spinal cord from the same regions. Look now at Fig. 2-38 and see the cell body of a motor nerve (an anterior horn cell) in the anterior horn of the grey matter. Its peripheral process or axon *(in blue)* leaves the cord in the ventral root, enters the spinal nerve and is distributed to the periphery in either a dorsal or ventral ramus (only the ventral distribution is shown in the diagram).

The peripheral process of a sensory nerve *(in red)* reaches the spinal nerve in either a dorsal or a

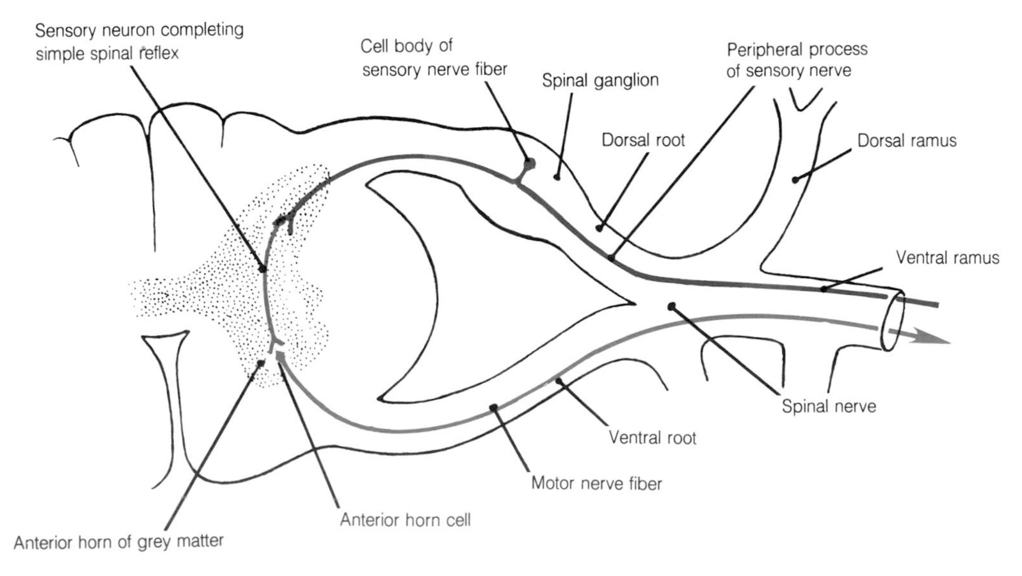

2-38 Diagram illustrating the pathways of sensory and motor nerve fibers between the spinal cord and a spinal nerve.

ventral ramus (again only an example in the ventral ramus is shown). The peripheral process passes from the spinal nerve to the dorsal root where the cell body of the fiber is situated in the dorsal root ganglion. From the ganglion the central process passes into the cord where it may synapse with the cell body of a second sensory fiber in the dorsal horn or ascend the cord to synapse at a higher level. A second sensory neuron is shown in the diagram *(in green)* that completes a spinal reflex between the sensory input and motor outflow of one segment.

The Autonomic Nervous System

Because of its close association with the spinal nerve it is useful to have a brief summary of the main features of the autonomic nervous system at this stage. In the peripheral nervous system the somatic motor and sensory fibers of the spinal nerves are concerned with the supply of skin and skeletal muscle. The autonomic or visceral nerves supply glands and smooth muscle and influence the muscle of the heart. They are divided into two complimentary parts, the **parasympathetic and the sympathetic systems**. These systems are commonly described as if they were isolated motor systems, however, both rely on sensory information received from the peripheral and central nervous systems for their function and despite their name have no real autonomy.

The two systems are also often said to be antagonistic. While they frequently stimulate diametrically opposite actions, there is no reason to believe that these are achieved as the stimulus of one system is overridden by the other.

In general terms the parasympathetic system tends to control and regulate vegetative functions and conserve energy while the sympathetic system responds to outside stimuli and prepares the body for action, be it fight or flight.

The Parasympathetic Nervous System

Efferent or motor fibers of the parasympathetic nervous system leave the central nervous system with the third, seventh, ninth, and tenth cranial nerves and the second and third or third and fourth sacral spinal nerves. Each of these pathways consists of two neurons. The first neuron has its cell body in the central nervous system (either brainstem or sacral portion of the spinal cord) and terminates by synapsing with the cell body of a

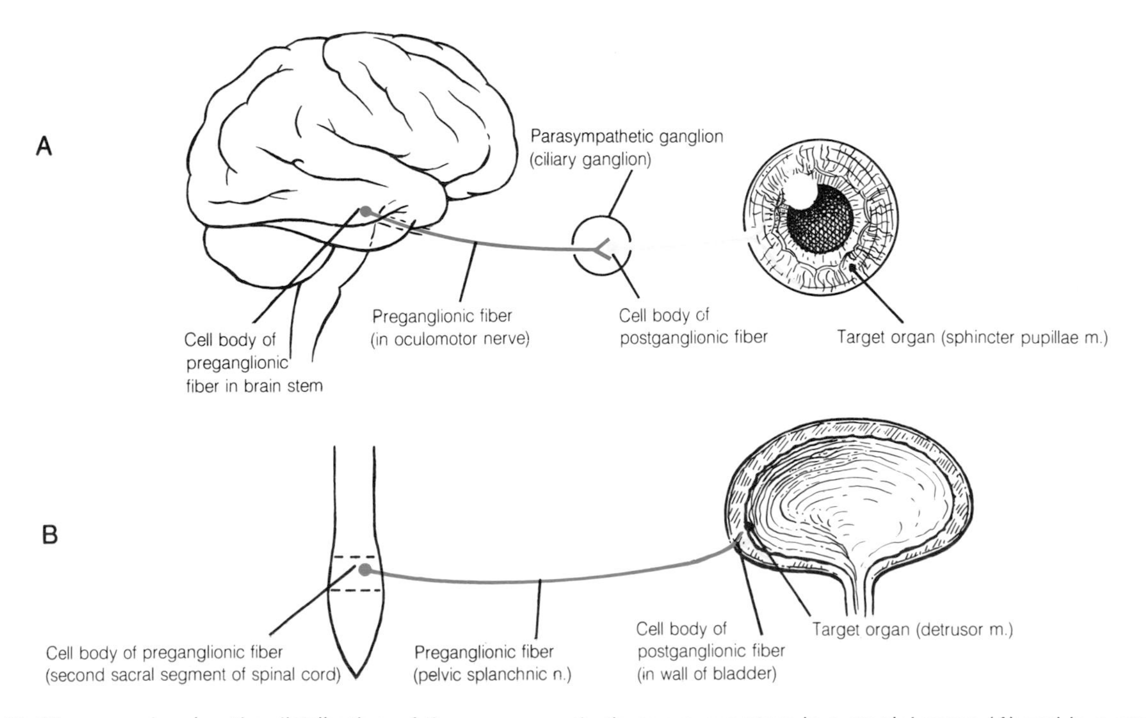

2-39 Diagrams showing the distribution of the parasympathetic nervous system in a cranial nerve (*A*) and in a sacral spinal nerve (*B*).

second neuron in a ganglion lying **close** to the organ to be supplied. The second neuron makes functional contact with the effector organ, i.e., smooth muscle or gland. Because of this arrangement the two neurons are described as preganglionic and postganglionic, respectively. Note that there is no distribution of parasympathetic fibers in branches of spinal nerves running to the body wall. The arrangement of preganglionic and postganglionic parasympathetic fibers and the peripheral ganglion is shown schematically in Fig. 2-39.

The Sympathetic Nervous System

The outflow of the sympathetic nervous system is segmentally arranged and its origin is limited to the spinal cord. It arises from the **first thoracic to the second lumbar segments.** Again, it is a two neuron motor system supplying smooth musculature and glands. The **preganglionic fibers** have their cell bodies in a small column of grey matter in the spinal cord which projects just dorsal to the anterior horn. It is known as the **lateral horn.** These fibers leave the cord with the ventral root of the spinal nerve, join the ventral ramus for a short distance, and then leave it in a fine nerve known as a **white ramus communicans** (it is white because the fibers are myelinated). The preganglionic fibers carried in this terminate by forming a synapse with the cell body of a **postganlionic fiber.** The synapse and cell body lie in a **sympathetic ganglion.** Unlike the parasympathetic ganglia, these lie near the spinal cord on each side of the vertebral column. Hence they are commonly called **paravertebral ganglia.** The ganglia are linked to each other by a **sympathetic trunk**. In Fig. 2-40 the two trunks can be seen extending from the base of the skull to the coccyx where they become united at the **ganglion impar**.* Above, the trunks enter the cranial cavity on the surface of the internal carotid arteries as the **internal carotid nerves**. Along this trunk are found three cervical, eleven thoracic, four lumbar, and four sacral ganglia. However, the inferior cervical and first thoracic ganglia are often fused to form the **stellate ganglion.** Note that both the ganglia and the trunks extend above and below the levels of sympathetic outflow from the cord.

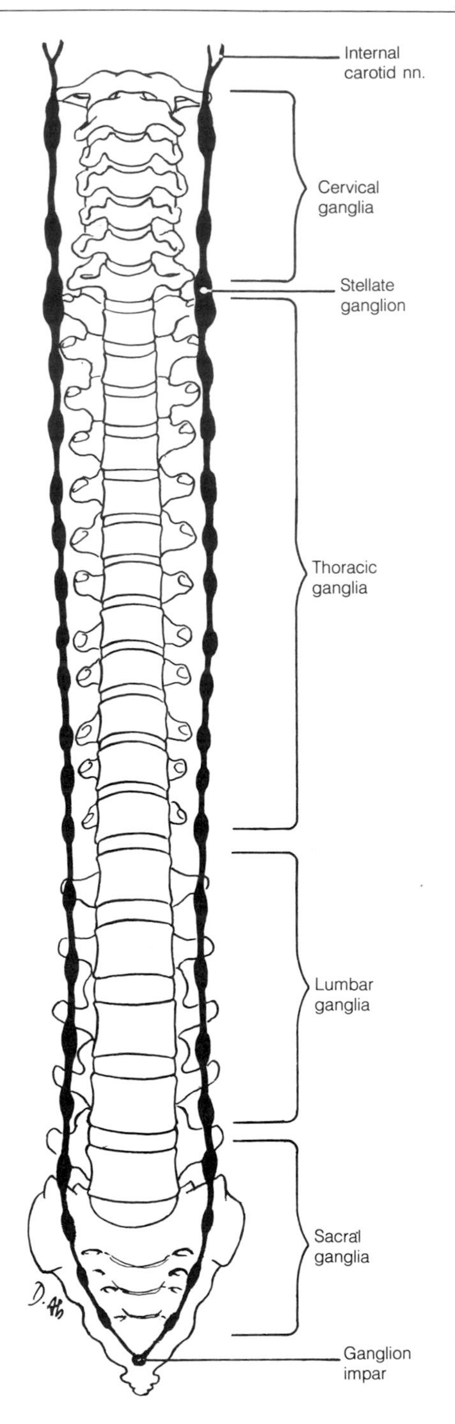

2-40 The sympathetic trunk and ganglia superimposed on the vertebral column.

The manner in which sympathetic fibers are distributed can now be studied in Fig. 2-41 A. In Fig. 2-41 A, first identify the cell body of a preganglionic fiber in the lateral horn of the grey mat-

* Impar, *L.* = unequal (or in this case no equal and therefore unpareid)

ter and then follow the fiber to the ganglion. On reaching the ganglion, the preganglionic fiber may synapse upon the cell body of a postganglionic fiber at that level, or ascend or descend the trunk to synapse in a ganglion at a higher or, as in the illustration, a lower level (i.e., lower lumbar and sacral). Thus, although the sympathetic outflow is limited to the thoracic and upper lumbar segments, its eventual distribution is extended to all segments.

Postganglionic fibers may be distributed in one of the following ways. In Fig. 2-41 B they can be seen to *(a)* leave the ganglion as a discrete and often specifically named visceral branch, e.g., a cardiac nerve; *(b)* join an adjacent blood vessel and share in its distribution; or *(c)* return to the ventral raumus of a spinal nerve in a **grey ramus communicans** to supply the smooth muscle of peripheral vessels (vasomotor fibers), hair follicles (pilomotor fibers), and the sweat glands of the skin (sudomotor fibers). Having established this pattern of distribution, it is necessary to say that an exception is found in some visceral nerves. In these many preganglionic fibers *(dotted red line)*

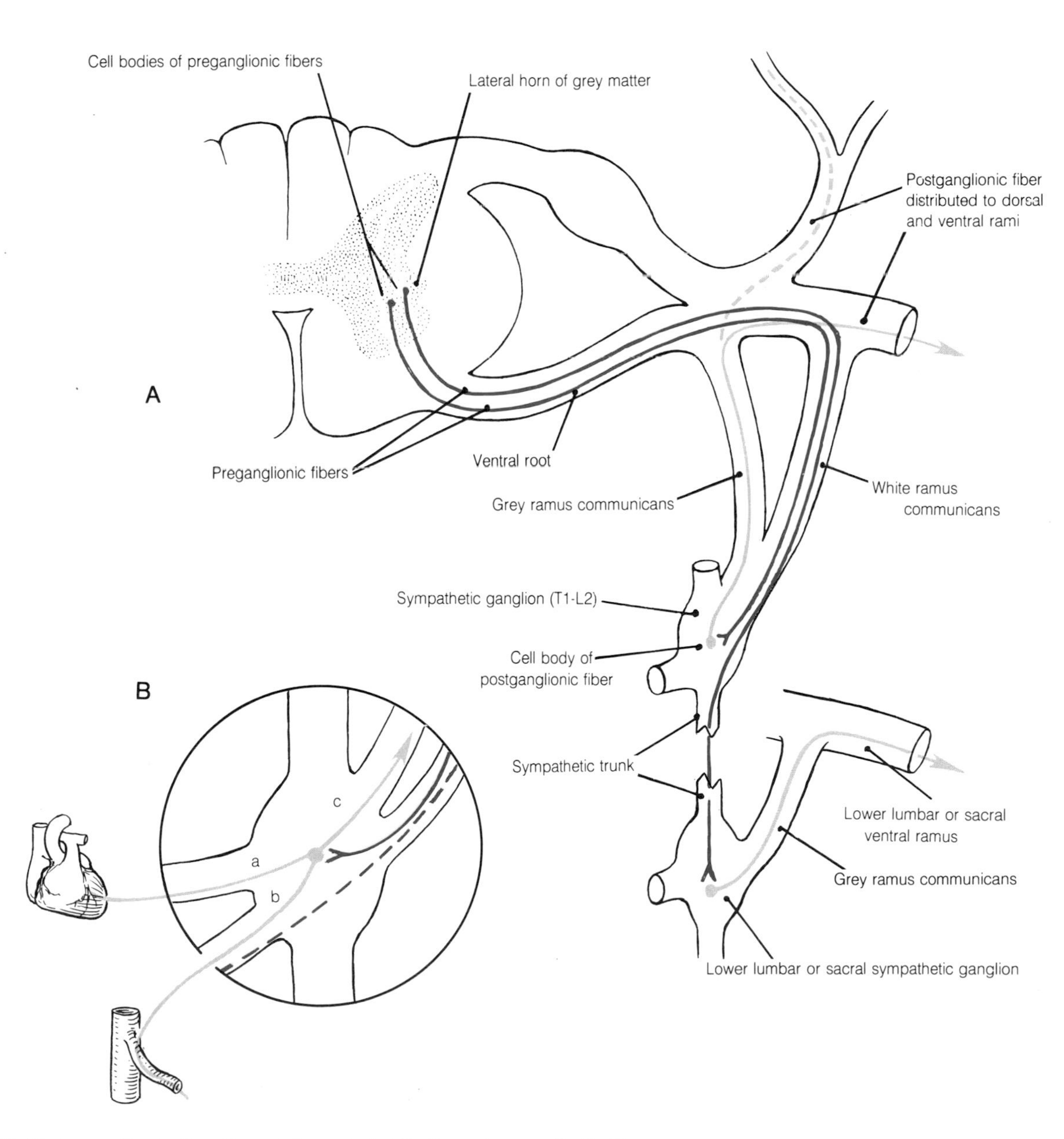

2-41 Diagrams illustrating the origin and distribution of sympathetic fibers.

(Fig. 2-41 B) pass straight through the ganglion into the nerve and do not synapse until they reach the organ they supply.

If these principles behind the distribution of sympathetic fibers have been understood it should be clear that white rami communicantes carrying preganglionic fibers are only found leaving the thoracic and upper lumbar spinal nerves while grey rami communicantes carrying postganglionic fibers leave every ganglion to join a spinal nerve. In the same manner fibers destined for the viscera have a wide distribution from the sympathetic chain. The thoracic and lumbar spinal cord segments from which this distribution to individual regions is made is summarized in Fig. 2-42.

This brief description belies the importance of the autonomic system in the control of almost all the vital homeostatic systems of the body. While it has long been possible to modify its effects by surgical means, for example by vagotomy in gastric surgery or sympathectomy to reduce excessive sweating, even more profound changes can now be produced by modern therapeutic agents.

The Mixed Peripheral Nerve

With this knowledge of the formation of a spinal nerve and its branches and their relationship to the sympathetic system, the term "mixed peripheral nerve" can now be appreciated and the effects of the division of such a nerve deduced, i.e., paralysis of the muscles supplied by motor nerve fibers and loss of sensory information from skin, muscle, tendon, and joints. In addition, the interruption

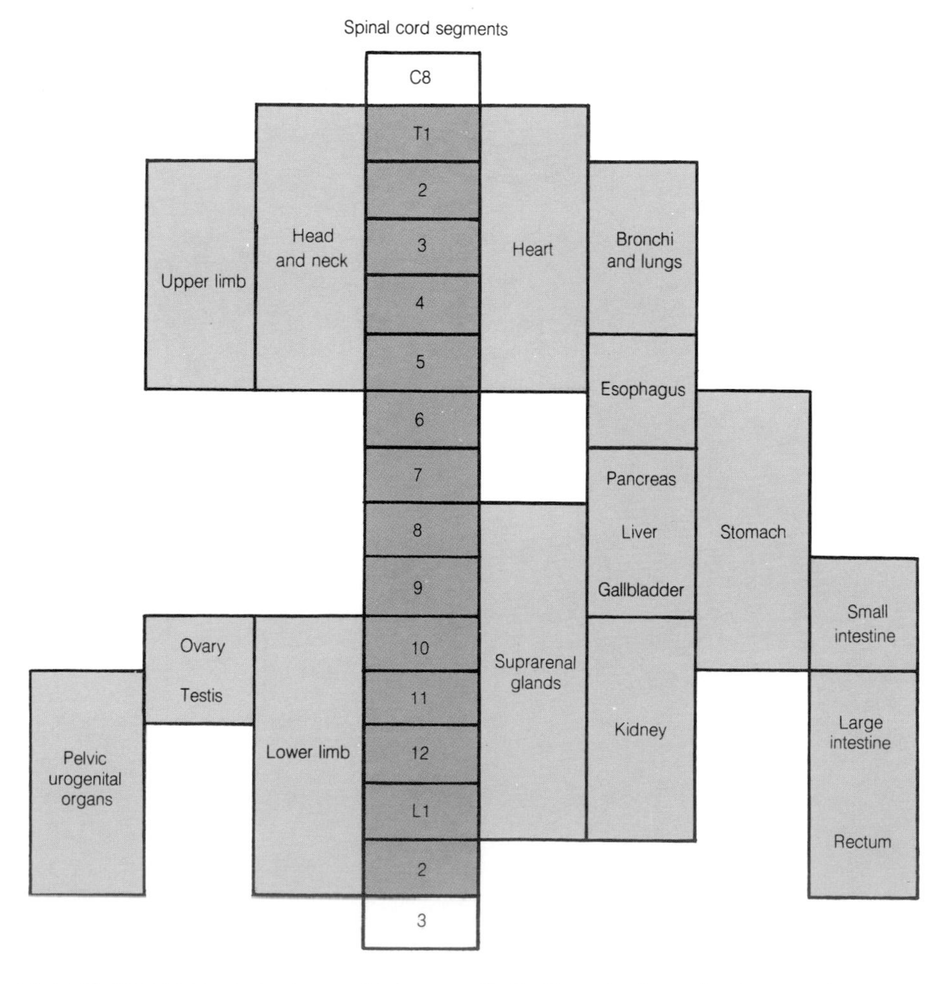

2-42 Showing the spinal cord segments from which sympathetic fibers to the main viscera arise.

of postganglionic vasomotor, sudomotor, and pilomotor sympathetic fibers will lead to a warm, red, and dry skin and a failure to erect hairs and produce "goose bumps."

The Herniated Intervertebral Disc

Much of the normal anatomy described in this chapter can be reviewed in a short description of the pathology of a herniated disc. This condition is most commonly found in the cervical and lumbar regions and herniation of the disc between the fourth and fifth lumbar vertebrae is taken as an example here.

In Fig. 2-43 the soft nucleus pulposus has herniated through the posterior margin of the annulus fibrosus and is projecting into the vertebral canal. At this level only the dorsal and ventral roots forming the cauda equina are present in the canal and it is these that become irritated and compressed by the herniation. In the illustration it can be seen that at the level of the herniation the fourth lumbar spinal nerve has already escaped through its intervertebral foramen. It is, then, the roots of the fifth nerve that will be compressed. This nerve makes a significant contribution to the sciatic nerve and as a result pain is felt along the course of the nerve down the back of the thigh. Injury to the dorsal root will give rise to numbness of the skin over the lateral side of the calf and between the first and second toe, and injury to the ventral root produces weakness of the muscles that lift the foot toward the front of the leg. To these neurological findings can be added the narrowing of the disc space between the vertebral bodies and a filling defect in the opacity produced by a myelogram.

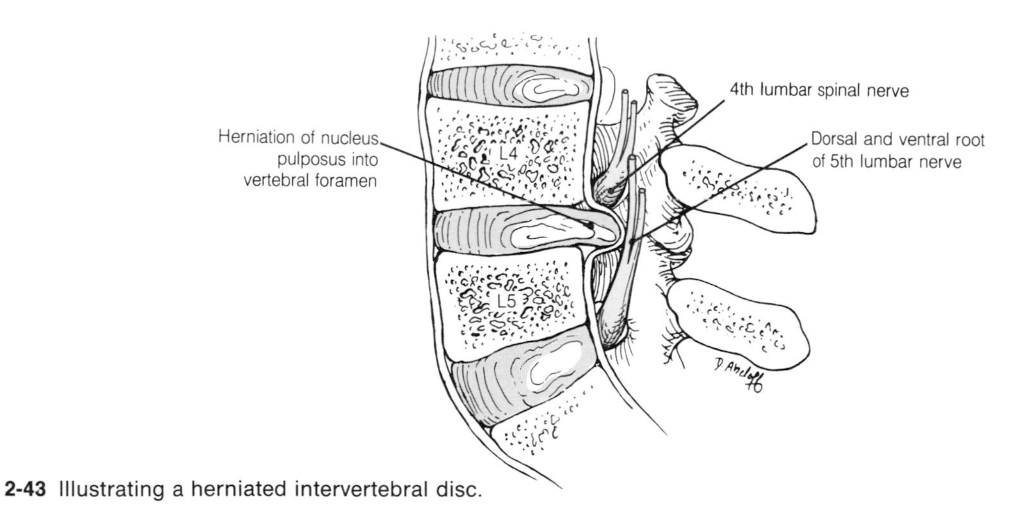

2-43 Illustrating a herniated intervertebral disc.

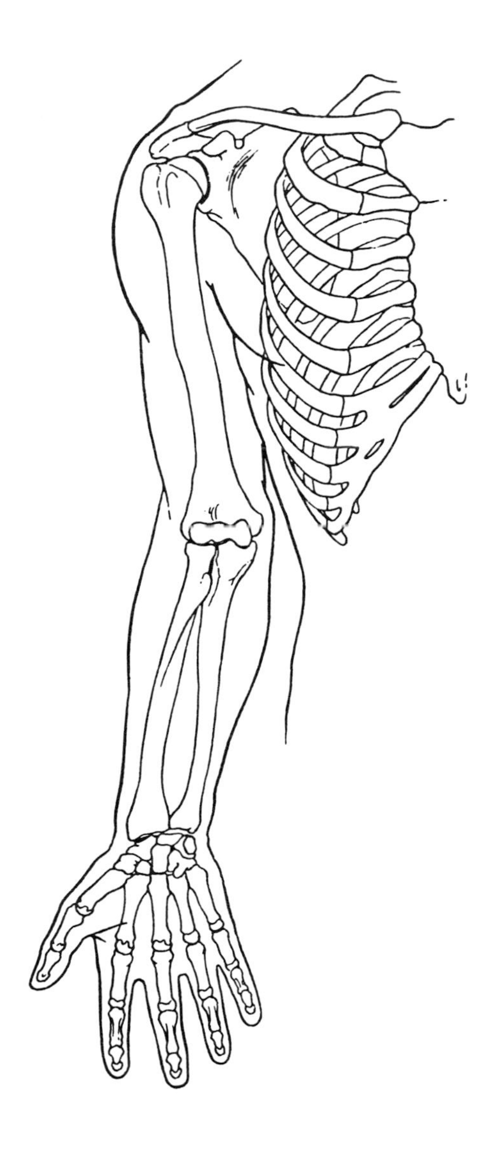

3 The Upper Limb

In man the upper and lower limbs possess many obvious similarities which stem from their phylogenetic history. Each is supported on a girdle which is linked to the axial skeleton. Each has a single proximal long bone and a pair of distal long bones on which articulate either a five-digit hand or foot. Evolutionary specialization has, however, led to profound differences between them but lit-

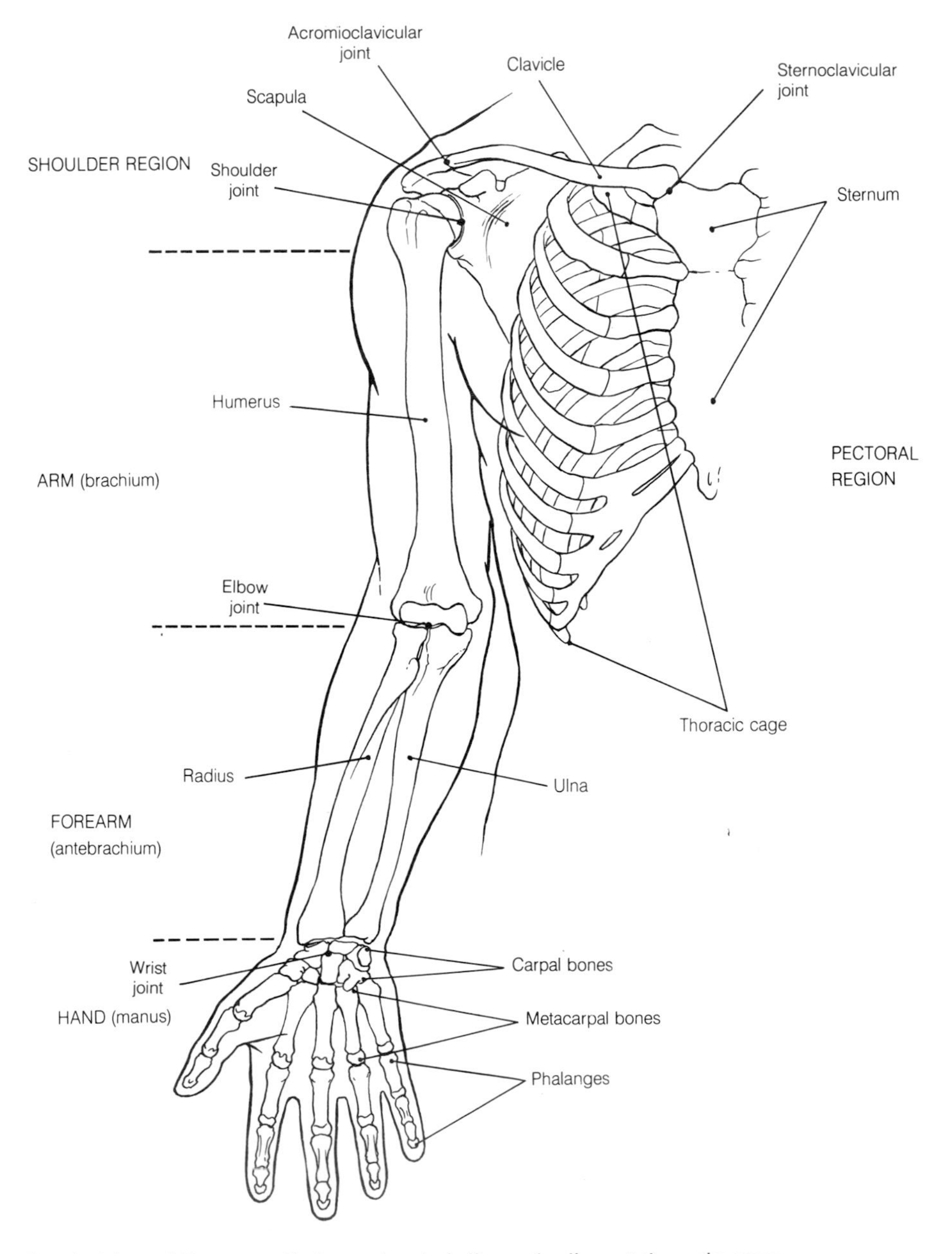

3-1 Showing the skeleton of the upper limb, pectoral girdle, and adjacent thoracic cage.

tle time need be spent on identifying homologous structures. What is of interest is to relate these differences to the functional demands now made upon each limb.

The assumption of the erect posture and a bipedal gait have freed the upper extremity from the task of supporting the body weight and it has become adapted for investigatory and manipulative skills. Contrary to common belief, this adaptation has not involved the anatomy of the hand which carries almost identical and unspecialized digits; it is the substantial representation of the hand in the cerebral cortex together with its lack of specialization that make it so useful. The exception to this is, of course, the opposable thumb which takes part in the particularly human "precision grip." As a result of this adaptation, the upper limb, which so often acts as an intermediary between man and his machines, is at risk both in the workplace and the home where crippling injuries and infections can occur. It may also be disabled by more generalized conditions such as the muscular dystrophies and rheumatoid arthritis.

The General Arrangement of the Upper Limb

In Fig. 3-1 note how the upper limb is suspended from the bony shoulder girdle which also links it to the trunk. The general region of the shoulder joint and the girdle is known as the shoulder region. The region between the shoulder joint and the elbow is known as the arm or brachium. The Latin word brachium, although uncommonly used by itself, is found in other words or in an adjectival form, e.g., the muscle brachialis or the brachial artery. Below the elbow is the forearm or antebrachium, which is joined to the hand at the wrist.

The main features of the skeleton of the upper limb are also illustrated in Fig. 3-1. The shoulder girdle is formed by the clavicle anteriorly and the scapula posteriorly. The clavicle articulates with the sternum at the sternoclavicular joint and with the scapula at the acromioclavicular joint (note that the scapula is not joined to the axial skeleton). The humerus, which lies in the arm, articulates with the girdle at the shoulder joint and with the radius and ulna of the forearm at the elbow joint. Of these bones, the radius articulates with the hand at the wrist joint.

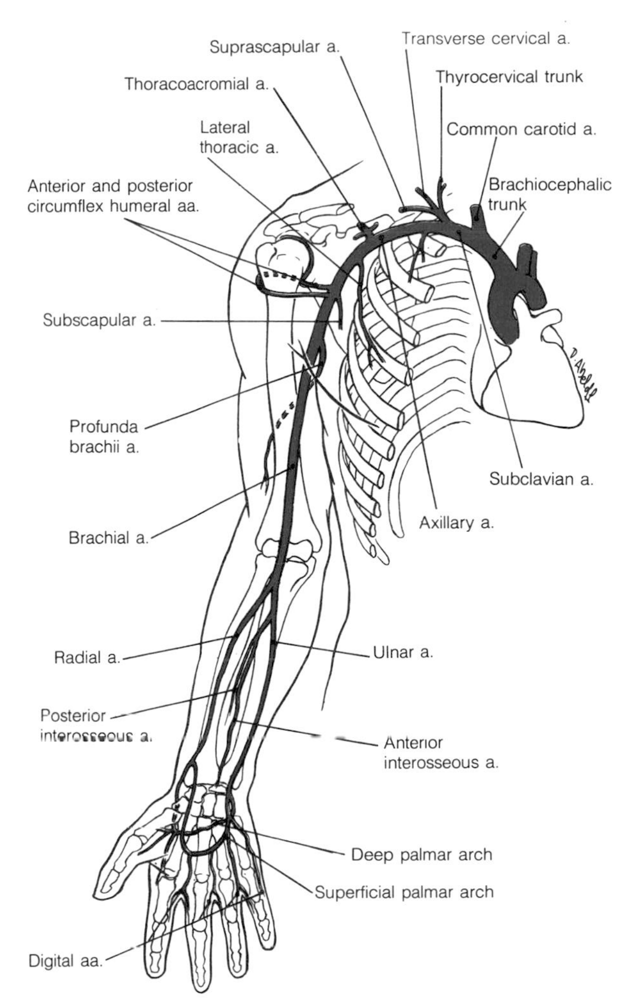

3-2 Showing the main arteries distributing blood to the upper limb. Parts of the thoracic wall and the clavicle have been removed.

In Fig. 3-2 the main vessels distributing blood to the upper limb are shown. On the right side a large vessel rises through the thorax from the aorta and divides behind the sternoclavicular joint into the common carotid artery, which supplies the head and neck, and the subclavian artery which supplies the upper limb. On the left side the subclavian artery is a direct branch of the aorta. The subclavian artery passes over the upper surface of the first rib into the armpit or axilla. Here it changes its name to the axillary artery. On leaving the axilla and entering the arm it becomes the brachial artery. Just below the elbow joint the brachial artery divides into the ulnar and radial arteries which run through the anterior part of the forearm. In the hand these vessels become united again by way of the superficial and deep palmar arches.

It can be appreciated that the main arterial vessels lie in the anterior part of the limb. The posterior part of the limb is supplied by a number of important branches. The scapular region receives branches from the subclavian and axillary arteries, the back of the arm is supplied by the profunda brachii branch of the brachial artery and the back of the forearm by the posterior interosseous artery. The latter is derived from the ulnar artery.

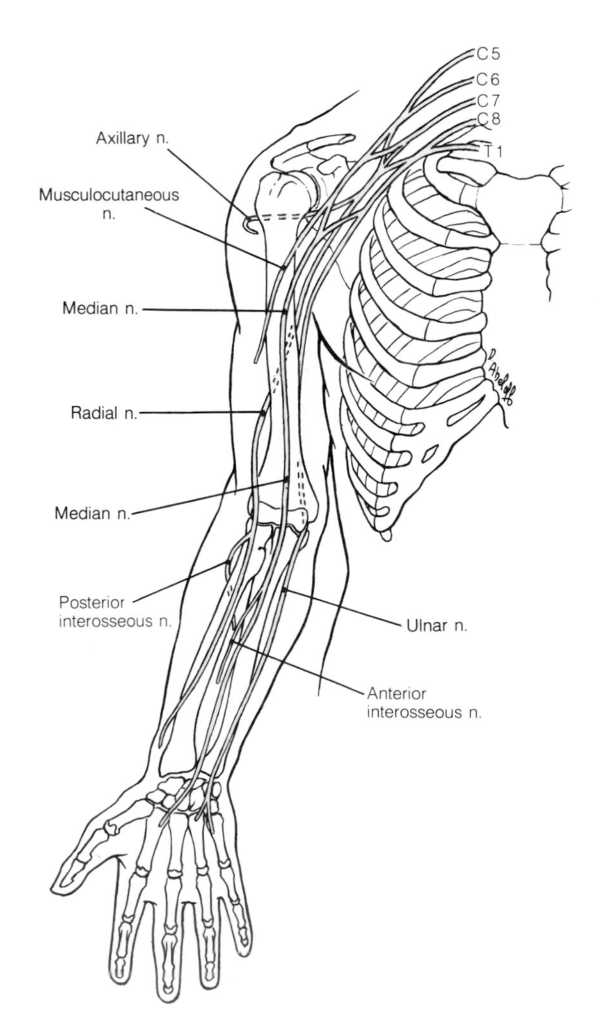

3-3 Showing the main nerves of the upper limb.

Blood is returned from superficial structures by a number of subcutaneous veins which empty into the axillary vein. Deep structures are drained by veins, often paired, which run alongside the arteries (venae comitantes) and eventually form or open into the axillary vein. The axillary vein becomes the subclavian vein as it enters the thorax.

While the blood supply of the limb ascends from the thoracic cavity its nerve supply descends from the cervical region to join the blood supply in the axilla. The skin over the shoulder region is supplied by the supraclavicular nerves (C,3,4) from the cervical plexus. Below this the fifth, six, seventh, and eight cervical and first thoracic ventral rami form the brachial plexus. From this plexus are derived nerves supplying the remainder of the limb. The more important of these are illustrated in Fig. 3-3. The median, musculocutaneous, and ulnar nerves supply the skin and muscles of the anterior or flexor part of the limb, whereas the axillary, radial, and posterior interosseous nerves supply skin and muscles of the posterior or extensor part.

Although great stress is not placed on the fact, do not forget that both arteries and nerves also supply the bones and joints of the limb.

Surface Anatomy of the Shoulder Region

Because the upper limb is intimately joined to the thoracic portion of the trunk some details of this region are included in this section.

Look at Fig. 3-4. In the midline and at the root of the neck identify the **suprasternal notch**. The bellies of the **two sternocleidomastoid muscles** of the neck converge on either side of it. Moving laterally from this notch an elevation formed by the medial end or the **clavicle** can be felt. From this the body of the clavicle extends outward to the **acromion** of the scapula. The subcutaneous anterior border of the clavicle is convex medially and concave laterally. Above and below the junction of its medial and lateral thirds are depressions known as the **supraclavicular and infraclavicular fossae**. Although the junction of the clavicle with the acromion of the scapula is not easily discernable, the lateral margin of the acromion can be felt, and when the arm is at the side of the trunk

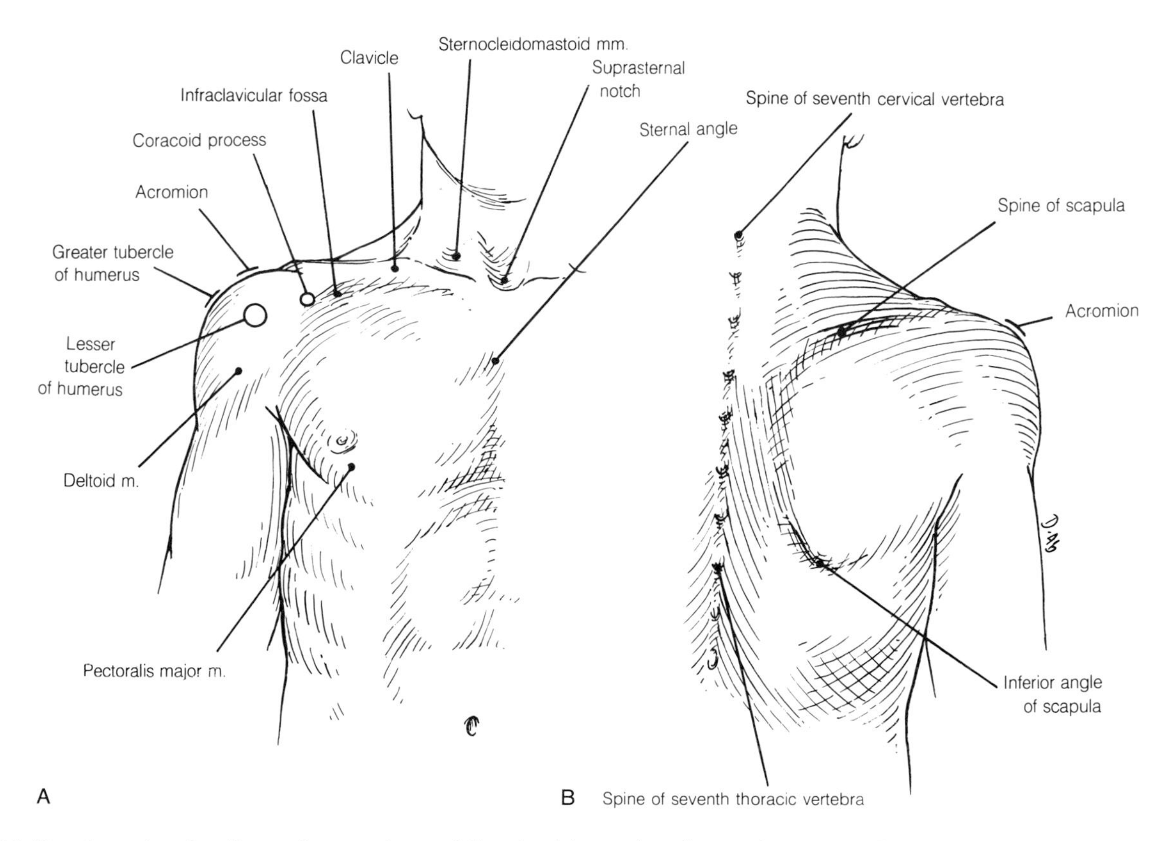

3-4 Drawings showing the surface anatomy of the shoulder region. *A*, anterior aspect; *B*, posterior aspect.

the **greater tubercle** of the humerus lies just below this and is the most lateral bony point of the shoulder region. When the shoulder is dislocated, the acromion becomes the most lateral point. The contour of the shoulder below the acromion is formed by the **deltoid muscle**. Having identified the greater tubercle of the humerus, the **lesser tubercle of the humerus** can be found medial to it and separated from it by a groove, the **bicipital groove**. Both of these tubercles move as the arm is rotated, but a third bony point just medial to the lesser tubercle will not. This is the tip of the **coracoid process** of the scapula.

Returning to the midline, the **manubriosternal joint** can be felt as an elevation below the suprasternal notch. This, called the **sternal angle**, marks the level of the **second costal cartilage and rib** and from here the ribs can be felt and counted. The male nipple usually lies over the fourth intercostal space about 10 cm from the midline.

In muscular subjects the lower border of the large fan-shaped **pectoralis major muscle** can be seen extending upward and laterally from below the nipple toward the shoulder. As this muscle leaves the chest wall, it and the overlying skin form the anterior wall of the armpit or **axilla**. The upper margin of this muscle descends from the medial part of the clavicle and can often be seen to be separated by a shallow **deltopectoral groove** from the deltoid muscle lying over the shoulder.

Posteriorly the margin of the acromion can be traced to the **spine of the scapula** which in turn leads to the **medial border of the scapula**. At the lower end of the medial border is the **inferior angle of the scapula** which lies opposite the **seventh thoracic vertebral spine**. This can be checked by counting down from the prominent spine on the seventh cervical vertebra or **vertebra prominens**.

The Superficial Structures of the Upper Limb

The superficial structures of the upper limb are best considered in a systemic rather than a regional manner, although the sources of cutaneous nerves, the terminations of superficial veins, and the lymph nodes to which superficial lymphatics drain will be encountered again as the underlying regions are described.

The Cutaneous Innervation

The skin of the upper limb is innervated by cutaneous branches derived from the ventral rami of the third to eighth cervical and first and second thoracic nerves. Look at Fig. 3-5 and see the **supraclavicular nerves** (cervical plexus, C3, 4) supply the skin of the shoulder region. The skin of the arm is supplied laterally by the **upper** (axillary nerve, C5, 6) and **lower** (radial nerve, C5, 6) **lateral cutaneous nerves of the arm**, and medially

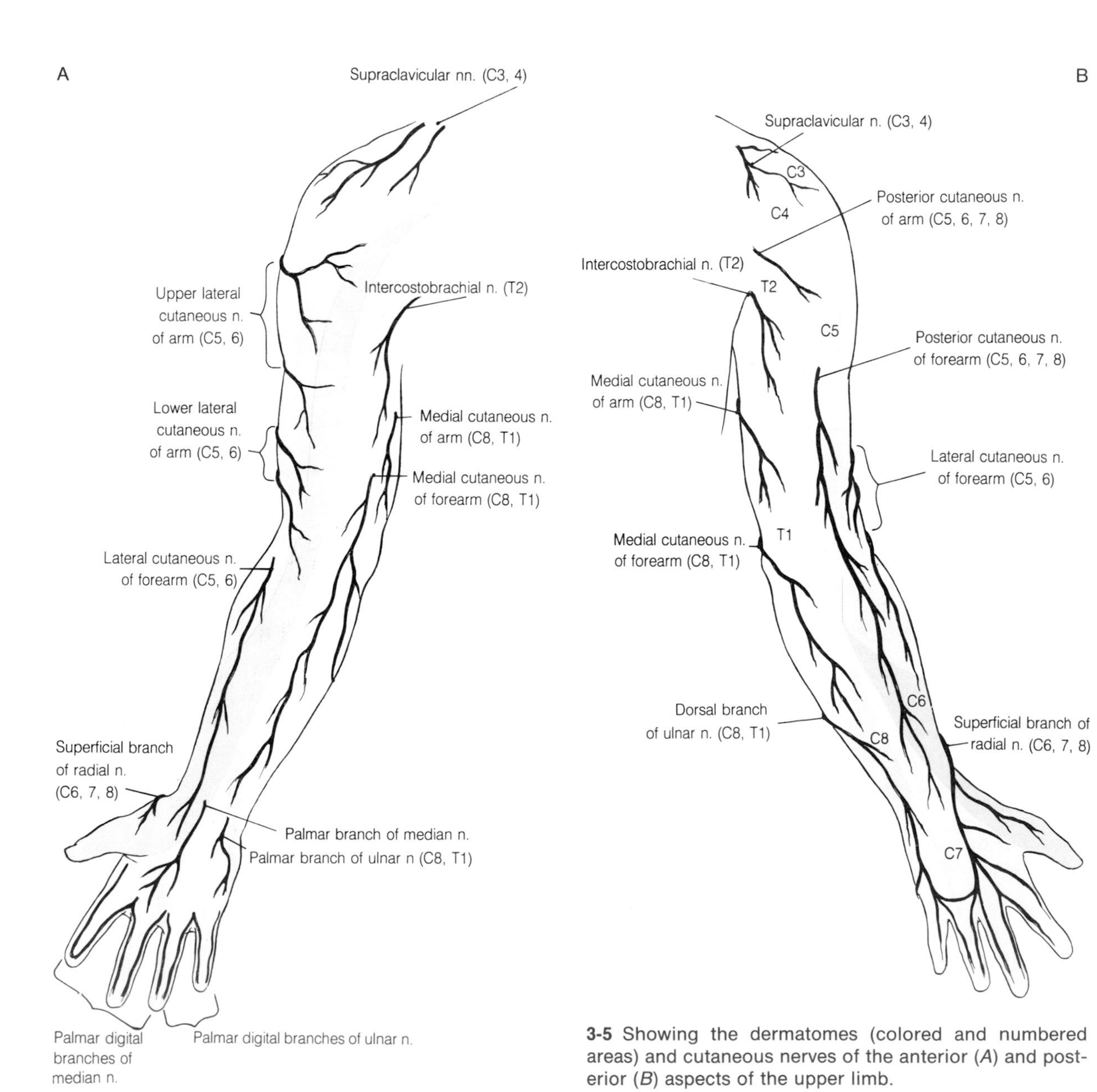

3-5 Showing the dermatomes (colored and numbered areas) and cutaneous nerves of the anterior (*A*) and posterior (*B*) aspects of the upper limb.

by the **intercostobrachial nerve** (second intercostal nerve, T2) and the **medial cutaneous nerve of the arm** (medial cord of brachial plexus, C8, T1). Posteriorly an area not reached by the medial and lateral cutaneous nerves is supplied by the **posterior cutaneous nerve of the arm** (radial nerve, C5, 6, 7, 8).

From the elbow to the wrist the skin of the anterior aspect of the forearm is supplied laterally by the **lateral cutaneous nerve of the forearm** (musculocutaneous nerve, C5, 6) and medially by the **medial cutaneous nerve of the forearm** (C8, T1). The posterior aspect of the forearm is supplied by the **posterior cutaneous nerve of the forearm** (radial nerve, C5, 6, 7, 8).

The skin of the palmar aspect of the wrist and hand is supplied laterally by cutaneous branches of the **median nerve** (C6, 7, 8) and medially by cutaneous branches of the **ulnar nerve** (C8, T1). The division of the territories of these two nerves lies along the middle of the ring finger. Note that the **superficial branch of the radial nerve** (C6, 7, 8) contributes to the supply of the skin of the lateral side of the ball of the thumb or thenar eminence.

The skin of the dorsal aspect of the wrist and hand is supplied by the **superficial branch of the radial nerve** laterally and the **ulnar nerve** medially. Note that the dorsal aspect of the skin of the nailbed and the distal phalanx of the fingers is supplied by branches of nerves on the palmar aspect of the hand.

In this description the spinal nerves contributing to the cutaneous branches are included in parentheses as well as their nerves of origin for later review. It can be seen that more than one spinal nerve is included in these cutaneous branches and the **dermatomes** they represent are also mapped out in Fig. 3-5. With the help of the diagram in Fig. 3-6, it can be understood how during development the dermatomes have been drawn down the limb and away from the trunk leaving the dermatomes of C5 and C6 adjacent to T1 and T2. The line separating these discontinuous dermatomes is called the **axial line**.

Some space has been spent in describing the cutaneous innervation of the upper limb because in it, although distorted, the segmental pattern of innervation can still be seen and use may be made of it in detecting the site or level of a spinal nerve or cord lesion.

The Superficial Venous Drainage

The superficial veins of the upper limb are illustrated in Fig. 3-7. The **dorsal and palmar digital veins** and the veins of the palm tend to drain to an irregular venous network on the dorsum of the hand. From the medial and lateral sides of this **dorsal venous network** two veins ascend the forearm and arm receiving tributaries along their course. On the medial side the **basilic vein** can be followed to the middle of the arm where it pierces the deep fascia and contributes to the formation of the axillary vein, the main deep vein of the upper

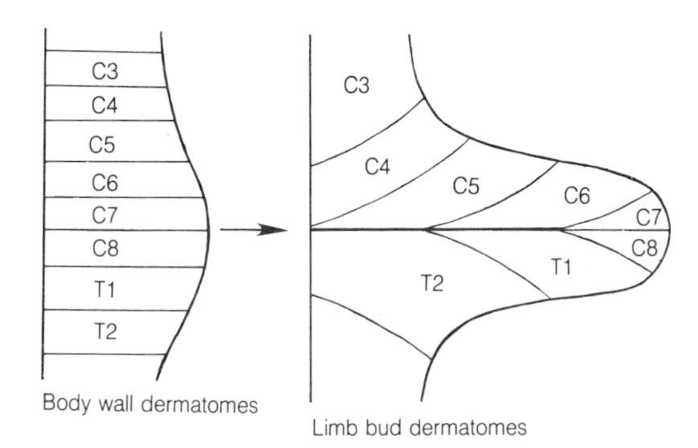

3-6 Diagram showing how the dermatomes supplied by segmental nerves are drawn down over the anterior surface of the developing limb bud of the embryo.

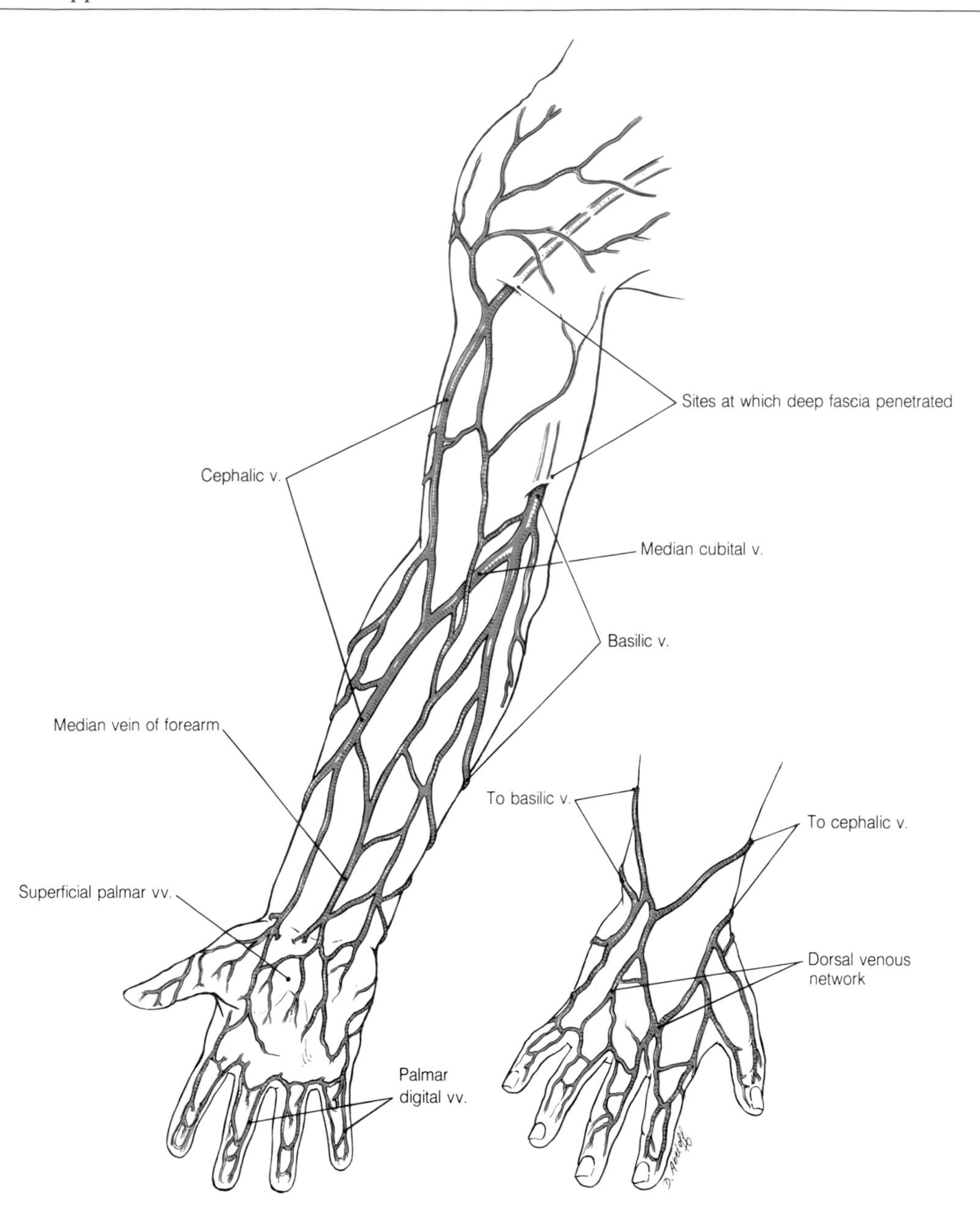

3-7 Showing the cephalic and basilic veins and their tributaries.

limb. Laterally the **cephalic vein** winds round onto the anterior aspect of the forearm and in front of the elbow joint communicates with the basilic vein through the **median cubital vein**. This short but constant, usually visible, and immobile vein is the one commonly chosen for venipuncture. Continuing up the lateral side of the arm the cephalic vein reaches the deltopectoral groove which carries it to the infraclavicular fossa. Here it pierces the deep fascia (**clavipectoral fascia**) and joins the **axillary vein**. A **median vein of the forearm** is also often prominent. It drains the central portion of the palm and joins the median cubital vein.

The Superficial Lymphatic Drainage

In the hand, digital vessels and those from the sides of the palm join vessels on the dorsum.

These in turn form trunks which ascend the forearm and arm along the course of the cephalic and basilic veins. The central portion of the palm is drained by vessels that run proximally in company with the median vein of the forearm. Superficial vessels from the shoulder region run toward the axilla passing round the anterior and posterior axillary folds to reach **axillary nodes**.

Vessels accompanying the cephalic vein end either in superficial **infraclavicular nodes** or pass with the vein through the deep fascia to reach axillary nodes. Some of the vessels alongside the basilic vein are interrupted at the elbow by a superficial **supratrochlear node**, but the main lymphatic drainage is again to deep axillary nodes.

It should be noted that the majority of vessels terminate in deep lymphatic nodes of the axilla. These will be described in more detail with that region.

The Skeleton of the Shoulder and Thoracic Wall

Before describing the muscles that link the upper limb to the pectoral girdle and trunk, the bones of the girdle, bony thoracic wall, and arm must be examined.

The Pectoral Girdle

The pectoral girdle supports the upper limb and the strut of the clavicle keeps it clear of the trunk. The girdle itself is in turn suspended from the head and neck by the upper fibers of the **trapezius muscle** which descend from the spines of cervical vertebrae and the skull and are attached to both the clavicle and the scapula.

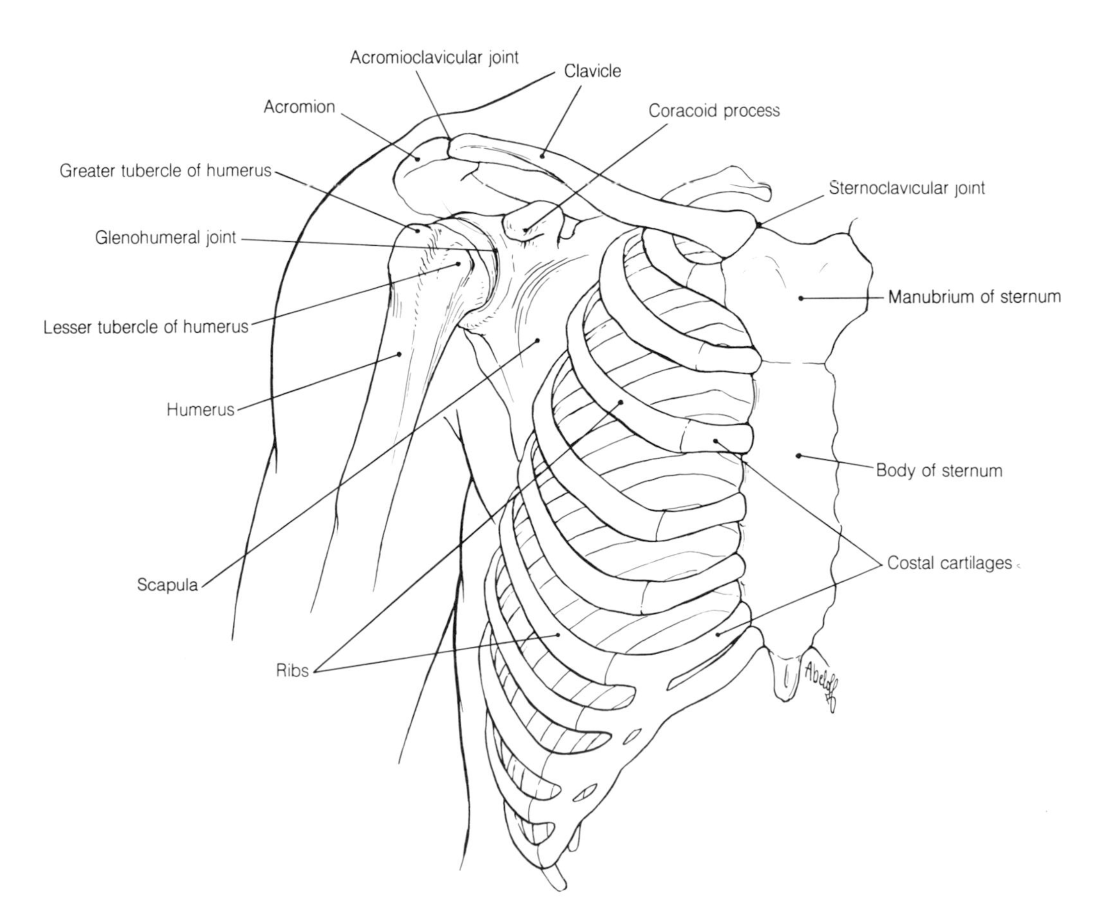

3-8 Showing the skeleton of the shoulder region and the adjacent thoracic wall.

The girdle is formed by the **clavicle** and **scapula**. Note how mobile it is in comparison with the rigid pelvic girdle. It only articulates with the trunk at the sternoclavicular joint. However, in functional terms there is what can be considered a gliding joint between the serratus anterior muscle deep to the scapula and the thoracic wall. Here the intervening space is filled with very loose areolar connective tissue and the scapula is able to sweep across the posterolateral aspect of the thoracic wall. The relationship of the girdle to the trunk is illustrated in Fig. 3-8. This figure should also be used to match the bony points with the surface anatomy shown in Fig. 3-4.

As always, the sections on osteology should be read, if possible, with the appropriate bones at hand and the important attachments of muscles and ligaments identified from the illustrations. Do not forget that many muscle attachments are well defined on bones by ridges, tubercles, and roughened areas, and that once both attachments of a muscle have been established, its actions can usually be deduced.

Clavicle

The clavicle* which is seen in Fig. 3-9*A* from above and *B* from below, is shaped like an open "S" or the f-hole in a member of the violin family. Medially its rounded **body** is convex anteriorly; laterally it is concave. The medial end is slightly expanded and articulates with the **sternum** and the **first costal cartilage** at the **sternoclavicular joint**.

* Clavis, *L.* = a key

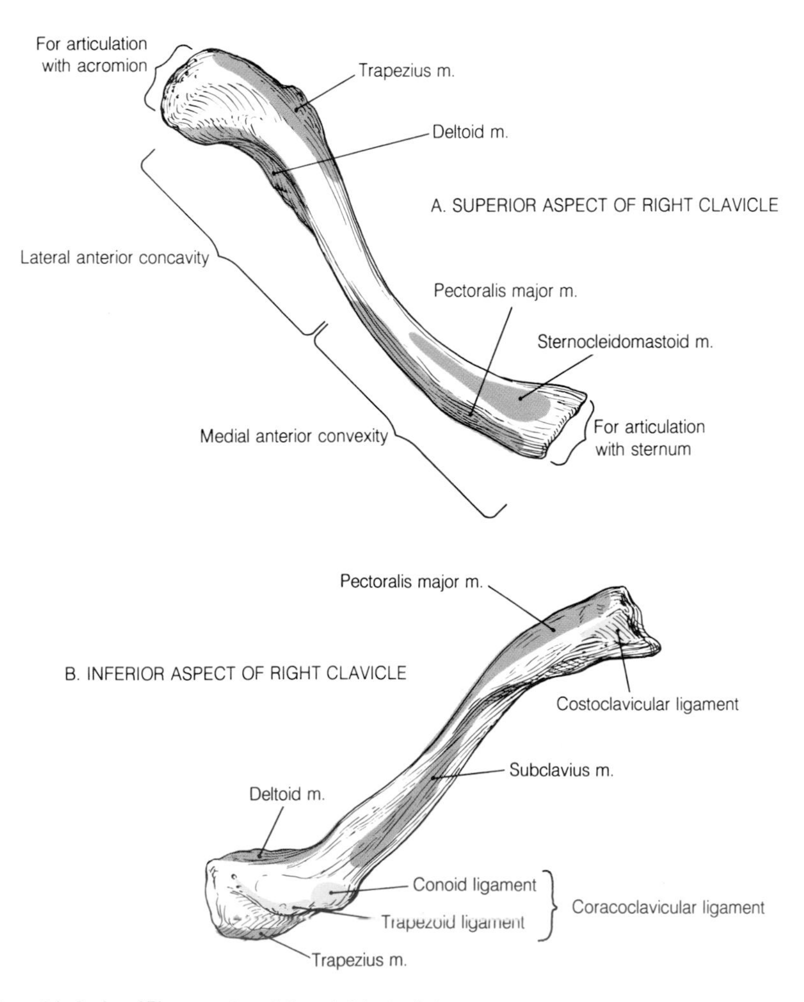

3-9 The superior (*A*) and inferior (*B*) aspects of the right clavicle.

The lateral end is flattened and articulates with the **acromion process** of the scapula at the **acromioclavicular joint**. The inferior surface is grooved and bears a tubercle and a ridge on its distal third. These are the **conoid tubercle** and the **trapezoid line**. Attached to these is the strong **coracoclavicular ligament** that links the clavicle of the coracoid process of the scapula. A roughened area also lies beneath the proximal end for the attachment of **the costoclavicular ligament**.

Important muscular attachments that should be found on the illustrations are those of the **pectoralis major** and **deltoid muscles** anteriorly, the **sternocleidomastoid muscle** superiorly, the **trapezius** posteriorly, and the **subclavius** inferiorly.

The development of the clavicle is unusual in that two primary centers of ossification appear in mesenchyme during the sixth intrauterine week before any other in the body. However, medial and lateral zones of mesenchyme do become transformed into cartilage and it is the growth of these that contributes to the length of the clavicle. A secondary center for the proximal end appears late at about 18 years and soon fuses with the body. The clavicle is thus partly ossified in membrane and partly in cartilage.

Fracture of the clavicle is a common injury and usually occurs at the junction of the middle and outer thirds. The distal fragment is displaced inferiorly by the weight of the limb.

The Scapula

The scapula is a large, triangular, flat bone lying on the posterolateral aspect of the upper chest wall. As can be seen in Fig. 3-10, it has a **medial, lateral,** and **superior border** and clearly defined **superior** and **inferior angles**. The **lateral angle** is lost in an expanded articular process called the **glenoid cavity** which is supported by the short **neck** of the scapula. While the superior and medial borders are thin and sharp, the lateral border contains a bar of thick bone which blends with the glenoid cavity. This bar acts as a lever which trans-

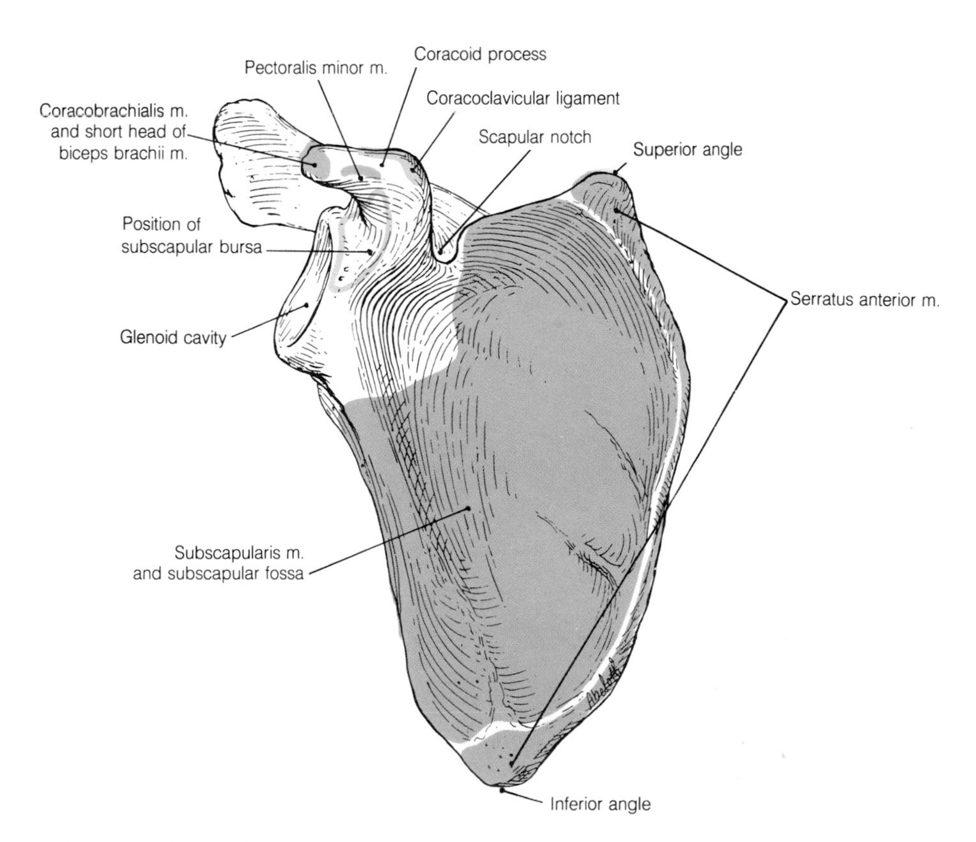

3-10 The ventral aspect of the right scapula.

mits the force of the lower four digitations of the **serratus anterior muscle** to the glenoid cavity.

From the upper surface of the neck of the scapula the **coracoid* process** projects upward and forward. It is the tip of this process that can be felt below the clavicle and lateral to the infraclavicular fossa. This small **scapular notch** is found where the superior margin joins the base of the coracoid process. The surface facing the chest wall is slightly hollowed and forms the **subscapular fossa** which shows ridges for muscular attachment.

The dorsal surface, which is seen in Fig. 3-11, is divided by the **spine** into a smaller upper part, the **supraspinous fossa**, and a larger lower part, the **infraspinous fossa**. The spine has a superior and inferior lip and, thus, presents a flattened surface posteriorly. If this is followed laterally, it is seen to become expanded into the flattened **acromion** which partly overhangs the glenoid cavity. The acromion has a small facet on its medial margin for articulation with the clavicle. Where the lateral border of the spine joins the neck of the scapula, the **spinoglenoid notch** is formed.

The shallow hollow of the glenoid cavity faces laterally and slightly anteriorly and articulates with the head of the humerus. Above it is found the **supraglenoid tubercle** and below the **infraglenoid tubercle**. This aspect of the scapula is shown in Fig. 3-12. Now identify the attachments of the following muscles and ligaments:

On the ventral surface of the scapula (Fig. 3-10):

1. The subscapularis muscle occupying the greater part of the surface and the bony ridges which show where intramuscular tendons are attached.
2. The serratus anterior muscle on the medial border. Note how this attachment is expanded at the inferior angle.

On the coracoid process (Fig. 3-10):

1. The short head of biceps brachii.
2. Coracobrachialis.
3. Pectoralis minor.
4. The coracoclavicular ligament which joins the coracoid process to the clavicle.

* Corax, *Gk.* = a crow; hence the idea of beak-like

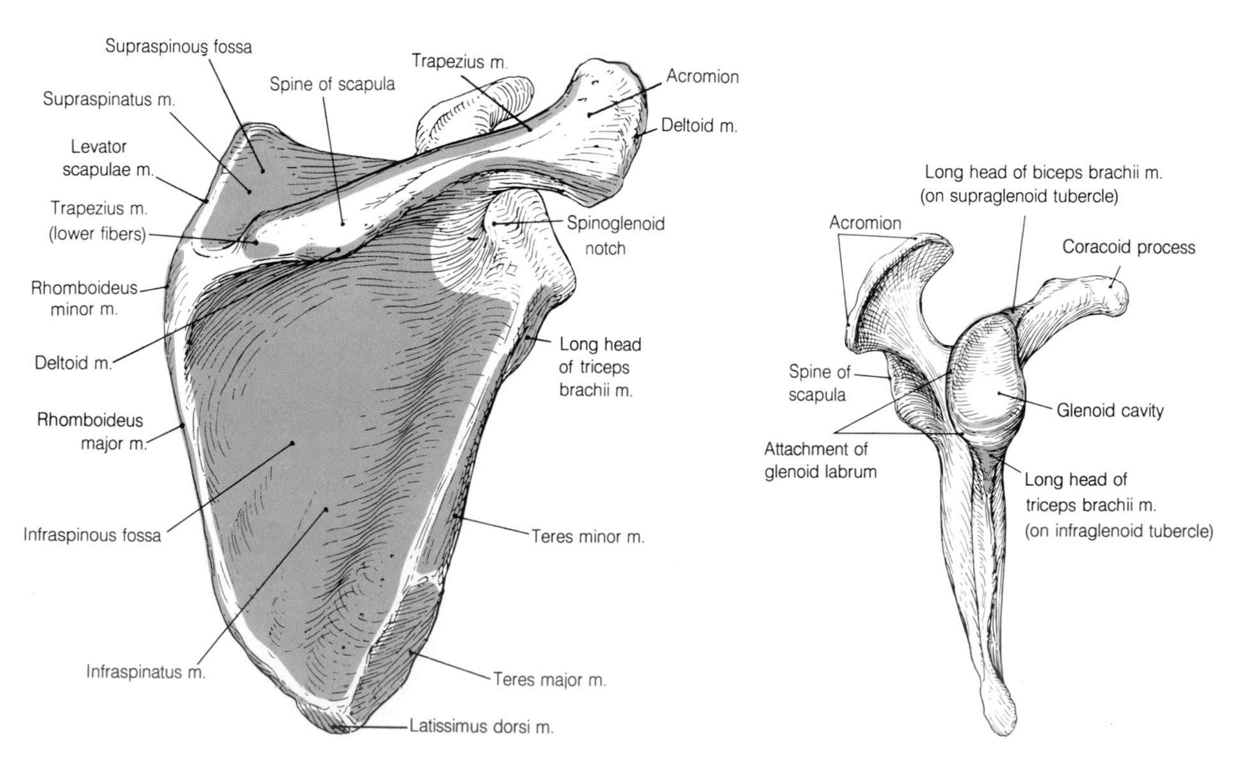

3-11 The dorsal aspect of the right scapula.

3-12 The lateral aspect of the right scapula.

5. Also note the position of the subscapular bursa.

On the glenoid process (Fig. 3-12):

1. The long head of biceps brachii to the supraglenoid tubercle.
2. The long head of triceps brachii to the infraglenoid tubercle.
3. Note also the attachment of the glenoid labrum to the margin of the glenoid cavity.

On the dorsal surface of the scapula (Fig. 3-11):

1. Supraspinatus filling the supraspinous fossa.
2. Infraspinatus in the infraspinous fossa.

On the medial border (Fig. 3-11):

1. Levator scapulae.
2. Rhomboideus major and minor.

On the lateral border (Fig. 3-11):

1. Teres minor.
2. Teres major.

On the acromion and spine (Fig. 3-11):

1. The deltoid muscle. The attachment of the deltoid muscle begins on the clavicle and is continued from the lateral margin of the acromion onto the lower lip of the spine.
2. Trapezius. This attachment to the upper lip of the spine is continued onto the medial margin of the acromion and thence to the clavicle. Compare the attachments of trapezius and deltoid.

The scapula develops in cartilage. A center of ossification for the body appears at the eighth week of intrauterine life and a center for the coracoid appears one year after birth. At puberty further centers appear for the coracoid, glenoid cavity, acromion, and one for the medial border and inferior angle. All the secondary centers fuse with the body by the twentieth year.

The Skeleton of the Thoracic Wall

A brief description of the skeleton of the chest wall is necessary at this point in order that the

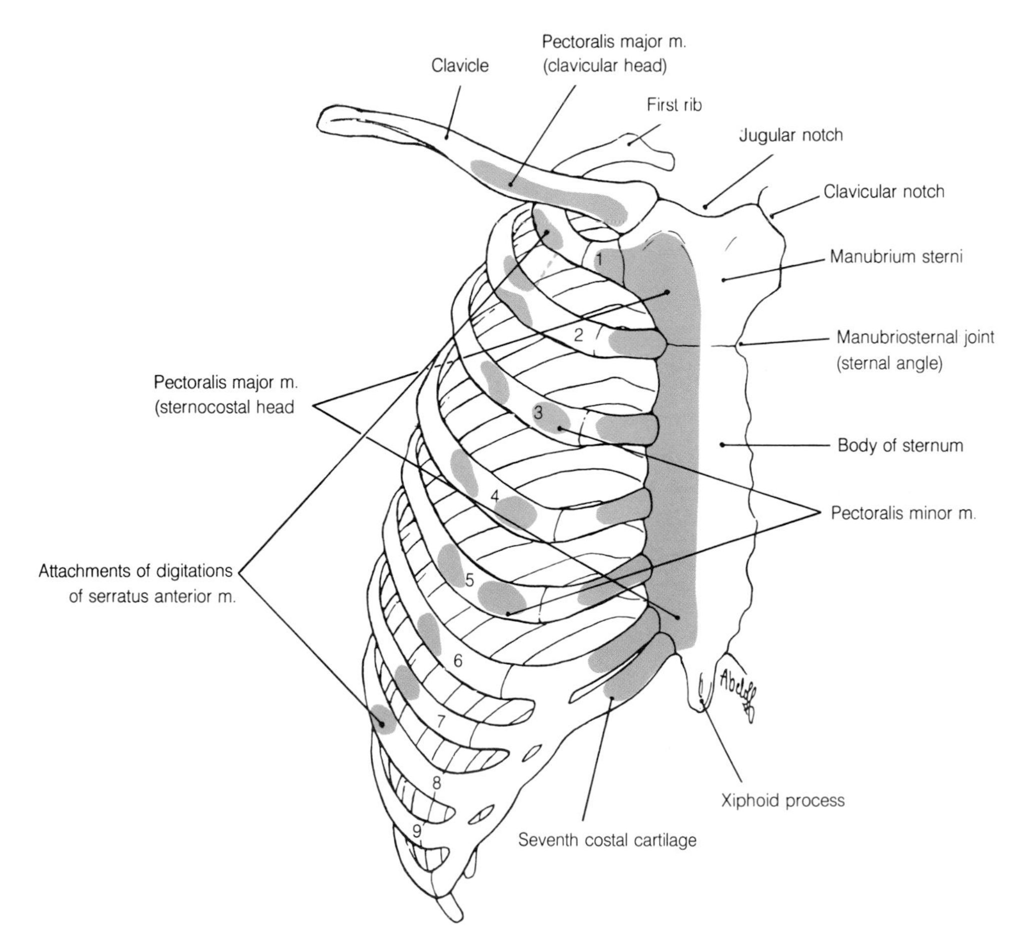

3-13 The skeleton of the thoracic wall.

proximal attachments of some upper limb muscles and the sternoclavicular joint may be understood.

The skeleton is illustrated in Fig. 3-13. It is made up of a series of **12 pairs of curved ribs** which articulate posteriorly with the thoracic vertebrae. From this point each slopes downward and forward. Anteriorly each bears a **costal cartilage**. The costal cartilages of the upper seven ribs articulate anteriorly with the **sternum**. The next three cartilages articulate with adjacent cartilages. The cartilages on the tips of the last two ribs lie free in the body wall musculature. The sternum consists of two parts. The most superior is the **manubrium** which articulates with the **body of the sternum** at the **manubriosternal joint**. At the lower end of the body lies the cartilaginous **xiphoid process**. The upper border of the manubrium is characterized by the midline **jugular notch** and on each side by the **clavicular notches**. Immediately below these lie the articular facets for the costal cartilages of the first rib.

In fig. 3-13 identify:

1. The attachment of pectoralis major to the manubrium and body of the sternum and to the costal cartilages of the upper seven ribs.
2. The attachment of pectoralis minor to the anterior extremities of the 3rd, 4th, and 5th ribs (or 2nd, 3rd, and 4th).
3. The attachments of the eight digitations of serratus anterior.

The Joints of the Pectoral Girdle

The Sternoclavicular Joint

At this joint the expanded proximal end of the clavicle articulates with the clavicular notch of the manubrium and the cartilage of the first rib below. It is a synovial joint characterized by the presence of an intraarticular disc which divides the joint cavity into two. The joint is surrounded by a fibrous capsule which is reinforced by **anterior** and **posterior sternoclavicular ligaments** and an **interclavicular ligament**. A more distant ligament which helps to stabilize the joint is the **costoclavicular ligament**, a bilaminar ligament joining the proximal end of the shaft of the clavicle to the first costal cartilage. The main features of this joint can be seen in Fig. 3-14.

The fact that it contains a fibrocartilaginous disc suggests a wide variety of movements and this is found to be so. The rotation, depression, elevation, protraction, and retraction of the clavicle which are associated with movements of the shoulder girdle are all reflected in movement at the sternoclavicular joint.

The Acromioclavicular Joint

The acromioclavicular joint which can also be seen in Fig. 3-14 is a small synovial joint between the flattened lateral end of the clavicle and the medial border of the acromion. The long axis of

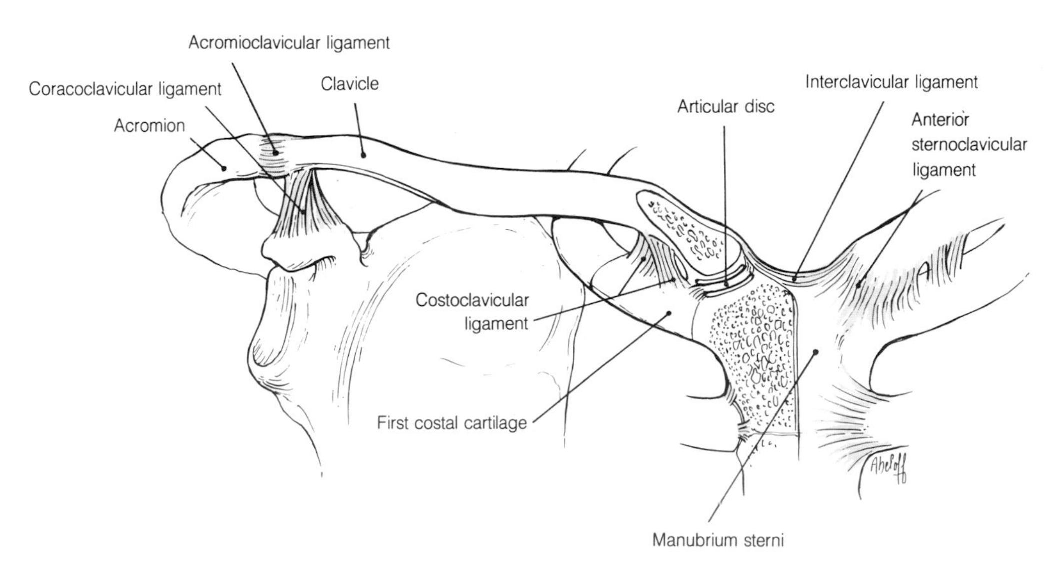

3-14 The right sternoclavicular and acromioclavicular joints.

the joint lies in an anteroposterior plane but its short axis is oblique as the clavicle overlaps the acromion. The joint is surrounded by a fibrous capsule which is reinforced superiorly by the **acromioclavicular ligament**. The range of movement of this joint is limited by the strong nearby **coracoclavicular ligament**. When this ligament becomes taut, movement of the scapula is transmitted to the clavicle. This joint presents a clinical problem as it is often dislocated in falls on the outstretched hand or on the point of the shoulder and may prove difficult to maintain in the reduced position.

The Skeleton of the Arm

The Humerus

The arm contains a single long bone called the **humerus**. This bone is illustrated in Fig. 3-15 where it can be seen that it possesses an expanded upper end or head whose medial aspect presents a rounded and smooth surface for articulation with the glenoid cavity of the scapula. The **body** or more commonly shaft is cylindrical in its upper part but below becomes flattened anteroposteriorly until at the lower end sharp medial and lateral borders are found. The rather irregular lower end has a **medial** and **lateral epicondyle** between which lie a spool-like **trochlea** for articulation with the ulna and a rounded **capitulum** for articulation with the radius.

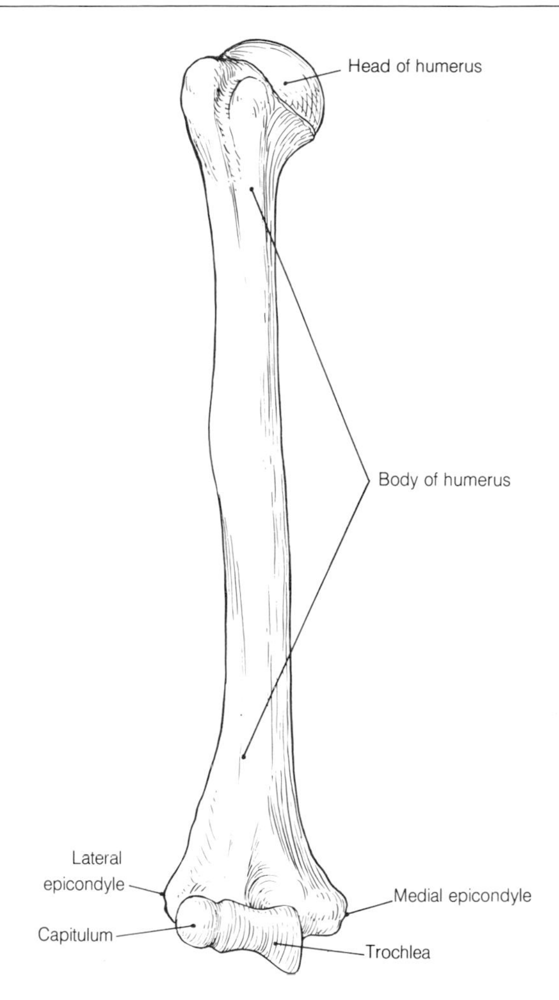

3-15 The anterior aspect of the right humerus.

Look now at the more detailed illustrations of the upper half of the bone in Fig. 3-16. Note again on the anterior view the rounded articular surface which forms rather less than half a sphere and is directed slightly upward and backward as well as medially. On the anterior aspect of the head is the **lesser tubercle**. This is clearly separated from the laterally placed **greater tubercle** by the **intertubercular sulcus** which contains the tendon of the long head of biceps brachii muscle.

The articular portion of the head is separated from the tubercles by a slight constriction. This is called the **anatomical neck** ($A-A$ in Fig. 3-16). The region where the combined head and tubercles join the body is called the **surgical neck** ($S-S$) and is a common site for fractures.

The intertubercular sulcus can be followed some way down the shaft. Its lateral lip leads to a roughened prominence about halfway down the shaft. This is the **deltoid tuberosity**.

Turn now to the posterior view. Two whole facets and a portion of a third which faces superiorly can be seen on the greater tubercle. On its posterior surface the body shows a roughened line running obliquely from medial to lateral border and below this a shallow **groove for the radial nerve**. Some important sites of attachments of muscles should now be examined.

On the anterior aspect of the shaft:

1. The deltoid muscle to the deltoid tuberosity.
2. The coracobrachialis muscle to the medial border of the shaft at the same level.
3. The pectoralis major to the outer lip, the latissimus dorsi to the floor, and the teres major to the inner lip of the intertubercular sulcus.

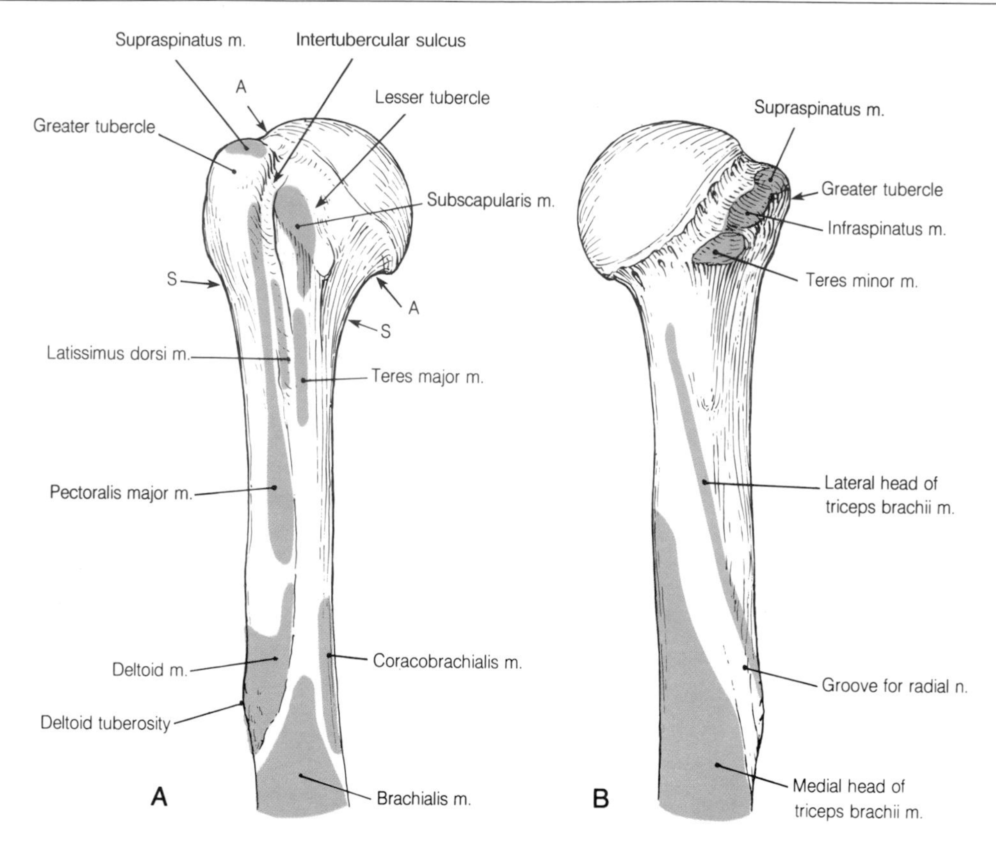

3-16 The anterior (*A*) and posterior (*B*) aspects of the upper end of the right humerus.

On the lesser tubercle:

1. The subscapularis muscle. Note that this attachment extends down onto the shaft below the tubercle and is in continuity with that of teres major muscle.

On the greater tubercle:

1. The supraspinatus muscle on the most superior facet.
2. The infraspinatus and teres minor muscles to the two facets on its posterior aspect.

On the posterior aspect of the body:

1. The linear origin of the lateral head of triceps brachii muscle above the groove for the radial nerve.
2. The medial head of triceps brachii muscle medial to and below the groove for the radial nerve.

The humerus develops in cartilage. A primary center of ossification for the body appears at the eighth week of intrauterine life. Secondary centers appear after birth. The center for the head appears at six months and is followed by those for the greater and lesser tubercles at two and five years. All three centers fuse to form a single epiphysis at six years before they join the body at the twentieth year. At the lower end secondary centers appear for the two epicondyles, for the capitulum, and for the trochlea between the first and twelfth years. The three lateral centers fuse with each other before joining the body at about sixteen years. That for the medial epicondyle becomes separated from the others by an extension of ossification of the body and does not join the body until the twentieth year.

Particular note should be made of the manner in which the secondary centers of ossification appear and fuse because the radiolucent cartilage that separates them from the body may be mistaken for fracture lines in radiographs.

Having considered the skeleton of the thoracic cage, the pectoral girdle, and the upper half of the humerus, these can now be covered by the muscles that connect the upper limb with the trunk and the girdle.

Muscles Connecting the Upper Limb to the Trunk

Muscles Arising from the Vertebral Column

A posterior group of muscles arising from the occipital bone, cervical, thoracic, lumbar, and sacral spines, and the iliac crest overlies the deeper intrinsic muscles of the back but is separated from them by a thick layer of deep fascia. Although these muscles arise from a region usually associated with the dorsal rami of spinal nerves, they are in fact innervated by ventral rami assisted by the (spinal) accessory cranial nerve.

Trapezius

Each trapezius muscle is an extensive flat triangular muscle lying over the posterior aspect of the neck and thorax on either side of the midline. Together the two muscles have the shape of a diamond or trapezium. The proximal attachment of the muscle is from the superior nuchal line and external occipital protuberance on the back of the skull, from the ligamentum nuchae (which is attached to the spines of cervical vertebrae), and from the spines and supraspinous ligaments of the twelve thoracic vertebrae. From this long, almost

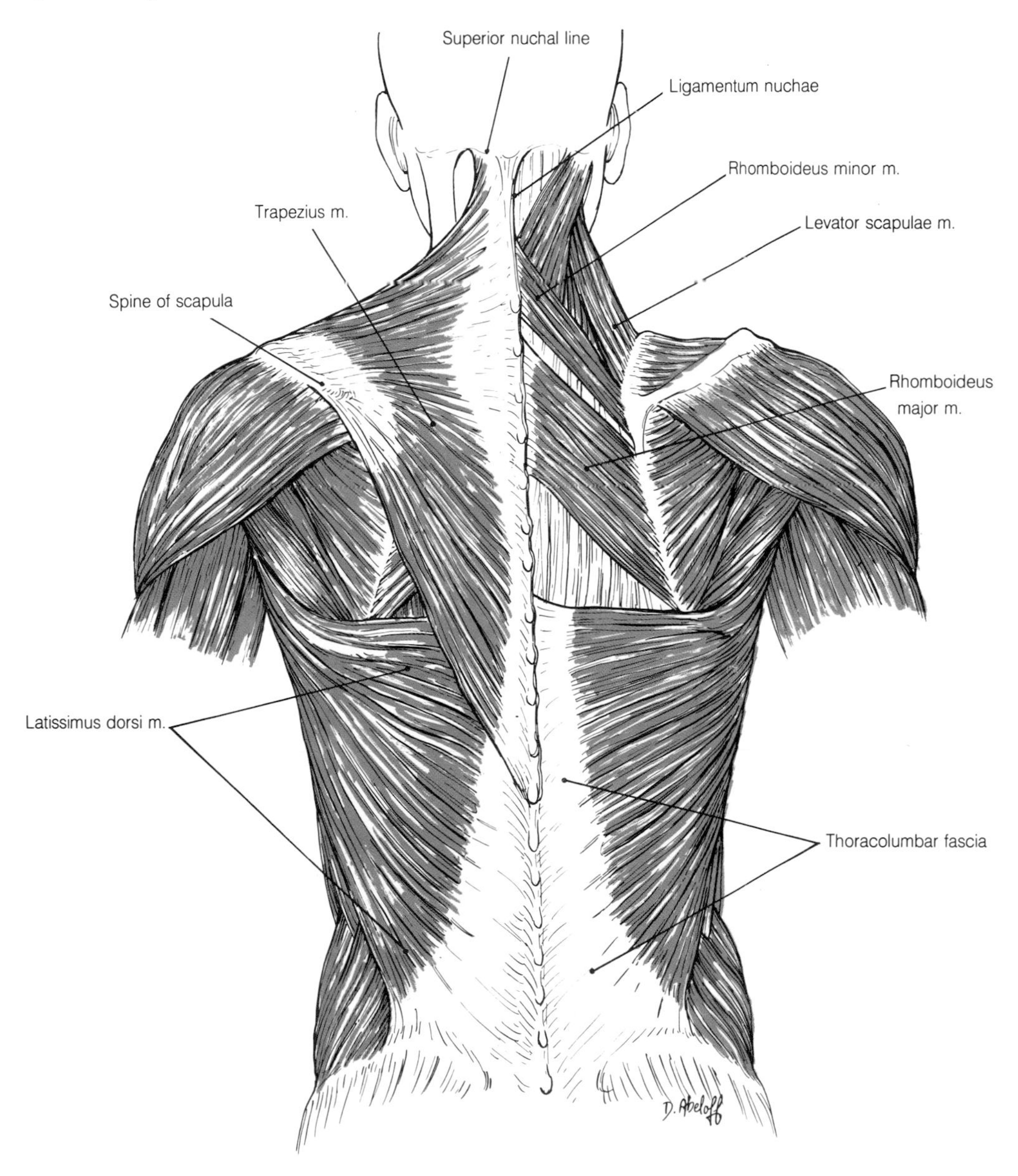

3-17 The superficial (*left* of figure) and deep (*right* of figure) muscles of the posterior aspect of the shoulder region and back.

linear attachment, the fibers of the muscle converge on the pectoral girdle and have a continuous insertion into the upper lip of the spine of the scapula, the medial border of the acromion and the posterior border of the lateral third of the clavicle. Look at Fig. 3-17 to see how the superior fibers roll over the shoulder on to the clavicle, the intermediate fibers run more horizontally to the acromion and spine, and the lower fibers converge on a small tubercle at the medial end of the spine.

The nerve supply of trapezius is unusual in that it receives its motor innervation from a cranial nerve, the **accessory nerve**, and its sensory fibers separately from ventral rami of the **third and fourth cervical nerves**. The accessory nerve approaches the deep surface of the muscle from the posterior triangle of the neck.

One of the most important actions of the trapezius is to rotate the scapula when the limb is raised above the head. This action will be discussed in more detail with the shoulder joint. It also helps to stabilize the shoulder in all movements of the upper limb and finally "square" and "shrug" the shoulder. The latter movement is used in a clinical examination to test the accessory nerve.

Levator Scapulae

When the trapezius muscle is reflected, the levator scapulae and rhomboid muscles are exposed. The levator scapulae descends as slips from the posterior tubercles of the transverse processes of the upper four cervical vertebrae to the medial border of the scapula between the superior angle and the spine. The muscle helps control the position of the scapula when the limb is used or loaded. In conjunction with the rhomboid muscles, it rotates the scapula in the opposite direction to trapezius, depressing the point of the shoulder and moving the inferior angle toward the midline.

The levator scapulae is supplied from the **third and fourth cervical nerves** and from a branch of the **dorsal scapular nerve**.

The Rhomboid Muscles

Rhomboideus minor and major have a continuous origin from the lower part of the ligamentum nuchae, the seventh cervical spine, and the upper five thoracic spines. The fibers slope downward and laterally to be inserted into the medial border of the scapula below levator scapulae. Both muscles are supplied by the **dorsal scapular nerve** from the brachial plexus. Their action is to stabilize and brace back the shoulder and with levator scapulae medially rotate the inferior angle.

Latissimus Dorsi

Latissimus* dorsi, like trapezius, is a large flat muscle. Although it is a muscle of the upper limb, its origin extends from the lower six thoracic spines, the lumbar spines, the sacral spines, and the iliac crest by means of the thoracolumbar fascia to which it is attached. In addition fleshy slips arise from the lower four ribs. The fibers converge toward the shoulder region where they wind around the muscles forming the posterior wall of the axilla. Here the muscle is replaced by a narrow flat tendon which is attached to the floor of the intertubercular sulcus of the humerus.

Latissimus dorsi is supplied by the **thoracodorsal nerve** from the posterior cord of the brachial plexus.

In conjunction with pectoralis major, latissimus dorsi is a powerful adductor of the arm, especially in climbing. Because it approaches the front of the humerus from its medial side, this muscle is also able to medially rotate the arm at the shoulder joint.

The general arrangement of these muscles is seen in Fig. 3-17.

Muscles Arising from the Thoracic Wall

Pectoralis Major

This is a large fan-shaped muscle having a extensive origin from the anterior surface of the medial half of the clavicle and the upper six costal cartilages and adjacent sternum. Additional fibers arise from the fascia covering the external oblique muscle of the abdomen. The clavicular and sternocostal portions can be distinguished by a shallow furrow between them. The muscle converges on the arm where it is inserted into the outer lip of the intertubercular sulcus of the humerus. In Fig. 3-18, which illustrates this muscle, note how

* Latus, *L.* = broad; latissimus is the superlative, i.e., broadest

the lower fibers curl beneath the upper to form the rounded lower border of the anterior wall of the armpit. If the flat tendon that attaches the muscle to the humerus is examined carefully, it can be seen to have two laminae. The clavicular and upper sternocostal fibers join the superficial lamina while the lower sternocostal fibers reach successively higher points on the deep lamina as they twist under the upper portion of the muscle.

Pectoralis major is supplied by both **medial** and **lateral pectoral nerves**. Note that these nerves are named after the cord of the brachial plexus from which they arise and not their anatomical relationships to each other. During dissection, it will be found that the lateral pectoral nerve reaches the deep surface of the muscle medial to the medial pectoral nerve which usually pierces pectoralis minor before reaching pectoralis major.

All fibers of pectoralis major are used in conjunction with latissimus dorsi to adduct the arm powerfully. This can easily be tested by pressing the palm against the thigh. However, if the arm is flexed by pushing against the edge of a table the clavicular fibers are brought into action while in pulling on the same table edge the sternocostal fibers are used. The muscle also assists in medial rotation of the arm.

Pectoralis Minor

Pectoralis minor is found beneath pectoralis major arising from the third, fourth, and fifth ribs (sometimes the second, third, and fourth) close to their costal cartilages. Its fibers converge on the coracoid process of the scapula.

It is supplied by the **medial pectoral nerve** and

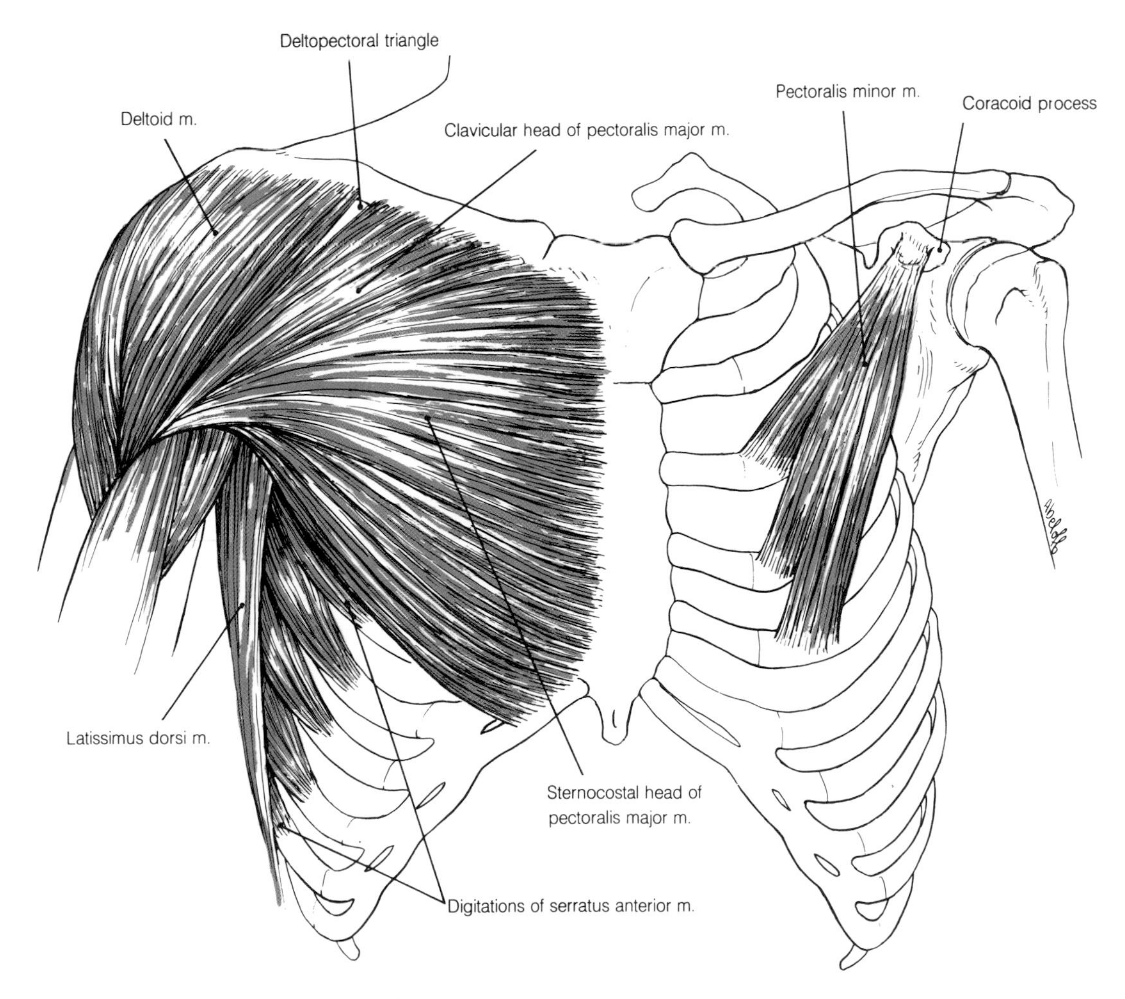

3-18 Showing the right pectoralis major muscle and the left pectoralis minor muscle.

its action is to help in pulling the scapula forward around the chest wall. In acute respiratory difficulty its action may be reversed and it is used to raise the ribs when the shoulder girdle is fixed. The left pectoralis minor muscle is illustrated in Fig. 3-18.

Serratus Anterior

Serratus anterior is a broad, flat muscle wrapped around the chest wall. It arises by eight fleshy digitations from the anterior extremities of the upper eight ribs and the intervening fascia covering the intercostal muscles. The digitations pass backward around the chest wall and beneath the scapula to be attached to the deep surface of its medial border. The first digitation which arises from the first and second ribs reaches the superior angle, the lower four digitations converge on the inferior angle, and the remainder are spread out along the medial border. This arrangment can be seen in Fig. 3-19 where the scapula has been lifted away from the chest wall.

Serratus anterior is supplied by the **long thoracic nerve** from the brachial plexus. The nerve passes through the axilla on the superficial surface of the muscle and may be damaged during excision of lymph nodes and radical operations for removal of the breast. Damage to the nerve and paralysis of the muscle lead to a "winged scapula." In this condition the medial border and inferior angle of the scapula protrude from the posterior thoracic wall. When efforts are made to use the muscle, as in pushing, the protrusion worsens.

Serratus anterior is used to draw the scapula and thus the pectoral girdle and arm forward around the chest wall in pushing and forward-reaching movements. Its powerful action on the inferior angle of the scapula rotates the angle laterally and in so doing turns the glenoid cavity upward. Use is made of this action each time the upper limb is raised above the head.

Subclavius

This is a small muscle arising from the junction of the first rib and costal cartilage and attaching to

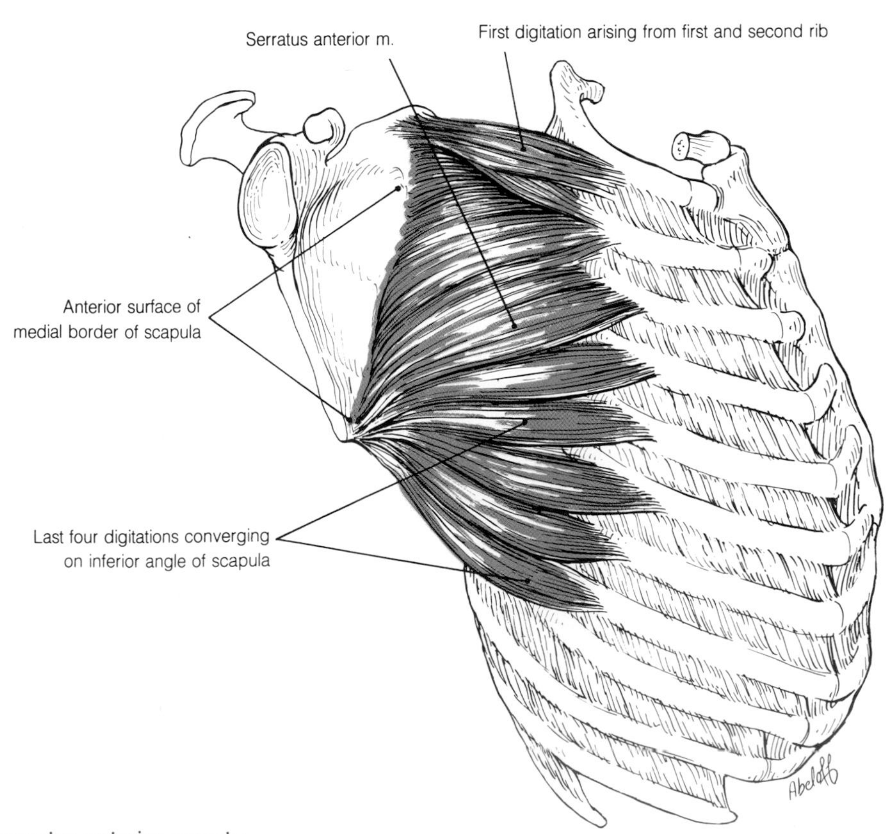

3-19 The right serratus anterior muscle.

the inferior surface of the clavicle. It is supplied by a **branch from the brachial plexus**. Its position suggest that it steadies the clavicle on the chest wall, but its action is difficult to test.

Sternocleidomastoid

Although situated in the neck, the sternocleidomastoid muscle is attached to both the pectoral girdle and the thoracic skeleton. Arising from the mastoid process of the temporal bone and the adjacent superior nuchal line, it descends obliquely downward and forward to be inserted into the manubrium sterni by a prominent tendon and to the medial third of the upper surface of the clavicle by a more fleshy attachment.

Like trapezius, the muscle is supplied by motor fibers from the **accessory nerve** and by sensory fibers from the **second and third cervical nerves**. Its actions are mainly concerned with tilting movements of the head.

The Clavipectoral Fascia

This is a thickening of the deep fascia that extends from the periosteum surrounding the clavicle to the axillary fascia in the floor of the axilla. Its arrangement is best understood from the diagram in Fig. 3-20 where it can be seen to enclose the subclavius muscle, fill in the gap between subclavius and pectoralis minor, surround pectoralis minor, and then reach and support the axillary fascia. The tough membrane it forms between subclavius and pectoralis minor is named the **costocoracoid membrane** and is pierced by the cephalic vein, the lateral pectoral nerve, and the thoracoacromial artery and vein with accompanying lymphatics. This region can be approached through the **deltopectoral triangle** between the clavicle and the adjacent borders of the deltoid and pectoralis major muscles below it.

Muscles of the Scapular Region

There is a group of short and mostly stout muscles between the scapula and upper humerus that are essential to the stability and movement of the shoulder joint. These are now described.

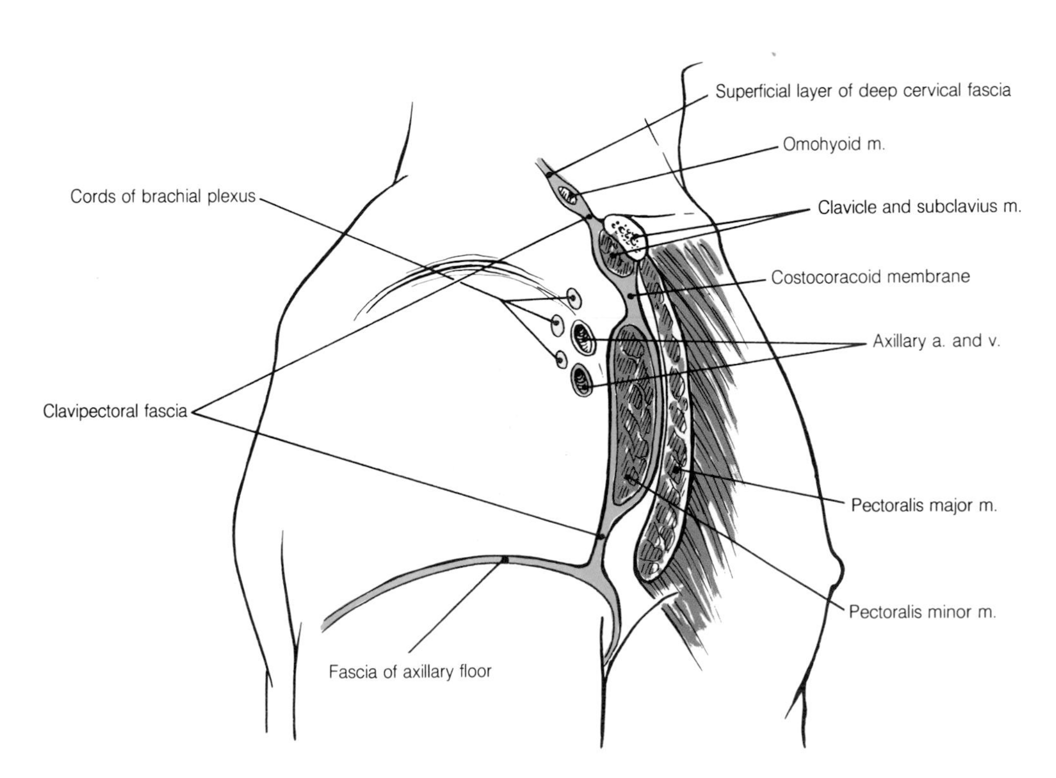

3-20 Diagram illustrating the clavipectoral fascia.

Subscapularis

In Fig. 3-21 subscapularis is seen to be a fanshaped muscle which arises from the greater part of the subscapular fossa. Its fibers pass laterally to converge on a tendon which is attached to the lesser tubercle of the humerus. It is sometimes difficult to picture the position of this muscle which clearly arises behind the chest but is attached in front of the humerus. If the scapula and upper humerus are considered as a unit and it is remembered that the subscapular fossa is on the anterior aspect of the scapula, the passage of the muscle from the anterior surface of one bone to the other becomes more understandable. The sketch seen in Fig. 3-22 should also help to make this clear. As will be seen later, this muscle forms the greater part of the posterior wall of the axilla.

As with a number of other short muscles around the shoulder which are grouped together as the **"rotator cuff" muscles**, the tendon of subscapularis becomes intimately fused with the anterior part of the fibrous capsule of the shoulder joint and in this way substantially reinforces it. In addition, a constant **subscapular bursa**, which communicates with the shoulder joint lies between the tendon and the neck of the scapula.

Subscapularis is innervated by the **upper and lower subscapular nerves**.

With the other short muscles to be described, subscapularis stabilizes the head of the humerus in the shallow glenoid cavity. When the arm is at the side, its attachment allows it to produce medial rotation of the humerus at the shoulder joint.

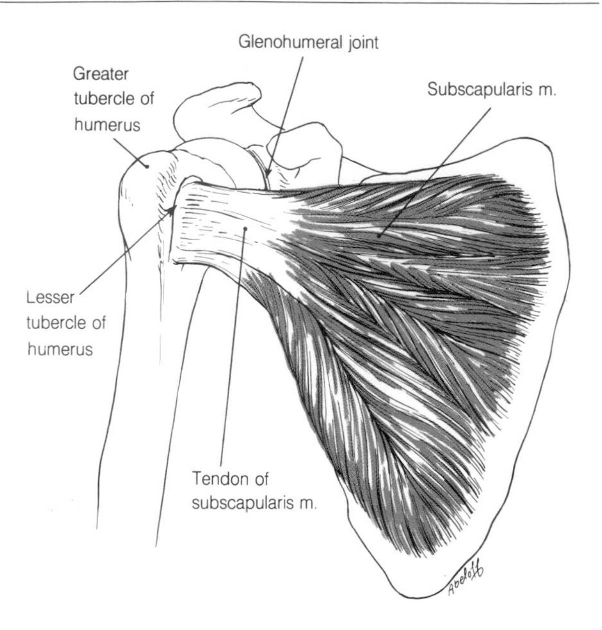

3-21 The right subscapularis muscle viewed from the front.

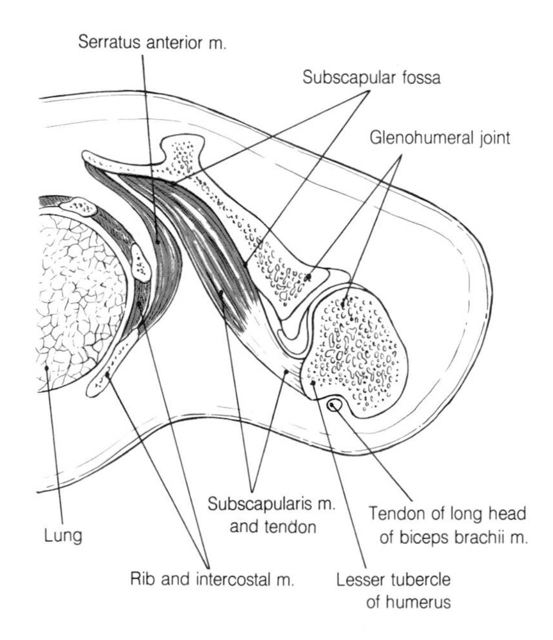

3-22 A horizontal section through the shoulder region to show the position of subscapularis.

Supraspinatus

Supraspinatus arises from the supraspinous fossa of the scapula and the tough overlying **supraspinous fascia**. The belly fills the fossa and almost hides it, but with the wasting of this muscle that is often associated with a painful shoulder joint the fossa can soon be recognized. The tendon of supraspinatus passes under the acromion and over the top of the shoulder joint to reach the highest of the facets on the greater tubercle of the humerus. The tendon fuses with the capsule of the shoulder joint and is separated from the overlying acromion by the **subacromial bursa**.

Supraspinatus is supplied by the **suprascapular nerve** and its action is to stabilize the shoulder joint, prevent the head of the dependent humerus slipping down off the glenoid cavity and, with the deltoid muscle, abduct the arm away from the body.

Infraspinatus

Infraspinatus arises from the infraspinous fossa and the overlying **infraspinous fascia**. Its fibers are directed up toward the shoulder joint and its tendon reinforces the capsule before becoming inserted into the middle facet on the greater tubercle. An **infraspinatus bursa** which may communicate with the joint cavity is sometimes present between the tendon and capsule.

Infraspinatus is supplied by the **suprascapular nerve** through a branch that reaches it through the spinoglenoid notch. Its action is to rotate the arm laterally and, with the other short muscles, stabilize the shoulder joint.

Teres Minor

Teres minor is a slender muscle arising from the upper two-thirds of the dorsal surface of the lateral margin of the scapula. Its fibers pass upward toward the posterior aspect of the shoulder joint where they are replaced by a tendon which becomes fused with the capsule of the joint. The tendon is inserted into the lowest of the three facets on the greater tubercle of the humerus and to the shaft for a short distance below this.

Teres minor is supplied by a branch of the **axillary nerve**. It joins the other short muscles in stabilizing the shoulder joint and, in addition, will produce lateral rotation of the dependent arm.

The Rotator Cuff

The subscapularis, supraspinatus, infraspinatus, and teres minor muscles are grouped together as the rotator cuff muscles. A look at the diagram in Fig. 3-29 explains the reason for this. Note how the tendons of these muscles surround the shoulder joint on all sides except inferiorly (the common site for dislocation). They are thus in an advantageous position to keep the large humeral head applied to the shallow glenoid cavity. Injuries to these muscles, which commonly occur in athletes and older people, severely limit the mobility of the joint.

Teres Major

Teres major arises from an area on the dorsal surface of the lateral border of the scapula below teres minor. It is helpful to think of this muscle as

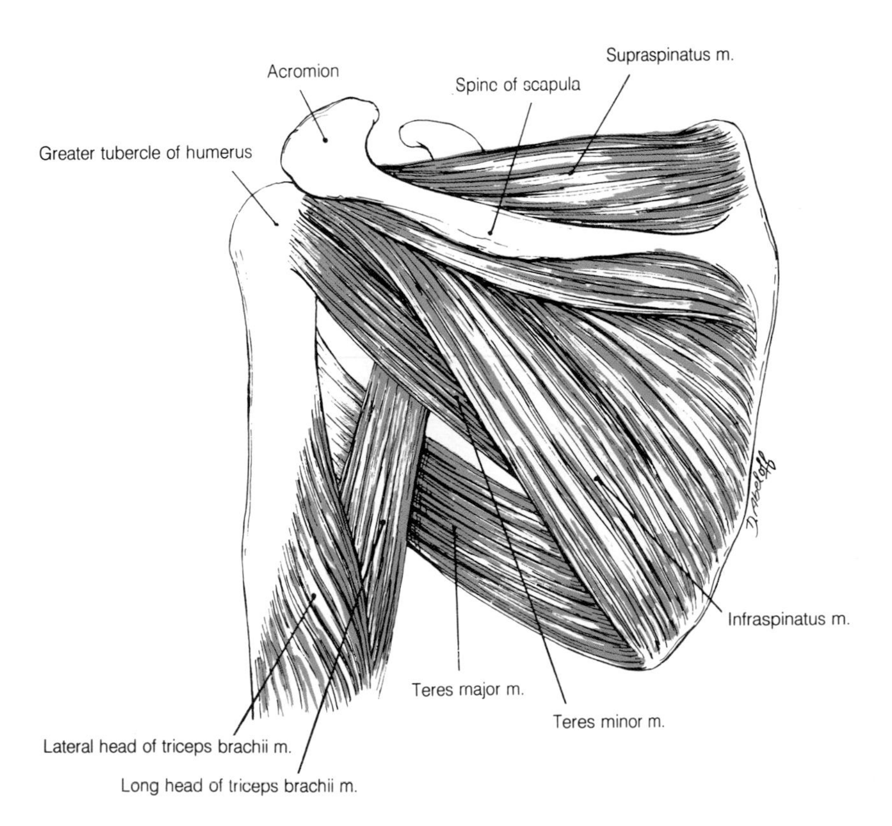

3-23 The posterior scapular muscles on the left side.

a portion of subscapularis that has migrated onto the dorsum of the scapula for it has a common nerve supply and its fibers, like those of subscapularis, pass upward and laterally onto the anterior surface of the humerus where they are attached to the medial lip of the intertubercular sulcus. Thus, its humeral attachment lies immediately below that of subscapularis to the lesser tubercle. It is supplied by the **lower subscapular nerve** and its action is to adduct and medially rotate the arm.

Look now at Fig. 3-23 and confirm the positions of each of these four scapular muscles on the posterior aspect of the shoulder.

The Quadrangular and Triangular Spaces

The configuration of the muscles below the shoulder is such that they allow a number of structures to pass from the anterior to the posterior regions of the axilla and arm. These "spaces" are best understood from an examination of Fig. 3-24. In this illustration note:

1. **A quadrangular space** bordered by:
   a. the surgical neck of the humerus
   b. teres minor and deep to it subscapularis
   c. the long head of triceps brachii
   d. teres major

   Through this space pass the axillary nerve and the posterior humeral circumflex artery.

2. **A triangular space** at the lateral border of the scapula bordered by:
   a. teres minor and subscapularis
   b. the long head of triceps brachii
   c. teres major

   Through this space passes the circumflex scapula branch of the subscapular artery.

3. **A triangular space** medial to the shaft of the humerus bordered by:
   a. the long head of triceps brachii
   b. teres major
   c. the shaft of the humerus

   Through this space pass the radial nerve and the profunda brachii artery.

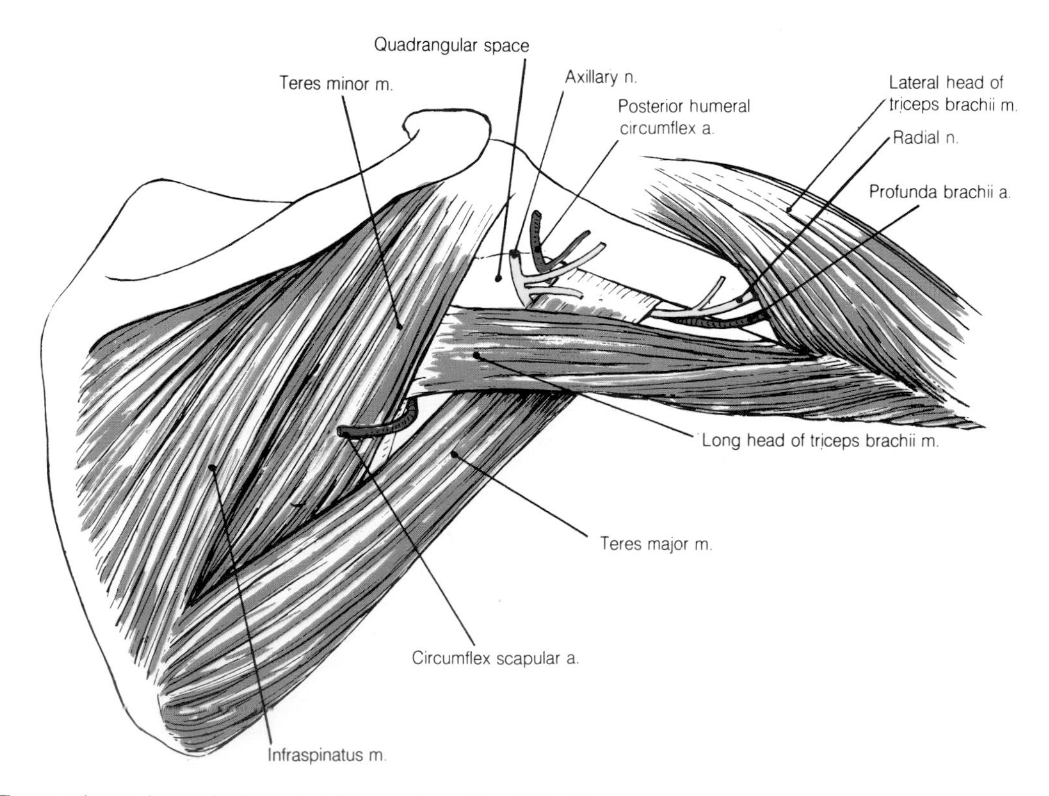

3-24 The quadrangular and triangulart "spaces" on the right side.

*Deltoid**

This powerful muscle is draped over the shoulder joint from its long V-shaped origin. This begins anteriorly over the anterior and superior aspect of the lateral third of the clavicle and continues around the outer margin of the acromion which carries it to the lower lip of the spine of the scapula. Its insertion into the humerus is at the deltoid tuberosity about half way down the lateral aspect of the shaft. The muscle is separated from the underlying shoulder joint by the **subacromial bursa** and the tendon of supraspinatus. The bursa is in fact more "subdeltoid" than subacromial.

The deltoid muscle is supplied on its deep surface by the **axillary nerve**, a branch of the posterior cord of the brachial plexus. It is this nerve that may be damaged when the shoulder joint is dislocated or while the dislocation is being reduced.

The V-shaped origin of the deltoid allows its fibers to approach the humerus anteriorly, laterally, and posteriorly. This fact, coupled with freedom of movement at the shoulder joint gives the muscle a wide range of actions. The anterior fibers are able to flex the arm, that is swing it forward, and the posterior fibers can extend it behind the body. The intermediate fibers arising from the acromion are powerful abductors of the arm. These fibers are arranged in a multipennate fashion and the intermuscular tendons to which they are attached arise from the several small tubercles on the lateral border of the acromion. The origin and these parts of the deltoid muscle are illustrated in Fig. 3-25.

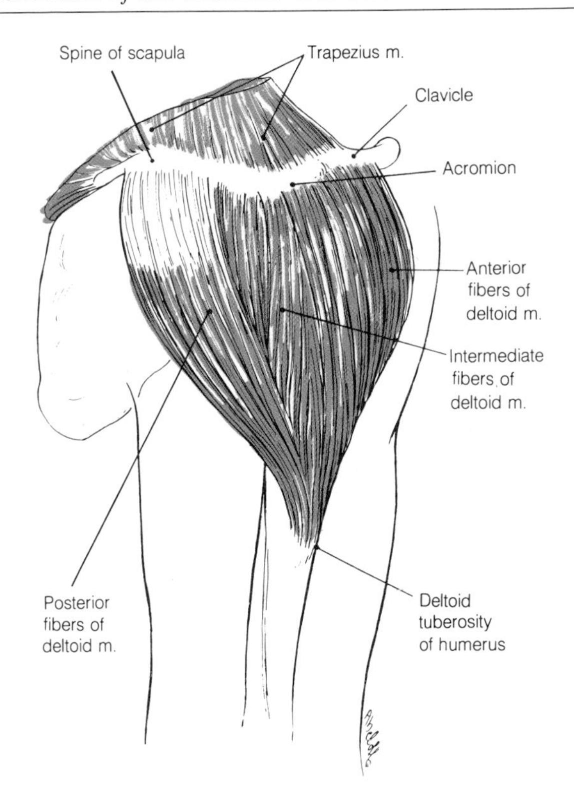

3-25 Showing the origin and parts of the right deltoid muscle.

The Shoulder Joint

The use made of the wide range of movements of the shoulder joint is not often appreciated until these are limited by disease or injury. Many everyday tasks, such as dressing or brushing the hair, require that the limb be elevated above the head and the details of this movement and the joints that it involves should be clearly understood.

The shoulder joint or **glenohumeral joint** is a multiaxial synovial joint of the ball and socket variety. However, the socket provided by the glenoid and surrounding glenoid labrum is not nearly as deep as the acetabulum of the hip joint and as a result the shoulder joint is prone to dislocation, especially among athletes. Some stability is provided by ligaments, but this joint depends largely on the muscles that surround it for its integrity. These do not span the joint inferiorly and it is, therefore, not surprising to find that this is a common site for a dislocation to occur.

The articular surfaces involved are the **glenoid cavity** of the scapula and the much larger **head of the humerus**. The depth of the glenoid cavity is slightly increased by a rim of fibrocartilage that surrounds it. This is the **glenoid labrum**.

The whole joint is surrounded by a fibrous capsule which is attached proximally to the glenoid outside the labrum. Superiorly it also includes the origin of the long head of biceps. Distally the capsule is attached to the margin of the articular surface of the head of the humerus. Because of the wide range of joint movement, the capsule is inevitably lax; nevertheless, the support it receives from the tendons of the surrounding "rotator

* Delta is the Greek letter D. In its capital form it is a triangle; hence deltoid or triangle-like.

cuff" muscles largely makes up for this deficiency. The capsule has a small opening anteriorly where the synovial cavity communicates with the subscapular bursa and another laterally that allows the tendon of biceps to pass through. Look at the illustration in Fig. 3-26 which demonstrates some details of the articular surfaces, the capsule, and the synovial sheath around the tendon of the long head of biceps.

Anteriorly the capsule is strengthened by three thickenings, the **superior, middle, and inferior glenohumeral ligaments**. Superiorly the **coracohumeral ligament** reinforces the capsule as it runs from the root of the coracoid process to the greater tubercle of the humerus. Arising from the supraglenoid tubercle within the joint capsule is the tendon of the long head of biceps. This arches over the head of the humerus to pierce the capsule and pass into the arm between the lesser and greater tubercles of the humerus. It is retained in the groove by the **transverse humeral ligament**. This tendon does not lie within the synovial membrane which lines the fibrous capsule, but is surrounded by a synovial sheath that is continued around the tendon into the intertubercular sulcus. While not directly concerned with the joint, the **coracoacromial ligament** forms an arch above it and prevents superior dislocation. Many of the features of the capsule and ligaments of the joint that have been described can be seen in Fig. 3-27.

Because of the multiaxial nature of this ball and socket joint the arm may be flexed, extended, abducted, adducted, circumducted, or rotated at the glenohumeral joint and the muscles that perform these movements are summarized below.

Flexion
- Pectoralis major (clavicular head when arm is in front of trunk, sternocostal head when behind)
- Anterior fibers of deltoid

Extension
- Posterior fibers of deltoid
- Teres major
- Latissimus dorsi

Abduction
- Deltoid
- Supraspinatus

Adduction
- Pectoralis major
- Latissimus dorsi
- Subscapularis
- Teres major and minor
- Infraspinatus

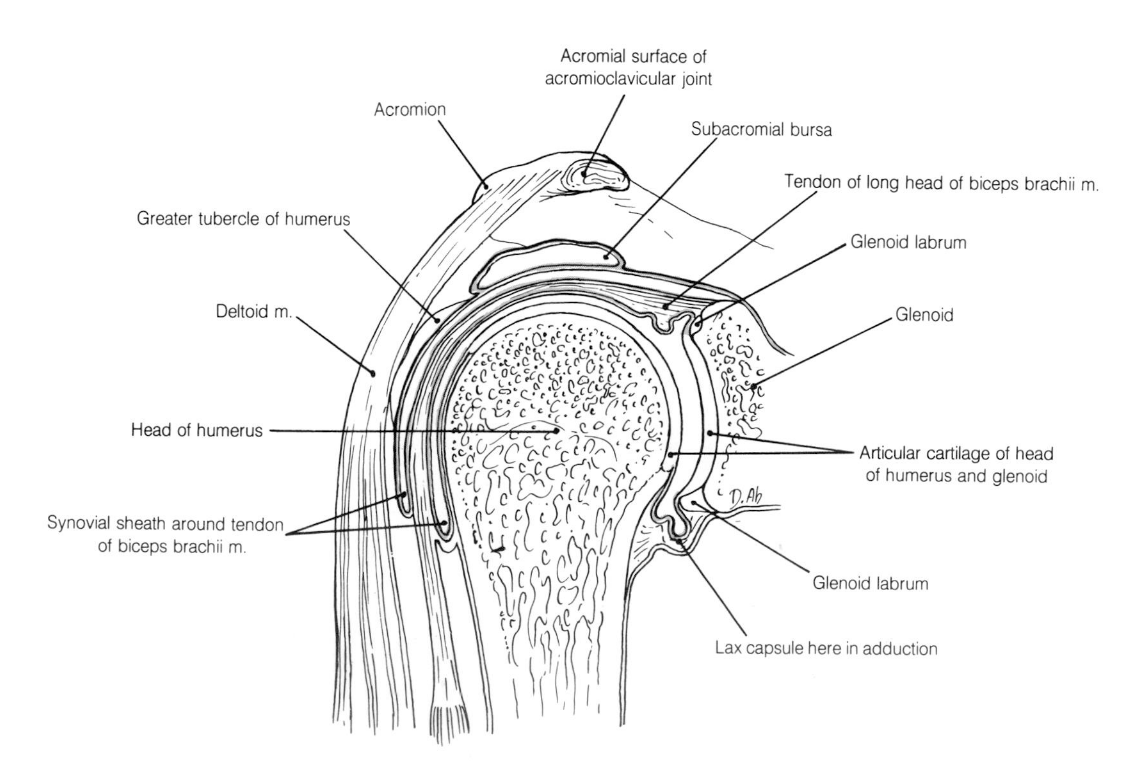

3-26 A coronal section of the shoulder joint.

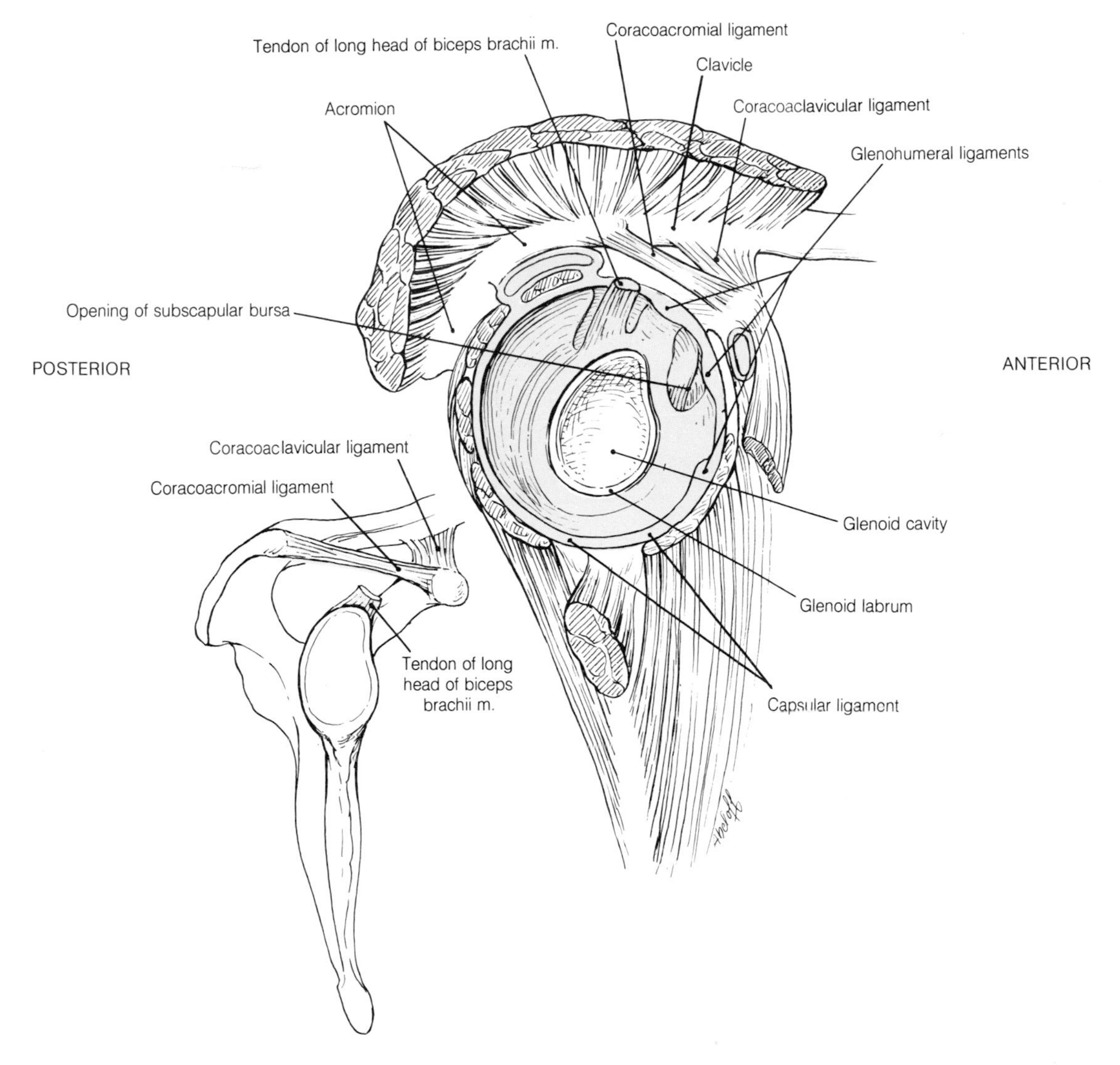

3-27 A section through the right shoulder joint made in a plane parallel to the glenoid cavity. The head of the humerus has been removed.

Lateral rotation
- Infraspinatus
- Teres minor
- Posterior fibers of deltoid

Medial rotation
- Pectoralis major
- Anterior fibers of deltoid
- Latissimus dorsi
- Teres major
- Subscapularis

Circumduction is produced by varying combinations of flexion, extension, abduction and adduction.

This list of muscles is not intended to be memorized. It should be used as a basis for reviewing the attachments of the muscles and from these, analyzing why the muscles are able to perform the movements.

The movement of abduction needs further examination. When the arm is at the side, the deltoid muscle can do little more than pull the head of the humerus upward. Patients with tears of their supraspinatus muscle find it impossible to start abduction, but, if helped over the first few degrees, can then complete the movement using their deltoid muscle. It is most likely, therefore, that supraspinatus is necessary to initiate the movement.

Further examination shows that a maximum of about 120° of abduction is possible at the glenohumeral joint. However, if the inferior angle of the scapula is felt as the arm is raised above the head, it becomes obvious that the movement of abduction does not occur solely at the glenohumeral joint. The remainder of the movement is achieved by rotating the inferior angle of the scapula laterally and forward so that the glenoid cavity is turned upward. If the inferior angle of the scapula is again felt it is clear that the rotation of the scapula begins soon after abduction of the arm is initiated and the two movements continue synchronously. As the scapula rotates, the coracoclavicular ligaments tighten and the rotation is transmitted to the clavicle and sternoclavicular joint. The mechanism whereby rotation of the scapula is achieved is illustrated diagrammatically in Fig. 3-28. The axis of rotation lies on the spine of the scapula and is indicated in the diagram by an asterisk. The powerful lower four digitations of serratus anterior, which are attached at the inferior angle of the scapula, swing this forward. The upper fibers of trapezius slightly raise the lateral extremity of the spine of the scapula while the lower fibers depress its medial end. In the erect position gravity will normally restore the position of the scapula but, if resistance is met, the rhomboid and levator scapulae muscles will be brought into play.

Fixation or arthrodesis of the glenohumeral joint is sometimes necessary and careful consideration of the angle of fixation allows useful scapular movement of the arm to be retained.

The important relations of the shoulder joint can best be appreciated from the illustration in Fig. 3-29 where a section has been made parallel to the plane of the glenoid cavity. A study of this will also serve as a useful review of the region.

Note above the joint:

1. the tendon of the long head of biceps brachii
2. the tendon of supraspinatus
3. the subacromial bursa
4. the acromion
5. the intermediate fibers of deltoid

Note anterior to the joint:

1. the opening of the subscapular bursa
2. subscapularis

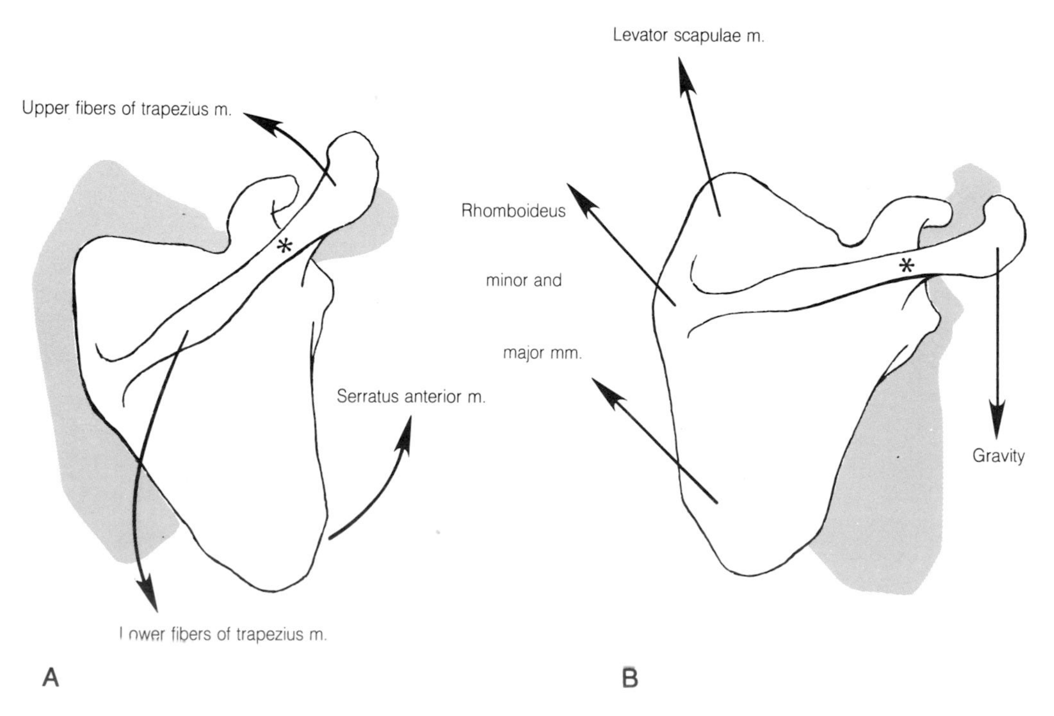

3-28 Diagrams showing how the inferior angle of the scapula is rotated laterally (*A*) and how its position is restored (*B*).

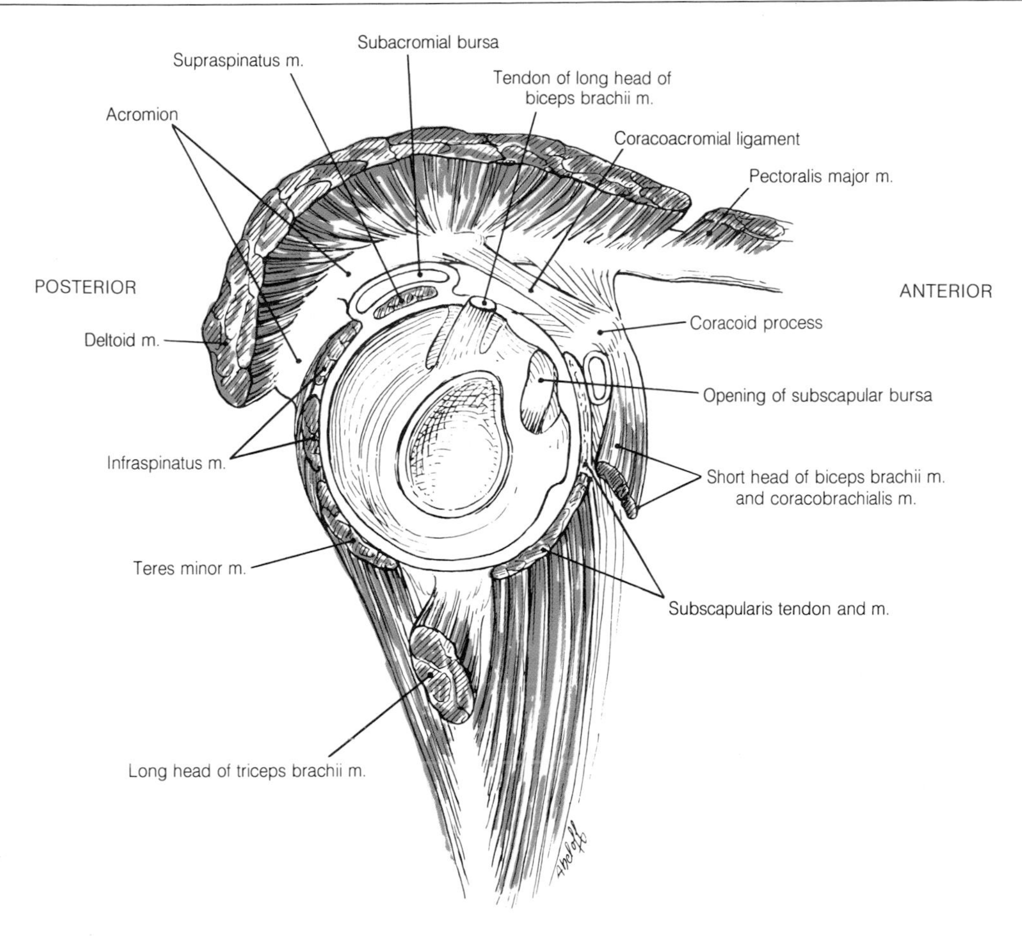

3-29 A section through the right shoulder joint made in a plane parallel to the glenoid cavity. The head of the humerus has been removed.

3. coracobrachialis and the short head of biceps brachii as they descend to the arm from the coracoid process
4. the anterior fibers of the deltoid muscle with pectoralis major below it would also lie anterior to the joint but have been removed.

Note below the joint:

1. the absence of muscles close to this aspect of the capsule of the joint except the long head of triceps brachii.

Note posterior to the joint:

1. infraspinatus
2. teres minor
3. the posterior fibers of deltoid.

The joint is supplied by the nearby **supra-scapular, axillary**, and **lateral pectoral nerves.**

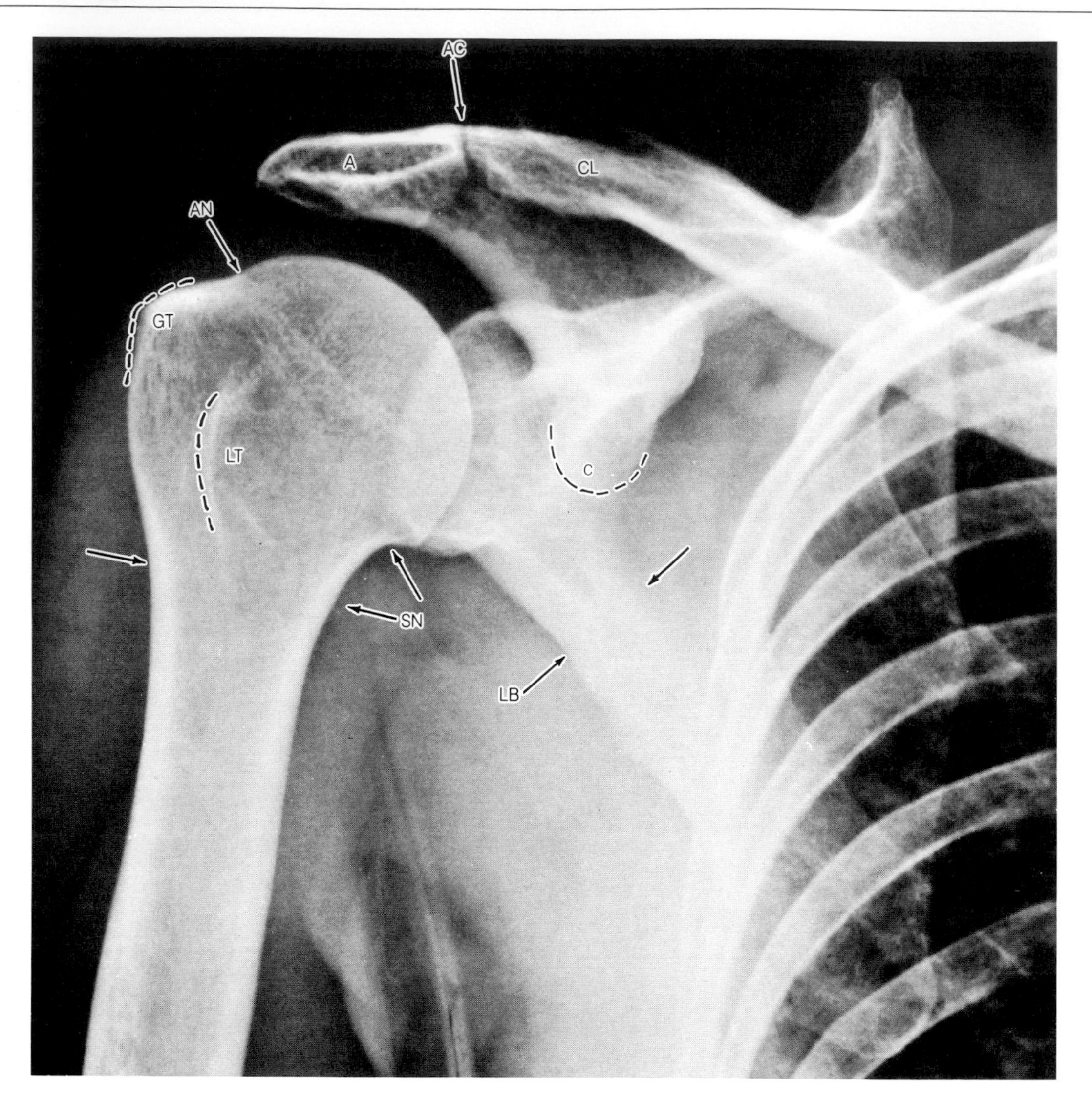

3-30 Radiograph of the right shoulder joint (a.p. projection) (Reproduced by permission from Wicke: *Atlas of Radiologic Anatomy*, 4th Ed, Urban & Schwarzenberg, Baltimore-Munich, 1987.)

The Radiological Anatomy of the Shoulder Joint

The shoulder joint can be demonstrated well by radiography and damage to the capsule can be revealed by the introduction of radiopaque material or air into the joint cavity. A study of the a.p. projection of the shoulder seen in Fig. 3-30 shows the normal relationship of the humerus to the scapula and serves as a review of the osteology of this region.

In the illustration note:

1. the head of the humerus and glenoid cavity
2. the acromion (*A*) and the clavicle (*CL*)
3. the acromioclavicular joint (*AC*) lying in an anteroposterior plane
4. the anatomical neck (*AN*) between the arrows
5. the greater (*GT*) and lesser tubercle (*LT*) with the intertubercular sulcus between
6. the surgical neck (*SN*) between the arrows
7. the thickened bone between arrows at the lateral border of the scapula (*LB*)
8. the tip of the coracoid process (*C*).

The Posterior Triangle of the Neck and the Axilla

The musculoskeletal arrangement of the shoulder region and upper arm have been described and a considerable number of nerves and vessels have been mentioned. It is now time to examine the source of these nerves and vessels.

The nerves are derived from ventral rami of cervical and thoracic spinal nerves and the plexiform arrangement into which they enter. This plexus is known as the brachial plexus and it approaches the limb from the posterior triangle of the neck above the level of the first rib.

The major vessel of the upper limb approaches it from the thoracic cavity and passing through the root of the neck enters the limb over the superior surface of the first rib.

The nerves and vessel join company at the root of the limb in a region known as the axilla. However, while there is little difficulty in understanding the source of the single artery and the termination of the large vein that accompanies it, to appreciate the formation of the brachial plexus a description of the posterior triangle of the neck is necessary.

The Posterior Triangle of the Neck

The term "posterior triangle" is a little misleading as it lies on the anterolateral aspect of the neck. It is also a very narrow triangle until its borders are drawn aside. Nevertheless, it serves as a good anatomical approach to a number of structures that pass into the upper limb.

This triangle is bounded below by a short length

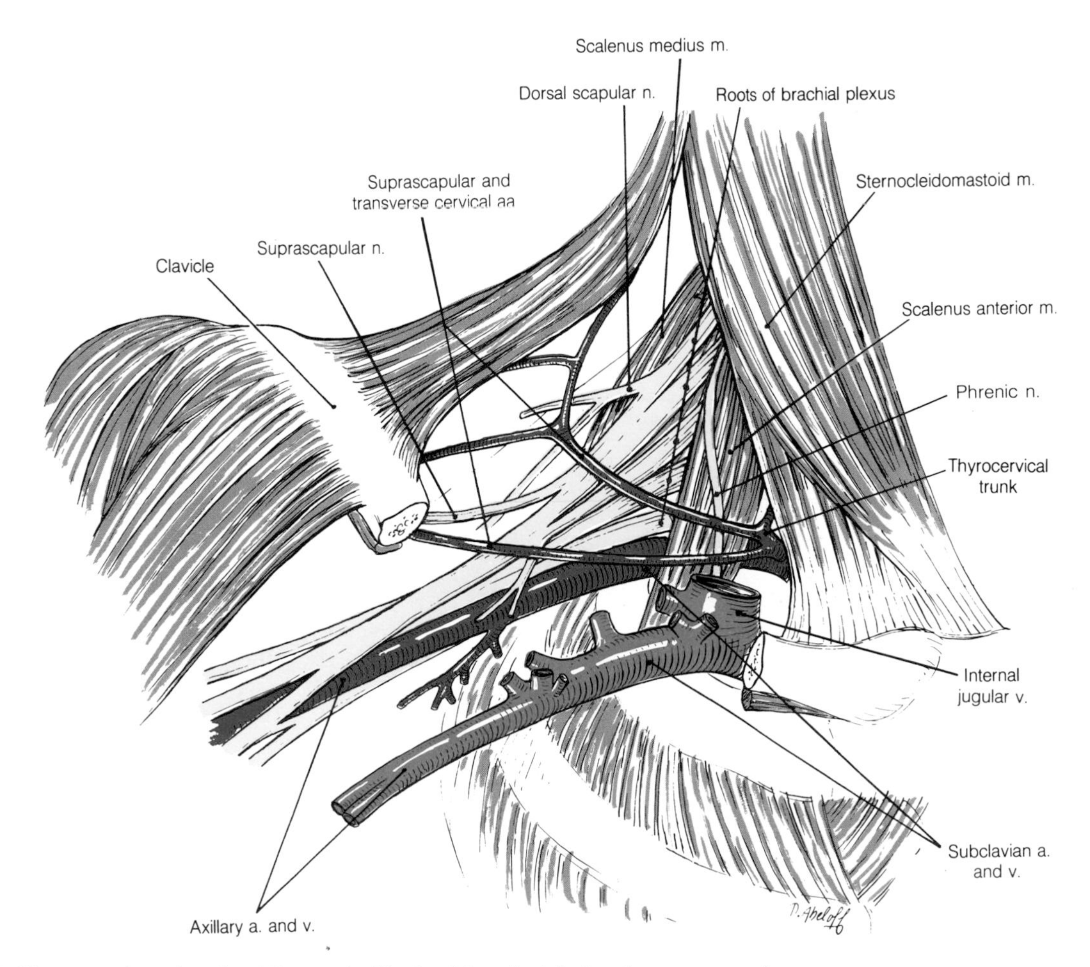

3-31 The posterior triangle of the neck. Much of the clavicle has been removed.

of clavicle which forms its base. The sides of the triangle are formed anteriorly by the lateral border of sternocleidomastoid and posteriorly by the anterior border of trapezius. The two muscular borders meet at the apex just posterior to the mastoid process. Enclosing the muscles and spanning the triangle is the investing deep fascia of the neck. This forms the roof of the triangle. More superficially lie the platysma muscle and skin. It is the contents of the triangle and that part of the root of the neck hidden by the clavicle that are of particular interest at this point. In Fig. 3-31 the borders of the triangle have been retracted and a portion of the clavicle removed to reveal the first rib. A number of veins and cutaneous nerves have also been eliminated.

In the illustration find the scalene muscles in the floor of the triangle. The **scalenus anterior** is the most anterior of these and close to the first rib this muscle separates the **subclavian vein** in front from the **subclavian artery** behind. (On the left side the subclavian artery arises directly from the **arch of the aorta** but on the right it is a branch of the **brachiocephalic trunk**. On both sides the subclavian vein joins the **internal jugular vein** to form the **left or right brachiocephalic vein**.) At the outer border of the first rib the subclavian artery changes its name to the **axillary artery** as it enters the axilla while the **axillary vein** leaving the axilla becomes the subclavian vein at the same point.

Behind and above the subclavian artery are seen the emerging roots of the **brachial plexus**. These unite and divide to form trunks and didions as they converge on the **axilla**. Branches of the plexus that should be noted in this region are the **dorsal scapular nerve**, the **long thoracic nerve**, and the **suprascapular nerve**.

At the medial border of scalenus anterior is found the **thyrocervical trunk** which is a branch of the subclavian artery. Two branches of this trunk cross the anterior surface of the muscle and the lower part of the triangle; these are the **transverse cervical** and **suprascapular arteries**.

The Axillary Fossa

As they leave the neck the brachial plexus and the axillary vessels enter a triangle bounded anteriorly by the clavicle, medially by the first rib and posteriorly by the scapula. This is the truncated apex of a pyramidal space called the axillary fossa or more simply, the axilla. The bony boundaries of this apex are illustrated in Fig. 3-32. The muscles forming the walls of the axilla have already been encountered and are summarized below:

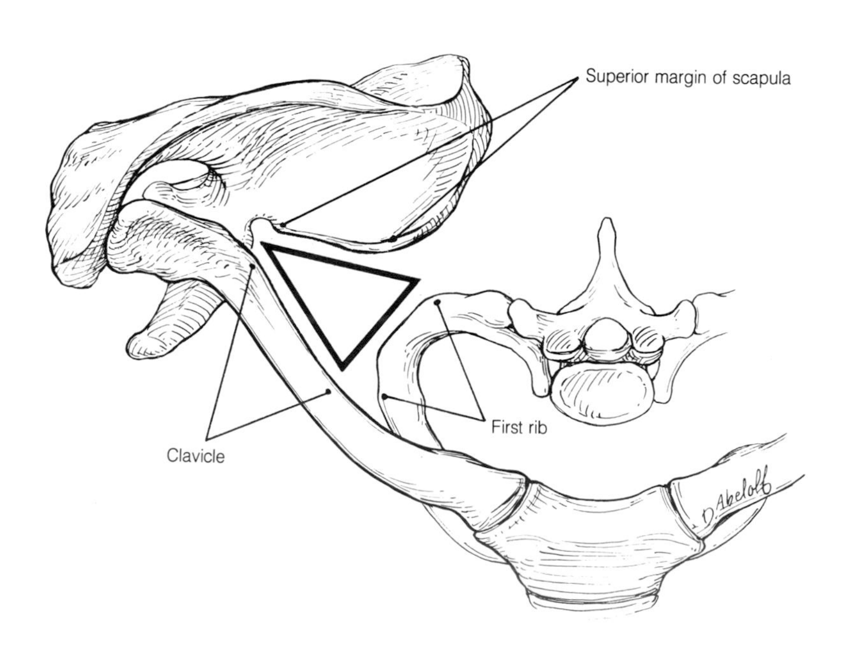

3-32 Diagram illustrating the bony boundaries of the apex of the axilla.

Medial wall:

Digitations of serratus anterior lying on the first four ribs and intervening intercostal muscles.

Anterior wall:

Pectoralis major
Pectoralis minor
Subclavius

Posterior wall:

Subscapularis
Teres major
Latissimus dorsi

The horizontal section of the axilla seen in Fig. 3-33 should help in the visualization of the walls of the axilla and also introduce the contents. Identify the muscles forming each of the walls (the section is too low to include the horizontally aligned subclavius muscle). At the lateral angle of the triangle are the long head of biceps, the upper portions of coracobrachialis, and the short head of biceps brachii. The greater part of the space is filled with fat, but within this lie the axillary artery and vein, axillary lymph nodes, and branches of the brachial plexus.

The lower borders of pectoralis major in the anterior wall and latissimus dorsi in the posterior

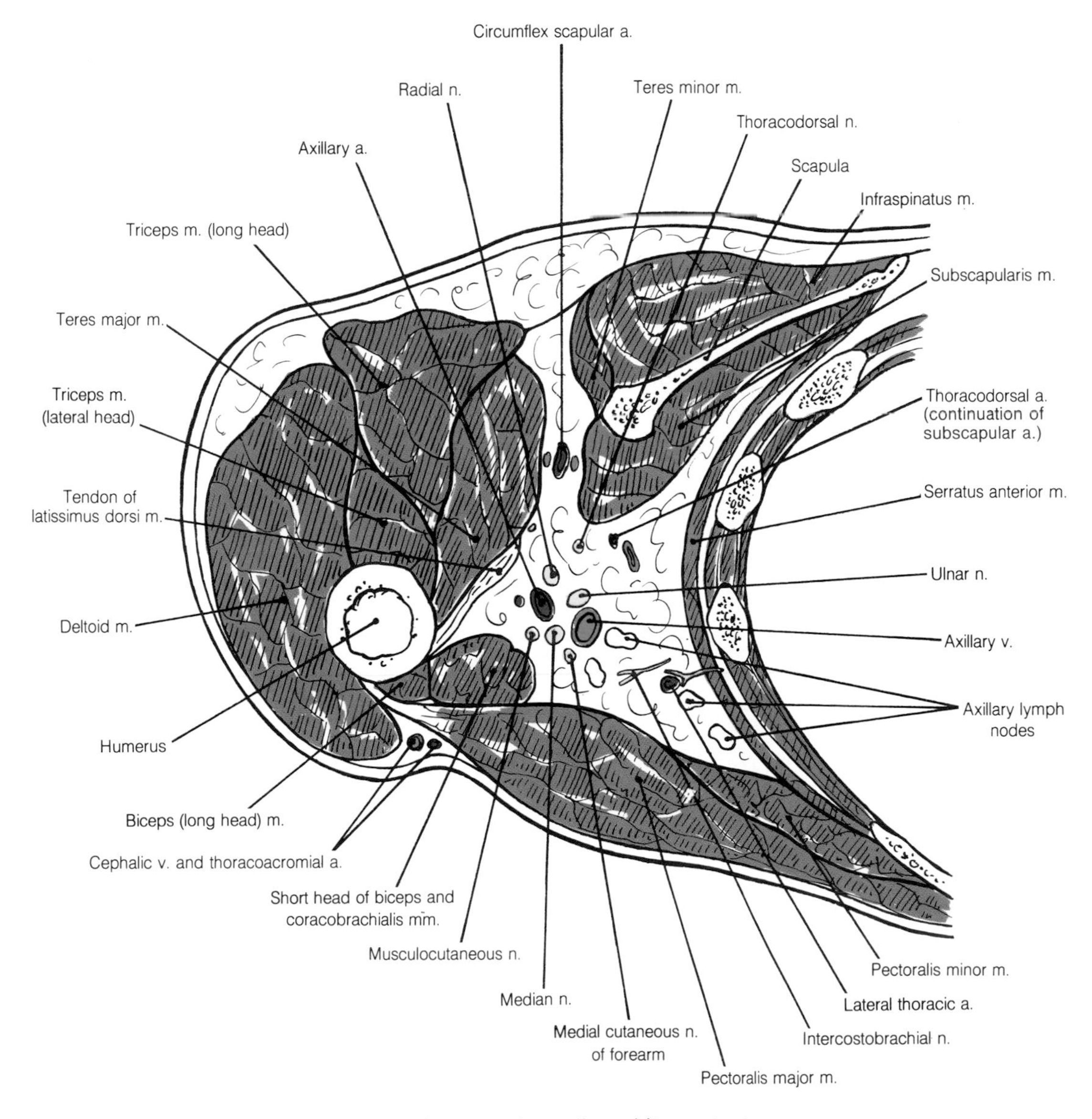

3-33 Horizontal section of the axilla to show its muscular walls and its contents.

wall are bridged by the tough **axillary fascia**. This forms the concave base of the axilla and the hollow of the "arm pit."

The contents of the axilla to be described are the brachial plexus, the axillary artery and vein, and the axillary lymph nodes.

The Brachial Plexus

The brachial plexus is a vulnerable structure prone to injury during an obstetric delivery, in road accidents, and in gunshot wounds. Its organization, therefore, needs to be known in some detail. There is no magic formula that makes this easy. Begin by understanding the basic pattern of its **roots, trunks, divisions** and **cords**. Then add the branches of the roots, the branches from the upper trunk, and finally those of the cords. The trunks are named after their relationship to each other while the cords are named after their relationship to the axillary artery.

The plexus is formed from a series of **roots**. These are in fact the ventral rami of the fifth to the eight cervical and the first thoracic spinal nerves. The use of the term "root" is confusing; the word implies no connection with the dorsal and ventral roots that form the spinal nerves. The roots lie within the prevertebral musculature and emerge into the posterior triangle between scalenus anterior and scalenus medius. Their subsequent rearrangement should be followed in Fig. 3-34. The upper two and lower two roots unite to form the **superior and inferior trunks**. The root from the seventh cervical nerve continues as the **middle trunk**. Behind the clavicle each trunk divides into an **anterior** and a **posterior division**. In the axilla the posterior divisions all unite to form the **posterior cord**. This supplies extensor structures on the posterior aspect of the limb. The two upper anterior divisions unite the form the **lateral cord**. The remaining anterior divison forms the **medial cord**. The two cords formed by anterior divisions supply flexor structures on the anterior aspect of the limb.

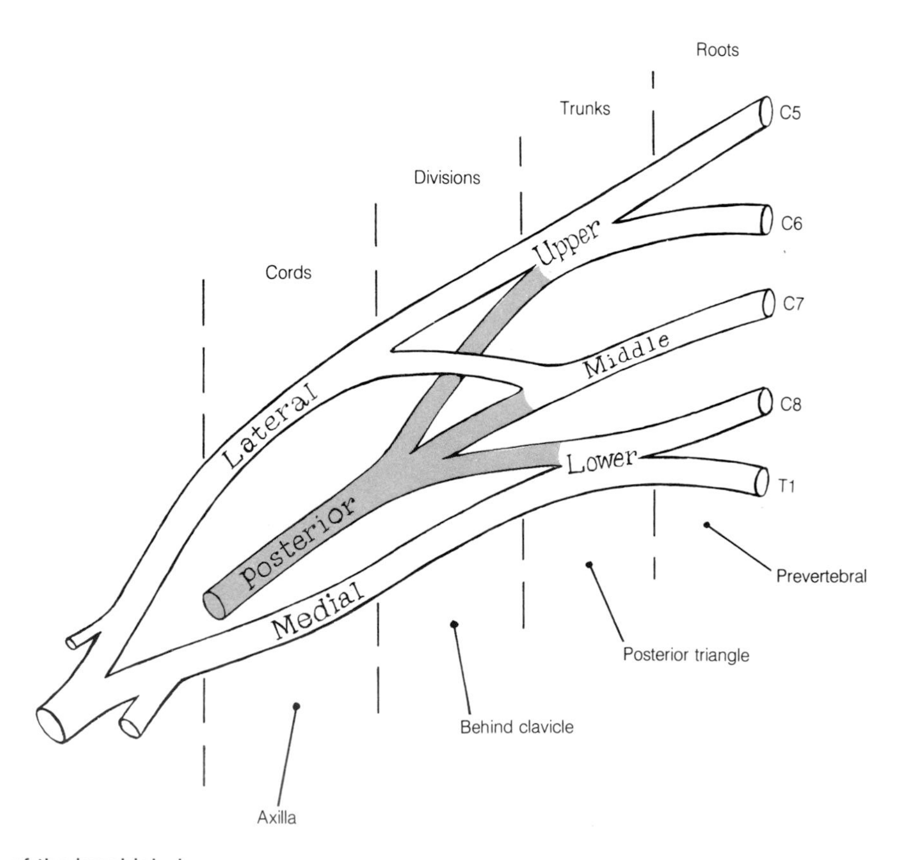

3-34 Diagram of the brachial plexus.

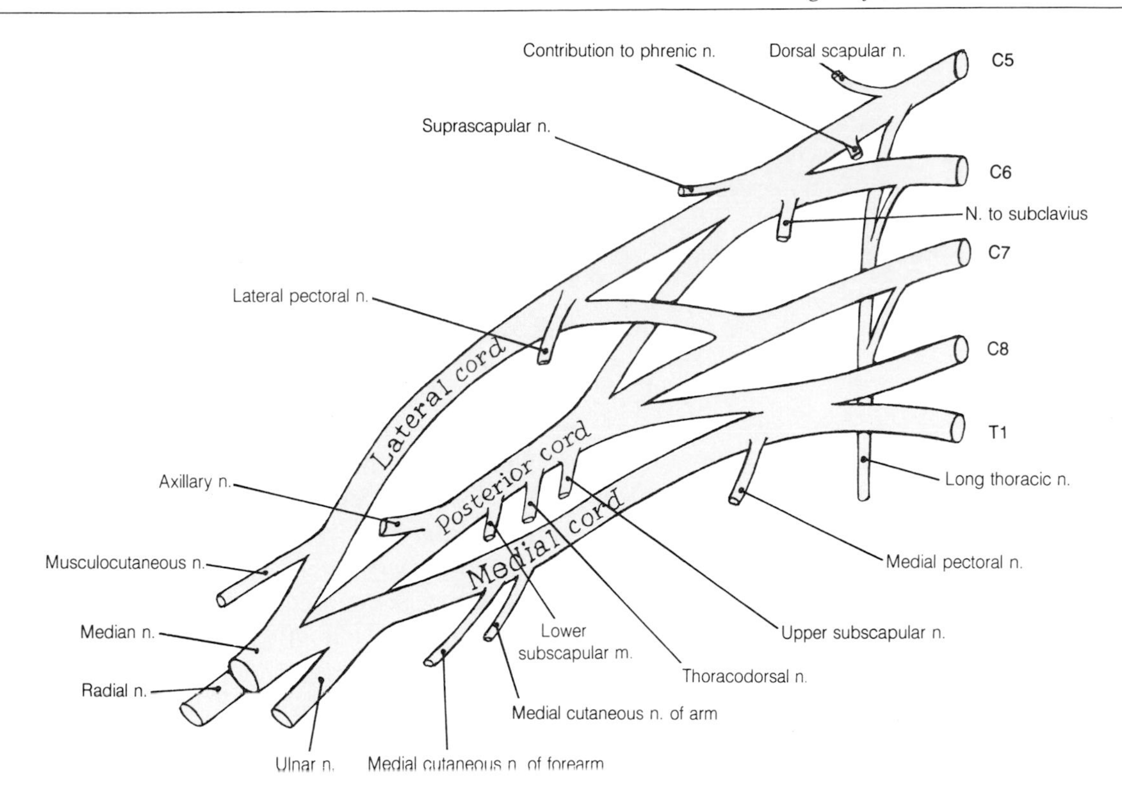

3-35 Diagram of the brachial plexus including its branches.

This plexiform arrangement allows neurons from several segmental levels of the cord to be distributed to the periphery in a single nerve. For example, a look at Fig. 3-34 shows that branches of the posterior cord could in theory contain neurons from all spinal nerves contributing to the plexus.

Having established the basic arrangement of the plexus it is now possible to add the branches of distribution. In Fig. 3-35 these have been added to the plan of the plexus shown in Fig. 3-34. The segmental origin of the neurons in each nerve are given in the text.

Branches from the Roots

The **dorsal scapular nerve** (C5) usually pierces scalenus medius to reach the deep surface of levator scapulae. Here it joins the deep branch of the dorsal scapular artery and descends over the deep surface of the rhomboid muscles which it supplies.

The **long thoracic nerve** (C5, 6, 7) is formed by branches from the fifth, sixth, and seventh cervical roots. The first and second branches unite within scalenus medius and pierce it to run over the chest wall posterior to the brachial plexus and axillary artery. Here the third root joins the nerve which continues inferiorly over the superficial surface of serratus anterior which it supplies. During operations performed for the removal of the breast and the excision of axillary lymph nodes, care is taken to avoid injury to the nerve as it lies on the surface of the muscle.

Branches from the roots also supply the scalene muscles and the longus colli, a prevertebral muscle in the neck.

It should be realized that the anterior primary ramus of the first thoracic nerve also contributes to the innervation of the chest wall by giving off the first intercostal nerve.

Branches of the Upper Trunk

The **suprascapular nerve** (C5, 6) is an important and large nerve which runs deep to trapezius to reach the supraspinous fossa by passing through the scapular notch below the ligament that bridges it. In the supraspinous fossa it supplies supraspinatus and then passes with the suprascapular artery through the spinoglenoid notch to supply infraspinatus in the infraspinous fossa. The scapular course of both nerve and artery are seen in Fig. 3-36.

A small **nerve to subclavius** is also given off from the upper trunk.

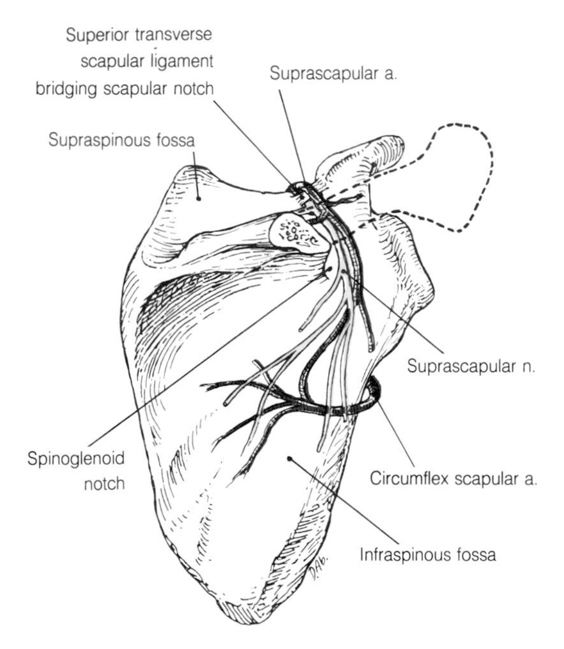

3-36 Showing the scapular course of the suprascapular nerve and artery.

Branches of the Cords

- The Lateral Cord
  - The musculocutaneous nerve (C5, 6, 7)
  - The lateral root of the median nerve (C6, 7)
    This joins with a similar contribution from the medial cord to form the median nerve
  - The lateral pectoral nerve (C5, 6, 7)
- The Medial Cord
  - The ulnar nerve (C8, T1)
  - The medial root of the median nerve (C8, T1)
  - The medial pectoral nerve (C8, T1)
  - The medial cutaneous nerve of the arm (C8, T1)
  - The medial cutaneous nerve of the forearm (C8, T1)
- The Posterior Cord
  - The radial nerve (C5, 6, 7, 8, T1)
  - The axillary nerve (C5, 6)
  - The upper subscapular nerve (C5, 6)
  - The thoracodorsal nerve (C6, 7, 8)
  - The lower subscapular nerve (C5, 6)

The musculocutaneous, median, ulnar, and radial nerves will all be described with the arm.

The **lateral pectoral nerve** crosses the axillary vessels and pierces the clavipectoral fascia to supply pectoralis major from its deep surface. Remember that on the chest wall it lies medial to the medial pectoral nerve.

The **medial pectoral nerve** passes forward between the axillary artery and vein to reach pectoralis minor. Some branches of the nerve supply pectoralis minor while other branches pierce it to reach and supply pectoralis major.

The medial cord gives off two cutaneous nerves, the medial cutaneous nerves of the arm and forearm. The **medial cutaneous nerve of the arm** supplies the medial aspect of the arm below the area supplied by the intercostobrachial nerve. The **medial cutaneous nerve of the forearm** pierces the deep fascia of the arm with the basilic vein and supplies skin along the course of the vein as far as the wrist.

The **intercostobrachial nerve**, although not a branch of the brachial plexus, contributes to the innervation of the skin of the upper limb. It is the lateral cutaneous branch of the second intercostal nerve and supplies skin of the axilla and upper medial aspect of the arm.

The **axillary nerve** is given off by the posterior cord and soon leaves the axilla by passing posteriorly through a quadrangular space bounded above by subscapularis, below by teres major, laterally by the surgical neck of the humerus and medially by the long head of triceps. Since teres minor overlies the posterior surface of subscapularis, the nerve appears at the back of the shoulder with this muscle above it. Now deep to the deltoid, it divides into an anterior and posterior branch. The former supplies the deltoid muscle and gives off a few perforating twigs to the overlying skin. The latter supplies both deltoid and teres minor before it appears at the posterior border of deltoid as the **upper lateral cutaneous nerve of the arm**. Note again the proximity of this nerve to the surgical neck of the humerus and the shoulder joint.

The **two subscapular nerves** arise from the posterior cord as it lies on subscapularis. They supply the subscapularis and teres major muscles.

The **thoracodorsal nerve** usually arises from the posterior cord between the subscapular nerves. It runs downward over the posterior wall of the axilla with the subscapular artery to reach lattisimus dorsi and supply it.

The relations of the brachial plexus in the axilla are so intimately connected with those of the axillary vessels that they will be described together.

The Axillary Artery

It has been seen already that the **axillary artery** is a continuation of the subclavian artery. The name axillary is assumed as the vessel passes into the axilla over the outer border of the first rib. In a similar manner the same vessel becomes the brachial artery as it leaves the axilla at the lower border of teres major and enters the brachium or arm. In a somewhat arbitrary manner, it is divided into three parts by pectoralis minor. The first part lies above the level of the muscle, the second behind it and the third below it. This scheme does have some merit in that the number of the part also indicates the number of named branches that it gives off. The artery is illustrated diagrammatically in Fig. 3-37, and this should be used to follow the branches described below.

Branches from the First Part

The **superior thoracic artery** is a small vessel that follows the upper border of pectoralis minor to supply both pectoral muscles and the thoracic wall. It also contributes to the blood supply of the breast.

Branches from the Second Part

The **thoracoacromial artery** is a short trunk that curls round the upper medial border of pectoralis minor, pierces the clavipectoral fascia, and breaks up into branches supplying the pectoral, acromial, clavicular, and deltoid regions.

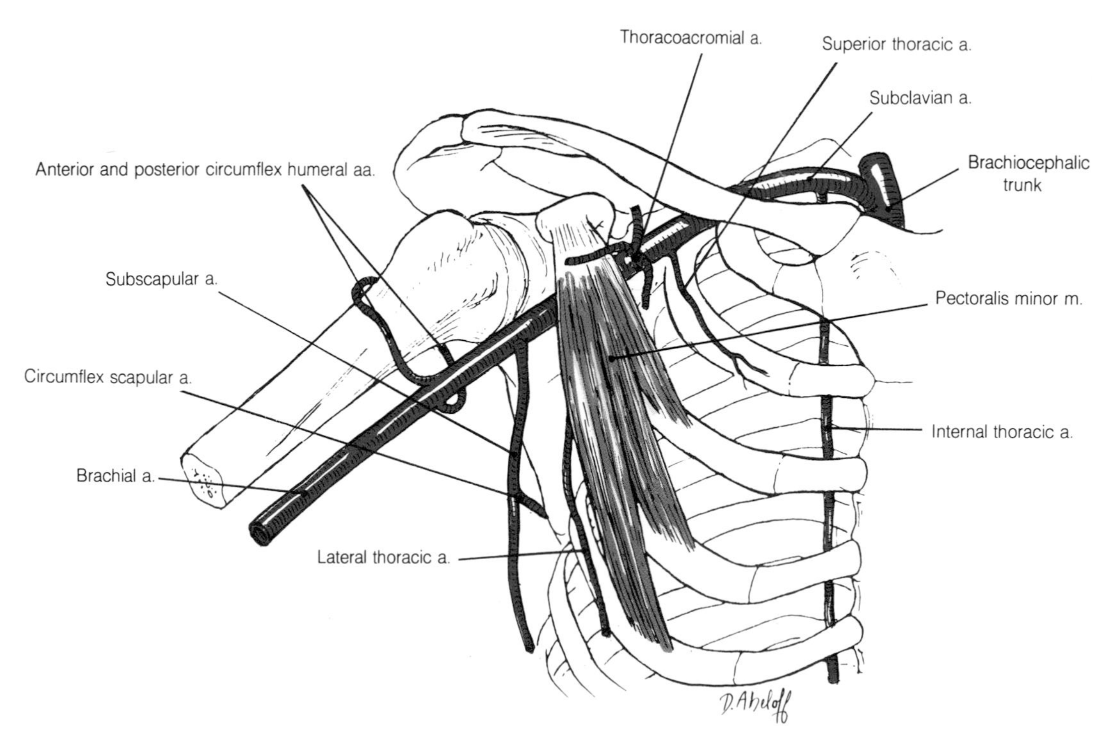

3-37 Diagram of the axillary artery and its branches.

The **lateral thoracic artery** appears at the lateral border of pectoralis minor and follows this onto the thoracic wall to supply the pectoral muscles and serratus anterior. It makes a substantial contribution to the blood supply of the breast and is intimately related to the pectoral group of lymph nodes that drain the breast.

Branches from the Third Part

The **subscapular artery** is the largest branch of the axillary and arises as that vessel crosses the lateral border of subscapularis. It follows the border of subscapularis to the inferior angle of the scapula where it passes onto the thoracic wall. Among its many muscular branches is the large **circumflex scapular artery** which turns posteriorly around the lateral border of the scapula to reach the infraspinous fossa. This and other branches anastomose with branches of the subclavian artery forming a potential collateral anastomosis that may bypass obstructions or injuries of the first and second parts of the axillary artery. This anastomosis is illustrated in Fig. 3-38.

The **anterior and posterior circumflex humeral arteries** form an anastomotic circle around the surgical neck of the humerus. The anterior is small. The posterior runs with the axillary nerve through the quadrangular space to supply the shoulder joint and surrounding muscles.

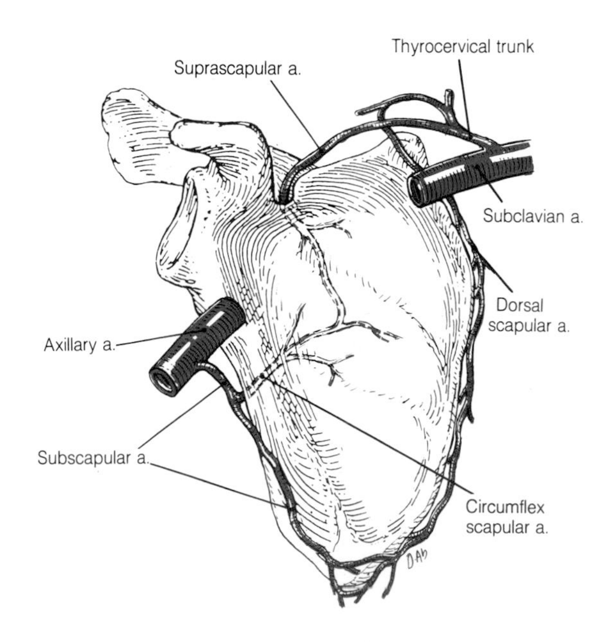

3-38 Showing the arterial anastomosis around the scapula.

The Axillary Vein

The **axillary vein** is a continuation of the basilic vein which is joined by the brachial veins or venae comitantes of the brachial artery. It begins at the lower border of teres major and ascends through the axilla medial to the axillary artery. At the outer border of the first rib it becomes the subclavian vein. It receives tributaries which correspond to the branches of the axillary artery and near its termination is joined by the cephalic vein.

The Relations of Nerves and Vessels in the Axilla

The mobility of the upper limb at the shoulder joint causes some changes in position, shape, and relationship of structures as different positions are assumed. When the limb is held at right angles to the trunk the course of the axillary artery is almost straight and can be represented on the surface by a line drawn from the center of the clavicle which passes just below the coracoid process to the shallow medial bicipital furrow. Look now at Fig. 3-39 and note that:

1. Throughout its course the axillary vein lies medial to the axillary artery.
2. The lateral and posterior cords of the brachial plexus lie lateral to the first part of the axillary artery while the medial cord lies posterior to it. At this level the vessels and cords of the plexus are surrounded by an extension of the prevertebral fascia called the axillary fascia.
3. The cords of the plexus become rearranged over the second part of the artery and assume around this the position implied by their names.
4. The medial cord now separates the axillary artery from the axillary vein.
5. The main terminal branches of the cords are related to the third part of the artery.
6. The musculocutaneous nerve and the lateral root of the median nerve lie lateral to the artery.
7. The medial cutaneous nerve of the forearm and the ulnar nerve lie medial to the artery and between it and the vein.

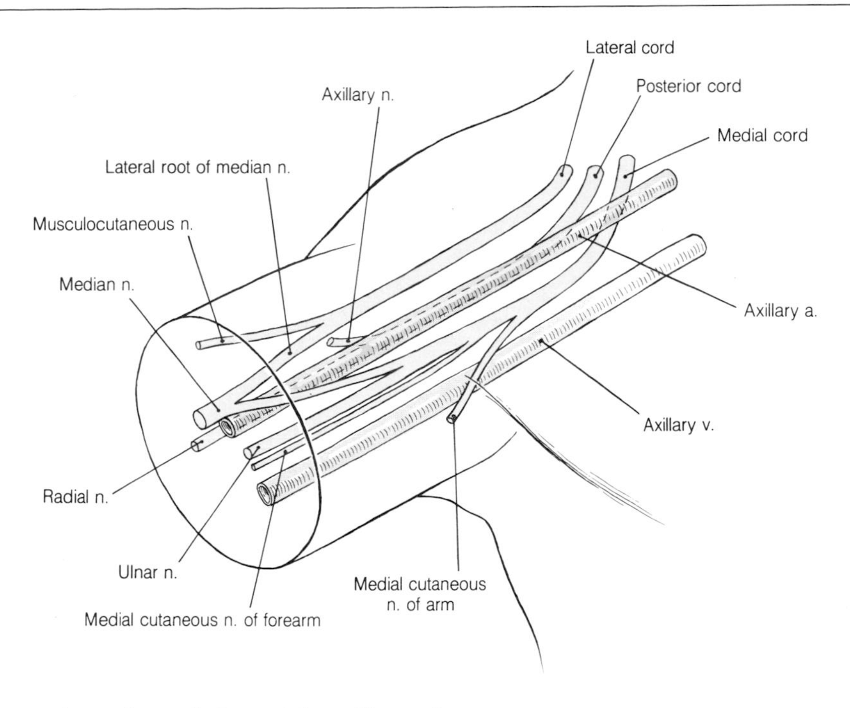

3-39 Diagram of the immediate relations of the axillary artery.

8. The medial root of the median nerve crosses in front of the artery.
9. The median nerve itself is formed anterior to or just lateral to the artery.
10. Posterior to the artery lie the radial nerve and part of the axillary nerve.

These contents of the axilla are under cover of the anterior and posterior axillary walls and are protected above and laterally by the shoulder. However, the second and third parts of the axillary artery have only the fascia and skin of the floor of the axilla covering them below. Here the pulsation of the artery can be felt.

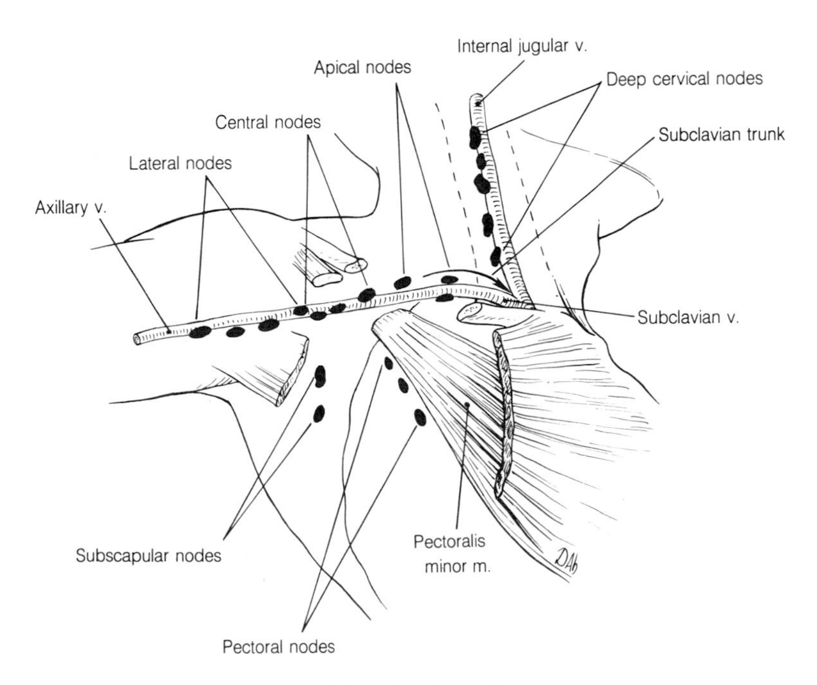

3-40 Diagram illustrating the arrangement of the axillary lymph nodes.

The Axillary Lymph Nodes

The axillary region contains 20-30 deep lymph nodes. These are arranged in five groups in the. manner shown in Fig. 3-40. They are:

1. **A lateral group** around the axillary vein which drains the deep and much of the superficial tissues of the upper limb.
2. **A subscapular group** arranged along the subscapular vessels. These drain the back of the shoulder, trunk, and lower neck.
3. **A pectoral group** lying at the lateral border of pectoralis minor which drain the anterior thoracic wall and breast.
4. **A central group** which receives afferents from all other groups and, in addition, afferents from the upper limb which accompany the cephalic vein.
5. **An apical group** which again receives afferents from all other groups and, in addition, afferents from the upper limb which accompany the cephalic vein.

The efferents from the apical group join to form the **subclavian trunk** which opens into the junction of the subclavian and internal jugular veins.

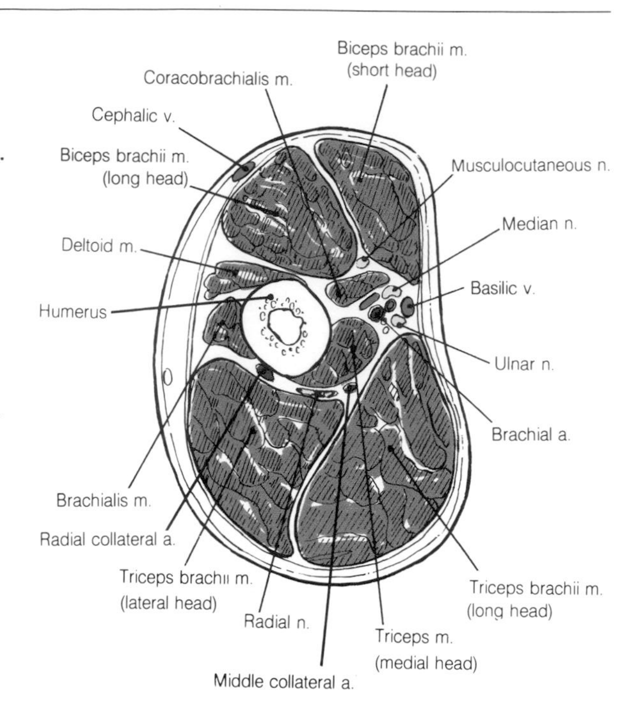

3-41 Cross-section of the upper arm.

The Arm

The anatomical arm lies between the shoulder and elbow joint. It consists of two well-defined muscle compartments – flexor and extensor, surrounding the shaft of the humerus. Many of the structures encountered in the axilla pass through the arm to cross the elbow joint and enter the forearm. The brachial fascia surrounding the arm is thin over the anterior aspect but thicker over the posterior. From the **brachial fascia** is given off a **lateral intermuscular septum** that extends to the body of the humerus from the lateral lip of the intertubercular sulcus to the lateral epicondyle. A **medial intermuscular septum** extends from the medial lip of the sulcus to the medial epicondyle. It is these septa that define the anterior flexor compartment and the posterior extensor compartment. Look now at Fig. 3-41. This cross-section of the arm shows the general arrangement of the two compartments and in it the position of the main structures listed below can be determined before they are described in more detail.

The Flexor Compartment

The biceps brachii and coracobrachialis muscles
The brachialis muscle
The brachial artery and veins
The musculocutaneous and median nerves
The proximal part of the ulnar nerve

The Extensor Compartment

The triceps brachii muscle
The terminal branches of the profunda brachii artery
The radial nerve
The distal part of the ulnar nerve

The musculocutaneous nerve is the motor nerve to the flexor muscles in the anterior compartment and the radial nerve is the motor nerve to the extensor muscles in the posterior compartment. The median and ulnar nerves have no branches in the arm. Note that the musculocutaneous nerve is derived from anterior divisions of the brachial plexus and the radial nerve from posterior divisions. In the forearm and hand it will be seen again that the median and ulnar nerves which are both formed from anterior divisions supply flexor muscles while the deep branch of the radial nerve formed from posterior divisions supplies extensor muscles.

Muscles of the Flexor Compartment of the Arm

Mention has already been made of the origins of biceps brachii, coracobrachialis, and the long head of triceps brachii from the pectoral girdle. Of these muscles, coracobrachialis is attached to the humerus and thus its action is only on the shoulder joint. The long heads of the other two traverse the length of the arm and are attached to the ulna or radius. They are, therefore, two joint muscles and act on shoulder and elbow joint. Brachialis and the medial and lateral heads of triceps arise from the humerus and, being attached to the ulna, act only on the elbow joint. Reference should be made to the illustrations of the humerus and forearm bones in Fig. 3-42 as the following descriptions are read.

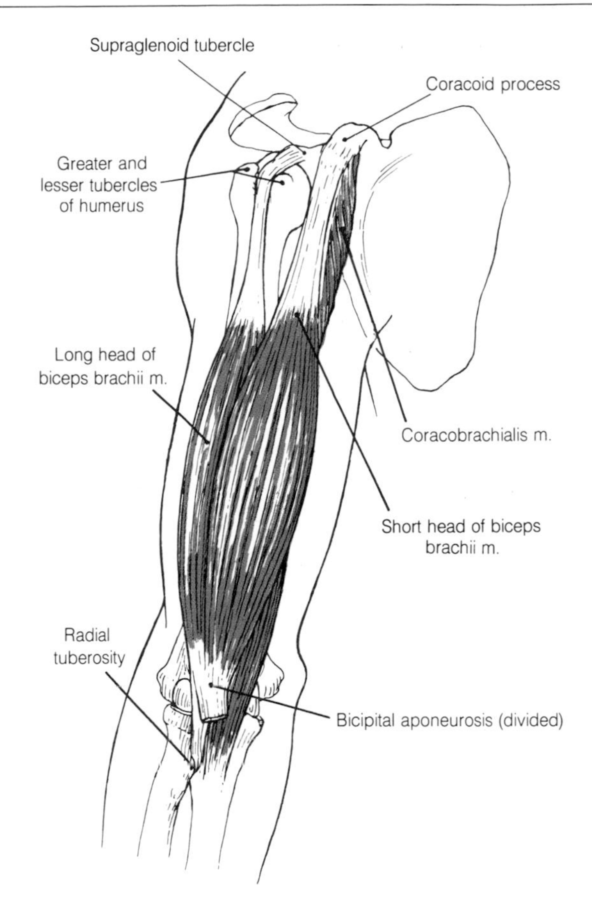

3-43 The biceps brachii muscle.

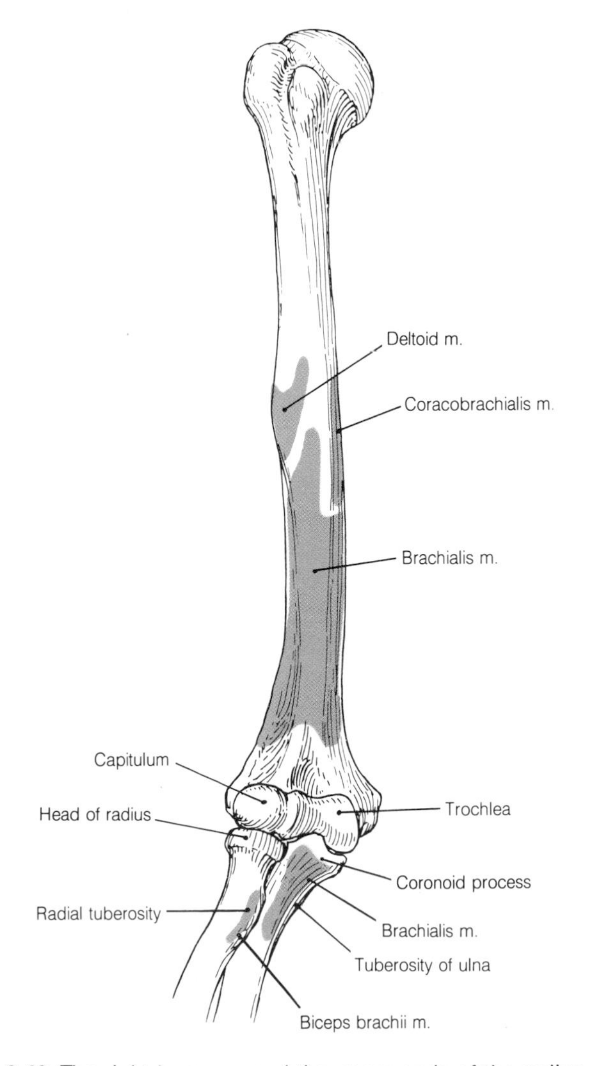

3-42 The right humerus and the upper ends of the radius and ulna seen from their anterior aspect.

Biceps brachii

This muscle has a long and a short head. The long head arises from the supraglenoid tubercle within the capsule of the shoulder joint. In Fig. 3-43 the tendon of origin can be seen to emerge from the capsule between the greater and lesser tubercles of the humerus and descend in the intertubercular sulcus. The short head lies medial to the long and arises from the tip of the coracoid process. The two muscular bellies continue through the anterior part of the arm fusing toward its lower extent. The flattened distal tendon of the muscle crosses the elbow joint and becomes attached to the posterior margin of the tuberosity of the radius. A bursa separates the tendon from the greater part of the tuberosity. As the tendon crosses the elbow joint an aponeurotic sheet, the **bicipital aponeurosis**, passes medially across the brachial artery to fuse with the deep fascia of the forearm. This fascia in turn fuses with the periosteum over the subcutaneous surface of the ulna thus providing a secondary attachment of the muscle to this bone.

Both heads of the muscle are supplied by the **musculocutaneous nerve** after it has pierced coracobrachialis.

Biceps is a flexor of the shoulder joint but its most powerful actions are on the forearm bones. It is a strong flexor at the elbow joint and a powerful supinator of the forearm. The latter movement is carried out as a right-handed person screws on a lid and involves lateral rotation of the radius. Biceps performs this rotation well in the semiflexed position as it pulls on the radial tuberosity. This action is described more fully with the radioulnar joints.

Coracobrachialis

Coracobrachialis arises from the tip of the coracoid process together with the short head of biceps. It descends into the arm to be attached to the medial margin of the humerus at the same level as the deltoid tuberosity. The brachial vessels and median nerve cross the muscle in the upper arm and the pulsation of the brachial artery can be felt in the groove formed between it and the anteriorly lying biceps.

Coracobrachialis is supplied by the **musculocutaneous nerve** which pierces it. Its action is to pull the arm forward and across the chest particularly from the extend position. It also helps steady the limb when it is abducted from the trunk.

Brachialis

Brachialis is a large muscle that arises from the anterior surface of the lower half of the shaft of the humerus. Its fibers converge on a flat tendon that crosses the elbow joint and is inserted into the tuberosity of the ulna and the adjacent surface of the coronoid process. It is supplied by the **musculocutaneous nerve** and is a powerful flexor of the elbow joint. Both brachialis and coracobrachialis are illustrated in Fig. 3-44.

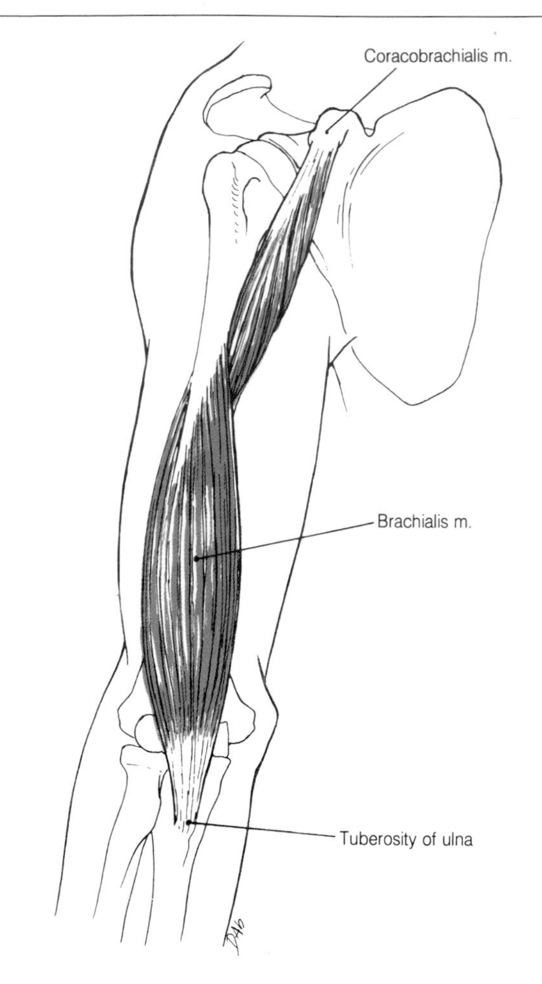

3-44 Coracobrachialis and brachialis muscles.

Note again that each of these muscles in the flexor compartment is supplied by the **musculocutaneous nerve**.

Muscles of the Extensor Compartment of the Arm

Triceps brachii

Triceps is the only large muscle in the posterior or extensor compartment of the arm. As its name suggests, it has three heads. A long head has already been seen to arise from the infraglenoid tubercle of the scapula. Medial and lateral heads arise from the humerus. Look at Fig. 3-45 and identify the radial groove as it spirals obliquely downward and laterally over the back of the

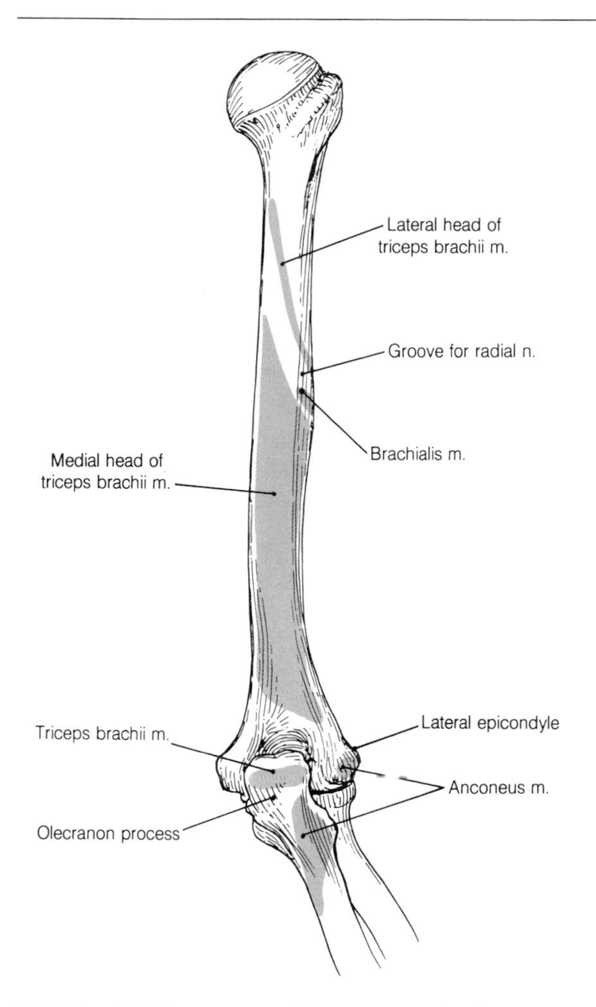

3-45 The right humerus and the upper ends of the radius and ulna seen from their posterior aspect.

humerus. Above and lateral to the groove is the linear origin of the lateral head of triceps. Below and medial to the groove is the much more extensive origin of the medial head. The bellies fuse and a common tendon is formed above the elbow joint which is inserted into the upper surface of the **olecranon process of the ulna**. The long and lateral heads of the muscle which cover the medial head are illustrated in Fig. 3-46. Near its distal attachment the tendon fuses with the **anconeus muscle**. This arises from the posterior aspect of the lateral humeral condyle and is inserted into the upper part of the posterior surface of the ulna.

Each head of triceps and anconeus are supplied independently by branches of the **radial nerve**.

The triceps is the only effective extensor of the elbow joint. Although under many circumstances extension is produced by the action of gravity, triceps is essential when resistance is met as in any form of pushing. The long head may also help to support the head of the humerus in the glenoid cavity when the arm is abducted.

Vessels of the Arm

The Brachial Artery

At the level of the lower margin of the posterior wall of the axilla where it is formed by teres major, the axillary artery becomes the brachial artery. From this point it traverses the length of the flexor compartment of the arm, crosses the elbow joint, and divides into its two terminal branches, the ulnar and radial arteries. Throughout its

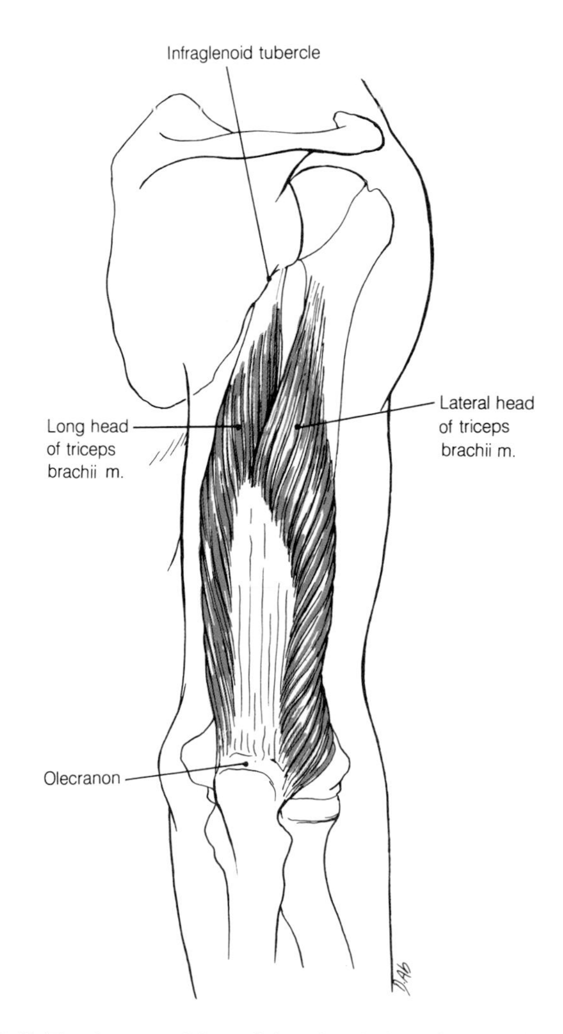

3-46 The long and lateral heads of the triceps brachii muscle.

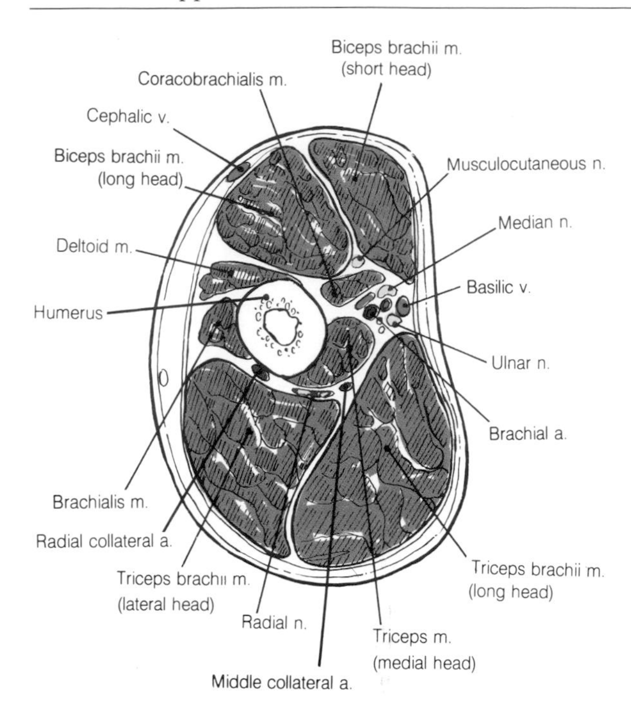

3-47 Horizontal section through the upper arm.

course it is superficial, being covered by skin and fascia only. Reference to the cross-section of the upper arm shown in Fig. 3-47 will confirm the superficial position of the artery and demonstrate biceps brachii lying anterior to it and the long head of triceps brachii lying posterior to it. In addition the median nerve can be seen crossing in front of it from its lateral side above the section to its medial side below. At this level the ulnar nerve lies posterior to it but at a lower level will leave the artery to enter the extensor compartment.

In its passage through the arm, the brachial artery gives off substantial muscular branches and a nutrient artery to the humerus. A **superior ulnar collateral artery** is given off near the middle of the arm and this accompanies the ulnar nerve. An **inferior ulnar collateral artery** leaves the brachial artery just above the elbow joint. Both these vessels take part in the arterial anastomosis around the elbow joint. The largest branch of the brachial artery is the **profunda brachii artery**. The course of these branches is summarized in Fig. 3-48.

The Profunda Brachii Artery

Only one major vessel reaches the arm from the trunk and this is confined to the flexor compartment. The extensor compartment is supplied by branches of the posterior humeral circumflex branch of the axillary artery and the profunda brachii artery which is given off from the brachial artery soon after it leaves the axilla. It runs posteriorly to join the radial nerve in the extensor compartment and accompanies this between the lateral and medial heads of triceps in the groove for the radial nerve. It gives off muscular and nutrient arteries, anastomoses with the posterior humeral circumflex artery, and ends by dividing into a **radial and a middle collateral artery**. The former accompanies the radial nerve into the anterior compartment of the forearm, while the latter passes posterior to the lateral condyle. These collateral arteries also enter the anastomosis around the elbow joint.

The Brachial Veins

The brachial artery is accompanied by a pair of large venae comitantes which may be described as brachial veins. They are joined by tributaries which correspond to the branches of the brachial artery. The veins end by joining with the **basilic vein** to form the **axillary vein**.

Nerves in the Arm

In the description of the axilla, the median, musculocutaneous, ulnar, and radial nerves were seen to be formed as terminal branches of the cords of the brachial plexus and to enter the arm. Their subsequent course through the arm will now be followed.

The Median Nerve

The median nerve, formed from both the lateral and medial cords, enters the arm on the lateral side of the brachial artery. In the arm it crosses in front of the artery and continues on its medial side to the elbow. Normally no muscular branches are given off until the nerve reaches the forearm.

The Musculocutaneous Nerve

A branch of the lateral cord, this nerve is the motor nerve to the muscles in the flexor compartment of the arm. It is not until it reaches the fore-

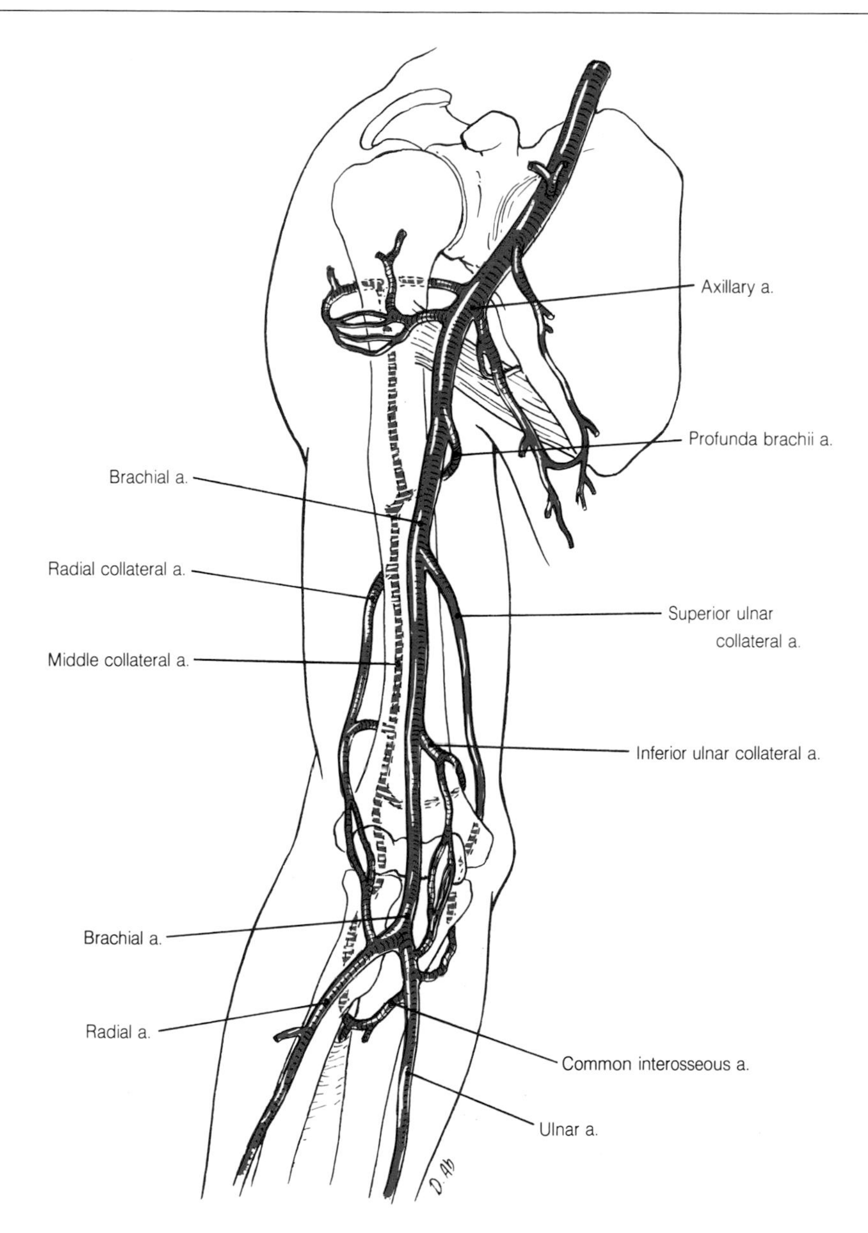

3-48 To show the brachial artery and its branches.

arm that it becomes a cutaneous nerve. Soon after its formation, it passes through coracobrachialis and then runs downward and laterally through the arm between biceps and brachialis to reach the lateral border of brachialis. Having supplied each of these muscles it pierces the deep fascia and becomes the **lateral cutaneous nerve of the forearm**.

The Ulnar Nerve

The ulnar nerve is a terminal branch of the medial cord of the brachial plexus. It enters the arm on the medial side of the brachial artery but half way down the arm it leaves the flexor compartment by piercing the medial intermuscular septum. Continuing between this and the medial head of triceps, it runs in a groove on the posterior aspect of the medial epicondyle of the humerus and enters the forearm between the humeral and ulnar heads of the muscle flexor carpi ulnaris. The tingling sensation produced by blows on the nerve as it passes behind the epicondyle has given rise to the term "funny bone." The ulnar nerve has no muscular branches in the arm.

The Radial Nerve

The radial nerve is a terminal branch of the posterior cord of the brachial plexus. Initially lying on the posterior wall of the axilla, it slips over the lower border of teres major through a triangular space formed between this muscle, the long head of triceps brachii, and the shaft of the humerus. Here it is joined by the profunda brachii artery and the two structures wind posteriorly round the humerus in the groove for the radial nerve between the attachments of the lateral (above) and medial (below) heads of triceps brachii. The nerve enters the flexor compartment in the lower part of the arm and passes into the forearm under cover of the brachioradialis muscle. In the axilla and arm it supplies muscular branches to each head of triceps brachii and a long branch to anconeus. Near the elbow joint it also supplies a branch to the most lateral fibers of brachialis and to two muscles of the forearm that arise from the lateral

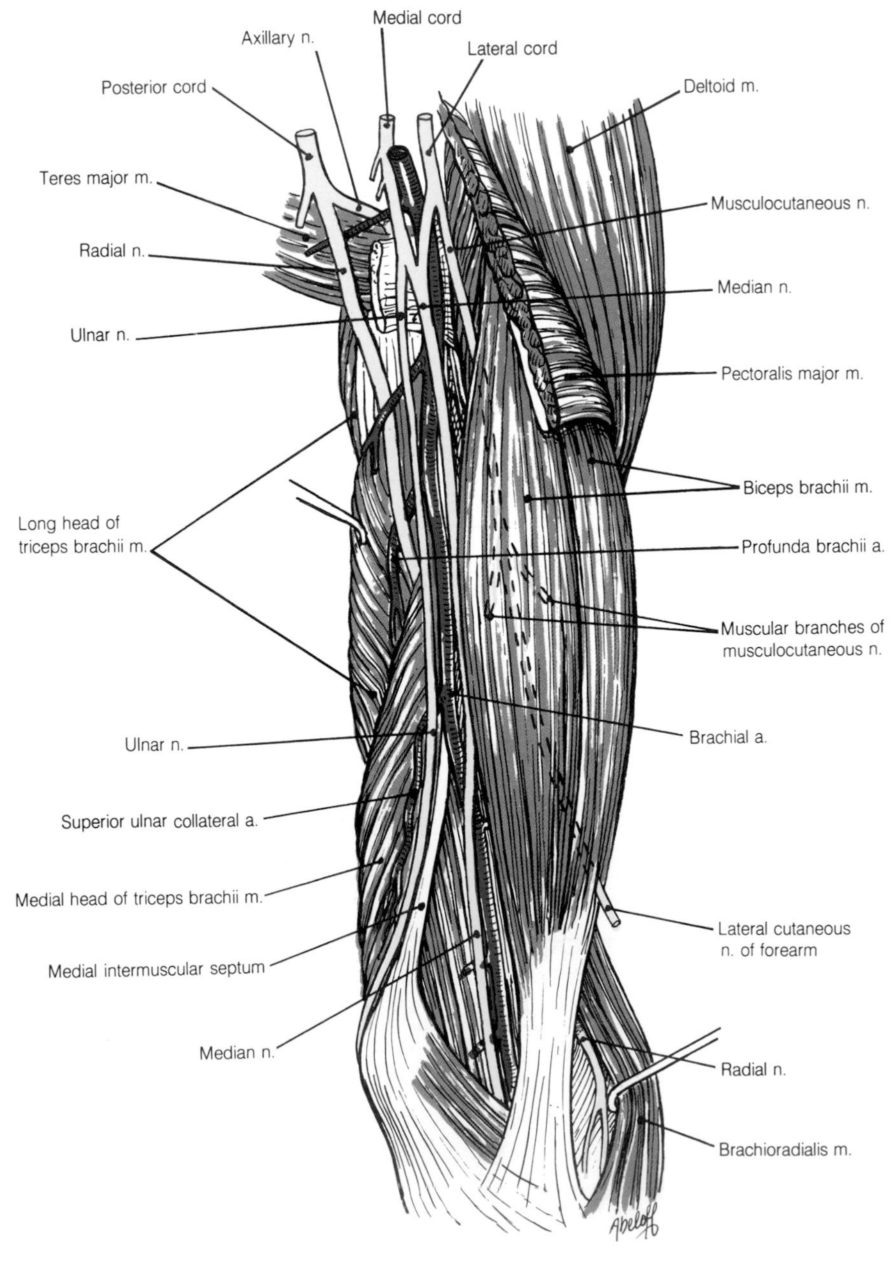

3-49 Showing the course of nerves through the flexor compartment of the arm.

supracondylar ridge. These are the brachioradialis and the extensor carpi radialis longus. Cutaneous branches of the radial nerve are the **posterior cutaneous nerve of the arm**, which leaves it in the axilla, the **lower lateral cutaneous nerve of the arm,** and the **posterior cutaneous nerve of the forearm**.

In fractures of the humerus the radial nerve may be damaged where it lies close to the bone or conduction in the nerve may be impaired if it is involved in scar tissue.

The course of these nerves through the arm can be followed in Fig. 3-49. Note again how each is formed from the cords of the brachial plexus. Follow the radial nerve from the posterior wall of the axilla (here formed by teres major) to the triangular space where it is joined by the profunda brachii artery. It can be seen again beneath the brachioradialis muscle at the lateral side of the elbow. The ulnar nerve can be seen to pass behind the medial intermuscular septum and the medial epicondyle. Note how the median nerve remains superficially placed and only partly overlapped by the biceps brachii muscle. The musculocutaneous nerve can be seen to disappear beneath biceps brachii and after supplying the flexor muscles reappears at the lateral border of biceps as the lateral cutaneous nerve of the forearm.

Much of the anatomy described in the arm can now be rapidly reviewed by comparing the cross-sections of the upper and lower arm seen in Fig. 3-50. Note the following:

1. In the upper section the three heads of triceps and the two heads of biceps can be distinguished but each muscle has fused to a single belly in the lower section.
2. In the upper section the radial nerve is seen separating the lateral and medial heads of triceps as it winds round the posterior aspect of the humerus. In the lower section it is about to enter the flexor compartment of the forearm with the muscle brachioradialis.
3. In the higher section the median nerve is seen as it begins to cross anterior to the brachial artery and the ulnar nerve is medial to the artery. In the lower section the median nerve is now medial to the artery and the ulnar nerve lies in the posterior compartment behind the medial inter muscular septum.
4. The musculocutaneous nerve in the upper section has become the lateral cutaneous nerve of the forearm in the lower.

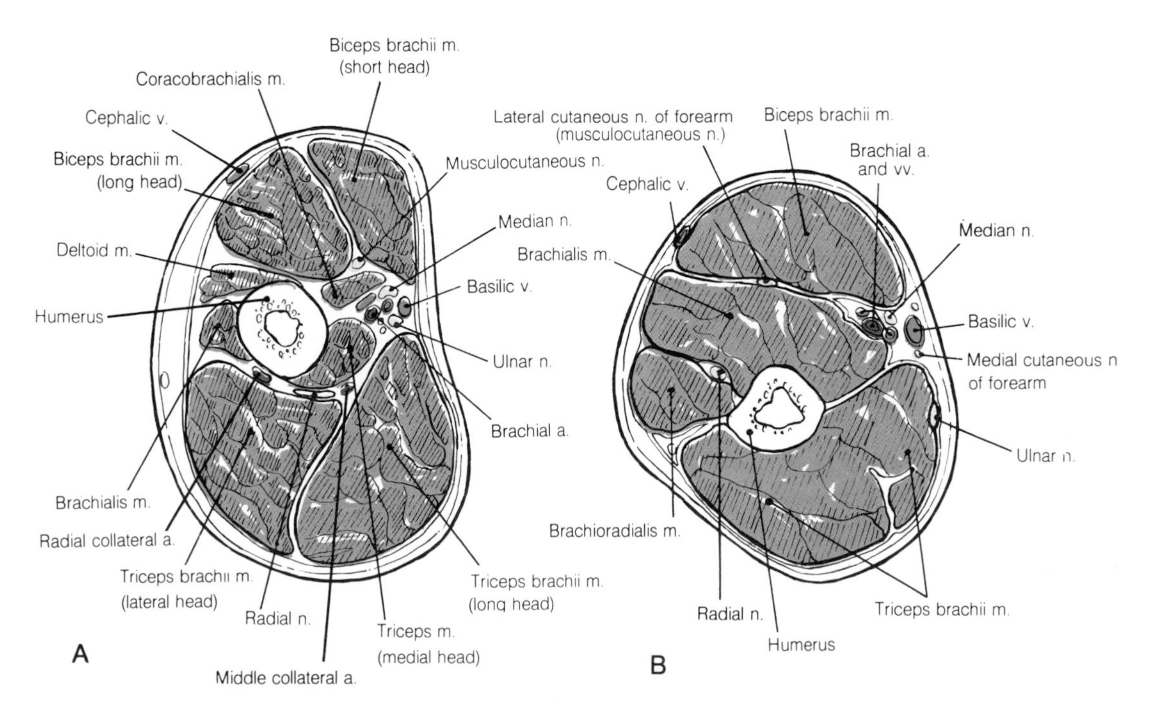

3-50 Cross-sections through the upper and lower arm.

5. The brachial artery, its venae comitantes, the median nerve, the basilic vein, and the medial cutaneous nerve of the forearm form a superficial neurovascular bundle lying in the groove between triceps and biceps in both sections and throughout the arm.

The Cubital Fossa

The cubital fossa is a triangular anatomical region lying anterior to the elbow joint. Its base is formed by an imaginary line drawn through the medial and lateral epicondyle of the humerus and below this its sides are formed by the lateral border of pronator teres and the medial border of brachioradialis. Its outlines are indicated in Fig. 3-51.

This region has more than descriptive significance. It includes most of the anterior relations of the elbow joint, a joint not infrequently subject to dislocation and fracture dislocation. It also forms a site of access to both venous and arterial blood. And lastly, with some reflection of its borders it is seen to contain all the important nerves and vessels running between the arm and forearm except the ulnar nerve which passes behind the medial epicondyle of the humerus at this level. It thus forms a useful point at which to establish the relationships of these structures and from which to follow them into the forearm.

The roof of the fossa is formed by skin and superficial and deep fascia. In the superficial fascia lie the median cubital vein running from the cephalic to the basilic vein and the medial cutaneous nerve of the forearm medial to it. The deep fascia is reinforced here by the bicipital aponeurosis.

When the roof of the fossa is removed and its margins reflected, the contents are exposed in the manner seen in Fig. 3-52. Examine this illustration carefully starting on the lateral or radial side, and note that:

1. The lateral cutaneous nerve of the forearm appears between biceps and brachialis and becomes superficial over brachioradialis.
2. Brachioradialis forms the lateral border of the fossa and is reflected to expose the radial nerve as it enters the flexor compartment of the forearm. The beginning of its deep branch which

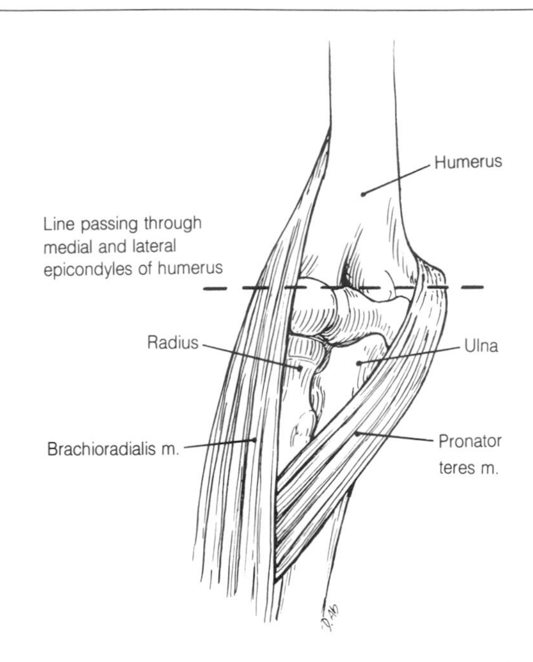

3-51 Diagram showing the borders of the cubital fossa.

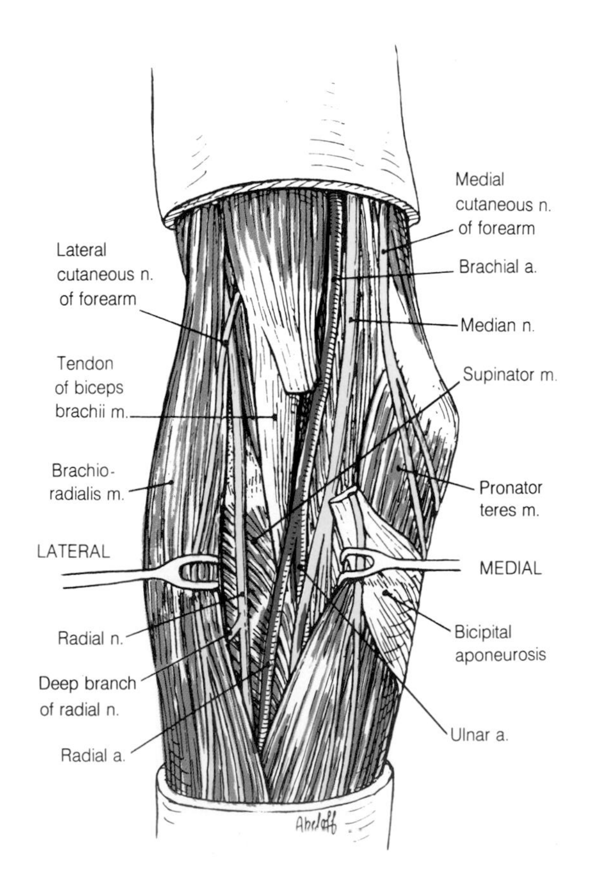

3-52 The contents of the cubital fossa and nearby structures.

will enter the extensor compartment is also seen.

3. Medial to this lies the lateral border of brachialis with biceps and its tendon lying anterior to the belly of brachialis. The bicipital aponeurosis has been divided.
4. Crossing over the tendon of biceps and deep to the bicipital aponeurosis is the brachial artery. This can be followed to its division into the radial and ulnar arteries.
5. The median nerve lies medial to the artery.
6. The medial cutaneous nerve of the forearm (or its branches) lies anterior to pronator teres as it forms the medial border of the fossa.

The floor of the fossa is formed largely by brachialis as it runs toward its attachment to the ulna. Below and lateral to this lies the supinator muscle which encircles the upper end of the radius.

The Skeleton at the Elbow

Before continuing with a description of the elbow joint and forearm, a more detailed examination of the lower end of the humerus and the upper ends of the radius and ulna must be made. Look at Fig. 3-53 as the following descriptions are read.

The Lower End of the Humerus

The lower end of the humerus is flattened anteroposteriorly and expanded medially and laterally. It shows a prominent and easily palpable **medial epicondyle** and a less prominent **lateral epicondyle**. Above each epicondyle lies a sharp **supracondylar ridge**. Between the epicondyles there is a rounded surface, the **capitulum**, for articulation with the radius and a spool or pulley-like surface, the **trochlea**, for articulation with the

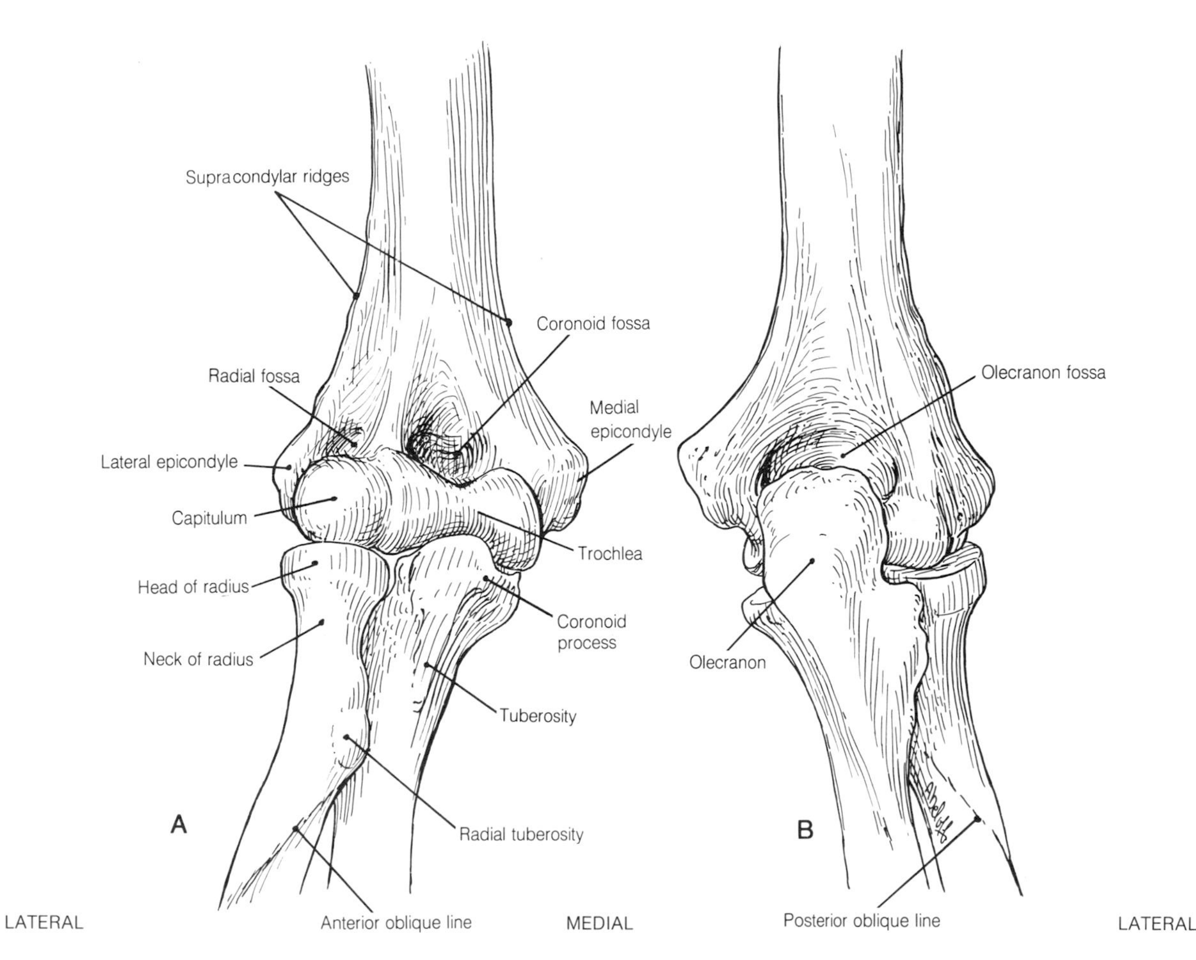

3-53 The anterior (*A*) and posterior (*B*) aspects of the lower end of the right humerus and the upper end of the ulna and radius.

ulna. Just lateral to this and separating the trochlea from the medial epicondyle is a small strip of shaft behind which is found a shallow groove for the ulna nerve. Both the capitulum and the trochlea have a fossa above them named the **radial and coronoid fossae**. The articular surface of the trochlea is carried on to the posterior aspect of the lower end and above this lies the **olecranon fossa** which houses the **olecranon process** in full extension of the elbow joint.

The Upper End of the Radius

The **head of the radius** is a short cylinder with a shallow depression on its superior surface which articulates with the capitulum. The smooth margin of the disc also articulates with the **radial notch of the ulna** (Fig. 3-54) and the encircling **annular ligament**. A slight constriction, called the **neck of the radius**, separates the head of the radius from the shaft below. Medially the upper part of the shaft bears the **radial tuberosity** and from the tuberosity a line can be followed downward and laterally. This is the **anterior oblique line**. A similar but much less well-marked line is sometimes found on the posterior aspect of the shaft.

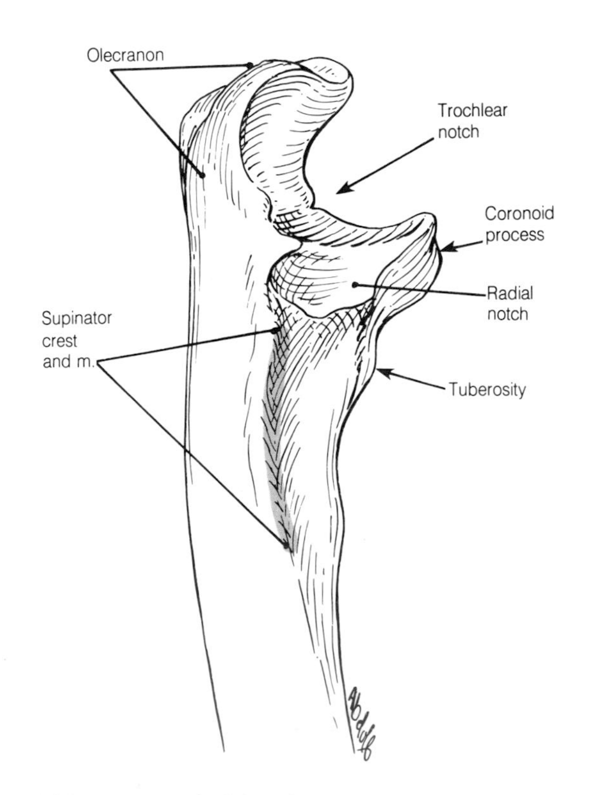

3-54 The upper end of the ulna.

The Upper End of the Ulna

The upper end of the ulna which is shown in more detail in Fig. 3-54 is characterized by a hook-like process called the **olecranon**. The inner aspect of this is articular. It grasps the trochlea of the humerus and is known as the **trochlear notch**. Continuous with this surface on the lateral aspect of the upper end is the **radial notch** which receives the margin of the head of the radius. Below the trochlear notch and forming its anterior prominence is the **coronoid process** and on the lower part of this the **tuberosity of the ulna**.

Below the radial notch the lateral surface of the ulna shows a shallow triangular depression which allows room for rotation of the radial tuberosity. The posterior margin of this triangle is formed by the **supinator crest**.

Having established the main bony features of this region some important muscular and ligamentous attachments can be added.

In Fig. 3-55 identify the attachments of brachioradialis and extensor carpi radialis to the lateral supracondylar ridge of the humerus. Below this is the attachment of the common extensor origin to the lateral epicondyle. The **radial collateral ligament of the elbow joint** also arises from the lateral surface of the epicondyle and the small origin of anconeus lies on its posterior aspect. Note how the capsular attachment encircles the radial, coronoid and olecranon fossae.

On the medial supracondylar ridge above the medial epicondyle is found a narrow triangular area for the humeral attachment of pronator teres. To the anterior aspect of the epicondyle are attached the common flexor origin and below this the **ulnar collateral ligament of the elbow joint**.

On the radius and ulna identify the attachments of biceps to the radial tuberosity and brachialis to the coronoid process and ulnar tuberosity. A small slip of pronator teres (the deep head) also arises from the ulna at the medial border of the coronoid process as does a slip of flexor digitorum superficialis. Note how the supinator muscle is attached to the upper part of the shaft of the radius above the anterior and posterior oblique lines. Posteriorly the triceps brachii is seen attached to the

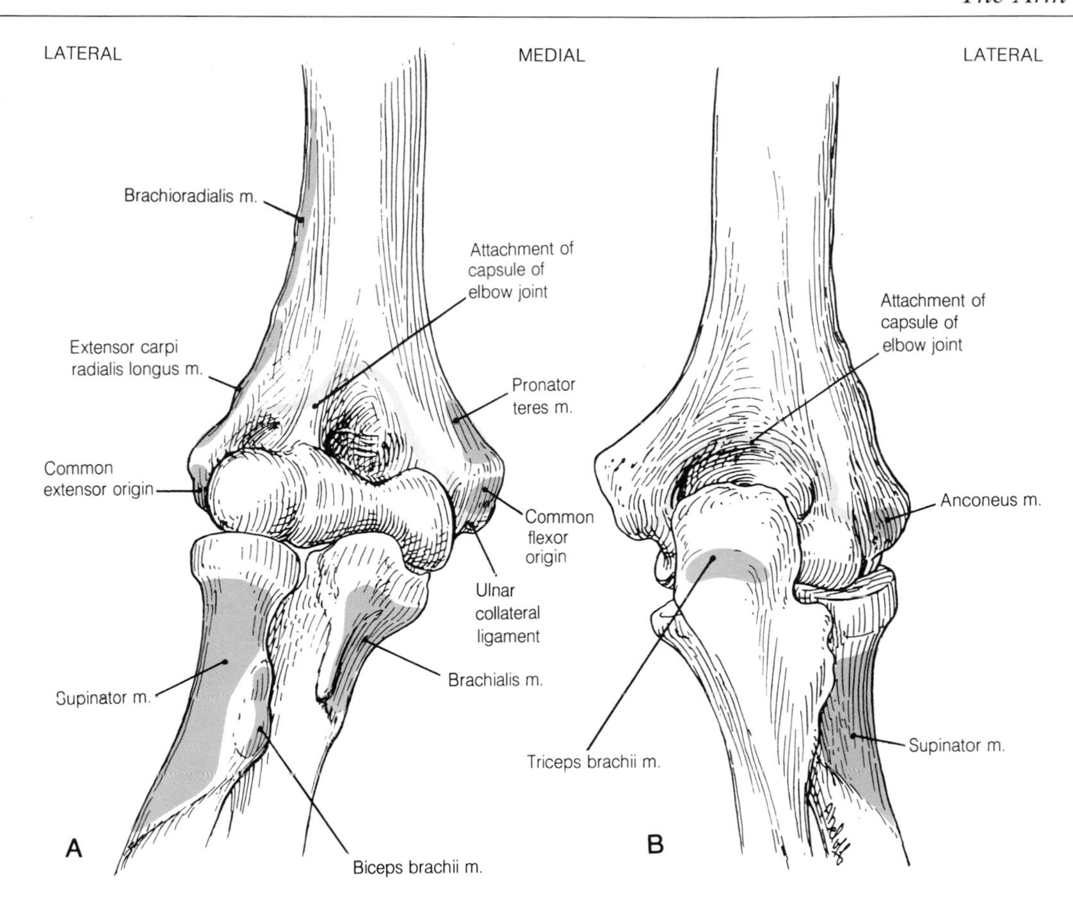

3-55 The anterior (*A*) and posterior (*B*) aspects of the lower end of the right humerus and the upper end of the radius and ulna showing muscular and ligamentous attachments.

superior surface of the olecranon. Below this is the distal attachment of anconeus and lateral to this area the deep head of supinator arises from the supinator crest of the ulna (Fig. 3-54).

The Elbow Joint

The elbow joint is a synovial hinge joint between the lower end of the humerus and the upper ends of the ulna and radius. Although its movements are almost entirely limited to flexion and extension, its continuity with the superior radioulnar joint introduces an unexpected complexity.

The bony surfaces concerned in the elbow joint are the **trochlear notch of the ulna** and **trochlea of the humerus** medially and the hollowed out **superior surface of the radius** and the **capitulum of the humerus** laterally. The shape of the trochlea and the almost congruent trochlear notch limit the movement at the humeroulnar portion of the joint to flexion and extension. However, if the alignment of the groove on the trochlea is examined carefully it can be seen that, while in flexion, the ulna and humerus will lie in the same sagittal plane but in extension the ulna will be abducted on the humerus. This accounts for the "carrying angle," a feature more marked in women than men and illustrated in the radiograph of the extended elbow in Fig. 3-56.

The hollowed superior surface of the radius and the spheroidal surface of the capitulum can be considered as a ball and socket joint. Clearly flexion and extension can occur with the movements of the ulna, but, in addition, the independent rotation of the radius at the proximal radioulnar joint

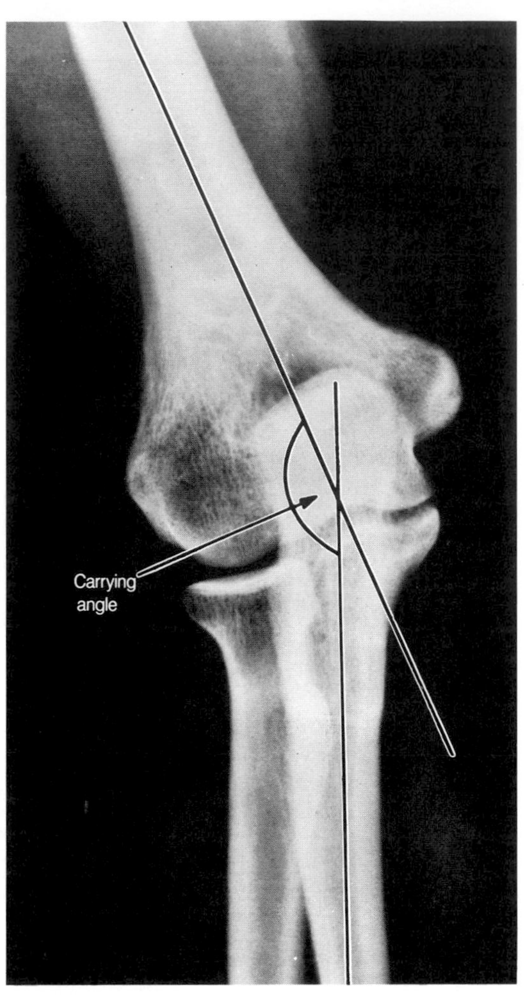

that is so necessary for pronation and supination of the forearm is also possible. With the help of the outline diagram almost all the articular surfaces at the elbow joint can be visualized in the radiograph seen in Fig. 3-57. This is a lateral projection of the left elbow.

The capsule, as might be expected in a hinge joint, is lax anteriorly and posteriorly but strengthened by collateral ligaments that lie medially and laterally. Its proximal attachment surrounds the radial, coronoid and olecranon fossae. Distally it is attached to the margins of the trochlear notch and to the annular ligament.

The **ulnar collateral ligament**, illustrated in Fig. 3-58, runs between the medial epicondyle and the ulna and has a triangular configuration with posterior and anterior bands extending to the olecranon and coronoid processes. An oblique

◀ **3-56** Radiograph of the elbow joint demonstrating the "carrying angle". (Reproduced by permission, from Wicke: *Atlas of Radiologic Anatomy*, 4th Ed, Urban & Schwarzenberg, Baltimore-Munich, 1987.)

3-57 Radiograph of the elbow joint (lateral projection). (Reproduced by permission, from Wicke: *Atlas of Radiologic Anatomy*, 4th Ed, Urban & Schwarzenberg, Baltimore-Munich, 1987.)
▼

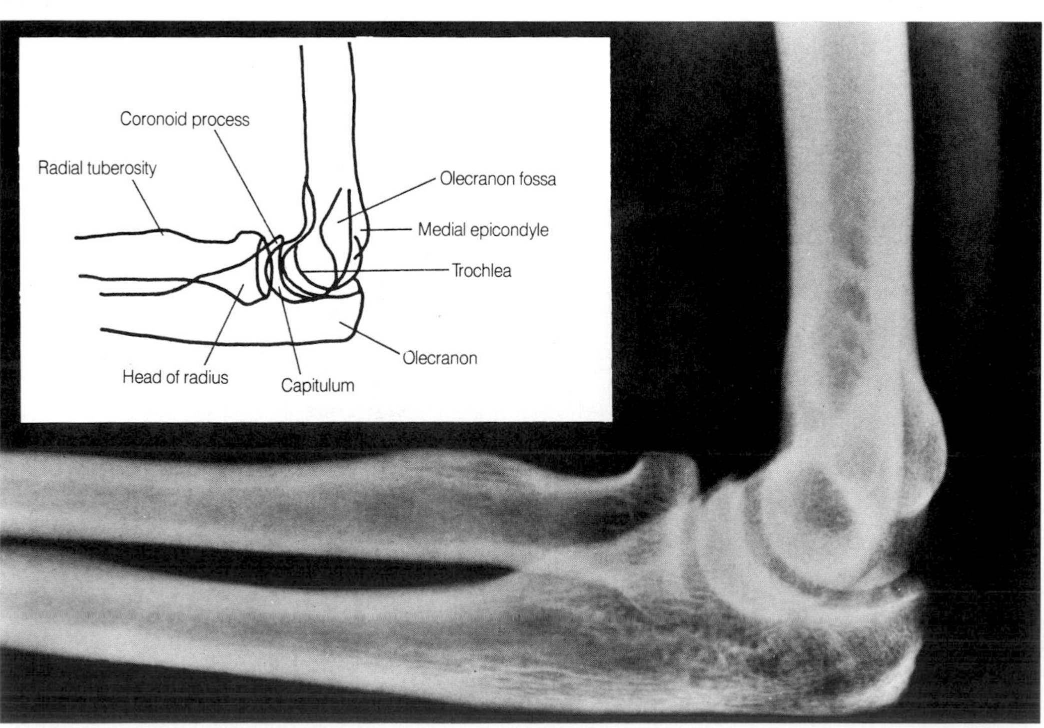

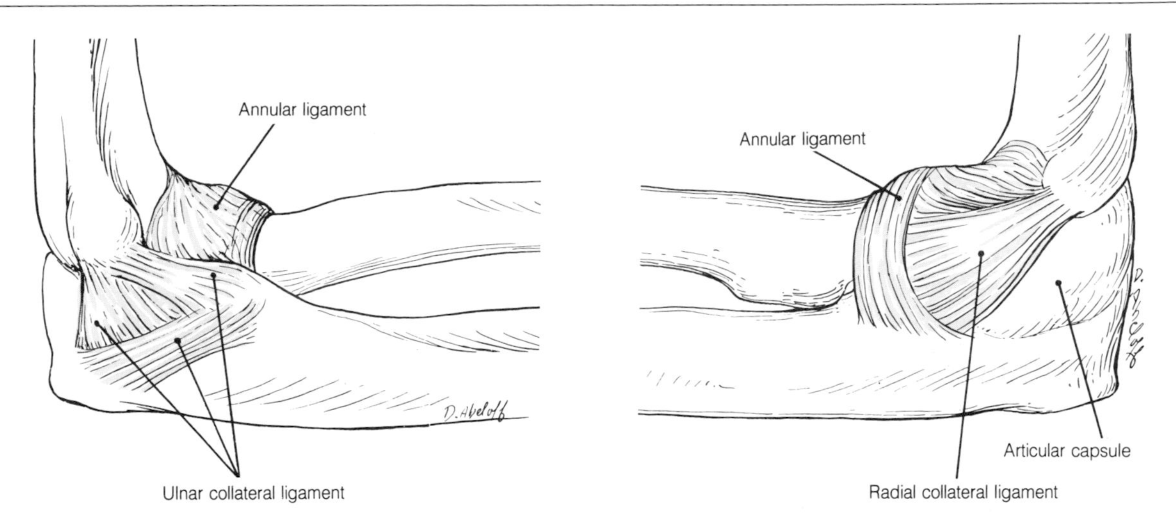

3-58 The ulnar collateral ligament of the elbow joint.

3-59 The radial collateral ligament of the elbow joint.

band running between these processes completes the ligament.

The **radial collateral ligament** seen in Fig. 3-59 arises from the lateral epicondyle but is attached to the stationary annular ligament and not the radius. In this fashion independent rotation of the radius is not impeded.

The joint is supplied by branches of the radial, median, ulnar and musculocutaneous nerves and its blood supply is derived from the network of vessels that surround the joint.

The movement of extension is produced by triceps brachii assisted by anconeus while flexion is produced by biceps and brachialis.

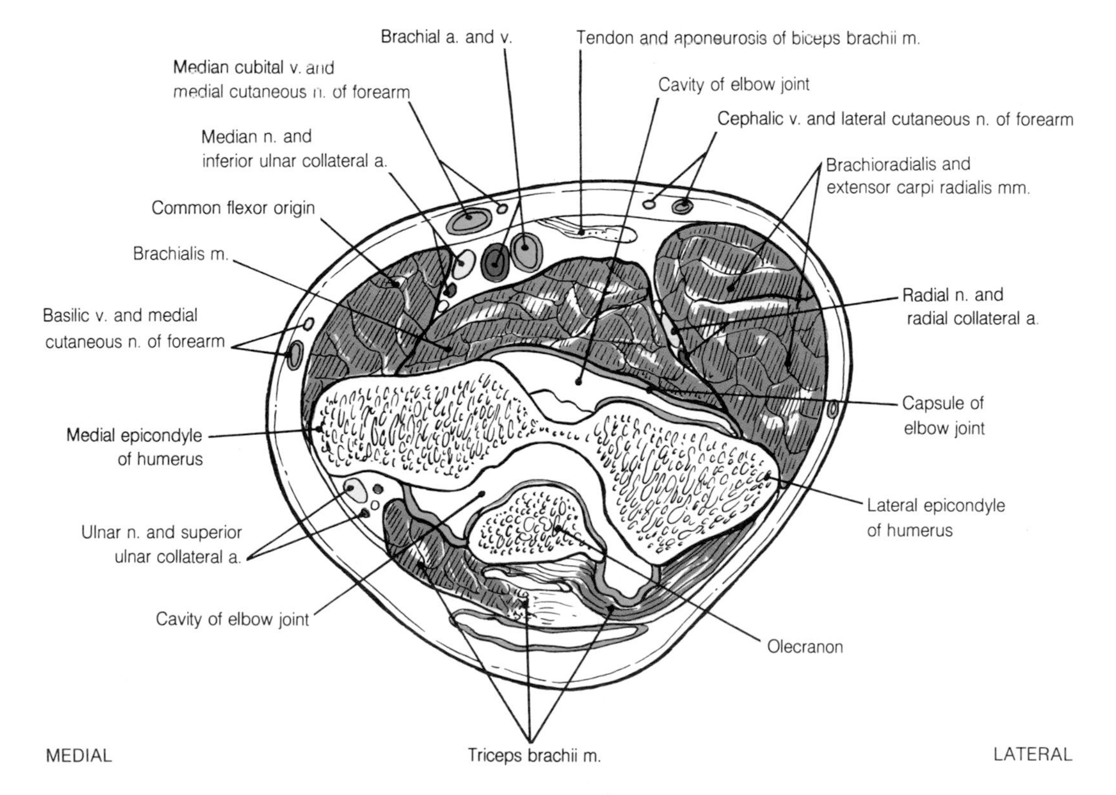

3-60 Cross-section through the elbow joint.

The elbow joint is liable to dislocation and a number of fractures occur close to the joint. Fractures across the thin, flattened lower end of the humerus carry with them the risk of damage to the brachial artery either by directed injury, sympathetic vasoconstriction or compression following swelling beneath an ill-planned cast. The anterior relations of the joint seen in the cubital fossa are thus important and should be reviewed again in the cross-section through the joint seen in Fig. 3-60.

The Forearm

The forearm, like the arm, is divided into flexor and extensor compartments. These lie anterior and posterior to a pair of parallel long bones, the ulna and radius. In practice, however, this anatomical position of the supinated forearm is neither comfortable nor very commonly used. Most manipulations are carried out in the mid-position between pronation and supination or in the fully prone position in which the thumb ap-

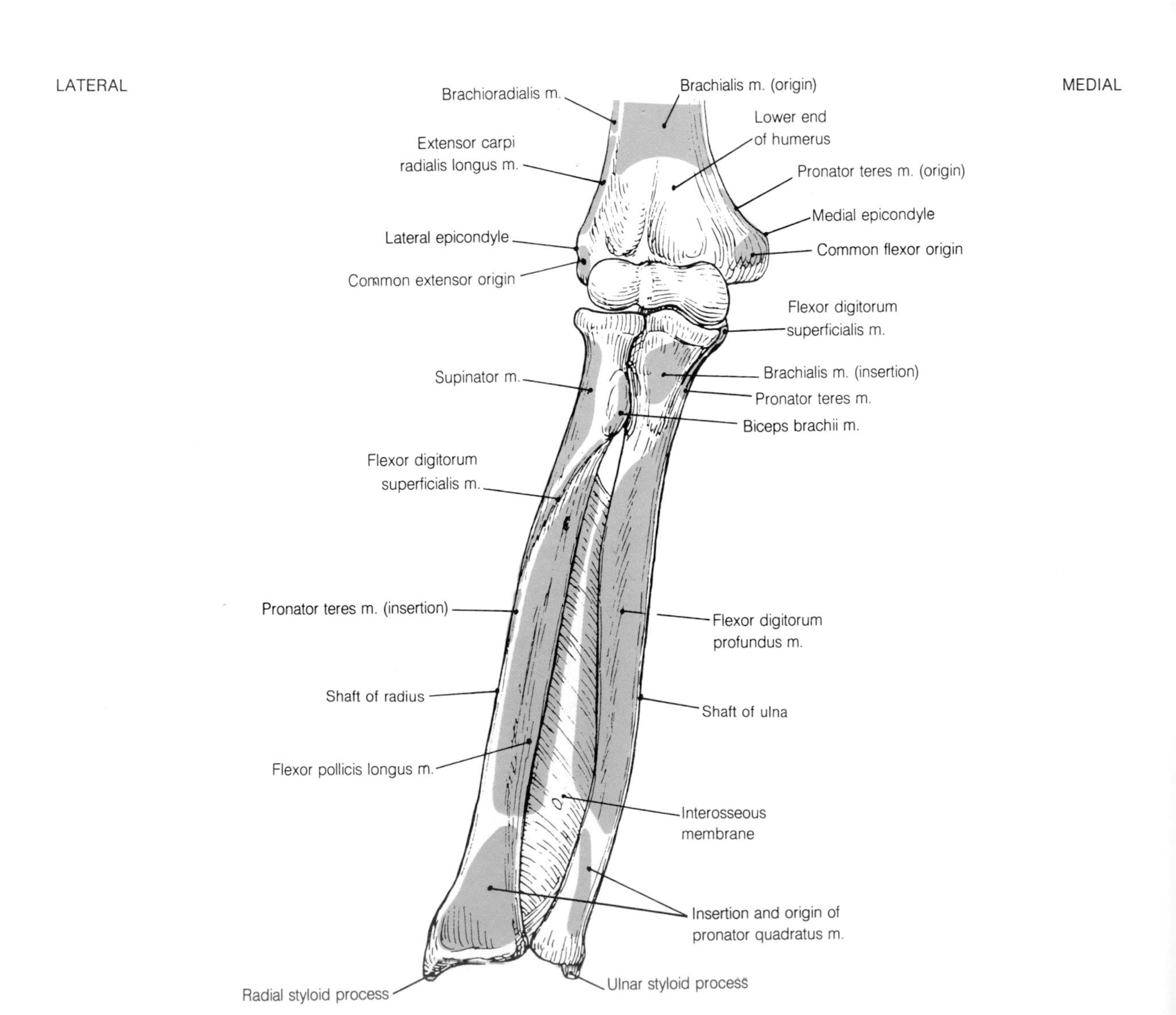

3-61 The anterior aspects of the right radius and ulna showing sites of muscular attachments.

pears to lie on the medial side of the hand. One of these positions is usually found to be assumed in the cadaver and to avoid confusion it is sometimes helpful to refer structures to the radial or ulnar sides of the forearm.

The bulk of the flexor musculature in the forearm arises from the medial epicondyle of the humerus and the anterior surfaces of the ulna and radius. Most of the muscles become tendinous at the wirst and continue thus into the hand. The extensor muscles have a similar arrangement except that their humeral origin is from the lateral epicondyle.

Before considering these muscles the shafts of the radius and ulna must be examined.

The Skeleton of the Forearm

The Radius and Ulna (Anterior Aspect)

The upper ends of these bones and the lower end of the humerus have already been examined in sufficient detail. In the supine or anatomical position, the shafts of the two bones lie approximately parallel to each other although, as can be seen in Fig. 3-61, each is bowed slightly away from the other. Distally the radius is markedly expanded while the ulna is only slightly so. Both bones show a **styloid process** and of these the radial styloid can be felt to extend about a quarter of an inch more distally than that of the ulna. This relationship is usually lost in the common "Colles" fracture of the lower end of the radius where the tips of both processes are found to lie at the same level.

Look now at Fig. 3-62. In cross-section the shafts of the two bones are triangular and the apices of the two triangles are formed by the opposing sharp **interosseous borders** which are united by the interosseous membrane. The same illustration shows the relationship to the flexor compartment of the anterior surfaces of both bones and the medial surface of the ulna. The posterior surfaces and the lateral surface of the radius are related to the extensor compartment. The extensor compartment appears to invade the territory of the flexor compartment but much of this is due to the muscle brachioradialis which, although supplied by the radial nerve has a rather indeterminate function. Note also the subcutaneous posterior margin of the ulna.

Look again at Fig. 3-61 and review the common

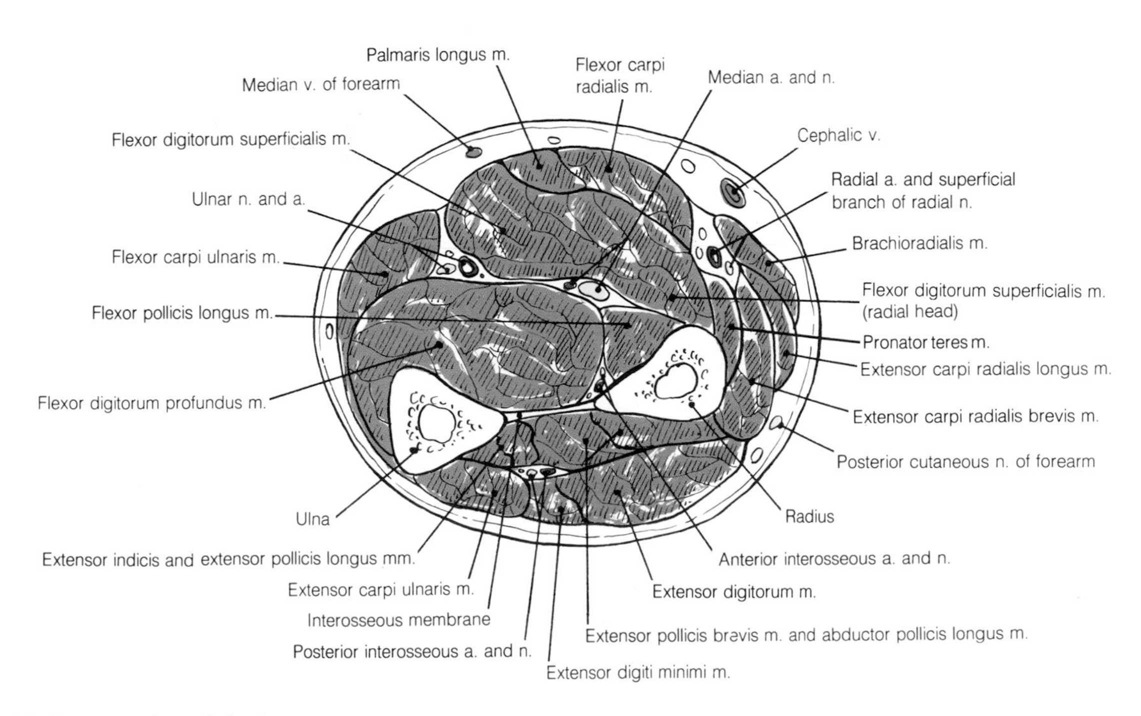

3-62 Cross-section of the forearm.

flexor and extensor origins from the humeral epicondyles and the origins of pronator teres, brachioradialis, and extensor carpi radialis longus from the supracondylar ridges. Find the additional sites of origin of pronator teres and flexor digitorum superficialis on the medial margin of the coronoid process and a further long linear origin of flexor digitorum superficialis from the shaft of the radius below the anterior oblique line. The distal attachment of pronator teres is found at the middle of the anterior border of the radius. The middle two-quarters of the shafts of both bones give origin to the flexor pollicis longus from the radius and flexor digitorum profundus from the ulna. Note how these attachments extend onto the adjacent interosseous membrane whose fibers are directed downward and medially from radius to ulna. The anterior surfaces of the lower quarter of both bones give attachment to the pronator quadratus which spans the interosseous membrane.

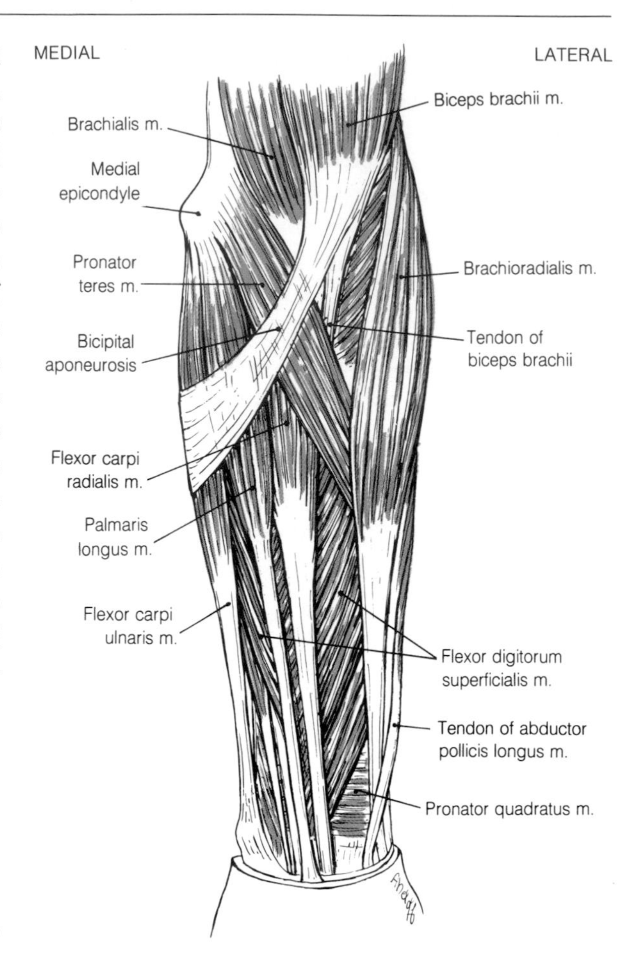

3-63 The superficial muscles of the flexor compartment of the left forearm.

The Muscles of the Flexor Compartment – The Superficial Group

From the medial supracondylar ridge and epicondyle of the humerus a group of five superficial muscles fan out over the flexor compartment like the five digits of the hand. From the medial to the lateral side these are:

Pronator teres
Flexor carpi radialis
Palmaris longus
Flexor digitorum superficialis
Flexor carpi ulnaris

The arrangement of these muscles is seen in Fig. 3-63 where brachioradialis, already found to intrude into the flexor compartment, is also included. The insertions of muscles reaching the hand will be described in more detail with that section.

Pronator Teres

This muscle arises from a narrow triangular region on the medial supracondylar ridge immediately above the medial epicondyle. Forming the medial margin of the cubital fossa it runs downward and laterally and passes beneath brachioradialis to become attached to the middle of the anterior border of the radius. It is joined on its deep surface by its ulnar head, a slip of fibers arising from the coronoid process. The median nerve passes between the two heads as it leaves the cubital fossa but the ulnar artery passes deep to both of them.

The action of pronator teres is to pull the shaft of the radius over that of the ulna in the movement of pronation.

Flexor Carpi Radialis*

This is a "flexor of the wrist of the radial side." It arises from the common flexor origin and becomes tendinous about halfway down the forearm. The tendon passes over the wrist beneath the **flexor retinaculum** and is inserted into the palmar sur-

* Karpos, *Gk.* = wrist joint. This is an unfortunate hybrid name for a muscle, but Greek words were Romanized in the same way they have been Anglicized.

face of the base of the second metacarpal bone with a slip to the third.

Looking ahead, there is a radial and ulnar flexor of the wrist and two radial and an ulnar extensor of the wrist. These may be thought of as the four guy ropes of a tent. When the two flexor muscles act together flexion of the wrist results, similarly the two extensors produce extension. If the two radial muscles act together they produce abduction or radial deviation of the wrist and the two ulnar muscles produce adduction or ulnar deviation. A graded combination of these actions will produce the movement of circumduction. If the right forearm is held between left thumb and index finger just above the wrist, it should be clear that this circumduction involves no element of rotation at the wrist joint.

Palmaris Longus

This muscle is probably becoming vestigal. It is not always present and, when it is, has a small short belly and a long thin tendon which is inserted into the **palmar fascia**. Its action is to aid in flexing the wrist joint. When necessary it also serves as a source of tendon for grafting.

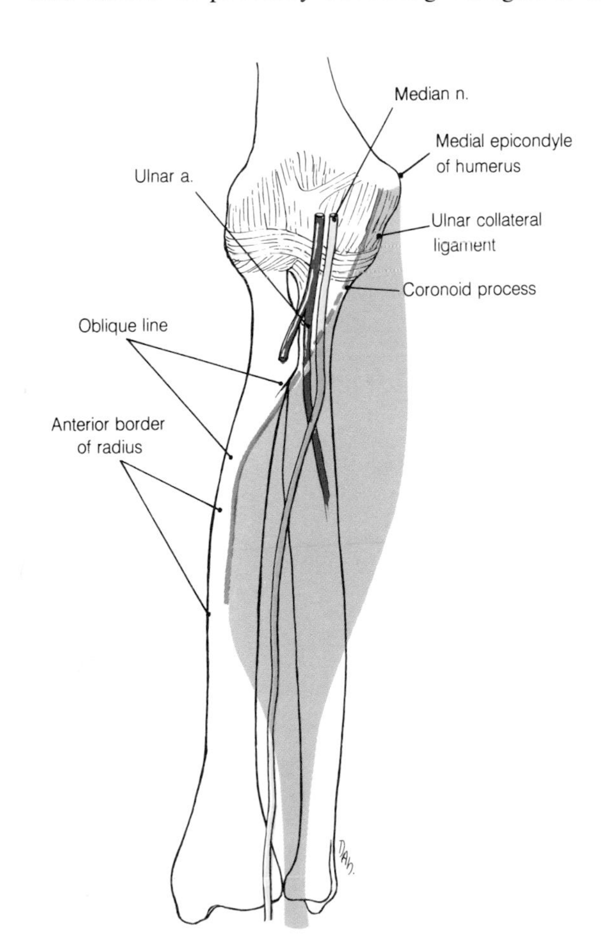

3-64 Showing the ulnar and radial attachments of the flexor digitorum superficialis muscle.

Flexor Digitorum Superficialis

The "superficial flexor of the fingers" is the largest of this superficial group of muscles and, in fact, lies partly beneath them on a slightly deeper plane. It has a long origin involving the common flexor origin, the ulnar collateral ligament of the elbow joint, the medial side of the coronoid process and the anterior border of the radius. The median nerve and the ulnar artery pass into the forearm between the humeroulnar and radial origins of this muscle. This relationship and the origin of the muscle are seen in Fig. 3-64. The median nerve remains deep to the muscle until near the wrist.

The somewhat flattened belly gives rise to four tendons just above the wrist. These pass deep to the flexor retinaculum to be inserted into the middle phalanges of the medial four digits.

Flexor digitorum superficialis is a flexor of the middle and proximal phalanges and the wrist.

Each of these four superficial muscles is supplied by **muscular branches of the median nerve**. The flexor carpi ulnaris, the fifth and most medial of the superficial muscles, is supplied by the **ulnar nerve**.

Flexor Carpi Ulnaris

This, the most medial of the superficial flexor muscles, has two heads. A humeral head arises

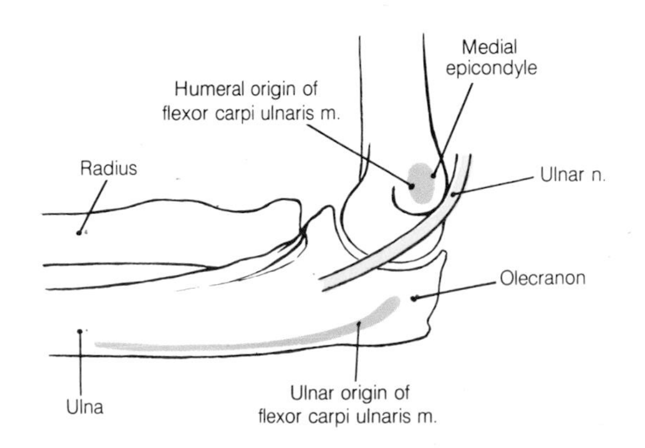

3-65 Showing the manner in which the ulnar nerve enters the forearm.

from the common flexor origin and an ulnar head from the medial aspect of the olecranon and the upper part of the posterior border of the ulna. The ulnar nerve passes into the forearm between the two heads as shown in Fig. 3-65. The belly extends almost vertically down the medial side of the forearm and the tendon formed in the lower half of the muscle is attached to the pisiform bone and by ligamentous extensions to the hamate and fifth metacarpal bones.

Flexor carpi ulnaris is a flexor and adductor of the wrist and is supplied by the **ulnar nerve**.

The Muscles of the Flexor Compartment — The Deep Group

A deep group of flexor muscles lies largely undercover of the superficial muscles described above.

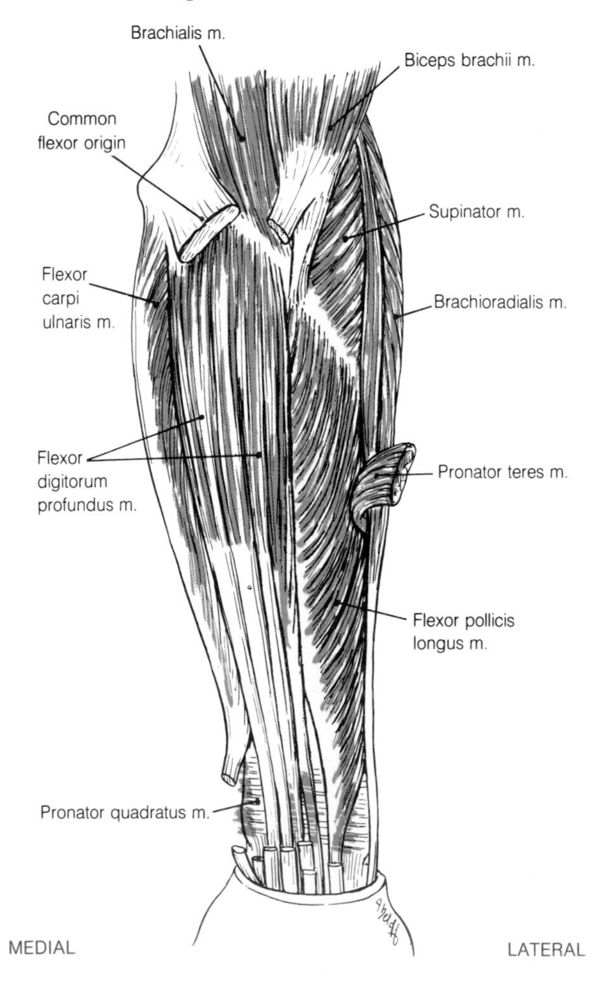

3-66 The deep muscles of the flexor compartment of the left forearm.

These are:
Flexor digitorum profundus
Flexor pollicis longus
Pronator quadratus

The arrangement of these deep muscles can be seen in Fig. 3-66.

Flexor Digitorum Profundus

The deep flexor of the fingers arises from the medial and anterior surfaces of the upper three-quarters of the shaft of the ulna and the adjacent interosseous membrane. The medial border of the muscle also gains origin from a tough aponeurosis attached to the posterior border of the ulna. This aponeurosis is shared with flexor carpi ulnaris and extensor carpi ulnaris.

About halfway down the forearm four tendons are formed, although only that to the index finger appears to be separate. These cross the wrist beneath the flexor retinaculum and run to the medial four digits to be inserted into their terminal phalanges. As will be seen later, the profundus tendon of each finger pierces the superficial tendon in order to reach its more distal attachment.

The innervation of this muscle is unusual in that its medial part is supplied by the **ulnar nerve** and its lateral part by the **anterior interosseous nerve** a branch of the **median nerve**. As a result, a lesion of the ulnar nerve is associated with weakness in flexion of the distal interphalangeal joint of the medial two digits. In fact, the individual axons distributed to the muscle by the ulnar nerve are probably derived from a communicating branch between median and ulnar nerves in the axilla.

The prime action of the muscle is to flex the distal interphalangeal joint, but it also aids in flexing the finger as a whole and the wrist.

Flexor Pollicis Longus

The long flexor of the thumb arises from the anterior surface of the shaft of the radius below the anterior oblique line and above the pronator quadratus muscle. Note that it does not extend onto the lateral surface of the radius which lies in the extensor compartment. However, like flexor digitorum profundus, it does arise from the interosseous membrane. It is also common to find a slip of fibers arising from the coronoid process and joining the main belly. The muscle ends in a single tendon which passes into the hand deep to the flexor retinaculum. It is inserted into the base of the distal phalanx of the thumb.

The muscle is supplied by the **anterior interosseous nerve** and its action is to flex the thumb.

Pronator Quadratus

Pronator quadratus is a flat quadrilateral muscle attached to and spanning the anterior surfaces of the lower ends of the radius and ulna.

It is supplied by the **anterior interosseous nerve** and it plays an important role in pronation of the forearm.

Brachioradialis

This muscle is not part of the flexor group of muscles and is supplied by the nerve to the extensor muscles. Nevertheless, despite much argument, it probably acts as a flexor of the elbow through most of the range of this joint. What is more to the point is that it lies in the front of the forearm and becomes an important relation when the vessels and nerves of the forearm are described.

It arises from the lateral supracondylar ridge of the humerus and descends the lateral margin of the forearm. It is inserted by a flat tendon into the base of the styloid process of the radius. As its name suggests, it extends from the arm to the radius and does not, like the other long muscles, enter the hand. It is supplied by the **radial nerve** before that nerve divides into its superficial and deep branch.

Vessels of the Flexor Compartment

The brachial artery enters the forearm through the cubital fossa where it divides into the **radial and ulnar arteries**.

The Radial Artery

From the cubital fossa the radial artery passes distally in the forearm at first undercover of the brachioradialis and then to the medial side of its tendon where it is covered by skin and fascia only. It is superficial to all other muscles attached to the underlying radius. These are supinator, pronator teres, flexor digitorum superficialis, flexor pollicis longus, and pronator quadratus. Where it lies on the lower end of the radius its pulsation is easily palpable. It is here also that a side-to-side anastomosis with the cephalic vein is formed prior to renal dialysis. After about six weeks the wall of the vein becomes hypertrophied, and needles leading to and from the dialysis machine can be repeatedly inserted. The superficial branch of the radial nerve joins the middle third of the artery in the forearm but leaves it again in the lower third. At the wrist the artery winds posteriorly to reach the dorsal surface of the hand.

Branches of the radial artery in the forearm are:

The radial recurrent artery
A palmar carpal branch
A dorsal carpal branch
A superficial palmar branch
Muscular branches

The **radial recurrent artery** arises soon after the radial artery is formed and its subsequent course will be described with the anastomosis around the elbow.

Small **palmar and dorsal carpal branches** anastomose with palmar and dorsal carpal branches of the ulnar artery at the wrist.

The **superficial palmar branch** is given off just before the radial artery turns posteriorly at the wrist. It contributes to the superficial palmar arch.

The Ulnar Artery

The ulnar artery arises together with the radial artery as a terminal branch of the brachial artery in the cubital fossa. It passes beneath both heads of pronator teres, flexor carpi radialis, and palmaris longus to reach the medial side of the forearm undercover of flexor carpi ulnaris. Continuing straight down the forearm it becomes superficial on the lateral side of the tendon of that muscle and medial to the tendons of flexor digitorum superficialis. For most of its course it lies on flexor digitorum profundus. At the wrist the artery passes through a superficial slip of the flexor retinaculum and, thence, into the hand.

As the ulnar artery passes beneath the pronator teres it is crossed superficially by the median nerve which passes between the heads of pronator teres. The artery is joined on its medial side by the ulnar nerve in the lower two-thirds of the forearm.

Branches of the ulnar artery in the forearm are:

- The anterior ulnar recurrent artery
- The posterior ulnar recurrent artery
- The common interosseous artery
- The anterior interosseous artery
- The posterior interosseous artery
- A palmar carpal branch
- A dorsal carpal branch
- Muscular branches

The **recurrent branches** are described with the anastomosis around the elbow joint. The **palmar and dorsal carpal branches** complete, with the corresponding branches of the radial artery, palmar and dorsal arches.

The **common interosseous artery** arises shortly after the ulnar artery is formed. It is a short branch which almost immediately divides into an anterior and posterior interosseous artery. The **posterior interosseous artery** passes into the extensor compartment of the forearm above the interosseous membrane. The **anterior interosseous artery** passes down the forearm on the anterior surface of the interosseous membrane. At the upper border of pronator quadratus it gives off a small branch which joins the palmar carpal arch and then pierces the membrane to anastomose with the posterior interosseous artery.

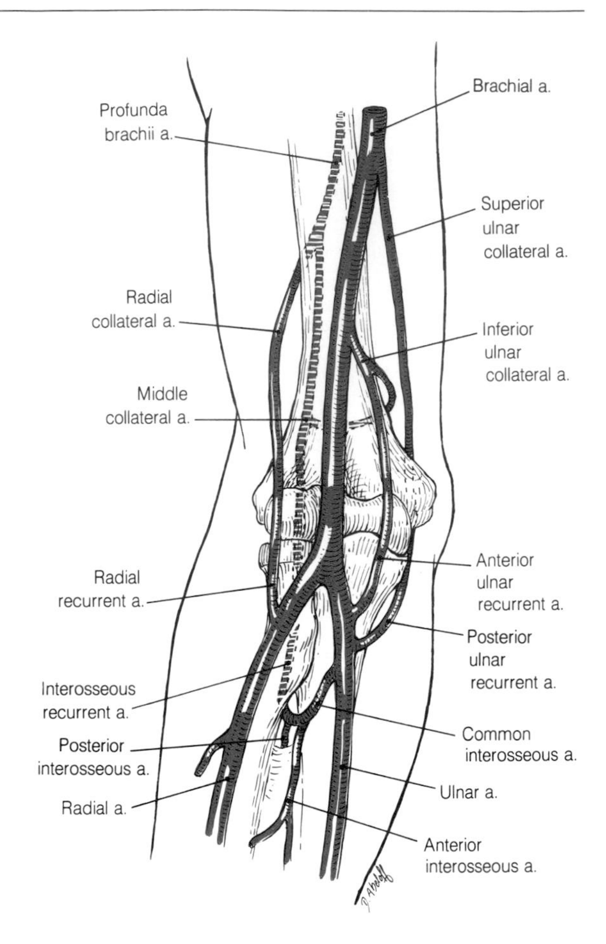

3-68 The arterial anastomoses around the elbow joint.

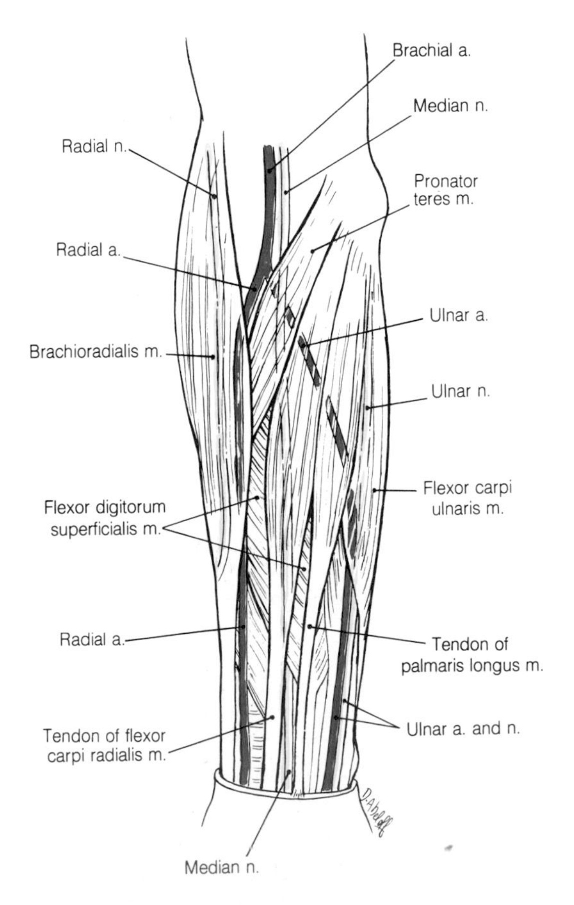

3-67 Semidiagrammatic illustration of the course of the ulnar and radial arteries through the forearm.

Review now the course and main relations of the radial and ulnar arteries in Fig. 3-67. Note:

1. That as it passes beneath pronator teres the ulnar artery is crossed superficially by the median nerve.
2. The radial artery and the superficial branch of the radial nerve lie alongside each other in their middle thirds and under cover of brachioradialis.
3. The ulnar artery and nerve lie alongside each other in their distal two-thirds.
4. The radial and ulnar nerves lie "outside" the arteries.

5. Both arteries are superficial in the lower third of their course.

Having described the brachial, radial, and ulnar arteries, it is now possible to consider together the collateral and recurrent branches that form an arterial anastomosis around the elbow joint. This is illustrated in Fig. 3-68. Note that an anastomotic vessel lies anterior and posterior to both humeral epicondyles. This figure also shows how the anterior and posterior interosseous arteries arise from the common interosseous branch of the ulnar artery.

Nerves of the Flexor Compartment

The **median and ulnar nerves** traverse the flexor compartment of the forearm and supply all the muscles it contains. The **superficial branch of the radial nerve** runs part of its course in this compartment but has no branches.

The Median Nerve

In the cubital fossa the median nerve lies on the medial side of the brachial artery. It enters the forearm by passing between the two heads of pronator teres and at the same time crosses the ulnar artery which lies deep to both heads. From pronator teres it passes onto the deep surface of flexor digitorum superficialis where it remains until just above the wrist. Here it appears between the tendons of flexor digitorum superficialis and flexor carpi radialis. Branches of the median nerve in the forearm are:

The anterior interosseous nerve.

Muscular branches to all the superficial flexors of the forearm except flexor carpi ulnaris.

A palmar cutaneous branch which is given off just above the wrist and runs into the hand to supply skin over the thenar eminence and the central part of the palm.

Articular branches to the elbow joint and proximal radioulnar joint.

The **anterior interosseous nerve** leaves the median nerve as it passes through pronator teres. It then joins the anterior interosseous artery and passes down the forearm on the surface of the interosseous membrane between flexor pollicis longus and flexor digitorum profundus. It supplies flexor pollicis longus, the lateral half of flexor digitorum profundus and pronator quadratus. It terminates by sending articular branches to the inferior radioulnar, wrist and carpal joints.

The Ulnar Nerve

The ulnar nerve enters the forearm by passing between the two heads of flexor carpi ulnaris as shown in Fig. 3-69. In this manner it comes to lie between this muscle and the underlying flexor digitorum profundus. It is joined on its lateral side by the ulnar artery and in the lower half of the forearm both structures become superficial on the lateral side of the tendon of flexor carpi ulnaris.

The branches of the ulnar nerve in the forearm are:

Muscular branches to flexor carpi ulnaris and the medial half of flexor digitorum profundus.

A palmar cutaneous branch which arises in the forearm and supplies skin of the medial part of the palm.

A dorsal branch that passes medially, deep to flexor carpi ulnaris, to reach the dorsal aspect of the hand.

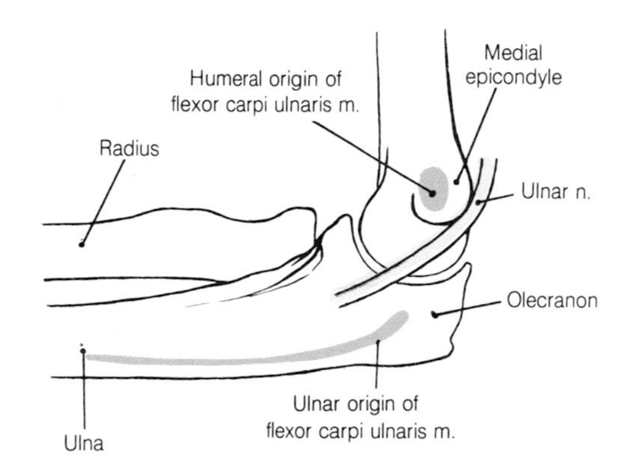

3-69 Showing the manner in which the ulnar nerve enters the forearm.

The Radial Nerve

The radial nerve enters the forearm between brachioradialis and brachialis. It immediately divides into a **superficial and deep branch**. The deep branch passes laterally around the radius be-

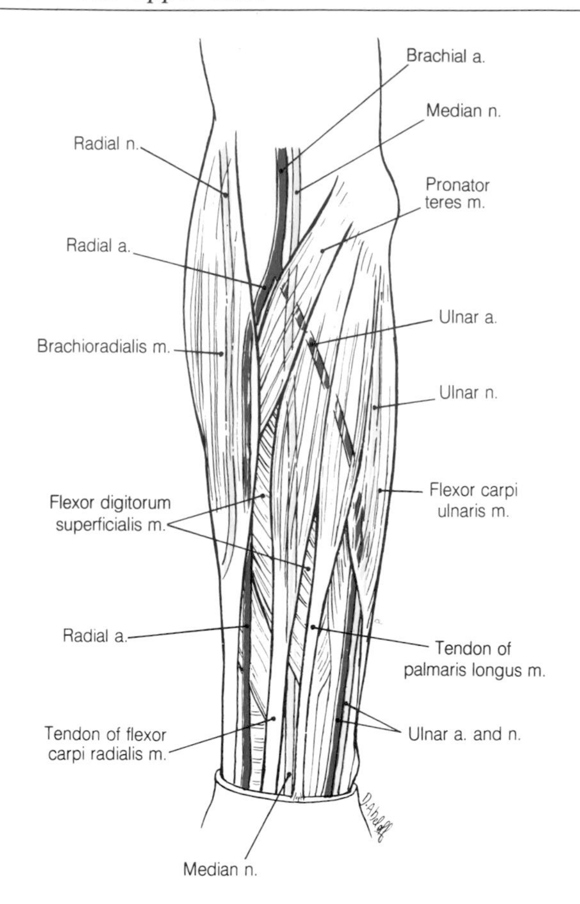

3-70 Semidiagrammatic illustration of the course of the radial, median, and ulnar nerves in the forearm.

tween the fibers of supinator and reaches the extensor compartment where it becomes the **posterior interosseous nerve**. The superficial branch continues down the forearm deep to brachioradialis. At the junction of the upper and middle thirds, it is joined on its medial side by the radial artery but in the lower third leaves the artery to pass from beneath brachioradialis onto the dorsum of the wrist and hand. The superficial branch of the radial nerve has no muscular or cutaneous branches in the forearm.

The course of each of these nerves in the forearm can now be reviewed in Fig. 3-70.

The Skeleton of the Forearm

Before describing the muscles of the extensor compartment, the dorsal aspects of the shafts of the ulna and radius must be examined.

The Ulna and Radius (Posterior Aspect)

Look at Fig. 3-71 and identify some bony features of the dorsal aspect of the ulna and radius. Note:

1. The olecranon process
2. The disc-like profile of the head of the radius and the posterior margin of the radial notch of the ulna
3. The supinator crest which hides the radial tuberosity
4. The subcutaneous border of the ulna descending from the olecranon process
5. The faint posterior oblique line of the radius
6. The interosseous borders of both bones and the interosseous membrane
7. The styloid process of the radius
8. The dorsal tubercle of the radius
9. The styloid process of the ulna; this projects from the posteromedial rather than medial side of the lower end of the ulna
10. The rounded lateral surface of the lower end of the ulna which articulates with the ulnar notch of the lower end of the radius.
11. In addition identify again the medial and lateral epicondyles and the lateral supracondylar ridge of the humerus. Each of these regions provides attachment for many of the muscles found in the forearm.

As in the flexor compartment of the forearm, a superficial group of muscles arises from the humerus. Attachment for these is provided by the lateral supracondylar ridge and the lateral epicondyle. In Fig. 3-72 identify the following attachments of extensor muscles of the humerus, ulna, and radius:

1. Brachioradialis and extensor carpi radialis longus to the lateral supracondylar ridge of the humerus
2. The common extensor origin on the anterior surface of the lateral epicondyle (Fig. 3-61)
3. The attachment of triceps brachii to the superior surface of the olecranon and that of anconeus below
4. The remainder of the attachment of the supinator muscle seen earlier on the ventral aspect of the neck and body of the radius above the anterior oblique line. Here it lies above the less obvious posterior oblique line.
5. The attachment of an aponeurosis common to flexor digitorum profundus, flexor carpi ulnaris

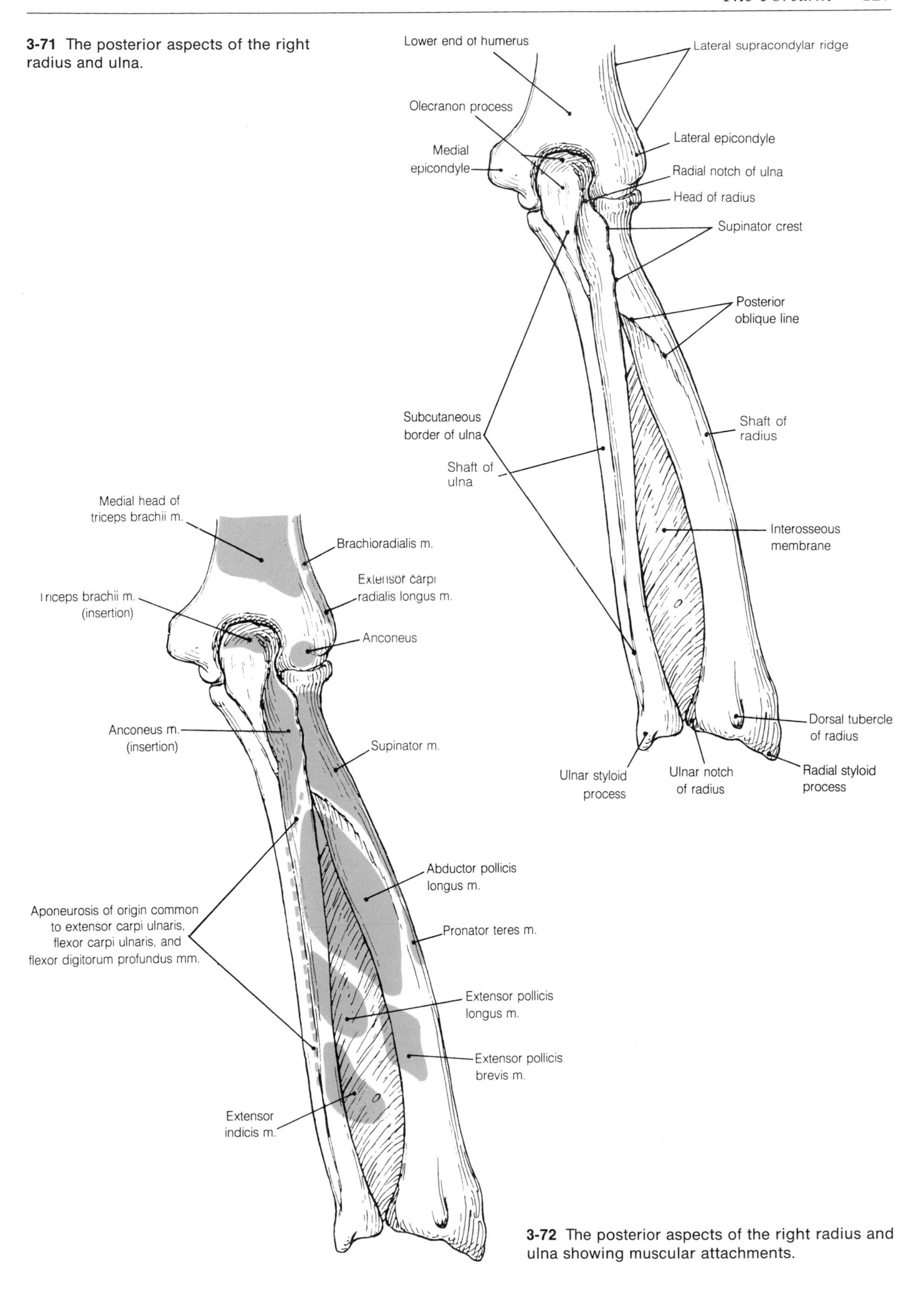

3-71 The posterior aspects of the right radius and ulna.

3-72 The posterior aspects of the right radius and ulna showing muscular attachments.

and extensor carpi ulnaris to the upper two-thirds of the posterior subcutaneous border of the ulna

6. The posterior surfaces of the bodies of the ulna and radius and the intervening interosseous membrane. This surface provides attachment for abductor pollicis longus, extensor pollicis longus, extensor pollicis brevis and extensor indicis.

The Muscles of the Extensor Compartment — Superficial Group

The muscles included in the superficial group are:

Brachioradialis
Extensor carpi radialis longus
Extensor carpi radialis brevis
Extensor digitorum
Extensor digiti minimi
Extensor carpi ulnaris

The arrangement of these muscles can be seen in Fig. 3-73.

Brachioradialis

This muscle was described with the flexor muscles. It arises from the lateral supracondylar ridge of the humerus and is inserted into the base of the styloid process of the radius. It is supplied by the **radial nerve** in the arm.

Extensor Carpi Radialis Longus

There are two radial extensors of the wrist. The longer arises from the lateral supracondylar ridge of the humerus between brachioradialis and the common extensor origin. It is initially covered by brachioradialis but in the middle of the forearm its tendon appears at the posterior border of brachioradialis. However it is soon hidden again by abductor pollicis longus and extensor pollicis brevis as they spiral round the radial side of the forearm to reach the base of the thumb. After crossing the dorsal surface of the lower end of the radius at the wrist, the tendon is inserted into the dorsal aspect of the base of the second metacarpal bone.

Extensor carpi radialis longus is supplied by the **radial nerve** before that nerve divides. Its action is to extend and abduct the wrist.

Extensor Carpi Radialis Brevis

The short radial extensor of the wrist arises from the common extensor origin just below and undercover of the long extensor. Its tendon appears in the middle of the forearm and lying medial to that of extensor carpi radialis longus accompanies it to cross the wrist and become inserted into the dorsal aspect of the base of the third metacarpal bone. The insertions of these two muscles can be compared with that of the single flexor carpi radialis that has a distal attachment to the palmar surface of both the second and third metacarpal bones.

Extensor carpi radialis brevis is supplied by the **posterior interosseous nerve** and is an extensor and abductor of the wrist.

Extensor Digitorum

Also arising from the common extensor origin, this muscle diverges from the lateral border of the forearm towards the midline. The belly divides into four tendons which pass together over the wrist and then separate to be attached to the dorsal aspects of the middle and distal phalanges of the medial four digits.

Extensor digitorum is supplied by the **posterior interosseous** nerve and its action is to extend the middle and distal phalangeal joints. By virtue of the position of its tendons on the dorsal aspect of the metacarpophalangeal joints and the extensor retinaculum that binds them to the wrist, the muscle will also extend these joints. Extension of all these joints occurs as the hand is opened to grasp an object.

Extensor Digiti Minimi

The extensor of the smallest or little finger arises from the common extensor origin and runs down the dorsal aspect of the forearm medial to the extensor digitorum. Its tendon divides into two fine slips after crossing the wrist joint and there, with a tendon from extensor digitorum, joins the extensor expansion of the little finger.

Extensor digiti minimi, like extensor digitorum, is supplied by the **posterior interosseous nerve** and extends the fifth finger and wrist.

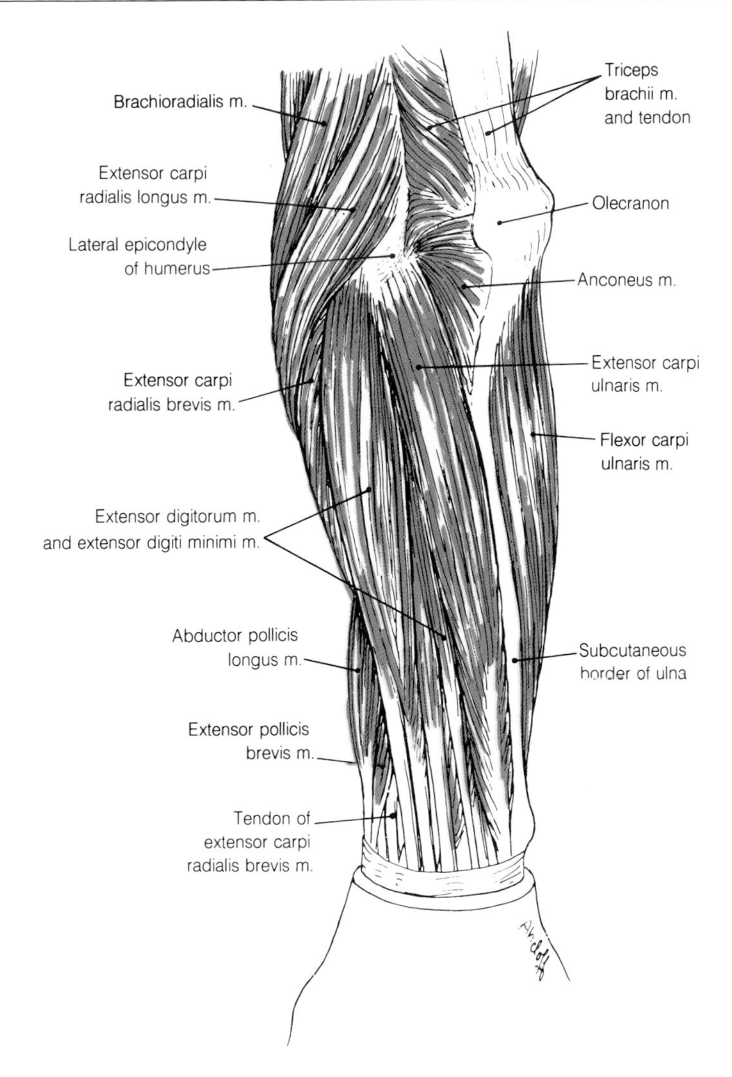

3-73 The superficial muscles of the extensor compartment.

Extensor Carpi Ulnaris

The ulnar extensor of the wrist arises from the common extensor origin and an aponeurosis attached to the posterior border of the ulnar. Its tendon is attached to a tubercle on the medial side of the base of the fifth metacarpal bone.

This muscle is supplied by the **posterior interosseous nerve** and its action is to extend the wrist and adduct it.

The Muscles of the Extensor Compartment – Deep Group

Beneath the superficial group of extensor muscles is a deep group. One of these, the supinator, acts only on the superior radioulnar joint, the remainder form long tendons which cross the wrist and enter the hand. The arrangement of these muscles is seen in Fig. 3-74 and they include:

Supinator
Abductor Pollicis Longus
Extensor Pollicis Longus
Extensor Pollicis Brevis
Extensor Indicis

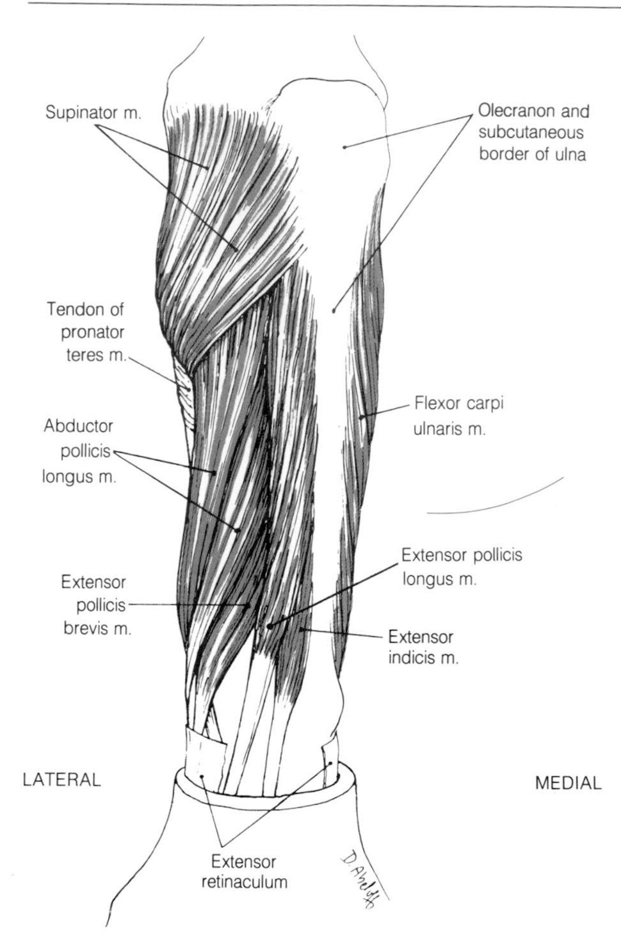

3-74 The deep muscles of the extensor compartment.

Supinator

Supinator is formed from a deep and superficial lamina. The deep lamina arises from the supinator crest of the ulna and its fibers pass posterior to the neck and upper shaft of the radius. A superficial lamina has a continuous origin from the lateral epicondyle of the humerus, the radial collateral ligament of the elbow joint and the annular ligament of the superior radioulnar joint. From these origins fibers wrap around the head, neck and upper shaft of the radius. Both laminae are inserted into the shaft of the radius above the anterior and posterior oblique lines. It is between the two laminae that the deep branch of the radial nerve passes lateral to the neck of the radius into the extensor compartment of the forearm.

The action of the muscle is to rotate the shaft of the radius laterally. This action forms an important part of the movement of supination and the arrangement of the fibers and their relationship to the deep branch of the radial nerve are illustrated in Fig. 3-75. Supinator is supplied by the **deep branch of the radial nerve**.

The remaining four muscles of the deep group arise from the posterior surfaces of the radius and ulna. Turn again to Fig. 3-72 and see how their origins are arranged.

Abductor Pollicis Longus

Abductor pollicis longus arises from the posterior surfaces of the ulna and radius and the interosseous membrane below the anconeus on the ulna and supinator on the radius. Its fleshy belly passes distally and laterally down the forearm to appear above the wrist between the superficial muscles extensor digitorum and extensor carpi radialis brevis and longus. Passing superficial to the latter two muscles and brachioradialis it crosses the wrist to reach the radial side of the base of the first metacarpal bone.

The muscle is supplied by the **posterior interosseous nerve** and its action is to abduct the thumb.

Extensor Pollicis Brevis

Extensor pollicis brevis arises from the posterior surface of the radius and interosseous membrane below the abductor pollicis longus. The muscle accompanies abductor pollicis longus through the forearm and over the wrist to become attached to the dorsal surface of the base of the proximal phalanx of the thumb.

Extensor pollicis brevis extends the proximal phalanx of the thumb and is supplied by the **posterior interosseous nerve**.

Extensor Pollicis Longus

Extensor pollicis longus arises from the posterior surface of the ulna and interosseous membrane below abductor pollicis longus. The belly passes directly down the forearm toward a deep groove on the dorsal aspect of the lower end of the radius. From this point the tendon passes laterally across the tendons of the two radial extensors of the wrist to reach the dorsal aspect of the base of the distal phalanx of the thumb.

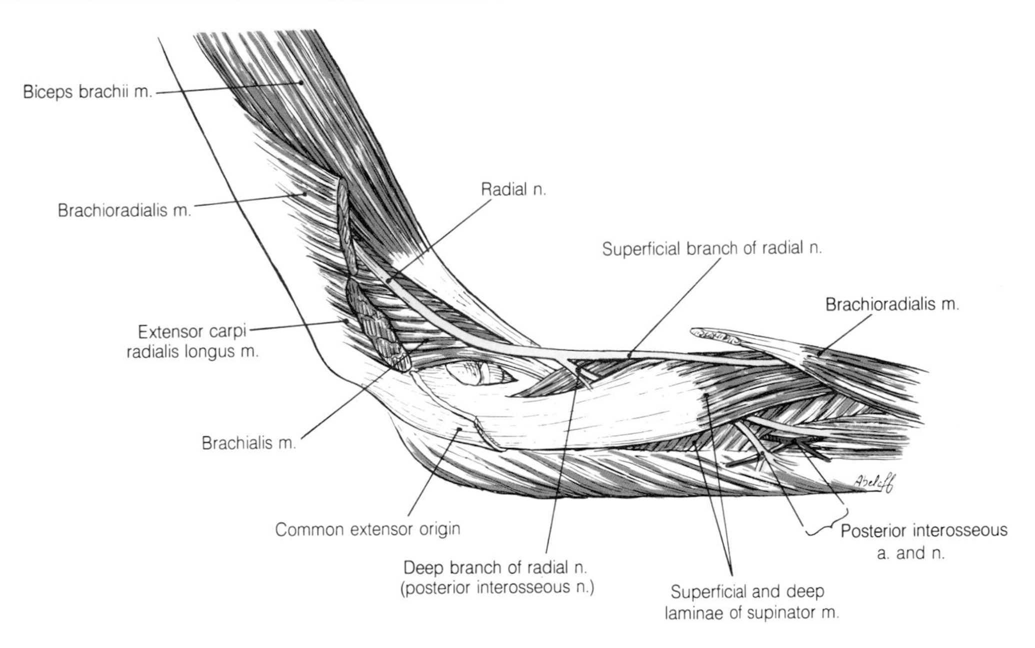

3-75 Showing the relationship of the supinator muscle to the deep branch of the radial nerve.

The muscle is supplied by the **posterior interosseous nerve** and serves to extend the distal phalanx of the thumb. In addition, the direction of its pull from the dorsum of the lower end of the radius allows it to adduct the thumb once it is fully extended.

When the thumb is actively extended a depression can be seen in the skin above the base of the first metacarpal bone. This is known as the "anatomical snuffbox." Its anterior margin is formed by the underlying tendons of abductor pollicis longus and extensor pollicis brevis and its posterior margin by the tendon of extensor pollicis longus. In its floor can be felt the radial artery as it winds round onto the dorsum of the hand. What is rather more important is that the scaphoid bone lies beneath the artery. Acute tenderness over this bone after a fall on the hand suggests a fracture even if no sign can be seen on an initial radiograph.

Extensor Indicis

A separate extensor muscle is provided for the index finger in addition to the tendon from the extensor digitorum muscle. This arises from the posterior surface of the ulna and interosseous membrane below extensor pollicis longus. Its tendon crosses the wrist to join that from the extensor digitorum longus.

The muscle is supplied by the **posterior interosseous nerve** and aids in extending the index finger.

To summarize the nerve supply of the superficial and deep muscles of the extensor compartment of the forearm, it can be said that with the exception of brachioradialis and extensor carpi radialis longus, which are supplied by the **radial nerve** before it divides, all are supplied by the **posterior interosseous nerve**.

Vessels of the Extensor Compartment of the Forearm

No major artery reaches the extensor compartment of the forearm from the arm. However, shortly after its formation in the flexor compartment, the **ulnar artery** gives off a short trunk called the **common interosseous artery**. This in turn divides into an **anterior and posterior interosseous artery**. Both these vessels are concerned with the blood supply of structures in the extensor compartment.

The Posterior Interosseous Artery

From its origin from the common interosseous artery, the small posterior interosseous artery passes posteriorly above the upper border of the interosseous membrane between ulna and radius. It appears between supinator and abductor pollicis longus and running distally between the deep and superficial extensor muscles, supples these.

The Anterior Interosseous Artery

During its course down the anterior surface of the interosseous membrane, the anterior interosseous artery provides branches which pierce the membrane to supply the underlying extensor muscles. At the upper border of pronator quadratus, it pierces the membrane to anastomose with the **posterior interosseous artery** and continue to the wrist where it joins the **dorsal carpal arch**.

Nerves of the Extensor Compartment of the Forearm

Two of the extensor muscles of the forearm, brachioradialis and extensor carpi radialis longus, are supplied by the radial nerve before it divides at the elbow into a superficial and deep branch. The deep branch, commonly called the posterior interosseous nerve, passes into the extensor compartment of the forearm to supply all the remaining extensor muscles.

The Posterior Interosseous Nerve

Formed at the elbow by the division of the radial nerve under cover of brachioradialis, the deep branch of the radial nerve passes laterally around the radius between the two laminae of supinator (Fig. 3-75). This muscle leads the nerve to the plane between the superficial and deep extensor muscles where it is now called the posterior interosseous nerve. Travelling distally in this plane, it crosses abductor pollicis longus and extensor pollicis brevis before passing onto the posterior interosseous membrane to terminate at the wrist. A muscular branch to the extensor carpi radialis brevis is given off before the nerve pierces supinator. The remaining muscles are supplied by branches arising soon after it leaves the supinator muscle. The nerve terminates as fine articular branches to joints at the wrist.

Now use the cross-section passing through the middle of the forearm shown in Fig. 3-76 to review the main relations of structures in the forearm. Note in this:

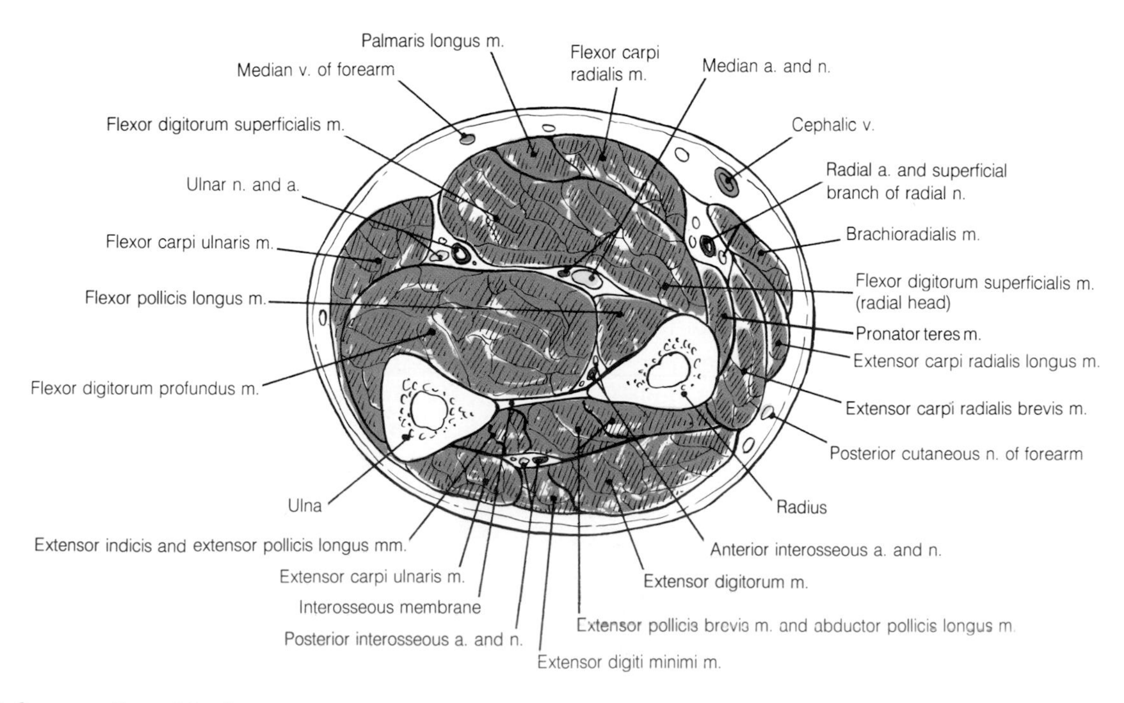

3-76 Cross-section of the forearm.

1. The ulna and radius with their sharp interosseous borders joined by the interosseous membrane.
2. Flexor pollicis longus on the anterior surface of the radius and flexor digitorum profundus on the anterior and medial surfaces of the ulna.
3. The anterior interosseous artery and nerve lying between these two muscles.
4. The flexor digitorum superficialis lying anterior to the deep flexor of the fingers and the long flexor of the thumb.
5. The median nerve on the deep surface of the flexor digitorum superficialis.
6. Flexor carpi ulnaris and its aponeurotic ulnar attachment.
7. The ulnar nerve and artery lying between flexor carpi ulnaris and flexor digitorum superficialis.
8. At the lateral surface of the radius pronator teres and its insertion into this bone.
9. Lateral to this a group of extensor muscles, extensor carpi radialis longus and brevis and brachioradialis.
10. Beneath brachioradialis the superficial branch of the radial nerve and the radial artery.
11. Most anteriorly palmaris longus with flexor carpi radialis on its lateral side.
12. On the posterior surfaces of the ulna, radius and posterior interosseous membrane, portions of extensor indicis, extensor pollicis longus, extensor pollicis brevis and abductor pollicis longus.
13. The posterior interosseous nerve and artery lying between the deep extensor muscles mentioned above and the more superficial extensor carpi ulnaris, extensor digiti minimi and extensor digitorum.

The Radioulnar Joints and Pronation and Supination

It is quite apparent that when the elbow is flexed to 90° so as to eliminate the effect of rotation of the humerus at the shoulder, the hand can be placed with the palm flat on a table in the prone position or with the dorsum flat on the table in the supine position. However, if the two forearm bones are grasped firmly above the wrist, this maneuver is not possible. The movements of pronation and supination that allow the hand to assume these positions must, therefore, be occurring in the forearm and not at the wrist joint. The elbow joint only allows flexion and extension: it is at the proximal and distal radioulnar joints that pronation and supination occur.

Some stress is laid on these movements because they add enormously to the manipulative ability of the hand and their loss after injuries to the forearm is a severe disability.

The Proximal Radioulnar Joint

At this joint the margin of the disc-like head of the radius rotates inside an osseofibrous ring formed by the **radial notch of the ulna** and the strong **annular ligament** that completes the ring. The joint and this ligament are illustrated in Fig. 3-77 which also shows how the lower margin of the ligament narrows around the neck of the radius thus resisting downward dislocation.

This is a synovial joint and its cavity is continuous with that of the elbow joint. The movement at the joint is a rotation of the head of the radius within the osseofibrous ring. This rotation is not limited by movement at the elbow joint because the hollowed superior surface of the radius ar-

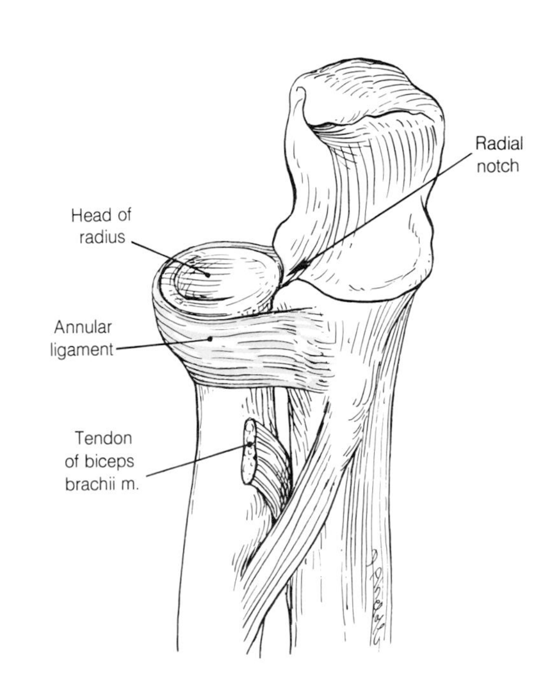

3-77 The proximal radioulnar joint.

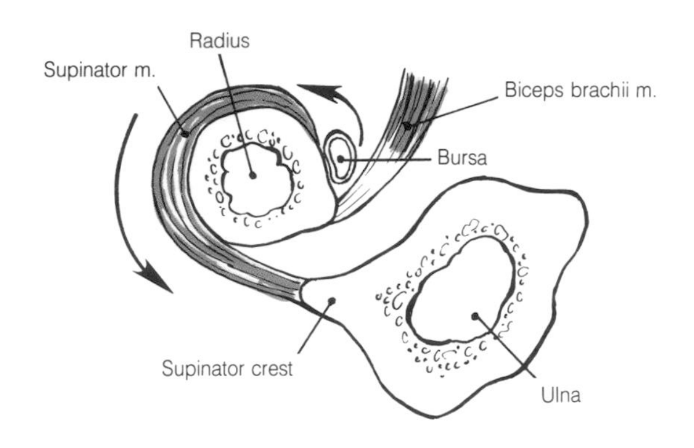

3-78 Diagram to show the actions of the biceps brachii and supinator muscles at the proximal radioulnar joint.

ticulates with the globular capitulum in the manner of a multiaxial ball and socket joint. Examine Fig. 3-78 to see how the actions of biceps brachii and supinator bring about supination at this joint. Rotation of the radius in the opposite direction is carried out by pronator teres and pronator quadratus.

The Middle Radioulnar Joint

The shafts of the ulna and radius are joined by the interosseous membrane whose fibers extend obliquely downward and medially from radius to ulna. If the radius is to move in relation to the ulna, tension in this membrane must be taken into account. In fact, it is found to be lax in the extremes of pronation and supination and only becomes tense in the commonly used intermediate position.

The Distal Radioulnar Joint

At the distal radioulnar joint the convex lower end of the ulna articulates with the ulnar notch of the radius. Although the proximal and distal joints have somewhat similar characteristics, at both joints the ulna remains stationary. Thus, at the distal joint the lower end of the radius turns around the ulna as indicated in Fig. 3-79. Again this is a synovial joint and the cavity extends over the inferior surface of the lower end of the ulna beneath a triangular fibrocartilage whose base is attached to the lower end of the radius and whose apex is attached to a pit between the styloid process of the ulna and its inferior articular surface. This fibrocartilage moves with the radius about its ulnar attachment. It also separates the ulna from the proximal row of carpal bones at the wrist or radiocarpal joint as can be seen in Fig. 3-80.

It is now possible to understand how the radius together with the wrist and hand can swing about an axis which passes through the center of the head of the radius and the center of the lower end of the ulna. This axis is illustrated in Fig. 3-81. In Fig. 3-81A the radius and ulna lie parallel as in the supinated or anatomical position, in Fig. 3-81B the forearm and hand have been pronated.

It is the muscles pronator teres and pronator quadratus that swing the radius over the ulna in pronation and the biceps and the supinator muscle that return the radius to the anatomical position in supination.

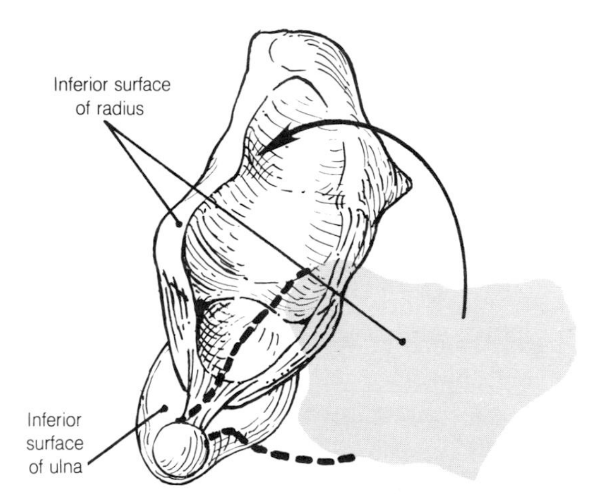

3-79 Diagram to illustrate the movement of the lower end of the radius about the ulna.

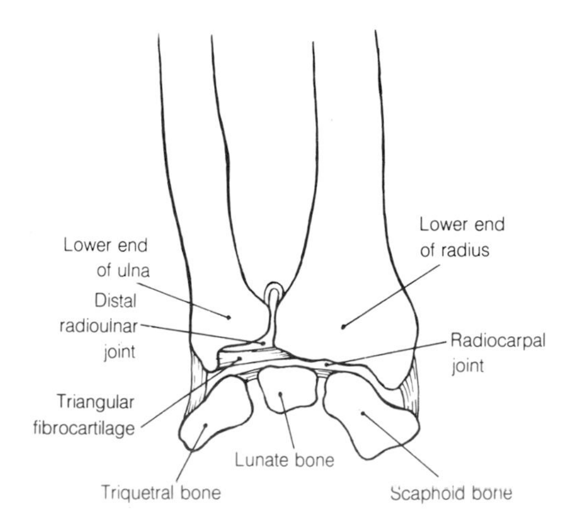

3-80 Diagram showing the relationship of the distal radioulnar joint to the radiocarpal joint.

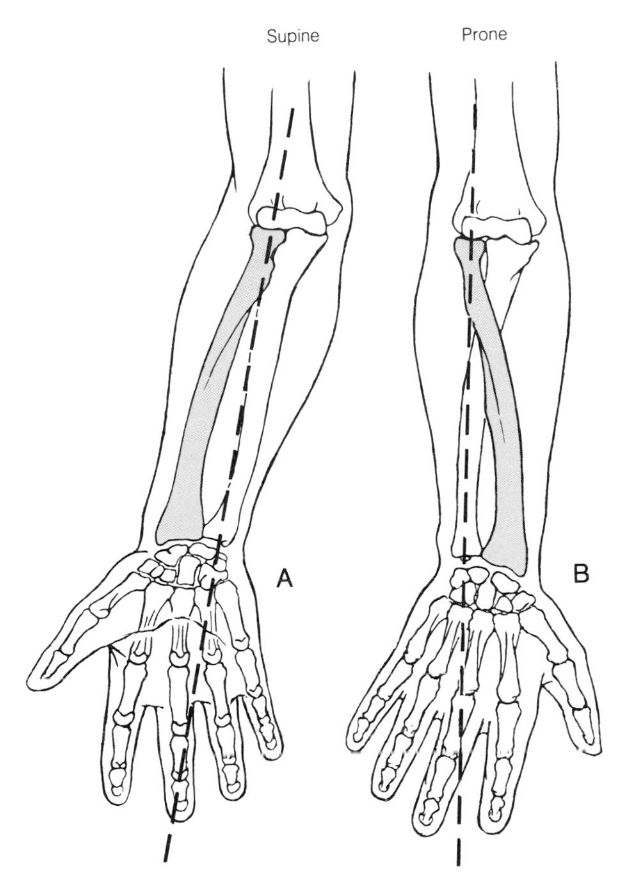

3-81 The ulna and radius shown in supination (*A*) and pronation (*B*).

The Dorsum of the Hand

The skin of the dorsum of the hand, unlike that of the palm is thin and not anchored to the underlying deep fascia and thus easily becomes edematous. As a result the most obvious outward sign of a deep infection of the **palm** may be a puffy swelling of the **dorsum** of the hand. The superficial veins show clearly, especially when the hand is dependent, and the irregular dorsal network and the commencements of the basilic and cephalic veins can be seen through the skin. Beneath the veins the extensor tendons can be made to stand out when the fingers and thumb are extended.

Except for the dorsal interosseous muscles, which are described with the palmar surface of the hand, there are no other muscles intrinsic to the dorsum of the hand and all tendons encountered in this region arise from muscles whose bellies lie in the extensor region of the forearm.

As they pass across the wrist joint all these tendons are surrounded by synovial sheaths and bound to its dorsal aspect by the **extensor retinaculum**.

Look now at Fig. 3-82, which illustrates the **synovial sheaths** and the **extensor retinaculum**. It also provides an opportunity to review the muscles

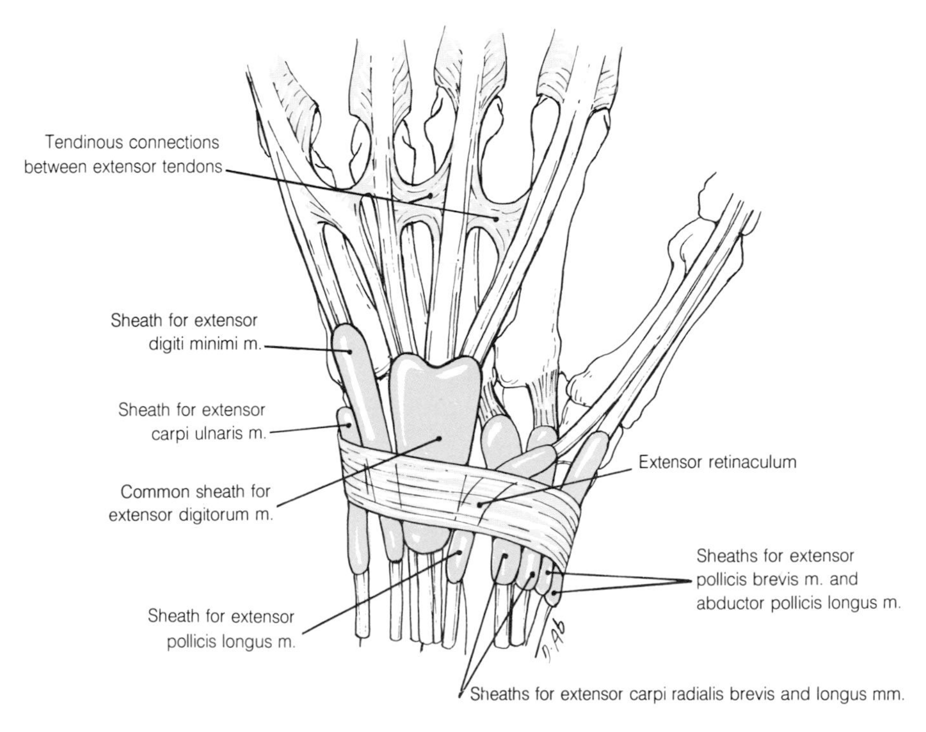

3-82 The tendons, their synovial sheaths, and the extensor retinaculum on the dorsal aspect of the wrist.

reaching the dorsum of the hand. Beginning on the radial side these are:

Abductor pollicis longus and extensor pollicis brevis which run in a groove on the lateral side of the radius. Remember that these two deep extensor muscles become superficial in the forearm between extensor digitorum and extensor carpi radialis longus and brevis. As they run over the carpus they form the lateral border of the anatomical snuffbox. Abductor pollicis is attached to the radial side of the first metacarpal bone.

Extensor carpi radialis longus and brevis which are attached to the dorsal surfaces of the bases of the second and third metacarpal bones.

Extensor pollicis longus which is separated from the preceding muscles by a well marked and palpable tubercle on the lower end of the radius. As the tendon of this muscle passes distal to this tubercle, it turns laterally across extensor carpi radialis longus and brevis to form the medial border of the snuffbox.

Extensor digitorum and extensor indicis which share a common synovial sheath.

Extensor digiti minimi which passes in its own sheath over the distal radioulnar joint.

Extensor carpi ulnaris which passes over the head of the ulna adjacent to its styloid process. It is attached to the medial side of the base of the fifth metacarpal.

The Extensor Retinaculum

This is a strong fibrous band extending from the lateral border of the radius and obliquely across the wrist to the triquetral and pisiform bones. Note in Fig. 3-82 how this retinaculum links the hand to the radius but not the ulna. In this way the unit of radius and hand are not limited in movement by the stationary ulna in the movements of pronation and supination.

The Long Extensor Tendons in the Hand

The distal attachments of the long extensor tendons of the digits have already been mentioned, but they now need to be described in greater detail. Note that because of their position on the dorsal surface of the hand, the long extensor tendons to the digits are also able to produce extension at the metacarpophalangeal and wrist joints.

Extensor pollicis longus is attached to the dorsal surface of the base of the distal phalanx of the thumb. Over the first metacarpophalangeal joints its tendon is joined by slips from the tendons of two palmar muscles; abductor pollicis brevis from the lateral side and adductor pollicis from the medial. The long extensor of the thumb extends the interphalangeal and metacarpophalangeal joints. It also extends and adducts the first metacarpal bone.

Extensor pollicis brevis is attached to the base of the proximal phalanx of the thumb and extends the metacarpophalangeal and carpometacarpal joints.

Extensor digitorum provides a tendon for each of the four fingers. Over each metacarpophalangeal joint the tendons become expanded to form the dorsal digital expansions or extensor hoods. These are triangular in shape with a base facing proximally and an apex facing distally. The dorsal digital expansions of the index and little fingers are joined by the tendons of extensor indicis and extensor digit minimi. It is to the sides of these expansions that the lumbrical and interosseous muscles are attached.

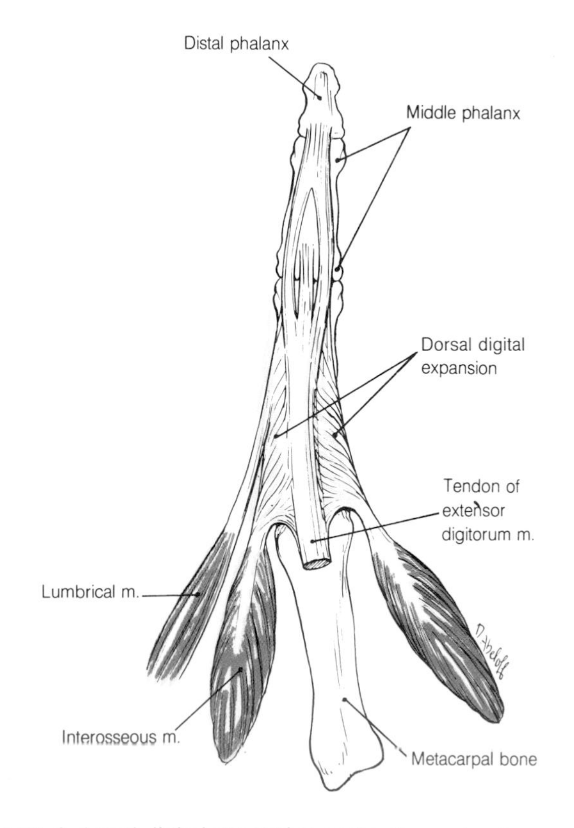

3-83 A dorsal digital expansion and extensor tendon.

The expansion narrows over the proximal phalanx and becomes partially separated into three slips. The central slip is inserted into the dorsal surface of the base of the middle phalanx and the lateral slips unite over the body of the middle phalanx and are inserted into the dorsal surface of the base of the distal phalanx.

Look now at Fig. 3-83 and confirm the features of the dorsal digital expansion and the long extensor tendon which have been described above. The independent functions of the slips inserted into the middle and distal phalanxes are well illustrated by two minor but nevertheless disabling and disfiguring injuries. When the two lateral slips are torn from their insertion into the distal phalanx a "mallet finger" is produced. This is illustrated in Fig. 3-84 where it can be seen that although the proximal interphalangeal joint can be extended there is an inability to extend the distal interphalangeal joint. In a "boutonnière"* lesion the lateral slips are torn from the central slip and slide down on either side of the proximal interphalangeal joint. The resulting deformity which is illustrated in Fig. 3-85 is flexion of the proximal joint and hyperextension of the distal joint.

The action of the long extensor tendons is to extend the interphalangeal, metacarpophalangeal and wrist joints. However, in some manipulations it is necessary to flex the metacarpophalangeal joints while at the same time extending the interphalangeal joints. This combination of movements is produced by the interosseous and lumbrical muscles. These muscles join the dorsal digital expansions by passing across the axes of the metacarpophalangeal joints on their palmar side and thus are able to flex them. Through the expansion and the long extensor tendon the pull of the muscles is transferred to the dorsal side of the axes of the interphalangeal joints and, thus, they are able to extend these. The diagram in Fig. 3-86 shows the relationship of these muscles to the axes of the metacarpophalangeal and proximal interphalangeal joints.

* Boutonnière, *Fr.* = buttonhole

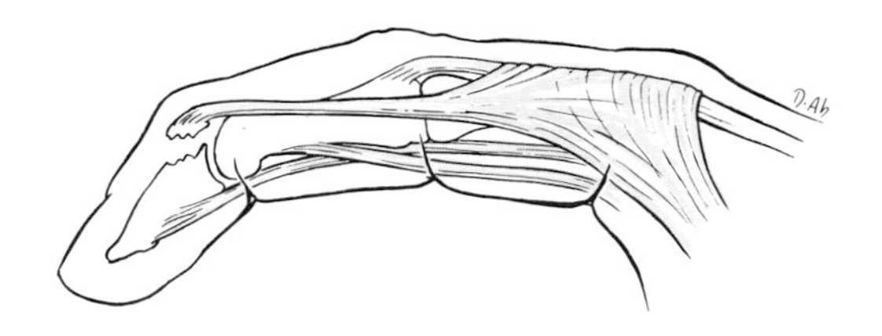

3-84 Showing the injury in a mallet finger.

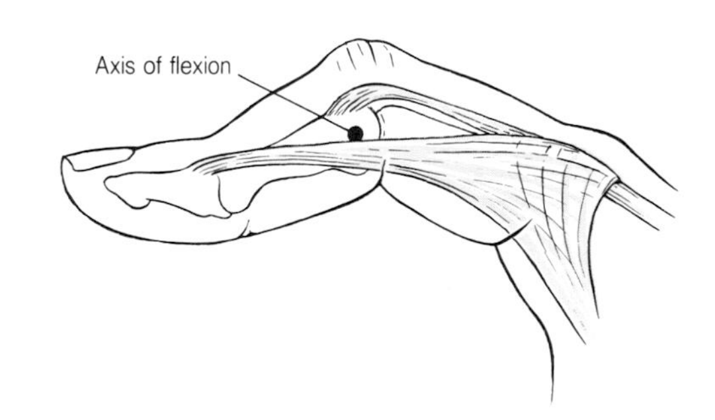

3-85 Showing a "boutonnière" injury.

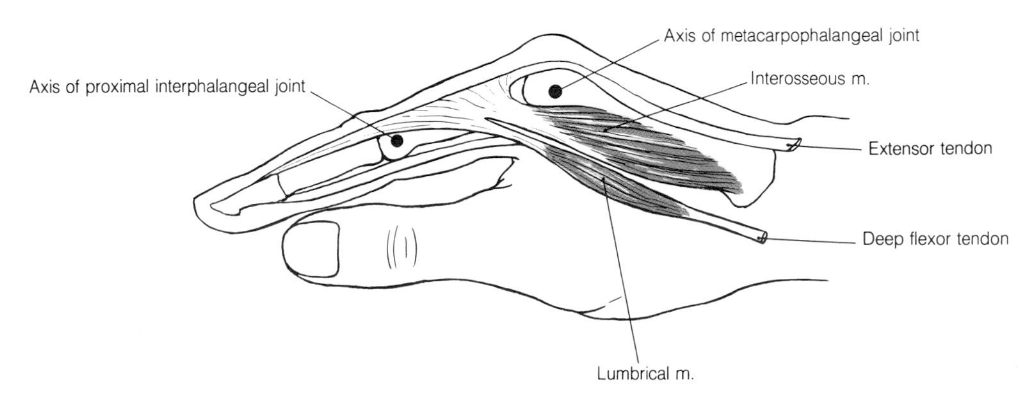

3-86 Showing the relationship of lumbrical and interosseous muscles to metacarpophalangeal and interphalangeal joint axes.

The Nerves of the Dorsum of the Hand

There are no intrinsic muscles to be supplied on the dorsum of the hand and branches of the radial and ulnar nerves reaching this region provide innervation for the skin of the dorsum of the hand and fingers.

The **Superficial branch of the radial nerve** leaves the lower third of the forearm by passing beneath the tendon of brachioradialis. It then winds round on to the back of the hand where it divides into dorsal digital nerves for the thumb, index, middle and often half the ring finger.

In a rather similar manner, the **dorsal branch of the ulnar nerve** winds around the ulnar side of the forearm just above the wrist to supply dorsal digital nerves to the ulnar half of the ring finger and both sides of the little finger. Note that:

1. The territory of the dorsal digital nerve to the radial side of the thumb includes a constant area of skin of the thenar eminence.
2. The territory supplied by the ulnar nerve may include the whole of the ring finger and half the middle finger.
3. The nail beds and at least the skin over the dorsum of the distal phalanges are supplied by dorsal branches of palmar digital nerves.

A typical pattern of the distribution of cutaneous nerves on the dorsum of the hand is illustrated in Fig. 3-87.

The Vessels of the Dorsum of the Hand

The radial artery (described in more detail with the palm of the hand) appears for a short distance on the dorsum of the hand. This vessel, its dorsal carpal branch, and the dorsal carpal branch of the ulnar artery form an arterial arcade over the back of the wrist from which arise dorsal metacarpal arteries. These in turn provide dorsal digital arteries to each side of the digits except for the thumb and the lateral side of the index finger. Here the dorsal digital arteries are direct branches of the radial artery which are given off before it returns to the palmar aspect of the hand.

A diagram of the distribution of these vessels is seen in Fig. 3-88.

In addition to these vessels, the anterior in-

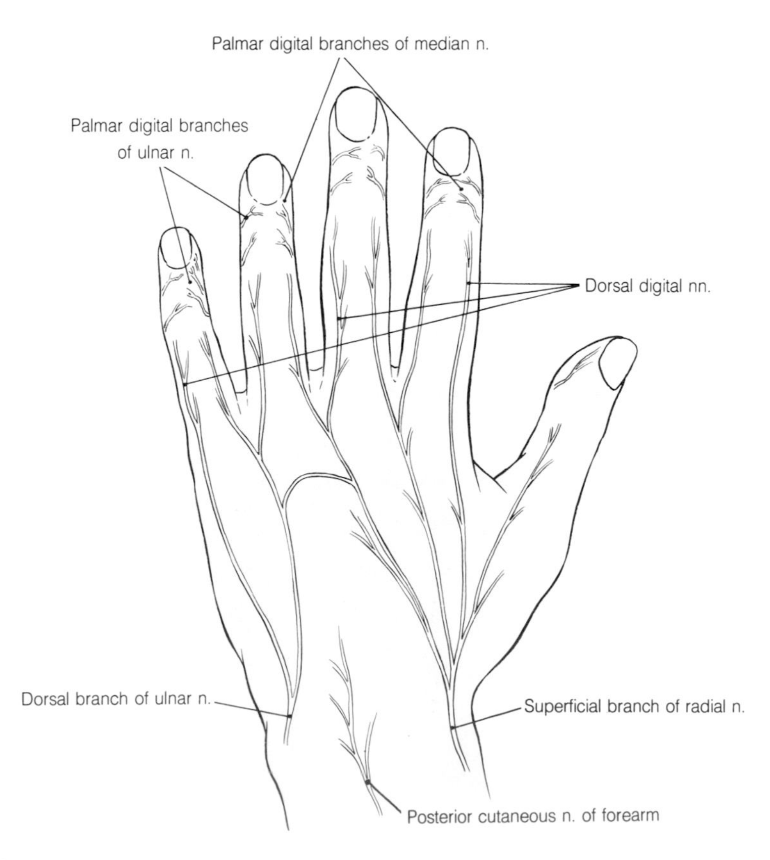

3-87 The distribution of cutaneous nerves on the dorsum of the hand.

terosseous artery which has pierced the interosseous membrane, contributes to the dorsal carpal arch and perforating branches of palmar metacarpal arteries join the dorsal metacarpal arteries. The dorsal arch and its branches lie deep to the extensor tendons.

The Wrist and Palmar Surface of the Hand

In addition to major industrial injuries, the hand is liable to much, apparently insignificant, minor trauma which may assume major importance when secondarily infected. While chemotherapy has reduced this risk considerably, a knowledge of the fascial planes and compartments and of the extent of the synovial sheaths of tendons, all of which determine the spread of infections, provides guidelines for treatment. Fractures of carpal bones, metacarpal bones, and phalanges also commonly occur.

Except for the position of the thumb and the movement by which it can be opposed to the fingers, the skeleton of the hand presents an arrangement of five similar and unspecialized digits and metacarpal bones. These are mounted on two rows of carpal bones which articulate with the forearm at the wrist joint. It is this lack of specialization coupled with an extensive motor and sensory cortical representation that allows the hand to perform so many differing tasks.

The outward appearance of the palm of the hand is well known. The thick epidermis with its profuse sweat glands and papillary ridges have already been described. The palmar surface is also bare of hair as is the skin of the dorsal aspects of the terminal phalanges; probably a characteristic

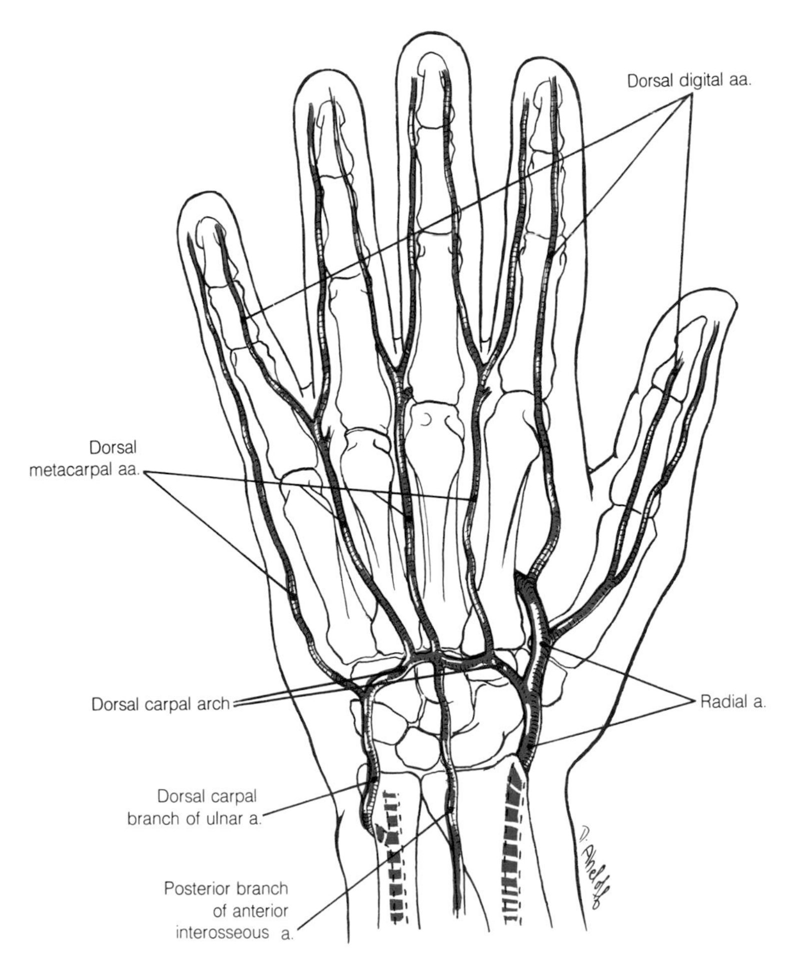

3-88 Showing the distribution of the radial artery on the dorsum of the wrist and hand.

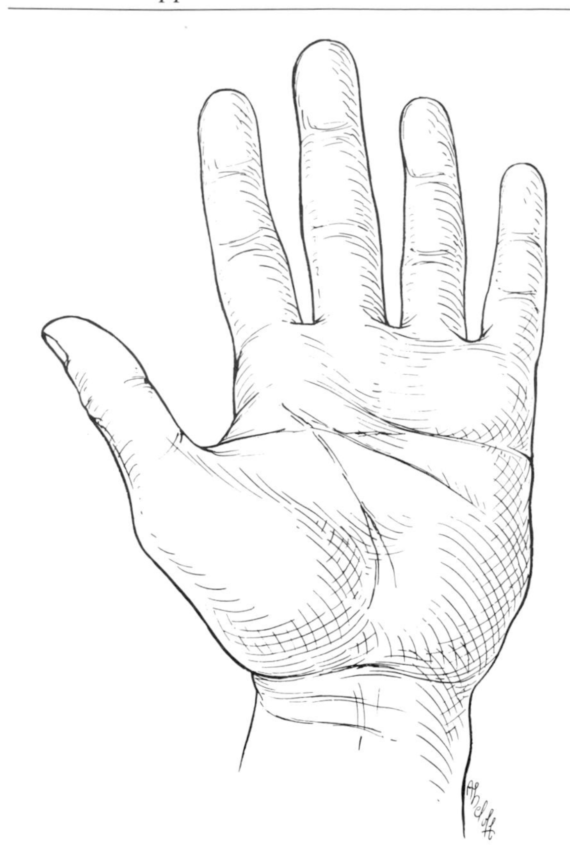

3-89 The flexure lines of the palm of the hand.

primate feature. The skin creases or flexure lines seen in Fig. 3-89, indicate sites at which the skin is anchored to the deep fascia. When incisions must be made on the palm or fingers, they are usually designed to follow these lines. Note that the distal interrupted crease across the palm lies just proximal to the level of the medial four metacarpophalangeal joints. These joints do not lie at the bases of the free fingers. Three transverse creases can also be seen at the palmar surface of the wrist. The most distal of these lies at the level of the proximal margin of the flexor retinaculum. An intermediate crease marks the radiocarpal joint and the bases of the radial and ulnar styloid processes, and the most proximal crease lies over the lower ends of the ulna and radius. The tubercle of the scaphoid and the pisiform bones can be felt laterally and medially just beyond the distal crease.

Lateral to a longitudinal palmar crease is a raised region called the thenar eminence. A similar but less well marked hypothenar eminence lies on the medial side of the palm. Finally note that the nail of the thumb lies at right angles to those of the fingers when the hand is at rest.

The soft tissues of the hand are supported on its skeleton and this must be considered next.

The Skeleton of the Hand

The skeleton of the hand is well illustrated in the radiograph shown in Fig. 3-90 and the key to the carpal bones that accompanies it (3-91).

Look first at the proximal row of the carpal bones and identify the **scaphoid,* lunate**, and **triquetral bones**. The **pisiform* bone** is usually included in this proximal row but it is really a sesamoid bone in the tendon of flexor carpi ulnaris and only articulates with the triquetral on whose palmar surface it lies. Note that the scaphoid and lunate bones articulate with the radius but the triquetrum is separated from the ulna by a substantial radiolucent space which is occupied by an articular disc. Additional features to recognize are the **tubercle of the scaphoid** and the **pisiform bone** superimposed on the triquetral bone.

Cupped in the three true proximal carpal bones is a distal row of four bones. In the radiograph identify these starting from the radial side. They are the **trapezium** which articulates with the thumb, the small **trapezoid bone**, the large **capitate bone**, and the **hamate* bone**. The trapezium is characterized by an elongated tubercle on its palmar surface. This cannot be seen on the radiograph but has been added to the key. A distinct groove lies on its medial side. The shadow of a hook-like process projecting from the palmar surface of the hamate can be made out on the radiograph. It is called the hook of the hamate.

Articulating with the distal row of carpal bones are the **five metacarpal bones**. The medial four are similar in appearance and each has a rounded head distally, a shaft and a rather irregular expanded base proximally. Note how crowded the proximal bases are. Little movement occurs here. The shorter and rather stouter first metacarpal also has a rounded head but its base presents a concavo-convex surface for articulation with a re-

* Skaphe, *Gk.* = a skiff, thus scaphoid or boat-shaped; Pisum, *L.* = a pea; Hamatus, *L.* = hooked.

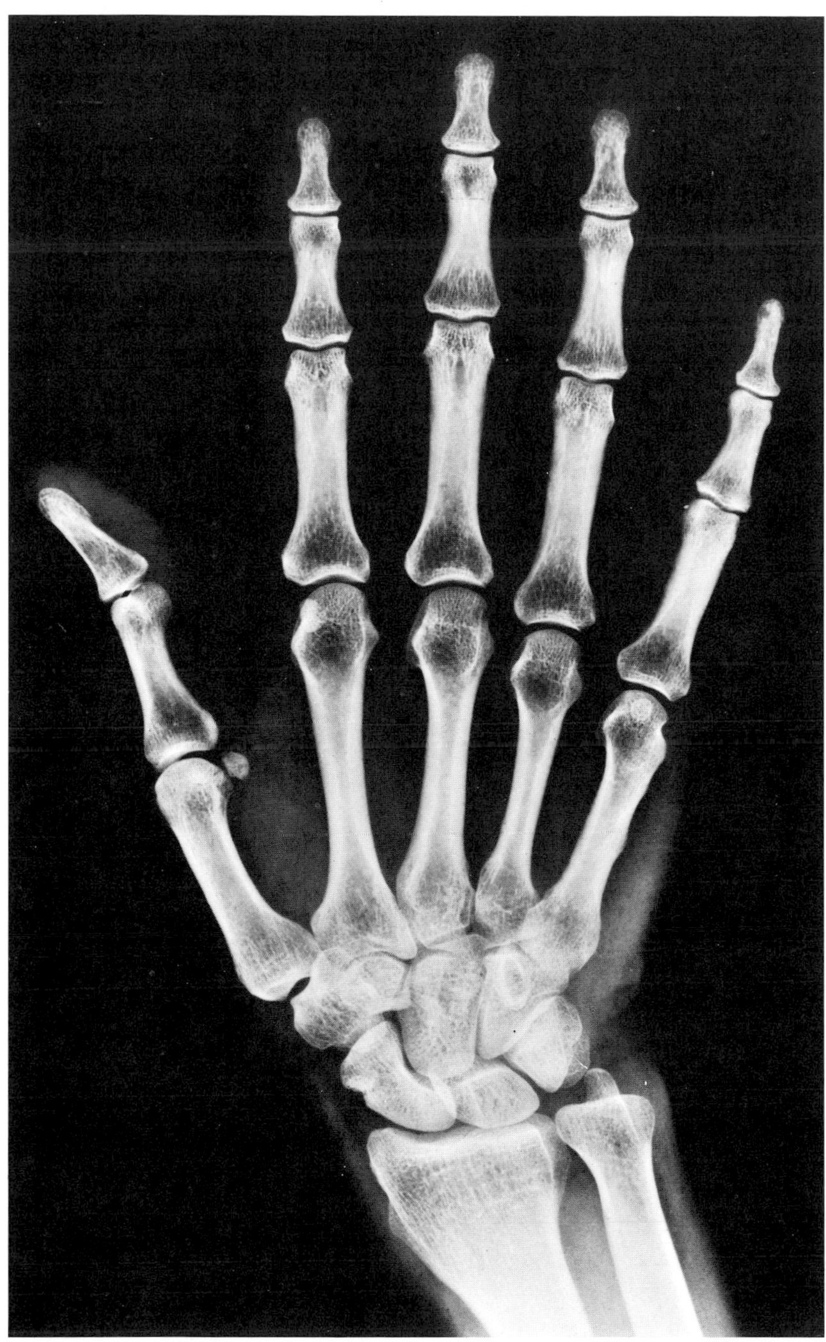

3-90 Radiograph of the bones of the hand and wrist. (Reproduced by permission, from Wicke: *Atlas of Radiologic Anatomy*, 4th Ed, Urban & Schwarzenberg, Baltimore-Munich, 1987.)

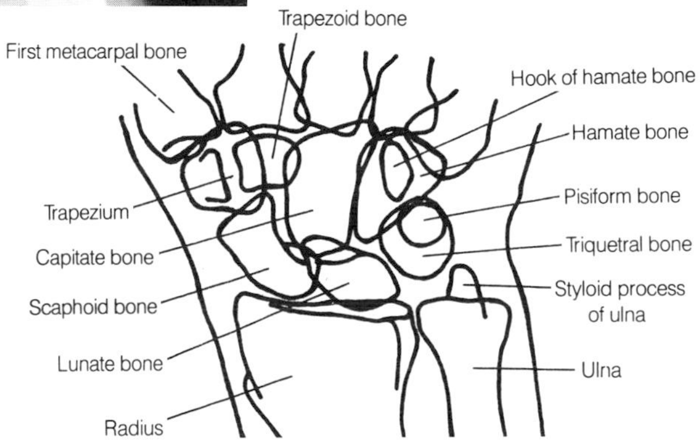

3-91 The carpal bones.

ciprocal surface on the trapezium. Perhaps the most important feature, however, is the fact that when the hand lies flat on the table (as in the radiograph) this metacarpal lies in a plane at 90° to the other four, i.e., its dorsal surface faces laterally. A sesamoid bone can be seen medial to the head of the first metacarpal bone and two more can be found superimposed over the heads of the second and fifth metacarpal heads. These lie in tendons on the palmar surfaces of the bones. They are not important in themselves and are not always present, but the possibility of their presence must be borne in mind before they are reported as fragments of bone.

Each finger has three phalanges, the thumb two. The proximal phalanx of the thumb and the proximal and intermediate phalanges of the fingers are similar in appearance. Their heads appear rounded but in fact have a shallow groove. The dorsal surface of the shafts are convex from side to side, but the palmar surface is flat. Faint ridges for attachment of the fibrous flexor sheaths can be felt at their borders. The bases of the proximal phalanges have concave oval facets for articulation with the heads of the metacarpal bones. The bases of the intermediate phalanges of the fingers fit the rounded and grooved surfaces with which they articulate. The distal phalanges are smaller flattened bones expanded both proximally and distally.

No centers of ossification are normally present at the wrist a birth. Centers for the capitate and hamate appear after two to three months. The centers for the triquetral, lunate, scaphoid, trapezium, and trapezoid appear between the third and fifth years. The pisiform is the last bone to begin to ossify during the tenth to twelfth years.

The metacarpal bones have primary centers for their shafts which appear at the ninth week **in utero**. Secondary centers appear for the heads of the metacarpals of the fingers between the first and second years. The first metacarpal, however, has a secondary center at its base but not at its head. Centers for the shafts of the phalanges appear **in utero** between the 8th and 12th week and secondary centers, which appear at the bases between the second and fourth years, fuse with the shafts by 18 years.

The Wrist or Radiocarpal Joint

The radiocarpal joint is a synovial joint which unites the hand to the forearm. The name radiocarpal is informative as it indicates immediately that the ulna takes no direct part in the articulation at the wrist.

The proximal articular surface is formed by the inferior surface of the **distal end of the radius** and a triangular fibrocartilaginous **articular disc** which extends from the medial side of the articular surface of the radius to a pit at the base of the ulnar styloid process. This surface is illustrated in Fig. 3-92. Note that its outline is roughly eliptical. It is also concave.

The distal articular surface is formed by the scaphoid, lunate, and triquetral bones as is seen in Fig. 3-93. Their proximal articular surfaces form a

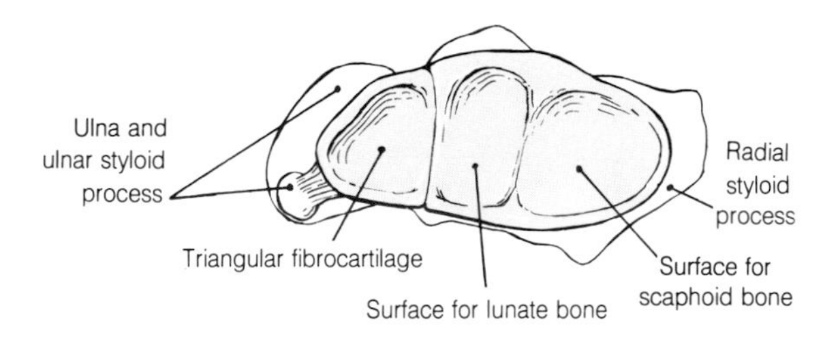

3-92 Showing the proximal articular surface of the radiocarpal joint.

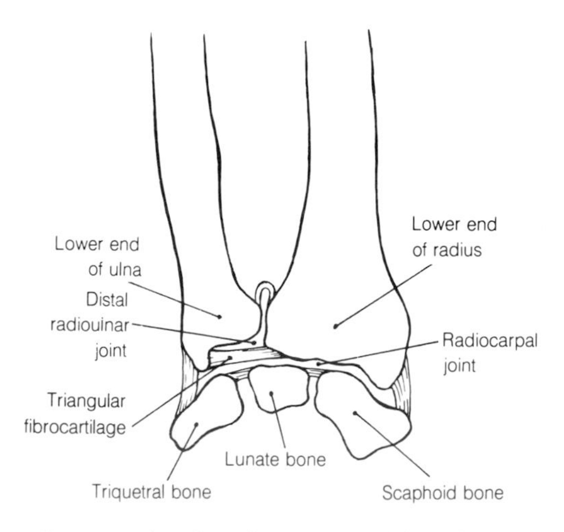

3-93 Diagram showing the relationship of the distal radioulnar joint to the radiocarpal joint.

convex surface which fits into the concavity formed by the radius and articular disc. The **capsule** of the joint is lined by **synovial membrane** and the **synovial cavity** does not normally communicate with the distal radioulnar joint or the intercarpal joints. The articular capsule is strengthened by **ulnar and radial collateral ligaments** and **palmar and dorsal radiocarpal ligaments**.

The collateral ligaments extend on the ulnar side from the ulnar styloid to the triquetral and pisiform bones and on the radial side from the radial styloid to the scaphoid. Neither of these ligaments is particularly strong.

The radiocarpal ligaments extend from the palmar and dorsal surfaces of the lower end of the radius to the proximal row of carpal bones. The fibers of both are directed obliquely downward and medially. Note that there is no major ligament attaching the ulna to the dorsal or palmar surface of the carpus. As a result the radius and hand are able freely to move as a unit about the ulna in pronation and supination.

The movements of the wrist joint are functionally inseparable from those occuring between the proximal and distal row of carpal bones. This is known as the midcarpal joint.

The Midcarpal Joint

Each carpal bone articulates with adjacent carpal bones and small movements occur at these synovial joints. A more distinct joint lies between the scaphoid, lunate, and triquetral bones proximally and the trapezium, trapezoid, capitate, and hamate bones distally. Here a more substantial amount of movement is possible.

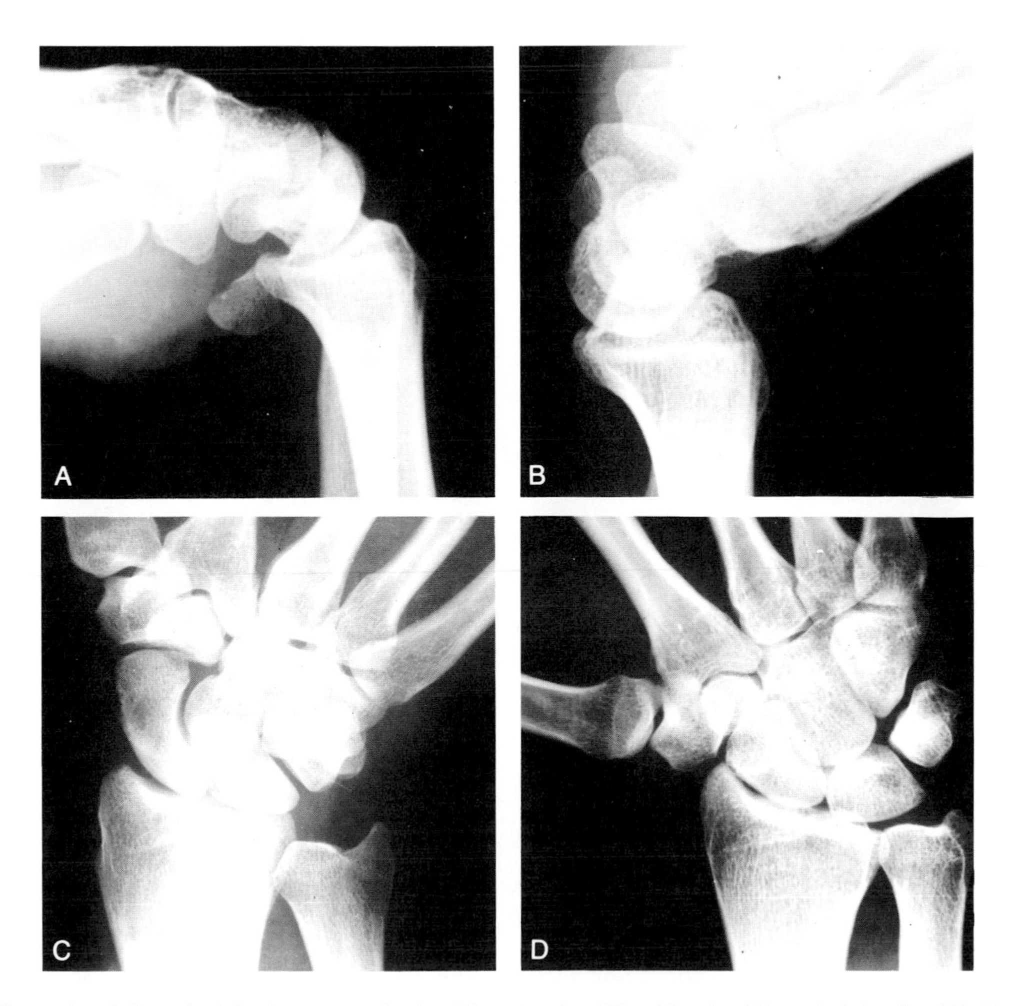

3-94 Radiographs of the wrist joint in extreme flexion (*A*), extension (*B*), adduction (*C*), and abduction (*D*). (Courtesy of Jeremy Young, M.D., Department of Diagnostic Radiology, University of Maryland School of Medicine.)

Movements of the Radiocarpal and Midcarpal Joints

Movements at the region of the "wrist" are flexion, extension, abduction, adduction, and, by combining these, circumduction. The full range of these movements results from movements at the radiocarpal and midcarpal joints together with small adjustments made at the intercarpal joints.

The contribution of each joint to these movements can be appreciated from examination of the radiographs seen in Fig. 3-94. In this note that:

1. The midcarpal joint contributes more to flexion than the radiocarpal joint.
2. The radiocarpal joint contributes more to extension.
3. In adduction, most of the movement occurs at the radiocarpal joint.
4. In abduction, which is more limited than adduction, the movement is almost entirely at the midcarpal joint.

The muscles performing these movements have been described with the forearm. In particular, recall the extensors and flexors of the carpus on the radial and ulnar sides of the wrist. The ulnar group produce adduction and the radial group abduction, while the radial and ulnar flexors perform flexion of the wrist and the radial and ulnar extensors perform extension. When appropriate, these muscles are aided by the long flexors and extensors of the digits and the long abductor of the thumb. Remember, however, that strong flexion of the digits is normally associated with an extended wrist.

The Carpometacarpal Joints

The medial four joints between the carpus and the metacarpal bones are irregular synovial joints which allow little movement. However, some flexion and extension of the fourth and fifth joint do occur when, in a "power grip," the hand is accommodated to a tool handle or rope. The movement can also be detected if the knuckles are watched when a tight fist is made.

The carpometacarpal joint between the trapezium and first metacarpal is a sellar or saddle-shaped multiaxial synovial joint. The reciprocal concavo-convex surfaces of this joint allow opposition of the thumb to the fingers in both a precision and power grip. This movement of opposition will be described later with the small muscles of the thumb that are involved.

The Metacarpophalangeal Joints

These can be described as condyloid synovial joints. A shallow concavity on the proximal surface of the phalanx articulates with the rounded surface of the metacarpal head which is partially divided into two condyles on its palmar aspect. Note that the articular surface on the metacarpal does not extend into its posterior aspect.

Each joint is surrounded by a fibrous capsule which is reinforced anteriorly by a **palmar ligament** and on each side by **two collateral ligaments**. Dorsally the extensor tendons serve as dorsal ligaments.

The palmar ligaments are tough plates of fibrocartilage firmly attached to the phalanges but only loosely so to the metacarpals. Between the palmar ligaments of the four fingers lie three **deep transverse metacarpal ligaments** which bind the palmar ligaments firmly together. No such ligament links the mobile thumb to the index finger.

The arrangement of the collateral ligaments can be seen in Fig. 3-95. Note how they run obliquely between the dorsal aspect of the metacarpal bone to the palmar aspect of the base of the phalanx. These and the palmar ligaments limit extension of the joint.

The movements possible at the metacarpophalangeal joints are flexion which is performed by the long and short flexors of the digits and by the interossei and lumbrical muscles, extension which is performed by the long extensors of the digits, and abduction and adduction performed by the interrosseous muscles. Slight rotation occurs when the fingers are flexed individually and a wider range of passive rotation can be elicited.

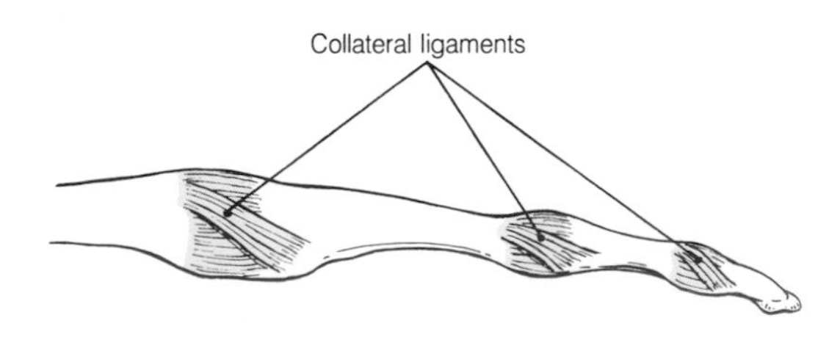

3-95 Showing the arrangement of the collateral ligaments of the metacarpophalangeal and interphalangeal joints.

The Interphalangeal Joints

The interphalangeal joints are very similar in structure to the metacarpophalangeal joints, but are only able to flex and extend. They are, thus, uniaxial hinge joints. Each has a strong **palmar ligament** and two **collateral ligaments**.

Flexion is performed by the long flexors of the finger and thumb. Extension is performed by the extensor digitorum and extensor pollicis longus muscles, but when combined with flexion at the metacarpophalangeal joints (which eliminates the long extensors), by the interossei and lumbrical muscles.

The Palm

With the description of the skeleton and joints completed, attention can now be turned to the soft tissues of the palm.

Beneath the tough and stable skin of the palm the fatty superficial fascia is broken up into small loculi by fibrous septa which anchor the skin to the underlying deep fascia. It is characteristic of cuts and incisions of the palm that the fat in these loculi springs out onto the surface. Over the central region of the palm the deep fascia is continuous with a sheet of fibrous tissue called the **palmar aponeurosis**.

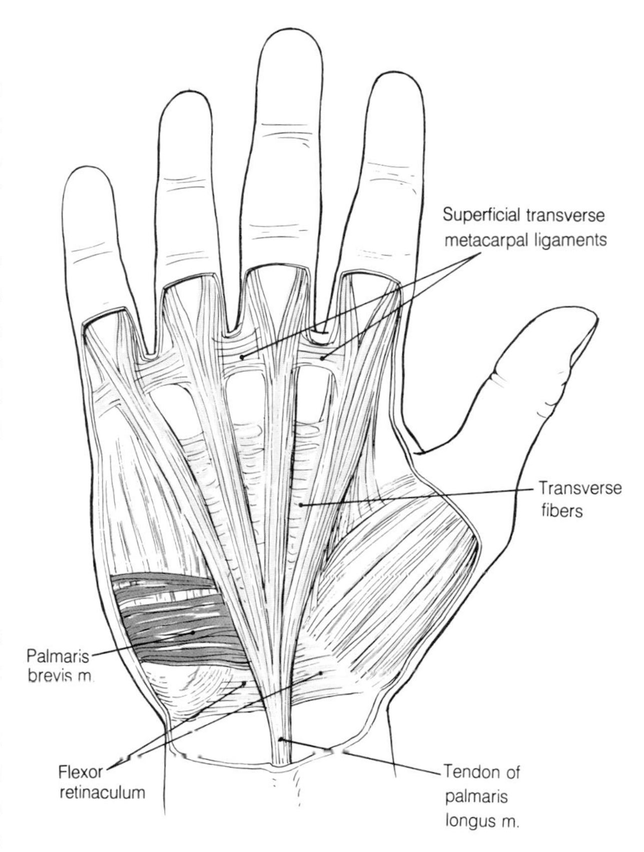

3-96 The palmar aponeurosis.

The Palmar Aponeurosis

This tough fibrous aponeurosis is illustrated in Fig. 3-96. It covers the palm between the thenar and hypothenar eminences. Proximally, it is continuous with the flexor retinaculum and the tendon of palmaris longus. Distally, it breaks up into four slips which are joined together by the **superficial transverse metacarpal ligament**. The four slips divide at the base of each finger. The divisions pass deeply on either side of the finger to fuse with the fibrous flexor sheath, the capsule of the metacarpophalangeal joint and the proximal phalanx. The palmaris longus and the palmar aponeurosis probably represent vestiges of a long flexor of the proximal phalanges. Contraction of the aponeurosis extending to the ring and little fingers leads to a permanent flexor deformity. This is known as a Dupuytren's contracture.

The deep fascia over the thenar and hypothenar eminences is continuous with the palmar aponeurosis but is much less thick.

The Flexor Retinaculum

It can easily be understood that if the long flexor tendons running through the wrist and palm and into the digits were not in some way retained close to the skeleton, they would spring away from each joint as it was flexed. At the wrist this retaining function is carried out by the flexor retinaculum. This is a strong band of fibrous tissue about the size of a commemorative postage stamp which spans the concave palmar aspect of the carpus. In this way an osseofibrous **carpal tunnel** is formed through which pass the median nerve and the long flexor tendons to the digits. The retinaculum is attached medially to the pisiform bone and the hook of the hamate and laterally to the tubercle of the scaphoid and a ridge on the trapezium. An additional septum descends to the medial lip of the groove on the trapezium forming a tunnel for the tendon of flexor carpi radialis. Look now at

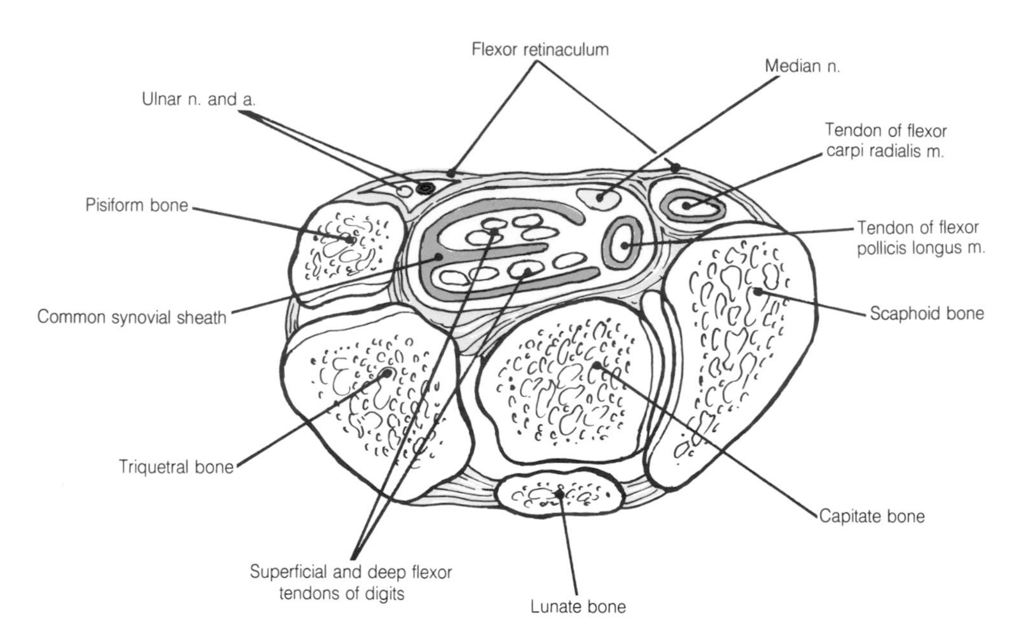

3-97 Cross-section of the carpal tunnel and its contents.

Fig. 3-97 which shows the formation of the carpal tunnel and the structures which traverse it.

Arising from the superficial surface of the flexor retinaculum and the adjacent carpal bones are the short muscles of the thumb and little finger. These form the **thenar** and **hypothenar eminences**.

The Thenar Eminence

This raised region between the wrist and the base of the thumb is made up of the abductor pollicis brevis, the flexor pollicis brevis, and the opponens pollicis muscles.

Before describing these muscles and the movements they perform, it is important to consider the relationship of the thumb to the four fingers. Examine the relaxed hand and see that the thumb nail and, thus, the rest of the digit lie in a plane at right angles to the nails of the fingers. As a result flexion of the thumb and first metacarpal brings them across the surface of the palm. This movement is illustrated in Fig. 3-98. In Fig. 3-99 the hand is viewed from the lateral aspect. In this illustration adduction can be seen to be a movement of the thumb towards the plane of the palm and abduction a movement away from it. The thenar muscles are illustrated in Fig. 3-100 and this should be followed as the muscles are described.

The **abductor pollicis brevis** arises chiefly from the flexor retinaculum and is the most superficial of the three muscles. Distally it is attached to the radial side of the base of the proximal phalanx of the thumb and to the tendon of flexor pollicis longus. This muscle draws the thumb away from the palm.

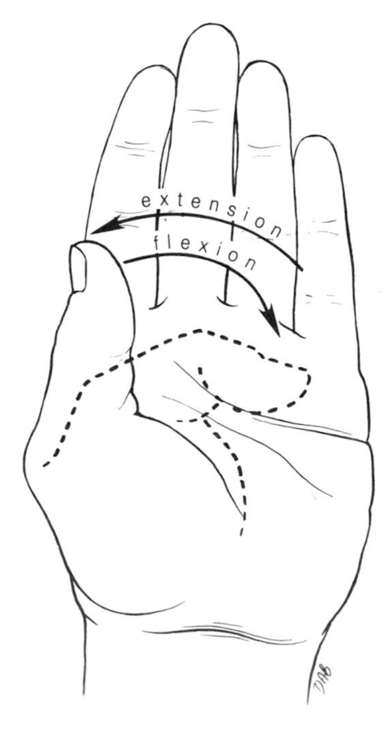

3-98 Showing the plane of flexion and extension of the thumb.

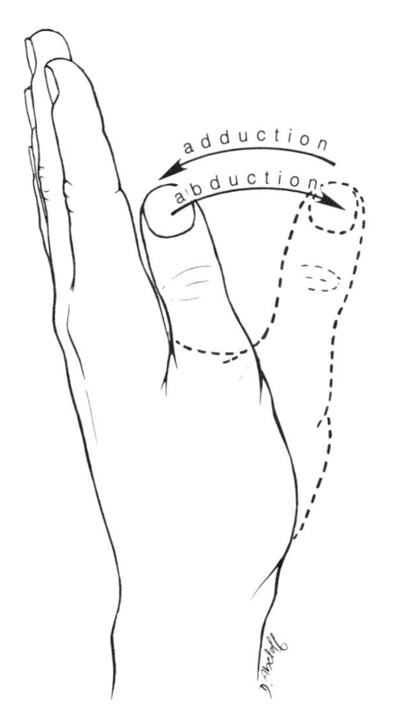

3-99 Showing the plane of abduction and adduction of the thumb

The **flexor pollicis brevis** also arises from the flexor retinaculum. It lies medial to and partly undercover of the abductor. It is attached to the radial side of the base of the proximal phalanx by a tendon which contains a small sesamoid bone. Contraction of this muscle flexes the proximal phalanx and the first metacarpal bone. It also contributes to rotation of the metacarpal bone at the carpometacarpal joint.

The **opponens pollicis** which lies deep to the abductor pollicis brevis arises from the trapezium and flexor retinaculum. It is attached to the lateral and palmar surfaces of the first metacarpal bone. When it shortens, it flexes the first metacarpal bone bringing it across the palm and also medially rotates it. It is this action aided by some rotation at the first metacarpophalangeal joint produced by the flexor pollicis brevis that is essential for the movement of opposition in which the pad of the thumb can be placed against that of any finger.

Each of the three thenar muscles is supplied by the **recurrent muscular branch** of the median

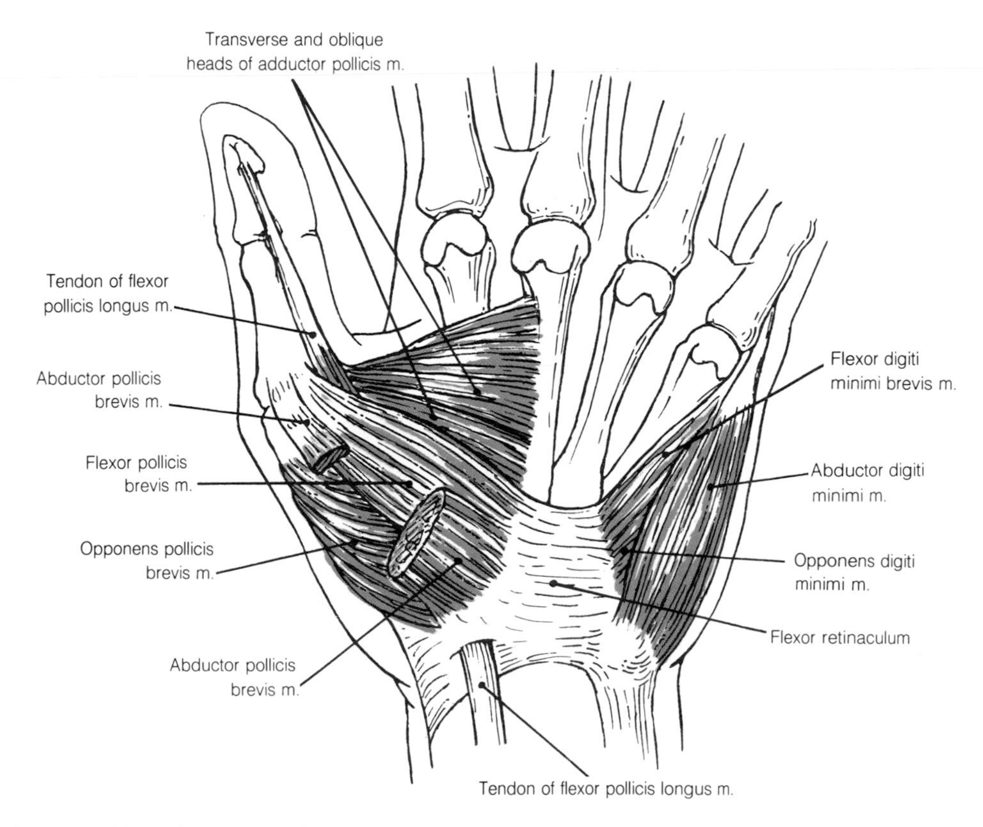

3-100 The thenar and hypothenar muscles.

nerve. A deep head of flexor pollicis brevis, if present, may be supplied by the **deep branch of the ulnar nerve**.

Although not strictly a thenar muscle, the **adductor pollicis** needs to be described at this point. It is a muscle with two heads. An oblique head arises from the capitate and bases of adjacent metacarpal bones and a transverse head arises from the palmar surface of the shaft of the third metacarpal bone. The two heads converge on the ulnar side of the base of the proximal phalanx of the thumb. This muscle draws the thumb towards the palm and provides the pinching or gripping force of the opposed thumb.

The adductor pollicis is supplied by the **deep branch of the ulnar nerve**.

The Hypothenar Eminence

This region which lies between the wrist and base of the little finger is much less prominent than the thenar eminence. It is composed of three hypothenar muscles which are also shown in Fig. 3-100.

The **abductor digiti minimi** arises from the flexor retinaculum and pisiform bone and is inserted into the ulnar side of the base of the proximal phalanx of the fifth finger. It abducts the fifth finger away from the fourth.

The **flexor digiti minimi** arises from the retinaculum and the hook of the hamate. It is inserted into the proximal phalanx with the abductor, and flexes the fifth metacarpophalangeal joint.

The **opponens digiti minimi** lies deep to both the preceding muscles. Arising also from the retinaculum and the hook of the hamate, it is inserted into the ulnar border of the fifth metacarpal bone. It slightly flexes and laterally rotates the metacarpal bones thus, helping to cup the palm.

All three hypothenar muscles are supplied by the **deep branch of the ulnar nerve**.

Before continuing to describe the hand, it is useful to review the structures that pass into its palmar aspect from the forearm. For this purpose, examine the cross-section made through the lower

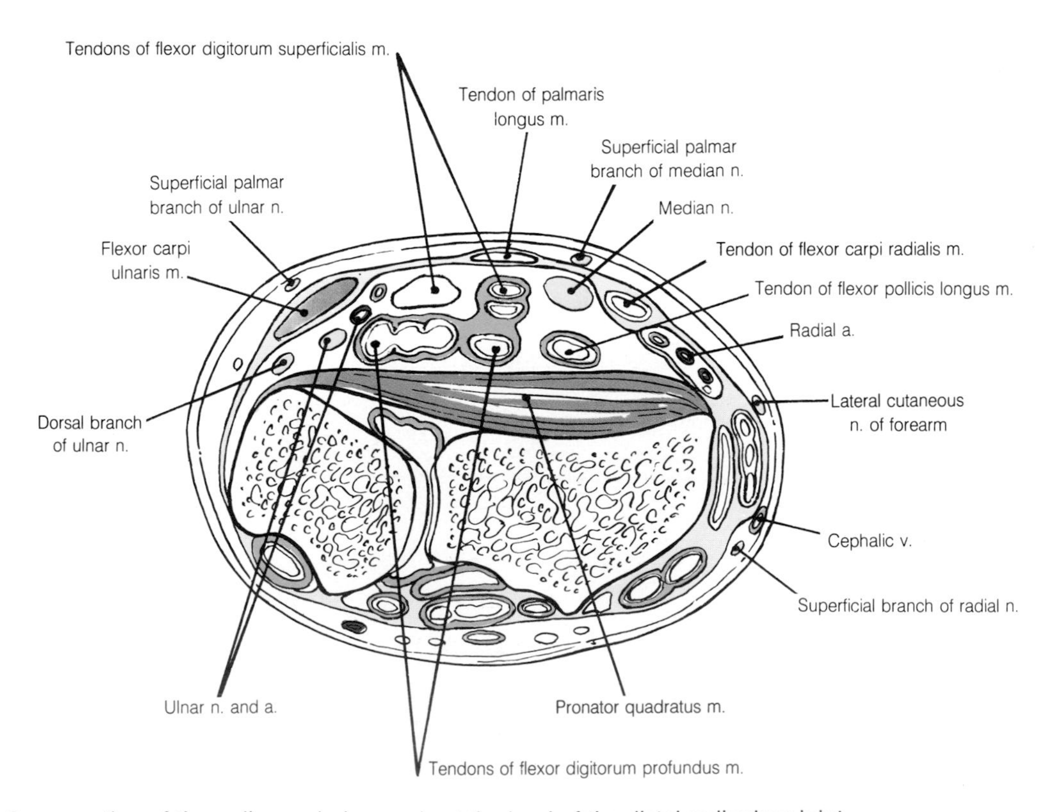

3-101 Cross-section of the radius and ulna made at the level of the distal radioulnar joint.

end of the radius and ulna seen in Fig. 3-101. Note the following:

1. That the section passes through the distal radioulnar joint and the muscle pronator quadratus. If the ulna remains stationary it can be seen how shortening of this muscle will rotate the lower end of the radius anteriorly over the ulna, thus bringing the forearm and hand into the prone position.
2. Lying on the anterior surface of pronator quadratus are the tendons of flexor digitorum profundus and more laterally that of flexor pollicis longus.
3. Superficial to these tendons is the flexor digitorum superficialis from which only two tendons have become clearly separated.
4. The deep and superficial tendons to the fingers are surrounded by a common synovial sheath *(blue)*, whereas that of flexor pollicis longus has its own sheath.
5. Of the more superficial muscles related to the deep fascia, flexor carpi ulnaris lies most medially. It covers the ulnar nerve and artery and the dorsal branch of the ulnar nerve which is about to escape from its medial border.
6. The tendon of palmaris longus lies in the midline with the tendon and synovial sheath of flexor carpi radialis on its lateral side.
7. Between these two tendons the median nerve lies deep to the fascia with its palmar branch superficial to it.
8. Between the flexor carpi radialis tendon and the radius lies the radial artery. It is at this point that its pulsation can be most easily felt.
9. As well as the palmar branch of the median nerve, the palmar branch of the ulnar nerve can be seen superficial to flexor carpi ulnaris. Terminal branches of the lateral cutaneous nerve of the forearm are also about to enter the palm on the lateral side.

The Long Flexor Tendons in the Hand

Having reviewed the positions of these structures as they pass over the flexor aspect of the wrist, the discription of the palm of the hand can be completed.

Of the structures seen in the section, the palmaris longus muscle has already been seen to fuse with the flexor retinaculum and become continuous with the palmar aponeurosis. The course in the hand of the remaining muscles must now be described.

Flexor Carpi Ulnaris

The tendon of this muscle can be followed to the pisiform at the medial margin of the flexor retinaculum. From here a small slip extends to the hamate bone (the **pisohamate ligament**) and the tendon continues to the base of the fifth metacarpal bone. Despite the fact that this extension of the tendon is called the **pisometacarpal ligament**, it does correspond to the metacarpal attachments of the other flexors and extensors of the carpus.

Flexor Carpi Radialis

The tendon of flexor carpi radialis passes beneath the most lateral part of the flexor retinaculum in a separate compartment already described. It is attached to the palmar surface of the base of the second metacarpal bone but also sends a slip to the base of the third. In this way it seems to balance the metacarpal attachments of the long and short radial *extensors* of the wrist.

Flexor Digitorum Superficialis and Profundus

The eight tendons derived from these muscles all pass beneath the flexor retinaculum. Note that the four profundus tendons lie as a single deep row while the superficialis tendons to the middle and ring fingers lie superficial to those to the index and little fingers.

Flexor Pollicis Longus

The tendon of flexor pollicis longus also passes beneath the flexor retinaculum lying lateral to the flexor tendons to the fingers. From here it runs between opponens pollicis and the adductor pollicis to be attached to the palmar surface of the base of the distal phalanx.

The Flexor Synovial Sheaths

Look at Fig. 3-102 and see that each of the tendons passing beneath the flexor retinaculum is surrounded by a synovial sheath. Note that those surrounding flexor carpi radialis and flexor pollicis longus are complete while the flexor tendons to the fingers have been invaginated from the radial side into an incomplete sheath. In Fig. 3-103, the extent of these sheaths in the hand can be seen.

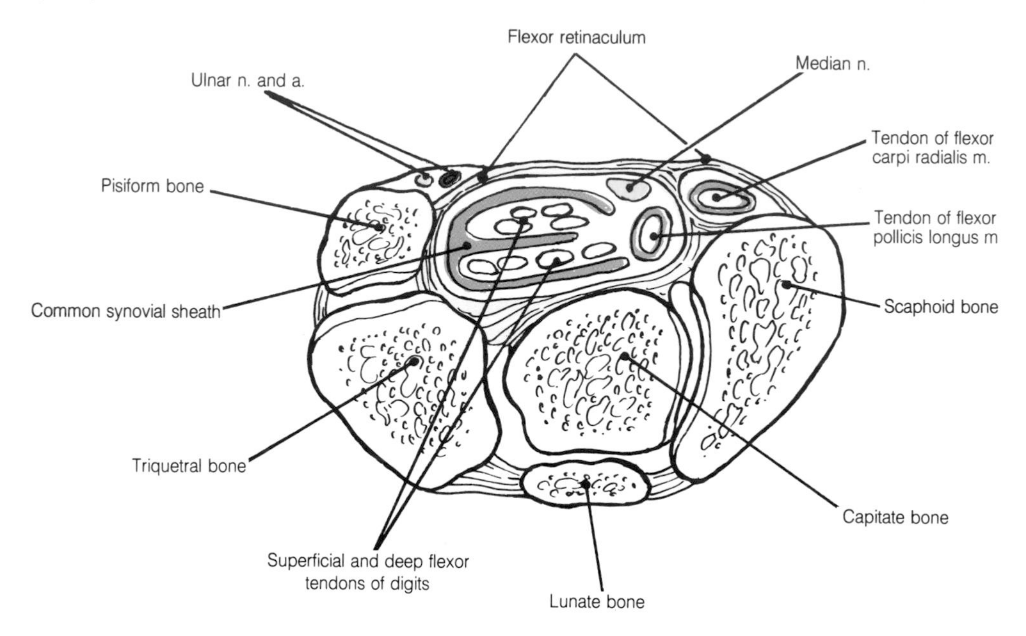

3-102 Cross-section of the carpal tunnel and its contents.

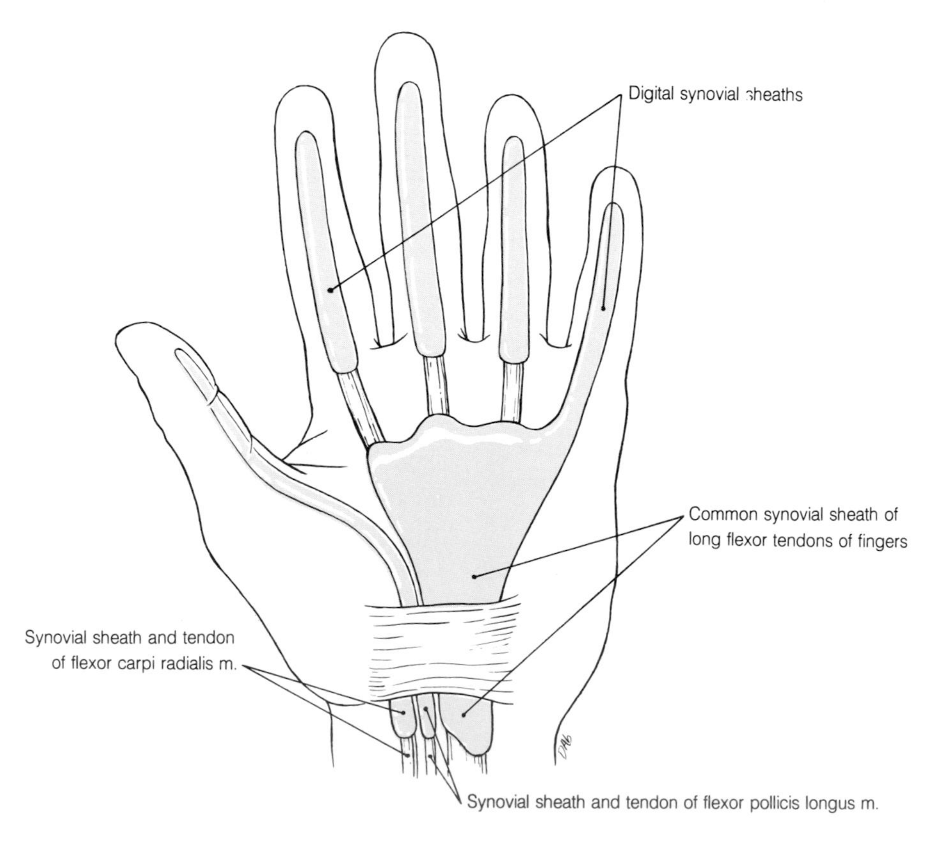

3-103 The flexor synovial sheaths in the hand.

Those for flexor carpi radialis and flexor pollicis longus extend close to the distal attachments of the muscles. The sheath around the tendons to the fingers terminates in the palm except that portion around the tendon to the little finger, which extends to the distal phalanx. Separate sheaths surround the tendons to the remaining digits as they lie in the free portions of the fingers.

Although these synovial sheaths perform a necessary function, they also provide easy pathways for the spread of infection. In this way infections of the sheath of the thumb or little finger can spread to the palm and beneath the retinaculum into the forearm.

The Long Flexor Tendons in the Digits

The single long flexor tendon to the thumb has been followed to the terminal phalanx of that digit. The pairs of superficial and deep tendons to the fingers need to undergo some rearrangement in order to reach their distal attachments because the deep tendons extend to terminal phalanges while the superficial tendons are attached to the sides of the middle phalanges. How this happens is best understood from the diagram in Fig. 3-104. Note how the superficialis tendon divides over the proximal phalanx and allows the profundus tendon to pass through it on its way to the base of the terminal phalanx. The two slips twist under the profundus tendon, join and partially decussate and then separate again to be attached to the sides of the middle phalanx.

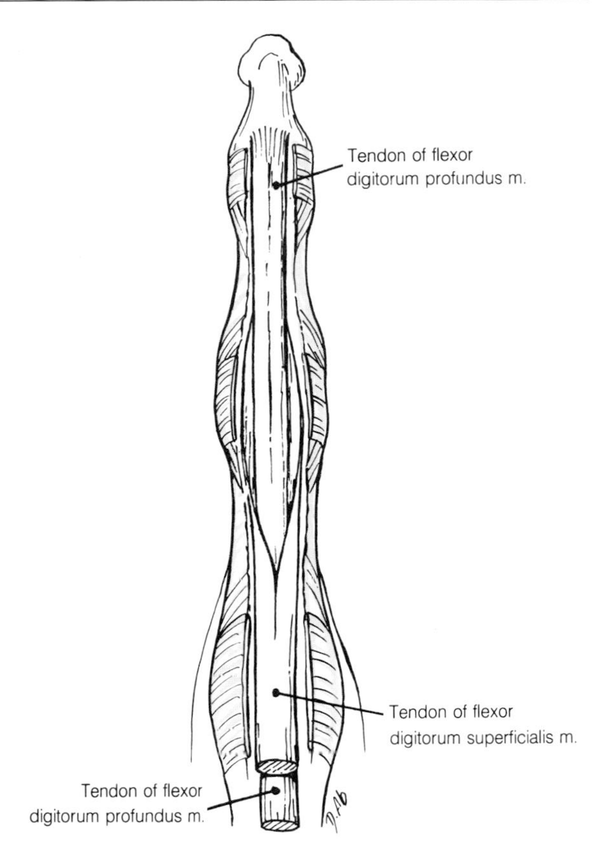

3-104 Diagram showing the manner in which the deep flexor tendon of a finger perforates the superficial tendon.

As these tendons pass through the fingers each pair is surrounded by a synovial sheath. If the tendons are lifted out of their sheaths from their pal-

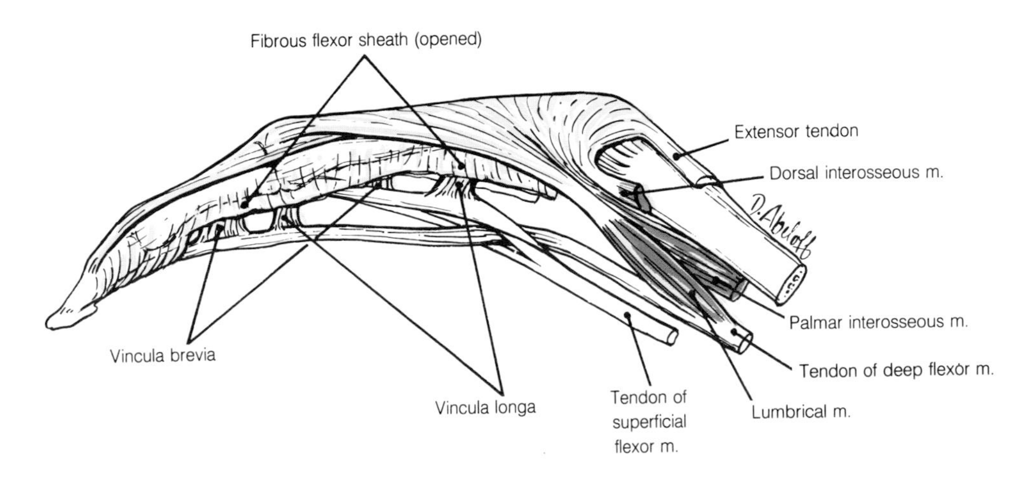

3-105 Showing the long flexor tendons of a finger and their vincula.

mar aspect, they will be seen to be attached to the dorsal aspect of the sheath by bands of synovial membrane. These bands are known as **vincula.*** A **vinculum breve** lies close to the attachment of each tendon and **vincula longa** are found more proximally in the finger. These vincula are illustrated in Fig. 3-105. This figure again shows the splitting of the superficialis tendon.

The Fibrous Flexor Sheaths

The importance of the flexor retinaculum in preventing the flexor tendons from springing away from the wrist has already been mentioned. A similar function is achieved in the thumb and fingers by the **fibrous sheaths of the fingers**. These are attached to the margins of the phalanges and arch over the flexor tendons forming, with the phalanges, osseofibrous tunnels. These are tough

* Vinculum, *L.* = a band or chain

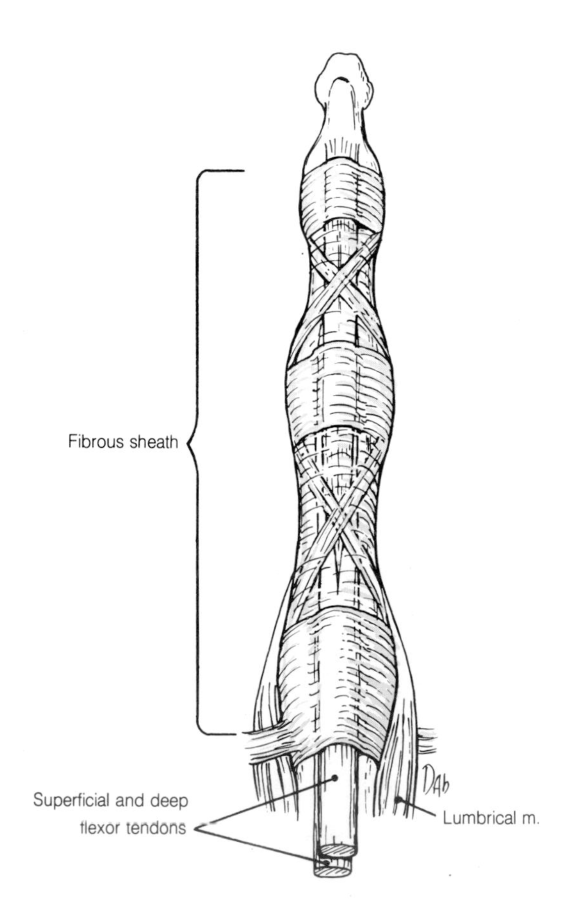

3-106 Illustrating a fibrous flexor sheath.

over the phalanges but are much thinner over the joints where their fibers have a cruciform arrangement. The fibrous sheath is illustrated on the middle finger in Fig. 3-106.

The Lumbrical* Muscles

These four small muscles are attached to the profundus tendons in the palm. The lateral two are unicipital and arise from the radial side of the tendons to the index and middle fingers. The medial two arise from the adjacent sides of the tendons to the middle and ring and ring and little fingers. One is attached to the radial side of the dorsal digital expansion of each of the fingers. They can be seen in Fig. 3-107.

The actions of the lumbricals are to flex the metacarpophalangeal joint and extend the interphalangeal joints. These actions have been described in more detail with the dorsal digital expansion.

The first and second lumbricals are supplied by the **median nerve** and the third and fourth by the **deep branch of the ulnar nerve**.

The Interosseous Muscles

Packed into the spaces between the metacarpal bones are a group of three palmar interosseous muscles and a group of four dorsal interosseous muscles.

The palmar interossei are unicipital and arise from the palmar surface of a metacarpal bone. Their distal attachment is by a small tendon to a dorsal digital expansion.

The dorsal interossei arise from the sides of adjacent metacarpal bones and are attached distally to the base of a proximal phalanx and its overlying dorsal digital expansion.

The part these muscles play in flexion of the metacarpophalangeal joint and extension of the interphalangeal joints has already been discussed. In addition to these actions, the palmar interossei adduct the fingers toward an imaginary line drawn down the center of the middle finger and the dorsal interossei abduct the index, middle and ring fingers away from this line. With this information

* Lumbricus, *L.* = earthworm

3-107 The lumbrical muscles.

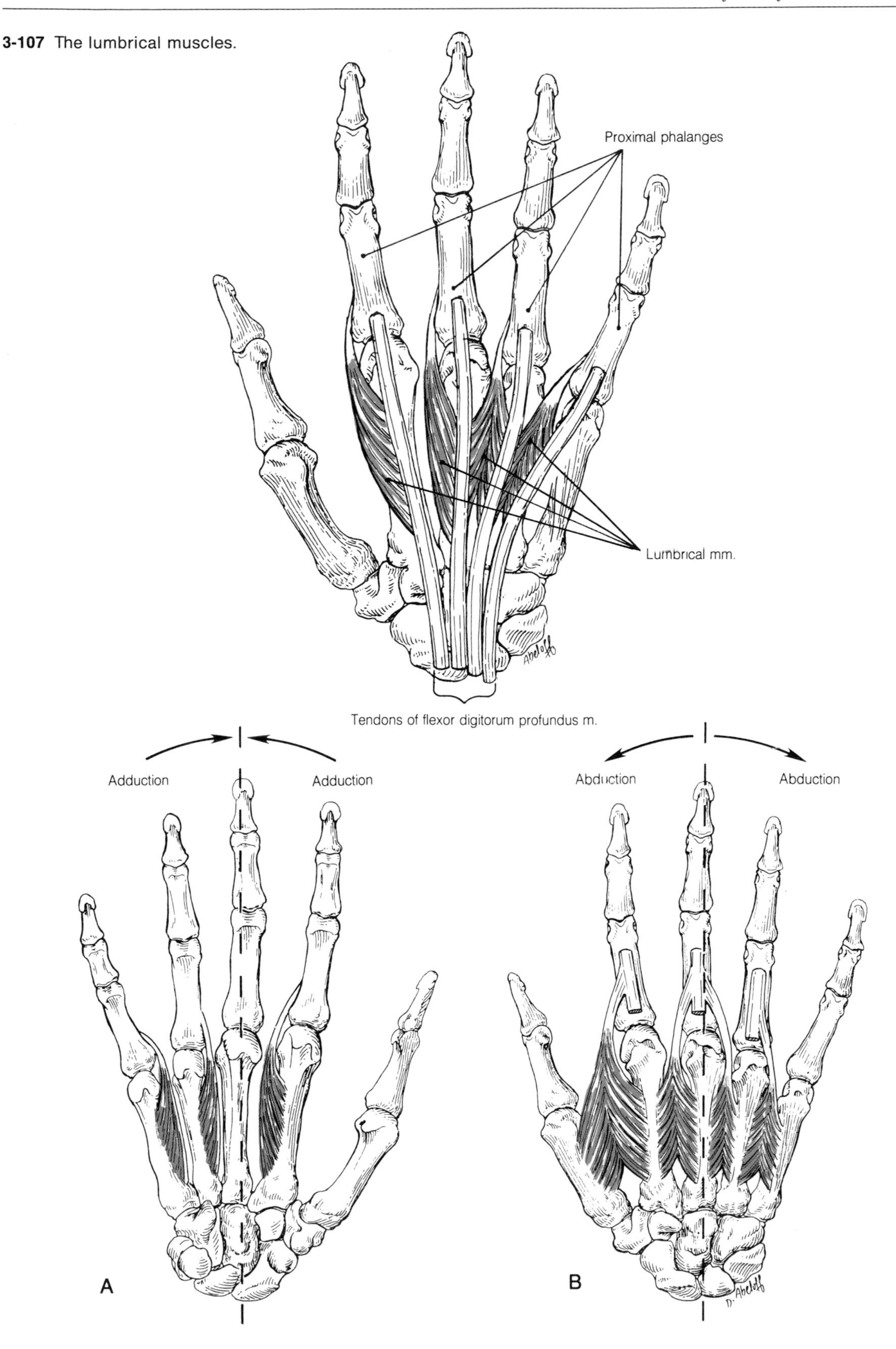

3-108 The palmar interosseous muscles (*A*) and dorsal interosseous muscles (*B*).

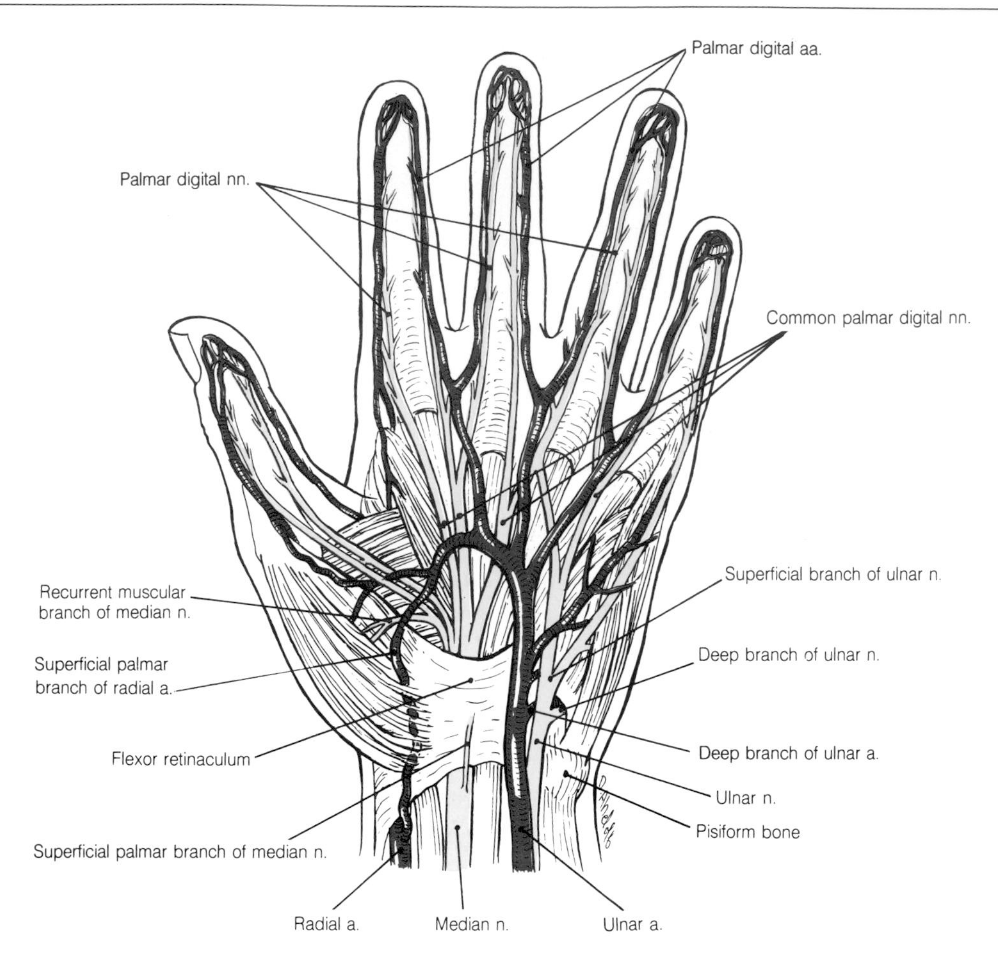

3-109 The median and ulnar nerves in the hand.

in mind and with the help of Fig. 3-108, it should be possible to determine the attachments and action of each individual muscle. Note that the middle finger can be abducted to either side of the imaginary line and, thus has two dorsal interossei attached to it. Also the thumb and little finger are not abducted by interossei but by abductor pollicis brevis and abductor digiti minimi. The thumb is adducted by adductor pollicis. The ability to separate the fingers is lost as they are flexed toward the palm.

All the interossei are supplied by the **deep branch of the ulnar nerve**. Testing the extended fingers will often provide early evidence of a lesion of the ulnar nerve as will wasting of the interossei between the metacarpal bones.

Nerves in the Hand

The Median Nerve

The median nerve has been seen to appear from beneath the deep surface of flexor digitorum superficialis above the wrist and then to lie between the tendon of that muscle and the tendon of flexor carpi radialis. As can be followed in Fig. 3-109, it passes from this point deep to the flexor retinaculum and through the carpal tunnel where it lies superficial to the long flexor tendons. It is here that pressure on or irritation of the nerve leads to a "carpal tunnel syndrome" in which sensory changes in the region supplied by the nerve and wasting of thenar muscles, which it innervates, occur.

Emerging from the retinaculum the nerve divides into a number of branches. The most lateral of these is an important **muscular branch** which curves laterally and proximally to gain the thenar eminence. Here it supplies flexor pollicis brevis, abductor pollicis brevis, and opponens pollicis.

Two **palmar digital nerves** to the lateral and medial sides of the thumb are given off next. These are followed by a palmar digital nerve to the lateral side of the index and two further branches (**common palmar digital nerves**) which divide to supply the adjacent sides of the index and middle and middle and ring fingers.

The first and second lumbrical muscles are supplied from the palmar digital nerve to the lateral side of the index and the common palmar digital nerve supplying the index and middle fingers.

Note that palmar digital nerves, including those derived from the ulnar nerve, give off dorsal branches which supply the nail bed and skin over the dorsal aspect of the terminal phalanges. It should also not be forgotten that the digital nerves, although ostensibly cutaneous in distribution, also carry sensory fibers from joints, secretor motor fibers to sweat glands, and vasomotor fibers to digital vessels.

The Ulnar Nerve

Although the ulnar nerve does not pass beneath the flexor retinaculum, it and the ulnar artery, which runs lateral to it, are bound down to its anterior surface by a superficial slip of fascia as they pass just lateral to the pisiform bone. Beyond the retinaculum the nerve divides into a deep and superficial branch and this can be seen in Fig. 3-109.

The **superficial branch** supplies the palmaris brevis muscle lying beneath the skin over the hypothenar eminence and then provides palmar digital nerves to the ulnar side of the little finger and to the adjacent sides of the ring and little fingers. A communication is often established with the median nerve in the palm. As in the other fingers, dorsal branches supply the nail bed and skin over the dorsum of the distal phalanx.

The **deep branch of the ulnar nerve** is another important motor nerve in the hand. In company with the deep branch of the ulnar artery, it passes through the muscles of the hypothenar eminence and supplies them. It then runs laterally across the palm beneath the long flexor tendons. Here it supplies the medial two lumbrical muscles, all the interosseous muscles, both heads of adductor pollicis and, finally, may send a branch to flexor pollicis brevis.

Having described the terminal branches of the radial, median, and ulnar nerves in the hand, the cutaneous innervation of its palmar and dorsal surfaces can now be reviewed. A picture of this

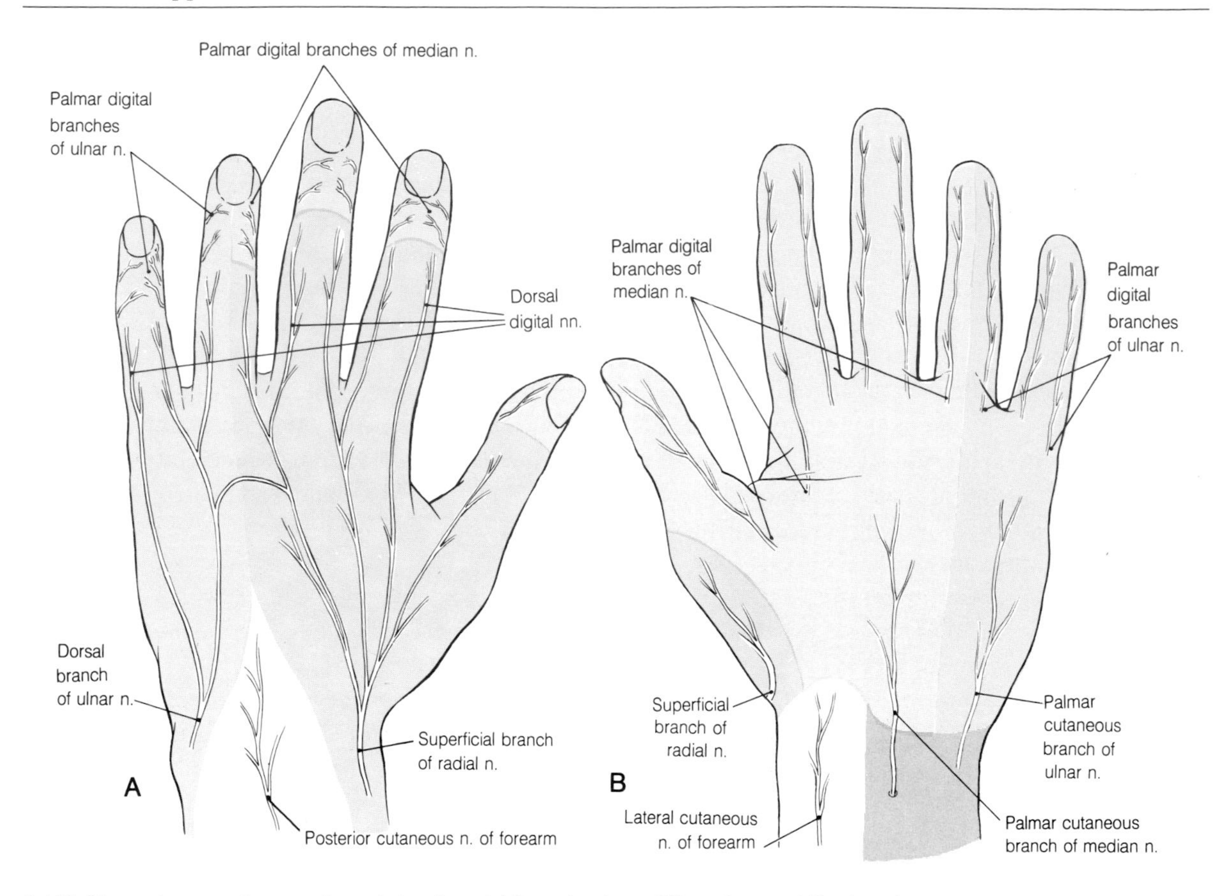

3-110 The cutaneous innervation of the dorsal (*A*) and palmar (*B*) surfaces of the hand.

innervation is seen in Fig. 3-110, although some variation does occur. Note on the palmar surface that:

1. A fairly constant region of the lateral aspect of the thenar eminence is supplied by the superficial branch of the radial nerve.
2. The palm is supplied by palmar branches of the median and ulnar nerves given off **superior** to the wrist.
3. The palmar digital nerves are terminal branches of the median and ulnar nerves in the hand.
4. The separation between the regions supplied by the median and ulnar nerves falls on a line passing through the ring finger.
5. The lateral cutaneous nerve of the forearm may contribute substantially to the supply of the palm.

Note on the dorsal surface that:

1. The separation between regions supplied by the dorsal branch of the ulnar nerve and the superficial branch of the radial nerve again passes through the ring finger. However, the dorsal branch of the ulnar nerve may supply the adjacent sides of the ring and middle fingers. In any event, the two nerves usually communicate on the dorsum of the hand.
2. The nail beds and the skin over the distal phalanges of all digits are supplied by branches of the palmar digital nerves.

Vessels in the Hand

The Radial Artery

The radial artery has been seen approaching the wrist between the tendon of flexor carpi radialis and the anterior border of the radius. From here it does not pass directly into the palm but winds around the lateral border of the carpus beneath the tendons of abductor pollicis longus, extensor pollicis brevis and extensor pollicis longus and over the surface of the scaphoid and trapezium. This is to say that it passes through the anatomical snuff box to reach the dorsal side of the space between the first and second metacarpal bones. Here it runs between the two heads of the first dorsal interosseous muscle to reach the palm. At first it runs deep to the oblique head of adductor pollicis and then, passing between this and the transverse head, crosses the palm deep to the long flexor tendons and anastomoses with the **deep palmar branch of the ulnar artery** to form the **deep palmar arch**.

Over its course from the wrist to the ulnar side of the palm the radial artery gives off a number of branches. These and the course of the radial artery in the hand can be followed in Fig. 3-111.

A **palmar carpal artery** anastomoses with a similar branch of the ulnar artery over the palmar surface of the carpus.

The **superficial palmar branch** is given off the radial artery just before it winds round the wrist. It passes through the thenar muscles and turns medially to anastomose with the ulnar artery. In this way it completes the **superficial palmar arch**.

A **small dorsal carpal branch** anastomoses with a dorsal carpal branch of the ulnar artery and the anterior and posterior interosseous arteries. From this dorsal carpal arch are derived three **dorsal metacarpal arteries** which provide **dorsal digital arteries** for the ulnar side of the index and the remaining fingers.

The **first dorsal metacarpal artery** arises directly from the radial artery and supplies dorsal digital arteries to the adjacent sides of the index

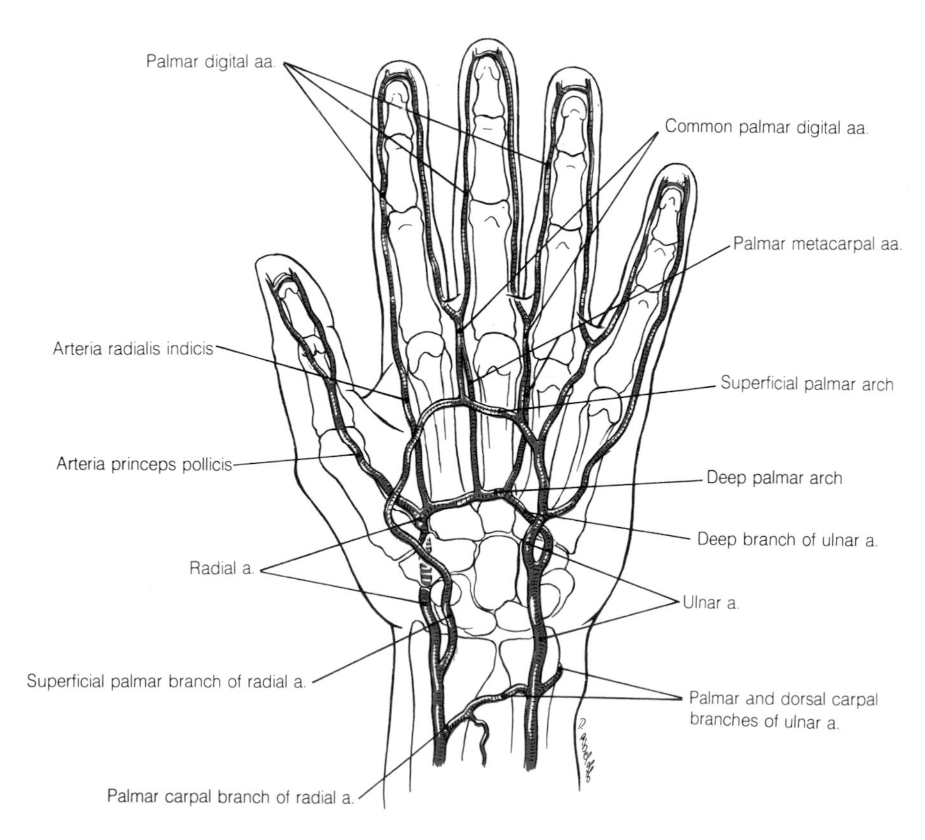

3-111 The course and distribution of the radial and ulnar arteries in the palm of the hand.

finger and thumb while that for the radial side of the thumb is given off from the radial artery.

On entering the palm the radial artery gives off the **arteria princeps pollicis**. This vessel divides to supply palmar digital arteries to the thumb on its medial and lateral sides.

The palmar digital artery of the radial side of the index finger is provided by the **arteria radialis indicis**, the last named branch of the radial artery.

The arteria princeps pollicis and the arteria radialis indicis can be thought of as palmar metacarpal arteries in series with those given off by the deep palmar arch. In fact, they not uncommonly arise from a common trunk which can be named the first palmar metacarpal artery.

The Ulnar Artery

The ulnar artery approaches the wrist between flexor digitorum superficialis and flexor carpi ulnaris and passes beneath a superficial layer of the flexor retinaculum with the ulnar nerve on its medial side. A **dorsal carpal branch** given off at this point winds posteriorly round the carpus to join with the radial and interosseous arteries in forming the dorsal carpal arch.

As it crosses the wrist the ulnar artery passes at first lateral to the pisiform bone and then to the medial side of the hook of the hamate. Here it gives off its **deep palmar branch** and then turns laterally across the palm beneath the palmar aponeurosis to form the **superficial palmar arch**. The arch is usually completed by the anastomosis of the ulnar artery with the **superficial palmar branch of the radial artery**.

The deep palmar branch passes through the hypothenar muscles to anastomose with the radial artery and complete the **deep palmar arch**.

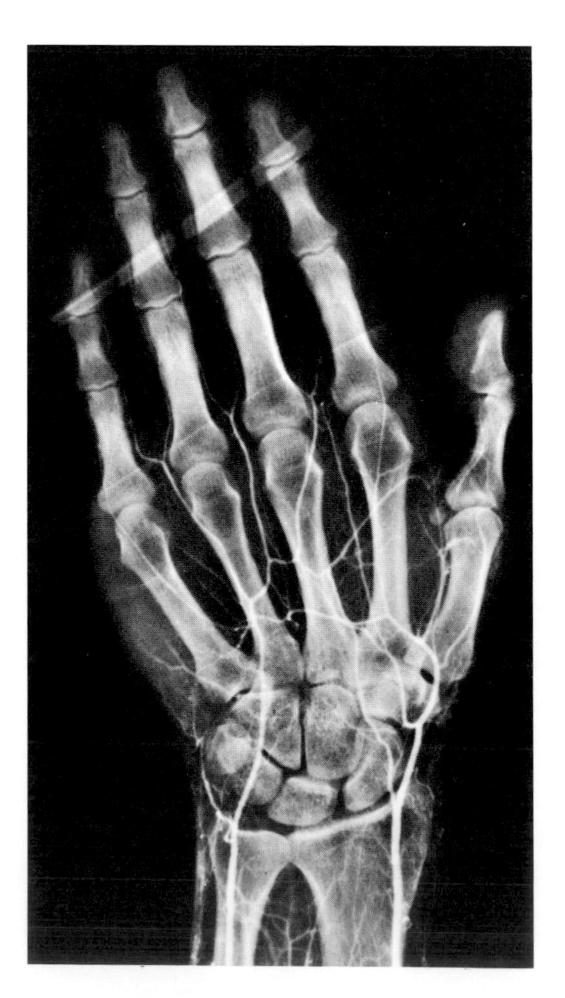

3-112 An arteriogram of the hand.

The Palmar Arches

Dorsal digital arteries have been seen to be branches of dorsal metacarpal arteries derived either directly from the radial artery or the dorsal carpal arch. The palmar digital arteries to the thumb and radial side of the index are derived from two branches of the radial artery, i.e., the princeps pollicis and radialis indicis arteries. The remaining palmar digital arteries are derived from the superficial and deep palmar arches. In Fig. 3-112 this rather complex arrangement can be seen as it exists in life. It also shows the relationships of the ulnar artery and its deep branch to the hook of the hamate bone. The course of these vessels can be followed more easily in Fig. 3-113 (overleaf).

The **deep arch**, which is formed by the radial artery and the deep palmar branch of the ulnar artery, lies deep to the long flexor tendons. It is also convex toward the fingers but lies about one centimeter nearer the wrist than the superficial arch. It gives off three **palmar metacarpal arteries** which join the three common palmar digital arteries at the clefts of the fingers. The vessels so formed divide into pairs of palmar digital arteries for the adjacent sides of the index and middle, middle and ring, and ring and little fingers.

The **superficial arch** is formed by the anastomosis of the ulnar artery and the superficial palmar branch of the radial artery. It lies just beneath

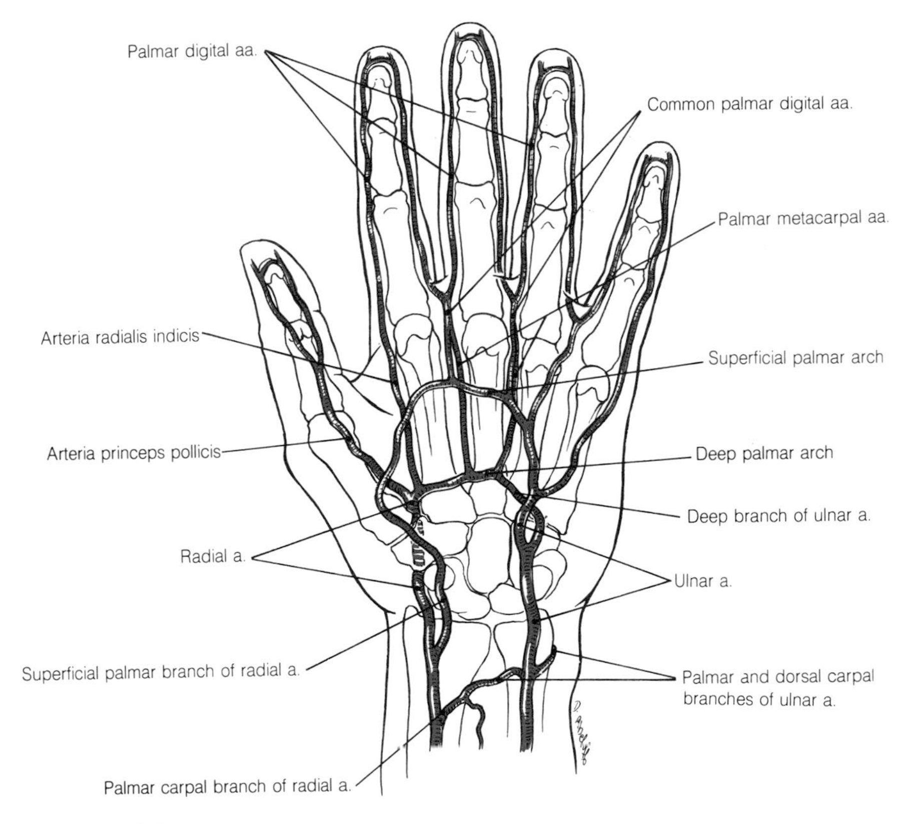

3-113 Showing the superficial and deep palmar arches.

the palmar aponeurosis, is convex toward the fingers, and its maximum convexity lies at the level of the distal border of the extended thumb. It gives off three **common palmar digital arteries** and the **palmar digital artery** to the medial side of the little finger.

The Palmar Spaces

The central part of the palm which lies deep to the palmar aponeurosis and between the thenar and hypothenar eminences is divided into two fascial spaces. Look at Fig. 3-114 and see how the **medial and lateral palmar septa** run dorsally from the palmar aponeurosis to the first and fifth metacar-

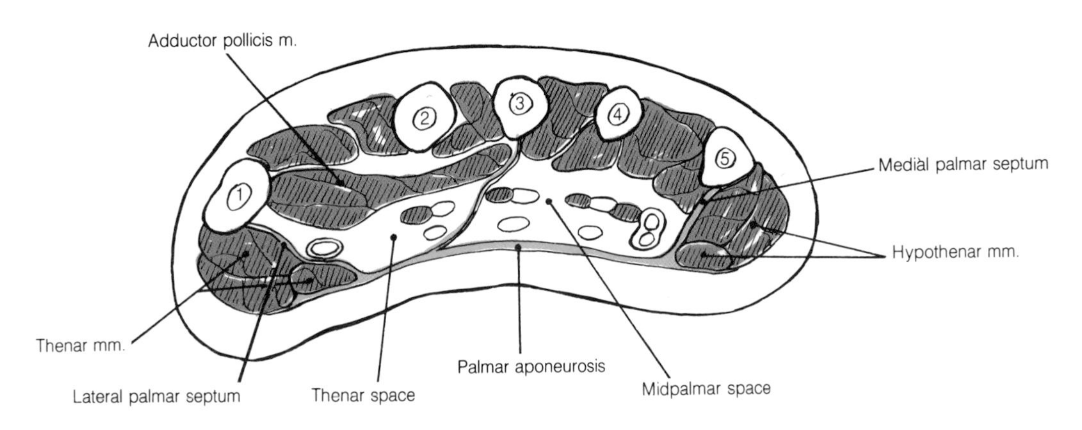

3-114 The palmar spaces.

pal bones. Between these septa lies a space which is divided into two by an **intermediate palmar septum**. The more lateral of these is called the **thenar space**. In its floor lies the fascia covering adductor pollicis and it contains the long flexor tendons to the thumb and index fingers and the first lumbrical muscle. The more medial space is the **midpalmar space**. Its floor is formed by the lateral three metacarpal bones and the interosseous muscles between them, and it contains the long flexor tendons and lumbrical muscles of the medial three fingers. The spaces communicate with the subcutaneous tissue of the webs between the fingers and deep infections of the midpalmar space often "point" at these sites.

The Nails

The nails are a modification of the **stratum lucidum and stratum corneum** of the epidermis. They appear on the dorsal surfaces of the distal phalanges from beneath the **nail fold**. The free margin of the **body of the nail** overhangs the tip of the phalanx at a recess of skin called the **hyponychium**. The **root of the nail** lies beneath the nail fold and is continuous with the pale **lunula** which is seen to a greater or lesser extent distal to the fold. The epidermis beneath the root and lunula is called the **germinal matrix** for it is here that the nail is being formed. Beyond the lunula the body of the nail glides over an inert surface of epithelium called the **sterile matrix**. These features of the nail can be seen in Fig. 3-115.

It is also worth pointing out that the subcutaneous tissue of the pulp of the finger tip is divided up into partitions by strong connective tissue septa radiating from the terminal phalanx to the skin. The tension developed by an infection here (a felon) may be sufficient to cause necrosis of the terminal phalanx.

The Finger in Cross-Section

The fingers are frequently involved in trauma and infections and, if it is carefully used, minor surgery can be performed under local anesthesia. For this reason, details of the anatomy of the nerves, the vessels, and the tendons with their associated sheaths are important. Much of this detail can be summarized in a cross-section of a digit and an example is shown in Fig. 3-116.

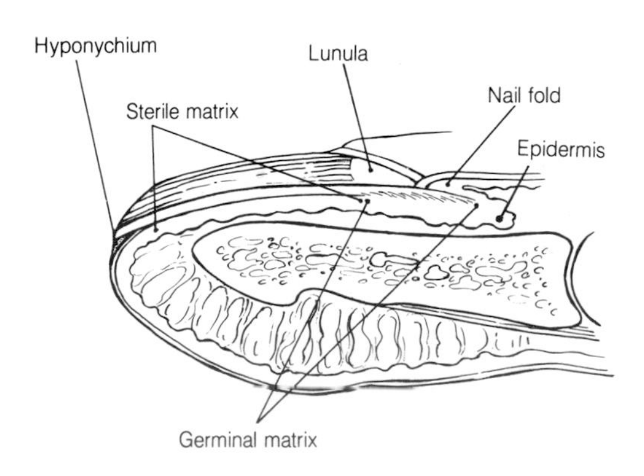

3-115 Showing the main features of a finger nail.

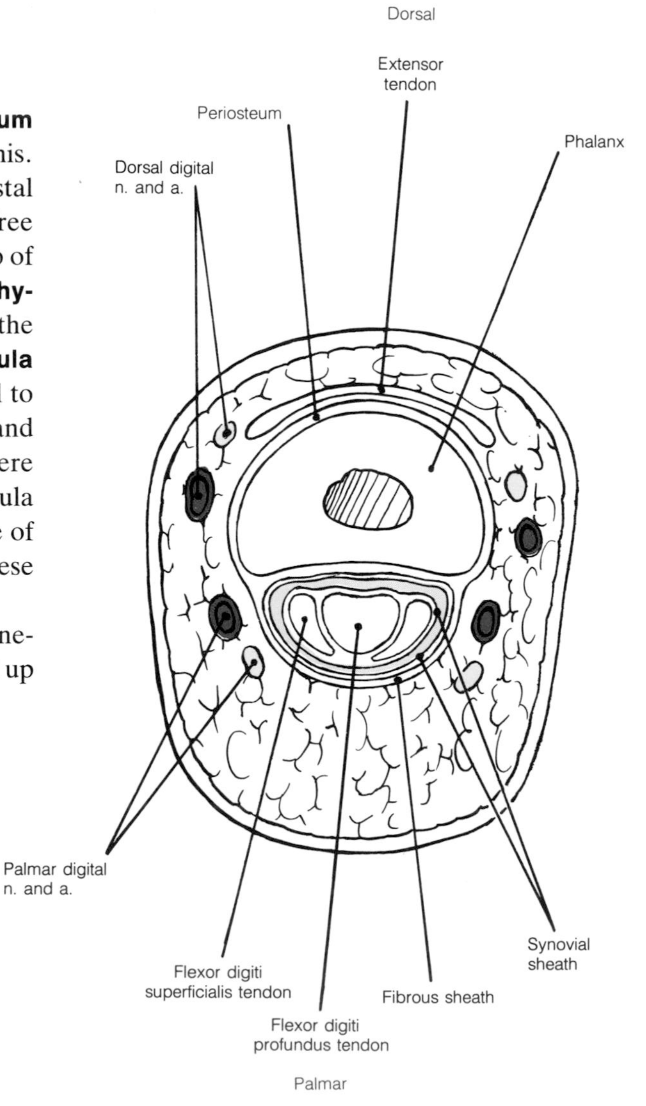

3-116 Cross-section through the proximal phalanx of a finger.

In this illustration note:

1. The flattened extensor tendon lying on the dorsum of the phalanx. This tendon has no synovial sheath.
2. The phalanx surrounded by periosteum.
3. The fibrous flexor sheath which, with the periosteum on the palmar surface of the phalanx, forms a protective tunnel for the long flexor tendon.
4. The tendon of flexor digitorum profundus flanked by the divided tendon of flexor digitorum superficialis.
5. The visceral and parietal layers of the synovial sheath surrounding the long flexor tendons.
6. The palmar digital nerve lying **anterior** to the palmar digital artery.
7. The dorsal digital nerve lying **posterior** to the dorsal digital artery.

Nerve Injuries in the Upper Limb

In the description of the axilla it was said that the brachial plexus was a vulnerable structure that is subject to injury particularly during childbirth and in road accidents. The major nerves in the limb are also liable to injury by penetrating wounds and in association with fractures. The discussion of some of these injuries that follows will serve as a useful review of the innervation of the upper limb. To aid in this the diagram of the brachial plexus is reproduced in Fig. 3-117.

Injuries of the Brachial Plexus

The brachial plexus is most commonly injured by excessive traction. This may involve the upper roots (C5, 6, and 7) when the shoulder is forced downward as may happen in a motorcycle accident or during a difficult delivery as the head is accidentally drawn away from an obstructed shoulder (Erb-Duchenne palsy). The lower roots (C8 and T1) become involved when the limb is pulled forcibly upward above the head. This injury may also occur during the delivery of a breech presentation (Klumpke paralysis).

A knowledge of the formation of the brachial plexus and the spinal nerves that supply the mus-

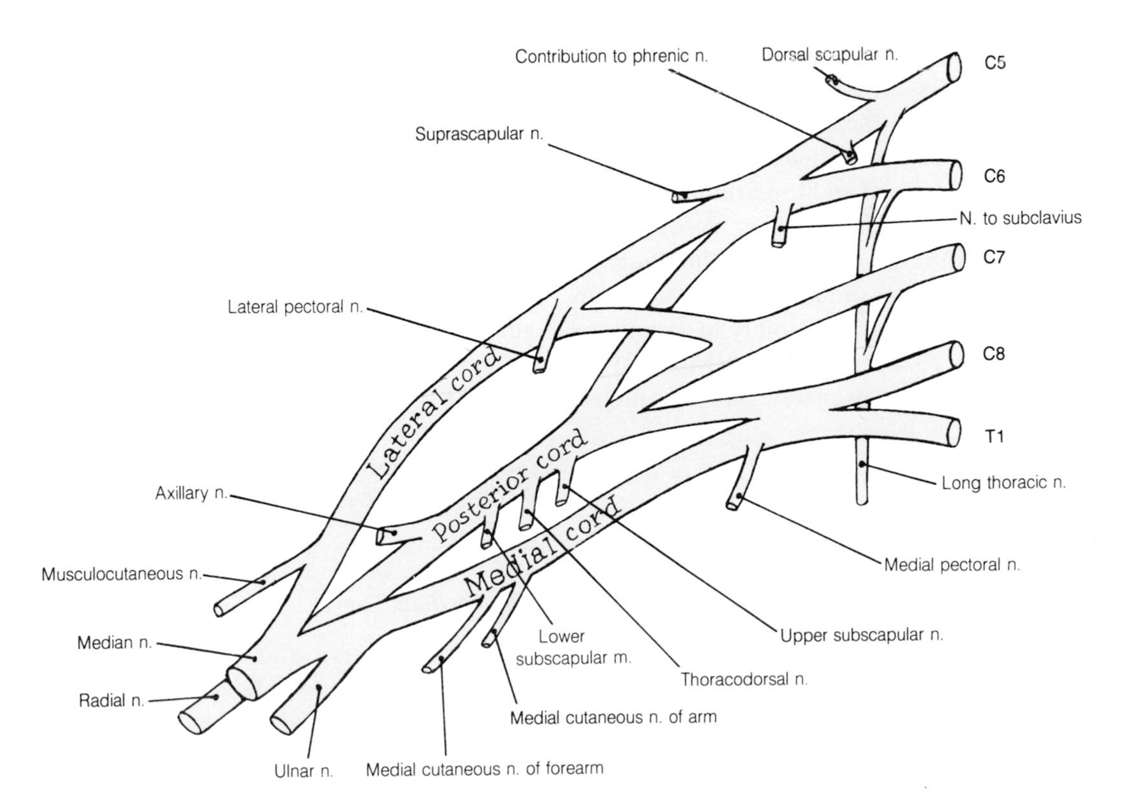

3-117 The brachial plexus.

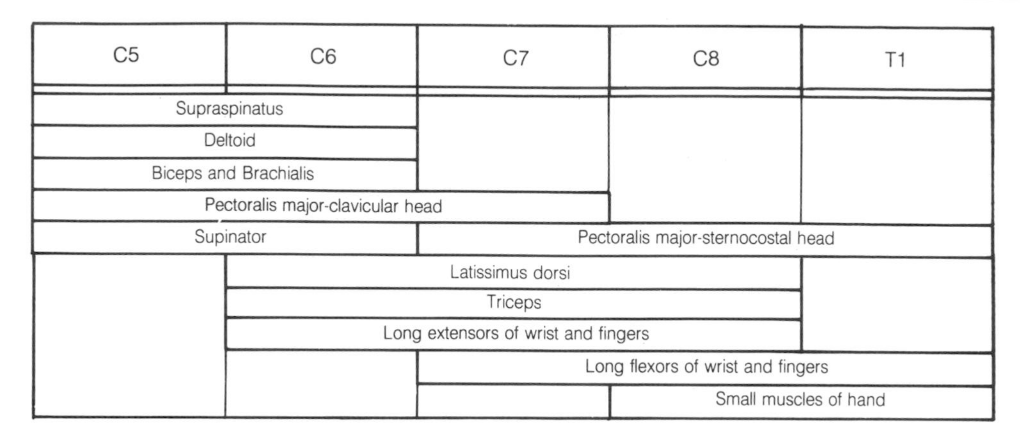

3-118 Diagram showing the contributions of spinal nerves to the innervation of some muscles of the upper limb.

cles and skin of the limb allow a diagnosis of the degree and extent of the injury to be made. The spinal nerves on which some important muscles rely for their innervation are shown in Fig. 3-118.

Injuries to the Upper Roots

Examination of Fig. 3-118 will show that injuries to the upper roots or trunk of the brachial plexus will produce a loss of abduction (deltoid), lateral rotation (spinati), and flexion (pectoralis major and anterior fibers of deltoid) at the shoulder, flexion at the elbow (biceps), and supination of the forearm (biceps and supinator). Skin over the lateral aspect of the arm and forearm may also be anesthetic. If the middle trunk is also involved adduction and internal rotation at the shoulder will be lost as well (latissimus dorsi and pectoralis major) and there will be an inability to extend the wrist and fingers. As a result the limb will lie at the side with the forearm pronated and the fingers flexed.

Injuries to the Lower Roots

When the lower roots or lower trunk are injured all the small muscles of the hand are paralyzed as well as the long flexor muscles of the fingers. This results in a characteristic claw-hand deformity. Anesthesia over the medial side of the forearm and hand is also found. Injury to the first thoracic root may also interrupt the sympathetic outflow to the head and neck and produce a **Horner's syndrome**. In this there is a drooping of the upper eyelid or ptosis, a constriction of the pupil, and dry, flushed, facial skin on the side of the injury.

Injuries to Individual Nerves

The Musculocutaneous Nerve

The musculocutaneous nerve (C5, 6 and 7) supplies the muscles in the arm that flex the elbow and injury will produce severe weakness of that movement as well as weakness of supination. Loss of sensation over the lateral aspect of the forearm is also found.

The Axillary Nerve

Injuries to the axillary nerve which are often associated with dislocations of the shoulder joint and fractures of the surgical neck of the humerus produce a paralysis of the deltoid muscle and an inability to abduct the arm at the shoulder beyond the first few degrees produced by supraspinatus (suprascapular nerve). An area of anesthesia over the lower part of the deltoid muscle is also present.

The Median Nerve

The median nerve and its anterior interosseous branch supply the long flexor muscles of the wrist and hand with the exception of the flexor carpi ulnaris and the medial part of the flexor digitorum profundus. The two pronator muscles in the forearm are also supplied as are the muscles of the thenar eminence which become typically flattened and wasted. Injury to the median nerve at the elbow, therefore, produces a severe disability in that the thumb and lateral two fingers cannot be flexed or the thumb opposed. Attempts to flex the wrist only produce ulnar deviation effected by the intact flexor carpi ulnaris. Anesthesia and trophic changes over the lateral part of the palm and palmar surfaces of the digits also present a serious problem. Injuries of the median nerve at the wrist or its involvement in the **carpal tunnel syndrome** lead to weakness or paralysis of the thenar muscles and a similar area of skin changes. Variations in the distribution of the median and ulnar nerves may, however, produce somewhat different pictures of paralysis and anesthesia.

The Ulnar Nerve

When the ulnar nerve is injured – usually at the elbow as it passes behind the medial epicondyle – all the small muscles of the hand are affected except those of the thenar eminence. As a result the ability to adduct the thumb and adduct and abduct the fingers is weak or absent. Ulnar deviation of the wrist is weak and because of the ulnar nerve's contribution to flexor digitorum profundus the medial two digits cannot be effectively flexed. Paralysis of the interossei also leads to an inability to flex the metacarpophalangeal joints while the interphalangeal joints are extended. The unopposed action of the long extensor tendons leads to the deformity known as a "claw hand." This analogy is suggested by a wasted and flattened palm and hypothenar eminence, hyperextended metacarpophalangeal joints, and flexed interphalangeal joints. Some anesthesia over the palmar and dorsal aspect of the medial side of the hand and of the little finger is to be expected.

The Radial Nerve

The radial nerve supplies all the extensor muscles of the arm and forearm and the degree of motor paralysis will depend upon the level at which the nerve is injured or severed. One of the commonest causes of injury is a fracture of the shaft of the humerus where the nerve lies close to the bone. At this level the triceps muscle may be spared paralysis but the inability to extend the wrist or fingers leads to the classical picture of "wrist-drop". Injury to the deep branch of the radial nerve will produce a similar picture but injury to the superficial branch only results in anesthesia over the dorsal aspect of the lateral part of the hand and bases of the lateral fingers and thumb (remember that the dorsum of the distal parts of the digits are supplied by the palmar digital nerves).

Because of anomalies of distribution and overlapping of fields of innervation, areas of sensory loss following nerve lesions do not exactly correspond to the "normal" distribution of cutaneous branches described in this text.

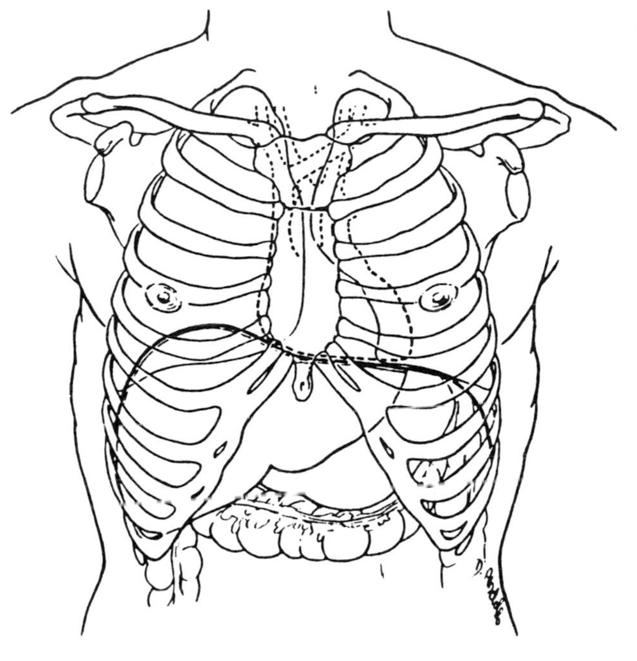

4 The Thorax

The thorax is the part of the body that lies between the neck and the diaphragm. The cavity of the thorax contains the lungs, the heart, a number of large blood vessels, the respiratory passages, and the esophagus which is continuous from the neck to the diaphragm. The cavity is surrounded by the thoracic wall which also surrounds and protects the upper part of the abdominal cavity. The reason for this is that the diaphragm, the fibromuscular septum* which separates the two cavities, is domeshaped with its convexity facing upward. It is clearly important, then, to be able to define the approximate level of the diaphragm on the surface of the thorax. Above the diaphragm lie the contents of the thoracic cavity. These can be examined by the simple but nonetheless useful methods of inspection, palpation, percussion, and auscultation (listening with a stethoscope) as in a routine medical examination, or by the more sophisticated methods of radiological and ultrasound techniques, echocardiography, computerized tomography, or magnetic resonance imaging. In none of these methods are the structures under investigation directly visualized, and it is therefore essential for the interpretation of the re-

* Septum, *L.* = barrier or wall

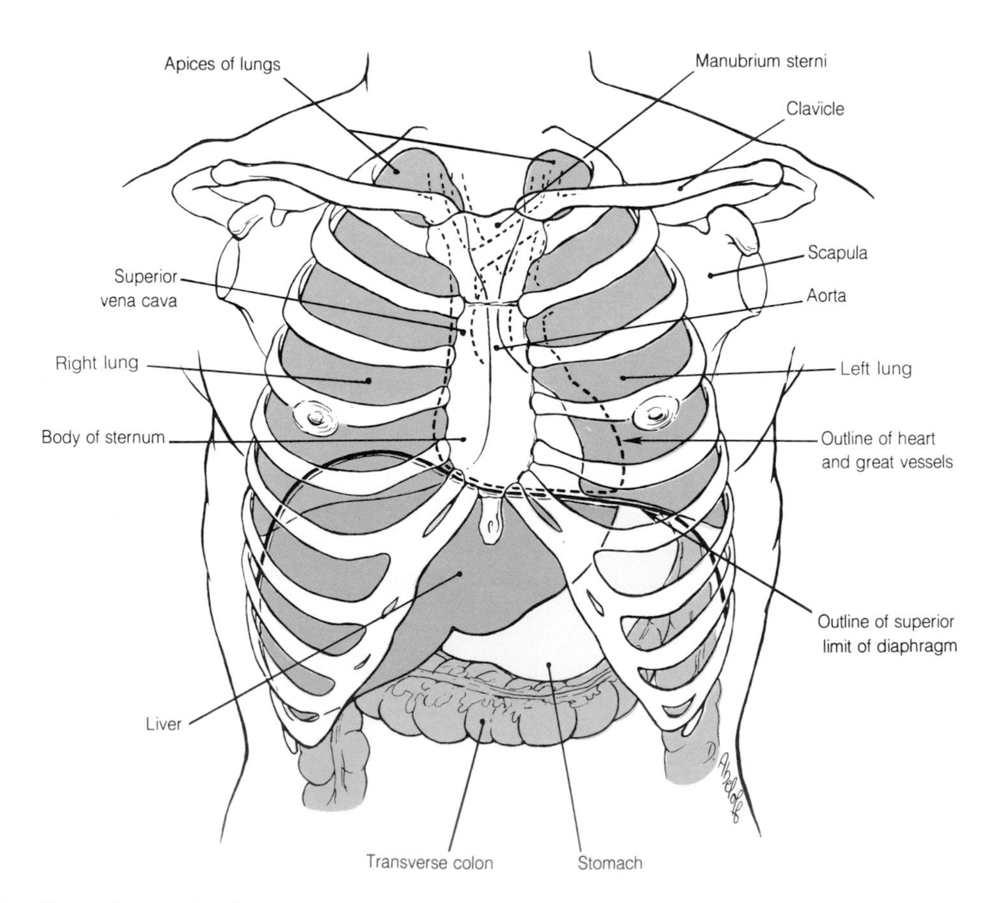

4-1 Showing the surface projections of some thoracic and abdominal organs and the thoracic cage.

sults obtained to have a clear three-dimensional picture of where the normal structures lie in relation to each other and to the surface. To stress the importance of this some of these relationships are illustrated in Fig. 4-1. Note the following features:

1. The outline of the heart and great vessels (*black broken line*) is deep to the anteromedial portions of the lungs.
2. The diaphragm extends superiorly as high as the fourth intercostal space on the right and to the fifth intercostal space on the left (*black solid line*).
3. The liver lies largely under cover of the thoracic cage immediately below the diaphragm.
4. The position of the superior vena cava and ascending aorta is superficial, just deep to the manubrium of the sternum above and a portion of the heart lying just deep to the body of the sternum below.
5. The apices of the lungs lie above the level of the clavicles and the anterior parts of the first ribs.

The Thoracic Wall

The thoracic wall is supported by the **thoracic cage.** The overlying skin is modified to form the **mammary papillae** or nipples which in thc male and immature female lie at the level of the fourth intercostal space. Each is surrounded by an **areola.** The female **breast** develops in the superficial fascia of the chest wall and its ducts open at the nipple. Beneath the skin and fascia of the chest wall lie a number of muscles that belong to and are discussed with the upper limb. They are innervated by nerves arising from the **brachial plexus** which is formed from the **fifth** to **eighth cervical** and **first thoracic nerves.** Similarly, over the lower part of the thorax are found muscles of the **anterior abdominal wall** which are supplied by **thoracic spinal nerves.**

Deep to this superficial group of muscles lie the bony thoracic cage and the **intercostal muscles** between adjacent **ribs.** Deeper still lies a thin and incomplete sheet of musculature lined by a layer of fascia and finally the **parietal pleura.** A needle passed through the thoracic wall to enter the pleural cavity must traverse each of these soft tissue layers.

The skin over the thoracic wall is innervated in a segmental fashion by the lateral and anterior cutaneous branches of underlying intercostal nerves. The exception to this is the skin over the first intercostal space which is supplied by the supraclavicular nerves.

The **lymphatic drainage of the superficial layers of the thoracic wall** is largely to the subscapular axillary nodes although some lymphatic vessels reach the pectoral nodes. Near the midline anteriorly, lymphatic vessels pierce the anterior ends of the intercostal spaces to reach the parasternal nodes which lie along the internal thoracic artery.

The Breasts or Mammary Glands

The breasts are present in both sexes. In the male they remain rudimentary throughout life although some enlargement is occasionally seen at birth, puberty, and following estrogen therapy. Maturation of the female breast begins at puberty and reaches a maximum during pregnancy and lactation. In other mammals more than the typical two human pectoral glands develop at sites along a

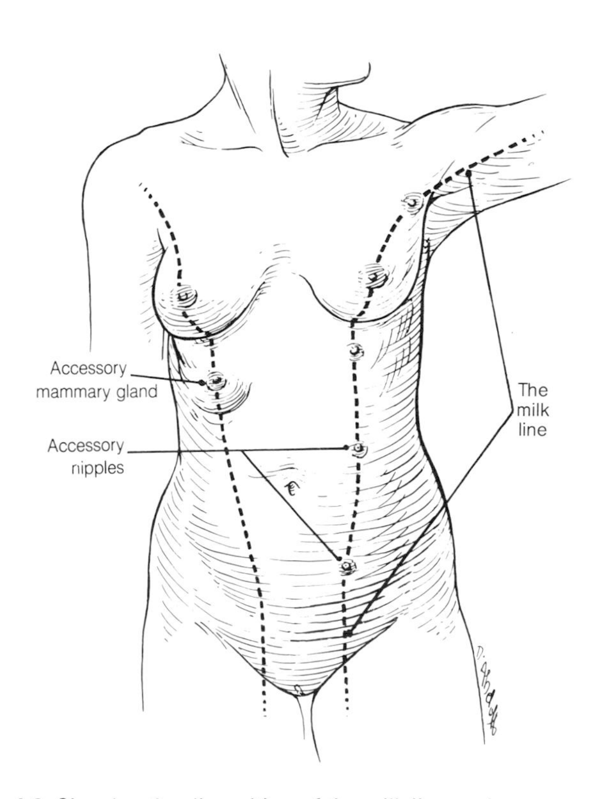

4-2 Showing the disposition of the milk line and an accessory mammary gland and nipples.

"milk line" extending from the axilla to the inguinal region and, in man, accessory nipples or glands are occasionally found along this line. Examples are illustrated in Fig. 4-2. The importance of the breast lies in its periodic phases of activity and its susceptibility to neoplastic change. However, its superficial position allows a chance for effective treatment by surgery and radiotherapy if tumors are recognized early. A "lump in the breast" should, therefore, always be taken seriously and the possible sites of lymphatic involvement should be known. The external appearance of the adult female breast is shown in Fig. 4-3A. Note that the nipple is centrally placed within the circular pigmented **areola**. **Areola glands** which lubricate the nipple during nursing can be seen as small surface elevations.

In the young adult female the breast extends between the levels of the second and sixth ribs. It lies **in the superficial fascia** and **on the deep fascia** covering the **pectoralis major** and **serratus anterior muscles.** It may extend over the superficial layer of abdominal muscles and into the axilla as the **lateral process** or "axillary tail" as is depicted in Fig. 4-3B. It is composed of 15 to 20 **lobes** containing a duct system, lobules of glandular tissue, supporting connective tissue, and surrounding fat. A single **lactiferous duct** from each lobe opens onto the nipple. These ducts become expanded deep to the nipple to form **lactiferous**

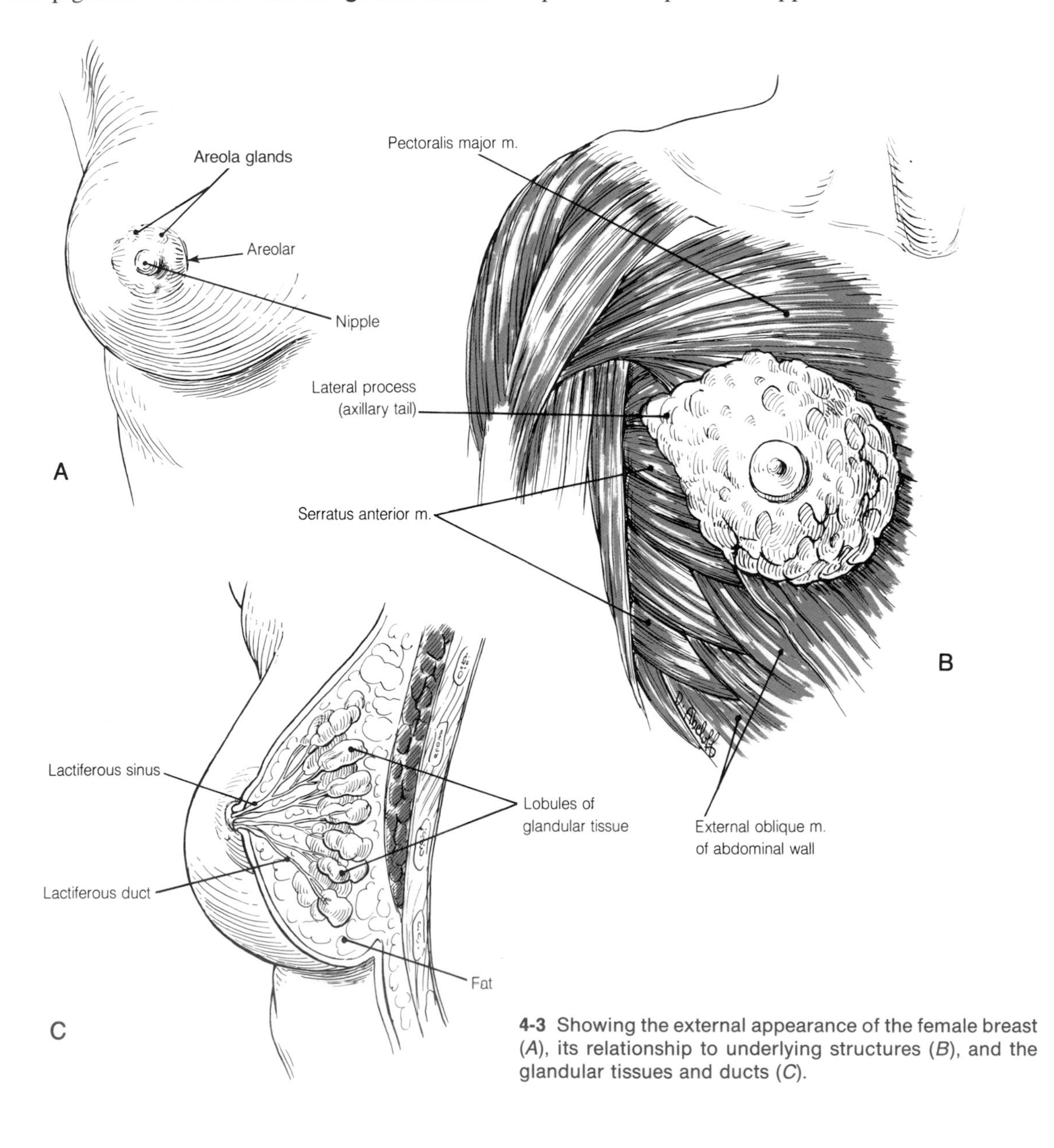

4-3 Showing the external appearance of the female breast (*A*), its relationship to underlying structures (*B*), and the glandular tissues and ducts (*C*).

sinuses. The arrangement of these structures is seen in Fig. 4-3C.

The nipple normally projects from the areola and smooth muscle fibers which it contains allow further erection. Occasionally the nipple remains in a prenatal retracted position thus causing difficulties in breast feeding. The surrounding areola is slightly pigmented but becomes much darker during the first pregnancy and never returns to its original color.

The **blood supply of the breast** is derived from branches of the **internal thoracic artery** which pierce the intercostal spaces, the **intercostal arteries,** and the **thoracic branches of the axillary artery.** Venous blood is drained from the breast by veins that accompany the arteries. Cells from malignant tumors of the breast may gain entry to these veins and account for widespread secondary tumors.

The **lymphatic drainage of the breast** assumes great importance in the treatment of malignant tumors and the assessment of their prognosis for it is along lymphatic channels that dissemination most commonly occurs.

Lymphatics from within the gland drain into a deep submammary or a superficial subareolar plexus. Further lymphatic channels radiate from these plexuses laterally to axillary nodes (pectoral, central, and apical), upward to infraclavicular and supraclavicular nodes, medially to the contralateral breast and nodes along the internal thoracic artery, and inferiorly to extraperitoneal tissues and thence to mediastinal nodes. These pathways are illustrated diagramatically in Fig. 4-4. Note the lymphatic communication with the contralateral breast. Of tumors arising in the lateral half of the breast more than 60% will involve the axillary nodes and some of these the internal thoracic nodes also. Tumors of the medial half are most likely to spread to the internal thoracic nodes.

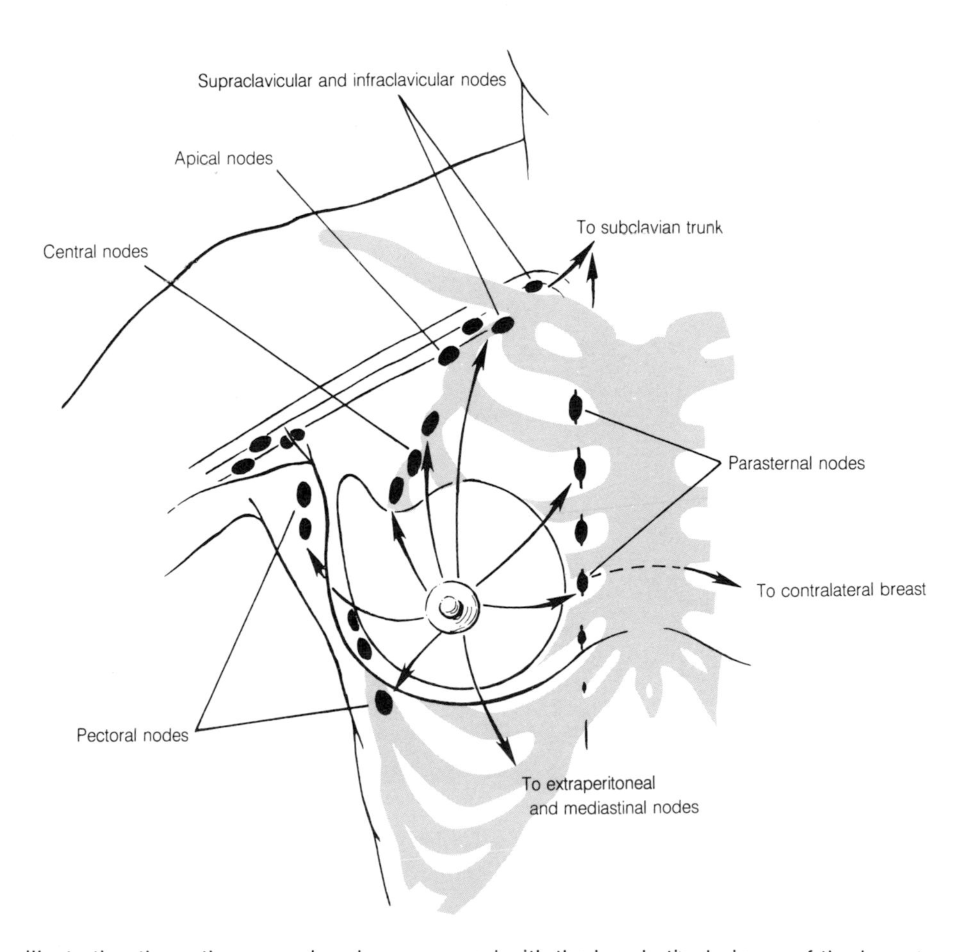

4-4 Diagram illustrating the pathways and nodes concerned with the lymphatic drainage of the breast.

The Thoracic Cage

The thoracic cage is formed by the **thoracic vertebral column, 12 pairs of ribs** and their **costal cartilages,** and the **manubrium** and **body of the sternum.** Posteriorly each rib typically articulates with the body and transverse process of the thoracic vertebra of the same number and with the body of the vertebra above. Anteriorly the upper seven pairs of ribs articulate through their costal cartilages with the sternum (vertebrosternal or true ribs); the eighth, ninth, and tenth pairs articulate with the costal cartilages of adjacent ribs (vertebrochondral or false ribs). The eleventh and twelfth pairs lie free anteriorly and are capped by only a small costal cartilage (vertebral or floating ribs). The anterior aspect of the thoracic skeleton is illustrated in Fig. 4-5. Note:

1. the small **thoracic inlet** lying superiorly and surrounded by the first thoracic vertebra, the two first ribs and their costal cartilages, and the manubrium sterni;
2. the large **thoracic outlet** lying inferiorly and surrounded by the twelfth thoracic vertebra, the costal margin, and the xiphisternal joint;
3. the plane in which each rib lies slopes anteriorly and downward and laterally and downward;
4. that while the first rib has a superior and inferior surface and a medial and lateral border, subsequent ribs are tilted so that the sixth rib clearly has superior and inferior borders and medial and lateral surfaces.

As well as protecting the underlying organs, the multiple bony elements and joints of the thoracic cage provide it with the ability to change in shape that is necessary for ventilation of the lungs.

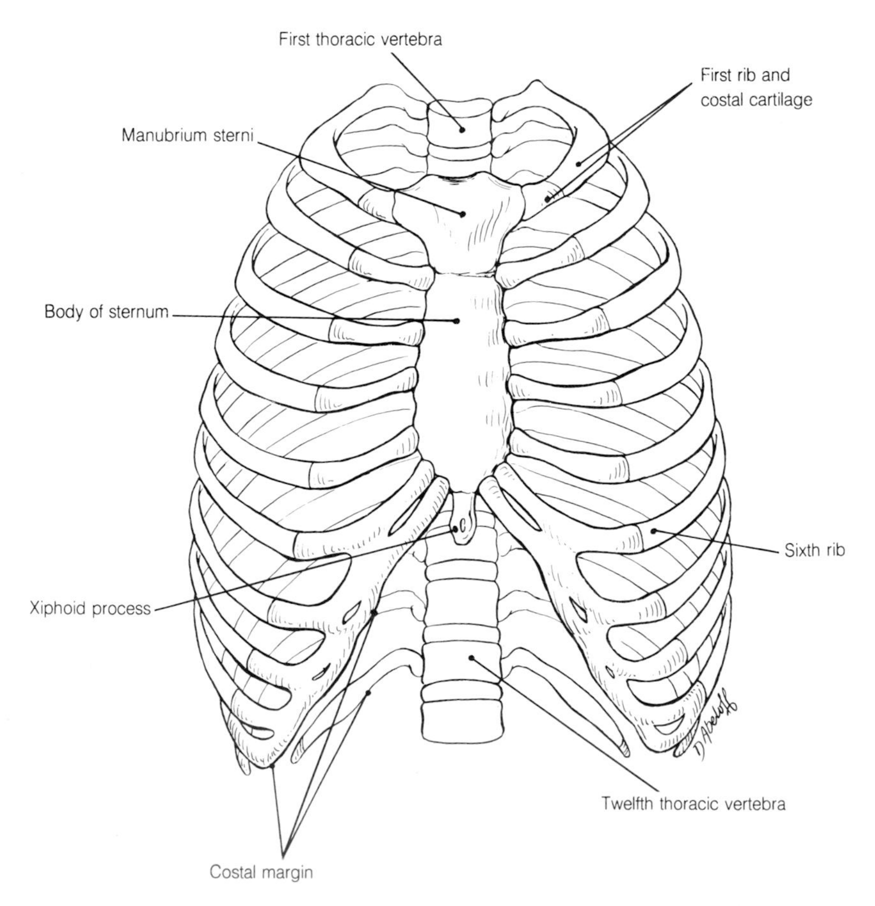

4-5 The thoracic cage.

The Thoracic Skeleton

A description of the thoracic vertebral column is found in Chapter 2, "The Back and Spinal Cord." The features of thoracic vertebrae, their joints, and the shape and movements of the column should now be reviewed.

The Ribs or Costae

A typical or **true rib** has a **body** which is flattened and curved. It has a rounded upper border, a sharp lower border, and an internal and external surface. Just inside the lower border is the **costal groove.** The body describes a curve of gradually increasing radius as the eye follows it forward. This curve is interrupted posteriorly by the **angle.** The anterior end of the rib is hollowed where it is attached to the **costal cartilage.** Posteriorly the body is joined to the **head of the rib** by a short **neck.** The details of the body and head of a rib are seen in Fig. 4-6, A and B, in which the eighth rib is seen from below and behind. Note:

1. the costal groove which houses the intercostal nerve and vessels,
2. the tubercle of the rib which bears a facet for articulation with the transverse process of the eighth thoracic vertebra, and
3. the two facets on the head for articulation with the bodies of the eighth and seventh thoracic vertebrae. The articulations of the tubercle and the head are synovial joints.

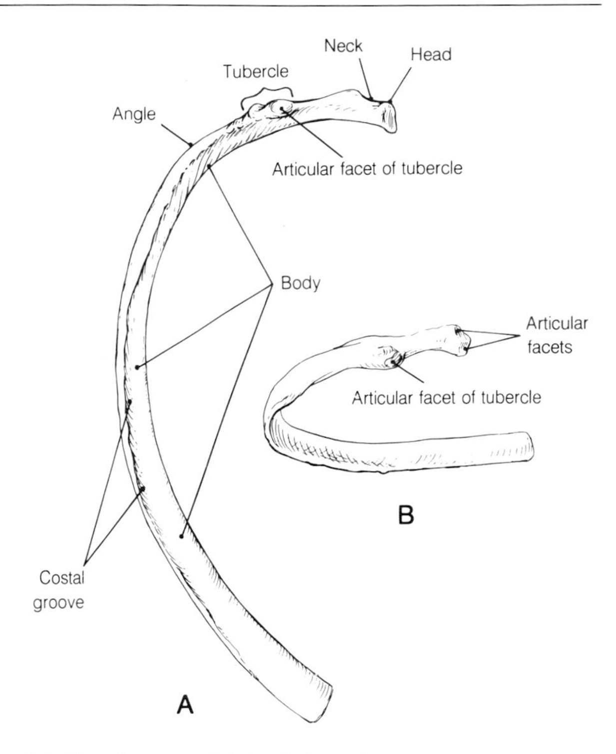

4-6 Showing the eighth rib from below (*A*) and from behind (*B*).

The **first rib** is atypical. The head shows only a single articular facet for the first thoracic vertebra. The shaft has an inner and outer border and a superior and inferior surface. The superior surface has a small elevation near its inner border called the **scalene tubercle,** and crossing the upper surface, anterior and posterior to the tubercle, are shallow grooves in which lie the subclavian vein and artery, respectively. These features of the superior surface of the first rib are illustrated in Fig. 4-7.

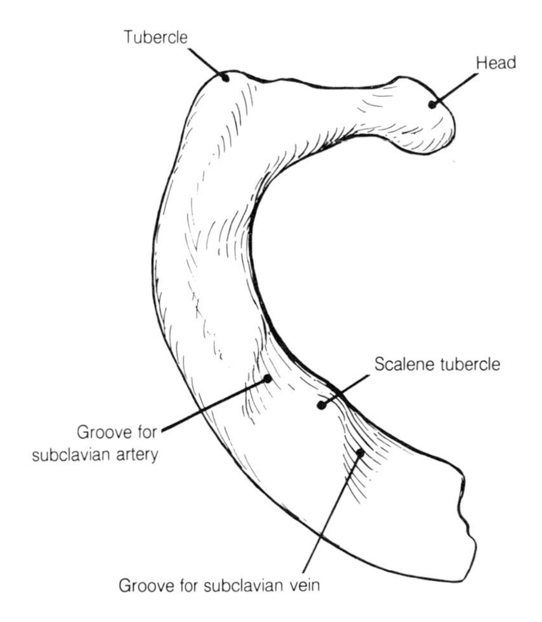

4-7 Showing the first rib from above.

The **second rib** is oriented in a position that is intermediate between that of the first and a typical rib. There are two articular facets on the head for the bodies of the first and second thoracic vertebrae.

The **eleventh and twelfth ribs** only articulate with the bodies of their corresponding vertebrae and, therefore, their heads only have a single facet. They also have no tubercles or articular facets for transverse processes. Other features are poorly marked and the tips are usually pointed. Their costal cartilages do not articulate with those of adjacent ribs.

Primary centers of ossification appear in the bodies of the ribs at the end of the second intrauterine month. Secondary centers for the head and, when present, for the tubercles, appear at puberty and fuse with the body around the twentieth year.

The Sternum

The sternum develops as a series of six **sternebrae** (compare with vertebrae). Normally all but the first and last sternebrae fuse to form the **body of the sternum.** The first remains separate and becomes the **manubrium sterni;** the last becomes the **xiphoid process.** Using Fig. 4-8, observe that the upper border of the manubrium has an easily palpable depression called the jugular notch. The lateral border articulates with the clavicle at the **clavicular notch,** the first costal cartilage, and part of the second costal cartilage. The lower border articulates at a slight angle with the body of the sternum by a secondary cartilaginous joint. This, the **manubriosternal joint,** is palpable and marks the level of the **second costal cartilage** and the surface elevation it produces is known as the **sternal angle.** The body of the sternum articulates with the second to seventh costal cartilages. At the lower end of the body is the **xiphoid process,** a narrow tongue of hyaline cartilage depressed below the level of the sternum.

One or two primary centers appear in each sternebra between the fifth and sixth months of intrauterine life. Fusion of the centers for the lower sternebrae begins at puberty.

The Intercostal Spaces

The intercostal spaces lie between the ribs. They are largely filled by the external and internal intercostal muscles. However, deep to each internal intercostal muscle and under cover of the costal groove lies a neurovascular bundle made up of an intercostal vein, artery, and nerve arranged in that order from above downward.

The Intercostal Muscles

The fibers of the **external intercostal muscles** run downward and forward from the lower border of the rib above to the upper border of the rib below. The muscle is deficient anteriorly being replaced by a fibrous membrane, the **external intercostal membrane.**

Fibers of the **internal intercostal muscles** run downward and backward deep to the external muscles. The fibers of this muscle are deficient posteriorly and are replaced by the **internal intercostal membrane.**

The **innermost intercostal muscles,** the **sub-**

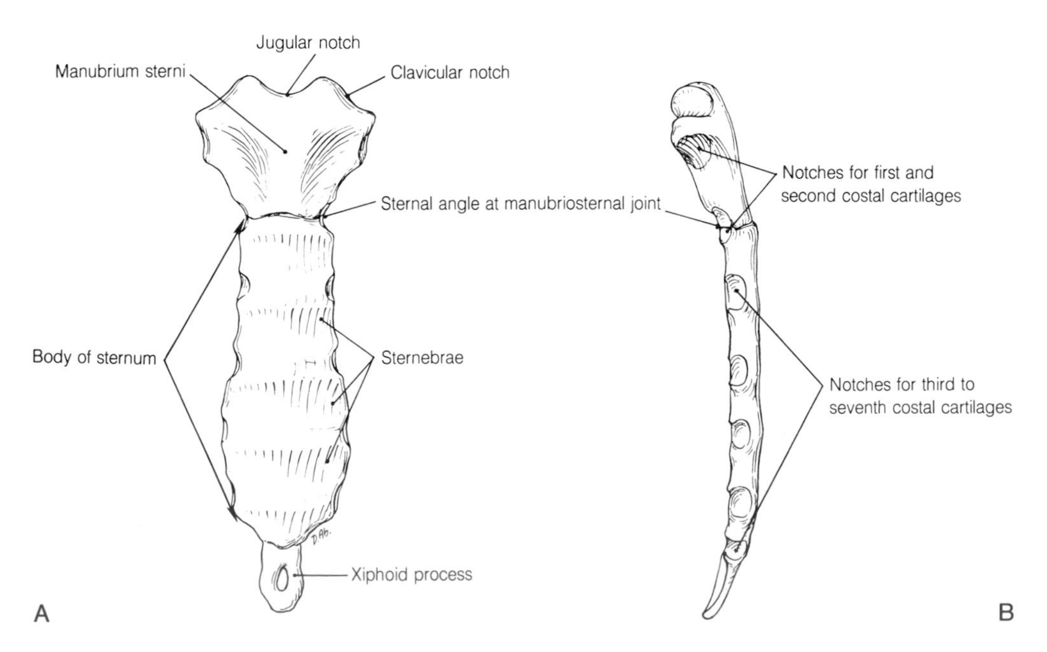

4-8 The sternum viewed from the front (*A*) and from the lateral aspect (*B*).

costal muscles, and the **transversus thoracis muscle** form a discontinuous layer on the deep surface of the thoracic cage linking ribs to ribs and costal cartilages to sternum.

The nerve supply of all these muscles is from the adjacent **intercostal nerves.**

The actions of the intercostal muscles are concerned with respiration and are discussed after the diaphragm has been described.

The Intercostal Nerves

The intercostal and subcostal nerves are the ventral rami of the twelve thoracic spinal nerves. Typically these nerves run forward in an intercostal space deep to the internal intercostal membrane and muscle and below the intercostal vessels. They give off a **lateral cutaneous branch** which pierces the intercostal and superficial muscles to reach and supply the skin of the lateral thoracic wall, and a **collateral branch** which runs forward along the lower border of the space. At the lateral border of the sternum the nerves pass anteriorly and terminate as **anterior cutaneous branches.** Along their course they supply the intercostal muscles. An intercostal nerve follows closely the distribution of the ventral ramus of a typical segmental nerve and this is illustrated in Fig. 4-9.

The **first intercostal nerve** provides a major contribution to the brachial plexus and only a small branch continues around the first intercostal space.

The lateral cutaneous branch of the **second intercostal nerve** is called the **intercostobrachial nerve** and helps supply the medial surface of the arm.

The **seventh to eleventh intercostal nerves** escape from the thoracic wall by passing from beneath the costal margin onto the anterior abdominal wall. They supply the abdominal muscles, including the rectus abdominis which they pierce to reach the skin on either side of the midline.

The **twelfth thoracic** or **subcostal nerve** has a similar distribution, but runs below the twelfth rib.

As a result of this distribution of the lower seven thoracic nerves, the greater part of the skin of the anterior and lateral abdominal walls is supplied by the anterior and lateral cutaneous branches of intercostal and subcostal nerves. As a guide to their distribution it should be remembered that **the umbilicus lies in the dermatome of the tenth thoracic nerve.** In addition, it should be noted that the skin over the first intercostal space may be supplied from the **fourth cervical segment** through the **supraclavicular nerves.** These facts all assume importance in a neurological examination when attributing areas of loss of skin sensation to a particular spinal cord level.

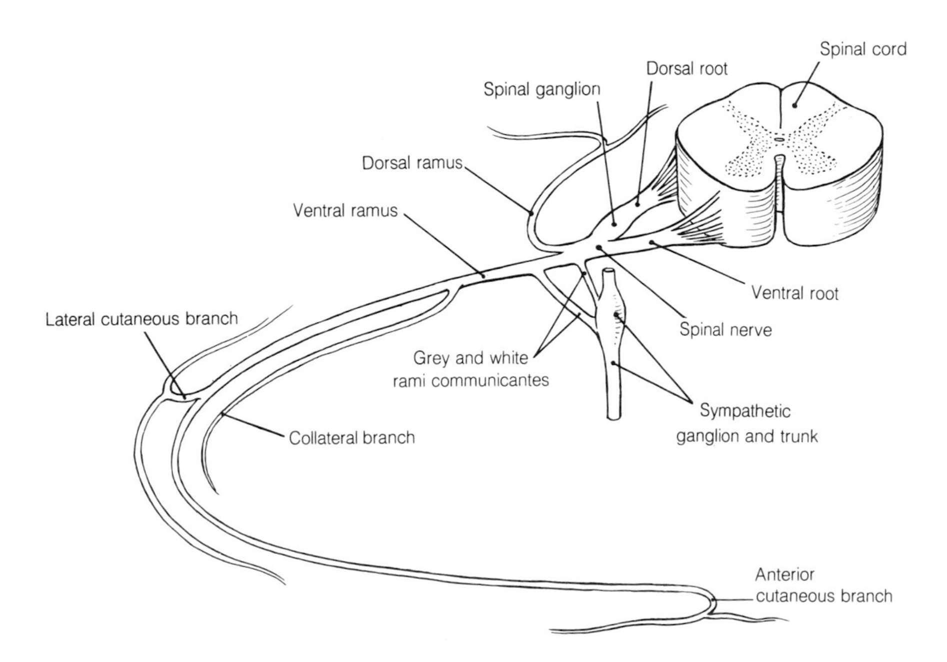

4-9 The distribution of a typical spinal nerve.

The Intercostal Arteries

The intercostal arteries lie between the intercostal veins and nerves in the costal grooves. Each space contains an **anterior intercostal artery** derived from the **internal thoracic artery** or one of its branches and a **posterior intercostal artery** typically derived from the **aorta.** The upper two posterior intercostal arteries are, however, branches of a **supreme intercostal artery** which is, in turn, a branch of the **costocervical trunk** given off by the **subclavian artery.** The lower anterior intercostal arteries are given off by the **musculophrenic artery,** one of the two terminal branches of the **internal thoracic artery.**

The anterior and posterior intercostal arteries anastomose with each other and supply branches to the intercostal muscles and the overlying superficial thoracic muscles, the breast, and skin.

Perhaps the most important branches of the posterior intercostal arteries are their **dorsal branches** which give rise to **spinal branches.** These enter the vertebral canal through thoracic **intervertebral foramina** to supply the vertebral bodies and the meninges and reinforce the blood supply of the spinal cord.

The Internal Thoracic Artery and Vein

The **internal thoracic artery** is a branch of the first part of the subclavian artery in the root of the neck. It descends anteriorly in the thorax deep to the cartilages of the upper six ribs and just lateral to the sternum. It divides at the level of the sixth intercostal space into the **superior epigastric artery** which will be followed later onto the anterior abdominal wall, and the **musculophrenic artery** which continues around the costal margin giving off the lower **anterior intercostal arteries** and supplying the diaphragm. Perforating branches pass anteriorly through each intercostal space and accompany the anterior cutaneous branches of the intercostal nerves. Those of the second, third, and fourth spaces also provide substantial branches to the breast. The **pericardiacophrenic artery** is a branch which accompanies the **phrenic nerve** and supplies the pleura, pericardium, and diaphragm.

Alongside the arteries and their branches lie the

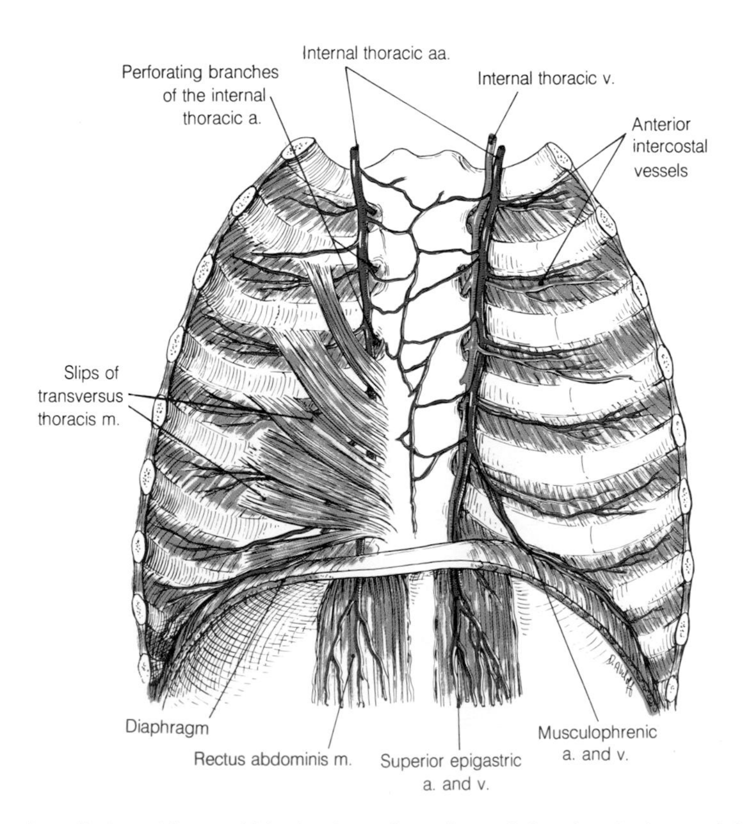

4-10 The anterior thoracic wall viewed from within the thoracic cavity and showing the internal thoracic artery and vein.

tributaries and main trunk of the **internal thoracic veins** which drain into the **left and right brachiocephalic veins.** The internal thoracic artery and vein are accompanied by an important chain of lymphatic nodes and vessels which frequently become involved in the spread of malignant tumors of the breast.

The course and distribution of the internal thoracic artery and vein can be followed in Fig. 4-10. The sternum and chest wall are viewed from inside the thoracic cavity. On the right side the slips of transversus thoracis have been removed to expose the whole course of the anterior intercostal arteries and veins.

The Intercostal Veins

The intercostal veins lie above the intercostal arteries in the subcostal groove. **Anterior intercostal veins** drain into the internal thoracic veins. **Posterior intercostal veins** from the first space ascend over the neck of the first rib and then turn forward over the pleura covering the lungs to drain into the **left and right brachiocephalic veins.** The veins from the second and third spaces on the left side form the **left superior intercostal vein** which crosses the arch of the aorta to reach the left brachiocephalic vein. The remaining posterior intercostal veins drain into the **azygos system of veins** which is discussed with the contents of the mediastinum.

The Endothoracic Fascia

The internal surface of the thoracic wall is lined by a layer of loose areolar tissue called the endothoracic fascia.

The Typical Intercostal Space

In order to review the features of a typical intercostal space, look at Fig. 4-11 and note that:

1. The external and internal intercostal muscles lie between the upper and lower borders of adjacent ribs.
2. The intercostal vein, artery, and nerve lie deep to both external and internal intercostal muscles.
3. The intercostal nerve is not under cover of the ribs and is thus suitably placed for infiltration with local anesthetic.
4. The only muscles lying deep to the intercostal

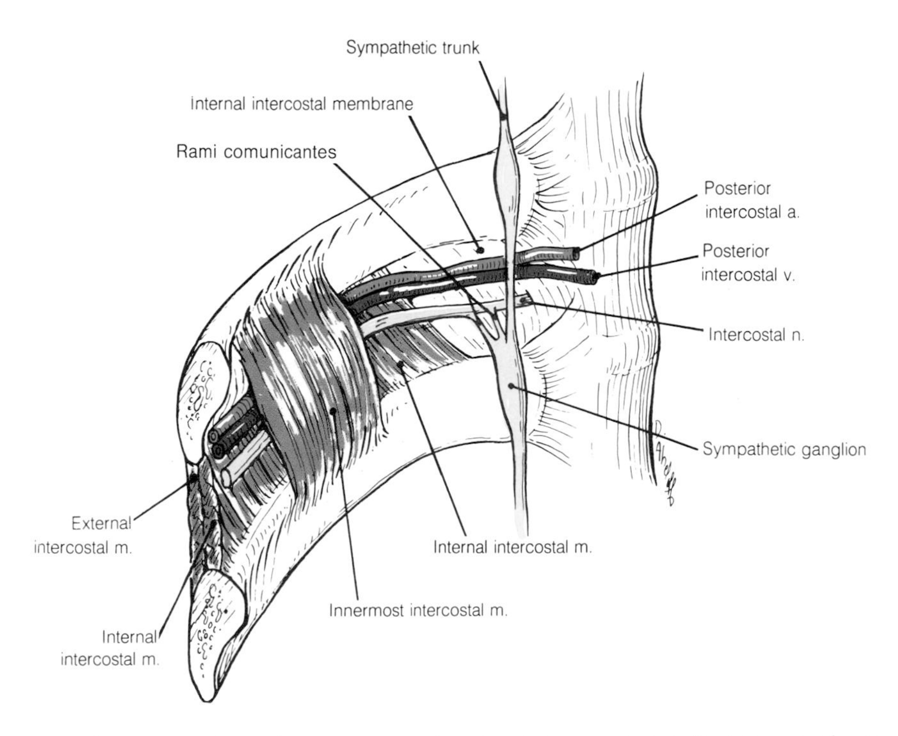

4-11 Semischematic diagram to show the arrangement of the intercostal vessels, nerve, and muscles in a typical intercostal space.

vessels and nerve are the discontinuous slips of the innermost intercostal muscles.

5. The sympathetic trunk passes over the heads of the ribs and the posterior intercostal membranes on a plane deep to all other intercostal structures.

The description of the thoracic wall has now been completed, but before the movements involved in ventilation of the lungs are considered, the diaphragm must be described.

The Diaphragm

The diaphragm is a fibromuscular partition which separates the thoracic cavity from the abdominal cavity. The majority of its muscle fibers arise from the inner aspect of the thoracic outlet and arch upward to reach a somewhat trefoil **central tendon.** In Fig. 4-12, the convex superior aspect of the diaphragm is viewed from within the thoracic cavity. Two slips of muscle arise from the deep surface of the **xiphoid process** and form the sternal part. The costal part arises from the inner aspects of the **lower six ribs and costal cartilages.** The lumbar part arises from the **lateral, medial** and **median arcuate ligaments** and as the **crura*** from the upper lumbar vertebrae. The details of the lumbar part of the diaphragm can be seen in Fig. 4-13. The lateral arcuate ligament spans the gap between the twelfth rib and the transverse process of the first lumbar vertebra. The medial arcuate ligament extends from this point to the lateral margin of a crus. The left crus arises from the bodies of the upper two lumbar vertebrae and the right from the upper three. The medial margins of the crura are joined across the midline by the median arcuate ligament. Returning to Fig. 4-12, it is sufficient to note at this point that the **descending aorta** passes behind the median arcuate ligament in front of the body of the first lumbar vertebra, the **esophagus** passes through fibers of the lumbar part of the diaphragm just anterior to the aorta, and at the level of the tenth thoracic vertebra and more anteriorly the **inferior vena cava** passes through an opening in the central tendon to the right of the midline at the level of the eighth or ninth thoracic vertebra. The supe-

* Crus (pl. crura), *L.* = leg

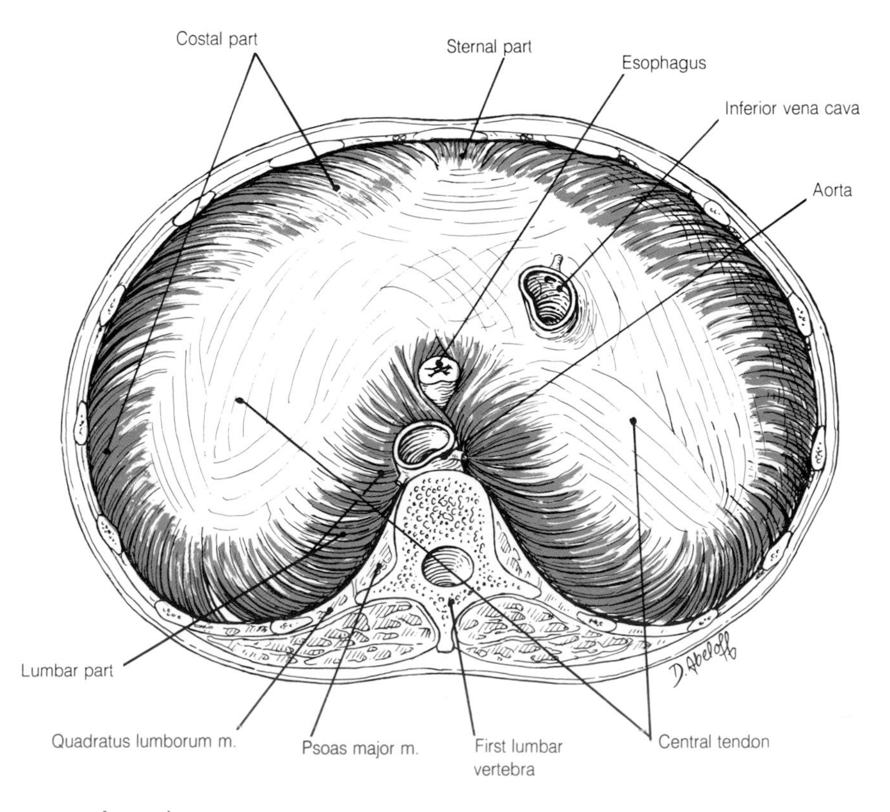

4-12 The diaphragm seen from above.

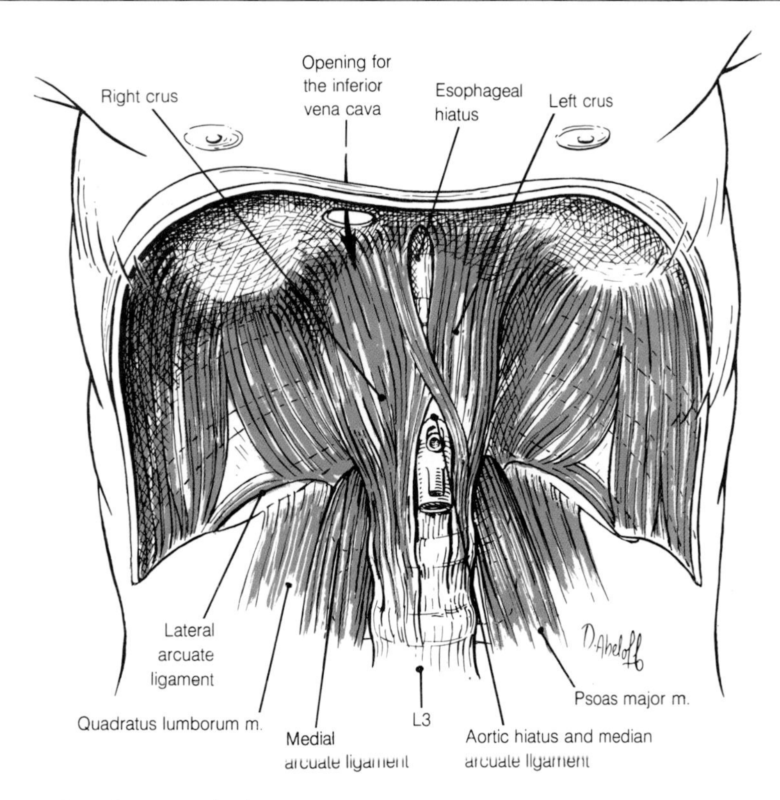

4-13 The diaphragm seen from its abdominal aspect.

rior surface of the diaphragm is covered by **parietal pleura** and by the **fibrous pericardium,** which is firmly attached to the central tendon. The inferior surface is covered by **parietal peritoneum** except for a small area where it is directly related to the liver.

The diaphragm is supplied with blood by the **pericardiacophrenic** and **musculophrenic branches** of the **internal thoracic artery,** the **lower posterior intercostal arteries,** the **superior phrenic arteries,** and the **inferior phrenic arteries** which are the first branches of the abdominal aorta.

Motor nerve fibers to the muscle of the diaphragm reach it in the **phrenic nerves. Sensory fibers** from the muscle and adjacent pleura and peritoneum are also carried in this nerve but those from the periphery run in the lower intercostal nerves.

It should be clearly understood that although its action may seem involuntary, the muscle of the diaphragm is striated skeletal muscle and will, therefore, be supplied by somatic motor nerve fibers. These are mainly contributed to the phrenic nerve by the **fourth cervical segmental nerve** but also by thc **third and fifth.** This apparently distant origin of the nerve is explained by the fact that the muscle of the diaphragm is largely developed from mesenchyme that has migrated from the fourth cervical myotome.

Contraction of the diaphragm when the costal margin is fixed causes it to descend and this movement combined with a contraction of the anterior abdominal wall muscles causes a rise in intraabdominal pressure which is used, for example, in expelling feces, urine, and in childbirth. However, its most important action is to increase the capacity of the thoracic cavity and, thus, draw in air through the respiratory passages. This action forms an important part of respiration.

Respiration

Respiration in the anatomical sense refers to the rhythmic expansion and contraction of the thoracic cavity that leads to the regular replacement of air within the lungs. This is achieved by inspiratory and expiratory movements of the thoracic wall and diaphragm. During inspiration the thoracic cavity is enlarged in volume, a negative pressure is created, and air is drawn into the lungs through the respiratory passages. This enlargement of the cavity can be achieved by increasing its vertical, anteroposterior, and transverse dimensions. The vertical dimension is increased when the diaphragm contracts and assumes a flatter profile. This movement must be coupled with a relaxation of the abdominal muscles to allow a downward displacement of abdominal viscera. The ribs slope downward and forward and as a result the anteroposterior diameter can be increased by elevating the anterior ends of the ribs about a transverse axis. The ribs also slope downward and laterally, and thus, when elevated about an anteroposterior axis (like a bucket handle) the transverse diameter is increased. Each of these movements can be more readily appreciated by reference to the diagrams in Fig. 4-14.

While the action of the diaphragm in producing an increase in vertical dimension is clear, the part played by the intercostal muscles and other accessory muscles of respiration in moving the ribs remains difficult to interpret despite a number of electromyographic studies.

During quiet inspiration there is little movement of the thoracic cage. However, some electrical activity can be recorded in the **scalene muscles** of the neck which probably fix the ribs, and in the intercostal muscles which contract sufficiently to resist deformation by the negative intrathoracic pressure. The change in volume is due to the contraction and descent of the diaphragm.

In quiet expiration no direct muscle action is involved. During inspiration the elastic tissue of the lungs has been stretched and its recoil is sufficient to drive out the inspired air and reduce the thoracic volume. Nevertheless, activity can again be recorded in the diaphragm and intercostal muscles. This is probably a paradoxical action as they relax in a controlled manner against the force of the elastic recoil.

Deep inspiration recruits more muscles acting with greater force. The scalene muscles with the help of the **sternocleidomastoid muscles** now raise the ribs producing some movement at the **manubriosternal joint.** The intercostal muscles become more active and may also help to elevate the ribs. The twelfth rib is fixed by the **quadratus lumborum** muscle which is attached to its lower border, thus allowing a more forcible downward movement of the diaphragm. As the need for or difficulty in inspiration increases, the back is arched by the **erector spinae muscles** and the pectoral girdle and upper limbs are stabilized by grasping a heavy object so that the **pectoral muscles** can assist in raising the ribs.

In forced expiration the elastic recoil of the lungs and deformed costal cartilages can be rein-

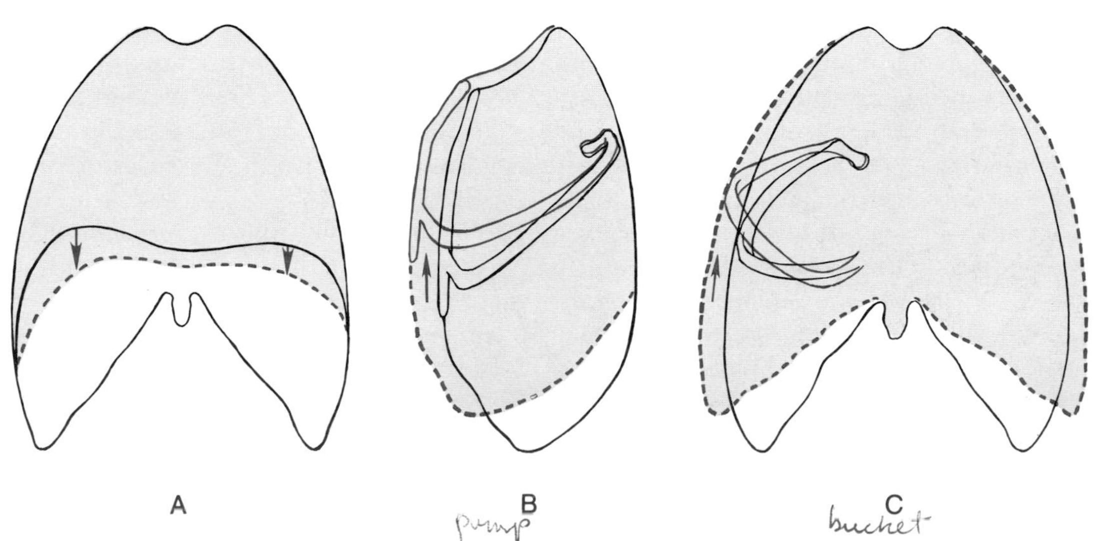

4-14 Diagram to illustrate the increase in the volume of the thoracic cavity during inspiration caused by (*A*) contraction of the diaphragm, (*B*) elevation of the ribs about a transverse axis, and (*C*) elevation of the ribs about an anteroposterior axis.

forced by strong contraction of the muscles of the anterior abdominal wall, which depress the ribs, and by increasing intraabdominal pressure force the diaphragm upward.

The Thoracic Cavity

The thoracic cavity is bounded by the thoracic wall, the diaphragm, and superiorly by the **supra-pleural membrane,** a layer of fascia that arches over each lung a little above the thoracic inlet.

On removing the anterior part of the thoracic wall during a dissection, it becomes apparent that there are two laterally lying spaces each containing a lung and its surrounding pleura, and a midline region in which the most prominent structures are the heart and great vessels. The midline region, or **mediastinum,** although firm and rigid in the cadaver, is extremely mobile in life. It changes shape with respiratory and cardiac movement and may become displaced by abnormal contents or pressure in the adjacent **pleural cavities.**

The Pleural Cavities

The pleural cavities surround each lung and it is of value, at this point, to understand how the serous membranes or **pleurae** that form these cavities are arranged around each lung in such a way that the cavities themselves remain empty. A similar relationship will be found between the heart and the serous pericardium, the intestines and the peritoneum, and a tendon and its synovial sheath. If, as in Fig. 4-15, a rather large but soft balloon can be imagined, plunge a lung into one side so that it becomes completely enveloped by a double layer of balloon except at the point at which the lung is held. The outer layer of the invaginated balloon is the parietal pleura lining the wall of the thoracic cavity, the superior surface of the diaphragm, and the mediastinum; the inner layer becomes the visceral pulmonary pleura covering the lung. Around the point at which the lung is held, the two layers become continuous with each other. This region is known as the root of the lung and allows for the passage of vessels and bronchi between the lung and the mediastinum. The balloon or pleural cavity remains empty except for a film of tissue fluid which assists in reducing the friction between the opposed smooth and shiny surfaces of pleura.

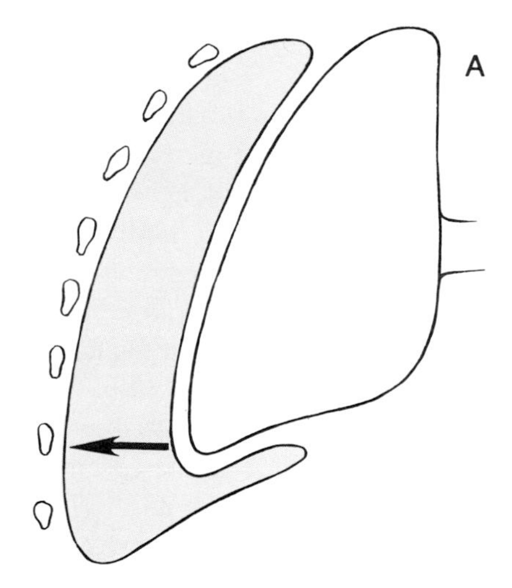

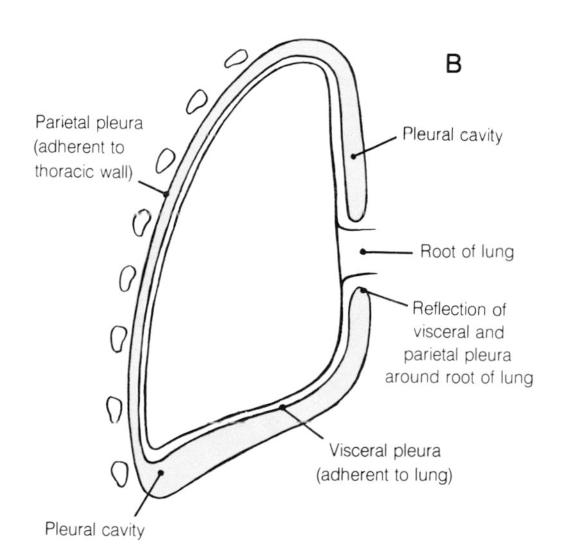

4-15 Diagrams illustrating how the pleural cavity envelops the lung.

Abnormal conditions within the pleural cavity give rise to many of the signs elicited during an examination of the chest. Inflammation of the opposed pleural surfaces gives rise to audible and sometimes palpable friction. Fluid within the cavitiy produces a dullness on percussion and a reduction of breath sounds on auscultation while air produces a resonance and an absence of breath sounds.

The Pleura

The **parietal pleura** lines the thoracic wall (**costal pleura**) and covers the superior surface of the diaphragm (**diaphragmatic pleura**) and the mediastinum (**mediastinal pleura**). At the thoracic inlet it arches over the lung as the **cervical pleura** or **cupula.** Here it is reinforced by a sheet of connective tissue called the **suprapleural membrane.** Because of the inclination of the thoracic inlet the cervical pleura lies above the level of the anterior part of the first rib and the clavicle. The parietal pleura is reflected at the root of the lung to become continuous with the pulmonary pleura. At this point it sags below the hilus, as a cuff hangs below a wrist, forming an unfilled, but potential space. This double layer of pleura is called the **pulmonary ligament.**

At the hilus and pulmonary ligament the pleura becomes the **visceral pleura** as it is reflected onto the surface of the lung to which it is closely adherent. It is invaginated into the fissures that divide each lung into lobes but at these sites the space between the two invaginating layers may be obliterated as a result of inflammatory disease.

In life the lungs do not extend to the limits of the pleural cavity even in a maximal inspiration. As a result, the surface markings of lung and parietal pleura do not coincide. This fact is of practical importance when an exploratory needle is introduced. However, the apex of each lung lies immediately adjacent to the cervical pleura and as a result, perforating wounds of the neck above the medial two-thirds of the clavicle should be assumed to have penetrated the pleural cavity and lung until shown otherwise. Where the parietal pleura is reflected off the diaphragm onto the thoracic wall a recess is formed that is never completely filled by the lung. This is known as the **costodiaphragmatic recess.** Similar recesses are found between the mediastinum and the diaphragm, and between the mediastinum and the thoracic wall at the anterior margin of the pleural cavity. These are the **phrenicomediastinal** and **costomediastinal recesses.** The surface anatomy of the pleural cavities which illustrates these points will be described with that of the lungs.

The parietal pleura, in common with the thoracic wall, is supplied with blood by intercostal arteries and branches of the internal thoracic artery and its venous and lymphatic drainage is similarly shared. It is innervated by intercostal nerves and the phrenic nerve (mediastinal and diaphragmatic pleura), and is sensitive to pain which may be referred to the thoracic and abdominal walls or in the case of the mediastinal pleura to the neck and shoulder, i. e., regions of skin whose segmental innervation is shared with the phrenic nerve.

The pulmonary pleura is supplied by bronchial arteries and its venous and lymphatic drainage is common with that of the lung. Its innervation is from the autonomic system and it is not sensitive to pain.

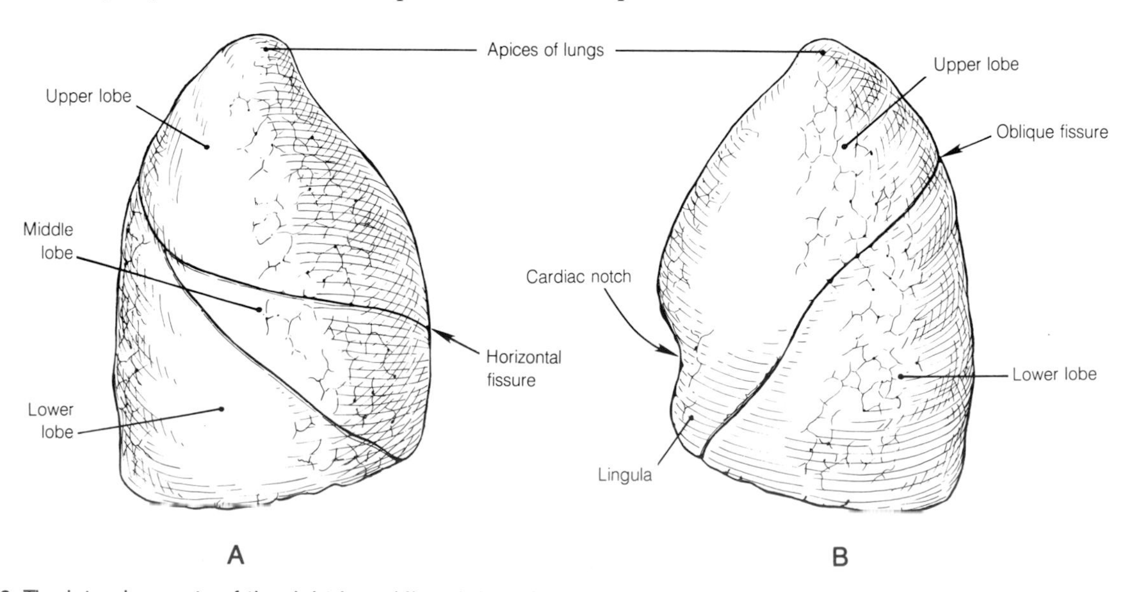

4-16 The lateral aspects of the right lung (*A*) and the left lung (*B*).

The Lungs

The lungs (pulmones) lie on either side of the mediastinum and are surrounded by the pleural cavities. In life they are light, spongy and will float in water. Their surface is mottled, the coloration varying from pink to black depending upon the amount of carbon particles deposited there. After fixation they are found to be heavier (with perfusion fluid) and more firm. As a result of this firmness they can be seen, after removal, to retain impressions of adjacent organs. The left lung is partially deficient anteriorly where it overlies the heart and pericardium. This deficiency is known as the **cardiac notch.**

Lateral views of the left and right lungs are shown in Fig. 4-16. Each can be seen to be divided by an **oblique fissure** into a more anterior **upper lobe** and a more posterior **lower lobe.** On the right side a **horizontal fissure** extends laterally to

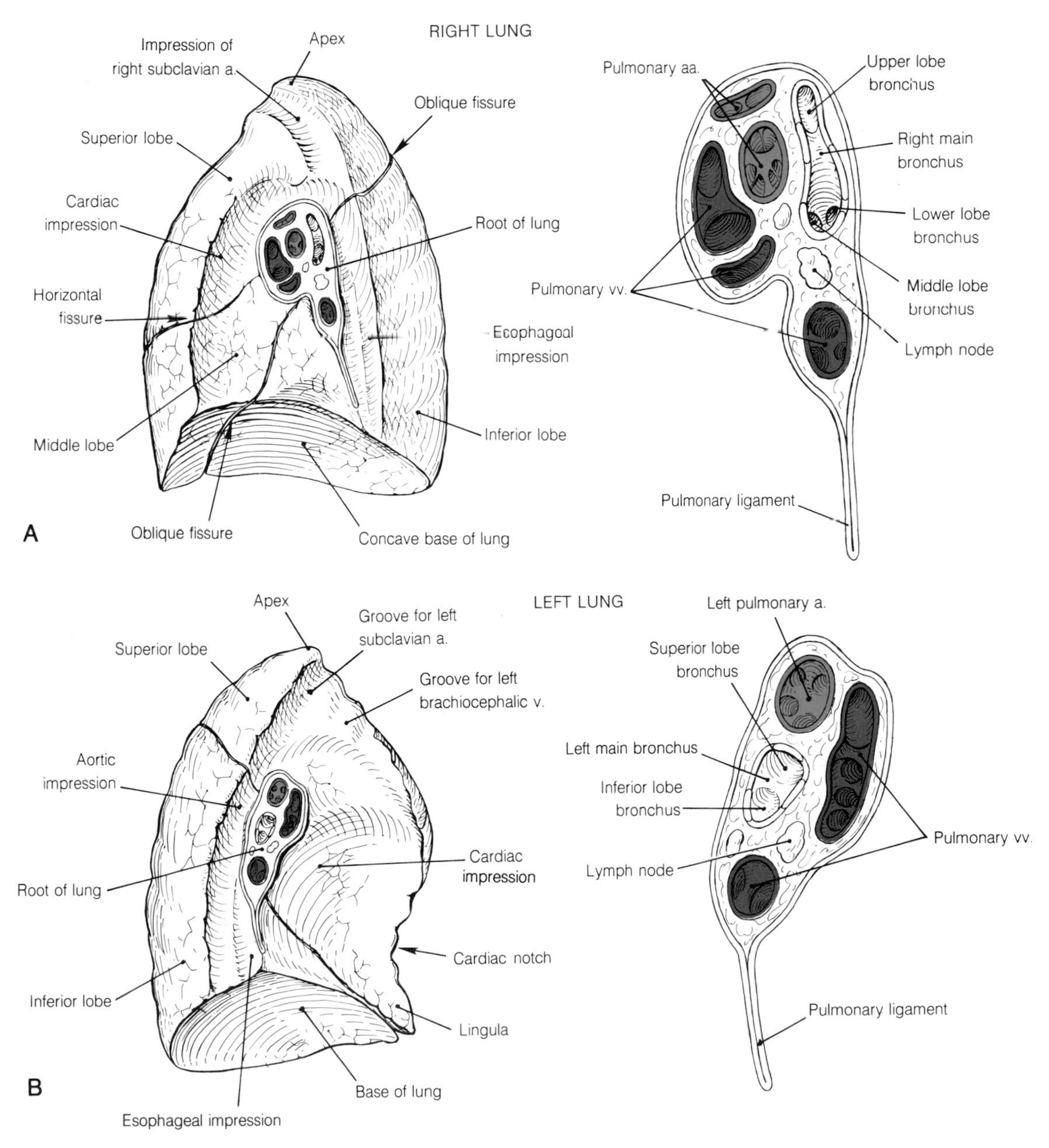

4-17 The medial aspect of the right lung (*A*) and the left lung (*B*). Detail of the root of each lung is alongside. The pulmonary veins which contain oxygenated blood are indicated in red and the pulmonary arteries which contain reduced blood are indicated in blue.

meet the oblique fissure and forms a **middle lobe** which lies between the upper and lower lobes. The outer surface of the lung is convex and follows the shape of the thoracic wall. The apex of each lung is rounded but the base is concave and molded to the dome of the diaphragm. Each anterior border is sharp and that of the left lung has a shallow notch, the cardiac notch. Below this, the **lingula*** forms the lowest part of the left upper lobe.

In Fig. 4-17, the lungs are seen from their medial aspects. Note again the rounded apex, concave base, and the fissures separating the lobes. The hilus lies centrally and is surrounded by impressions made by the mediastinal structures to which the lung has been applied. Of these identify the deep left and shallow right **cardiac impression,** grooves formed by the great vessels and on the right by the esophagus. The structures entering and leaving the lung at the hilus are known as the root of the lung.

The Roots of the Lungs

The roots of the lungs which are also illustrated in Figs. 17A and B should now be examined. Each is surrounded by the cut edge of the pleura as it is reflected off the lung onto the mediastinum. The extension of this cut edge below the root that forms the **pulmonary ligament** can now be seen. Within this boundary lie the **pulmonary arteries, pulmonary veins, bronchi, bronchial vessels,** and **lymph nodes.** The arteries, veins, and bronchi branch in the root and the number of cross-sections of each may vary depending upon the level of the section. The **two pulmonary veins** (depicted in *red* because they carry oxygenated blood) lie anteriorly and inferiorly. In the left root the **pulmonary artery** lies most superiorly and above the **left bronchus.** On the right, the **pulmonary artery** lies between the vein and the bronchus. The latter branches early, sending an "eparterial" bronchus to the upper lobe. The lymph nodes (which are often quite black in city dwellers) form part of the **bronchopulmonary** group of lymph nodes.

* Lingua, *L.* = tongue; lingula is the diminuative, i.e., little tongue

The Bronchial Arteries

The small bronchial arteries supply the tissue of the lungs, the bronchial tree and the pulmonary pleura.

Two left bronchial arteries usually arise directly from the thoracic aorta near the level of the left main bronchus. A single right bronchial artery arises from the third right posterior intercostal artery. The bronchial veins terminate either in pulmonary veins or in the azygos system of veins.

Advances in the medical and surgical treatment of pulmonary disease coupled with the more accurate localization of lesions has increased the need for a detailed knowledge of the internal structure of the lung, including the distribution of the bronchi and blood vessels.

The Bronchi

The main bronchi have been seen entering each lung at its root. Both are terminal branches of the **trachea** which has passed into the upper part of the mediastinum from the neck. In structure, they are fibromuscular tubes reinforced by incomplete cartilaginous rings or plates and lined by respiratory mucous membrane. Within the lung they divide in an orderly manner into secondary or **lobar bronchi** for each lobe. These, in turn, divide into tertiary or **segmental bronchi** running to discrete regions of lung known as **bronchopulmonary segments.** The details of the bronchial tree can be followed in Fig. 4-18.

The **right main bronchus** is short and diverges little from the trachea. For this reason inhaled foreign bodies are more likely to pass into the right lung. As it enters the lung a secondary bronchus for the upper lobe (eparterial bronchus) leaves it. Continuing through the lung the primary bronchus terminates by dividing into secondary bronchi for the middle and lower lobes. The upper lobe bronchus divides into three tertiary bronchi, the middle into two, and the lower into five.

The **left main bronchus** is longer than the right and leaves the trachea at a distinct angle. Passing into the left lung it divides into secondary bronchi for the upper lobe (including the lingula) and the lower lobe. Typically, the upper lobe bronchus divides into three tertiary bronchi for the upper lobe and two for the lingula. The lower lobe bronchus divides into five tertiary bronchi.

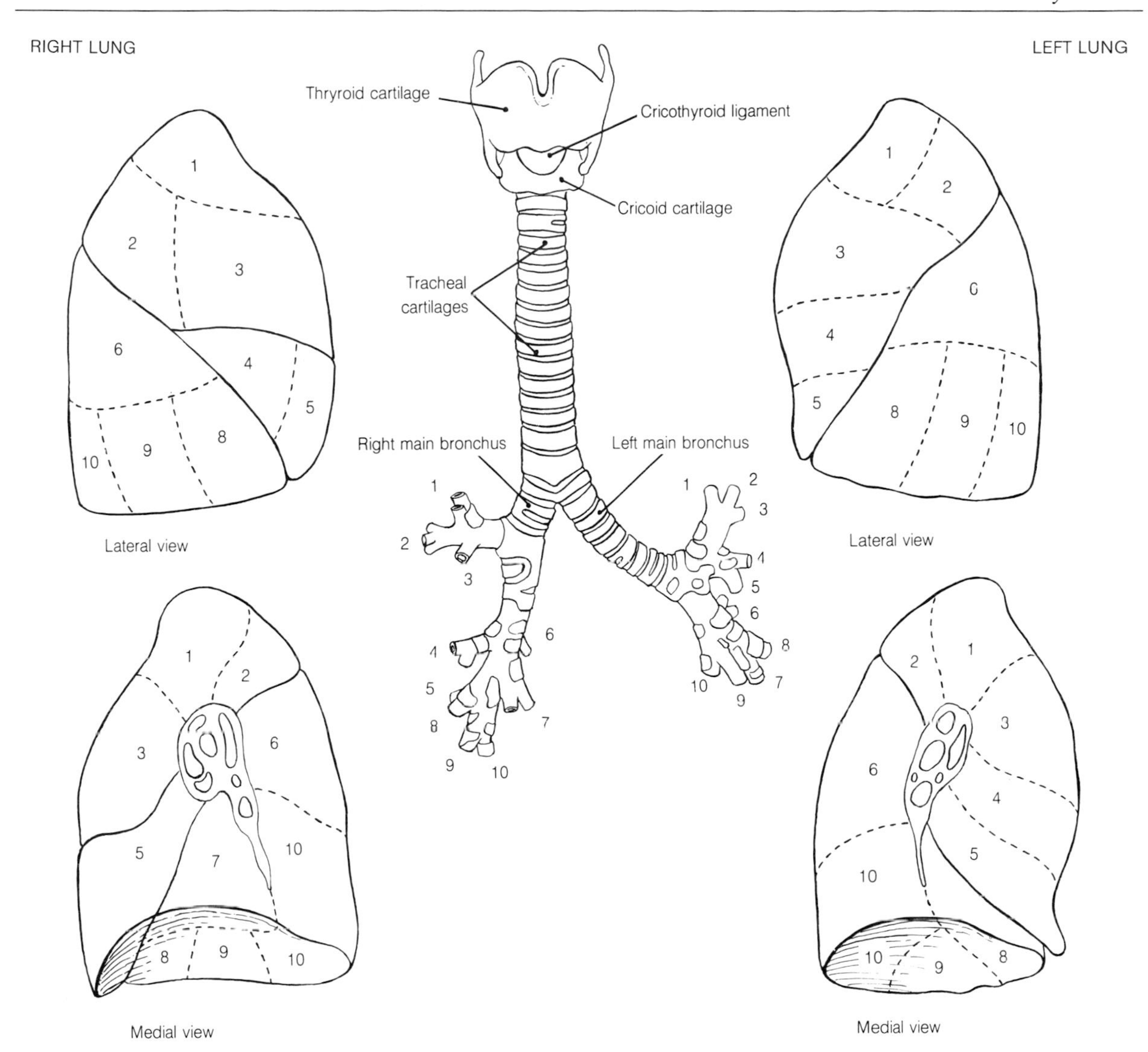

4-18 The trachea and the distribution of the main, lobar, and segmental bronchi. The segmental bronchi are numbered to correspond with the outlines of the bronchopulmonary segments seen on the accompanying lateral and medial views of the lungs. The left medial basal segment (7) is frequently absent.

The Bronchopulmonary Segments

Each tertiary bronchus and an accompanying branch of the pulmonary artery supplies a bronchopulmonary segment. This is an independent functional unit of pulmonary tissue that can be defined radiographically, and on occasion, excised surgically. Its location and relevant bronchus is also of interest to the physical therapist who may be required to position the body so as to drain a particular segment of lung. In Fig. 4-18, the surface projections of the segments are shown and numbered to correspond with the numbered tertiary bronchi.

Within a bronchopulmonary segment, tertiary bronchi repeatedly subdivide and become reduced in caliber. At a point where the cartilage in their walls is no longer present, the air passages are known as bronchioles and these, after further division, end by breaking up into the terminal bronchioles of a lung lobule. The ultimate distribution of these bronchioles to the alveoli where respiratory exchanges occur is shown in Fig. 4-19.

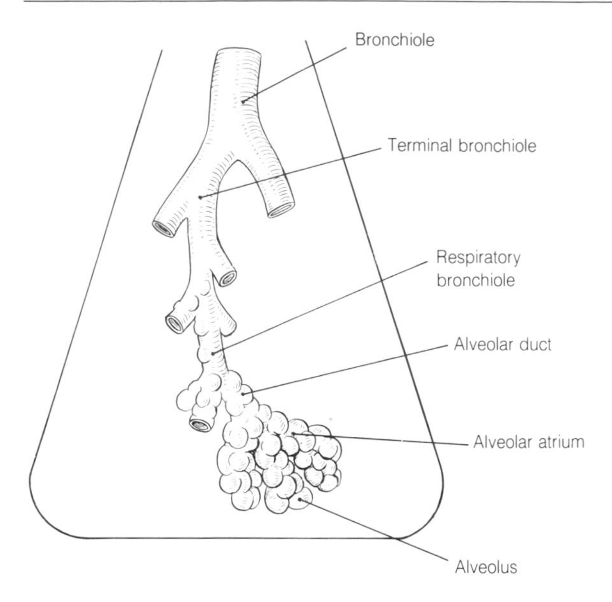

4-19 Diagram of the distribution of a bronchiole in a lobule of the lung.

The Pulmonary Arteries and Veins

The **right pulmonary artery** divides early in the root of the lung where it gives off a branch to the superior lobe; its continuation supplies the middle and lower lobes. The **left pulmonary artery** supplies branches to the upper lobe and lingula and thereafter to the lower lobe. In both lungs these branches follow bronchi to supply the bronchopulmonary segments.

A **superior and inferior pulmonary vein** are formed at the hilus of each lung. The tributaries of the superior veins arise from the upper and middle lobe on the right and the upper lobe and lingula on the left. The tributaries forming the inferior veins arise from the lower lobes. These tributaries do not conform to the pattern of the bronchopulmonary segments and many are intersegmental and drain adjacent segments.

The Nerve Supply of the Lungs

The lungs are supplied by both the sympathetic and parasympathetic nervous systems and branches from the **anterior and posterior pulmonary plexuses** lying in front of and behind the roots of the lungs supply the smooth muscle of the bronchial tree, the vessels, and the mucous membrane. The muscles of the bronchi are supplied by parasympathetic fibers which, on stimulation, produce bronchial constriction, and by sympathetic fibers which, on stimulation, produce bronchial relaxation.

The Lymphatic Drainage of the Lungs

The currently high incidence of cancer of the lung places extreme importance on lymphatic drainage. The lung is drained by a **superficial lymphatic plexus** lying beneath the pulmonary pleura and a **deep plexus** that follows the bronchial tree. The two plexuses communicate at the periphery and finally unite at the hilus where they end in the **bronchopulmonary group of lymph nodes** at the root of the lung. Efferents from these pass successively through the **tracheobronchial and paratracheal nodes** and eventually empty into the **left or right subclavian vein** through the **left and right bronchomediastinal trunks.** The disposition of these almost continuous groups of nodes is seen in Fig. 4-20. The parietal pleura is drained by vessels running in the intercostal spaces to parasternal or intercostal lymph nodes.

The Examination of the Lungs

A number of techniques are available for examination of the bronchial tree and the lungs. A limited direct examination of the trachea and bronchi can be made using a bronchoscope, and CT scanning now allows the accurate localization of intrapulmonary lesions. However, a radiograph of the chest can provide a wealth of information and may, perhaps, be the only feasible imaging technique possible in a very ill patient. Fig. 4-21 is an example of a normal chest radiograph. The subject has been positioned so that the film is against the chest and the X-rays are penetrating the body from behind. This is therefore a p. a. (posteroanterior) projection and is the one commonly used. Note that the subject is well centered as the medial ends of the clavicles (*C*) each lie just outside the outline of the vertebral column. Follow the outlines of the ribs starting at that of the first rib (*1*). Note that the posterior parts of the shafts are horizontal. This means that, as is usual, the radiograph has been taken in inspiration. The costal cartilages are translucent and are not seen here, but calcification, which is radiopaque, may occur with age. Now follow the outline of the diaphragm (*D*) from the right costodiaphragmatic recess

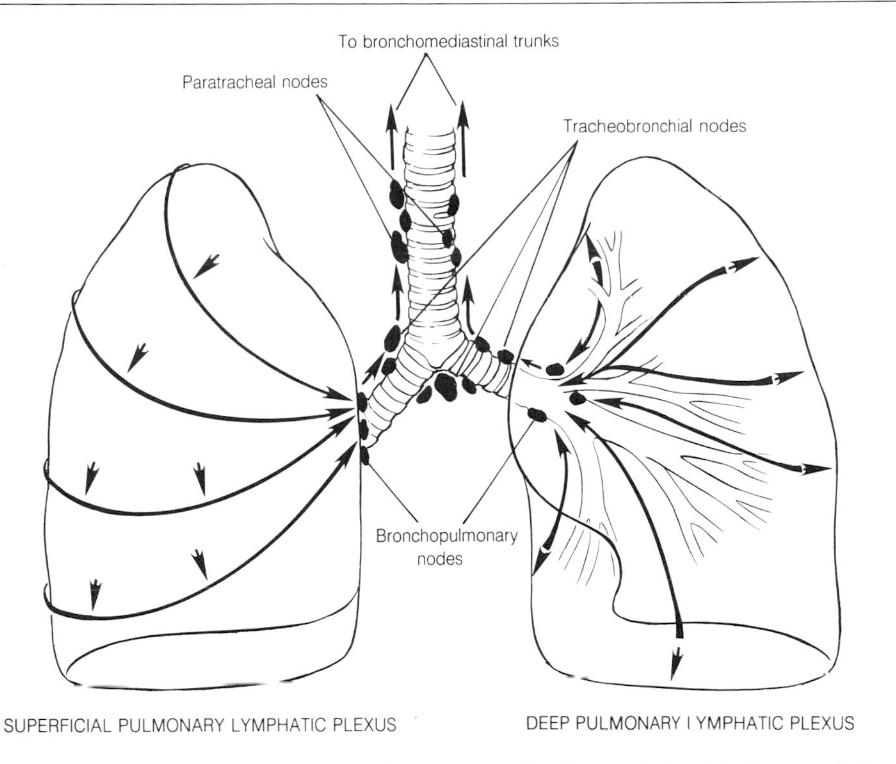

4-20 Diagram illustrating the drainage of the superficial lymphatic plexus of the right lung and the deep lymphatic plexus of the left lung.

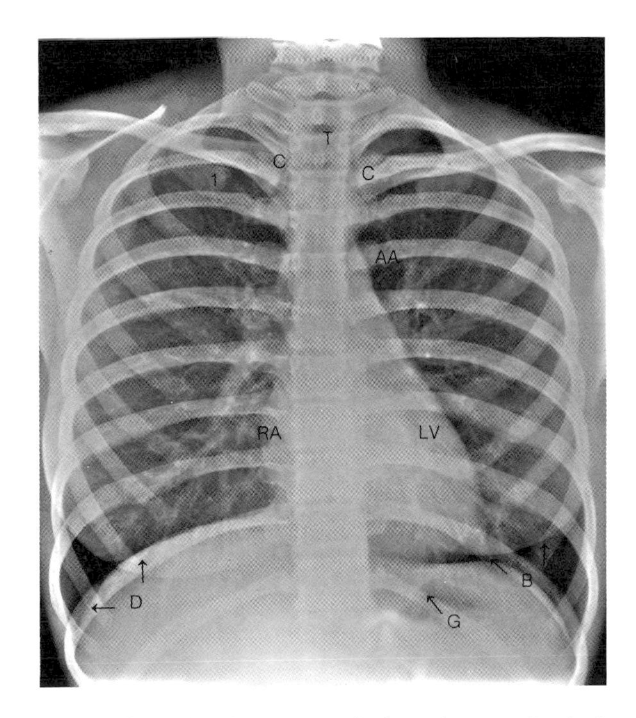

4-21 Radiograph of the normal chest (p.a. projection). Reproduced by permission, from Wicke: *Atlas of Radiologic Anatomy* 4th Ed, Urban & Schwarzenberg, Baltimore–Munich, 1987.

across the dense shadow of the liver, with which it blends, to the right margin of the heart. The level of the right dome is low because of the inspiration but is still higher than the left which is often more clearly outlined by gas (*G*) in the underlying stomach. The upper part of the mediastinal shadow shows a midline translucency (*T*) formed by the air in the trachea; the lower part is formed by the heart. For the present, note the outline of the right atrium (*RA*), the left ventricle (*LV*) and the aortic arch (*AA*). On either side of the heart the hilar shadows are seen. These are produced largely by the pulmonary arteries and veins but the translucency of air in a bronchus seen in transverse section can sometimes be recognized. Outside the hilar shadows are the lung fields which have an overall translucency, but superimposed upon this is a branching network of fine vascular markings radiating from the hilus. In many good films a fine line can be seen extending from the right hilus to intersect the sixth rib. This marks the horizontal fissure between the right upper and middle lobes. Finally, note the sometimes confusing outlines of the breast shadows (*B*).

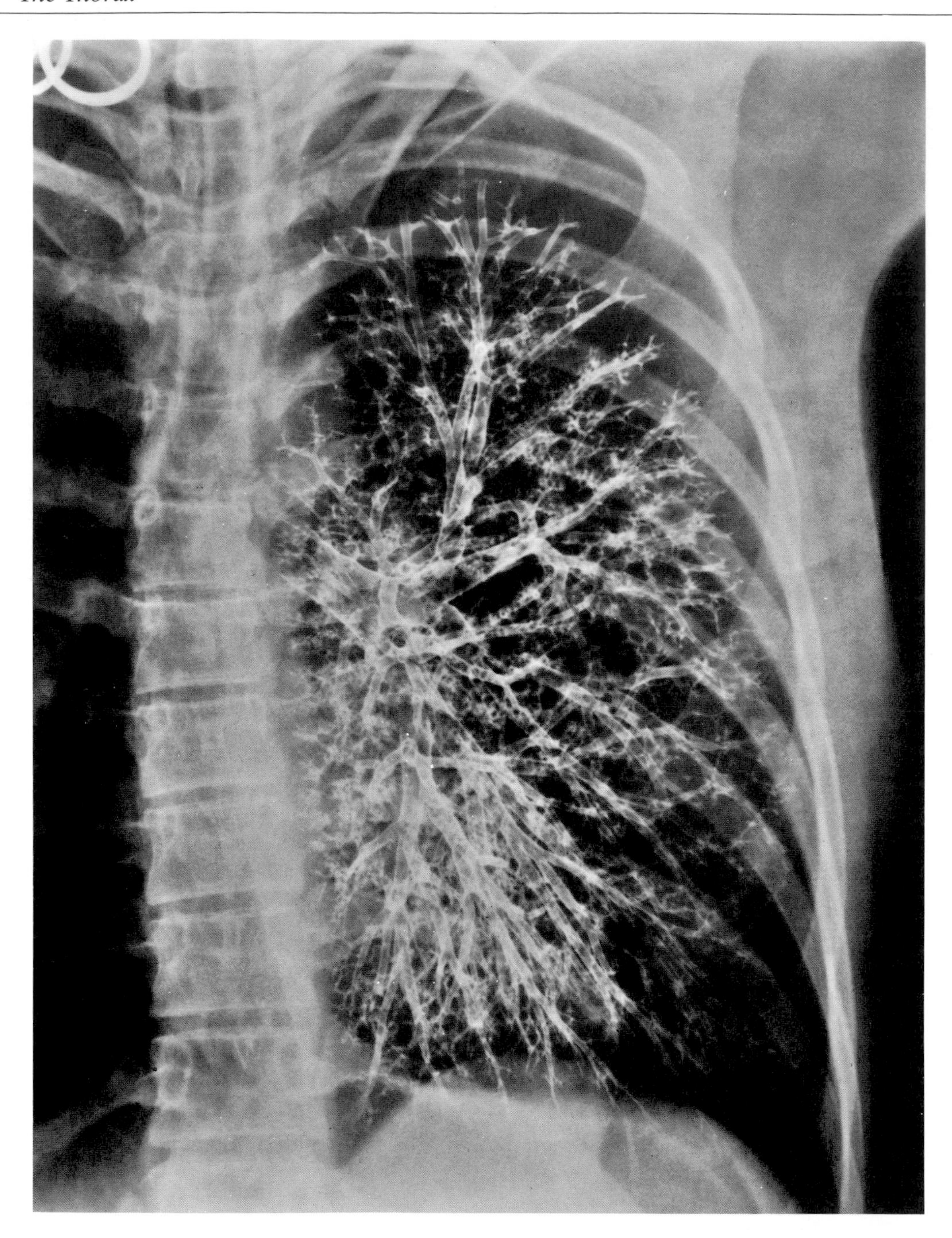

4-22 Normal left bronchogram (a. p. projection). (Reproduced by permission, from Wicke: *Atlas of Radiologic Anatomy*, 4th Ed, Urban & Schwarzenberg, Baltimore–Munich, 1987.)

A more complete radiographic picture can be gained from oblique and lateral views and arteriograms and bronchograms. In the last procedure an oily radiopaque substance is introduced into the trachea and the subject positioned so that the walls of the bronchi to be examined are coated. An example of a normal bronchogram is given in Fig. 4-22, and should be compared with the diagram of the bronchial tree in Fig. 4-18. In addition, a CT scan of the thorax is shown in Fig. 4-23.

This horizontal section has been made at the level of the sixth thoracic vertebra and passes

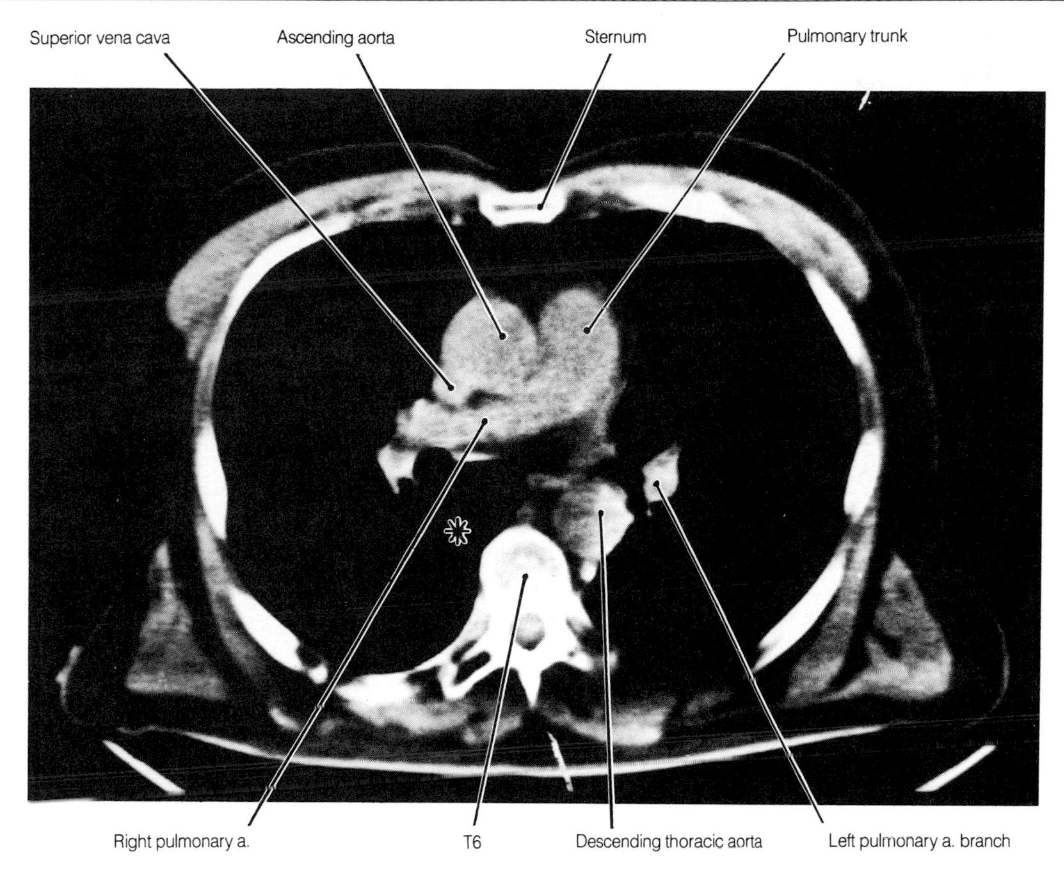

4-23 A CT scan of the thorax made at the level of the sixth thoracic vertebra. (Photograph courtesy of Nancy Whitley, M. D., Dept. of Diagnostic Radiology, University of Maryland School of Medicine.)

through the great vessels just above the level of the heart. Remember that it is presented as if viewed from below and that the right side of the subject appears on the left side of the figure in this scan. Now identify the profiles of the ascending aorta and descending aorta and imagine the aortic arch above the section. Beneath the arch lies the right pulmonary artery which can be seen and followed back to the pulmonary trunk. A branch of the left pulmonary artery appears in the left lung field. The superior vena cava can be identified to the right of and posterior to the ascending aorta. Finally note the area indicated by the *asterisk*. This region is normally filled by lung but it may be encroached upon by mediastinal tumors. It is almost impossible to visualize these tumors through the mediastinal shadow of a radiograph but they can be easily recognized in a CT scan.

The Surface Anatomy of the Lungs

The description of the pleural cavities and lungs now makes it possible to map out in more detail the relationship of these structures to the surrounding thoracic wall. A knowledge of the pleural reflections and the extent to which the parietal and pulmonary pleurae coincide is important in any physical examination of the chest and in the accurate placing of a needle for, say, aspiration or biopsy purposes.

Looking at Fig. 4-24, which illustrates the surface projection of the parietal pleura (*in blue*) and the margins of the lungs and their lobes (*black dashed line*), start with the anterior projection (*A*) at the level of the second costal cartilage. On the right side both the parietal pleura and anterior margin of the lung run together down to the level of the sixth costal cartilage. Here they diverge as

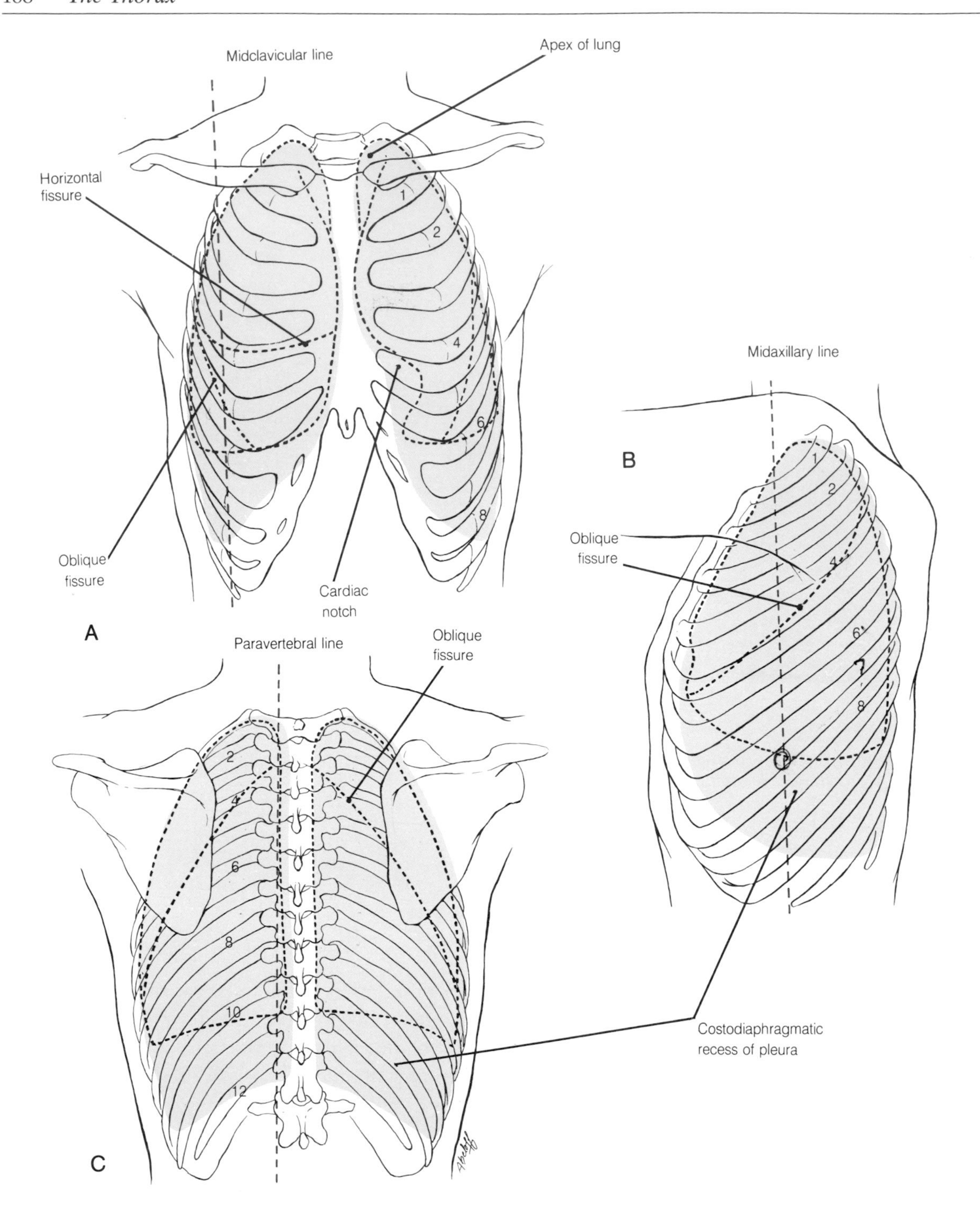

4-24 Surface projections of the lungs (*black dashed lines*) and pleural cavities (*blue*) on the anterior (*A*), lateral (*B*), and posterior (*C*) aspects of the thoracic wall.

the lower margin of the lung crosses the midclavicular line at the sixth costal cartilage and the parietal pleura at the eighth. Moving to the lateral view (*B*), the lower margin of the lung is seen to cross the eighth rib at the midaxillary line and the parietal pleura the tenth. Passing around to the posterior view (*C*), the lower margin of the lung meets its posterior margin at the level of the eleventh thoracic vertebra while the lower margin of the parietal pleura actually passes outside the

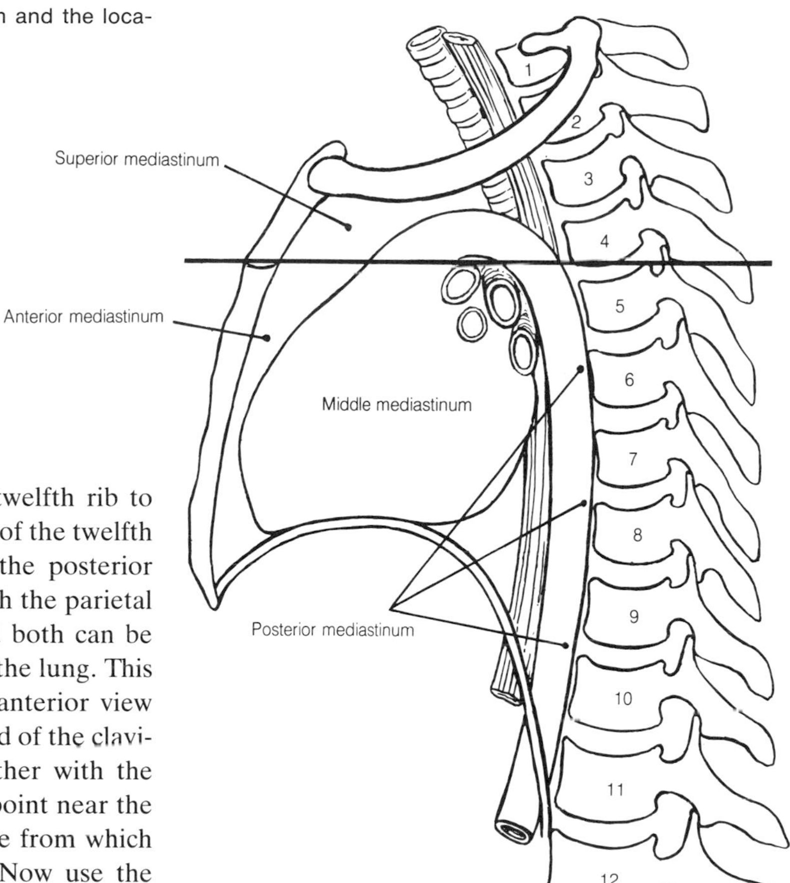

4-25 Diagram illustrating the mediastinum and the location of its four parts.

thoracic cage as it runs below the twelfth rib to meet its posterior border at the level of the twelfth thoracic vertebra. The position of the posterior border of the lung now coincides with the parietal pleura at the paravertebral line and both can be followed up to and over the apex of the lung. This line can be picked up again in the anterior view and seen to lie **above** the medial third of the clavicle. The margins of the lung, together with the parietal pleura, now converge on a point near the midline opposite the second cartilage from which their course was originally traced. Now use the figures to determine how the projection of the lung and pleural cavity on the left differs from that of the right and, in addition, note:

1. that both lung and pleural cavity lie above the level of the clavicle;
2. that on the left side, the lung, and to a lesser extent, the parietal pleura diverge from the midline over the heart, producing the cardiac notch in the lung;
3. that the lower borders of the lung and pleura are widely separated around the costal margin. This represents the extent of the **costodiaphragmatic recess** between parietal pleura lining the thoracic wall and that covering the diaphragm. It is a potential space only, but will be partly filled by the expanding lung in deep inspiration or may contain pathological fluids such as pleural effusions or blood;
4. that the costodiaphragmatic recess extends below the level of the twelfth rib posteriorly and is in danger when the classical incision for exposure of the kidney is made; and
5. that the surface projection of the horizontal fissure lies over the fourth rib and costal cartilage.

The Mediastinum and its Contents

Between the two lungs and their sorrounding pleural cavities lies the region known as the **mediastinum.** It forms a mobile midline septum and contains, among other structures, the heart, great vessels, trachea, esophagus, portions of the thoracic duct and the phrenic and vagus nerves, and lymph nodes. These structures are loosely bound together by connective tissue, a fact of importance when infection spreads either from the neck or from a perforation of the esophagus. An understanding of the close relationships of these structures is also crucial to the interpretation of symptoms because disease of one may initially become manifest through the dysfunction of another; for example, an aneurysm of the aorta may produce a hoarse voice, or a tumor of the lung may cause an eyelid to droop.

As a descriptive and topographical aid and for clinical localization, the mediastinum may be divided into a **superior** and **inferior mediastinum** and the inferior mediastinum further divided into an **anterior, middle,** and **posterior mediastinum.** These regions are illustrated in Fig. 4-25. Note that the superior mediastinum lies above a plane passing through the manubriosternal joint anteriorly and the lower border of the fourth thoracic vertebra posteriorly. The anterior mediastinum lies between the body of the sternum and the pericardium surrounding the heart, the middle mediastinum is the region taken up by the heart and the roots of the great vessels, and the posterior mediastinum lies between the middle mediastinum and the vertebral column. These three portions of the inferior mediastinum are bounded inferiorly by the diaphragm.

The Superior Mediastinum

The horizontal section through the thorax at the level of the body of the fourth thoracic vertebra (seen in Fig. 4-26) illustrates the shape and the contents of the superior mediastinum. Note from before backward: (1) the attachments of two strap muscles of the neck, (2) the remnants of the **thymus,** (3) the left and right **brachiocephalic veins** which will unite to form the **superior vena cava,** (4) the **brachiocephalic trunk** and the **left common carotid** and **left subclavian arteries** close to the **trachea,** (5) the **left vagus** between the latter two arteries and the **right vagus** on the right side of the trachea, (6) the **left recurrent laryngeal nerve** between the trachea and the **esophagus,** (7) the **thoracic duct** immediately to the left of the esophagus, and (8) the many **lymph nodes** packed between these structures.

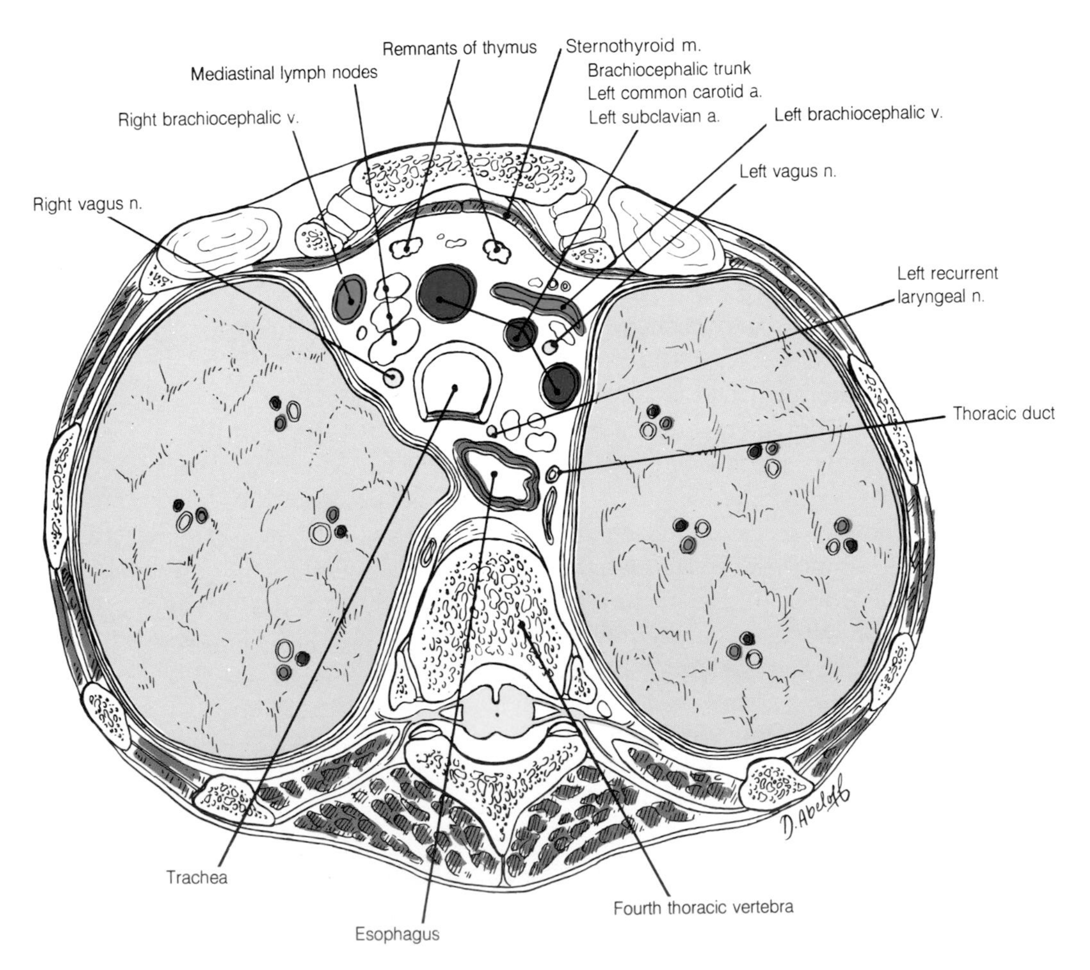

4-26 Horizontal section of the thorax at the level of the fourth thoracic vertebra viewed from below. The section passes just above the arch of the aorta.

The large arteries have arisen from the **arch of the aorta** which, with the upper part of the superior vena cava, formed by the union of the two brachiocephalic veins, lies in the lower part of the superior mediastinum. Only the left recurrent laryngeal nerve is seen as the right nerve leaves the right vagus in the root of the neck to run superiorly to the larynx.

Using Fig. 4-27 the main contents of the inferior mediastinum can be determined. The section shown has been made at the level of the eighth thoracic vertebra and should be referred to as the anterior, middle, and posterior mediastina are described.

The Anterior Mediastinum

This narrow portion of the inferior mediastinum lies between the body of the sternum and the pericardium. Its only contents of note are **lymph nodes** which may be involved in the metastatic spread of tumors or the site of a lymphoma. However, the region may also become invaded by an enlarging tumor of the thymus or by a teratodermoid tumor. The **costomediastinal recess** is also well shown in this section.

The Middle Mediastinum

The middle mediastinum is largely filled by the **heart** and the surrounding **pericardium,** but also contains the **ascending aorta,** the **pulmonary trunk** and its division into **left and right pulmonary arteries,** portions of the **superior vena cava** and **pulmonary veins** as they enter the heart, the termination of the **azygos vein,** the **phrenic nerves,** the bifurcation of the **trachea,** and the **left and right main bronchi.** This is a formidable list of important structures and gives some idea of the possible gravity of perforating wounds of the

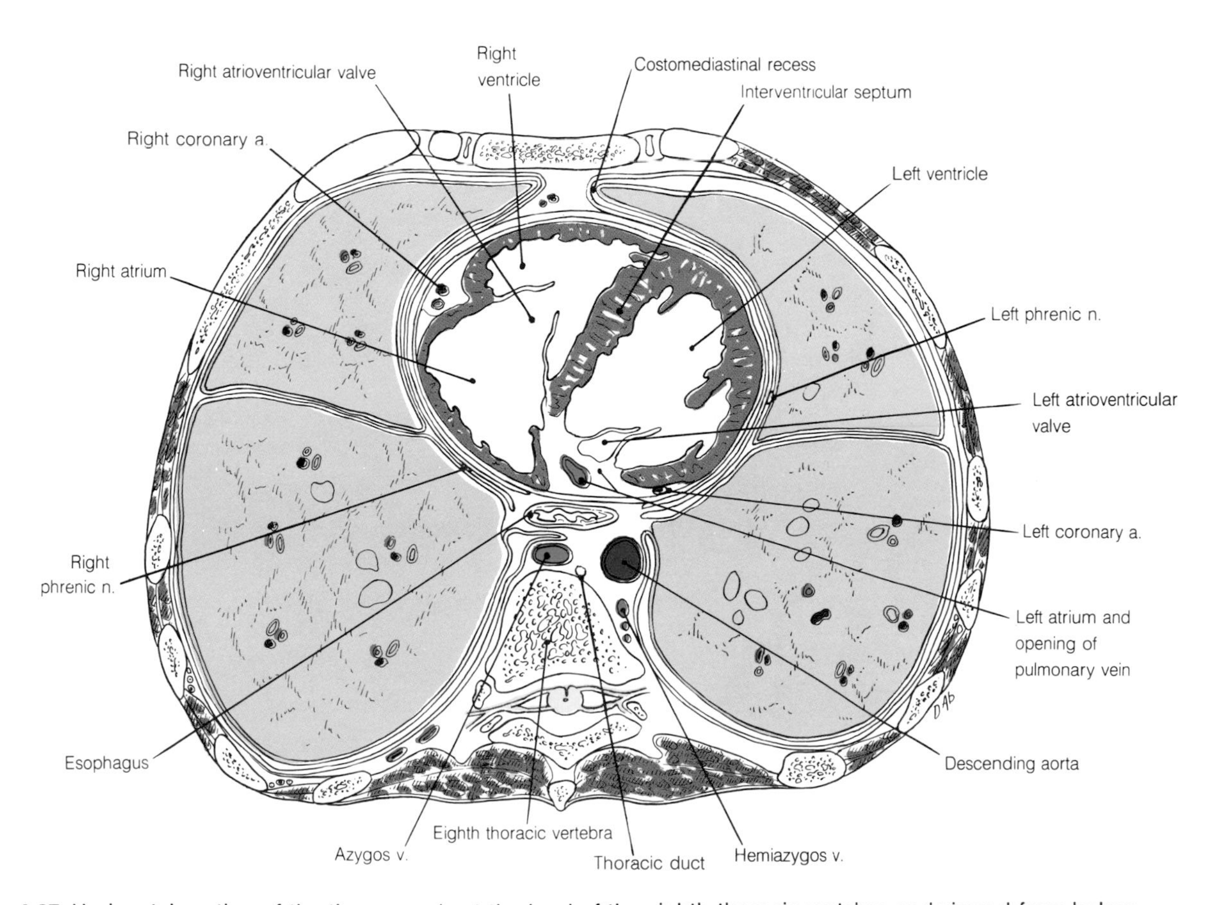

4-27 Horizontal section of the thorax made at the level of the eighth thoracic vertebra and viewed from below.

chest. The description of each structure will, however, turn the list into a logical grouping of closely associated organs, vessels, and nerves.

In the cross-section note the four chambers of the heart and the **left and right coronary arteries** within the pericardium and the **phrenic nerves** between the pericardium and mediastinal pleura. The **pulmonary veins** can be seen entering the left atrium. These lead from the root of each lung where sections of **secondary bronchi, branches of the pulmonary arteries,** and **lymph nodes** can be identified.

The Posterior Mediastinum

The posterior mediastinum lies below the posterior part of the superior mediastinum and behind both the middle mediastinum and the sloping posterior part of the diaphragm. The majority of the structures lying in it are oriented in a vertical manner and run through its whole extent.

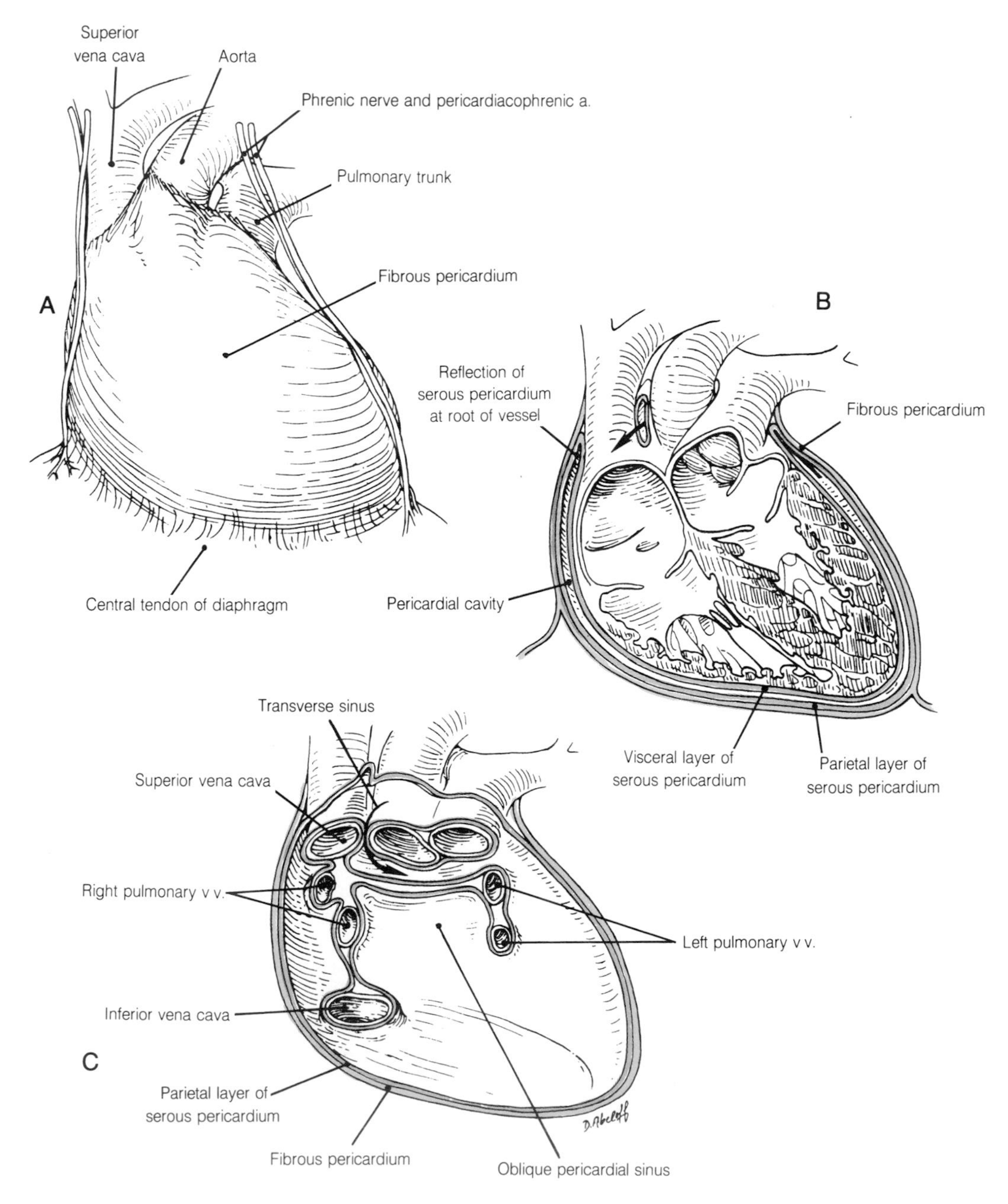

4-28 Illustrating the fibrous pericardium (*A*), the serous pericardium (*B*), and the pericardial sinuses (*C*).

In the section pick out the thick-walled **esophagus** with the **descending aorta** behind and to its left. Between the aorta and the vertebral body the **thoracic duct** and the **azygos* and hemiazygos veins** can be seen.

This brief summary of the contents of the mediastinum illustrates how many of these structures pass through more than one of its parts. For this reason the more detailed descriptions of individual structures are given over their entire course.

The Pericardium

The Fibrous Pericardium

The heart is surrounded by a tough fibrous sac known as the fibrous pericardium and this is illustrated in Fig. 4-28A. At the roots of the great vessels this sac blends with the connective tissue or tunica adventitia covering them. Elsewhere, the heart is free to move inside it. The sac is also fused to the **central tendon of the diaphragm.** On its surface lie the **phrenic nerves** and the **pericardiacophrenic branches** of the **internal thoracic arteries.** where it is adjacent to the pleural cavities, it is covered by parietal pleura. Constriction of the fibrous pericardium by inflammatory processes can severely impede the filling of the heart.

The Serous Pericardium

The fibrous pericardium is separated from the wall of the heart by the serous pericardium. Examine Fig. 4-28B and see that the serous pericardium, like the pleura, is arranged as an outer **parietal layer** which lines the inner surface of the fibrous pericardium and an inner **visceral layer** which covers the heart. At the roots of the great vessels the two layers become continuous with each other and thus form the closed and empty **pericardial cavity.** A potential space is, however, present and this may become filled with blood or other fluids. As can be imagined, the fibrous pericardium will not distend rapidly and such collections of fluid within the pericardial cavity may interfere with the function of the heart.

* Azygos, *Gk.* = a, without; Zygon, yoke; hence, unpaired

If this arrangement is now clear, look again at Fig. 4-28B and see that the serous pericardium is carried up as a separate sleeve over the superior vena cava on the one hand and the aorta and pulmonary trunk on the other. This leaves a communication of the pericardial cavity between the two sleeves called the **transverse pericardial sinus.** Its position is indicated by the *arrow*.

Now look at Fig. 4-28C and see that when the heart and visceral pericardium are removed the sleeve of pericardium that encloses the superior vena cava also encloses the openings of the inferior vena cava and the pulmonary veins. The irregular outline of this sleeve surrounds a portion of the pericardial cavity called the **oblique pericardial sinus.**

The Heart

Although apparently a single organ, the heart is functionally a pair of muscle pumps linked to each other by the pulmonary circulation. The right pump is fed by the venous return which reaches it by way of the superior and inferior venae cavae. The left pump empties into the aorta.

The heart consists of four chambers, a **right atrium and ventricle** (the right pump) and a **left atrium and ventricle** (the left pump). It is important to be able to orient these so as to understand their relationship to the great vessels reaching and leaving the heart and also their projection onto the body surface.

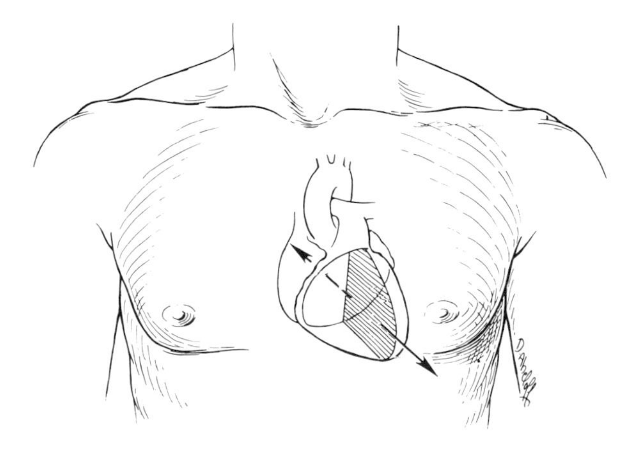

4-29 Showing the orientation of the heart and the vertical plane separating its left and right sides.

The vertical plane between the left and right sides of the heart lies obliquely in the body. As can be seen in Fig. 4-29, the result of this orientation is that the anterior or **sternocostal surface** of the heart is composed almost entirely of the right atrium and ventricle and only a small part of the left ventricle appears at its left-hand margin. In Fig. 4-30 follow the margin of the heart from the superior vena cava on the right to the left auricle (a portion of the left atrium) on the left. The right border is formed by the right atrium, the inferior, by the right and left ventricles, and the left by the left ventricle and left auricle. Where the inferior and left borders meet there is a blunt point known as the **apex.** As will be seen later, a knowledge of the structures forming the margins of the heart is essential for the interpretation of the shadow it produces in a radiograph.

The inferior or **diaphragmatic surface** of the heart rests on the diaphragm and is made up of portions of the walls of the two ventricles. The posterior aspect or **base of the heart** is illustrated in Fig. 4-31, and can be seen to be formed almost entirely by the left atrium but also a small part of the right atrium. The separation between the two ventricles is marked on the surface of the heart by the **anterior and posterior interventricular sulci** which meet just to the right of the apex. The sulci, however, are normally obscured by their content of fat and vessels. Posteriorly the atria are separated by a shallow and less obvious **interatrial sulcus.** The point at which the interventricular and interatrial sulci cross the coronary sulcus is known as the **crux.** The ventricles are separated from the atria by the **coronary sulcus.** It is worth pointing out at this point that the word coronary is derived from the Latin word *corona,* which means a crown and, hence, the fanciful name for this sulcus. The coronary arteries are so named for their relationships to this sulcus rather than to the heart they supply. The confusion perhaps arises from the fact that the Latin word for the heart is *cor* but the stem is *cord-* (see later "venae *cordis* minimae"). For some reason the Greek word for heart (*kardiakos*) has passed into the English language in preference to the Latin.

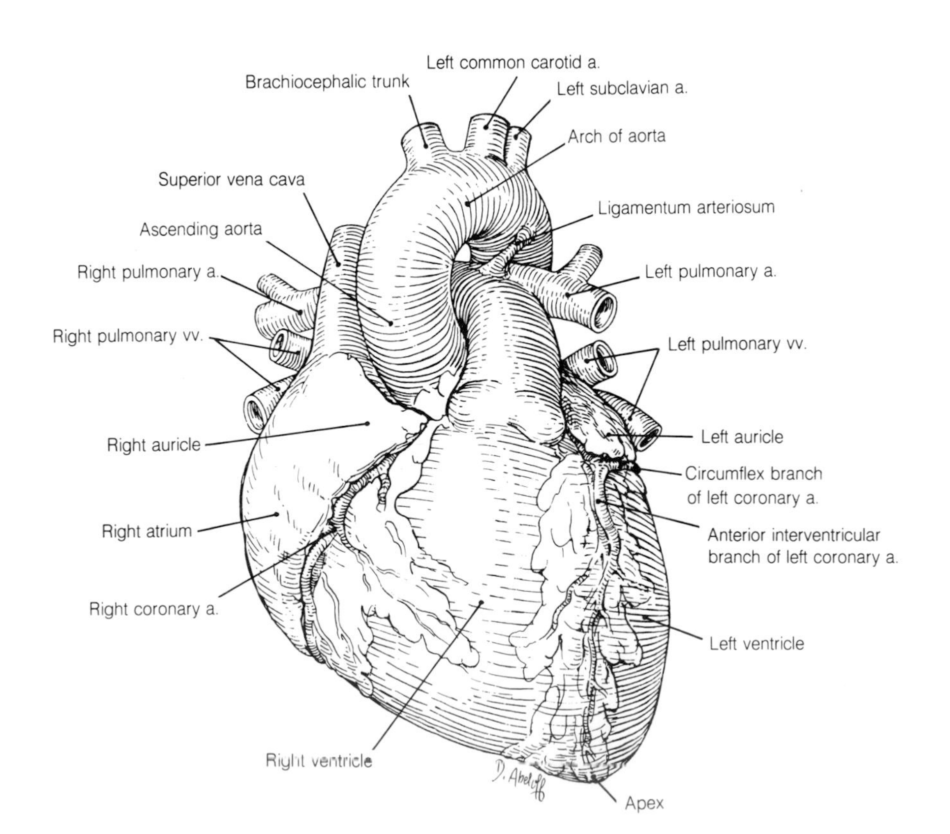

4-30 Anterior view of the heart and great vessels.

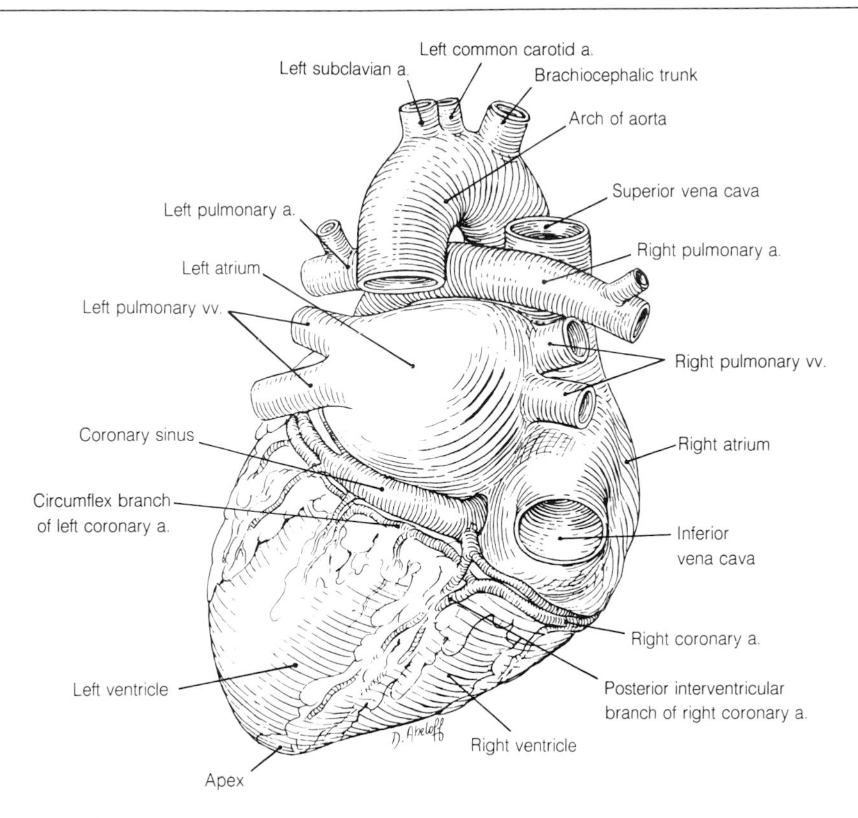

4-31 Posterior view of the heart and great vessels.

Using the same illustrations, the relationships of the great vessels to the chambers can be seen. In the anterior view, the superior vena cava empties superiorly into the right atrium and the pulmonary trunk leaves the right ventricle. The position of the ascending aorta is misleading but its origin from the left ventricle is concealed behind the pulmonary trunk. The posterior view of the heart shows the four pulmonary veins draining into the upper part of the left atrium and the inferior vena cava emptying into the right atrium.

The Structure of the Heart

The greater part of the wall of the heart is made up of cardiac muscle fibers and is called the **myocardium.** As has been already seen, it is covered externally by the visceral layer of the serous pericardium and this, together with a thin subserous layer of connective tissue, is the **epicardium.** The chambers of the heart are lined by the **endocardium,** which also covers the valves and is continuous with the endothelium and underlying connective tissue of the vessels entering and leaving the chambers.

Within the wall of the heart is a connective tissue skeleton. This is composed of four firmly connected rings of fibrous tissue, one around each atrioventricular and arterial orifice. This provides a relatively rigid attachment for the valves and the myocardium.

The myocardium is made up of two separate, rather complex systems of spiraling and looping bundles of fibers, one for the atria and one for the ventricles. The two systems are nowhere continuous with each other, hence, the need for a specialized atrioventricular conducting system to bridge the electrical gap.

The Chambers and Valves of the Right Side of the Heart

The thin-walled **right atrium*** receives the venous return from the systemic circulation through the superior and inferior venae cavae which open into its posterior part, superiorly and inferiorly. No

* Atrium, *L.* = a hall or lobby

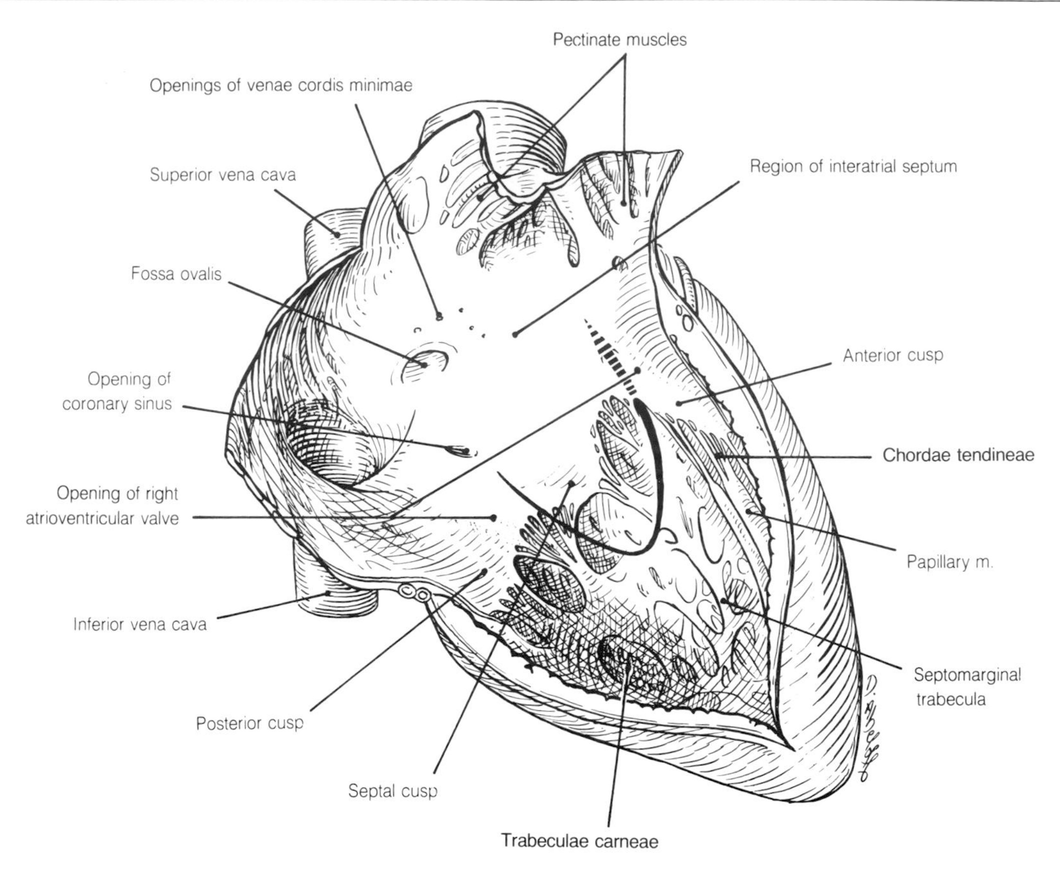

4-32 The right side of the heart opened to show the internal appearance of the right atrium and ventricle.

valves restrict the flow of blood in or out of these openings in the adult, although in the fetal heart, valve-like projections from the wall do serve to direct the flow of blood.

When the right atrium is opened, as is demonstrated in Fig. 4-32, the posterior wall receiving the venae cavae is seen to be smooth, whereas the wall of the anterior portion, including the auricle* is thrown into ridges by thickenings in the underlying muscle. These thickenings are called **pectinate* muscles** after their resemblance to a comb. The junction between these two areas is called the **crista terminalis** and its position can be identified on the outer surface of the right border of the atrium by a shallow groove, the **sulcus terminalis.** The practical importance of the latter structure is that its superior end surrounds the **sinuatrial node** or "pacemaker" of the heart.

The wall of the right atrium that separates it from the left is called the **interatrial septum** and it shows a shallow oval depression, the **fossa ovalis.** This indicates the site of the flap-like valve that allowed venous blood from the inferior vena cava to enter the left atrium and bypass the pulmonary circulation in fetal life. It sometimes remains patent after birth and may give rise to symptoms. Below the fossa ovalis is the **opening of the coronary sinus.** It is through this that the majority of the blood supply of the heart returns to the right atrium. Smaller veins, the **venae cordis minimae,** also open directly into the atrium. Blood from the right atrium passes through the **right atrioventricular** or tricuspid valve into the right ventricle.

The **right ventricle** has a considerably thicker muscular wall than the right atrium and its internal surface shows irregular projections of muscle called **trabeculae carneae.** One of these, the **septomarginal trabecula** or moderator band, lies partly free in the ventricle and carries the **right**

* Auriculus, *L.* = a little ear; Pecten, *L.* = a comb

branch of the atrioventricular bundle of the conducting system. On closure of the tricuspid valve, contraction of the ventricular muscle forces blood upward through the smooth-walled **infundibulum** to the **pulmonary valve.** In Fig. 4-32 this region is hidden by the anterior cusp of the right atrioventricular valve.

The **right atrioventricular** or tricuspid valve allows the free flow of blood from the right atrium during the period of ventricular relaxation (diastole) but is designed to close as contraction of the right ventricle (systole) raises the intraventricular pressure. The ostium* of the valve lies almost on the median plane and it is larger than the left atrioventricular ostium and can accommodate the tips of three fingers. The ostium is closed during systole by three **cusps.** Of these, one lies on the anterior margin of the orifice, one on the posterior, and one, the **septal cusp,** on the margin adjacent to the septum between the two ventricles. The free margins of the cusps are irregular and to them are attached fine strands of connective tissue called **chordae tendinae.** These, in turn, are attached in groups to the **papillary muscles** which project from the wall of the ventricle. These features of the valve should now be identified in Fig. 4-32. As the intraventricular pressure rises the cusps are forced toward the orifice and, thus, close it. They are prevented from evaginating into the right atrium by the chordae tendinae and papillary muscles attached to their margins. In pathological conditions eversion of a cusp does sometimes occur.

The design of the **pulmonary valve** is somewhat different from that of the tricuspid valve. It lies at the junction of the ventricular wall and the pulmonary trunk and consists of three delicate cup-shaped valvules whose cavities face away from the direction in which blood is forced from the ventricle during systole. In the adult heart one valvule lies anteriorly and the other two on the left and right posteriorly. At the end of systole, when intraventricular pressure falls, there is a tendency for blood to return to the ventricle. In so doing it fills the valvules and approximates their free margins, thus preventing further blood flow. This is illustrated diagrammatically in Fig. 4-33. In Fig. 4-34 the free margin of each valvule can be seen to be thickened and that this thickening is most pronounced at the **nodule** in the center. The region adjacent to the free margin is particularly thin and formed of little more than a double layer of endothelium. This is called the **lunule.** The valvules are normally avascular.

These chambers and valves are concerned with pumping the venous return from the systemic circulation to the lungs for oxygenation.

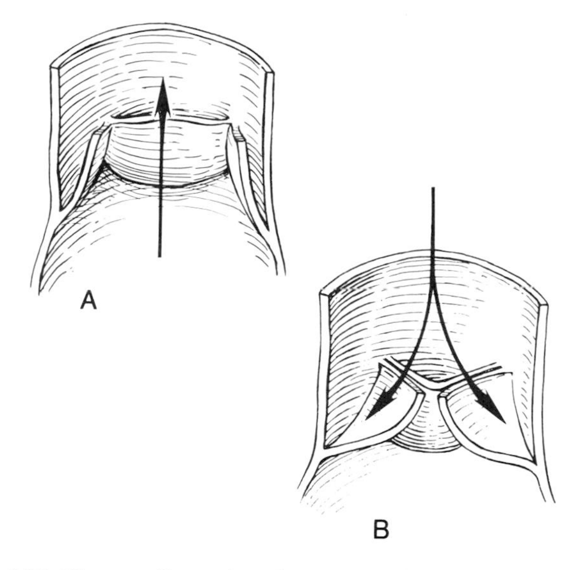

4-33 Diagram illustrating the manner in which the pulmonary valve opens (*A*) and closes (*B*) as the direction of the blood flow changes.

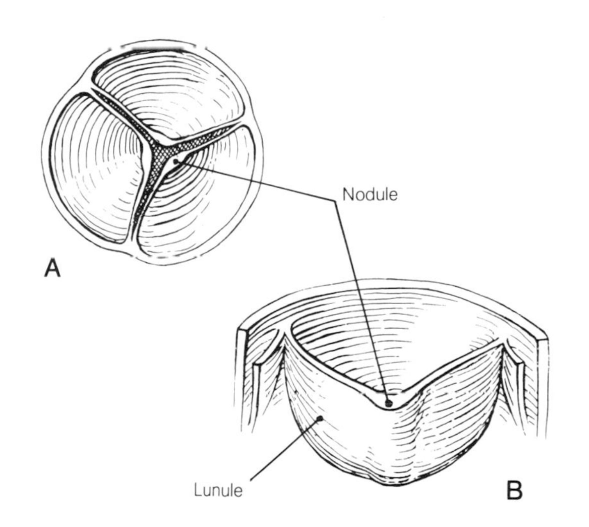

4-34 The pulmonary valve viewed from above (*A*) and a pulmonary valvule (*B*).

---

* Ostium, *L.* = a door or opening

The Chambers and Valves of the Left Side of the Heart

Oxygenated blood returns from the lungs in the four pulmonary veins which enter the upper posterior aspect of the **left atrium.** The position of the atrium, the pulmonary veins, and the left auricle can be seen in Fig. 4-35. The left atrium lies largely behind the roots of the pulmonary trunk and aorta and to the right it is separated from the right atrium by the **interatrial septum.**

The interior of the atrium is smooth and the ridges raised by musculi pectinati are only found in the auricle. There are no valves at the orifices of the pulmonary veins. Blood passes from the left atrium through the **left atrioventricular valve** into the **left ventricle.**

The **left ventricle** lies to the left of and behind the right ventricle and forms the apex and the greater part of the diaphragmatic and posterior surfaces of the heart. Its wall is three times as thick as that of the right and its cavity is round in cross-section. As a result the **interventricular septum** which separates the ventricles bulges into the cavity of the right ventricle making this crescentic in shape. This can be seen in Fig. 4-36. As in the right ventricle, the muscle wall displays trabeculae carnae which here are even more pronounced. Contraction of this ventricle forces the now-oxygenated blood through the aortic valve

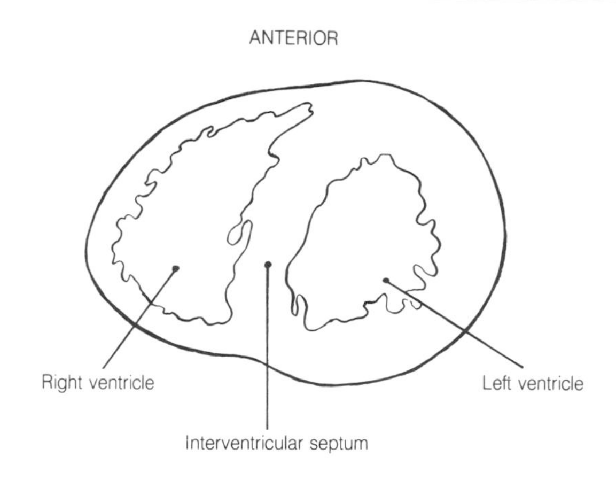

4-36 Horizontal section through the left and right ventricles to show the relative thickness of their walls and the interventricular septum. The section is viewed from below.

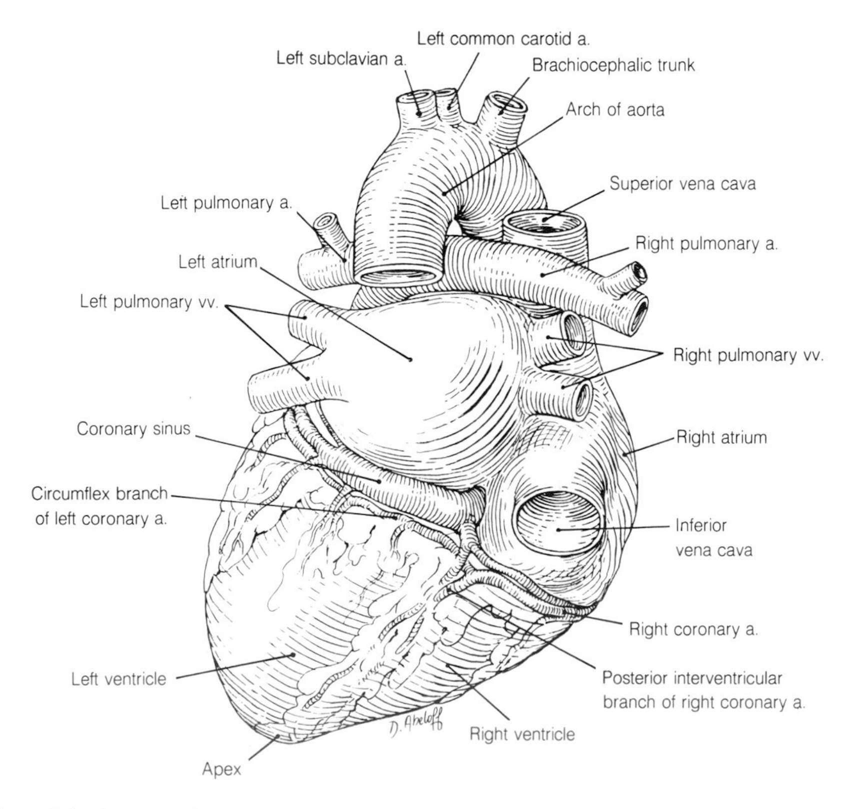

4-35 Posterior view of the heart and great vessels.

into the aorta and thus returns it to the systemic circulation.

The **left atrioventricular, bicuspid,** or **mitral*** valve has many similarities with the right atrioventricular valve although it is smaller and admits the tips of only two fingers. It has only two cusps and of these the more anterior is larger and lies between the orifices of the left atrioventricular and aortic valves; the smaller lies posteriorly. To the margin of each are attached chordae tendinae which blend with papillary muscles projecting from the ventricular wall. The position of this valve between the left atrium and ventricle can be seen in Fig. 4-37.

The **aortic valve** lies in the upper part of the ventricle in front of the left atrioventricular valve and is separated from it by the anterior cusp of the latter. It has three valvules which are similar to those of the pulmonary valve, but arranged with one lying posteriorly behind a left and right anterior cusp. Above each valve is an **aortic sinus** and it is from two of these that the left and right coronary arteries arise.

* Mitral, from the English word "mitre," used to describe the tall cleft cap worn by a Bishop and which characterizes the chess piece. The word is intended to indicate the two cusps of this valve and is commonly used in clinical medicine.

The **interventricular septum** is a thick muscular partition separating the left from the right ventricle. A small region of its upper part is thin and fibrous and is called the **membranous part.** Separate mention of this septum is made because in a number of congenital conditions it is found more or less deficient and, in addition, it carries the conducting bundle which propagates the signal for muscular contraction from the atria to the ventricles.

Considerable space has been devoted to a description of the chambers and valves of the heart in preparation for an understanding of the pathophysiology of heart disease. However, no illustrations can replace the careful examination of a postmortem or preserved specimen.

The Conducting System of the Heart

A conducting system is essential for the orderly sequence of atrial and ventricular contraction and to establish a common rate at which these chambers will beat. The stimulus that causes the con-

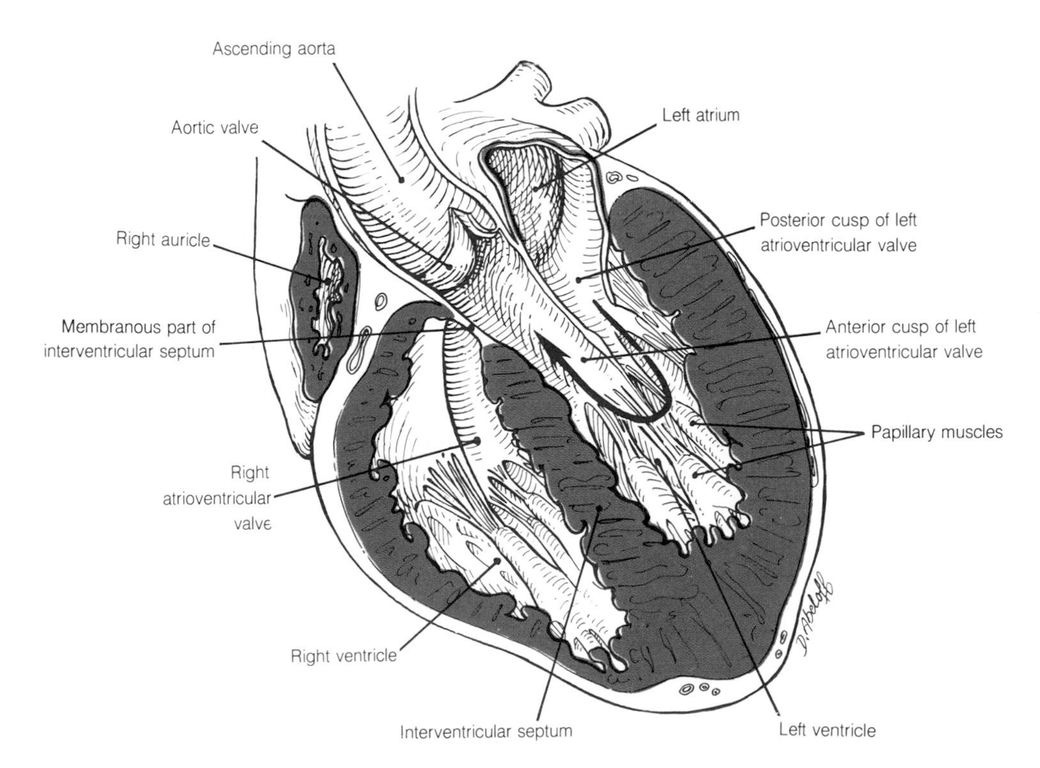

4-37 The heart sectioned to show the left ventricle and the left atrioventricular valve. The *arrow* indicates the direction of the flow of blood around the anterior cusp of the left atrioventricular valve.

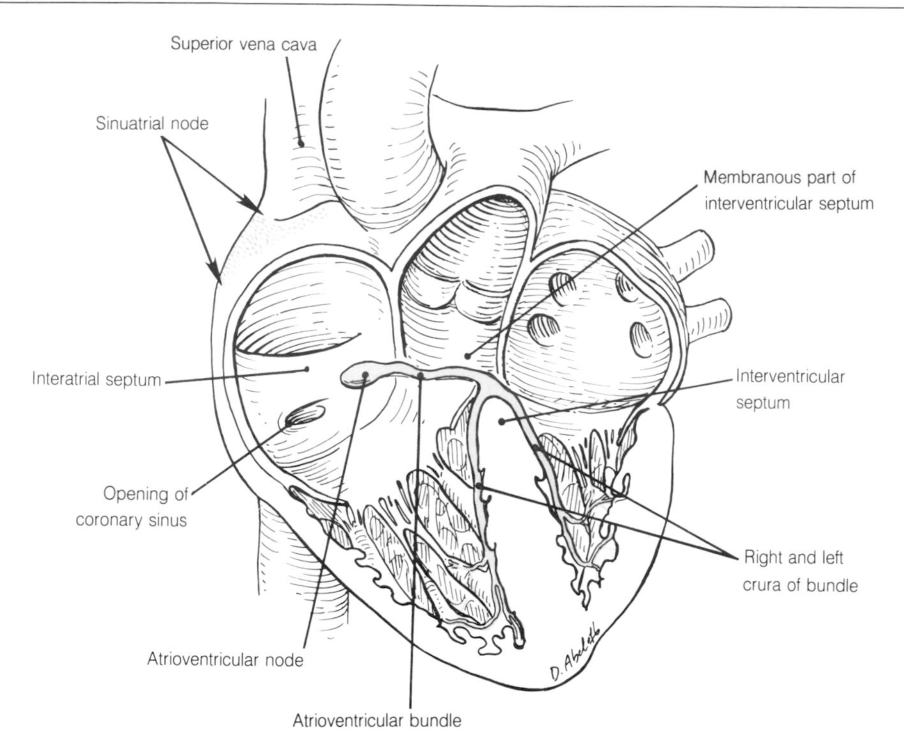

4-38 Semidiagrammatic illustration of the conducting system of the heart.

traction is initiated in a region of specialized muscle fibers known as the **sinuatrial node** or pacemaker which can be seen in Fig. 4-38 to lie in the wall of the upper part of the right atrium and partially surround the opening of the superior vena cava. From this site the impulse is spread throughout the two atria by the electrical coupling of their muscle fibers. However, there is now good electrophysiological and histological evidence for the existence of preferential pathways for the more rapid conduction of the impulse between the atria and to the atrioventricular node. These are three in number and are called **internodal pathways.**

The **atrioventricular node** lies in the interatrial septum close to the opening of the coronary sinus. This node receives the impulse from the atrial muscle fibers and passes it to the **atrioventricular bundle.** This bundle runs to the membranous and then the muscular interventricular septum where it divides into a **right and left crus** (known by clinicians as bundle branches). The right crus runs toward the apex beneath the endocardium, enters the septomarginal trabecula to supply the papillary muscles, and then breaks up into fine fibers which supply the remainder of the right ventricle. The left crus has a similar superficial course down the left surface of the septum but breaks up earlier into two or more strands for the supply of left papillary and ventricular muscles. Like the sinuatrial node, the atrioventricular node and bundle are formed from specialized cardiac muscle fibers.

The Blood Supply of the Heart

The continuous beating of the heart throughout life and the ability to increase its output by many times during periods of intense physical exercise requires a constant and reliable source of oxygen and other nutrients. No stores of energy are maintained within heart muscle cells and it is common knowledge that any major sudden reduction of the blood supply caused by the thrombosis* of coronary vessels leads to the rapid necrosis* of the region of heart muscle involved. This blood supply is provided by the **left and right coronary arteries.**

The coronary arteries are branches of the ascending aorta and arise from the left and right

* Thrombos, *Gk.* = a clot; Nekros, *Gk.* = dead

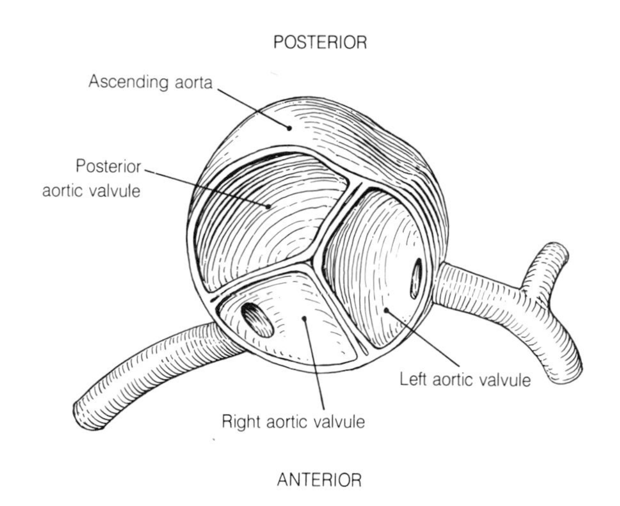

4-39 Diagram to illustrate the origin of the left and right coronary arteries from the ascending aorta. The aortic valve and sinuses are viewed from above.

aortic sinuses which lie above the valvules of the aortic valve. The diagram in Fig. 4-39 should make this arrangement clear.

The **left coronary artery** arises from the **left anterior sinus** and runs forward hidden between the pulmonary trunk and the left auricle. Its course can be followed in Figs. 4-40 and 4-41. On reaching the atrioventricular sulcus it divides into a **circumflex branch** and an **anterior interventricular branch.** The circumflex branch turns to the left in the atrioventricular sulcus which leads it round the left margin of the heart (margo obtusus) toward the posterior interventricular sulcus. In its course it gives off atrial branches, a left marginal, and posterior ventricular branches to the left ventricle.

The anterior interventricular branch descends

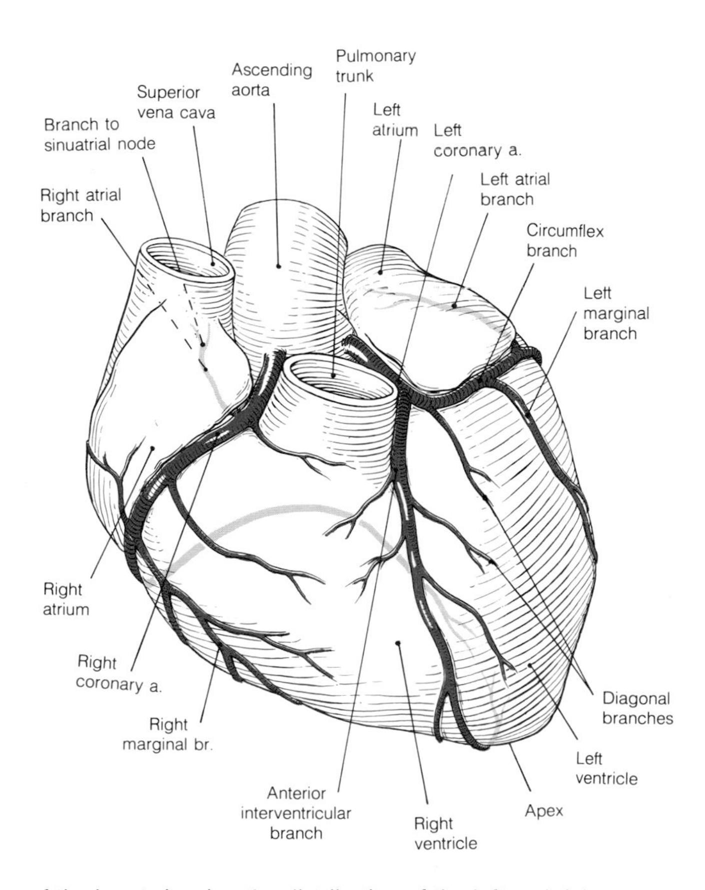

4-40 The anterior surface of the heart showing the distribution of the left and right coronary arteries.

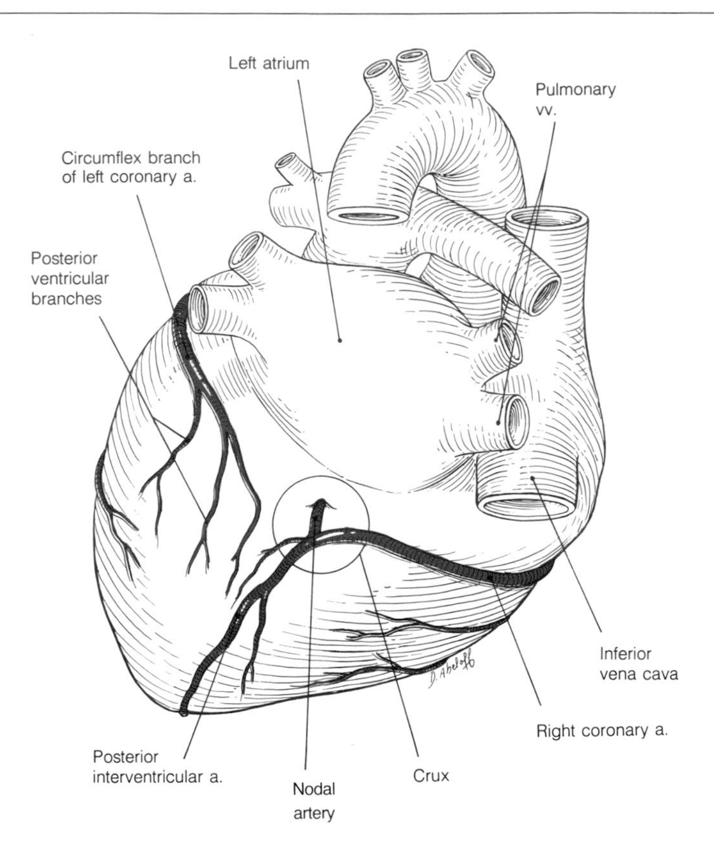

4-41 The base and diaphragmatic surface of the heart showing the distribution of the right coronary artery and the circumflex branch of the left coronary artery.

in the anterior interventricular sulcus toward the apex of the heart. It gives off a number of "diagonal" branches to the left ventricle and a series of shorter branches to the right ventricle. From its deep surface and anchoring it to the myocardium arise septal branches for the supply of the interventricular septum.

The **right coronary artery** arises from the **right anterior aortic sinus** and, referring again to Figs. 4-40 and 4-41, it can be seen to pass downward and to the right in the coronary sulcus. This leads it round the inferior margin of the heart (margo acutus) towards the posterior interventricular sulcus. Here it forms an anastomosis with the circumflex branch of the left coronary artery before running forwards to the apex in the posterior interventricular sulcus where it is known as the posterior interventricular branch.

The right coronary artery gives off atrial branches to the left atrium and ventricular branches to the left ventricle including a constant right marginal branch. In its passage over the posterior interventricular sulcus it also gives off perforating septal branches and small ventricular branches to the left ventricle.

Two branches of the coronary arteries need particular attention because they supply important regions of the conducting system. The sinuatrial node is supplied by a sinuatrial node branch of the right coronary artery in 60-70% of cases or by a sinuatrial node branch of the left coronary artery (30-40% of cases). The **atrioventricular node artery** is given off at the crux as the right coronary artery turns down into the posterior interventricular sulcus. The atrioventricular node artery and posterior and anterior septal branches of the interventricular branches supply the atrioventricular bundle and its branches.

The distribution of the coronary arteries which has been described is that most commonly found. However, the left or right coronary may make a preponderant or dominant contribution to the posterior and inferior aspects of the ventricles or the contributions may be approximately balanced.

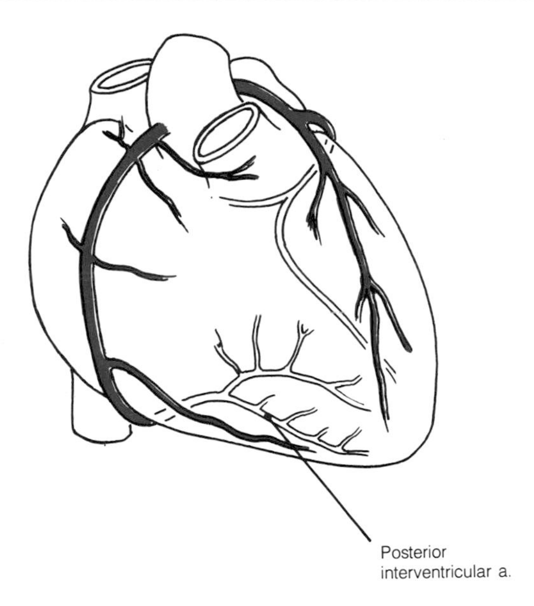

4-42 Right coronary artery dominance.

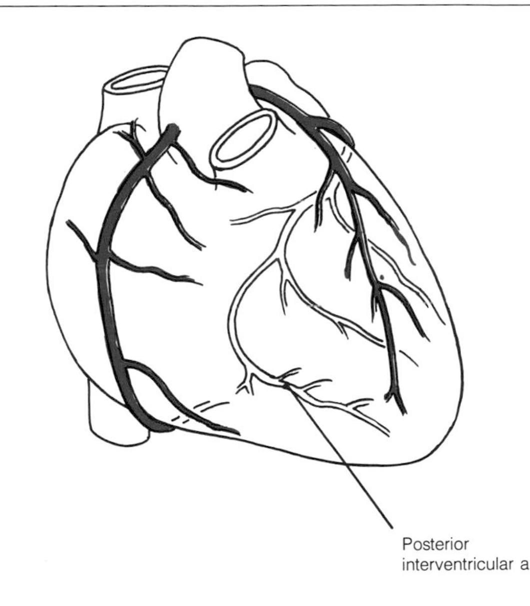

4-44 Left coronary artery dominance.

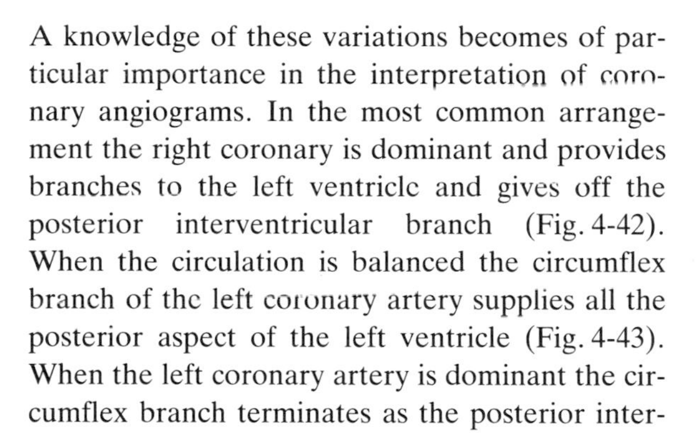
A knowledge of these variations becomes of particular importance in the interpretation of coronary angiograms. In the most common arrangement the right coronary is dominant and provides branches to the left ventricle and gives off the posterior interventricular branch (Fig. 4-42). When the circulation is balanced the circumflex branch of the left coronary artery supplies all the posterior aspect of the left ventricle (Fig. 4-43). When the left coronary artery is dominant the circumflex branch terminates as the posterior interventricular artery (Fig. 4-44). While damage to the cardiac musculature following thrombosis of coronary arteries or their branches is in itself a life threatening condition, its impact becomes more serious if the function of any part of the conducting system is also impaired.

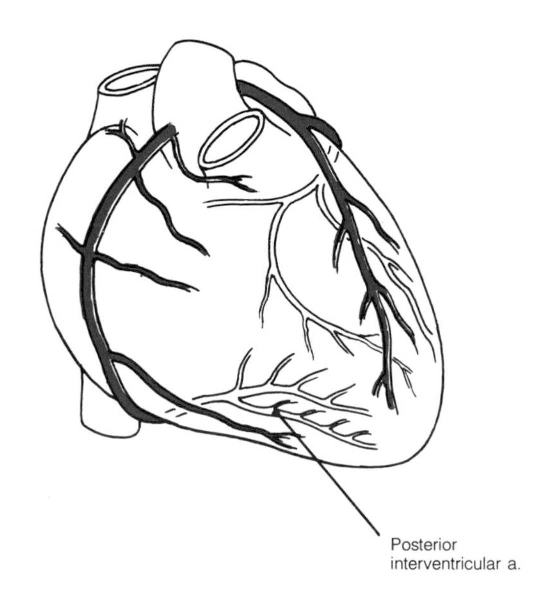

4-43 Balanced coronary circulation.

The ability to demonstrate the coronary arteries in life and the modern surgical techniques for alleviating their obstruction have generated a terminology now commonly used among clinicians. Before its division the left coronary artery (LCA) is known as the **left main common trunk** (LMCT). Its two branches are the **circumflex artery** (Cx) and the **left anterior descending artery** (LAD). The right coronary (RCA) usually terminates as the **posterior descending artery** (PDA). The left marginal branch is known as the **obtuse marginal artery** and the right as the **acute marginal artery.**

The Venous Drainage of the Heart

The greater part of the venous blood leaving the heart is collected by a system of veins that converge on the **coronary sinus.** This lies in the posterior part of the coronary sulcus and opens into the right atrium between the orifices of the inferior vena cava and the tricuspid valve.

The larger veins follow the course of the arteries and like these are often hidden by fat. How-

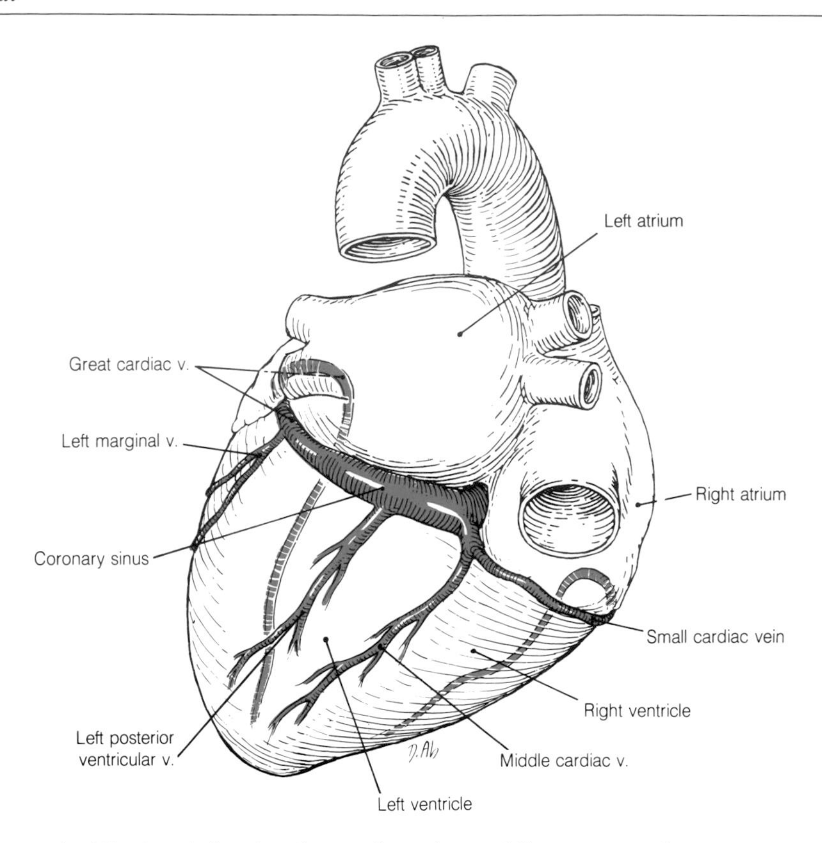

4-45 The posterior aspect of the heart showing the cardiac veins and the coronary sinus.

ever, their nomenclature differs and this and their arrangement can be seen in Fig. 4-45. Note that in this illustration the heart is viewed from behind. The **great cardiac vein** lying in the anterior interventricular sulcus is joined by a **left marginal vein** and the **posterior vein of the left ventricle** to form the coronary sinus. This is joined near its termination by the **small** and **middle cardiac veins.** In addition to these veins draining into the coronary sinus, two or three **anterior cardiac veins** draining the right ventricle pass from the surface of the ventricle into the right atrium anteriorly and a number of very small veins (**venae cordis minimae**) drain directly into the atria from the myocardium and do not appear, as the others, on the surface of the heart. It should be noted that the coronary sinus may be covered by a thin layer of muscle fibers as it approaches the right atrium.

The Examination of the Heart

A considerable amount of information about the action and size of the heart can be obteined using the traditional techniques of inspection, palpation, percussion, and auscultation. To carry these out and interpret the findings requires a knowledge of the surface projection of the heart and where particular sounds are best heard, and these will be discussed with the surface anatomy. To this information can be added radiological findings, and initially these will come from an examination of a p. a. projection of the heart. An example is shown in Fig. 4-46. With the normal anatomy of the heart now in mind a more detailed examination of the borders of its shadow can now be made. Beginning at the upper end of the right margin, identify the superior vena cava (*SVC*) and

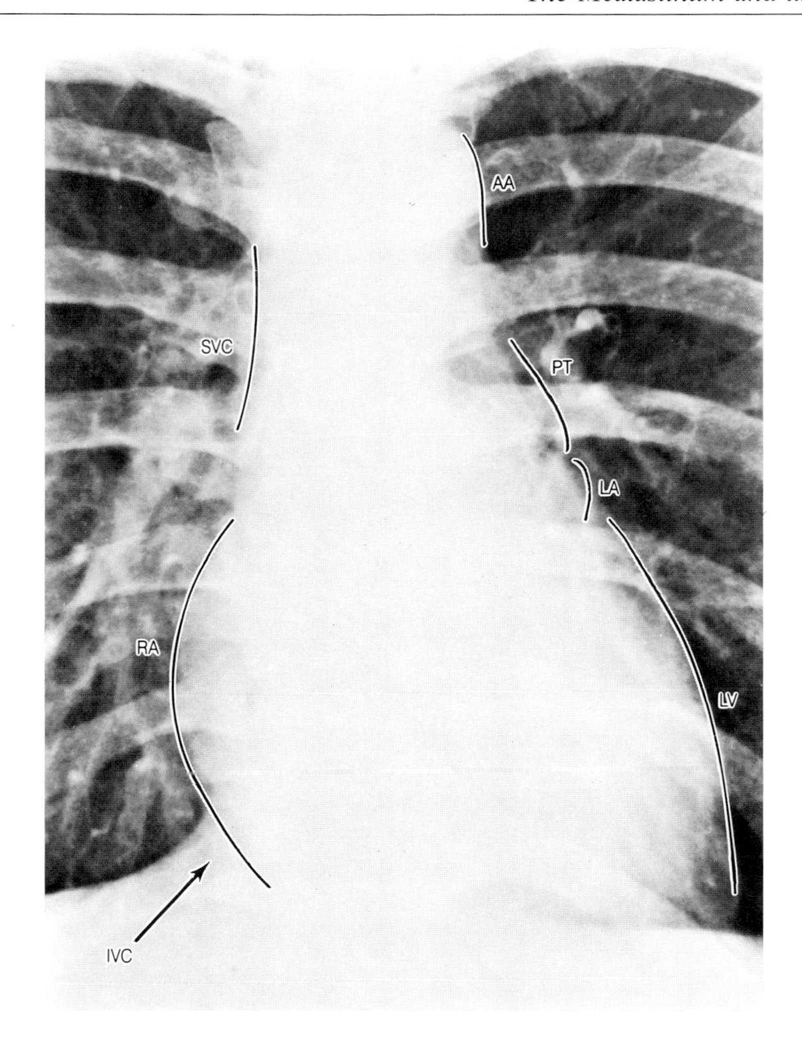

4-46 Radiograph of the heart (p. a. projection). (Reproduced by permission, from Wicke: *Atlas of Radiologic Anatomy*, 4th Ed. Urban & Schwarzenberg, Baltimore–Munich, 1987.)

the right atrium (*RA*). The inferior border is lost in the shadow of the diaphragm and underlying liver, but the edge of the inferior vena cava (*IVC*) can be seen filling in the angle between the right atrium and the diaphragm. On ascending the left border, the margin is formed by the left ventricle (*LV*), the left auricle (*LA*), the pulmonary trunk (*PT*) and the arch of the aorta (*AA*). The prominence formed by the last is often termed the aortic bulb or knuckle.

The shape of the heart shadow varies with body shape. The tall, thin individual will have a long narrow shadow, but in the shorter, broad individual it will be shorter and more rounded. However, a rough estimate of the heart size can be obtained by adding the maximum distance to the left and right borders from the midline. In a normal heart the total is less than half the transverse diameter of the thorax.

By introducing radiopaque material through a cardiac catheter the chambers of the heart, the nearby great vessels, and the coronary arteries can also be visualized. An example of a left coronary angiogram is given in Fig. 4-47. This investigation is an important preliminary to coronary artery surgery.

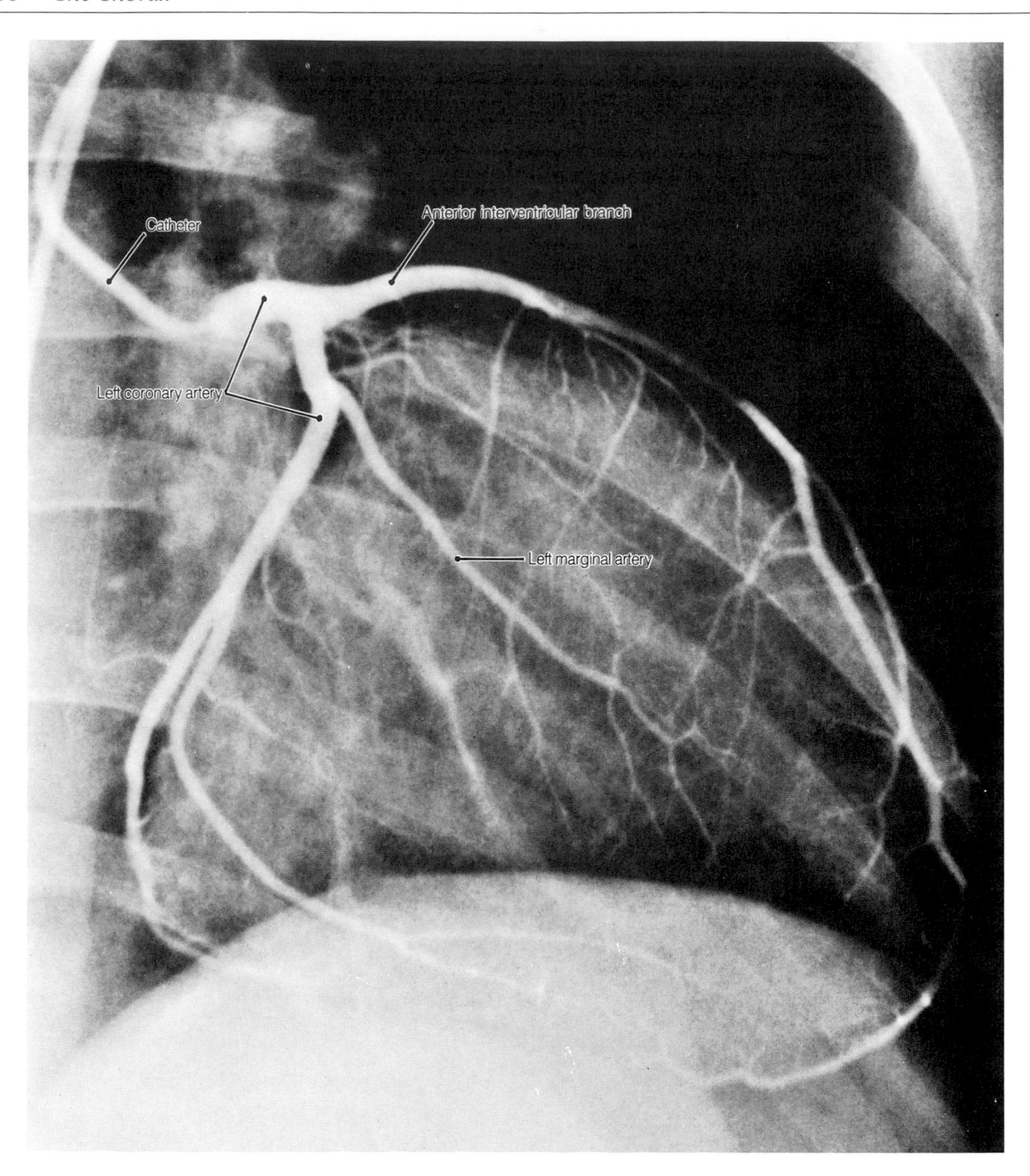

4-47 Left coronary angiogram (right anterooblique projection). (Reproduced by permission, from Wicke: *Atlas of Radiologic Anatomy,* 4th Ed, Urban & Schwarzenberg, Baltimore–Munich, 1987.)

The Surface Anatomy of the Heart

Figs. 4-48 and 4-49 illustrate the relationships of the heart, its valves, the superior vena cava, and the aorta to the thoracic cage. By identifying the sternal angle and thus the level of the second costal cartilages, these structures can be transferred to the surface by palpating and counting ribs and intercostal spaces. This projection should not be considered as rigid, for normal variations in the shape of the heart, the position of the body, and respiratory movement will modify the generalized picture presented here.

Returning to Fig. 4-48, note that the upper border of the heart lies opposite the second intercostal space, that the right border lies 3 to 5 cm from

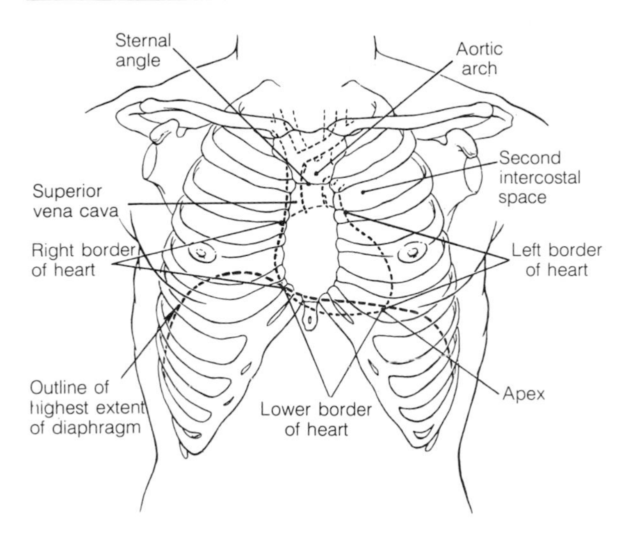

4-48 Showing the relationships of the heart, the superior vena cava, and the aorta to the overlying thoracic cage.

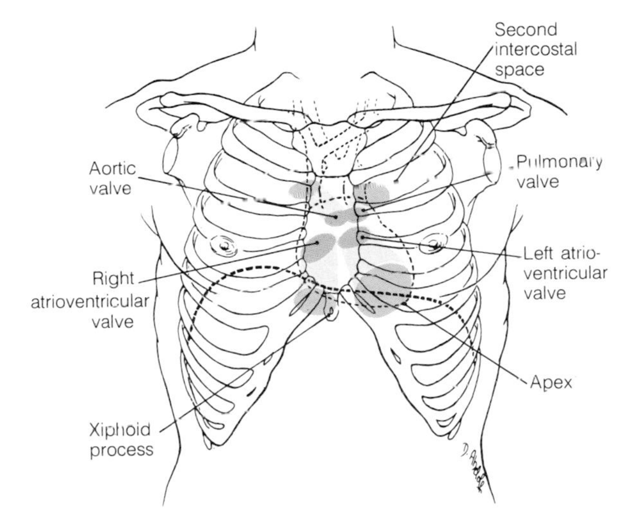

4-49 Showing the surface projections of the valves of the heart and the regions over which the sounds that they make can best be heard.

the midline, and that the left border slopes laterally to the apex which lies opposite the fifth intercostal space, 8 to 9 cm from the midline. The **apex beat** which may be both visible and palpable lies a centimeter or two medial to the anatomical apex. A line drawn from the apex beat through the xiphisternal sternal joint indicates the lower border and completes the outline of the heart. Note also that the right margin of the superior vena cava is an extension of the right border of the heart and that the aortic arch lies behind the manubrium sterni and extends as high as its midpoint.

Percussion of the thorax reveals a variable degree of dullness over the outline of the heart. This is most pronounced over the area not covered by lung, i. e., at the cardiac notch.

In Fig. 4-49 the surface projections of the pulmonary, aortic, and atrioventricular valves are shown. However, the sites at which the normal or abnormal sounds produced by the valves can best be heard do not correspond to their surface projections. The aortic valve is best heard at the second right intercostal space close to the sternum, and the pulmonary valve at the second left space. The sound of the mitral valve is transmitted to the region of the apex beat and the tricuspid valve can be heard over the lower end of the sternum.

The Great Vessels

Having studied the heart as a muscular pump and followed the course of the blood through its chambers and valves, the great vessels which connect it with the systemic and pulmonary circulations must now be described. The venous return from the systemic circulation has been seen to enter the right atrium through the superior and inferior venae cavae. The superior vena cava carries blood returning from the thorax, the upper extremities, and the head and neck, while the inferior vena cava carries blood from the abdomen, pelvis and lower extremity.

The Inferior Vena Cava

The greater part of the inferior vena cava lies within the abdominal cavity and its intrathoracic course is very short. After piercing the central tendon of the diaphragm just to the right of the midline, a small portion of it lies outside the pericardium and, as has been mentioned, it fills in the cardiodiaphragmatic recess in a radiograph of the heart (Fig. 4-46). The remainder extends the short distance between the fibrous pericardium and the inferior surface of the right atrium into which it opens.

The Superior Vena Cava

In Fig. 4-50 the superior vena cava can be seen to be formed by the union of the **left and right brachiocephalic veins.** Each of these begins be-

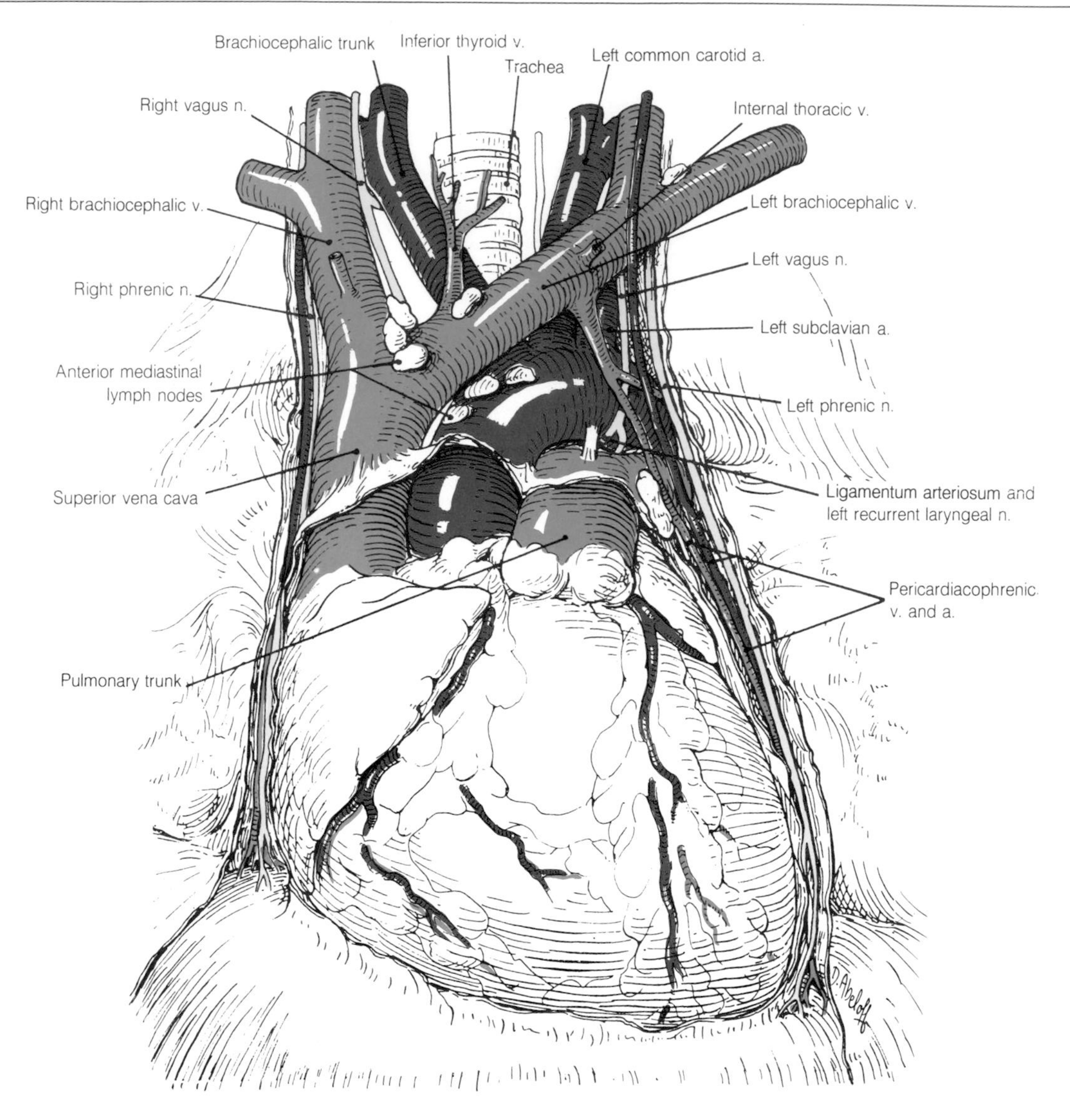

4-50 The great vessels in the superior mediastinum.

hind the corresponding sternal end of the clavicle and they unite below the first costal cartilage just to the right of the sternum. Note the short vertical course of the right vessel and the longer, oblique course of the left as it runs deep to the manubrium of the sternum. In a child with acute respiratory obstruction, an engorged left brachiocephalic vein may be at risk if a tracheostomy incision is extended too low. Note also the **right phrenic nerve** running along the lateral margin of the right vein just beneath the mediastinal pleura and the **right vagus** lying behind and medial to the vein. In its course toward the left, the left brachiocephalic vein crosses the **left phrenic** and **vagus nerves,** and **left subclavian and common carotid arteries,** the **trachea,** and the **brachiocephalic trunk.** The **internal thoracic veins** and the **inferior thyroid vein** drain into the brachiocephalic veins and both are closely associated with **anterior mediastinal lymph nodes.** Close to the point at which the superior vena cava enters the right atrium it is joined by the **azygos vein** although this cannot be seen in the illustration.

The Pulmonary Trunk

Deoxygenated blood is passed from the right ventricle to the lungs through the **pulmonary trunk** and by the **left and right pulmonary arteries** which are formed by its bifurcation. In Fig. 4-50 the trunk can be seen to partially obscure the root of the **aorta** as it runs superiorly, but passing to the left it reaches the **arch of the aorta** where it divides into the left and right pulmonary arteries. Of these, the right passes laterally beneath the arch of the aorta to reach the hilus of the right lung. The left pulmonary artery follows a rather shorter course as it runs to the hilus of the left lung.

Close to the bifurcation of the pulmonary trunk, the left pulmonary artery is joined to the concavity of the arch of the aorta by a band of fibrous tissue called the **ligamentum arteriosum** and behind and to the left of this can be seen the left recurrent laryngeal nerve as it loops beneath the arch. The significance of the ligamentum arteriosum will be discussed in the section describing the fetal circulation at the end of this chapter.

The pulmonary trunk and pulmonary arteries are sites of fatal obstruction by massive emboli. The clot of blood responsible may break away from the wall of the right side of the heart or, in some instances, a clot in the form of a cast of the deep veins of the leg may be found.

The Pulmonary Veins

Oxygenated blood returns from the lungs in the paired left and right pulmonary veins which open into the left atrium. Turn back now to Fig. 4-17 and review the relationships of the pulmonary arteries and veins in the root of the lung. In Fig. 4-35 the entry of the pulmonary veins into the left atrium can be seen.

The Aorta

Through the aorta, blood from the left ventricle is distributed throughout the body. The first portion of this vessel is the **ascending aorta.** When viewed from in front it is largely hidden by the infundibulum of the right ventricle and the pulmonary trunk. Just above the aortic valve its wall shows three dilatations known as the **aortic sinuses.** Two of these, the left and right, mark the origins of the **coronary arteries;** the posterior sinus is not related to a coronary artery.

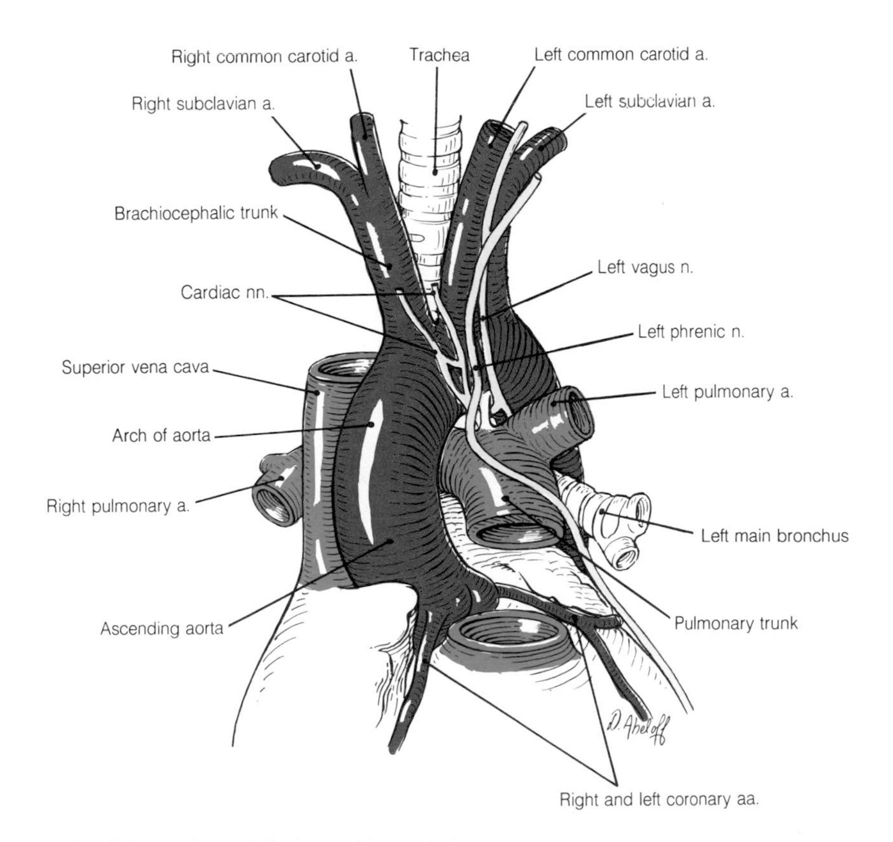

4-51 Showing the arch of the aorta and its immediate relations.

The ascending aorta blends with the **arch of the aorta.** Although this arch frequently appears in diagrams to lie in a coronal plane, it does, in fact, lie almost in a sagittal plane diverging only slightly to the left as it arches backward over the **right pulmonary artery** and the **left main bronchus.** Refer to Fig. 4-51 and see that its convexity gives rise to three large vessels, the **brachiocephalic trunk,** and the **left common carotid and subclavian arteries.** These can be seen embracing the thoracic portion of the trachea. Note also the **left phrenic nerve,** some fine **cardiac nerves,** and the **left vagus nerve** crossing its anterolateral aspect. The arch lies in the superior mediastinum and terminates at the level of the lower border of the fourth thoracic vertebra where it becomes continuous with the thoracic part of the **descending aorta.**

A number of important thoracic structures remain to be described. Most lie close to the midline and pass from the superior to the middle or posterior mediastinum, others lie in one mediastinum only or outside the mediastinum in the paravertebral region. These are grouped as other mediastinal and paravertebral structures.

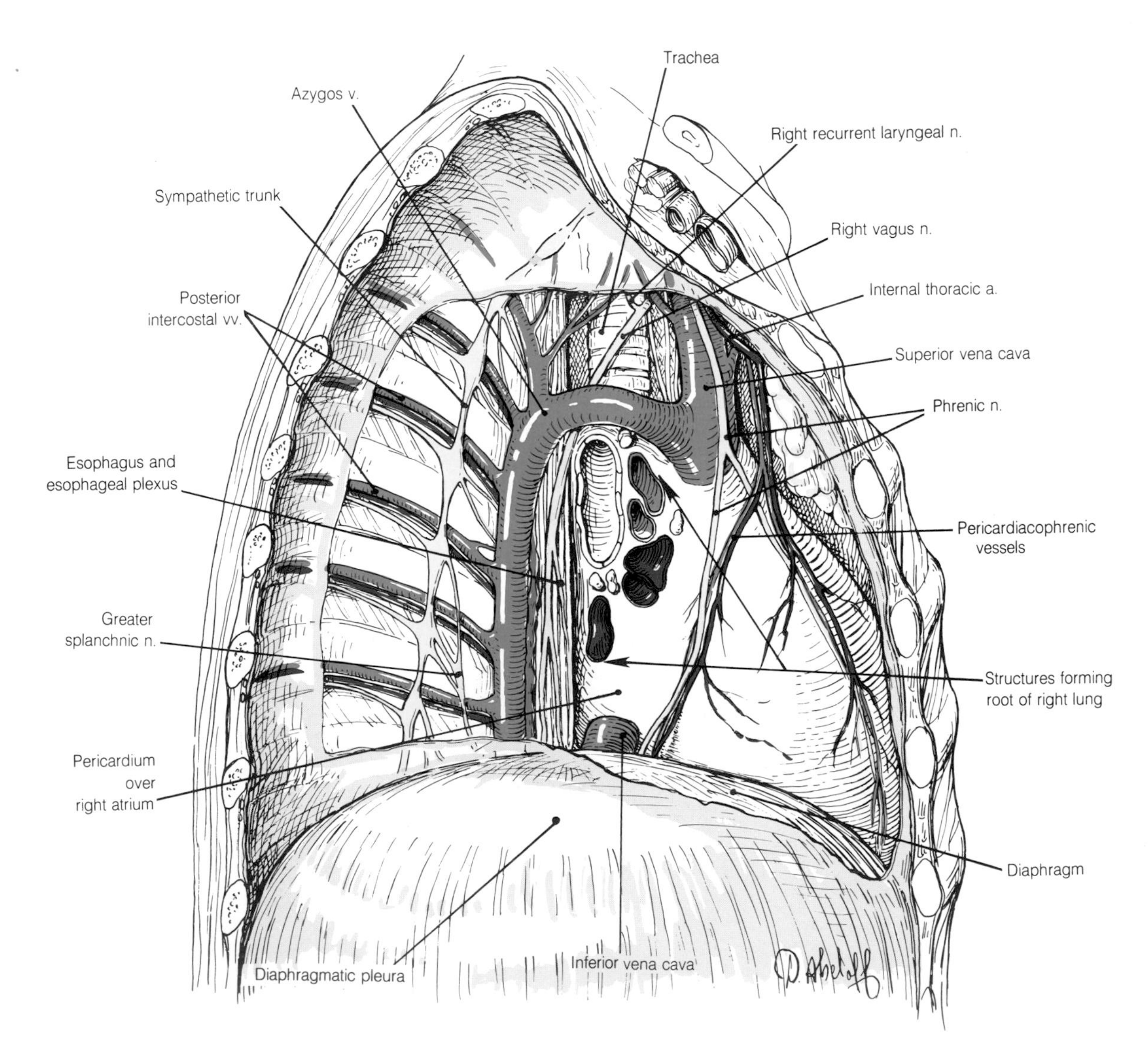

4-52 Showing the right surface of the mediastinum and the right posterior thoracic wall. Parietal pleura has been removed from the mediastinum, thoracic wall, and diaphragm to expose underlying structures.

Other Mediastinal and Paravertebral Structures

The Phrenic Nerves in the Thorax

The phrenic nerves are formed in the neck by branches of the **ventral rami of the third, fourth, and fifth cervical nerves.** They alone carry motor fibers to the striated muscle of the diaphragm and the division of one leads to total paralysis of the corresponding half of the diaphragm. In addition, they carry sensory fibers from the mediastinum, the pericardium, the diaphragm, and its covering layers of pleura and peritoneum.

The course of the nerves can be followed in Figs. 4-52 and 4-53. Each nerve approaches the thoracic inlet on the superficial surface of the muscle scalenus anterior which is attached to the upper surface of the first rib. Slipping off the medial border of this muscle both nerves enter the thoracic cavity. The **left nerve** descends between the left common carotid and left subclavian arteries and deep to the left brachiocephalic vein. Crossing the arch of the aorta it reaches the pericardium and continues over the surface of this to the diaphragm. The **right nerve** lies lateral to **venous structures** in its course to the diaphragm. These are the right brachiocephalic vein, the superior vena cava, and the pericardium over the right atrium and the inferior vena cava. On reaching the diaphragm, sensory branches are given off to the parietal pleura and peritoneum. On the left, the nerve pierces the muscle while giving off its mus-

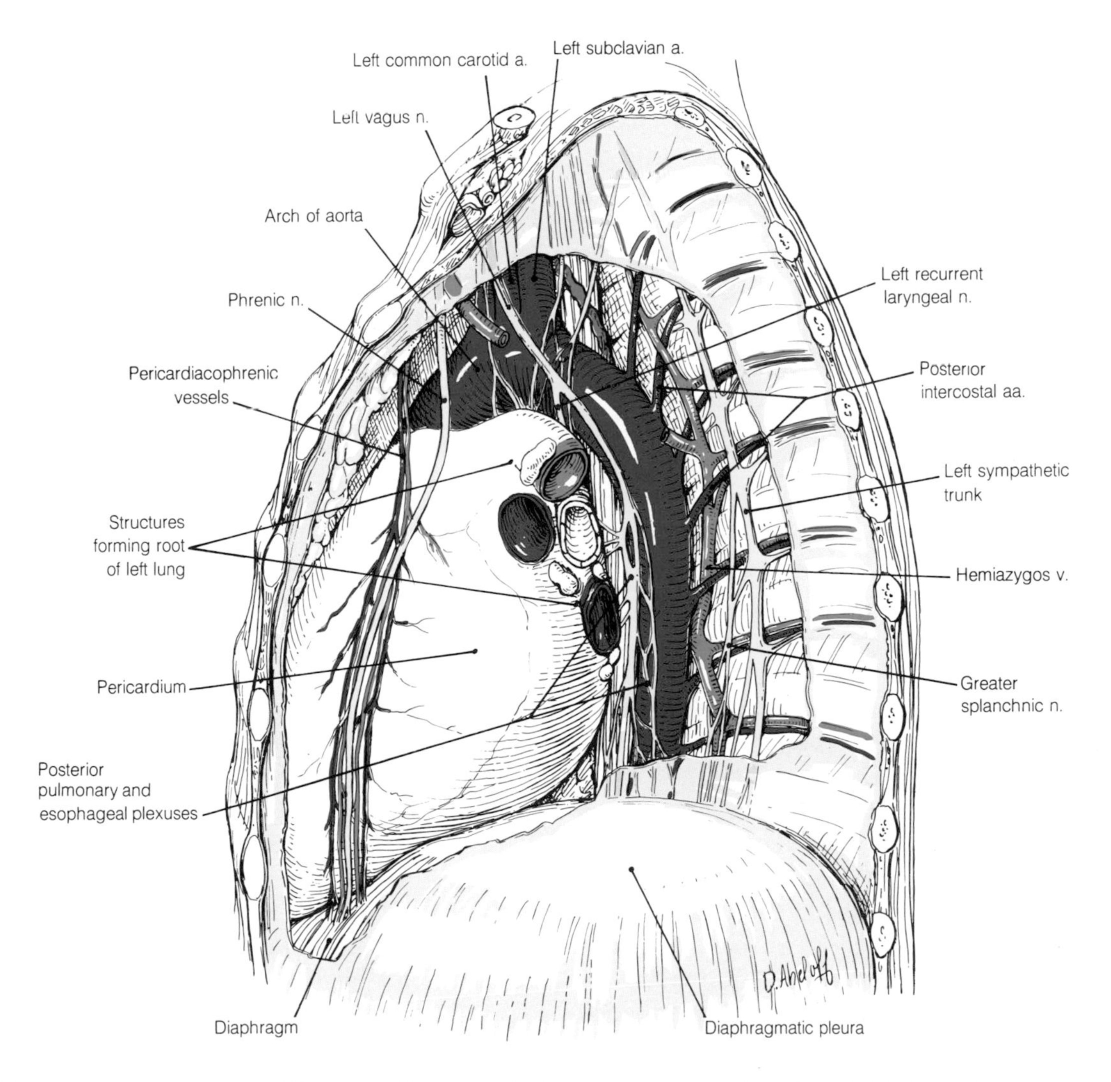

4-53 Showing the left surface of the mediastinum and the left posterior thoracic wall. Parietal pleura has been removed from the mediastinum, thoracic wall, and diaphragm to expose underlying structures.

cular branches. The right nerve pierces the central tendon in company with the inferior vena cava before breaking up into its muscular branches. Both nerves are joined by a **pericardiacophrenic branch** of the internal thoracic artery on the surface of the pericardium.

Because of its origin from cervical spinal nerves, irritation of the pleura or peritoneum covering the diaphragm (which is supplied by the phrenic nerve) may give rise to pain referred to the shoulder or neck which are supplied by the same spinal nerves.

The Thoracic Aorta

The descending limb of the arch of the aorta becomes the **thoracic aorta** at the level of the lower border of the fourth thoracic vertebra und thus at the lowest level of the superior mediastinum. At this point the thoracic aorta lies just to the left of the vertebral column but as it descends through the posterior mediastinum it gradually moves to the right. As a result, when it leaves the thoracic cavity by passing between the **two crura of the diaphragm,** it lies anterior to the twelfth thoracic vertebra.

In Fig. 4-53, the left lung has been removed and most of the course of the thoracic aorta can be seen. Lying anteriorly are the **root of the left lung** and below this the pericardium and underlying **left atrium.** Behind lie the **hemiazygos veins** and the **vertebral column,** and to the right, but not seen, are the **azygos vein** and the **thoracic duct.** The **esophagus** lies initially to the right of the thoracic aorta but as both structures descend it comes to lie anterior to the aorta. Laterally, of course, lie mediastinal and pulmonary pleura and the left lung, all of which have been removed. Note that the **left vagus** passes off the arch of the aorta to form the posterior pulmonary plexus and then continues as a number of communicating trunks called the **esophageal plexus.**

The **bronchial arteries** arising from the thoracic aorta have already been mentioned. In addition, nine pairs of **posterior intercostal arteries** supplying the lower nine intercostal spaces, and a pair of **subcostal arteries** are given off. The adjacent esophagus, pericardium, and diaphragm also receive branches.

The Sympathetic Trunk

In Fig. 4-53, the left sympathetic trunk can be seen to run a paravertebral course on the posterior wall of the thoracic cavity. The right trunk is similarly placed. At the junction of the neck and thorax, the trunk overlies the **heads of the ribs** but inferiorly it approaches the **bodies of the vertebrae** before leaving the cavity by passing behind the **medial arcuate ligament** of the diaphragm on the surface of the muscle **psoas major.** Along its course can be seen small swellings which are the thoracic sympathetic ganglia. One of these is usually associated with each intercostal space but the first is often fused with the inferior cervical ganglion to form the **cervicothoracic** or stellate* ganglion. The ganglia communicate with the intercostal nerves by means of **white and gray rami communicantes.** The general arrangement of the nerve fibers within the trunk and ganglia and their relationship to the spinal cord and peripheral nerves is described in Chapter 2. Two well marked branches, the **greater and lesser splanchnic nerves,** which contain a substantial number of myelinated nerve fibers, can be seen to be derived from the fifth to ninth ganglia and ninth to eleventh ganglia, repectively. The nerves of each side leave the cavity by passing through the **crura of the diaphragm.** A **least splanchnic nerve** leaves the lowest ganglion to pass with the sympathetic trunk beneath the **medial arcuate ligament.** These splanchnic branches, although arising in the thorax, are destined to join the **abdominal autonomic plexuses.**

Finer medially running branches of the upper ganglia run to join the **thoracic aortic plexus** and the **cardiac, pulmonary and esophageal plexuses.**

The Vagus Nerves in the Thorax

The vagi represent the parasympathetic nervous system in the thorax. As well as preganglionic parasympathetic fibers, they also carry sensory fibers from the thoracic viscera to the brainstem and, in addition, the left carries branchial motor fibers which are distributed to the larynx via the **left recurrent laryngeal nerve.** Parasympathetic fibers

* Stella, *L.* = a star

also reach the heart by way of cardiac branches of the vagi which are given off in the neck. The two vagi do not follow identical courses through the thorax. The **right vagus** enters the thorax close to the trachea and passes behind the right bronchus and the hilum of the right lung. Here, branches are given off to the **anterior and posterior pulmonary plexuses.** From the root of the lung the nerve passes onto the esophagus where, together with the left vagus and sympathetic fibers, it breaks up to form the **esophageal plexus.** As the esophagus passes through the diaphragm the vagi are reformed on its surface and enter the abdominal cavity as the **anterior and posterior vagal trunks.**

The **left vagus** is separated from the trachea by the arch of aorta which it leaves to reach the posterior aspect of the hilus of the left lung and then the esophagus. As the left vagus crosses the arch of the aorta a substantial branch, the **left recurrent laryngeal nerve,** is given off. This nerve as its name suggests, hooks backward and upward beneath the ligamentum arteriosum and arch of the aorta, to reach the groove between the trachea and esophagus. In this groove, it ascends into the neck. (The right recurrent laryngeal nerve carries out a similar maneuver but around the subclavian artery in the root of the neck.)

The Thymus

The thymus is a most intriguing organ whose function until recent years was completely unknown, although on empirical grounds it had been associated with the disease myasthenia gravis. This association has now been confirmed by the discovery that the thymus plays an important part in the immune response.

This "gland" is large at birth when its two lobes lie in the superior mediastinum beneath the sternum and anterior to the great vessels and upper part of the heart. It may also extend upward into the root of the neck and downward into the anterior mediastinum (Fig. 4-54). Although it continues to grow until puberty, it becomes relatively smaller over this period. From puberty onward there is a real reduction in size which is probably associated with the establishment of an adequate population of T cells throughout the body. In later life this organ is extremely difficult to find.

The thymus is supplied by branches from the **internal thoracic** and **inferior thyroid** arteries and drained by veins accompanying these vessels.

The Trachea in the Thorax

The trachea begins in the neck at the lower border of the **cricoid cartilage** and its anatomy and cervi-

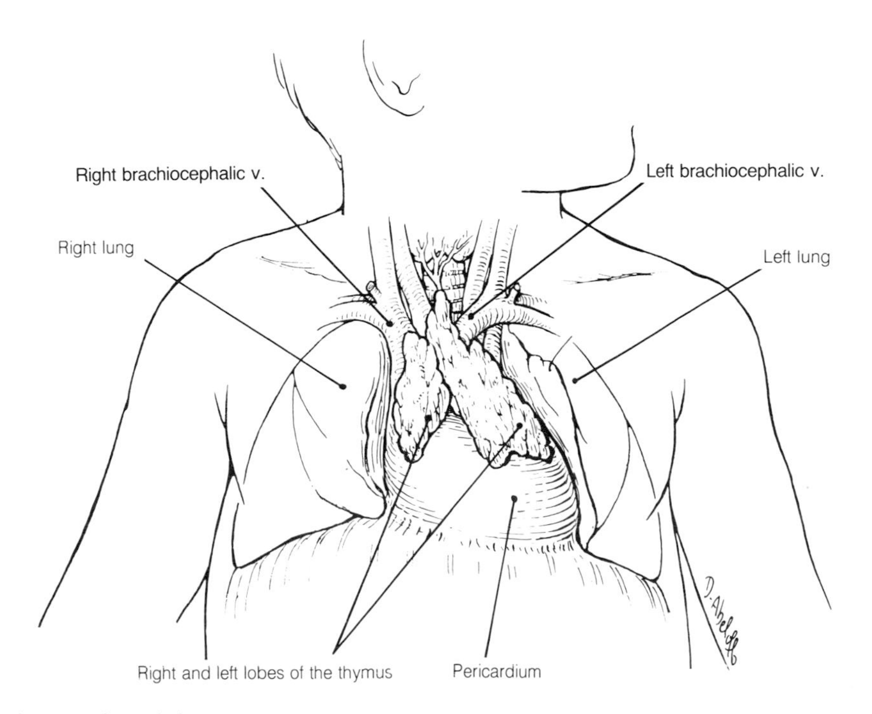

4-54 The thymus in a newborn infant.

cal course are described with that region. It enters the superior mediastinum through the thoracic inlet close to the midline where it lies anterior to the **esophagus;** a relationship that it maintains until its bifurcation into the two main bronchi at the level of the **fifth or sixth thoracic vertebra** (respiratory movements make this level imprecise). Anteriorly it is related to the **brachiocephalic trunk,** the **left common carotid artery,** and the **arch of the aorta.** The deep portion of the **cardiac plexus** lies between the aortic arch and the trachea. These anterior relationships are shown in Fig. 4-55. In this it can be seen that the brachiocephalic trunk moves to the right side of the trachea at the thoracic inlet and the left common carotid artery joins company with the left subclavian artery on the left side. The tracheobronchial and paratracheal lymph nodes are very closely related to the walls of the trachea. The trachea is also related to the **right vagus** which descends on its right side and the **left recurrent laryngeal nerve** which ascends at first between the trachea and the arch of the aorta and then between the trachea and the esophagus. The latter relationships can be best seen in the horizontal section through the superior mediastinum shown in Fig. 4-56. This also serves as a review of the region. The important structures that should be identified from before backward are:

1. the remnants of the thymus,
2. the left and right brachiocephalic veins,
3. the brachiocephalic trunk in front of the trachea,
4. the left common carotid and subclavian arteries to the left of the trachea with the left vagus between them,
5. the right vagus on the right side of the trachea and the left recurrent laryngeal nerve between the trachea and the esophagus, and
6. the thoracic duct on the left side of the esophagus.

The trachea can be felt as it enters the thorax and its central position confirmed. Its intrathoracic course can frequently be determined in

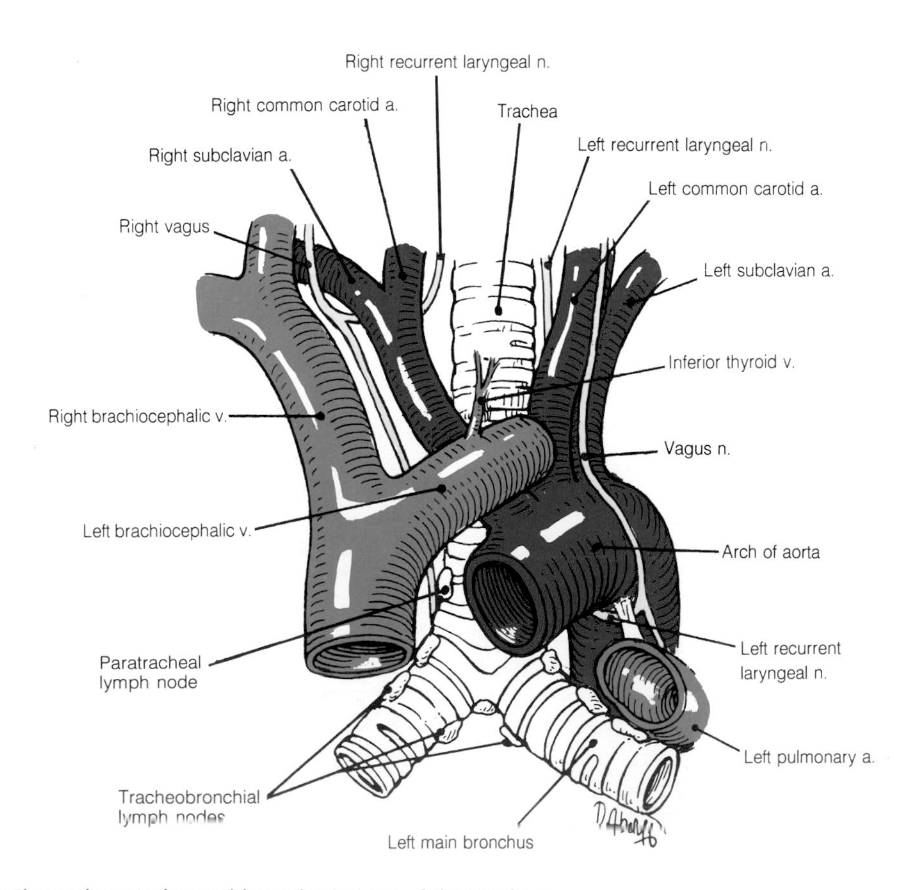

4-55 Showing the main anterior and lateral relations of the trachea.

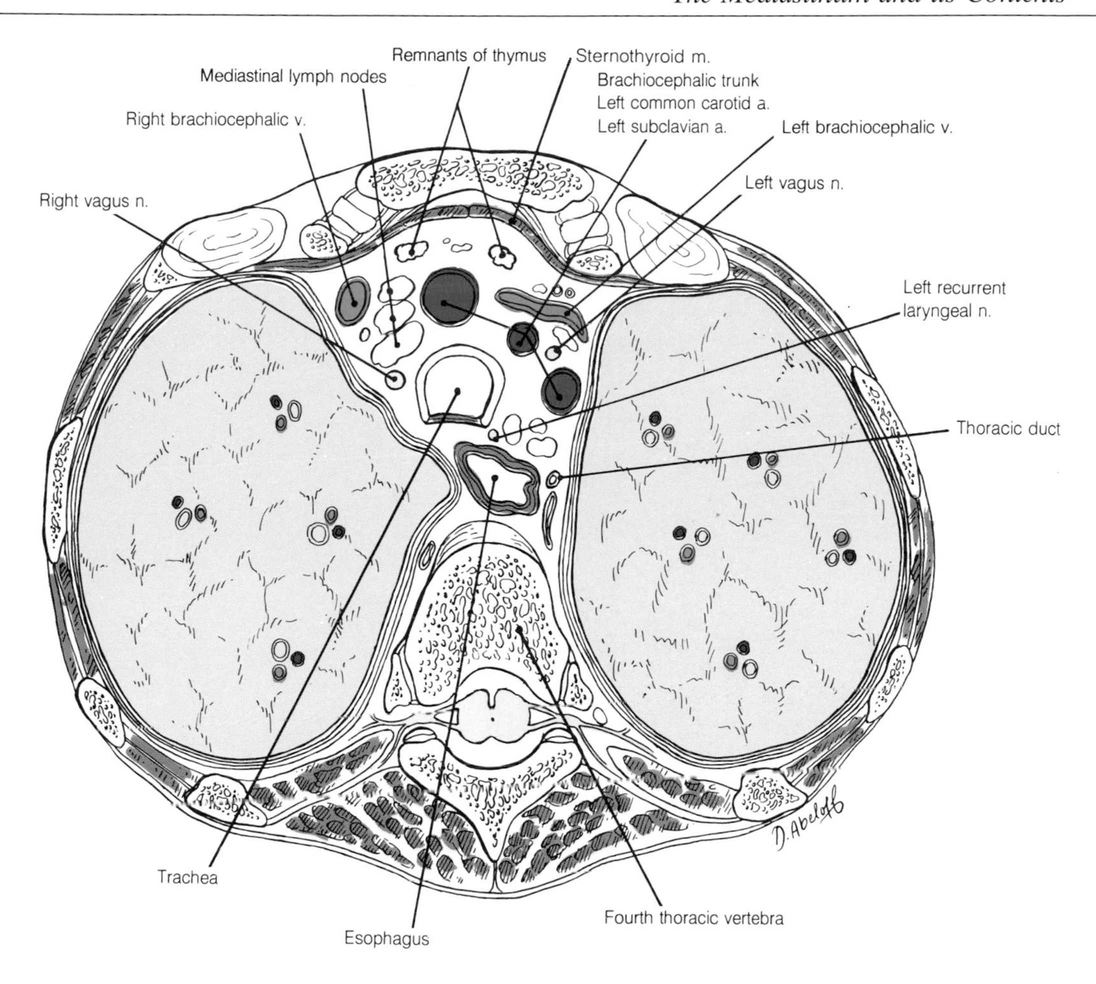

4-56 Horizontal section of the thorax at the level of the fourth thoracic vertebra viewed from below. The section passes just above the arch of the aorta.

a chest radiograph by the translucency produced by the air it contains. The interior of the trachea can be examined with the bronchoscope and it is possible to take biopsies of enlarged tracheo-bronchial lymph nodes through this instrument. For this reason it is useful to know that the **carina***, the projection of the last tracheal cartilage between the openings of the main bronchi, lies about ten inches or 25 centimeters from the incisor teeth.

The Esophagus

The esophagus is a muscular tube lined by a stratified squamous but not keratinized epithelium. It begins in the neck at the level of the cricoid cartilage where it is continuous with the pharynx. It passes into the thorax just to the right of the midline and between the vertebral column and the trachea. In Fig. 4-57, note how it lies behind both the trachea and aorta in the upper part of the thorax. However, it gradually moves forward and to the left to leave the thoracic cavity by passing through the muscular part of the diaphragm at the level of the tenth thoracic vertebra. Other structures lying anterior to the esophagus but not seen in the illustration are the pericardium and left atrium and finally, the diaphragm.

The blood supply of the esophagus is provided by the **inferior thyroid artery** from above, by branches from the **thoracic aorta,** and by branches of the **left gastric artery** ascending from the

* Carina, *L.* = a keel

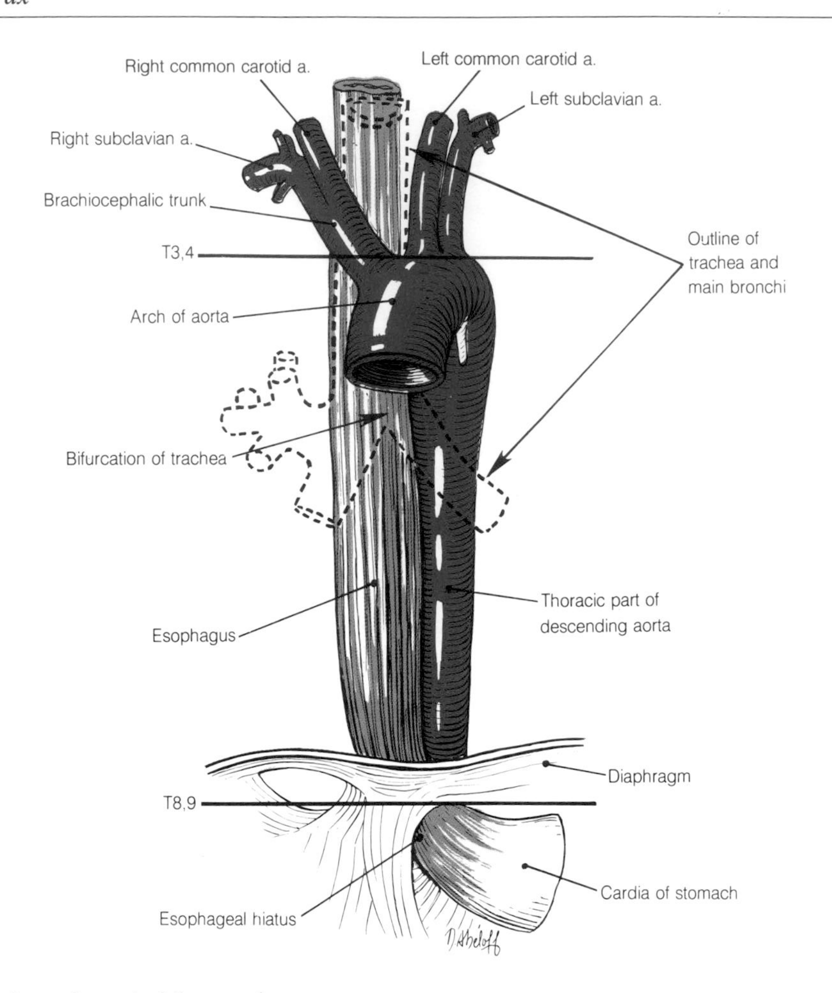

4-57 Showing the thoracic part of the esophagus.

abdominal cavity. Blood returns to the **inferior thyroid veins** and the **azygos system of veins** both of which are part of the systemic venous system. Blood from the lower part of the esophagus drains into the **left gastric vein** which is part of the **hepatic portal system.** This is, then, a site at which the two venous systems are confluent and is known as a **portocaval anastomosis.** The importance of these anastomoses is considered with the description of the portal vein in Chap. 5. The nerve supply is derived from both the vagi and branches from the sympathetic trunks which together form the esophageal plexus.

The Examination of the Esophagus

By means of an esophagoscope, the mucous membrane lining the esophagus can be observed directly. It is pale pink in color and thrown into vertical folds which converge on each other as the lumen is followed through the diaphragm. The pulsation of the nerby arch of the aorta and respiratory and cardiac movements can be seen during this examination.

The lumen of the esophagus can also be outlined by a barium swallow and the passage of the opaque material followed on a screen. In Fig. 4-58, the entire lumen has been filled and a distinct narrowing caused by the arch of the aorta can be seen at *AA*. The region *LA* is related to the left atrium and may become narrowed or displaced if the left atrium enlarges. Note also the effect produced by the longitudinal folds in this region. The projection used in this radiograph is the right anterooblique. Identify the structures *C, S,* and *D.* They are the medial ends of the clavicles, the inferior angle of the scapula, and the dome of the diaphragm.

The point at which the esophagus passes through the diaphragm is of some interest as, al-

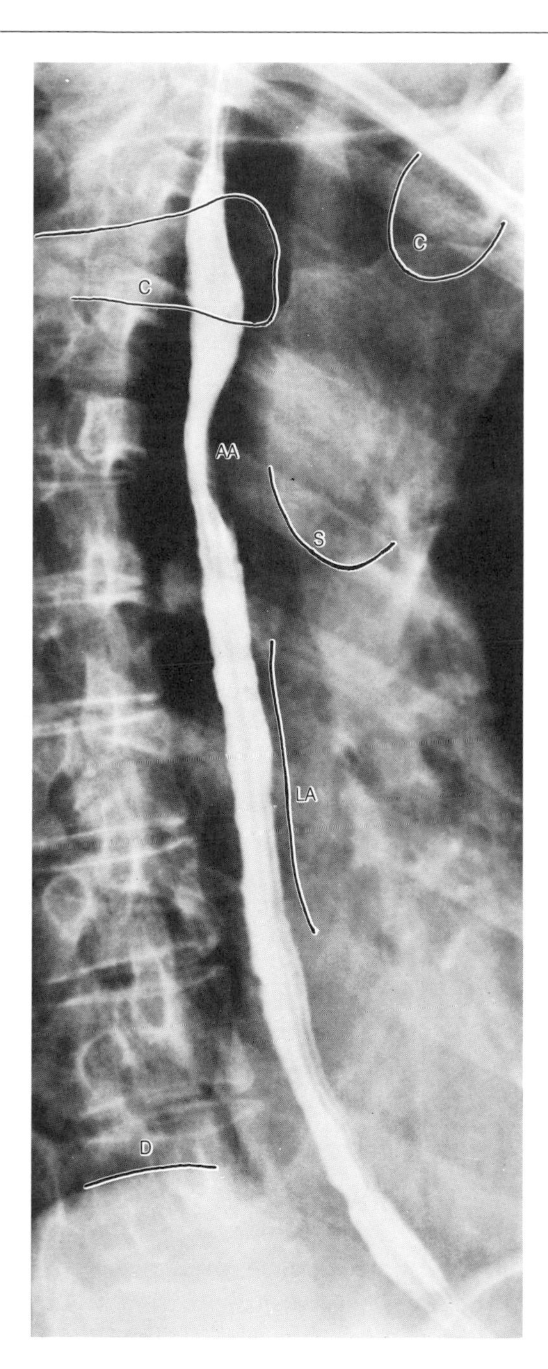

4-58 The esophagus during a barium swallow (right anterooblique projection). (Reproduced by permission, from Wicke: *Atlas of Radiologic Anatomy*, 4th Ed, Urban & Schwarzenberg, Baltimore–Munich, 1987.)

though a functional spincter is present here, no convincing morphological correlate can be demonstrated. Congenital and acquired herniae in which the stomach or other abdominal structures pass into the thorax are also found here.

The Azygos System of Veins

The posterior intercostal veins were earlier described as draining into the azygos system of veins. Like the venae cavae, the main trunk of this system lies on the right-hand side of the body and is called the **azygos vein.** As can be seen in Fig. 4-59, its most inferior tributaries are the right ascending lumbar and subcostal veins and often a lumbar azygos vein. These unite in an inconstant manner to form the azygos vein which enters the thoracic cavity on one or other side of the right crus of the diaphragm. The azygos vein ascends in the posterior mediastinum on the right side of the vertebral column and on reaching the level of the fourth thoracic vertebra arches forward over the root of the right lung to join the superior vena cava. Note that the right posterior intercostal veins of all but the first space drain into it. On the left, there is a similar but smaller trunk, called the **hemiazygos vein,** which crosses the midline at a variable level to join the azygos vein. The lower posterior intercostal veins drain into the hemiazy-

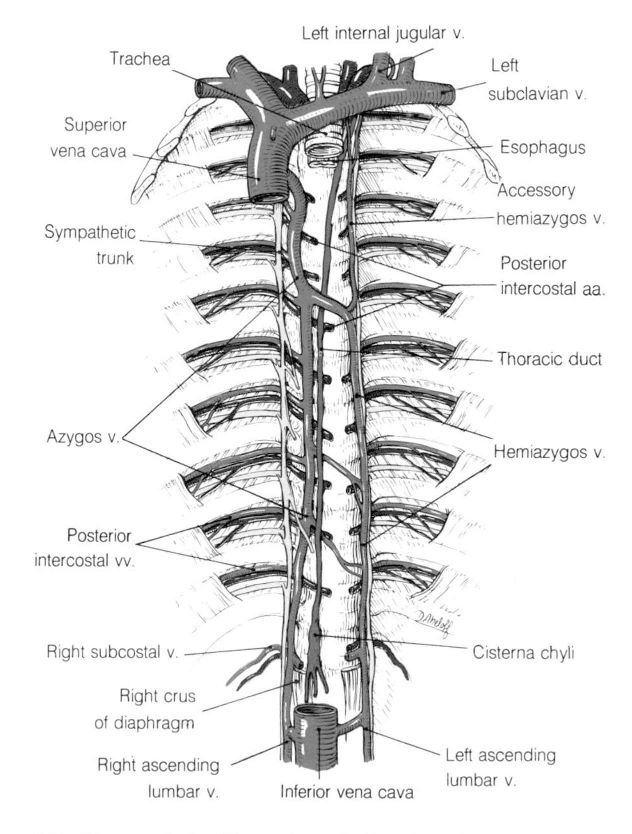

4-59 The posterior thoracic wall showing the azygos system of veins and the thoracic duct.

gos vein and further cross communications may exist between the hemiazygos and azygos veins. The upper posterior intercostal veins can be seen draining into the **accessory hemiazygos vein** which descends to join the hemiazygos vein. Note the relationships of these vertically running venous trunks. Lateral to them are the sympathetic trunks, the thoracic duct lies between them, posterior to them are the posterior intercostal arteries, and the esophagus and aorta, which have been removed, lie anterior to them.

The azygos system forms a potential anastomosis between the inferior and superior vena cava, and in the event of obstruction of either will form an important alternative circulation.

The Descending Aorta

The thoracic part of the descending aorta is a continuation of the aortic arch. It begins at the level of the lower border of the fourth thoracic vertebra, traverses the posterior mediastinum, and leaves the thoracic cavity by passing through the aortic hiatus of the diaphragm at the level of the twelfth thoracic vertebra. Here it becomes continuous with the abdominal part. Initially lying to the left of the vertebral column it gradually moves to the midline and in front of the vertebral column. At the beginning of its course the root of the left lung lies anterior to it and the esophagus lies on its right side. As the aorta descends through the posterior mediastinum it passes behind the esophagus and at the level of the diaphragm this is situated to its left.

Branches are given off to the mediastinum and include those to the bronchi, pericardium, and esophagus. Paired lateral branches are the posterior intercostal arteries, the subcostal arteries, and the superior phrenic arteries for the supply of the diaphragm. The large paired ostia of the intercostal arteries which lie on the posterior aspect of the aorta are often met with surprise by those watching, for the first time, the evisceration of the mediastinum at an autopsy.

The Thoracic Duct

The thoracic duct is a large valved lymphatic vessel which is formed in the upper part of the abdominal cavity and ascends through the thoracic cavity to reach the root of the neck. Here it terminates by opening into the junction of the **left subclavian and internal jugular veins.** Refer again to Fig. 4-59 and find the **cisterna chyli.** This is a dilatation at the lower end of the thoracic duct lying anterior to the **body of the first lumbar vertebra.** It usually lies behind the right crus of the diaphragm and to the right of the aorta. Into the cisterna chyli drain the intestinal trunks from the intestinal canal and the lumbar trunks which receive lymph from the lower limbs, the deep lymph vessels of the abdominal and pelvic walls, and vessels from the abdominal and pelvic organs not associated with the intestinal tract. Leaving the cisterna chyli the thoracic duct enters the thorax alongside the aorta and lying on the right of the bodies of the thoracic vertebrae and anterior to the right posterior intercostal arteries. At about the level of the fifth vertebra it crosses the midline and ascends into the neck on the left side of the esophagus. During its course through the thorax, the thoracic duct receives lymphatic vessels from the lower right intercostal spaces, the left intercostal spaces, and from the posterior mediastinal lymph nodes. Despite the fact that it receives lymph from the greater part of the body, obstruction of the thoracic duct is symptomless because alternative pathways can be found. Occasionally, however, it may be ruptured by trauma or a malignant tumor. If its contents enter the pleural cavity, a **chylous effusion** is formed.

The Lymphatic Drainage of the Thorax

Superficial lymphatic vessels of the thoracic wall drain to nearby groups of lymphatic nodes. Posteriorly, vessels converge on the subscapular group of axillary nodes and anteriorly on either side of the midline, they pierce the intercostal spaces to reach the parasternal nodes lying along the internal thoracic vein. Between these two regions vessels run to the pectoral group of axillary nodes. The lymphatic drainage of the breast has been described earlier in this chapter and the axillary nodes are described with the upper limb.

The **drainage of the deep surfaces of the thoracic wall and the thoracic viscera** appear at first sight rather complicated, but if the diagram in Fig. 4-60 is followed, the flow of lymph from tis-

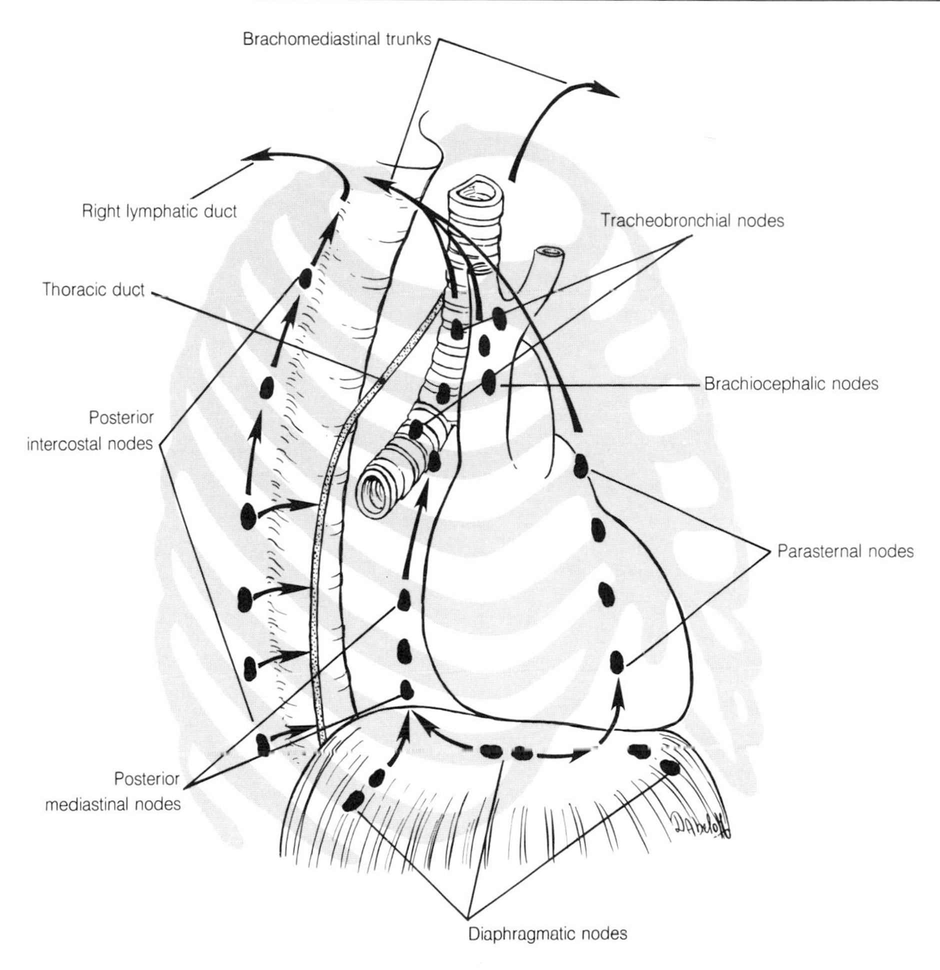

4-60 Diagram illustrating the pathways and nodes concerned with the lymphatic drainage of the thoracic cavity.

sues to groups of nodes and then to the larger vessels which empty into the confluence of the subclavian and internal jugular veins becomes more logical. The deeper tissues of the thoracic wall and parietal pleura drain to a ring of peripheral nodes. These lie behind the sternum (**parasternal nodes**), on the diaphragm (**diaphragmatic nodes**), and in the paravertebral region (**posterior intercostal nodes**).

The thoracic viscera drain to nodes lying between the heart and esophagus (**posterior mediastinal nodes**), those around the main bronchi and trachea (**tracheobronchial nodes**) and a group associated with the brachiocephalic veins (**brachiocephalic nodes**).

Efferent vessels from these three groups and those from the parasternal nodes form the **bronchomediastinal trunks** which directly or indirectly join the systemic venous system in the root of the neck. Efferents from the diaphragmatic nodes drain to the nearby posterior mediastinal or parasternal nodes and those from the posterior intercostal nodes drain to the thoracic duct (lower right nodes and left nodes) or into the right lymphatic duct (upper right nodes).

The Autonomic Nervous System in the Thorax

The function of the autonomic system in the thorax is of great importance to the physician for it is concerned with the control and regulation of the heart, the blood pressure, and the caliber of the bronchi. The manner in which parasympathetic fibers from the sympathetic trunks reach the thorax has already been described. It remains to discuss the contribution of these to the thoracic autonomic plexuses.

These plexuses consist of ill-defined collections of pre- and postganglionic fibers of both sympathetic and parasympathetic systems. They can be isolated to some extent by careful dissection but tend to be dismissed as rather tough connective tissue surrounding the vascular structure to which they are often closely adherent. The contributions of each system to the plexuses is summarized below.

Sympathetic Contributions

Preganglionic sympathetic fibers for distribution to the thorax arise from cell bodies in the lateral columns of the upper four of five thoracic segments of the spinal cord. They leave the cord in the ventral roots of the corresponding spinal nerves and join the sympathetic trunk. Synapse with the cell bodies of postganglionic fibers occurs either in upper thoracic sympathetic ganglia or in cervical sympathetic ganglia to which the preganglionic fibers have ascended in the trunk. From these ganglia the postganglionic fibers are distributed in the following manner:

Cardiac branches of the superior, middle and inferior cervical ganglia to the cardiac and coronary plexuses

Branches from the upper five thoracic ganglia to the aortic, pulmonary, and esophageal plexuses

Parasympathetic Contributions

Preganglionic parasympathetic fibers arising centrally in the dorsal nucleus of the vagus are distributed to the thoracic plexuses by the vagus and its branches. Synapse between pre- and postganglionic fibers occur in the plexuses or in the wall of the organ supplied. The branches of the vagi concerned are:

1. **superior cardiac branches** arising high in the neck and joining the cervical cardiac branches of the sympathetic ganglia,
2. **inferior cardiac branches** arising in the root of the neck and running directly to the cardiac plexuses,
3. **cardiac branches of vagus and recurrent laryngeal nerves** in the thorax,
4. **anterior and posterior pulmonary branches** to the anterior and posterior pulmonary plexuses, and
5. **esophageal branches** to the esophageal plexus.

The Thoracic Autonomic Plexuses

The **superficial cardiac plexus** lies below the arch of the aorta and recieves contributions from the cardiac branch of the left superior cervical sympathetic ganglion and the inferior cardiac branch of the left vagus. This plexus communicates with the deep cardiac plexus and the left coronary plexus.

The **deep cardiac plexus** lies between the arch of the aorta and the bifurcation of the trachea and receives contributions from the cardiac branches of the cervical sympathetic ganglia except that of the left superior ganglion, the cardiac branches of the upper thoracic sympathetic ganglia, and the cervical and thoracic cardiac branches of the vagi and recurrent laryngeal nerves. Branches of this plexus reach the atria of the heart, the coronary plexuses and the pulmonary plexuses.

The **coronary plexuses** are extensions of the cardiac plexuses along the coronary arteries and they carry autonomic fibers to the atria and ventricles of the heart.

The **anterior and posterior pulmonary plexuses** lie in front of and behind the root of the lung. On each side they receive contributions from the vagus, the cardiac plexuses, and upper thoracic sympathetic ganglia, and are distributed to the bronchi and pulmonary vessels.

The **esophageal plexus** is formed by several branches of the vagus that leave the posterior pulmonary plexuses and break up into filaments on the surface of the esophagus. These reform at the lower end of the esophagus to pass through the diaphragm as the **anterior and posterior vagal trunks.** The esophageal plexus also receives fibers from the upper five sympathetic ganglia.

The Fetal Circulation

During the description of the thorax and the abdomen a number of vestigial vascular structures are mentioned whose presence in the adult body can

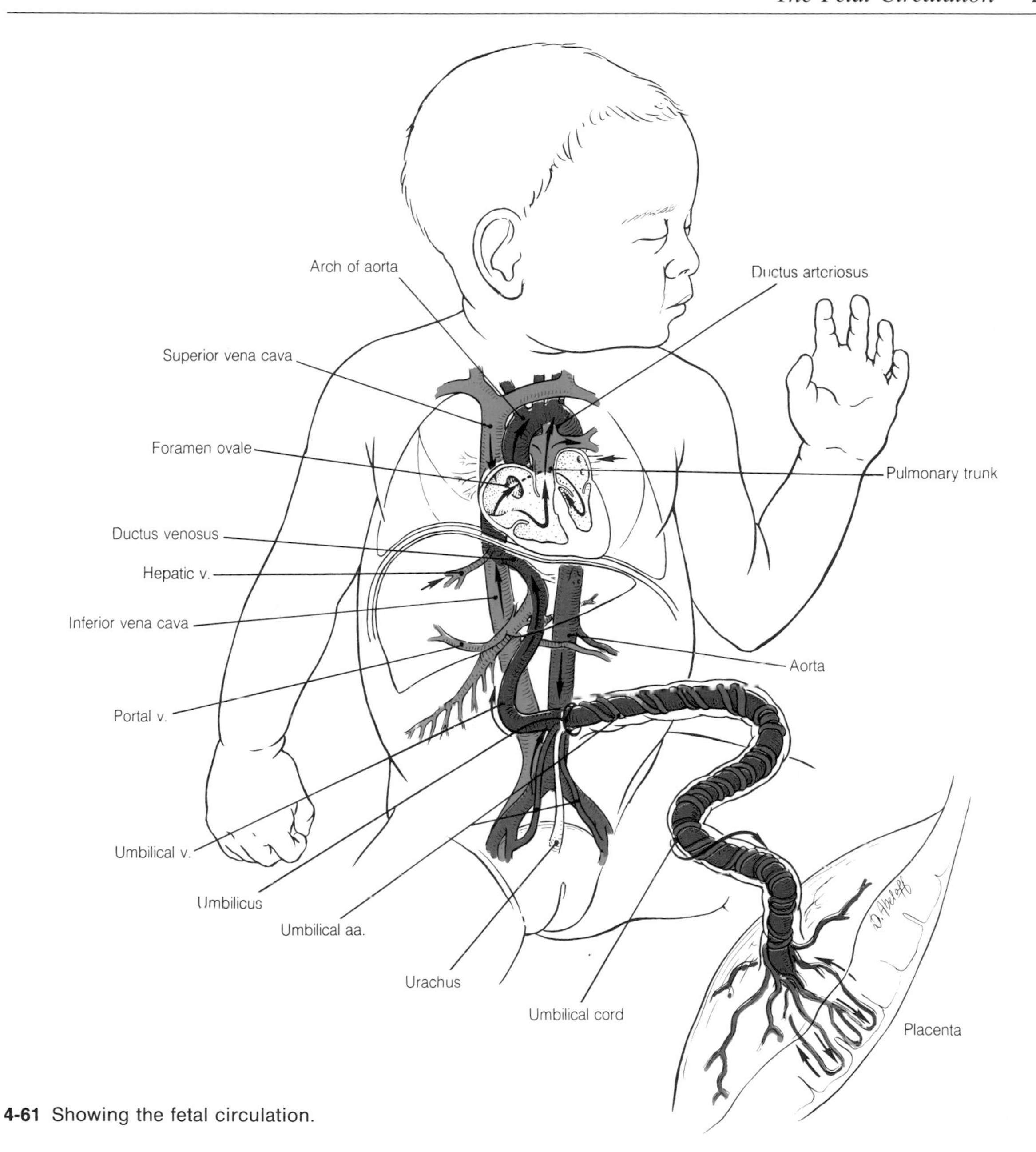

4-61 Showing the fetal circulation.

only be explained in the light of the circulation of blood in the fetus immediately before birth. At this time the **placenta** serves to oxygenate the fetal blood and provide it with nutrients. Look now at Fig. 4-61 and follow the course of this enriched blood as it leaves the placenta in the **umbilical vein** to reach the fetus at the umbilicus. From this point the umbilical vein runs in the free border of the falciform ligament to the porta hepatis. Here it gives off branches to the left lobe of the liver before joining the left branch of the portal vein. Near the point at which these two veins join a large vessel called the **ductus venosus** is formed. This runs across the visceral surface of the liver in the hepatic margin of the gastrohepatic ligament to join the left hepatic vein and thence the inferior vena cava. In this way a small quantity of placental blood reaches the liver either directly or through the portal vein. The greater part, however, bypasses the liver and, traveling in the ductus venosus, reaches the inferior vena cava.

In the vena cava the placental blood is mixed with some deoxygenated blood returning from the lower part of the fetus but, nevertheless, remains from 65-70% saturated with oxygen. On entering the right atrium the blood from the vena cava is directed by the **valve of the inferior vena cava** toward the **foramen ovale** through which it passes to reach the left atrium. Here it is mixed with a small amount of blood returning from the pulmonary circulation. Blood in the left atrium flows into the left ventricle and is pumped from this chamber into the aorta for distribution to the head, neck, and upper limbs through the large vessels which leave the arch.

Returning from the head and neck the deoxygenated blood enters the right atrium in the superior vena cava and from here is directed with some oxygenated blood from the inferior vena cava to the right ventricle. Contraction of the right ventricle forces blood into the pulmonary trunk. From here most of the blood flows in the large **ductus arteriosus** which joins the aorta after the branches of the arch have been given off. Only a small quantity reenters the pulmonary circulation through the relatively small pulmonary arteries. Poorly oxygenated blood in the descending aorta now passes to the lower part of the fetus and to the **umbilical arteries** which leave the internal iliac arteries to return to the umbilical cord and the placenta. Note that because of the manner in which the placenta is inserted into the fetal circulation, the umbilical veins carry oxygenated blood and the umbilical arteries deoxygenated blood.

At birth a rapid adjustment of this circulation is necessary as the flow of blood from the placenta is cut off and ventilation of the lungs commences. These events are briefly summarized below:

1. Thrombosis and eventual obliteration of the umbilical vein to form the **ligamentum teres**;
2. Closure and obliteration of the ductus venosus to form the **ligamentum venosum**;
3. A fall in pressure in the pulmonary trunk and arteries as the resistance in the pulmonary vascular bed falls;
4. The closure of the valve-like foramen ovale as a relative increase in left atrial pressure forces the valve against the septum. Fusion of this valve with the septum gives rise to the **fossa ovalis**;
5. Contraction and eventual obliteration of the ductus arteriosus. The vestige of this is the **ligamentum arteriosum**; and
6. Obliteration of the umbilical arteries from their vesical branches in the pelvis to the umbilicus. The remnants of these vessels form the **medial umbilical ligaments**.

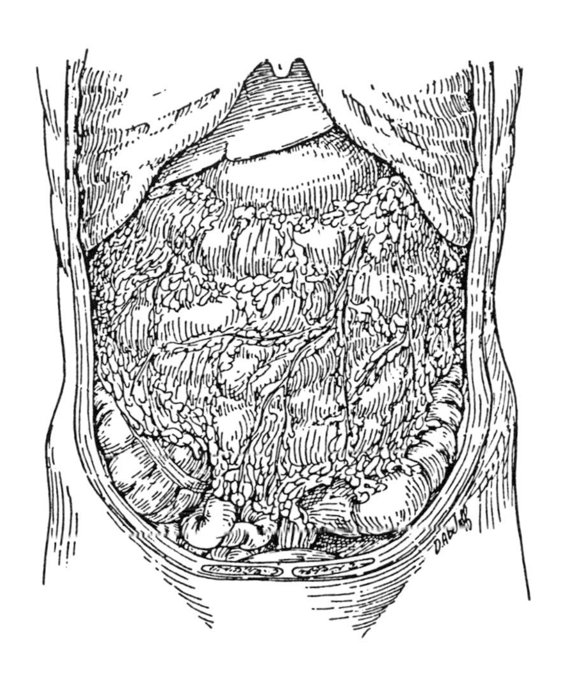

5 The Abdomen

The study of the abdomen follows naturally on that of the thorax. The two are separated by no more than the musculature of the diaphragm through which a number of major structures pass and which, because of its domed shape, allows some abdominal organs to lie within the protection of the thoracic cage.

The Abdominal Wall

The abdominal cavity is to a large extent well protected by musculoskeletal walls. Its upper part lies within the thoracic cage and its lower part within the greater pelvis and between the wings or alae of the iliac bones. Posteriorly, the intervening gap is filled in by a thick layer of back muscles and the vertebral column. Anteriorly and laterally, the walls are made up of a relatively thin expansile muscular sheet, but even here rapid reflex contraction of muscle and the resultant narrowing of the vulnerable area affords some protection.

It is through the thin relaxed, anterior abdominal wall that a routine abdominal examination is made and once more it is evident that the normal position of organs must be known. This norm is, however, somewhat less clear-cut than elsewhere in the body. Even relatively fixed organs show a normal variation while others move considerably with respiration or are suspended from a sheet of mesentery and are thus extremely mobile.

Surgical approaches are also made through the anterior and lateral abdominal walls. Incisions for these are designed to provide the most effective exposure, but at the same time do the minimum damage to the abdominal wall. Each of these criteria depends on anatomical details of muscle fiber direction, fascial sheaths, and the course of the motor nerves. Except in the lowermost part of the abdominal wall these nerves are the terminations of the intercostal nerves which escape from the costal margin to continue their downward and forward course. As a landmark for these, remember that the dermatome around the umbilicus is related to the tenth intercostal nerve.

The abdominal wall possesses areas of potential weakness through which the abdominal contents may herniate. The region most commonly involved in this condition is the inguinal canal. This canal allows the descent of the testicle from the posterior abdominal wall to its postnatal scrotal position, and a number of conditions in this region – congenital indirect hernia, undescended testicle, hydrocele of the cord – can be traced directly back to faulty embryological development.

As well as having a protective function, the muscles of the anterior and lateral abdominal wall also serve to move the trunk and, by altering the intraabdominal pressure, take part in respiration, coughing, and defecation.

Superficial Structures of the Abdominal Wall

Even in the cadaver it is worthwhile examining the skin of the abdominal wall. Are there any scars? Do midline pigmentation or striae gravidarum indicate previous pregnancies? Is there a hernia present or has one been repaired? Observations of this kind may well influence the course of your dissection of the abdominal contents and are, of course, a normal part of a clinical examination.

The skin is supplied by the cutaneous branches of the **seventh to twelfth intercostal nerves** and by the **first lumbar nerve** in the form of the **iliohypogastric nerve**. Note in Fig. 5-1 how these become superficial over the lateral abdominal wall and near the midline and behave as the terminal branches of the lateral and anterior cutaneous branches of a typical ventral ramus. Also in Fig. 5-1 can be seen the free communication which exists between **thoracoepigastric, intercostal**, and **superficial epigastric veins**. This network forms

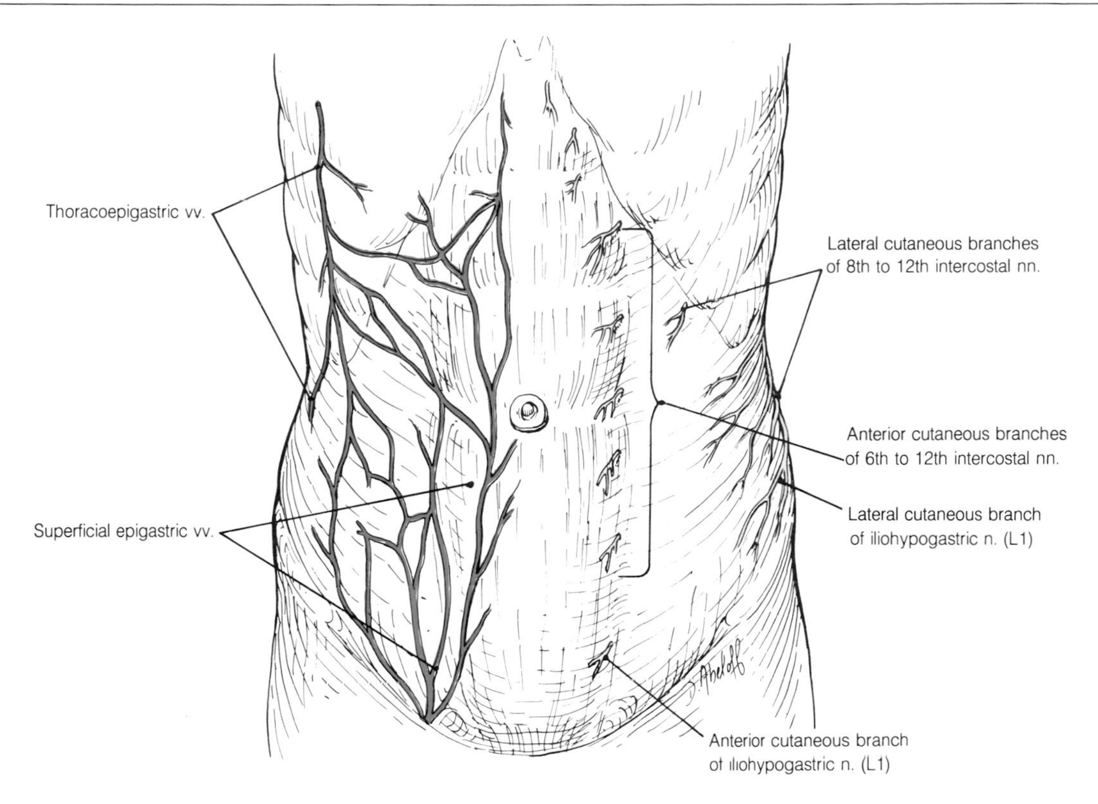

5-1 The superficial veins and nerves of the anterior abdominal wall.

an alternative venous pathway if the inferior vena cava becomes obstructed.

Beneath the skin the **superficial fascia** exists as a layer of adipose tissue of very variable thickness. However, over the lower part of the anterior abdominal wall its deeper stratum is condensed into a thin but definite sheet of fibrous connective tissue. Hence, the superficial fascia is said to have a superficial and a deep layer. The deep or membranous layer will be seen to be continuous with a similar layer in the perineum. The deep fascia amounts to the fascial sheaths of the abdominal muscles and except in the case of the rectus sheath is thin and does not inhibit their freedom of movement.

Superficial lymphatic vessels draining the anterior abdominal wall above the level of the umbilicus drain upward to the axillary nodes. Below this level, vessels drain to the superficial inguinal nodes in the groin where they are joined by those sweeping round from the region of the buttocks. Lymphatic vessels from deeper structures in the wall drain to the nodes lying along the main vessels within the abdominal cavity, i.e. iliac and lumbar nodes.

The Skeleton of the Abdominal Wall

Before the muscles of the abdominal wall can be usefully discussed some new osteology must be introduced and some reviewed.

The formation of the costal margin by the fusion of the costal cartilages of the seventh to tenth ribs and by the free-lying tips of the eleventh and twelfth costal cartilages should be reviewed as should the lumbar spine.

The **fifth lumbar vertebra** articulates with the **sacrum** at the brim of the **greater pelvis**. Posteriorly the two pelvic bones articulate with the sacrum on either side at the **sacroiliac joints** and anteriorly with each other at the **symphysis pubis**, a secondary cartilaginous joint. Before puberty the pelvic bone is formed from three phylogenetically independent bones which are fused together by hyaline cartilage. These are the **iliac bone**, the **pubic bone**, and the **ischium**. The iliac bone presents a large **ala** or wing on either side of the greater pelvis which serves largely for muscle attachment and protection. Its abdominal surface gives origin to the strong iliacus muscle. The free upper border is known as the **iliac crest**. This crest is limited posteriorly by the **posterior superior iliac spine** and anteriorly by the **anterior superior iliac spine**. Posteriorly, the crest provides attachment for the **thoracolumbar fascia**, the **latissimus dorsi**, and **quadratus lumborum muscles**. From the intermediate and anterior part of each crest arise the three flat sheets of muscle that form the greater part of the lateral and anterior abdominal wall. About two inches behind the anterior superior iliac spine the lateral margin of the crest bears a projection called the **iliac tubercle**.

Below the anterior superior spine lies the less well marked **anterior inferior spine**. Beyond this the ilium fuses with the pubic bone and its superior ramus. This bears a linear marking, the **pectineal line**. This line, when followed medially, leads toward the **pubic tubercle** and **pubic crest**.

Turn now to Fig. 5-2 in which many of the features described above are illustrated. Identify the costal margin, the xiphoid process, the lumbar vertebrae and their transverse processes, the promontory of the sacrum, the iliac bones and iliac crests, the anterior superior and inferior iliac spines, the pectineal line or pecten pubis, pubic tubercle and pubic crest, and in the midline the pubic symphysis. Note that in the correctly oriented pelvic girdle the coccyx articulating with the lower margin of the sacrum can be seen above the level of the symphysis pubis.

The pelvic bone is described in greater detail with the pelvic region and the lower extremity.

The Surface Anatomy of the Anterior Abdominal Wall

Having established some bony landmarks, the anterior abdominal wall can be divided by horizontal and vertical planes into a number of regions which are of use to the clinician when describing the site of pain felt by a patient or of abnormal physical signs such as areas of tenderness or tumors. In Fig. 5-3, two transverse and two vertical planes are indicated which divide the anterior abdominal wall into three midline, three left, and three right regions. The level of the diaphragm is also in-

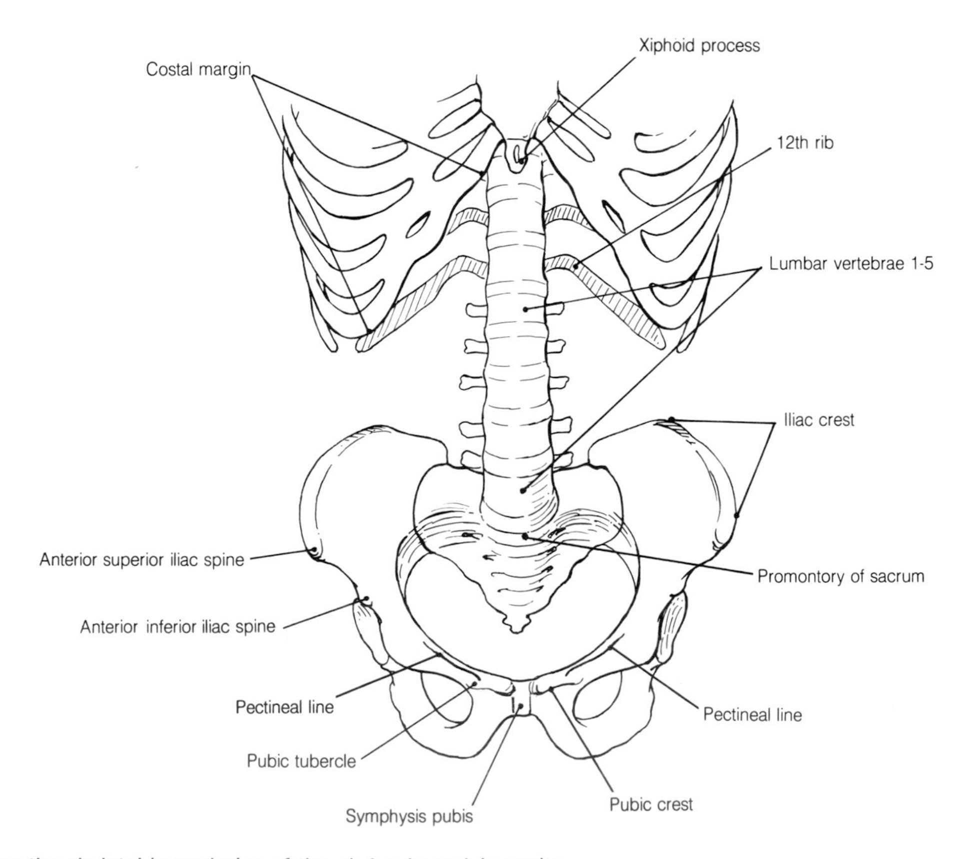

5-2 Showing the skeletal boundaries of the abdominopelvic cavity.

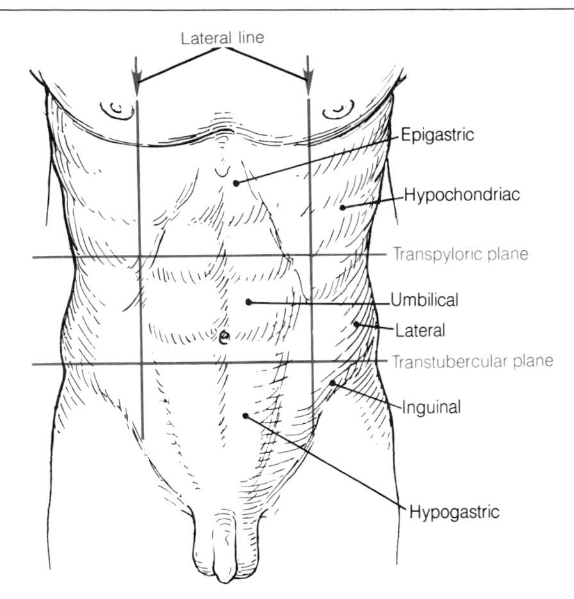

5-3 The anterior abdominal wall showing descriptive regions and planes.

cluded to show the upper limit of the abdominal cavity. Note in the illustration that:

1. The **transpyloric plane** passes through a point midway between the jugular notch and symphysis pubis (or more simply, between the xiphisternum and umbilicus).
2. The **transtubercular** plane passes through the tubercles of the iliac crests.
3. The **vertical right and left lateral planes** almost correspond to the midclavicular planes of the thorax and pass through the midpoint of a line joining the anterior superior iliac spine and the symphysis pubis (the midinguinal point).
4. The midline regions are called the **epigastric, umbilical**, and **hypograstric regions**.
5. The lateral regions are called the **hypochondriac** and the **lateral (lumbar)** and **inguinal (iliac)** regions.

In practice these descriptive regions are used without precise reference to the planes that define them.

As will be seen, the transpyloric plane which also passes through the body of the first lumbar vertebra, is of use to the anatomist to both describe and picture the relationships of several abdominal organs clustered about it.

An **intercristal plane** passing through the highest points of the iliac crests is of value in determining the level of the spine of the fourth lumbar vertebra prior to performing a spinal tap.

The surface projections of individual abdominal organs will be given as each is described.

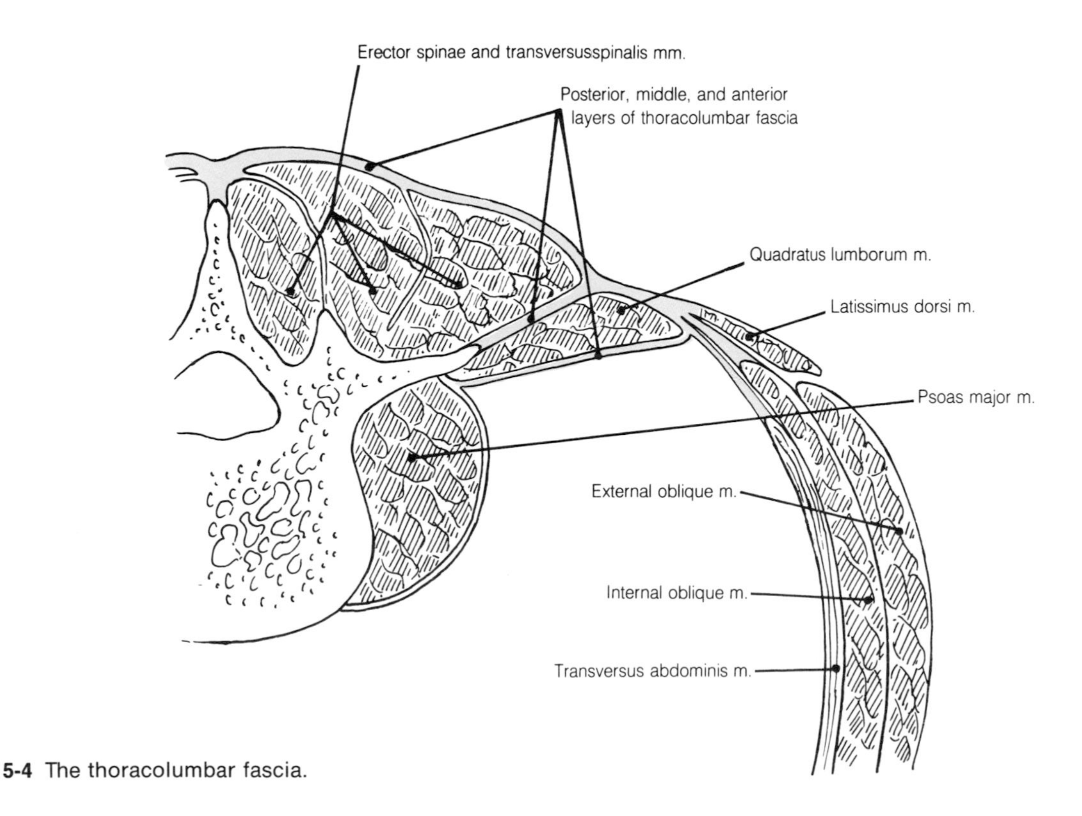

5-4 The thoracolumbar fascia.

The Thoracolumbar Fascia

Although the lumbar part of the thoracolumbar fascia is an integral part of the posterior abdominal wall, it forms an important site of origin for muscles of the lateral and anterior abdominal walls. It must, therefore, be given some consideration at this point. Look at Fig. 5-4 and see that sheets of fascia arise from:

1. The tips of the lumbar spines
2. The tips of the lumbar transverse processes
3. The anterior aspects of the lumbar transverse processes.

The anterior and middle sheets enclose the quadratus lumborum muscle and the middle and posterior sheets enclose the rector spinae muscle. All three sheets fuse lateral to the lumbar transverse processes and provide attachment for the **internal oblique** and **transversus abdominis muscles**.

The Muscles of the Anterior Abdominal Wall

The musculature of the anterior and lateral walls of the abdomen is made up of a trilaminar sheet on either side of a pair of vertically oriented muscles. The thin aponeurotic tendons of the three lateral muscles form a sheath around each vertical muscle before fusing in the mid-line at the **linea alba**.* The trilaminar sheet is composed of:

1. The external oblique muscle (obliquus externus abdominis)
2. The internal oblique muscle (obliquus internus abdominis)
3. The transversus abdominis muscle.

The vertically oriented muscles are the rectus* abdominis muscles.

The External Oblique Muscle

The external oblique muscle arises as digitations from the outer surfaces of the lower eight ribs. The fleshy fibers fan out downward and medially over the anterior abdominal wall. There is a **free posterior margin** to the muscle where its most posterior fibers run from the twelfth rib to the anterior half of the outer margin of the iliac crest. The remaining more obliquely running fibers become an aponeurotic sheet which contributes to the anterior sheath of the rectus muscle before fusing with its fellow at the **linea alba** in the mid-line. The lower free margin of the aponeurosis extends from the anterior superior iliac spine to the pubic tubercle and is called the **inguinal ligament**. The arrangement of the fibers and aponeurosis is seen in Fig. 5-5. Note that there are no fleshy fibers below a line joining the umbilicus to the anterior superior iliac spine, a point to remember when the layers through which an abdominal incision passes are identified.

The Internal Oblique Muscle

The fibers of the internal oblique muscle which are shown in Fig. 5-6 arise from the thoracolumbar fascia, the anterior two-thirds of the iliac crest deep to the attachment of the external oblique, and from the lateral two-thirds of the inguinal ligament. The fibers fan out from this origin. The uppermost run upward and medially to become attached to the costal margin. The intermediate fibers become aponeurotic and help in the forma-

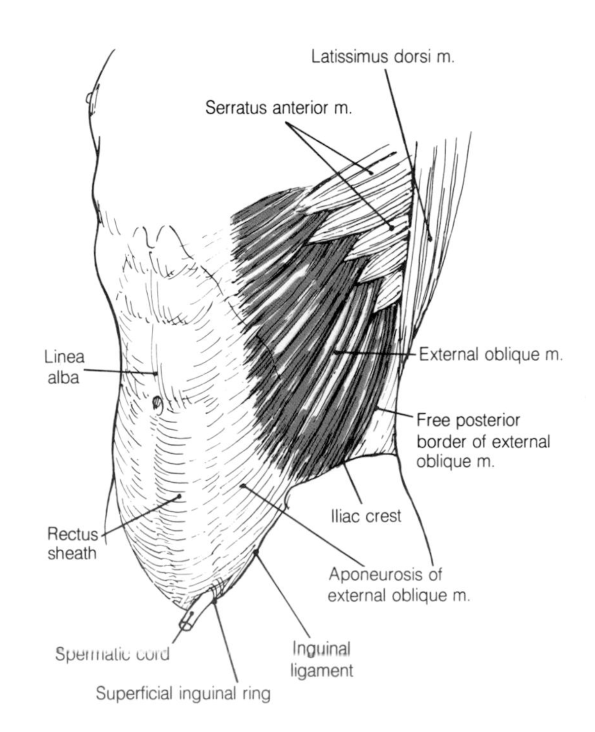

5-5 The external oblique muscle of the abdomen.

* Linea alba, *L.* = white line; Rectus, *L.* = straight or upright

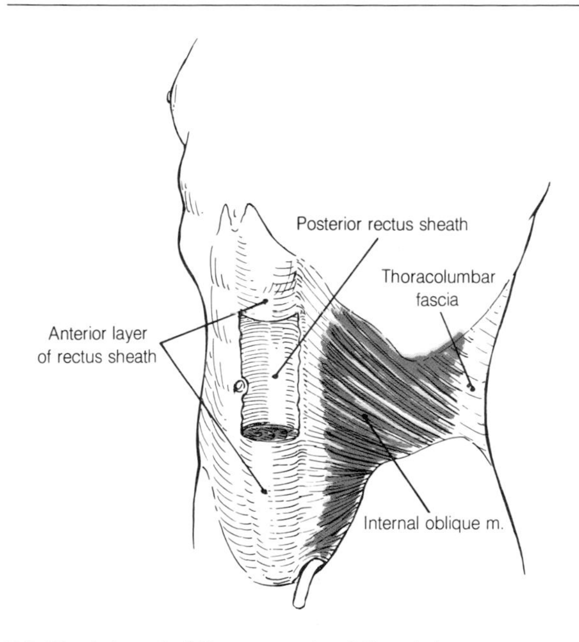

5-6 The internal oblique muscle of the abdomen.

tion of the **rectus sheath** before joining the linea alba. The lowermost are attached by a flattened tendon to the **pectineal line** on the superior pubic ramus. This tendon is fused with a similar attachment of the transversus abdominis muscle and is known as the **conjoint tendon**.

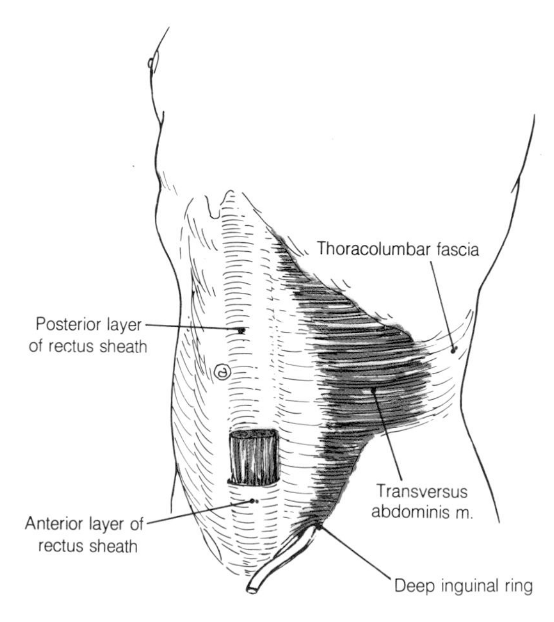

5-7 The transversus muscle of the abdomen.

The Transversus Abdominis Muscle

The fibers of transversus abdominis arise from a long origin which extends from the deep surface of the costal margin, the thoracolumbar fascia, the anterior two-thirds of the medial margin of the iliac crest, and the outer half of the inguinal ligament. Running approximately transversely across the abdominal wall, the fibers also become aponeurotic and contribute to the rectus sheath before joining the linea alba. The lowermost fibers are attached to the pectineal line by the conjoint tendon. The arrangement of the fibers of this muscle can be seen in Fig. 5-7.

The Rectus Abdominis

The two rectus abdominis muscles form the vertical component of the anterior abdominal musculature and lie on either side of the linea alba. The muscles are broad superiorly and narrow inferiorly. Each is attached to the fifth, sixth, and seventh costal cartilages above and below by tendinous and fleshy insertions to the pubic crest and the symphysis pubis. As can be seen in Fig. 5-8, the fibers are separated into segments by **fibrous intersections**. One of these lies at the level of the umbilicus and two are above. The intersections are adherent to the anterior wall of the sheath of the muscle and must be freed from this before the muscle can be displaced laterally, as in making a paramedian surgical incision. A small triangular muscle called **pyramidalis** covers the lower fibers and tendons of attachment of each rectus muscle.

The Rectus Sheath

Each rectus abdominis muscle is enclosed in a fibrous sheath formed by the aponeurotic tendons of the three lateral muscles.

The external oblique contributes to the anterior layer of the sheath over its whole extent. Below the costal margin the internal oblique aponeurosis splits around the rectus muscle contributing to anterior and posterior layers and the aponeurosis of the transversus abdominis passes into the posterior layer.

Midway between the umbilicus and the symphysis pubis, the posterior wall of the sheath becomes deficient since all aponeuroses pass anterior to the rectus abdominis. The posterior wall of the sheath

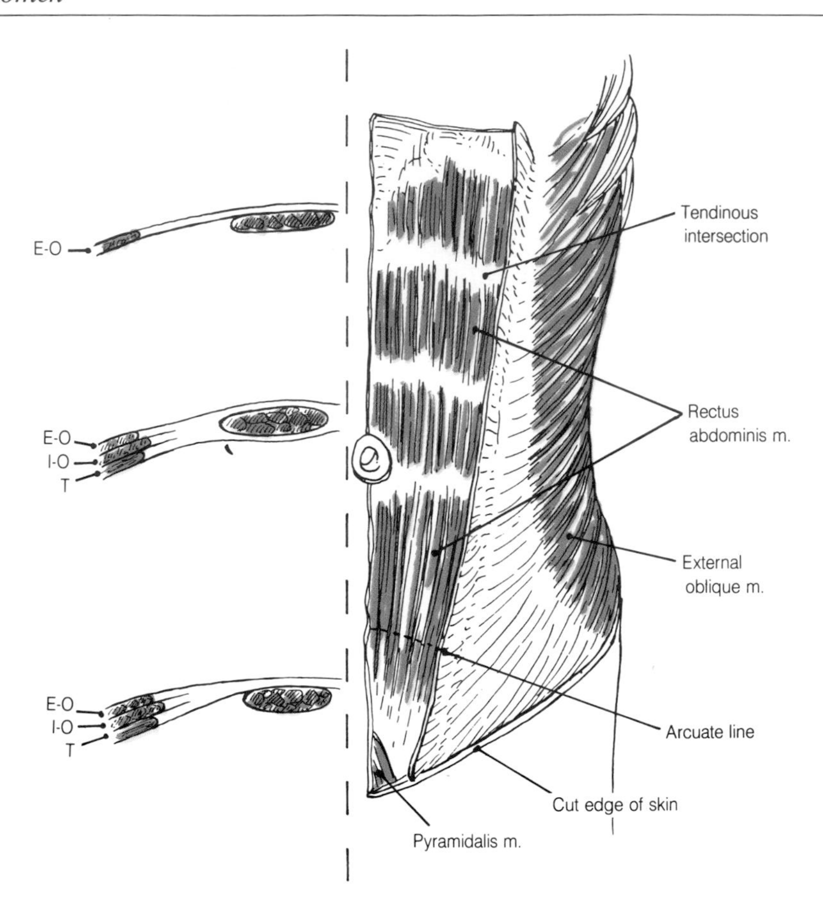

5-8 The rectus abdominis muscle and the contributions of the three lateral abdominal muscles to its sheath.

is also deficient above the costal margin where the muscle lies directly on the underlying costal cartilages. The composition of the rectus sheath at different levels is shown diagrammatically in Fig. 5-8 alongside an illustration of the muscle.

At the level at which the aponeuroses of all three lateral muscles fuse to form only the anterior layer of the sheath, the posterior sheath terminates at a free margin called the **arcuate line**. It is here that the **inferior epigastric artery** enters the sheath to run superiorly on the deep surface of the rectus abdominis muscle. The artery anastomoses with the **superior epigastric artery** which has entered the sheath from above by passing deep to the costal margin. Below the level of the arcuate line the rectus abdominis lies on the **transversalis fascia**.

The muscles of the anterior abdominal wall are supplied by the **lower six thoracic** and **first lumbar segmental nerves**. The thoracic nerves emerge beneath the costal margin and run downward and forward around the abdominal wall between the internal oblique and transversus abdominis muscles. This is called the **neurovascular plane** as the nerves are accompanied by branches of the **musculophrenic** or the **first lumbar artery**. The nerves, but not the arteries, pierce the rectus sheath to supply the muscles within. To thoracic nerves are added the **iliohypogastric** and **ilioinguinal nerves** which are derived from the first lumbar nerve. These supply the lower fibers of the external oblique, internal oblique, and transversus abdominis muscles.

In addition to branches of the musculophrenic and lumbar arteries, which supply the lateral muscles, the superior and inferior epigastric arteries supply the rectus abdominis muscle.

As has already been mentioned, the muscles of the anterior abdominal wall act in a protective capacity and by their contraction are able to increase intraabdominal pressure. The rectus abdominis, an antagonist of erector spinae, flexes the trunk against gravity or resistance and the lateral muscles produce rotation and lateral flexion of the trunk to their own side.

The Transversalis Fascia

Beneath the muscles of the anterior abdominal wall lies a thin layer of transversalis fascia which separates the deep surfaces of transversus abdominis and rectus abdominis from the underlying extraperitoneal fat and peritoneum.

The Inguinal Region

It is through the lower part of the anterior abdominal wall that the testis passes in its descent into the scrotum. In so doing it sets the scene for congenital and acquired defects that allow the protrusion of a sac of peritoneum through the abdominal wall. Such protrusions commonly contain intestine and are known as **herniae**. The repair of herniae or the treatment of the acute conditions of obstruction or necrosis of any contents they may contain, forms an appreciable part of general surgical practice. For this reason some emphasis is given to what is known as the inguinal region. Strictly speaking, the region is confined to a triangular area of the anterior abdominal wall which lies below the transtubercular plane, lateral to the lateral line and medial to the iliac crest and inguinal ligament. In a discussion of the inguinal canal and its contents, the adjacent part of the hypogastric region is included.

The Inguinal Ligament

The inguinal ligament is not a free-lying structure but a thickening in the lower border of the aponeurosis of the external oblique muscle. It extends from the anterior superior iliac spine to the pubic tubercle. Its lower edge is not sharp but incurved forming a shallow trough. To the outer convexity of the trough is attached the **fascia lata** of the thigh whose tension keeps the ligament bowed downward. The simple diagram seen in Fig. 5-9 should make this clear. If the inguinal ligament is superimposed on the pelvis, it is separated from the pectineal line on the superior ramus of the pubic bone by a triangular space. In life this space is filled in by the **lacunar ligament**. If the pelvis is held correctly oriented, i.e., with the symphysis pubis and the anterior superior iliac spines in the same vertical plane, the lacunar ligament can be seen to lie in a horizontal plane and, thus, is able to form part of the floor of the **inguinal canal** and support its contents. The diagram in Fig. 5-14 should help to demonstrate this point.

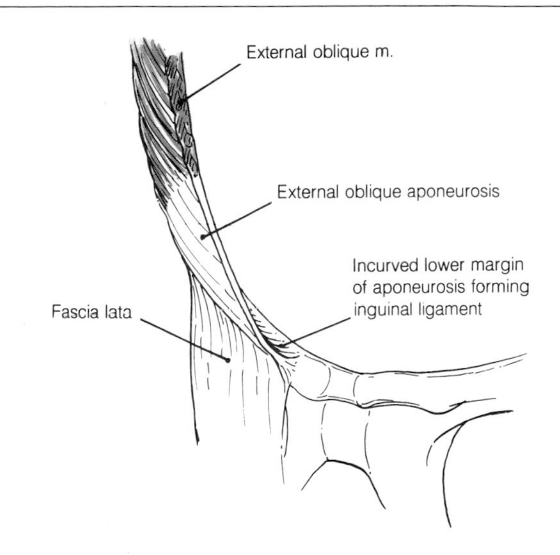

5-9 To show the shape of the inguinal ligament and its attachment to the fascia lata of the thigh.

The Superficial Inguinal Ring

Just above the pubic tubercle the fibers of the external oblique aponeurosis split to leave a triangular gap with its apex above and laterally. The lateral side of the triangle called the **lateral crus** descends to the pubic tubercle and the medial side, the **medial crus**, to the symphysis pubis. Bone of the **pubic crest** forms the short base. This triangular gap is the **superficial inguinal ring** through which the **spermatic cord** in the male and the **round ligament of the uterus** in the female leave the inguinal canal. This triangular space is commonly made more "ring-like" by intercrural fibers which obliterate its apex. Look at Fig. 5-10 to be sure that the form and positon of the superficial ring is understood.

The Inguinal Canal

The word canal brings to mind a sizeable and patent structure. As far as the inguinal canal is concerned this is far from a true picture for its lumen

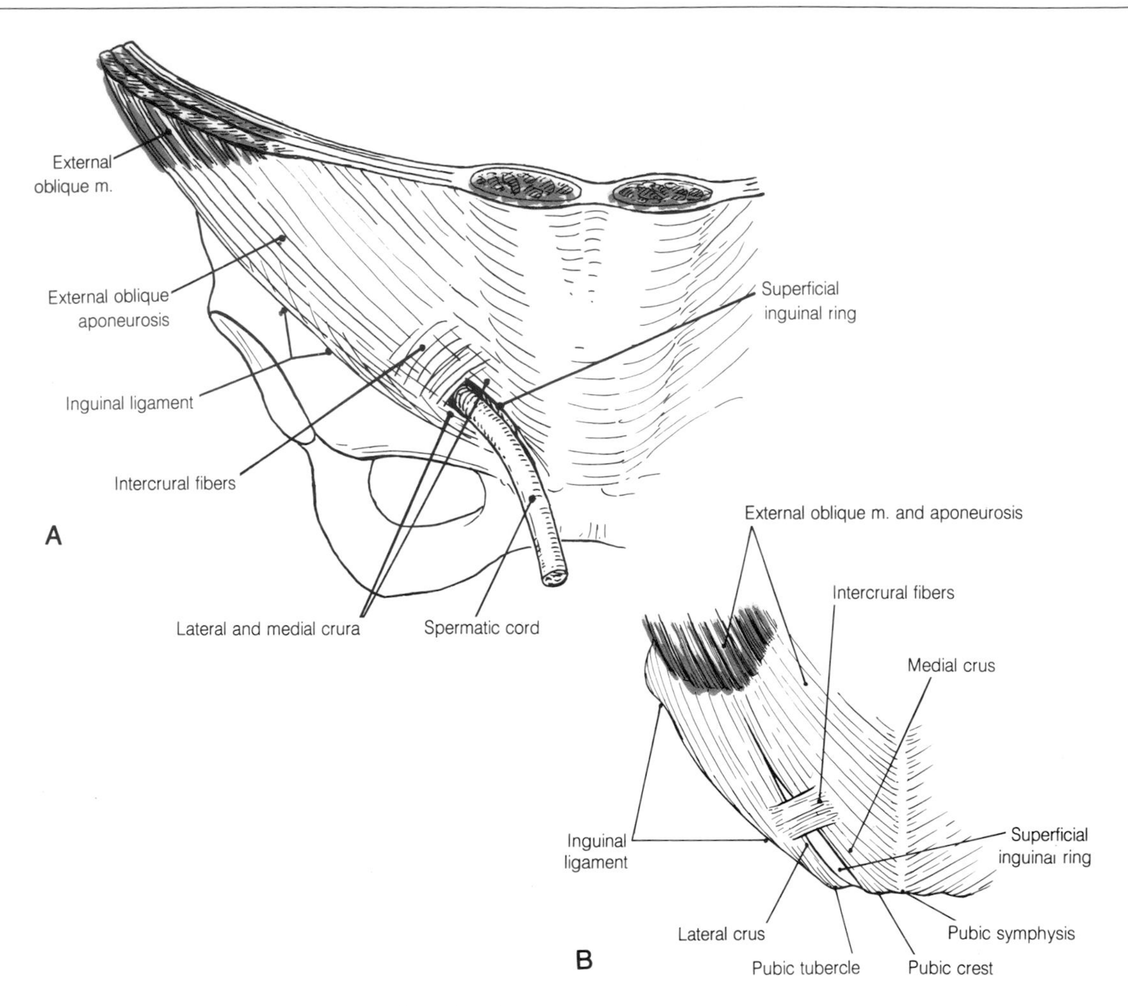

5-10 Showing the relationship of the superficial inguinal ring to the anterior abdominal wall and spermatic cord (*A*) and detail of the ring itself (*B*).

is a narrow, vertically compressed slit. It is also quoted as a region of weakness in the anterior abdominal wall. However, except when a preformed and abnormal passage between the peritoneal cavity and the scrotum exists, no structure other than the spermatic cord is ever found to pass along its full length in a healthy adult.

When reading the ensuing description it is well worthwhile to have a pelvic bone or, better still, an articulated pelvis to refer to. This should be correctly oriented with the anterior superior iliac spines and the symphysis pubis in the same vertical plane.

The inguinal canal lies above and parallel to the inguinal ligament. It extends medially for about 4 cm from the deep to the superficial inguinal ring and can be said to have an anterior and posterior wall and a narrow roof and floor. Its contents are the spermatic cord and its coverings and the ilioinguinal nerve which joins the cord in the canal to pass with it out of the superficial ring. Bearing in mind the slit-like nature and the orientation of the canal, its walls can now be described.

The Anterior Wall. As the superficial ring lies in the external oblique aponeurosis and the deep inguinal ring lies both deep and lateral to this, the aponeurosis must lie in the anterior wall of the canal. In addition, the more medial fibers of the internal oblique muscle, which arise from the lateral two thirds of the inguinal ligament, will also contribute. Look at Figs. 5-10 and 5-11 to confirm this description.

The Roof. Fibers of the internal oblique and transversus muscles arise from the inguinal ligament in a plane superficial to that of the canal. They arch over the canal in an oblique manner

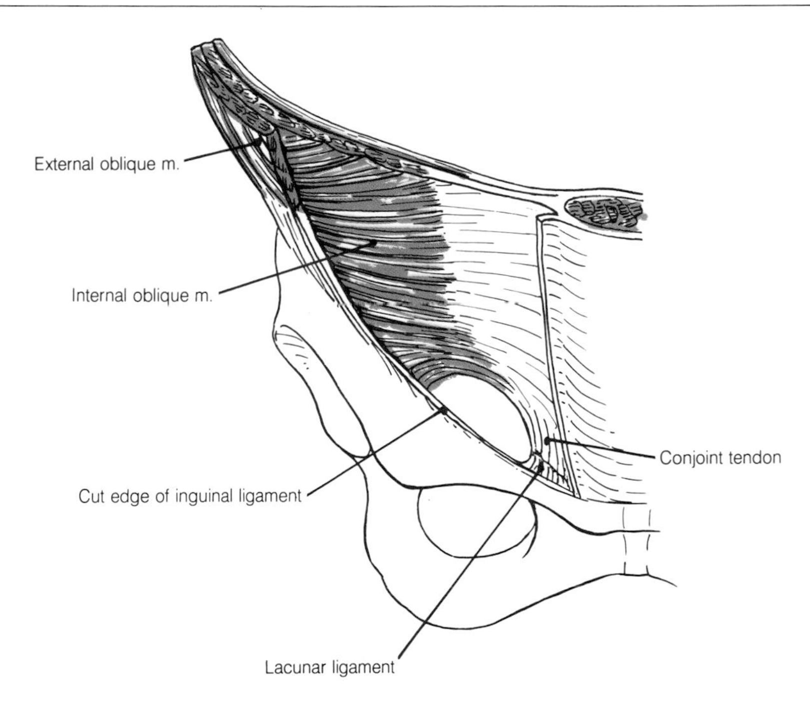

5-11 The internal oblique muscle and the inguinal canal.

from their lateral origin to their medial insertion through the conjoint tendon. In so doing they form the narrow roof of the slit-like canal.

The Posterior Wall. The same muscles that formed the roof now descend behind the canal as the conjoint tendon and, thus, also form its posterior wall. Deep to these muscles lies a rather indefinite and loose layer of fascia, the transversalis fascia. This also contributes to the posterior wall. It is through this fascia that the structures forming the spermatic cord enter and leave the canal at the deep inguinal ring. In Fig. 5-12 the entire anterior wall of the canal has been removed to show the composition of its posterior wall.

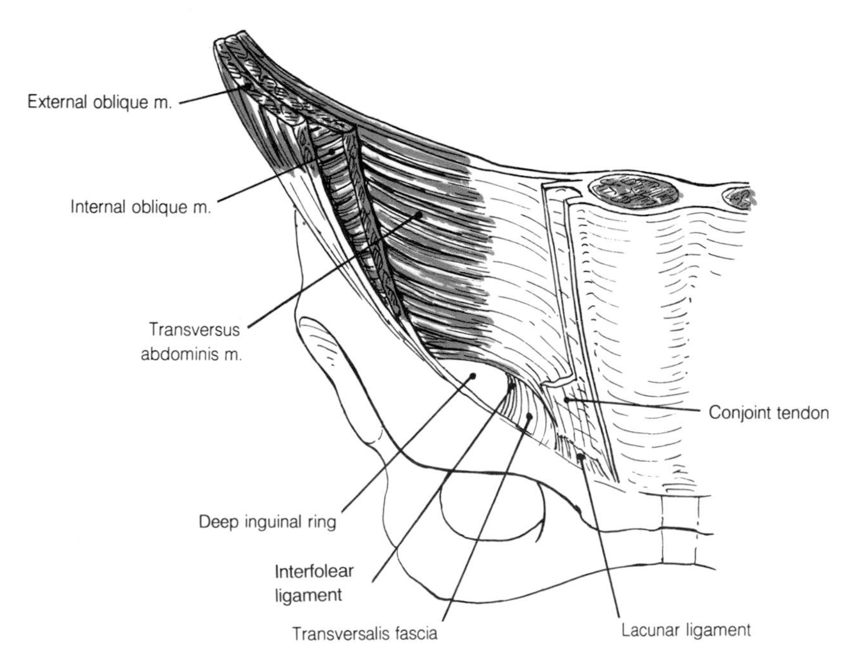

5-12 Showing the internal oblique muscle and the posterior wall of the inguinal canal.

The Deep Inguinal Ring

The deep inguinal ring lies at the mid-inguinal point just above the inguinal ligament which forms its inferior margin. Medially lies transversalis fascia which is sometimes thickened at this point to form the **interfoveolar ligament**. On its lateral side and above lie the lower fibers of transversus abdominis which arch over it. In Fig. 5-13 the deep inguinal ring is viewed from within the abdominal cavity. Note:

1. how the inferior epigastric artery curves round the medial margin of the ring on its way to pass beneath the arcuate ligament.

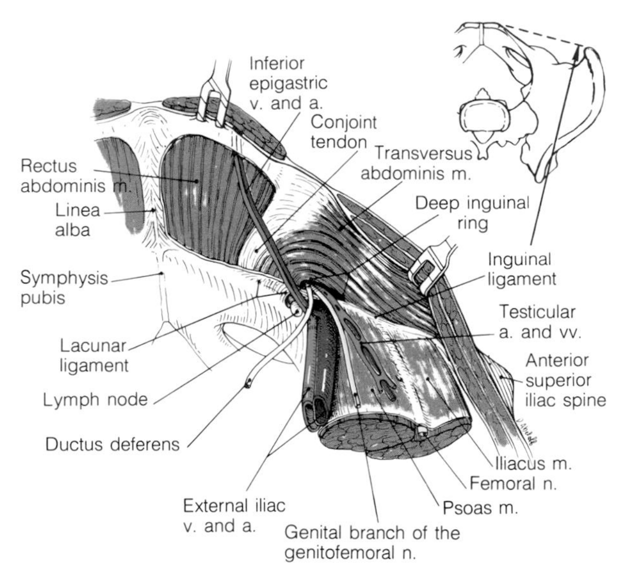

5-13 The deep inguinal ring and surrounding structures viewed from within the abdominal cavity.

2. the ductus deferens as it leaves the ring and crosses the inferior epigastric vessels before passing inferiorly into the lesser pelvis.
3. the testicular artery as it passes off the surface of psoas major on the posterior abdominal wall to enter the ring. It is accompanied by testicular veins.
4. the genital branch of the genitofemoral nerve entering the deep ring. It comes to lie on the dorsal aspect of the cord.

The Floor. This feature is perhaps the hardest to picture. Remember it is the floor of a narrow slit and will itself be narrow. Laterally, a longer portion is formed by the incurved lower edge of the inguinal ligament. A short medial portion is formed by the **lacunar ligament** which bridges the triangular gap between the medial half-inch of the inguinal ligament and the adjacent pectineal line.

The Spermatic Cord

Entering or leaving the male inguinal canal at the deep inguinal ring are:

the **ductus deferens**,
the **artery of the ductus deferens**,
the **testicular artery**,
the **cremasteric artery**,
the **testicular vein or veins**,
the **genital branch of the genitofemoral nerve**,
and autonomic nerves.

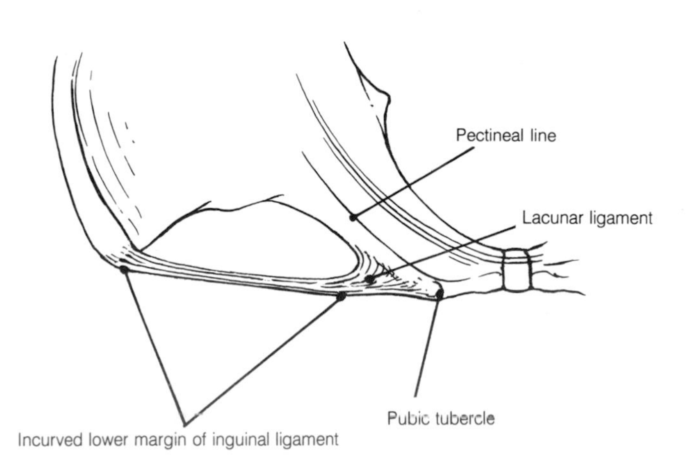

5-14 Showing the contribution of the inguinal and lacunar ligaments to the floor of the inguinal canal when it is viewed from above.

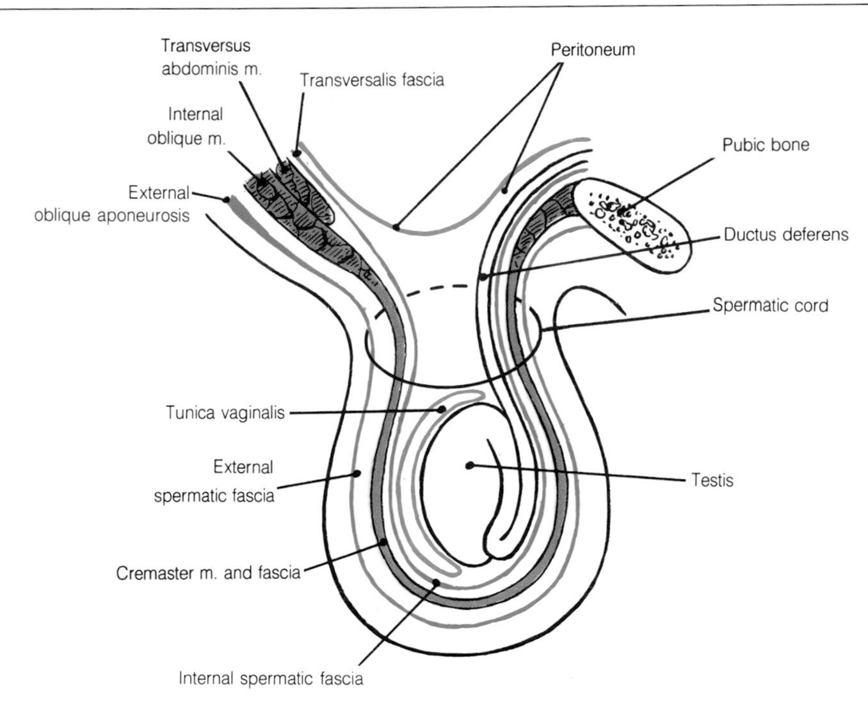

5-15 Diagram illustrating the formation of the sheaths of the spermatic cord.

All these structures pass through the whole length of the inguinal canal but at the superficial ring they are found enclosed in a series of sheaths. These sheaths, as can be followed in Fig. 5-15, are derived from the layers of the anterior abdominal wall through which the structures have successively passed. A deep fascial layer, the **internal spermatic fascia**, is derived from the transversalis fascia, a fasciomuscular layer called the **cremasteric fascia** and containing the **cremaster muscle**, is derived from the internal oblique muscle, and an outer fascial layer, the **external spermatic fascia**, is derived from the aponeurosis of the external oblique muscle. These sheaths and the structures which they surround form the **spermatic cord**. Although the spermatic cord is said to pass along the inguinal canal, it can now be seen not to be fully formed until it leaves the superficial ring. From this point it passes over the pubic tubercle to enter the **scrotum** and here descends vertically to the **testis**.

In the female the round ligament follows a similar course but terminates in the fibrofatty tissue of the **labium majus**.

Some of the components of the spermatic cord will now be described in more detail.

The Cremaster Muscle

In man the cremaster muscle is incomplete and formed from discrete bundles of muscle fibers derived from the internal oblique muscle which run an open spiral course around the spermatic cord and testis and return to the pubic tubercle or parent muscle. The bundles are linked by loose connective tissue, **the cremasteric fascia**. These fibers are supplied by the **genital branch of the genitofemoral nerve** and on contraction raise the testis toward the superficial inguinal ring.

The Pampiniform Plexus of Veins

Between the testis and the superficial inguinal ring the testicular veins form a plexus around the cord. This is known as the **pampiniform plexus**. The plexus condenses to three or four vessels which ascend through the inguinal canal to reach the **inferior vena cava** on the right side and the **left renal vein** on the left.

The Ilioinguinal Nerve

The ilioinguinal nerve is a branch of the first lumbar segmental nerve. It sweeps around the posterior abdominal wall to enter the neurovascular plane between the internal oblique and transversus abdominis muscles and reach a point just above the deep inguinal ring. Here it perforates the internal oblique and comes to lie between that muscle and the overlying aponeurosis of the external oblique. It runs downard below the spermatic cord which it leaves at the superficial ring to supply skin of the scrotum (or mons pubis) and adjacent thigh.

Because of the close association of the testis and spermatic cord with the anterior abdominal wall and the condition of inguinal hernia, both the testis and the scrotum will be described here rather than with the male reproductive organs.

The Scrotum

The scrotum is a pouch-like structure lying inferior to the penis. Its walls are composed of thin rugose skin which is relatively free of hair but contains many sweat and sebaceous glands. Beneath the skin is a layer of membranous superficial fascia which is continuous with that of the penis and lower anterior abdominal wall. In the superficial facia there is a layer of smooth muscle called the **dartos muscle**. This muscle responds to cold by contracting, thus drawing the testis nearer the body and wrinkling the overlying skin to which it is attached. The superficial fascia also extends deeply in the midline dividing the scrotum into two compartments within which the testes lie.

The Testes

Each testis with its associated epididymis lies within the scrotum at the lower end of the spermatic cord. It is partially invaginated into a serous sac of peritoneal origin known as the **tunica vaginalis**. Between the visceral layer of the sac, which clothes the front and sides of the testis and the parietal layer, there is a potential space. Following injury or disease of the testis, this space may be filled with watery fluid (hydrocele) or blood (hematocele). Examine Fig. 5-16 to be sure that the relationship of this sac to the testis is understood. The almond-shaped testis is about 5 × 3 × 2.5 cm in dimension. Its tough outer fibrous coat, called the **tunica albuginea**, sends many fibrous septula into the interior of the gland which divide it into small **testicular lobules**.

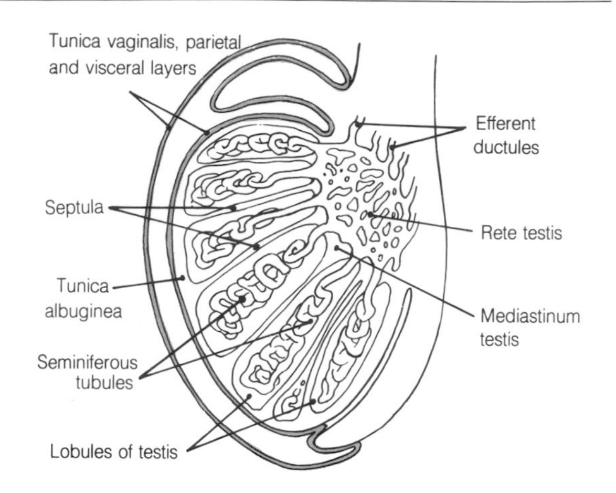

5-16 Semidiagrammatic section of the testis and tunica vaginalis.

These contain the **seminiferous tubules**. The septa and tubules converge on the posteriorly lying **mediastinum testis** where the tubules form a network called the **rete testis**. Further tubules, the **efferent ducts**, open into the **head of the epididymis** which is attached to the upper pole of the testis. These features of the testis should be identified in Fig. 5-16.

The testis is supplied with blood by the **testicular artery**. This artery is a direct branch of the abdominal aorta and arises just below the renal arteries. It descends in the spermatic cord to the posterior aspect of the testis. From here branches embrace the testis and perforate the tunica albuginea to reach the interior. Venous blood returns via the **pampiniform plexus** and **testicular vein** to reach the inferior vena cava (right vein) and the left renal vein (left vein). Obstruction of the left renal vein by a renal tumor may produce a dilatation of the veins around the left testis and epididymis (a varicocele). Occasionally the testis may rotate within the scrotum and, in so doing, twist and obstruct the testicular artery. This causes acute pain and necrosis of the testis may follow if the condition is not rapidly relieved.

Lymphatics from the testis run back with the testicular artery to reach lymph nodes alongside the aorta.

The Epididymis

The epididymis is a long (6 m), highly convoluted tube attached to the posterior border of the testis. Its **head** lies at the upper pole of the testis where it is joined by the efferent ducts. It descends to the

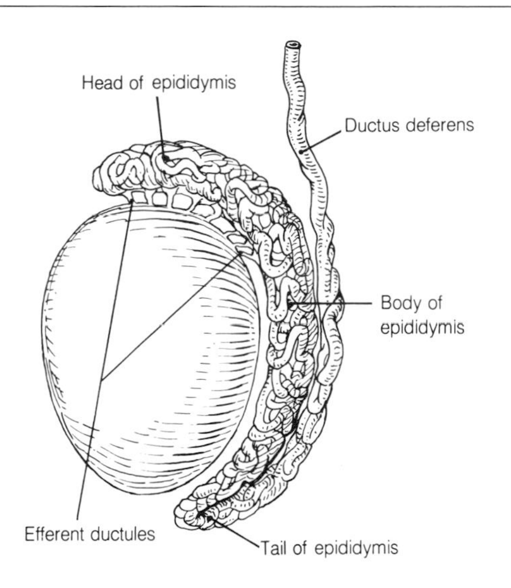

5-17 The epididymis.

lower pole as the **body of the epididymis** where it becomes continuous with the **ductus deferens**. This region is called the **tail of the epididymis**. The relationship of the epididymis to the testis can be seen in Fig. 5-17.

The Ductus Deferens

The ductus deferens is a firm muscular tube which can be easily palpated in the living spermatic cord. Continuous with the epididymis it ascends on its medial side to become incorporated into the spermatic cord above the testis. It eventually joins the urinary tract at the prostatic part of the urethra into which it delivers spermatozoa carried from the testis and epididymis.

The Descent of the Testis

In order that the surgical anatomy of the inguinal region, spermatic cord, and testis may in due course be understood, a brief account of the descent of the testis and the manner in which normal development may fail is now given. A diagrammatic illustration of this process can also be seen in Fig. 5-18.

The testis develops on the posterior abdominal wall of the embryo. However, it subsequently migrates downward and leaves the abdominal cav-

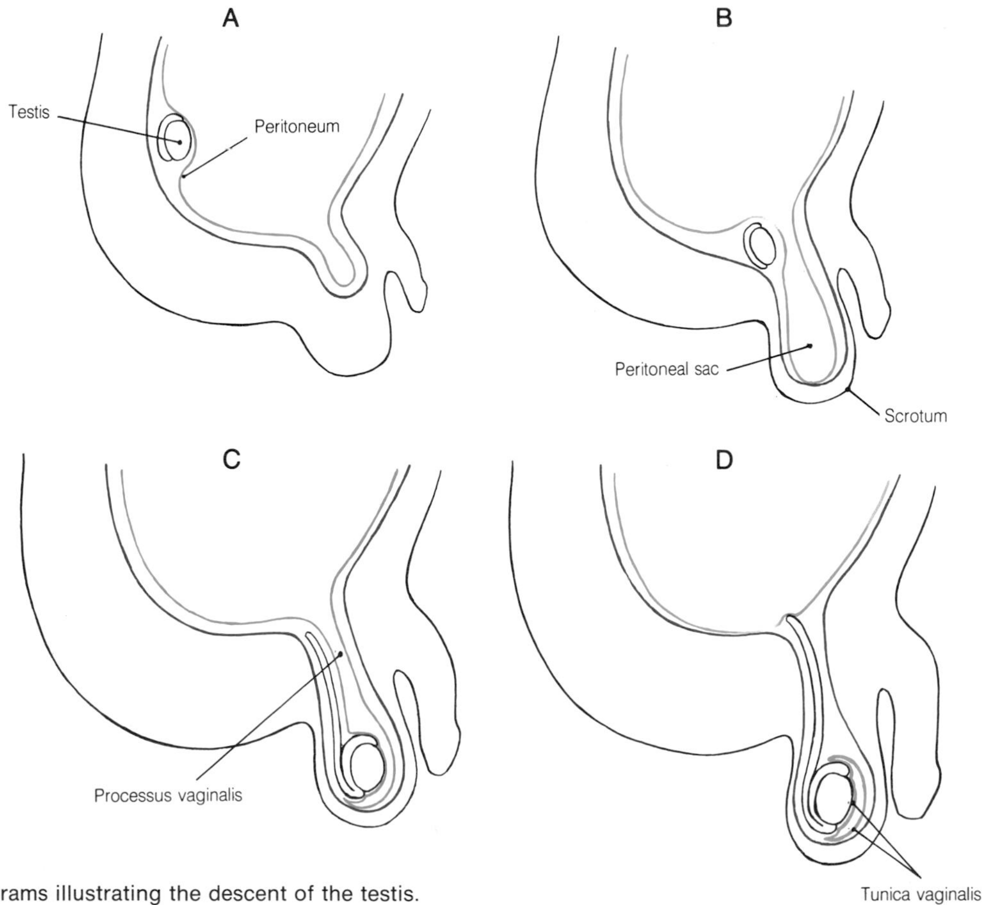

5-18 Diagrams illustrating the descent of the testis.

ity through the inguinal canal to reach the scrotum at about the time of birth. The course of this migration seems to be determined by the presence of the **gubernaculum testis**, a fibromuscular cord extending from the lower pole of the testis to the developing scrotal swellings. It also seems that the slightly lower temperature prevailing in the scrotum is necessary for the normal maturation of spermatozoa.

As it descends, the testis is preceded by a sac of peritoneum which lines the scrotum. This sac is known as the **processus vaginalis**. On reaching the scrotum, the testis invaginates this sac from behind and in this way becomes partially clothed by a visceral and parietal layer of peritoneum. Communication of the sac with the main peritoneal cavity becomes obliterated and that portion left around the testis forms the tunica vaginalis, which has already been described. The descending testis draws with it its duct, the ductus deferens, and its vessels, nerves, and lymphatics. The origin of the testicular arteries from the abdominal aorta provides some evidence of the site of origin and subsequent course of the testis.

Failure of these developmental processes to follow a normal course leads to a number of relatively common abnormalities. Some of these are now described.

Abnormalities of the Testis

The testis normally lies in the scrotum at birth although descent may occur within the first nine months of life. Failure of the testis to reach the scrotum may be due to incomplete descent in which case it is arrested within the abdomen or along the course of the inguinal canal. Alternatively the testis may descend through the canal but be guided to an ectopic site in the perineum or the anterior aspect of the thigh. Where possible the testis should be surgically transferred to the scrotum. However, if this fails, surgical removal may be necessary because a testis retained in the abdominal cavity or inguinal canal is sterile and also liable to develop malignant tumors.

Abnormalities of the Processus Vaginalis

This communication between the peritoneal cavity and the tunica vaginalis surrounding the testis is normally completely obliterated. Partial failure of this process at some point along the course of the processus vaginalis may lead to the development of a fluid-filled sac called a **hydrocele of the cord**. If the processus vaginalis remains totally patent, a pathway exists through which abdominal structures can pass into the scrotum. This is called a congenital or oblique hernia and is described now under the general heading of herniae.

Herniae

Herniae involve the passage of a peritoneal sac with or without abdominal contents through a site of congenital or acquired weakness in the abdominal wall. Common sites of herniae are at the umbilicus, the inguinal region, and the femoral canal. Less commonly, they are found in the linea alba, at the obturator foramen, and through a triangular area behind the free posterior border of the external oblique muscle where only the internal oblique and transversus abdominis muscles contribute to the lateral abdominal wall.

Inguinal Herniae

An inguinal hernia is a common but disabling condition. In addition it carries the threat of becoming an acute surgical emergency if the intestine it may contain becomes obstructed or ischemic. Inguinal herniae are of two varieties, namely congenital and acquired, and the anatomical features of each are described below.

Congenital Herniae

Complete failure of the processus vaginalis to close results in the presence of a peritoneal sac continuous with the abdominal peritoneal cavity. This passes from the deep inguinal ring, along the inguinal canal within the coverings of the spermatic cord, out of the superficial inguinal ring and down to the scrotum where it is continuous with the tunica vaginalis. Although present from birth, the existence of such a hernial sac does not become obvious until a portion of intestine (usually small intestine) is forced through it. The condition is, therefore, most commonly found in boys and young men. Because congenital herniae do not pass directly through the abdominal wall, they are often described as indirect herniae.

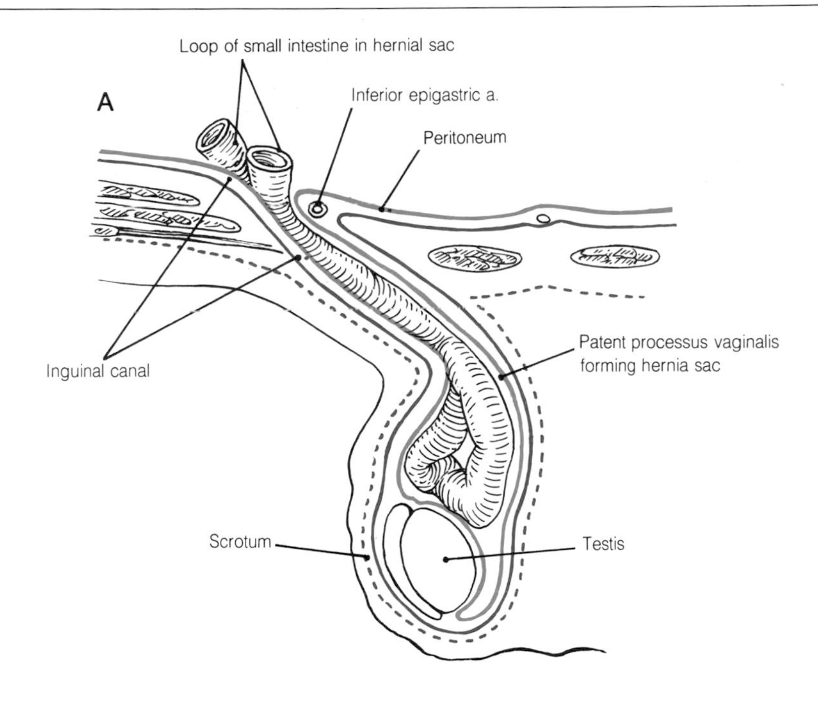

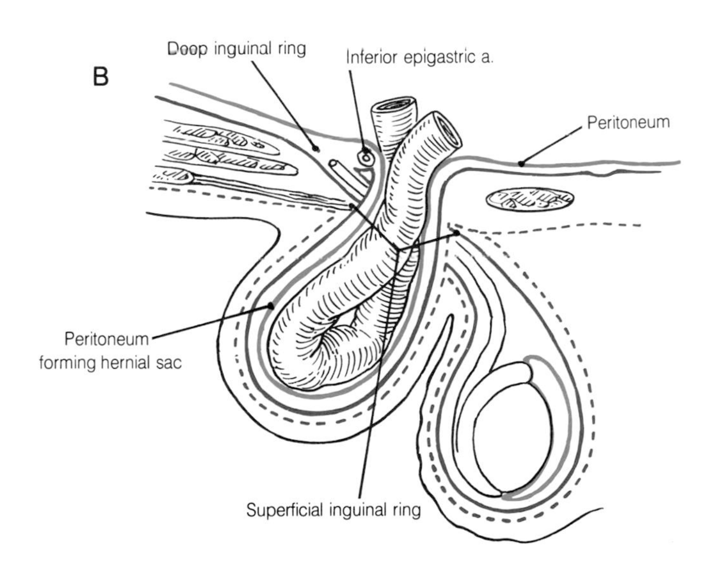

5-19 Congenital (*A*) and acquired (*B*) inguinal herniae.

Acquired Herniae

This variety of hernia is usually found in older men. The hernial sac enters the inguinal canal through a weakness in its posterior wall and does not lie within the coverings of the spermatic cord. Because the site of the weakness and the superficial ring through which it emerges are superimposed, this type of hernia is referred to as a direct hernia. Again because there is no preformed pathway to the scrotum and the hernia lies outside the spermatic cord it is directed superficially onto the anterior abdominal wall. Any doubt about the variety of hernia can often be settled by feeling the pulsation of the inferior epigastric artery. If this is found lateral to the path of the hernia, the hernia is direct; if felt medially, the hernia is congenital and indirect.

Now compare the anatomy of congenital and acquired herniae in Fig. 5-19. Note in Fig. 5-19A that the loop of bowel in the hernial sac passes through the inguinal canal from the deep to the superficial ring, and continues into the scrotum with the other contents of the spermatic cord. In Fig. 5-19B the hernial sac passes through the posterior wall of the inguinal canal medial to both the deep inguinal ring and the inferior epigastric artery. It emerges onto the surface of the abdominal wall through a dilated superficial inguinal ring.

The Inguinal Canal in the Female

The impression that the inguinal canal is of no importance in the female should not be gained from the above description. The arrangement of structures forming the boundaries of the canal is identical to that in the male, but the gonad, i.e., the ovary, descends only to the pelvis and does not traverse the canal. However, the gubernaculum is retained in the form of the **round ligament**. This extends from the ovary, along the inguinal canal, and is anchored in the fibrofatty tissue that makes up the **labium majus**. Not only is the gubernaculum retained, but in fetal life a processus vaginalis is formed and normally obliterated. For this reason congenital indirect herniae do occur in women, although much less frequently than in men. As in the male, an isolated portion of the processus vaginalis may form a hydrocele known by gynecologists as a **hydrocele of the canal of Nuck**.

The Deep Aspect of the Anterior Abdominal Wall

When the anterior abdominal wall is reflected to expose the peritoneal cavity and the abdominal organs, its deep aspect should be carefully examined. The entire surface is covered by parietal peritoneum but through this a number of features can be seen or felt. These are illustrated in Fig. 5-20. Using this figure, identifiy the region of the umbilicus and, above it, the **falciform ligament** leading to the liver. The **ligamentum teres** can be felt as a firm cord in the free border of the falciform ligament. Extending inferiorly from the umbilicus to the pelvis, three folds of peritoneum

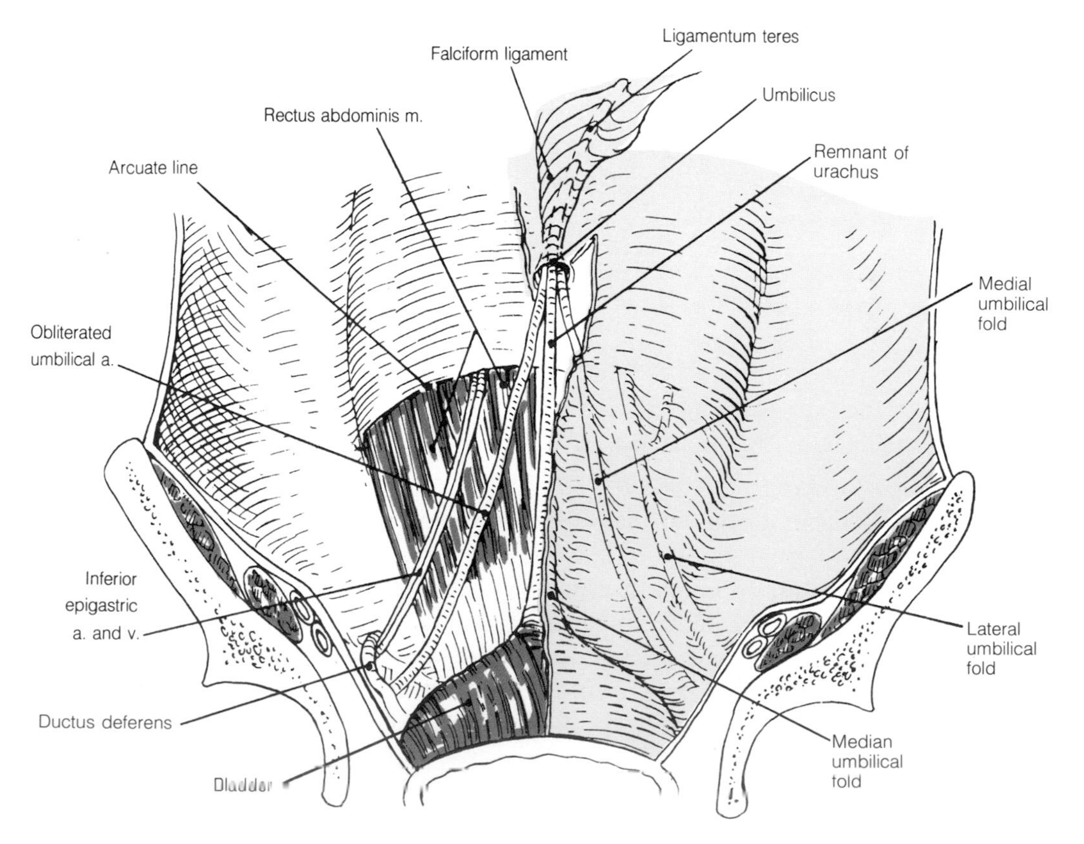

5-20 The deep aspect of the anterior abdominal wall. The peritoneum has been removed from its left side.

can be seen. A **median umbilical fold** contains the remains of the **urachus** and leads to the bladder. On either side two **medial umbilical folds** can also be followed into the pelvis and are found to approach the internal iliac artery. These folds are formed by the **obliterated umbilical arteries**. Two **lateral umbilical folds** can be seen outside the medial folds and they are formed by the **inferior epigastric vessels**. These folds can be followed from the external iliac vessels to the **arcuate line**. At this point the posterior sheath of the rectus abdominis becomes deficient and the vessels can pass superiorly on the deep surface of the muscle. In the iliac fossa the inferior epigastric vessels lie medial to the **deep inguinal ring** and the **ductus deferens** hooks laterally around them to enter the **inguinal canal**.

The Alimentary Tract

The alimentary tract spans the regions of the head, neck, thorax, abdomen, pelvis, and perineum, and before embarking on a description of its abdominal portion an overview of its whole course is given in order to maintain the continuity of this important system.

The alimentary tract begins at the **oral cavity** where the ingested food is prepared for swallowing by mastication and salivation. From here the food is forced back into the oral part of the **pharynx**, which is shared with the respiratory tract. From this point the passage of ingested material is continued onward by a series of reflex mechanisms. The **nasopharynx** is closed by the **soft palate** and the laryngeal opening is narrowed and raised beneath the posterior third of the tongue as the bolus of food is propelled from **oropharynx** through the **larynopharynx** into the **esophagus**. The esophagus begins in the lower neck and is continuous through the thoracic cavity. At the diaphragm it passes through diaphragmatic muscle at the level of the eighth or ninth thoracic vertebra and after a short course within the abdominal cavity joins the **stomach** at its cardiac region just to the left of the midline.

The greater part of the remainder of the alimentary tract lies within the **abdominopelvic cavity**. The bag-like stomach is suspended from the diaphragm and posterior abdominal wall by peritoneal ligaments or mesenteries whereas the **duodenum** with which it is continuous is, except for its first half-inch, applied to the posterior abdominal wall. It can thus be seen that the stomach lies free within the abdominal cavity and is anchored only at its proximal and distal extremities. The stomach terminates distally at the **pyloric sphincter** which controls the egress of food into the duodenum. The **gastroduodenal junction** lies just to the right of the midline at the level of the first lumar vertebra and is the beginning of the **small intestine**. The fixed duodenum describes a C-shaped course to the right of the vertebral colum. The lower limb of the "C" crosses the midline to terminate at the **duodenojejunal junction**. The long **jejunum and ileum** complete the small intestine. While it is usually possible to distinguish proximal jejunum from distal ileum, these two segments of the small intestine are continuous and their characteristics change rather imperceptibly along their course. The ileum terminates in the right iliac region at the **ileocecal junction**.

At the ileocecal junction the ileum joins the **large intestine**. Below the junction lies the blindly ending sac of the **cecum**,* above is the **ascending colon** passing upward on the right side of the posterior abdominal wall. The greater part of the large intestine is characterized by features which help in its identification. It is usually of larger caliber than the small intestine, is sacculated and small tags of fat called **appendices epiploicae** are found hanging from its surface. In addition, its outer longitudinal layer of muscle is condensed into three visible bands called **taeniae*** **coli**. If these are followed proximally they lead back to the base of the **vermiform*** **appendix** to the left of the cecum just below the ileocecal junction. The ascending colon becomes the **transverse colon** just below the liver at the **right colic flexure**. Unlike the ascending and descending colon, the transverse colon is freely suspended by a mesentery attached across the posterior abdominal wall. At the **left colic flexure** on the left side of the abdominal cavity, the colon continues downward as the **descending colon** attached again to the posterior abdominal wall. Near the

* Cecus, *L.* = blind; Taenia, *L.* = flat band; Vermis, *L.* = worm

brim of the pelvis the colon again becomes freely suspended by mesentery and is known as the **sigmoid or pelvic colon**. The sigmoid colon joins the fixed **rectum** in the pelvic cavity.

The somewhat sinuous rectum lies within the pelvic cavity but only its upper two-thirds are related to the peritoneal cavity. As it passes through the muscular floor of the pelvic cavity it becomes the short terminal portion of the intestinal tract, the **anal canal**. This is surrounded by a sphincter of smooth muscle and a sphincter of striated muscle and terminates at the **anus**.

Associated with the alimentary tract and, in fact, derived as diverticula from the duodenum are the **liver** and **pancreas**.* Their respective alimentary functions are the secretion of bile and digestive enzymes. The liver occupies the upper right quadrant and part of the upper left quadrant of the abdominal cavity and lies almost entirely within the protection of the thoracic cage. It is suspended between the anterior abdominal wall, diaphragm, and stomach. On its inferior surface lies the **gallbladder**. This can be considered as a diverticulum of the duct of the liver. It temporarily stores and concentrates the bile. When suitably stimulated it passes the bile back into the duct of the liver which carries it to the duodenum. Just as the duct enters the duodenum it is joined by the **pancreatic duct**. The pancreas itself lies on the posterior abdominal wall and its head nestles in the concavity of the duodenum. Because of its common blood supply and venous drainage and perhaps because the products of the aging red cells it destroys are carried to the liver, the **spleen** is normally considered with the alimentary tract. It lies beneath the diaphragm in the upper left quadrant of the abdominal cavity and is suspended by mesentery between the stomach and the posterior abdominal wall.

* Pancreas, *Gk.* = all flesh

The Abdominopelvic Cavity

In describing the walls of the abdominal cavity it has already been pointed out that this cavity extends up into the thoracic cage and down into the pelvis. While it is largely protected in its upper and lower extents, it should not be forgotten that perforating wounds in these regions may involve the cavity and its contents. When the abdomen is opened through the anterior abdominal wall, the inner aspect of its walls and the exposed organs are seen to be covered by a shiny serous membrane called the peritoneum.

The Peritoneum

Like the pleura, the peritoneum is a serous membrane lined by mesothelial cells. It also can be thought of as an empty balloon into which the alimentary canal and its associated glands have to a greater or lesser extent been invaginated. Thus, it comes about that there is a *parietal layer* of peritoneum lining the abdominal walls, the inferior surface of the diaphragm, and the walls of the pelvic cavity, and a **visceral layer** covering those structures which have become invaginated. The two layers are separated by a potential space, the **peritoneal cavity**. During the early stages of development, the abdominal part of the alimentary tract is suspended in the peritoneal cavity by a **dorsal mesentery** (Fig. 5-21 A) and the more proximal part by a **ventral mesentery** as well (Fig. 5-21 B). A mesentery is a double sheet of visceral peritoneum which is reflected off the sides of the structure it surrounds and which becomes continuous with the parietal peritoneum where it reaches the abdominal wall. Unfortunately, this simple arrangement has become complicated by the fact that during development mesenteries may, in places, become adherent to each other or, structures originally suspended by a mesentery may fall back onto the posterior abdominal wall and become adherent to it in the manner illustrated in Fig. 5-22 A and B. Such structures become only partially covered by peritoneum and are described as being **retroperitoneal**. Organs which remain surrounded by peritoneum are termed **intraperitoneal**. It should be realized that the adjective intraperitoneal is misleading as no part of the

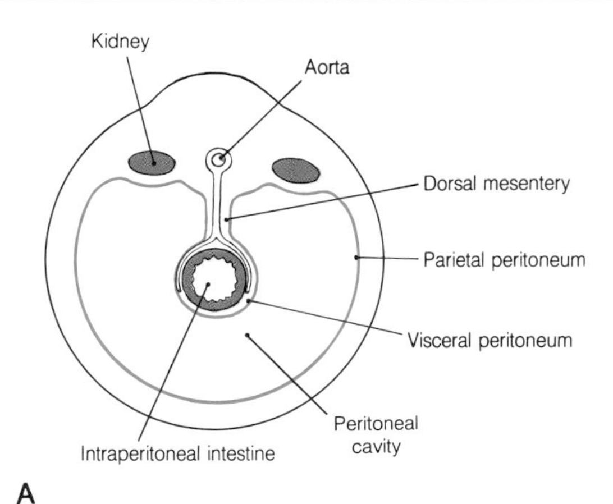

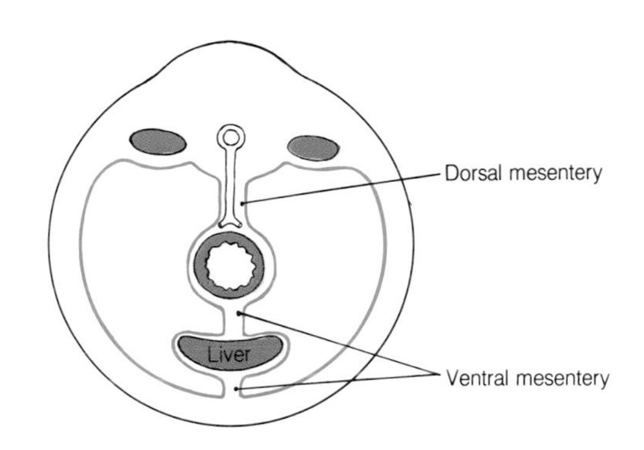

5-21 Diagrams to show the embryonic dorsal (*A*) and ventral (*B*) mesenteries.

alimentary tract actually lies inside the peritoneal cavity and for the same reason none is completely covered by peritoneum as this is missing at the site of the attachment of the mesentery.

Examine now in more detail the diagrams in Figs. 5-21 and 5-22 and note the following:

1. In Fig. 5-21B a section of the upper part of the developing abdominal cavity is shown. The intestine, in this case the stomach, is suspended between a dorsal and a ventral mesentery. Both mesenteries are formed by a double layer of peritoneum and each layer is reflected off either the anterior or posterior abdominal wall. The dorsal mesentery carries a branch of the dorsal aorta to the intestine and also veins and lymphatics. In the ventral mesentery, the liver can be seen to be developing.
2. In Fig. 5-21A the ventral mesentery is no longer present and the intestine hangs freely from the dorsal mesentery. This arrangement is retained in the jejunum and ileum.
3. In Fig. 5-22A the intestine and dorsal mesentery have fallen back onto the abdominal wall and in Fig. 5-22B the adjacent layers of parietal and visceral peritoneum have fused and disappeared. The intestine is now only partially covered by peritoneum and is said to be retroperitoneal. Examples of this arrangement are found in the duodenum and the ascending and descending colon.

The Nerve Supply of the Peritoneum

The parietal peritoneum is supplied by the segmental nerves that innervate the overlying skin and muscles. The peritoneum covering the inferior surface of the diaphragm is supplied by intercostal nerves peripherally, but by the phrenic nerve centrally. The visceral peritoneum is not thought to have a sensory nerve supply.

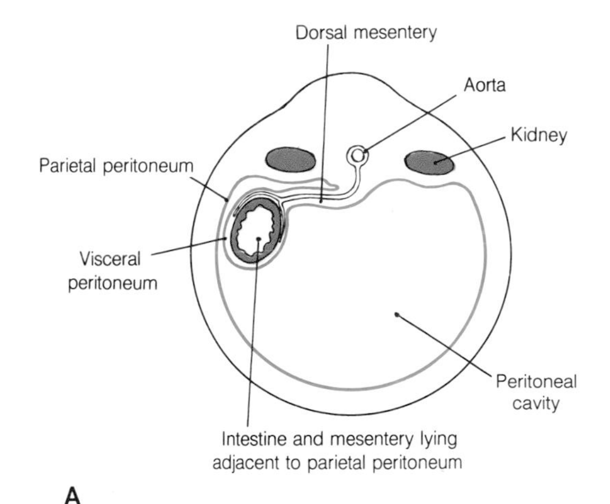

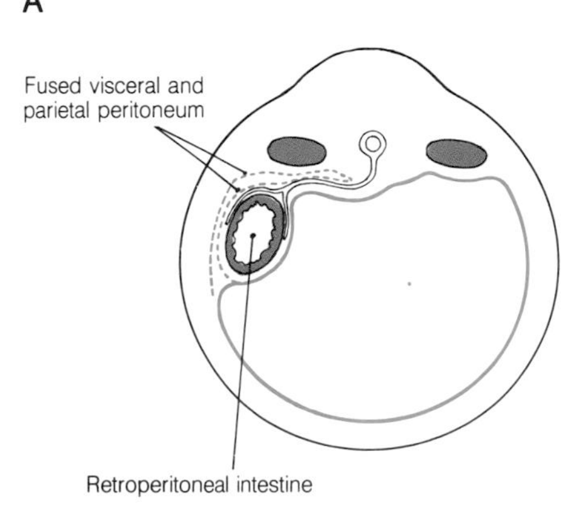

5-22 Diagrams to illustrate the formation of a retroperitoneal part of the intestine.

Irritation of the parietal peritoneum gives rise to fairly well localized pain. On the other hand, visceral peritoneum may be painlessly incised. (Pain from abdominal organs probably arises from distention or contraction of their intrinsic musculature.) Irritation of the diaphragmatic parietal peritoneum sometimes gives rise to pain in the shoulder region. This is known as "referred pain" and is related to the common segmental nerve supply of the two areas (C4 and 5).

The Greater and Lesser Peritoneal Sacs

The greater part of the peritoneal cavity forms what is known as the **greater sac**. However, a pouchlike evagination of the dorsal mesentery of the stomach forms a much smaller compartment and is called the **lesser sac**. The lesser sac communicates with the greater sac through a narrow opening called the **epiploic foramen**. In the adult the lesser sac lies posterior to the stomach and extends to a variable extent into the **greater omentum** and its relationships will become more clear when surrounding structures are described. For the time being some idea of how it is formed can be gained from a study of the diagrams in Fig. 5-23. In this, part A is very similar to part B of Fig. 5-21, except that the posterior part of the dorsal mesentery is now swinging to the left and the stomach and ventral mesentery are swinging to the right. The site of the developing spleen is also indicated in the dorsal mesentery. In Fig. 5-23B the dorsal mesentery has become partially fused to the posterior abdominal wall leaving a portion between the kidney and spleen – **the lienorenal ligament*** – and a portion between spleen and stomach – **the gastrosplenic ligament**.* The opening into the lesser sac narrows to form the **epiploic foramen**. The portion of the ventral mesentery between the stomach and liver will become the **lesser omentum** and that between liver and anterior abdominal wall the **falciform ligament**.

* Splen, *Gk.* = spleen; Lien, *L.* = spleen. The use of these two words with the same meaning is unfortunate in this situation, but less so if that meaning is understood.

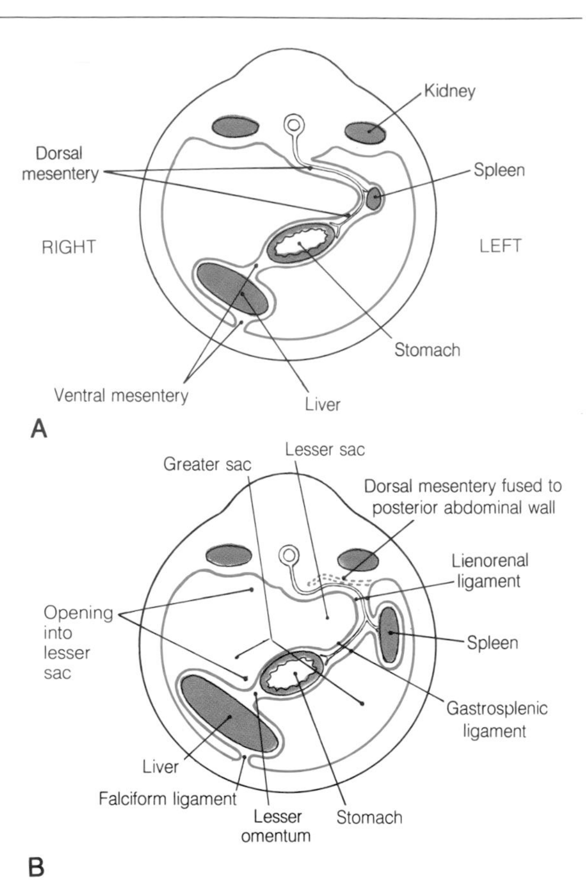

5-23 Diagrams to illustrate the formation of the lesser sac.

The General Arrangement of the Abdominal Contents

On opening the peritoneal cavity, it is normal to find the contents covered by an "apron" of fat-filled peritoneum. This is the **greater omentum*** and it hangs from the lower border or **greater curvature** of the stomach. Fused to the posterior aspect of the omentum are the transverse colon and its mesentery or **mesocolon**.

The greater omentum has the remarkable property of moving toward areas of inflammation or perforation of the intestinal tract and, by taking part in the inflammatory process, helps to seal off the region from the general peritoneal cavity. The appearance of the undisturbed omentum and abdominal contents is seen in Fig. 5-24. The attach-

* Omentum, *L.* = fatty skin

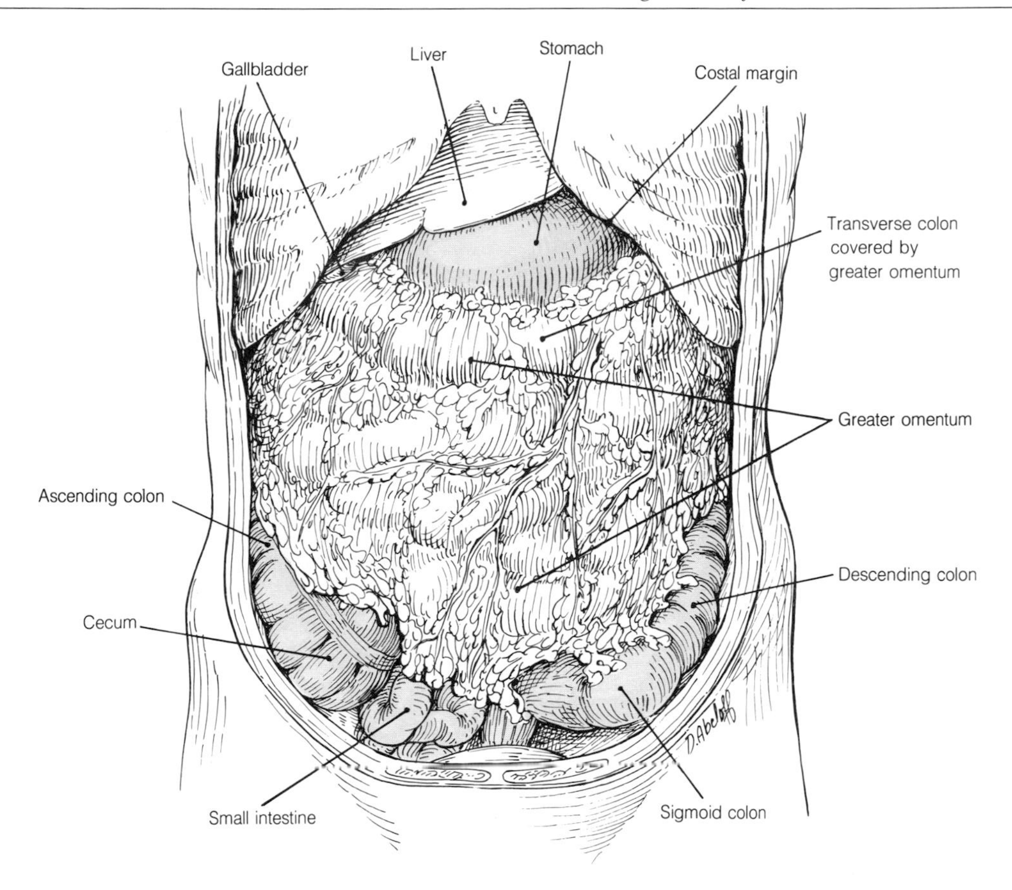

5-24 To show the undisturbed greater omentum.

ments of the greater omentum and underlying mesocolon divide the abdominal cavity into a **supracolic** and an **infracolic compartment**.

The Supracolic Compartment

When the greater omentum lies in its normal position, the examining hand is directed up over the **stomach** to the supracolic compartment, which is limited superiorly by the inferior surface of the diaphragm. Little of the contents of this compartment can be seen with ease as they lie beneath the cover of the costal margin. As a result, examination must be made by palpation in the manner used by a surgeon. Using the diagram in Fig. 5-25, it can be seen that when the hand is passed to the right, the **liver** is encountered and the fingers may be slipped between the right lobe and costal margin. If the fingers are moved across, toward the midline, they are obstructed by the **falciform ligament**. This is a remnant of the ventral mesentery in which the liver developed. During fetal life it contained the **left umbilical vein**, and the remnant of this vein, the **ligamentum teres**, can be followed inferiorly in the free border of the falciform ligament to the umbilicus. On the inferior or visceral surface of the liver will be found the **gallbladder** with its tip or **fundus** close to the costal margin (typically at the ninth costal cartilage).

In the midline, the hand will pass up onto the anterior surface of the stomach. If this is followed to the right it will lead the hand to the **first part of the duodenum** and if upward and to the left, to the last half-inch of the **esophagus** as it penetrates the diaphragm to join the stomach. The upper border of the stomach is called the **lesser curvature** and this is attached to the liver by a double layer of peritoneum known as the **lesser omentum**, which like the falciform ligament, is part of

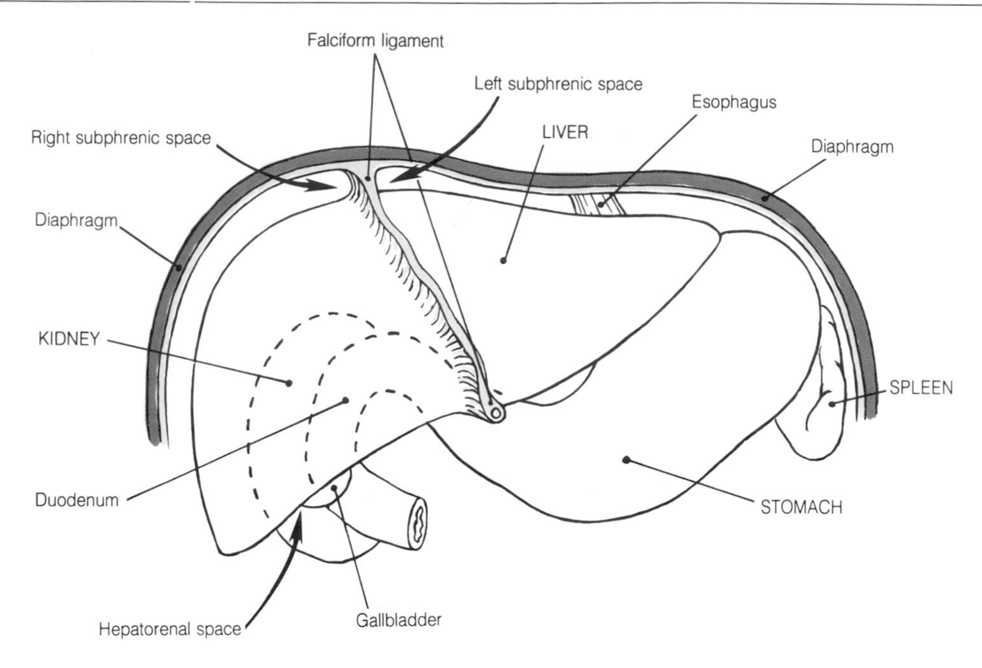

5-25 Diagram showing the main organs in the supracolic compartment.

the ventral mesentery. When the hand is passed to the left over the inferior surface of the diaphragm, the tip of the **spleen** will be felt.

Any infected fluid within the peritoneal cavity tends, in the supine body, to gravitate either to the pelvis or the subphrenic (i.e., subdiaphragmatic) or subhepatic (i.e., beneath the liver) regions where it may be sealed off and an abscess formed. Two **subphrenic spaces** are described. A right subphrenic space lies between liver and diaphragm to the right of the falciform ligament and a left, beneath the diaphragm to the left of the falciform ligament. A further **subhepatic or hepatorenal space** lies between the inferior visceral surface of the liver and the right kidney which it overlies. These spaces are also illustrated in Fig. 5-25. The lesser sac is also sometimes described as a **left subhepatic space.** The presence of these spaces must always be borne in mind when seeking the cause of an unexplained fever, because infections they may contain produce few localizing signs.

The Infracolic Compartment

If the greater omentum is folded back and upward over the costal margin, as in Fig. 5-26, the infracolic compartment is exposed. On each side there is a distinct **paracolic gutter**. On the right the gutter lies between the lateral abdominal wall and the **ascending colon**, on the left it lies between the lateral abdominal wall and the **descending colon**. The right gutter communicates with the supracolic compartment and, in particular, with the right subhepatic space. The left gutter is limited superiorly by a fold of peritoneum between the diaphragm and the left colic flexure called the **phrenicocolic ligament**. The ascending colon becomes continuous with the **transverse colon** at the **right colic flexure**. The transverse colon lying on the posterior surface of the greater omentum joins the descending colon at the **left colic flexure**. At the brim of the pelvis another mobile length of large intestine is found suspended from a mesentery. This is the **pelvic or sigmoid colon** which links the descending colon to the **rectum** in the pelvic part of the abdominal cavity. Within the arch formed by the colon lies the greater part of the **small intestine**. Although up to twenty feet in length, the mesentery which tethers it to the posterior abdominal wall is only about 15 cm in length at its attachment. This attachment lies obliquely across the infracolic compartment from the upper left to lower right corner where the terminal part of the small intestine joins the ascending colon.

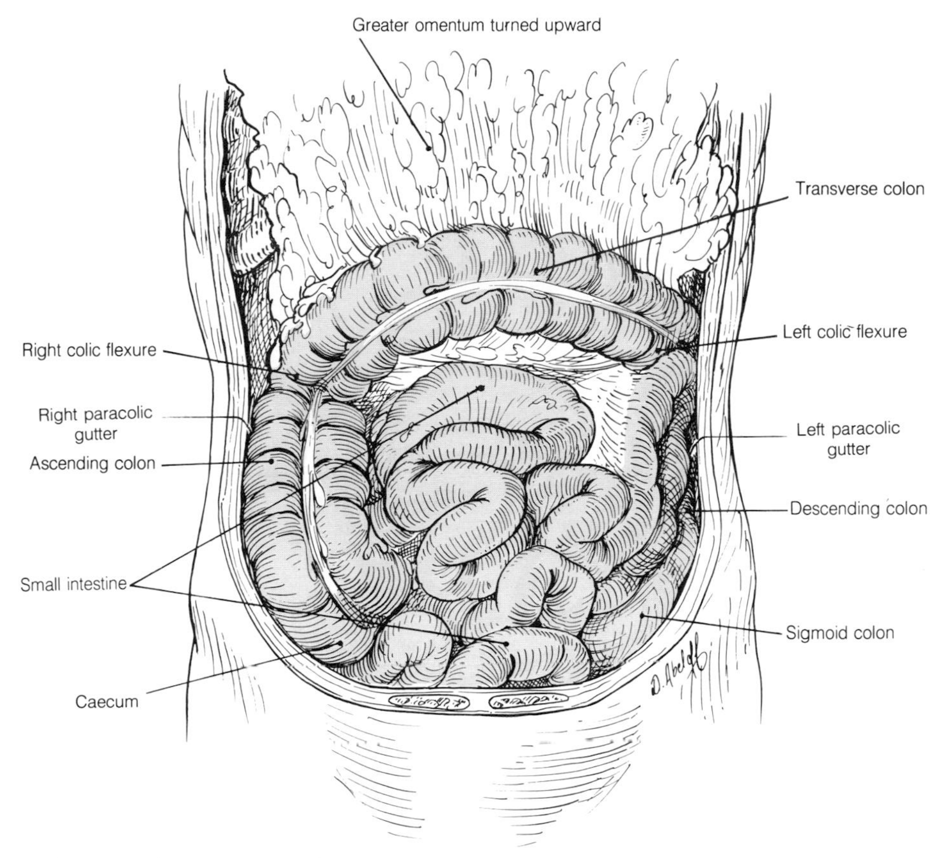

5-26 To show the infracolic compartment.

The stomach is joined to the mobile portion of the small intestine by the **duodenum**. The duodenum is almost entirely retroperitoneal. It begins in the supracolic compartment and passes behind the attachment of the greater omentum and the transverse mesocolon to the posterior abdominal wall to become continuous with the jejunum in the infracolic compartment.

With its general arrangement now in mind the abdominal portion of the alimentary tract and its associated organs can now be described.

The Abdominal Organs

The Esophagus

The esophagus passes through the muscular part of the diaphragm in company with the anterior and posterior vagal trunks at the level of the ninth or tenth thoracic vertebra. The opening lies to the left of the midline and is surrounded by a loop of muscle fibers derived from the **right crus** of the diaphragm. After a very short intraabdominal course, the esophagus joins the **cardiac region** of the stomach at the **gastroesophageal junction**.

The importance of the last inch or two of the esophagus is considerable. Although difficult to demonstrate anatomically, this region forms a functional sphincter controlling the entry of food to the stomach and preventing the reflux of stomach contents. Dysfunction of this sphincter may lead to the accumulation of food in the thoracic esophagus and its subsequent dilatation. Also, parts of the stomach or other mobile abdominal organs may pass into the thoracic cavity through defects in the esophageal diaphragmatic opening or hiatus. These hiatal defects may be congenital or acquired.

The abdominal esophagus is supplied by a branch of the left gastric artery and drained by the left gastric vein. These vessels and their anastomoses will be described with the blood supply of the foregut. Its autonomic nerve supply is received from the two **vagi** (parasympathetic) and the **splanchnic nerves** (sympathetic).

The Stomach

The stomach is a mobile, muscular, and distensible organ lying between the esophagus proximally and the duodenum distally. Its anterior and posterior surfaces are covered by peritoneum and at its **greater and lesser curvatures** this is reflected onto the **greater and lesser omenta**, respectively. The **gastroesophageal and gastroduodenal junctions** are relatively fixed where the esophagus passes through the diaphragm and where the first part of the duodenum becomes retroperitoneal. The stomach is commonly described as "J-shaped" and may assume this form when empty. However, its shape will vary with the degree to which it is filled and with the general body build. In Fig. 5-27 it can be seen that tall, thin people may have a long "J-shaped" stomach and short, robust people a more horizontal "steerhorn" stomach.

Look now at Fig. 5-28 and identify:

1. The right concave border of the stomach called the **lesser curvature** to which the lesser omentum is attached
2. The left convex border called the **greater curvature** to which the greater omentum is attached
3. A notch in the lesser curvature called the **incisura angularis**
4. The **fundus** which is that part of the stomach lying above the entry of the esophagus
5. The **cardiac part** surrounding the cardiac orifice at the junction of esophagus and stomach
6. The **body of the stomach**
7. The funnel-shaped **pyloric part** which begins at the incisura and may be divided into the **pyloric antrum** and the **pyloric canal**
8. The **pyloric sphincter** which terminates at the **pyloric orifice**.

The Pyloric Sphincter

The pyloric sphincter, which controls the flow of gastric contents into the duodenum, can be recognized by a thickening in the intestinal wall and in life by the small prepyloric vein which shows through the overlying peritoneum. An interesting "hyperplasia" of the sphincter may affect infants

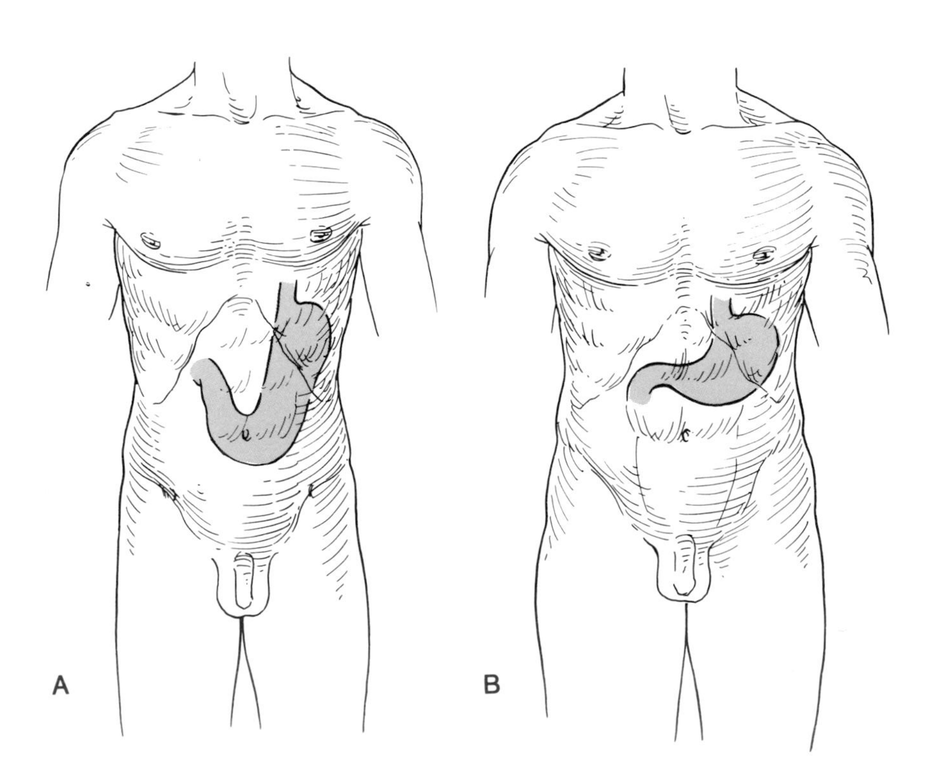

5-27 Illustrating varieties of stomach shape.

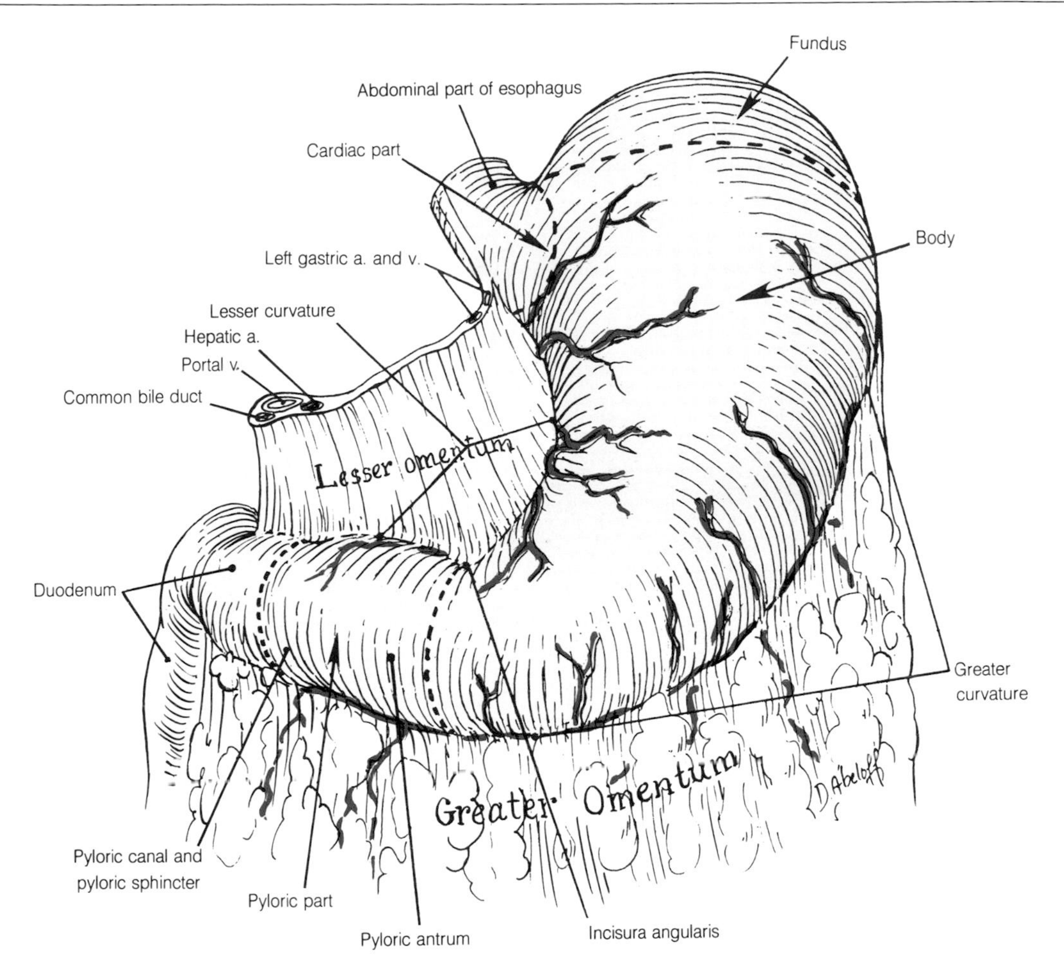

5-28 The anterior aspect of the stomach.

in the first six weeks of life. The narrowing it causes leads to potentially fatal obstruction and vomiting, but when recognized can be treated by anticholinergic drugs or by an incision that divides the muscle fibers but does not penetrate the underlying mucous membrane.

The Interior of the Stomach

When the stomach is opened, the lining mucosa is found to be thrown into folds. As is shown in Fig. 5-29 these are arranged in a longitudinal manner along the lesser curvature and radiological examination reveals that swallowed fluids initially follow this pathway. In life the interior of the stomach can be directly examined with the gastroscope or by radiographic methods.

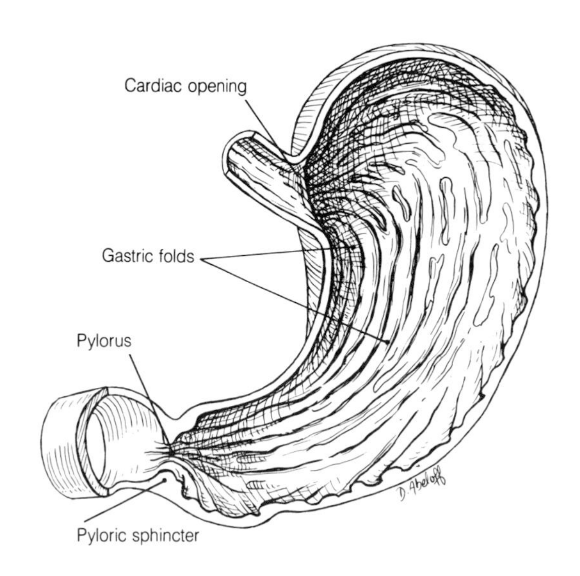

5-29 Showing the interior of the stomach.

Important Relations of the Stomach

Anteriorly and above lie the anterior abdominal wall, the left costal margin, and the attached diaphragm, which arches up and over the stomach. To the left, contact is made with the spleen and, to the right, the quadrate and left lobes of the liver lie above and in front of the stomach.

Posteriorly the stomach is again related to the diaphragm and also to the left suprarenal gland, the upper pole of the left kidney, the pancreas, and the left colic flexure. These structures together with the spleen are said to form the **bed of the stomach**, but it should be remembered that all are separated from the stomach by the peritoneum forming the lesser sac. Examine Fig. 5-30 to confirm these relationships. A study of the horizontal section illustrated in Fig. 5-31 will also show some of these relationships but in addition it demonstrates the position of the lesser sac.

The Vessels of the Stomach

The stomach has an extremely rich blood supply (its mucous membrane produces about 2.5 liters of secretion each day). The vessels contributing to this supply are the **left and right gastric arteries**, the **short gastric arteries**, and the **left and right gastroepiploic arteries**, and these are illustrated in Fig. 5-32. Each of these vessels is directly or indirectly a branch of the **celiac trunk**, the first midline branch of the **abdominal aorta**. The veins which accompany these arteries drain into the **portal vein** or its tributaries. All these vessels will be described in more detail with the distribution of the celiac trunk and the regions drained by the portal vein.

The Innervation of the Stomach

The stomach is supplied by both sympathetic and

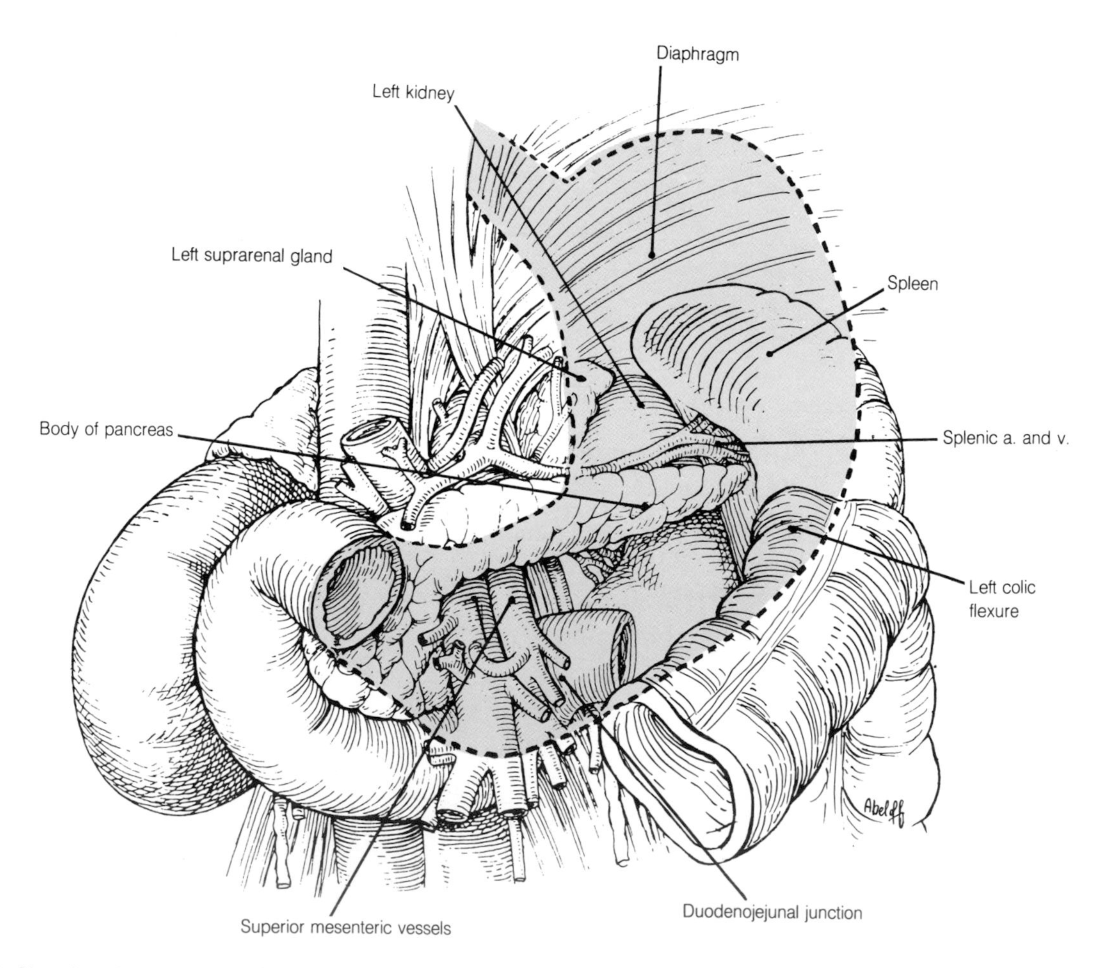

5-30 Showing the structures forming the bed of the stomach.

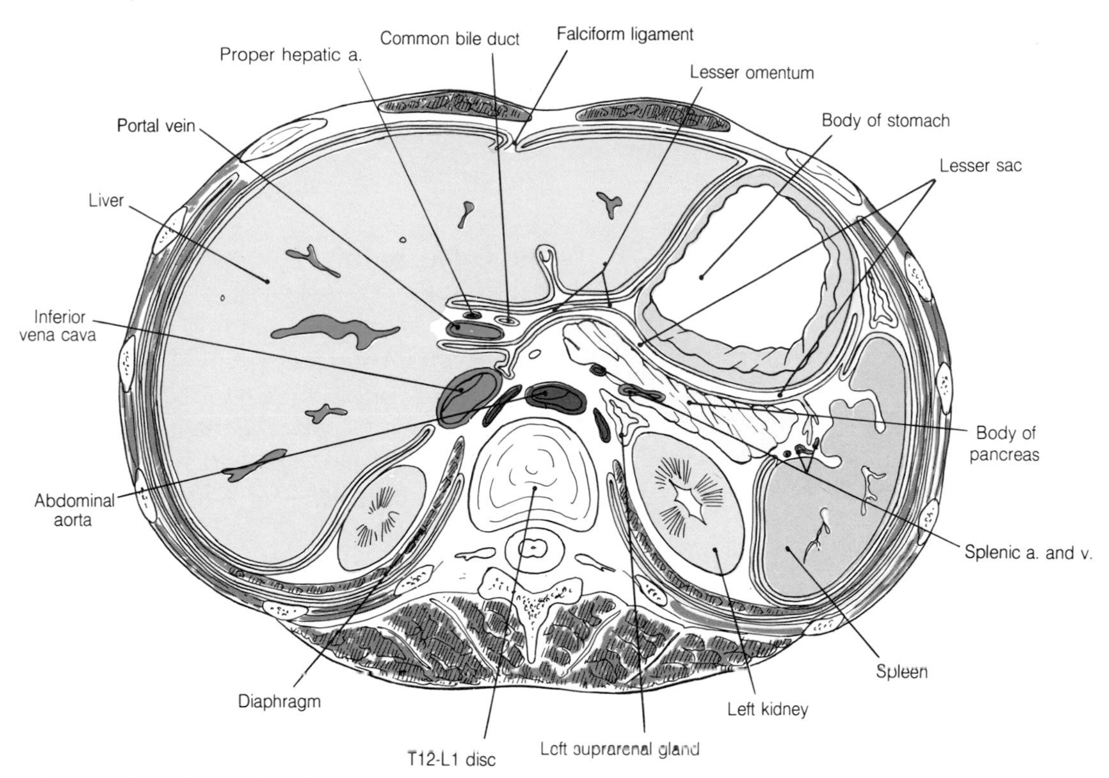

5-31 Horizontal section of the abdomen made at the level of the disc between the twelfth thoracic and first lumbar vertebrae and viewed from below.

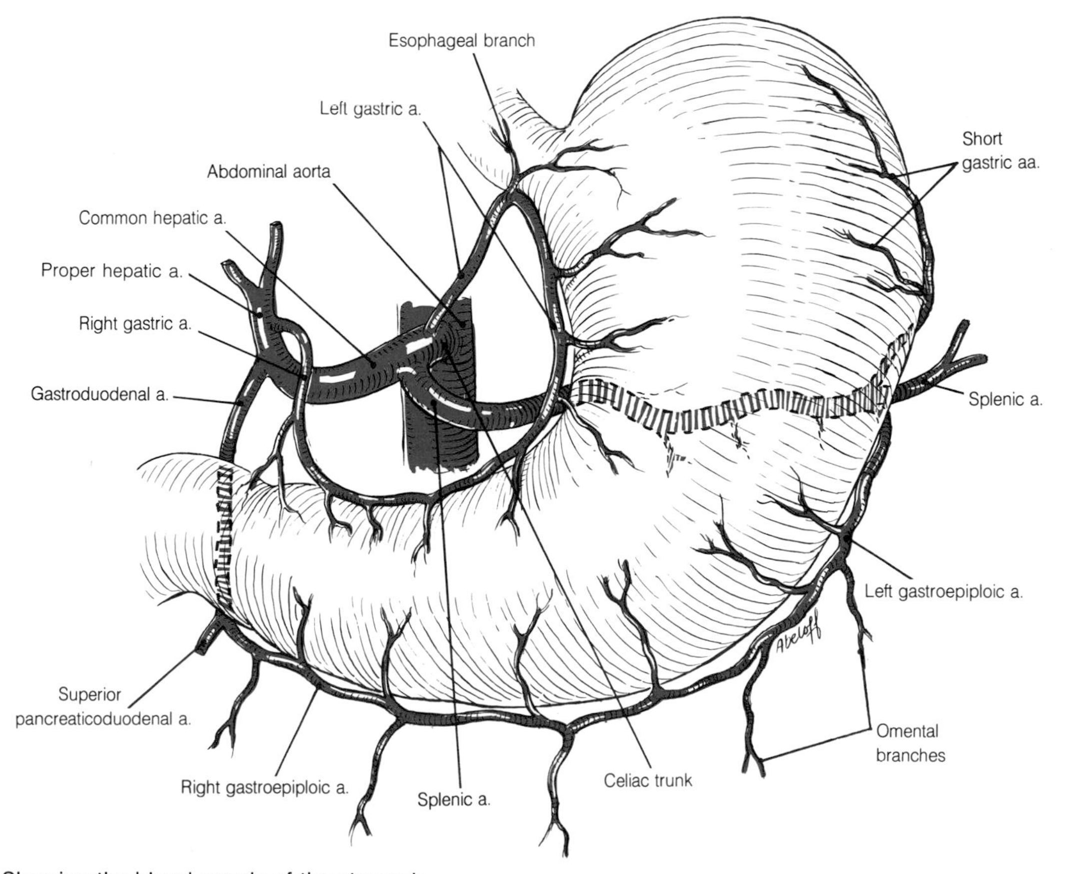

5-32 Showing the blood supply of the stomach.

parasympathetic nerves. Sympathetic nerves reaching the stomach are derived from autonomic plexuses on nearby arteries, the largest of which is the **celiac plexus**. These nerves are vasomotor to the gastric blood vessels and carry pain fibers from the stomach. The parasympathetic supply is provided by the vagus. **Anterior and posterior vagal trunks** derived from the esophageal plexus pass through the diaphragm with the esophagus and break up into **anterior and posterior gastric branches** on the anterior and posterior surfaces of the stomach. The posterior gastric branches also contribute to the celiac plexus. The gastric branches themselves form plexuses in the muscular and submucous coats of the stomach and postganglionic fibers control muscular activity and secretion.

The Lymphatic Drainage of the Stomach

The incidence of malignant growths of the stomach places considerable importance on its lymphatic drainage. The involvement of the surrounding groups of lymph nodes in the spread of such growths determines the prognosis of the condition and often the symptoms with which a patient presents.

The lymphatic vessels largely follow the course of the arteries and are represented diagrammatically in Fig. 5-33, in which the arteries they follow and the nodes to which they drain can be traced as follows:

Artery	Nodes
L. gastric a.	L. gastric nodes
Short gastric a. L. gastroepiploic a. Splenic a.	Pancreaticosplenic nodes
R. gastroepiploic a.	R. gastroepiploic and pyloric nodes
R. gastric a.	Pyloric, hepatic, and L. gastric nodes

Lymph from each of these groups of nodes drains to the preaortic celiac nodes before passing to the cisterna chyli.

It will be realized that the proximity of pyloric and hepatic nodes to the biliary system may lead, if they become involved in the spread of a tumor, to the obstruction of the common bile duct and therefore to jaundice.

In this and subsequent discussions the importance of the physiological role of the lymphatics in removing products of digestion from the intestine should not be lost.

The Duodenum

The duodenum* forms the first part of the small intestine. It is continuous with the stomach at the pylorus and with the jejunum at the duodenojejunal junction. The first centimeter or so is mobile and is attached to the lesser omentum. The remainder is retroperitoneal and firmly attached to the posterior abdominal wall. In passing from its commencement at the pylorus just to the right of the midline at the level of the body of the first lumbar vertebra to its termination to the left of the midline at the level of the body of the second lumbar vertebra, it describes a C-shaped curve. The **superior part** passes posteriorly and to the right at the side of the vertebral column. Thus, when the duodenum is viewed in a p.a. radiograph, the first part is observed down its length and an oblique projection is needed to reveal the typical picture of the **duodenal bulb** (see Fig. 5-48). The **descending part** descends alongside the vertebral column. The **horizontal part** crosses the vertebral column at the level of the third lumbar vertebra and the **ascending part** ascends to the duodenojejunal junction at the level of the second lumbar vertebra. At this point the dependent and free jejunum produces the **duodenojejunal flexure**.

At its junction with the jejunum the duodenum is anchored to the right crus of the diaphragm by the **suspensory ligament of the duodenum**. This

* Duodenum, Mediaeval, *L.* = twelve, i.e. the length of the duodenum is twelve finger breadths

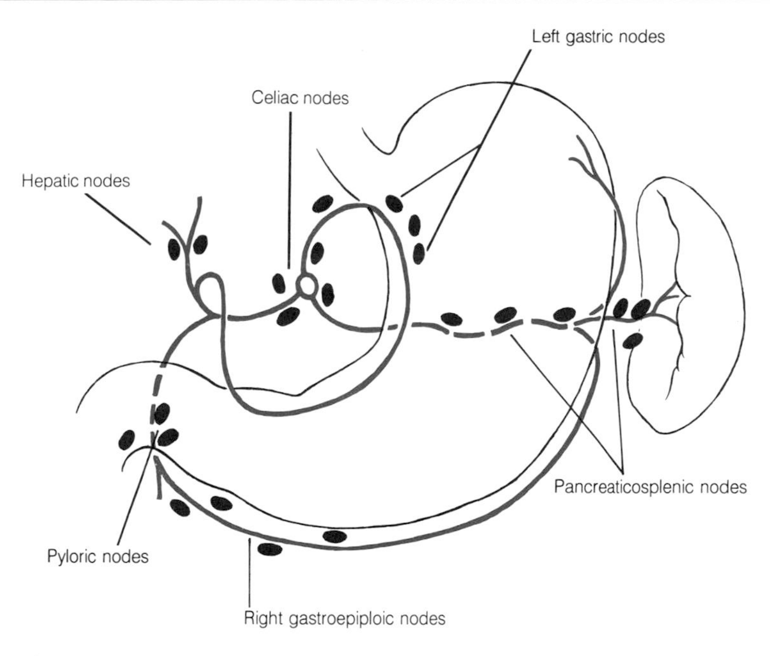

5-33 Diagram illustrating the lymphatic drainage of the stomach.

fibrous band may also contain smooth muscle fibers. The ligament is palpable and, during a laparotomy, it may be used to identify the proximal end of a loop of jejunum.

The Interior of the Duodenum

Although the duodenum has a thin muscular wall, the thick mucous membrane is thrown into irregular circular folds called **plicae circulares** which give it a firm feeling when palpated. On the posteromedial aspect of the descending part can be found a papilla surmounted by a hood-like fold of mucous membrane. This is the **duodenal papilla** and marks the site of entry of the combined **biliary and pancreatic ducts**.

Look now at Fig. 5-34 and identify the parts of the duodenum and their vertebral levels. Note also the position of the duodenal papilla and its relationship to the biliary and pancreatic ducts.

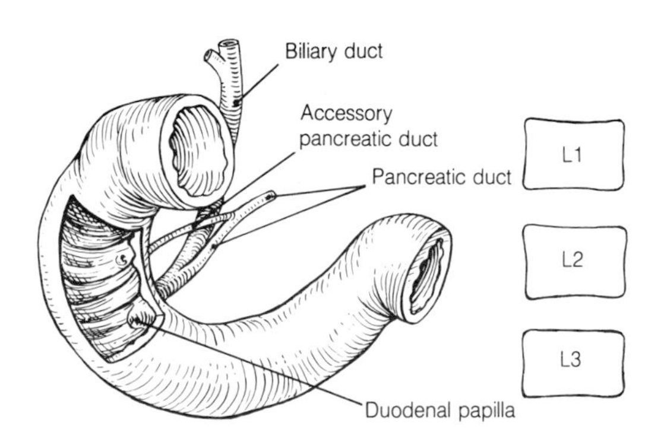

5-34 Showing the shape of the duodenum and the interior of its descending part.

Important Relations of the Duodenum

The curve of the duodenum surrounds the head and neck of the pancreas and lies on the hilus and vessels of the right kidney. The biliary duct and the portal vein lie behind the superior part of the duodenum. The biliary duct, after joining with the pancreatic duct, curves to the right to enter the posteromedial aspect of the descending part. More deeply the superior part is separated from the inferior vena cava by the epiploic foramen and the horizontal part crosses both the inferior vena cava and the aorta. The **superior mesenteric vessels**, however, pass in front of the horizontal part. Also crossing the descending part from right to left is the attachment of the transverse mesocolon. Overhanging the duodenum anteriorly is the visceral surface of the liver and the attached gallbladder. Look carefully at Fig. 5-35 and confirm these relationships.

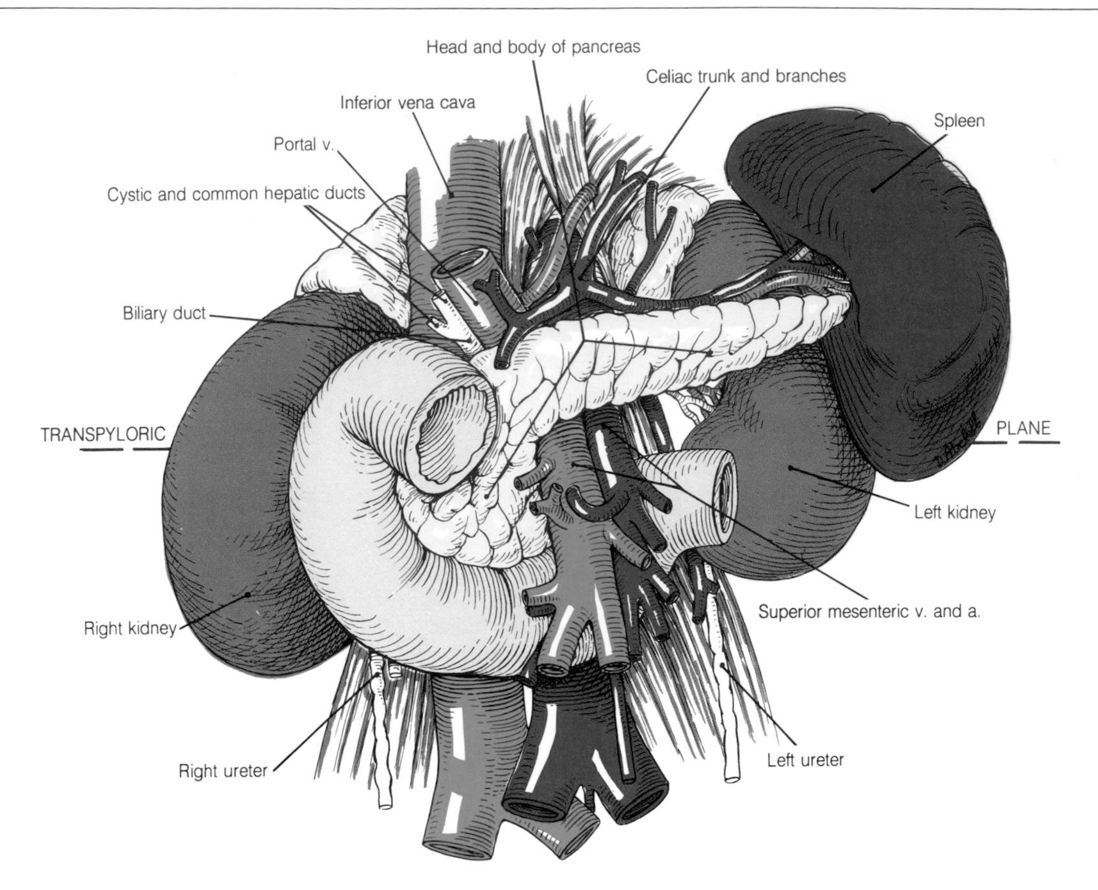

5-35 Showing the relations of the duodenum.

Duodenal Fossae

Folds of peritoneum associated with the ascending part of the duodenum and duodenojejunal junction may form small recesses or fossae. The importance of these lies only in that very occasionally a piece of small intestine may become trapped in one and form an **internal hernia**. The two fossae most commonly found are the **superior and inferior duodenal fossae**.

The Blood Supply of the Duodenum

The duodenum is supplied by **duodenal** and **superior pancreaticoduodenal branches** from the **gastroduodenal branch** of the **hepatic artery** and by **inferior pancreaticoduodenal branches** of the **superior mesenteric artery**. Venous drainage is to the portal and superior mesenteric veins. Lymph from the duodenum passes proximally to pyloric, hepatic, and celiac nodes and distally to superior mesenteric nodes.

During the description of the stomach and duodenum, the liver, gallbladder, pancreas, and spleen have been mentioned as having close relationships to them. All but the last share with them a common origin from the embryonic foregut and all have a common blood supply from the celiac trunk. It is, therefore, proposed to complete the description of these organs before continuing with that of the small and large intestines.

The Liver

The liver is a large solid organ which normally weighs about 1,500 g. It is roughly wedge-shaped and mostly lies in the upper right portion of the abdominal cavity although the thin end of the wedge extends across to the left. Its **anterior, superior, lateral**, and **posterior** surfaces which form a continuous convexity are molded to the inferior surface of the diaphragm and, thus, called the **diaphragmatic surface**. A posteroinferior or **visceral surface** slopes forward and downward to the sharp **inferior border**. Both these surfaces and the inferior border are illustrated in Fig. 5-36. The inferior border lies deep to the right costal margin but as it passes to the left it lies immediately deep to the anterior abdominal wall before passing undercover of the left costal margin. In the epigastrium the normal liver may be demonstrated by a dullness on percussion although it cannot usually be felt here.

The liver is traditionally divided into a **left and right lobe** by the attachment of the falciform ligament. This division leads to the inclusion of the **quadrate** and **caudate lobes** (seen on the visceral surface) with the right lobe. However, it is more important to understand that functionally, the quadrate and part of the caudate lobe belong to

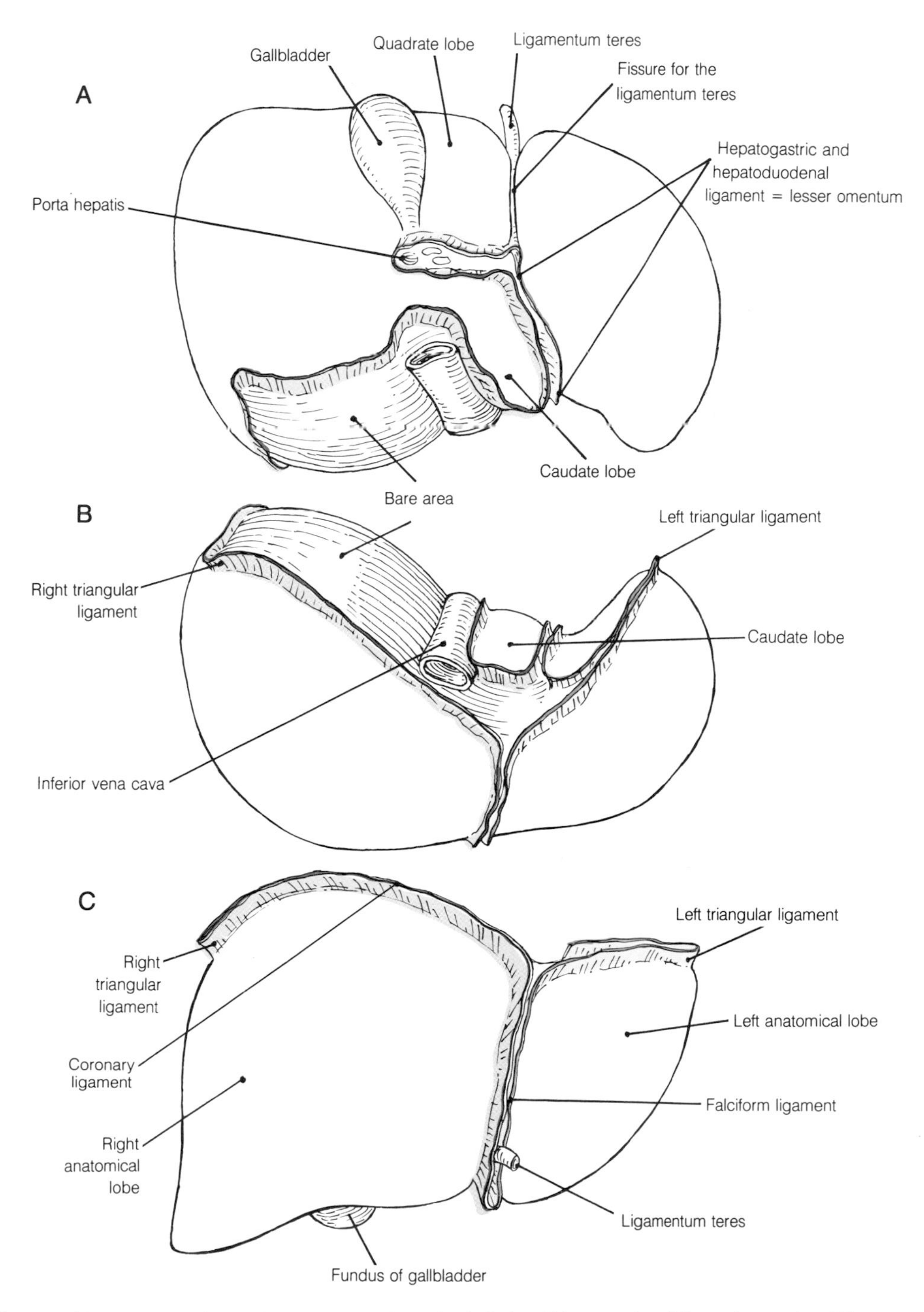

5-36 The liver and its peritoneal attachments seen from its inferior (A), superior (B), and anterior (C) aspects.

the left lobe of the liver. Except over what is called the **bare area** and beneath the gallbladder, the liver is covered by visceral peritoneum derived from the embryonic ventral mesentery in which it has developed. This mesentery is still represented by the **falciform ligament** which links the liver to the diaphragm and the anterior abdominal wall and the **lesser omentum** which links the liver to the lesser curvature of the stomach. Look now at Fig. 5-36 and see how the peritoneum is reflected off the liver. Begin with the lowest diagram and follow the left peritoneal layer of the falciform ligament (*blue*). It runs vertically up the anterior surface of the liver, but on reaching the superior surface it makes a temporary diversion to the left and then returns. In this manner it forms the **left triangular ligament**. When the right layer is followed (*red*), it is seen to make a diversion to the right on the superior surface of the liver forming the **coronary ligament** whose right extremity is the **right triangular ligament**. However, before returning, this layer encircles a substantial part of the posterior surface of the liver and inferior vena cava, leaving them uncovered by peritoneum and immediately adjacent to the diaphragm hence, the **bare area**.

The right and left layer join company again and descend over the visceral surface of the liver between left and caudate lobes forming now the **hepatogastric and hepatoduodenal ligaments**, both of which are part of the lesser omentum. Between the caudate and quadrate lobes the two layers turn to the right, encircle the hepatic artery, portal vein, and biliary ducts of the **porta hepatis**, and at this point become continuous with each other at the free border of the lesser omentum.

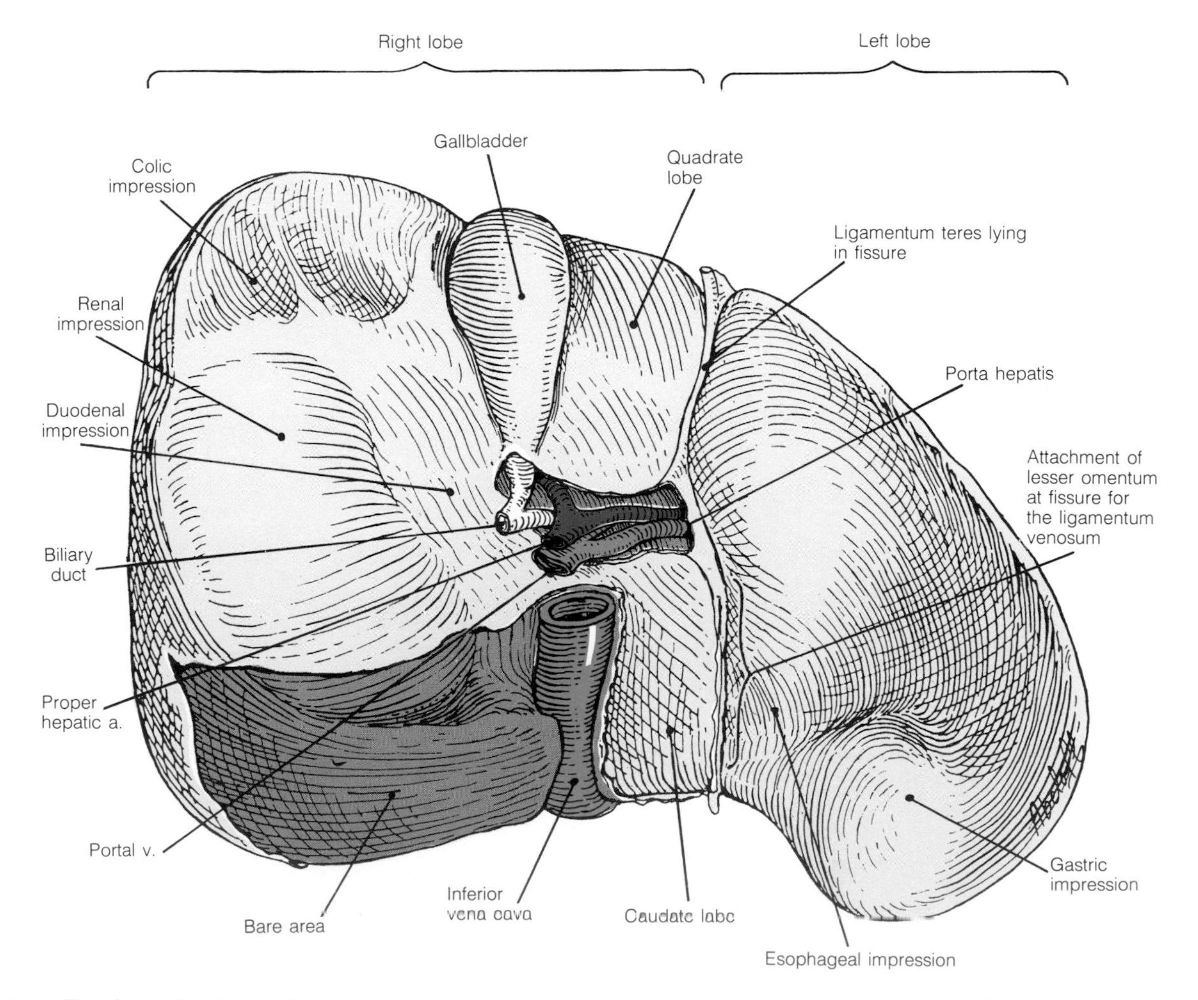

5-37 The visceral and part of the posterior surfaces of the liver seen after lifting the inferior border in a cranial direction.

The arrangement of these peritoneal reflections and ligaments should be confirmed in the cadaver at the earliest opportunity.

A more detailed examination of the surfaces of the liver can now be made using Fig. 5-36 and Fig. 5-37, in which it is viewed from its visceral aspect. At the same time its relationships with nearby structures can be considered. Note then the following features:

1. The rounded anterior, lateral, superior, and posterior parts of the diaphragmatic surface are smooth and related to the muscle and central tendon of the diaphragm.
2. The costal impressions often seen in the hardened liver from a cadaver are not present in life but a shallow cardiac depression near the center of the superior surface corresponds to the inferior ventricular surface of the heart.
3. The sharp inferior margin shows a deep notch which becomes the **fissure for the ligamentum teres** on the visceral surface. A further shallow notch is often present to the right of this near the fundus of the gallbladder.
4. The posterior surface is deep on the right side but narrows to a sharp edge on the left.
5. The bare area on the right side of the posterior surface. This is related to the diaphragm and the inferior vena cava which lies here in a deep groove in the liver just before it pierces the central tendon of the diaphragm. Below, the bare area is related to the right suprarenal gland.
6. Between the inferior vena cava and the attachment of the lesser omentum lies the caudate lobe. This lobe limits the superior extent of the lesser sac.
7. Between the two peritoneal layers of the lesser omentum is a **fissure for the ligamentum venosum**. This ligament is the remnant of the **ductus venosus** which, in fetal life, carried blood from the portal and umbilical veins to the hepatic veins.
8. To the left of this fissure the posterior surface shows an impression for the abdominal portion of the esophagus.
9. To the right, the visceral which is continuous with a larger impression related to the stomach surface overlies the right kidney and below this is related to the right colic flexure.
10. To the left of this region there is a shallow impression in which lies the gallbladder. Its superior end joins the porta hepatis which is surrounded by the free margin of the lesser omentum.
11. Between the gallbladder and the fissure for the ligamentum teres lies the quadrate lobe of the liver.

The Porta Hepatis

A particular note should be made of the **porta*** **hepatis**. It is a deep fissure on the visceral surface of the liver lying between the caudate and quadrate lobes. It is here that the left and right branches of the hepatic artery and portal vein enter the liver and the right and left hepatic ducts leave to form the common hepatic duct. The anterior and posterior layers of the lesser omentum are reflected around the lips of the fissure and it is in the free border of this ligament that these vessels and duct travel.

The Venous Drainage of the Liver

Venous blood leaves the liver by a number of short wide **hepatic veins** that join the inferior vena cava as it lies in a deep groove in the posterior surface of the liver. These veins, therefore, have no course outside the liver. They are best demonstrated when the inferior vena cava is opened.

The Lymphatic Drainage of the Liver

Lymphatics from the liver may pass to mediastinal nodes in the thorax by means of vessels accompanying the inferior vena cava and esophagus or to hepatic and, thence, to celiac nodes.

The Blood Supply of the Liver

The hepatic artery, a branch of the celiac trunk, runs in the free border of the lesser omentum. It divides into left and right branches to supply the left and right lobes of the liver. It has, in addition, a cystic branch which supplies the gallbladder. An accessory left hepatic artery, often arising from the left gastric artery, is not uncommonly found. Blood also reaches the liver by way of the portal

* Porta, *L.* = gate or door

vein. This venous blood is by no means desaturated and contributes substantially to the oxygen supply of the liver.

The Gallbladder

The gallbladder forms part of the excretory apparatus of the liver and acts as a temporary reservoir for bile. Upon the appropriate physiological stimulus this is released into the duodenum. The gallbladder has already been seen to lie applied to the visceral surface of the liver and to be partly covered by peritoneum reflected off that surface. It has a **fundus** which lies anteriorly at the inferior margin of the liver and close to the anterior abdominal wall. The **body** narrows to **a neck**, which turns medially toward the porta hepatis. The neck blends with the **cystic duct**, which joins the **common hepatic duct** to form the **biliary duct**. The mucous membrane lining the neck and cystic duct is thrown into spirally arranged folds. These form the **spiral valve**.

The relations of the gallbladder can largely be deduced from its position on the visceral surface of the liver. Above and in front of the body lies the liver and, anterior to the fundus as it projects below the liver, is the anterior abdominal wall. To the right is the transverse colon and crossing behind it and lying on its right-hand side is the duodenum.

The **cystic artery**, a branch of the hepatic artery, supplies the gallbladder. Veins draining it pass directly to the liver either from its deep surface or along the cystic duct to the porta hepatis. Lymphatics pass to the hepatic nodes.

The gallbladder is subject to both acute and chronic inflamation; the latter being often associated with the formation of gallstones, a condition known as cholelithiasis.* Such stones may leave the gallbladder, become impacted in the biliary tract, and cause both acute pain and obstruction to the flow of bile. This in turn can lead to jaundice. These conditions may require surgical intervention under difficult conditions when a clear picture of the anatomy of this organ and the biliary tract becomes essential.

* Chole, *Gk.* = bile; Lithos, *Gk.* = a stone

The Biliary Tract

The arrangement of the biliary tract can most easily be understood from an examination of Fig. 5-38. Note the **left and right hepatic ducts** which join to form the **common hepatic duct**. The **cystic duct** runs for a variable distance alongside the hepatic duct before joining it at an acute angle to form the **biliary duct**. This is carried in the free margin of the lesser omentum to the first part of the duodenum behind which it passes to become embedded in the head of the pancreas. Turning to the right in company with the **pancreatic duct**, it reaches the descending part of the duodenum. Here the two ducts unite at the **hepatopancreatic ampulla** which opens into the duodenum at the **duodenal papilla**. The circular muscle around the terminations of the bile and pancreatic ducts and around the hepatopancreatic ampulla is thickened to form the **sphincter of the ampulla** which controls the flow of bile and pancreatic secretions to the duodenum.

Variations of the Biliary Tract

Cholecystectomy, that is removal of the gallbladder, is a frequently performed operation. However, after repeated episodes of acute inflammation the biliary tract and the nearby branches of the hepatic artery become difficult to define. For this reason awareness of the common variations in the courses of the ducts and vessels becomes particularly important.

Fig. 5-39 shows how the cystic duct may join the common hepatic duct very close to its formation at the porta hepatis (left). Alternatively the cystic duct may join the hepatic duct quite close to the duodenum (center) and even cross over the hepatic duct to achieve this (right).

While the right hepatic artery usually passes behind the common hepatic duct as it nears the porta hepatis, it may pass anterior to the duct. Furthermore, the cystic artery, which normally arises from the right hepatic artery after the latter has passed beneath the common hepatic duct, may arise to the left of the duct from the right or left hepatic artery and pass either anterior or posterior to the duct.

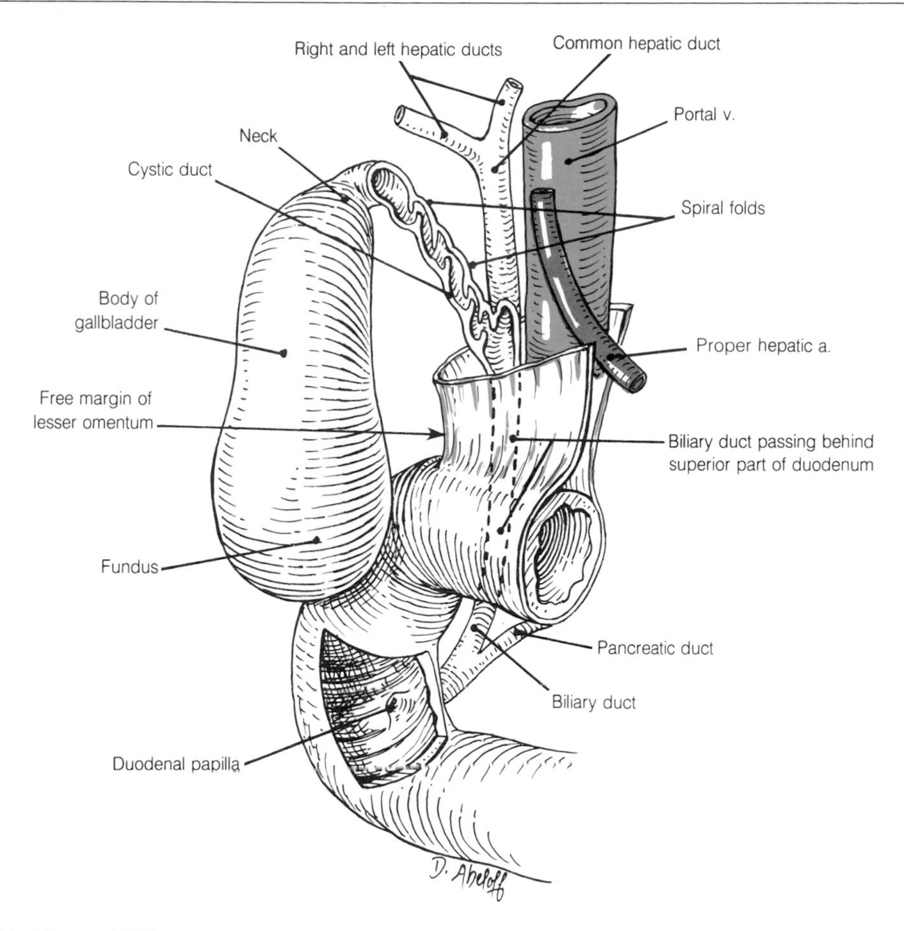

5-38 The gallbladder and biliary tract.

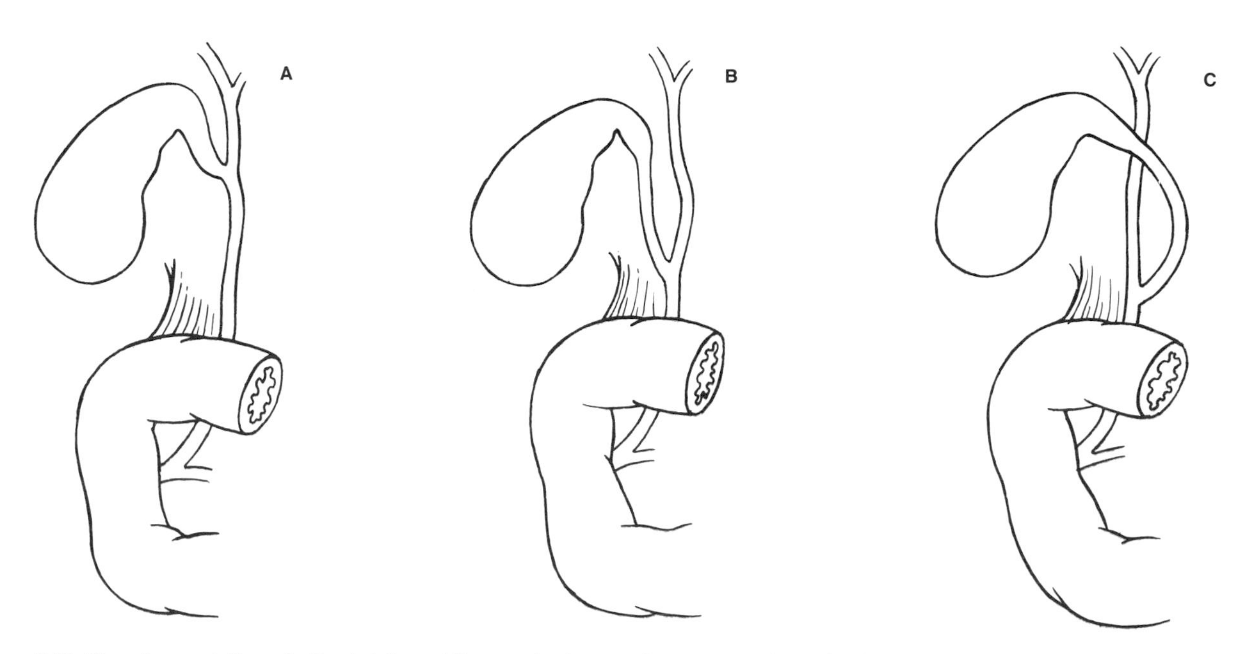

5-39 Showing variations in the joining of the cystic duct to the common hepatic duct.

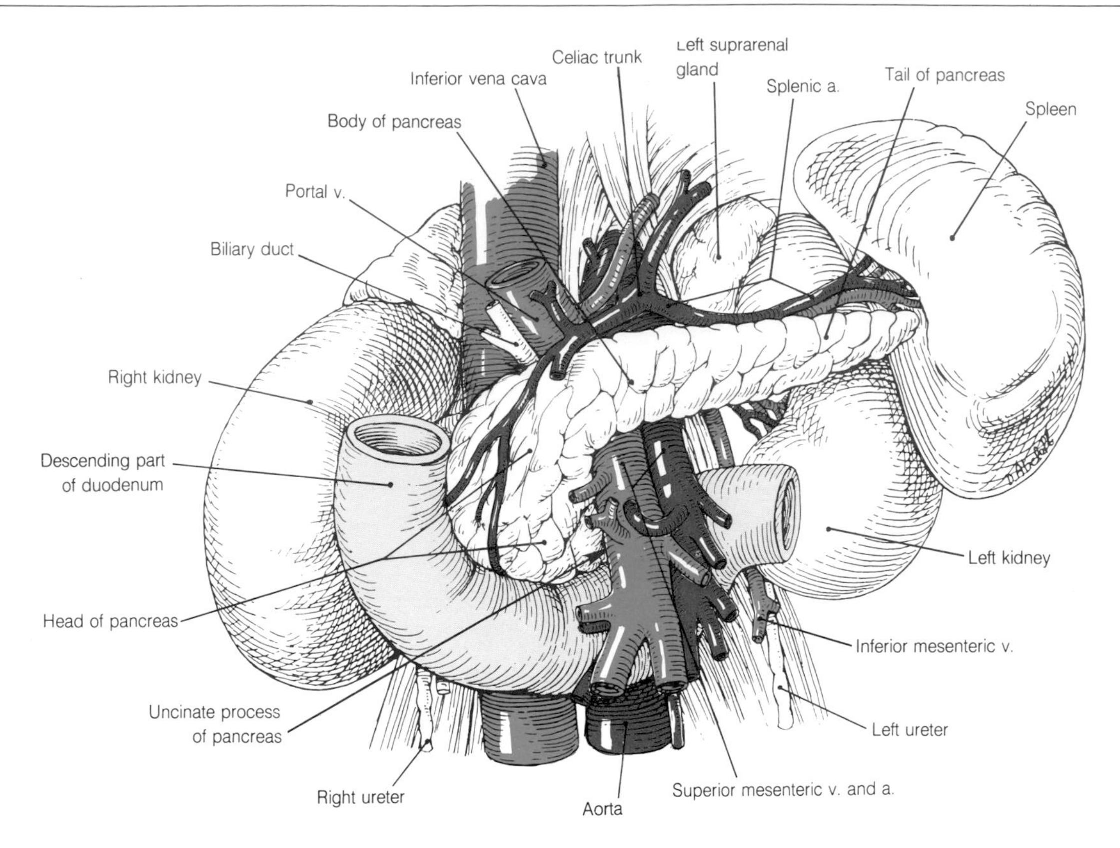

5-40 The pancreas and its nearby relationships.

The Pancreas

The pancreas is both an endocrine and an exocrine gland and as might be expected has both a substantial blood supply and a duct system. As this organ is described its various parts and their nearby relationships should be followed in Figures 5-40 and 5-41.

The pancreas lies on the posterior abdominal wall aligned roughly about a line drawn through the body of the first lumbar vertebra. In cross-section its triangular and posterior, anterior, and inferior surfaces can be recognized (Fig. 5-41). The posterior surface which is applied to the posterior abdominal wall has no peritoneal relationship. The pancreas is therefore said to be retroperitoneal. The anterior and inferior surfaces are covered by peritoneum, but, at the border where these surfaces join, the peritoneum is reflected off each to form the fused peritoneal layers of the posterior wall of the lesser sac and the transverse mesocolon. It is at this point that the middle

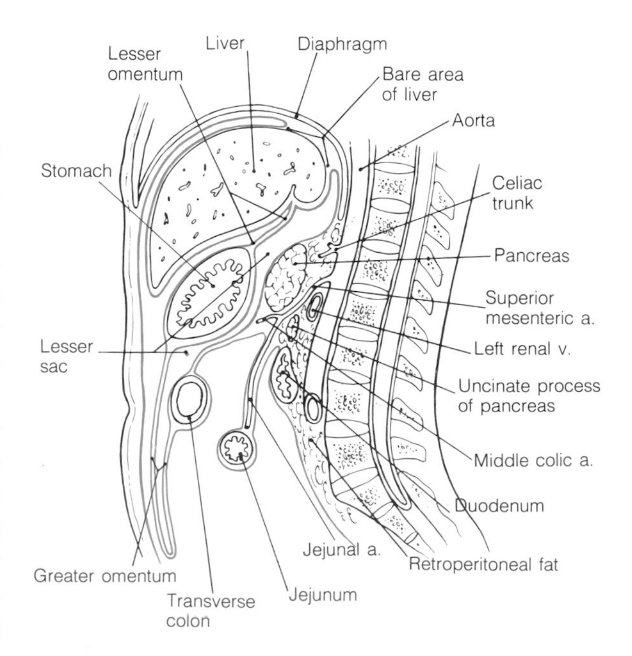

5-41 Semidiagrammatic sagittal section through the pancreas and the origin of the superior mesenteric artery.

colic artery passes between the two layers to reach the transverse colon. As a result of this arrangement of the mesentery the superior surface of the pancreas lies posterior to the lesser sac. The inferior surface on the other hand faces the infracolic compartment of the greater sac.

The pancreas has a **head** which nestles in the curve of the duodenum and lies anterior to the inferior vena cava and the left renal vein, and through which travels the biliary duct. Below, a small portion of the head is tucked beneath the superior mesenteric vein and is known as the **uncinate process**. The head is joined to the **body** by the **neck** which overlies the superior mesenteric vessels and the portal vein. The body extends to the left as far as the hilus of the left kidney and overlies the aorta, the left renal vein, the splenic vessels, and the termination of the inferior mesenteric vein. Anteriorly, it is crossed by the attachment of the transverse mesocolon. The **tail** of the pancreas leaves the posterior abdominal wall in the lienorenal ligament to end at the hilus of the spleen. Figs. 5-40 and 5-41 should be used to confirm all these relationships. The main **pancreatic duct** traverses the organ to open into the second part of the duodenum in company with the bile duct. An **accessory duct** may drain the lower part of the head or **uncinate process** and open independently into the duodenum **above** the level of the main duct.

The head of the pancreas is supplied by both the **superior and inferior pancreaticoduodenal arteries**. The remainder is supplied by many branches of the splenic artery with which it is intimately related. Veins draining the pancreas join the portal, splenic, and superior mesenteric veins. Lymphatics from the pancreas follow the course of its blood vessels to preaortic nodes around the celiac and superior mesenteric arteries. Intermediate pancreaticosplenic nodes are found along the splenic artery as it traverses the upper border of the pancreas.

The Spleen

The spleen forms part of the reticuloendothelial system and is concerned with hematopoiesis in fetal life, and, in the adult, with the reutilization of iron from the hemoglobin of destroyed red blood cells. Although not functionally part of the intestinal canal, it has a common blood supply and venous drainage and such close topographical relations that its description is included with neighboring abdominal organs.

The normal spleen, which is solid and firm, lies comfortably in the palm of the hand. The convex surface that fills the palm lies in the left hypochondrium against the diaphragm with its long axis aligned with the tenth rib (however, the pleural cavity separates spleen and diaphragm from the rib). A border, which is notched anteriorly, separates this surface from a somewhat concave visceral surface. Here is found the hilus where vessels enter and leave the organ. Look at Fig. 5-42 and note that the visceral surface is related to the left kidney, the stomach, and the splenic flexure of the colon. Each surface is covered with visceral peritoneum which is reflected as a double layer onto the left kidney as the **lienorenal ligament** in which lies the tail of the pancreas and onto the stomach as the **gastrosplenic ligament**.

The spleen is supplied with blood by the **splenic artery** and blood drains from it in the **splenic vein**. This is a tributary of the portal vein and, thus, blood from the spleen is carried to the liver. Lymphatic drainage is to nodes at the hilus and, thence, to celiac nodes.

The normal spleen is completely undercover of the thoracic cage and is not palpable. However, in a number of conditions enlargement occurs and in

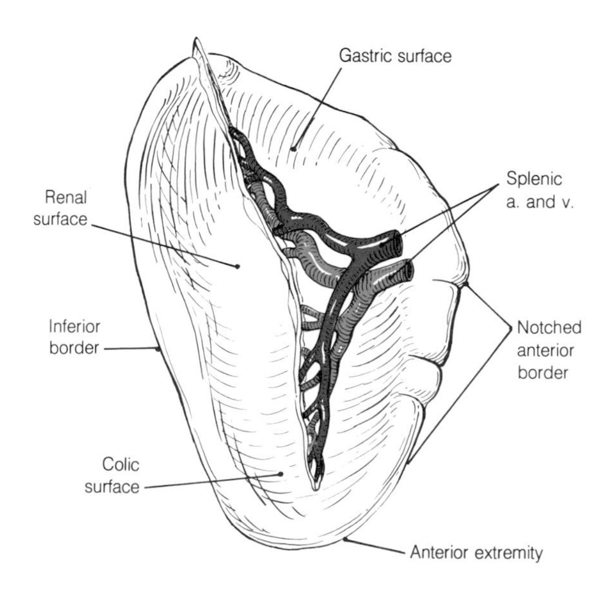

5-42 The visceral surface of the spleen.

due course its tip becomes palpable immediately beneath the anterior abdominal wall at the left costal margin. Because no hollow viscera intervene, an area of dullness to percussion will develop. Further enlargement occurs in a diagonal direction toward the right iliac region. The spleen is also particularly liable to rupture in falls and automobile accidents. Bleeding from this friable organ is difficult to control and splenectomy may be necessary.

The Celiac Trunk and its Distribution

The abdominopelvic portion of the intestinal canal is supplied by three large midline branches of the abdominal aorta. The first of these is the **celiac trunk** which supplies both the abdominal part of the canal derived from the embryological foregut and the spleen. These structures have been described and their blood supply mentioned; it is now time to follow these vessels from their common source in more detail.

The celiac trunk is a short, wide, midline vessel arising from the anterior aspect of the abdominal aorta at the level of the twelfth thoracic vertebra. Here it is closely invested by the celiac plexus of autonomic nerves and lies behind the cavity of the lesser sac. The latter relationship means that its branches must cross the posterior abdominal wall to the margins of the lesser sac before finding a suitable point from which they can reach the structures they supply. Look now at Fig. 5-43 and follow each of the three branches as they are described. These are the **splenic, left gastric**, and **common hepatic arteries**.

The Splenic Artery

The splenic artery follows a sinuous course across the posterior abdominal wall at the upper border of the pancreas. Reaching the lienorenal ligament,

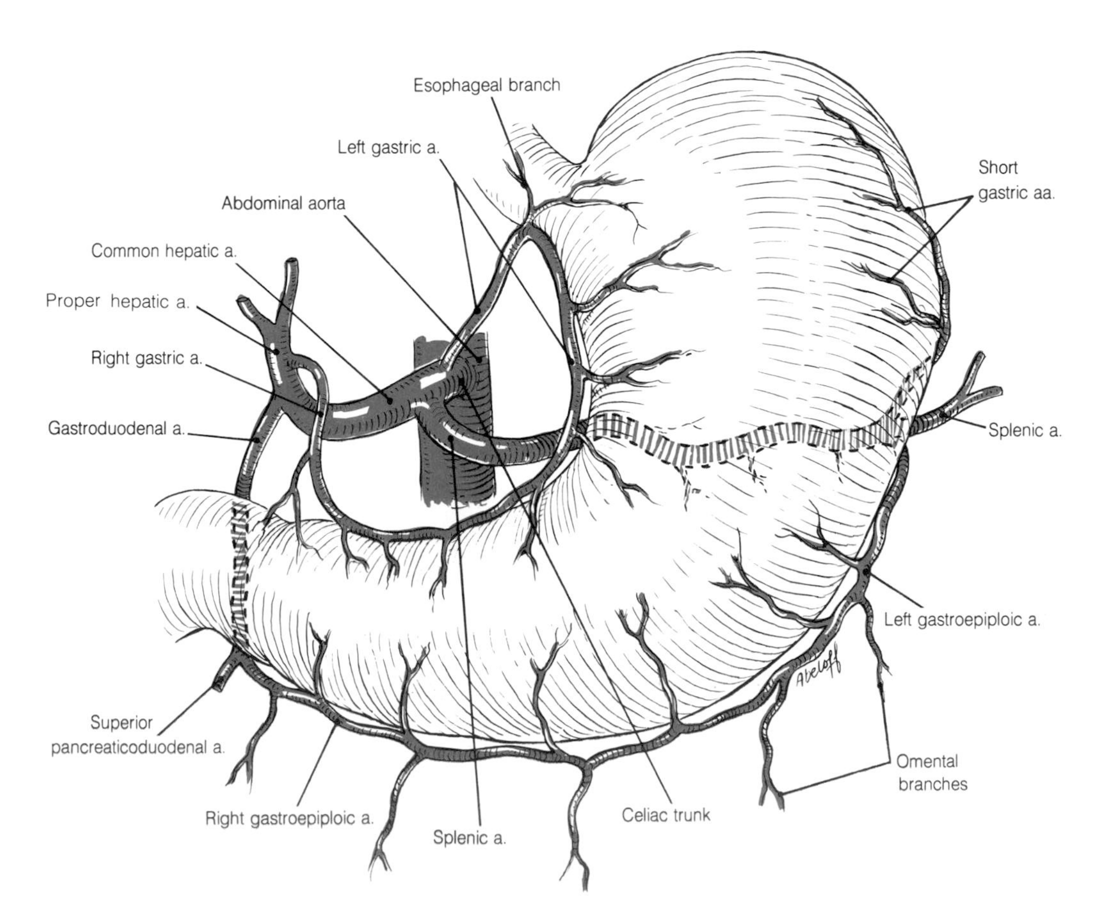

5-43 Showing the branches and distribution of the celiac trunk.

it is carried in this to the spleen. As well as supplying the spleen, it gives off **short gastric arteries** and the **left gastroepiploic artery** which are carried to the greater curvature of the stomach in the gastrosplenic ligament. The splenic artery also gives off many **pancreatic branches**.

The Left Gastric Artery

The left gastric artery ascends to the left across the posterior abdominal wall to reach the gastroesophageal junction close to the diaphragm. After giving off **esophageal branches** it descends the lesser curvature of the stomach to supply it and anastomose with the **right gastric artery**.

The Common Hepatic Artery

The common hepatic artery is slightly more complex. It descends to the right across the posterior abdominal wall until it lies above the retroperitoneal first part of the duodenum. Here it gives off the **right gastric and gastroduodenal arteries**. From the duodenum, the artery, now called the **proper hepatic artery**, is carried to the liver in the free border of the lesser omentum. Here it lies to the left of the common hepatic duct and anterior to the portal vein. As it nears the liver the artery divides into a **left and right hepatic branch**. The **cystic artery** supplying the gallbladder is usually given off by the right hepatic branch.

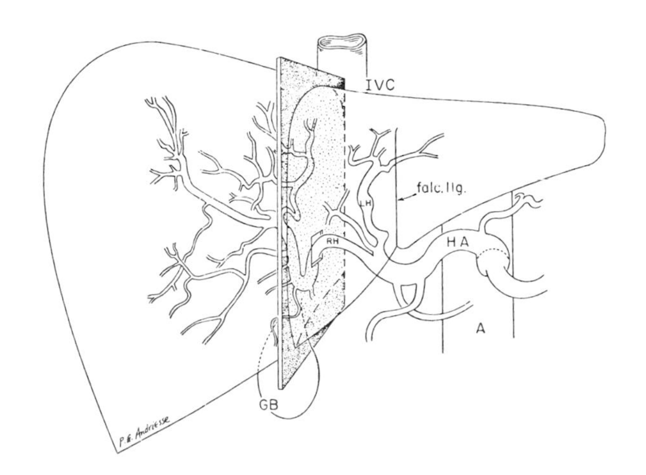

5-44 Tracing of a hepatic arteriogram to show the distribution of the left and right branches of the hepatic artery. (Reproduced by permission, from Nebesar, et al.: *Celiac and Superior Mesenteric Arteries: A Correlation of Angiograms and Dissections.* Little, Brown and Co., Boston, 1969.)

Recent advances have made possible more radical surgery of the liver in the form of partial resection and transplantation. As a result, a knowledge of the intrahepatic distribution of the hepatic artery and variations in its extrahepatic course has become important.

In the tracing of the angiogram seen in Fig. 5-44, it can be seen that the left branch of the hepatic artery supplies not only the classical left lobe but also the regions of the caudate and quadrate lobes. The right branch supplies the region to the right of a plane passing through the gallbladder and the groove for the inferior vena cava.

Variations in the hepatic artery include its origin from the superior mesenteric artery or the presence of accessory hepatic arteries arising from either the left gastric artery or the superior mesenteric artery.

Continue now to follow the distribution of the celiac trunk in Fig. 5-45.

The Cystic Artery

As has already been mentioned, the cystic artery is commonly a branch of the right branch of the hepatic artery beyond the point at which this passes behind the common hepatic duct. As ligation of this artery forms part of the operation of cholecystectomy, the surgeon must be aware of the fact that it may arise sooner or even from the left branch and pass either in front of or behind the common hepatic duct.

The Right Gastric Artery

The right gastric artery, which may be a branch of the common hepatic or of the hepatic artery proper, also reaches the lesser omentum to pass along the lesser curvature of the stomach and anastomose with the left gastric artery.

The Gastroduodenal Artery

Arising from the common hepatic artery at the first part of the duodenum, the gastroduodenal artery passes behind the duodenum and divides after a shourt course into the **right gastroepiploic artery** and the **superior pancreaticoduodenal ar-**

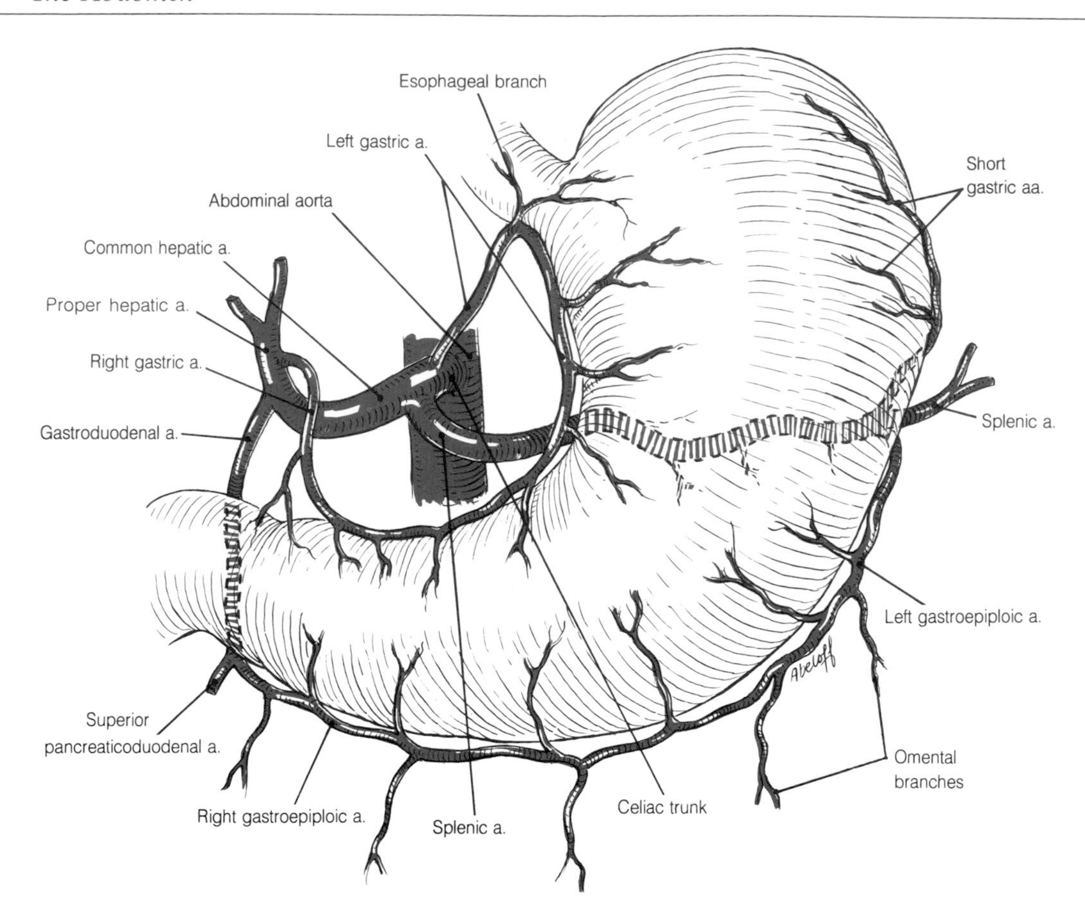

5-45 Showing the branches and distribution of the celiac trunk.

tery or arteries. In addition, it supplies branches to the pyloric end of the stomach and the duodenum.

The Superior Pancreaticoduodenal Arteries

Arising from the gastroduodenal artery as a single or two vessels, the anterior and posterior superior pancreaticoduodenal arteries descend in front of and behind the head of the pancreas. They anastomose with anterior and posterior divisions of the **inferior pancreaticoduodenal artery** and supply both pancreas and duodenum.

The Gastroepiploic Arteries

The left and right gastroepiploic arteries, arising from the splenic and gastroduodenal arteries, respectively, form an arcade in the greater omentum where it is attached to the greater curvature of the stomach. From it, branches ascend over both surfaces of the stomach and descend into the greater omentum.

The Surface Anatomy of the Upper Abdomen

Before looking at any guidelines for the surface projections of abdominal viscera, it must be remembered that a number of factors may modify them. Among these factors should be included:

1. Respiratory movement both of the diaphragm and anterior abdominal wall
2. The position of the subject, i.e., erect or supine
3. The general bodily build or somatotype of the subject, i.e., short and broad or tall and thin
4. The degree of distention of hollow organs by solid, liquid, or gaseous contents
5. Intrinsic movements of which the organ is capable.

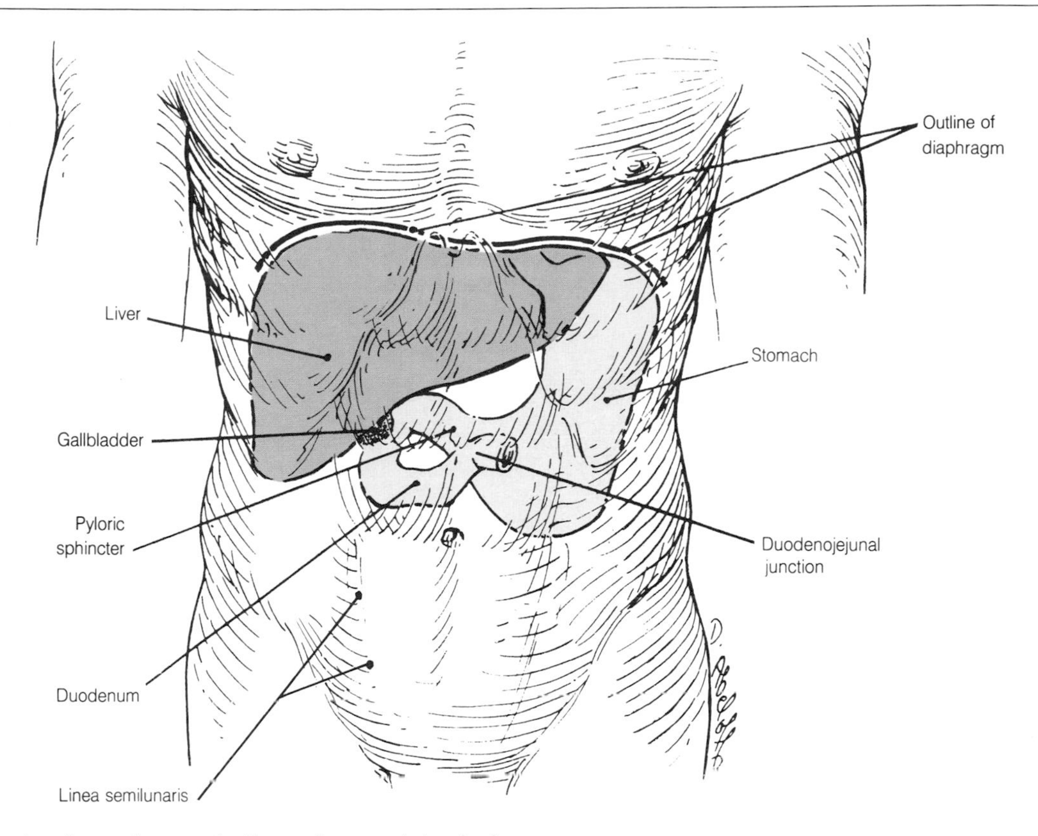

5-46 Showing the surface projections of some abdominal organs.

Examine now the outlines of structures projected onto the anterior thoracic and abdominal walls in Fig. 5-46 and note that:

1. The convex margin of the liver follows the outline of the diaphragm and may normally reach as high as the fifth rib and costal cartilage in the midclavicular line on the right, dip toward the xiphisternum in the midline, and then rise again and end near the fifth rib in the left midclavicular line.
2. From the right the inferior border of the liver follows the costal margin to the tip of the ninth rib where the **linea semilunaris** at the lateral border of rectus abdominis transects the margin. From here, ascending less acutely than the costal margin, it crosses the infrasternal angle, and runs beneath the left costal margin at the tip of the eighth costal cartilage to join with the superior surface.
3. The fundus of the gallbladder lies opposite the tip of the right ninth costal cartilage where it is crossed by the linea semilunaris.
4. The relatively fixed gastroesophageal junction lies opposite the tip of the left eighth costal cartilage and the curved fundus of the stomach is immediately adjacent to the diaphragm at the level of the left fifth or sixth costal cartilage.
5. In the erect subject the pylorus usually lies a little below the transpyloric plane, two to three centimeters to the right of the midline.
6. The duodenojejunal junction lies at the level of the body of the second lumbar vertebra about three centimeters to the left of the midline.
7. The normal spleen lies so deeply under the costal margin that it has no useful surface projection on the anterior abdominal wall. However, on the back it may be projected over the posterior parts of the ninth, tenth, and eleventh ribs with its posterior margin about five centimeters from the midline.

The importance of this exercise lies in acquiring the ability to recognize a change in position or size of an organ when examined at the bedside or radiographically. This change may be intrinsic to the organ or due to pathology of adjacent structures.

The Radiological Anatomy of the Upper Abdominal Organs

The Stomach, Duodenum, Liver, and Spleen

The mucous membrane of the stomach may be examined directly through a flexible gastroscope or the cavity filled with a swallowed radiopaque material (usually an emulsion of barium sulphate) and examined by fluoroscopy.

Swallowed barium rapidly begins to enter the stomach at its cardiac end and then streams toward the pylorus before filling the body. The fundus remains outlined by air. In the radiograph seen in Fig. 5-47, note:

1. Barium outlining the esophagus and leading to the cardia
2. The translucent air in the fundus
3. The lesser and greater curvatures
4. The incisura angularis on the lesser curvature (I. A.)
5. A wave of muscular contraction at the pyloric end (M. C.)
6. That this is a "J-shaped" stomach
7. The partially filled duodenum and jejunum showing the plicae circulares
8. The outlines of the lower borders of the breasts.

Because the superior part of the duodenum runs posterolaterally, an oblique projection provides a better view of its junction with the stomach. In Fig. 5-48, the narrow pylorus can be seen to lead into the duodenal bulb. The flexure between the superior and descending parts of the duodenum can also be seen.

The vessels supplying the stomach, liver, spleen, and pancreas can also be visualized by celiac arteriography. An example is seen in Fig.

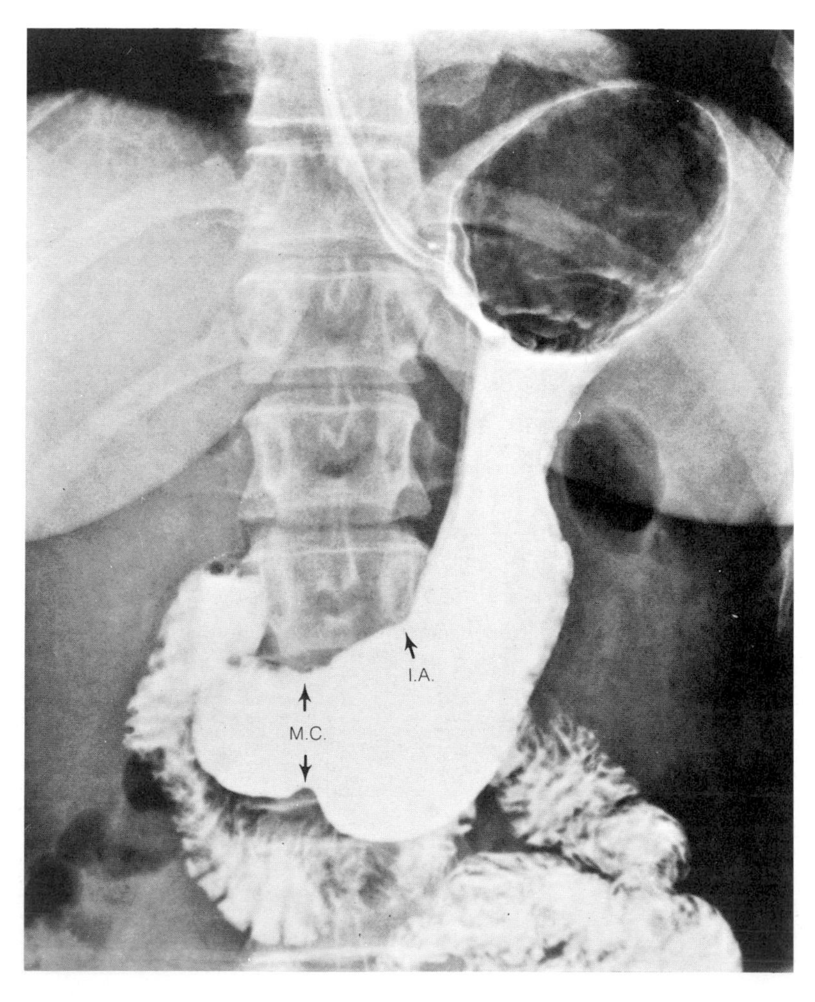

5-47 The stomach filled during a barium meal (p.a. projection). (Reproduced by permission, from Wicke. *Atlas of Radiologic Anatomy,* 4th Ed, Urban & Schwarzenberg, Baltimore-Munich, 1987.)

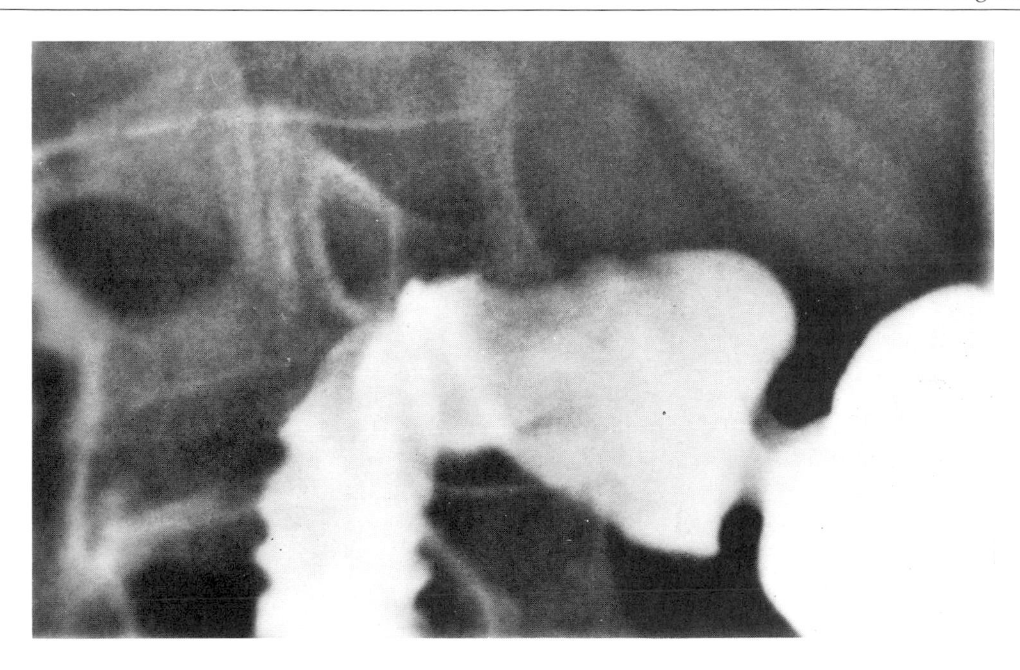

5-48 The superior part of the duodenum filled with barium and showing the pylorus and duodenal bulb (right antero-oblique projection). (Reproduced by permission, from Wicke: *Atlas of Radiologic Anatomy,* 3rd Ed, Urban & Schwarzenberg, Baltimore-Munich, 1983.)

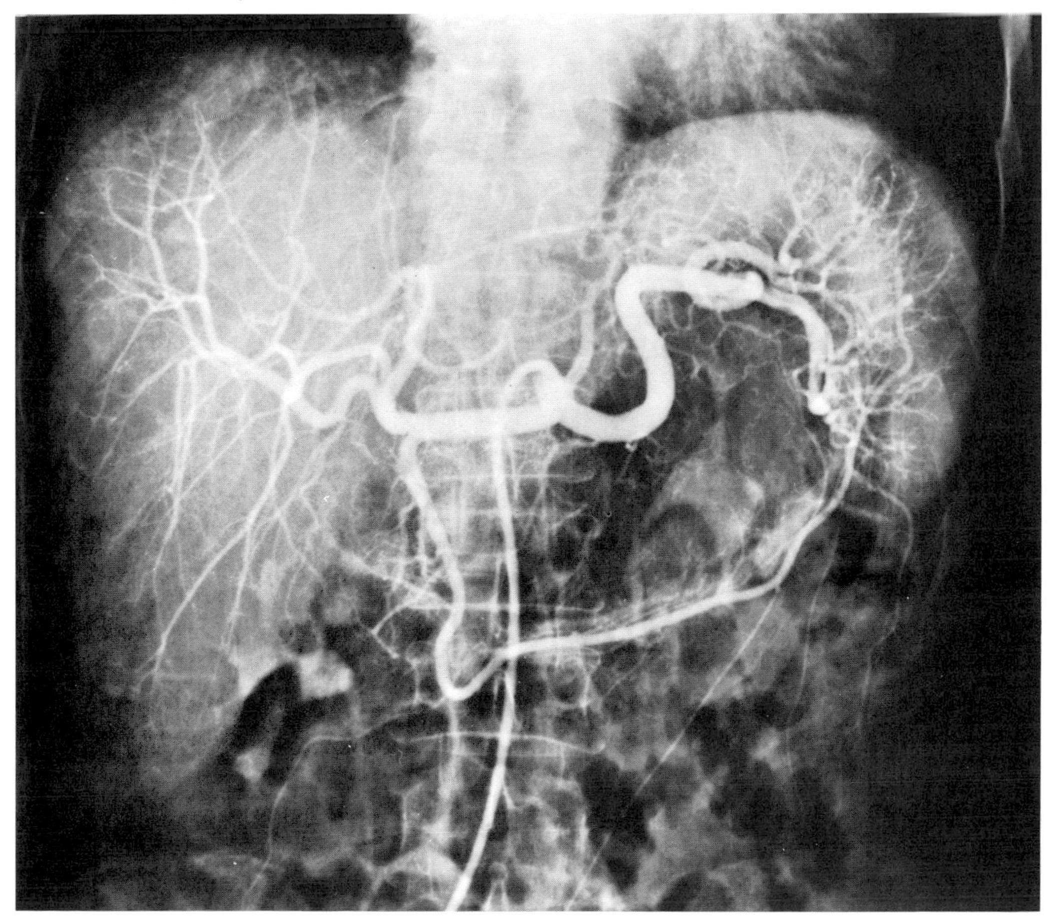

5-49 A celiac arteriogram. (Reproduced by permission, from Wicke: *Atlas of Radiologic Anatomy,* 4th Ed, Urban & Schwarzenberg, Baltimore-Munich, 1987.)

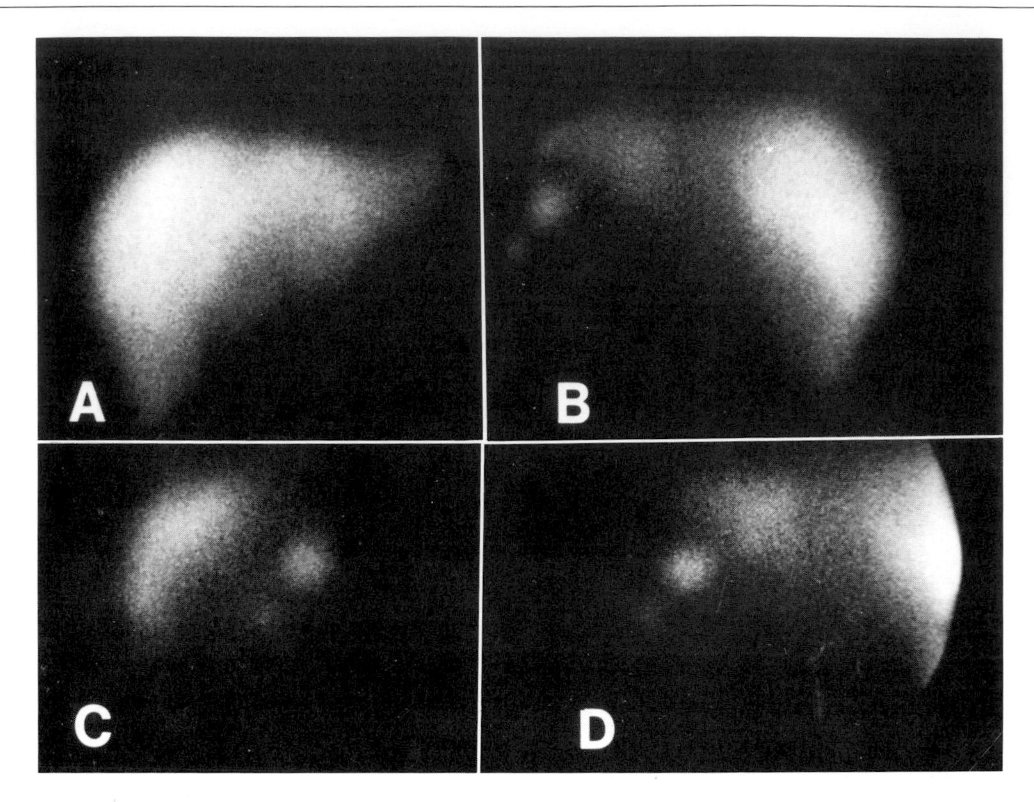

5-50 A nuclear scan of the liver and spleen. (Courtesy of Gerald S. Johnston, M.D., Division of Nuclear Medicine, University of Maryland School of Medicine.)

5-49, in which the branches of distribution of the celiac trunk should be identified.

Although not strictly speaking a radiological examination, the liver and spleen can be displayed by scanning after the reticuloendothelial cells which they contain have taken up a radioactive tracer (Technetium-99m sulfur colloid). An example of this technique is shown in Fig. 5-50. In this, anterior (*A*), posterior (*B*), left lateral (*C*), and left posterior oblique (*D*) views are presented. An interesting feature of this scan is the presence of multiple splenic images caused by splenunculi (little spleens) rather than a single organ.

The Gallbladder

In cholecystography advantage is taken of the ability of the gallbladder to concentrate the bile secreted by the liver. Thus, a relatively dilute opaque medium introduced by mouth or intravenously becomes, after excretion from the liver, sufficiently concentrated to outline its cavity. Fig. 5-51 illustrates a normal gallbladder 15 hours after the contrast medium has been taken by mouth. Note that:

1. The body lies at the level of the third lumbar vertebra.
2. The cavity is evenly filled and the outline is smooth.
3. The rounded fundus lies inferiorly and the tapering neck lies superiorly.

Some idea of the functioning of the gallbladder can be obtained by comparing the picture seen before and after a fatty meal which normally stimulates it to empty. Radiopaque stones can be recognized in the plain film that should precede this investigation and radiolucent stones can be seen to produce a filling defect in the contrast medium.

The Biliary and Pancreatic Ducts

It is now possible to cannulate the biliary and pancreatic ducts using a flexible endoscope and directly introduce an opaque medium. The procedure is known as an endoscopic retrograde cholan-

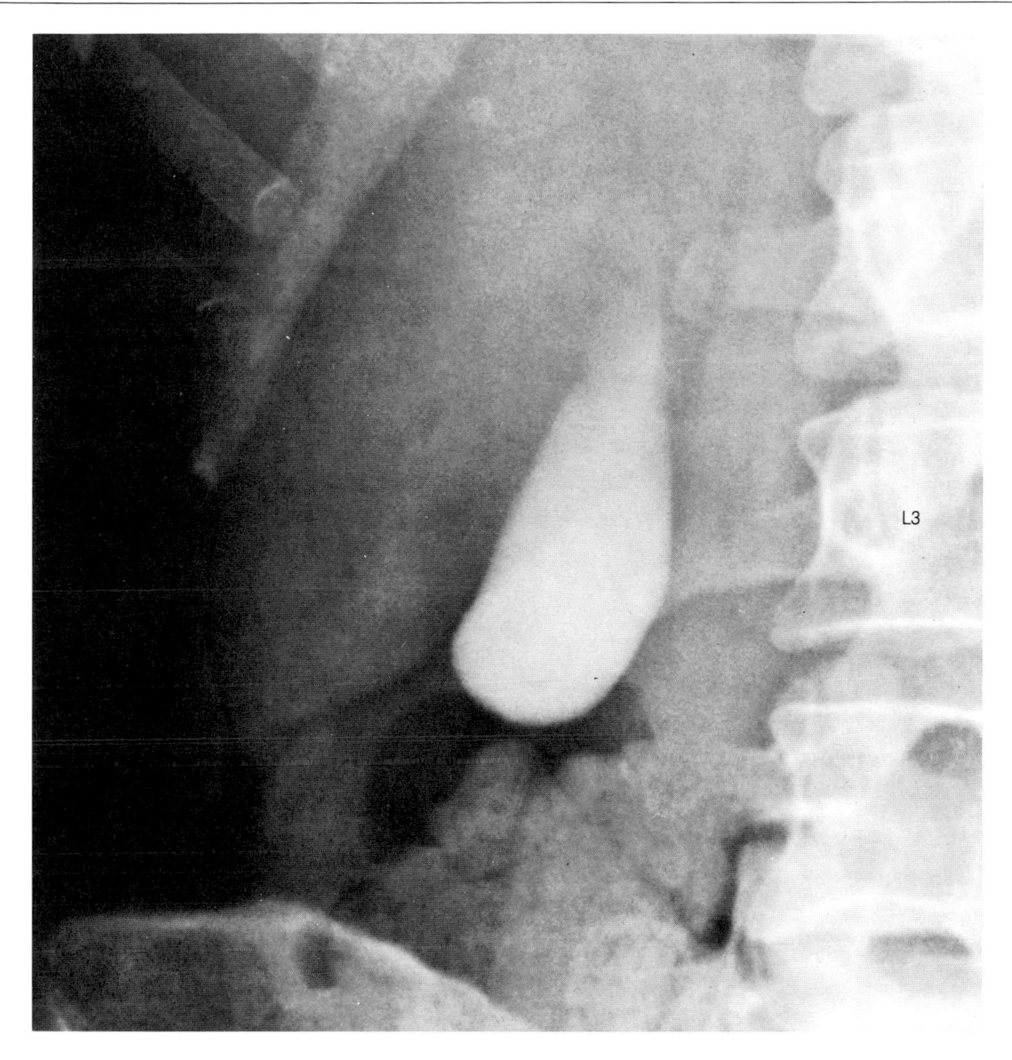

5-51 The gallbladder 15 hours after the oral administration of contrast medium. (Reproduced by permission, from Wicke: *Atlas of Radiologic Anatomy,* 4th Ed, Urban & Schwarzenberg, Baltimore-Munich, 1987.)

giopancreatogram (ERCP) and an example is shown in Fig. 5-52. A fine cannula can be identified by a small opacity at its tip. Radiating upward from this are extra- and intra-hepatic biliary ducts and, passing at first upward and then to the left, the pancreatic duct.

The Small Intestine

The small intestine consists of the duodenum which has already been described, the **jejunum**, and the **ileum**. Unlike the duodenum the twenty feet or more of jejunum and ileum are mobile and fill any space available in the abdominopelvic cavity. They are also the portions of the intestinal tract most commonly found in hernial sacs. The jejunum is fixed to the posterior abdominal wall at the duodenojejunal junction; the ileum at the ileocolic junction. Between these two points the intestine is attached to the posterior abdominal wall by an extensive mesentery. Despite the great length of this mesentery at its attachment to the bowel, its attachment to the posterior abdominal wall is only about 15 cm in length. As a result it and the attached bowel are thrown into a series of deep folds and coils. The bowel is entirely surrounded by peritoneum except at the attachment

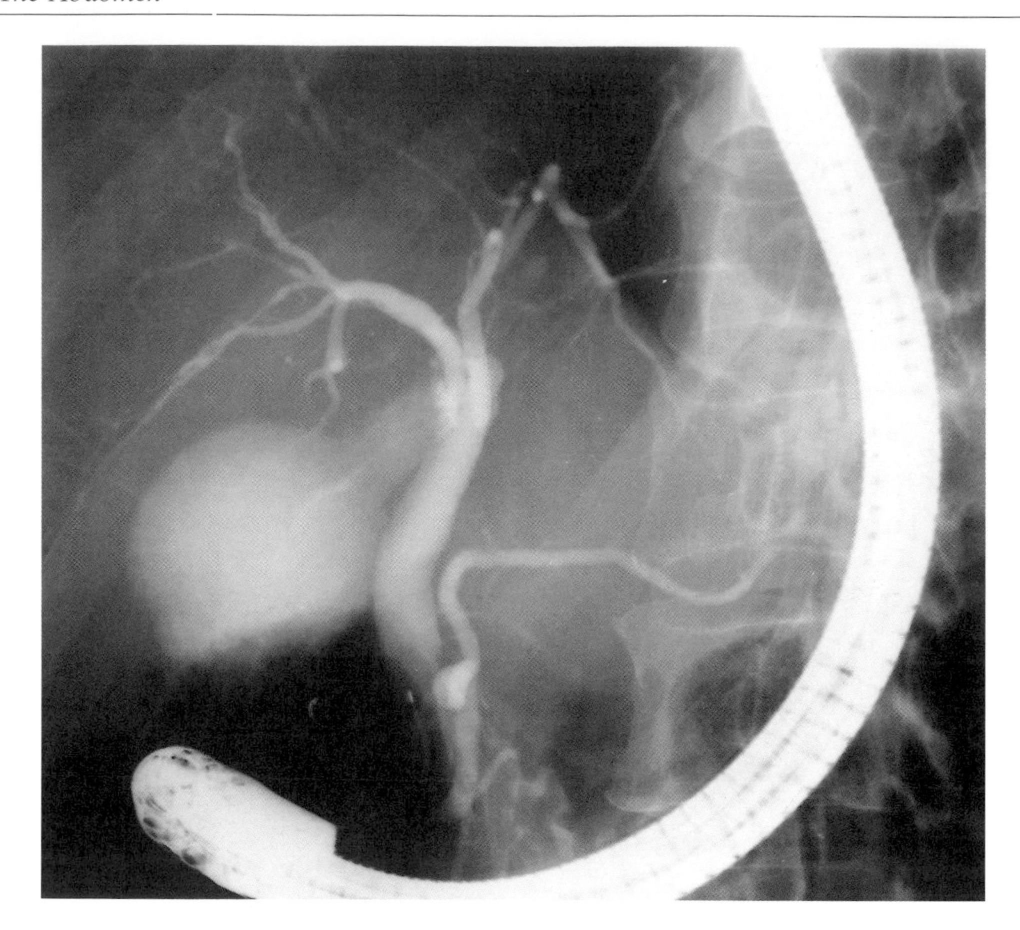

5-52 An endoscopic retrograde cholangiopancreatogram showing biliary and pancreatic ducts (Courtesy of M. A. Hall-Craggs, B. A., M. R. C. P., F. R. C. R., and the E. R. C. P. Unit, The Middlesex Hospital, London).

of the mesentery, and this fact, together with its good blood supply, allows relatively easy and safe surgical resection and anastomosis.

As has already been mentioned, the small intestine is divided into the jejunum and the ileum. However, while the beginning of the jejunum may be distinguished from terminal ileum, the characteristics of the two portions blend imperceptibly with each other. Typically the jejunum is of greater caliber, has a thicker wall, its lymphoid tissue is diffuse, and the jejunal arteries are closely packed and united by only one or two arcades as they approach the intestine. The ileum, on the other hand, is thinner walled, has well marked aggregations of lymphoid tissue at its antimesenteric border (called Peyer's patches), and the more widely spaced ileal arteries are linked by multiple arcades. Some of these features are illustrated in Fig. 5-53.

The Large Intestine

The large intestine is larger in caliber but is considerably shorter in length than the small intestine. It begins in the right iliac fossa at the **ileocecal junction** and terminates at the **anus** and includes the **cecum**, the **vermiform appendix,**

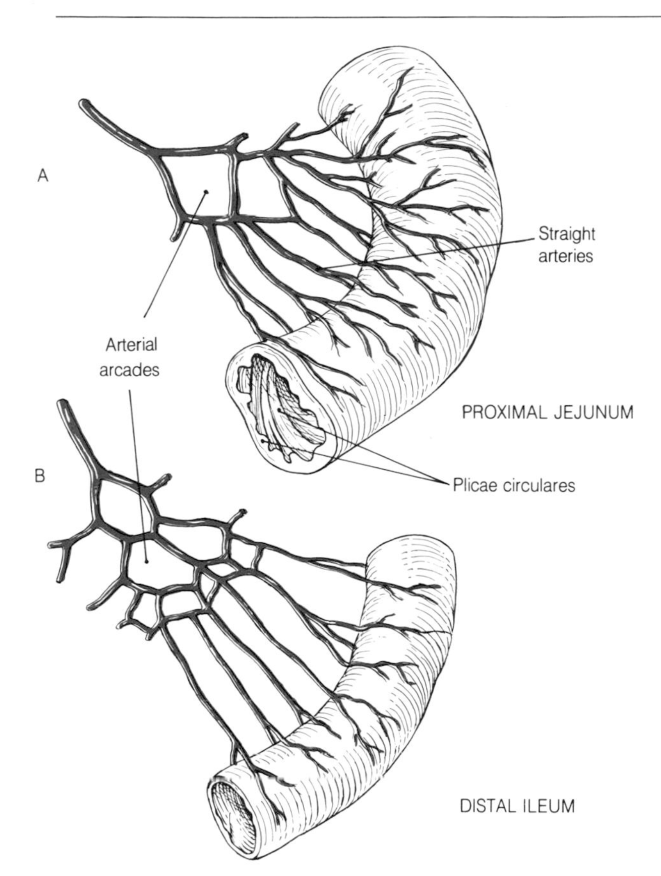

5-53 Semidiagrammatic drawings illustraing the features of the proximal jejunum (*A*) and the distal ileum (*B*).

colon, rectum, and **anal canal**. While the small intestine presents a generally smooth tubular appearance the wall of the cecum and colon shows a series of sacs or **haustra**. These are well demonstrated radiographically by an opaque medium introduced through the rectum as in Fig. 5-54. Haustra are present because the outer longitudinal coat of muscle fibers in the wall of the intestine is concentrated into three bands called **taeniae*** **coli**. These span the more voluminous circular fibers and throw them into pouch-like folds.

The large bowel is disposed in a somewhat rectangular manner around the base of the mesentery of the small intestine and further description of its various regions now follows. The rectum and anal canal, however, will be described with the pelvis and perineum to which they are more intimately related.

In Fig. 5-55 the appearance and arrangement of the large intestine in the abdominal cavity is illustrated and this should be referred to as the following descriptions are read.

* Taenia, *L.* = flat band

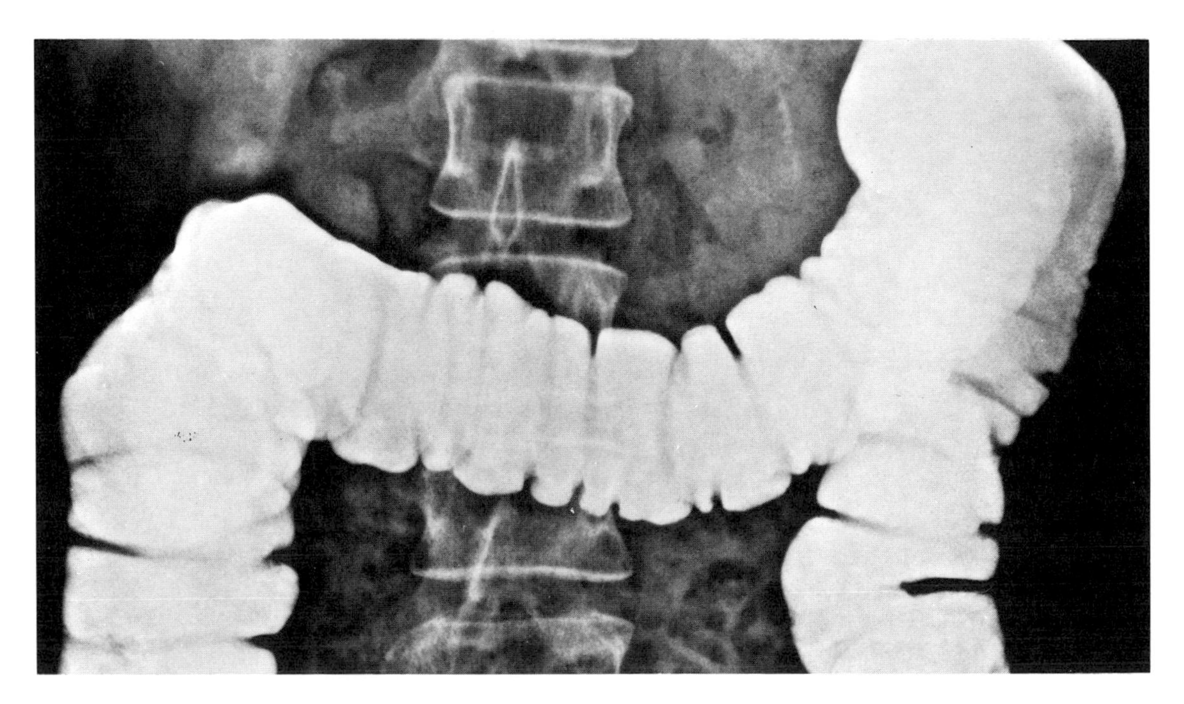

5-54 Haustra of the transverse colon displayed by a barium enema (Reproduced by permission, from Wicke: *Atlas of Radiologic Anatomy,* 3rd Ed, Urban & Schwarzenberg, Baltimore-Munich, 1983.)

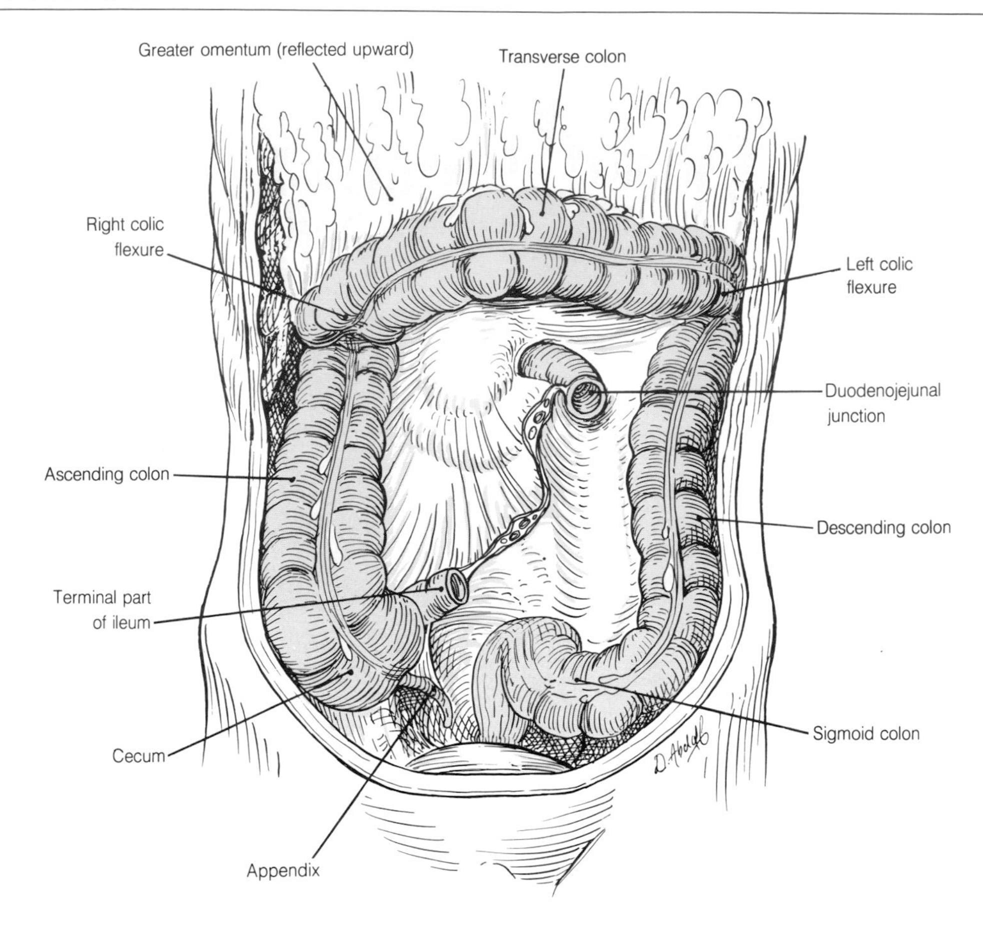

5-55 The abdominal cavity exposed to show the arrangement of the large intestine. The small intestine has been removed.

The Cecum and Vermiform Appendix

The cecum and vermiform* appendix lie in the right illiac fossa below the level of the ileocecal junction and there form a large and a small cul-de-sac. The cecum is usually completely covered with peritoneum and thus lies free in the peritoneal cavity. It has the external appearance of large intestine. The base of the appendix lies at the posteromedial aspect of the cecum. The organ itself is a narrow muscular tube containing large aggregations of lympoid tissue in its wall. It is suspended from the terminal ileum by a **mesoappendix**. This mesentery allows considerable freedom of movement to the appendix which may be found lying over the pelvic brim or, more often, tucked behind the cecum or ascending colon. When it is inflamed, a retrocolic position may give rise to misleading symptoms. The presence of inflammation may also cause the surgeon difficulty in finding the base of the appendix. This dilemma can often be solved by following a taenia coli proximally because all three meet at this point. Although variable in position, the surface projection of the base of the appendix is classically described as lying at the junction of the lateral and middle third of a line joining the anterior superior iliac spine with the umbilicus. This is known as **McBurney's point**. In practice, patients with acute appendicitis often describe a localized pain very close to this point and it also serves as a guide to a suitable appendectomy incision.

At the ileocecal junction a slit-like valve with lips pointing into the cecum is found. However,

* Vermis, *L.* = a worm; hence, vermiform or wormlike

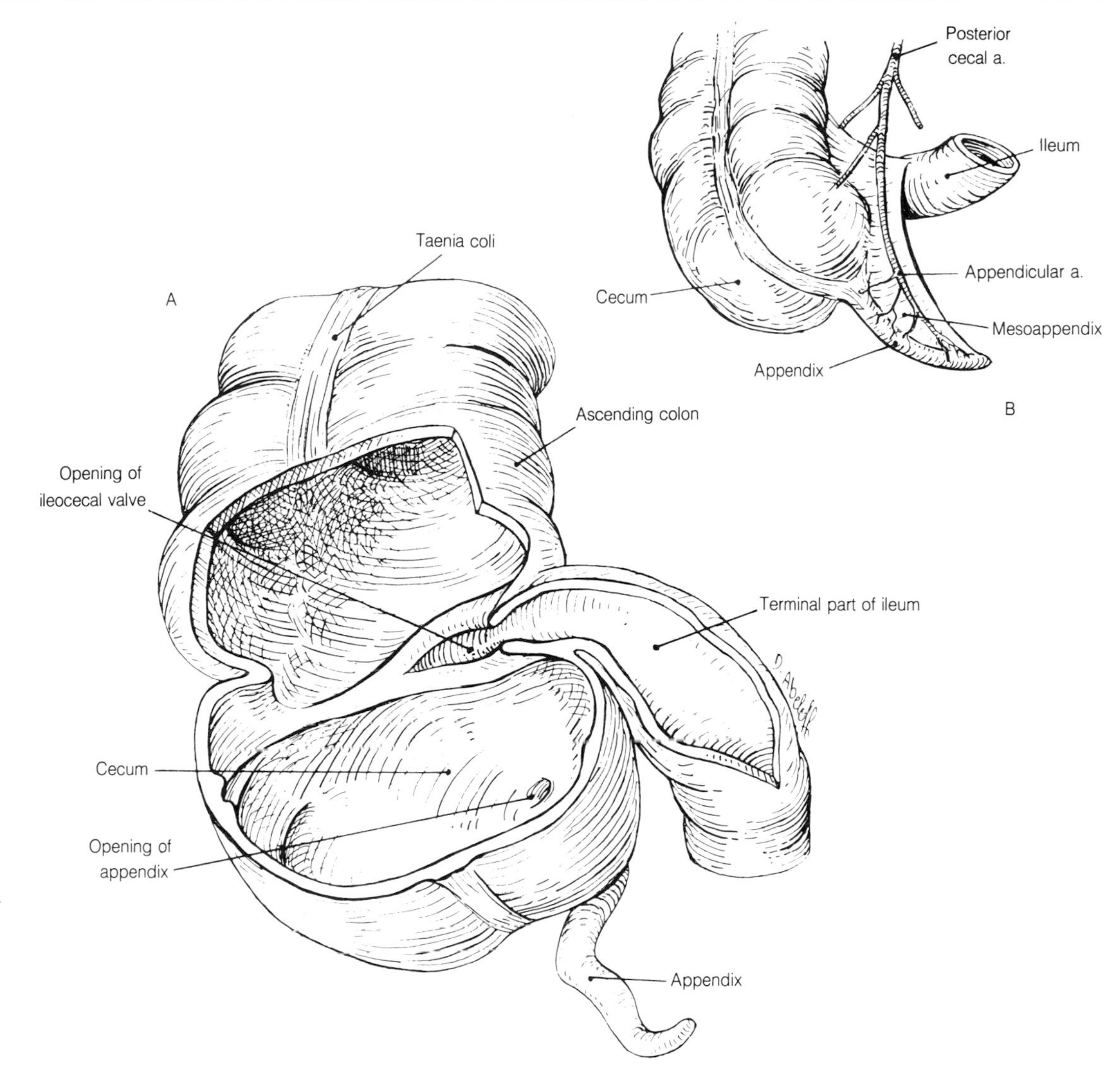

5-56 The cecum opened to show the ileocecal junction and valve (*A*) and detail of the mesoappendix and appendicular artery (*B*).

there is no strong evidence that this valve has much control over the passage of the intestinal contents from small to large intestine. Look at Fig. 5-56, A and B, and confirm these features of the cecum and appendix.

The Colon

The **ascending colon** is only partially covered by peritoneum and lies on the right side of the abdomen extending from the right iliac fossa to the right colic flexure just below the liver. Its bare posterior surface is related to the muscles of the posterior abdominal wall and the lower pole of the right kidney. Just below the liver the ascending colon bends sharply toward the left at the **right colic flexure**.

The **transverse colon** extends from the right colic flexure to the **left colic flexure**, which lies at a slightly higher level on the left side of the abdomen. The transverse colon hangs from the **transverse mesocolon** and may descend to a variable degree towards the pelvis.

The **descending colon** begins at the left colic flexure where it is tethered laterally to the nearby diaphragm by a distinct peritoneal fold, the **phrenicocolic fold**. The descending colon like the ascending colon is only partially covered by

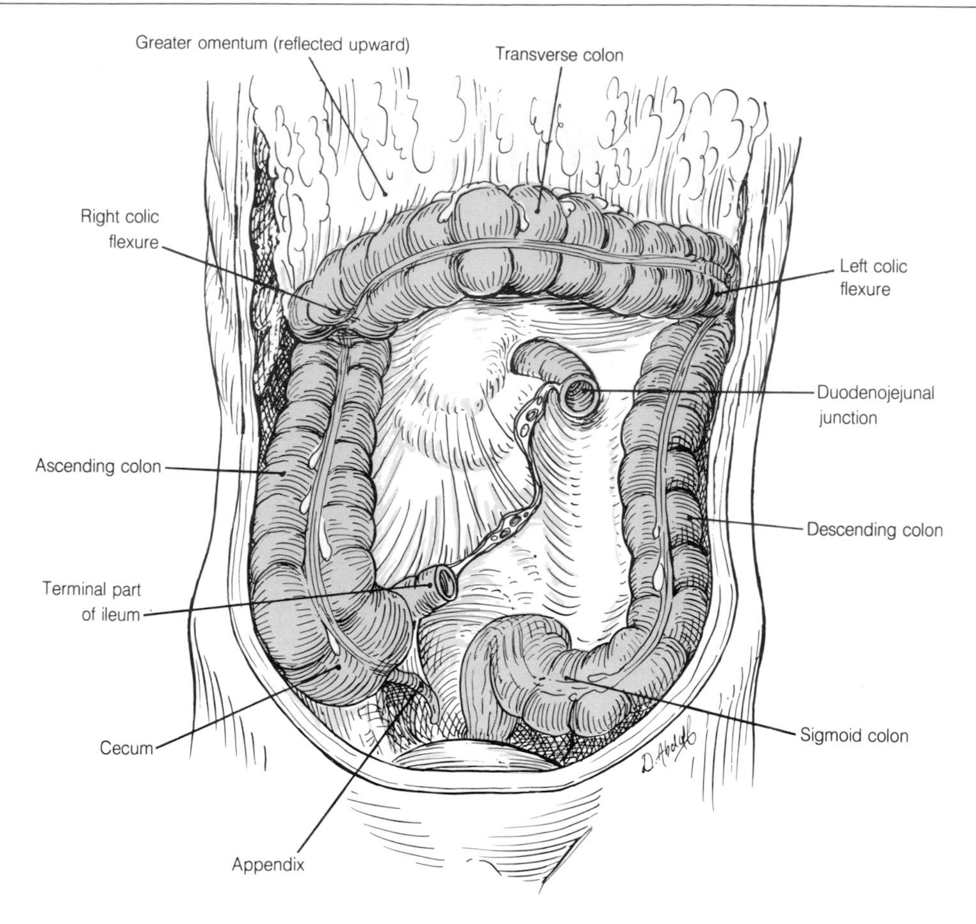

5-57 The abdominal cavity exposed to show the arrangement of the large intestine. The small intestine has been removed.

peritoneum and lies on the left side of the abdominal cavity. Its bare posterior surface is related to the lower pole of the left kidney and below this to the quadratus lumborum, psoas, and iliacus muscles. It ends near the brim of the lesser pelvis where it becomes continuous with the sigmoid or pelvic colon.

The **sigmoid or pelvic colon**, like the transverse colon, is completely covered by peritoneum and is suspended by a **pelvic mesocolon** whose inverted V-shaped attachment is to the pelvic brim and posterior wall of the pelvis. Opposite the third piece of the sacrum the pelvic colon becomes continuous with the **rectum**. The attachment of the mesocolon crosses the left common iliac vessels, the left gonadal vessels, and the left ureter.

The Blood Supply of the Small and Large Intestine

The small and large intestine are supplied with blood by the **superior and inferior mesenteric arteries**. These are both single midline branches of the abdominal aorta. The territory of the superior mesenteric artery extends from the distal half of the duodenum to the latter third of the transverse colon. The intestine from this point to the upper part of the anal canal is supplied by the inferior mesenteric artery. These two regions correspond to the embryological midgut and hindgut. A substantial proportion of the cardiac output is diverted to the vessels supplying the alimentary canal following a meal; the need being not only to supply the tissue itself but to provide fluid for secretions and blood for the removal of absorbed products of digestion.

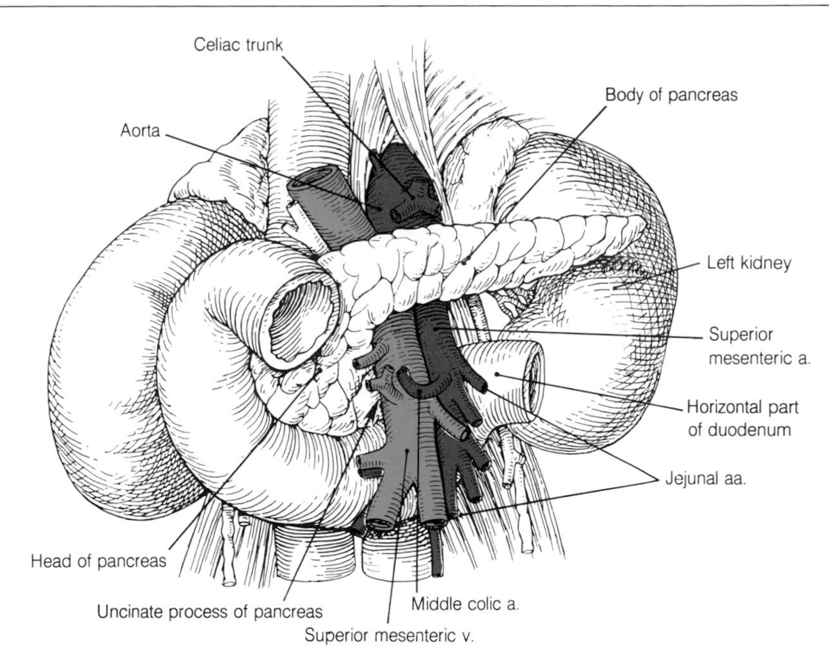

5-58 The superior mesenteric artery as it appears from behind the body of the pancreas.

The Superior Mesenteric Artery

The superior mesenteric artery arises at an acute angle from the aorta just below the celiac trunk at the level of the first lumbar vertebra. Here it is covered by the pancreas and crossed by the splenic vein which is traveling to join the portal vein. Behind it and separating it from the aorta is the left renal vein. Passing downward and slightly forward the artery appears between the body and uncinate process of the pancreas and crosses in front of the horizontal part of the duodenum. Look at Fig. 5-58 in which these relationships are summarized.

On emerging from the pancreas, it descends in the root of the mesentery of the small intestine toward the last part of the ileum where it forms an anastomotic arcade with an earlier branch, the **ileocolic artery**. Its first branch is the **inferior pancreaticoduodenal artery** whose anterior and posteror branches embrace the head of the pancreas and anastomose with the corresponding branches of the superior pancreaticoduodenal artery. During its course through the root of the mesentery many **jejunal and ileal branches** are given off from its left side. From its right side arise the **ileocolic, right colic**, and **middle colic arteries**.

The ileocolic artery gives off the **anterior and posterior cecal arteries**, of which the latter usually supplies the appendix. The ileocolic and right colic arteries cross the posterior abdominal wall behind the peritoneum to reach the colon. The middle colic artery is given off where the superior mesenteric artery is crossed by the root of the transverse mesocolon and is distributed in this to the transverse colon. The distribution of these vessels is vividly illustrated in the superior mesenteric arteriogram seen in Fig. 5-59 and they should be identified using the diagram seen in Fig. 5-60.

The Inferior Mesenteric Artery

The inferior mesenteric artery is the artery of the embryonic hindgut and supplies the remainder of the large intestine, including the upper part of the anal canal. It arises from the aorta 3 or 4 cm above its bifurcation and descends beneath the peritoneum of the posterior abdominal wall. Its

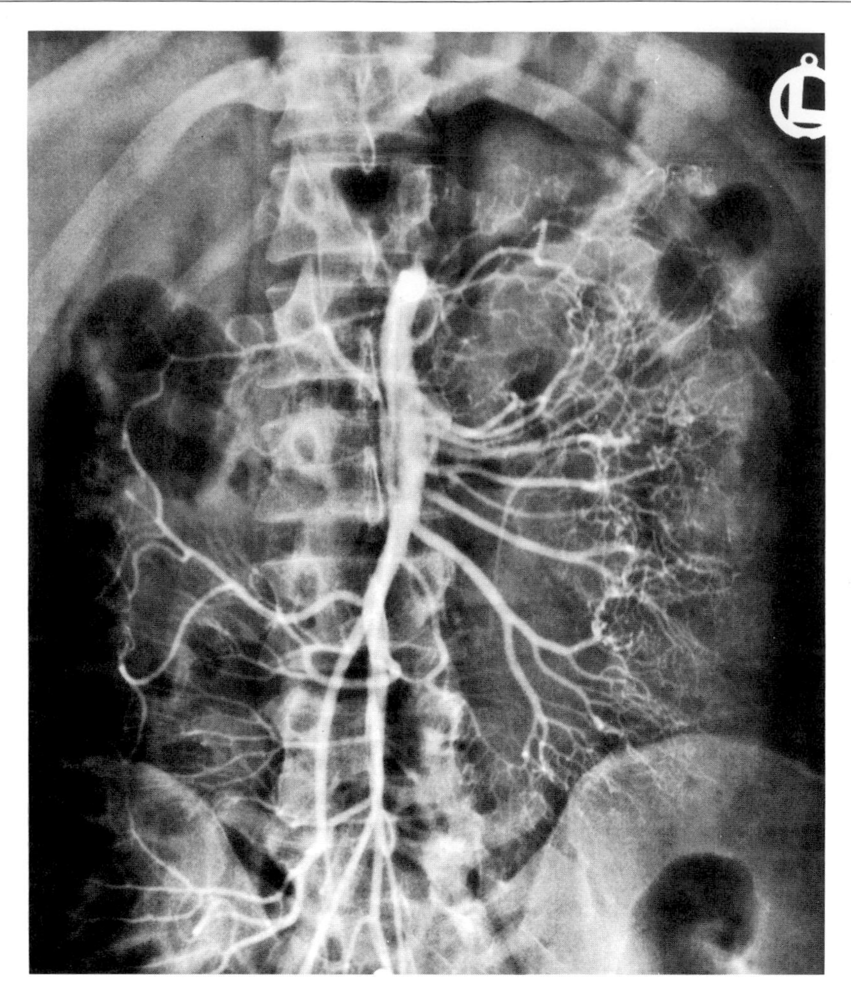

5-59 Superior mesenteric arteriogram. (Reproduced by permission, from Wicke: *Atlas of Radiologic Anatomy,* 4th Ed, Urban & Schwarzenberg, Baltimore-Munich, 1987.)

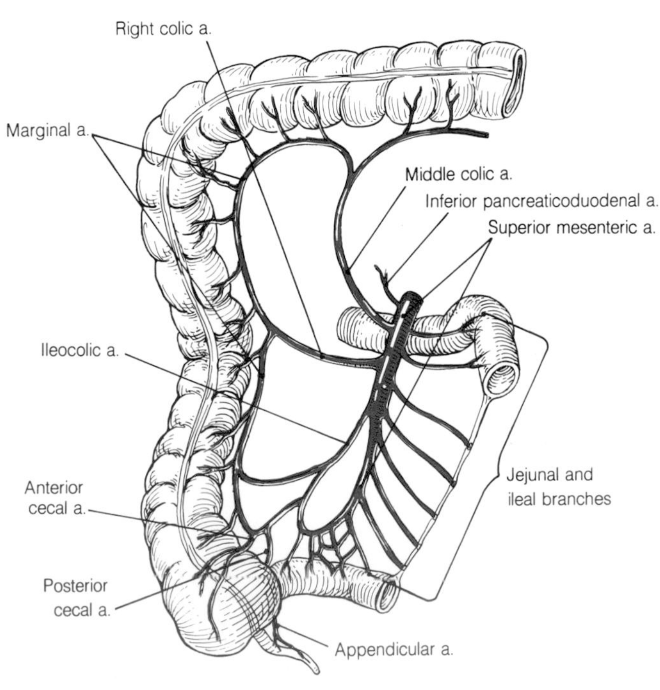

5-60 Showing the branches of distribution of the superior mesenteric artery.

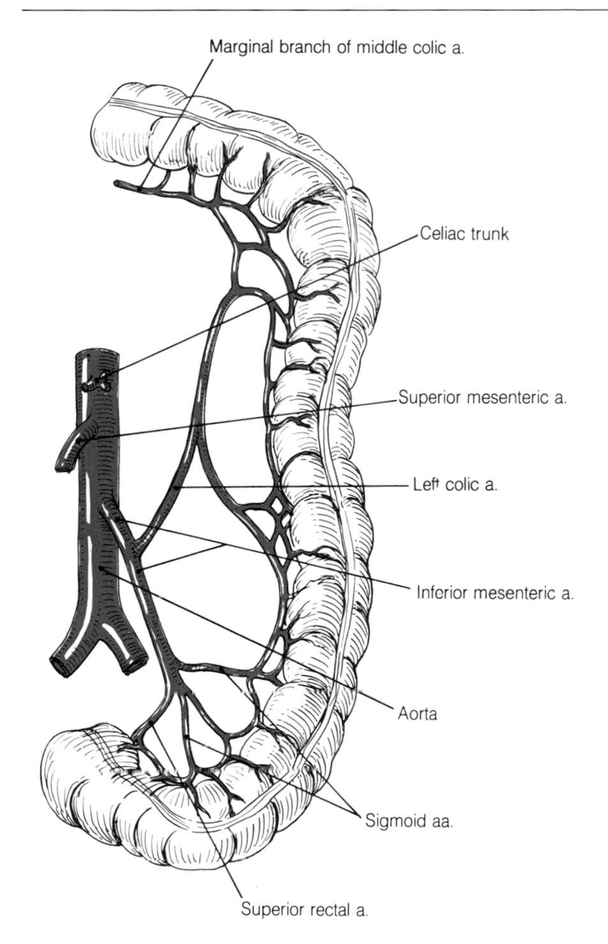

5-61 Showing the branches of distribution of the inferior mesenteric artery.

terminal branches are given off in a variable manner but a typical distribution is shown in Fig. 5-61. A **left colic artery** runs upward and to the left, supplying the last part of the transverse colon, left colic flexure, and upper descending colon; a **descending branch of the left colic artery** supplies the remainder of the descending colon. A series of **sigmoid arteries** supplies the sigmoid colon. The continuation of the inferior mesenteric artery is the **superior rectal artery** which divides into two or three branches within the pelvis to supply the rectum. In addition, the rectum and anal canal are supplied by the **middle rectal arteries** from the two internal iliac arteries and by two **inferior rectal arteries** arising from the internal pudenal arteries. These are described with the perineum.

The abundant blood supply to the small intestine has been commented upon. That to the large intestine is less generous and greater care on the part of the surgeon is needed when choosing sites for excision and anastomosis. Some communication between the main vessels of supply is maintained near the mesenteric border of the intestine by a **marginal artery**.

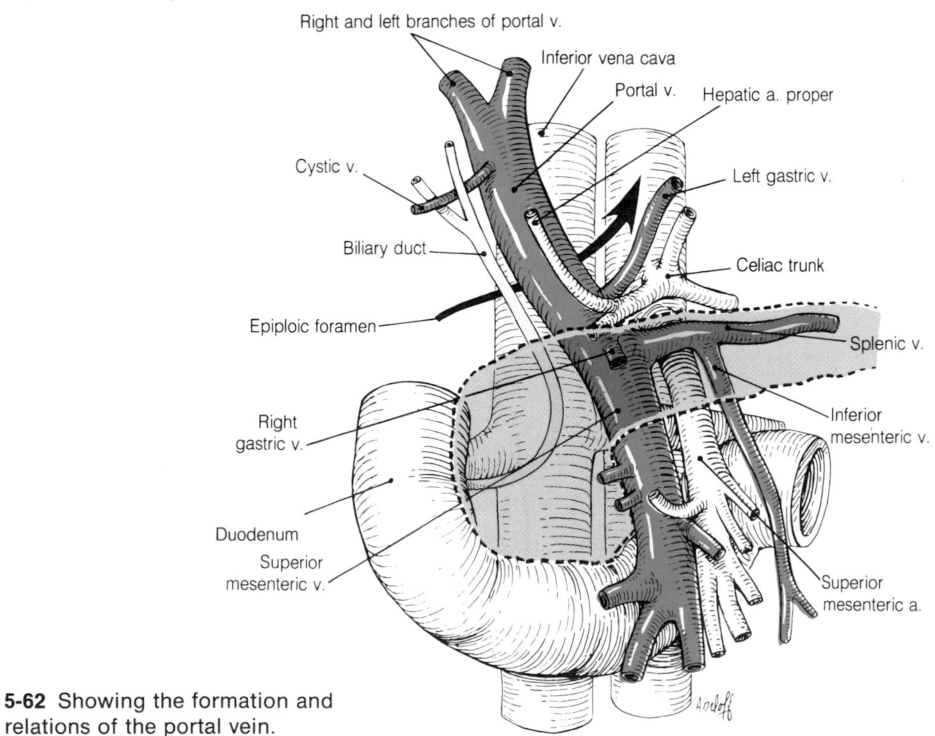

5-62 Showing the formation and relations of the portal vein.

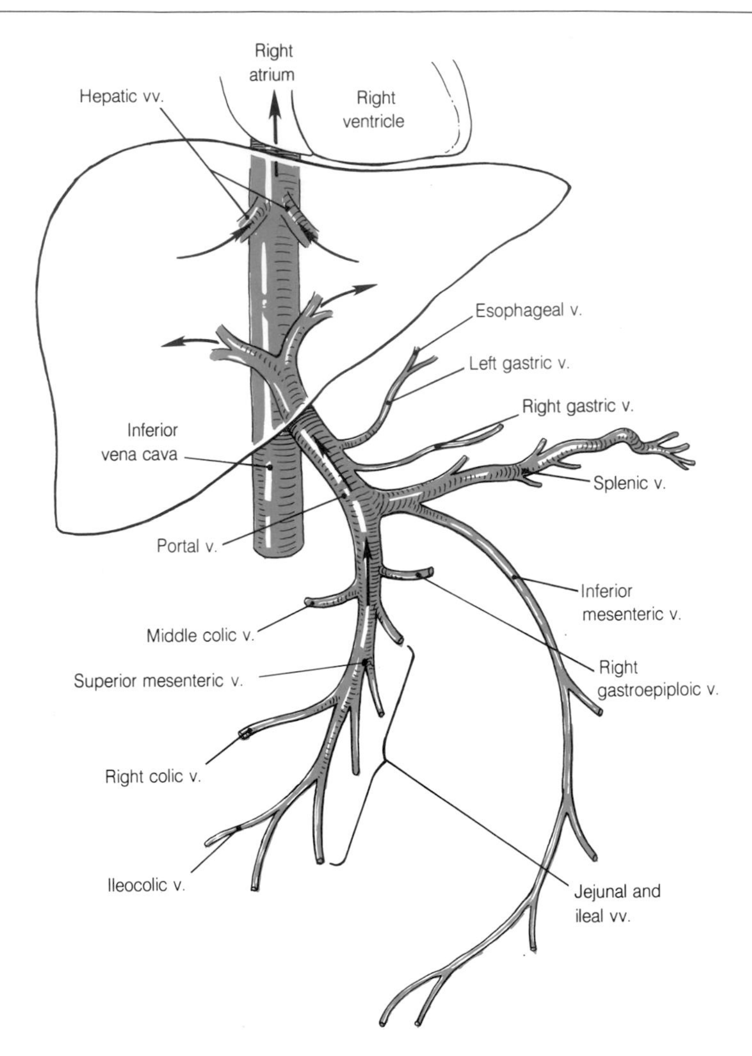

5-63 Semidiagrammatic illustration of the hepatic portal venous system.

The Venous Drainage of the Intestine

The venous blood returning from the intestinal canal and spleen has been enriched with nutrients absorbed from the intestinal lumen. This blood is carried directly to the liver for the storage and metabolism of these nutrients by the **hepatic portal system** of veins. These veins, unlike those of the systemic venous system, do not return directly to the heart, but join to form the portal vein which breaks up into a further system of veins and fine sinusoids in the liver. The blood from these is collected up again by hepatic veins which empty into the inferior vena cava. In this way the blood is eventually returned to the heart. The hepatic portal system consists of the **portal vein** and its tributaries, the **splenic vein**, the **superior mesenteric vein**, and the **inferior mesenteric vein**. Additional smaller tributaries are the **left and right gastric veins** and the **cystic vein**.

The Portal Vein

The portal vein is formed posterior to the neck of the pancreas by the junction of the splenic and superior mesenteric veins. It ascends posterior to the superior part of the duodenum to enter the free border of the lesser omentum. Here the epiploic foramen separates it from the inferior vena cava on the posterior abdominal wall. Blood from the lower end of the esophagus, the stomach, and the gallbladder drains into the portal vein through the **left and right gastric veins** and **the cystic vein**. Its formation, course, and relations can be seen in Fig. 5-62. The tributaries of the vein are shown in Fig. 5-63 and this should be followed as they are described.

The Splenic Vein

The splenic vein is formed at the hilus of the spleen by the union of a number of splenic tributaries and veins from the greater curvature of the stomach. Along its course behind the upper border of the pancreas it receives many small tributaries from that organ. The splenic vein is normally joined by the **inferior mesenteric vein**.

The Superior Mesenteric Vein

The superior mesenteric vein is formed by tributaries which correspond to the many branches of the superior mesenteric artery. It unites with the splenic vein to form the portal vein.

The Inferior Mesenteric Vein

The inferior mesenteric vein drains an area similar to that supplied by the inferior mesenteric artery. The vein, however, leaves the artery to ascend over the left side of the posterior abdominal wall and slip beneath the lower border of the pancreas where it joins the splenic vein.

Portal Obstruction

For a number of reasons the return of blood through the liver to the right side of the heart may become restricted. Among these can be mentioned failure of the right side of the heart to efficiently pass on the venous blood reaching it, compression of portal venules in the liver by fibrosis, and thrombosis of the portal vein itself. As a result, the hydrostatic pressure in the portal system rises and the vessels become distended. Alternative pathways for the return of this blood to the heart are found at sites at which the venous drainage is shared by both hepatic portal and systemic venous systems. These alternative pathways are known as **portocaval** or **portosystemic anastomoses**. They are found at the following sites:

1. **At the lower end of the esophagus:**
   Here esophageal veins drain via the left gastric vein to the portal vein **or** to the azygos vein in the thorax.
2. **At the anal canal:**
   Here rectal veins drain via the inferior mesenteric vein to the portal vein **or** to the internal iliac veins via the middle and inferior rectal veins.
   The distended veins in the anal canal are known as hemorrhoids. However, it should be made clear that portal obstruction is a rare cause of this common condition.
3. **At the umbilicus:**
   This region may be drained by small veins traveling in the falciform ligament to the portal vein **or** by systemic veins draining the anterior abdominal wall.
   The pattern of distended veins radiating from the umbilicus that results from portal hypertension is known as a "caput medusae" after the snake-like hair of the mythical gorgon Medusa.
4. **At regions where the intestine is retroperitoneal:**
   Here veins in the wall of the intestine may drain via portal vein tributaries **or** to veins of the adjacent abdominal wall.
   Examples of these sites occur at the surfaces of the duodenum or the ascending and descending colon that are not covered by peritoneum.

While the occurrence of these anastomoses is of largely diagnostic importance, the distended veins under the mucous membrane of the lower end of the esophagus (*esophageal varices*) may rupture and produce an extremely severe hemorrhage.

The Radiological Anatomy of the Small and Large Intestine

The small intestine begins to fill shortly after the ingestion of a barium meal, but it may be two to four hours before this is seen to reach the ileocecal valve. The radiological appearances of the jejunum and the ileum are well contrasted in Fig. 5-64. The fine indentations into the lumen of the jejunum are caused by the plicae circulares. These

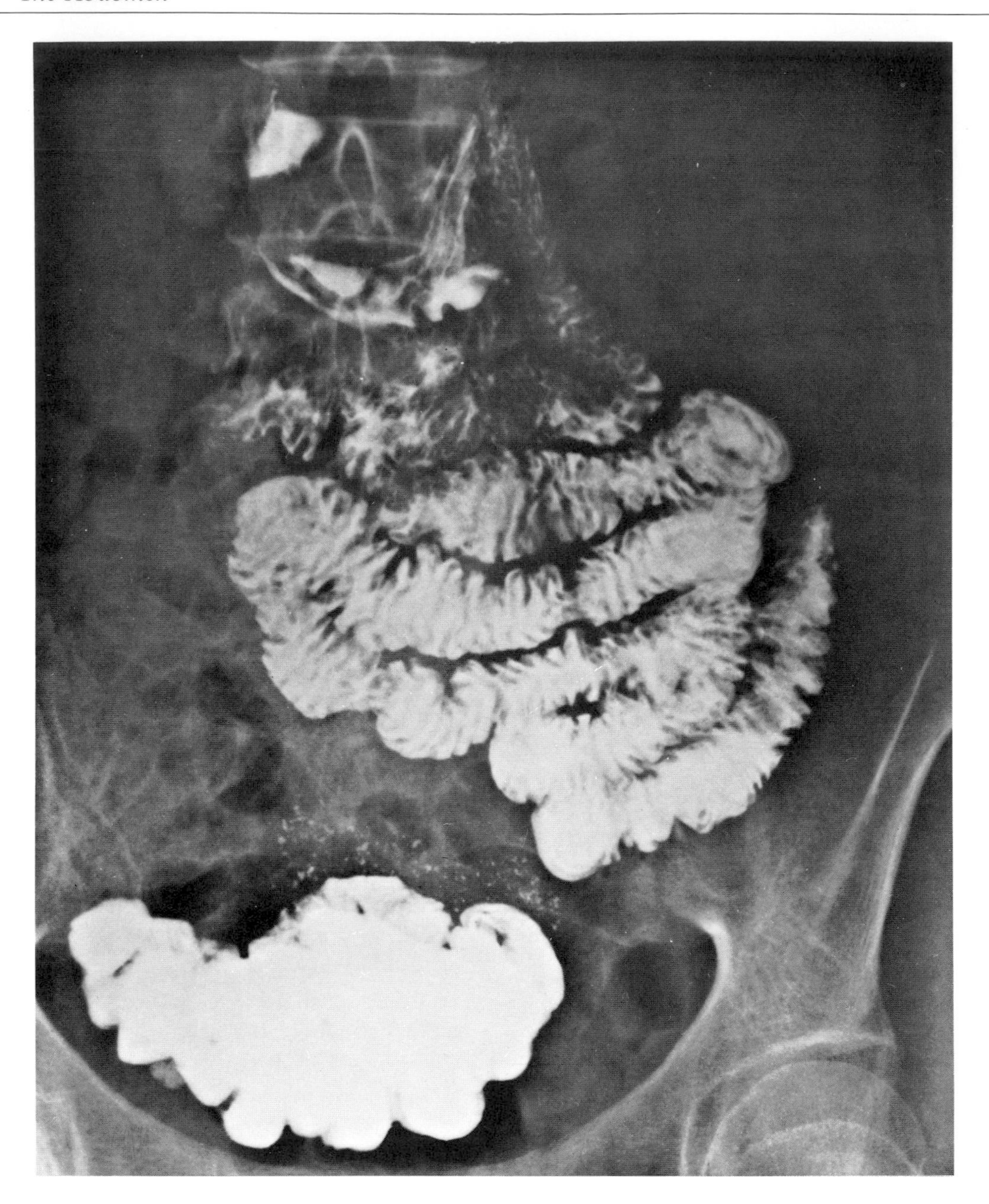

5-64 The jejunum and ileum partially filled by a barium meal. (Reproduced by permission, from Wicke: *Atlas of Radiologic Anatomy,* 4th Ed, Urban & Schwarzenberg, Baltimore-Munich, 1987.)

are absent from the distal portion of the ileum which shows more complete filling with only occasional indentations. Using fluoroscopy, rhythmic segmentation and peristalsis can be observed.

The large intestine is best examined after a barium enema in which the opaque material is introduced through the anal canal. The typical appearance of haustration has already been mentioned and was seen in Fig. 5-54. In Fig. 5-65 the enema has been evacuated and replaced by air giving a double-contrast between the wall and the lumen. In this identify the different regions of the

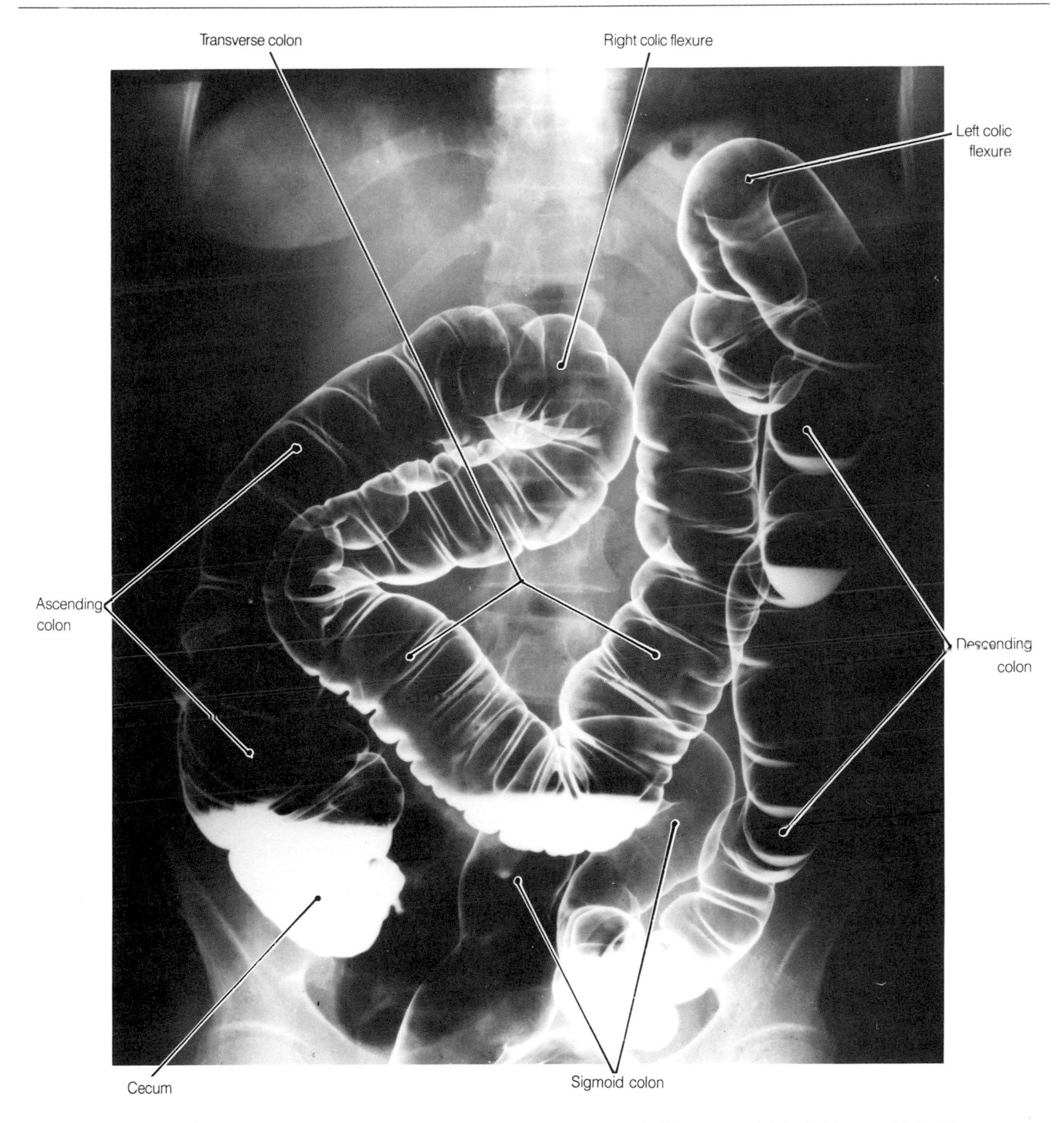

5-65 The colon (double-contrast enema; p.a. projection in erect position). (Courtesy of Jakob Altaras, M. D., Department of Radiology, Giessen University.)

colon and note the high and acute left colic flexure. In many instances the transverse colon will be found to loop down toward the brim of the pelvis. The vermiform appendix does not always fill or is hidden behind the cecum.

The Transpyloric Plane

This hypothetical plane passes through the lower part of the body of the first lumbar vertebra and, anteriorly, through a point halfway between the jugular notch and the symphysis pubis. While it

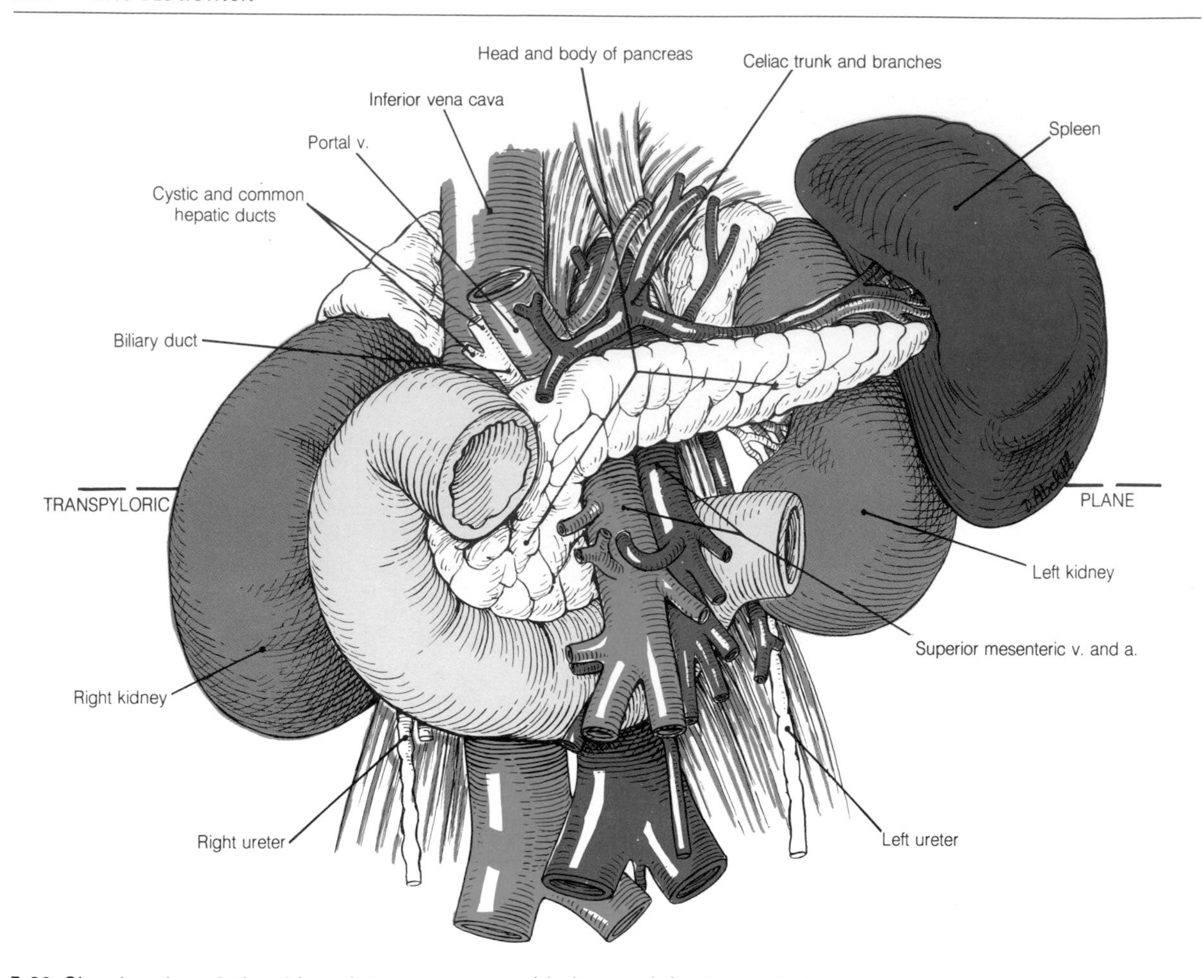

5-66 Showing the relationships of structures assembled around the transpyloric plane.

has some value in establishing the surface relationships of a number of relatively fixed abdominal structures it can also be of great help in reviewing many of the close relationships of structures described in the previous pages. Look at Fig. 5-66 and note that:

1. The pancreas lies diagonally across the plane.
2. The hilum of the right kidney lies just below it and that of the left just above.
3. The pylorus lies on or just below it and leads to the superior part of the duodenum.
4. The descending part of the duodenum is related to the hilum of the right kidney and the head of the pancreas whose positions have already been established.
5. The spleen lies at the tail of the pancreas.
6. The celiac trunk arises from the aorta just above the pancreas.
7. The superior mesenteric artery which arises from the aorta behind the neck of the pancreas, appears between the body and uncinate process of the pancreas.
8. The superior mesenteric vein passes deep to the neck of the pancreas and, with the splenic vein, forms the portal vein.
9. The portal vein together with the biliary duct ascends behind the superior part of the duodenum to be joined by the hepatic artery in the free border of the lesser omentum.
10. Although not seen, the renal vessels must lie deep in this region. The left renal vein passes in front of the abdominal aorta just below the

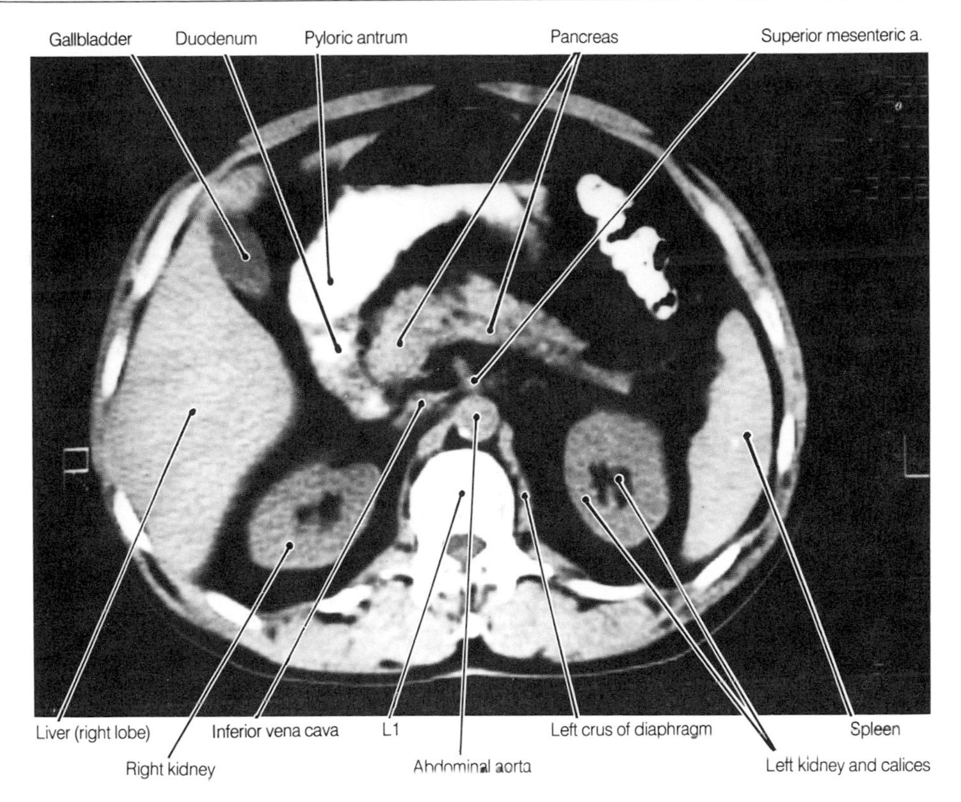

5-67 CT scan made through the trunk at the level of the first lumbar vertebra. (Courtesy of Nancy Whitley, M. D., Department of Diagnostic Radiology, University of Maryland School of Medicine.)

origin of the superior mesenteric artery. The renal arteries lie at a deeper level still; the right artery passing behind the inferior vena cava.

Having established the arrangement of structures around the transpyloric plane it is now possible to interpret the CT scan made through the body of the first lumbar vertebra that is shown in Fig. 5-67. In this the cavity of the stomach and the first part of the duodenum are made more obvious by the presence of contrast medium. Note that because the subject is supine a fluid level and a gas bubble can be seen in the ventral part of the stomach. Compare the scan with Fig. 5-66 and see why each labeled structure appears in it.

The Posterior Abdominal Wall

Unlike the thin, elastic, and vulnerable anterior abdominal wall, the posterior abdominal wall is dense, somewhat rigid, and protective. It is composed of the **bodies of the five lumbar vertebrae** and their **intervertebral discs**, the **psoas major** and **quadratus lumborum muscles**, and the **iliacus muscle** which covers the ala of the **iliac bone**. The bodies of the vertebrae project forward into the abdominal cavity forming a paravertebral gutter on either side. In a thin supine subject the lumbar lordosis brings the lumbar vertebrae and the overlying abdominal aorta close to the anterior abdominal wall and through this they are easily palpable.

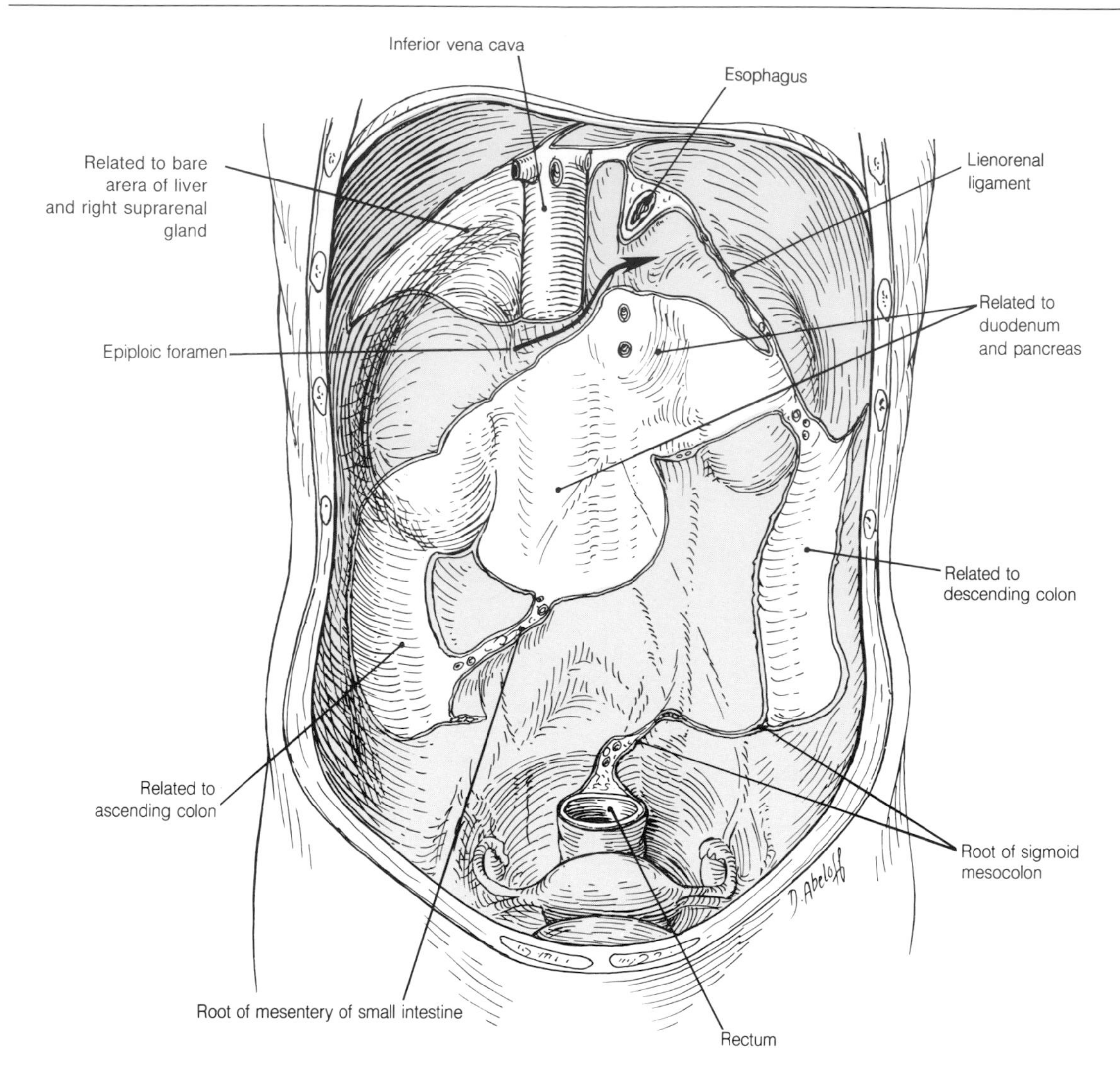

5-68 Showing the regions of the posterior abdominal wall exposed by the removal of the abdominal organs.

Following the removal of the intestinal canal and its associated glands a number of structures remain applied to the posterior abdominal wall which are partially or completely covered by peritoneum. The extent of this peritoneal cover depends upon whether the organs removed were intraperitoneal and had a mesentery or were retroperitoneal and, thus, kept the peritoneum from the posterior abdominal wall. The relationships of these mesenteries and sites of organ attachments can now be reviewed in Fig. 5-68. In this illustration identify where the bare area of the liver was applied to the diaphragm, the large area exposed by the removal of the duodenum and pancreas, and the areas underlying the ascending and descending colon. Note also the roots of the mesenteries of the small intestine and sigmoid colon. The root of the transverse mesocolon does not appear because it lies across the ventral surface of the pancreas.

The structures exposed after the removal of all the remaining peritoneum from the posterior abdominal wall are the **abdominal aorta** and the **inferior vena cava**, the **kidneys**, and the **suprarenal glands** which cap them, the **ureters** which are the excretory ducts of the kidneys, the abdominal portion of the **sympathetic trunk**, the **subcostal nerve**, and **branches of the lumbar plexus**. Behind these lie the muscles of the posterior abdominal wall. The general position of these structures can be seen in Fig. 5-69.

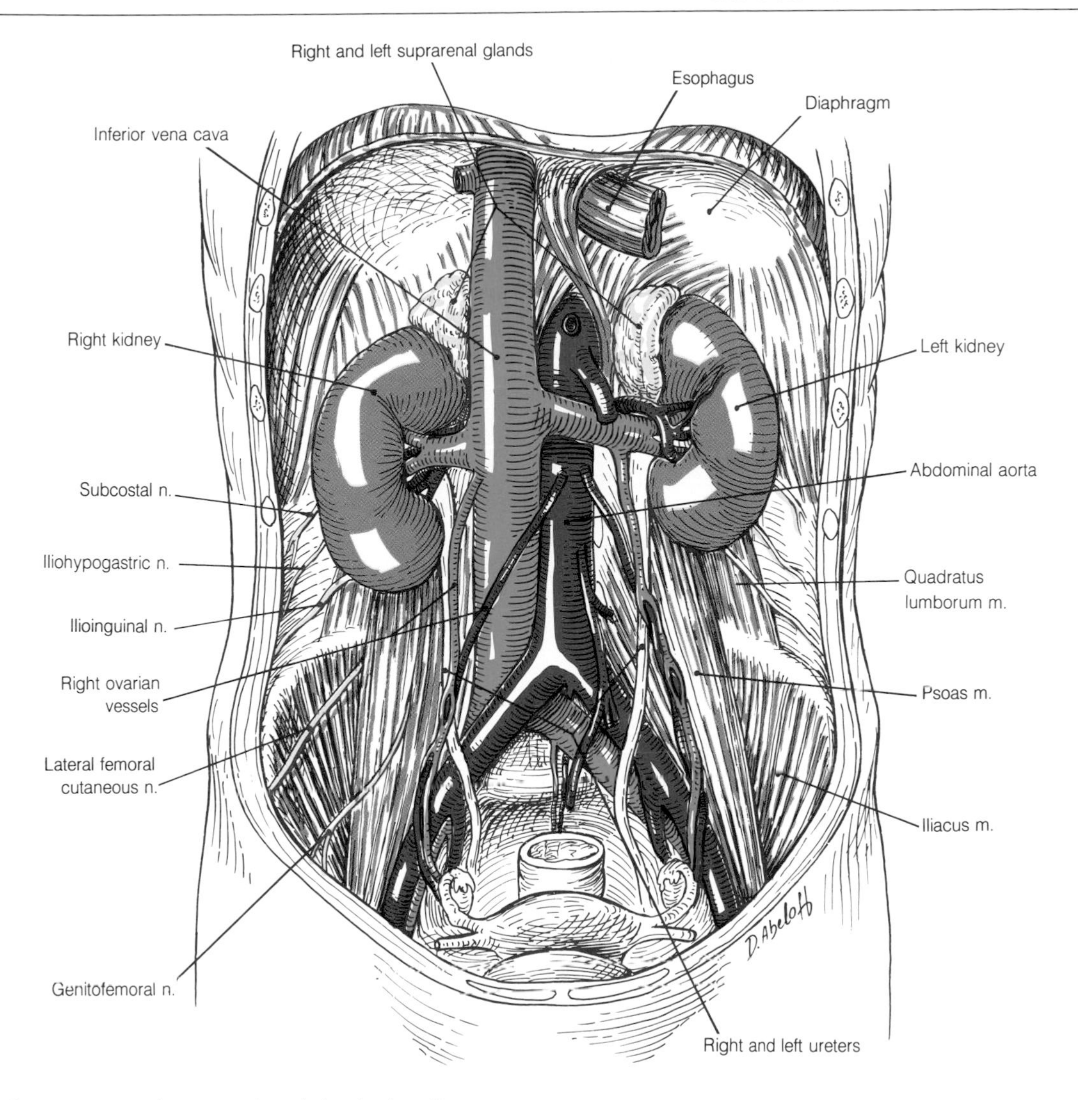

5-69 Structures on the posterior abdominal wall.

The Muscles of the Posterior Abdominal Wall

The Psoas and Iliacus Muscles

The **psoas major muscle** lies in the paravertebral gutter at the side of the bodies of the lumbar vertebrae and arises from the lumbar intervertebral discs and adjacent areas on the lumbar and the twelfth thoracic vertebral bodies. Between these sites of attachment the concavities of the bodies are spanned by tendinous arches. The muscle also arises from these arches and the roots of the transverse processes. Beneath the tendinous arches **lumbar arteries and veins** and **sympathetic rami communicantes** gain access to the neurovascular plane of the abdominal wall. The fan-shaped **iliacus muscle** arises from the internal aspect of the ala of the iliac bone and converges on the lateral border of the psoas muscle with which it passes beneath the **inguinal ligament**. These muscles share a common attachment to the **lesser trochanter of the femur** and the adjacent shaft. Both are covered by a common sheet of fascia called the **iliac fascia**. The most important action of the "iliopsoas" muscle is to flex the hip joint. A psoas minor muscle is sometimes encountered on the surface of the major and this is attached distally to the pectineal line. The psoas muscle is supplied by branches of the **ventral rami of the first, second, and third lumbar segmental nerves** and the iliacus by branches of the **femoral nerve**.

Quadratus Lumborum

This flat muscle lies immediately lateral to the upper part of psoas major. It extends from the lower border of the twelfth rib and the tips of the lumbar transverse processes to the **iliolumbar ligament** which spans the gap between the fifth lumbar transverse process and the iliac crest and to an adjacent region of the iliac crest itself. The muscle is covered by the anterior layer of the lumbar fascia which provides origin for the transversus abdominis and internal oblique muscles at its lateral border. The quadratus lumborum stabilizes the twelfth rib in inspiration and is a lateral flexor of the trunk. It is supplied from the **ventral rami of the twelfth thoracic and upper three lumbar segmental nerves**.

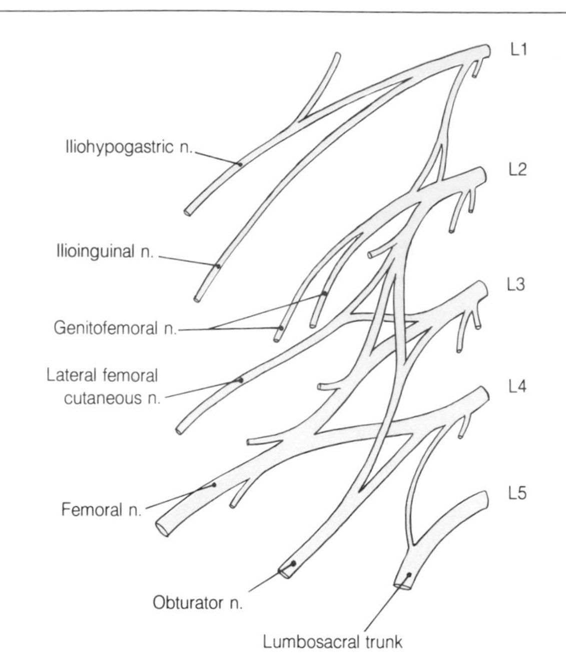

5-70 Diagram of the lumbar plexus.

The Nerves of the Posterior Abdominal Wall – the Lumbar Plexus

The twelfth thoracic and the five lumbar segmental nerves leave intervertebral foramina to gain access to the posterior abdominal wall but the subsequent course and distribution of their ventral rami varies. The ventral ramus of the twelfth thoracic or **subcostal nerve** continues as a single nerve. The upper four lumbar rami give rise to the **lumbar plexus** which is largely concerned with the innervation of the extensor and adductor compartments of the thigh. Part of the fourth and all the fifth lumbar ventral rami contribute to the **sacral plexus** and through this also supply structures of the lower limbs. Fig. 5-70 shows a diagram of the formation of each nerve. In Fig. 5-71 the course of each nerve can be followed as it is described below.

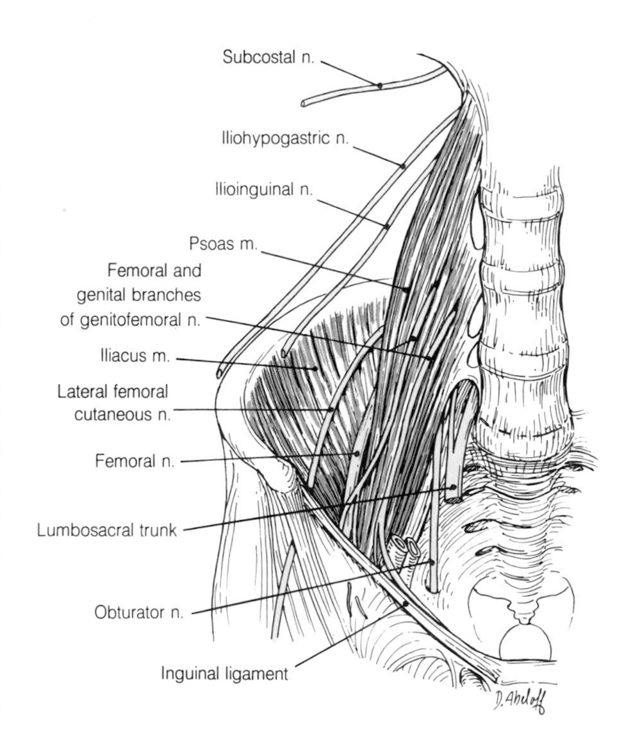

5-71 The nerves of the posterior abdominal wall.

The Subcostal Nerve

This is the ventral ramus of the twelfth thoracic nerve. It appears on the posterior abdominal wall from beneath the **lateral arcuate ligament of the diaphragm** and crosses the surface of quadratus lumborum below the twelfth rib. It is carried obliquely around the abdominal wall between the transversus and internal oblique muscles and terminates by supplying the lower part of the rectus abdominis muscle. During its course it also supplies the lateral abdominal muscles and by lateral and anterior cutaneous branches, a strip of overlying skin.

The Iliohypogastric and Ilioinguinal Nerves

These two nerves are derived from the first lumbar nerve. They appear singly or together at the lateral border of the psoas muscle, run over the posterior abdominal wall, and pierce the transversus abdominis muscle to reach the neurovascular plane. The **iliohypogastric nerve** divides into lateral and anterior cutaneous branches which become superficial to supply skin over the upper lateral surface of the thigh and the lower abdomen above the pubis. The **ilioinguinal nerve** pierces the internal oblique muscle and joins the spermatic cord as it passes through the superficial inguinal ring. It is distributed to skin of the external genitalia and an adjacent region of the thigh. Both nerves give off muscular branches to the transversus and internal oblique muscles.

The subcostal, iliohypogastric and ilioinguinal nerves are posterior relations of the kidney as it lies on the posterior abdominal wall.

The Lateral Femoral Cutaneous Nerve

Formed from the second and third lumbar nerves, the lateral femoral cutaneous nerve emerges from the lateral border of psoas and runs over the iliacus muscle to enter the thigh beneath the lateral end of the inguinal ligament. At this point it may become "entrapped" and give rise to pain. Becoming superficial below the ligament, it supplies the skin of the lateral aspect of the thigh.

The Femoral Nerve

This large nerve is formed from the second, third, and fourth lumbar nerves. It appears in the groove between psoas and iliacus and is thus guided beneath the inguinal ligament. It supplies the extensor muscles of the thigh and overlying skin.

The Genitofemoral Nerve

The small genitofemoral nerve is formed from the first and second lumbar nerves. It appears on the surface of the psoas muscle and descends over this before it divides into a genital and a femoral branch. The **genital branch** joins company with the spermatic cord or round ligament of the uterus. It supplies some skin of the external genitalia in both sexes and in the male it supplies the cremaster muscle. The **femoral branch** passes deep to the inguinal ligament on the lateral side of the **external iliac and femoral arteries** to supply a small area of skin below the ligament.

The Obturator Nerve

This nerve, like the femoral nerve, is derived from the second, third, and fourth lumbar nerves. It appears at the medial border of psoas and its course through the pelvis and thigh will be described later.

The Lumbosacral Trunk

A portion of the fourth lumbar nerve joins the fifth nerve to form a stout trunk which enters the pelvis to take part in the formation of the **sacral plexus**.

The Lumbar Sympathetic Trunk

This portion of the sympathetic trunk appears beneath the **medial arcuate ligament of the diaphragm** on the surface of the psoas muscle. Running down the medial border of the muscle and largely covered by the vena cava or aorta, the trunk passes over the pelvic brim posterior to the common iliac vessels and enters the pelvic cavity. It usually bears four ganglia. These are linked to the upper two lumbar nerves by both grey and white rami and to the lower by grey rami only. In addition, four **lumbar splanchnic nerves** pass medially to join the abdominal autonomic plexuses.

The Vessels of the Posterior Abdominal Wall

The Abdominal Aorta

The abdominal portion of the descending aorta passes through the diaphragm at the level of the twelfth thoracic vertebra and runs down in front of the anterior longitudinal ligament covering the lumbar vertebrae. It divides into the two **common iliac arteries** over the body of the fourth lumbar vertebra. The many branches of the abdominal aorta can be usefully divided into: (1) paired arteries to the body wall, (2) arteries to paired viscera, and (3) unpaired midline branches to the

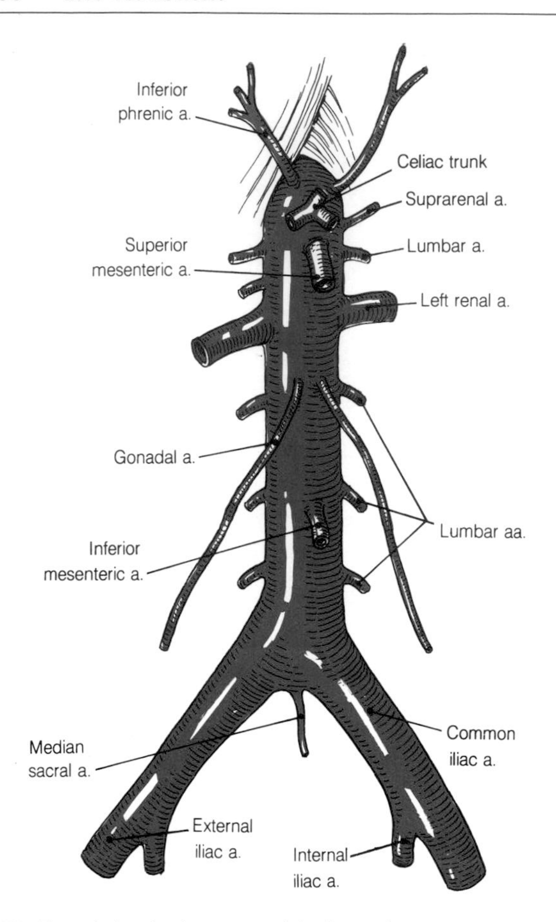

5-72 The abdominal aorta and its branches.

alimentary canal. These are summarzied below and illustrated in Fig. 5-72.

Branches to the body wall:
1. The inferior phrenic arteries
2. Four pairs of lumbar segmental arteries

Branches to paired viscera:
1. The middle suprarenal arteries
2. The renal arteries
3. The gonadal arteries to testis or ovary

Unpaired branches to the alimentary canal:
1. The celiac trunk
2. The superior mesenteric artery
3. The inferior mesenteric artery

The unpaired branches to the alimentary canal have already been described. The body wall branches are segmental arteries in series with the posterior intercostal branches of the thoracic aorta. The **inferior phrenic arteries** supply the diaphragm and often send a branch to the suprarenal gland. The **lumbar arteries** pass laterally behind the sympathetic trunks and, on the right, behind the inferior vena cava. Running deep to the tendinous arches which give origin to the psoas muscle, they reach either the anterior or posterior surface of quadratus lumborum and, thence, the neurovascular plane. Like the intercostal arteries, the lumbar arteries give off dorsal and spinal branches. The **middle suprarenal arteries** supply the suprarenal glands directly from the aorta. The **renal arteries** are short and of considerable caliber and arise just below the level of the superior mesenteric artery. The right artery passes **behind** the inferior vena cava and right renal vein to reach the right kidney. The left artery, which is shorter than the right, passes posterior to the left renal vein. Both arteries give off small **inferior suprarenal branches**. The renal arteries transmit about 20% of the resting cardiac output to the kidneys.

During development the gonads have migrated from the posterior abdominal wall to the pelvis or scrotum. For some reason, and unlike the kidneys which have migrated in the opposite direction, they have retained their earlier arterial supply from the abdominal aorta. The small **testicular** or **ovarian arteries** arise below the renal arteries and descend obliquely into the pelvis just beneath the peritoneum of the posterior abdominal wall.

The abdominal aorta terminates at the level of the fourth lumbar vertebra by dividing into the left and right **common iliac arteries**.

The Inferior Vena Cava

The inferior vena cava is formed just to the right of the bifurcation of the aorta by the junction of the two **common iliac veins**. This large vein ascends the posterior abdominal wall immediately to the right of the aorta, passes posterior to the liver in a deep groove on its posterior surface, and enters the thoracic cavity by piercing the **central tendon of the diaphragm** and the attached **fibrous pericardium**. It then enters the **right atrium of the heart**.

The tributaries of the inferior vena cava do not all correspond to the branches of the abdominal aorta. Note, therefore, that:
1. There are no alimentary veins – these all join the hepatic portal system.

2. There are lumbar and inferior phrenic veins which follow the course of their corresponding arteries and join the inferior vena cava.
3. There are two renal veins which lie anterior to the renal arteries. The left renal vein has to cross the aorta to reach the inferior vena cava and is rather longer than the right.
4. Only the right gonadal vein drains directly into the inferior vena cava. The left joins the left renal vein.
5. The short hepatic veins drain directly into the inferior vena cava as it passes posterior to the liver and do not retrace the course of the hepatic artery.

The Autonomic System in the Abdomen

The components contributing to this system in the abdomen are:
1. The vagi (parasympathetic)
2. The sympathetic trunks and their rami communicantes
3. The thoracic splanchnic nerves (sympathetic)
4. The lumbar splanchnic nerves (sympathetic)
5. The pelvic splanchnic nerves (parasympathetic).

The Vagi

The **vagal trunks** which enter the abdomen on the surface of the esophagus are named **anterior** and **posterior**. By **anterior gastric branches** the anterior trunk supplies the anterior surface of the stomach and the pylorus and by **posterior gastric branches** the posterior trunk supplies the posterior surface of the stomach. The posterior trunk also supplies fibers to the nearby **celiac plexus**. Beyond this point the vagi cannot be distinguished as discrete nerves although their distribution in the form of perivascular plexuses extends to the latter part of the transverse colon. The left part of the transverse colon, the descending and the pelvic colon receive their parasympathetic innervation from the **pelvic splanchnic nerves**.

The Sympathetic Trunks

On entering the abdomen beneath the medial arcuate ligaments the **sympathetic trunks** become slightly convergent and pass through the lumbar region on the surface of the vertebrae just medial to the psoas muscle. The number and size of the associated ganglia are irregular but four on each side are usually described. They receive preganglionic fibers in **white rami communicantes** from the first and second lumbar ventral rami only. Postganglionic fibers in the form of **grey rami communicantes** pass from ganglia to all the lumbar nerves for the supply of the body wall and lower limb. Visceral branches also leave the ganglia and these are called the **lumbar splanchnic nerves**.

The Thoracic Splanchnic Nerves

The thoracic splanchnic nerves are formed by medially running visceral branches from the lower seven thoracic sympathetic ganglia. Those from the fifth to ninth ganglia form the **greater splanchnic nerves**, those from the tenth and eleventh the **lesser splanchnic nerves** and those from the lowest ganglion, be it eleventh or twelfth, the **least splanchnic nerves**. These nerves enter the abdomen by piercing the crura of the diaphragm and join the celiac plexus. It should be noted that unlike the grey rami which carry sympathetic fibers to the peripheral nerves, these splanchnic nerves contain a high proportion of preganglionic fibers.

The Lumbar Splanchnic Nerves

Visceral branches also leave the four lumbar sympathetic ganglia. These are **lumbar splanchnic nerves**. Again, these contain a substantial number of preganglionic fibers.

The Pelvic Splanchnic Nerves

The **pelvic splanchnic nerves** form the pelvic **parasympathetic** outflow and arise from the second and third or third and fourth sacral ventral rami. This outflow is mainly concerned with the parasympathetic innervation of pelvic viscera. However, fibers also pass upward into the abdominal cavity to supply the large intestine beyond the middle third of the transverse colon. Travelling by way of the **hypogastric plexus** these preganglionic fibers reach the **inferior mesenteric plexus** for distribution along that vessel.

The Autonomic Plexuses of the Abdomen

An extensive prevertebral autonomic plexus lies on the anterior surface of the abdominal aorta and around the origins of its branches. This plexus is in continuity with similar plexuses in the thorax above and the pelvis below. However, the abdominal portion is divided into a number of parts named after the vascular structures to which they are related, hence the **celiac, superior mesenteric, intermesenteric, and inferior mesenteric plexuses**. Further plexuses, subsidiary to the celiac plexus, are the **hepatic, gastric, splenic, suprarenal, renal, and gonadal** plexuses through which its fibers are distributed to those organs. The gastrointestinal tract receives its autonomic innervation from the celiac, superior mesenteric, and inferior mesenteric plexuses along the branches of the similarly named arteries.

Contributing to these plexuses are the vagi (preganglionic fibers) and the thoracic splanchnic nerves (mostly preganglionic). The lower part of the plexus is reinforced by the lumbar sphlanchnic

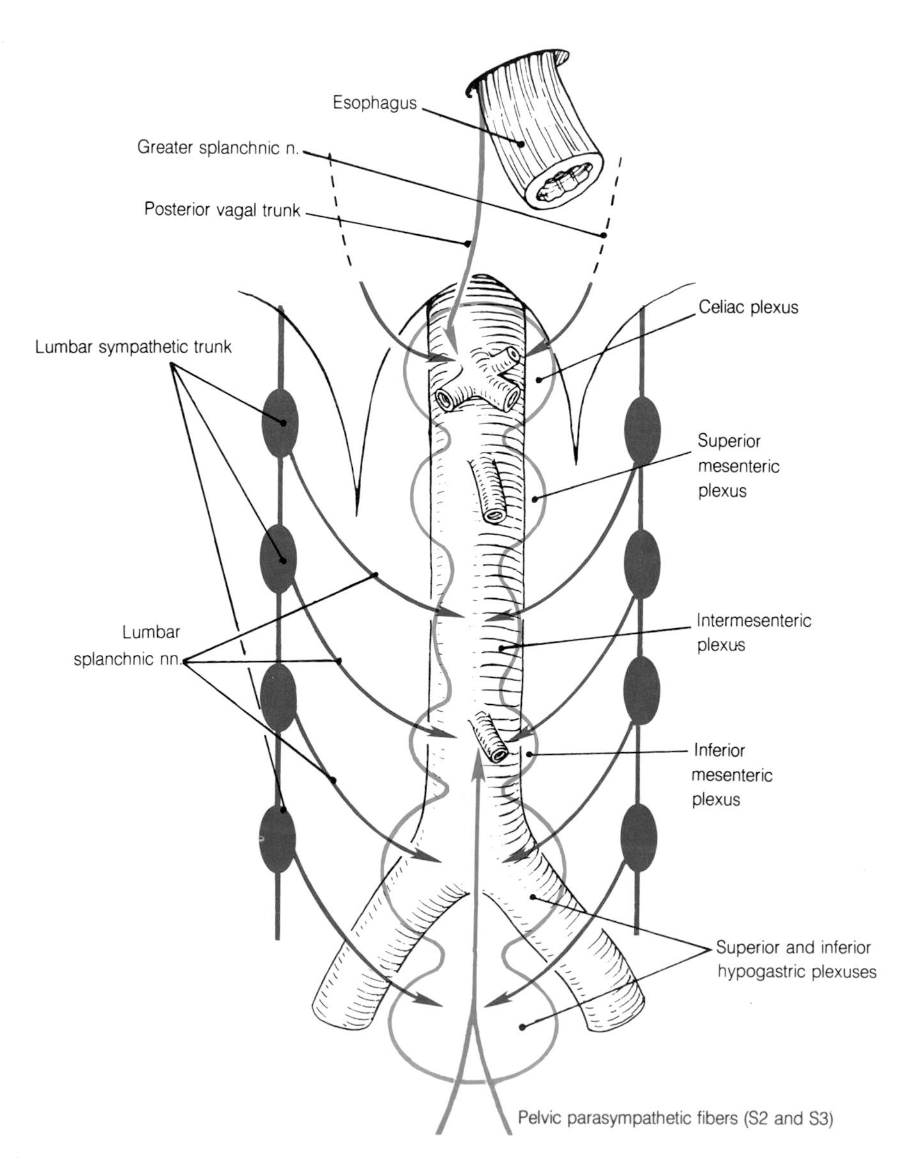

5-73 Diagram of the autonomic nerves and plexuses of the abdomen.

nerves from the lumbar sympathetic ganglia. As has been noted, the inferior mesenteric plexus receives its parasympathetic fibers not from the vagus but from the pelvic parasympathetic outflow known as the pelvic splanchnic nerves (preganglionic). Preganglionic vagal and pelvic splanchnic fibers pass through the plexuses without interruption and synapse with postganglionic fibers in the walls of the organs they supply. Most of the preganglionic sympathetic fibers synapse in the plexuses but the exceptions to this are sympathetic fibers destined for the suprarenal medulla whose cells correspond developmentally to postganglionic fibers. The arrangement of the main abdominal plexuses is illustrated diagrammatically in Fig. 5-73.

Before leaving the subject of the autonomic system in the abdomen it should not be forgotten that both the vagal and sympathetic contributions also carry afferent fibers back to the central nervous system. These fibers do not, of course, synapse in autonomic ganglia, but like other sensory nerves have their cell bodies in spinal or cranial ganglia. They play an essential part in autonomic reflexes which do not reach the conscious level and are also responsible for the sensation of pain in abdominal organs.

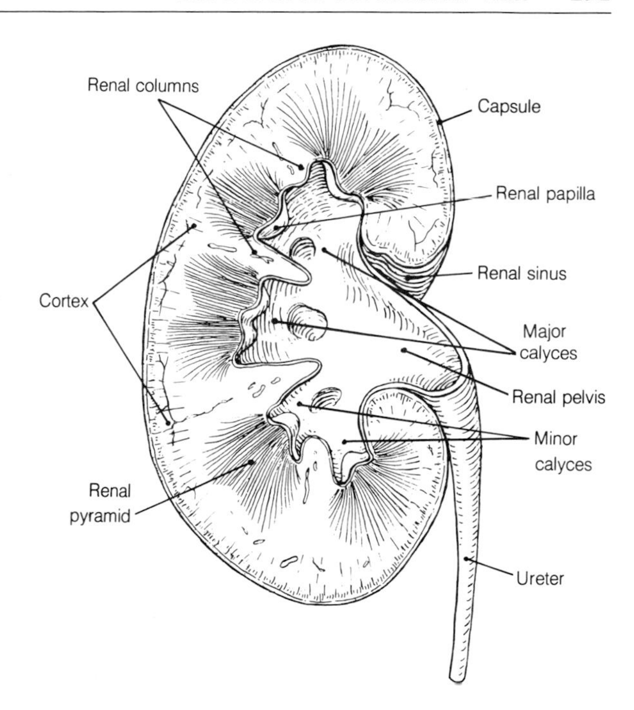

5-74 Coronal section of the kidney.

The Kidneys

Whether kindeys are bean-shaped or beans are kidney-shaped is a moot point. Most people are aware of their shape. However, at birth the kidneys still retain the surface appearance of **fetal lobulation**. This lobulation bears no relation to the underlying structure and normally is lost by the fourth or fifth year of age. On occasion it persists into adult life. Each kidney is about 12 cm in length, 6 cm in breadth, and 3 cm in thickness and lies on the posterior abdominal wall in the paravertebral gutter. The exact position of the kidneys varies with respiration and the position of the body, but they lie approximately opposite the upper three lumbar vertebrae. The left lies a little higher than the right, with its upper pole overlapping the diaphragm and extending to the level of the eleventh rib. A posterior surface faces posteromedially and an anterior surface anterolaterally. A convex border faces outward and a concave border inward. It is important to remember this oblique position when radiographs are interpreted. The middle of the concave border of the kidney is hollowed out to form the **renal sinus** and it is here that its vessels and excretory duct, the ureter, enter and leave the organ. Each kidney is covered by a **fibrous capsule** which in health may be stripped off easily. When sliced open, across its borders, some indication of its internal structure can be seen, and this is illustrated in Fig. 5-74, where the following features should be noted:

1. A lighter **cortex** lies immediately beneath the capsule and extends inward as the **renal columns** between roughly triangular and darker areas known as the **renal pyramids** which form the **medulla**.
2. The apices of the pyramids are "grasped" by the hollow **minor calyces**.
3. The minor calyces are branches of three or four divisions of the **renal pelvis**. These divisions are the **major calyces**.
4. The renal pelvis is continuous with the **ureter**.

Collecting ducts within the kidney open at the apices of the pyramids and, via the calyces, renal pelvis, and ureter, urine is carried to the bladder.

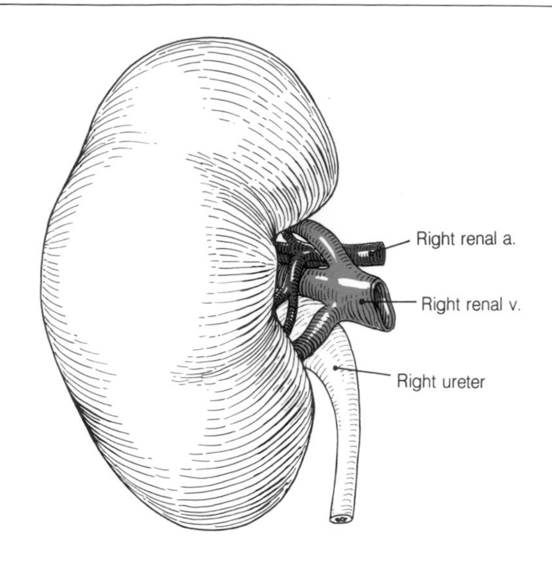

5-75 Anterior view of the right kidney.

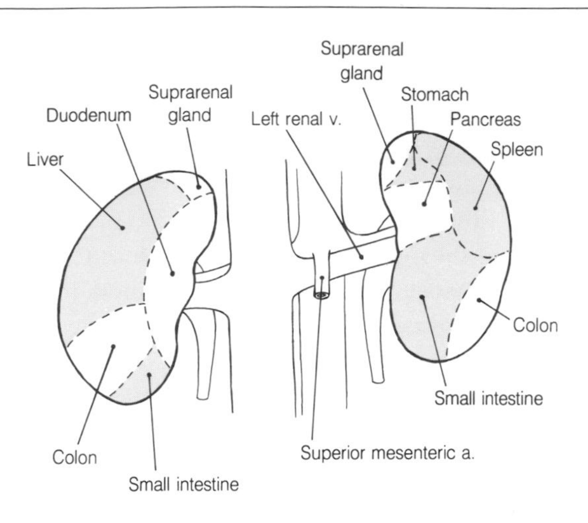

5-76 Diagram of the anterior aspects of the kidneys and their immediate relations.

Look at Fig. 5-75 to see that as it nears the hilus the renal vein lies in front of the renal artery and that both artery and vein lie in front of the ureter. Terminal branches of the renal artery may, however, pass in front of the vein. It should be remembered that the surgeon usually approaches the kidney from the back and, therefore, encounters the renal pelvis and ureter first.

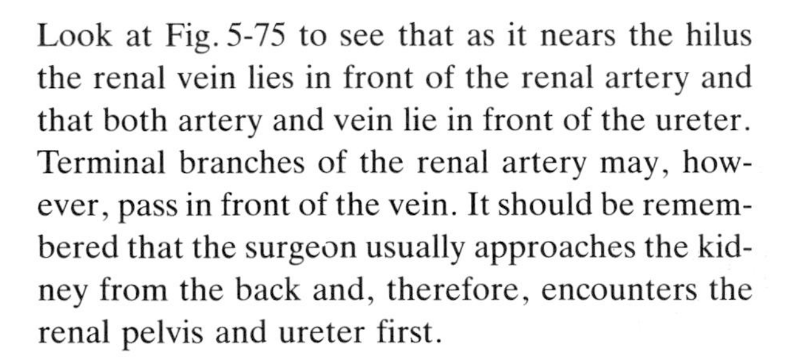

The Main Relations of the Kidneys

The anterior aspects of the kidneys are shown in Fig. 5-76 and in this the structures immediately related to their surface are labeled. In some regions (indicated in *blue*) the kidneys are covered by **parietal peritoneum** of the posterior abdominal wall. This means that they are separated by this layer and the peritoneal cavity from related structures which themselves will be covered by **visceral peritoneum**. In other regions the relationship is to retroperitoneal structures which lift the peritoneum off the kidney. With this in mind compare the anterior relations of each kidney and note that:

1. The medial aspect of both upper poles is directly related to a suprarenal gland.
2. The hilar region of the left kidney is directly related to the pancreas and the right to the duodenum.
3. A substantial part of the lower pole of the right kidney is directly related to the right colic flex-

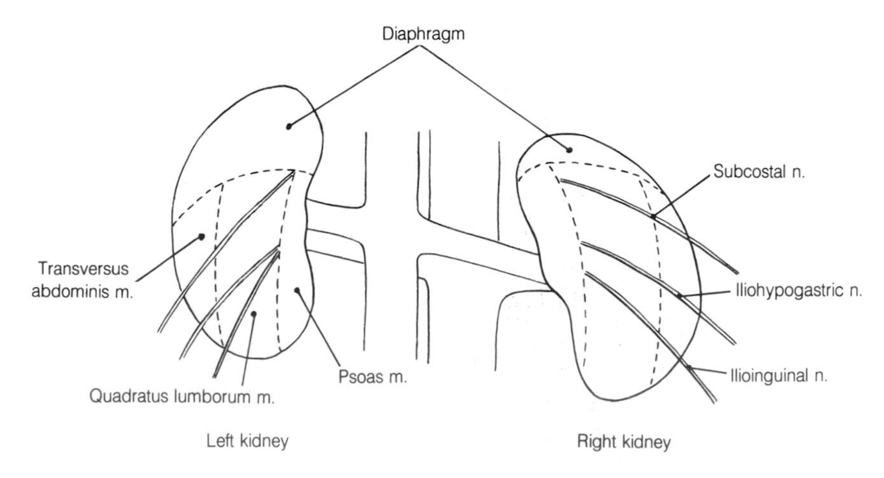

5-77 Diagram of the posterior aspects of the kidneys and their immediate relations.

ure and similarly a lesser part of the left to the left colic flexure.

4. The left lower pole is covered with peritoneum and a similar but smaller area is present over the right. These areas will be in contact with overlying small intestine.
5. The remainder of both kidneys is also covered by peritoneum and related to the overlying liver on the right and to the stomach and spleen on the left.

Except for the slight difference in levels, the posterior relations of the kidneys to the diaphragm, the muscles of the posterior abdominal walls, and the subcostal, iliohypogastric, and ilioinguinal nerves are similar on both sides and can be seen in Fig. 5-77 without further explanation. It should, however, be remembered that the pleural cavity extends below the level of the twelfth rib and that it may be accidentally opened during a posterior approach to the kidney.

The Renal Fascia

Each kidney and suprarenal gland is surrounded by adipose tissue called **perirenal fat**. This is in turn enclosed by the anterior and posterior layers of the **renal fascia**. These layers fuse above and lateral to the kidney. Below they remain separate around the ureter and blend with adjacent retroperitoneal tissue. Some connection between these layers probably exists at the medial margin of the kidney as fluid effusions within the fascia do not usually extend across the midline.

The Blood Vessels of the Kidney

The blood reaching the kidney is needed not only to supply the renal parenchyma but to take part in the process of renal filtration. As has already been pointed out, the renal arteries carry about 20% of the cardiac output and of this about 90% is delivered to the cortex where the **glomeruli**, the structures responsible for filtration, lie. At the hilus of the kidney each renal artery divides into a number of branches which supply five segments of the kidney. The main artery at first divides into **anterior** and **posterior branches** which pass anterior and posterior to the renal pelvis. The anterior branch breaks up into an upper, middle and lower segmental artery, the last supplying the inferior pole. The posterior branch typically divides into posterior segmental arteries. The superior pole or apical segment is usually supplied by a branch of the upper segmental artery. These segments and a typical pattern of distribution of the segmental branches is shown in Fig. 5-78.

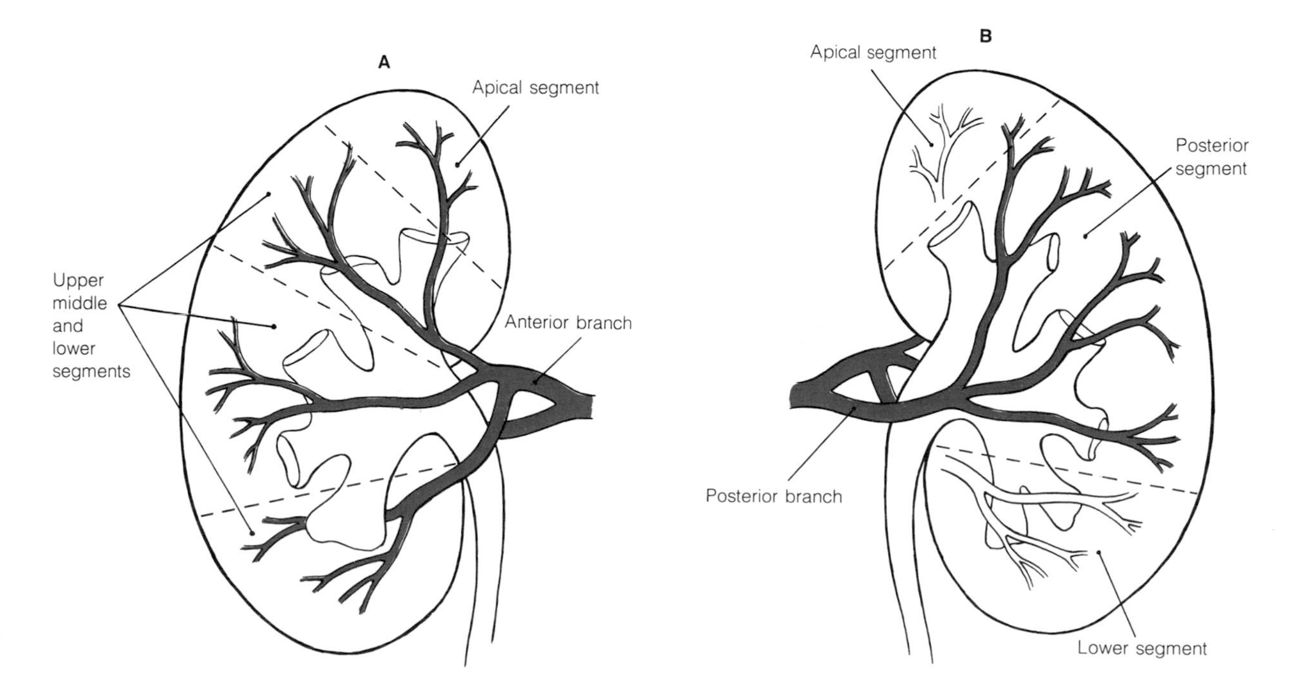

5-78 The segments of the right kidney and their blood supply seen from in front (left) and behind (right).

The segmental arteries are probably functional end arteries. When temporarily ligated an obvious area of ischemia is produced in the segments they supply and it is impossible to fill these segments when postmortem angiography is attempted through the remaining segmental arteries. This arrangement of the segmental arteries influences the incisions made in the kidney for partial nephrectomy or the removal of calculi through the renal parenchyma.

The segmental arteries in turn branch into **interlobar arteries**. At the junction of cortex and medulla the interlobar arteries, which lie at right angles to the borders of the kidney, give off the **arcuate arteries** which run parallel to the borders. From these are given off **interlobular arteries** which run into the cortex and provide the **afferent glomerular arterioles** to the **glomerular capillary network** or glomerulus. **Efferent glomerular arterioles** leaving the glomeruli form capillary networks around cortical renal tubules or descend as **vasa recta** to supply the medulla. The rather complex arrangement of these vessels is illustrated schematically in Fig. 5-79.

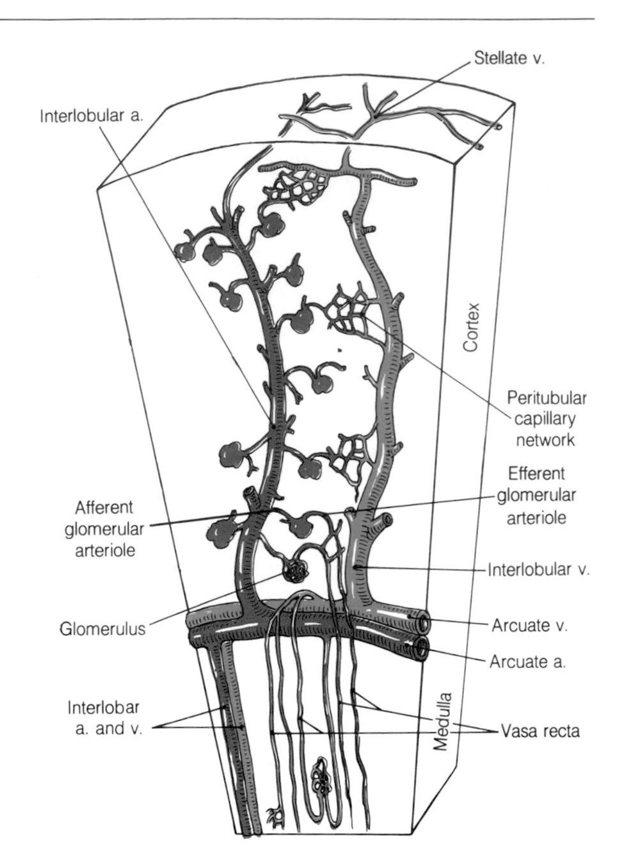

5-79 Diagram illustrating the blood vessels of the renal cortex and medulla.

Blood returns from the kidneys to the inferior vena cava through the large **renal veins** which run medially and anterior to the renal arteries. Because the inferior vena cava lies on the right hand side of the posterior abdominal wall the right renal vein is short. Note in Fig. 5-76 that the longer left vein passes anterior to the aorta just below the origin of the superior mesentric artery to reach the inferior vena cava.

The Innervation of the Kidney

Autonomic fibers from the celiac plexus, the least splanchnic nerves and the first lumbar splanchnic nerve reach the kidney through the renal plexus. Despite this profuse autonomic supply innervation appears to have little effect on the normal function of the kidney.

The Radiological Anatomy of the Kidney

The kidneys may be examined radiographically. The intravenous pyelogram or urogram is performed by introducing a radiopaque substance into the blood for filtration and concentration by the kidney. If renal tubular function is present, concentration occurs and the kidney itself forms a soft opacity. The calyces and ureter, however, are clearly outlined. In an alternative method, the opaque material is introduced via a cystoscope and a fine catheter which is passed into a ureteric orifice of the bladder. This is the retrograde pyelogram and only displays the ureter and calyces. Both investigations are preceded by plain radiographs. Radiopaque stones are not uncommonly found in the calyces and renal pelvis and may form elaborate casts of these cavities known as "staghorn calculi". An example of the detail shown by an intravenous urogram is seen in Fig. 5-80. In this note:

1. The faint opacity of the kidney
2. That the upper pole of the kidney overlies the eleventh and twelfths ribs
3. That the lower poles extend inferiorly to the level of the third lumbar vertebra
4. The outline of the minor and major calyces and the renal pelvis.

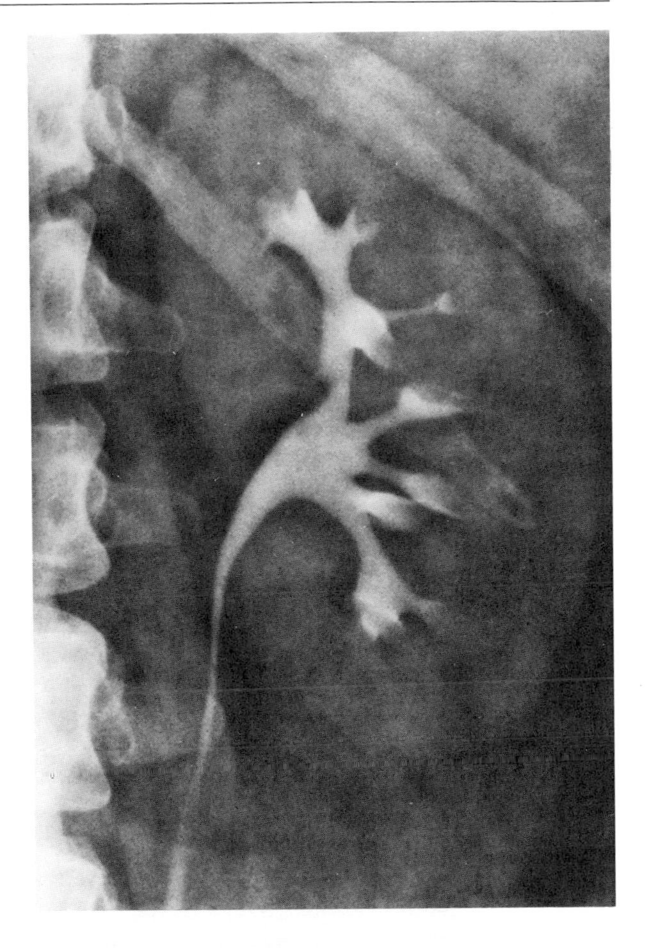

5-80 Intravenous urogram showing detail of the left kidney. (Reproduced by permission, from Wicke: *Atlas of Radiologic Anatomy,* 4th Ed, Urban & Schwarzenberg, Baltimore-Munich, 1987.)

The presence (one may be congenitally absent), size, and relationships of the kidneys can also be demonstrated by sonography. In the example given in Fig. 5-81 the left kidney can be clearly made out. The relationship of the left renal vein to the superior mesenteric artery which lies anterior to it and to the aorta which lies posterior to it can also be seen in this horizontal section.

The Ureters

Each ureter is a continuation of the renal pelvis and leaves the hilus of the kidney posterior to the blood vessels. It descends over the posterior abdominal wall behind the peritoneum and on the surface of psoas major. At the pelvic brim it crosses the common iliac artery near its bifurcation and enters the pelvis. At the level of the ischial

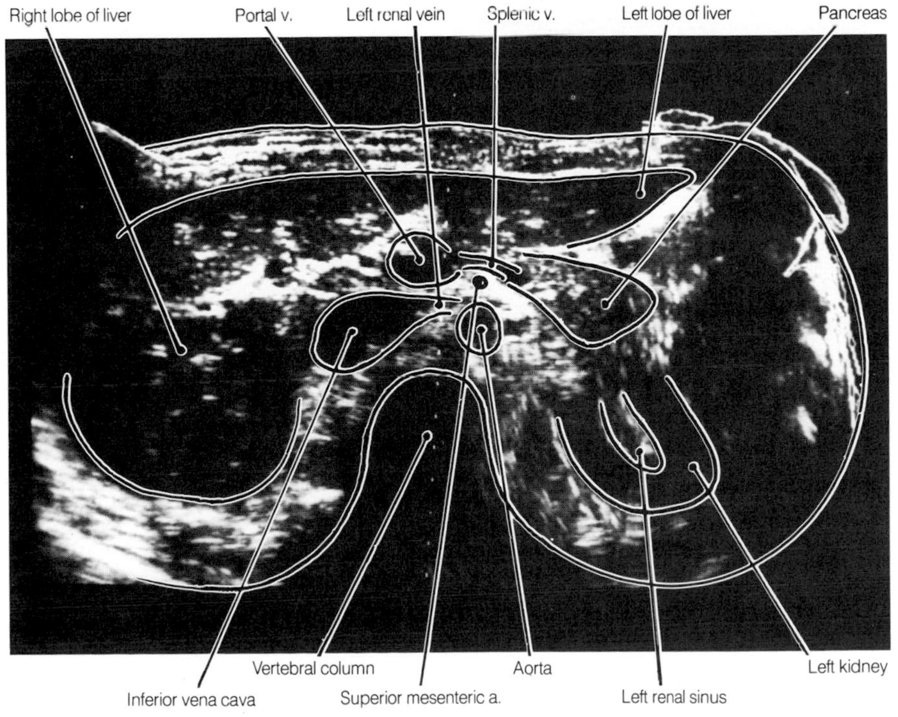

5-81 Sonogram of the abdominal cavity. Horizontal section close to the transpyloric plane viewed from below. (Courtesy of Morgan G. Dunne, M. D., Department of Diagnostic Radiology, University of Maryland School of Medicine.)

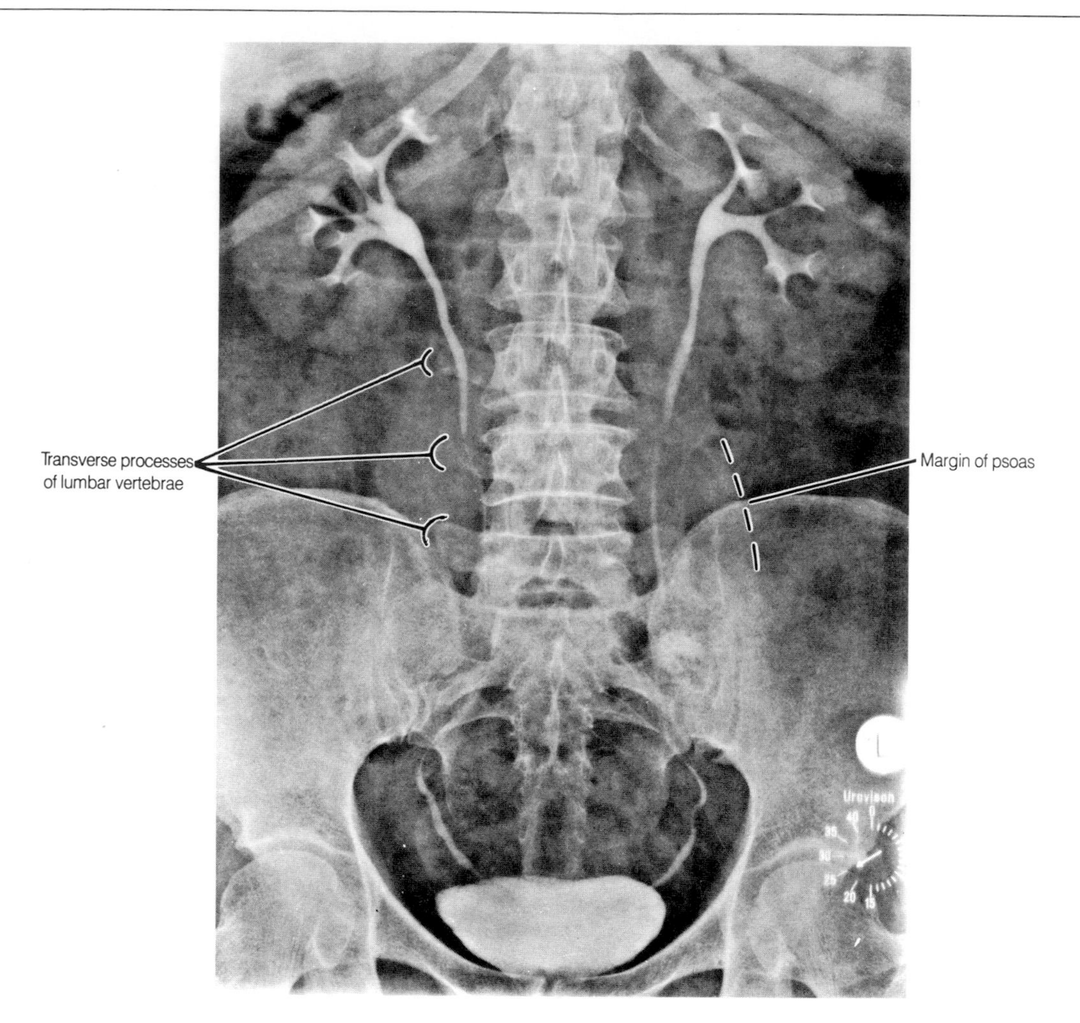

5-82 Intravenous urogram. (Reproduced by permission, from Wicke: *Atlas of Radiologic Anatomy,* 4th Ed, Urban & Schwarzenberg, Baltimore-Munich, 1987.)

spine it turns medially into the bladder. The ureters have muscular walls and exhibit active peristalsis while transmitting urine from the kidney to the bladder. For this reason an absence of contrast medium over a short length is often seen during intravenous urography.

The blood supply of the ureters is derived from both renal and vesical arteries but some reliance is placed on vessels reaching them from overlying peritoneum. Care should, therefore, be taken not to "strip" them unnecessarily from peritoneum during surgery. Where this is inevitable, as in a renal transplant, every effort is made not to disturb the supply from the renal artery.

Refer now to the urogram seen in Fig. 5-82 and note:

1. The faint shadows of the kidneys and the dense shadows of the calyces and renal pelvis
2. The course of the ureters over the faint shadows produced by the psoas muscles
3. The relationship of the ureters to the lumbar transverse processes
4. The relationship of the ureters to the sacroiliac joints
5. The pelvic course of the ureters to enter the partially filled bladder
6. The filling defects in the ureters caused by peristalsis.

The Suprarenal Glands

On the medial aspect of the superior pole of each kidney lies a suprarenal or adrenal gland. These are endocrine glands which are essential for life but have no immediate functional relationship to the kidney. They are separated from the kidneys by a layer of perirenal fat by which they are also surrounded. The left is semilunar in shape, the right more pyramidal. In cross-section, the gland is triangular and can be seen to be divided into an outer cortex which is yellow in color and an inner, more vascular medulla. The blood supply of the suprarenal glands comes from branches of the inferior phrenic and renal arteries and from the suprarenal branches of the aorta. Blood drains to the inferior vena cava on the right and to the left renal vein on the left. Lymph from both the kidneys and the suprarenal glands drains to paraaortic nodes. The suprarenal glands receive numerous preganglionic sympathetic fibers from the greater splanchnic nerves. These are distributed to the medulla where they stimulate the liberation of epinephrine and norepinephrine from the chromaffin cells. The cortex of the gland is under the control of the adrenocorticotrophic hormone secreted by the anterior lobe of the pituitary gland. The appearance of the right suprarenal gland and its blood supply are illustrated in Fig. 5-83.

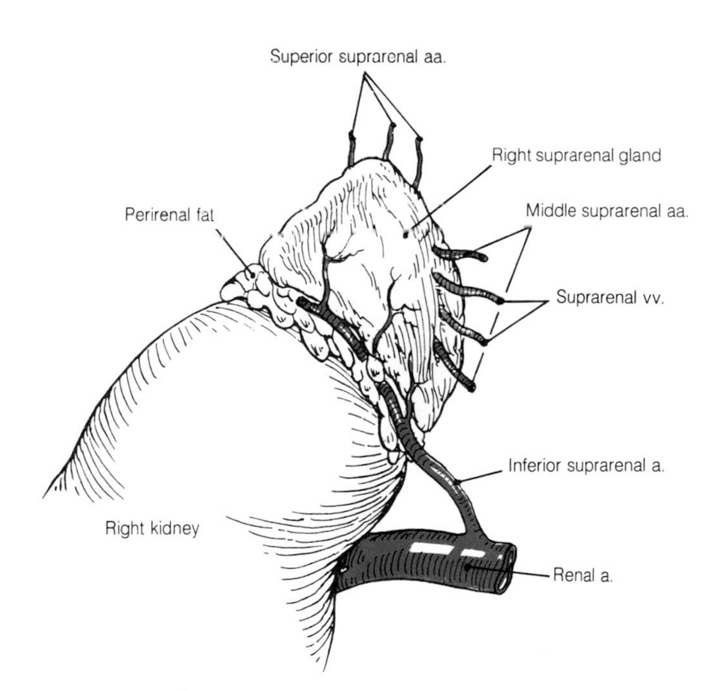

5-83 The right suprarenal gland and its blood supply.

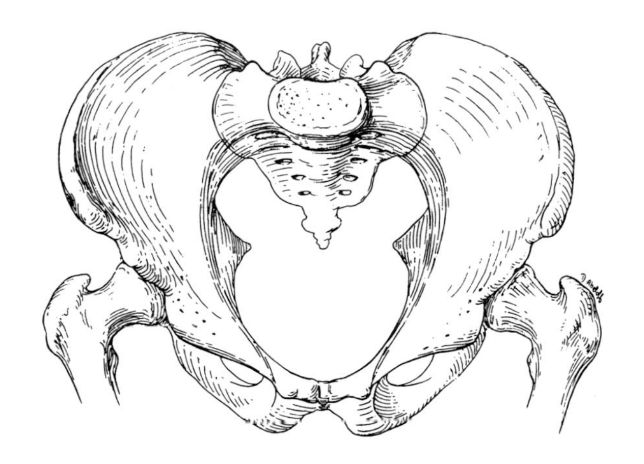

6 The Pelvis

While the word pelvis* is confidently used to describe a part of the body, its use becomes somewhat confusing when details of this region are presented. The word pelvis is also used to describe the complete, articulated, bony ring or girdle formed by the sacrum and the two hip bones. The bony pelvis thus formed is divided into an upper part called the **greater pelvis** which is flanked by the wings or alae of the iliac bones and a lower part, the **lesser pelvis**, which, although structurally continuous with the greater pelvis, is descriptively separated from it by the **superior pelvic aperture** or pelvic brim.

* Pelvis, *L.* = a basin

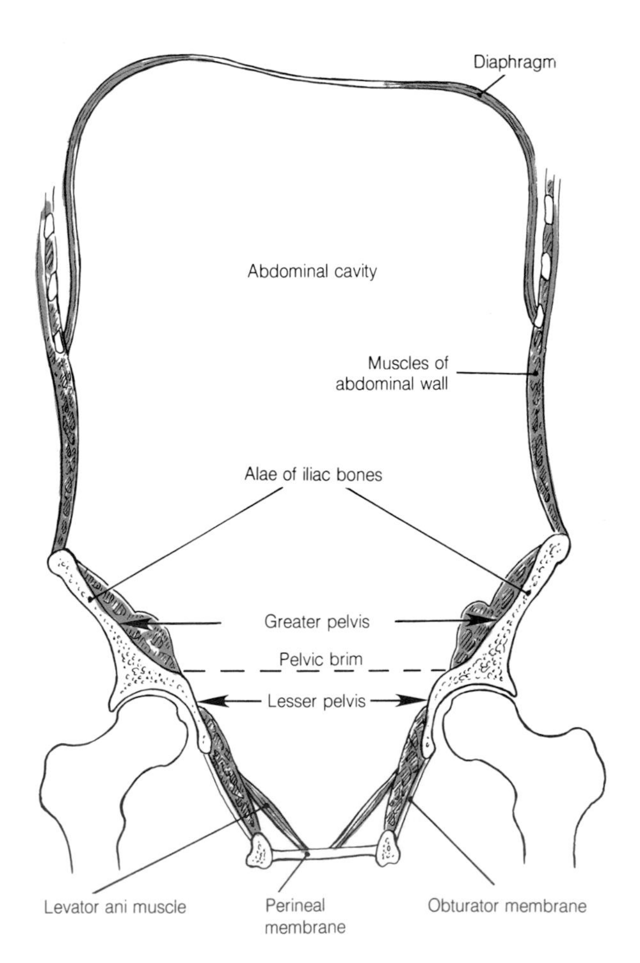

6-1 Coronal section through abdomen and pelvis.

The cavity of the greater pelvis, which is part of the abdominal cavity, is lined by the greater peritoneal sac. This dips down into the lesser pelvis to partially cover the terminal portion of the alimentary tract, the urinary bladder, and the internal reproductive organs of the female. This extension of the peritoneal cavity may also contain any sufficiently mobile abdominal organ able to reach it. The cavity of the lesser pelvis or pelvic cavity as it is more commonly called, is completed by muscular walls and a muscular floor which is perforated by the urinary, reproductive, and alimentary tracts in both sexes. Look now at Fig. 6-1 and establish the relationships of the abdominal cavity to the pelvic cavity and the greater pelvis to the lesser pelvis.

Finally the word pelvis is used, as it is at the top of this page, to describe an ill-defined region where the trunk joins the lower limbs.

The Bony Pelvis or Pelvic Girdle

In the anterosuperior view of the pelvis shown in Fig. 6-2, the pelvic girdle can be seen to be formed by the **two hip bones** which articulate with each other anteriorly at the **pubic symphysis** and with the unpaired **sacrum** posteriorly at the **sacroiliac joints**. The sacrum is a single bony mass formed by the fusion of five sacral vertebrae. Superiorly, it articulates with the fifth lumbar vertebra and inferiorly with two or three fused and vestigial coccygeal vertebrae. The complete ring thus formed surrounds the cavity of the pelvis. This has a **superior aperture**, brim or inlet, and an **inferior aperture** or outlet. The superior aperture is bounded by the **promontory of the sacrum** and the **linea terminalis**. The latter line includes the **arcuate line** of the ilium and the **iliopectineal line**

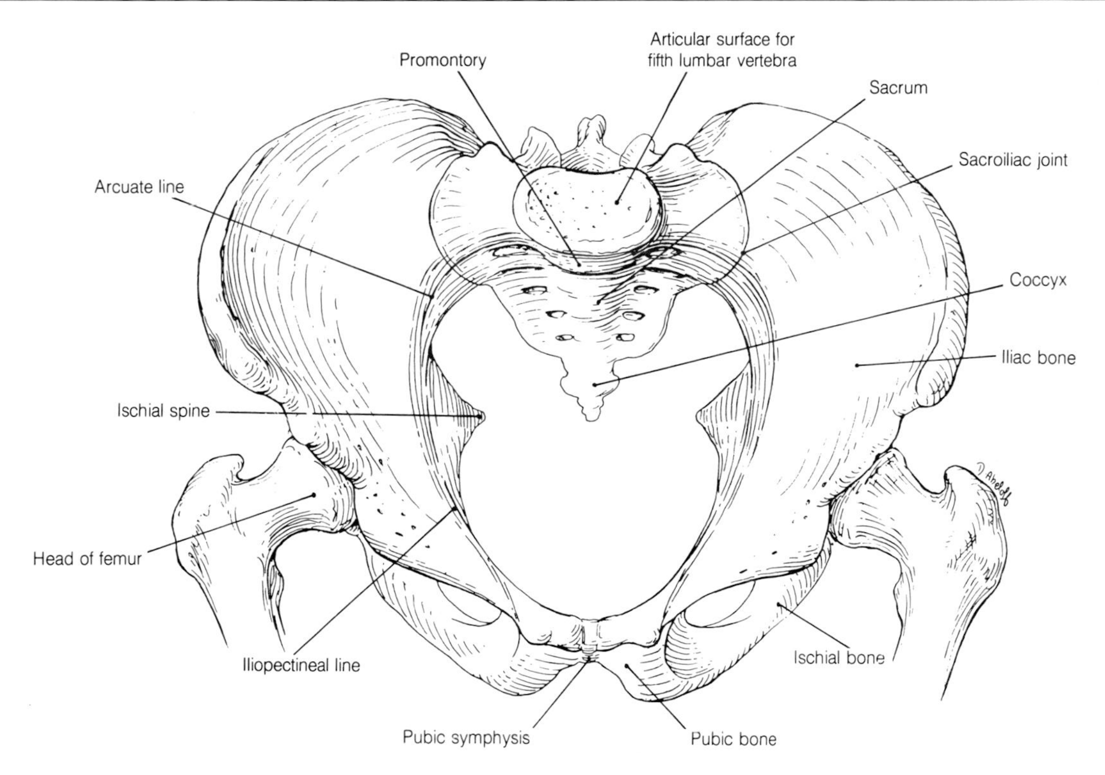

6-2 Anterosuperior view of the pelvic girdle.

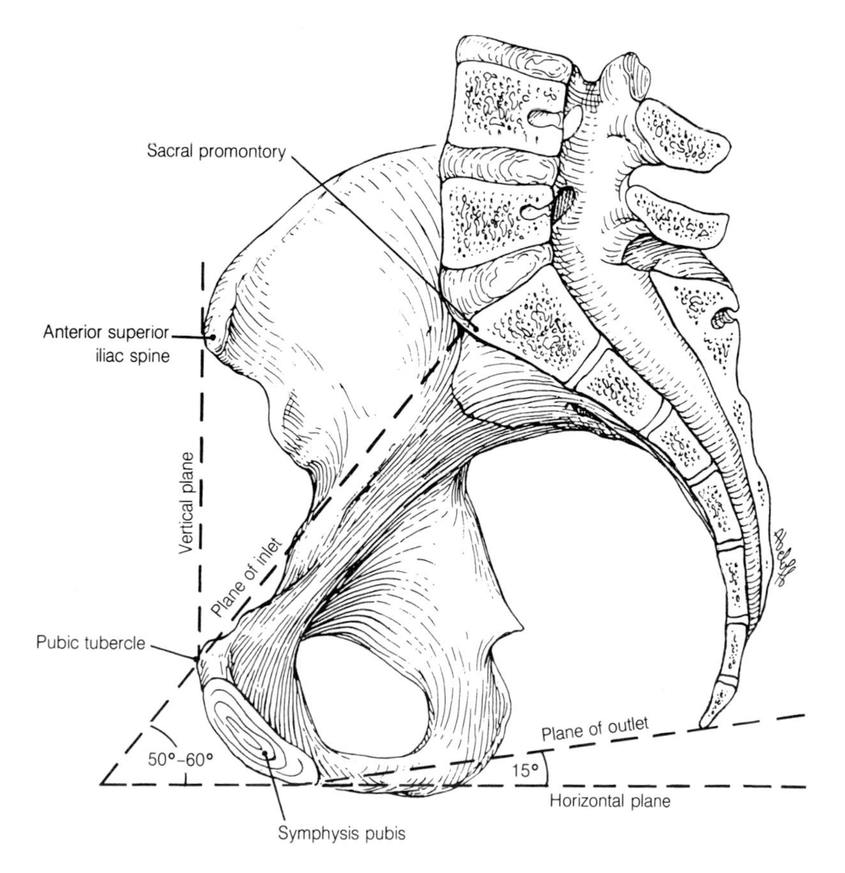

6-3 The planes of the pelvic inlet and outlet.

of the pubis. Whereas the outlet is almost horizontal, the inlet or brim is tilted well forward so that the inner aspects of the pubic bones face upward and help to support the contents of the cavity. The angles formed with the horizontal plane by the inlet and outlet can be seen in Fig. 6-3. To orient an articulated specimen of the pelvis in this position, it is necessary to hold it so that the pubic tubercles and the anterior superior iliac spines are all in the same vertical plane. The shape of the inlet is of particular importance in the female for it is through this aperture that the fetal head enters the pelvic cavity during labor.

The Sacrum

The five sacral vertebrae are fused into a solid mass. **Anterior and posterior sacral foramina** allow the ventral and dorsal rami of the sacral nerves to leave the vertebral column. There is an earshaped **auricular surface** on each side for articulation with the hip bones and behind these, a roughened area for the very strong interosseous ligament of the sacroiliac joint. The fused spines appear as a narrow, irregular mid-line ridge called the **median sacral crest**. The lower laminae are normally defective, leaving a posterior hiatus cal-

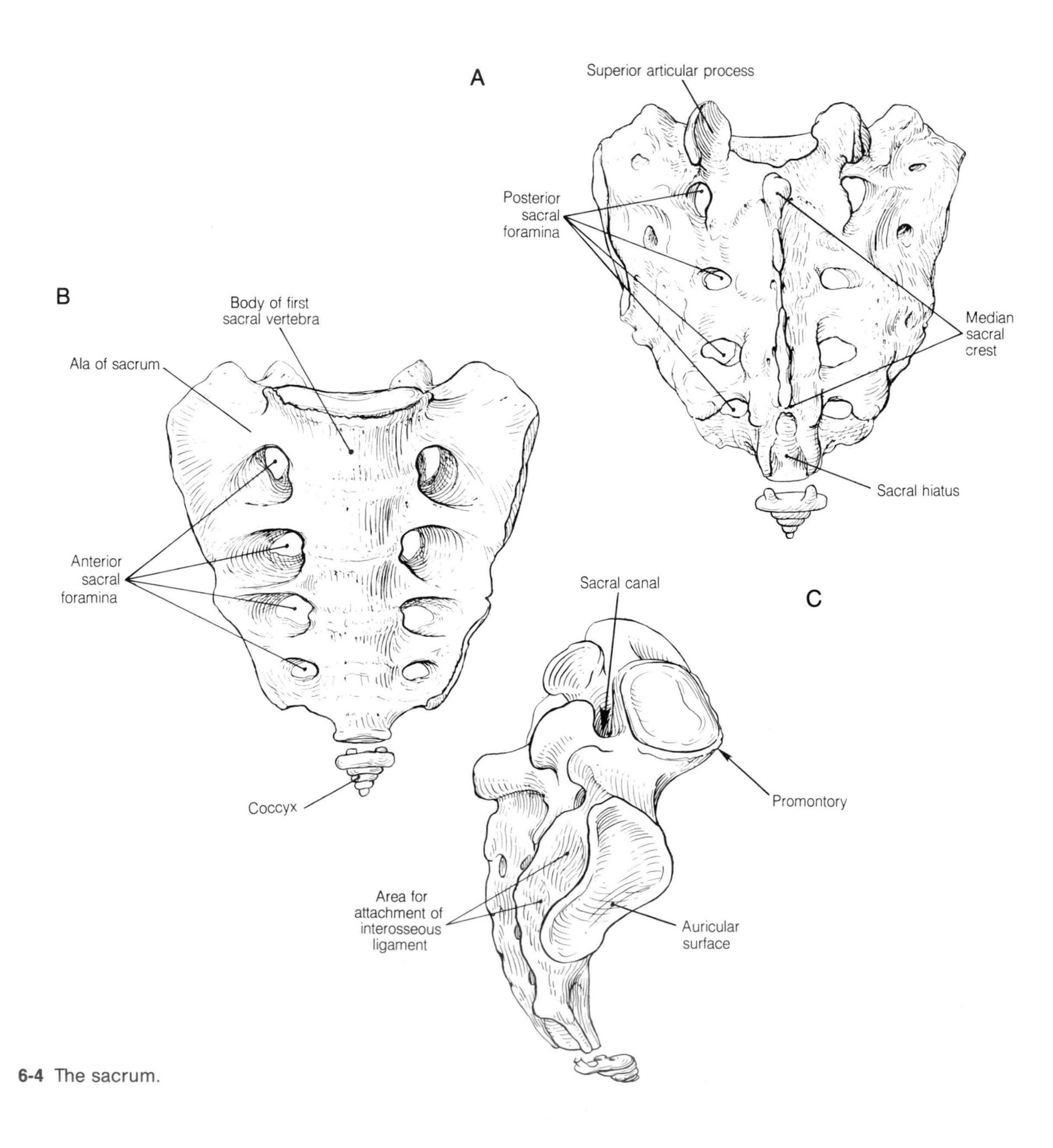

6-4 The sacrum.

led the **sacral hiatus** through which anesthetic solutions may be introduced. These features of the sacrum are illustrated in Fig. 6-4 and the bone is described in more detail with the lower limb.

The Hip Bone

Both developmentally and phylogenetically the hip bone is formed from three bones: the **iliac bone, pubic bone**, and the **ischium**. At puberty the three bones are separated at the acetabulum by a Y-shaped piece of cartilage, but shortly afterward, ossification of this cartilage is completed and the three bones form a single unit called the **hip bone**. In this chapter only the medial aspect of the hip bone is considered in detail.

The Iliac Bone

The upper part of the iliac bone presents a large wing-like surface or **ala** from whose inner aspect, called the **iliac fossa**, arises the iliacus muscle. Behind this can be seen an **auricular surface** for the sacroiliac joint. A roughened area posterior to this gives attachment to the **interosseous sacroiliac ligament**. The bone is bounded above by the **iliac crest** and its anterior border bears the **superior and inferior anterior iliac spines**. The posterior border bears **superior and inferior posterior iliac spines**.

The lower extremity of the bone is continuous with the pubic bone and ischium. It contributes to the **acetabulum** and below the level of the **linea terminalis** it also contributes to the bony wall of the lesser pelvis and to the margin of the **greater sciatic notch**. The features on the pelvic surface of the hip bone should be identified in Fig. 6-5.

The Pubis and Ischium

The description of the pubis and ischium should also be followed in Fig. 6-5. The **pubic bone** has a

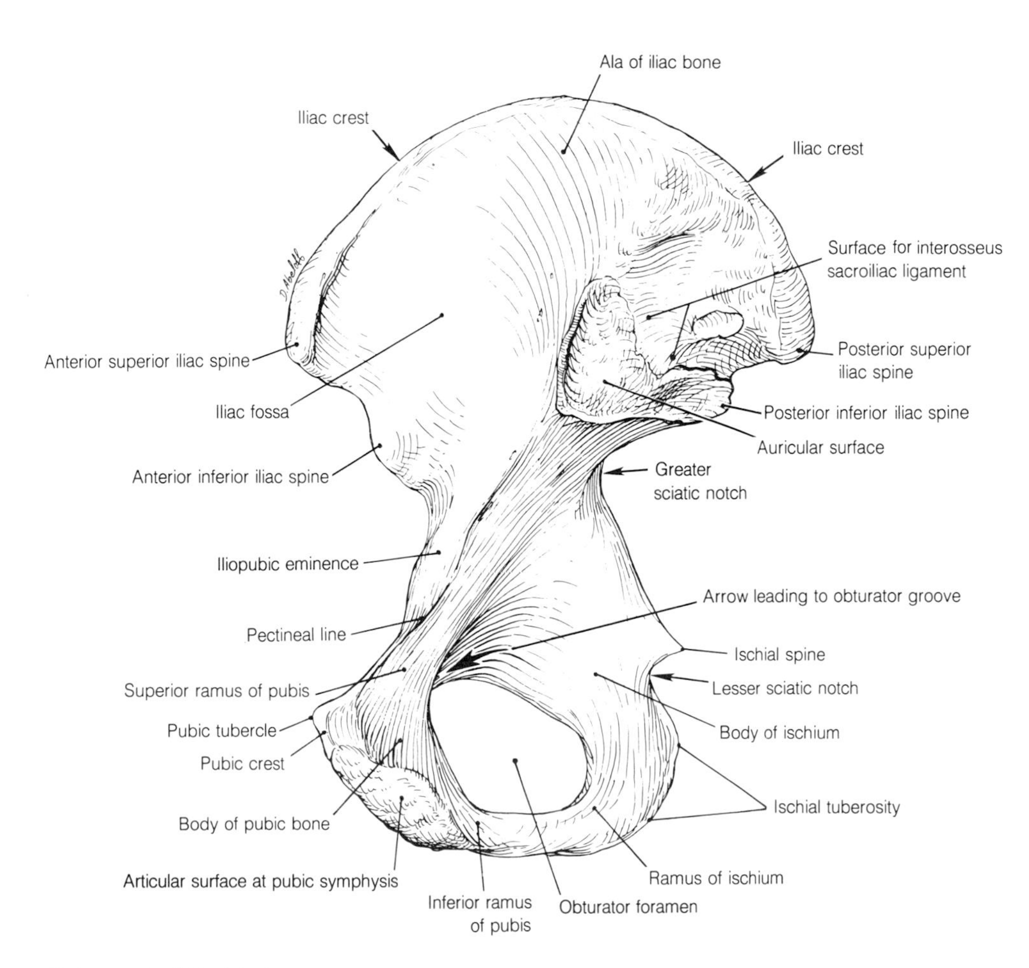

6-5 The medial aspect of the hip bone.

body which articulates with its fellow on the opposite side at the pubic symphysis and a **superior and inferior ramus**. On the upper surface of the body the **pubic crest** should be identified and followed laterally to the rounded prominence called the **pubic tubercle**. The superior ramus extends laterally from the upper part of the body toward the acetabulum. Here it forms the superior margin of the **obturator foramen** which is occluded in life by the **obturator membrane** (see Fig. 6-6). This border is grooved laterally by the **obturator groove** which carries the obturatar nerve and vessels from the pelvis to the thigh. The posterior border of the superior surface of the ramus is sharp and forms the **pectineal line** which can be followed to the **iliopubic eminence**.

The inferior ramus links the pubis to the **ischium** and with the contralateral inferior ramus forms the **pubic arch** and the **subpubic angle**. As the inferior ramus of the pubis is followed backward, it joins the **ramus of the ischium** at an indistinct point. Hence, these rami are often known together as the ischiopubic ramus. The ramus of the ischium leads to the **body of the ischium** and the large and roughened **ischial tuberosity** at its postero-inferior border. Above, the posterior border of the ischium contributes to a large upper notch challed the **greater sciatic notch** and a smaller and lower notch called the **lesser sciatic notch**. The two notches are separated by the **ischial spine**; in life they are transformed into two foramina by the **sacrotuberous and sacrospinous ligaments** in the manner seen in Fig. 6-6.

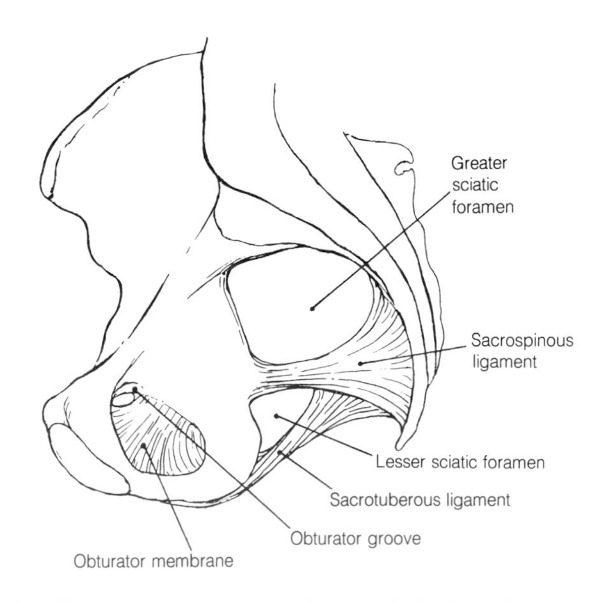

6-6 To show the greater and lesser sciatic foramina.

The Male and Female Pelvis

Male and female pelves may differ markedly in a number of respects and it can sometimes be quite easy to assign a pelvis to its correct sex, an exercise often performed by the anthropologist or forensic pathologist. However, the examination of skeletal material in an anatomy laboratory will soon show that this sexual dimorphism is not always clear cut and that a number of features must be taken into account before a decision can be reached.

Some of these features depend upon the primary function of the pelvis as a weight-bearing structure, others are related to the adaptation of the pelvic cavity to the passage of a fetal head. Useful points of distinction between the male and female pelvis are tabulated below:

1. Males are normally larger and more muscular and their pelves are more stoutly built and have more pronounced markings at muscle attachments.
2. The joint between the fifth lumbar vertebra and the sacrum is broader in the male and spans one-third of the breadth of the sacrum.
3. The margins of the ischiopubic rami are more roughened and everted in the male where they provide attachment for the crura penis and ischiocavernosus muscles.
4. The acetabulum is larger in the male and its diameter is equal to the distance between its edge and the pubic symphysis. In the female this distance is considerably greater than the breadth of the absolutely smaller acetabulum.
5. The male pelvis is deeper and more conical than the female and with this generalization are associated:
   a. a greater intercristal dimension in the male,
   b. a subpubic angle of less than 70° in the male,
   c. a subpubic angle of greater than 80° in the female,
   d. a smaller distance between the ischial spines which are also inverted in the male,
   e. a less curved and more backwardly tilted sacrum in the female.

f. a broader greater sciatic notch in the female, and

g. a shallower anterior pelvic wall in the female and a more triangular shaped obturator foramen.

Perhaps the most important difference between male and female pelves is the shape of the superior pelvic aperture.

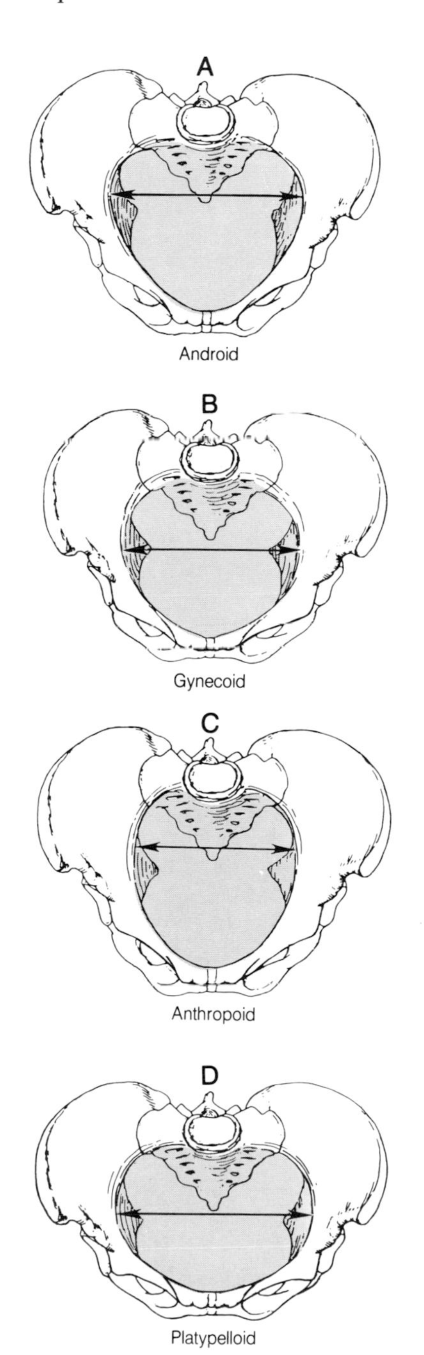

6-7 Showing the variety of shapes of the pelvic inlet.

In Fig. 6-7 A the shape of the typical male superior aperture can be seen to be heart-like with a maximum transverse diameter close to the sacrum. This is the **android pelvis**. The typical female or **gynecoid pelvis** (Fig. 6-7 B) has a much more rounded profile with a maximum transverse diameter well in front of the sacrum. An oval-shaped aperture with a long anteroposterior diameter is said to be **anthropoid** (Fig. 6-7 C) and a rare type with a wide transverse diameter is said to be **platypelloid** (Fig. 6-7 D).

For anthropological purposes, the superior pelvic aperture can be described using the following three measurements which are illustrated in Fig. 6-8:

1. the **conjugate diameter** between the upper border of the symphysis pubis and the sacral promontory,
2. the maximum **transverse diameter**, and
3. the **oblique diameter** between one iliopubic eminence to the opposite sacroiliac joint.

Similar measurements can also be made of the inferior pelvic aperture:

1. the **anteroposterior diameter** between the tip of the coccyx and the lower border of the symphysis pubis, and
2. the **transverse diameter** between the medial surfaces of the ischial tuberosities.

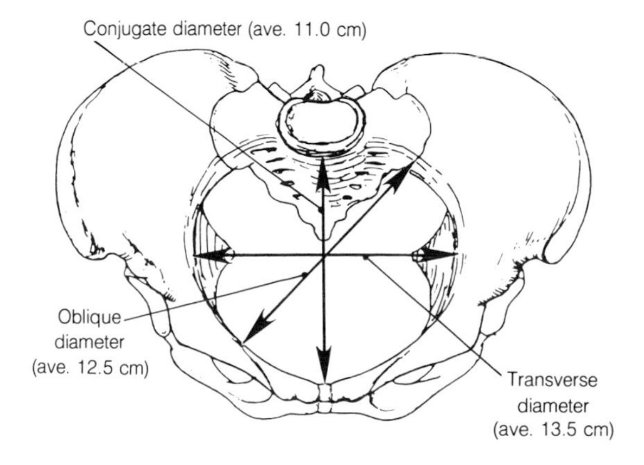

6-8 Showing the diameters of the superior pelvic aperture.

The Pelvis in Obstetrics

Successful labor and delivery depend largely on pelvic dimensions and shape that are compatible with those of the fetal head. In the past these were frequently assessed by the technique of X-ray pelvimetry. However, because of the risks to both mother and child this practice has been discontinued except in special circumstances. As a result, careful manual examination of the bony pelvis has assumed greater importance in eliminating clearcut instances of disproportion and where some doubt still exists, a trial labor is allowed to proceed with all in readiness for a Caesarian section. The fetogram seen in Fig. 6-9 illustrates the relationship of the fetus to the abdominopelvic

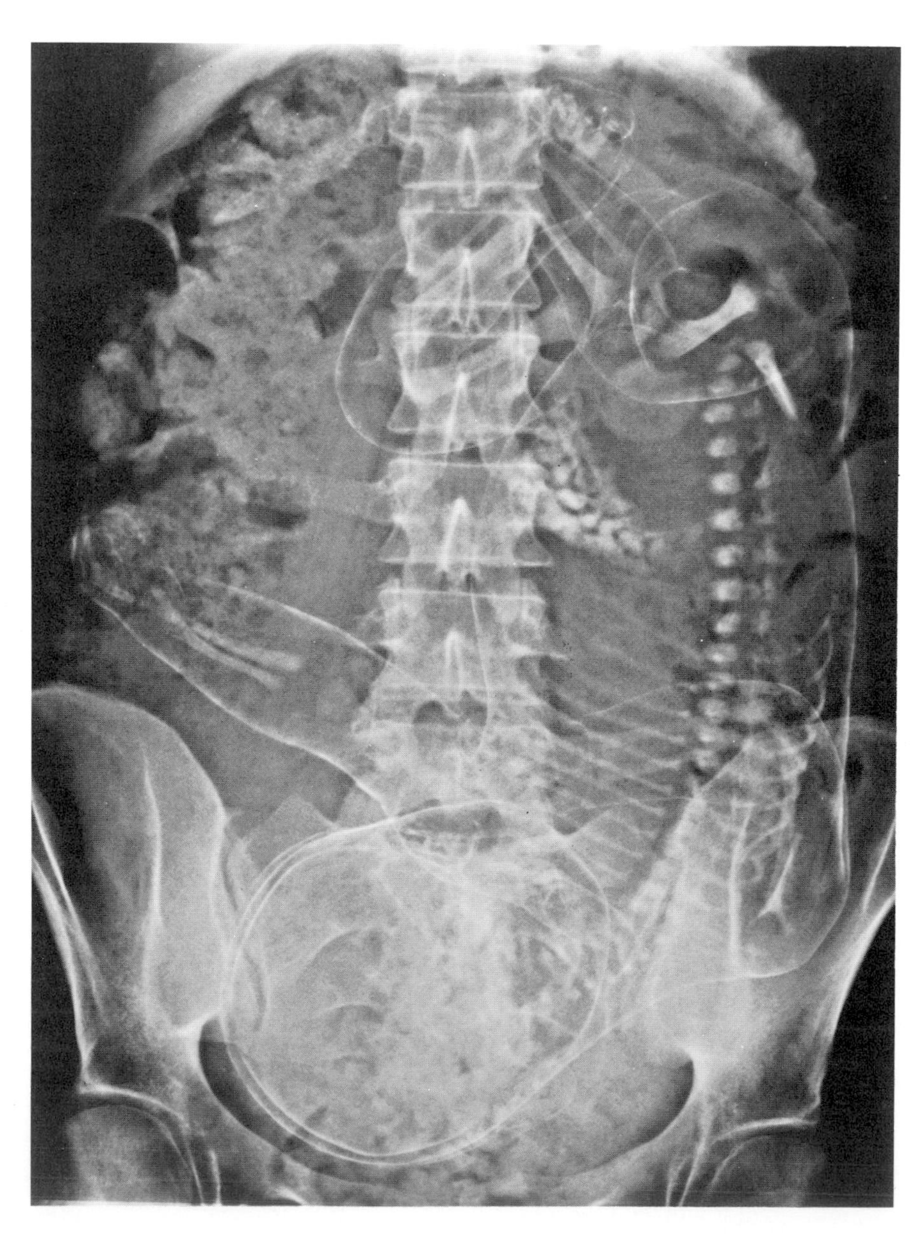

6-9 A fetogram. (Reproduced by permission, from Wicke: *Atlas of Radiologic Anatomy,* 4th Ed, Urban & Schwarzenberg, Baltimore-Munich, 1987.)

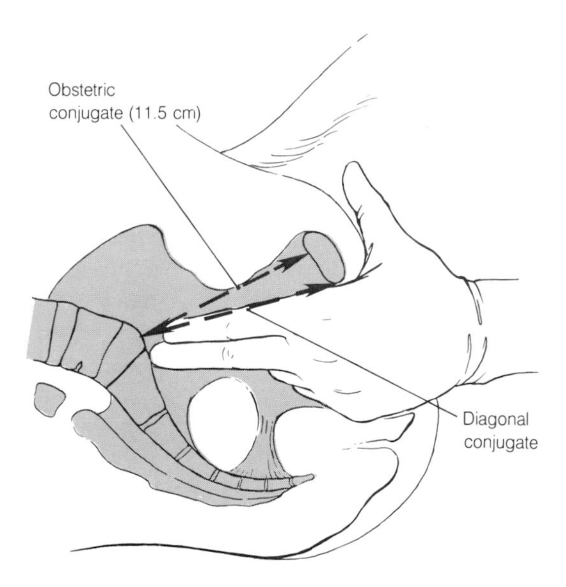

6-10 Showing the manner in which the diagonal conjugate is estimated.

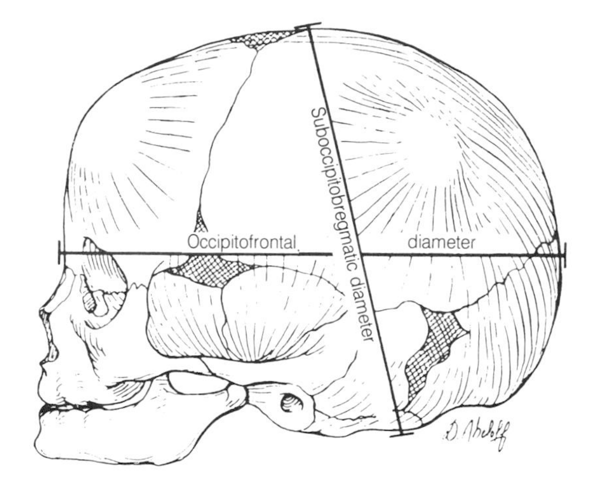

6-11 Showing the important diameters of the fetal skull.

cavity and of the fetal head to the pelvis before the onset of labor.

In making an obstetrical assessment of the pelvis it is the minimum dimensions that are important; at the superior aperture this is normally the anteroposterior diameter. This is estimated in a vaginal examination by feeling the sacral promontory with the tip of the middle finger. The measurement obtained is called the **diagonal conjugate** and is about 2.0 cm greater than the minimum diameter of the aperture or **obstetrical** or **true conjugate**. These measurements are illustrated in Fig. 6-10. The midpelvic diameter between the ischial spines can also be judged in this examination. It is usually the smallest diameter of the pelvis and lies at the point at which most difficulty in labor occurs. The smallest diameter of the outlet is that between the ischial tuberosities.

It will be noticed that the maximum diameter of the superior aperture lies transversely while that of inferior aperture lies anteroposteriorly. As the fetal head enters the superior aperture its maximum diameter lies across the pelvis. However, during its subsequent passage through the birth canal, it rotates through 90° so that its maximum diameter now lies anteroposteriorly at the outlet with the occiput appearing beneath the pubic arch. In the mid-pelvis the anteroposterior diameter of the fetal head is greatly reduced by flexion of the neck. As a result, the smaller suboccipitobregmatic diameter (approximately 9.5 cm) is exchanged or the occipitofrontal diameter (approximately 11.0 cm). These diameters can be compared in Fig. 6-11. A further reduction in diameters also occurs as the flat bones of the skull overlap each other slightly.

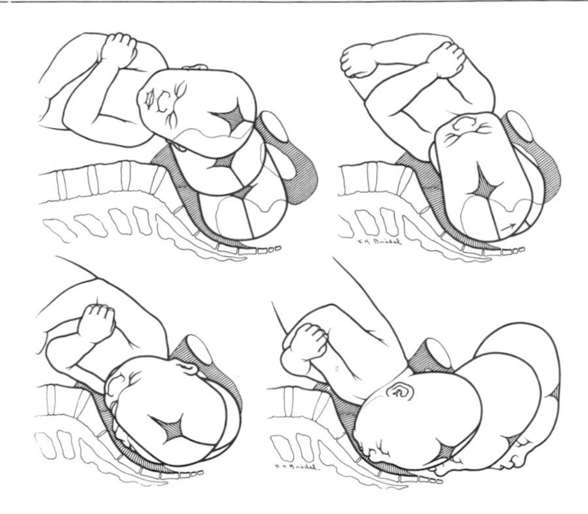

6-12 Diagrams illustrating the movements of the fetal head during delivery. (Reproduced by permission, from Steele and Javert: The mechanism of labor for transverse positions of the vertex. *Surg. Gynecol Obstet* 75:480, 1942.)

The progress of the fetal head through the pelvis during a normal delivery is illustrated in Fig. 6-12. In this the cardinal movements of the fetal head can be appreciated by following the movement of the anterior fontanelle. These are: **engagement, descent, flexion, internal rotation**, and **extension**. The final movement of the head is **external rotation** as the shoulders come to lie across the greatest diameter of the outlet.

The Joints of the Pelvis

The Pubic Symphysis

The two pubic bones articulate with each other anteriorly at the **pubic symphysis**. This is a secondary cartilaginous joint and in the young the two articulating surfaces are covered by a thin layer of hyaline cartilage but separated from each by a midline disc of fibrocartilage. The joint is immovable except in the latter months of pregnancy when the ligaments become lax. The **superior pubic ligament** above and the **arcuate pubic ligament** below reinforce the joint.

The Sacroiliac Joint

The auricular surface on the ilium has already been examined. It articulates with a similarly shaped surface on the lateral aspect of the sacrum. The joint is a synovial joint of the "plane" variety, so these surfaces are covered by hyaline cartilage and separated by synovial fluid. However, although roughly planar, the two surfaces show gross but congruent irregularities which reduce movement to a minimum. Movement is still further restricted by the very short and strong **interosseous ligament** which is attached to the irregular bony surfaces behind the auricular surfaces. Weaker **anterior** and **posterior ligaments** also help to stabilize the joint. More distant and extrinsic ligaments are the **sacrotuberous** and **sacrospinous ligaments** which were illustrated earlier in Fig. 6-6.

The **sacrotuberous ligament** extends from the posterior iliac spines and the dorsal surface of the sacrum and coccyx to the medial margin of the ischial tuberosity.

The **sacrospinous ligament** extends from the medial margin of the lower part of the sacrum to the ischial spine.

The Pelvic Cavity

The Walls of the Pelvis

The walls of the lesser pelvis have an outer bony skeleton formed by the hip bone anteriorly and laterally and by the sacrum posteriorly. The obturator foramen is occluded by the **obturator membrane** and much of the bony inner surface is lined by two muscles, the **obturator internus** anterolaterally and the **piriformis muscle** posteriorly. These two muscles are illustrated in Fig. 6-13 and described below.

Obturator Internus

This muscle arises from the bone surrounding the obturator foramen and the thick obturator membrane which almost fills it. Essentially a muscle of the hip joint, it is directed toward the lesser sciatic foramen where its tendon makes a 90° turn to run laterally out of the pelvis and gain insertion on the medial surface of the greater trochanter of the femur. The muscle is supplied by the **nerve to obturator internus** which is derived from the fifth lumbar and first and second sacral nerves. This nerve is formed in the pelvis, leaves it through the greater sciatic foramen, and re-enters through the lesser sciatic foramen to supply the muscle.

Piriformis

Piriformis arises by digitations from the anterior surfaces of the middle three pieces of the sacrum. The digitations surround the anterior sacral foramina and allow the exit of the ventral rami of sacral nerves. The muscle passes laterally through the greater sciatic notch largely filling it. Its tendon is inserted into the tip of the greater trochanter of the femur. Its nerve supply is from the **first and second sacral nerves**.

The Floor of the Pelvis

The muscular pelvic floor forms a rounded sling or gutter on which is supported the terminal part of the rectum and the prostate and urethra in the male, and the rectum, the vagina, and the urethra in the female. When the pelvis is correctly

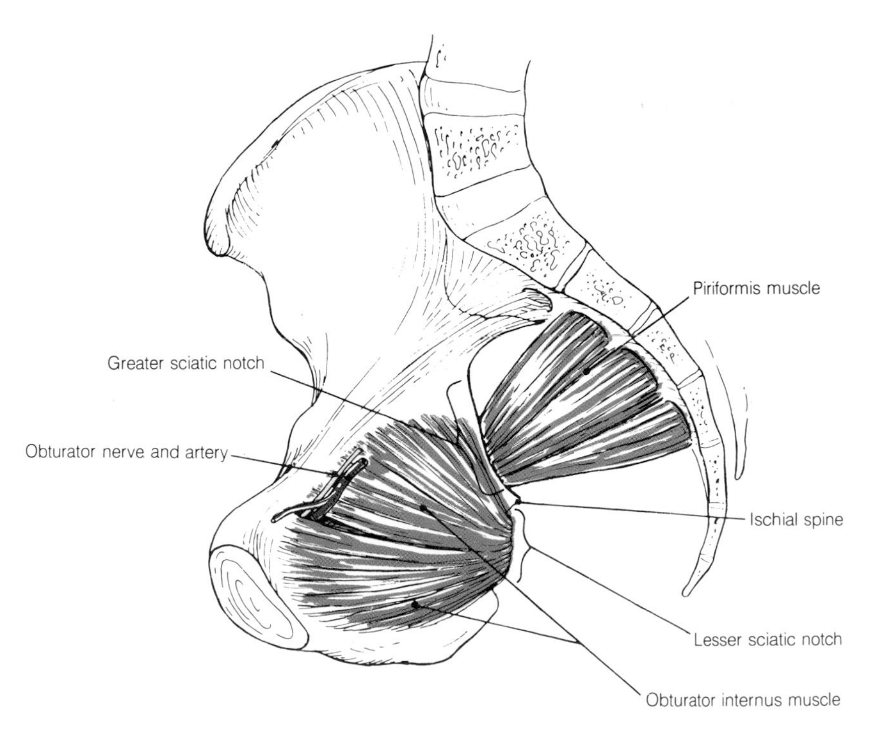

6-13 Showing the muscles of the pelvic wall.

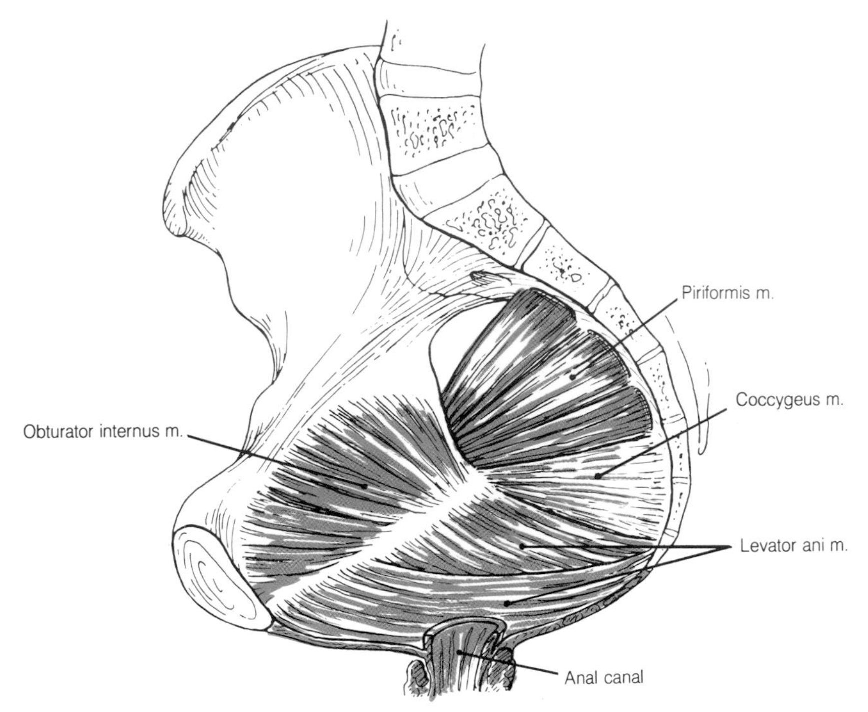

6-14 Showing the muscles of the pelvic floor and wall on the right side.

oriented, as in Fig. 6-14, the gutter can be seen to slope downward and forward. The sheets of muscle forming the floor are the **coccygeus muscle** posteriorily and the **levator ani muscle** anteriorly. The combined muscles are referred to as the **pelvic diaphragm** and their relationship to the muscles forming the walls of the pelvis can be determined by comparing Fig. 6-13 with Fig. 6-14.

Coccygeus

The coccygeus muscle arises from the ischial spine and fans out to be inserted into the lateral margin of the lower sacrum and coccyx. Immediately posterior to the muscle and almost inseparable from it is the sacrospinous ligament. The muscle is supplied by the **perineal branch of the fourth sacral nerve**.

Levator Ani

The levator ani is the more important muscle and the integrity of the pelvic floor depends upon its function. It is particularly liable to damage during difficult deliveries and this damage may be followed by urinary incontinence, prolapse of the bladder, and prolapse of the uterus through the vagina.

The muscle has a linear origin from the pelvic wall. This origin starts anteriorly on the inner aspect of the body of the pubis, extends across the surface of the obturator fascia from an overlying condensation of pelvic fascia called the **arcus tendineus**, and terminates at the spine of the ischium where its most posterior fibers lie parallel with the lower fibers of coccygeus. The levator ani can be described in two parts according to the origin of their fibers. Those fibers arising from the pubis are known as **pubococcygeus** and those from the arcus tendineus and the spine of the ischium as **iliococcygeus** (Fig. 6-15).

Pubococcygeus

In the male the most medial fibers of pubococcygeus pass backward around the prostate (which contains the first part of the urethra) and are inserted into the **perineal body**. This portion of levator ani is known as the **levator prostatae**. In the female similar fibers pass around the lower end of the vagina forming a sphincter for it and

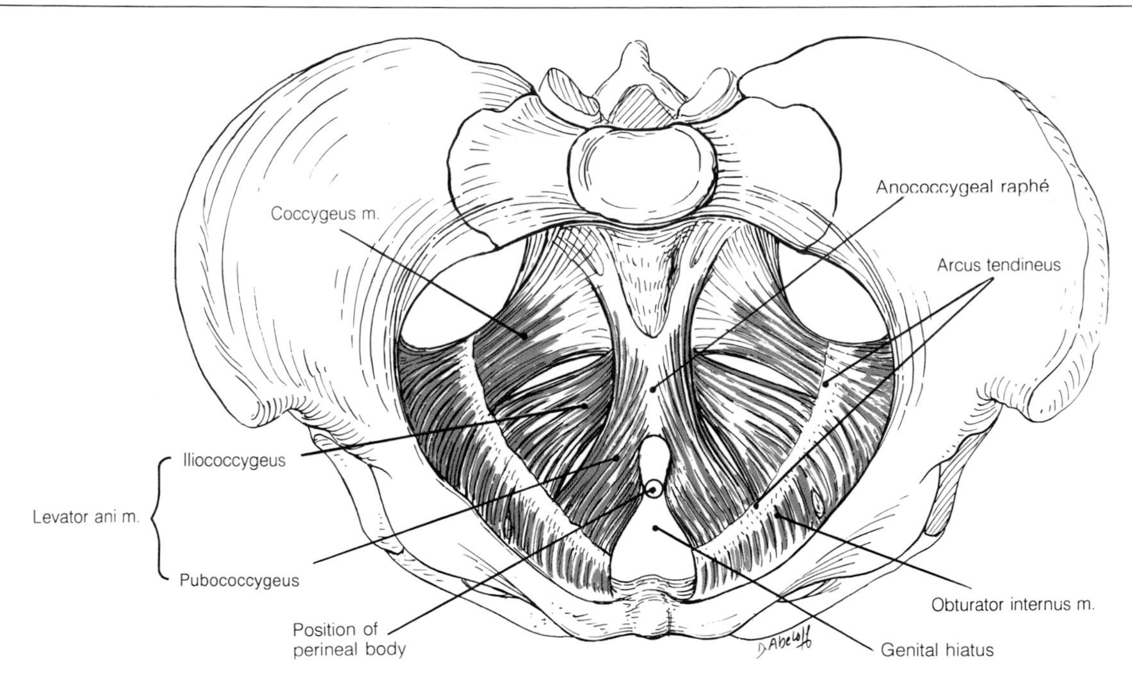

6-15 The pelvic diaphragm viewed from above.

again terminate at the perineal body. These fibers are known as the **pubovaginalis muscle**.

Succeeding fibers of pubococcygeus pass backward to the side of the anorectal junction and unite with those of the opposite side at the **anococcygeal raphé** which extends from the anorectal junction to the coccyx. Deeper fibers of this group form a sling around the junction and become intimately mixed with those of the deep portion of sphincter ani externus. These fibers, called the **puborectalis muscle**, play an important part in maintaining rectal continence and must relax during defecation. The effect of this "rectal sling" can be gauged from Fig. 6-16.

Iliococcygeus

The remaining fibers of levator ani which arise from the surface of obturator internus and the ischial spine form a sheet of muscle which passes deep to pubococcygeus and is inserted into the anococcygeal raphé and the tip of the coccyx. The reader should, however, be aware that in the elderly subject it is often difficult to dissect the parts described from what is little more than a thinned out fibromuscular sheet.

Both parts of levator ani are supplied by the **fourth sacral nerve** and by branches of the **pudendal nerve** which reach its inferior surface.

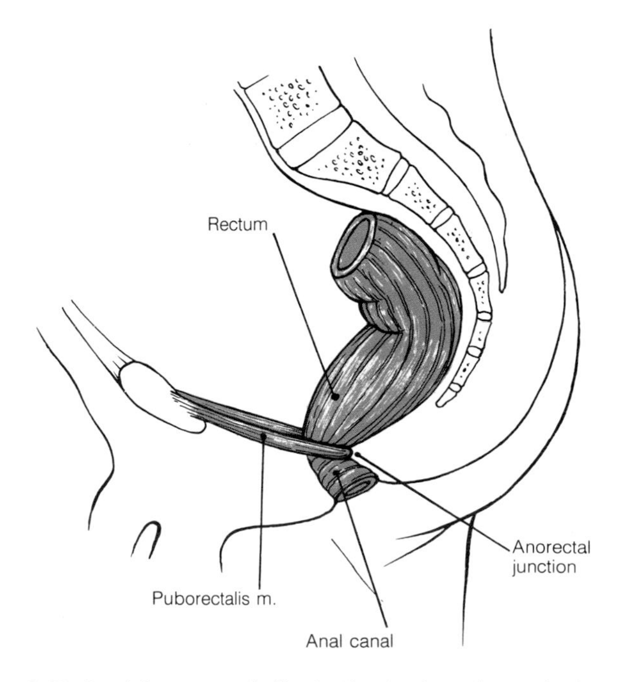

6-16 Semidiagrammatic illustration to show the occlusion of the lower end of the rectum that follows contraction of puborectalis.

The Central Tendon of the Perineum

In describing the levator ani muscle, the **central tendon of the perineum** or, more commonly the perineal body, has been mentioned. This is a midline knot of fibrous tissue lying posterior to the prostate or vagina. To it are fused portions of the levator ani and the external sphincter of the anal canal and also several perineal muscles. It is this structure that may be torn through during delivery of the fetal head. If repair is not made, subsequent contraction of the surrounding fibers of levator ani increases the defect in the pelvic floor and allows the bladder, vagina, and occasionally the uterus to prolapse into the perineum below.

The Pelvic Fascia

The pelvic floor is covered with loose areolar tissue which allows its mobility. Similar areolar tissue surrounds the distensible organs (e. g., bladder and rectum) but the prostate, which does not distend, has a distinct membranous fascia around it. The areolar tissue surrounding the distensible organs, unfortunately, provides a pathway for the spread of infection in the pelvis (pelvic cellulitus). It is in this loose fascia that the pelvic venous plexuses lie. Anteriorly, the gap between the two medial borders of levator ani, the **genital hiatus**, is completed by thickenings in the superior fascia of the pelvic diaphragm known as the **puboprostatic** (male) or **pubovesical** (female) ligaments. Further thickenings in the fascia form various ligaments which will be described with the viscera. Where the walls of the cavity are formed by the obturator internus muscle the lining fascia is thick, but over piriformis it is less well differentiated.

The Pelvic Organs

Structures Entering the Pelvis

The greater peritoneal sac of the abdominal cavity dips down into the lesser pelvis and provides a partial covering of peritoneum for some of the organs contained in the pelvic cavity. This extension of the peritoneal cavity may contain mobile abdominal structures such as the small intestine, the appendix, the pelvic colon, or the greater omentum, but these are separated from the cavity by their own covering of visceral peritoneum.

In discussing the abdomen and its contents, a number of structures have been described up to the point at which they enter the pelvic cavity at the superior pelvic aperture. A brief review of these now follows.

The Intestinal Tract

The retroperitoneal **descending colon** becomes continuous with the **sigmoid colon** at the brim of the pelvis. The sigmoid colon is surrounded by peritoneum and has a sigmoid mesocolon which is attached to the lateral wall of the cavity. This allows considerable mobility of the sigmoid loop. The sigmoid colon becomes directly anchored to the posterior wall of the pelvis at its junction with the rectum.

The Urinary Tract

The **two ureters** cross the pelvic brim behind the peritoneum covering the bifurcation of the **common iliac arteries**.

The Reproductive Tract

In the female the reproductive organs lie within the pelvis. In the male the **ductus deferens** of the testis pierces the anterior abdominal wall through the inguinal canal, passes upward along the course of the external iliac artery, and then turns slightly downward across the pelvic brim immediately beneath the peritoneum.

The Nervous System

The **lumbosacral trunk** formed from the ventral rami of the fourth and fifth lumbar nerves crosses the pelvic brim anterior to the sacroiliac joint and at the medial margin of the psoas muscle. In the pelvis it joins the ventral rami of sacral nerves to form the **sacral plexus**.

The **obturator nerve**, which is derived from the ventral rami of the second, third, and fourth lumbar nerves, appears at the medial border of the psoas muscle at the brim of the pelvis. Here it lies behind the common iliac vessels. As these vessels bifurcate, the nerve follows the internal iliac vessels into the pelvis on their lateral side.

The **sympathetic trunks** enter the pelvis on either side of the intervertebral joint between the fifth lumbar vertebra and the sacrum. The **hypogastric plexus**, which contains contributions from both the sympathetic and parasympathetic systems, also spans the pelvic brim over the promontory of the sacrum.

The Vascular System

Anterior to the fourth lumbar vertebra, the abdominal aorta divides into the **left and right common iliac arteries**. Each of these divides into an **external and internal iliac artery**. The external iliac artery passes around the brim of the pelvis at the medial border of the psoas muscle which guides it beneath the inguinal ligament into the lower limb. The internal iliac artery enters the pelvic cavity posteriorly in front of the sacroiliac joints. The **internal iliac veins** run proximally alongside the arteries.

In both male and female, the **left and right gonadal arteries** arise from the abdominal aorta. In the female these enter the pelvic cavity to reach the ovaries by crossing the common iliac vessels on their pelvic surface. Gonadal veins lie alongside the arteries.

The **superior rectal artery**, the terminal branch of the **inferior mesenteric artery**, enters the pelvis in the root of the sigmoid mesocolon as it crosses the left common iliac vessels.

The Embryological Structures

On the deep aspect of the anterior abdominal wall, a midline and two medial and lateral folds of peritoneum are found. The midline fold overlies the remnants of the **urachus** which formed a communication between the developing bladder and the extraembryonic allantois. This can be followed as the **median umbilical ligament** into the pelvis anteriorly where it joins the apex of the bladder. Two **medial umbilical folds** cover the remnants of the **umbilical arteries** and these can be followed out of the pelvis from the **superior vesical branches** of the **internal iliac arteries** and over the anterior abdominal wall to the umbilicus. The two **lateral folds** overlie the inferior epigastric arteries and are not related to the pelvic cavity.

The organs lying within the pelvic cavity are the **rectum**, the **bladder**, and the **distal parts of the ureters** of both sexes; the **ovaries**, the **uterine tubes**, the **uterus and vagina** of the female; and the **ductus deferens**, the **seminal vesicles**, and the **prostate** of the male.

The Rectum

The rectum occupies a similar pelvic position in both sexes but its anterior relations differ considerably. It is a continuation of the mobile sigmoid colon but is itself fixed and has no mesentery. It begins at the level of the middle piece of the sacrum and despite its name (rectus = straight), follows a curved course over the lower sacrum and coccyx onto the pelvic floor where it is supported by the levator ani muscle. The curved course is itself sinuous in that the middle portion of the rectum or **ampulla** shows a bend to the left. At the border of the levator ani muscle (formed here by puborectalis) it pierces the pelvic floor to become the **anal canal**. This point is referred to as the **anorectal junction**. In the undistended state, the mucous membrane of the rectum is thrown into longitudinal folds. There are also usually three permanent **transverse folds**, containing submucosa and muscularis, which become more marked in distension. The lateral curvatures of the rectum and the transverse folds are illustrated diagrammatically in Fig. 6-17.

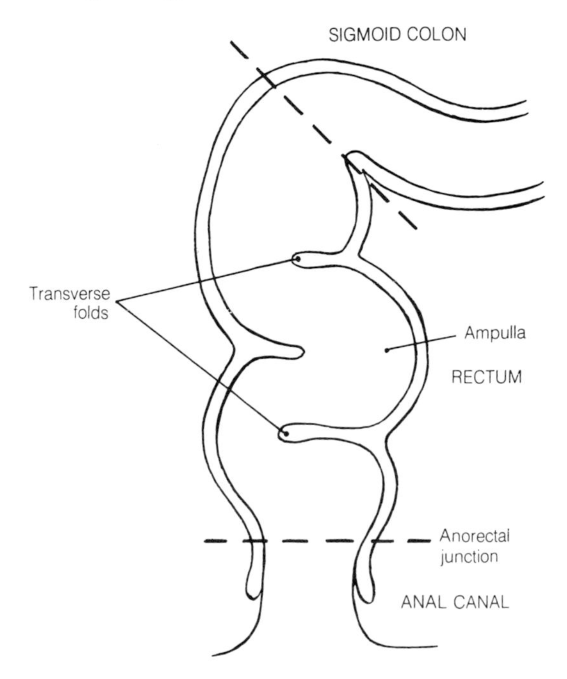

6-17 Diagrammatic coronal section through the rectum.

The upper one-third of the rectum is covered by peritoneum anteriorly and laterally. The middle third is only covered anteriorly and the lower third has no peritoneal relationship. Beneath the peritoneum the three taeniae coli typical of the large intestine fuse again into a continuous longitudinal muscle coat.

In Fig. 6-18 it can be seen that posteriorly, the rectum is related successively to the sacrum, the coccyx, and the pelvic floor. The sympathetic trunks, lateral sacral vessels and lymph nodes also lie posteriorly. Between the upper part of the rectum and the lateral pelvic walls lie mobile small bowel or pelvic colon in the **pararectal fossa** of the pelvic parietal peritoneum. Anteriorly, the relations differ between the sexes. In the male, the **rectovesical pouch** containing coils of small intestine separates the rectum from the bladder. Below this the rectum is related to the bladder, the seminal vesicles, and the prostate without the intervention of peritoneum. In the female, the **rectouterine pouch** separates the upper part of the rectum from the uterus and the posterior fornix of the vagina. Below the pouch the rectum is directly related to the vagina.

The Blood Vessels of the Rectum

The rectum is primarily supplied by the **superior rectal branch** of the **inferior mesenteric artery**. This branch breaks up into two or three branches on the surface of the rectum before piercing its muscular wall. The superior rectal artery is reinforced by the **middle rectal arteries** and finally communicates with branches of the **inferior rectal arteries**.

A **rectal plexus of veins** is drained by the **superior rectal vein**, a tributary of the **inferior**

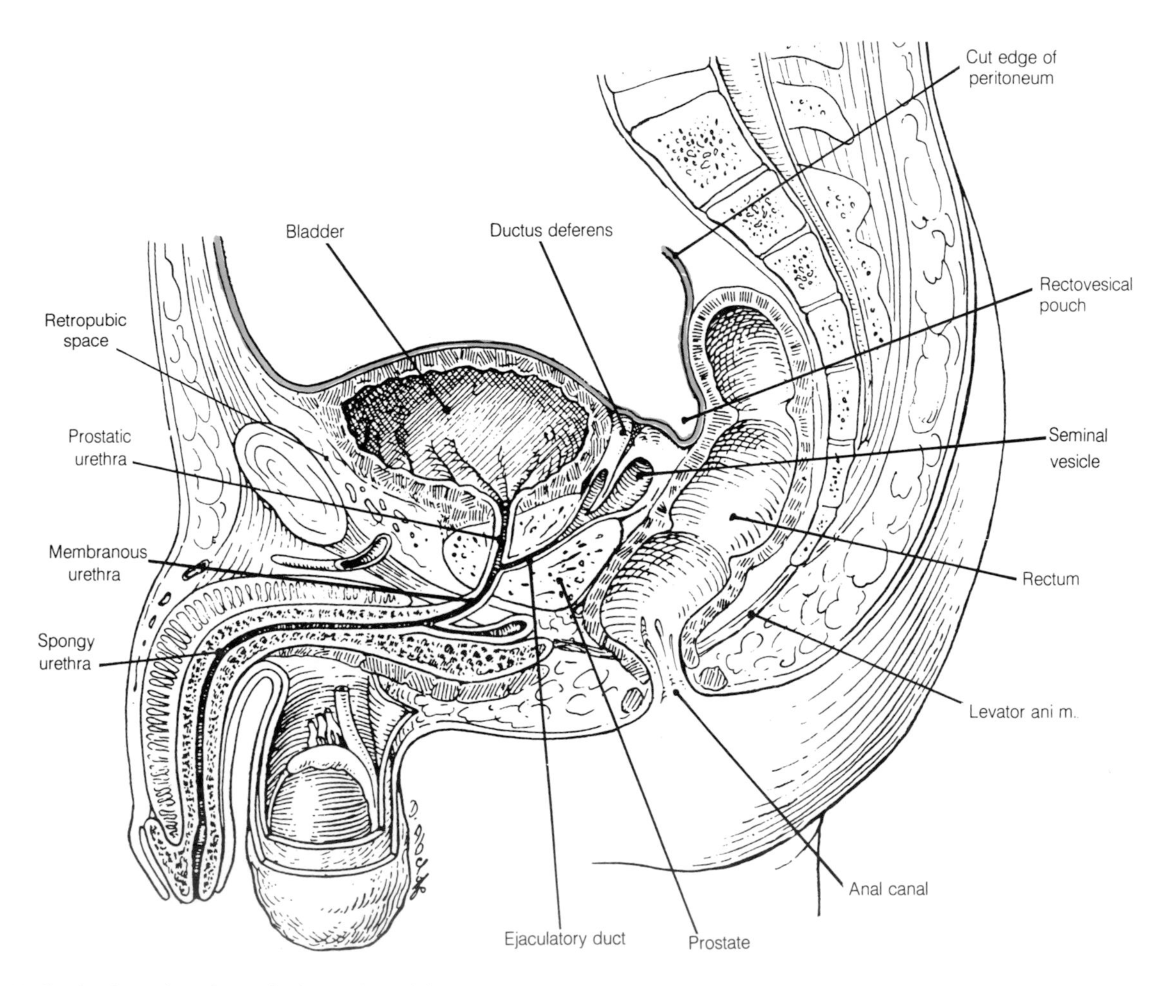

6-18 Sagittal section through the male pelvis.

mesenteric vein. The rectal plexus, which extends into the anal canal, is also drained by the **middle and inferior rectal veins**. It is the longitudinally running venous channels of the rectal plexus that become dilated to form hemorrhoids or piles. This condition is common and inconvenient. However, hemorrhoids may also be a manifestation of portal obstructions because blood which is prevented from draining through the superior rectal vein passes through the middle and inferior rectal veins to join the systemic venous system. This is an example of a **portocaval anastomosis**.

The blood supply and venous drainage of the rectum is discussed in more detail with the perineum (Chapter 7).

The Examination of the Rectum

The anal canal and lower rectum may be palpated by a finger inserted into the anal canal; this simple procedure should form part of every full physical examination. In addition to the anal and rectal walls, it allows the prostate to be felt in the male. In the female, rectal examination provides an alternative to vaginal examination, particularly during a delivery when the degree of dilatation of the uterine cervix may be estimated. Pathological conditions of the seminal vesicles and ovaries and other pelvic disease may also be detected in this manner. The rectum can be viewed directly with the proctoscope and this examination may be extended to the sigmoid colon with the sigmoidoscope.

The Ureters

The course of the ureters has been followed from the kidney over the posterior abdominal wall to the pelvic brim. Each crosses the brim beneath the peritoneum covering the bifurcation of the common iliac artery. In Fig. 6-19 the right ureter can be seen to continue downward into the lesser pelvis toward the ischial spine and cross, on their

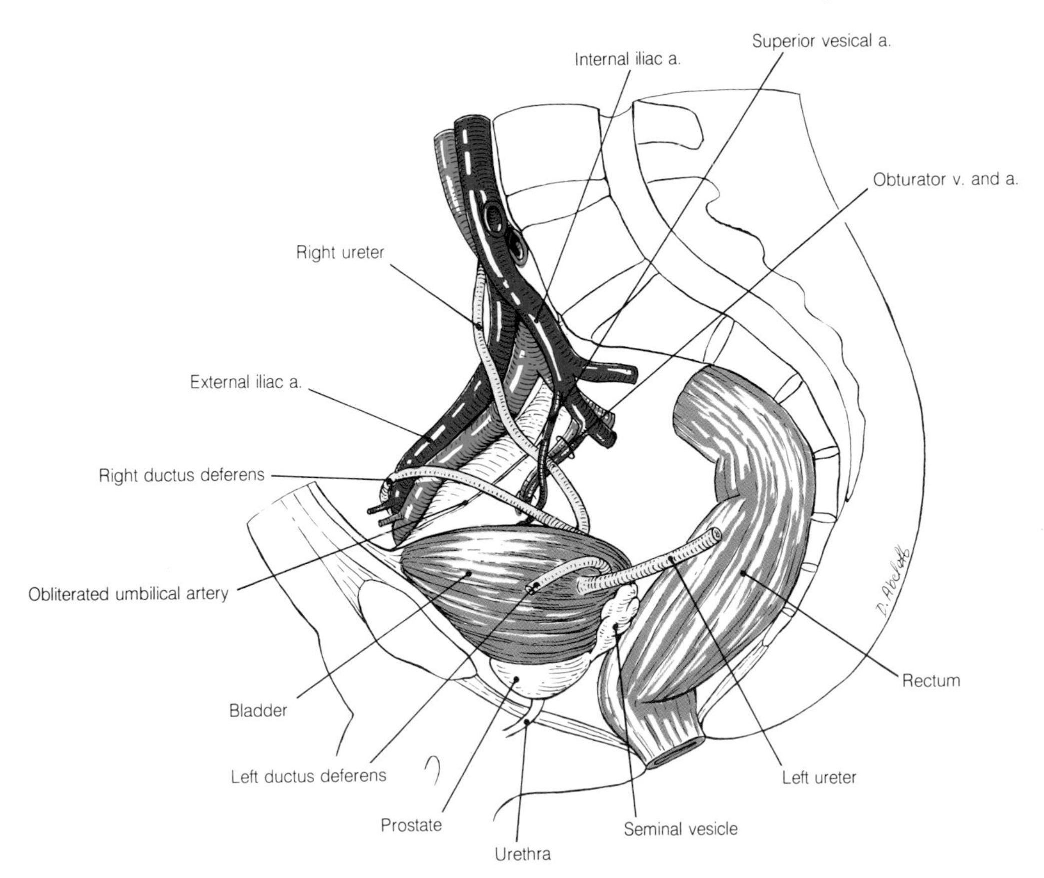

6-19 Section through the male pelvis to show the course of the ureter.

medial side, the obturator nerve and vessels and the obliterated umbilical artery. On reaching the pelvic floor, it runs forward to enter the **base of the bladder** near its superior angle. Its course through the muscle of the bladder wall is oblique and thus it opens into the cavity of the bladder at a lower level, i.e., at the upper border of the **trigone**. In its pelvic course the ureter is crossed superiorly by the **ductus deferens** in the male and the **uterine artery** in the female. This latter relationship is of extreme importance, for it is possible to damage the ureter when the uterine artery is ligated prior to removal of the uterus during a hysterectomy.

The innervation of the ureter is of some interest. Although supplied by both sympathetic and parasympathetic nerves, the nature of the motor supply to the thick muscular wall of the ureter is unclear. What is certain is that this undergoes active peristalsis and that spasm of the muscle, usually related to the presence of a stone, produces acute pain. This is referred to the skin innervated by the last two thoracic and first two lumbar spinal cord segments. Typically the pain starts in the loin and radiates toward the scrotum and penis or to the labium majus.

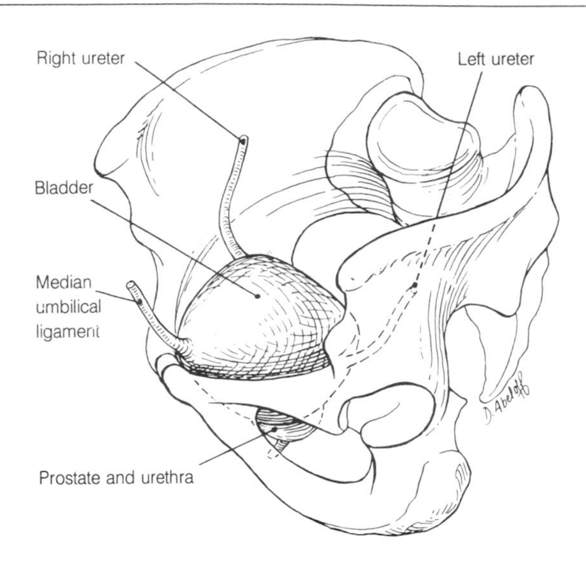

6-20 To show the shape of the bladder and its relationship to the bony pelvic walls.

The Urinary Bladder

The urinary bladder or **vesica*** **urinaria** is a highly distensible muscular organ which, when empty, lies in the pelvis and rests on the symphysis pubis and the floor of the pelvis. It is partially covered by peritoneum, its muscular walls are of smooth muscle and it is lined by transitional epithelium; a lining layer that can adapt to the large changes in its volume. As it is filled with urine from the ureters, it enlarges upward into the abdominal cavity stripping peritoneum off the anterior abdominal wall as it ascends. In cases of prostatic obstruction to the flow of urine from the bladder, it may extend to the umbilicus. The absence of peritoneum over its anterior aspect allows percutaneous surgical drainage of the bladder above the pubis. In the infant the pelvic cavity is very flat and shallow and the bladder normally occupies an intraabdominal position.

Look at Fig. 6-20 and see how the angle between the two pubic bones and the floor of the pelvis mold the moderately full bladder into a three-sided pyramid. There is a superior surface related to intraabdominal contents (usually small bowel) and in the female to the overhanging uterus, two inferolateral surfaces which become continuous with each other at the **retropubic space** (Fig. 6-18), and a base which faces backward. This surface is related to the seminal vesicles and the rectum in the male and to the vagina and uterus in the female.

To the apex of the pyramid is attached the median umbilical ligament, a remnant of the urachus. Occasionally this may remain patent a birth and urine may discharge at the umbilicus. At the two superior angles of the base, the ureters enter the bladder and, at the inferior angle, the urethra leaves it. It is here that the smooth muscle of the bladder wall becomes circularly arranged around the commencement of the urethra and forms the **sphincter vesicae**. This region is firmly adherent to the prostate in the male and in both sexes is known as the **neck of the bladder**. The neck of the bladder is relatively immobile, being fixed by puboprostatic and lateral vesical ligaments, all thickenings of the pelvic fascia.

* Vesica, *L.* = a bladder; hence vesical artery, pubovesical ligament, etc.

In the male the superior surface of the bladder is entirely covered by peritoneum. This is reflected on either side into the **paravesical fossae** and onto the upper part of the base where it forms the anterior wall of the rectovesical pouch.

The Interior of the Bladder

The bladder can be distended with water and examined directly, using a cystoscope. Under these circumstances the wall appears smooth and yellowish and at the base can be seen a triangular, more reddish region, the **trigone**. In Fig. 6-21 a less distended bladder is shown and the mucous membrane of the wall is thrown into irregular folds by the trabeculae of underlying muscle fiber bundles. The **internal urethral orifice** can be seen at the lower angle of the base, and toward its upper angles, the slit-like **ureteric orifices**. These are joined by a fold of mucous membrane called the **interureteric fold**. The three orifices lie at the angles of the smooth surface of the trigone. In addition to being more red than the rest of the bladder wall, the trigone is less mobile, does not wrinkle as the bladder contracts, and is the region most sensitive to painful stimuli (particularly with bladder stones).

The Vessels and Nerves of the Bladder

The bladder is supplied by the **superior and inferior vesical branches** of the **internal iliac arteries**. Veins which drain a **vesical and, in the male, a prostatic venous plexus** terminate in the **internal iliac veins**.

The bladder receives both motor and sensory innervation. The motor fibers are both parasympathetic and sympathetic. The parasympathetic fibers are motor to the smooth muscle of the bladder wall, which is called the **detrusor*** **muscle**, but inhibit the sphincter vesicae. The sympathetic fibers on the other hand are said to be inhibitory to the detrusor muscle and motor to the sphincter vesicae. The sensory fibers give rise to the conscious sensation of a full bladder and also pain resulting from disease. The fact that these pain fibers return to both sacral and lumbar segments of the cord makes it difficult to surgically eliminate the intractable pain of terminal disease in this region. More will be said of the innervation of the bladder when the autonomic system in the pelvis is discussed.

* Detrudere, *L.* = to push down

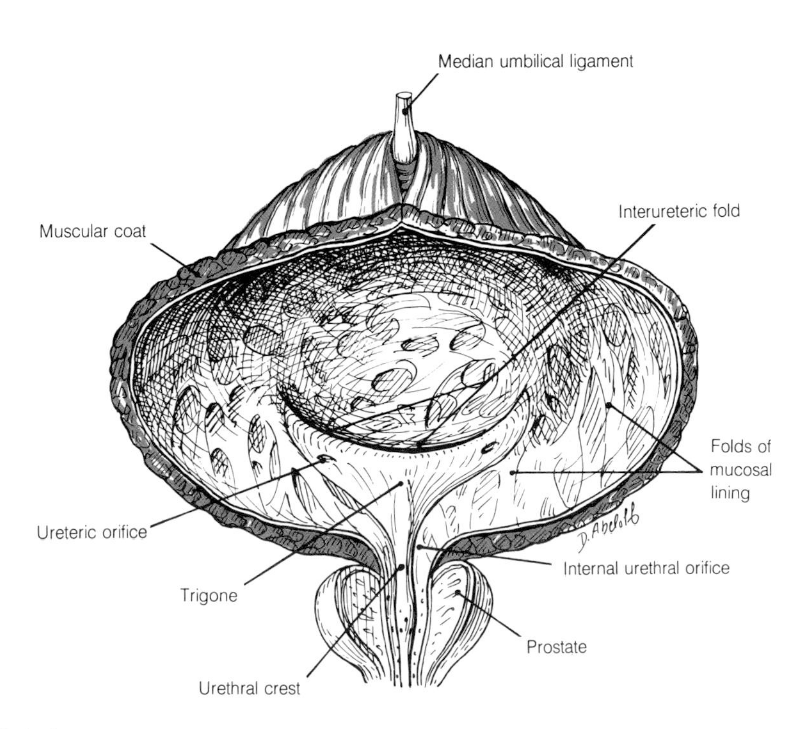

6-21 The interior of the bladder and urethra exposed anteriorly.

The Male Urethra

The urethra is much longer in the male than in the female (20 cm as opposed to 4 cm). Beginning at the neck of the bladder it passes through the prostate, the floor of the pelvis, the perineal membrane, and the penis, to terminate as the external urethral orifice at the tip of the glans penis. It is divided into **prostatic, membranous**, and **spongy parts** and the sinuous course that it follows when the penis is dependent can be seen in Fig. 6-18. In Fig. 6-22 the whole length of the urethra is shown exposed and this should be followed as each part is described.

The Prostatic Urethra

The unfilled prostatic urethra is somewhat semilunar in cross-section with the convexity facing forward. Its posterior wall shows an elevated central region, the **urethral crest**, flanked by the **prostatic sinuses**. The crest becomes expanded to form the **seminal colliculus**. At the summit of this elevation there lies a blindly ending diverticulum, the **prostatic utricle**, and on either side of this open the **ejaculatory ducts**. These details of the prostatic urethra are demonstrated more clearly in Fig. 6-23.

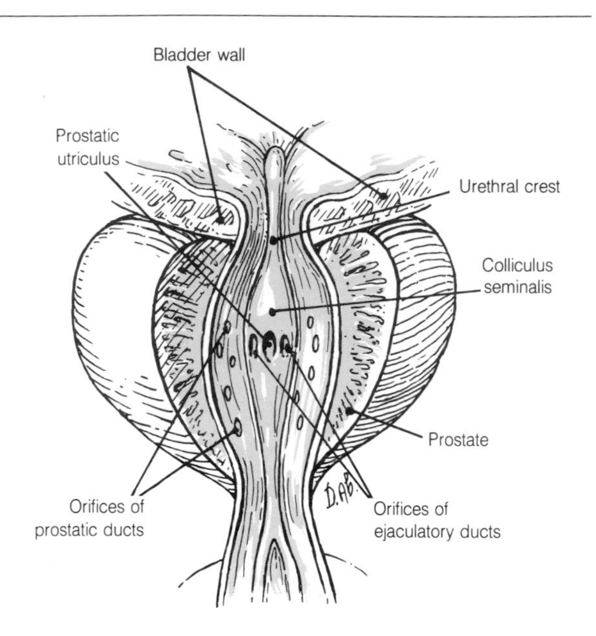

6-23 Showing details of the prostatic urethra.

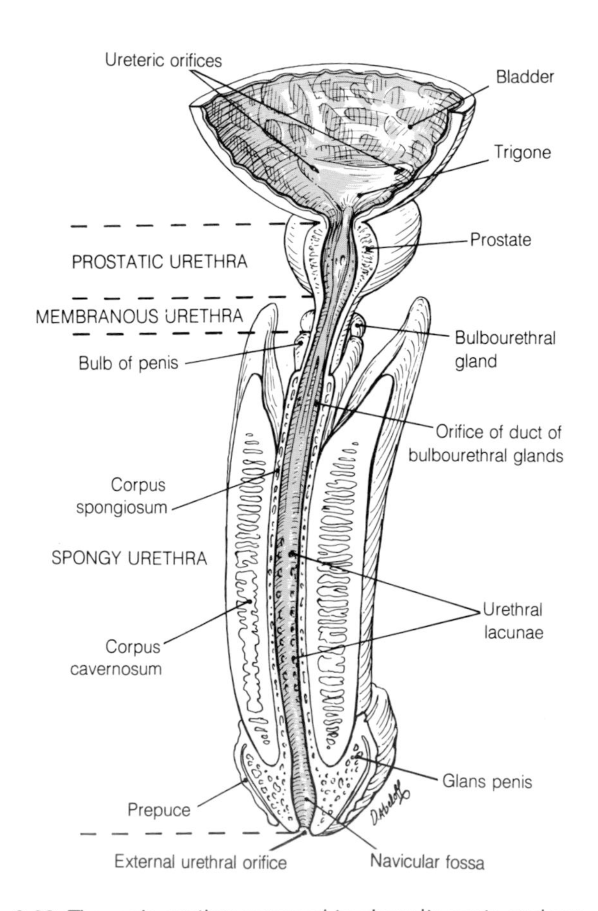

6-22 The male urethra exposed to show its parts and surrounding structures.

The Membranous Urethra

Lying between the apex of the prostate and the bulb of the penis, this, the shortest part of the urethra, is surrounded by the **sphincter urethrae** and the **perineal membrane**. The **bulbourethral glands** lie on either side of it. The rigid perineal membrane anchors this portion of the urethra to the bony pelvis and renders it liable to injury. This may be brought about by faulty instrumentation or trauma. In the latter case, often as a result of an automobile accident, a full bladder may be torn from the urethra or it may be ruptured as the left and right halves of the pelvis are separated by a fracture.

The Spongy Urethra

This, the longest portion of the urethra, traverses the bulb, the corpus spongiosum, and the glans of the penis. Just before its termination at the **external urethral opening**, the urethra is expanded to from the **navicular fossa**. The ducts of the bulbourethral glands open near its commencement and over the rest of its course it is joined by the ducts of many **urethral glands** which may open into small out-pouchings of the mucous membrane called **lacunae**. These glands are prone to gonococcal infections which may result in scarring and stricture formation. Such strictures may be dilated by instruments known as bougies. To achieve this successfully, the course of the urethra and its narrow fixed membranous portion must be borne in mind as the instrument is introduced.

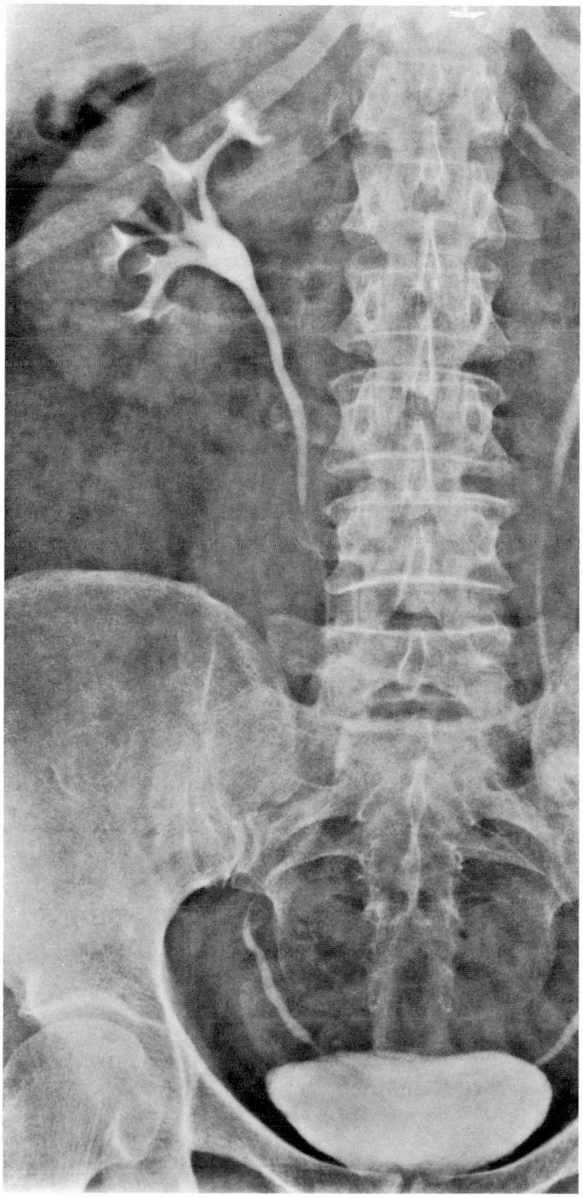

6-24 An intravenous urogram showing the pelvic course of the ureters and the bladder. (Reproduced by permission, from Wicke: *Atlas of Radiologic Anatomy,* 4th Ed, Urban & Schwarzenberg, Baltimore-Munich, 1987.)

The Female Urethra

The female urethra, which is only about 4.0 cm in length, runs from the neck of the bladder through the floor of the pelvis and perineal membrane to open into the vestibule just anterior to the opening of the vagina. Over its course it is firmly bound to the anterior wall of the vagina.

The Radiological Anatomy of the Lower Urinary Tract

The abdominal and pelvic course of the ureters is well demonstrated by intravenous pyelography, especially if a series of radiographs is compared. The normal ureter can be followed over the posterior abdominal wall at the tips of the lumbar transverse processes and then into the pelvis over the sacroiliac joint toward the ischial spine. At this level it turns medially to join the bladder. Peristaltic waves are frequently seen and persistent narrowing of the lumen is seen at its junction with the renal pelvis, as it crosses the pelvic brim and as it passes through the bladder wall. Look at the radiograph in Fig. 6-24 to check these details. Using the same technique, the filling of the bladder can also be observed. Note in Fig. 6-24 that the bladder lies at a rather higher level than might be imagined and that its neck just overlies the upper margin of the pubic symphysis. The bladder may also be visualized by filling it with an opaque medium introduced through a urethral catheter.

The Male Pelvic Reproductive Organs

In the male pelvic cavity are found the termination of each **ductus deferens**, the **seminal vesicles** and their **ducts**, the **ejaculator ducts** formed as the deferent ducts and the ducts of the seminal vesicles join, and finally the **prostate gland** in which the ejaculator ducts join the prostatic urethra.

The Ductus Deferens

The ductus deferens has been seen earlier to ascend from the epididymis in the spermatic cord and enter the abdominal cavity through the inguinal canal. On reaching the deep inguinal ring, it hooks around the inferior epigastric artery, crosses over the external iliac vessels, and enters the pelvic cavity. Passing posteriorly beneath the peritoneum lining the wall of the cavity it now crosses over the obturator nerve and vessels and the ureter to reach the base of the bladder. Its course in the pelvis can be seen in Fig. 6-19. Here the terminal portion becomes dilated at the **ampulla** and joins with the duct of the seminal vesicle to form the **ejaculator duct**. This duct enters a cleft between the neck of the bladder and the posterior surface of the prostate and opens into the prostatic urethra on the colliculus.

The Seminal Vesicles

These are two lobulated sacs each lying lateral to the ampulla of a ductus deferens. The tapering lower end of each vesicle joins the ductus deferens to form the ejaculatory duct. The seminal vesicles lie in front of the anterior wall of the rectum and, if enlarged, may be palpated through the rectum.

Look now at Fig. 6-25 in which the seminal vesicles are viewed from their posterior aspect. Their relationship to the bladder, the ureters, and the prostate gland can be seen.

The Prostate Gland

This glandular organ, supported by firm fibromuscular connective tissue, lies inferior to the male bladder and surrounds the urethra. It measures

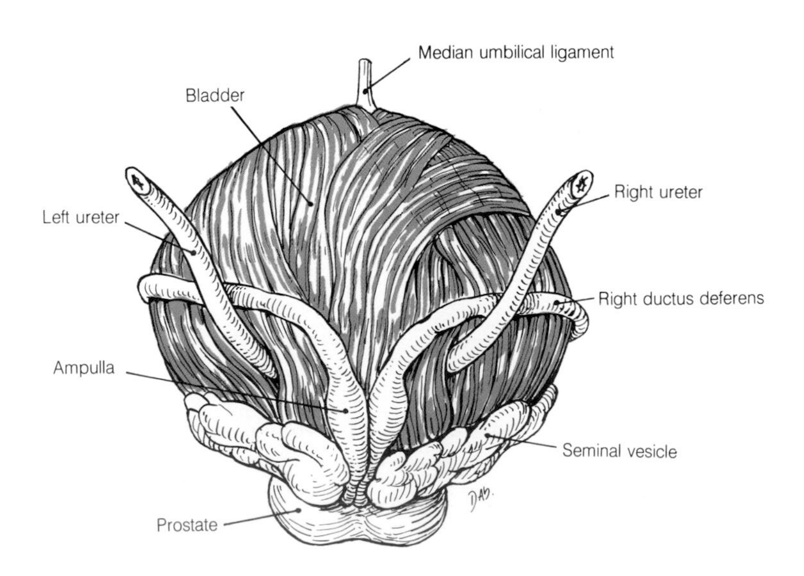

6-25 The base of the bladder showing the ureters, ductus deferentes, seminal vesicles, and prostate.

approximately 4 × 3 × 2 cm. It is conical in shape and has a **base** applied to the bladder and a blunt **apex** directed inferiorly and toward the urogenital diaphragm. The connective tissue stroma of the gland is in direct continuity with that of the bladder wall, thus intimately fusing the two structures. Although not apparent on the surface, except for an indistinct posterior groove, the gland consists of a left and right **lateral lobe** united in front of the urethra by a fibromuscular strip called the **isthmus**. A **middle lobe** of variable size lies above the lateral lobes and the point of entry of the ejaculatory ducts. Enlargement of this lobe rapidly leads to urethral obstruction at the bladder neck. The sphincter urethrae which surrounds the urethra immediately beyond the prostate also embraces the apex of the prostate. Awareness of this fact is important as the preservation of these fibers in prostatic surgery leads to optimum urinary continence.

The prostate has its own connective tissue **capsule** and, in addition, it is surrounded by a further thick sheath derived from the pelvic fascia. The capsule and sheath are separated by the **prostatic plexus of veins**. It is supplied by the inferior vesical artery and venous blood drains to the prostatic plexus and, thence, to the vesical plexus and internal iliac veins.

The prostate is influenced by circulating male sex hormones and undergoes a rapid increase in size at puberty due to the development of secretory follicles from a preexisting duct system. A further increase in size or hypertrophy normally occurs from the fourth decade onwards. This is due to the formation of nodules of hyperplastic glandular and connective tissue. These may eventually produce urinary obstruction especially if they project into the bladder around the internal meatus or compress the prostatic urethra.

Semen

To the spermatozoa delivered to the urethra through the deferent and ejaculatory ducts is added the **semen**. This is secreted by the seminal vesicles (about 60%), the prostate (about 20%), the bulbourethral glands, and the urethral glands. The volume of semen produced at a single ejaculation is approximately 3.0 ml with a sperm count of a hundred million per milliliter.

The Female Pelvic Reproductive Organs

The female reproductive organs in the pelvic cavity include the ovaries, the uterine tubes, the uterus, and the greater part of the vagina. Each have, to a lesser or grater extent, a peritoneal covering and, therefore, a surface related to the peritoneal cavity. Before describing the peritoneal coverings in detail, it is useful to understand that the peritoneum of the pelvic cavity is thrown up in a horizontal fold over the midline uterus and the uterine tubes as these extend laterally from the upper part of the uterus toward the wall of the cavity. The anterior and posterior layers of this fold are separated by the uterus but on either side where they "hang" from the uterine tubes, they come together to form the broad ligaments of the uterus. In this way the female pelvic cavity is divided into an anteroinferior and posterosuperior compartment, the former containing the bladder, the latter, the rectum.

The Ovaries

The ovaries are the female counterpart of the testes and during reproductive life are the source of a single ovum shed at monthly intervals. As a result of this process, the surface of the adult ovary appears puckered and scarred. Look now at Fig. 6-26 in which the uterus and the broad ligament are viewed from **behind**. Each ovary lies on the posterior aspect of the **broad ligament** near the lateral wall of the pelvis. It is almond-shaped being about 3 cm long and 1.5 cm thick and its long axis is aligned almost vertically. It is attached to the broad ligament by a fold of peritoneum called the **mesovarium**. Its upper pole is closely related to the fimbriae surrounding the abdominal opening of the **uterine tube** and from it extends the **suspensory ligament of the ovary**. This is really an extension of the broad ligament to the lateral wall of the pelvis which carries the ovarian vessels in its upper free border. The lower pole of the ovary is attached to the uterus near its junction with the uterine tube by a fibromuscular cord called the **ligament of the ovary**. This lies within the layers of the broad ligament and its continuous with the **round ligament of the uterus**. The exposed surface of the ovary is covered by a layer of cuboidal cells which becomes continuous with the sur-

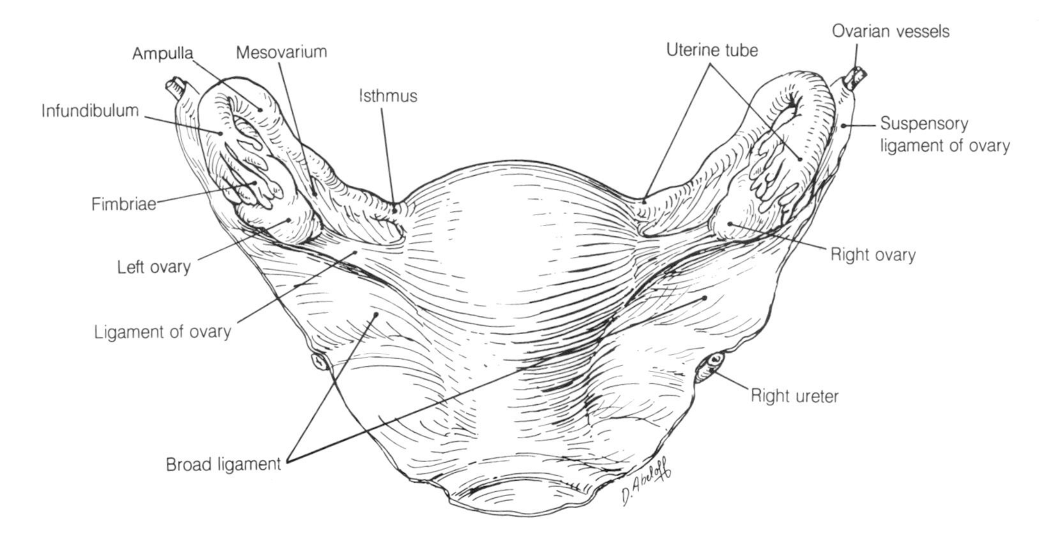

6-26 The ovaries, uterine tubes and uterus seen from their posterior aspect.

rounding peritoneum. Thus, when the **Graafian follicle** ruptures through this, the ovum is liberated and enters the peritoneal cavity. It is probably immediately guided to the nearby infundibulum of the uterine tube, but there is good evidence that if this is occluded or has been removed, the ovum may reach the contralateral tube.

The ovaries are supplied by the **ovarian arteries** which descend into the pelvis from the **abdominal aorta**. Veins leave the ovary as a network around the artery and this is called the **pampiniform plexus**. From such a plexus the two **ovarian veins** are formed of which the left drains into the **left renal vein** and the right into the **inferior vena cava**.

The Uterine Tubes

The uterine tubes extend laterally from the junction of the **fundus** and **body of the uterus** and lie in the upper border of the broad ligament. Their position and relationship to the ovary can also be seen in Fig. 6-26. The lumen of the tube communicates with the cavity of the uterus by its most medial narrowed part called the **isthmus**. Running laterally, it becomes more expanded as the **ampulla**. This curls posteriorly over the back of the broad ligament to end close to the superior pole of the ovary as the trumpet-like **infundibulum**. The border of the infundibulum bears finger-like processes termed **fimbriae** of which one, the **ovarian fimbria**, is attached to the ovary.

When the ovum is shed into the peritoneal cavity, it is engulfed by the infundibulum and passed to the uterus. Fertilization takes place in the tube. Occasionally the fertilized ovum is arrested in the tube and a tubal pregnancy results. Rupture of the tube with severe hemorrhage follows within about six weeks.

The uterine tubes are supplied by branches of both the ovarian and uterine arteries.

The Uterus

The uterus is a thick-walled, muscular structure having a narrow cavity which is continuous with the peritoneal cavity through the uterine tubes and with the perineum through the vagina. During reproductive life its lining mucous membrane, called the endometrium, is prepared every month for the reception of a fertilized ovum. If this event does not occur the endometrium becomes necrotic and is shed as the menstrual flow. If fertiliziation occurs, the ovum becomes embedded in the wall of the uterus through which it acquires its nutrition for growth to a full-term fetus. At this time contraction of the muscular wall delivers the fetus from the uterine cavity through the dilated cervix and vagina and into the perineum.

Despite the fact that the uterus may enlarge to enclose a full-term fetus, in the virgin state it measures only 3 × 2 × 1 inches. In shape it is somewhat like an inverted flattened pear. It is joined on either side by the uterine tubes which open into its slit-like **cavity**.

The uterus is divided into three parts, a muscular **fundus** above the level of the uterine tubes, a tapering **body** which contains the cavity, and a **cervix** or neck, the narrowest part, which projects into the vagina. The cavity of the uterus is continuous with the **canal of the cervix** through the **isthmus** and the canal opens into the vagina through the **external os**. These parts of the uterus should be identified in the diagram shown in Fig. 6-27.

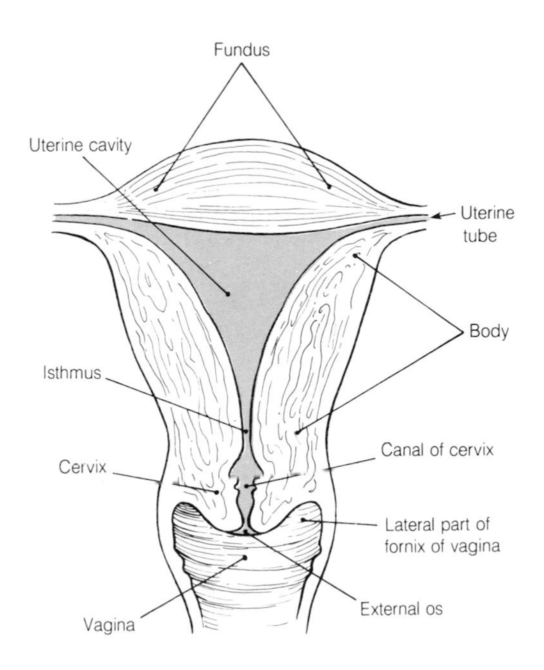

6-27 Semidiagrammatic coronal section of the uterus.

In Fig. 6-27 it can also be seen that the upper anterior surface, the superior surface, and the posterior surface of the uterus are covered by peritoneum. That covering the anterior and posterior surfaces continues onto the sides of the uterus and from these is reflected laterally to the pelvic wall as the two layers of the broad ligament.

The Position of the Uterus

Although the lower part of the body of the uterus is firmly adherent to the base of the bladder, above this point it is mobile and free to overhang the superior surface of the empty bladder. This is the normal **anteverted** position. The uterus is also slightly bent forward along its own axis, and this is known as **anteflexion**. Occasionally, the uterus is found to slope backward. It is then said to be **retroverted**. These features of the uterus should be confirmed in Fig. 6-28, and in the same illustration the following relationships should also be noted:

1. The uterus lies above the bladder and is separated from it by the **vesicouterine pouch**.

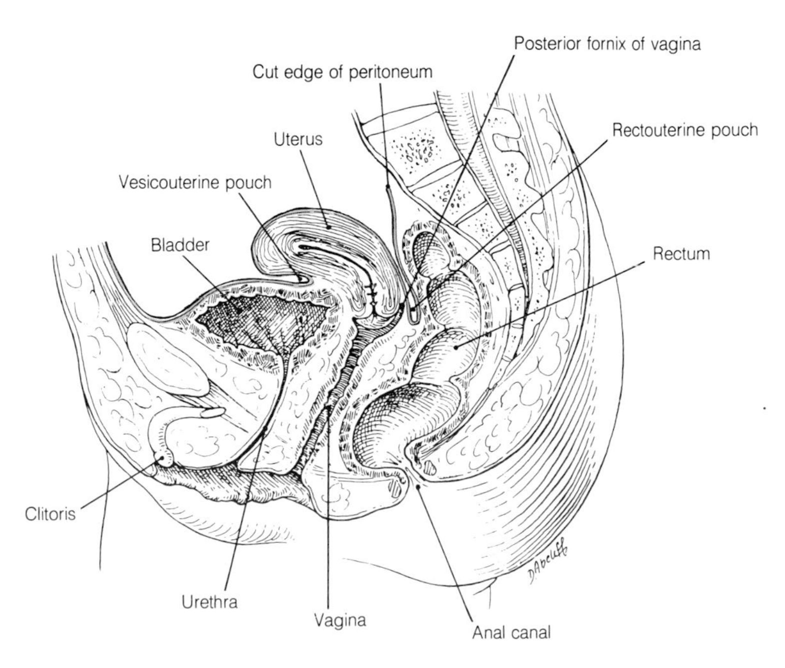

6-28 Sagittal section through the female pelvis.

2. The middle third of the rectum lies behind the uterus and is separated from it by the **recto-uterine pouch**.
3. The cavity of the anteverted uterus lies at approximately 90° to the cavity of the vagina.

The uterus is supported in the pelvis by the muscular pelvic diaphragm and retained in position by a number of ligaments. Some of these ligaments are little more than folds of peritoneum; others have a more certain functional value.

The Ligaments of the Uterus

The Broad Ligaments

The uterine tubes and ovarian vessels extend from the uterus towards the pelvic brim and have peritoneum hanging down both in front and behind them like a sheet on a clothesline. In this way are formed the broad ligaments of the uterus and the suspensory ligaments of the ovaries. In practice, the broad ligaments and the uterus divide the female pelvic peritoneal cavity into an anteroinferior and a posterosuperior compartment. Fig. 6-29, which is a sagittal section through the ligament made between the uterus and the ovary, shows this arrangement and the structures lying within the ligament namely the uterine tube, the ovarian vessels, the ligament of the ovary and the round ligament. Note also the uterine vessels and the ureter at the base of the broad ligament.

The Round Ligaments

The round ligaments of the uterus extend from the body of the uterus, just below the uterine tube, to the pelvic brim and thence through the inguinal canal to end in the fibrofatty tissue of the **labium majus**. They are probably remnants of gubernacula comparable to those that seem to guide the testes to the scrotal sac. In fact, on occasion the round ligament may be accompanied through the inguinal canal by a patent **processus vaginalis** and thus prepare the ground for the appearance of an indirect inguinal hernia. The round ligament is composed of fibrous tissue and smooth muscle fibers.

The Transverse Cervical Ligaments

These are sometimes known as the cardinal ligaments. They are fibromuscular thickenings of the pelvic fascia surrounding the uterine vessels in the base of the broad ligament. They are attached to the side of the cervix of the uterus and the lateral fornix of the vagina. These ligaments probably play a part in stabilizing the uterus and vagina.

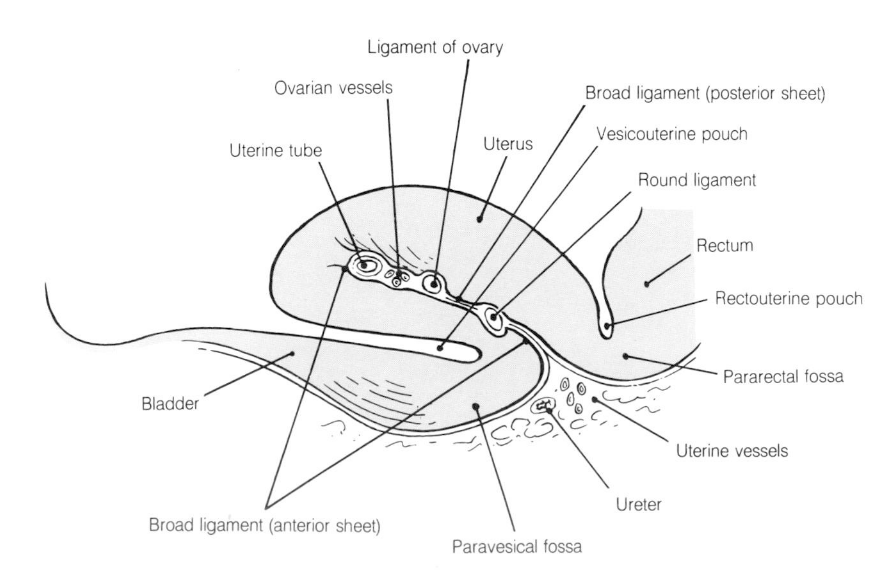

6-29 Semidiagramatic sagittal section through the broad ligament.

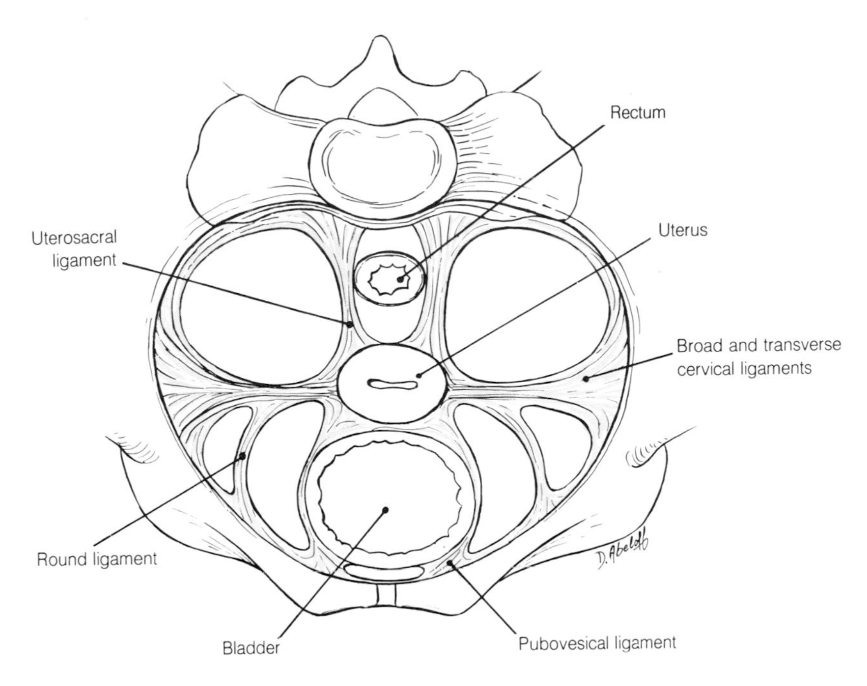

6-30 Semidiagrammatic illustration of the ligaments of the uterus.

The Anterior and Posterior Ligaments

These are little more than alternative names for the uterovesical and rectouterine folds of peritoneum forming the margins at the vesicouterine pouch and the rectouterine pouch.

The Uterosacral Ligaments

As they run backward from uterus to rectum, the margins of the rectouterine pouch are raised as peritoneal folds. These folds contain fibromuscular thickenings which pass on either side of the rectum to become attached to the sacrum. These are the uterosacral ligaments.

The ligaments described act to a lesser or greater extent to stabilize the uterus in the pelvis. In addition, the uterus obtains support from its attachment to the vagina, bladder, and rectum. The importance of the integrity of the muscular pelvic diaphragm and the perineal body in this respect has already been mentioned. The arrangement of the ligaments between the uterus and the parietal pelvic fascia is illustrated diagrammatically in Fig. 6-30.

The Vessels of the Uterus

The uterus is supplied by the **uterine artery**, a branch of the **internal iliac artery**. Leaving its parent vessel on the wall of the pelvic cavity, the uterine artery crosses the floor in the base of the broad ligament. It is here that it passes over the ureter, which may be at risk when the artery is ligated. On reaching the cervix the artery runs up the lateral wall of the uterus to end by anastomosing with the **ovarian artery**. Along its course it sends branches to the vagina, the cervix, and the walls of the uterus.

Blood draining from the uterus forms a venous plexus in the broad ligament which empties into the **internal iliac veins**.

The Vagina

The vagina is the terminal portion of the female genital tract and its cavity is continous proximally with that of the uterus at the **external os of the cervix**. Its distal opening into the perineum is known as the **vaginal orifice**. The anterior and posterior walls of the vagina are normally apposed and thus obliterate the cavity but this accepts the erect penis during coitus and becomes widely distended during parturition.

Look again at Fig. 6-28 and see that the long axis of the vagina lies at approximately 90° to that of the uterus. As a result, the cervix of the uterus projects into its anterior wall which is, therefore, shorter than the posterior wall. At its point of entry, the cervix is surrounded by a recess of invaginated vaginal wall called the **vaginal fornix**. Again, as a result of the alignment of the two organs, the posterior part of the fornix is the deepest.

Anteriorly, the vagina is related to the cervix and is separated from the bladder by loose connective tissue. Below the bladder the vagina is firmly adherent to the urethra which lies embedded in its anterior wall. Posteriorly, the posterior fornix is related to the rectouterine pouch and below this to the rectum.

The fact that the posterior vaginal fornix is related to the peritoneum of the rectouterine pouch, means that straight instruments inserted into the vagina for the purpose of procuring an abortion, penetrate the fornix rather than enter the cervix and may give rise to both hemorrhage and peritonitis. It is at this site also that pelvic abscesses forming in the rectouterine pouch may rupture or be drained surgically.

Blood reaches the upper part of the vagina from the uterine arteries. Vaginal branches of the internal iliac, inferior vesical and middle rectal arteries and branches of the internal pudendal artery supply the remainder. Venous blood reaches the uterine and vesical plexuses. The blood supply of the ovaries, the uterus and vagina are summarized in Fig. 6-31.

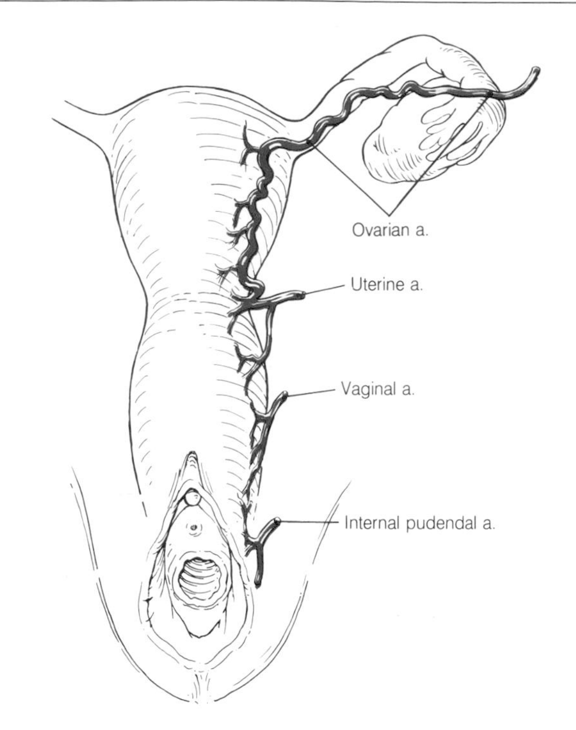

6-31 Showing the arteries supplying the uterus and vagina.

The Examination of the Female Reproductive Tract

The vagina and cervix may be examined by digital palpation and through the wall of the vagina the normal uterus may be felt, especially if this is steadied by a second hand exerting pressure on the lower anterior abdominal wall (a bimanual examination). Tumours and regions of tenderness can also be detected in this manner. Direct observation of the vaginal wall and cervix can also be made when the vagina is distended by a speculum. The uterine cavity and uterine tubes can be demonstrated radiographically by introducing a radiopaque contrast medium into the cervix. Patency of a uterine tube is shown as the medium spills from the infundibulum into the peritoneal cavity. This procedure, known as hysterosalpingography,* has been used in the radiograph seen in Fig. 6-32. Note in this the cervical canal, the shape of the uterine cavity, the tortuous course of the uterine tubes, and their fimbriated extremities. The slightly less dense opacities above the uterus and below the end of the left tube are formed by contrast medium in the peritoneal cavity.

* Hystera, *Gk.* = uterus; Salpinx, *Gk.* = a tube

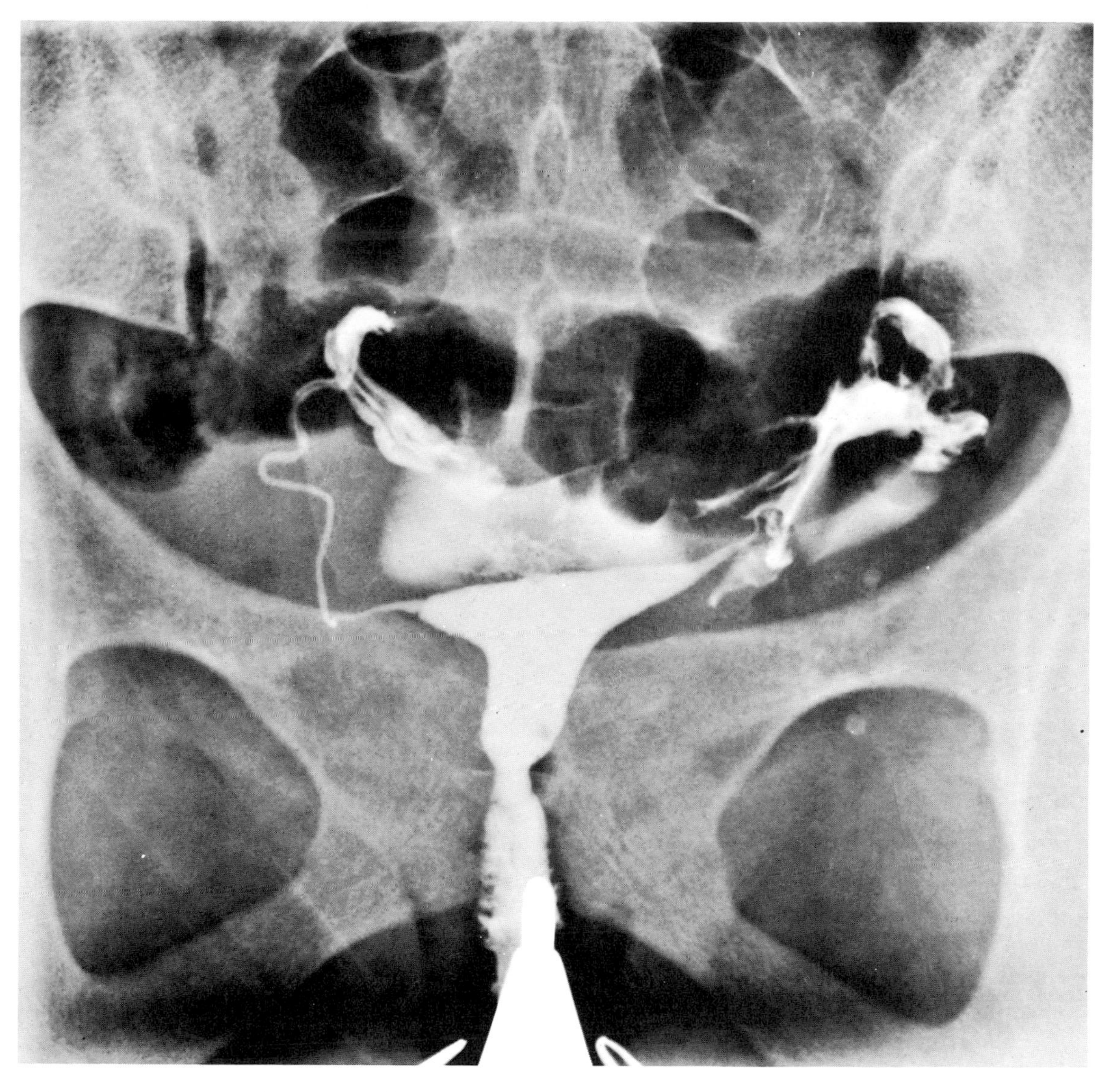

6-32 Hysterosalpingogram. (Reproduced by permission, from Wicke: *Atlas of Radiologic Anatomy,* 4th Ed, Urban & Schwarzenberg, Baltimore-Munich, 1987.)

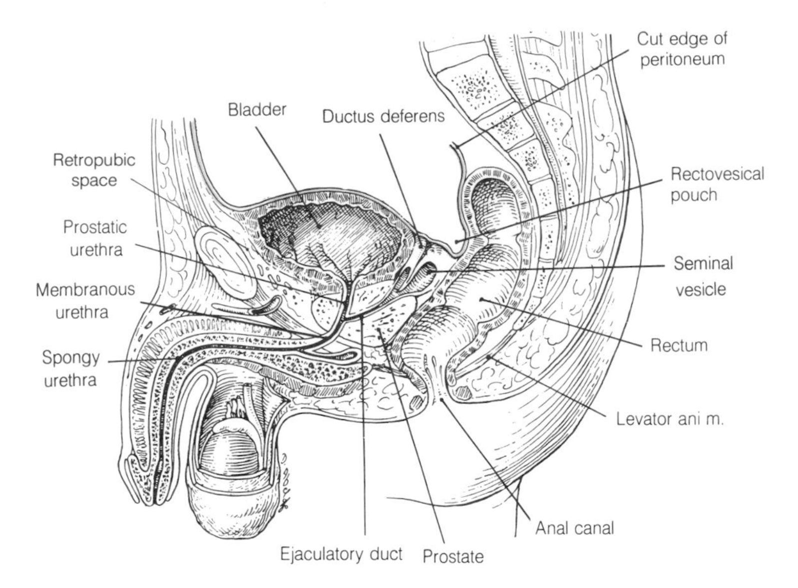

6-33 A sagittal section through the male pelvis.

Sagittal and Horizontal Sections of the Pelvis

A study of the median sagittal sections through the male and female pelves illustrated in Figs. 6-33 and 6-34, provides a useful review of the peritoneal reflections in each sex and the relationships of the major organs to each other and to the surrounding bony lesser pelvis.

First, follow the peritoneum down over the anterior abdominal wall and across the upper surface of the bladder. From here in the male it descends a short distance over the fundus of the bladder and then up onto the anterior wall of the middle third of the rectum. In this way the rectovesical pouch is formed. In the female the peritoneum is reflected forward off the superior surface of the bladder and onto the anterior wall of the uterus. This is the vesicouterine pouch. Note that the shallow anterior vaginal fornix receives no peritoneal covering. The peritoneum can now be followed over the fundus and down the posterior wall of the body of the uterus onto the posterior fornix of the vagina. As the peritoneum is reflected from the vagina onto the rectum the rectouterine pouch is formed. In the same illustration, note some further features:

1. Most of the cavity of the lesser pelvis lies below the curve of the sacrum and not below the abdominal cavity proper.
2. The empty or partially filled bladder rests on the symphysis pubis.
3. The prostate lies immediatly in front of the lower third of the rectum where it may be palpated.
4. The female urethra lies between the vagina and the symphysis pubis.
5. The lower third of the rectum lies behind the vagina.

In Fig. 6-35 a CT scan of the male pelvis is shown as an example of a horizontal section of the pelvis. The level of the section is indicated by the inset sketch and it can be seen from this that it passes through the head of the femur and the tip of the sacrum. Identify these bony landmarks on the scan and then note the acetabulum and the underlying obturator internus muscles forming the lateral walls of the pelvic cavity. Between the lateral walls lie the bladder, the seminal vesicles, and the rectum. The anterior part of the bladder must be lying in the abdominal cavity as profiles of the two rectus muscles can be seen anterior to it. Although less precise detail is revealed, the simpler sonogram can often help in the diagnosis of pelvic conditions. An example of a horizontal section through the female pelvic cavity using this technique is shown in Fig. 6-36.

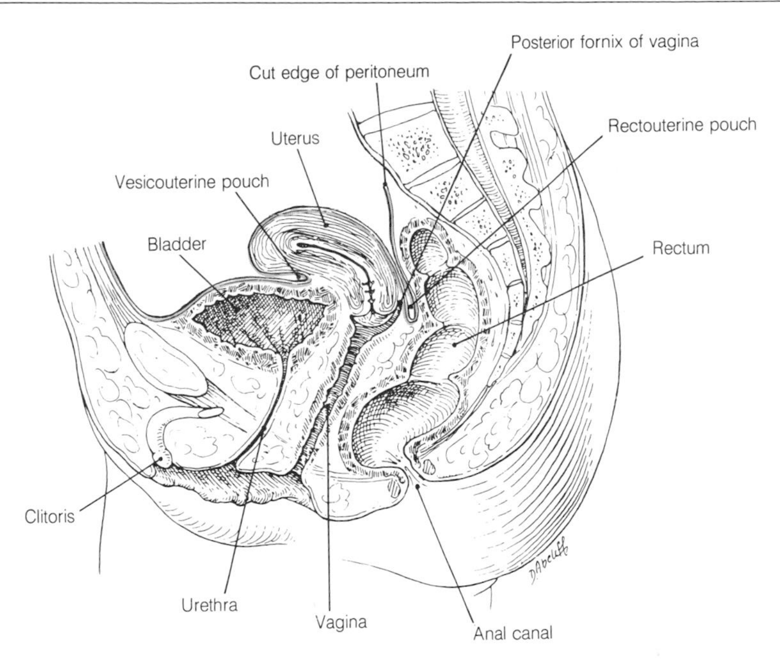

6-34 A sagittal section through the female pelvis.

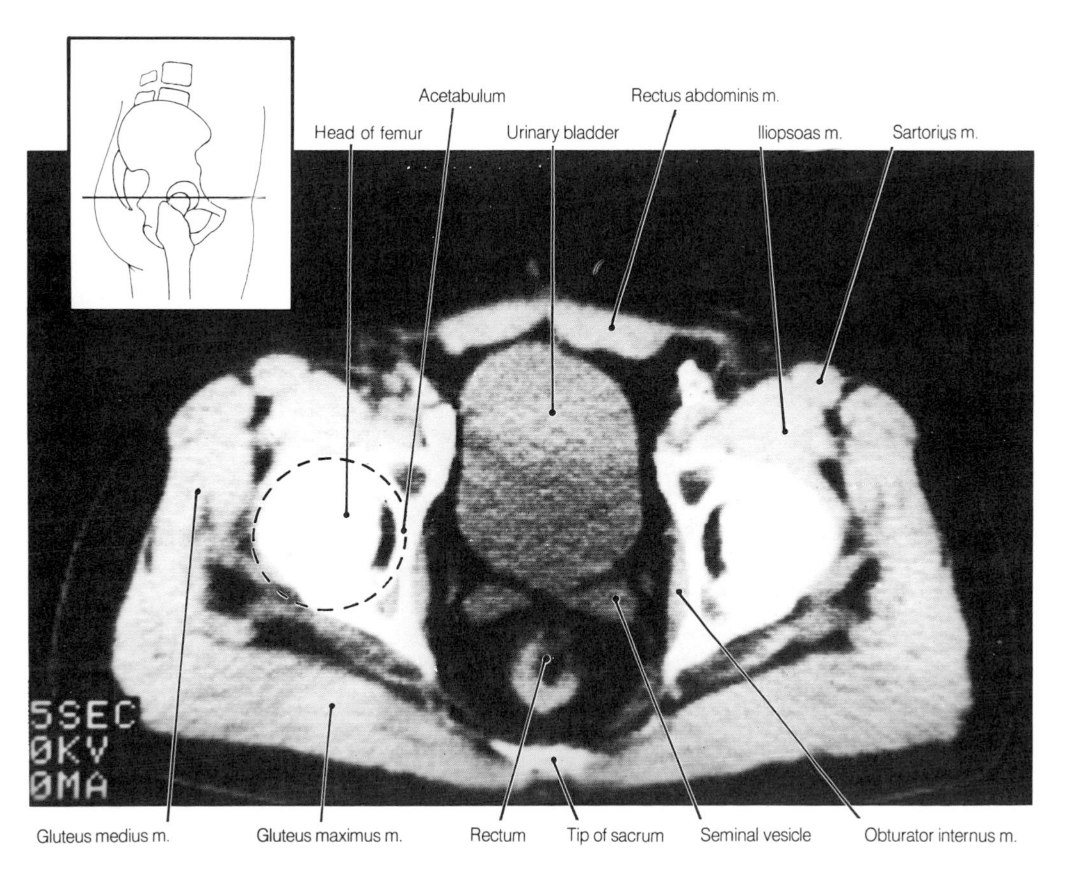

6-35 CT scan through the male pelvis at the level indicated by the outline drawing. (Courtesy of Nancy Whitley, M. D., Dept. of Diagnostic Radiology, University of Maryland School of Medicine.)

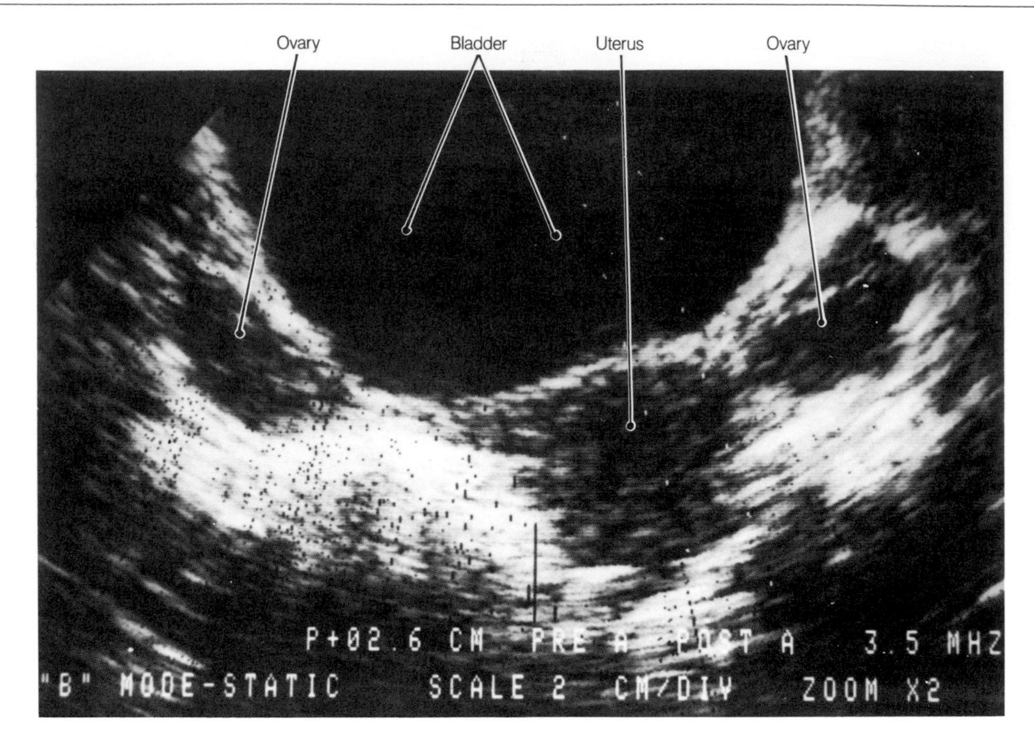

6-36 Sonogram of the female pelvic cavity (horizontal section), (Courtesy of Morgan Dunne, M. D., Dept. of Diagnostic Radiology, University of Maryland School of Medicine.)

The Vessels of the Pelvis

The Internal Iliac Artery

The **abdominal aorta** divides over the fourth lumbar vertebra into the two **common iliac arteries**. These can be followed to the pelvic brim where they, in turn, bifurcate into an **internal and external iliac artery** over the sacroiliac joint. The external iliac artery continues around the brim and enters the lower extremity beneath the inguinal ligament as the **femoral artery**. The internal iliac artery descends into the pelvic cavity and there terminates in a number of branches running either to or through the pelvic wall (parietal branches) or to the pelvic viscera (visceral branches). The internal iliac artery is only about 4.0 cm in length. It is crossed anteriorly by the ureter and lies on the corresponding internal iliac vein and the lumbosacral trunk. The artery divides into an anterior and posterior trunk but beyond this point the pattern of distribution is irregular and branches often share a common origin. There is also a variation according to sex. For this reason the pattern shown in Fig. 6-37 must be considered as one example of a variety of possiblities. The branches are summarized below and their pelvic course should be followed in the illustration.

Parietal Branches of the Anterior Trunk

The **obturator artery** runs around the lateral wall of the pelvis beneath the peritoneum and on the medial surface of obturator internus. It leaves the pelvis through the **obturator canal** in company with the **obturator nerve**. It is crossed medially by the ureter and in the male by the ductus deferens.

The **internal pudendal artery** crosses the pelvic surface of the sacral plexus and leaves the pelvis through the **greater sciatic foramen** below the piriformis. It then curves behind the ischial spine to enter the perineum through the **lesser sciatic foramen**. Its distribution in the perineum will be described later.

The **inferior gluteal artery** also leaves the pelvis through the greater sciatic foramen below the

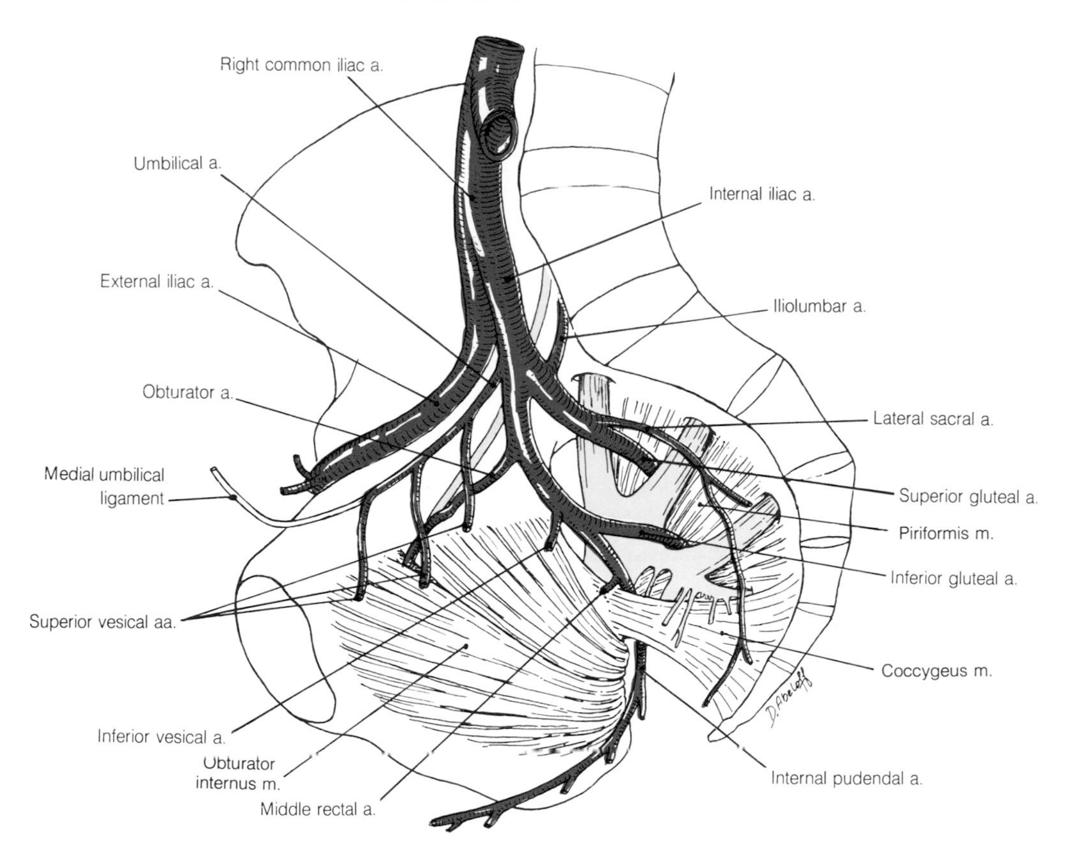

6-37 Showing the branches of the internal iliac artery.

piriformis. Its subsequent course and distribution are described with the gluteal region of the lower limb.

Visceral Branches of the Anterior Trunk

Several **superior vesical arteries** usually arise from the patent origin of the **umbilical artery** and supply the bladder. The rest of the umbilical artery is obliterated and forms the **medial umbilical fold** seen on the deep surface of the anterior abdominal wall.

An **inferior vesical artery** runs across the floor of the pelvic cavity to supply the fundus of the bladder. It also supplies the lower part of the ureter and, in the male, the prostate and seminal vesicles. In the female it may provide a vaginal artery.

The small **artery to the ductus deferens** arises from either a superior or inferior vesical artery.

The **middle rectal artery** may arise in common with the inferior vesical artery and run to the rectum where it anastomoses with the superior and inferior rectal arteries.

The **vaginal artery** in the female replaces or is a branch of the inferior vesical artery. On reaching the vagina it anastomoses with its fellow and forms longitudinal vessels which in turn anastomose with vaginal branches of the uterine artery.

The **uterine artery** crosses the pelvic floor in the base of the broad ligament to reach the junction of the cervix and body of the uterus above the lateral fornix of the vagina. Just before reaching the uterus it crosses above and anteriorly to the ureter. This structure is, therefore, at risk when the uterine artery is ligated prior to a hysterectomy. The artery follows a tortuous course up the body of the uterus and finally turns laterally between the walls of the broad ligament to anastomose with the **ovarian artery**. It supplies many branches to the uterus and, as already described, it anastomoses below with the vaginal artery.

Branches of the Posterior Trunk

These are all distributed to the pelvic wall. The **iliolumbar artery** ascends to the iliac fossa and supplies iliacus and psoas. The **lateral sacral ar-**

teries pass medially and supply spinal branches which enter the anterior sacral foramina. The **large superior gluteal artery** passes backward to leave the pelvic cavity through the greater sciatic foramen above the piriformis muscle.

The Ovarian Arteries

Each ovarian artery arises from the **abdominal aorta** just below the renal artery. It passes over the pelvic inlet to enter the suspensory ligament of the ovary which carries it to the mesovarium of the broad ligament. Here it supplies the ovary and uterine tube and sends a branch to anastomose with the **uterine artery** on the lateral wall of the uterus.

The Veins of the Pelvis

Each pelvic viscus is surrounded by a plexus of veins to which it drains. Which the exception of the gonadal veins, which drain to the inferior vena cava on the right and left renal vein on the left, the venous plexuses drain into the internal iliac veins. As has already been mentioned, the **pelvic venous plexuses** also communicate with the **external and internal vertebral plexuses**. These communications play an important part in the spread of disease from the pelvis. It should also be remembered that the **middle rectal veins** communicate with the **superior rectal veins** which form part of the portal venous system. The veins accompanying the parietal branches of the internal iliac arteries also drain into the internal iliac veins.

Nerves in the Pelvis

The nerves in the pelvis include the **obturator nerve** derived from the **lumbar plexus**, the **sacral plexus** including its parasympathetic branches, the terminal portion of the **sympathetic trunk**, and the **pelvic autonomic plexuses**.

The Obturator Nerve

This nerve is formed from the **ventral primary rami of the second, third**, and **fourth lumbar nerves**. It appears in the pelvis at the medial border of psoas and, now joined by the obturator artery, runs across the lateral wall of the pelvis to leave through the obturator foramen and supply muscles of the lower limb. It does not supply obturator internus. It may be irritated by disease of the ovary to which it is closely related and if this is so, pain may be referred to the cutaneous distribution of the obturator nerve on the medial aspect of the thigh.

The Sacral Plexus

The sacral plexus is formed by the fourth and fifth lumbar ventral rami (the **lumbosacral trunk**) and the **first four sacral ventral rami**. The latter nerves enter the pelvis through the anterior sacral foramina. The fifth sacral ventral ramus joins with coccygeal nerves to form the small coccygeal plexus. The sacral plexus lies largely on the piriformis muscle in the pelvis and is covered by fascia of the pelvic wall. Much of this plexus is concerned with the nerve supply of the lower limb but a number of branches supply structures within the pelvis and perineum. There is in addition a pelvic parasympathetic outflow. Because of its protected position, injury to individual roots or divisions is unusual. A diagram of the whole plexus is given in Fig. 6-38. At first sight this appears extremely complex but some order can be made of this if it is carefully examined.

Look first at the nerves shown in *blue*. These supply structures concerned with the gluteal region and lower limb. Identify the sciatic nerve and note that it is formed partly by posterior divisions of the ventral rami L4 to S2 (*dark blue*) and partly by anterior divisions of the ventral rami L4 to S3 (*light blue*). As in the upper limb posterior divisions supply extensor muscles and in the sciatic nerve these contributions remain discrete and are contributed to the **common peroneal nerve**. The anterior divisions supply flexor muscles and are contributed to the **tibial nerve**. The posterior divisions of this part of the plexus also give off the **superior and inferior gluteal nerves** and the anterior divisions the **nerves to obturator internus, quadratus femoris**, and the **gemelli**. Piriformis is supplied by the posterior division of S2 in the pelvis. One cutaneous branch, the **posterior femoral cutaneous nerve** which is derived from posterior

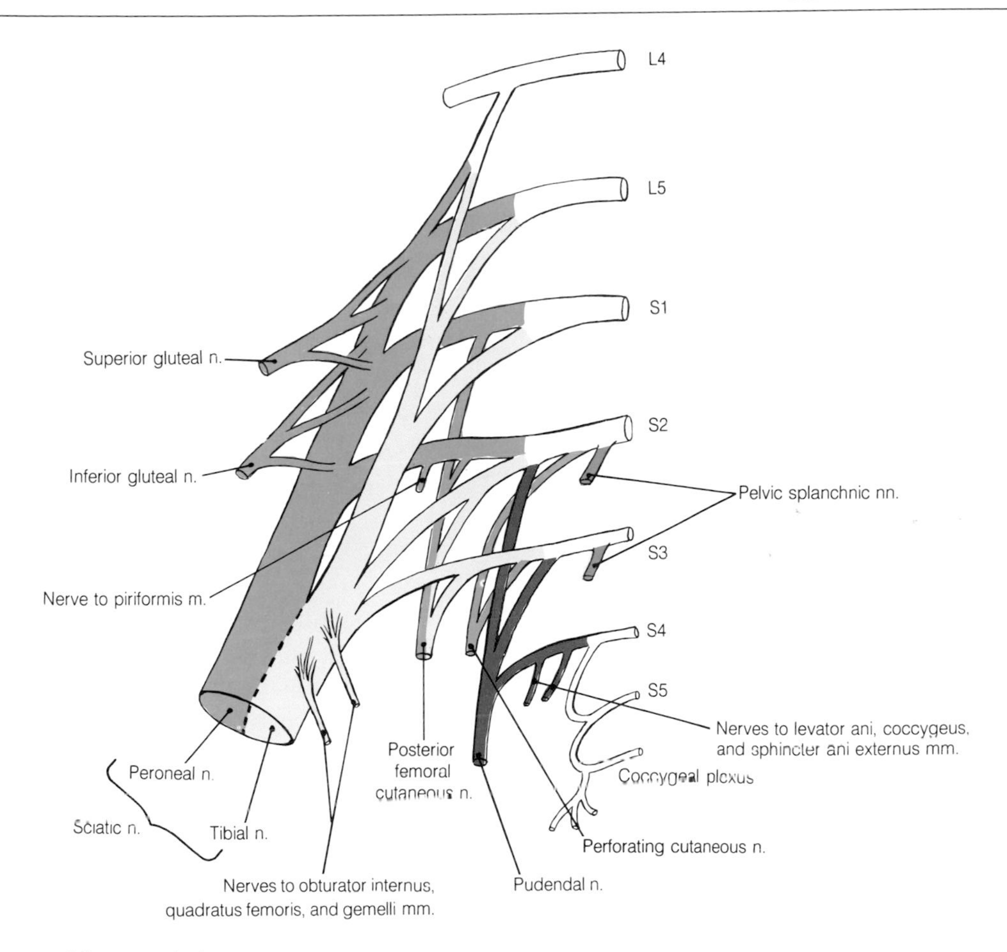

6-38 Diagram of the sacral plexus.

and anterior divisions runs to the lower limb and a second, the **perforating cutaneous nerve** (when present it perforates the sacrotuberous ligament), helps supply the skin of the buttock.

The nerves indicated in *red* are concerned with the supply of the muscular floor of the pelvis, i.e., coccygeus and levator ani, and the skeletal muscle and skin of the perineum and external genitalia through the **pudenal nerve**. The important parasympathetic outflow arising from S2 and S3 is shown in *green*. This will be described later.

Note that with the exception of the nerve to the piriformis and the perforating cutaneous nerve, all the nerves to the gluteal region and lower limb leave the pelvis through the greater sciatic foramen and to these must be added the pudenal nerve.

The Autonomic System in the Pelvis

The autonomic system in the pelvis plays an essential part in the control of defecation, micturition,* and sexual intercourse and its dysfunction has a deep effect on the social life and behavior of a patient. It is, therefore, important that the physician be able to provide good advice and support when such a situation arises.

As in other visceral regions of the body, nerves and plexuses of both the sympathetic and parasympathetic systems are encountered in the pelvis. Before considering the plexuses and their functions, it is important to understand the source of the contribution made by each part of the autonomic system.

* Micturire, *L.* = to urinate

The Sympathetic Contribution

At the pelvic inlet the **lumbar sympathetic trunks** are continuous with the **sacral sympathetic trunks** which lie anterior to the sacrum and medial to the anterior sacral foramina. The two trunks unite over the coccyx at a small enlargement known as the **ganglion impar**. There are usually four **ganglia** along the trunk. **Grey rami communicantes** carry postganglionic fibers to sacral nerve roots for distribution to the lower limbs and perineum. Visceral branches known as **sacral splanchnic nerves** join the pelvic autonomic plexuses. These, like the lumbar splanchnic nerves, are largely formed by preganglionic fibers.

Fibers of lumbar splanchnic nerves, i.e., visceral branches of the lumbar sympathetic trunk, also find their way down into the pelvis, in company with fibers from the aortic plexus. The leash of nerves thus formed is called the **superior hypogastric plexus** or the presacral nerve.

The Parasympathetic Contribution

The contribution from this system arises as several rootlets from the second and third or third and fourth sacral nerves. These are the **pelvic splanchnic nerves** (nervi erigentes) which pass forward to join the pelvic plexuses. The fibers in these nerves are preganglionic. It will be remembered that the gastrointestinal tract relies largely on the vagus for its parasympathetic supply, but that this ceases at the terminal part of the transverse colon. From this point onward the remaining intestine is supplied by preganglionic fibers from the pelvic splanchnic nerves that pass upward into the abdominal cavity through the hypogastric plexuses.

Preganglionic parasympathetic and postganglionic sympathetic fibers from the sources described above, join in complex networks situated in the lower lumbar and pelvic regions for the supply of pelvic organs. These networks are the pelvic

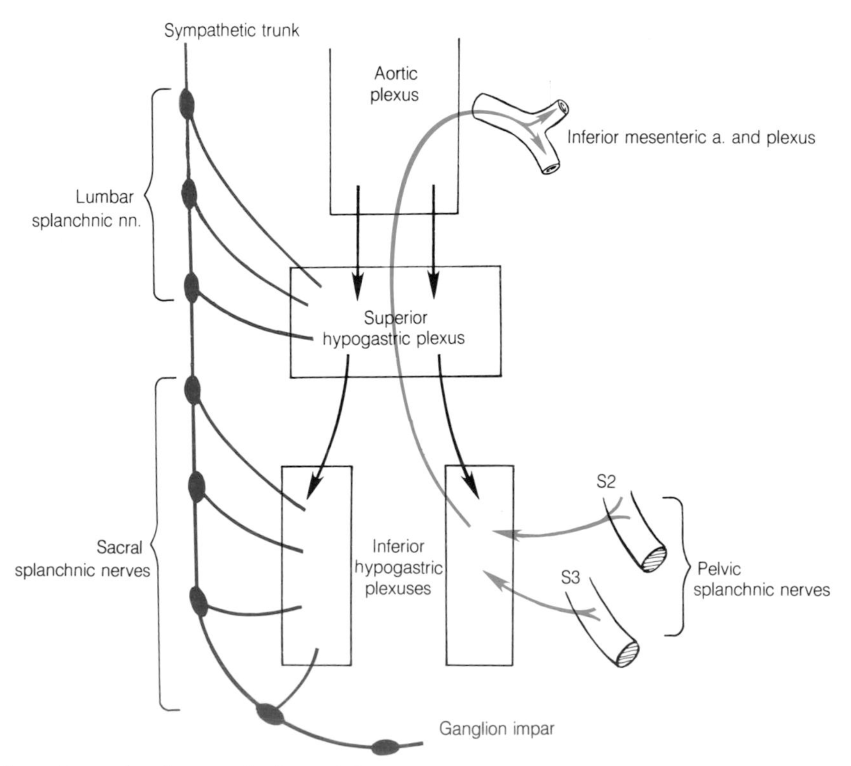

6-39 Diagram of the autonomic plexuses in the pelvis.

autonomic plexuses. They are illustrated diagrammatically in Fig. 6-39 and described below.

Pelvic Autonomic Plexuses

The **superior hypogastric plexus** is formed over the fifth lumbar vertebra by lumbar splanchnic nerves and fibers from the aortic plexus. It also contains preganglionic parasympathetic fibers ascending from the pelvic splanchnic nerves. The superior hypogastric plexus descends into the pelvis and divides into a left and right inferior hypogastric plexus.

The **left and right inferior hypogastric plexuses** lie on either side of the rectum and receive contributions from the superior hypogastric plexus, **sacral splanchnic nerves** of the sacral sympathetic trunk, and parasympathetic fibers from the **pelvic splanchnic nerves**. These plexuses further divide according to the organ that they supply and in this way the **rectal, vesical, and uterovaginal** or **prostatic plexuses** are formed.

Visceral afferent or sensory fibers arising in pelvic and perineal structures also join these plexuses and reach the spinal cord via the pelvic splanchnic or the lumbar splanchnic nerves. It is for this reason that it is difficult to alleviate by surgical means intractable pain from pelvic organs.

The ovary, although postnatally a pelvic organ, receives autonomic fibers from the ovarian plexus which descends from the aortic plexus on the surface of the ovarian artery.

Function and the Pelvic Autonomic System

Micturition

Urine is continually passed into the bladder from the ureters. With filling, the tension in the bladder wall increases and at a certain point, somewhat dependent on the rate of filling, a micturition reflex is initiated. As a result the smooth muscle of the bladder wall contracts and the smooth and striated muscle of the vesical and urethral sphincters is relaxed and urine is voided. In humans this reflex may be consciously stimulated or inhibited.

Sensory nerve fibers in the bladder wall are stimulated by a sufficient rise in tension and these pass back to the central nervous system in the pelvic splanchnic nerves. Efferent parasympathetic fibers also running in the pelvic splanchnic nerves are in turn stimulated and cause the bladder wall muscle to contract. A reciprocal relaxation of the vesical sphincter and a voluntary relaxation of the urethral sphincter, which is supplied by the pudendal nerve, allow urine to pass through the urethra. Emptying is aided by the relaxation of the pelvic floor musculature and by contraction of the abdominal wall. The spinal reflex is subject to the facilitatory and inhibitory influence of centers in the brainstem and emptying may also be inhibited by conscious control of the urethral sphincter.

Transection of the spinal cord initially abolishes the reflex and the bladder overfills and overflows. The reflex then returns without the unconscious or conscious control of higher centers. However, with training it is possible to devise alternative methods of voluntarily stimulating the reflex at appropriate times.

Defecation

Distension of the rectum initiates a defecation reflex. The contraction of the sphincter ani internus, which is maintained by its sympathetic nerve supply, is inhibited by its parasympathetic nerve supply and with the voluntary relaxation of the sphincter ani externus and muscles of the pelvic floor, defecation occurs. This function may also be aided by contraction of the abdominal wall. The filling of the rectum usually follows the taking of food, which through a gastrocolic reflex sets off activity of the colon. In small children this leads to defecation after meals but with training and habit the response to the reflex becomes less frequent.

Lymphatic Drainage in the Pelvis

As has been found elsewhere, the lymphatic drainage of an organ in the pelvis is closely related to that organ's blood supply whether this is from a branch of the internal iliac artery or directly from the aorta. In the pelvis, lymphatic vessels also follow the fascial and peritoneal coverings of organs which reach the iliac fossae and hence the nodes around the external iliac vessels. Groups of lymphatic nodes are found around the internal, external, and common iliac arteries and along the aorta. Additional nodes are found in the hollow of the sacrum. Connections between these groups, however, allow the spread of lymph and, therefore, tumors, from one organ or group to another in an unpredictable manner. It must also be realized that many pelvic structures extend into the perineum where the lymphatic drainage is largely to inguinal lymph nodes. Bearing these facts in mind, the common lymphatic drainage of the pelvic organs listed in chart may be more easily understood.

Organ	Nodes
Rectum	Pararectal nodes
	Inferior mesenteric nodes
	Preaortic nodes
	Internal iliac nodes
Anal Canal	Superficial inguinal nodes
Ovary	Aortic nodes
Uterus and Uterine Tubes	External iliac nodes
	Internal iliac nodes
	Sacral nodes
Vagina	Internal and external iliac nodes
	Superficial inguinal nodes
Bladder	Internal and external iliac nodes
Urethra	Internal iliac nodes
	Superficial inguinal nodes

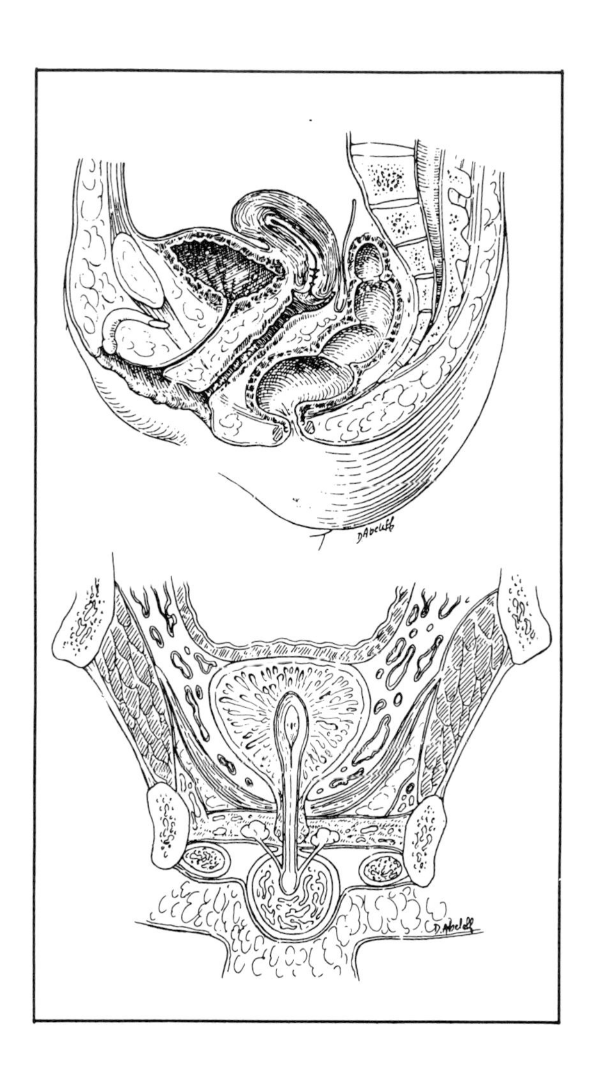

7 The Perineum

The perineum is the region bounded by the **pelvic outlet** that lies below the **pelvic diaphragm**. It follows from this that while the pelvic diaphragm forms the floor of the pelvic cavity, it also forms the roof of the perineum. Also, because the floor of the pelvic cavity slopes downward and medially, the roof of the perineum will do likewise. Thus, the perineum will be relatively deep laterally and shallow as the midline is reached. Furthermore, a number of structures must either pass through the pelvic floor to reach the perineum (alimentary, urinary, and genital systems) or circumnavigate it (pudendal nerves and vessels).

The Boundaries of the Perineum

Use Fig. 7-1 to define the boundaries of the pelvic outlet and thus the boundaries of the perineum. Most anteriorly is the **arcuate pubic ligament** lying below the **symphysis pubis**. This boundary is continued laterally along the **ischiopubic ramus** to the **ischial tuberosity** and is completed between the tuberosity and the coccyx by the **sacrotuberous ligament**. It is conventional to divide this diamond shape into an anteriorly situated **urogenital triangle** and a posteriorly situated **anal triangle** by drawing an imaginary line through the

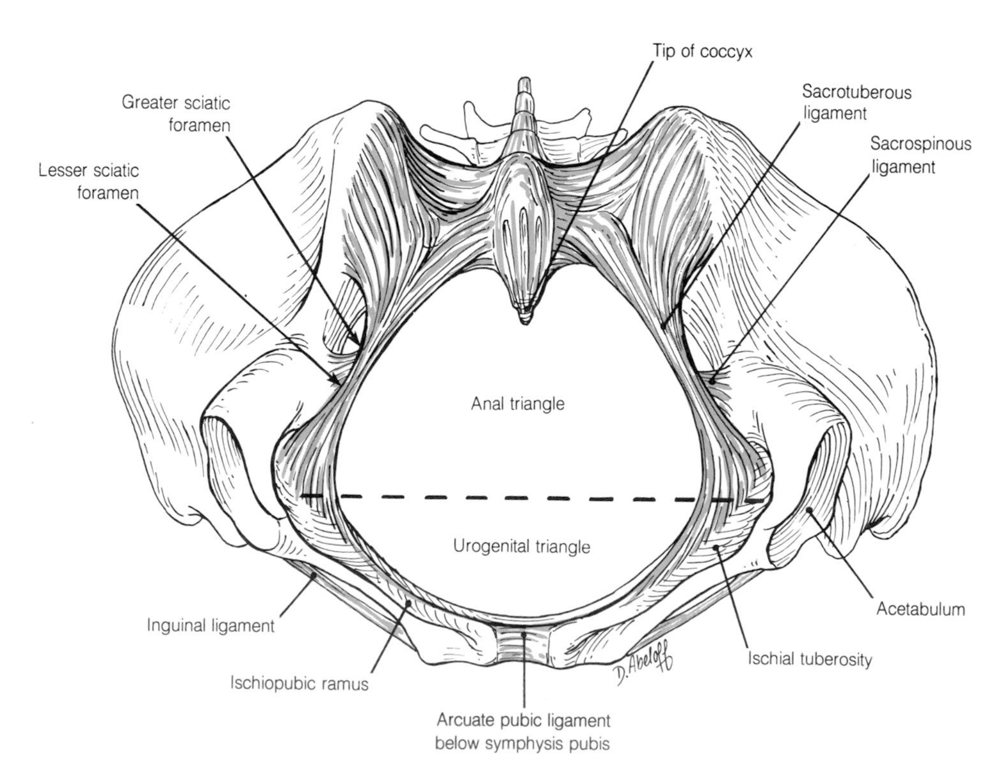

7-1 Showing the bony and ligamentous boundaries of the perineum.

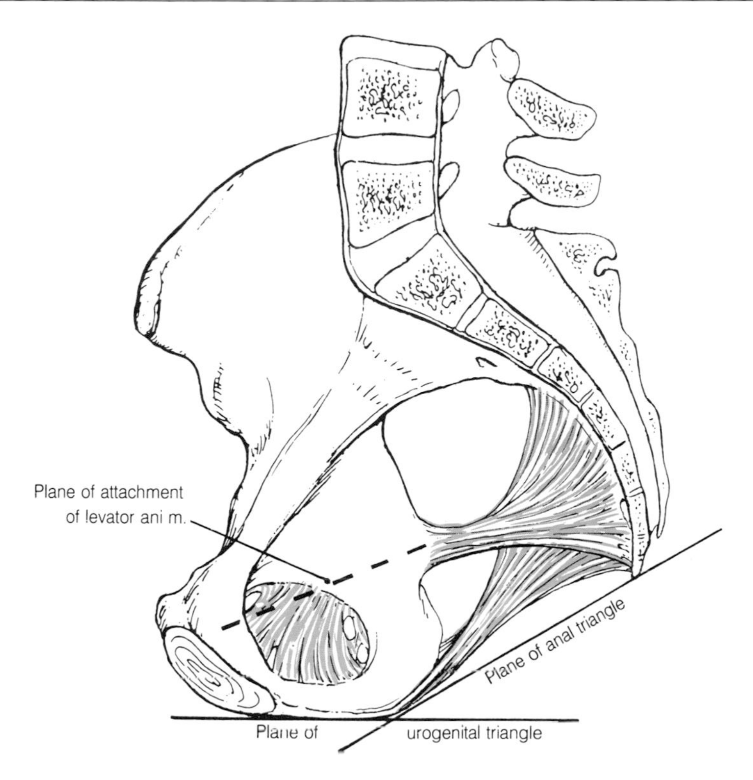

7-2 To show the angle formed between the planes of the urogenital and anal triangles.

ischial tuberosities. Although this procedure simplifies description, it should not be forgotten that the two triangles communicate freely.

Next observe the boundaries of the fossa when seen from the lateral aspect and as illustrated in Fig. 7-2. Note how the diamond is folded about the ischial tuberosities and that the plane of the urogenital triangle forms an angle with the plane of the anal triangle. The attachment of the roof of the perineum, i.e., the levator ani, to the lateral wall of the pelvis is also indicated in the illustration. It can now be seen that the depth of the perineum is minimal anteriorly and posteriorly and maximal in the intermediate region. The perineum is limited inferiorly by skin and fat-filled superficial fascia.

The Anal Triangle

The anal triangle contains the two **ischiorectal fossae** which are separated by the **anal canal** and adjacent midline connective tissue (the fossae have been more accurately named "ischioanal" in the most recent edition of *Nomina Anatomica*). The midline connective tissue comprises the **anococcygeal ligament** posteriorly and the **perineal body** anteriorly.

The Anal Canal

The rectum ends at the floor of the pelvic cavity where it is partly encircled by the **puborectalis** portion of the levator ani muscle. From this point the anal canal extends downward and backward for about one and a half inches to terminate at the **anus**. In this way quite a sharp angle is formed at

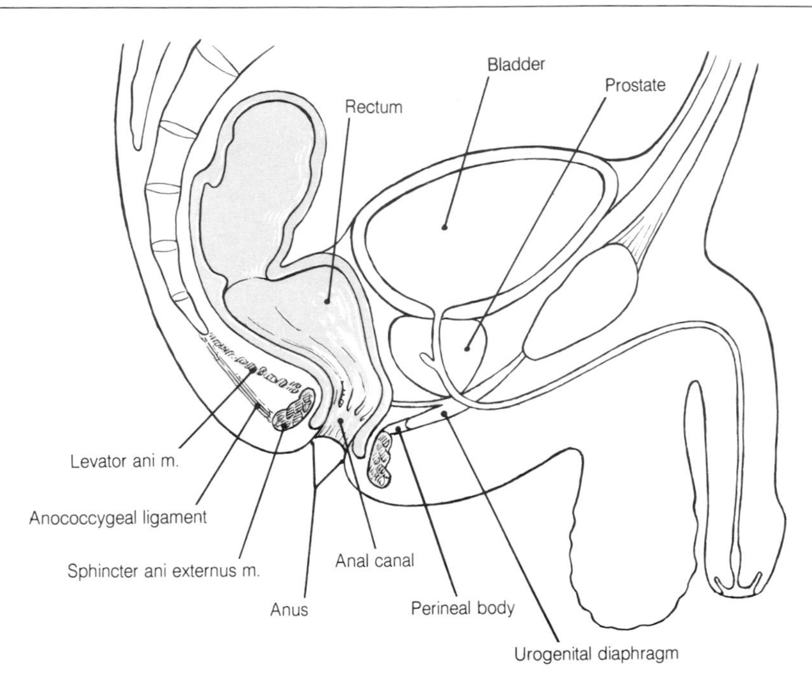

7-3 A sagittal section of the pelvis and perineum to show the position and immediate relationships of the anal canal.

the **anorectal junction** and this can be appreciated from the drawing in Fig. 7-3. The canal is normally empty and flattened from side to side. On filling with feces it may expand by displacing fat in the ischiorectal fossae which lie lateral to it.

The upper part of the canal is derived from the endoderm-lined cloaca and the lower part from the ectoderm-lined proctodeum, and this difference in origin is reflected in differences in the epithelial lining, blood supply, lymphatic drainage, and nerve supply of each part.

The Lining of the Anal Canal

A number of features of the lining of the anal canal that are described can only be clearly recognized in the living young person and are often difficult to demonstrate in the older laboratory cadaver. In its upper part the canal is lined by a columnar epithelium containing mucus-secreting goblet cells and is similar to that found in the rectum. Look at Fig. 7-4 and see that the mucous membrane in this region is thrown into vertical folds called the **anal columns** and that the columns are joined at their lower ends by horizontal folds which form the **anal valves**. The recesses which lie between the columns and the valves are called the **anal sinuses**. Into the sinuses open a number of **anal glands** which lie in the submucosa or surrounding muscle. Obstruction and infection of these glands may produce fistulae and painful abscesses.

The lower margins of the anal valves together form the **pectinate line** and below this there is a zone of transitional stratified epithelium. This is limited below by the white or **anocutaneous line** where the lining of the canal becomes true skin.

The Muscular Wall of the Anal Canal

The mucous membrane of the canal is surrounded by a muscular wall composed of the **sphincter ani internus** and the **sphincter ani externus**. The arrangement of these two sphincters is illustrated in Fig. 7-4.

The **sphincter ani internus** is a somewhat thickened continuation of the circular smooth muscle of the rectum. It surrounds the upper three-quarters of the anal canal and terminates at the level of the white line.

The **sphincter ani externus** is formed from three parts named **subcutaneous, superficial**, and **deep**. Unlike the internal sphincter, all are striated muscles and, therefore, under voluntary control. Look at Fig. 7-4 and see that the subcutaneous part is indeed subcutaneous and sur-

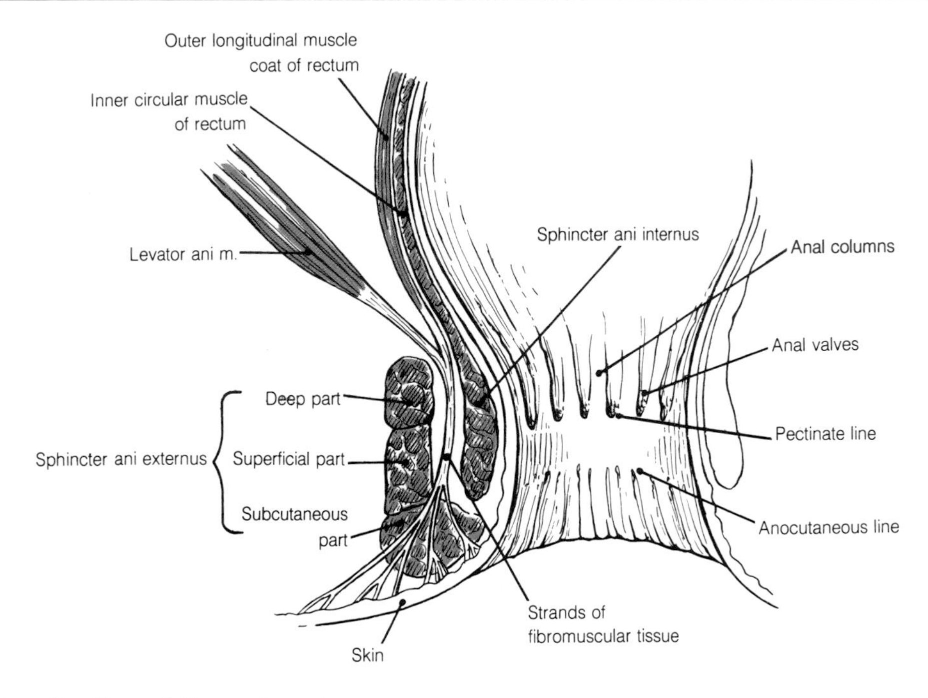

7-4 A coronal section through the anal canal.

rounds the anal opening. The superficial and deep parts, which lie above it, are separated from the anal mucosa by the internal sphincter and a layer of fibroelastic tissue continuous above with the outer longitudinal smooth muscle coat of the rectum. Below, this coat breaks up into a number of septa which pass through the subcutaneous part to become attached to the skin of the anal margin.

The superficial part is attached posteriorly to the coccyx by means of the **anococcygeal ligament** and anteriorly to the **perineal body**. The deep part is fused with the **puborectalis portion of levator ani**.

The external sphincter is supplied by the **perineal branch of the fourth sacral spinal nerve** and by the **inferior rectal branch of the pudendal nerve** (S2 and S3). Voluntary contraction of the external sphincter is used to delay defecation initiated by the filling of the rectum with feces. Damage to the important "anorectal ring" formed by the deep part of the external sphincter, puborectalis, and the internal sphincter, either surgically or in childbirth, may lead to incontinence of feces.

The Blood Vessels of the Anal Canal

In keeping with its origin from the hind gut, the mucous membrane of the upper part of the anal canal is supplied by the **inferior mesenteric artery** through its continuation the **superior rectal artery**. Terminal branches of this vessel pass from the rectum to the anal canal between the muscular and mucous coats to anastomose with branches of the **middle rectal artery** and supply the muscular coat of the upper part of the canal. The middle rectal artery is a branch of the **internal iliac artery**.

The **inferior rectal artery** arises from the **internal pudendal artery** on the lateral wall of the ischiorectal fossa, crosses the fossa and, reaching the anal canal, supplies both muscle and the skin which lines the lower part of the canal. It anastomoses with its fellow and with the middle and superior rectal arteries. Look at Fig. 7-5 in which the origin and distribution of the rectal arteries is illustrated.

The venous drainage of the anal canal is to an internal venous plexus in the submucosa and an

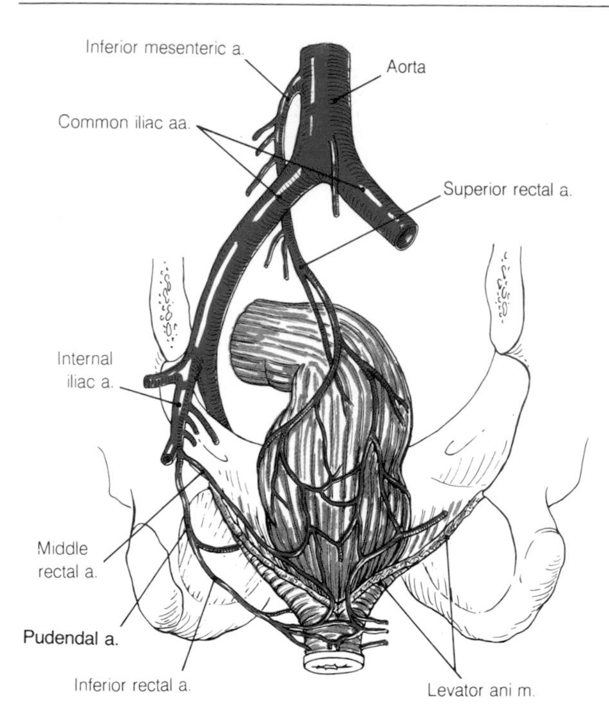

7-5 Semischematic diagram of the posterior aspect of the rectum and anal canal and the arteries supplying them.

external plexus outside the muscular coat. While the internal plexus drains mainly to the **superior rectal veins** and thence to the **inferior mesenteric veins**, both plexuses are also drained by the **inferior rectal veins** which are tributaries of the **internal pudendal veins. Middle rectal veins** drain the muscular coat of the upper part of the anal canal. The internal venous plexus is, therefore, a site of a portocaval venous anastomosis and dilatation of the veins of the plexus may accompany portal hypertension. However, perhaps because of little support from surrounding tissue or the absence of valves in the inferior mesenteric veins, dilatation of these veins, known as hemorrhoids or piles, occurs much more frequently in the absence of any general rise in portal venous pressure.

The Lymphatic Drainage of the Anal Canal

The lymphatic vessels draining the upper part of the anal canal pass upward to nodes alongside the rectum and sigmoid colon, which in turn drain to preaortic nodes. Vessels from the lower part, but above the mucocutaneous junction, drain to internal and common iliac nodes. The skin lining the lower part drains to the most medial superficial inguinal nodes and an enlarged or painful node here should always prompt a rectal examination.

The Innervation of the Anal Canal

As might be suspected, the lining of the upper part of the anal canal is supplied by the **autonomic system** while the skin of the lower part is supplied by somatic spinal nerves in the form of the **inferior rectal nerves** which are branches of the **pudendal nerve**.

The Examination of the Anal Canal

The anal canal cannot be left without mention of the importance of carrying out a digital examination whenever it is indicated. Not only will gross local lesions be revealed, but the condition of the sphincters can be assessed and in the male the prostate, whether enlarged or not, can be felt through the anterior rectal wall. In the female this examination may be used as an alternative to a vaginal examination. The cervix can be palpated through the anterior wall of the rectum and other enlarged, abnormal, or painful pelvic structures detected. The examination can be completed by direct observation through a proctoscope.

The Ischiorectal Fossae

The ischiorectal fossae lie on either side of the midline of the perineum. Each is shaped rather like a segment of an orange having a sharp linear upper border and a curved lower surface which narrows as it approaches the upper border anteriorly and posteriorly. In Fig. 7-6, which illustrates a coronal section through the deepest part of the fossae, note the following features:

1. The sharp upper border is formed where the levator ani muscle and the **inferior fascia of the pelvic diaphragm** meet the **obturator internus muscle** and overlying **obturator fascia**.
2. The sloping roof and medial wall are formed by the levator ani muscle and the anal canal which is surrounded by the internal and external sphincters.
3. The lateral wall is formed largely by obturator internus and the obturator fascia as they span the obturator foramen.

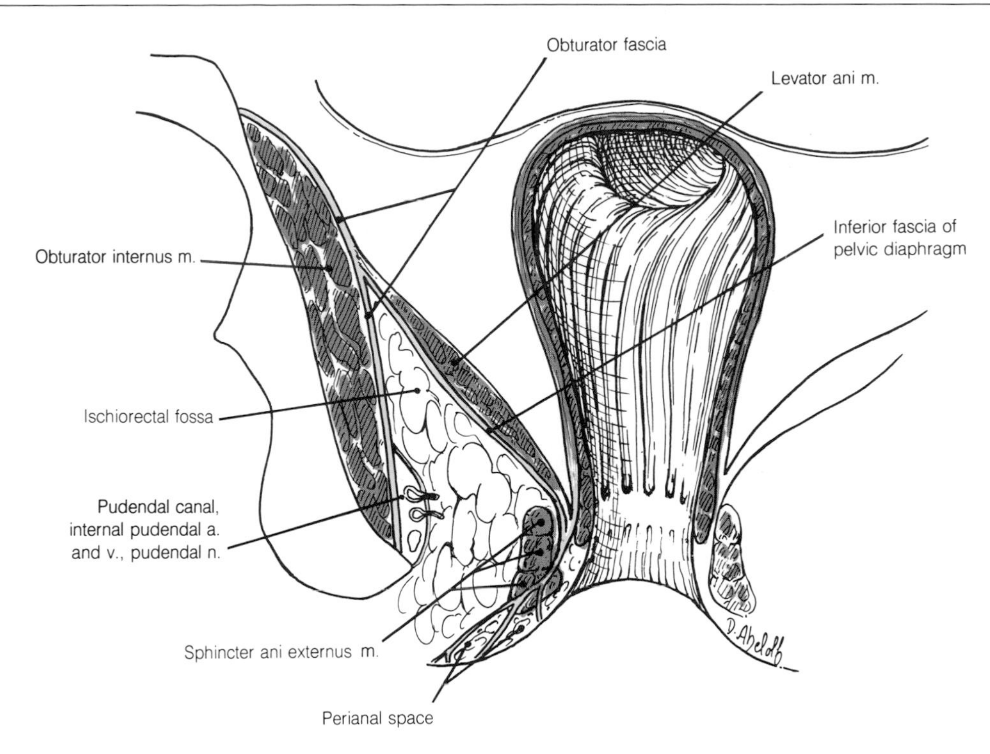

7-6 A coronal section through the anal canal and ischiorectal fossa.

4. Below obturator internus, the lateral wall is completed by the medial surface of the ischial tuberosity.
5. Between obturator internus and the ischial tuberosity, the **pudendal canal** containing the **internal pudendal vessels** and the **pudendal nerve** has been sectioned transversely.
6. The space is filled with coarsely loculated fatty tissue, but note that the fibroelastic septa derived from the external longitudinal smooth muscle coat of the rectum separate this region from the typical subcutaneous fat underlying the skin around the anus. This is the **perianal space** which contains the subcutaneous portion of the external sphincter and the external anal venous plexus.

Posterior to the section illustrated, the two fossae are separated by the anococcygeal ligament and anterior to the section by the perineal body.

Both posteriorly and anteriorly the fossae narrow as the floor meets the upper border and an anterior extension, which extends into the urogenital triangle, will be encountered again when that region is described.

The Pudendal Canal

The **internal pudendal vessels** and the **pudendal nerve** leave the pelvis through the **greater sciatic foramen** and enter the perineum through the **lesser sciatic foramen** below the level of the levator ani. The manner in which this occurs is illustrated in Fig. 7-7. On leaving the lesser sciatic foramen these structurs come to lie on the medial surface of obturator internus below the level of the attachment of levator ani where they are enclosed by a discrete sheath of fascia. This is the **pudendal canal** and it carries the vessels and nerves forward toward the urogenital triangle at the posterior margin of the **perineal membrane**. The inferior rectal artery and nerve leave the internal pudendal artery and pudendal nerve as the pudendal canal passes across the lateral wall of the ischiorectal fossa.

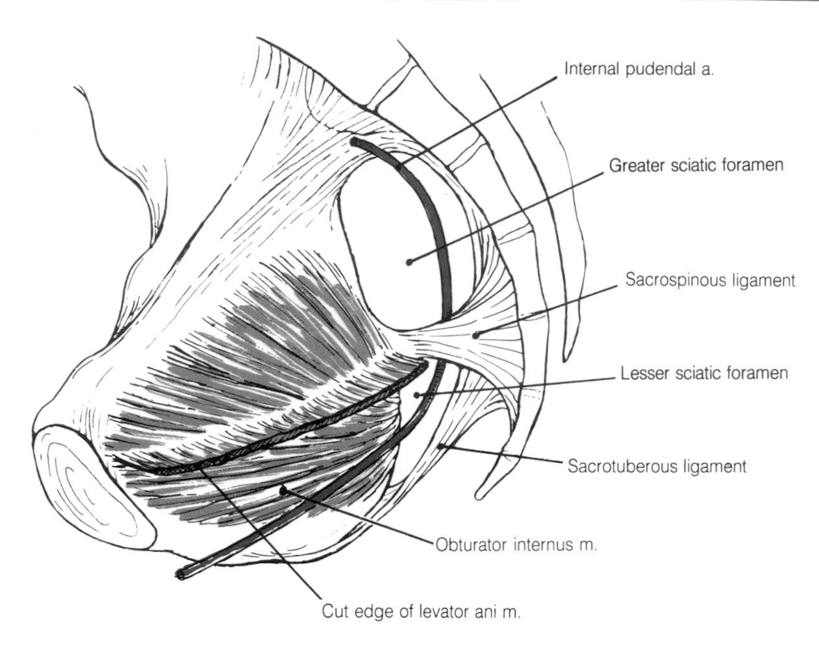

7-7 To show the passage of the internal pudendal artery from the pelvis to the perineum.

The Urogenital Triangle in the Male

Because of differences in the postnatal anatomy of the urinary and genital tracts and the external genitalia of the male and female, it is necessary to describe the urogenital triangle in the two sexes separately. However, a study of the embryology of this region shows that the differences are not as profound as appear at first sight. The urogenital triangle includes the urogenital diaphragm.

The Urogenital Diaphragm

The urogenital diaphragm lies below the anterior part of the pelvic floor and the genital hiatus between the medial margins of the levator ani muscle. The diaphragm consists of a layer of striated muscle sandwiched between two fascial layers. The deep layer or **superior fascia of the urogenital diaphragm** is an insubstantial layer which blends posteriorly with the perineal body and perineal membrane. The muscular layer consists of the **deep transverse perineal muscles** and the **sphincter urethrae**. The more superficial fascial layer is known as the **inferior fascia of the urogenital diaphragm** or more commonly the **perineal membrane**. Through the diaphragm passes the membranous urethra.

The Perineal Membrane

The perineal membrane is a strong triangular fascial sheet spanning the space between the two ischiopubic rami. When the pelvis is correctly oriented the membrane faces downward and slightly forward. Its appearance when observed from below can be seen in Fig. 7-8 and its position and alignment in Fig. 7-9.

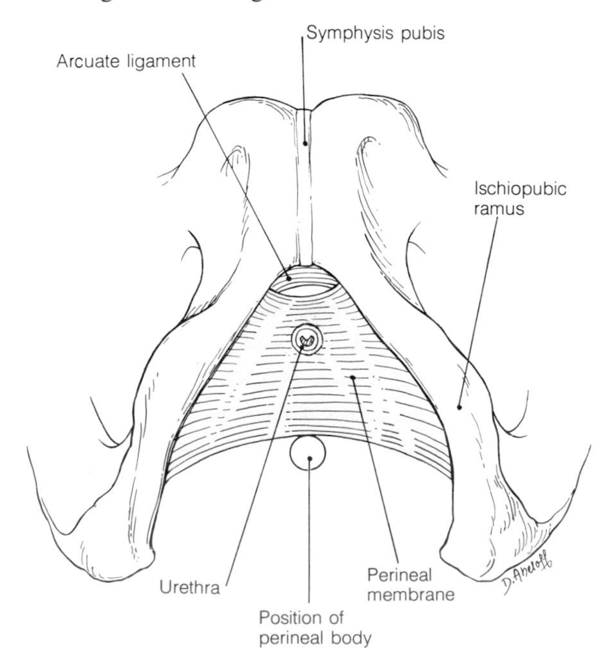

7-8 The perineal membrane as seen from below.

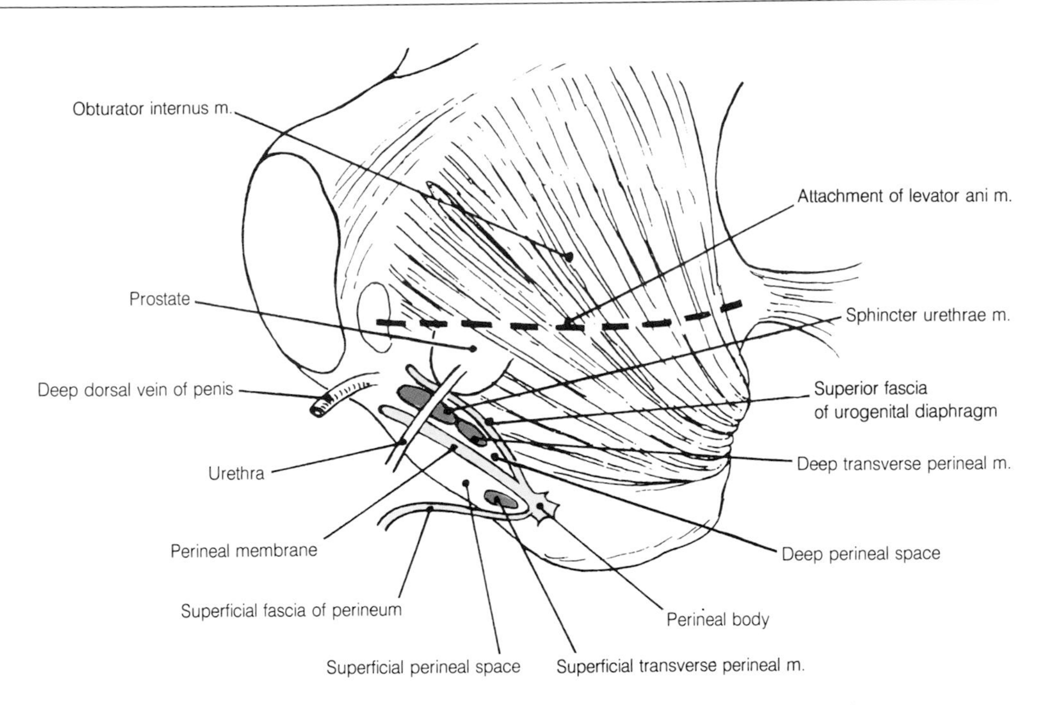

7-9 Semidiagrammatic illustration to show the position and alignment of the perineal membrane.

The apex of the membrane is truncated and does not reach the arcuate pubic ligament; a space is left for the passage of the **deep dorsal vein of the penis**. The sides of the membrane are attached to the ischiopubic rami and its base fuses posteriorly with the **perineal body** and gives attachment to the **superficial fascia of the perineum**.

Perforating the membrane in the midline is the **urethra**. This leaves the bladder and passes into the prostate where it is joined by the ejaculatory ducts. On leaving the prostate it is immediately surrounded by the sphincter urethrae muscle which lies deep to and above the perineal membrane. It is over this short course (1.5-2.0 cm), between the prostate and perineal membrane, that it is called membranous. Lying superior to the perineal membrane and in the same plane as the sphincter urethrae are the deep transverse perineal muscles.

The Deep Perineal Space

The region between the superior and inferior fascias of the urogenital diaphragm is commonly known as the **deep perineal space**. Its position can be seen in sagittal section in Fig. 7-9, and in coronal section in Fig. 7-10, where its contents and immediate relations are shown in more detail. As well as the deep perineal and sphincter urethrae muscles, it contains the **membranous urethra**, the **bulbourethral glands**, the **pudendal vessels**, and the **dorsal nerves of the penis**.

The Sphincter Urethrae

This striated muscle surrounds the membranous urethra in the male. Its deeper fibers encircle the urethra while its superficial fibers pass backward on either side of it from the anterior margin of the perineal membrane to the perineal body.

Relaxation of this sphincter is necessary before micturition can occur. It also provides voluntary control over a reflexly initiated desire to micturate. It is supplied by the **perineal branch of the pudendal nerve**.

The Deep Transverse Perineal Muscles

These striated muscles arise from fascia of the lateral wall of the perineum and passing deep to the

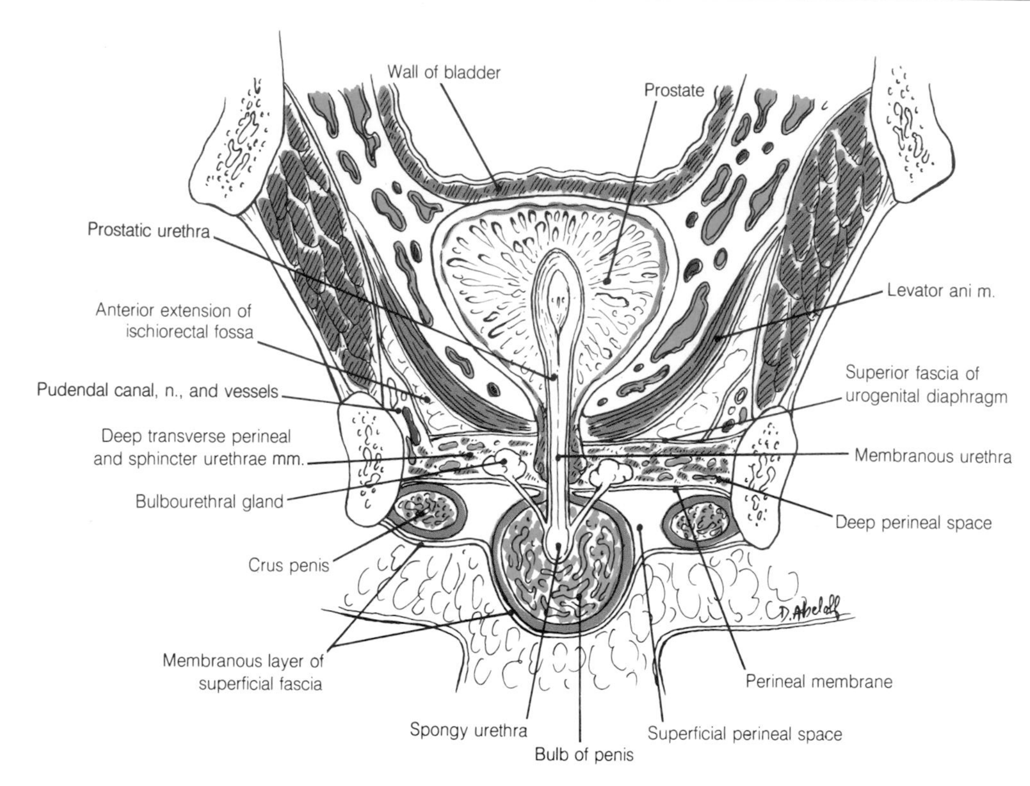

7-10 Coronal section through the male perineum.

perineal membrane they fuse with each other at a tendinous raphe and with the perineal body. Their attachment to the perineal body probably contributes to the support of the pelvic floor. These muscles are supplied by the **perineal branch of the pudendal nerve**.

The Bulbourethral Glands

These two small glands lie on either side of the membranous urethra and are embedded in the fibers of the sphincter urethrae and deep transverse perineal muscles. Their relatively long ducts pierce the perineal membrane to join the **spongy part of the urethra** about 2.5 cm beyond the membrane. The secretion of the glands is contributed to the seminal fluid.

The arrangement of the structures in the deep perineal space is shown in Fig. 7-11. Superficial to and below the perineal membrane lies the superficial perineal space.

The Superficial Perineal Space

The superficial fascia of the perineum consists of relatively thick areolar tissue containing a variable

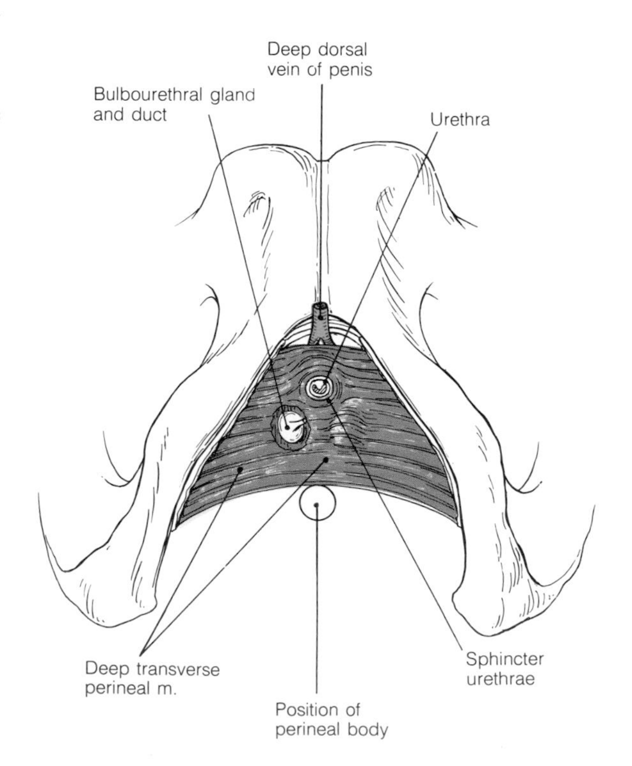

7-11 The contents of the deep perineal space after removal of the perineal membrane.

amount of fat. Over the urogenital triangle there is a deeper but thin and strong membranous layer. The extent and attachments of the membranous layer are of clinical importance because they determine the direction of the spread of urine which may leak from a rupture of the spongy urethra. These attachments are illustrated in Fig. 7-12.

Using Fig. 7-12A, note the posterior attachment of the fascia to the posterior border of the perineal membrane. On each side its attachment is continued forward and upward along the ischiopubic rami and then laterally over the thigh where it fuses with the fascia lata. Extending from these attachments, the fascia surrounds the scrotum and the penis and then becomes continuous above with the membranous layer of the superficial fascia of the anterior abdominal wall (Fig. 7-12B). In this way the superficial perineal pouch is formed. It can now be understood that urine leaking from the urethra will not pass backward into the anal triangle or laterally into the thigh, but after distending the scrotum and penis, will pass up onto the anterior abdominal wall. Contained within the superficial perineal space are the male external genitalia which are described next.

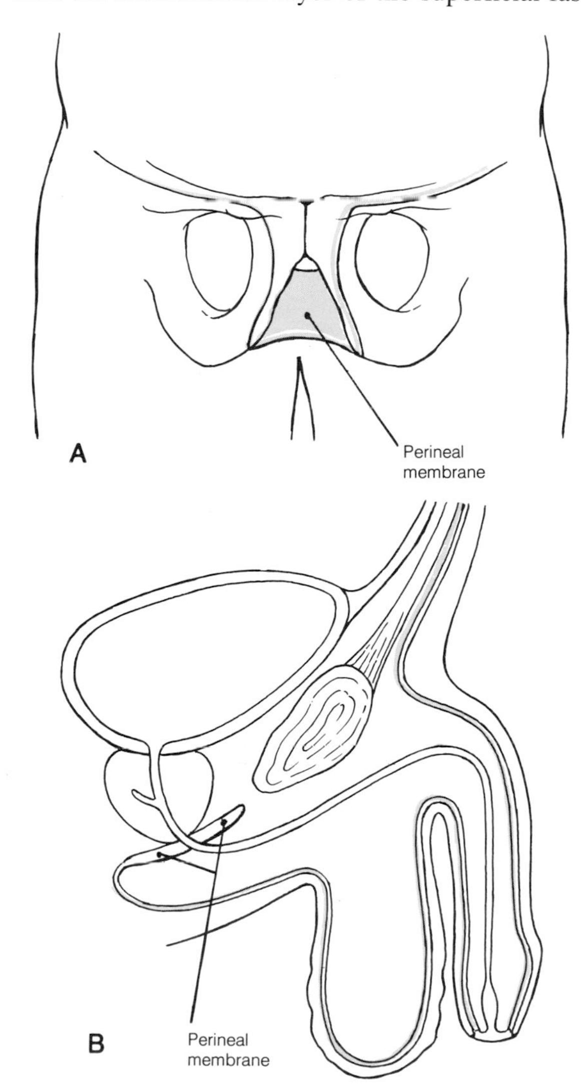

7-12 To show the extent and attachments of the membranous layer of the superficial perineal fascia.

The Male External Genitalia

The male external genitalia are the **penis** and **scrotum**. A description of the scrotum is given in association with the anterior abdominal wall (Chap. 5).

The Penis

The penis is described in two parts, namely the **root** which lies in the perineum and the free **body** which is enveloped in skin.

The root of the penis is formed by three masses

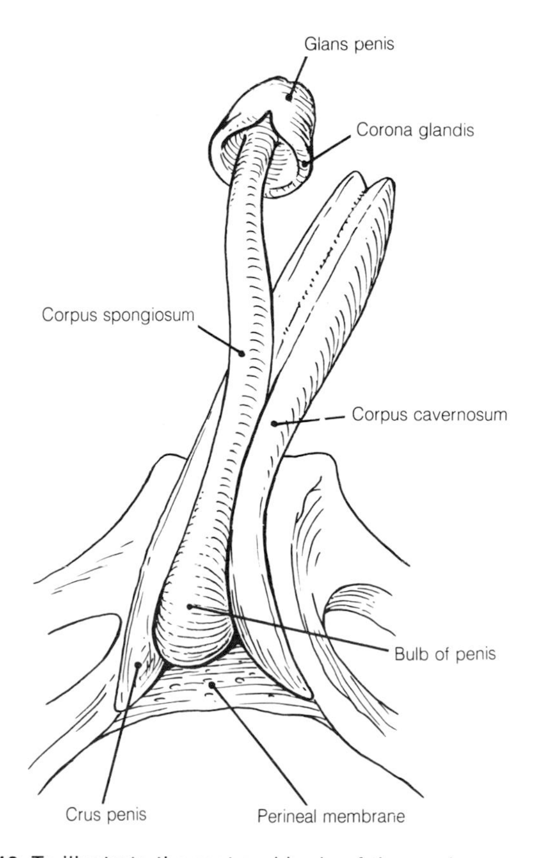

7-13 To illustrate the root and body of the penis.

of erectile tissue which are surrounded by fibrous tissue and lie below the perineal membrane. These are the two **crura** and the **bulb of the penis**.

Look at Fig. 7-13 and see that the crura are attached to the everted margins of the ischiopubic rami and that the bulb is attached to the perineal membrane. It is here that the urethra passes into the erectile tissue of the bulb and continues in it to the external urethral orifice.

These three structures merge to form the body of the penis and in Fig. 7-14 a transverse section of this is shown. The crura, now called the **corpora cavernosa**, are each surrounded by a fibrous sheath and the medial surfaces of these sheaths fuse in the midline to form the **septum penis**. The bulb, now called the **corpus spongiosum**, narrows and comes to lie in a groove on the ventral surface of the fused corpora cavernosa. The distal end of the corpus spongiosum becomes expanded to form the **glans*** **penis** whose hollowed proximal surface accepts the blunt terminations of the corpora cavernosa. The slightly everted margin of the glans is called the **corona glandis** and narrowing of the body proximal to this is called the **neck of the penis**.

It is necessary here to point out that the surfaces of the penis are described as if it were suspended, as in quadupeds, from the ventral surface of the body or in the position it assumes when erect. As a result, the anterior surface of the flaccid dependent penis is in fact its dorsal surface.

* Glans, *L.* = an acorn

The bulb and the two crura of the penis are partially surrounded by the **bulbospongiosus** and **ischiocavernosus muscle**. A further pair of muscles, called the **superficial transverse perineal muscles**, are closely associated with these.

The **bulbospongiosus muscle** arises from the perineal body and a midline raphe lying over the ventral surface of the bulb. Its fibers fan out laterally to be attached in sequence to the perineal membrane on either side of the bulb, to encircle the bulb, and finally encircle all three erectile masses to be attached to an aponeurosis on the dorsum of the penis.

The **ischiocavernosus muscle** arises from the ischial ramus on each side of the crus and initially covers it. Its fibers extend forward to end in an aponeurosis on the deep surface of the crus.

Both muscles assist in producing an erection of the penis and the bulbospongiosus helps to empty the urethra at the end of micturition.

The **superficial transverse perineal muscle** is a small slip that arises from the medial aspect of the ischial tuberosity and crosses the posterior margin of the perineal membrane to meet its fellow at the perineal body.

Each of these superficial perineal muscles is supplied from the **perineal branch of the pudendal nerve**.

The body of the penis is surrounded by thin loose skin. At the neck of the penis the skin is reflected upon itself to form the **prepuce** or foreskin which covers the glans. The prepuce is attached to the ventral surface of the glans by a fold

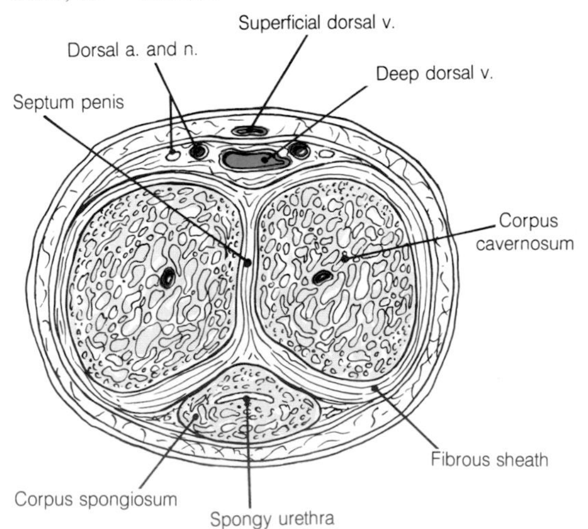

7-14 A transverse section of the body of the penis.

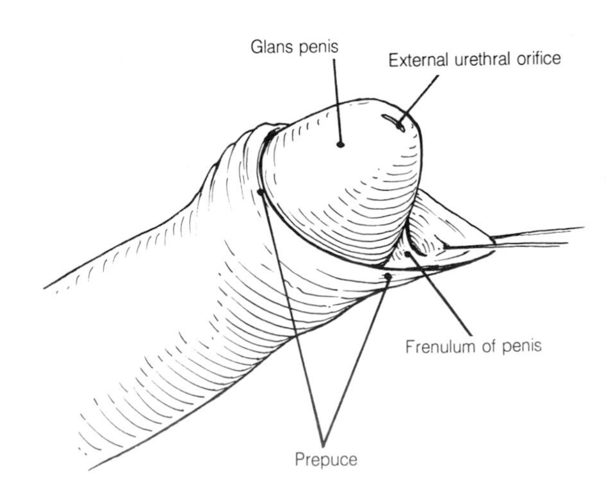

7-15 Illustrating the prepuce of the penis.

of skin called the **frenulum of the prepuce** (Fig. 7-15).

Beneath the skin the superficial fascia contains no fat, but its deepest layer is membranous and is continuous with the superficial perineal fascia.

The proximal part of the body of the penis is supported by the **fundiform** and **suspensory ligaments** arising from the linea alba and the pubic symphysis.

The Vessels of the Penis

The crura and corpora cavernosa of the penis are supplied by the **deep arteries of the penis** and by branches of the **dorsal artery**. The bulb and corpus spongiosum are supplied by the **artery of the bulb** and the **dorsal artery**. The dorsal artery also supplies the skin and superficial layers of the penis.

Branches of these vessels that enter the erectile tissue either open directly into the cavernous spaces or continue as the tortuous **helicine arteries**. It is these arteries that provide for the rapid distension of the cavernous spaces that is needed to achieve an erection of the penis.

Venous drainage is through a deep and a superficial dorsal vein.

The Nerve Supply of the Penis

Vasodilator parasympathetic fibers and sensory fibers reach the penis through the pudendal nerves and their terminal branches, the dorsal nerves of the penis.

View Fig. 7-14 to see the position of the paired dorsal arteries and nerves and the single midline superficial and deep veins.

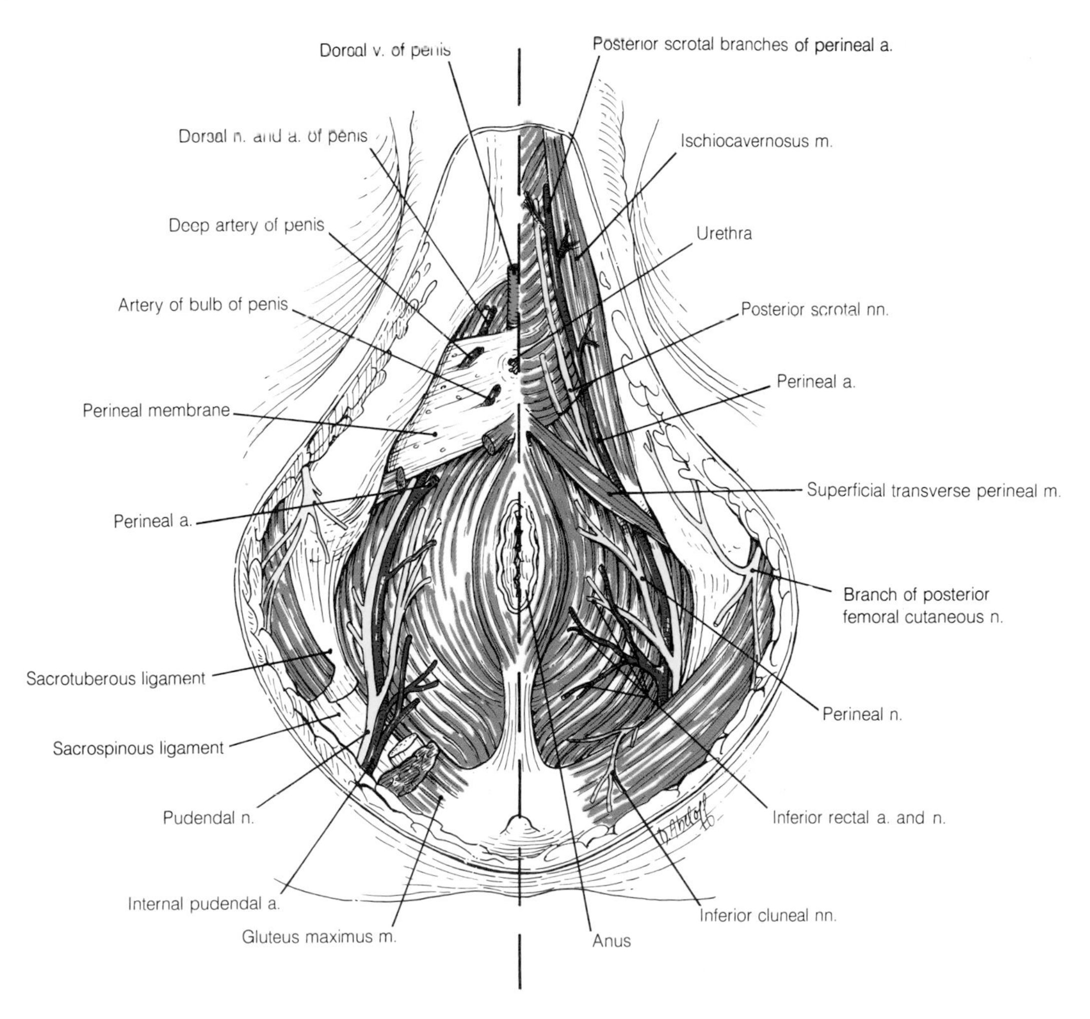

7-16 Showing a deep (*left side*) and superficial (*right side*) exposure of the distribution of the pudendal nerve and internal pudendal artery in the male. The pudendal veins have been omitted.

Vessels in the Urogenital Triangle in the Male

The Internal Pudendal Artery

The internal pudendal artery has been followed through the anal triangle in the pudendal canal which lies in the lateral wall of the ischiorectal fossa. The canal leads the artery to the deep perineal pouch where it continues forward on the deep surface of the perineal membrane and medial to the ischiopubic ramus. It ends by dividing into the deep and dorsal arteries of the penis. Its branches include:

- the inferior rectal artery,
- the perineal artery,
- the artery of the bulb of the penis,
- the deep artery of the penis,
- the dorsal artery of the penis, and
- branches to perineal muscles and the urethra.

The distribution of these vessels can be followed in Fig. 7-16.

The **inferior rectal artery** has been described with the anal triangle.

The **perineal artery** leaves the pudendal artery before that artery passes deep to the perineal membrane. Entering the superficial perineal space, it gives branches to the superficial perineal muscles and the scrotum.

The **artery to the bulb of the penis** runs medially across the deep surface of the perineal membrane and pierces it near the midline. Here it enters the bulb where it supplies the erectile tissue of the bulb and corpus spongiosum. A small branch also supplies the bulbourethral gland.

The **deep artery of the penis** pierces the perineal membrane and passes into the overlying crus to supply the erectile tissue of this and the corpus cavernosum.

The **dorsal artery of the penis** passes into the superficial perineal pouch between the pubic symphysis and the anterior border of the perineal membrane. Reaching the dorsum of the penis it continues distally between the fascia penis and the fibrous sheath of the corpus cavernosum. Its relationships to its fellow and the deep dorsal vein of the penis can be seen in Fig. 7-17. This artery supplies the cavernous tissue and the skin and superficial layers of the penis.

A small **urethral artery** pierces the perineal membrane to reach and supply the urethra in its penile course.

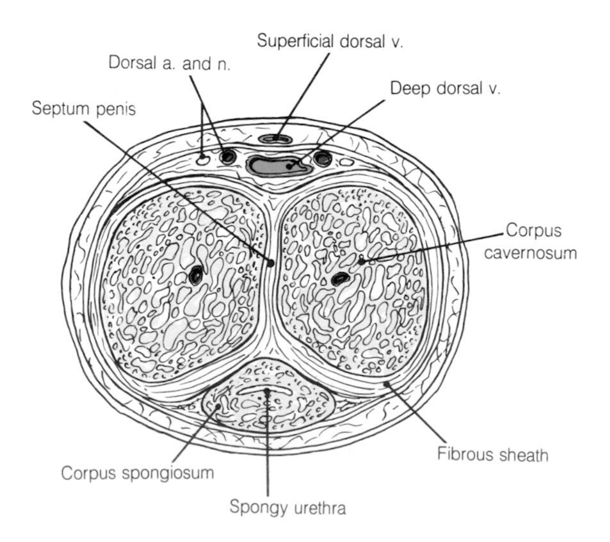

7-17 A transverse section of the body of the penis.

The Internal Pudendal Veins

The internal pudendal veins, which are the venae comitantes of the pudendal arteries, drain into the internal iliac vein. Their tributaries correspond to the branches of the pudendal artery. Note, however, that the single **deep dorsal vein of the penis** (Fig. 7-17) drains predominantly into the prostatic venous plexus which surrounds the prostate deep to the symphysis pubis and that the **superficial dorsal vein of the penis** drains to the left or right **superficial pudendal vein** and thence to the femoral vein.

Lymphatic Vessels

Vessels from the scrotum and superficial parts of the penis pass to the medial group of superficial inguinal lymph nodes. Those from deeper structures, including the penile urethra, may reach the deep inguinal nodes or, in company with the deep dorsal vein of the penis, run directly to internal iliac nodes.

Nerves in the Urogenital Triangle in the Male

The Pudendal Nerve

The pudendal nerve contains fibers of the second, third, and fourth spinal nerves. Like the internal pudendal artery, it enters the perineum by passing through the lesser sciatic foramen and it accompanies the artery across the lateral wall of the ischiorectal fossa in the pudendal canal. While in the canal it gives off the **inferior rectal nerve** which crosses the fossa to supply the sphincter ani externus, the mucous membrane, and skin of the lower part of the anal canal and the skin around the anus.

Near the posterior border of the perineal membrane the pudendal nerve divides into the **perineal nerve** and the **dorsal nerve of the penis**.

The **perineal nerve** passes below and, therefore, superficial to the perineal membrane. It gives off **posterior scrotal branches** and **muscular branches** to all the remaining striated muscles of the perineum. The anterior scrotal skin is supplied by branches of the ilioinguinal and genitofemoral nerves.

The **dorsal nerve of the penis** accompanies the internal pudendal artery over the deep surface of the perineal membrane and runs with the dorsal artery of the penis through the gap between the anterior border of the membrane and the arcuate pubic ligament. The nerve continues over the dorsum of the penis on the lateral side of the artery (Fig. 7-17) and terminates at the glans.

To review the urogenital triangle in the male, look now at the illustration of a coronal section through this region in Fig. 7-18 and identify the following features:

1. The prostatic, membranous, and spongy portions of the urethra in the midline. The spongy portion lies in the bulb of the penis and passes out of the section as it turns forward.
2. The levator ani and obturator internus muscles and the superior fascia of the urogenital diaphragm. These enclose an anterior extension of the ischiorectal fossa.
3. The termination of the pudendal canal carrying the pudendal vessels and the dorsal nerve of the penis into the deep perineal space.
4. The perineal membrane attached to the ischiopubic rami and forming the inferior boundary of the deep perineal space.
5. The deep transverse perineal and sphincter urethrae muscles and the bulbourethral glands within the deep space.
6. The midline bulb of the penis lying in the in-

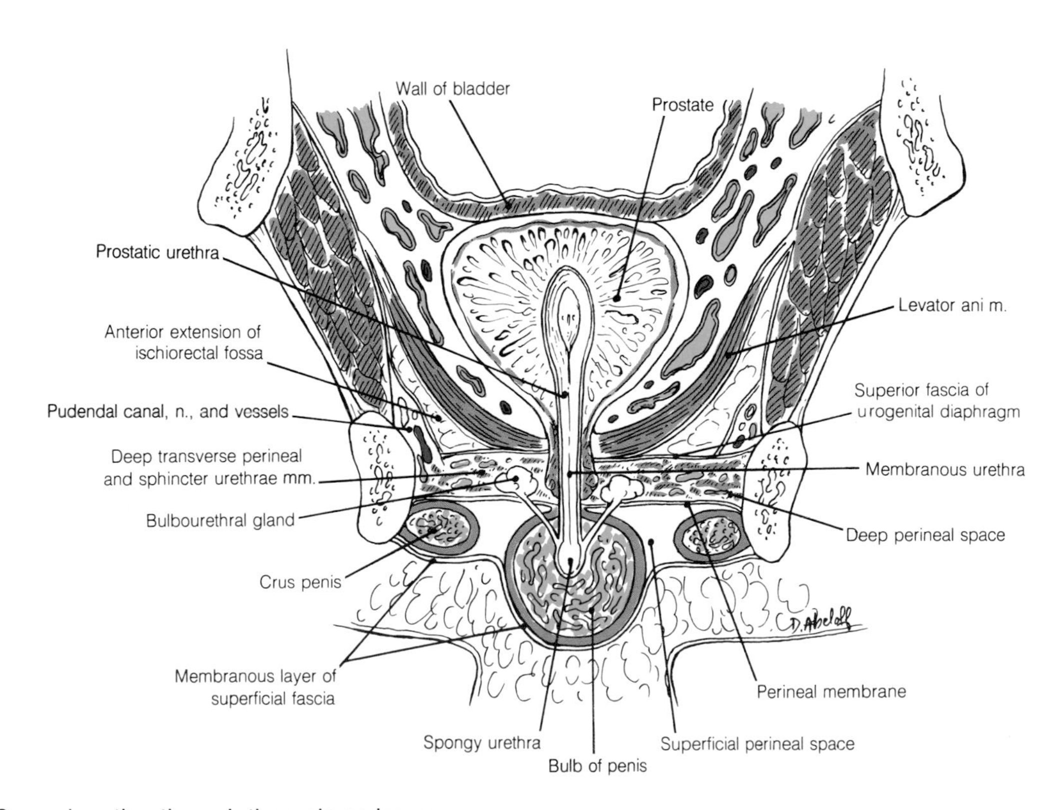

7-18 Coronal section through the male perineum.

ferior and superficial surface of the perineal membrane.
7. The crura of the penis attached on each side to the membrane and adjacent ischiopubic ramus.
8. The membranous layer of the superficial fascia of the perineum.

The Urogenital Triangle in the Female

Using the male as the model, a pattern for the anatomy of the urogenital triangle has been established. This pattern can be seen to be repeated in the female although modified by the presence of the vaginal canal, the absence of extraabdominal gonads, and the fact that the urethra does not traverse the clitoris, the homologue of the penis.

The Perineal Membrane

The presence of the vagina and its need to dilate during childbirth are associated with a reduction in size of the perineal membrane. This is little more than a fibrous arch which provides attachment for the crura of the clitoris at its lateral attachments to the ischiopubic rami and continues across the midline anteriorly. Behind this arch the urethra and vagina divide the membrane which becomes fused with their outer fascial covering.

The Vagina

The vagina has been seen to pass through the pelvic floor partly surrounded by the most anterior fibers of levator ani called in the female the **pubovaginalis muscle**. Below this level it lies in the urogenital triangle. The urethra is embedded in its anterior wall and posteriorly it is separated from the anal canal by the perineal body. Distally it opens into the **vestibule** at the **vaginal orifice** or **introitus** which in the virgin is partially occluded by the **hymen**. The pelvic and perineal course of the vagina and its main relations can be now reviewed in Fig. 7-19.

The lower part of the vagina is supplied by the pudendal nerve and local anesthesia can be produced by infiltration of this nerve via a needle passed through the vaginal wall. The blood supply is provided by the internal pudendal artery. Lymphatics from the lower vagina pass to the internal iliac nodes or, from the region of the orifice, to the superficial inguinal nodes.

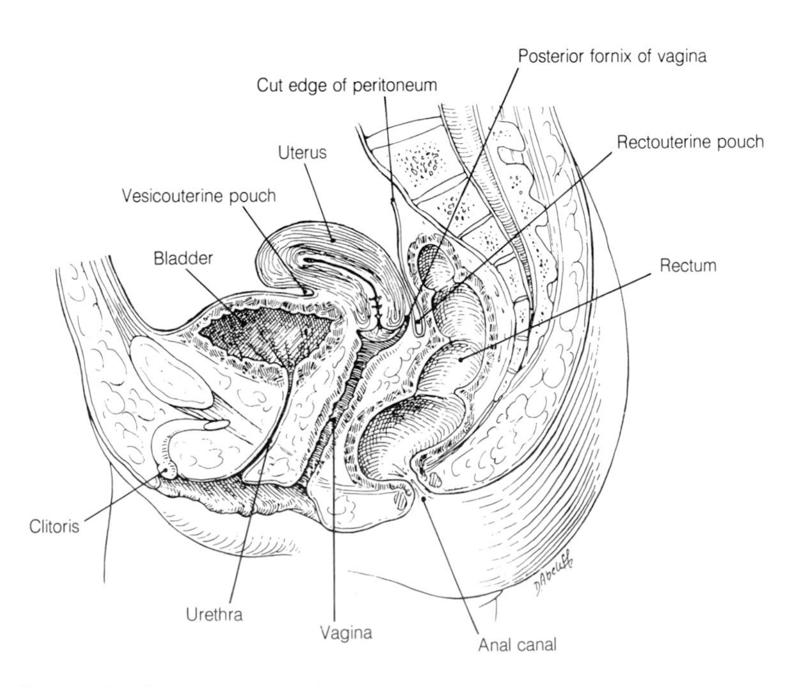

7-19 Sagittal section through the female pelvis and perineum.

The Hymen of the Vagina

The hymen* is a thin membrane that partially occludes the virgin vagina. Occasionally it completely occludes the canal, a fact not usually noticed until the onset of menstruation. After rupture the only signs of its presence are small nodules at the periphery of the orifice called **carunculae hymenales**.

The Urethra

The short female urethra descends into the urogenital triangle from the neck of the bladder and becomes embedded in the anterior wall of the vagina. Just above the level of the perineal membrane it is surrounded by the **sphincter urethrae muscle**. The urethra opens into the vestibule just anterior to the vagina. Its orifice can usually be recognized by a slightly raised margin.

The Clitoris

Like the penis the clitoris is a sensitive erectile structure. Two crura are attached to the perineal membrane adjacent to the ischiopubic rami. Anteriorly these become two corpora cavernosa which are bound together by fascia to form the body of the clitoris. The body which is directed inferiorly toward the vestibule is surmounted by a small rounded glans. The glans is connected by two slender bands of erectile tissue to the bulbs of the vestibule.

The Bulbs of the Vestibule

In the female the erectile tissue homologous with the bulb of the penis is divided into two masses which lie on either side of the vaginal orifice. These structures are called the **bulbs of the vestibule**. Their anterior extremities narrow and become united in front of the urethral opening. Their expanded posterior ends overlie the greater vestibular glands.

The Greater Vestibular Glands

These small glands lie deep to the bulbs of the vestibule on each side of the vaginal orifice. They are "greater" only when compared with the many lesser vestibular glands which open onto the surface of the vestibule. The ducts of the two greater vestibular glands open at the margin of the vaginal orifice.

The Muscles of the Female Urogenital Triangle

Each of the muscles of the male urogenital triangle has its counterpart in the female and all are supplied by the **pudendal nerve**. The **ischiocavernosus muscles** are somewhat smaller than in the male and the two portions of the **bulbospongiosus muscle** surround the vagina and cover the bulbs of the vestibule rather than meet in a midline raphe.

The Perineal Body in the Female

This midline node of fibromuscular tissue forms an essential feature in the support and integrity of the female pelvic floor. It lies between the anal canal and vagina. To it are attached the superficial and deep transverse perineal muscles, the pubovaginalis portion of the levator ani, the bulbospongiosus muscle, and the superficial part of the sphincter ani externus.

Tears of the perineal body and surrounding muscles occur during childbirth. If these are not repaired, the subsequent weakness may allow protrusion of the bladder wall through the pelvic floor and prolapse of the uterus into the vaginal canal.

Vessels and Nerves

The **internal pudendal artery**, although smaller in the female, has a course and distribution similar to that in the male. **Posterior labial branches** replace posterior scrotal branches and the three branches to the penis are represented by the **artery of the bulb of the vestibule** and **deep and dorsal arteries of the clitoris**.

Each artery is accompanied by a vein draining back to the internal pudendal veins. There is also a single **deep dorsal vein of the clitoris** which passes back benath the symphysis pubis to reach the **vesical plexus of veins**.

The **pudendal nerve** is also similar to that of

* Hymen, *L.* and *Gk.* = the god of marriage

the male in course and distribution. All skeletal muscle of the anal and urogenital triangles is supplied by either the inferior rectal nerve or the perineal nerve. The dorsal nerve of the clitoris which runs deep to the perineal membrane is small.

The Vestibule and Labia

The female urethra and the vagina have been followed to the vestibule. This is the midline cleft lying between the **two labia minora**. It is illustrated in Fig. 7-20 where the labia minora have been separated. In this illustration note the relative positions of the vaginal and external urethral orifices, the **vestibular fossa** behind the vagina and the openings of the ducts of the greater vestibular glands.

In the same illustration identify the labia minora. These are the two folds of skin that surround the vestibule. The skin is devoid of hair and subcutaneous fat. Follow the folds anteriorly and see that they divide into two. The upper folds enclose the glans clitoris in a **prepuce**. The lower folds unite beneath the clitoris to form the **frenulum of the clitoris**. When followed posteriorly, the labia are seen to unite in the nulliparous woman at the **frenulum of the labia minora** or **fourchette**. This fold of skin is usually ruptured during the first delivery.

The **labia majora** are two rounded and raised folds of skin lying outside the labia minora. The outer surface of each bears hairs but the inner surface is smooth and, although hairless, has many sebaceous follicles. Posteriorly the labia majora blend with the perineal skin. Anteriorly they fuse with the **mons pubis**, a rounded elevation lying over the symphysis pubis. The shape of the labia majora and mons is provided by underlying fatty connective tissue and, in the case of the labia majora, some smooth muscle. The **round ligament of the uterus** ends in this tissue and the peritoneal sac of an oblique congenital hernia may extend along the course of the round ligament to reach the labium majus.

Lymphatic vessels from the vagina below the hymen, the vestibule, and the labia drain to the **inguinal lymph nodes**.

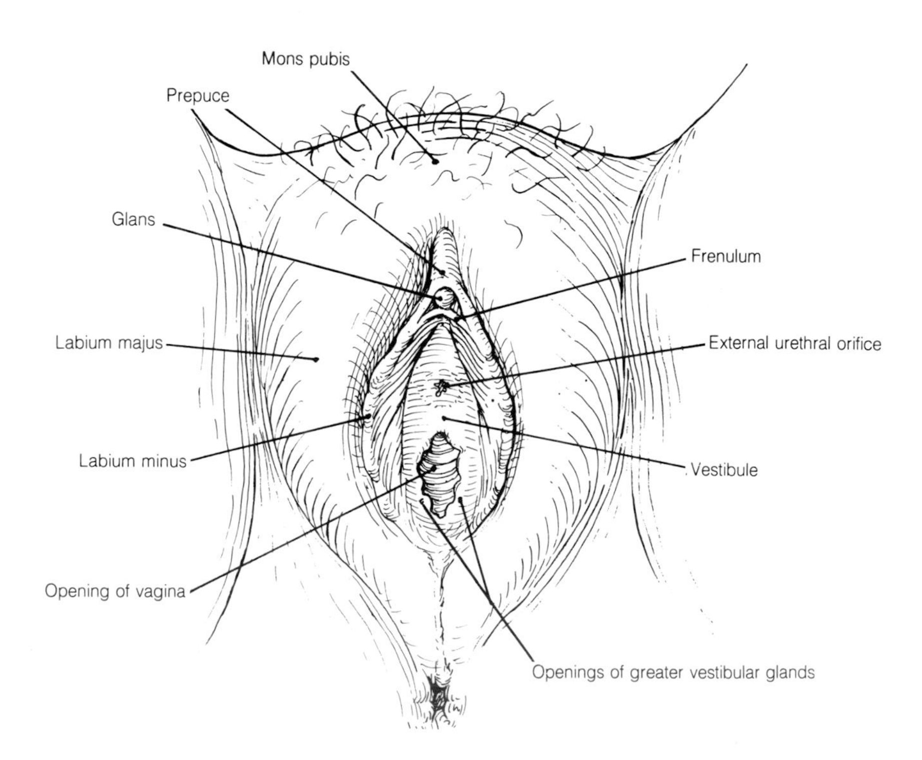

7-20 Showing the female external genitalia.

The Sexual Response

The sexual response is a complex process which involves the psychological state of the subject, the somatic sensory and motor systems, and the autonomic nervous system.

Both visual and tactile sensation may contribute to the preparatory phase. Tumescence of erectile tissue in both sexes is controlled by the parasympathetic outflow in the pelvic splanchnic nerves (S2 and S3). This causes arterial dilatation and increased blood flow. In the presence of a restrictive connective tissue envelope, the increased hydrostatic pressure leads to a reduction in the venous return and erection occurs. In the female this is coupled with lubrication of the introitus and vagina by glandular secretions. Sympathetic fibers derived from the twelfth thoracic and first lumbar spinal cord segments initiate ejaculation by causing contraction of the ductus deferens, seminal vesicles, prostate, and ejaculatory ducts. At the same time sympathetic fibers also inhibit the bladder wall musculature and stimulate the sphincteric action of the bladder neck. In this way micturition and the reflux of semen are prevented. Ejaculation is completed by the rhythmic contraction of the muscles of the corpora and pelvic floor while similar contractions of the pelvic musculature occur in the female. The stimulus for these is carried by the pudendal nerves.

8 The Lower Limb

The General Arrangement of the Lower Limb

While the erect posture allows the upper limb to be devoted to manipulative functions, it has left the lower limbs to serve as supporting and locomotor structures. Their "design" can be seen to reflect the demands of these functions. The skeleton of the pelvic girdle and lower limbs is shown in Fig. 8-1. The weight of the trunk can be seen to be transmitted from the sacrum, through the almost immobile sacroiliac joints to the pelvic bones and from these through the stable but mobile hip joints to the femora. Note how each femur is directed downward and medially through the thigh and toward the knee joint. Here it articulates with the tibia (but not with the fibula) and the kneecap or patella. Between the knee joint and the ankle joint the weight is transmitted by the tibia. The slender fibula only serves for muscle attachment in its upper part but its lower end is firmly bound to the tibia and forms an important part of the ankle joint. The tarsal and metatarsal bones of the foot form a stable but flexible unit. Although movement is limited at most of the intertarsal joints, the joints between the calcaneus and talus allow the rotation of the forefoot that is called inversion and eversion. These movements make it possible to adapt the sole of the foot to the surface on which it is planted.

In order that the body should remain balanced in the upright position its center of gravity must lie within an area delimited by the heels and the balls of the feet, i.e. the points at which the weight is finally transferred to the ground. This position can be maintained with the expenditure of very little muscular energy because the center of gravity lies approximately above the axes of the hip joints and knee joints. However, the center of gravity of the trunk and limbs lies a little in front of the axes of the ankle joints and about half way between the heels and the balls of the feet. As a result a consid-

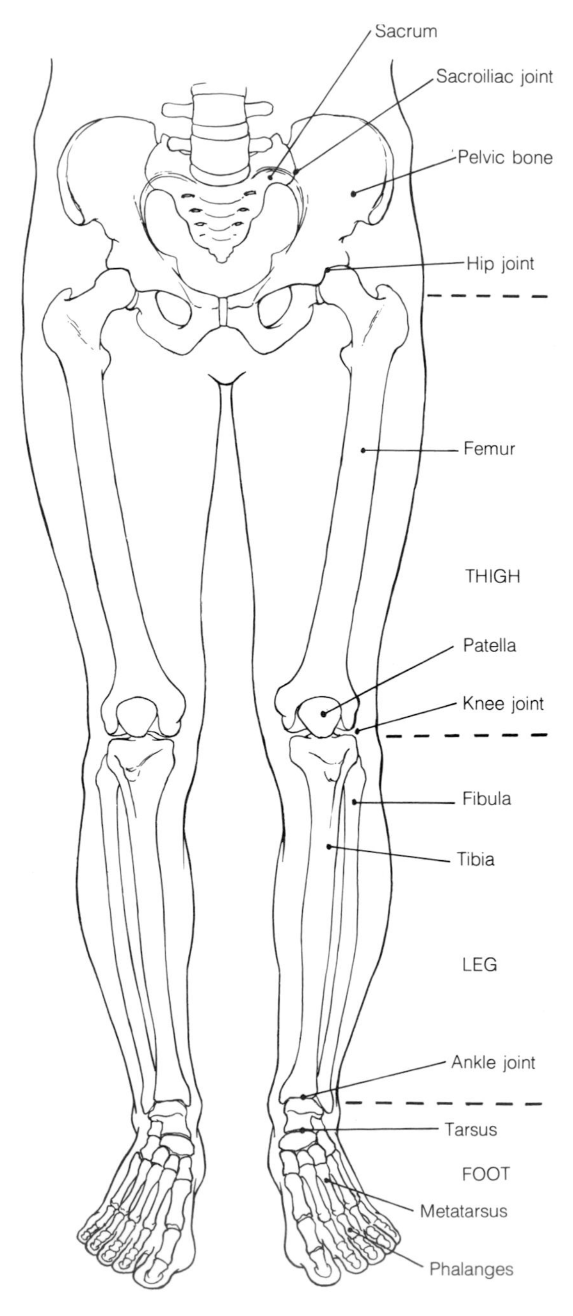

8-1 The skeleton of the lower limb seen from in front.

erable amount of forward and backward swaying is possible without loss of balance.

The spectrum of trauma and disease to which the limbs are subject is broadly speaking related to their functions. The osteoarthritic hip is related to the wear and tear of both weight-bearing and locomotion and numerous varieties of fracture of the lower limb result from the human penchant for active locomotor sports.

The blood supply of the lower limbs is derived from the **abdominal aorta** through the **common** and **external iliac arteries**. As is shown in Fig. 8-2B each external iliac artery passes into the anterior part of the thigh beneath the inguinal ligament. It is at this point that the vessel changes its name to the **femoral artery**. This artery supplies the great mass of musculature of the front of the thigh and, through the **circumflex** and **profunda femoris arteries**, the medial and posterior regions of the thigh. In the lower third of the thigh the femoral artery passes onto the posterior aspect of the limb and over its course behind the knee the vessel is known as the **popliteal artery**.

In the upper part of the leg the popliteal artery divides into the **anterior** and **posterior tibial arteries**. The anterior tibial artery passes between the tibia and fibula to supply the anterior part of the leg and, by the **dorsalis pedis artery** and its branches, the dorsum of the foot. The posterior tibial artery supplies the calf and, through the **peroneal artery**, the lateral part of the leg. At the ankle the posterior tibial artery divides into a **medial** and a **lateral plantar artery** and it is these vessels that supply the sole of the foot. Note in Fig. 8-2 how the dorsalis pedis artery on the dorsum of the foot pierces the first intermetatarsal space to anastomose with the **plantar arch** in the sole of the foot.

The region of the buttock is supplied by the **superior** and **inferior gluteal arteries** which are branches of the **internal iliac artery**. In the thigh these gluteal arteries form important anastomoses with branches of the femoral artery and in so doing provide an alternative or collateral circulation to more distant parts of the limb.

The innervation of the lower limb is provided by all the lumbar and the majority of the sacral ventral rami of the lumbosacral plexus. The main contribution of the lumbar part of the plexus is the **femoral nerve** (L 2, 3, 4) which passes in front of the pelvic girdle to supply the anterior part of the thigh and the extensor muscles of the knee joint. Note, however, that a cutaneous branch, the **saphenous nerve**, extends beyond this region to supply skin of the medial side of the leg and foot. The medial part of the thigh and the adductor muscles it contains are supplied by the **obturator nerve** (L 2, 3, 4). This nerve is also a branch of the lumbar plexus and leaves the pelvic girdle through the obturator foramen. The remainder of the limb, that is the posterior part of the thigh, the leg, and the foot, is supplied by the **sciatic nerve** (L 4, 5, S 1, 2, 3). This nerve appears from the posterior aspect of the pelvic girdle as it passes through the greater sciatic notch. After supplying the posterior part of the thigh the sciatic nerve

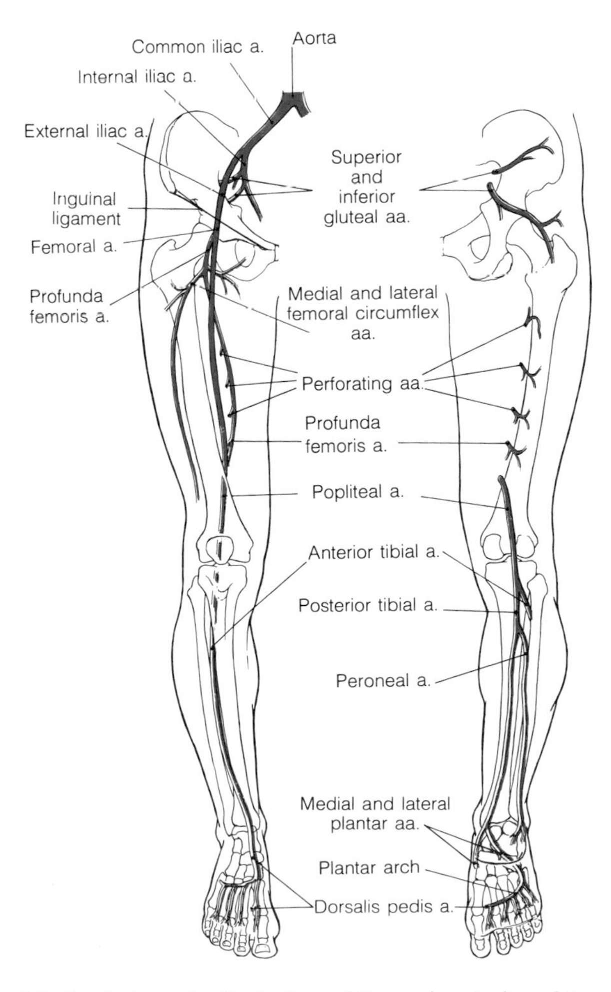

8-2 Semischematic illustration of the main arteries of the right lower limb viewed from in front (A) and from behind (B).

divides into the **common peroneal** and **tibial nerves**. Through its branches, the **superficial** and **deep peroneal nerves**, the common peroneal nerve innervates the front of the leg and the dorsum of the foot. The tibial nerve innervates the back of the leg and through the **medial** and **lateral plantar nerves** the sole of the foot.

The association of anterior and posterior divisions of ventral rami with flexor and extensor musculature which was encountered in the upper limb is also present in the lower limb. However, it has become partly obscured during the evolution of the quadripedal mammal. In this process the primitive hind limb has rotated so as to bring extensor musculature to its front. With the musculature have travelled the femoral and peroneal nerves both of which are formed from posterior divisions of ventral rami.

Two further important branches of the sacral plexus supply the muscles of the buttock or gluteal region. These are the **superior** and **inferior gluteal nerves** which leave the pelvis through the greater sciatic notch.

Using Fig. 8-3 identify each of the nerves that has been mentioned and the regions of the limb which it supplies.

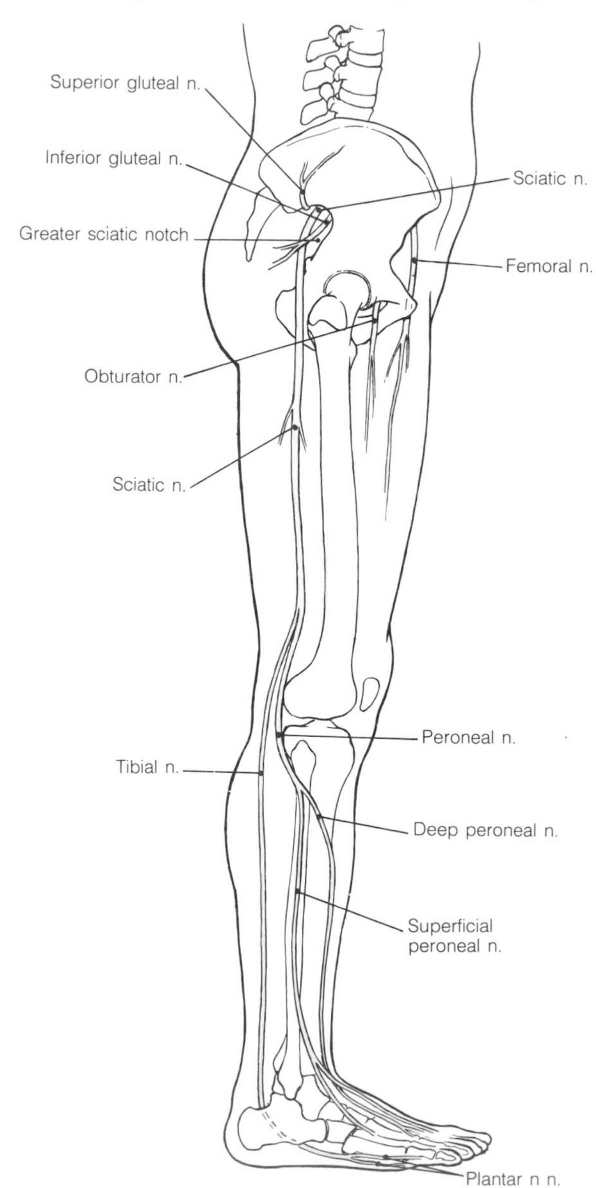

8-3 Semidiagrammatic illustration of the main nerves supplying the rightlower limb.

Superficial Structures of the Lower Limb

Attention should be paid to the superficial structures of the lower limbs for their details have real clinical value and significance. The segmental level of a spinal cord lesion can often be accurately determined by recording the distribution of pain or of regions of skin in which the sensation is altered (paresthesia) or absent (anesthesia). Knowledge of the area drained by a lymph node allows the cause of its enlargement to be discovered. The problem of varicose superficial veins is a common one and its correct treatment depends upon the relationship of the superficial veins to the deep venous drainage of the limb.

The Superficial Veins

The venous drainage of the foot whether from plantar or dorsal regions, is directed to a **dorsal venous arch**. Into this drain **dorsal metatarsal veins** and superficial veins from the sole of the foot that reach the dorsal venous arch through **medial and lateral marginal veins**. The dorsal arch is drained medially by the **great saphenous vein** and laterally by the **small saphenous vein**.

The Great Saphenous Vein*

The great saphenous vein passes anterior to the medial malleolus and continues proximally in the

* Saphenes, *Gk.* = obvious, i.e., clearly visible

subcutaneous tissue of the medial side of the leg and thigh until it reaches the **saphenous opening** in the fascia lata, the deep fascia of the thigh. Here it perforates the **cribriform fascia** which covers the opening and joins the femoral vein. In its course it receives tributaries from the anterior and medial aspects of the leg and from the thigh and lower anterior abdominal wall through the **lateral accessory saphenous veins**, the **superficial circumflex iliac, superficial epigastric**, and **external pudendal veins**. The course and termination of the saphenous vein are illustrated in Figs. 8-4 and 8-5. There are a number of features of this vein which should be carefully noted. These are:

1. Its position in front of the medial malleolus is constant and it can be found and used for intravenous infusions when other superficial veins are collapsed and invisible.
2. It possesses numerous "one way" valves which maintain blood flow in a centripetal direction.
3. It communicates with the deep veins of the limb particularly in the leg by **perforating veins**.
4. These communications possess valves which ensure that blood passes from superficial to deep veins. Failure of these valves may cause varicose veins.

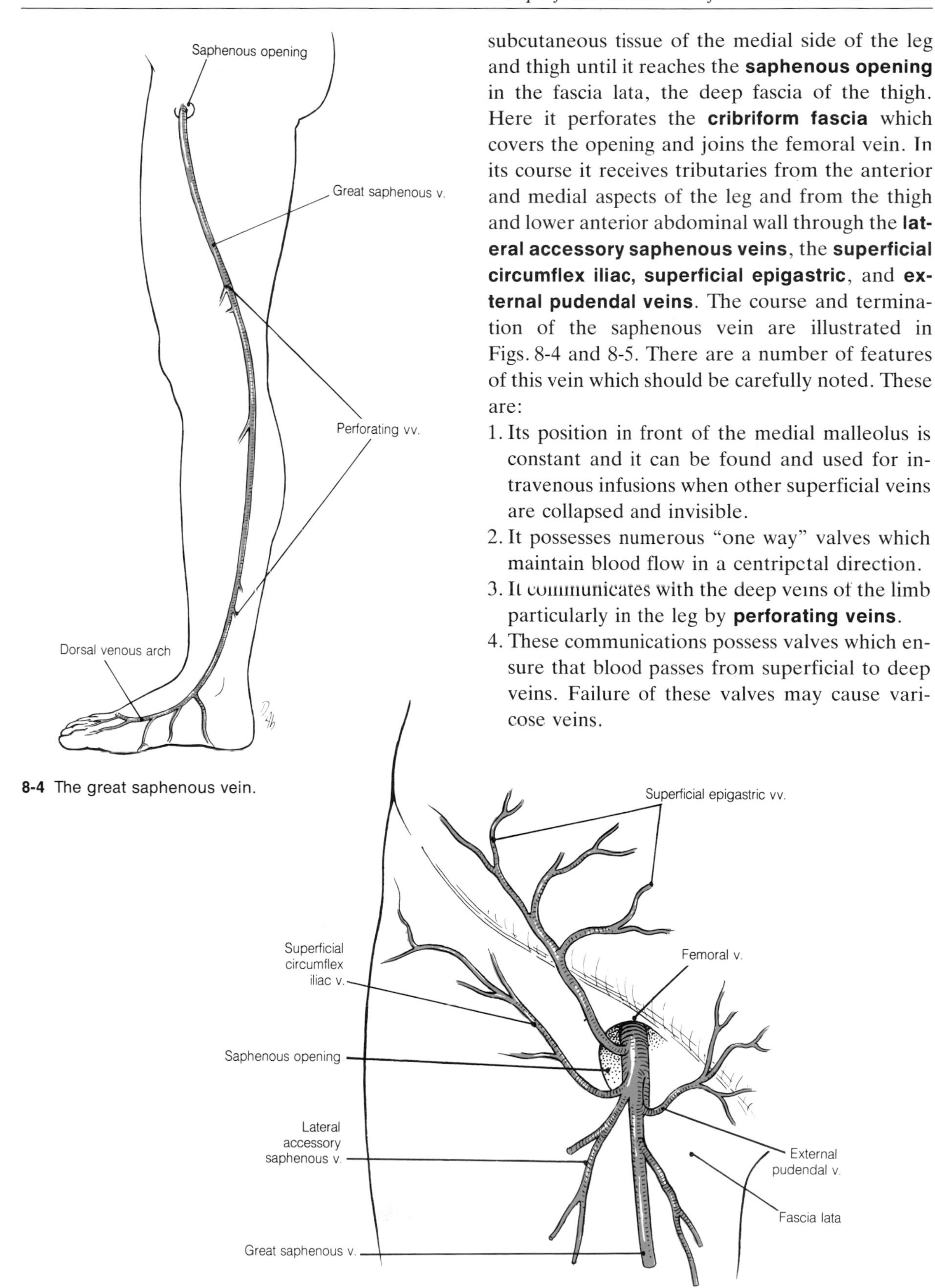

8-4 The great saphenous vein.

8-5 The saphenous opening and tributaries of the great saphenous vein.

5. Communications between the **superficial epigastric veins** on the anterior abdominal wall and veins draining the thoracic wall form a link between the superior and inferior venae cavae. These may enlarge and serve to bypass an obstruction of the inferior vena cava.

The Small Saphenous Vein

The small saphenous vein passes behind the lateral malleolus and winds upward and posteriorly around the calf to pierce the deep fascia of the popliteal fossa behind the knee. Here it joins the popliteal vein. Its tributaries drain the lateral and posterior aspects of the leg. Its course is seen in Fig. 8-6. Like the great saphenous vein it also communicates with deep veins through perforating veins.

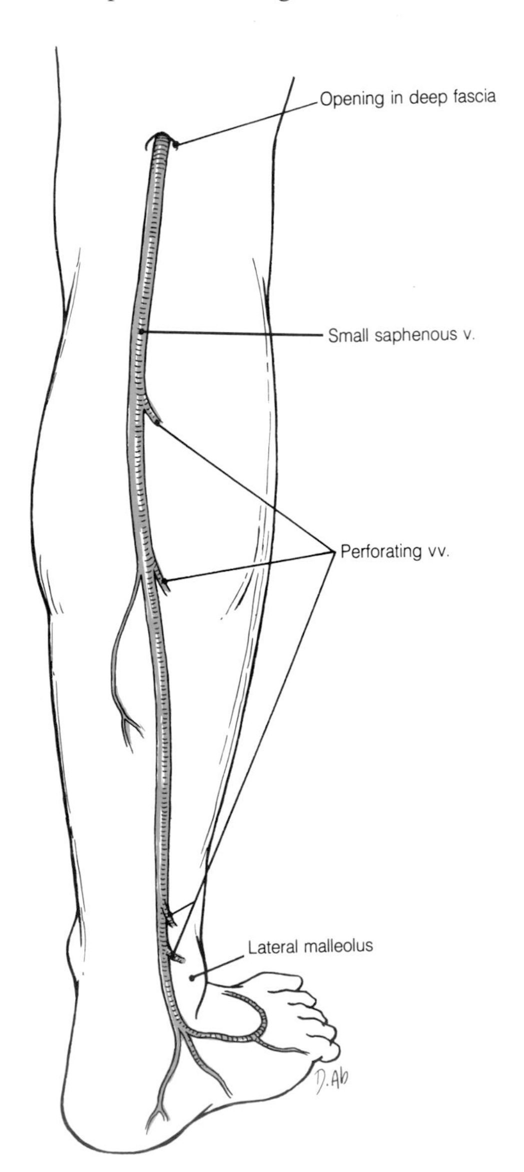

8-6 The small saphenous vein.

Lymphatic Drainage

As elsewhere, the superficial lymphatic vessels of the lower limbs tend to follow the superficial veins and the deep lymphatic vessels follow the main blood vessels. Medial superficial lymphatic vessels, therefore, follow the course of the great saphenous vein to end in the **superficial inguinal nodes** which lie just below the level of the inguinal ligament. Lateral vessels travel proximally with the small saphenous vein to end in **popliteal nodes** beneath the deep fascia of the popliteal fossa or pass medially to reach the superficial inguinal nodes. Deep vessels end either in the popliteal nodes or in **deep inguinal nodes**. Efferents from both the superficial and deep inguinal nodes pass to nodes lying along the external iliac vessels.

The Inguinal Lymph Nodes

The inguinal lymph nodes lie at the root of the thigh just below the **inguinal* ligament**. They are arranged in two groups. A group of **superficial nodes** lies superficial to the deep fascia and a group of **deep nodes** beneath it. It is important to remember that lymph from the abdominal wall and perineum drains to these nodes as well as that from the lower limbs.

The Superficial Nodes

Look at the diagram in Fig. 8-7 and see that the superficial nodes lie immediately below the inguinal ligament and alongside the termination of the great saphenous vein. Note also that the superficial nodes receive lymph from the buttock, the adjacent abdominal wall, the external genitalia, and the lower part of the anal canal and vagina. Clinical enlargement of these nodes should, therefore, suggest an examination of the perineum. The

* Inguen, L. = groin

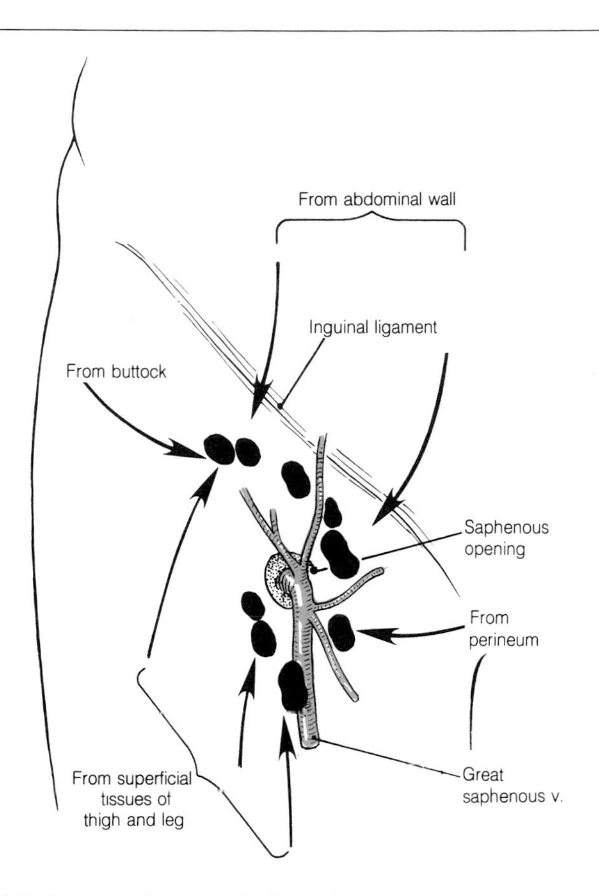

8-7 The superficial inguinal lymph nodes.

inferior nodes around the vein receive lymph from the superficial tissues of the lower limb.

The Deep Nodes

The deep nodes which lie alongside the femoral vein drain the deep tissues of the limbs by afferent vessels accompanying the main blood vessels.

Deep and superficial nodes communicate with each other and their efferent vessels run beneath the inguinal ligament to nodes lying alongside the external iliac vessels.

Cutaneous Innervation

Segmental Innervation

Whereas the pattern of the distribution of dermatomes in the upper limb is fairly well organized around axial lines, those of the lower limb have become distorted by rotation of the primitive limb to a plantigrade position and the fact that skin of the anterior abdominal wall has slipped down over the ventral aspect of the thigh. As can be seen in Fig. 8-8, the first, second, third, and fourth lumbar segments are well represented on the front of the thigh and spiral medially onto the medial surface of the thigh and leg. The fourth and fifth lumbar segments supply the leg medially and laterally and to these is added the first sacral segment on the dorsal and plantar surfaces of the foot. A strip of skin up the posterior aspect of the limb is supplied by the first and second sacral segments and the third and fourth sacral segments complete the innervation of buttock and perineum.

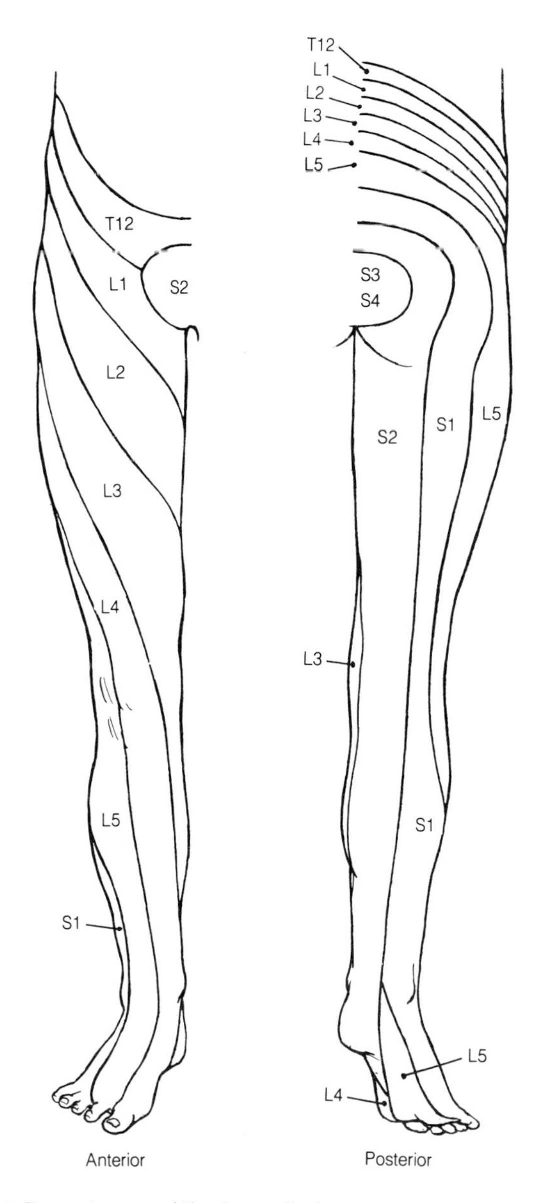

8-8 Dermatomes of the lower limb.

Cutaneous Nerves

Again it should be made clear that anatomical cutaneous nerves contain neurons derived from more than one segmental nerve. For example, the skin of the anterior aspect of the thigh is supplied in vertical strips by the lateral, intermediate, and medial femoral cutaneous nerves. Each of these nerves contains neurons from the second and third lumbar segmental nerves and these are distributed to the obliquely arranged second and third lumbar dermatomes. As a result the pattern of the distribution of segments does not correspond to the distribution of the nerves carrying neurons to their dermatomes.

The **subcostal, genitofemoral**, and **ilioingui-**

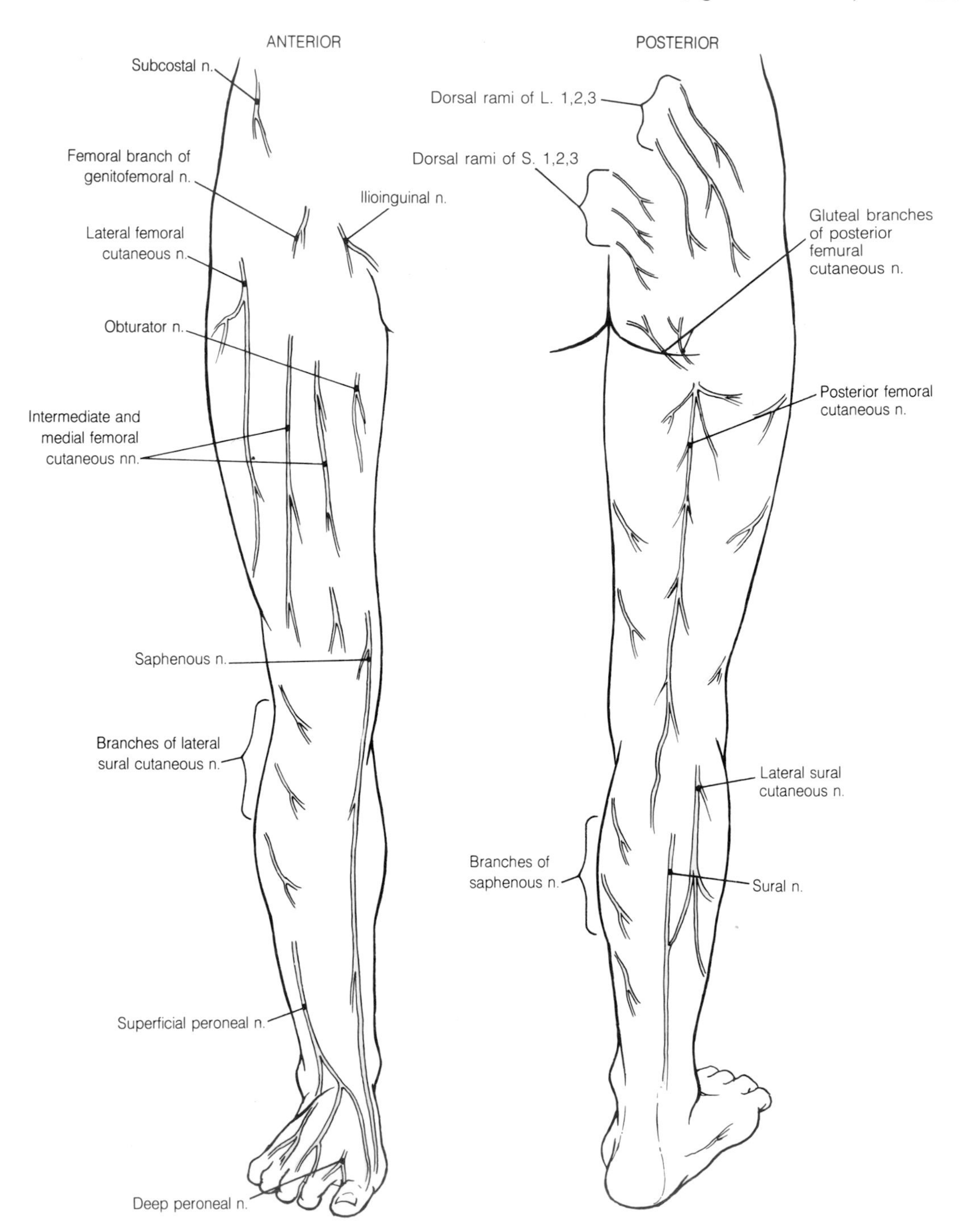

8-9 Cutaneous nerves of the lower limb.

nal nerves supply a strip of skin just below the inguinal ligament. The front and sides of the thigh are supplied by the **lateral, intermediate**, and **medial femoral cutaneous nerves** and cutaneous branches of the **obturator nerve**, while the posterior aspect is supplied by the **posterior femoral cutaneous nerve**. At the medial side of the knee the **saphenous nerve** appears superficially and extends down to supply the medial side of the leg and heel. The lateral aspect of the leg is supplied by the **sural*** **nerve** and the **superficial peroneal nerve**. The latter nerve also supplies the dorsum of the foot and gives dorsal digital branches to most of the toes. However, the cleft between the hallux and the second toe is supplied by the **deep peroneal nerve** and the lateral border of the fifth toe by the **sural nerve**. The sural nerve also supplies skin over the calf and lateral side of the foot. The sole of the foot is largely supplied by the **medial and lateral plantar nerves**. The skin over the gluteal region is supplied by lateral branches of the **dorsal rami** of the first, second, and third lumbar nerves, and the first, second, and third sacral nerves. It is also supplied by branches of the posterior femoral cutaneous nerve of the thigh.

The points at which these cutaneous nerves pierce the deep fascia are illustrated in Fig. 8-9 and more will be said of them as their deeper parent nerves are described.

* Sura, *L.* = calf

The Thigh and Gluteal Region

The Fascia of the Thigh

Removal of the skin and superficial fascia of the thigh reveals a tough layer of deep fascia which surrounds the muscles like a stocking. This is the **fascia lata**. Superiorly, it is attached to the inguinal ligament, the iliac crest, the thinner fascia covering the gluteus maximus of the buttock, the ischium and pubis. Below and anteriorly it is attached to the patella and tibial condyles, but posteriorly it extends into the leg over the popliteal fossa and into the calf. Laterally, it splits to enclose the belly of the muscle **tensor fasciae latae** which arises from the anterior part of the lateral lip of the iliac crest. This muscle pulls upon a band-like thickening of the fascia lata called the **iliotibial tract** which extends down the lateral aspect of the thigh to be attached to the lateral condyle of the tibia.

Just below the medial end of the inguinal ligament the fascia lata is deficient at the **saphenous opening**. Here the **great saphenous vein** passes deeply to reach the **femoral vein**. It is accompanied by lymphatics which penetrate the thin fascia which fills in the opening and give it its name, the **cribriform*** **fascia**.

A tough **lateral intermuscular septum**, which

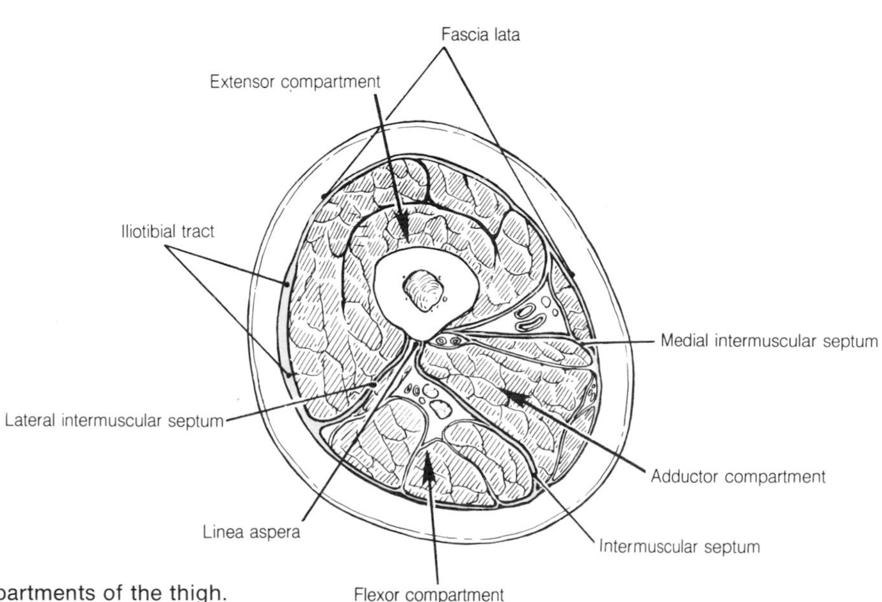

8-10 The compartments of the thigh.

clearly separates the anterior extensor muscles from the posterior flexor or hamstring muscles is attached to the deep surface of the fascia lata and to the **linea aspera** of the femur. It is through its attachment to this septum that the greater part of the **gluteus maximus muscle** exerts its action on the femur. A less robust **medial intermuscular septum** which also extends from the fascia lata to the linea aspera separates the extensor from the adductor muscles on the medial side of the thigh. Finally an intermediate septum separates the adductor from the flexor muscles. The presence of these septa allows the thigh to be divided into extensor, flexor, and adductor compartments and these are illustrated in Fig. 8-10.

Before embarking on a description of the deeper structures of the thigh and gluteal regions, a description of the hip bone and femur is given.

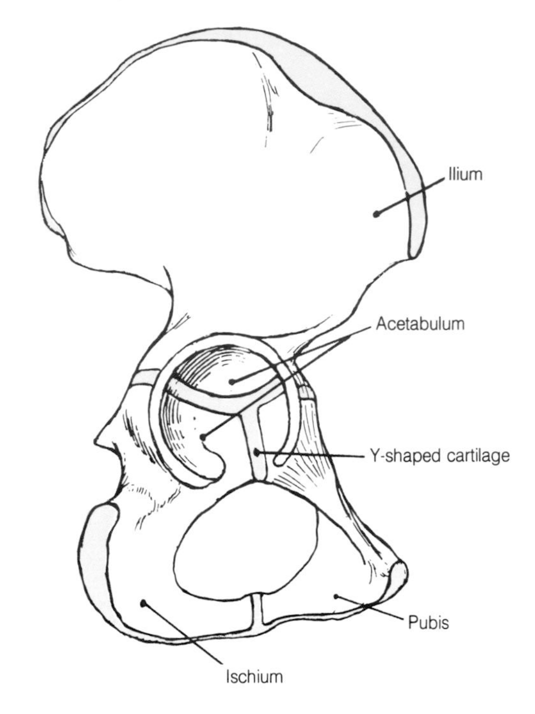

8-11 The parts of the hip bone.

The Skeleton of the Hip and Thigh

The Pelvic Girdle

The bones of the pelvic girdle form the protective bony walls of the pelvic cavity and a part of the birth canal. The girdle also functions as a stable platform which supports a substantial part of the body weight through its link with the sacrum and transmits this weight to the mobile lower extremities. The pubic symphysis and the sacroiliac joints which firmly unite the bones of the girdle are described with the pelvic cavity but should be reviewed here. Additional features of the borders and outer aspects of the pelvic bones must now be described together with the sacrum and femur.

The Hip Bone

The hip bone consists at birth of three separate bones joined together by hyaline cartilage. These are the **ilium**, the **ischium**, and the **pubis**, and each has its own primary center of ossification. By puberty ossification has progressed to a point at which the bones are only separated by a Y-shaped area of cartilage centered in the acetabulum, the cup that accepts the head of the femur. The three bones and the cartilage can be seen in Fig. 8-11. The fact that the bones are separated in this manner during childhood is of importance when viewing radiographs and to illustrate this a radiograph of an infant's pelvis and hip joint is shown in Fig. 8-12. Secondary centers appear within the Y-shaped cartilage and complete fusion occurs after puberty. Additional secondary centers (also shown in Fig. 8-11) appear in the cartilage for the iliac crest, pubic symphysis, ischial tuberosity, and anterior inferior iliac spine at about the same time. A knowledge of the mode of ossification of the pelvic bone is essential for the interpretation of radiographs in both disease and injury. It is also important to orient the bone with the anterior superior iliac spine and the pubic tubercle in the same vertical plane.

With the help of Figs. 8-13 and 8-14, and ideally with the pelvic bone itself available, further detail of the component bones can be now considered.

The Pubis

Medially the flattened **body** of the pubis articulates with its fellow at the symphysis pubis and from it radiate its **superior and inferior rami** which partly surround the **obturator foramen**. The superior ramus passes laterally to join with the **ilium** and **ischium** at the **acetabulum**. The inferior ramus passes laterally, downward and

* Cribra, *L.* = a sieve

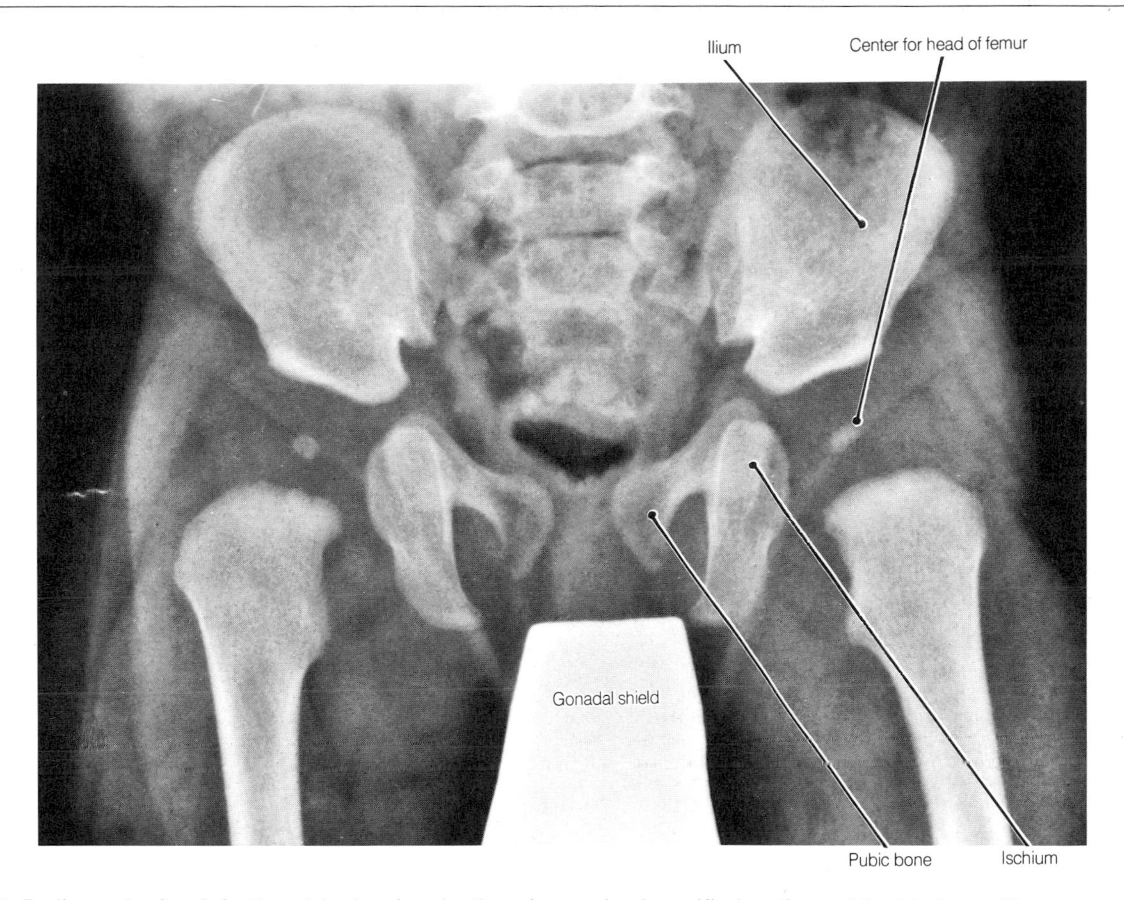

8-12 Radiograph of an infant's pelvis showing the three incompletely ossified portions of the hip bone. (Reproduced by permission, from Wicke: *Atlas of Radiologic Anatomy,* 4th Ed, Urban & Schwarzenberg, Baltimore-Munich, 1987.)

backward to join the **ramus of the ischium** at a point which can no longer be determined after fusion of these bones. The upper border of the body forms the **pubic crest** and at the lateral extremity of this is found the **pubic tubercle**. The posterior margin of the superior ramus shows a sharp raised ridge called the **pectineal line** or **pecten pubis**. This forms part of the pelvic brim. The inferior ramus has an everted medial border more marked in the male than female. The outer aspect of the pubic bone is largely covered by the proximal attachments of the adductor group of muscles. The obturator foramen is occluded in life by the thick **obturator membrane** which together with surrounding bone provides attachment for the **obturator externus muscle**. A small opening in the membrane allows the **obturator nerve** and **vessels** to leave the pelvis through the **obturator groove**. This lies on the inferior surface of the superior ramus close to the acetabulum.

The Ilium

The ilium presents a wing-like or **alar** posterolateral surface which provides attachment for the **three gluteal muscles** and the **tensor fasciae latae muscle**. Its borders present a number of features. Anteriorly are the **anterior superior iliac spine** and below this the **anterior inferior iliac spine** to which the **rectus femoris muscle** is attached. From the superior spine the **iliac crest** extends backward to a more expanded region called the **iliac tubercle**. Continuing beyond this the crest terminates at the **posterior superior iliac spine** and below this lies a **posterior inferior iliac spine**. This spine marks the upper end of a deep indentation in the posterior margin of the bone called the **greater sciatic notch**.

The lateral surface of the ala shows three roughened lines called the **posterior, anterior**, and **inferior gluteal lines**. Identify each of them in

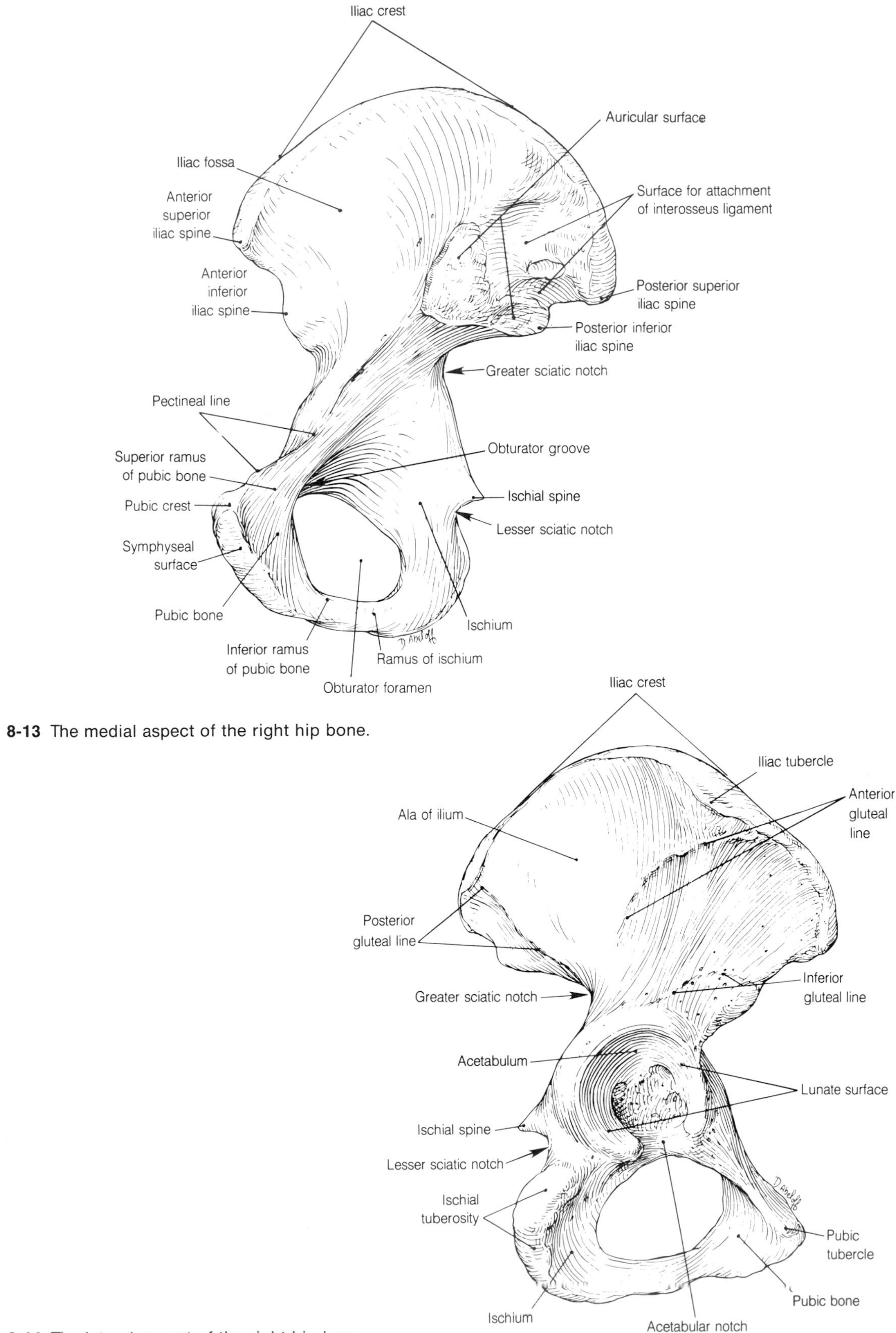

8-13 The medial aspect of the right hip bone.

8-14 The lateral aspect of the right hip bone.

Fig. 8-14. These lines separate the sites of attachment of the important gluteal muscles.

Medially the ala shows a shallow concavity in its anterior part. This is the **iliac fossa** and gives attachment to the fan-shaped muscle **iliacus**. Posteriorly the medial surface shows an ear-shaped **auricular surface** for articulation with a corresponding surface on the sacrum and behind this an irregular surface for the attachment of strong interosseous ligaments which join the sacrum and ilium.

The Ischium

The **body of ischium** is continuous with that of the ilium at the acetabulum and its **ramus** is fused to the inferior ramus of the pubis forming the **ischiopubic ramus**. Its posterior border contributes

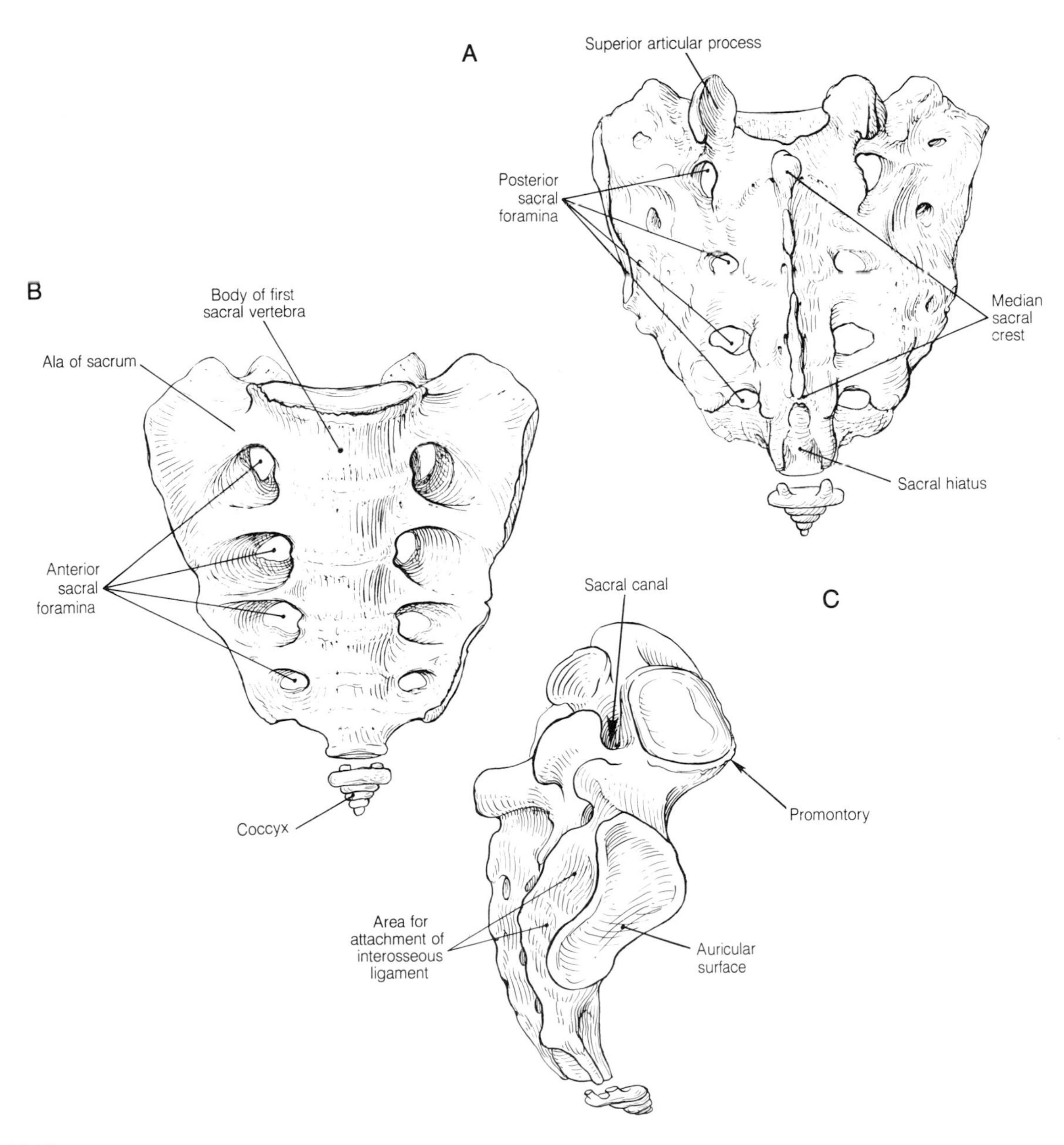

8-15 The sacrum.

to the greater sciatic notch and presents the **ischial spine** at the lower margin of the notch. To this the **sacrospinous ligament** is attached. Below the spine lies the **lesser sciatic notch**. Most inferiorly is the **ischial tuberosity**, whose roughened surface gives attachment to the "hamstring" muscles which extend the hip and flex the knee.

*The Acetabulum**

The acetabulum lies at the site of fusion of the three parts of the hip bone. It is a cup-shaped cavity facing laterally and slightly downward and forward. Its margin is deficient inferiorly at the **acetabular notch**. A similar deficiency is seen in the smooth articular or **lunate surface** that lines the inner surface of the cup. The floor of the cup is, however, not articular.

If the wing of the ilium is felt between the articular surface of the sacroiliac joint and the upper margin of the acetabulum, it will be found substantially thickened. It is through this bar of bone that the weight of the trunk is transmitted to the head of the femur as it lies in the acetabulum.

The Sacrum

Refer to Fig. 8-15 as the following description is read. The five sacral vertebrae ar fused into a solid wedge-shaped mass. The superior surface or **base** is formed by the stout **alae** which lie on each side of the body of the first sacral vertebra. Anteriorly the superior margin of the first sacral vertebra is called the **promontory**. There are two superior articular processes on the first sacral vertebra and these together with its body articulate with the fifth lumbar vertebra. It is not uncommon to find these two vertebrae partially fused.

A concave **pelvic surface** faces anteriorly and downward. Transverse bony ridges indicate the junctions of the individual vertebrae and two vertical rows of four **anterior sacral foramina** allow the exit of the sacral ventral rami. A **dorsal surface** shows an irregular **median sacral crest** flanked by four **posterior sacral foramina**

* Acetabulum, *L.* = vinegar cup

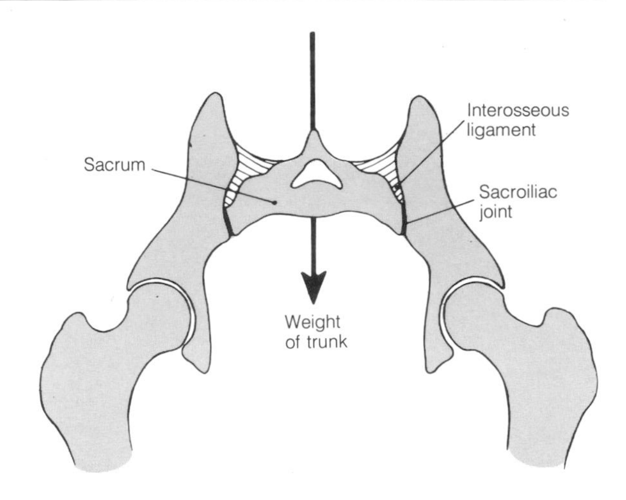

8-16 Diagrammatic coronal section through the sacroiliac joints.

through which dorsal rami pass. Both anterior and posterior sacral foramina lead inward to intervertebral foramina which are occupied by the sacral spinal nerves. On either side of the posterior sacral foramina lie indistinct bony ridges called the **intermediate** and **lateral sacral crests**. Below, the neural arch of the fifth sacral vertebra is absent and the resultant defect in the dorsal surface is called the **sacral hiatus**. It is through this that local anesthetic may be introduced into the **sacral canal**. The pelvic and dorsal surfaces meet inferiorly at the **apex** of the sacrum where it articulates with the coccyx.

On the upper part of each **lateral surface** there is an ear-shaped **auricular surface** for articulation with the iliac bone and behind this a roughened area, the **sacral tuberosity**, to which the short and strong **sacroiliac interosseous ligament** is attached.

From the description given above it might be imagined that the weight of the trunk that is transmitted through the sacrum to the pelvic girdle would tend to force the iliac bones apart. However, if the sacrum is correctly oriented, the long axes of its auricular surfaces lie anteroposteriorly and examination of Fig. 8-16 shows how, in fact, the wedge is reversed and the weight of the trunk is "suspended" between the iliac bones by the interosseous ligaments. Note also in this section the thick bars of bone which transmit the body weight to the head of the femur on each side.

The Femur

The femur is the single long bone of the thigh. It has a **body or shaft**, a **head**, and a lower end formed by two **condyles**. The head is globular and articulates with the acetabulum. Distal to the head is a stout **neck** which is directed laterally and slightly backward. It forms an angle of about 125° with the long axis of the shaft. Where the neck joins the shaft there are two substantial projections. These are laterally the **greater trochanter** and medially the **lesser trochanter**. Medially the greater trochanter overhangs a hollow in its base called the **trochanteric fossa**. The place where the neck joins the shaft is marked anteriorly by a roughened ridge called the **intertrochanteric line**. A similar but smoother elevation joins the trochanters posteriorly. This is the **intertrochanteric crest**. On the crest lies a rounded elevation, the **quadrate tubercle**. The body is smoothly rounded over the greater part of its length, but

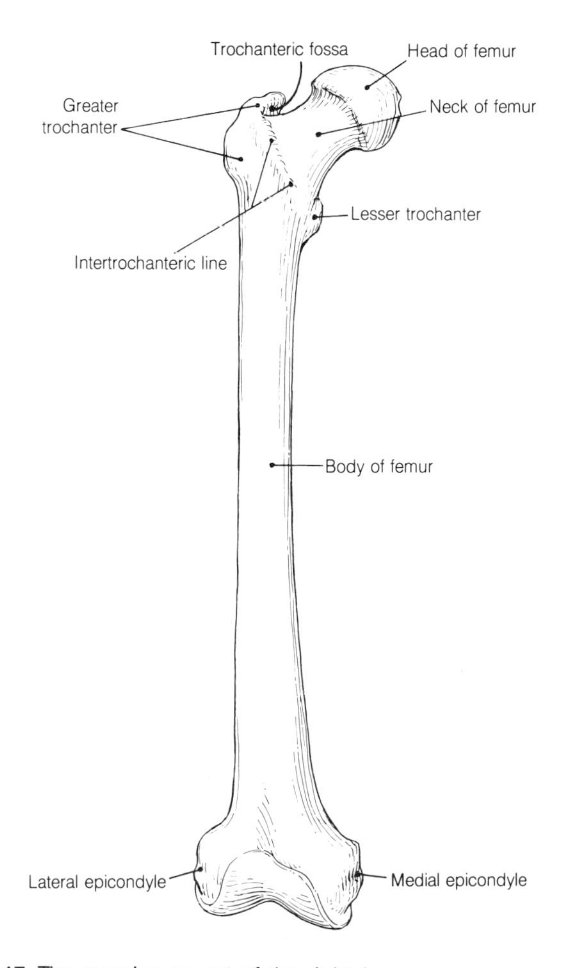

8-17 The anterior aspect of the right femur.

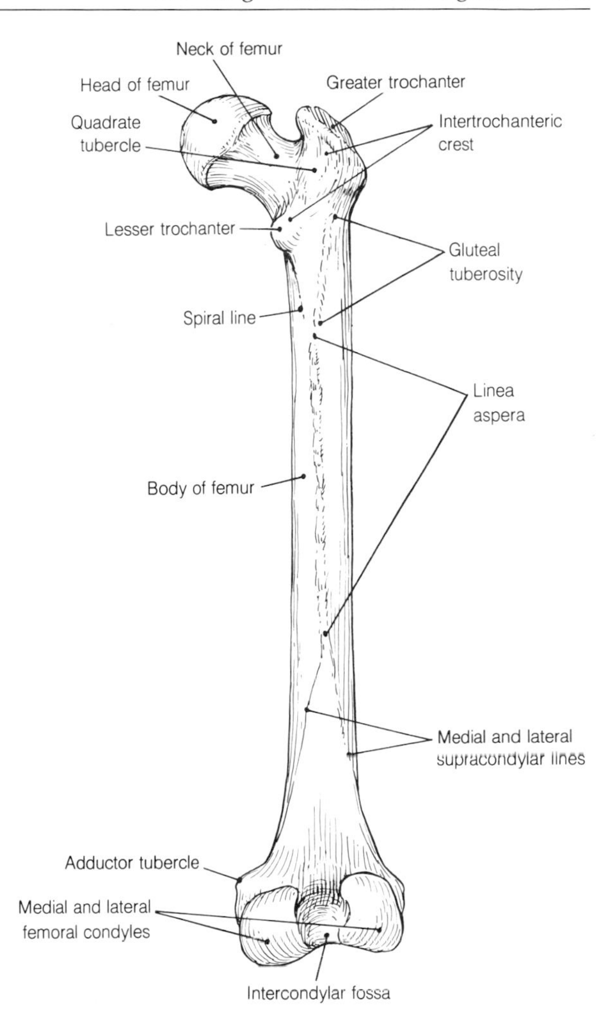

8-18 The posterior aspect of the right femur.

posteriorly there is a roughened vertical line, the **linea aspera**. Superiorly this expands near the greater trochanter and becomes the **gluteal tuberosity** and a less well marked **spiral line** extends toward the lesser trochanter. Inferiorly, the linea aspera divides into the **lateral and medial supracondylar lines**. These lead to the two femoral condyles which make up the lower end of the bone. The two condyles are separated by an **intercondylar fossa** inferiorly and posteriorly. The femoral condyles articulate below with the condyles of the tibia. Anteriorly the condyles merge at a shallow depression to articulate with the patella. The lateral surface of the lateral condyle shows a central elevation, the **lateral epicondyle**, and a well-marked groove for the tendon of the muscle **popliteus** below this. The medial surface of the medial condyle also shows an **epicon-**

dyle above which lies a further elevation, the **adductor tubercle**.

All the features described above should now be confirmed in Figs. 8-17 and 8-18 or on a specimen of the bone.

The primary center of ossification for the body and the anatomical neck, appears during the eight week of intrauterine life. A secondary center for the lower ends is often present at birth and is evidence of full term. The center for the head appears at one year, that for the greater trochanter at three years, and that for the lesser trochanter at ten years. The centers for the greater and lesser trochanters join the shaft individually, unlike those for the greater and lesser tubercles of the humerus which fuse with the center for the head and then together fuse with the shaft.

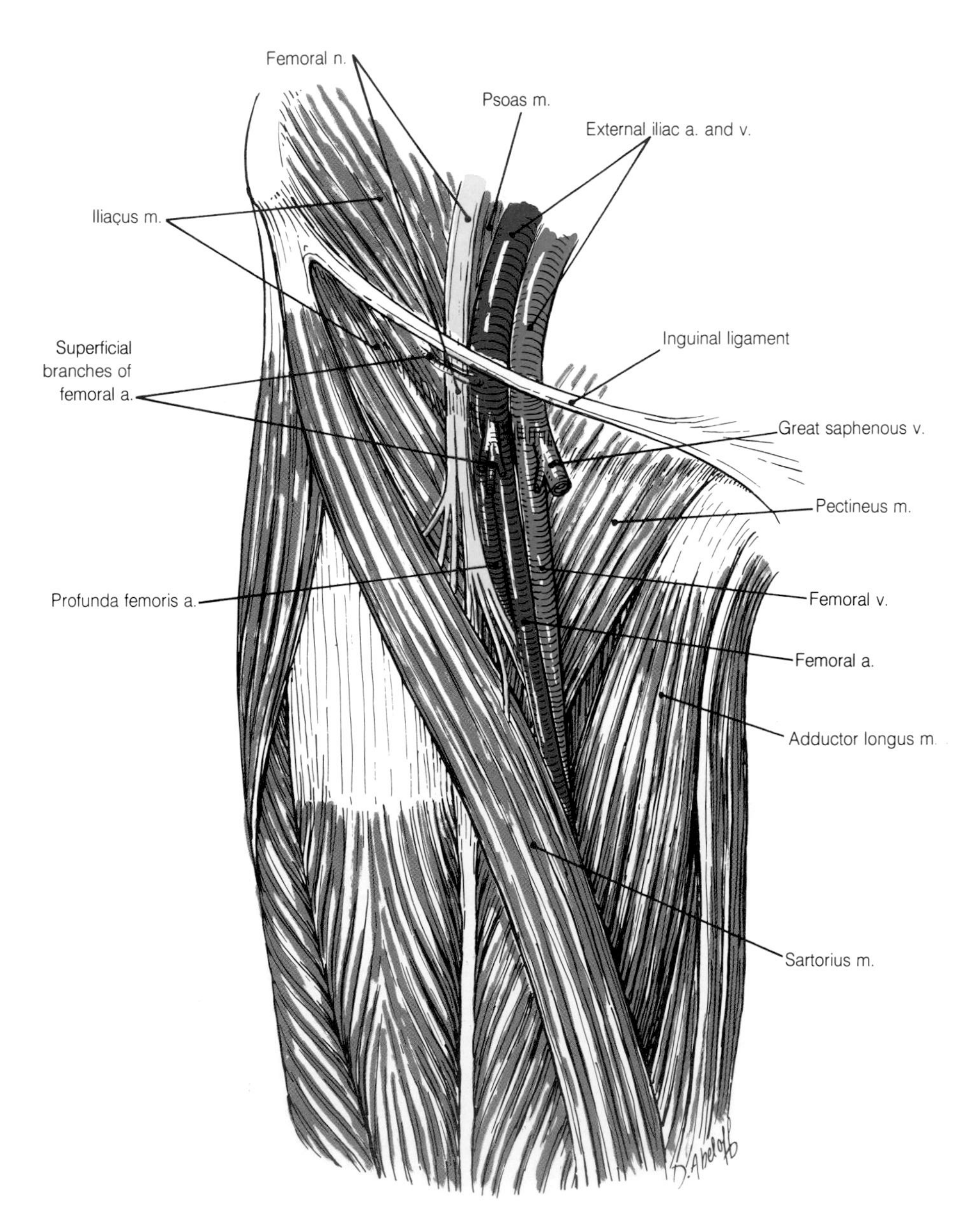

8-19 The femoral triangle and its contents.

The Extensor Compartment of the Thigh

This is a well defined fascial compartment limited anterolaterally by the fascia lata and separated from the adductor compartment by the medial intermuscular septum and from the flexor compartment by the lateral intermuscular septum. It contains the massive **quadriceps femoris muscle**, the **femoral nerve**, and the **femoral vessels**. The nerve and vessels lie at first in a region called the **femoral triangle**.

The Femoral Triangle

The femoral triangle has importance as a junctional region between the trunk and lower limbs. It contains the main vessel supplying the limb and the nerve supplying the muscles of the extensor compartment. These structures are covered by little other than skin and fascia and are, therefore, vulnerable. Wounds involving the vessels are not uncommon and occasionally lead to the formation of an arteriovenous fistula between the femoral artery and vein.

The **base** of the triangle is formed by the inguinal ligament and its **lateral border** by the medial margin of the sartorius muscle. The **medial border** is variously described but in this text it is taken to be the lateral margin of the adductor longus muscle. The two borders meet below at the **apex**. The triangle is roofed over by the fascia lata and the cribriform fascia. The muscles iliopsoas and pectineus lie in its floor.

Look at Fig. 8-19 to confirm this description of the triangle and see also that, of the structures passing beneath the inguinal ligament into the triangle, the femoral nerve is the most lateral. The nerve breaks up almost immediately into its terminal branches. Moving medially this is followed by the femoral artery, the femoral vein, and finally, the femoral canal. The latter three structures are found initially to be bound together by thickened fascia known as the **femoral sheath**. This is derived from extensions of the **transveralis fascia** anteriorly and the **iliac fascia** posteriorly. Note that the femoral nerve is separated from the sheath by the iliac part of the **iliopectineal fascia**. In the diagram in Fig. 8-20 the region between the inguinal ligament and the hip bone is viewed from below and the arrangement of the structures passing beneath the ligament and into the femoral triangle can be seen. Note the **femoral canal** lying between the femoral vein and the lacunar ligament. The canal is closed at its abdominal end by

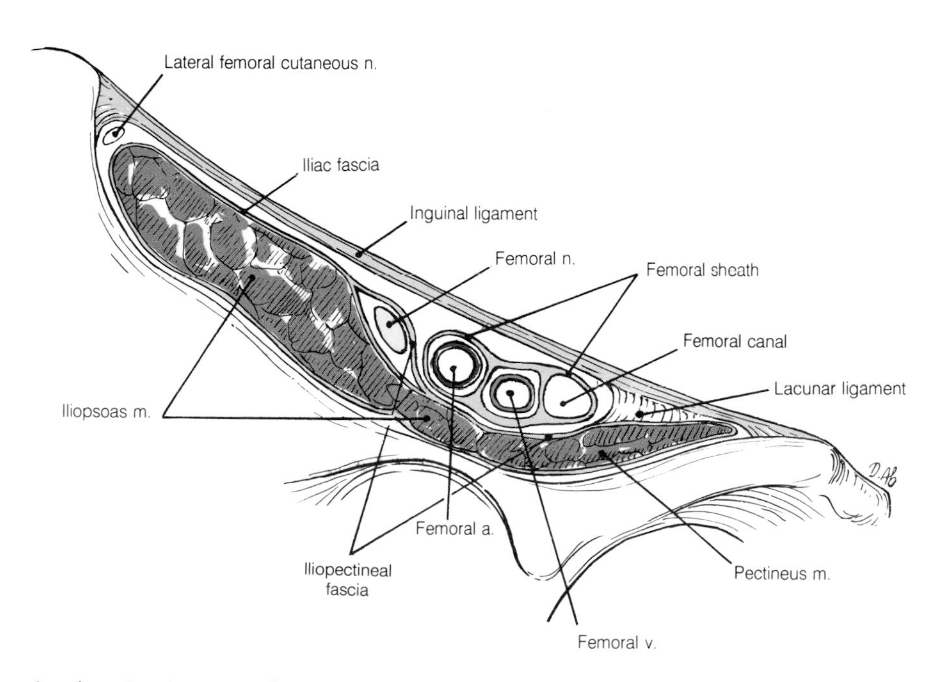

8-20 Diagram showing structures passing deep to the inguinal ligament.

peritoneum. On occasion this may be pushed or drawn through the canal to form a **femoral hernia**. The triangle also contains deep lymphatic nodes lying alongside the femoral vein and it is in the triangle that the femoral artery gives off its deep branch, the **profunda femoris artery**. At the apex of the triangle the femoral vessels pass into the **adductor canal**.

The Adductor Canal

At the apex of the femoral triangle the sartorius muscle overlaps the adductor longus muscle and at this point the femoral vessels disappear beneath sartorius. The vessels follow this muscle as they wind round to the medial aspect of the thigh in a channel called the **adductor canal**. Initially, the canal is bounded laterally by the vastus medialis and medially by the adductor longus and then by the vastus medialis and the adductor magnus. In Fig. 8-21 the canal is seen in cross-section in the middle of the thigh. Note that the roof, formed by sartorius, is now medial to the canal and the borders are anterior and posterior.

Having defined these two regions, the structures of the extensor compartment can be considered in more detail.

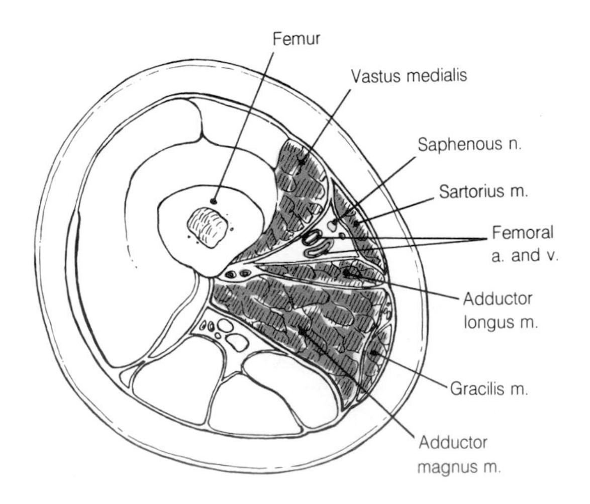

8-21 Cross-section of the thigh showing the adductor canal.

The Muscles of the Extensor Compartment

The main mass of muscle in this compartment is made up of the four heads of **quadriceps femoris** which extend the knee joint. Also included is the slender **sartorius muscle** and two muscles that have been seen in the floor of the femoral triangle. These, **iliacus** and **psoas**, are only found in the upper part of the thigh and are primarily flexors of the hip joint. Because it acts as an extensor of the knee joint, the **tensor fasciae latae** is described here although, unlike the other muscles of this compartment, it is supplied by a branch of the sciatic nerve.

Iliacus

Iliacus is a large, fan-shaped muscle arising from the upper part of the iliac fossa in the posterior wall of the abdominal cavity. It descends over the superior pubic ramus and beneath the inguinal ligament to join the lateral side of the tendon of **psoas major**. In common with psoas it is separated from the lower part of the iliac fossa and the hip joint over which it passes by a large bursa called the **psoas bursa**. The surface of the muscle is covered by the iliac fascia and the femoral nerve passes toward the lower limb in a shallow gutter between it and psoas. It is supplied by branches of the **femoral nerve** (L2, 3).

Psoas Major

The psoas arises as slips from the lower part of the twelfth thoracic vertebra, the upper and lower parts of the lumbar vertebrae, and the adjacent intervertebral discs. The muscular arches formed over the middles of the vertebral bodies allow the passage of lumbar segmental arteries and the rami connecting the sympathetic chain to the lumbar ventral rami. The muscle so formed descends alongside the lumbar vertebrae and following the brim of the pelvis passes deep to the inguinal ligament. A tendon is formed on its lateral aspect which receives the fibers of iliacus. This tendon is inserted into the lesser trochanter of the femur.

The combination of psoas with iliacus (often known as the iliopsoas muscle) is a powerful flexor of the hip joint. Any part played in rotation of the femur is probably inconsiderable but it should be

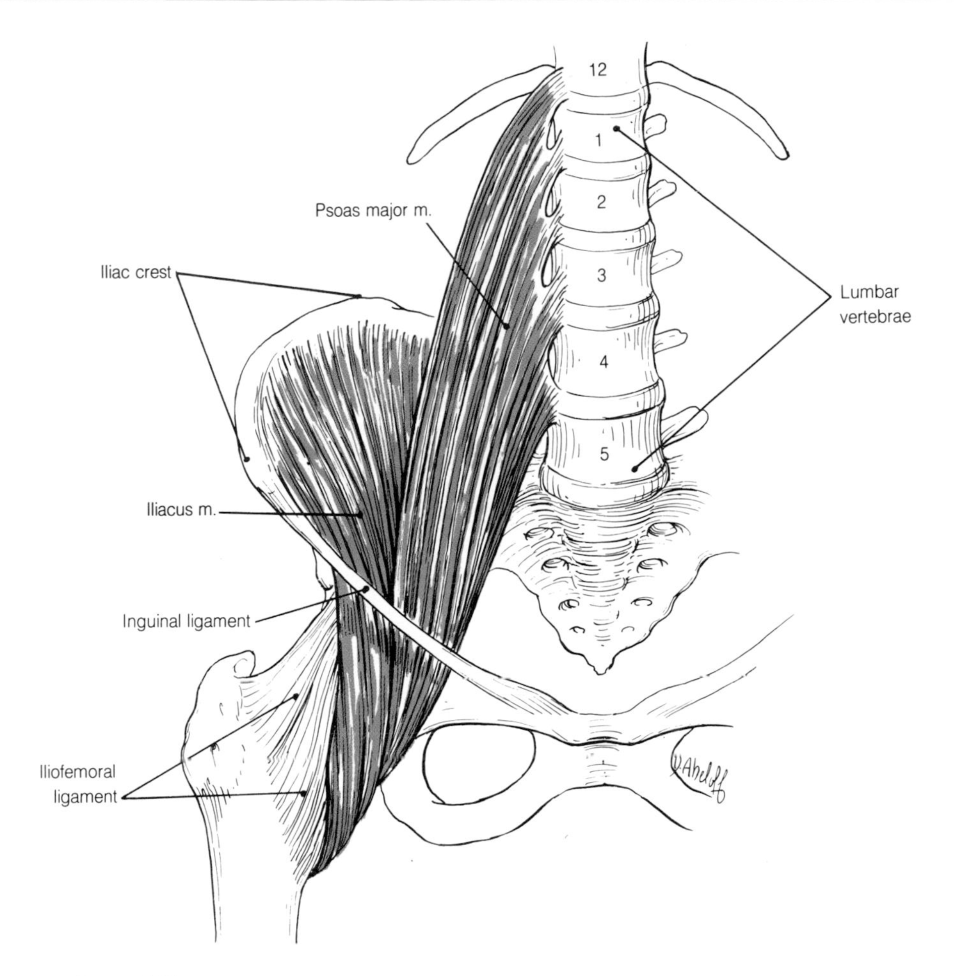

8-22 The psoas major and iliacus muscles.

remembered that when the neck of the femur is fractured and movement is possible between the fragments, the iliopsoas becomes a lateral rotator of the lower fragment. The iliopsoas may also act as a sensitive indicator of disease in or around the lumbar spine, which may be reflected as a flexion deformity of the hip joint. The psoas muscle is supplied by branches of the **second and third lumbar ventral rami**.

The arrangement of these two muscles in relation to the hip bone and femur are seen in Fig. 8-22.

A **psoas minor muscle** is sometimes found on the anterior surface of psoas major. Its flat tendon is inserted into the pectineal line.

Sartorius

The sartorius* or tailor's muscle is a long narrow strap muscle which arises from bone just below the anterior superior iliac spine. It descends the front of the thigh obliquely from lateral to medial side and is inserted into the shaft of the tibia below its medial condyle. The muscle crosses both the hip and knee joint and is a weak flexor of the thigh on the trunk and of the leg on the thigh. These two actions are combined in the traditional tailor's crossed-leg sitting position. The muscle is supplied by the **femoral nerve** (L2, 3).

* Sartus, *L.* = patched or repaired

Tensor Fasciae Latae

This small muscle arises from the anterior part of the outer lip of the iliac crest. It is enveloped by the fascia lata from which it also arises and it becomes continuous with the iliotibial tract. This means that its distal attachment is to the lateral condyle of the tibia. The action of this muscle is not easy to determine but it probably plays a part in the last few degrees of extension of the knee when its line of action commes to lie in front of the axis of the knee joint.

The muscle is supplied by the **superior gluteal nerve** (L4, 5).

Quadriceps Femoris

The quadriceps* femoris makes up the main bulk of the muscles of this compartment. Three of its heads arise from the femur, one from the hip bone. **Rectus femoris**, the pelvic head, is attached to the anterior inferior iliac spine and to the upper lip of the acetabulum by means of a bifurcate tendon. It can easily be recognized by the bipennate arrangement of its fibers. The **three vasti muscles**, the femoral heads of quadriceps femoris, are named **lateralis, medialis**, and **intermedius**. The vastus lateralis has a long linear origin from the intertrochanteric line, the lateral lip of the linea aspera and the lateral supracondylar ridge of the femur. Fibers also arise from the lateral intermuscular septum. Vastus medialis arises from the spiral line and the medial lip of the linea aspera. Vastus intermedius arises from the anterior and lateral aspects of the upper two-thirds of the shaft of the femur. The relationships of the bellies of these muscles to one another are seen in the horizontal section of the thigh illustrated in Fig. 8-23. In this, note how they are wrapped around almost the entire body of the femur leaving only the linea aspera for other muscles to gain attachment. Their appearance on the anterior aspect of the thigh is shown in Fig. 8-24 and their origins from the femur in Fig. 8-25.

* Quadr. *L.* = a prefix indicating four; Caput, *L.* = head, i.e., a four-headed muscle

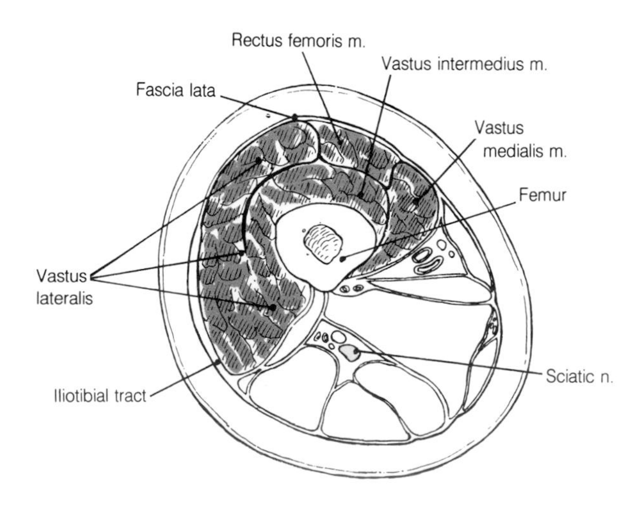

8-23 Cross-section of the thigh showing the quadriceps femoris muscle.

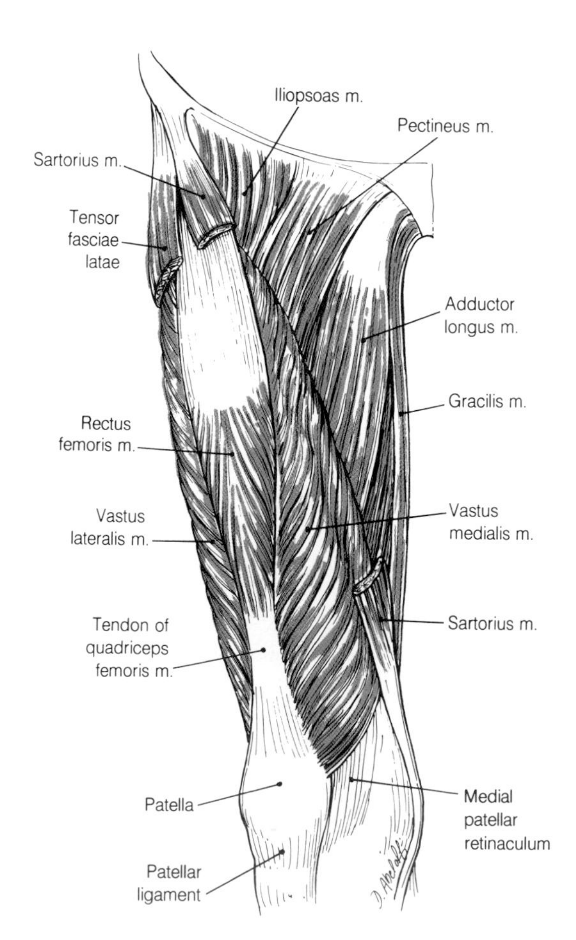

8-24 The anterior aspect of the thigh showing the quadriceps femoris muscle.

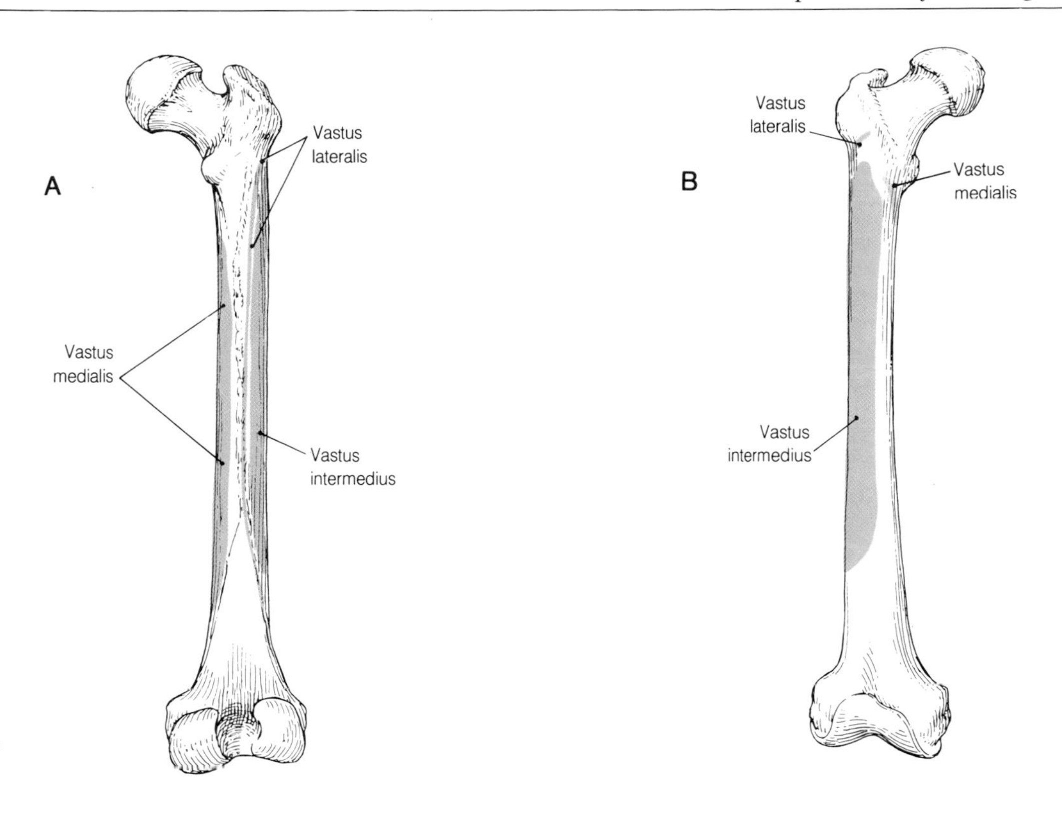

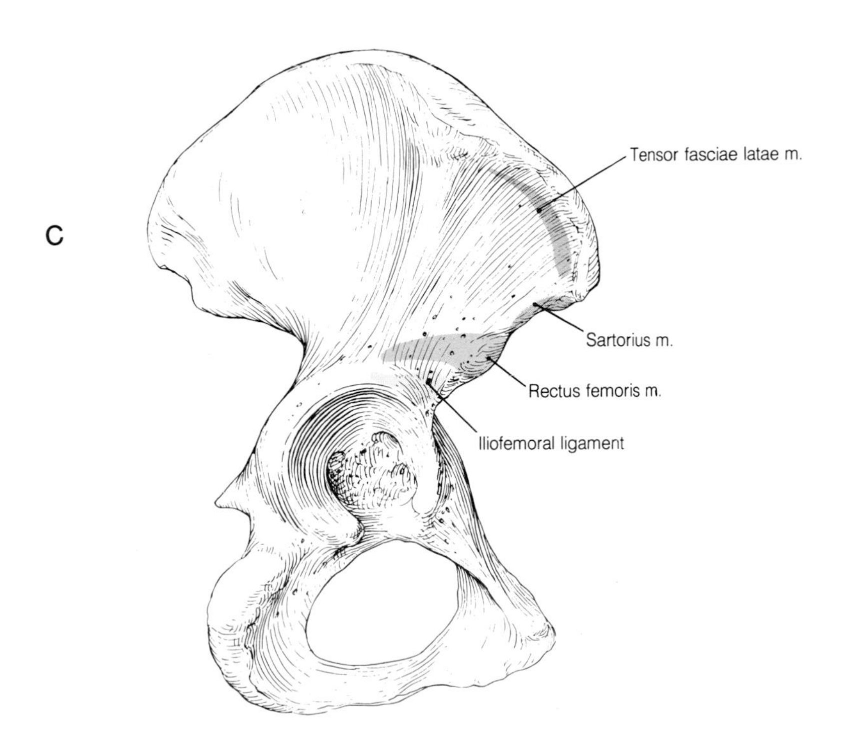

8-25 The posterior (A) and anterior (B) aspects of the femur and the hip bone (C) showing the origins of muscles of the extensor compartment of the thigh.

The four bellies of the muscle fuse at the **quadriceps tendon**. In the midline the tendon is interrupted by the presence of the **patella**, a sesamoid bone. The continuation of the tendon below the patella is known as the **patellar ligament** which is attached to the tibial tuberosity. In addition, the vasti send fibrous expansions over the front of the knee joint to the tibial condyles and thus reinforce its capsule. These are called the **medial and lateral patellar retinacula**.

Some space has been taken to describe this muscle, as its action as an extensor of the knee joint is required for walking and, more particularly, for rising, running, climbing. In addition, the activity of the lower oblique fibers of vastus medialis is essential for the stability of the patella into which they are inserted directly. Also the whole extensor apparatus acts as a ligament of variable length to maintain the stability of the knee joint. The muscle rapidly wastes when immobilized and this initiates a vicious circle when coupled with an injury to the knee joint. The orthopaedic surgeon and physical therapist go to great lengths to prevent this wasting. Unlike the three vasti, the rectus femoris is attached to the pelvis and is, therefore, a "two-joint" muscle and will help to flex the thigh on the trunk.

The four bellies are each supplied by branches of the **femoral nerve** (L2, 3, 4).

Vessels of the Extensor Compartment

The Femoral Artery

The femoral artery is a continuation of the **external iliac artery**. It appears in the femoral triangle beneath the inguinal ligament at the mid-inguinal point, i.e., halfway between the anterior superior iliac spine and the pubic symphysis. The **femoral vein** lies immediately medial to it and both structures are enclosed here in the femoral sheath. Deep to the artery lie the iliopsoas muscle and the head of the femur. The arterial pulse can normally be felt at this point, but it appears weak or absent if the head of the femur is dislocated. The accessability of the femoral artery in the femoral triangle makes it an ideal vessel through which catheters

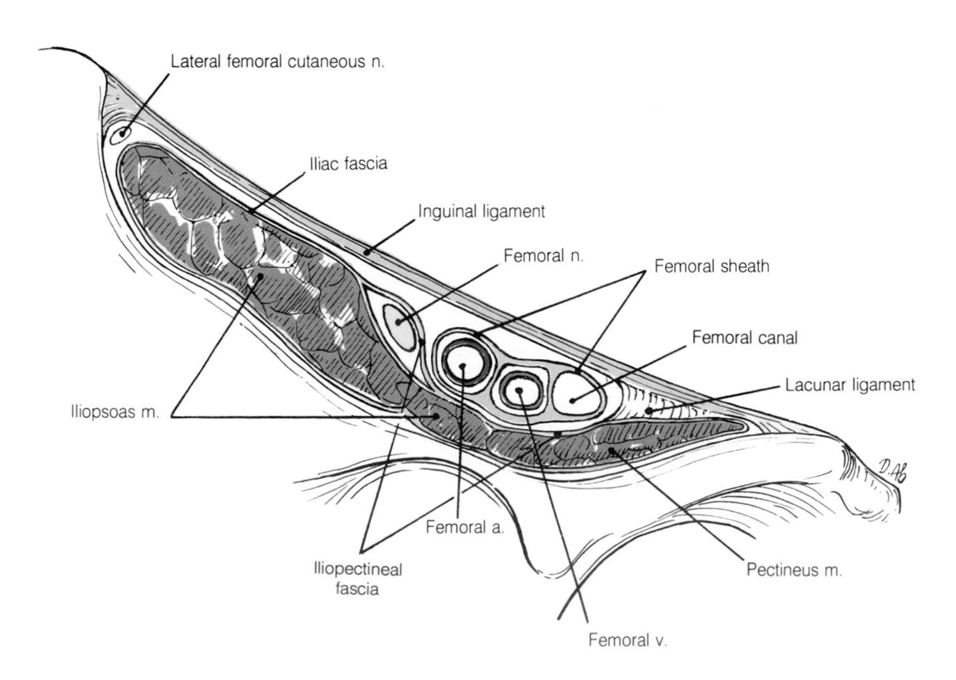

8-26 Diagram showing structures passing deep to the inguinal ligament.

may be introduced into the aorta and coronary arteries for diagnostic and therapeutic purposes. The relations of the artery as it passes beneath the inguinal ligament are illustrated in Fig. 8-26.

The artery descends superficially through the femoral triangle to its apex where it passes into the adductor canal anterior to the femoral vein. In the canal the artery winds round the medial side of the thigh until it reaches the **adductor hiatus** in the attachment of adductor magnus to the linea aspera. On passing through the hiatus into the popliteal fossa the name of the vessel changes to the **popliteal artery**. It is accompanied in its course by the **femoral vein** which initially lies medial to the artery but gradually assumes a posterior relationship so that on reaching the popliteal fossa behind the knee joint the vein is more posterior and, therefore, the more superficial of the two vessels.

The femoral artery is the only large axial artery of the thigh and its branches must, therefore, supply all structures within this region. Some alternative collateral circulation is, however, made possible by anastomoses between its branches and vessels in the gluteal region above and popliteal region below.

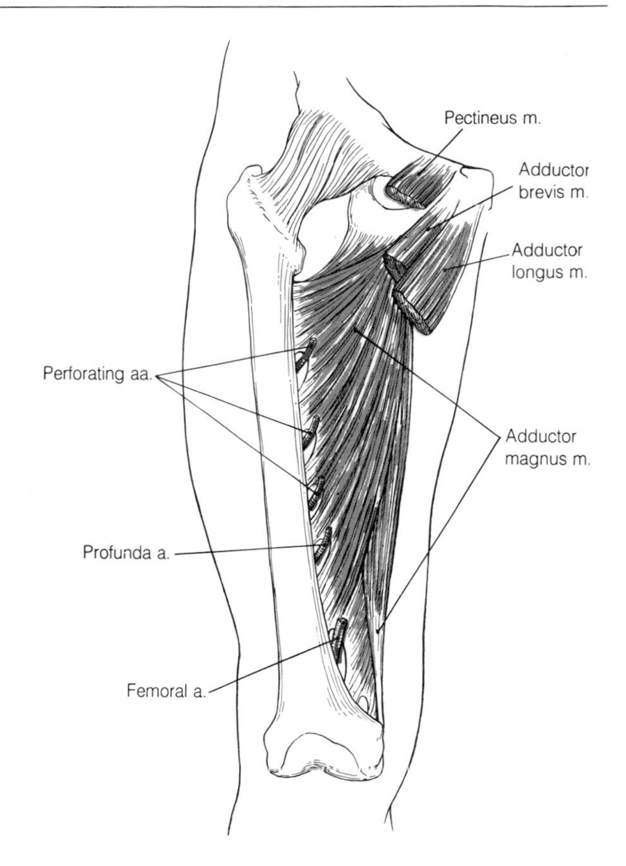

8-27 Semidiagrammatic drawing to show the vessels that perforate adductor magnus.

Superficial Branches of the Femoral Artery

Soon after appearing beneath the inguinal ligament three superficial branches are given off. These are the **superficial epigastric, superficial circumflex iliac**, and **external pudendal arteries**. They radiate out from the femoral triangle reaching the anterior abdominal wall, the region of the anterior superior iliac spine, and the external genitalia.

Deep Branches of the Femoral Artery

A **deep external pudendal artery** runs medially across pectineus and adductor longus beneath the fascia lata. After piercing the fascia it supplies the skin of the perineum. The **profunda femoris artery** is a large branch given off in the femoral triangle. It arises laterally, passes behind the femoral artery, and leaves the triangle through the floor between the **pectineus** and **adductor longus muscles**. Now in the adductor compartment it descends between **adductor longus and magnus**. Finally it pierces adductor magnus to end by anastomosing with muscular branches of the **popliteal artery**. During its course through the adductor compartment, the profunda femoris artery supplies these muscles and gives off a number of **perforating arteries**. These branches all pierce **adductor magnus** near its attachment to the linea aspera and supply muscles in the flexor compartment. The manner in which these arteries, the termination of the profunda femoris artery, and the femoral artery pass through adductor magnus is illustrated in Fig. 8-27.

The **medial and lateral circumflex femoral arteries** are usually given off by the profunda femoris artery in the femoral triangle but both may be direct branches of the femoral artery. The **medial circumflex femoral artery** leaves the floor of the triangle between psoas and pectineus and winds round the femur. An **ascending branch** anastomoses with the **inferior gluteal artery** and

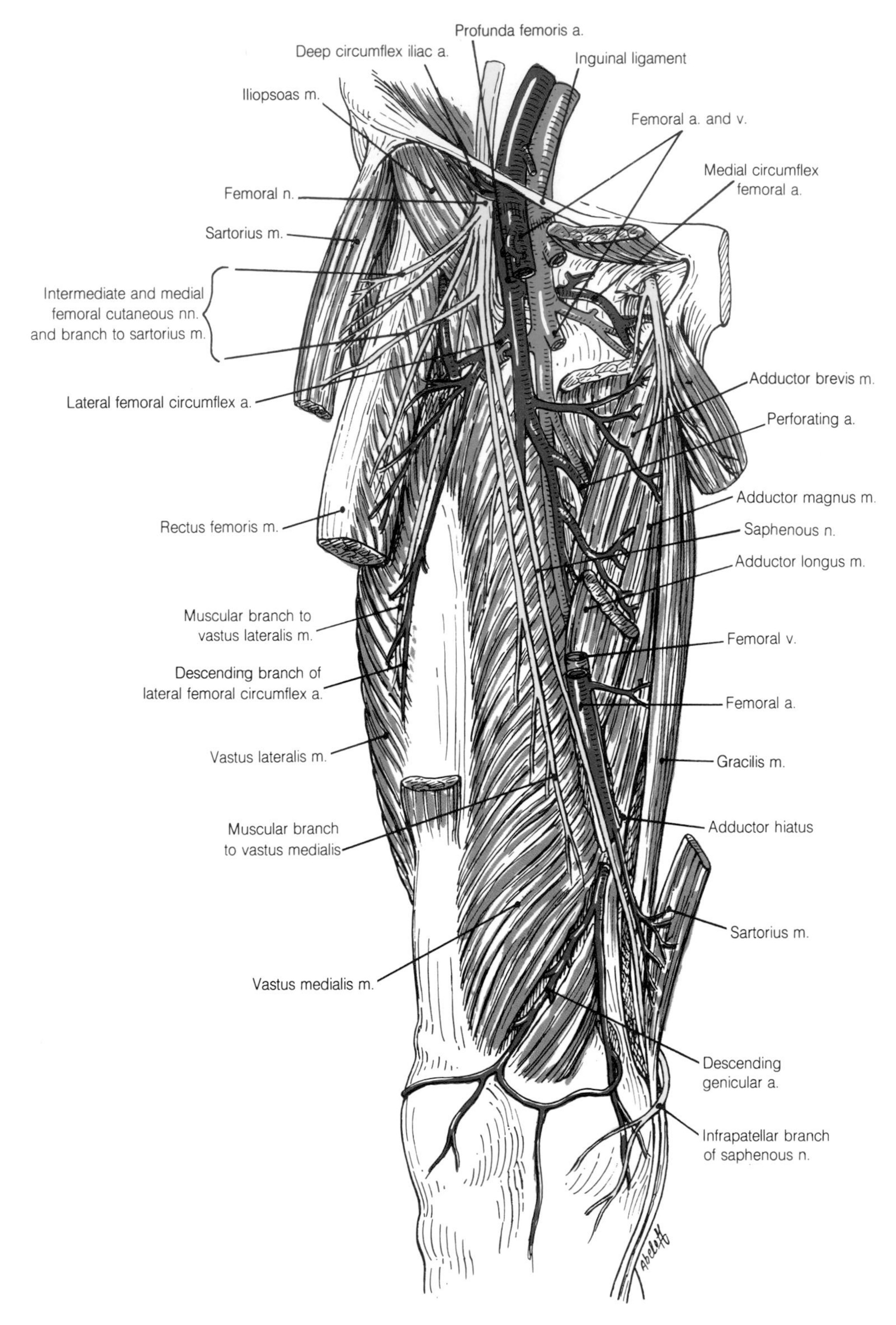

8-28/29 The course and distribution of the femoral artery and nerve in the thigh.

also helps supply the neck and head of the femur. A **transverse branch** anastomoses with a corresponding branch of the lateral artery. The **lateral circumflex femoral artery** passes deep to sartorius and rectus femoris and divides into **ascending, transverse**, and **descending branches**. The ascending branch supplies the **hip joint** and anastomoses with the **superior gluteal artery**. The transverse branch anastomoses with the transverse branch of the medial circumflex femoral artery and the descending branch passes down behind the rectus femoris to anastomose with **genicular arteries**.

A **descending genicular artery** leaves the femoral artery just before it passes through the adductor hiatus.

The femoral, the profunda femoris, and circumflex femoral arteries all give off important but unnamed **muscular branches**. The distribution of the deep branches of the femoral artery can be followed in Fig. 8-28. In this the sartorius muscle is shown divided and drawn aside to expose the whole length of the adductor canal.

The Cruciate Anastomosis

Reference has already been made to the existence of a collateral circulation which may serve as an alternative pathway for blood flow after obstruction to the femoral artery. This is known as the **cruciate anastomosis**. In this the gluteal branches of the internal iliac artery anastomose with both the circumflex femoral arteries which in turn anastomose with the first perforating artery. The perforating arteries all anastomose with each other and with muscular branches of the popliteal artery. In this way the internal iliac artery is linked to the popliteal artery.

Nerves of the Extensor Compartment

The Femoral Nerve

The femoral nerve, the largest nerve of the lumbar plexus, is formed, from the **ventral rami of the second, third, and fourth lumbar nerves**. Descending through the fibers of psoas major, it appears at the lateral border of that muscle and continues between it and iliacus to the inguinal ligament. The femoral nerve passes beneath the inguinal ligament into the femoral triangle just lateral to the femoral artery. After a very short course in the thigh it divides into an anterior and posterior division.

Both divisions of the femoral nerve give off muscular and cutaneous branches. The distribution of these branches is confined to the region of the thigh except for the cutaneous branch of the posterior division, the saphenous nerve, whose distribution extends from the knee to the medial side of the foot. The branches of the femoral nerve can be followed in Fig. 8-28 as they are described.

Branches of the Femoral Nerve

The **nerves to iliacus and pectineus** are given off by the femoral nerve during its course over the posterior abdominal wall.

The **branches of the anterior division** are the **intermediate and medial femoral cutaneous nerves** and a muscular branch to sartorius. Branches of the two cutaneous nerves are found piercing the deep fascia along a line drawn by the medial border of the sartorius. Their cutaneous distribution was discussed earlier.

The **branches of the posterior division** include muscular branches to all heads of the quadriceps femoris muscle and care must be taken to preserve these if the femur is surgically exposed. A particularly vulnerable branch is that to vastus lateralis which joins the descending branch of the lateral femoral circumflex artery to supply the muscle in its lower third.

The **saphenous nerve** is the cutaneous branch of the posterior division. It descends through the thigh in the adductor canal in company with the femoral vessels. It does not, however, pass through the adductor hiatus but appears at the medial border of sartorius and then turns forward below the knee. Here it gives off an **infrapatellar branch** before continuing down the medial side of the leg to reach the medial side of the foot. Its branches supply the skin over the medial half of the leg and the medial side of the foot but excluding the heel.

Branches of both the divisions of the femoral nerve supply articular branches to the hip and knee joints.

The Lateral Femoral Cutaneous Nerve

Like the femoral nerve, the lateral femoral cutaneous nerve is formed in the lumbar plexus but from only the second and third lumbar nerves, It appears on the posterior abdominal wall at the lateral border of psoas, sweeps downward and laterally over the surface of iliacus and passes beneath the inguinal ligament close to the anterior superior iliac spine. In the thigh it divides into two branches to supply the skin of the lateral aspect of the thigh. As it passes beneath or sometimes through the inguinal ligament, this nerve may become entrapped by surrounding connective tissue and give rise to pain over its distribution.

The Adductor Compartment of the Thigh

The adductor compartment of the thigh, although fairly clearly delineated from the anterior compartment is not so clearly separated from the posterior compartment. In fact, one of the muscles it contains (adductor magnus) functions and is innervated as a member of both compartments. The compartment contains the **adductor muscles** which arise from the ischiopubic part of the pelvic bone and the **obturator nerve** by which they are supplied.

The Muscles of the Adductor Compartment

As the muscles of the adductor compartment are described, reference should be made to the illustrations of their attachments to the pelvis and femur seen in Figs. 8-30 and 8-31.

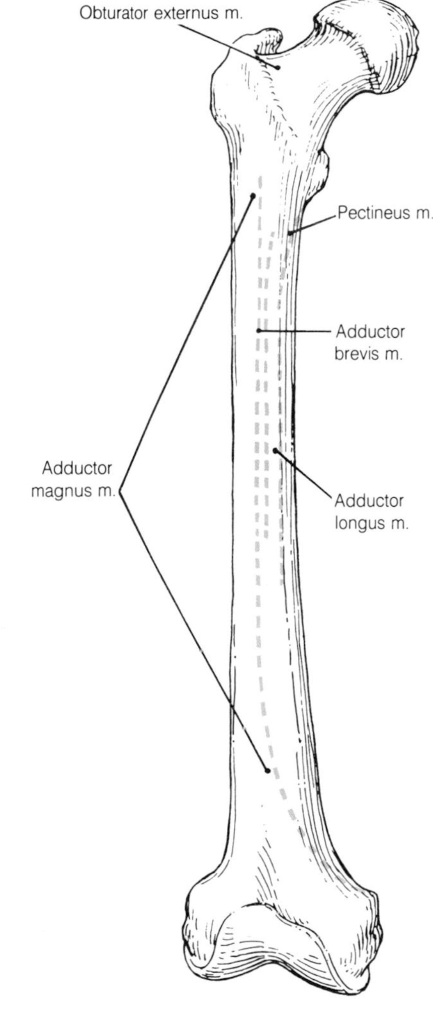

8-31 The femoral attachments of the adductor muscles **seen through the femur,** for comparison with Fig. 8-30.

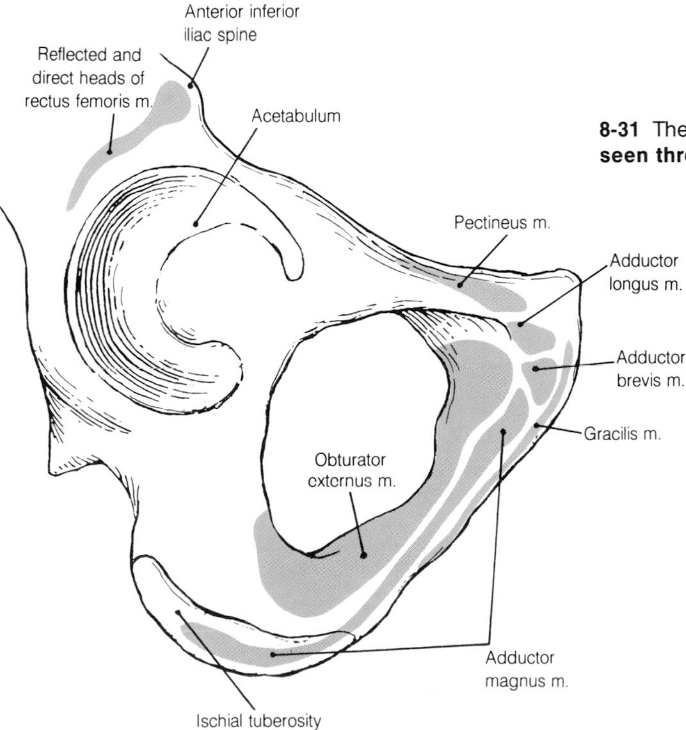

8-30 The ischium and pubis showing the origins of the adductor muscles.

Pectineus

Pectineus has already been seen in the floor of the femoral triangle. It arises from the **pectineal line** of the pubis and is attached to the **spiral line** of the femur which leads from the lesser trochanter to the linea aspera. Its attachment to the femur is, therefore, continuous with that of iliopsoas alongside which it lies. Its action is to adduct and flex the thigh. Unlike the remaining adductor muscles, pectineus is normally supplied by the **femoral nerve**. However, branches from the **obturator nerve** are occasionally found to supply its deeper fibers.

Adductor Longus, Brevis, and Magnus

Look at Fig. 8-30 and see the proximal attachments of the adductor muscles extending from the anterior aspect of the pubic bone, along the ischiopubic ramus and reaching the ischial tuberosity. From their origins these muscles run backward, laterally, and downward to the femur. In so doing they form an overlapping sheet of muscle that, becoming aponeurotic, finds a continuous attachment to the spiral line, the medial lip of the linea aspera, the medial supracondylar line, and the adductor tubercle. The detail of the insertion of each muscle is shown in Fig. 8-31.

The **adductor longus** lies in the floor of the femoral triangle and the medial wall of the upper part of the subsartorial canal. Together with **pectineus** it overlies **adductor brevis**. Adductor longus arises from the body of the pubic bone just below the pubic crest and is inserted into the middle third of the femur at the linea aspera.

Adductor brevis arises from the ramus of the pubis below the origin of adductor longus. Its belly therefore lies deep to this muscle and pectineus and is attached distally to the upper part of the linea aspera.

Adductor magnus is a much more bulky muscle. Those fibers arising from the ischiopubic ramus fan out to their attachment to the **linea aspera** and their aponeurotic tendon shows a deficiency in its lower part called the **adductor hiatus** which transmits the **femoral vessels**. In fact, smaller openings in the aponeurosis above this can also be found transmitting the **profunda femoris artery** and its perforating branches.

The fibers of adductor magnus that arise from the ischial tuberosity form belly that is attached by a rounded tendon to the **adductor tubercle of the femur**.

The action of all three muscles is to adduct the thigh. Although normal activity does not often require this as a prime movement, these muscles also play a synergistic part in locomotion and posture. The ischiocondylar portion of adductor magnus is also a weak extensor of the thigh. All three muscles are supplied by the **obturator nerve** and the ischiocondylar portion of adductor magnus by the **tibial division of the sciatic nerve**. The segments concerned are L2, 3, and 4.

Gracilis

This is a thin sheet of muscle lying superficially on the medial aspect of the thigh. It arises from the body of the pubis and the ischiopubic ramus and descends vertically to be attached behind sartorius to the shaft of the tibia below the medial condyle. It is a feeble flexor and medial rotator of the leg and may produce some adduction of the thigh. It is supplied by the **obturator nerve** (L2, 3).

Obturator Externus

Although this muscle has little function as an adductor, it is so closely associated at its origin with the adductors of the thigh and their nerve supply that it will be described with them.

Obturator externus arises from the outer face of the **obturator membrane** and the surrounding pelvic bone. Its fibers converge on a tendon which passes beneath the hip joint and behind the neck of the femur to end in the **trochanteric fossa**. Its action is to laterally rotate the thigh and with other short muscles it helps to stabilize the hip joint. Obturator externus is supplied by the **obturator nerve** (L3, 4).

Look now at Fig. 8-32 in which the adductor muscles are shown and correlate their positions and the way in which they overlap with their proximal and distal attachments which were illustrated in Figs. 8-30 and 8-31.

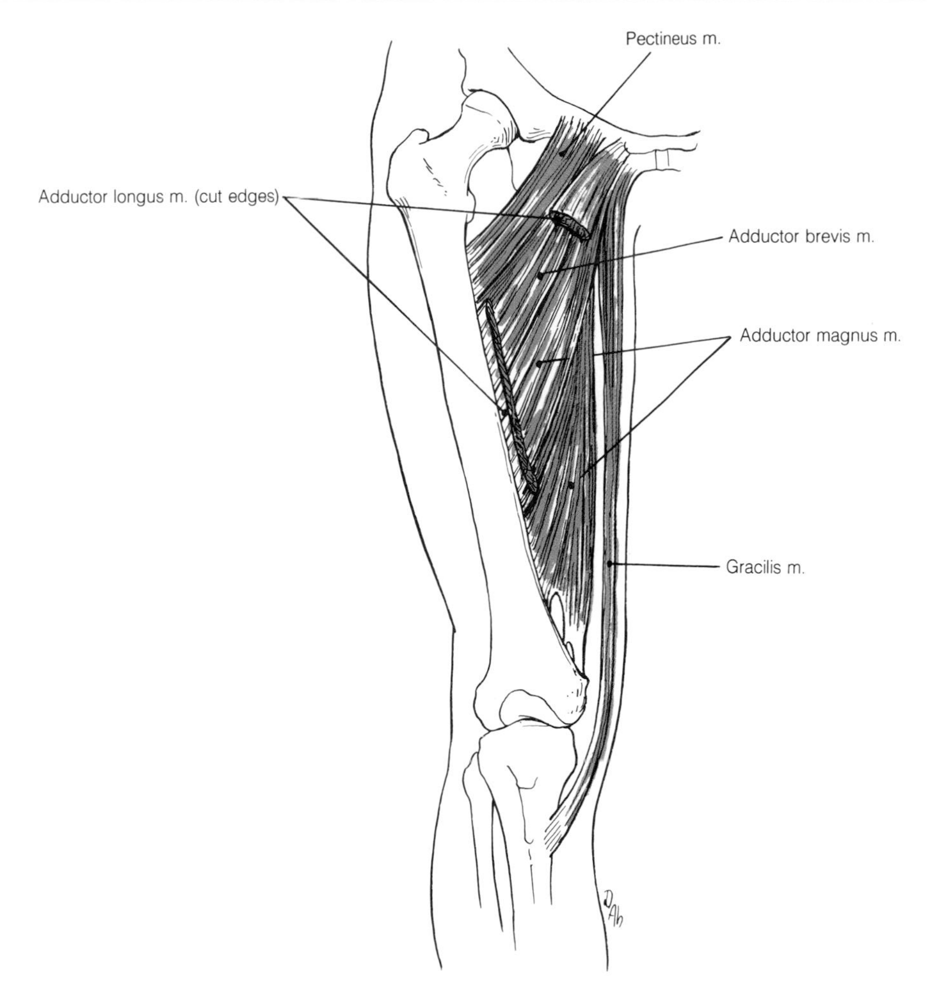

8-32 Showing the adductor muscles. The adductor longus muscle has been removed in part.

Vessels and Nerves of the Adductor Compartment

The greater part of the blood supply to the adductor muscles is obtained from the perforating branches of the profunda femoris artery and from muscular branches of the femoral artery in the subsartorial canal. Some supply is gained from the obturator artery. However, the obturator nerve plays the major part in the innervation of the muscles of this compartment.

The Obturator Artery

The obturator artery is a branch of the **internal iliac artery** in the pelvis. It enters the thigh through the **obturator canal** in company with the **obturator nerve**. It immediately divides into two branches which encircle the obturator foramen. Small branches supply the obturator externus muscle and the adjacent adductor and extensor muscles. An **acetabular branch** gives rise to the small artery that runs in the **ligament of the head of the femur**.

The Obturator Nerve

The obturator nerve is derived from the **second, third, and fourth lumbar ventral rami** in the lumbar plexus. It divides into anterior and posterior branches before passing with the obturator artery through the obturator canal. Here it supplies **obturator externus** and sends a small branch to the **hip joint**.

The anterior branch continues in front of adductor brevis to supply **adductor brevis, adductor longus**, and **gracilis**. It also supplies the skin over the medial aspect of the thigh.

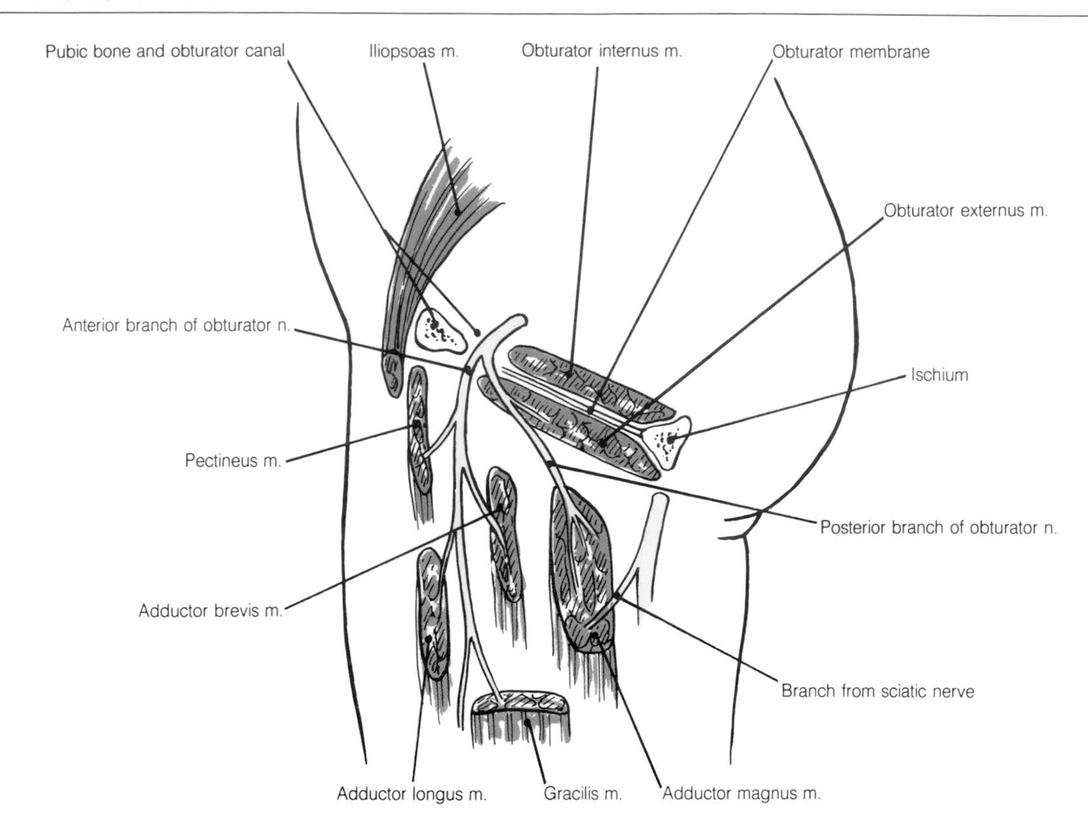

8-33 Diagram illustrating the distribution of the obturator nerve in the thigh.

The posterior branch passes behind adductor brevis, sometimes supplying it, and ends by supplying **adductor magnus** and the **knee joint**. The course and distribution of these branches is illustrated in Fig. 8-33.

Note that the femoral and obturator nerves supply both the hip joint and the knee joint. Pain generated by disease of the hip joint may be mistakenly referred by a child to the knee joint. A negative examination of this joint should not lead the physician to neglect the true cause of the pain.

The Profunda Femoris Artery

The obturator artery contributes little to the blood supply of the adductor compartment. However, this deficiency is made good by the **profunda femoris artery** and its **perforating branches**. These were described with the parent **femoral artery** and reference should now be made to this description and the accompanying illustration.

The Gluteal Region

The gluteal region underlies the skin of the buttock although it should be noted that the skin fold that forms the lower limit of the buttock does not follow the border of the large muscle gluteus maximus. The skin of the upper part of the buttock is supplied by the subcostal and iliohypogastric nerves and by cutaneous branches of lumbar dorsal rami. Below, it is supplied by branches of the posterior cutaneous nerve of the thigh. An underlying thick layer of fat and fascia gives the buttock much of its characteristic shape. The buttock is commonly used as a site for intramuscular injections and to elimiate any risk of damaging the sciatic nerve, the upper outer quadrant should be used for this purpose. Within the gluteal region there are three large muscles, the gluteus maximus, gluteus medius, and gluteus minimus, and a number of smaller muscles mainly concerned with lateral rotation of the thigh. The main nerve is the sciatic nerve. There is no major axial vessel accompanying the sciatic nerve and the region is supplied by branches of the internal iliac artery which reach it, like the sciatic nerve, through the greater sciatic notch. The gluteal region leads naturally into the posterior or flexor compartment of the thigh.

The Muscles of the Gluteal Region

A detailed illustration of the pelvic bone and the upper end of the femur is given in Fig. 8-34, and this should be followed as the muscles of the gluteal region are described. These include the three gluteal muscles and piriformis, obturator internus, the gemelli, and quadratus femoris muscles.

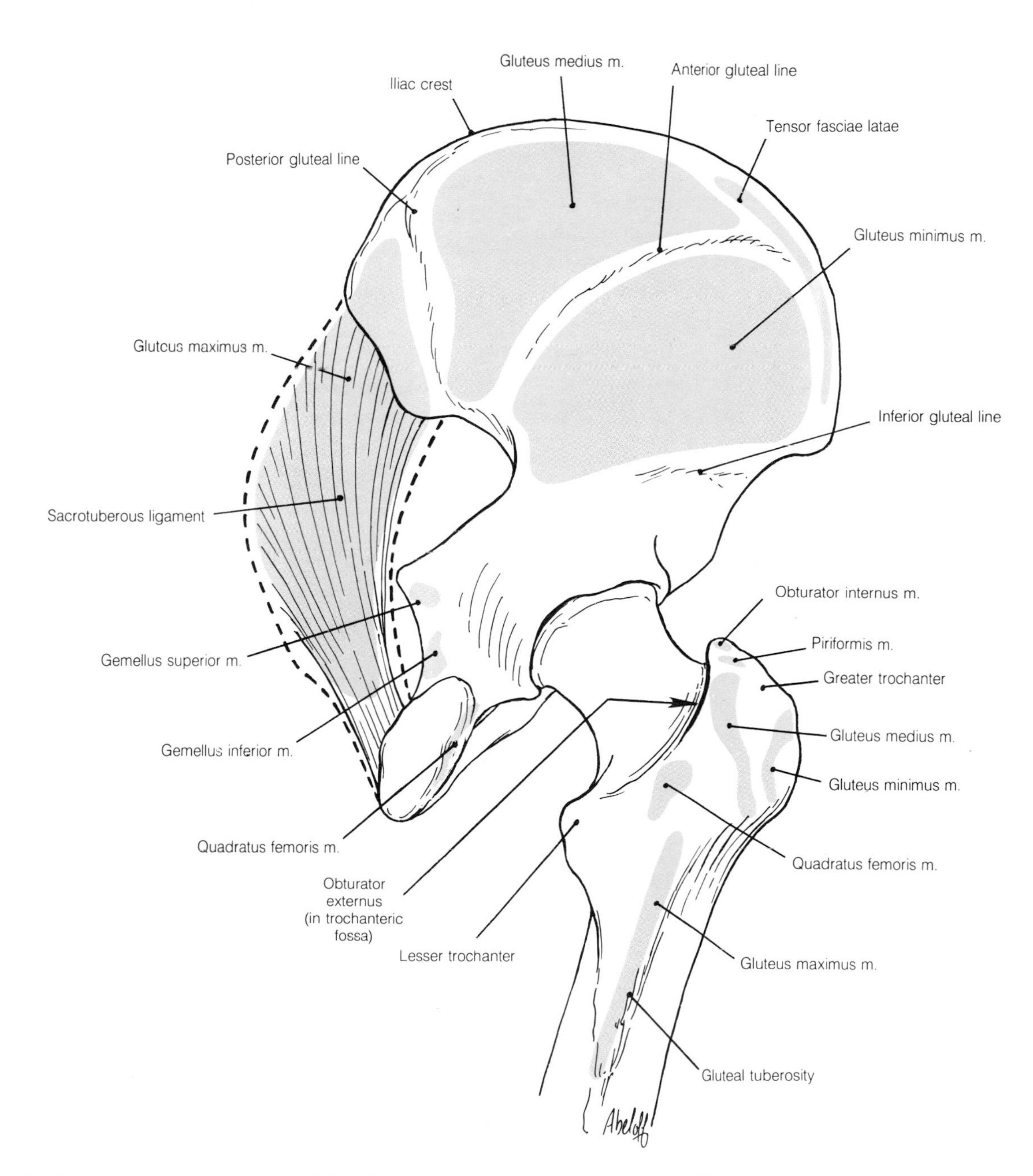

8-34 The posterolateral aspects of the right hip bone and the upper part of the femur showing the attachments of muscles in the gluteal region.

Gluteus Maximus

Gluteus maximus is a large muscle that arises from the **ala of the iliac bone** behind the **posterior gluteal line**. This relatively small bony origin from the pelvis is expanded by inclusion of the fascia covering the **erector spinae muscle**, the dorsum of the **sacrum** and **coccyx** below this, and the **sacrotuberous ligament**. The deeper fibers of the muscle are inserted into the **gluteal tuberosity**. The remaining fibers are attached to the **fascia lata** and, hence, via the **lateral intermuscular septum**, to the **linea aspera** below the gluteal tuberosity. As the muscle crosses the greater trochanter it is separated from this by a large bursa.

The gluteus maximus is a powerful extensor of the thigh on the trunk, a uniquely human action. It is little used in walking, but brought into play in running and climbing, or, when the limb is fixed, raising the trunk as from a ceremonial bow. Gluteus maximus is supplied by the **inferior gluteal nerve** (L5, S1, 2).

Gluteus Medius

The gluteus medius arises from the ilium between the **posterior and anterior gluteal lines**. It is largely overlapped by the gluteus maximus posteriorly. Its fibers converge on a tendon which is attached to the lateral aspect of the greater trochanter of the femur. Here a bursa separates the tendon from the upper part of the trochanter. It is supplied by the **superior gluteal nerve** (L5, S1).

Gluteus Minimus

The gluteus minimus, despite its name, is no small muscle. It arises from the ilium between the **anterior and inferior gluteal lines** and lies deep to **gluteus medius**. Its tendon also finds attachment to the lateral aspect of the **greater trochanter of the femur** and again a bursa lies between the tendon and the underlying trochanter. It is also supplied by the **superior gluteal nerve** (L5, S1).

Both gluteus medius and minimus act as abductors of the thigh from the trunk. This is not, however, a common action. Their importance lies in the stabilization of the pelvis on one leg when the opposite leg is raised from the ground. This is a function continuously required in walking and running, and paralysis of these muscles leads to a characteristic waddling gait as the subject throws the trunk and center of gravity away from the side of the raised leg. This problem in the mechanics of walking is illustrated in Fig. 8-35. In this the pelvis is first shown balanced evenly on the two lower limbs with the center of gravity lying between the feet. In the second drawing the right leg has been raised. The body is tilted to the left in order to place the center of gravity over the left foot. There is a tendency for the pelvis to rotate downward about the axis of the left hip. However, shortening of the left gluteus medius and minimus normally prevents this from occurring. If these muscles are paralyzed on the left side the gluteal fold on the right will fall. This is the anatomical basis for the Trendelenberg test.

Piriformis

The piriformis arises from the **middle three pieces of the sacrum** and leaves the pelvis through the **greater sciatic notch** still as a fleshy belly. It is inserted by a tendon into the upper border of the **greater trochanter**. It acts as a lateral rotator of the thigh and is supplied in the pelvis by branches of the **first and second sacral nerves**.

Obturator Internus

The obturator internus arises from the inner aspect of the pelvic bone and here has important relations to the pelvic cavity and perineum which are described with those regions. Its fleshy belly is replaced by a tendon as it passes over the **lesser sciatic notch**. Here the tendon makes a rightangled turn to reach the medial surface of the **greater trochanter** above the trochanteric fossa. Where this sudden change in direction is made the tendon is separated from the bone by a bursa and a layer of hyaline cartilage. Obturator internus is supplied by the **nerve to obturator internus** (L5, S1) which reaches the perineal surface of the muscle through the lesser sciatic foramen.

The Gemelli

The gemelli muscles are two small bellies which arise from the ischium above and below the **lesser**

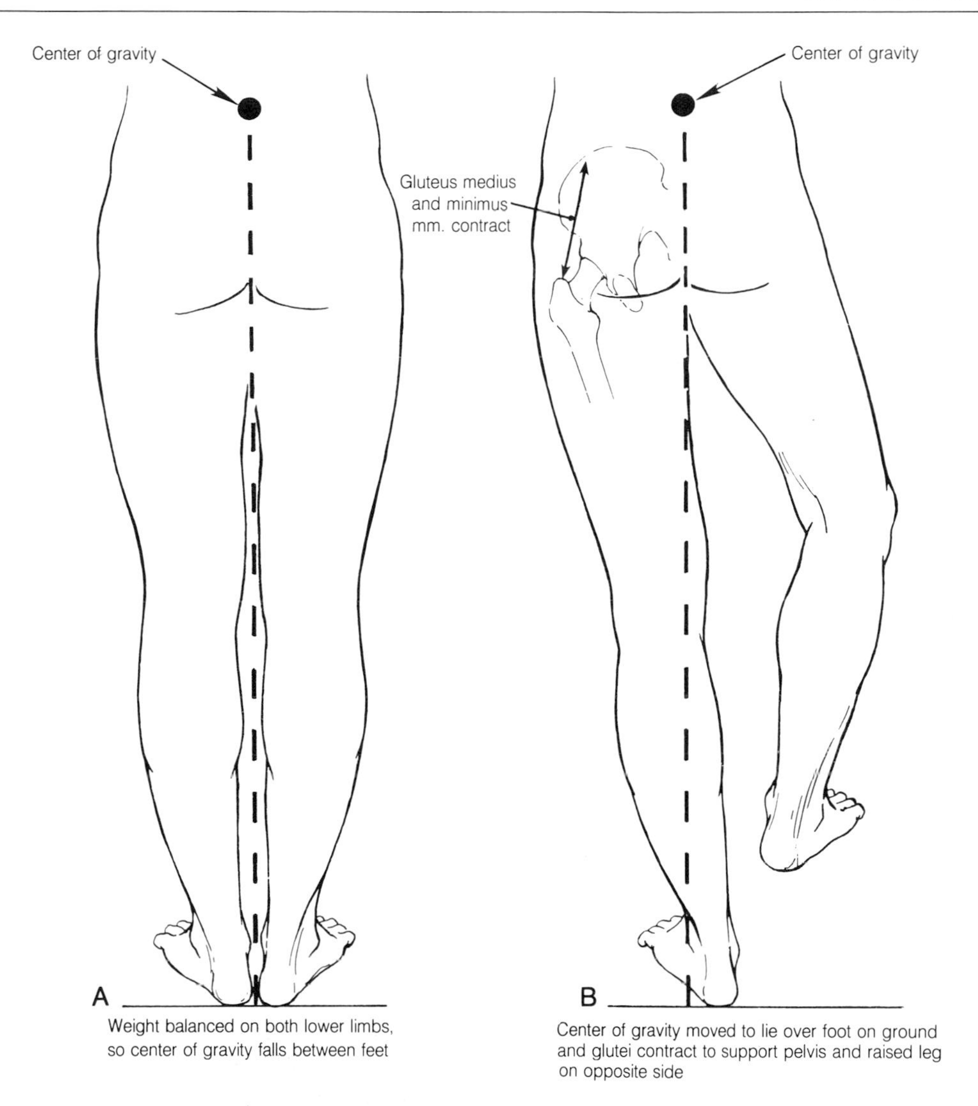

8-35 Diagram illustrating the actions of gluteus medius and minimus when the weight of the body is borne by one leg.

sciatic notch and accompany, and often hide, the extrapelvic course of the **tendon of obturator internus** with which they gain insertion.

The action of obturator internus and the gemelli is to rotate the thigh laterally. The superior gemellus is supplied by the **nerve to obturator internus** and the inferior by the **nerve to quadratus femoris**.

Quadratus Femoris

Quadratus femoris arises from the **ischium** just anterior to the ischial tuberosity and extends to the **intertrochanteric crest** and **quadrate tubercle of the femur**. It is a lateral rotator of the thigh and is supplied by the **nerve to quadratus femoris** (L5, S1).

Obturator Externus

This muscle, which is supplied by the **obturator nerve**, was described with the adductor compartment. Remember that its tendon passes below and behind the hip joint to be inserted into the trochanteric fossa. It is a lateral rotator of the thigh.

The Vessels and Nerves of the Gluteal Region

During the dissection of the pelvis, the piriformis muscle is seen to leave the cavity through the greater sciatic notch. With it pass the gluteal vessels and nerves, the superior above and the inferior below. Also formed on the pelvic wall is the sciatic nerve. This together with the pudendal vessels and nerve leaves the pelvic cavity through the greater sciatic notch and below the piriformis muscle. Reflection of the gluteus maximus exposes the piriformis in the gluteal region and the subsequent course of these structures can now be described and followed in Fig. 8-36.

The Superior Gluteal Artery

The superior gluteal artery appears at the **upper border of piriformis**, gives muscular branches to gluteus maximus, and then runs forward between **gluteus medius** and **minimus** to supply them. The superior gluteal artery anastomoses with the inferior gluteal artery and takes part in the **cruciate anastomosis**.

The Superior Gluteal Nerve (L4, 5, S1)

The superior gluteal nerve appears with its companion artery and follows it beneath gluteus medius. It supplies **gluteus medius, gluteus minimus**, and **tensor fasciae latae**.

The Inferior Gluteal Artery

The inferior gluteal artery appears **below the piriformis** and medial to the sciatic nerve. Its branches supply the gluteal muscles and muscles

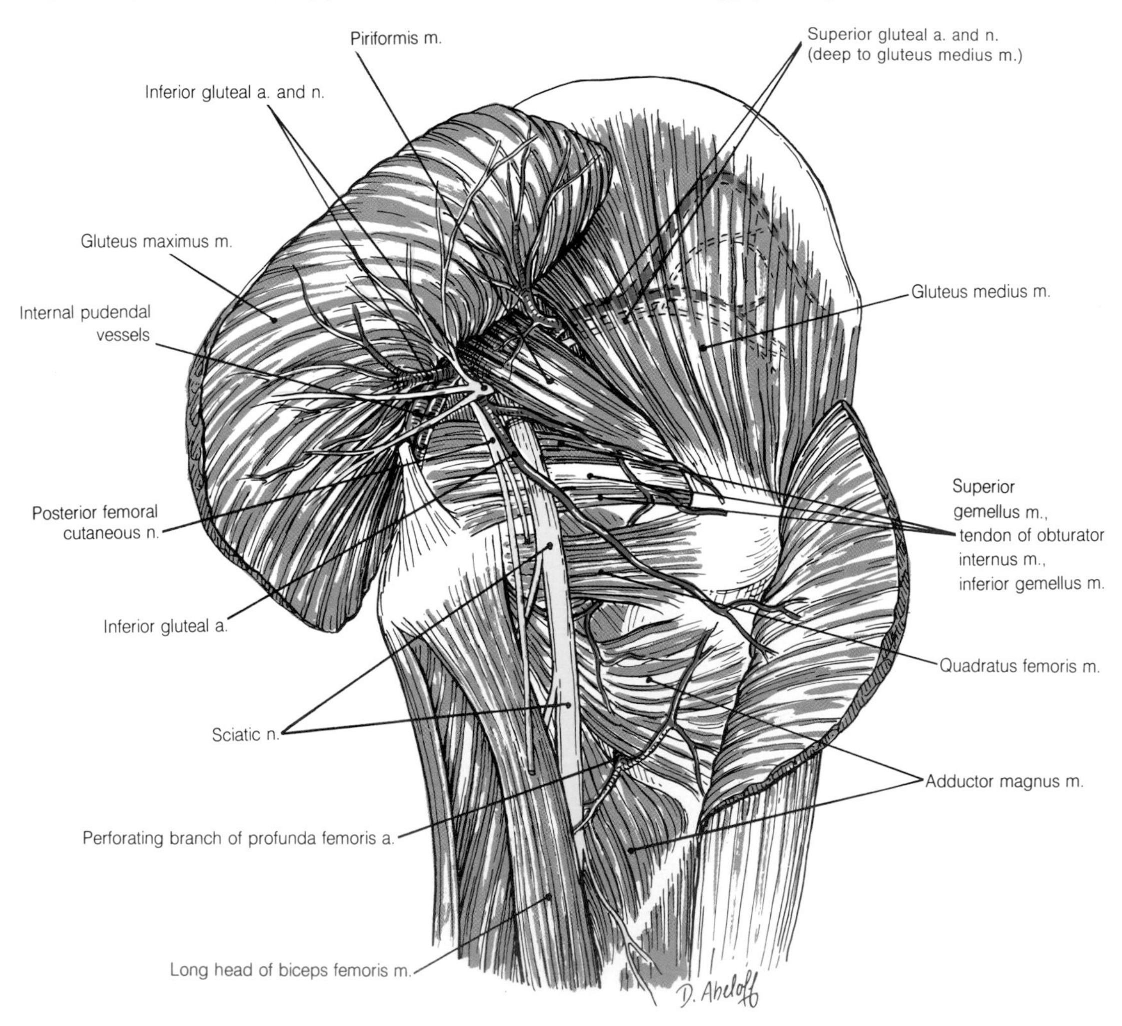

8-36 The gluteal region.

of the flexor compartment of the thigh, and it also takes part in the cruciate anastomosis.

The Inferior Gluteal Nerve (L5, S1, 2)

The inferior gluteal nerve appears **below piriformis** and supplies the gluteus maximus. That such a small nerve supplies such a large muscle can be accounted for by the fact that the number of muscle fibers in each motor unit, i.e., those supplied by one axon, is very large.

The Gluteal Veins

These accompany the gluteal arteries and, through their connections with veins in the thigh, form an alternative venous return from the lower limbs after obstruction of the femoral vein.

The Pudendal Vessels, Pudendal Nerve, and Nerve to Obturator Internus

On entering the gluteal region through the greater sciatic notch **below piriformis**, the **internal pudendal vessels** and **pudendal nerve** cross the ischial spine and leave the region through the **lesser sciatic notch**. The **nerve to obturator internus** follows a similar course lying lateral to the pudendal vessels as they cross the ischial spine.

The Sciatic Nerve

The sciatic nerve is the most important structure in the gluteal region as its branches supply the flexor muscles in the thigh, and all the muscles of the leg and foot. These branches also convey sensation from much of the skin over the same regions. Formed in the pelvic cavity from the **fourth and fifth lumbar nerves** and the **first three sacral nerves**, the sciatic nerve leaves the pelvis through the **greater sciatic notch**. Appearing as a flattened band in the gluteal region at the **lower border of piriformis**, it descends into the thigh crossing successively the **ischial wall of the acetabulum** (and thus the hip joint), the **obturator internus and adjacent gemelli**, and the **quadratus femoris muscle**. It is covered superficially by **gluteus maximus** in this part of its course. Check these anterior relations of the nerve in Fig. 8-36.

Because of the proximity of the nerve to the hip joint, it may be injured in posterior dislocations and fracture dislocations. The danger of damaging the nerve by giving intramuscular injections into the buttock has already been mentioned.

The sciatic nerve is in fact two nerves wrapped together by connective tissue. These nerves might be called the tibial and peroneal divisions after the terminal branches to which they ultimately give rise. The medial tibial division arises from anterior divisions of ventral rami (L4, 5; S1, 2, 3) and supplies flexor and plantar flexor muscles (comparable to the median and ulnar nerves of the upper limb). The peroneal division formed from posterior divisions of ventral rami (L4, 5; S1, 2) supplies extensor muscles (comparable to the radial nerve). The reality of the dual nature of the nerve can often be seen when the sciatic nerve divides high in the thigh or appears in the gluteal region as two separate nerves of which the common peroneal portion perforates the piriformis muscle.

In the gluteal region only articular branches to the hip joint are given off by the sciatic nerve.

The Posterior Femoral Cutaneous Nerve

The posterior femoral cutaneous nerve (S1, 2, 3) appears **below piriformis** and usually on the superficial surface of the sciatic nerve. It descends through the thigh just beneath the fascia lata which it pierces at the popliteal fossa. Branches which become superficial supply skin of the buttocks (inferior cluneal nerves), the external genitalia and the back of the thigh, popliteal fossa, and upper calf.

The Nerve to Quadratus Femoris

The nerve to quadratus femoris enters the gluteal region through the greater sciatic foramen **deep to the sciatic nerve**. It supplies the hip joint and quadratus femoris.

The Flexor Compartment of the Thigh

The flexor compartment of the thigh contains a group of muscles called the "hamstring" muscles after their tendons around the knee joint. (The hamstrings are hidden by the chef and cut by the duelist.) They arise from the ischial tuberosity and are inserted below the knee into the tibia or fibula. They are, therefore, "two-joint" muscles and will both extend the hip and flex the knee. Because of this, appropriate antagonists must be brought into use if only one of these movements is needed. They are supplied by the sciatic nerve which passes through the compartment. There is no major axial vessel in this compartment and the muscles are supplied by the gluteal arteries and perforating branches of the profunda femoris artery. The region can be logically followed down into the popliteal fossa which lies behind the knee joint.

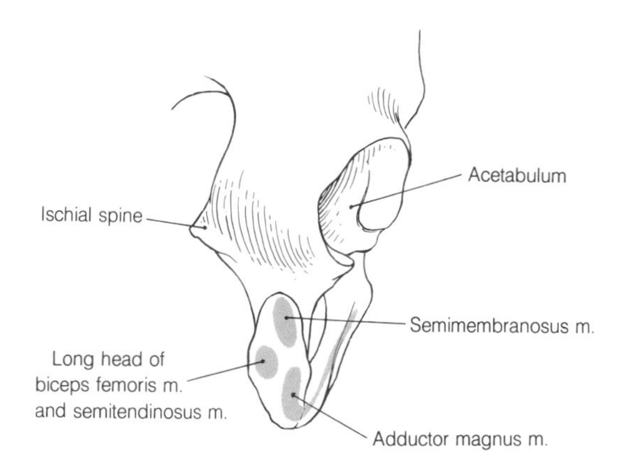

8-37 The ischial tuberosity showing the pelvic attachments of the "hamstring" muscles.

The Muscles of the Flexor Compartment

The muscles of this compartment are the **long head of biceps femoris, semitendinosus, semimembranosus**, and the "hamstring" portion of **adductor magnus** which was described with the muscles of the adductor compartment. In Fig. 8-37 the proximal attachments of these muscles on the ischial tuberosity are shown. Three of them pass downward toward the **medial side** of the knee joint. Of these the "hamstring" portion of adductor magnus ends at the adductor tubercle of the femur. The distal attachments to the tibia of the other two muscles are seen in Fig. 8-38. The biceps femoris passes **laterally** to be inserted into the head of the fibula. In its course it first covers and then crosses to the lateral side of the sciatic nerve. The arrangement of these muscles on the back of the thigh is illustrated in Fig. 8-39.

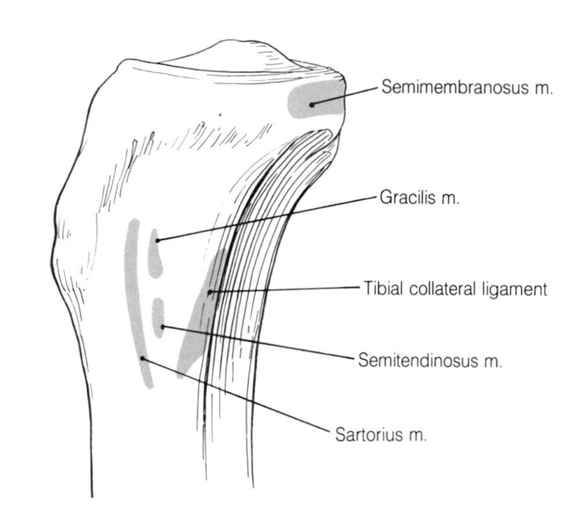

8-38 The medial aspect of the upper end of the tibia showing the tibial attachments of semimembranosus and semitendinosus muscles.

Biceps Femoris

Biceps femoris has a long head arising in common with **semitendinosus** from the ischial tuberosity. A short head (originally a separate muscle) arises from the lateral lip of the linea aspera. The long head crosses the sciatic nerve from the medial to the lateral side of the thigh, fuses with the short head, and is inserted into the head of the fibula. At its insertion, the tendon bifurcates around the fibular collateral ligament of the knee joint. The peroneal nerve lies just beneath the tendon as it travels from the popliteal fossa to the neck of the fibula.

Semitendinosus

Semitendinosus, as its name suggests, possesses a long tendon. It arises from the ischial **tuberosity** in common with the **long head of biceps femoris** and descends medially down the thigh to become tendinous over its lower third. It is inserted into the upper part of the medial aspect of the shaft of the tibia. Three muscles have now been described as inserting at this site, one from each compartment. Their tendons lie superficial to the tibial collateral ligament of the knee joint from which they are separated by a bursa. Look again at Fig. 8-38 and observe the attachments of **semitendinosus, gracilis, and sartorius**.

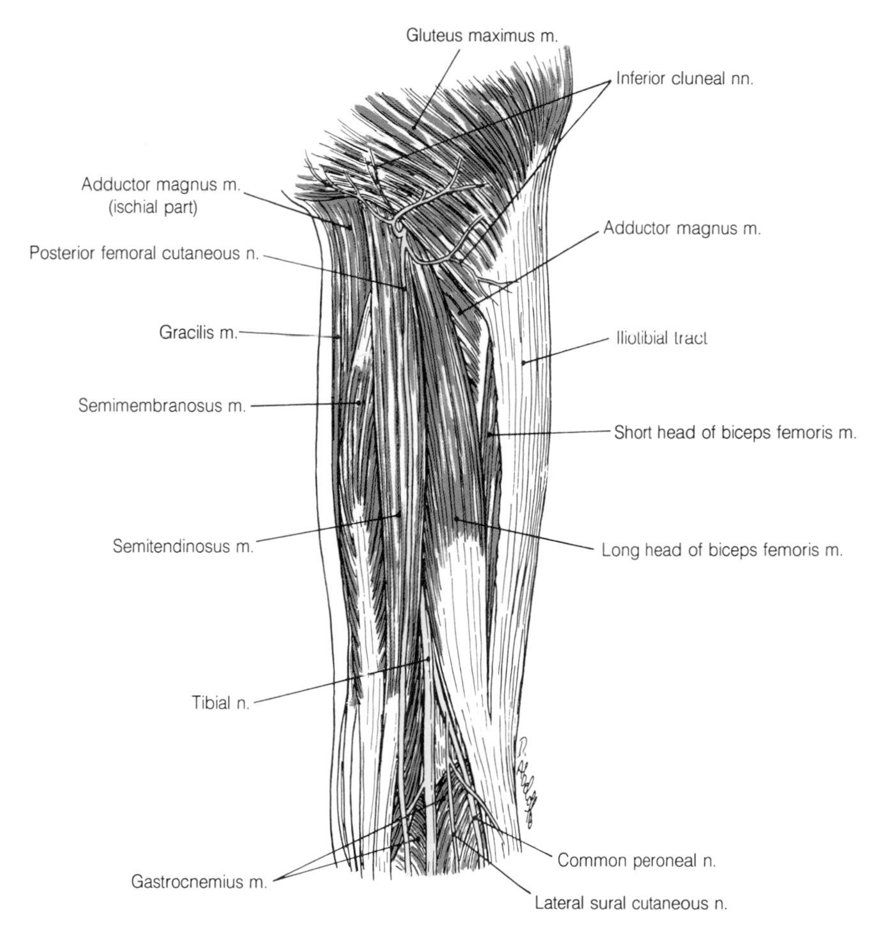

8-39 The posterior aspect of the thigh showing the "hamstring" muscles.

Semimenbranosus

Semimembranosus arises as a broad flat tendon from the ischial tuberosity and, hence, its membranous nature. Its origin is covered by the two preceding muscles and its belly and distal tendon (also flat) remain largely deep to semitendinosus until it finds attachment to a tubercle on the posterior aspect of the rim of the medial tibial condyle. A number of additional slips extend from the tendon and of these the most substantial is the **oblique popliteal ligament** of the knee which extends upward and laterally across the back of the joint to the lateral femoral condyle.

The muscles of the flexor compartment extend the thigh at the hip joint and flex the leg on the thigh at the knee. When the leg is flexed, biceps femoris can rotate the leg laterally on the thigh, and the semimembranosus and semitendinosus rotate it medially. The hamstring muscles, including the hamstring portion of adductor magnus, are supplied by branches of the **tibial division of the sciatic nerve** (L5; S1, 2). The short head of biceps femoris is supplied by a branch from the **peroneal division** of the sciatic nerve.

The Vessels and Nerves of the Posterior Compartment

The sciatic nerve enters the posterior compartment from beneath the lower border of gluteus maximus by passing off the surface of quadratus femoris on to adductor magnus. Superficially it is covered by the long head of biceps femoris and it continues beneath this muscle until it appears at the apex of the popliteal fossa. At this point it divides into the **tibial** and **common peroneal nerves**. Its branches in the thigh are muscular to the hamstring muscles. The branches to the long

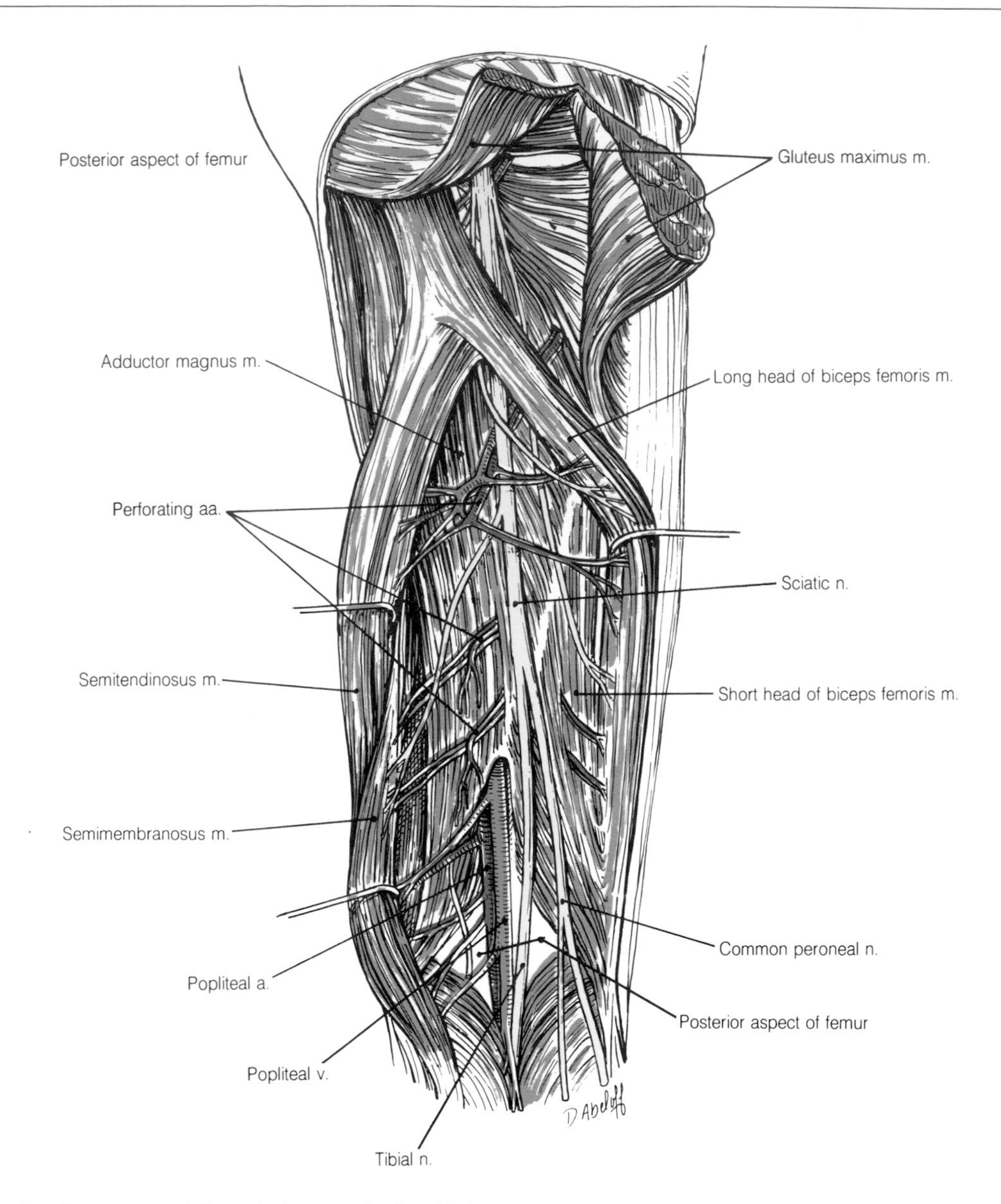

8-40 Showing the course of the sciatic nerve in the thigh.

muscles are from the tibial division of the nerve whereas that to the short head of biceps is derived from the common peroneal division, thus, indicating the different origins of the long and short heads. While the nerve normally divides into its terminal branches just before entering the popliteal fossa, division at a higher level is not uncommon.

The blood supply of structures in the posterior compartment is obtained from **perforating branches of the profunda femoris artery** and to some extent from the **inferior gluteal artery**. Look now at Fig. 8-40 which shows the course of the sciatic nerve through the flexor compartment. This illustration also shows the perforating branches of the profunda femoris artery and the adductor hiatus which carries the femoral vessels into the popliteal fossa.

With each of the compartments now described, the complete cross-section of the thigh seen in Fig. 8-41 can be used to review the region. Note again:

1. The fascia lata, the iliotibial tract, and the intermuscular septa
2. The four heads of quadriceps femoris in the extensor compartment

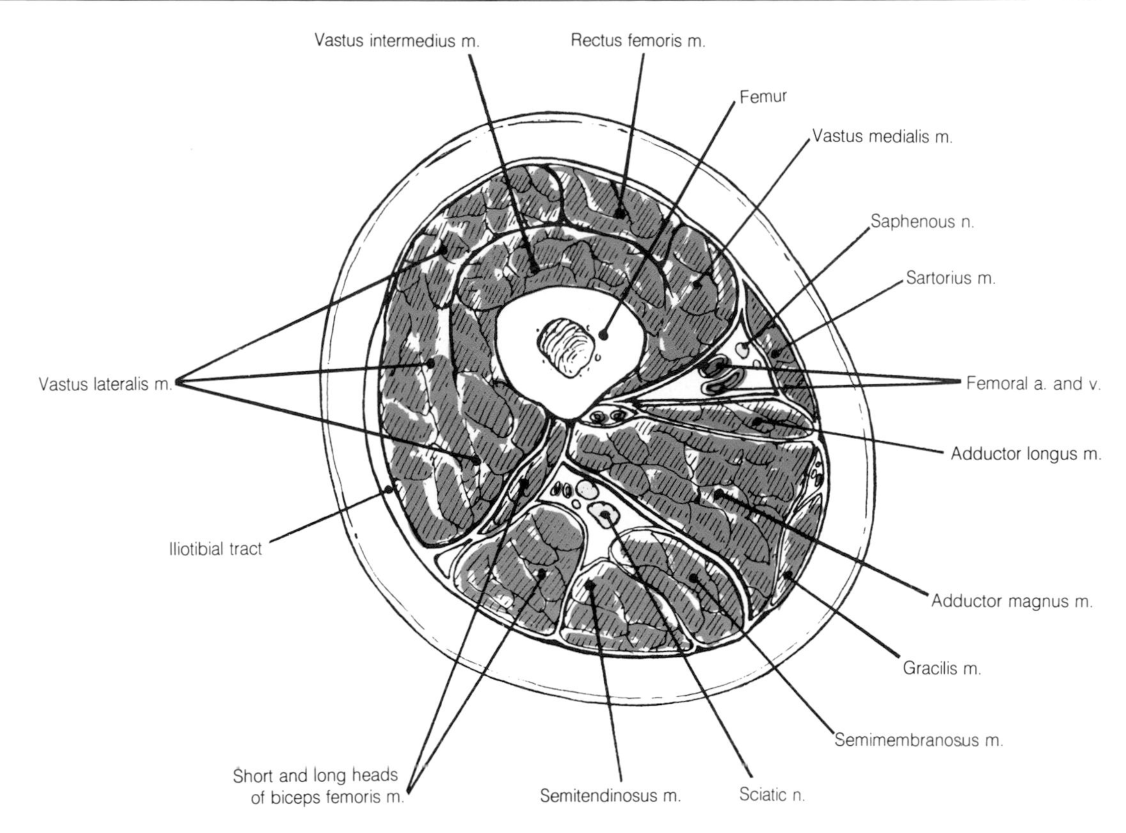

8-41 Cross-section through the middle of the thigh.

3. The adductor canal and the muscles that form its walls and roof at this point
4. The femoral vessels and the saphenous nerve travelling in the canal
5. That at this level only three of the adductor muscles are seen, i.e., the section lies below the level of the attachment of adductor brevis to the linea aspera
6. The hamstring muscles that arise from the ischial tuberosity in the flexor compartment with the sciatic nerve lying deep to them
7. That the section has passed through the upper end of the short head of biceps where it arises from the lateral lip of the linea aspera.

The Hip Joint

The hip joint is a synovial ball and socket joint which combines a wide range of movements with exceptional stability. Dislocation is rare except when extreme forces such as those encountered in an automobile accident are involved. The joint is, however, subject to both inflammatory and degenerative disease and its importance is underlined by the fact that it was one of the first joints for which replacement prostheses were designed.

The surfaces involved are the cartilage-covered, globular head of the femur and an incomplete ring of cartilage lining the acetabulum over the **lunate surface**. The defect in the cartilaginous ring is opposite the downward-facing acetabular notch but the articular surface here is completed by the **transverse ligament of the acetabulum**. The depth of the socket is increased by the **labrum acetabulare**, a complete rim of fibrocartilage surmounting the edge of the acetabulum and the transverse ligament. The floor of the acetabulum which is not articular contains a pad of fat.

The fibrous capsule is attached proximally to the margin of the acetabulum and to the transverse acetabular ligament. It extends over the head and neck of the femur to reach the bone close to the intertrochanteric line anteriorly but not reaching the crest posteriorly. From here it is reflected back toward the joint margin on the surface of the neck. Over the neck the capsule is thrown into a number of longitudinal folds called **retinacula**. These contain vessels supplying the neck and head of the femur. The fibers of the capsule are reinforced by circularly running fibers, the **zona orbicularis**, which clasp the free portion of the capsule close to the neck of the femur.

The **synovial membrane** lines the capsule including that part reflected over the neck. It also covers the fat in the floor of the acetabulum.

The capsule is reinforced by three strong ligaments extending from the pelvic bone to the femur. These are the **pubofemoral, ischiofemoral, and iliofemoral ligaments**. Of these, the strongest and most important is the iliofemoral ligament which is illustrated in Fig. 8-42. This arises from the **anterior inferior iliac spine** and nearby **lip of the acetabulum**. Its fibers diverge in two broad bands to be attached to the **intertrochanteric line**. All three ligaments take a spiral course around the neck of the femur and are arranged so that they become taut in full extension. In the relaxed standing position the center of gravity of the body lies behind the axis of the hip joint and tends to extend it. The iliofemoral ligament passively supports the weight of the trunk by preventing further extension of the joint. An insignificant **ligament of the head of the femur** runs from the floor of the acetabular fossa to a pit in the head of the femur. It carries a small branch of the obturator artery to the head of the femur.

The movements of the joint are described as those of the thigh on the trunk rather than those between the head of the femur and the acetabulum. The two are not the same because of the angle formed between the neck and shaft of the femur (approximately 125°). Thus, rotation about the long axis of the neck is translated into extension and flexion of the shaft of the femur (or the thigh) on the trunk. Using this convention the movements of the hip joint are flexion, extension, abduction, adduction, circumduction, and rotation, i.e., turning the thigh about its long axis. When disease of the hip joint is suspected these movements and their range must be very carefully examined and compared with those of the opposite joint. A flexion deformity (or limitation in extension) can easily be hidden by an increased lumbar lordosis. The important muscles producing movement at the joint are:

Flexion	Iliopsoas
Extension	Gluteus maximus The hamstring muscles

(These extensors are also used to control flexion in the common movement of stooping over to pick up an object.)

Abduction	Gluteus medius Gluteus minimus
Adduction	The adductor muscles
Medial rotation	Tensor fasciae latae The anterior fibers of gluteus medius and minimus
Lateral rotation	Both obturator muscles Piriformis Quadratus femoris

(Lateral rotation is a much more powerful movement than medial rotation.)

The **nerve supply** of the joint is from branches of the femoral and obturator nerves and by branches of the sacral plexus.

The **blood supply** of the joint is provided by branches of the medial circumflex femoral, superior, and inferior gluteal and obturator arteries.

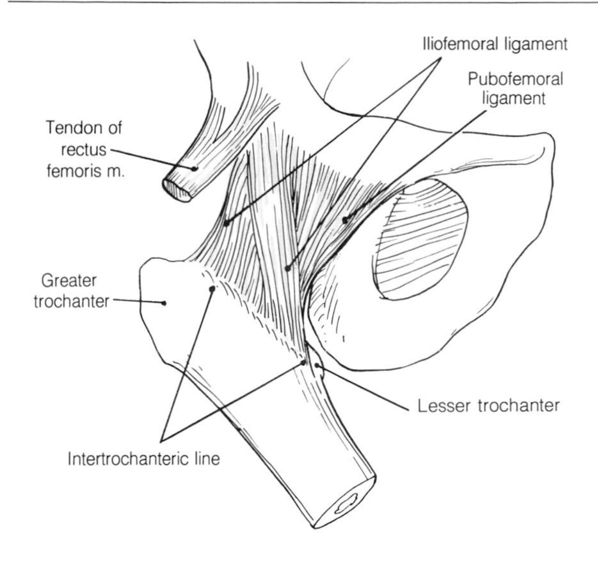

8-42 Showing the ligaments of the hip joint.

It should be noted that vessels supplying the head of the femur are bound to the neck of the femur by the retinacula and are easily damaged in fractures of the neck. The result may be an avascular necrosis of the proximal fragment.

The **relations of the hip joint** are important and also serve as a useful review of this region. In the femoral triangle anterior relations have been seen to be the muscles pectineus, iliopsoas and rectus femoris, and the femoral artery and nerve. Above lie the gluteus minimus, medius, and overlying gluteus maximus. Posteriorly is the group of short lateral rotators of the hip, piriformis, obturator internus, the gemelli, and quadratus femoris. The sciatic nerve is nearby as it runs over the posterior aspect of the acetabulum and all are covered by the gluteus maximus. Below the joint is found the tendon of obturator externus as it winds round to the back of the femur. In Fig. 8-43, a sagittal section through the head of the femur, a number of these relationships can be identified.

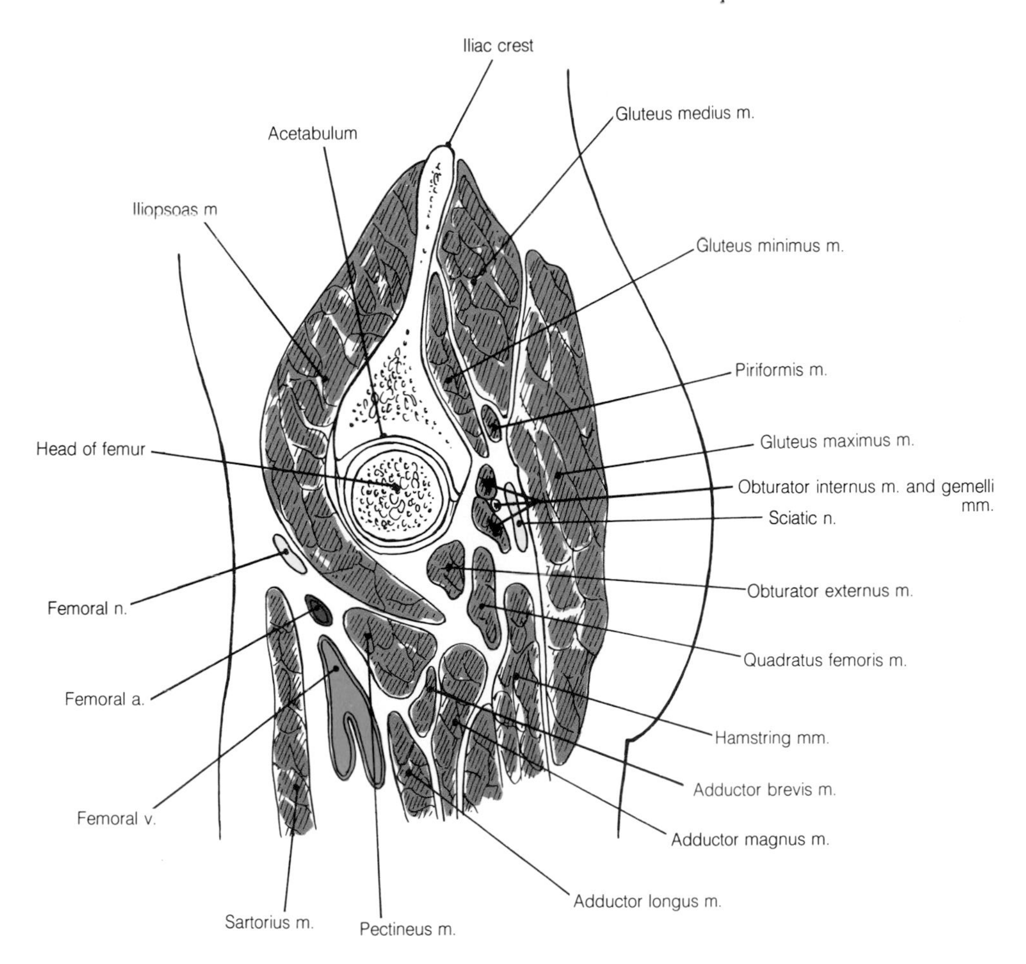

8-43 Sagittal section through the hip joint showing its relations.

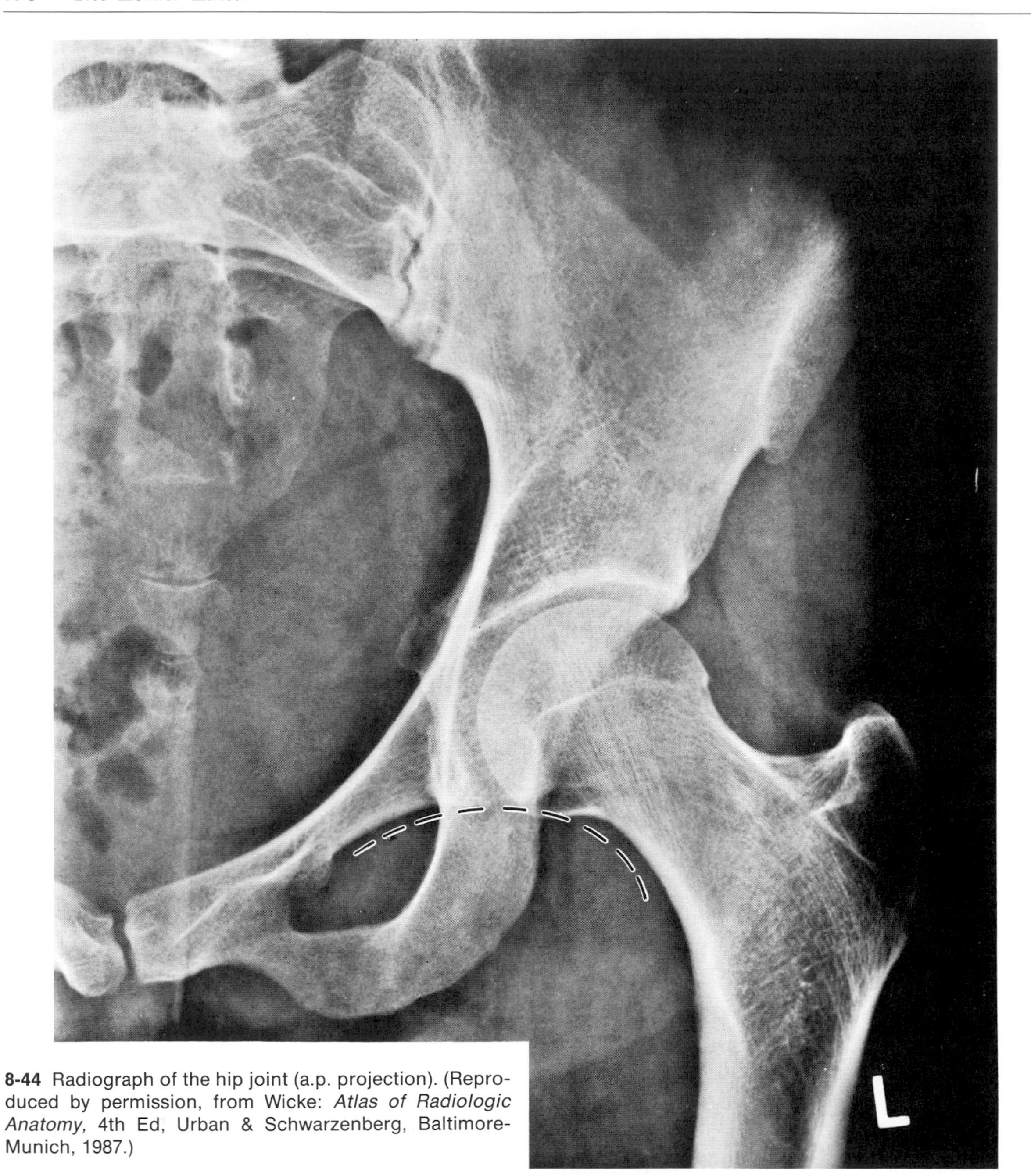

8-44 Radiograph of the hip joint (a.p. projection). (Reproduced by permission, from Wicke: *Atlas of Radiologic Anatomy,* 4th Ed, Urban & Schwarzenberg, Baltimore-Munich, 1987.)

The Radiological Anatomy of the Hip Joint

Early recognition of disease of the hip joint is of vital importance to prognosis and, in this, radiological examination plays a large part. A radiograph of the normal adult hip is seen in Fig. 8-44. Note:

1. The smooth unblemished curve of the femoral articular surface,
2. The articular surface of the acetabulum,
3. The "joint space" filled with radiolucent hyaline cartilage,
4. The architecture of the trabecular bone of the head, neck, and upper shaft. The strength of this angled bone largely depends upon this feature; the osteoporosis (rarefaction of bone) that accompanies old age severely weakens the neck of the femur.
5. The smoothly curved line produced by joining the medial margin of the neck to the lower mar-

gin of the superior pubic ramus (*dotted line*). Changes in the relationship of the head of the femur to the neck due to fracture or epiphyseal disease distorts this line (often known as Shenton's line),

6. While viewing this radiograph identify also the anterior superior iliac spine, the ischial spine, the ischial tuberosity, the superior pubic and ischiopubic rami, and the pubic symphysis.

The Popliteal Fossa

Little else but tendons and ligaments link the thigh with the leg anteriorly, medially, and laterally. All the important nerves and vessels pass from the thigh to the leg posteriorly through the region known as the **popliteal fossa**. It, therefore, deserves attention.

The margins of the fossa form a diamond shape and they are illustrated in Fig. 8-45. The upper lateral border is formed by the **biceps femoris muscle** as it nears its attachment to the head of the fibula. The upper medial border is formed by **semimembranosus** which is attached to a groove on the medial condyle of the tibia and outside this by the tendon of **semitendinosus** which is attached to the medial aspect of the shaft of the tibia below the condyle. Appearing from between biceps femoris and semimembranosus are the two heads of the **gastrocnemius muscle**. Their adjacent borders form the two lower borders of the diamond. The fossa is roofed by deep fascia which is pierced by the **small saphenous vein** and lymphatics from the back of the leg. The floor is formed by the posterior aspect of the **lower end of the body of the femur**, the **oblique popliteal ligament** which separates it from the knee joint, and

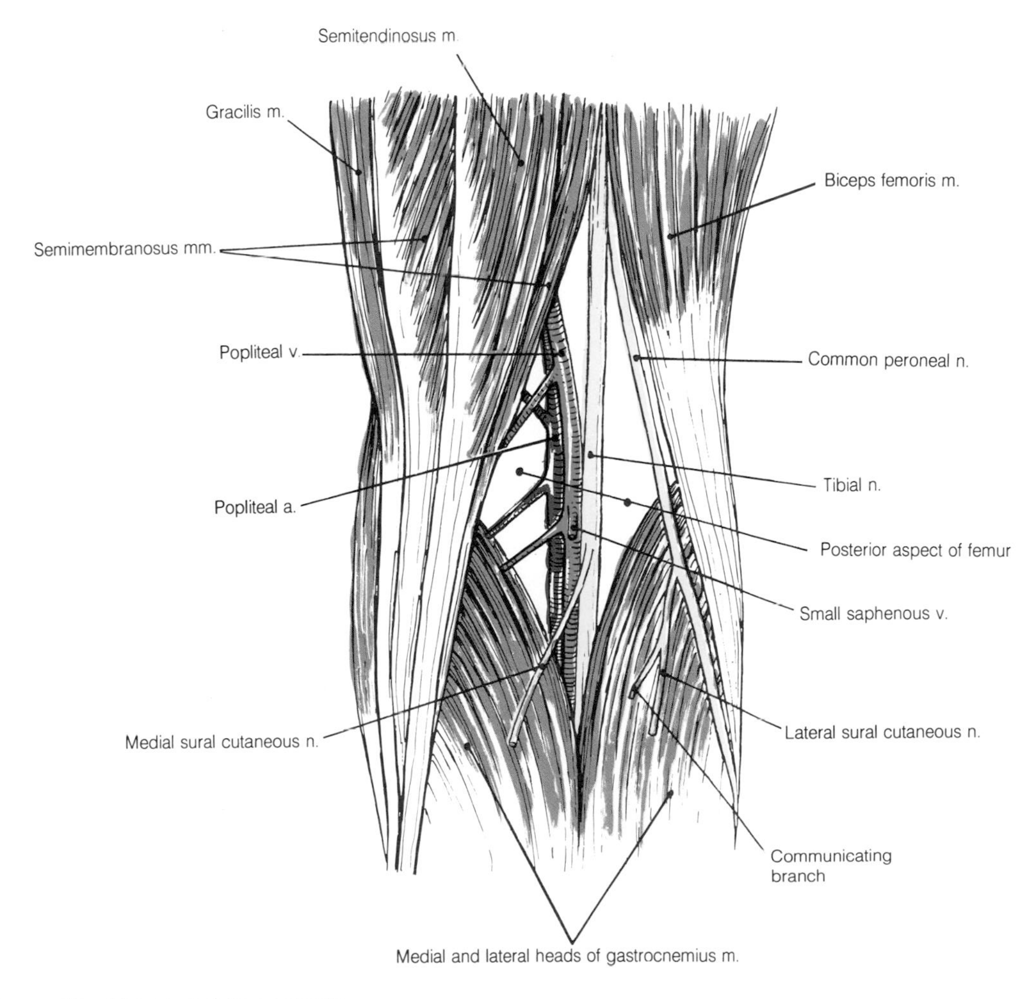

8-45 The popliteal fossa and its contents.

the **popliteus muscle**. In truth to display the fossa as described, the borders require some artificial separation.

The Contents of the Popliteal Fossa

Repeated reference to Fig. 8-45 should be made as the following structures are described.

The Popliteal Artery

The popliteal artery is, of course, a continuation of the femoral artery. The name of the vessel changes as it passes through the **adductor hiatus** and enters the upper part of the fossa from beneath is medial border. The artery runs across the floor of the fossa to terminate at the lower border of the **popliteus muscle** where it divides into the **anterior and posterior tibial arteries**. As it passes through the fossa, two superior, a middle, and two inferior **genicular arteries** are given off as well as muscular branches. Anastomoses of these genicular vessels with descending branches of the femoral and profunda femoris arteries and ascending branches from the anterior and posterior tibial arteries form an important potential collateral circulation which may serve to bypass the main vessels when these are narrowed or occluded.

The Popliteal Vein

The popliteal vein runs upward through the popliteal fossa at first medial to, then superficial to, and finally posterolateral to the popliteal artery before it enters the **adductor hiatus**. Its tributaries are the **small saphenous vein** and veins corresponding to the branches of the popliteal artery.

The Common Peroneal and Tibial Nerves

The common peroneal and tibial nerves are the terminal branches of the **sciatic nerve**. Their component fibers are discretely organized as the sciatic nerve enters the thigh in the gluteal region and they may become separate entities at any level of the thigh. The **common peroneal nerve** is found beneath the lateral border of the fossa which is formed by biceps femoris. It follows this muscle toward the **head of the fibula**. While in the popliteal fossa, a **lateral sural cutaneous nerve** to the skin of the calf and a communicating branch to the **medial sural cutaneous nerve** are given off. Branches also supply the knee joint. The **tibial nerve** roughly bisects the fossa as it travels from upper apex to lower apex of the diamond. While in the fossa, branches supply the **soleus, gastrocnemius, plantaris, and popliteus mucles** and there are also **articular branches**. A **medial sural cutaneous nerve** is also given off which, when joined by the communicating branch from the lateral sural cutaneous nerve, becomes the **sural nerve**. This nerve and its branches supply the lateral side of the calf and by **calcaneal branches** the heel. It terminates on the lateral aspect of the fifth toe.

The remaining space in the fossa is filled with fat in which a number of lymph nodes are embedded. These receive afferent vessels from the deep structures of the leg, and via vessels running with the short saphenous vein receive lymph from superficial tissues of the lateral aspect of the calf, heel, and foot.

The Knee

The thigh articulates with the leg at the knee joint and almost all the muscles that operate this joint have been considered with the thigh. Before discussing the knee joint, the adjacent bony parts must be described in more detail.

The Skeleton of the Knee Joint

The Lower End of the Femur

Look at Fig. 8-46 and see that the lower end of the femur is expanded and bears a pair of somewhat disc-like **articular condyles** united anteriorly, but elsewhere separated by an **intercondylar fossa**. The united part of the anterior surface articulates with the patella. Below this surface the separated condyles articulate with the medial and lateral condyles of the tibia. The lateral condyle bears an **epicondyle** below which lies a shallow pit for the attachment of the popliteus muscle and a groove posterior to this which houses the tendon of this muscle when the knee joint is in full flexion. The medial condyle also has an **epicondyle** and projecting from the condyle superiorly is the **adductor tubercle**. Posteriorly the shaft of the femur

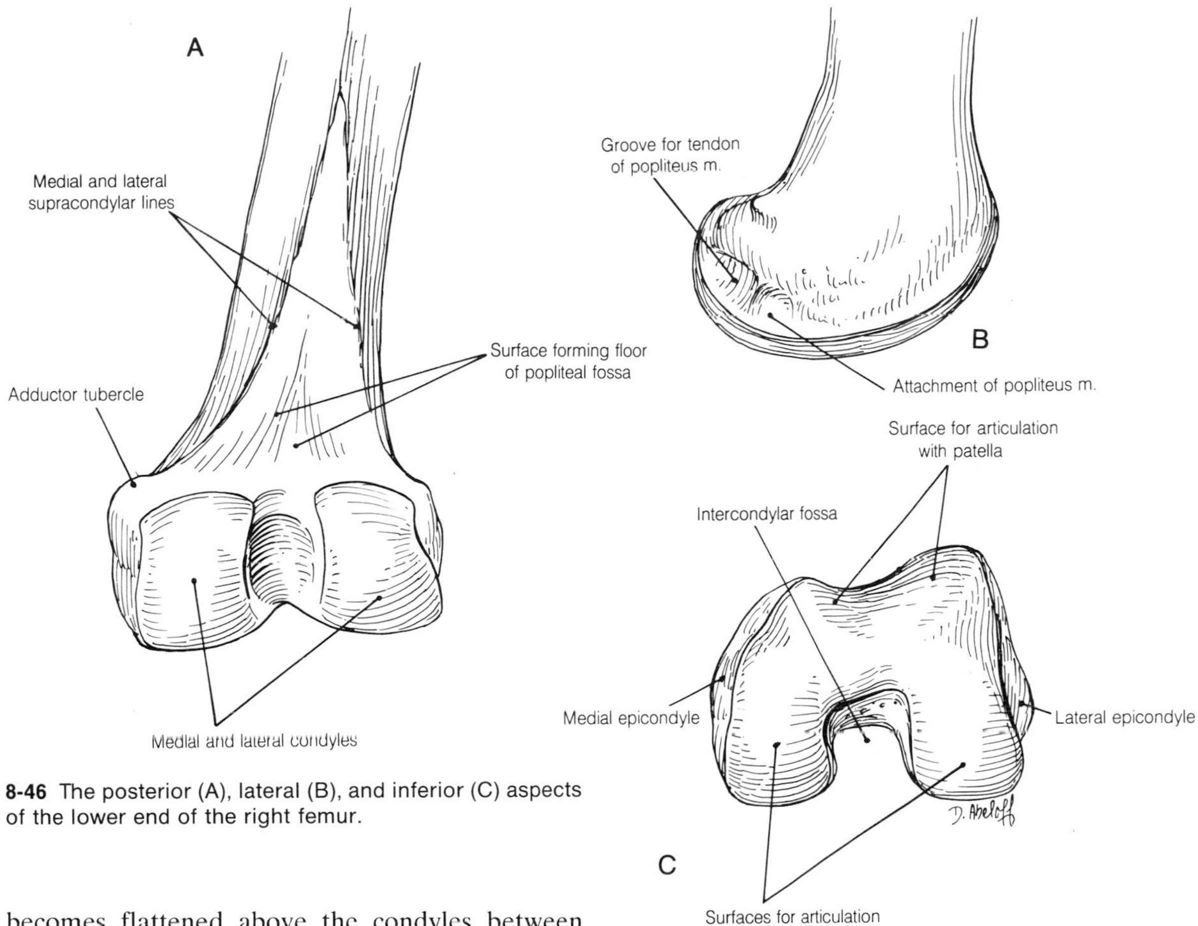

8-46 The posterior (A), lateral (B), and inferior (C) aspects of the lower end of the right femur.

becomes flattened above the condyles between the **medial and lateral supracondylar lines**.

The secondary center of ossification for the lower end of the femur appears just before birth. It is usually the only secondary center present at birth and indicates, in medicolegal terms, a full-term child. Fusion with the shaft occurs between the eighteenth and twentieth years.

The Patella

The patella is a sesamoid bone in the tendon of quadriceps femoris. It has a convex upper border and is triangular in shape below this. The **anterior surface** is roughened, while the deep **articular surface** is smooth. This surface displays two facets of unequal size. The larger lateral facet articulates with the larger and more prominent facet on the lateral femoral condyle, the smaller with the medial. Both surfaces are shown in Fig. 8-47. The patella is ossified by centers appearing and fusing between the third and sixth years.

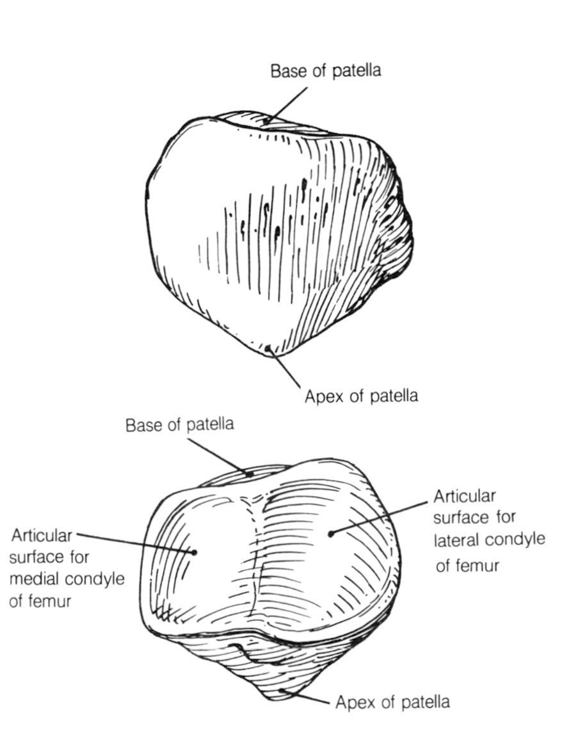

8-47 The anterior and posterior aspects of the patella.

The Upper End of the Tibia

In Fig. 8-48 it can be seen that the expanded upper end of the tibia is formed by the **medial and lateral condyles**. These show two slightly hollowed **superior articular surfaces** and are separated by a roughened **intercondylar area** which bears a **medial and a lateral intercondylar tubercle**. Anteriorly the **tibial tuberosity** lies at the junction of the **body or shaft** with the upper end, and posterolaterally, below the overhanging lateral condyle, there is an **articular surface for the head of the fibula**.

The secondary center of ossification for the upper end of the tibia appears soon after birth and joins the shaft between the sixteenth and eighteenth years. This epiphysis also includes the greater part of the tibial tuberosity and this portion may become partially separated and painful during adolescence.

The Upper End of the Fibula

The upper end or **head of the fibula** is only slightly expanded. The **apex** or styloid process of the head extends upward from it. On the superomedial aspect of the head there is a circular articular facet for the tibia. Note that the fibula does not take part in the knee joint. The secondary center for ossification of the head appears much later than that for the tibia, at about the fifth year, and fuses with the shaft at about the eighteenth year.

The Knee Joint

The knee joint is a synovial joint of the condylar variety. Although it does allow some rotation, it is primarily a hinge joint allowing flexion and extension. It is in many ways a remarkable structure.

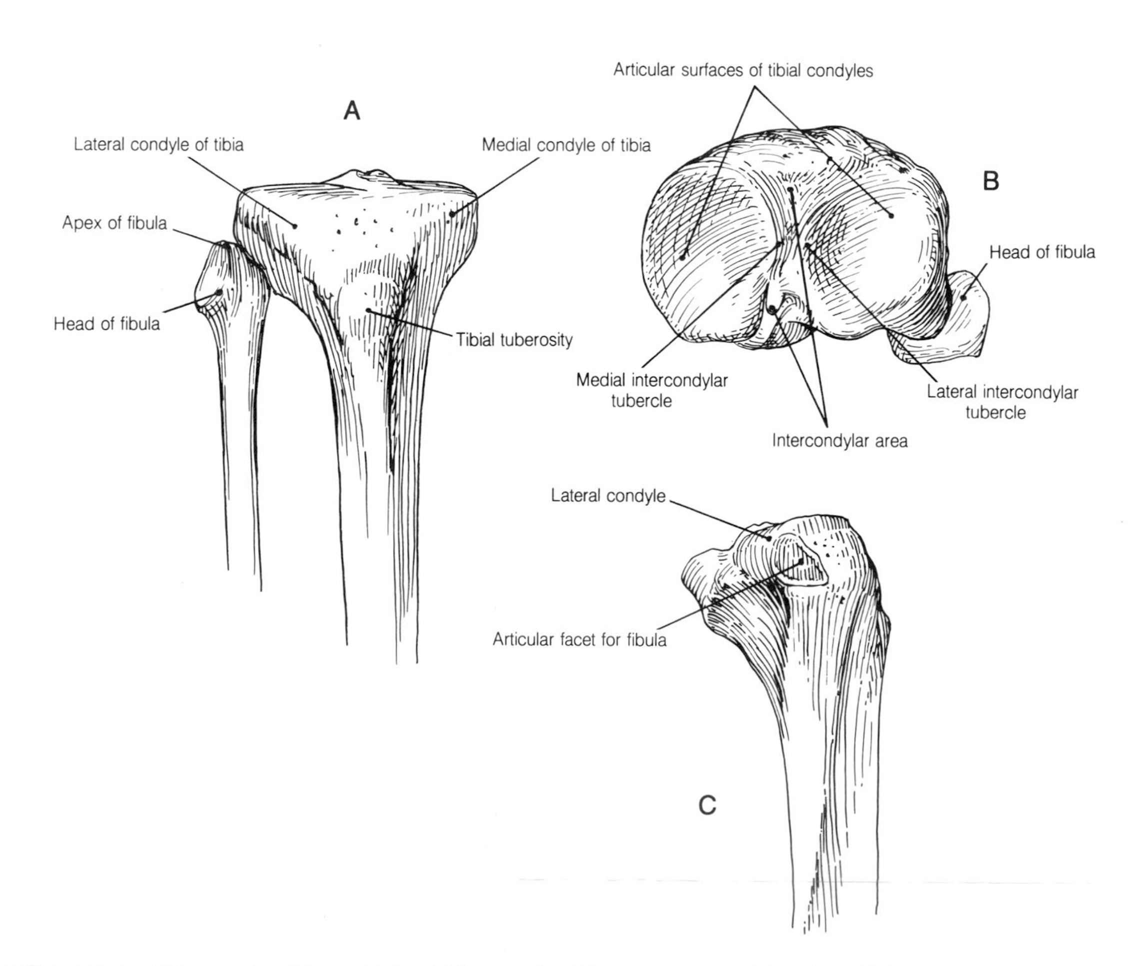

8-48 The anterior (A), superior (B), and lateral (C) aspects of the upper end of the right tibia.

Frequently bearing the full weight of the body it does not, at first sight, display the features normally associated with a stable joint. The bony articular surfaces are far from congruent but dislocation is extremely unusual. Stability of the joint depends upon the strength of the ligaments connecting the femur to the tibia and the actions of the muscles whose tendons of attachment surround the joint. Nevertheless, the joint is prone to serious injuries to its ligaments and intra-articular discs which more often than not result from the strains placed on it by recreational and professional sports.

The **bony articulating surfaces** covered, of course, by hyaline cartilage consist of the margins of the two disc-like femoral condyles and intervening bony bridge, the patella, and the superior surfaces of the tibial condyles. Note that the patella only articulates with the femur. Here the articular surface for the patella is delineated by two faint ridges which separate it from the surfaces for articulation with the tibial condyles. The extent of these articular surfaces on the femur, tibia, and patella can be seen in Figs. 8-46, 47, and 48.

The **capsule** of the knee joint is attached just outside the articular margins of the femoral condyles medially and laterally and above the intercondylar notch posteriorly. Anteriorly, it is defective in that it allows communication between the articular cavity of the joint and the **suprapatellar bursa** which lies between the lower end of the shaft of the femur and the **quadriceps muscle and tendon**. From the femur the capsule extends to the margins of the upper end of the tibia with a posterior defect allowing entry to the tendon of the muscle **popliteus**. The femoral attachment of this muscle lies within the capsule but the tendon itself lies outside the synovial membrane. Anteriorly, the capsule is replaced by the articular surface of the patella. Generally, the capsule is thin and that part of it surrounding the joint margins is often known as the **coronary ligament**. The capsule is reinforced anteriorly by the **patellar retinacula** which are expansions of the medial and lateral vasti muscles and by the quadriceps tendon, patella, and **patellar ligament**.

The **synovial membrane** lines the fibrous capsule, but a number of particular features need to be noted. Above the patella it escapes from the capsule to become continuous with the synovial

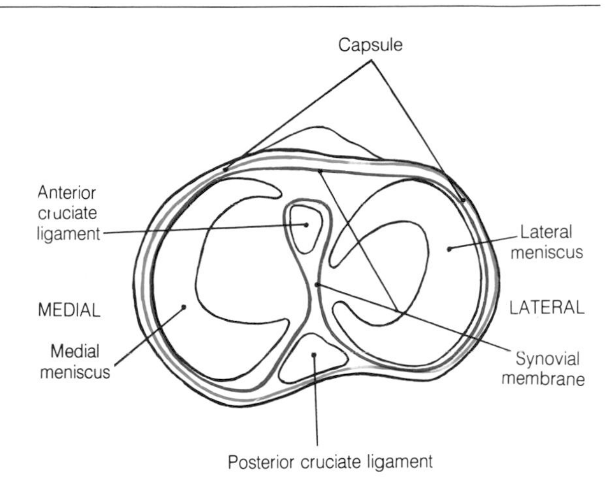

8-49 Diagram showing the tibial attachments of the capsule and synovial membrane of the knee joint.

lining of the suprapatellar bursa. Below the patella it is separated from the patellar ligament by a pad of fat which raises it to form the **infrapatellar fold**. It is separated from the capsule by the tendon of popliteus and posteriorly it is invaginated forward between the condyles of both the femur and tibia by the **cruciate ligaments** which thus lie outside the articular cavity. Check this feature in Fig. 8-49.

The stability of the knee joint depends largely on its ligaments which are shown in Figs. 8-50, 8-51, and 8-52. Among these can be included the **patellar ligament** which is in reality the termination of the quadriceps tendon. It extends from the lower border and **apex** of the patella to the upper part of the **tibial tuberosity**. Its tension is control-

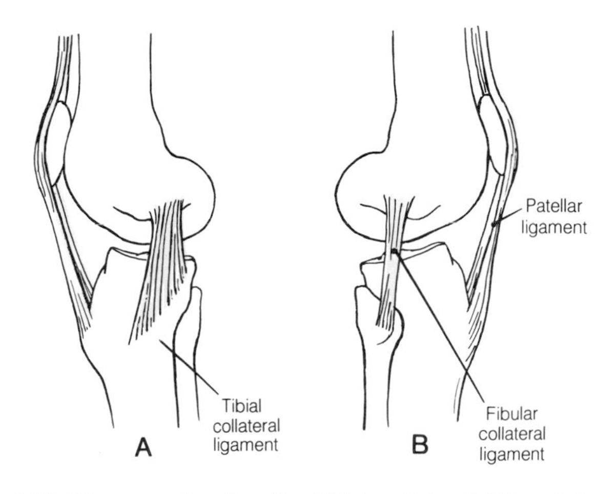

8-50 Diagrams showing the tibial collateral (A) and the fibular collateral (B) ligaments of the knee joint.

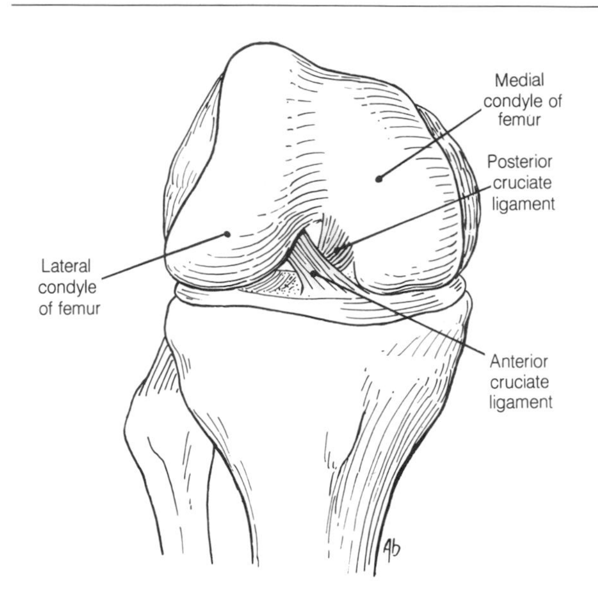

8-51 The cruciate ligaments of the knee joint seen from the front when the joint is flexed.

led by the quadriceps muscle and it is able, therefore, to help stabilize the joint throughout its range of movement.

The **tibial collateral ligament** (Fig. 8-50 A) is a broad, flattened ligament extending from the **medial femoral epicondyle** to the medial aspect of the **shaft of the tibia** below the medial condyle. A short deep portion extends as a thickening in the capsule from the femur to the **medial meniscus**.

The **fibular collateral ligament** (Fig. 8-50 B) is a strong rounded ligament extending from the **lateral femoral epicondyle** to the **head of the fibula**. It has no attachment to the **lateral meniscus**.

The **oblique popliteal ligament**, an expansion of the tendon of semimembranosus, gives some reinforcement to the capsule posteriorly.

The **cruciate ligaments** lie largely within the intercondylar notch. The word cruciate* indicates that they form a cross and this can be seen in Fig. 8-51. They are named **anterior** and **posterior** after their **tibial** attachments. The **anterior cruciate ligament** is attached to the anterior part of the in-

* Crux, L. = a cross

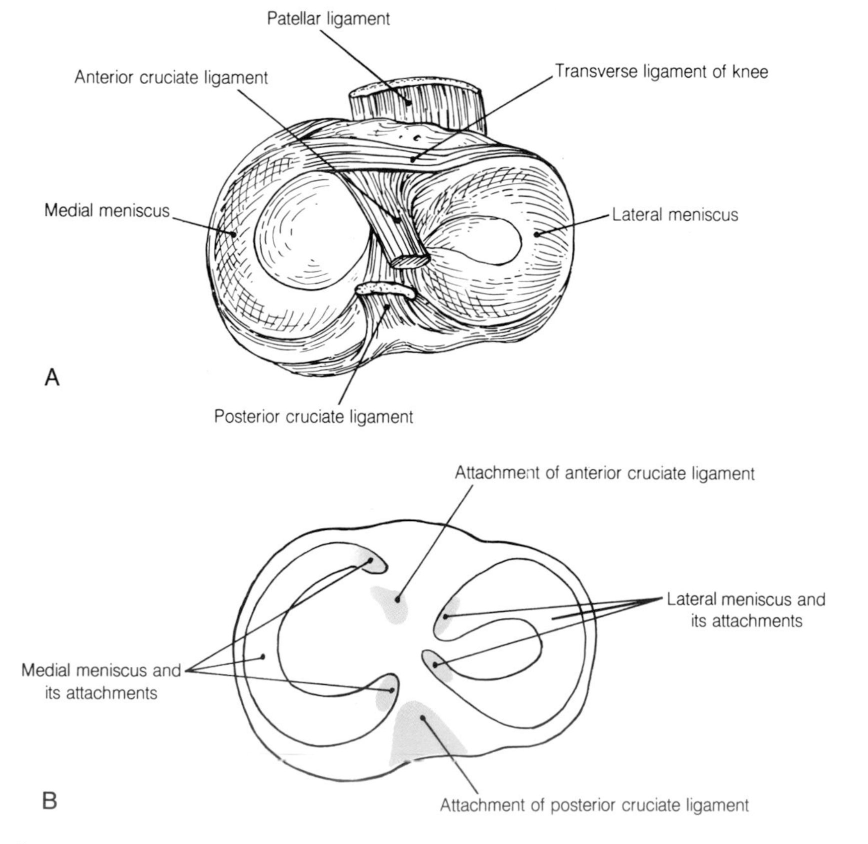

8-52 The cruciate ligaments and the menisci seen from above (A) and a diagram indicating their sites of attachment (B).

tercondylar area of the tibia and is directed backward to the medial surface of the lateral femoral condyle. The **posterior cruciate ligament** is attached to the posterior part of the intercondylar region of the tibia and is directed anteriorly to the lateral surface of the medial femoral condyle. The tibial attachments of both ligaments are shown in Fig. 8-52. Both ligaments are concerned with keeping the articular surfaces applied to each other throughout the joint's range of movement. The anterior ligament also prevents forward gliding of the tibia on the femur and the posterior prevents backward gliding. Their rupture can be detected by testing these functions when the knee is flexed. Abnormal anteroposterior movement in this position is called the "drawer" sign. As has already been described, these ligaments are excluded from the synovial cavity by a posterior invagination of the synovial membrane.

The **menisci** or semilunar cartilages of the knee joint are two crescentic plates of fibrocartilage having a wedge-shaped profile when cut in transverse section. In Fig. 8-52 A it can be seen that the lateral meniscus has the form of a closed "C," the medial, a more open "C." The horns of the menisci are attached to the intercondylar area of the tibia and at the periphery to the loose **coronary ligament** of the capsule. In addition, the medial meniscus is attached to the **deep portion** of the **tibial collateral ligament**. The inner margins and upper and lower surfaces project freely into the joint cavity. While the menisci are not normally weight-bearing, they do increase the congruity of the opposing articular surfaces and may aid lubrication and the "locking" of the joint in the slightly hyperextended position. The sites of attachment to the tibia of both the menisci and the cruciate ligaments are shown in Fig. 8-52B.

The **main movements of the knee joint** are flexion and extension. Extension is accomplished by the quadriceps femoris muscle and is limited as the cruciate and collateral ligaments become taut and the menisci compressed. At this point the joint is slightly hyperextended. Flexion, performed by the hamstring muscles, is limited by contact between calf and thigh. When the knee is partially flexed, medial and lateral rotation are possible and performed by the medial or lateral "hamstring" muscles. Rotation also occurs during the last part of extension. This is a passive or conjunct movement imposed upon the joint by the geometry of the articular surfaces and the sequential tightening of the joint ligaments. As the last 30° of extension are reached, the medial condyle of the femur and its surrounding meniscus begin to slide backward on the more extensive tibial articu-

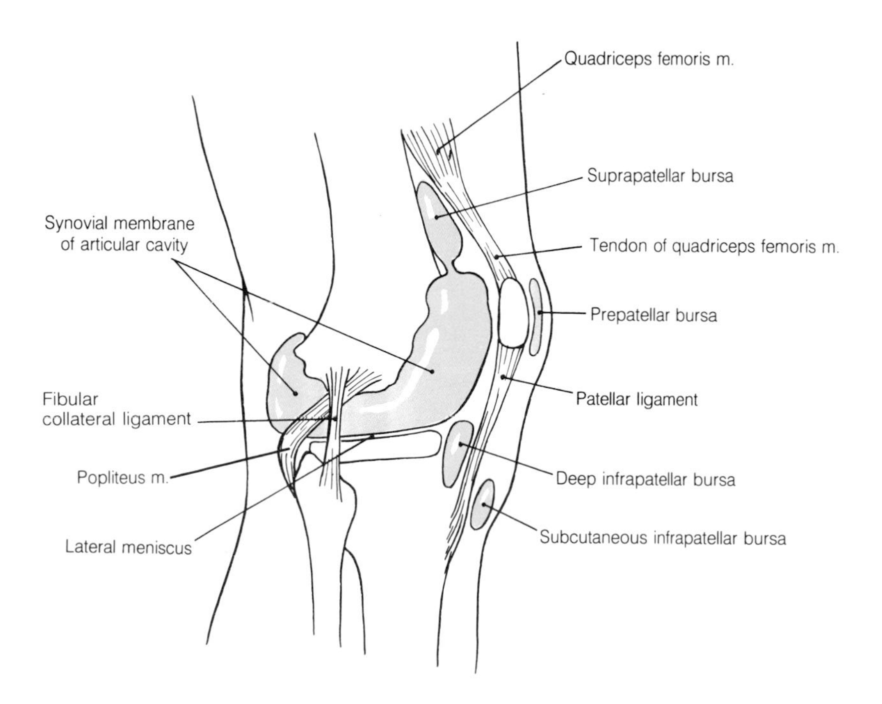

8-53 Showing the synovial membrane of the knee joint and associated synovial bursae.

lar surface until the locked position is reached. Lateral rotation of the femur on the tibia occurs as the joint is flexed again.

When the knee joints are thus locked in a few degrees of hyperextension and the center of gravity lies anterior to the transverse axes of the joints (as it does when standing comfortably), the position can be maintained with the minimum expenditure of muscular energy.

The **blood supply of the knee joint** is derived from the **genicular branches of the popliteal artery**. It is innervated by branches of the **sciatic, femoral**, and **obturator nerves** (remember that pain from the hip may be incorrectly referred to the knee).

The knee joint is surrounded by a number of bursae. The **suprapatellar bursa** has been mentioned. Do not forget that it communicates with the knee joint and that perforating injuries may introduce infection into the joint. A **prepatellar bursa** lying over the superficial surface of the patella and a **subcutaneous infrapatellar bursa** over the patellar ligament frequently become inflamed after prolonged kneeling. Additional bursae are the **deep infrapatellar bursa**, and bursae associated with the attachments of the gastrocnemii, gracilis, sartorius, semitendinosus, and the tendon of semimembranosus. The latter bursa may also communicate with the synovial cavity. The patellar bursae are shown in Fig. 8-53.

Injuries of the knee joint are commonly encountered and range from "sprains" of the collateral ligaments to rupture of the cruciate ligaments or tears in the menisci. Nearly all are associated with an effusion of fluid into the synovial cavity and some wasting of the quadriceps muscle. Tears of the menisci occur when the flexed and weight bearing knee is forcibly rotated. A partially separated fragment may slip between the articular surfaces and "lock" the joint. Because they are virtually avascular, the menisci do not repair and excision is, therefore, often needed. Of the two, it is the medial that is more frequently damaged, perhaps, because it is bound to the deep portion of the tibial collateral ligament and, therefore, less mobile. Tears of this meniscus are also found in association with rupture of the anterior cruciate ligament and the tibial collateral ligament. This is the "unhappy triad" and follows blows on the lateral side of the knee as in "clipping."

The Radiological Anatomy of the Knee Joint

The knee joint is frequently radiographed and the normal appearance of the joint should be studied in Fig. 8-54. Identify the femoral and tibial condyles and the shadow of the patella overlying the femur but not the tibia. Note the joint space into which project the intercondylar tubercles of the tibia. Although not seen in this radiograph, small sesamoid bones are occasionally present in the tendons of the calf muscle gastrocnemius. These may appear behind the femoral condyles in a lateral projection and be mistaken for "loose bodies" in the joint cavity.

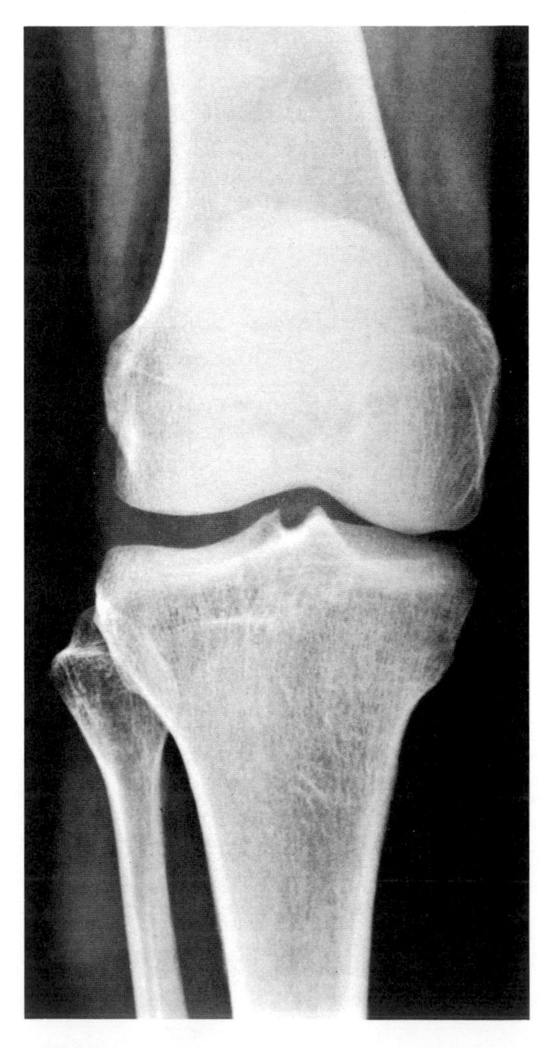

8-54 Radiograph of the right knee (a.p. projection) (Reproduced by permission, from Wicke: *Atlas of Radiologic Anatomy,* 4th Ed, Urban & Schwarzenberg, Baltimore-Munich, 1987.)

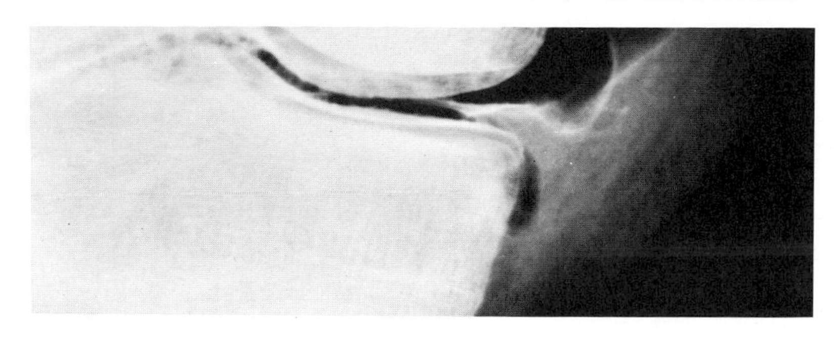

8-55 An arthrogram showing a medial meniscus. (Reproduced by permission, from Wicke: *Atlas of Radiologic Anatomy*, 4th Ed, Urban & Schwarzenberg, Baltimore-Munich, 1987.)

The menisci can be demonstrated by introducing air or opaque fluids into the joint cavity and an example of such an arthrogram is shown in Fig. 8-55.

The Superior Tibiofibular Joint

The superior tibiofibular joint is formed between the head of the fibula and the lateral condyle of the tibia and is a plane synovial joint surrounded by a fibrous capsule. Its synovial cavity does not communicate with that of the knee joint. No movement of importance occurs at this joint.

The Leg and Dorsum of the Foot

The upper ends of the tibia and fibula were described before a discussion of the knee and superior tibiofibular joints. It is now necessary to examine the remainder of these long bones before continuing with the description of the soft parts of the leg.

The Skeleton of the Leg

The Shaft and Lower End of the Tibia

The shaft and lower end of the tibia are illustrated in Fig. 8-56. Note that the shaft is triangular in section. Its sharp **anterior border** can be easily palpated as can its subcutaneous **medial surface**. A **lateral surface** lies between the anterior and **interosseus borders** and a **posterior surface** between this and its **medial border**. The upper part of the posterior surface is marked by an oblique ridge running downward and medially to blend with the medial border. This is the **soleal line**.

The lower end of the tibia is expanded and more quadrangular on section than the shaft. Its medial side can be easily identified by a downward projection called the **medial malleolus**. There is no sharp anterior border and the smooth anterior surface blends with the lateral aspect of the shaft. The medial surface and medial malleolus can be felt to be subcutaneous. Between the medial malleolus and the posterior surface is a **malleolar groove** that houses the tendon of tibialis posterior. The lateral surface is triangular and hollowed out to form a **fibular sulcus** that accepts the lower end of the fibula. Its upper part is roughened for the attachment of the **interosseous ligament** of the inferior tibiofibular joint. The inferior surface and the adjacent surface of the medial malleolus are smooth and form part of the articular surface of the ankle joint.

The Shaft of the Fibula

As seen in Fig. 8-56, the shaft of the fibula is somewhat trapezoidal in section with each surface associated with a muscle group. The narrow **medial surface** between the **anterior** and **interosseus borders** is related to dorsiflexors and the **lateral surface** to peroneal muscles. The **posterior surface** is divided by a vertical crest and related to plantarflexor muscles. The shaft of the fibula has no weight-bearing functions but serves

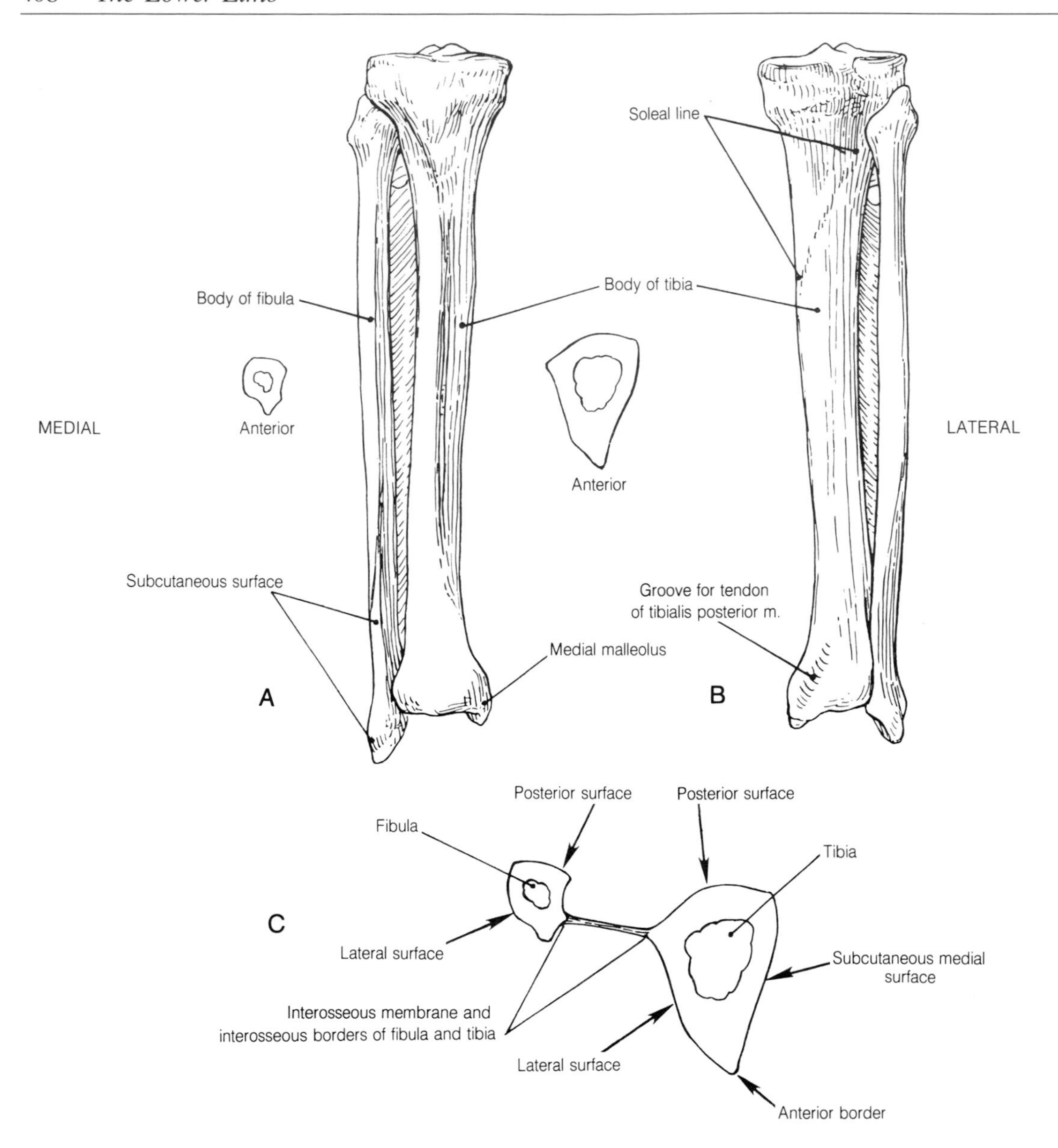

8-56 The anterior (A) and posterior (B) aspects of the right tibia and fibula. In (C) the relationship of the two bones and the interosseous membrane is shown in cross-section.

rather to give attachment to muscles. Portions of the shaft can be removed for bone grafting without loss of function.

The Lower End of the Fibula

The lower end of the fibula is also slightly expanded and projects downward below the level of the inferior surface of the tibia, forming the **lateral malleolus**. The medial surface of the malleolus shows a facet covered by articular cartilage which completes the proximal articular surface of the ankle joint. Just behind this facet lies a **malleolar fossa**. The anterior surface is rough, the lateral surface smooth and subcutaneous and the

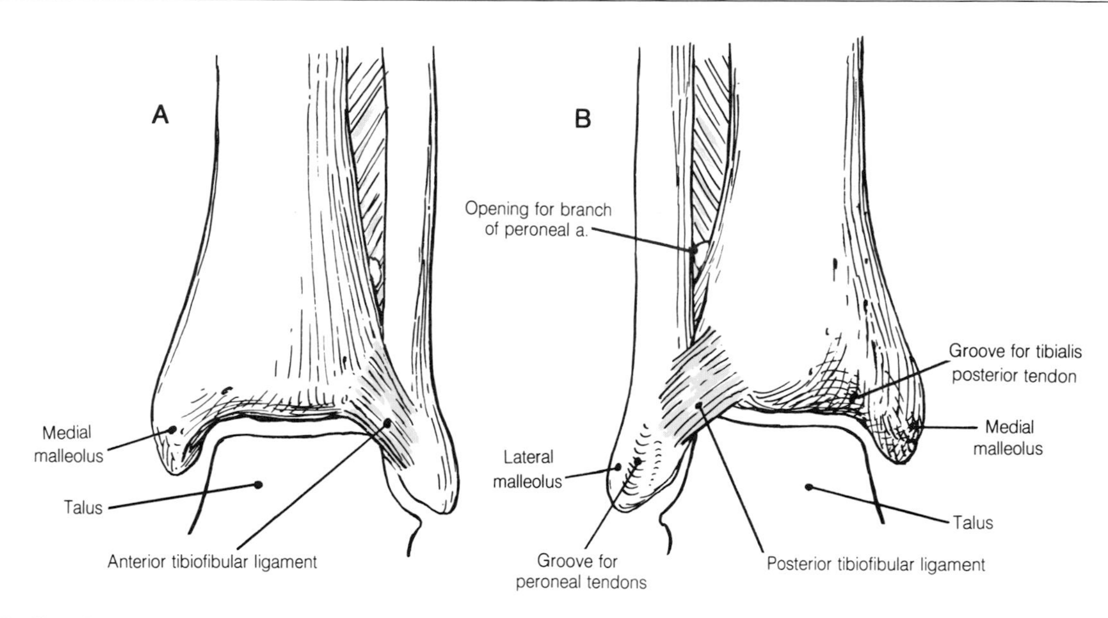

8-57 Showing the anterior and posterior tibiofibular ligaments.

posterior surface is marked by the **malleolar groove** in which lie the peroneal tendons.

Primary centers of ossification for the shafts of the tibia and fibula appear between the seventh and eighth weeks of intrauterine life. The secondary center for the upper end of the tibia appears at or soon after birth. Secondary centers for the lower ends of both bones appear during the first postnatal year and fuse with the shaft between the sixteenth and eighteenth year, the tibial preceding the fibular center. Note that the secondary center for the upper end of the fibula appears unusually late (fifth year), but nevertheless, also fuses late (eighteenth year). It is an exception to the general rule that lower limb epiphyses adjacent to the knee joint appear first.

The Interosseous Membrane

Look at Fig. 8-56 and see that the interosseous borders of the tibia and fibula are linked by a tough membrane called the **interosseous membrane**. The fibers of this membrane pass downward and laterally from tibia to fibula. The membrane has an opening at its upper end for the passage of the **anterior tibial artery** and is perforated by branches of the **peroneal artery** at its lower end. Its fibers are continuous with those of the **interosseous ligament of the inferior tibiofibular joint**.

The Inferior Tibiofibular Joint

The interior tibiofibular joint is formed between the roughened adjacent areas at the lower end of the tibia and fibula. It is a fibrous joint and permits only very slight movement. Its stability is maintained by a strong **interosseous ligament**, a continuation of the interosseous membrane, and **anterior and posterior tibiofibular ligaments** which are illustrated in Fig. 8-57. The integrity of the inferior tibiofibular joint is essential for the stability of the ankle joint as it keeps the lateral malleolus clasped against the lateral surface of the talus.

The Compartments of the Leg

The leg, which is that part of the lower limb between the knee and ankle, is conveniently organized into compartments by bony and fibrous structures. Look at the diagram in Fig. 8-58 and note the following:

1. An anterior compartment is formed by the lateral surface of the tibia, the strong **anterior crural* fascia**, the interosseous membrane, the **anterior intermuscular septum**, and the narrow anterior surface of the fibula. The compartment contains extensor and dorsiflexor muscles.
2. A lateral compartment containing the peroneal muscles lies between the **anterior and posterior intermuscular septa** attached to the anterior and posterior borders of the lateral surface of the fibula. Note here that the fibula lies on a much more posterior plane than the tibia.
3. The remaining posterior compartment is enclosed superficially by the **posterior crural fascia** and contains flexor muscles. These are separated into a deep and superficial group by a further septum called the **deep transverse crural fascia**. The anterior crural fascia is much thicker than the posterior. It is difficult to strip it off the underlying muscles of the extensor and peroneal compartments for which it provides attachment. In fact, the anterior compartment is so tightly encased that an increase in tissue fluid following unusual exercise may lead to vascular obstruction, acute pain, and possibly necrosis of muscles.

* Crus, *L.* = a leg

It will also be seen that each compartment has associated with it a major nerve and artery although in the case of the lateral compartment, the artery does not run in the compartment.

The Anterior Compartment

The Muscles of the Anterior Compartment

The course and distal attachments of the extensor muscles of the leg will be described in more detail with the dorsum of the foot which immediately

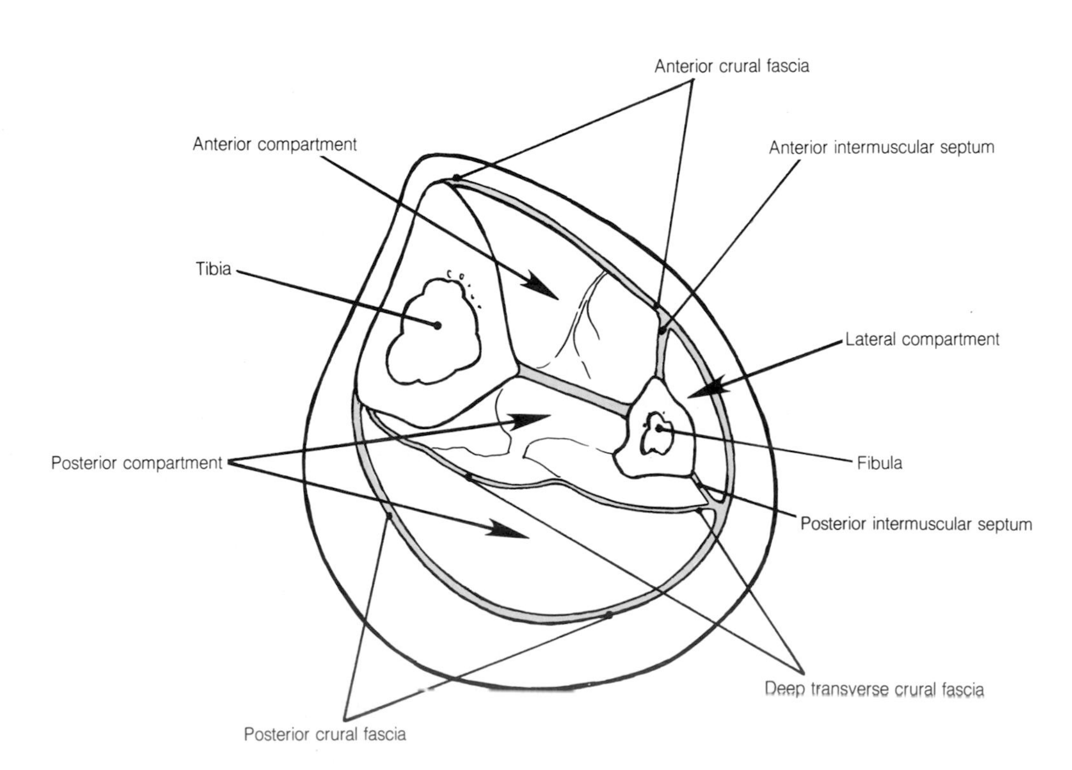

8-58 Diagram showing the compartments of the leg and their boundaries.

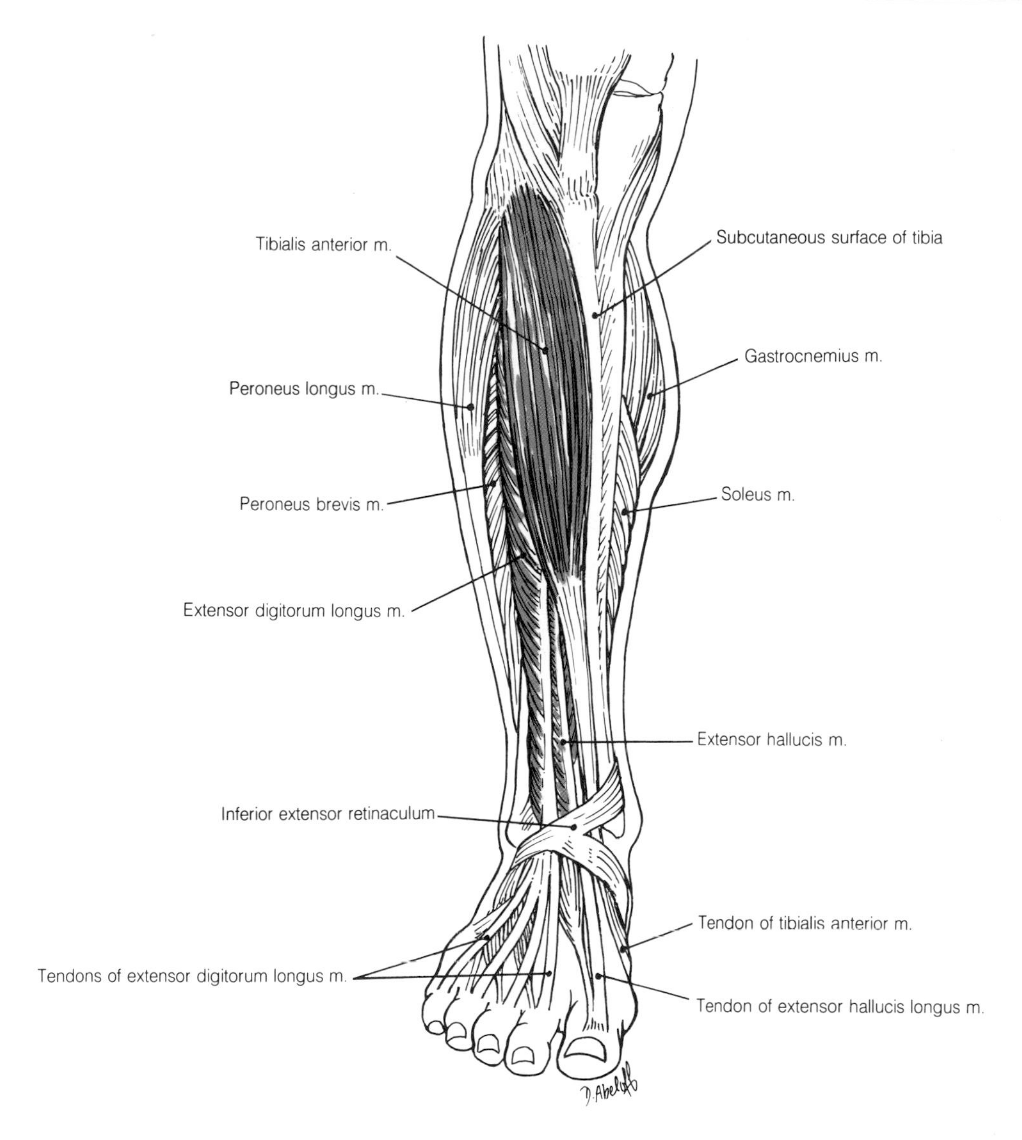

8-59 Showing the extensor muscles of the anterior compartment of the leg.

follows the description of this compartment. The appearance and position of the extensor muscles are illustrated in Fig. 8-59 and their attachments to the tibia and fibula in Fig. 8-60.

Tibialis Anterior

Tibialis anterior arises from the lateral condyle and upper half of the lateral surface of the tibia, from the adjacent interosseous membrane and the overlying crural fascia. It becomes tendinous in the lower third of the leg and passes over the ankle joint beneath the extensor retinacula to pass on to the medial side of the foot.

Extensor Digitorum Longus

Extensor digitorum longus arises from the lateral condyle of the tibia, the upper three-quarters of the medial surface of the fibula, the adjacent interosseous membrane, the anterior intermuscular septum, and the overlying fascia. Its tendon passes beneath the extensor retinacula dividing here into four slips for the lateral four toes.

Extensor Hallucis Longus

Extensor hallucis longus arises from the middle of the medial surface of the fibula and adjacent in-

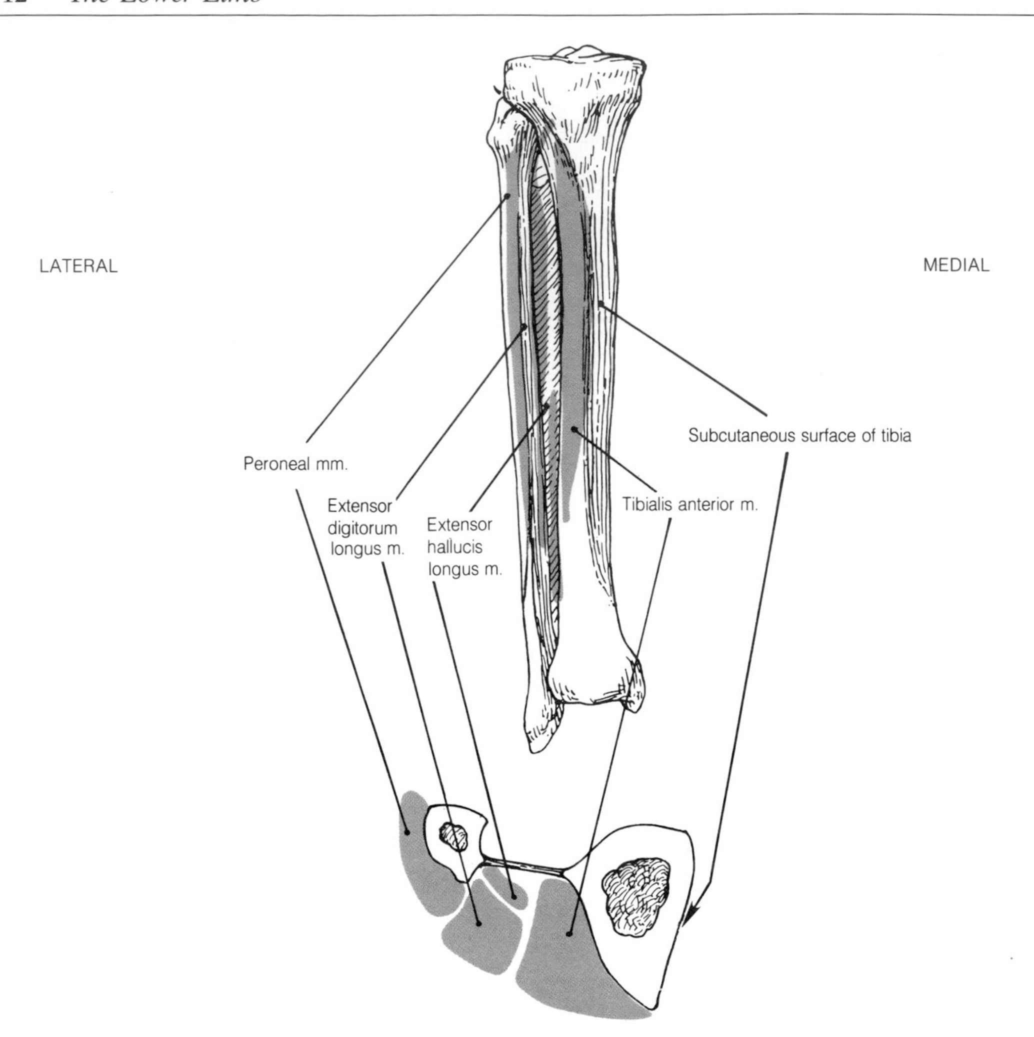

8-60 The anterior aspect of the tibia and fibula showing the attachments of the extensor muscles.

terosseous membrane. Initially, under cover of extensor digitorum longus its tendon becomes superficial in the lower part of the leg, passing beneath the retinacula lateral to the tendon of tibialis anterior on its way to the great toe.

Peroneus Tertius

Despite its name, peroneus tertius is not a true peroneal muscle but a portion of extensor digitorum longus with which it shares a common origin. Its tendon is attached to the dorsum of the fifth metatarsal bone. The muscle is not always found.

This group of extensor muscles in the anterior tibial compartment is supplied by branches of the **deep peroneal nerve** (L4, 5; S1) as it traverses the compartment.

The Nerves and Vessels of the Anterior Compartment

The Common Peroneal Nerve (L4, 5; S1, 2)

This nerve has been seen to leave the popliteal fossa under cover of the tendon of biceps femoris. It subsequently winds forward around the lateral aspect of the neck of the fibula to divide in the lateral compartment into a **deep and superficial peroneal nerve**. The deep peroneal nerve enters the anterior compartment while the superficial remains in the lateral compartment. The common peroneal nerve can be felt as it passes around the fibula and is vulnerable at this point. For example, the upper edge of a below-knee plaster cast may, by repeated percussion of the nerve, produce a classical picture of "foot drop" due to paralysis of the extensor muscles that the nerve supplies. A further problem is that this may only become apparent when the cast is removed.

The Deep Peroneal Nerve

The deep peroneal nerve arises in the lateral compartment and, passing deep to peroneus longus and extensor digitorum longus, enters the anterior compartment. This it descends lying close to the interosseous membrane between **extensor digitorum longus** and **extensor hallucis** laterally, and **tibialis anterior** medially. Here it is joined by the **anterior tibial artery**. In the lower part of the leg this neurovascular bundle is crossed by the extensor hallucis longus so that when the nerve and artery become more superficial over the ankle joint the tendons of extensor hallucis longus and tibialis anterior lie medially and the tendons of extensor digitorum longus lie laterally. The nerve itself is at this point lateral to the vessel. As well as muscular branches to the extensor compartment, articular branches are supplied to the ankle joint. The subsequent course of the nerve will be described with the foot.

The Anterior Tibial Artery

The anterior tibial artery is one of the two terminal branches of the popliteal artery formed at its bifurcation at the lower border of the **popliteus muscle**. From this point it passes forward through an opening in the interosseous membrane between tibia and fibula to reach the anterior compartment. It descends through this compartment in company with the deep peroneal nerve to appear over the ankle joint on the medial side of the nerve. The vessel contributes to the arterial anastomosis around the knee by an **anterior tibial recurrent branch** and supplies the muscles of the anterior tibial compartment. Its continuation on to the dorsum of the foot is called the **dorsalis pedis artery**.

Now examine the cross-section of the middle of the leg seen in Fig. 8-61. Identify again the boundaries of the anterior compartment and note the position of the anterior tibial vessels and the deep peroneal nerve which lie close to the interosseous membrane and between tibialis anterior lying medially and the extensor muscles of the digits lying laterally. The extensive subcutaneous medial surface of the tibia is also well shown.

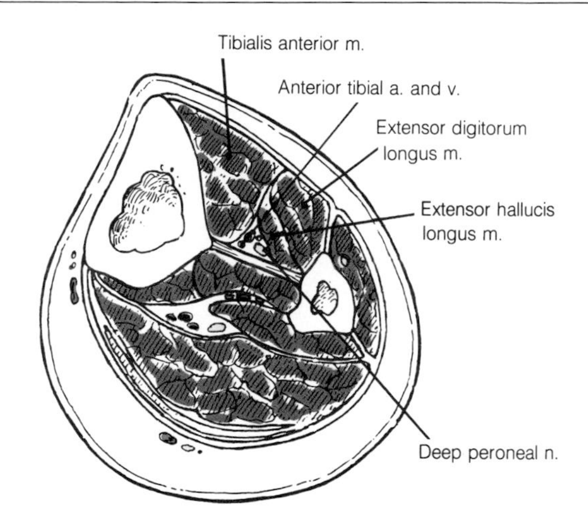

8-61 Cross-section through the middle of the leg showing the anterior compartment.

The Dorsum of the Foot

From the description of the structures lying in the anterior tibial compartment it should be clear that as these pass down on to the dorsum of the foot they will form important anterior relations of the ankle joint.

Note that in the foot the deep peroneal nerve supplies only one small muscle and a small area of skin between the great and second toe. Additional nerves contributing to the cutaneous innervation of this region are the **superficial peroneal nerve** and the **saphenous nerve** medially and the **sural nerve** laterally.

Superficial structures on the dorsum of the foot include the dorsal venous arch, the great and small saphenous veins, and the cutaneous nerves supplying the skin. The venous drainage has already been described. Details of the cutaneous innervation are given below.

The Cutaneous Innervation

The **superficial peroneal** nerve pierces the deep fascia in the lower part of the leg and, dividing into two branches, supplies the greater part of the dorsum of the foot. The medial branch provides a dorsal digital nerve to the medial side of the great toe and dorsal digital nerves to the second cleft. The lateral branch divides to supply the third and fourth clefts.

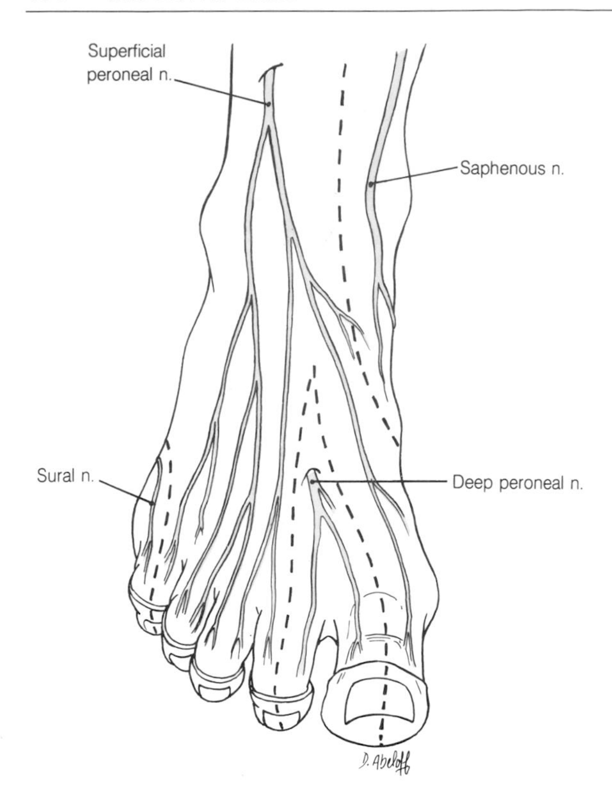

8-62 Showing the cutaneous innervation of the dorsum of the foot.

The **deep peroneal** nerve enters the foot between the tendons of extensor hallucis longus and extensor digitorum longus. It supplies the extensor digitorum brevis and then, becoming superficial, provides dorsal digital nerves for the first cleft. The oddness of this arrangement should make it memorable.

The **sural nerve**, appearing from behind the lateral malleous, supplies the lateral side of the foot and little toe.

The **saphenous nerve**, running in front of the medial malleolus, supplies the medial side of the foot up to the base of the great toe.

As in the hand, the dorsum of the terminal phalanges and the nail beds are supplied by ventral or in this case plantar digital nerves. This cutaneous innervation is summarized in Fig. 8-62.

The Muscles of the Dorsum of the Foot

The Extensor Retinacula

As the tendons of the muscles of the anterior compartment of the leg cross the ankle joint, they are prevented from "bowstringing" by two thickenings of the **anterior crural fascia** called the **superior and inferior extensor retinacula**. The superior is a simple band extending from the anterior border of the tibia to the anterior border of the fibula. The inferior is Y-shaped and lies immediately over the ankle joint. The stem of the "Y" is attached laterally to the **calcaneus**, while the two limbs are attached medially to the **medial malleolus** above and the **plantar fascia** below. Below the inferior retinaculum, the deep fascia of the foot is thin. A synovial sheath surrounds each tendon as it passes beneath the lower of these retinacula and that for tibialis anterior also extends beneath the superior retinaculum. As the tendons spread out over the dorsum, they cover a small muscle, the extensor digitorum brevis. These structures are illustrated in Fig. 8-63.

Extensor Digitorum Brevis

Extensor digitorum brevis arises from the upper surface of the calcaneus. It is directed medially across the dorsum of the foot, and gives off four

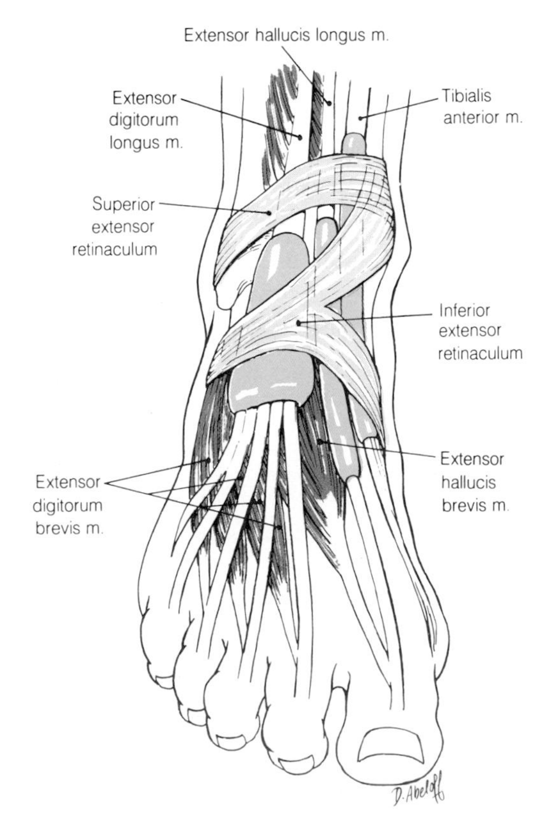

8-63 The anterior aspect of the ankle showing the extensor retinacula and extensor tendons and sheaths.

tendons for the medial four toes. The most medial part of the muscle may appear as a discrete belly and be known as **extensor hallucis brevis**. Its tendon is attached to the dorsal aspect of the proximal phalanx of the great toe. The remaining tendons join those of **extensor digitorum longus**. The muscle is supplied by the **deep peroneal nerve** (S1, 2).

The Tendons of the Long Extensor Muscles

The **tendon of tibialis anterior** passes medially to be attached to the medial aspect of the medial cuneiform bone and an adjacent region on the first metatarsal. The muscle is a dorsiflexor of the foot at the ankle joint and as will be seen later produces inversion of the foot.

The **tendon of extensor hallucis longus** passes toward the great toe to be inserted into the dorsal surface of the terminal phalanx. The tendon from extensor digitorum brevis is attached to the dorsal surface of the proximal phalanx. The action of extensor hallucis longus is to extend the great toe and dorsiflex the ankle joint.

The **extensor digitorum longus** breaks up into four tendons of which the three medial ones are joined by tendons of extensor digitorum brevis. Over the proximal phalanx an extensor expansion is formed that is similar to that on the dorsum of the fingers as is the attachment of the extensor tendon to the middle and distal phalanges. The muscle's action is to extend the toes and dorsiflex the foot.

The **tendon of peroneus tertius**, when present, runs to the base of the fifth metatarsal bone and is often found extending forward onto its shaft. The muscle is a dorsiflexor and evertor of the foot.

The Nerves and Vessels of the Dorsum of the Foot

The contributions made by the sural, superficial and deep peroneal, and saphenous nerves to the cutaneous innervation of this region have already been discussed. It remains to complete the distribution of the **deep peroneal nerve** and the **anterior tibial artery**. Both of these structures enter the dorsum of the foot beneath the inferior extensor retinaculum and at this point lie lateral to the tendons of extensor hallucis longus and tibialis anterior. The course and distribution of this nerve and artery are shown in Fig. 8-64. In this illustration the long extensor tendons and the extensor digitorium brevis have been removed.

The Deep Peroneal Nerve

This nerve passes deep to the extensor digitorum brevis muscle to which it gives off a branch. It terminates as a digital branch which through two dorsal digital nerves supplies the skin of the adjacent sides of the cleft between the first and second toes.

The Dorsalis Pedis Artery

As it enters the foot the anterior tibial artery becomes the dorsalis pedis artery. This vessel runs downward over the dorsum of the foot to the proximal end of the first intermetatarsal space. Here it passes through the space to anastomose with the **plantar arch** on the sole of the foot. The

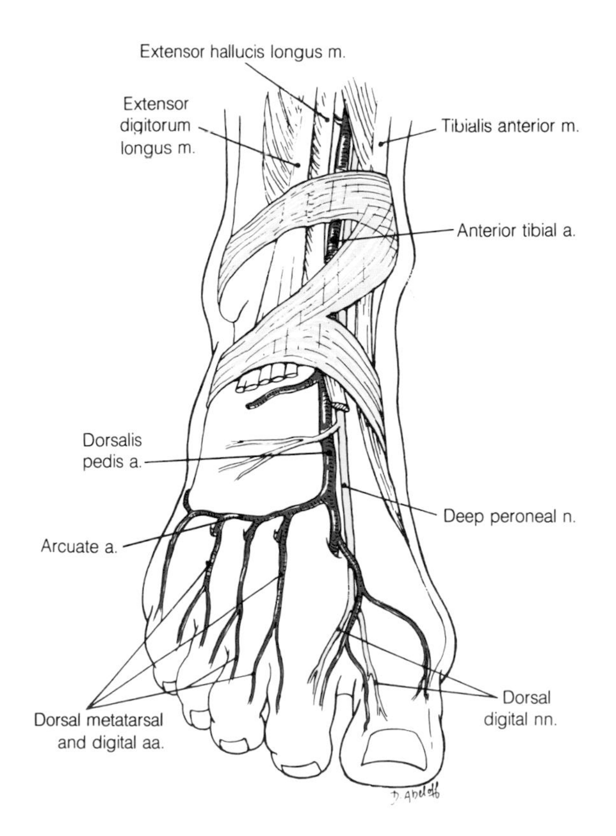

8-64 The anterior aspect of the ankle and dorsum of the foot showing the distribution of the dorsalis pedis artery and the deep peroneal nerve.

dorsalis pedis artery lies superficial to other structures on the dorsum of the foot and its pulsation can be felt and compared with that of the opposite side.

The artery gives off the **medial and lateral tarsal arteries**, a **first dorsal metatarsal artery**, and the **arcuate artery**. Branches of the latter two arteries provide dorsal digital arteries to the toes and their distribution can be seen in Fig. 8-64.

The Lateral Compartment of the Leg

The lateral compartment contains two muscles and the superficial peroneal nerve. Its blood supply is received from branches of the peroneal artery, but this vessel does not itself lie in the lateral compartment.

The Peroneal Muscles

Peroneus longus arises from the head and lateral aspect of the upper part of the shaft of the fibula. In addition, a much more extensive origin is obtained from the surrounding fascial walls of the lateral compartment. Its tendon passes in a groove behind the **lateral malleolus** where it is held, in company with that of peroneus brevis, by the **superior peroneal retinaculum**. The tendon continues forward and downward across the lateral aspect of the calcaneus to which it is held by the **inferior peroneal retinaculum**. It then curls beneath the cuboid where it lies in an osseofibrous tunnel. Continuing obliquely across the sole of the foot, it becomes attached to the lateral side of the base of the first metatarsal bone and to an adjacent area on the medial cuneiform bone. This attachment should be compared with that of tibialis anterior to similar regions on the medial sides of the same bones.

Peroneus brevis arises from the lower part of the lateral aspect of the fibula deep to peroneus longus. It also gains additional origin from surrounding fascia. Its tendon passes behind the lateral malleolus and over the calcaneus above that of peroneus longus to reach the tuberosity at the base of the fifth metatarsal bone. The peroneal tendons are surrounded by a common synovial sheath as they pass beneath the superior peroneal retinaculum. This sheath divides to surround the individual tendons as they pass beneath the inferior peroneal retinaculum and above and below the **peroneal tubercle** on the calcaneus. This can be seen in Fig. 8-65 in which peroneus longus and brevis are illustrated.

The tendons of both peroneal muscles pass on the lateral side of the axis about which inversion and eversion of the foot take place and are, therefore, evertors of the foot. Passing just behind the axis of dorsiflexion and plantarflexion at the ankle joint, they are probably also weak plantarflexors at that joint.

Both muscles are supplied by the **superficial peroneal nerve** (L4, 5; S1, 2).

The Superficial Peroneal Nerve

This nerve is formed in the substance of the peroneus longus by the division of the common peroneal nerve. Passing downward between the

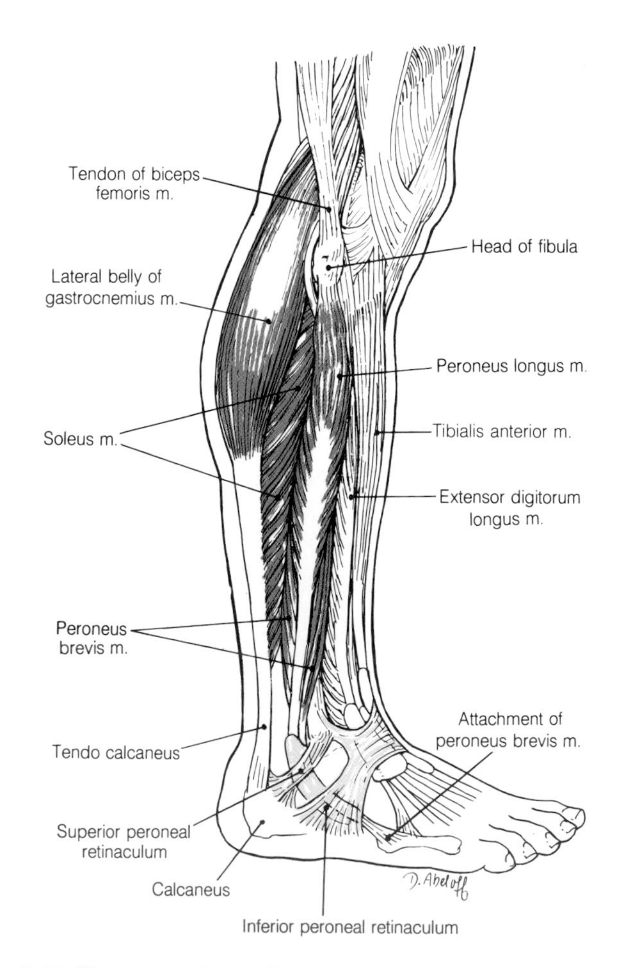

8-65 The peroneal muscles.

peroneal muscles and the anterior intermuscular septum, it supplies peroneus longus and brevis and pierces the crural fascia in the lower leg to become cutaneous. Here it divides into two branches that descend on the front of the leg and cross anterior to the ankle to reach the dorsum of the foot.

The Posterior Compartment of the Leg

By far the largest of the three compartments of the leg, this contains a superficial group of powerful muscles which plantarflex the foot, and a deeper group which comprises the long flexors of the digits and the tibialis posterior muscle. The tibial nerve and the posterior tibial artery run through the whole extent of this compartment.

The Superficial Muscles of the Posterior Compartment of the Leg

This group of muscles comprises the gastrocnemius, soleus, and plantaris muscles. In Fig. 8-65 they are viewed from the lateral side and in Fig. 8-66 from behind. The sites of their bony origins are shown in Fig. 8-67.

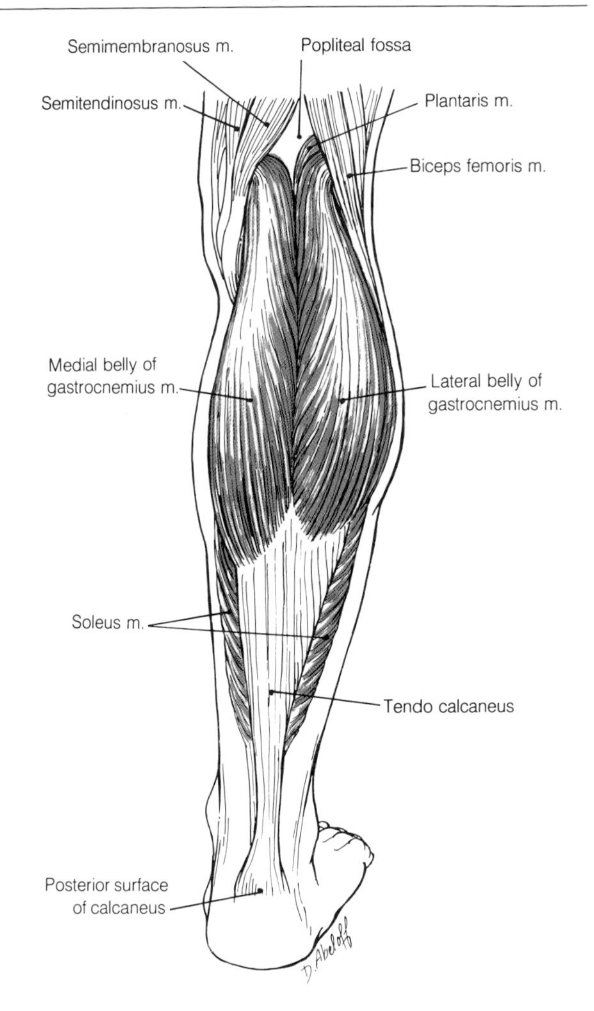

8-66 The posterior aspect of the superficial group of the plantar flexor muscles.

Gastrocnemius

Gastrocnemius* is the most superficial of the calf muscles and gives it its shape. It has two bellies which arise from the posterior surface of the femur just above the femoral condyles and from the adjacent capsule of the knee joint. The bellies unite at their attachment to the upper expanded end of the Achilles tendon or **tendo calcaneus**.

Soleus

Soleus lies deep to the gastrocnemius. It is a large flat muscle arising from the posterior aspect of the head and upper shaft of the fibula, the soleal line of the tibia, the middle third of the medial border of the tibia and from a fibrous band bridging its fibular and tibial attachments. The main nerve and vessels of this compartment pass deep to this bridge. From their extensive origin the fibers converge on a short tendon which joins the deep surface of the tendo calcaneus.

Plantaris

Plantaris is a vestigial muscle with a small short belly and a long thin tendon (compare with palmaris longus in the forearm). It arises close to the lateral head of gastrocnemius, descends obliquely between gastrocnemius and the underlying soleus and appears at the medial border of the tendo calcaneus with which it has a common insertion. The only value of this muscle may lie in its use for tendon grafting.

* Gaster, *Gk.* = stomach; Kneme, *Gk.* = leg, hence the "belly" of the calf

Both the gastrocnemius and soleus muscles act as powerful plantarflexors of the foot. However, it is probable that the gastrocnemius is more concerned with phasic activities such as running, jumping, and walking, while the soleus more with the tonic activity required for the maintenance of posture, particularly standing erect. By virtue of its attachment to the femur above the knee joint, gastrocnemius is also a flexor of the knee.

Each of the superficial muscles is supplied by branches of the **tibial nerve** (S1, 2) in or just below the popliteal fossa.

The Tendo Calcaneus

Flat and wide at its junction with the two bellies of gastrocnemius, the tendo calcaneus narrows and thickens as it approaches its attachment at the heel. Here it is inserted into the middle of the posterior aspect of the **calcaneus**. A smooth area of the bone above this is separated from the tendon by a **bursa** and some fatty tissue. This bursa may be subject to chronic inflammation.

Despite the fact that the tendo calcaneus is the strongest tendon in the body, it may rupture suddenly during active but often unaccustomed exercise. It is worth looking carefully at the arrangement of the fibers of the tendon. These can be seen to spiral downward as the more medial fibers become posterior and the posterior more lateral. This anatomical fact determines the manner in which incisions are made to lengthen a pathologically shortened tendon.

The superficial muscles described above are separated from the deep group by the **deep transverse fascia of the leg** which was illustrated in Fig. 8-58.

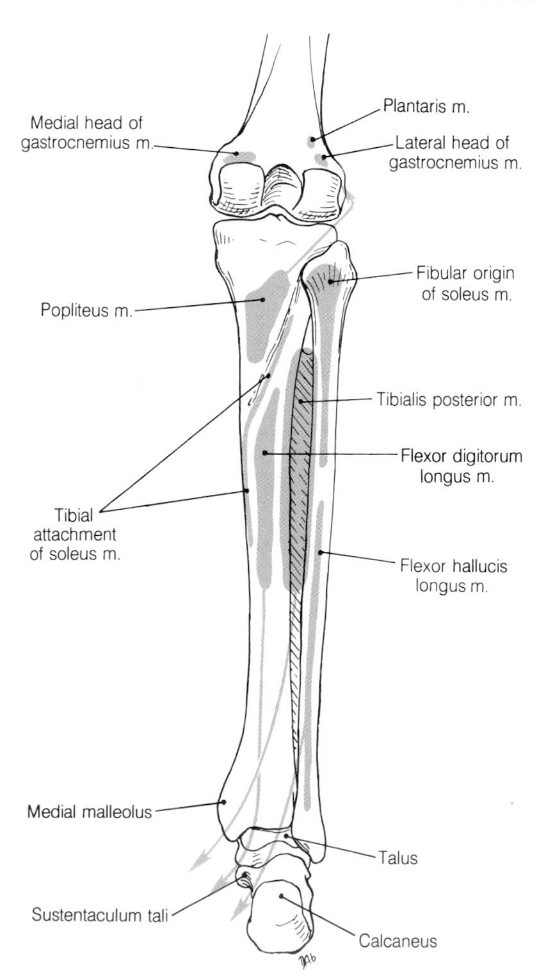

8-67 The posterior aspect of the femur, tibia, fibula, and tarsus showing the origins of the muscles of the posterior compartment of the leg.

The Deep Muscles of the Posterior Compartment

The deep muscles of the calf undergo a substantial rearrangement of their bellies and tendons between their origins from the tibia and fibula and their insertions in the foot. For this reason the schematic illustration in Fig. 8-67 should be followed as each muscle is described.

Flexor Hallucis Longus

Flexor hallucis longus is a large and powerful muscle which arises from the lower two-thirds of the posterior aspect of the fibula and the adjacent interosseous membrane and posterior intermuscular septum. It passes obliquely downward and medially. Its tendon lies in successive grooves on the lower end of the **tibia, talus, and sustentaculum tali of the calcaneus**. Passing through the sole of the foot, it is attached to the distal phalanx of the great toe.

Flexor Digitorum Longus

Flexor digitorum longus arises from the shaft of the tibia below the soleal line. It is rather smaller than the flexor hallucis longus. It descends the leg almost vertically and its tendon passes behind the **medial malleolus** in company with, but lateral to, that of tibialis posterior. In the sole, the tendon breaks up into four slips for the lateral four toes. These slips correspond to the flexor profundus tendons in the hand and, like them, perforate the superficial or, in this case, the flexor brevis tendons before inserting into the distal phalanx.

Both of these muscles are clearly flexors of the toes and perforce plantarflexors of the ankle joint. This action becomes of importance when combined with plantarflexion in the "take-off" movement of walking. Note that the muscle flexing the most medial toe arises laterally from the fibula and that for the lateral four toes arises medially from the tibia. It is evident that their tendons must cross. This crossing occurs in the sole of the foot and it will be described with that region.

Tibialis Posterior

Tibialis posterior is the most deeply placed muscle in the calf. It arises from both tibia and fibula and the intervening interosseous membrane. Its tendon appears at the medial border of flexor digitorum longus and passes immediately behind the medial malleolus with the tendon of flexor digitorum longus lying posterior to it. Its main attachment is to the tubercle of the navicular bone. Numerous additional slips extend to other bones of the tarsus and metatarsus. The muscle's action is to plantarflex and invert the foot.

Each of the deep flexor muscles described above is supplied by the **tibial nerve** (L4, 5 – tibialis posterior) (S2, 3 – flexor hallucis and digitorum longus).

Popliteus

The popliteus muscle does not quite fit in with the other muscles of the calf. It has been seen in the floor of the popliteal fossa arising from the posterior aspect of the tibia above the soleal line. Triangular in shape, it converges on a strong tendon which passes within the capsule of the knee joint to become attached to the lateral condyle of the femur. Additional attachment is made to the **lateral meniscus**. The muscle is able to laterally rotate the femur on the tibia when the latter is fixed and also withdraw the meniscus from the path of the backwardly moving lateral femoral condyle. It is supplied by the **tibial nerve** (L4, 5; S1).

The Flexor Retinaculum

This retinaculum extends from the medial malleolus to the calcaneus and plantar fascia. Its up-

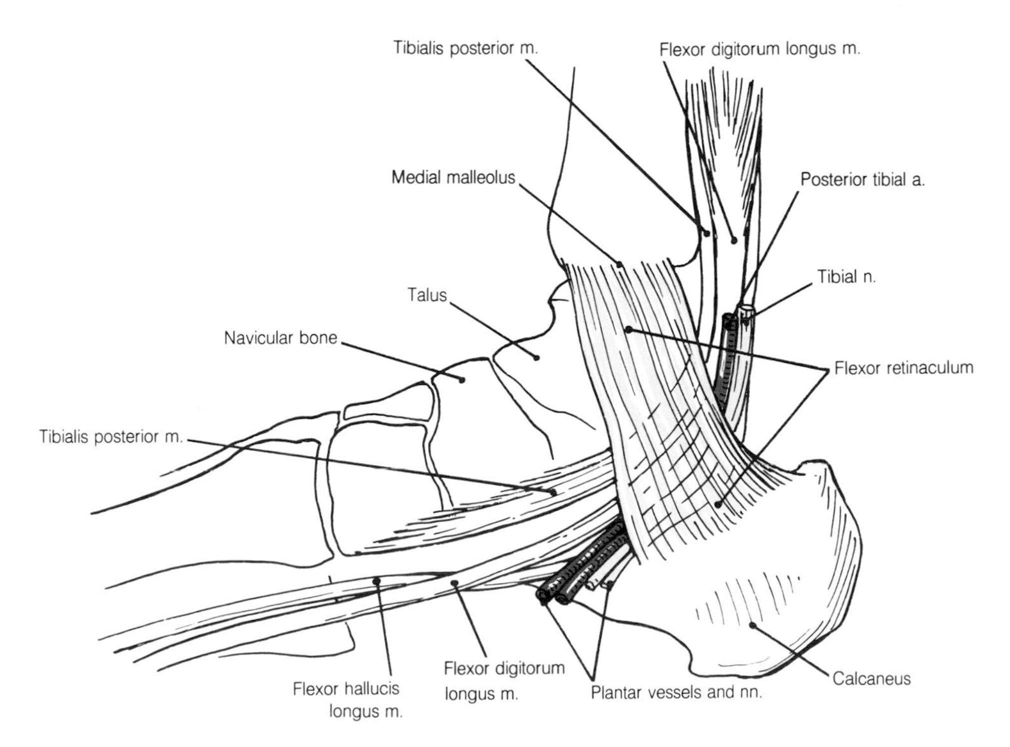

8-68 The flexor retinaculum at the ankle and subjacent structures.

per and lower margins are not clear cut as these blend with the deep transverse fascia of the leg above and with the plantar fascia below. Beneath the retinaculum pass the deep flexor muscles surrounded by synovial sheaths, the tibial nerve, and the posterior tibial artery. At a point just above the retinaculum the pulsation of the artery can be felt. The retinaculum and the structures passing beneath it can be seen in Fig. 8-68. Note that both the nerve and artery divide into medial and lateral plantar branches beneath the retinaculum. The illustration also shows the tendon of flexor hallucis longus as it crosses that of flexor digitorum longus in the sole of the foot.

The Nerves and Vessels of the Posterior Compartment

The Tibial Nerve

The tibial nerve (L4, 5; S1, 2, 3) is the larger of the two terminal branches of the **sciatic nerve**. It leaves the popliteal fossa to enter the posterior compartment of the leg between the two heads of gastrocnemius. In company with the **posterior tibial artery**, it passes deep to the soleus muscle and descends the leg between flexor digitorum longus and flexor hallucis longus. At the ankle it passes deep to the flexor retinaculum and divides into the **medial** and **lateral plantar nerves**. Articular branches supply the knee and ankle joints and muscular branches supply all the muscles of the posterior compartment.

The Posterior Tibial Artery

This artery is formed with the **anterior tibial artery** by the bifurcation of the **popliteal artery** at the lower border of popliteus. Here it gives off the **circumflex fibular artery**. It then passes beneath the fibrous arch between the tibia and fibula from which soleus arises and, on the medial side of the tibial nerve, runs downward and medially toward the medial malleolus. While deep to the flexor retinaculum, it divides into the **medial and lateral plantar arteries**. It gives off muscular branches and, at the ankle, malleolar and calcaneal branches. Its largest branch is the **peroneal artery**.

The Peroneal Artery

The peroneal artery arises from the posterior tibial artery shortly after the latter is formed. It descends through the posterior compartment close to the fibula between tibialis posterior and flexor hallucis longus. Its branches are muscular and these include branches which pass through the posterior intermuscular septum to the peroneal muscles in the lateral compartment. It terminates as calcaneal and lateral malleolar branches and a perforating branch which passes through the interosseous membrane to support the anterior tibial artery.

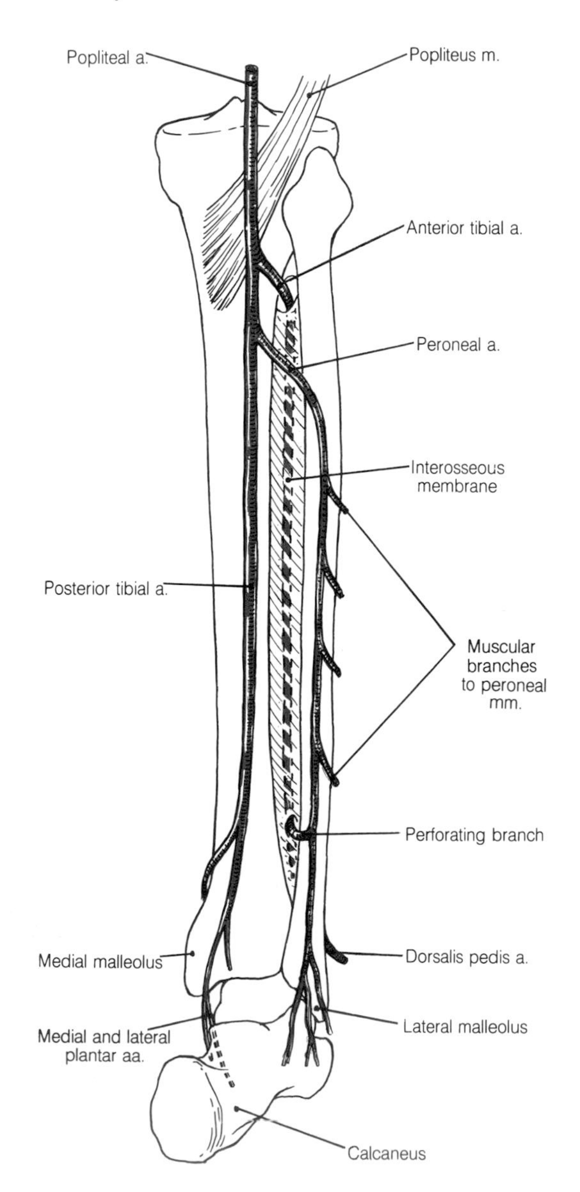

8-69 Semidiagrammatic illustration of the distribution of the anterior and posterior tibial arteries.

The distribution of the popliteal artery and its branches in the leg are summarized in Fig. 8-69. Note the division of the artery at the lower border of popliteus, the opening in the interosseous membrane through which the anterior tibial artery passes, a similar opening at the lower end of the membrane through which a perforating branch of the peroneal artery anastomoses with the anterior tibial artery or one of its branches, and the terminal malleolar and calcaneal branches of the posterior tibial and peroneal arteries.

The Deep Veins

The posterior tibial and peroneal arteries are accompanied by two or more venae comitantes. It is with these and other deep veins that the superficial veins communicate through perforating veins and it is the valves that they contain that allow them to function as a muscle pump.

Having described the main arterial trunks of the thigh and leg, the branches of each that contribute to an anastomosis around the knee joint can be summarized:

Femoral artery
  Descending genicular artery
Profunda femoris artery
  Descending branch of lateral circumflex artery
Popliteal artery
  Medial and lateral superior genicular arteries
  Medial and lateral inferior genicular arteries
Anterior tibial artery
  Anterior tibial recurrent artery
Posterior tibial artery
  Circumflex fibular artery

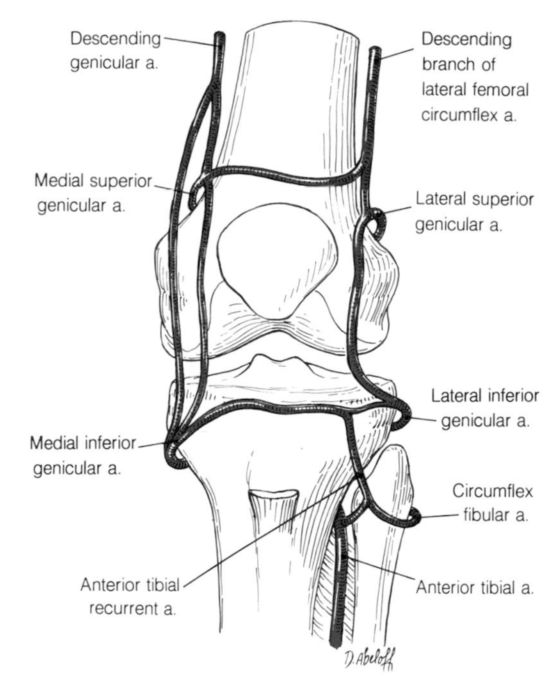

8-70 The arterial anastomosis around the knee joint.

Each of these vessels can be identified in Fig. 8-70.

As a final review of the leg examine again the cross-section made through the middle of the leg which is shown in Fig. 8-71. Identifiy each com-

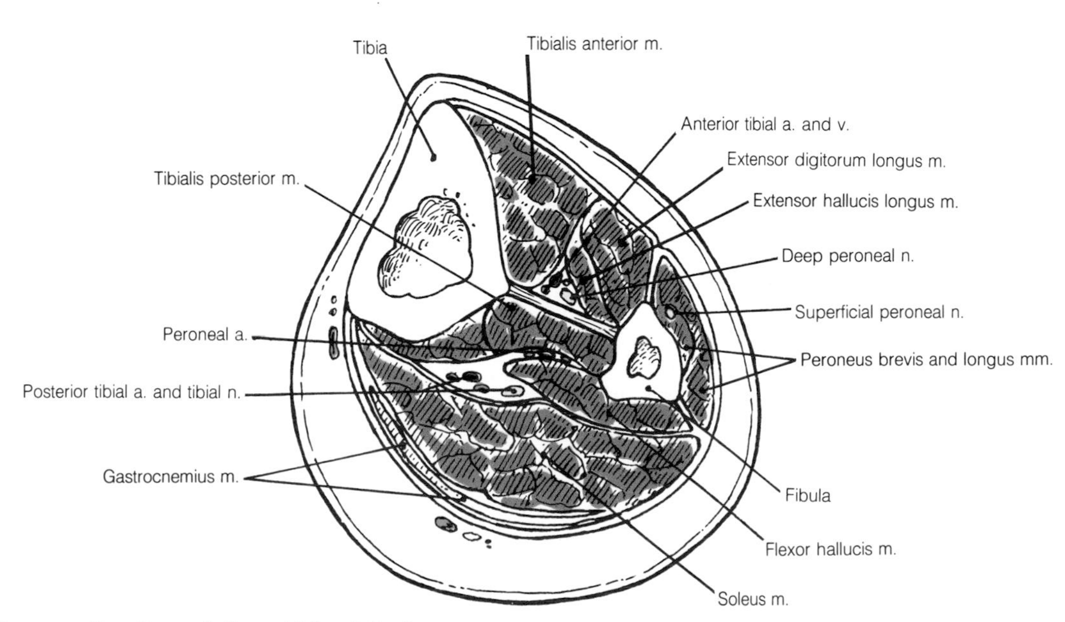

8-71 Cross-section through the middle of the leg.

partment, the muscle groups that it contains, and the nerve and artery that supplies each muscle group. Note that the section passes through the tendinous part of gastrocnemius and as a result the soleus muscle forms the main bulk of the calf at this level.

The Ankle and Sole of the Foot

Before describing the ankle joint, the soft tissues of the sole, and the functions of the foot, the bones involved must be described in some detail.

The Skeleton of the Ankle and Foot

The adaptation of the skeleton of the foot to the erect position and to bipedal plantigrade locomotion was mentioned earlier. It is now time to look at this skeleton in greater detail. The general arrangement of the bones, of the foot can be seen in Fig. 8-72. Note that the individual bones form two series which radiate from the calcaneus. A medial series consists of the calcaneus, talus, navicular and three cuneiforms, three metatarsals, and three sets of phalanges. The lateral series consists of the calcaneus, the cuboid, two metatarsals, and two sets of phalanges. These series correspond to the medial and lateral plantar arches but with the addition of the skeletons of the toes. While this may be a helpful concept when remembering the arrangement of the bones, the foot in life functions as a single pliable unit. The body weight is transmitted to the talus and calcaneus and, thence, across the remaining tarsal and metatarsal bones. It is applied to the ground at the tuber calcanei and the heads of the metatarsals. The irregularity in shape and arrangement of the tarsal bones tends to deter further examination, but the importance of their anatomy to pain-free feet should make this worthwhile. As before, this examination can be greatly simplified if the bone in question is examined while its description is read.

The Lower Ends of the Tibia and Fibula

The lower ends of the tibia and fibula both contribute to the ankle joint and the mortise-like configuration given to the crural articular surface by the

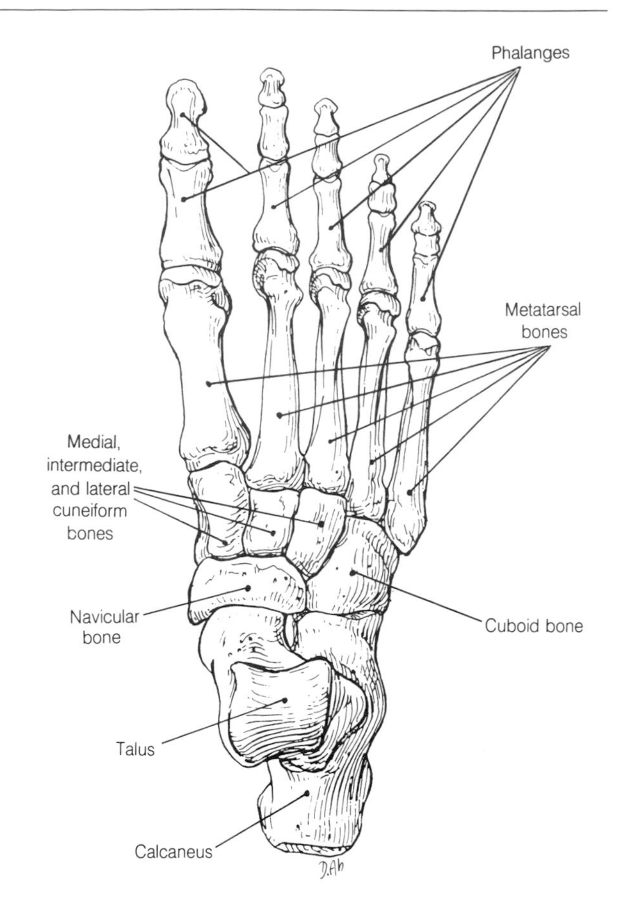

8-72 The bones of the foot.

two malleoli adds substantially to the stability of the joint. In fact even partial dislocations of the joint are nearly always associated with fractures of the medial malleolus or the lower third of the fibula. The tibia and fibula are firmly united above the ankle at the inferior tibiofibular joint and the strong interosseous and anterior and posterior tibiofibular ligaments. Confirm these features in Fig. 8-73 and note also the groove on the back of the lateral malleolus for the peroneal tendons and on the back of the medial malleolus for the tendon of tibialis posterior.

The Talus

It is by means of the talus that the foot articulates with the leg at the ankle joint. The **body of the talus** is roughly cuboidal. Using the bone and the illustration in Fig. 8-74 note the following features:

1. The superior aspect of the body is covered by an **articular surface** for the lower end of the **tibia**.

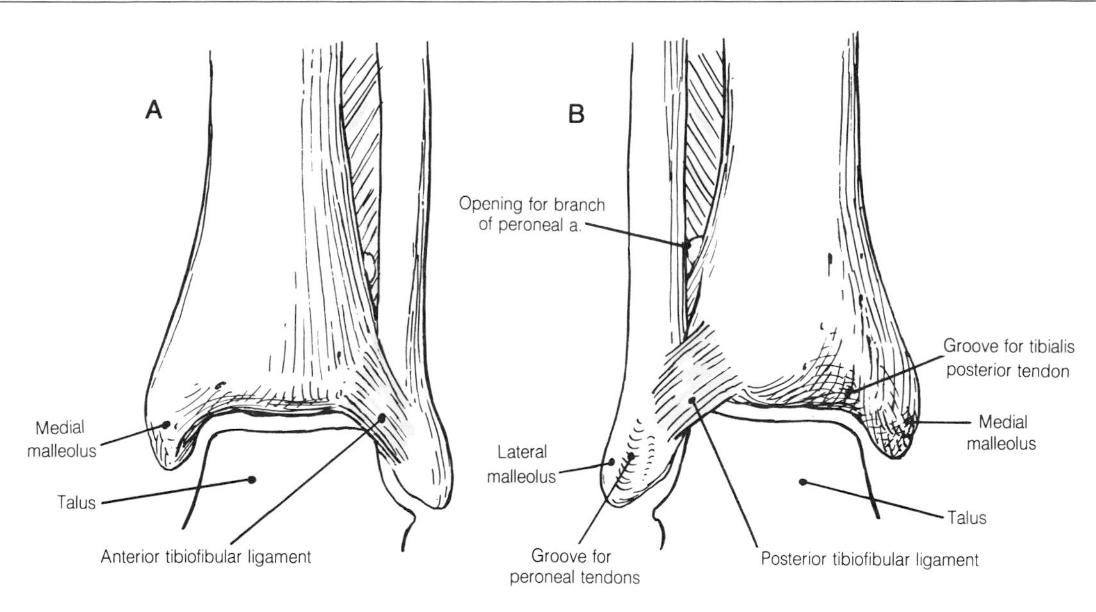

8-73 The anterior (A) and posterior (B) aspects of the lower end of the left tibia and fibula. The anterior and posterior tibiofibular ligaments are also shown.

2. This surface is extended on the medial and to a greater extent on the lateral surface for articulation with the **medial and lateral malleoli**.
3. The superior articular surface is convex from front to back and slightly concave from side to side. It is also slightly wider anteriorly.
4. The posterior surface shows a larger **lateral** and smaller **medial tubercle** between which is a groove for the tendon of **flexor hallucis longus**.
5. From the anterior surface projects a stout **neck** upon which lies the **head**. This is covered by an articular surface for the **navicular bone**, the next bone in the medial arch series.
6. This anterior articular surface is continued onto the inferior aspect of the neck as two further articular facets which form an **anterior articulation** with the **calcaneus**.
7. The inferior aspect of the body has an extensive

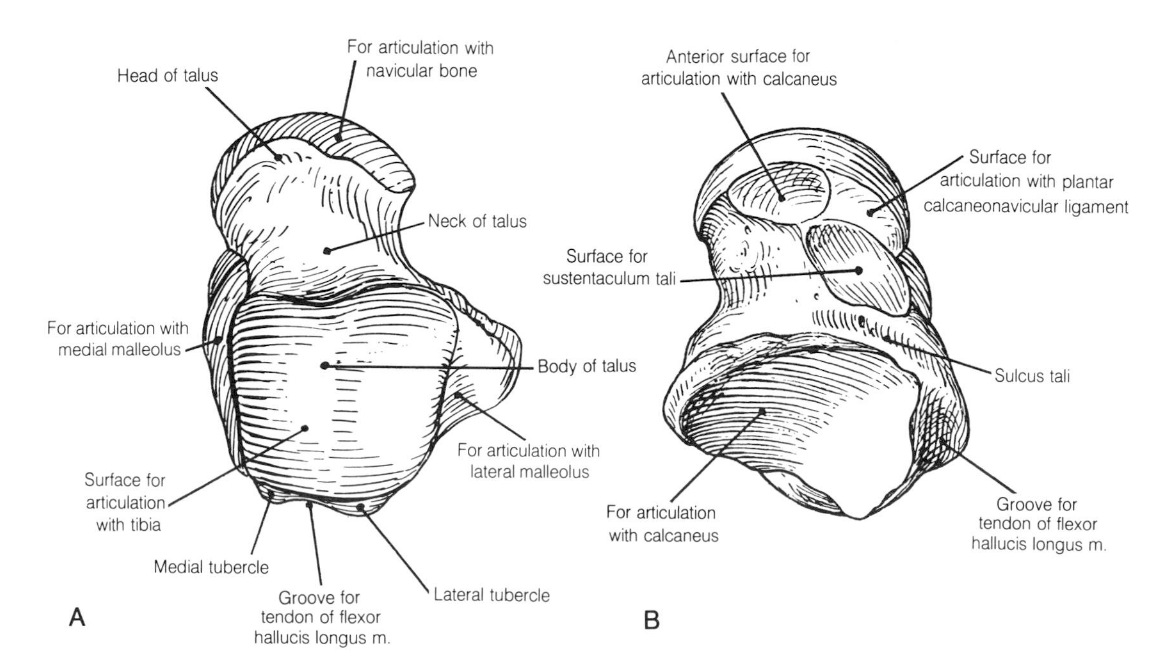

8-74 The right talus viewed from above (A) and below (B).

concave articular surface which forms a **posterior articulation with the calcaneus**. The depression between the anterior and posterior calcaneal articulations is the **sulcus tali**.

The Calcaneus

The calcaneus, the largest of the tarsal bones, is also roughly cuboidal. It lies beneath the talus and projects posteriorly in the foot forming the heel. Using the illustration in Fig. 8-75, note the following features:

1. The superior surface of the **body** bears a large central convex articular facet for the posterior articulation with the talus.
2. Medially there is a shelf-like projection of bone called the **sustentaculum tali**.
3. The upper surface of the sustentaculum tali and the adjacent part of the body bear two further facets for an anterior articulation with the talus. Note that these anterior facets form a slightly concave surface.
4. Between the anterior and posterior facets for articulation with the talus is the **sulcus calcanei**. When the sulcus tali and the sulcus calcanei are approximated in life a space called the **sinus tarsi** is formed. This space houses the strong **interosseous talocalcaneal ligament**.
5. The anterior surface of the bone is almost completely covered by an articular surface for the **cuboid bone** – the second bone in the lateral arch.
6. The posterior surface is smooth above where it is separated from the **tendo calcaneus** by fat and a bursa. An intermediate rough area gives attachment to the tendon. Below this the posterior surface curves downward and forward as the **tuber calcanei**.
7. The inferior surface bears the **tuber calcanei** posteriorly and this can in turn be divided into a **lateral and medial process**.
8. The inferior surface of the sustentaculum tali shows a groove for the tendon of flexor hallucis longus.

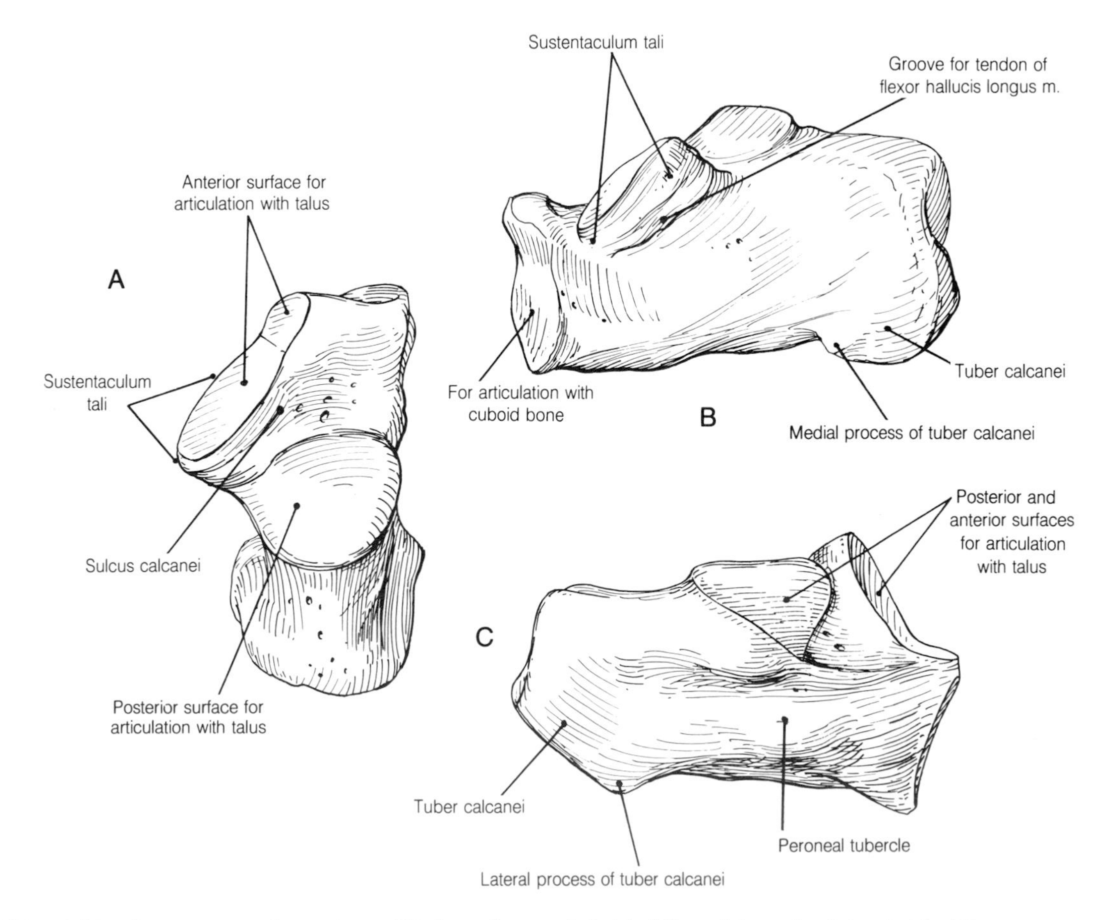

8-75 The right calcaneus seen from above (A), from the medial side (B), and from the lateral side (C).

9. The main feature of the lateral surface is the **peroneal tubercle**. It is at this point that the tendons of peroneus longus and brevis diverge.

The Navicular Bone

The navicular bone (Fig. 8-76) lies between the talus and the cuneiform bones. The proximal surface is oval and concave for articulation with the talus. The distal surface shows three facets for articulation with the three cuneiform bones. A prominent feature of the medial surface is the **tuberosity** to which the main part of the tibialis posterior tendon is attached.

The Cuneiform Bones

The three wedge-shaped cuneiform bones (Fig. 8-76) articulate between the navicular bone proximally and the three medial metatarsal bones distally. The lateral and intermediate bones have the broad surface of the wedge superiorly and the medial and lateral bones project distally on either side of the intermediate and thus form a socket for the base of the second metatarsal bone.

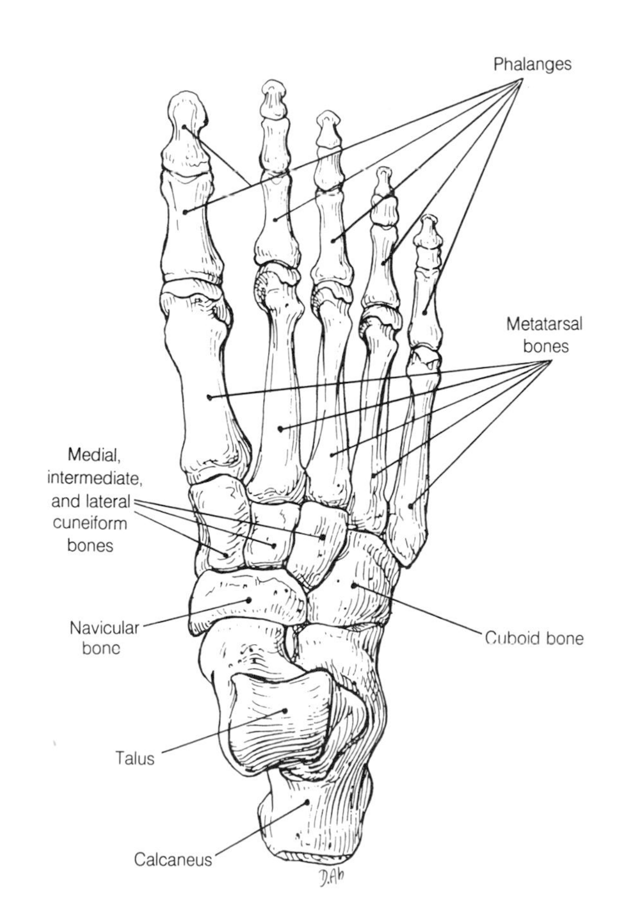

8-76 The bones of the foot.

The Cuboid Bone

The cuboid bone (Fig. 8-76) articulates with the calcaneus proximally and the lateral two metatarsal bones distally. Its plantar surface shows a groove for the **tendon of peroneus longus**.

The Metatarsal Bones

The five metatarsal bones (Fig. 8-76) each have a body, a proximal base, and a distal head. The first metatarsal is shorter and stouter than the others. The base of the fifth has a laterally projecting **tuberosity** to which the tendon of **peroneus brevis** is attached.

The Phalanges

The phalanges of the foot (Fig. 8-76) are similar in shape and arrangement to those of the hand, although they are smaller in size. The first digit or hallux has a proximal and distal phalanx while the remaining digits possess proximal, middle, and distal phalanges.

The three large tarsal bones have primary centers of ossification before birth. That for the calcaneus appears at the fifth month, the talus at the sixth month, and the cuboid at the ninth month. The primary centers for the four smaller tarsal bones appear within the first three years of infancy. Primary centers for the metatarsal and phalangeal shafts appear before birth and their secondary centers between the third and fifth years. Note that the secondary centers appear at the heads of the lateral four metatarsals but that for the hallux appears at the base as do those for all the phalanges.

The Ankle or Talocrural Joint

The ankle joint or talocrural joint is a synovial joint of the hinge variety. This means that it has a single axis and a single plane of movement. The axis passes transversely between the malleoli and through the body of the talus. The plane of movement allows flexion and extension or, as they are more commonly called, **plantarflexion** and **dorsiflexion**. The movements of inversion and eversion in which the sole of the foot is turned to face medially and laterally do not and cannot occur at this joint because the body of the talus is held firmly between the closely fitting malleoli by medial and lateral ligaments. The crural articular surface involves the **lower end of the tibia**, its **medial malleolus**, and the **lateral malleolus of the fibula**. These surfaces articulate with the upper surface of the talus which is curved convexly from before backward and is slightly concave from side to side. It is also slightly wider anteriorly, thus giving greater stability in dorsiflexion, the push-off position for most active movements. As the foot is plantarflexed the narrower portion of the talar articular surface becomes engaged between the malleoli and a small amount of abduction, adduction, and rotation are possible. Additional articular surfaces on its medial and lateral sides are for the malleoli. The joint has a fibrous capsule which is thin and lax anteriorly and posteriorly allowing for movement. The **capsule** is lined by synovial membrane and reinforced by strong **medial and lateral ligaments**. These ligaments are illustrated in Figs. 8-77, 8-78, and 8-79, and described below.

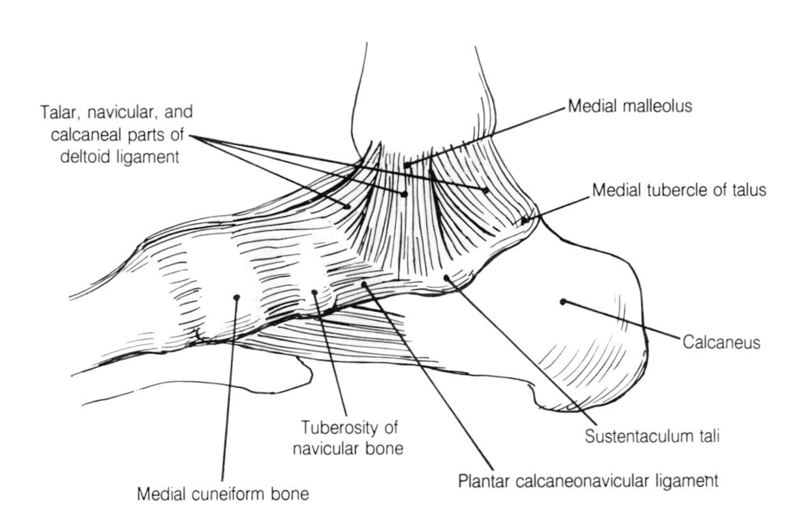

8-77 The medial ligament of the ankle joint.

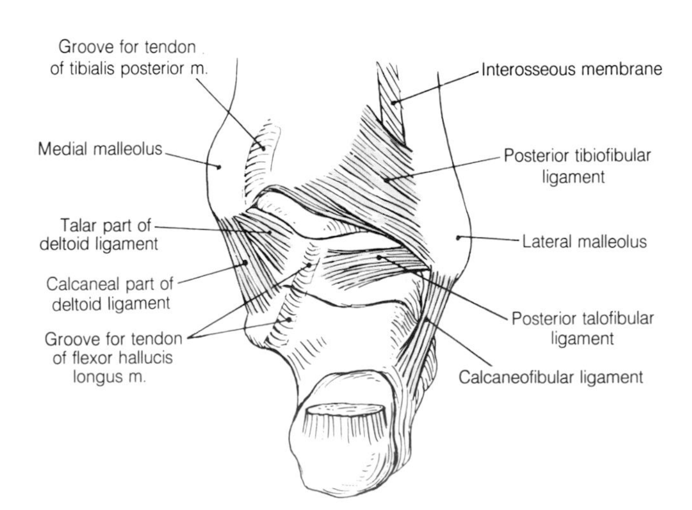

8-78 The ligaments of the ankle joint seen from behind.

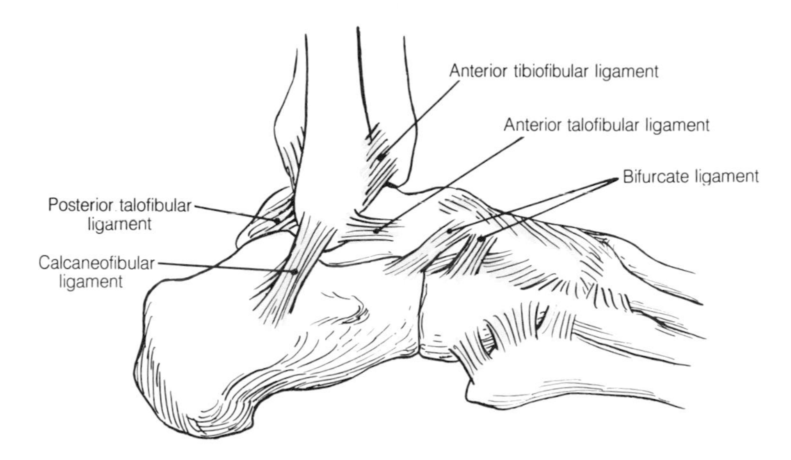

8-79 The lateral ligament of the ankle joint.

The Medial Ligament

The medial or deltoid ligament (Figs. 8-77 and 8-78) is a strong triangular ligament which arises from the **medial malleolus** and fans out to the **tuberosity of the navicular bone**, the **plantar calcaneonavicular ligament**, the **sustentaculum tali**, and the **medial tubercle of the talus**. If on the articulated skeleton the plantar calcaneonavicular ligament is, in imagination, placed in the gap between the tuberosity of the navicular and the sustentaculum, this will be seen to be a continuous attachment. A strong deep slip also extends from the malleolus to the medial surface of the talus.

The Lateral Ligament

The lateral ligament (Figs. 8-78 und 8-79) has three distinct bands arising from the **lateral malleolus**. An **anterior talofibular ligament** extends horizontally forward to the neck of the talus. A **calcaneofibular ligament** extends downward and backward to the calcaneus and a **posterior talofibular ligament** extends horizontally and posteriorly to the **lateral tubercle of the talus**.

The movements of the joint are **dorsiflexion** and **plantarflexion**. Dorsiflexion is produced primarily by tibialis anterior but any muscle whose tendon passes in front of the axis of the joint, for example, extensor digitorum longus or extensor hallucis longus, will assist in this movement. Similarly, any muscle whose tendon passes behind the axis will be a plantarflexor. Gastrocnemius and soleus are the most powerful of these, but are aided by tibialis posterior, flexor digitorum longus, and flexor hallucis longus. The tendons of the peroneal muscles probably pass too close to the axis to have much effect as plantarflexors.

The joint is supplied by branches of the **tibial** and **deep peroneal nerves**.

The Relations and Surface Anatomy of the Ankle Joint

A study of the relations and surface anatomy of the ankle joint provides a good opportunity to review the structures passing from the leg to the foot. This can be most usefully done by examining the cross-section that passes through the joint seen in Fig. 8-80. In this note the following features:

1. Anterior to the ankle joint and from its medial to lateral side are the tendons of tibialis anterior, extensor hallucis longus, and extensor digitorum longus. Each of these tendons is palpable, especially when the appropriate movement is made to tense them.
2. Between the tendons of extensor hallucis longus and extensor digitorum longus lie the anterior tibial artery and deep peroneal nerve. These were crossed by the extensor hallucis longus

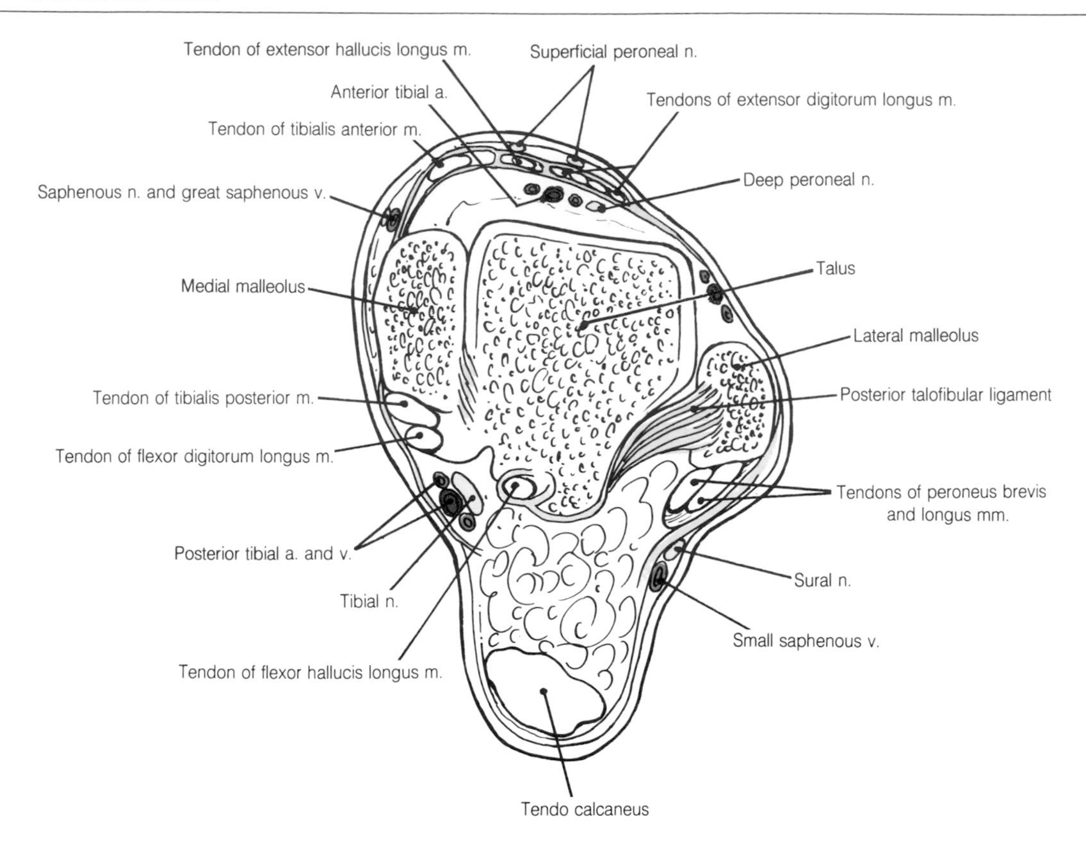

8-80 Cross-section through the ankle joint.

tendon just above this level as it moved from the fibular to the tibial side of the leg. Below the level of the section, the anterior tibial artery becomes superficial as the dorsalis pedis artery and the pulsation of this superficial vessel can be felt.

3. On either side of the talus lie the medial (tibial) malleolus and the lateral (fibular) malleolus. Both of the malleoli are easily palpable and the lateral malleolus can be felt to extend to a slightly lower level than the medial. Portions of the anterior talofibular and posterior talofibular ligaments can be seen attached to the lateral malleolus and, to the medial malleolus, the posterior tibiotalar part of the deltoid ligament.
4. Immediately behind the lateral malleolus are the two peroneal tendons and behind the medial malleolus the tendons of tibialis posterior and flexor digitorum longus.
5. The tendon of flexor halucis longus can be seen lying in the groove between the medial and lateral tubercles of the talus.
6. In a gutter between the talus and calcaneus lie the posterior tibial vessels and the tibial nerve. Here again the pulse may be felt over the posterior tibial artery. The vessels and nerve will shortly divide into medial and lateral plantar structures.
7. Posterior to the calcaneus the attachment of the tendo calcaneus is just included in the section.
8. Superficially the sural nerve and small saphenous vein can be seen **posterior to the lateral malleolus**, the terminal divisions of the superficial peroneal nerve anteriorly, and the saphenous nerve and great saphenous vein **in front of the medial malleolus.**

The Radiological Anatomy of the Ankle Joint

In Figs. 8-81 and 8-82 anteroposterior and lateral projections of the ankle are shown. Identify the two malleoli and the difference between the levels

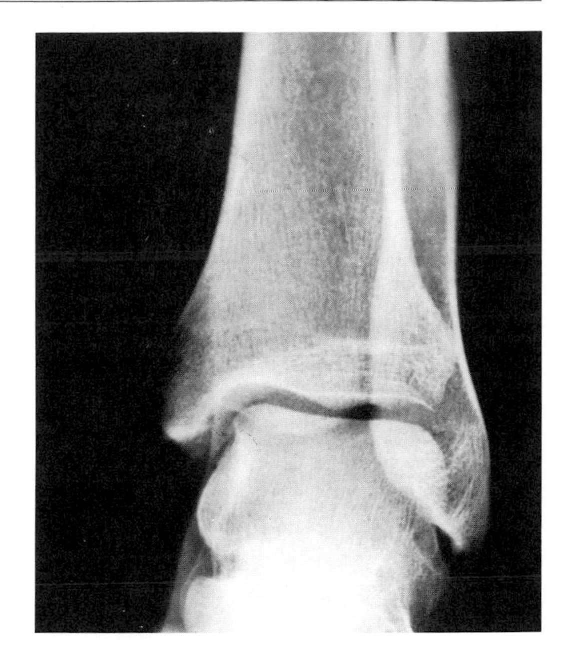

8-81 Radiograph of the ankle joint (a.p. projection). (Reproduced by permission, from Wicke: *Atlas of Radiologic Anatomy,* 4th Ed, Urban & Schwarzenberg, Baltimore-Munich, 1987.)

of their tips. Between them is grasped the body of the talus. Note also the "joint space." In the lateral projection the body, neck, and head of the talus can be identified as can the articulation of the head with the navicular bone. Below this lies the calcaneocuboid joint. The joints between the talus and calcaneus cannot be clearly seen in this view but the **sinus tarsi** that separates them can. Fractures around the ankle joint are extremly common and the picture of the normal joint should be well known.

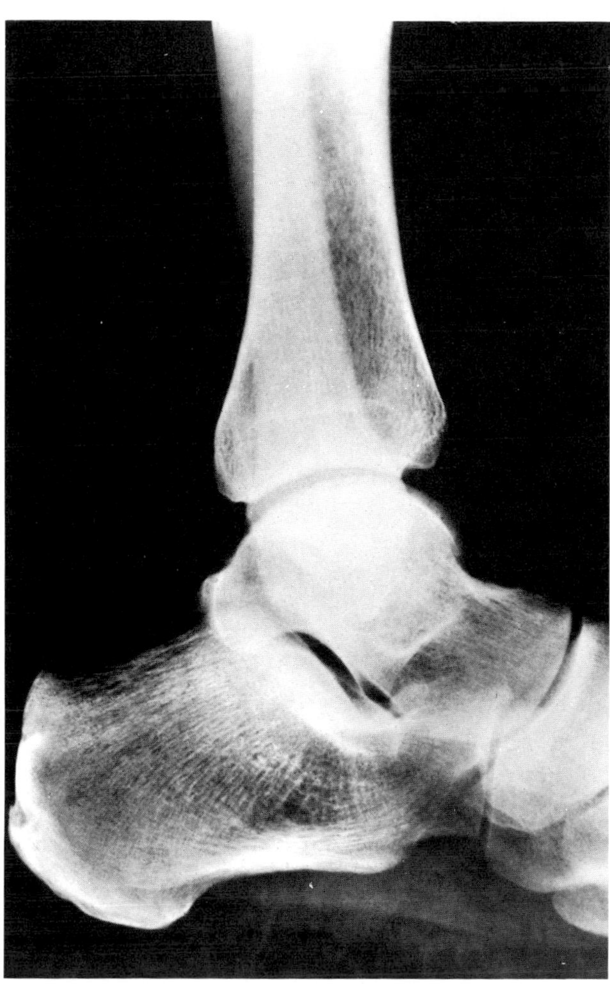

8-82 Radiograph of the ankle joint (lateral projection), (Reproduced by permission, from Wicke: *Atlas of Radiologic Anatomy,* 4th Ed, Urban & Schwarzenberg, Baltimore-Munich, 1987.)

The Sole of the Foot

The sole of the foot bears the weight of the body and in the natural state must withstand the wear and tear of the surface of the ground. It is covered by a thick layer of skin whose thickness is due to the depth of the superficial keratinized layer of the epidermis. Beneath the skin the fat of the subcutaneous tissue is divided up into small loculi by fibrous septa. These septa also serve to anchor the skin to the underlying deep fascia here called the **plantar aponeurosis** which is illustrated in Fig. 8-83. This is a strong layer of fibrous tissue extending from the calcaneal tuberosities to the bases of the toes. Its central portion is extremely thick and tough. It is perforated by cutaneous nerves and vessels, but the musculature and all-important nerves and vessels lie deep to this protective sheet. The cutaneous innervation of the sole of the foot is also shown in Fig. 8-83. The region of the heel is supplied by calcaneal branches of the tibial and sural nerves while the remainder, including the plantar surfaces of the digits, is supplied by cutaneous branches of the medial and lateral plantar nerves. The medial three-and-a-half digits receive plantar digital nerves from the medial plantar nerves and the lateral one-and-a-half receive these from the lateral plantar nerve.

The remaining structures of the sole of the foot are classically described in relation to four layers of muscles. As long as too much attention is not paid to the details of the attachments of the small muscles, this descriptive technique still remains the best way to organize the region.

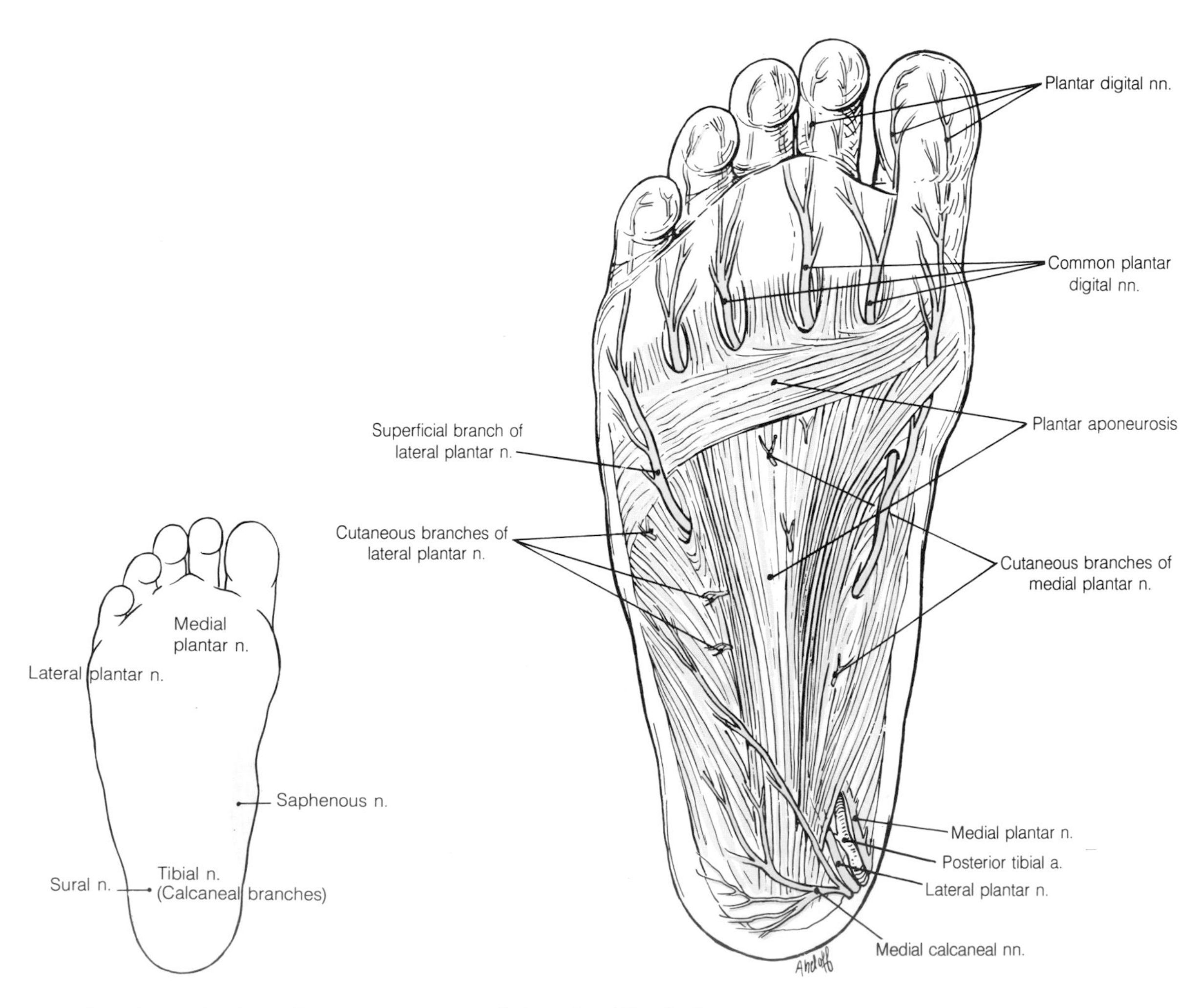

8-83 The plantar fascia and cutaneous nerves of the sole of the foot.

The Muscles, Nerves and Vessels of the Sole of the Foot

The First Layer of Muscles

This layer consists of three superficial muscles and they should be identified in Fig. 8-84:

1. An abductor of the great toe, **abductor hallucis**
2. A short flexor muscle for the lateral four toes (comparable to flexor superficialis in the hand), **flexor digitorum brevis**
3. An abductor of the little toe, **abductor digiti minimi**.

These muscles arise in sequence across the tuber calcanei. The abductors are inserted into the outer sides of the bases of the appropriate proximal phalanges and the short flexor tendons are perforated by the long flexor tendons before becoming attached to the intermediate phalanges.

Between the first and second layers of muscles lie the greater part of the course of the **medial and lateral plantar nerves and vessels**. Their distribution is illustrated in Fig. 8-85.

The Medial Plantar Nerve and Artery

The **medial plantar nerve** (L4, 5) is formed with the lateral plantar nerve by the division of the **tibial nerve** beneath the **flexor retinaculum**. Both nerves run forward into the sole beneath the **abductor hallucis muscle**. The medial plantar nerve continues forward and, between the abductor hal-

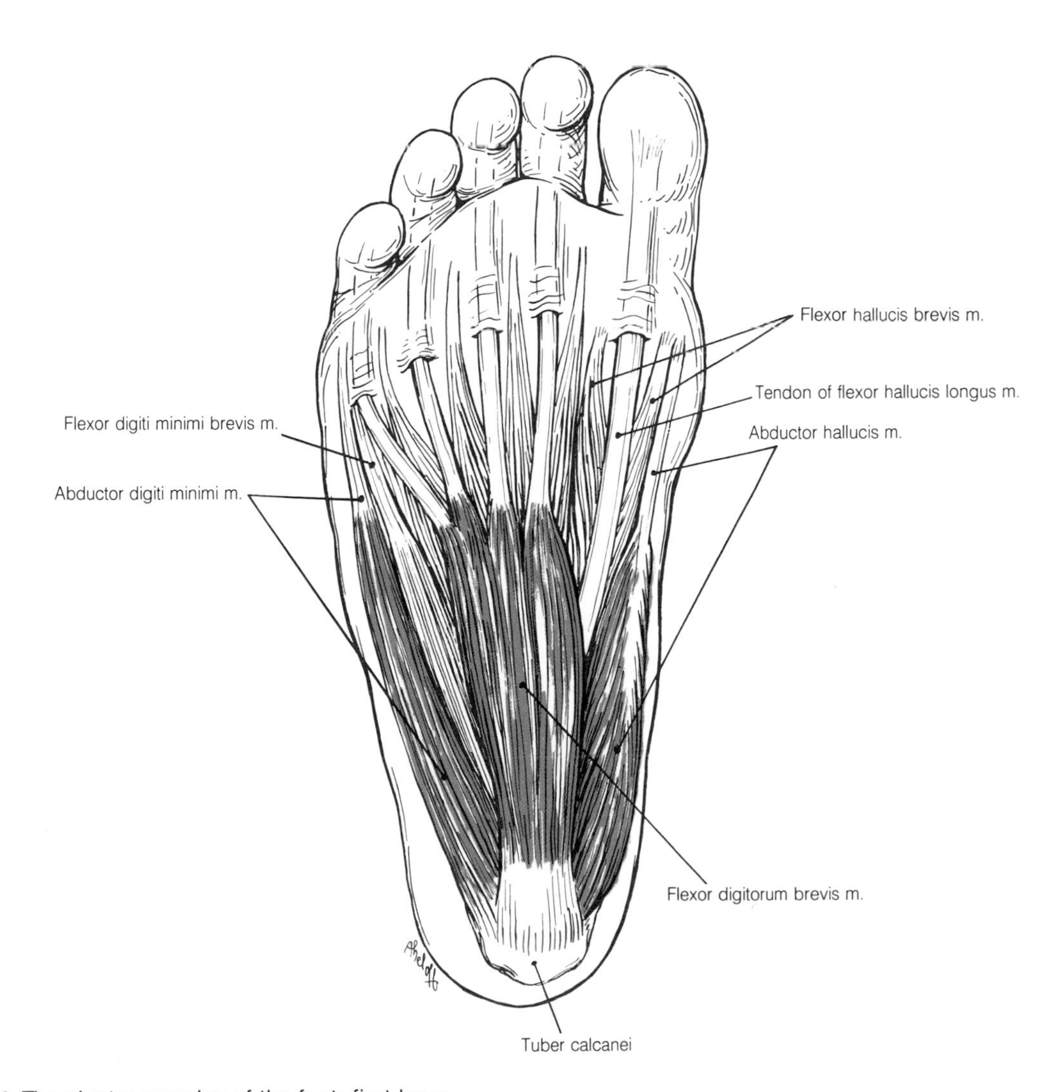

8-84 The plantar muscles of the foot; first layer.

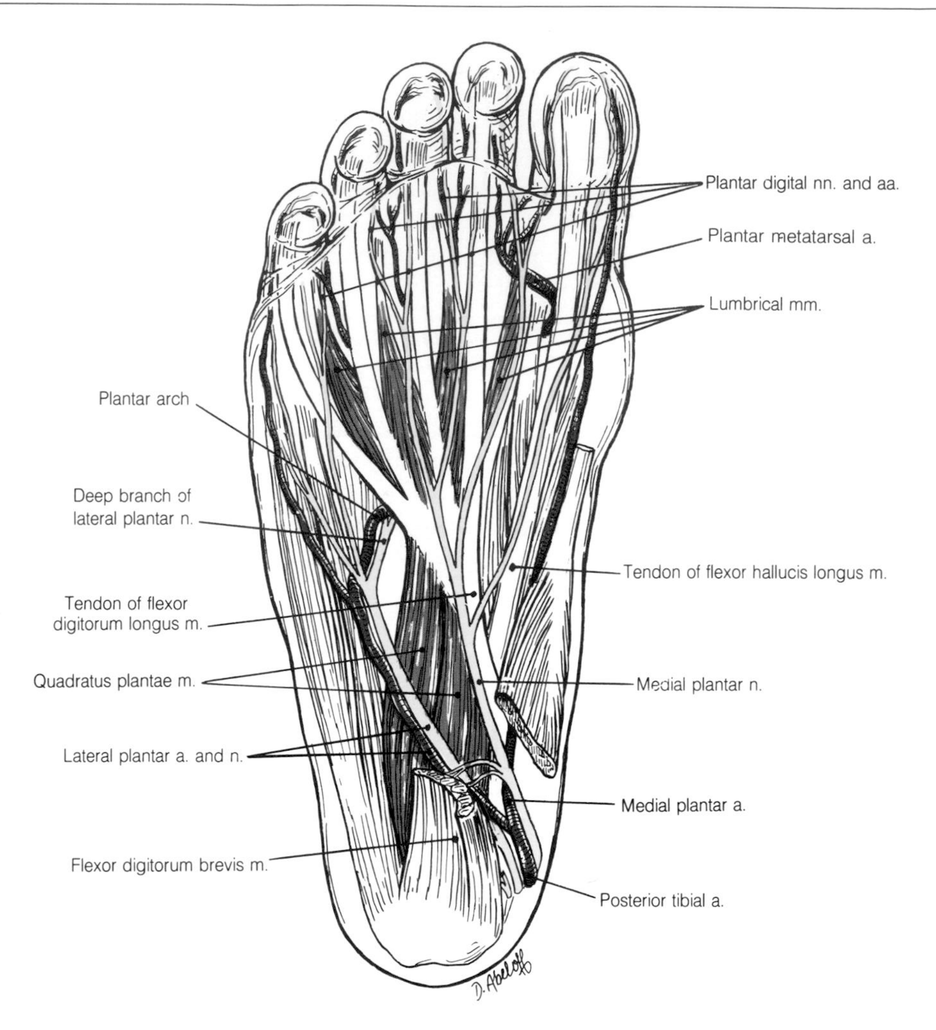

8-85 The plantar muscles of the foot: second layer. The medial and lateral plantar nerves and arteries are also shown.

lucis and the flexor digitorum brevis, breaks up into **plantar digital branches** for the medial three-and-a-half toes. While it has an extensive cutaneous distribution to the sole and digits, it only supplies four muscles, the **abductor hallucis**, the **flexor digitorum brevis**, the **flexor hallucis brevis**, and the **first lumbrical muscle**. All other muscles intrinsic to the sole of the foot are supplied by the **lateral plantar nerve and its deep and superficial branches**.

The **medial plantar artery** accompanies the nerve forward on its medial side. After giving off muscular branches, it terminates as a plantar digital branch to the medial side of the great toe and further branches which join the metatarsal branches of the plantar arch.

The Lateral Plantar Nerve and Artery

The **lateral plantar nerve** (S1, 2) and **artery** cross the sole of the foot obliquely just deep to the first layer of muscles to reach its lateral border. Both give off muscular branches and supply overlying skin. The muscular branches of the nerve which are given off before it divides are to **quadratus plantae and abductor digiti minimi**. At the level of the base of the fifth metatarsal, the artery runs medially, deep to the oblique head of the **adductor hallucis muscle** (third layer) to reach the first intermetatarsal space (Fig. 8-86). Here it anastomoses with the **dorsalis pedis artery** and completes the **plantar arch** (note that there is only one arterial arch in the foot). From the arch, **plantar**

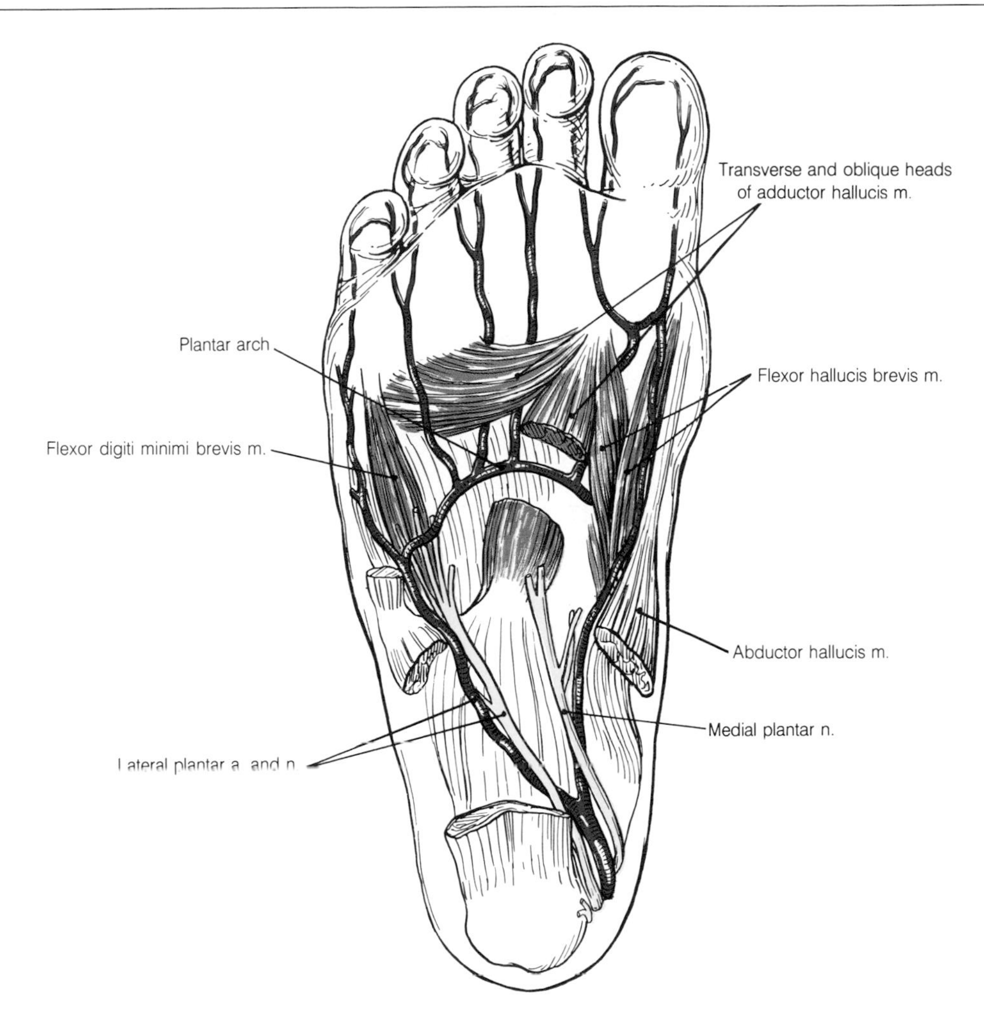

8-86 The plantar muscles of the foot: third layer. The oblique head of adductor hallucis has been partly removed to display the plantar arch.

metatarsal arteries run forward to divide into **plantar digital arteries** for the toes. The **lateral plantar nerve** divides into a superficial and a deep branch. The superficial branch supplies **plantar digital branches** to the lateral one-and-a-half toes and muscular branches to the **flexor digiti minimi brevis** and the **interossei** of the fourth intermetatarsal space. The **deep branch** follows the plantar arch and supplies the **remaining interossei**, the **lateral three lumbrical muscles**, and the **adductor hallucis**.

Plantar digital veins and medial and lateral plantar veins accompany the arteries as they drain the sole of the foot. They join to form posterior tibial veins alongside the posterior tibial artery.

The Second Layer of Muscles

This layer is composed of **quadratus plantae** (flexor digitorum accessorius) and the **lumbrical muscles**. The **tendons of the long flexor muscles** can also be included as they are intimately associated with each other and the intrinsic muscles of this layer. These are also illustrated in Fig. 8-85. The quadratus plantae arises from the calcaneus by a large medial head and a smaller lateral head. This muscle is inserted into the tendon of flexor digitorum longus. The four lumbrical muscles arise from the long flexor tendons given off to the lateral four digits and are inserted into the dorsal digital expansions as in the hand. The tendons of the long flexors enter the foot on its medial side with that of the long flexor of the digits lying medial to that of the great toe. In the sole, the tendon to the digits crosses the tendon to the great toe superficially and receives a susbstantial slip from it. An old name for the quadratus plantae muscle is included as it indicates that the muscle may aid the long flexors of the toes either by aligning their tendons or substituting for them when the foot is plantarflexed and the long flexors are unable to shorten sufficiently to flex the toes.

The Third Layer of Muscles

The third layer of muscles which is illustrated in Fig. 8-86 includes the **flexor hallucis brevis**, the **transverse and oblique heads of adductor hallucis**, and the **flexor digiti minimi brevis**. The flexor hallucis brevis divides into two heads which are attached to the base of the proximal phalanx. The tendons of attachment contain two constant sesamoid bones which articulate with the base of the phalanx. The long flexor tendon lies, possibly protected, between these two small bones. The adductor hallucis is attached with the lateral head of flexor hallucis brevis to the base of the proximal phalanx. The flexor digiti minimi brevis is attached to the lateral side, of the base of the proximal phalanx of the fifth digit.

The Fourth Layer of Muscles

This layer consists of the **four dorsal and three plantar interossei** arising from the metatarsal bones. As in the hand, the dorsal muscles abduct and the plantar adduct the digits. However, in the foot these actions occur about an axis which passes through the second digit rather than the middle digit. Running across the sole of the foot from lateral to medial side is the tendon of peroneus longus. Approaching from the lateral side of the cuboid bone it turns under this bone into an osseofibrous groove on its plantar surface. From here it runs obliquely forward and medially to be inserted into the lateral aspect of the base of the first metatarsal and the adjacent medial cuneiform bone.

The Innervation of the Intrinsic Muscles of the Sole

The intrinsic muscles of the sole are supplied by the **medial and lateral plantar nerves** (S2, 3), the two terminal branches of the tibial nerve. The distribution of these nerves to the muscles is summarized below.

The medial plantar nerve
- Abductor hallucis
- Flexor digitorum brevis
- Flexor hallucis brevis
- First lumbrical muscle

The lateral plantar nerve before division
- Quadratus plantae
- Abductor digiti minimi
- **Superficial branch**
  - Flexor digiti minimi brevis
  - The interossei of the fourth space
- **Deep branch**
  - The remaining interossei
  - Lateral three lumbrical muscles
  - Adductor hallucis

The names of the intrinsic muscles of the foot imply theoretical actions which are not altogether borne out in practice. Individual movements cannot be demonstrated except in the case of flexion and extension of the digits, but it is difficult to move one toe without the others. The intrinsic muscles seem to play little part in the support of the body in standing but do come into action as the heel is raised and the foot pushes off the ground in locomotion. This is combined with flexion of the toes at the proximal interphalangeal joints while the distal interphalangeal joints and metatarsophalangeal joints are held extended.

The Joints of the Foot

The Subtalar Joint

In the section on the talus and the calcaneus, complementary anterior and posterior articular surfaces were described on the lower surfaces of the talus and the upper surface of the calcaneus. Strictly speaking, these surfaces form part of the **talocalcaneal joint** posteriorly and the **talocalcaneonavicular joint** anteriorly (the calcaneonavicular joint has a joint cavity that is common to the anterior talocalcaneal joint). However, the two talocalcaneal articulations function together as a single unit and are described together as the **subtalar joint**. Examination of these concave and convex joints (one is a reverse of the other) shows, as seen in Fig. 8-87, that a line join-

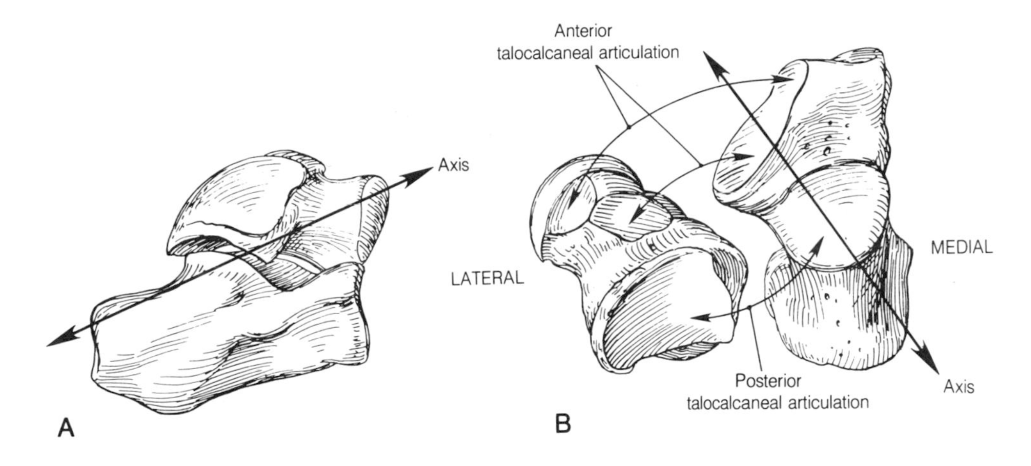

8-87 Showing the axes and articular surfaces of the subtalar joint.

ing the centers of rotation of each is directed forward, upward, and medially. It is about this combined axis that the two joints allow the important movements of inversion and eversion.

The Talocalcaneonavicular Joint

The rounded head of the talus articulates with the calcaneus at the anterior talocalcaneal joint (as described above) and with the navicular bone and the superior surface of the **plantar calcaneonavicular ligament** which bridges the gap between the sustentaculum tali and navicular bone. The arrangement of these articular surfaces and the position of the plantar calcaneonavicular ligament can be seen on the calcaneus and navicular bone in Fig. 8-88.

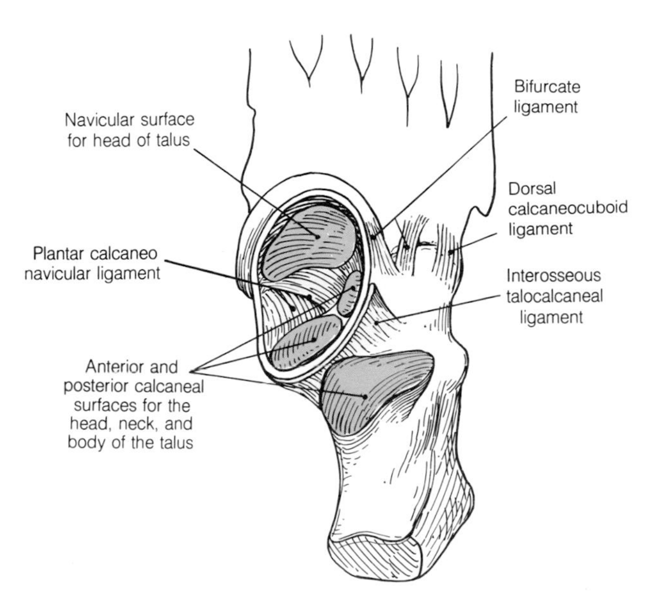

8-88 Showing the distal articular surfaces of the talocalcaneonavicular joint.

The Calcaneocuboid Joint

Lying in almost the same transverse plane as the talonavicular joint is the calcaneocuboid joint. This saddle-shaped synovial joint has limited movement and its capsule is closely reinforced inferiorly by the plantar calcaneocuboid ligament (short plantar ligament) and more superficially by the long plantar ligament. On the dorsal aspect of the joint the bones are bound together by the **calcaneocuboid ligament**, a part of the **bifurcate ligament**, and more laterally by the **dorsal calcaneocuboid ligament** (Fig. 8-88). The calcaneocuboid and talonavicular articulations lie almost in the same plane and are together often called the **midtarsal joint**.

The remaining intertarsal and tarsometatarsal joints are stable synovial joints strongly reinforced by dorsal, plantar, and interosseous ligaments. Except for the first tarsometatarsal joint which allows some elevation and depression, all show minimum gliding movements. The metatarsophalangeal and interphalangeal joints are synovial joints and very similar to those of the hand. The former are condyloid and allow flexion, extension, abduction, and adduction. The latter are hinge joints and only allow flexion and extension.

Important Ligaments of the Foot

As can be realized from the descriptions of the joints of the foot, only small amounts of movement are possible between individual tarsal bones and, thus there is a remarkable degree of overall stability. This is largely maintained by short strong ligaments linking the bones together. A number of these have already been mentioned, but the important ones are now summarized.

The Bifurcate Ligament

The strong bifurcate (Fig. 8-88) ligament is attached proximally to the upper surface of the calcaneus. As it passes forward it divides into two portions. The lateral portion is attached to the cuboid bone and becomes the **calcaneocuboid ligament**. The medial portion passes to the navicular bone as the **calcaneonavicular ligament**. In so doing it contributes to the lateral articular surface of the talocalcaneonavicular joint.

The Interosseous Talocalcaneal Ligament

This ligament lies between the anterior and posterior talocalcaneal articulations and extends from the **sulcus tali** to the **sulcus calcanei**. It binds the two bones together and becomes tight as the limit of eversion is reached.

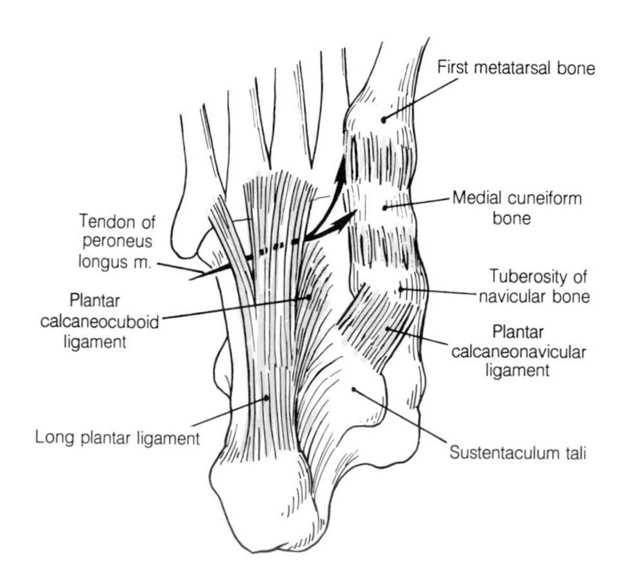

8-89 Showing the long and calcaneocuboid plantar ligaments. The course traveled by the tendon of peroneus longus and its attachments are also indicated.

The Plantar Calcaneonavicular Ligament

This ligament lies on the plantar aspect of the talocalcaneonavicular joint and bridges the gap between the sustentaculum tali and the navicular bone. In so doing it provides an additional articular surface for the head of the talus. This can be seen in Figs. 8-88 and 8-89. It was once known as the spring ligament but the absence of any elastic fibers makes this name inappropriate.

The Plantar Calcaneocuboid Ligament or Short Plantar Ligament

The extremely strong short plantar ligament lies on the plantar aspect of the calcaneocuboid joint and extends from the anterior tubercle of the calcaneus to the adjacent surface of the cuboid bone (Fig. 8-89).

The Long Plantar Ligament

The long plantar ligament overlies the short plantar ligament. It is attached posteriorly to the plantar surface of the calcaneus anterior to its medial and lateral processes. Anteriorly it is attached to the plantar surface of the cuboid bone and the bases of the lateral four metatarsal bones. As it crosses the deep surface of the cuboid bone, it turns the **groove for the tendon of peroneus longus** into an osseofibrous tunnel which in turn is lined by a synovial sheath for the tendon (Fig. 8-89).

Functions of the Foot

The feet serve to support the body weight and maintain stability and balance in a wide range of body postures. A perpendicular line dropped from the body's center of gravity to the ground must fall within the quadrangle limited by the margins of the feet if the body is to remain stationary and upright. The size of this quadrangle can, of course, be increased for added stability by separating the feet. The feet also act as propulsive levers in the acts of walking and running. Each of these functions can be performed more effectively if the sole of the foot can be readily adapted to the irregularities in the surface to which it is applied. This adaptation is largely secured by the move-

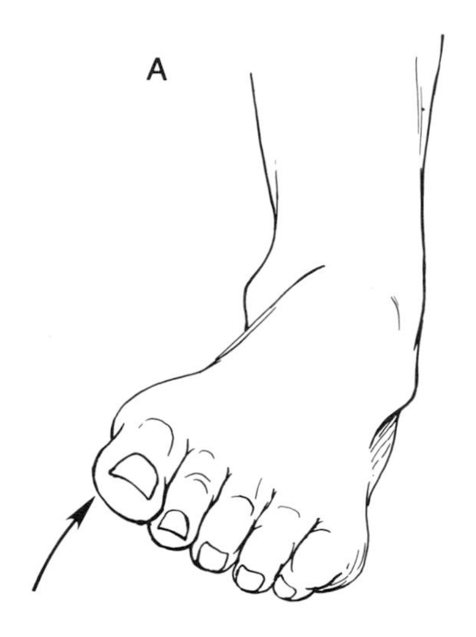

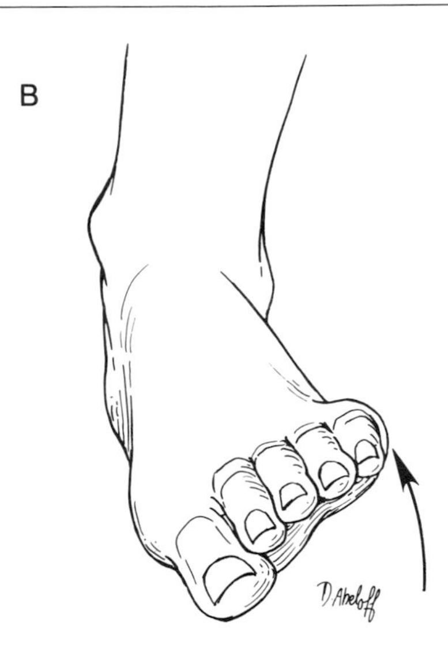

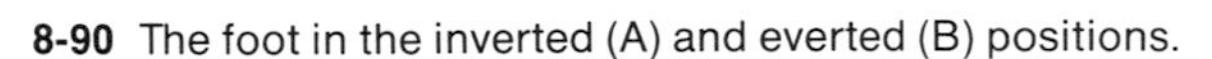

8-90 The foot in the inverted (A) and everted (B) positions.

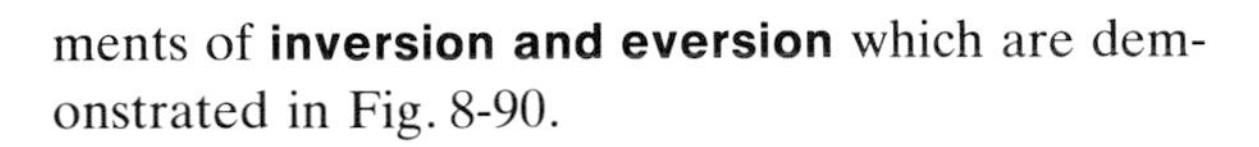

ments of **inversion and eversion** which are demonstrated in Fig. 8-90.

Inversion and Eversion

These movements occur to a limited extent at the "midtarsal joint" which comprises the **talonavicular and calcaneocuboid joints**. The small amount of movement here allows some rotation of the forefoot and thus the elevation of either its medial or lateral border. By far the greater part of inversion and eversion occurs at the **subtalar joint**. Movements about the sloping and oblique axis of this joint which were illustrated in Fig. 8-87 produce:

1. Inversion –
   - elevation of the medial border of the foot
   - adduction
   - plantarflexion
2. Eversion –
   - elevation of the lateral border of thc foot
   - abduction
   - dorsiflexion

When the medial and lateral borders of the foot are raised, it will be found that the other movements comfortably follow. Muscles whose tendons pass medial to this subtalar axis will be invertors. The best examples are **tibialis anterior and posterior** (note that their actions on the ankle joint differ because the tendon of one passes behind and the other in front of its axis). Muscles whose tendons pass lateral to the subtalar axis will be evertors. The most effective are **peroneus longus and brevis**.

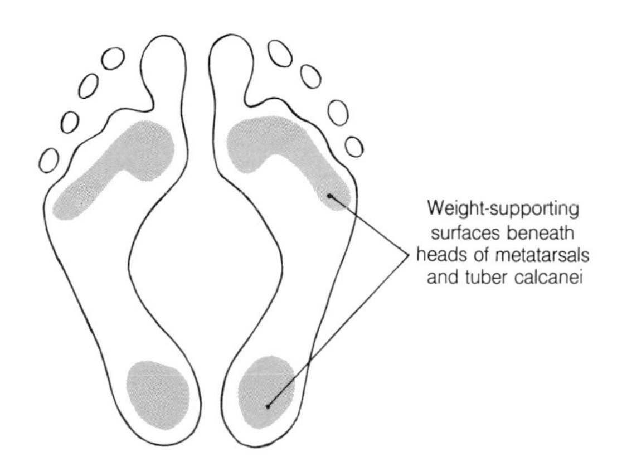

8-91 Footprints illustrating the weight-bearing regions of the feet.

The Arches of the Foot

An examination of the adult foot shows that it is arched longitudinally and the footprints seen in Fig. 8-91 suggest that the medial side of the arch is higher than the lateral. As a result, the longitudinal arch is somewhat arbitrarily divided into a medial and lateral arch. The skeleton of the medial arch consists of the calcaneus, talus, navicular and cuneiform bones, and the medial three metatarsals, whereas the lateral arch consists of the calcaneus, cuboid, and the lateral two metatarsals. When the two feet are placed together a transverse arch also becomes apparent. The skeletal elements of these arches can be identified in Fig. 8-92. The important thing to remember is that the foot is not a rigid lever and that the multiple elements of the arches give it resilience and the ability to change shape. However, these dynamic features require that the arches be supported and controlled. It is here that anatomy (i.e., the discipline) fails, for a number of theories as to how this is achieved have been put forward but none clearly proven. As long as the supporting extremities of the arches do not move apart, the congruity of the articulating surfaces of the component bones help to stop it collapsing. Beyond this the support must depend on plantar ligaments and muscles. In standing, electromyography indicates that there is very little activity in the intrinsic muscles of the foot. Support is almost certainly maintained by ligamentous structures linking the plantar surfaces of the bones. Among these are the strong, long and short plantar ligaments, the plantar calcaneonavicular ligament, and the plantar fascia. Directly, the foot is used to apply force to the ground, the intrinsic muscles become active, and the tension in the tendons of the long flexor muscles ties the extremities of the arches together taking the strain off the ligaments.

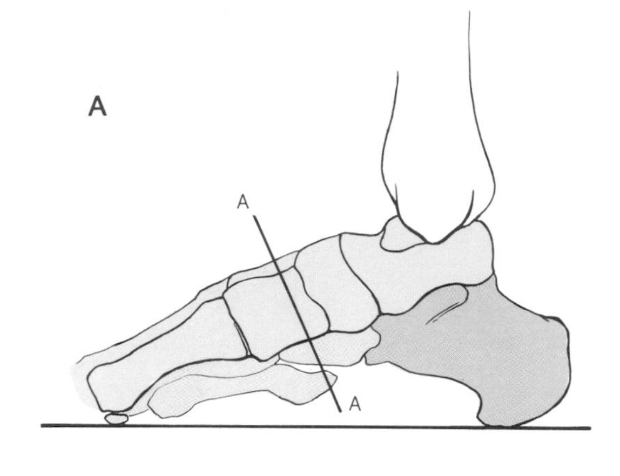

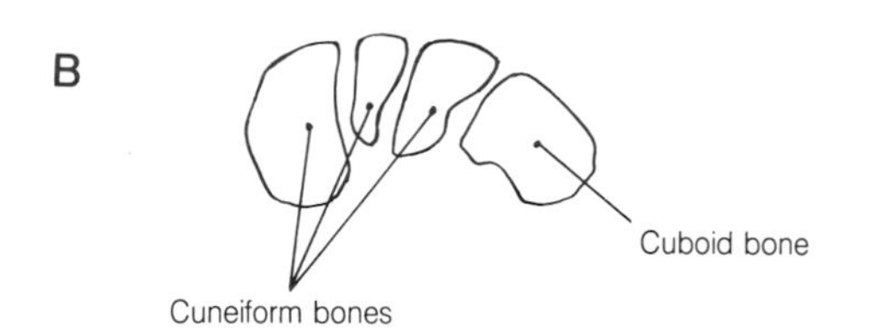

8-92 Illustrating the longitudinal (A) and transverse (B) arches of the foot.

Walking

Walking is an extremely complex function which we take very much for granted. However, quite small defects in the neurological, skeletal, or muscular structures involved give rise to a new gait pattern, often of a compensatory or trick nature. Some knowledge of the normal gait and the ability to recognize the abnormal is important to the physician.

The act of walking can be broken down into a **stance or supporting phase** and a **swing phase** for each limb and these are illustrated in Fig. 8-93. Of each complete limb cycle three-fifths is stance

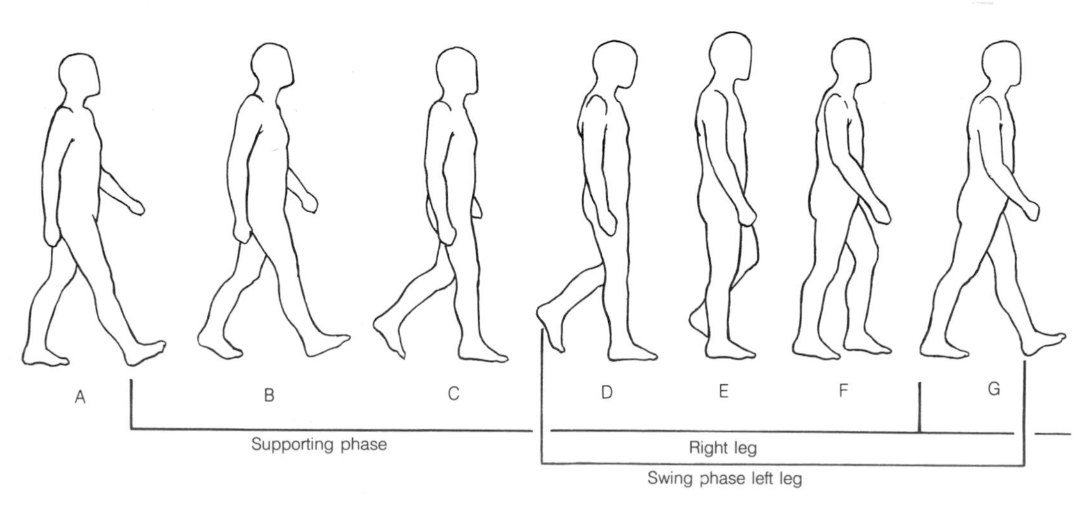

8-93 Illustrating the supporting and swing phases of walking.

phase and two-fifths swing and, thus, for a period both limbs are on the ground at the same time. The stance phase of one limb starts as the heel touches the ground (Fig. 8-93 A, right leg). The weight of the body is transferred to this heel (Fig. 8-93 B and C) and the opposite limb is free to go into its swing phase (Fig. 8-93 D). In this, the body's center of gravity moves forward over the supporting limb which is extended at hip and knee and the weight is transferred forward along the outer border of the foot to the heads of the metatarsals and the great toe. Finally, the heel is raised, the weight transferred to the opposite limb (now in the stance phase) and by flexing the hips and knee, the limb is swung forward. The knee is then extended and the cycle completed as the heel strikes the ground again (Fig. 8-93 G). As the weight is transferred from limb to limb, the gluteus medius and minimus of the weight-bearing side must contract to prevent the pelvis from tilting to the opposite side. All of these actions, together with those of the trunk and arms, are blended into a fluid motion in which the minimum muscular effort is used. Of this, most is expended in raising the center of gravity and giving it some forward momentum. When walking at a normal frequency the swing phase is largely achieved by a pendulum-like movement of the free limb.

The Segmental Innervation of Lower Limb Muscles

In previous sections the spinal cord segments concerned in the innervation of the lower limb muscles have been given in parentheses after the nerves supplying the muscles. A detailed knoweldge of these is not required but with the help of the table in Fig. 8-94 some information which may prove useful in the clinical examination of a weak limb may be gained. In the table the main flexors and extensors of the hip, knee, and, ankle joints can be picked out and their segments determined:

Hip joint	Flexors	L2 and L3
	Extensors	L5 and S1
Knee joint	Extensors	L3 and L4
	Flexors	L5 and S1
Ankle joint	Dorsiflexors	L5 and S1
	Plantarflexors	S1 and S2

From this analysis it can be seen that the more distal the joint the lower the segments that are concerned with its movements. The movements of adduction of the hip involve the segments L3 and 4, and abduction L4 and 5. Also, inversion involves L4 and 5, and eversion L5 and S1.

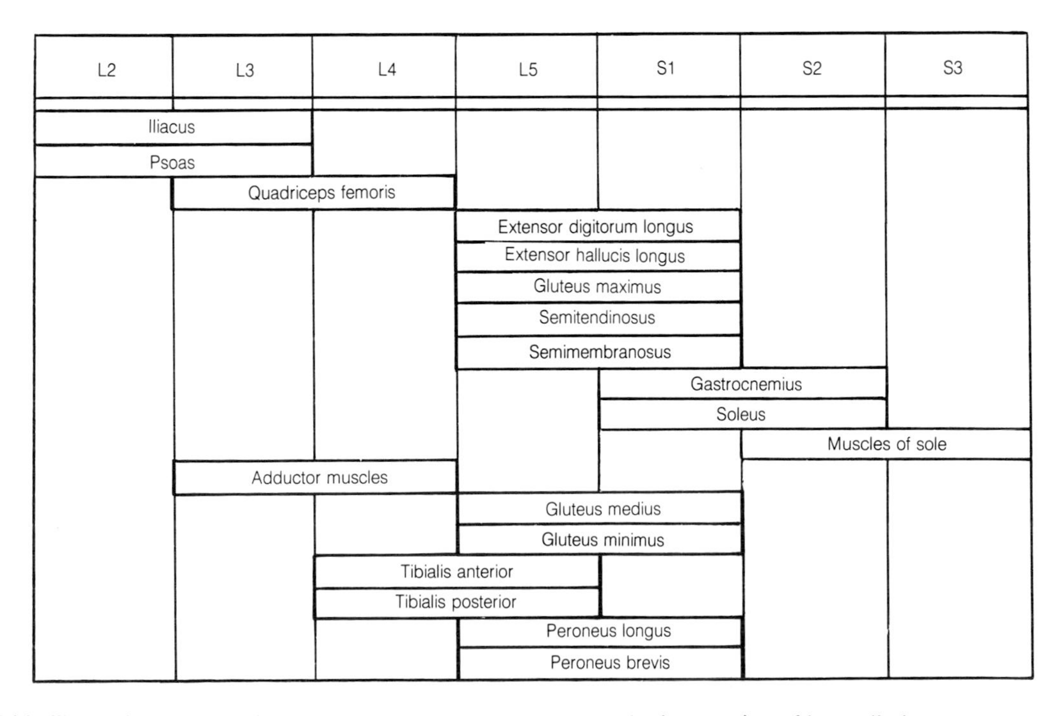

8-94 Table illustrating the contributions of spinal cord segments to the innervation of lower limb muscles.

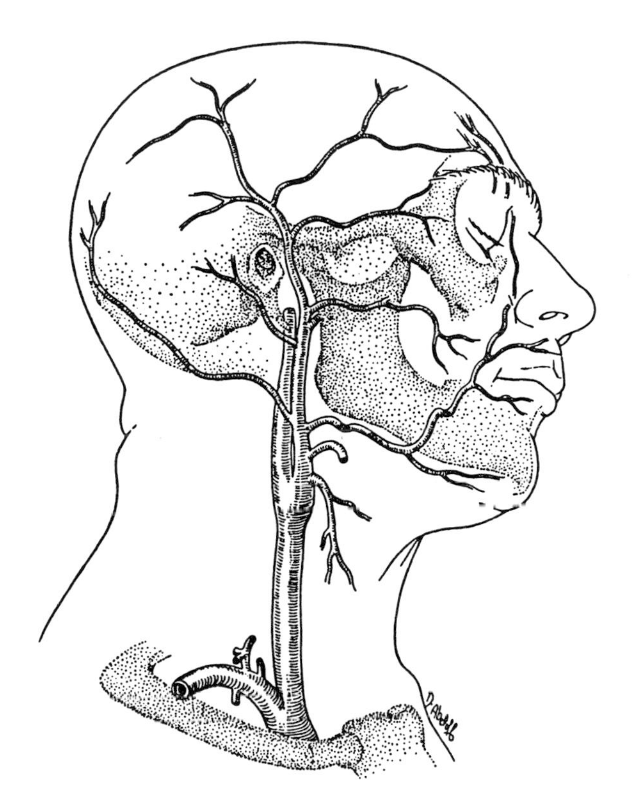

9 The Head and Neck

The study of the head and neck provides a formidable learning task in gross anatomy and even if the brain is assigned to "neurosciences" this region remains packed with small but important structures. Many of these are associated with the localization of organs of special sensation in the head (encephalization) and the presence of the proximal parts of the alimentary and respiratory systems which share, in part, a common channel. The head and neck presents their own particular problem as an exercise in dissection because many superficial structures are encountered some time before their parent structures are displayed at a deeper level. This fact demands a certain amount of courage and trust from the student for, if the early details can be accepted in isolation, as dissection proceeds they will be found to assemble into an understandable whole.

The osteology of the skull is worth mastering not only for the clinical importance of some regions but because it serves both literally and metaphorically as a skeleton on which to hang the soft parts and associate them logically with each other. For this reason the description of the head is begun with a study of the bones of the skull.

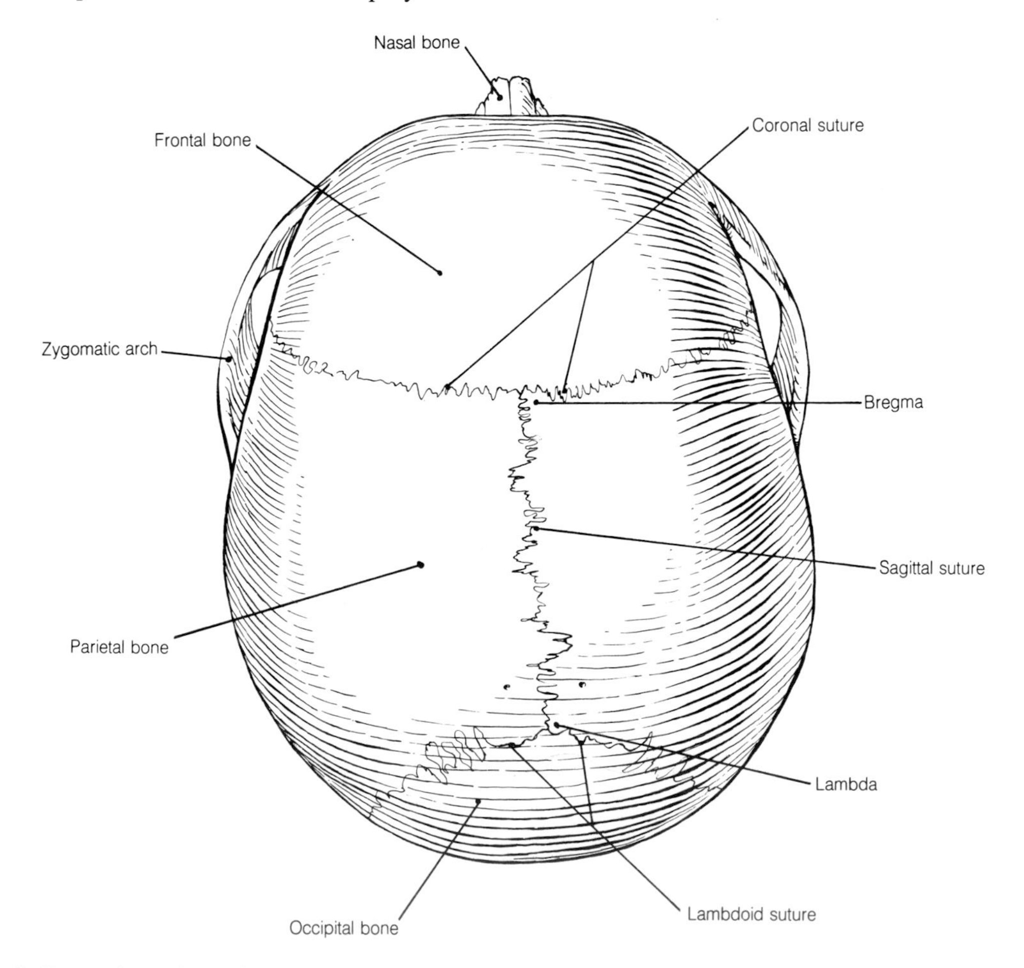

9-1 The skull seen from above (norma verticalis).

The General Osteology of the Skull—External Appearance

The bones of the skull can be divided into those that form the walls of the cranial cavity enclosing the brain, and those that form the face including the separate mandible. In man, unlike other primates or mammals, the facial skeleton lies directly below the anterior half of the cranial cavity. Thus, when the skull is viewed from above, little or nothing of it is seen. This separation of cranial and facial skeleton is useful for descriptive purposes but it has no strict morphological basis and does not preclude cranial bones contributing in part to the facial skeleton.

Norma Verticalis

A view of the skull from above (norma verticalis) is illustrated in Fig. 9-1. It can be seen that it is formed by a single **frontal bone** (occasionally a midline **metopic*** **suture** bisects it), two **parietal bones**, and an **occipital bone**. The parietal bones are joined to the frontal bone by the **coronal suture** and to each other by the midline **sagittal suture**. The point at which these two sutures meet is called the **bregma**. Posteriorly the two parietal bones are joined to the single **occipital bone** at the **lambdoid suture** and the point at which the

* Metopion, *Gk.* = forehead

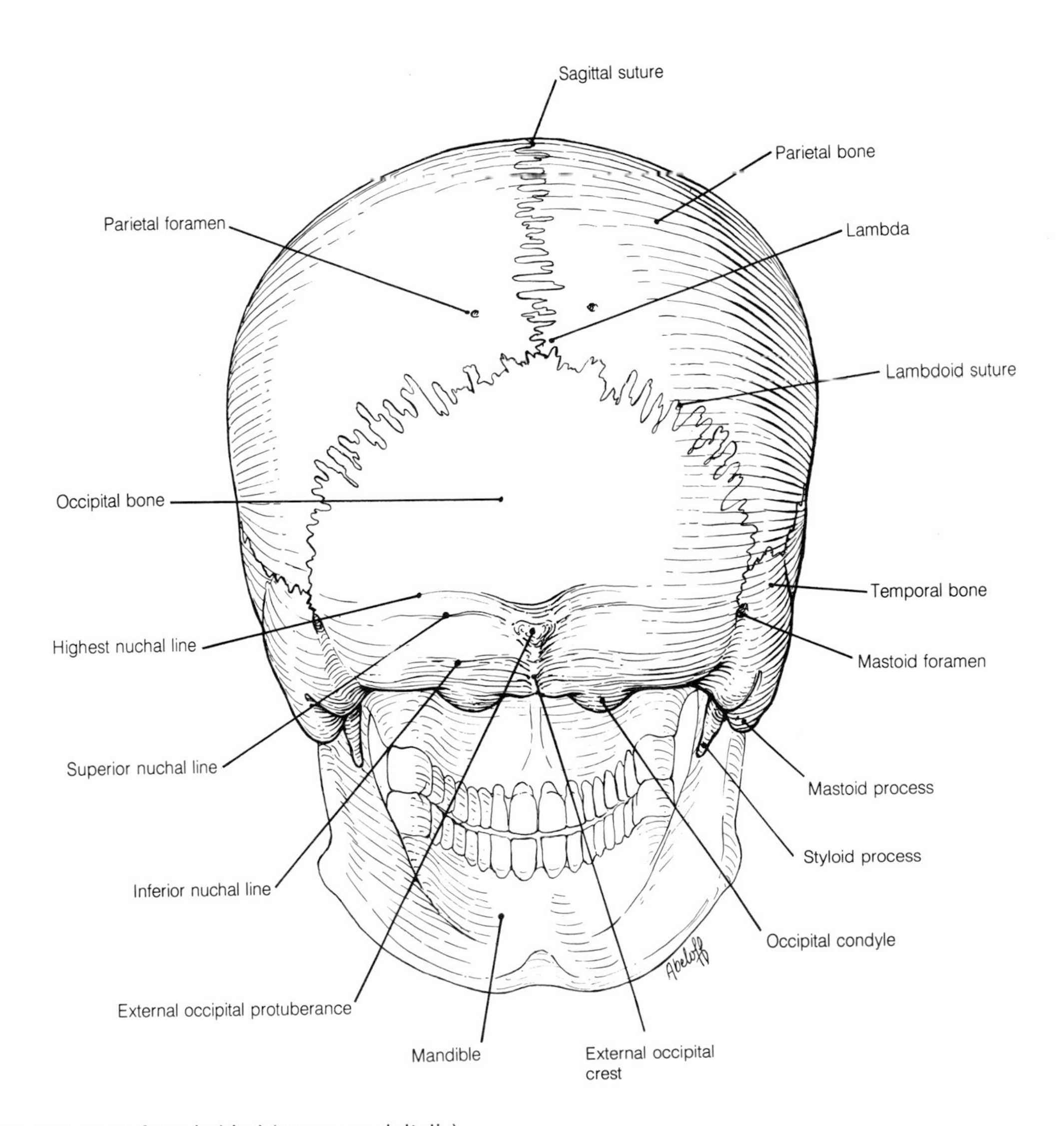

9-2 The skull seen from behind (norma occipitalis).

sagittal suture meets this is the **lambda***. Close to the lambda two small foramina can be seen in the parietal bones. These **parietal foramina** transmit emissary veins between the scalp and the venous sinus lying deep to the bone.

The **zygomatic arch** is a feature of the lateral aspect of the skull but its arching profile can be seen in this view. Note that of the facial skeleton little more than the **nasal bones** can be seen.

Norma Occipitalis

The skull is viewed from behind (norma occipitalis) in Fig. 9-2. The posterior portions of the parietal bones are seen again as well as the whole of the squamous part of the occipital bone which forms the posterior wall of the cranial cavity. On either side of this lie the **mastoid parts** of the **temporal bone** from which the **mastoid processes** project inferiorly. The temporal bones are joined to the occipital bone by the **occipitomastoid sutures** and near the superior end of these is seen a **mastoid foramen**. Again these foramina carry emissary veins. The **styloid processes** of the temporal bones can also be seen in this view as their tips project below the mastoid parts.

The most prominent point on the midline of the occipital bone is the **external occipital protuberance**. From it the **superior nuchal lines** extend laterally. Below the protuberance the **external occipital crest** leads to the **inferior nuchal lines**. A faint **highest nuchal line** can sometimes be made out. The elevations projecting from the lower margin of the occipital bone are the **occipital condyles** whereby it articulates with the first cervical vertebra or **atlas.**

Norma Frontalis

Looking at the skull from the front (norma frontalis) as in Fig. 9-3, the frontal bones are found to

* Lambda, *Gk.* = the letter λ of the Greek alphabet

9-3 The skull seen from the front (norma frontalis).

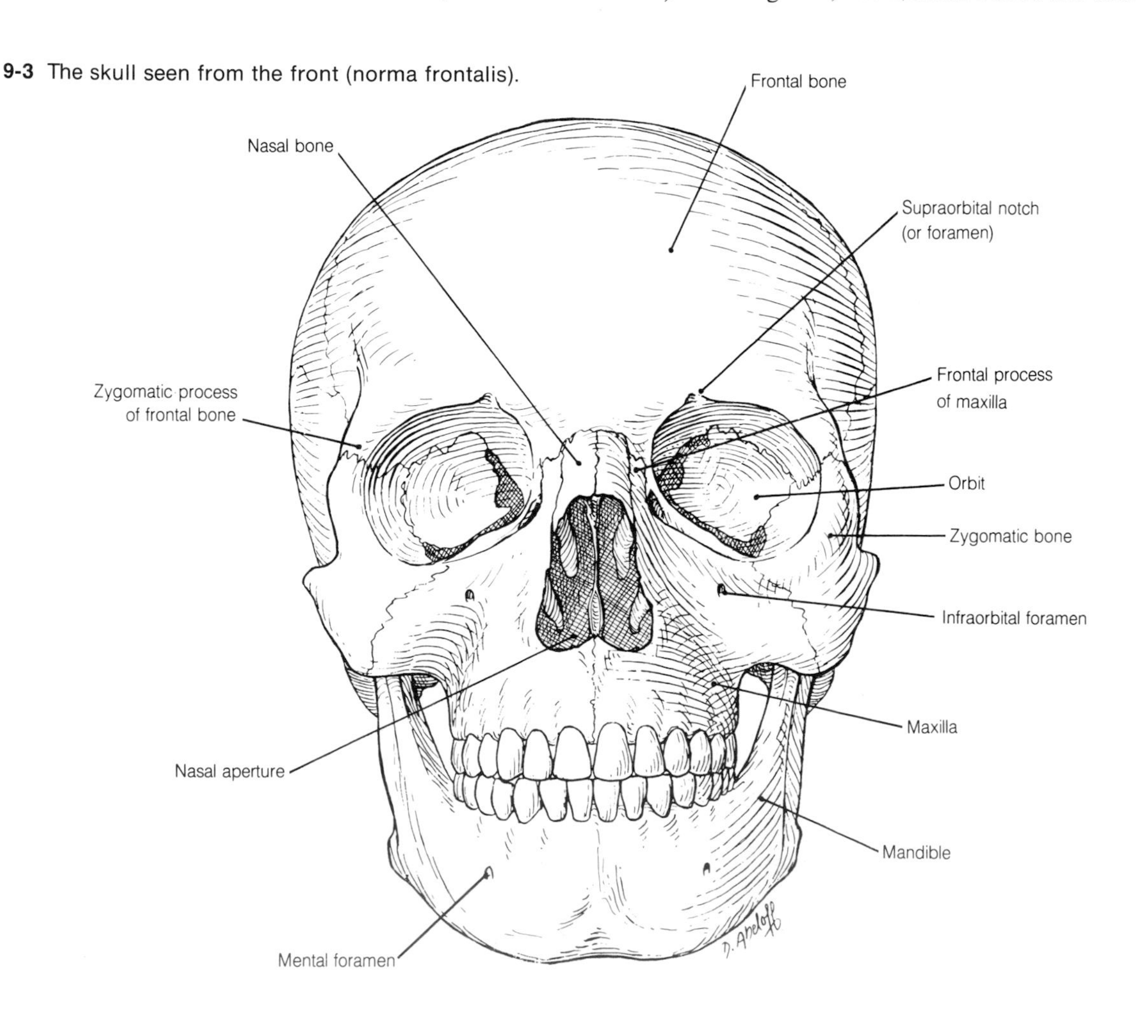

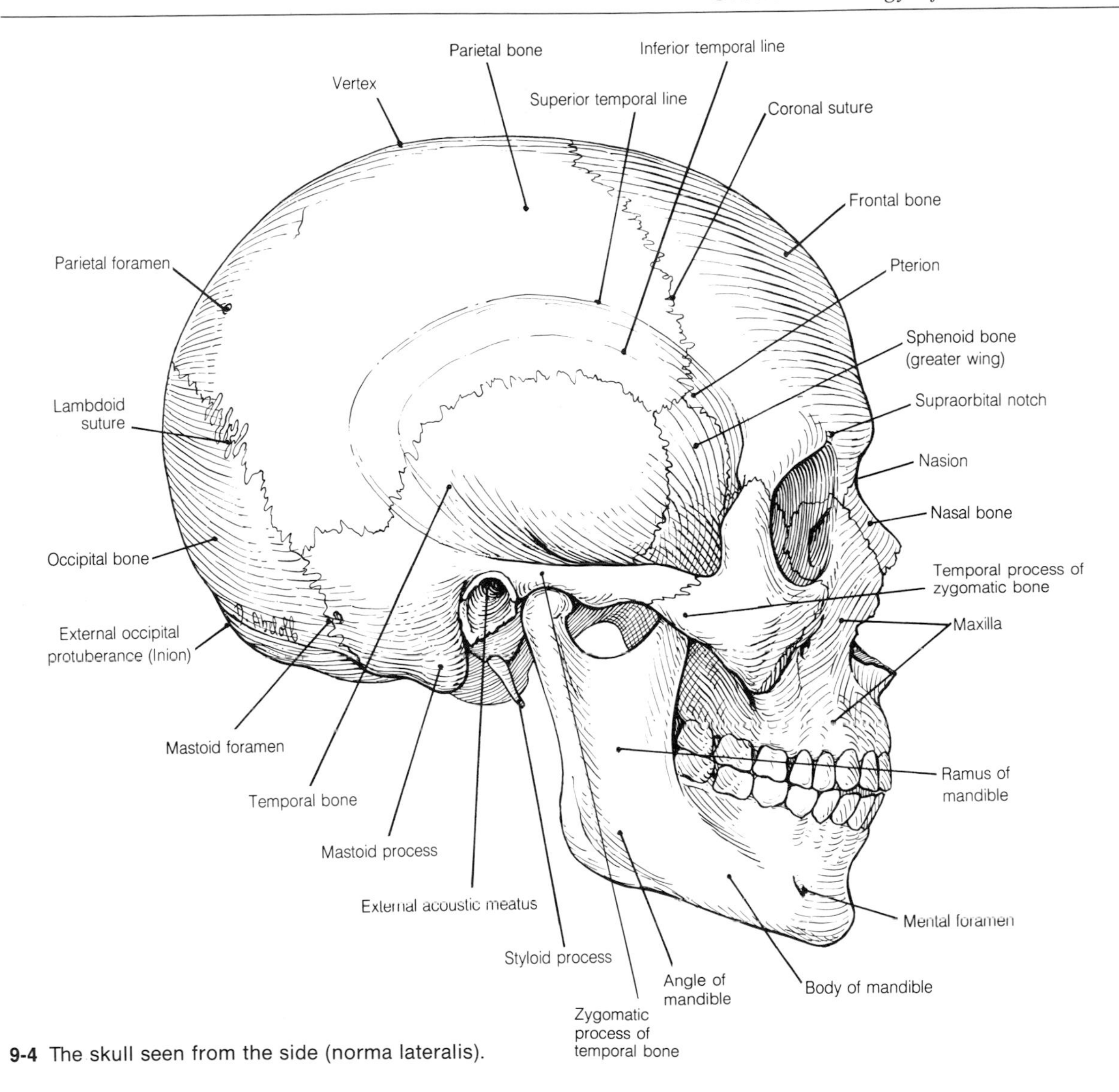

9-4 The skull seen from the side (norma lateralis).

descend to form the forehead and the upper borders of the orbits. Below and on either side of the midline the frontal bone articulates with a pair of **nasal bones** and the **frontal processes of the maxillae.** The latter form the medial margins of the orbits. By its **zygomatic process** the frontal bone articulates laterally with the **zygomatic bone**. The zygomatic bones form the lateral and part of the lower margins of the orbits and these margins are completed by the **maxillae.** The maxillae also complete with the nasal bones the margins of the nasal cavity and bear the upper teeth. Below the maxillae lies the **mandible** bearing the teeth of the lower jaw. A **supraorbital foramen or notch** is found in the frontal bone at the upper margin of each orbit. Below the lower margins of the orbits the **infraorbital foramina** are seen in the maxillae.

Norma Lateralis

In Fig. 9-4 the skull is viewed from the side (norma lateralis). The **parietal bone** is seen to articulate with the **temporal bone** and its mastoid portion posteriorly and with the **greater wing of the sphenoid bone** anteriorly. The latter also articulates with the frontal bone and zygomatic bone. The proximity of the articulations of frontal, parietal, sphenoid, and temporal bones gives rise to a landmark called the **pterion** (the *p* is silent). Now note the **superior** and **inferior temporal lines** near the lower margin of the parietal bone. These lines are named respectively after the **temporal fascia** and **temporal muscle** which are attached to them. Further landmarks that can be appreciated from this aspect are the **nasion** at the junction of the frontal and nasal bones, the **inion**, the most prominent posterior projection of the occipital bone, and the **vertex**, the highest point of the skull.

In the figure find the **zygomatic process of the temporal bone**. This extends forwards to articulate with the **temporal process of the zygomatic bone** and thereby forms the **zygomatic arch**. Through the space that lies between the skull and this arch passes the powerful temporalis muscle which can be felt above the arch when the jaws are clenched. Below the posterior end of this arch lies a bony opening, the **external acoustic meatus** which leads to the middle ear. A sharp spine of bone, the **styloid process** of the temporal bone can be seen below the meatus although on a deeper plane.

At the root of the zygomatic process of the temporal bone there is a hollow called the **mandibular fossa** and an elevation in front of this called the **articular tubercle**. These two bony features articulate with the **head of the mandible**.

The **mandible**, which is also illustrated in Fig. 9-4, consists of a horizontal **body** fused with its fellow in the midline and a vertical **ramus**. The body and the ramus are continuous with each other at the **angle of the mandible**. Anteriorly the body shows a small **mental foramen** and the alveolar bone along its superior margin bears the lower teeth. The upper border of the ramus bears the **coronoid process** anteriorly and the **condylar process** posteriorly. The two processes are separated by the **mandibular notch**. The condylar process is made up of the **neck** and the **head of the mandible** which articulates with the skull just anterior to the external acoustic meatus.

Further details of the cranial and facial skeleton will be given as the surrounding regions are described.

Norma Basalis

Finally, the skull is viewed from below (norma basalis) in Fig. 9-5. The palate of the facial skeleton is seen anteriorly and is formed by the **palatine process** of the **maxilla** in front and the **horizontal plate** of the **palatine bone** posteriorly. Here note the greater palatine foramen. Partially surrounding the palate is the **alveolar process** of the maxilla which carries the teeth. The remaining bones form the base of the skull and thus the floor of the cranial cavity. Behind the palate and the **posterior nasal apertures** lies the **body of the sphenoid bone** and its two **greater wings** which extend laterally. Extending inferiorly on either side of the body are two plates of bone, the **medial and lateral pterygoid plates**. The medial of these bears a hook-like process, the **hamulus**.

Separated from the body of the sphenoid by a fused or ill-defined suture is the **basal part** of the occipital bone. This is continuous, on either side of the **foramen magnum**, with the **condylar parts** of the bone. These bear the occipital condyles which articulate with the **lateral masses** of the **atlas** or first cervical vertebra. Posterior to the foramen magnum is the flattened **squamous portion** of the occipital bone which has already been seen from behind. Wedged between the greater wing of the sphenoid and condylar part of the occipital bone is the **petrous*** **portion of the temporal bone** containing the middle and inner ear.

Finally identify a bilateral series of foramina in the base of the cranial skeleton. The most anterior of these is the **foramen ovale** which is closely followed by the **foramen spinosum**. At the tip of the petrous part of the temporal bone is found the irregular **foramen lacerum** and in this bone but more laterally lies the rounded opening of the **carotid canal**. Very close to the canal but between the petrous part of the temporal bone and the occipital bone is the **jugular foramen** and nearby is the external opening of the **hypoglossal canal**. This canal cannot be seen in the figure but its course through the base of the occipital condyle is indicated by the arrow. Rather more laterally and between the base of the styloid process and the mastoid process the **stylomastoid foramen** can be seen. Each of these foramina will later be discussed in more detail with the important nerves and vessels which they transmit.

* Petrosus, *L.* = stony

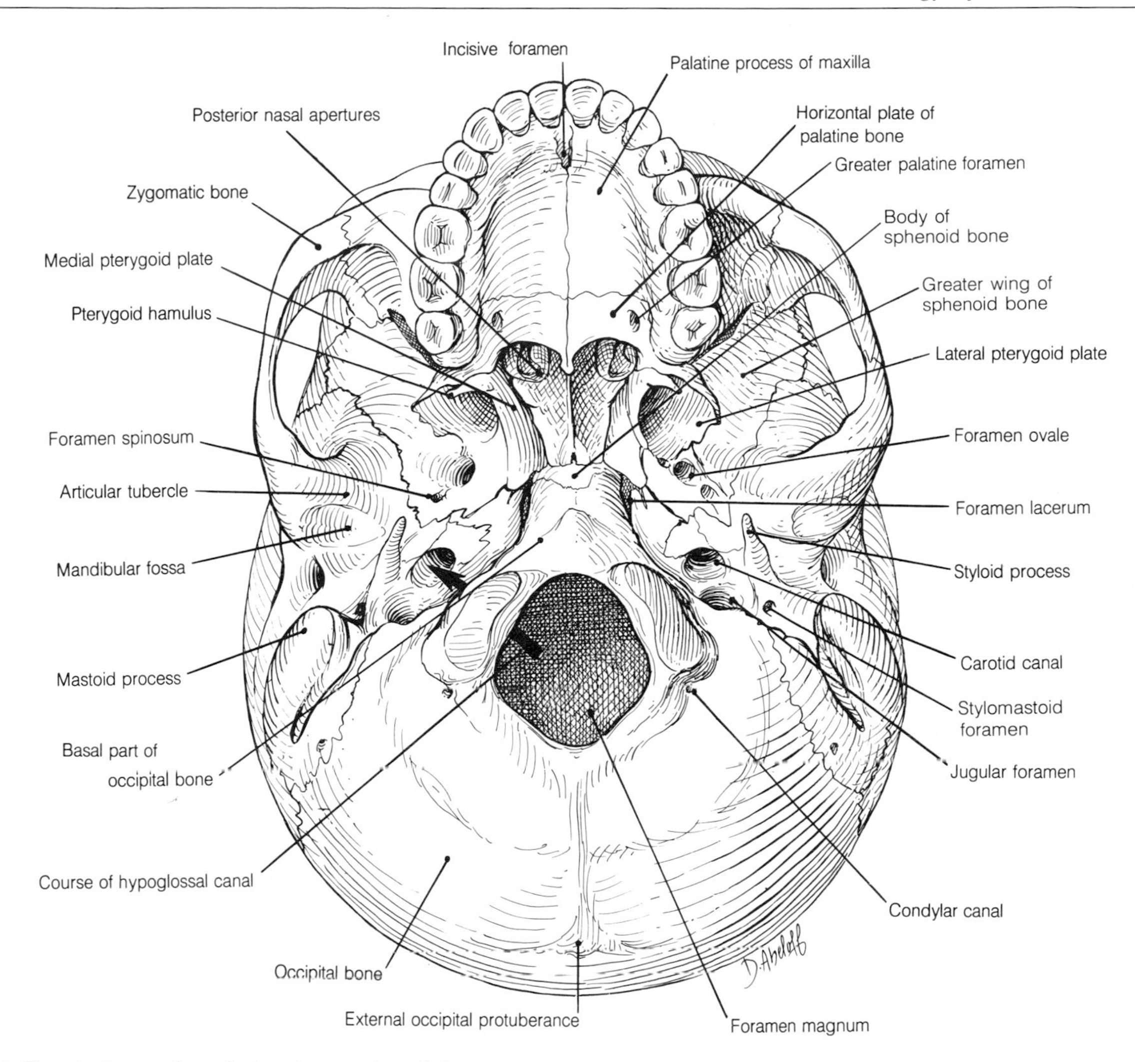

9-5 The skull seen from below (norma basalis).

The Development of the Skull

Development of the skull begins as a chondrification (cartilage formation) in condensations of mesenchyme in the future basioccipital and sphenoidal regions. This is followed by chondrification around the otocyst in the region of the future petrous portion of the temporal bone. To this mass of cartilage at the base of the skull is added a thinner membranous covering at the sides and roof.

The base of the skull ossifies in cartilage but, in the bones of the sides and roof, ossification occurs without prior chondrification, i. e., it is intramembranous. Most of the bones of the facial skeleton are also ossified in this manner. Centers of ossification appear between the second and fifth months of intrauterine life and at birth the majority of the bones of the skull are at least partly ossified. The ossified regions include the two parietal tuberosities which lie on the widest transverse diameter of the newborn skull, a dimension of importance during childbirth. At this time, also, the sagittal suture is not rigid and some overlapping of the parietal bones may occur without harm during passage of the head through the mother's pelvis.

At birth the skull shows some features which are distinctly different from those of the adult. Ossification of the membrane bones is not com-

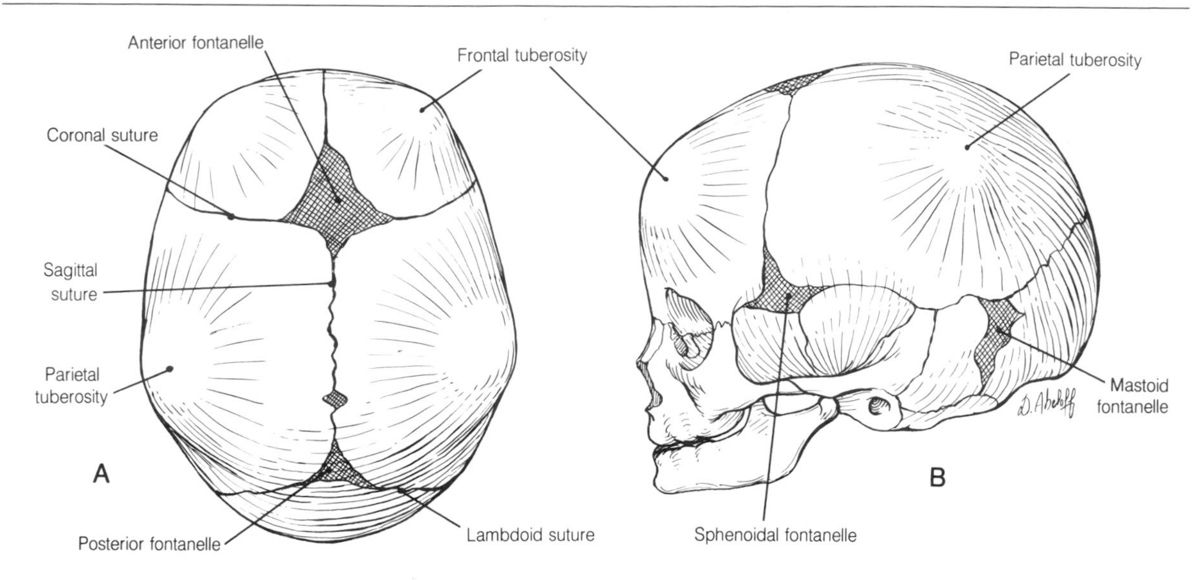

9-6 The skull of a newborn infant seen from above (*A*) and from the side (*B*).

plete and deficiencies remain that are covered only by a thin layer of fibrous tissue or membrane. These areas are known as the **fontanelles** and are seen in Fig. 9-6. Two of them, the **anterior and posterior fontanelles**, are widely patent at birth and are of clinical value to the pediatrician.

The anterior of these fontanelles lies at the bregma and the posterior at the lambda. The posterior fontanelle closes fairly soon after birth but the anterior remains patent well into the second year of life.

Pulsation of the brain can be felt through the fontanelles and they are obviously vulnerable areas. However, they also provide access to blood in the sagittal sinus and to the ventricles of the brain for purposes of investigation. At this site a practiced hand can also detect a rise in intracranial pressure or estimate the degree of hydration.

Another striking feature of the skull at birth is the disproportion between the cranial and facial skeleton which is clearly illustrated in Fig. 9-6B. Despite the fact that at birth the cranial capacity is about a quarter of that of the adult, the facial skeleton is proportionally even smaller. The proportions of facial to cranial skeleton are about 1:8 at birth, 1:5 at five years, and at maturity 1:2. These changes in proportion as age advances are largely due to the appearance of the dentition and growth of the air sinuses in the facial bones.

The mastoid process and tympanic plates are also poorly developed at birth leaving the facial nerve in a very exposed position and thus vulnerable to the inexpertly applied obstetrical forceps.

Despite its close association with the cartilage of the first or mandibular branchial arch **the mandible is ossified in membrane** which surrounds the cartilage. This ossification begins at a center near the mental foramen during the sixth week of fetal life and gradually surrounds and invades the cartilage. At birth the mandible is in two halves united by the **symphysis menti**. There is no development of alveolar bone which will bear the teeth and the ramus forms a large angle with the body (175°). The two halves of the mandible fuse after a year or so and with the eruption of the primary dentition and the development of alveolar bone in both upper and lower jaws, the angle between the ramus and body lessens to keep the teeth in apposition. In the adult this angle is reduced to about 120°.

As old age creeps on, the jaws may become edentulous and some return to the new-born state occurs. The alveolar bone is resorbed and the angle increases again. These alterations in the profile of the mandible are shown in Fig. 9-7 where the change in position of the mental foramen relative to the margin of the mandible can also be seen.

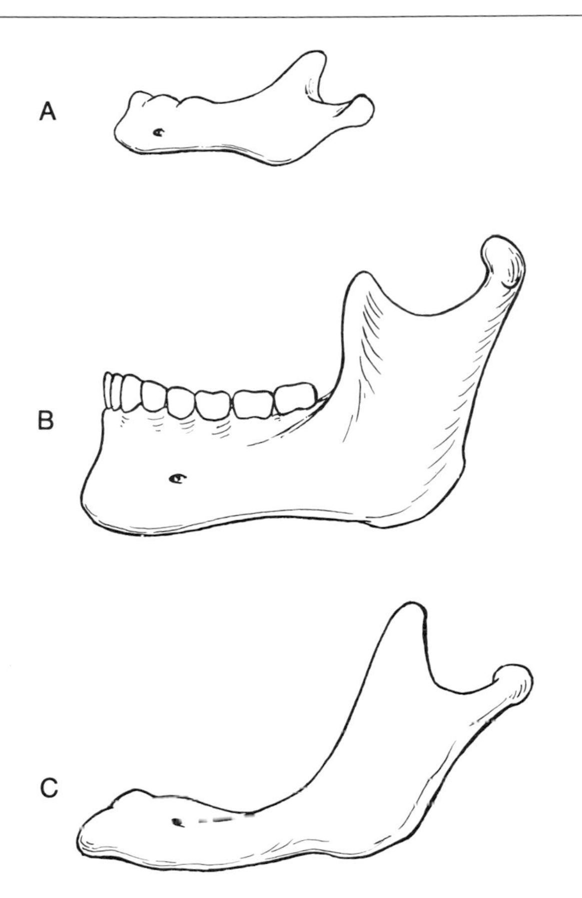

9-7 Showing the shape of the mandible at birth (*A*), at maturity (*B*), and in an aged edentulous individual (*C*).

The Face and Scalp

The observant physician can gain much information from the facial appearance or "facies" of his or her patient. This appearance may be due to the state of the capillary bed and its circulating blood, excessive amounts of tissue fluid (edema), pigmentation, or the functioning of the facial musculature. Both the face and scalp have a very rich blood supply derived from the large external carotid artery and the many anastomoses of its branches across the midline. Wounds of both regions bleed freely and repair well. Details of the many small muscles of the face are not important but those involved in the sphincters of the eye and mouth are. All these muscles are innervated by the seventh cranial or facial nerve and injuries to this nerve have serious consequences. Socially, a facial paralysis is a severe disability because an expressionless half of the face is paired with a grimace due to the unbalanced contraction of the unaffected muscles of the other half. Of more importance functionally is the inability to close the eye or blink and thus the loss of an essential protective mechanism. Paralysis of the sphincter of the mouth leads to difficulties in eating, drinking, speaking, and a failure to control the escape of saliva. Despite its name the facial nerve does not provide innervation for the skin of the face. This and the anterior part of the scalp are supplied in well defined areas by the three divisions of the trigeminal nerve, a fact of which considerable use is made in a neurological examination.

The Muscles of the Face and Scalp

The muscles of the human face allow a wide range of communicative expressions not displayed by other animals. They lie in the subcutaneous tissue and most are attached to the skin. The majority of these are illustrated and labeled in Fig. 9-8, but only four of them are described in any detail. It should be noted, however, that many of the small muscles converge on and blend with the lips, the most mobile and expressive features of the face.

Orbicularis oculi

This is an extensive flat sheet of muscle surrounding the **palpebral* fissure**. An outer orbital part is attached medially to the maxilla and frontal bone and its fibers loop around the orbit. Its use allows the eyelids to be closed tightly. An inner palpebral part lies in the upper and lower lid. These fibers are attached to the **medial palpebral ligament** and to a **lateral palpebral raphe**. They allow the lids to be closed gently or to blink. (The upper lids are raised by the **levator palpebrae superioris**, a muscle which will be described with the orbit.) A small slip of orbicularis oculi also passes deep to the **lacrimal sac**. These fibers are thought to dilate the lacrimal sac and enhance the drainage of tears.

* Palpebra, *L.* = eyelid

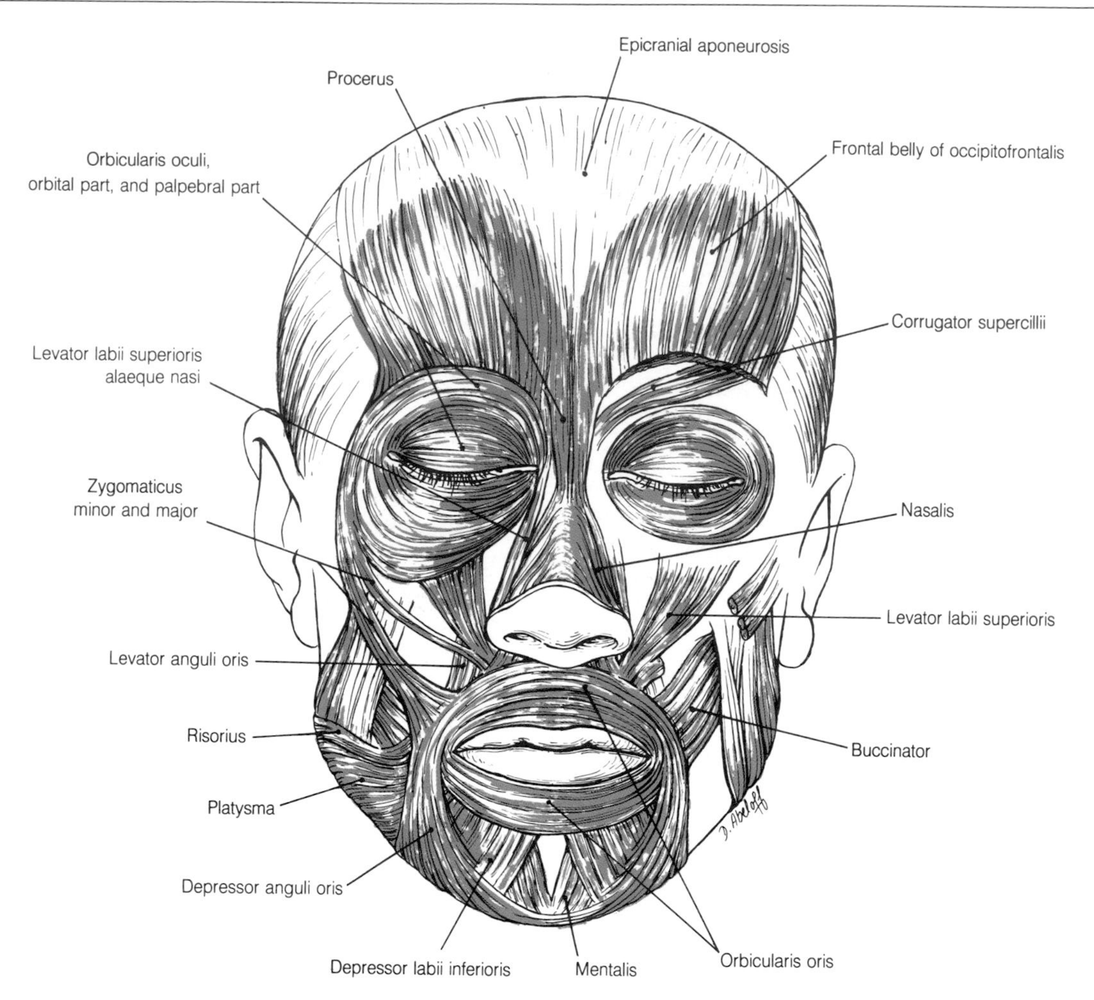

9-8 The muscles of facial expression.

Orbicularis oris

The details of the arrangement of orbicularis oris are best understood from Fig. 9-8 and 9-9. Lying in the upper and lower lip and being attached to bone in the midline, the fibers of this muscle sweep around the mouth and provide for its closure. Additional fibers related to elevation and depression intermingle with the sphincter at the angles of the mouth and when acting together dilate the mouth. This action is, of course, normally coupled with depression of the lower jaw. Also fusing with the orbicularis oris and the angle of the mouth is the **buccinator muscle**.

Buccinator

Buccinator* lies at a deeper level than other facial muscles. It is a flat sheet of horizontally running fibers lying in the cheek deep to the mucous membrane of the oral cavity. Posteriorly it is attached to the **pterygomandibular raphe** and portions of the maxilla and mandible adjacent to the upper and lower third molar teeth. Its fibers extend forward to blend with those of orbicularis oris as is seen in Fig. 9-9. While it controls the cheeks in blowing and sucking and is, therefore, a true facial muscle, it also serves to empty the gutter between the teeth and cheek during mastication.

---

* Buccinator, *L.* = trumpeter

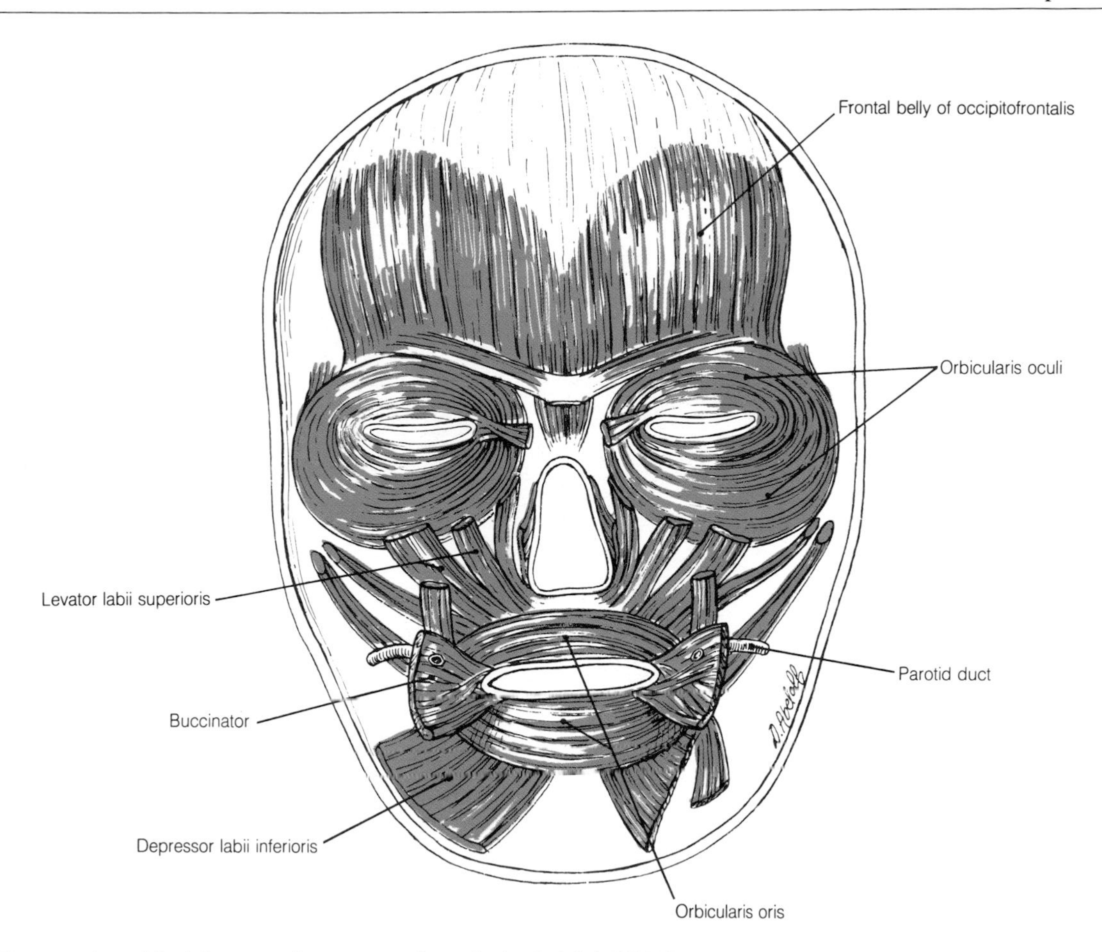

9-9 The muscles of facial expression as seen from their skeletal aspect.

Occipitofrontalis

The **occipital belly** of this muscle arises as a flat quadrilateral sheet from the occipital bone and mastoid part of the temporal bone. A similar **frontal belly** arises from the complex of muscles over the eyebrows and bridge of the nose. The importance of these two parts lies in the sheet of aponeurotic tendon that joins them. This, the **epicranial aponeurosis**, forms a layer of the scalp which will be described shortly.

The Platysma

One superficial sheet of muscle, the platysma, extends from the upper part of the thorax, over the clavicles, and through the **superficial fascia** of the neck to blend with the skin and muscles near the margin of the mandible. Its action is typically associated with the expression of horror or exhaustion. The muscle forms a distinct layer in the neck and can be used in the closure of surgical incisions to improve the cosmetic result.

Each of these muscles is supplied by the **facial nerve**.

The Facial Nerve in the Face

The facial or seventh cranial nerve leaves the base of the skull through the **stylomastoid foramen** deep to the **parotid gland** where this is wedged between the mastoid process and the ramus of the

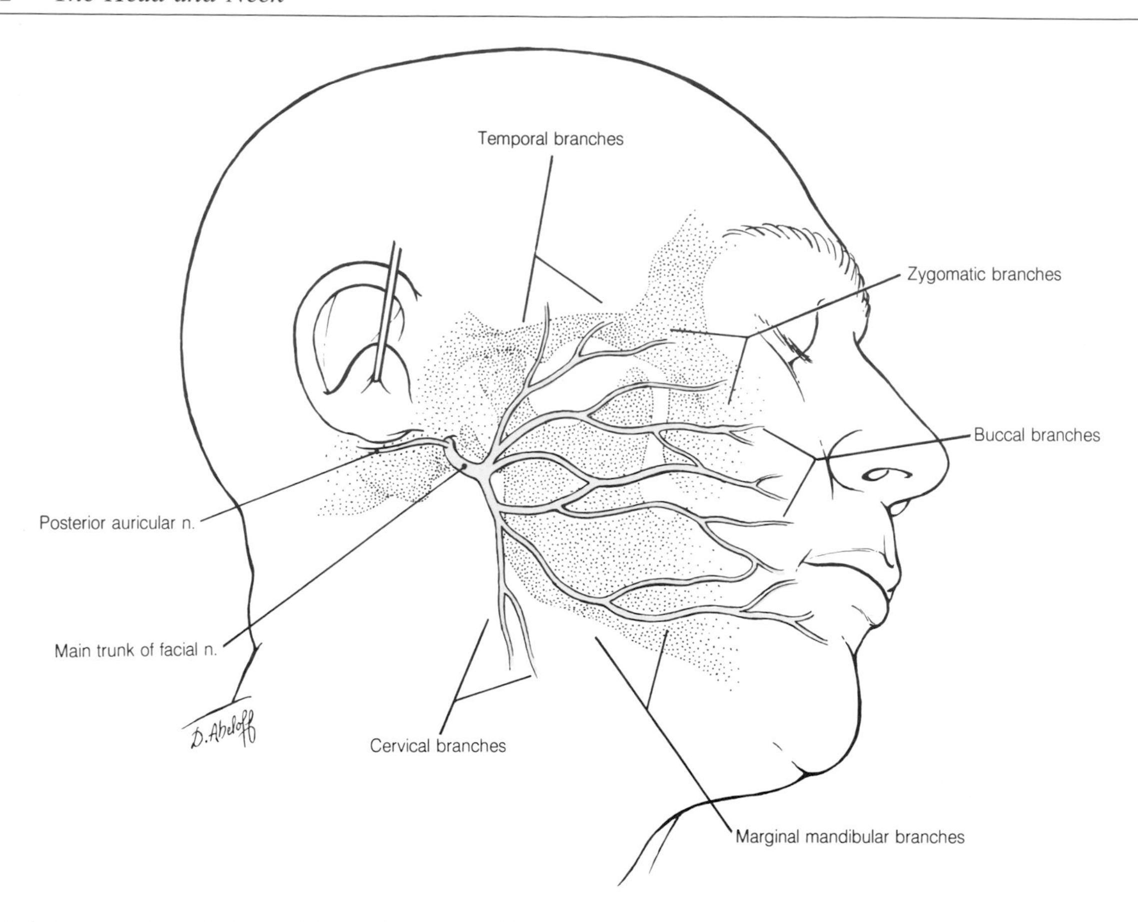

9-10 Showing the distribution of the facial nerve to the face and neck.

mandible. Before entering the gland it gives off the **posterior auricular nerve** whose branches supply the **occipital belly of occipitofrontalis**, the **stylohyoid muscle**, and the **posterior belly of the digastric muscle.** The nerve then enters the parotid gland and breaks up into five groups of branches which fan out like the digits of the hand to supply the remaining muscles of the face. These branches are named **temporal, zygomatic, buccal, marginal mandibular**, and **cervical** after the regions toward which they are directed. Their course and names are summarized in Fig. 9-10. Of these it is the zygomatic, buccal, and mandibular branches that supply the muscles around the eye and mouth and great care must be taken not to damage them during surgical procedures near the parotid gland.

The Skin of the Face and the Scalp

The Skin of the Face

The skin of the face is highly sensitive and well represented in the sensorimotor areas of the cerebral cortex. It is also very vascular and its color may reflect the emotions, the body temperature, and the degree of oxygenation and quantity of circulating hemoglobin.

The sensory innervation of the face is provided by the **trigeminal or fifth cranial nerve.** This nerve leaves the skull in three divisions through three separate foramina. The three divisions supply three distinct areas of skin by cutaneous branches which appear on the face at further foramina. This complication must be temporarily accepted until the deeper course of these branches is described. The three divisions are named the **ophthalmic nerve, the maxillary nerve** and the **mandibular nerve**. Their names indicate the regions of the face which they supply and these regions with the terminal cutaneous branches that

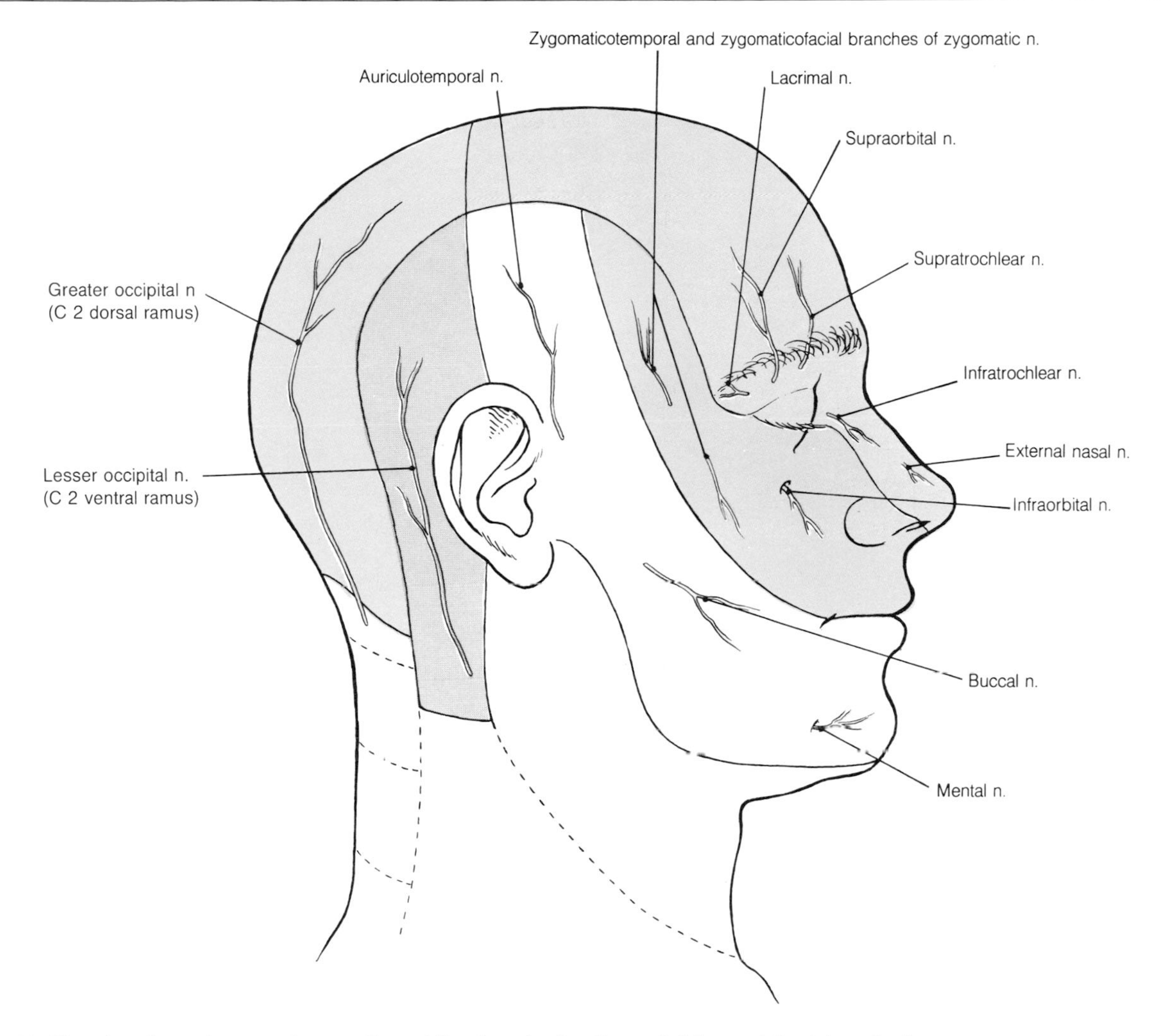

9-11 Showing the cutaneous innervation of the face by the three divisions of the trigeminal nerve.

supply them are seen in Fig. 9-11. Using the listing that follows, identify each of these branches and the division of the trigeminal nerve from which they are derived.

Division	Cutaneous Branch
Ophthalmic nerve	Supraorbital nerve
	Supratrochlear nerve
	Lacrimal nerve
	Infratrochlear nerve
	External nasal nerve
Maxillary nerve	Infraorbital nerve
	Zygomaticotemporal branches
	Zygomaticofacial branches
Mandibular nerve	Auriculotemporal nerve
	Buccal nerve
	Mental nerve

Note that the cornea and conjuctiva of the eyeball are supplied by branches of the ophthalmic nerve arising in the orbit. The lower lid, the lateral part of the nose, and the upper lip fall into the territory of the maxillary nerve. The skin over the angle of the mandible, however, falls outside the territory of the mandibular nerve and is supplied by ventral rami of cervical spinal nerves.

The Scalp

The skin of the scalp is normally covered by coarse hairs. Beneath the skin lies the fatty superficial fascia. It is very vascular and like that of the palm and sole is broken up into fatty loculi by fibrous septa. These septa anchor the skin to the underlying frontalis and occipitalis muscles and the epicranial aponeurosis which joins them. Beneath the aponeurosis is a potential space filled with loose

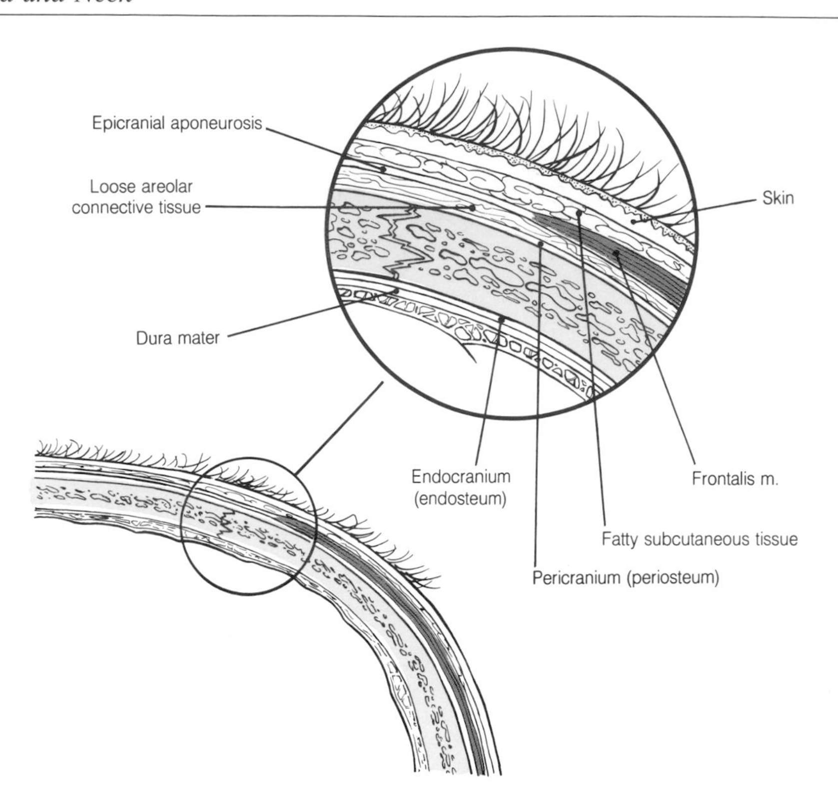

9-12 Showing the layers of tissue comprising the scalp.

areolar tissue. Bleeding into this space often becomes manifest as a bilateral black eye as the blood tracks forward beneath frontalis and orbicularis oculi and into the upper lid. The bones of the skull are covered by periosteum, which in this region is known as the pericranium. These features of the scalp are summarized in Fig. 9-12.

Refer again to Fig. 9-11 and note that the anterior part of the scalp is supplied by the **supratrochlear** and **supraorbital nerves** and the skin of the temporal region by **zygomaticotemporal branches** of the maxillary nerve and by the **auriculotemporal branch** of the mandibular nerve. The skin of the posterior part of the scalp is supplied by **cervical spinal nerves**. The **greater occipital nerve**, derived from the **dorsal** ramus of the second cervical nerve, ascends from the suboccipital triangle to supply the scalp up to the vertex. The **lesser occipital nerve** derived from the **ventral** ramus of the second cervical nerve supplies the region behind the ear.

The Blood Supply of the Face and Scalp

The blood supply of the face and scalp is largely provided by branches of the **external carotid artery.** However, branches of the **ophthalmic branch of the internal carotid artery** supply the forehead, scalp, upper lid, and nose.

The **facial artery**, a branch of the external carotid artery, enters the face by curling around the body of the mandible just anterior to the masseter muscle. From this point it runs a tortuous course past the angle of the mouth to the medial angle of the palpebral fissure of the eye. Near the angle of the mouth it gives off **a superior and an inferior labial branch** which communicate directly with those of the opposite side across the midline. As a result, deep wounds or incisions of the lip bleed copiously from either side.

The **superficial temporal artery**, a terminal branch of the external carotid artery supplies the temple and scalp and by the **transverse facial ar-**

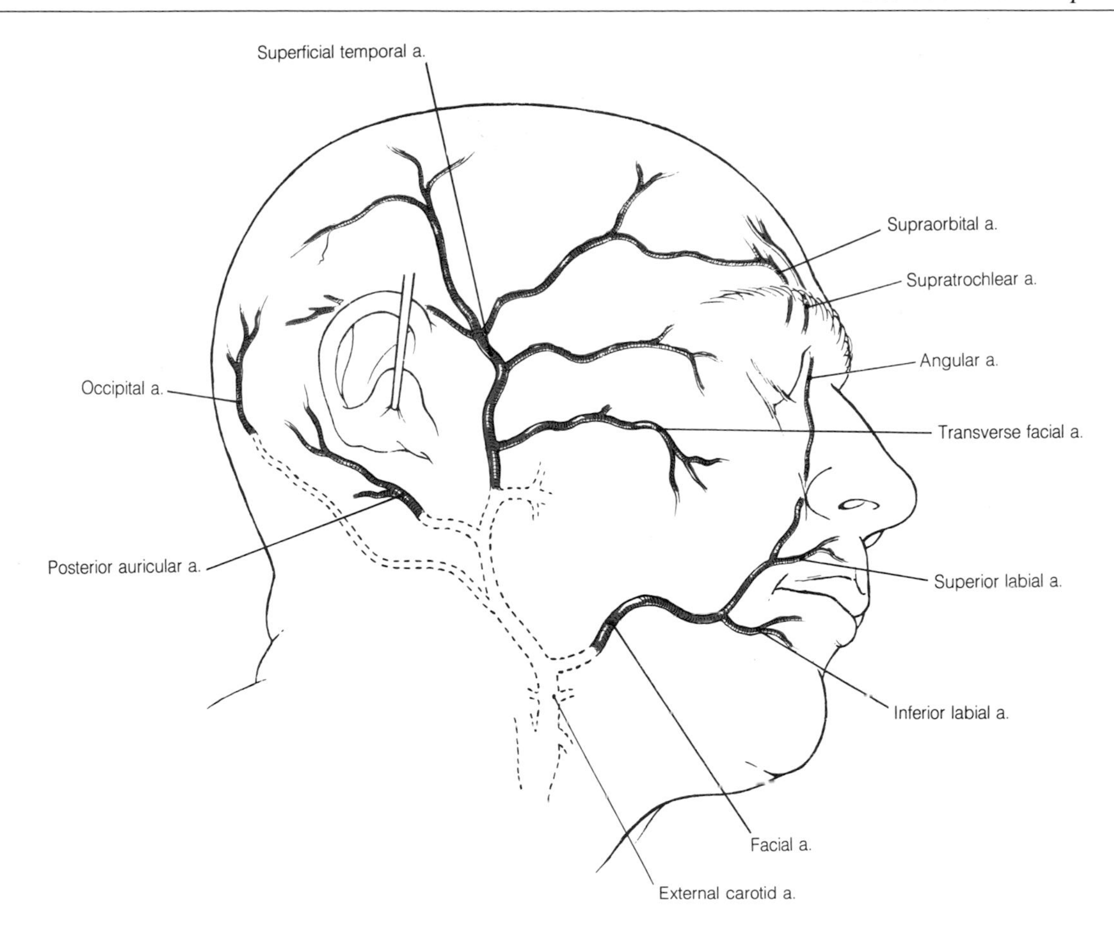

9-13 Showing the arterial supply of the face and scalp.

tery, the cheek. The forehead and anterior part of the scalp are supplied by the **supraorbital and supratrochlear arteries**, both branches of the ophthalmic artery which is itself a branch of the internal carotid artery. The posterior part of the scalp is supplied by two further branches of the external carotid artery. These are the **posterior auricular** and **occipital arteries**. The course and distribution of these vessels are illustrated in Fig. 9-13.

These vessels supplying the face and scalp communicate freely with each other across the midline and are thus able to establish an effective collateral circulation following obstruction or ligation of one external carotid artery.

Blood from the face and scalp is drained by veins accompanying the arteries. However, their important communications and their termination must await the description of deeper structures.

The Lymphatic Drainage of the Face and Scalp

The face and scalp are surrounded by a ring of regional lymph nodes into which the adjacent superficial tissues drain. The nodes of this ring in turn drain into the deep cervical lymph nodes that are distributed over the course of the internal jugular vein. The regional nodes are illustrated in

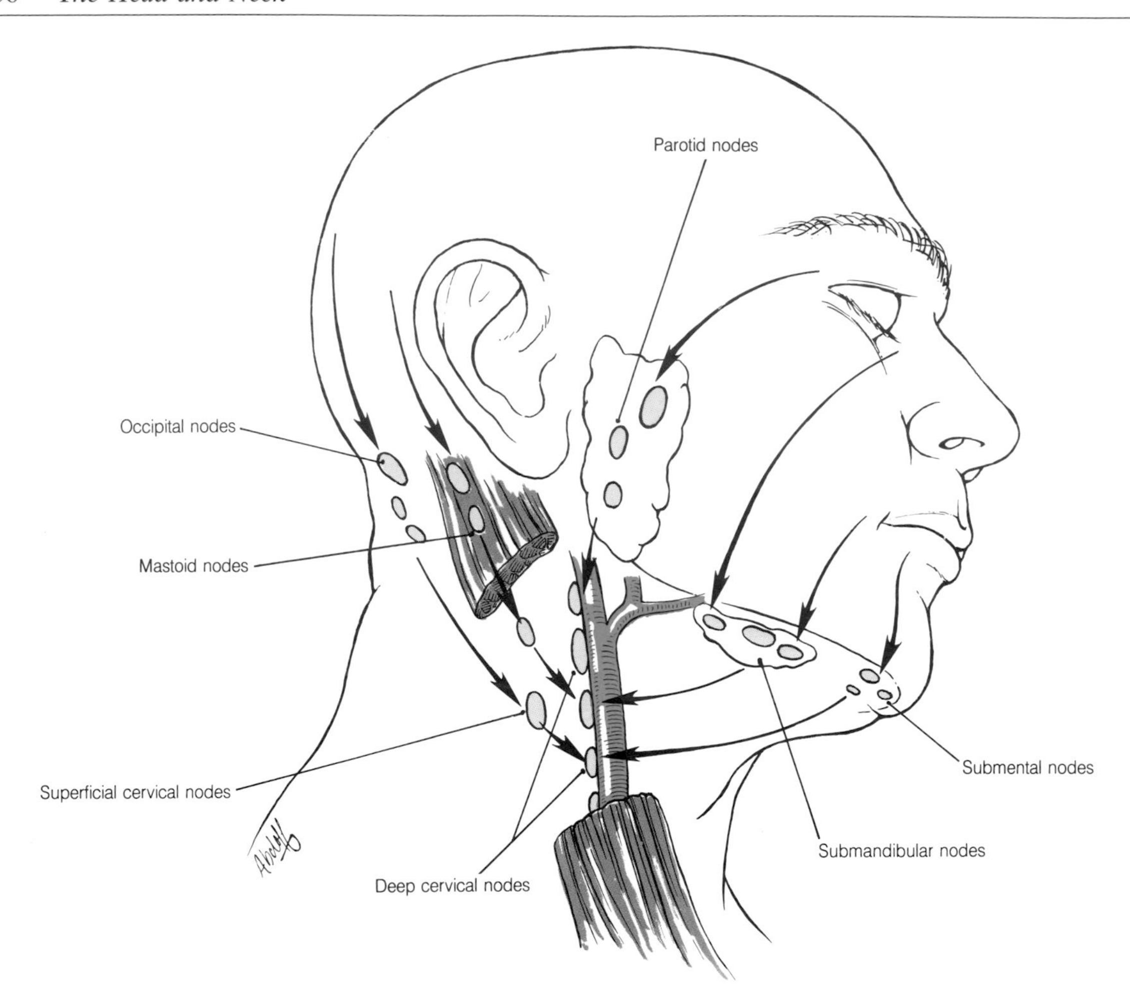

9-14 Showing the lymphatic drainage of the face and scalp.

Fig. 9-14 where the following groups can be identified:

Occipital
Mastoid
Superficial cervical on the surface of the sternocleidomastoid muscle
Parotid
Submandibular
Submental

Vessels from the scalp, ear, and temporal region drain into the occipital, mastoid, superficial cervical, and parotid nodes. The lids drain laterally to the parotid nodes and medially to the submandibular nodes. The remainder of the face drains to the submandibular nodes except the central part of the lower lip which drains to the submental nodes.

The Eyelids, Conjunctiva, and Lacrimal Apparatus

Although the corneal epithelium is similar in many respects to epidermis its superficial cells are alive and contain nuclei and no keratin. Because of this it must be kept protected and continually moist. In addition it is, like the cornea itself, translucent. This property so essential for function may be lost if infection or trauma breaches the epithelium. The importance of the sensory innervation of the cornea has already been mentioned. This innervation forms part of the protective corneal reflex. The effector pathway of this reflex is the innervation of the orbicularis oculi muscle which rapidly closes the lids, i. e., the eye blinks. The cornea is kept moist by the secretions of the lacrimal apparatus of which the conjunctival sac and lids form an important part.

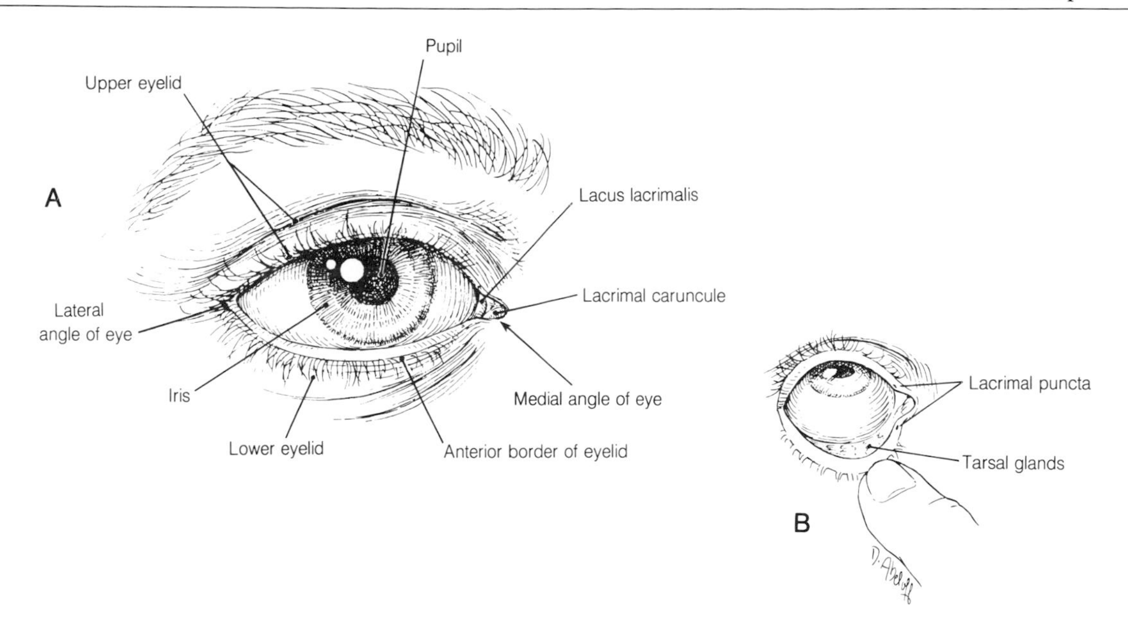

9-15 Showing the eyelids, the palpebral fissure, and the exposed portion of the eyeball.

The Eyelids

The eyelids or **palpebrae** are two soft mobile folds of skin lying in front of the eye and separated by a transverse gap called the **palpebral fissure**. The upper lid may be independently raised by the **levator palpebrae superioris**. Both may be closed by the **orbicularis oculi muscle**. In Fig. 9-15 note the following features:

1. The triangular space at the wider medial angle of the eye – this is the **lacus lacrimalis** and contains an elevation called the **caruncula lacrimalis**;
2. On the margin of each lid, above and below the base of this triangle, lies a minute orifice called a **lacrimal punctum** which can be seen more clearly in Fig. 9-15B;
3. The eyelashes, which are attached to the outer edge of the lid margin. Although not visible, the lashes are associated with the openings of modified sweat or **ciliary glands**;
4. Little or no **sclera** shows between the lower lid and the iris;
5. The upper lid covers about half the width of the iris and the lower lid lies just below it. Changes in these relationships produce the clinical signs of drooping of the upper lid or ptosis if the lid's elevator is paralysed or proptosis if for some reason the eyeball protrudes forward.

Look now at Fig. 9-16 and see that beneath the skin of the upper lid lies a layer of loose areolar tissue which may easily become edematous. Deep to this lie the fibers of the orbicularis oculi which are separated by a further layer of areolar tissue from the tarsi. It is into this loose tissue that blood may spread from the subaponeurotic space of the scalp producing the black eye often associated with severe blows to the head. The **tarsi** are two thin plates of fibrous tissue which provide support for the lids. They are attached to the margins of

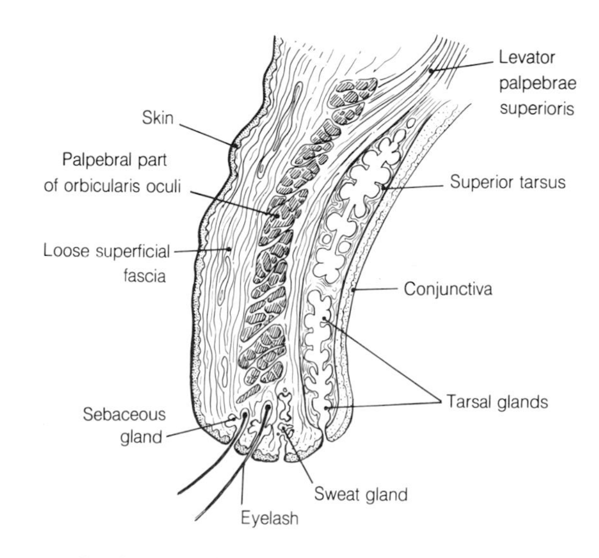

9-16 Semischematic section of the upper lid.

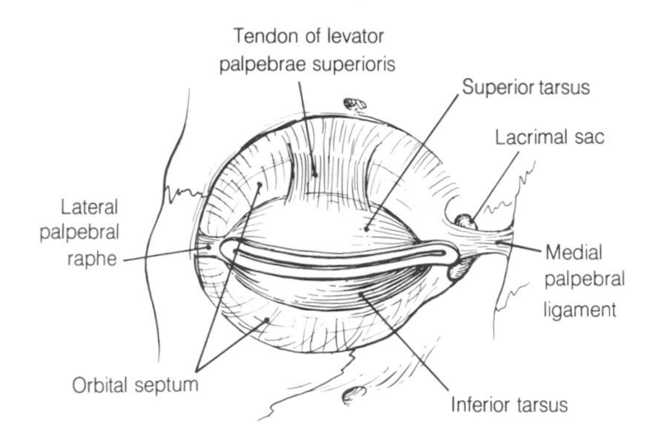

9-17 Showing the orbital septum, the tarsi, and the palpebral ligaments.

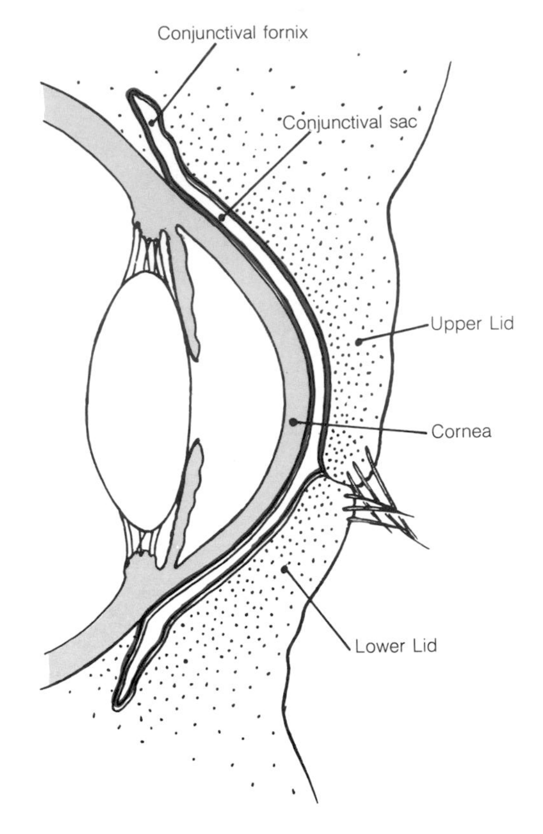

9-18 Diagrammatic sagittal section through the conjunctival sac and fornices.

the orbit by a thin fascial sheet called the **orbital septum**. This is thickened to form the **medial palpebral ligament** and the **lateral palpebral raphe**. To the superior tarsus is attached the levator palpebrae superioris. The arrangement of the tarsi and septum can be seen in Fig. 9-17.

Embedded in the tarsi and often visible through the conjuctiva of the everted lid are the **tarsal glands**. These open onto the lid margin.

The deep surfaces of the lids are covered by conjunctiva.

The Conjunctiva

The inner surface of the lids is covered by a mucous membrane called the **conjunctiva**. Above and below, the conjunctiva is reflected off each lid onto the sclera of the eyeball. The angles formed by these reflections are known as the **conjunctival fornices** and, as can be seen in Fig. 9-18, when the lids are closed a complete **conjunctival sac** is formed. At the margin of the cornea, often called the **limbus**, the conjunctiva becomes continuous with the **corneal epithelium**.

The Lacrimal Apparatus

Strictly speaking the lacrimal apparatus consists of the **lacrimal gland** which secretes the tears, and the ducts by which they are drained to the nasal cavity. However, to the secretion of the lacrimal gland is added those of **mucous glands** in the conjunctiva and the **tarsal glands**. A film of secretion is continually distributed over the exposed conjunctiva and corneal epithelium by the periodic blinking of the lids. Tears are formed and drained away continuously and only when in excess, as in crying, do they spill over the lids. The importance of the eyelids in spreading the tears is well seen in facial nerve lesions where the orbicularis oculi is paralyzed. In these, if special precautions are not taken to protect the cornea, it becomes damaged and scarred.

The lacrimal gland has a larger **orbital part** which is lodged in a fossa just inside the upper lateral angle of the orbital margin. The orbital part is continuous with a smaller **palpebral part** which lies behind the upper lid. The gland is connected to the superior conjunctival fornix by about a dozen ducts.

The tears drain downward over the eye and on reaching the lower lid margin are forced by blinking toward the **lacus lacrimalis**. Here they are drawn through the **lacrimal puncta** into the **lacrimal canaliculi** and, thence, to the **lacrimal sac**. The lacrimal sac lies in a hollow formed by the lacrimal and maxillary bones at the lower inner margin of the orbit. It communicates by means of

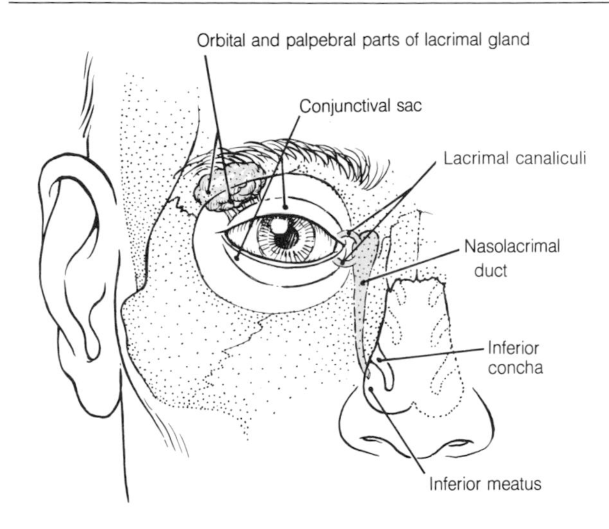

9-19 Semischematic illustration of the lacrimal apparatus.

the **nasolacrimal duct** with the **inferior meatus of the nasal cavity**. In Fig. 9-19 the lacrimal apparatus can be seen in relation to the lids, the orbit, and the nasal cavity.

Postganglionic secretomotor fibers innervating the lacrimal gland have their cell bodies in the **pterygopalatine ganglion**. These fibers pass to the maxillary nerve and its zygomatic branch. In the orbit they are transferred to the lacrimal nerve and carried in this to the lacrimal gland. The details of this innervation are given more fully in a discussion of the pterygopalatine ganglion.

The Cranial Cavity

The cranial cavity contains the brain and its surrounding coverings, the meninges. Although a detailed study of the brain is carried out in a neuroscience course, some idea of the disposition of its parts is necessary in order to appreciate the significance of the cranial nerves that leave it, its blood supply and venous drainage, and its very close association with the internal aspect of the bony cavity within which it lies.

The brain develops as three swellings at the proximal end of the neural tube of the embryo (the longer distal portion of the neural tube gives rise to the spinal cord). These primary vesicles develop into the **forebrain, midbrain**, and **hindbrain**. From the forebrain develop large bilateral secondary vesicles, the **cerebral hemispheres**. In the adult brain these dwarf in size the remaining portions and occupy a very large part of the cranial cavity. The two hemispheres are separated by a deep **longitudinal fissure** and in the depths of this they are joined together by a thick band of communicating nerve fibers (a commissure) called the **corpus callosum**. The remainder of the forebrain and the midbrain are hidden by the hemispheres. The midbrain is short in extent. It is characterized anteriorly by two stout pillars leading from the forebrain to the hindbrain. These are

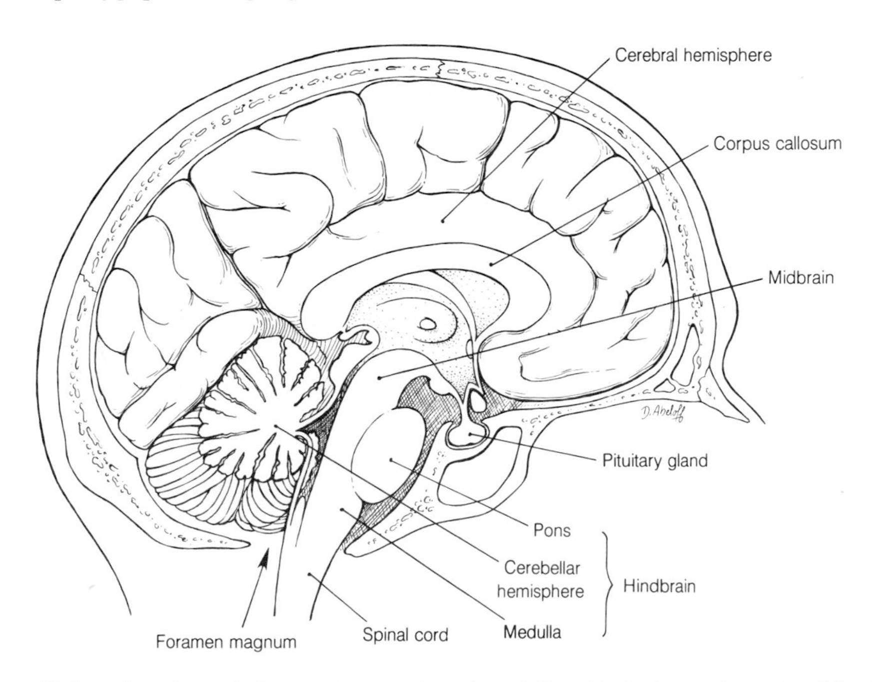

9-20 A median sagittal section through the cranium to show its relationship to the main parts of the brain.

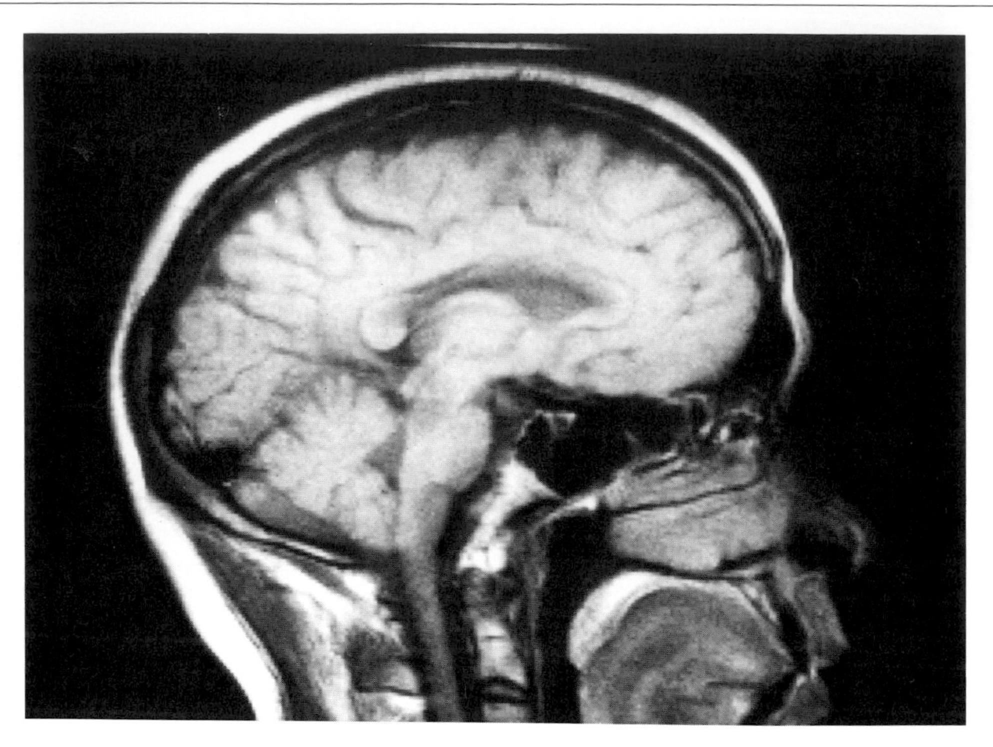

9-21 A magnetic resonance image scan of the median plane of the cranial cavity. (Photograph courtesy of Technicare Corporation.)

called the **cerebral peduncles**. Posteriorly, the midbrain shows four small elevations called the **superior and inferior colliculi** (little hills). The hindbrain which consists of the **pons** and **medulla** with the **cerebellar hemispheres** joined to their posterior aspect, appears below the cerebral hemispheres. The medulla is continuous with the **spinal cord** at the foramen magnum. The manner in which these parts of the brain are related to the cranial cavity that encloses them is illustrated in Fig. 9-20. The drawing in Fig. 9-20 was made from a median section of the skull and brain of a cadaver. It should be compared with the remarkable image seen in Fig. 9-21 which is an MRI scan of the head of a living subject, also made in the median plane.

The cavity of the neural tube is retained in the adult central nervous system. It is represented by the fine **central canal** of the spinal cord and, within the cerebral hemispheres, the remainder of the forebrain, and in the hindbrain, the cavity is dilated to form the **ventricles**. These are continuous with each other and the central canal. Within the ventricles are the **choroid plexuses**. At these sites the lining of the ventricles (ependyma) and the overlying vascular pia are specialized for the formation of **cerebrospinal fluid.** The fluid leaves the ventricular system through small apertures in the roof of the ventricle of the hindbrain and comes to circulate in the subarachnoid space. It is absorbed by further specialized regions called arachnoid granulations where arachnoid mater lies adjacent to the endothelial lining of venous sinuses.

To examine the interior of the cranium it is usual to remove the skull cap or **calva** (literally that area of the scalp that becomes bald) and this will be considered first. The remainder of the cranium or cranial base forms the three cranial fossae.

The outer surface of the cranial bones has been seen to be covered by periosteum, here called **pericranium**. The inner surface is similarly clothed by the **endocranium**. These two layers become continuous with each other at foramina that penetrate the bones and are also continuous with the sutural ligaments between the bones.

When the cut edges are examined after the skull cap is removed, the cranial bones are seen to be made up of two thin layers of compact bone separated by an irregular layer of cancellous bone. The compact bone forms the **inner and outer tables of the skull** and the cancellous bone which contains red bone marrow is called the **diploë**.

The Interior of the Skull Cap

The main features of the interior of the skull cap are seen in Fig. 9-22. Identify the inner aspects of the frontal, parietal, and occipital bones and the sutures that join them. Branching grooves are found particularly over the parietal bones. In life these house meningeal blood vessels. In the midline a shallow **sagittal groove** is related to the **superior sagittal venous sinus** and on either side of this further depressions are shaped around the **lacunae laterales** and more irregular excavations around **arachnoid granulations**. The latter structures are concerned with the reabsorption of cerebrospinal fluid and are described later. Anteriorly a spur of bone is seen projecting from the frontal bone. This ist the **frontal crest** which gives attachment to the falx cerebri, a partition of dura mater which separates the two cerebral hemispheres.

The Cranial Fossae

The base of the skull is arranged in a series of three step-like fossae that accommodate the parts of the brain that have been described. These fossae are perforated by either foramina or fissures (usually spaces between bones) through which pass cranial nerves, arteries, and venous structures. Examine Figs. 9-23 and 9-24 as the following descriptions are read.

The Anterior Cranial Fossa

The walls of the anterior fossa are formed by the frontal bone. The floor is largely formed by the **orbital plate** of the frontal bone (orbital because it forms the roof of the orbit). Also taking part is the

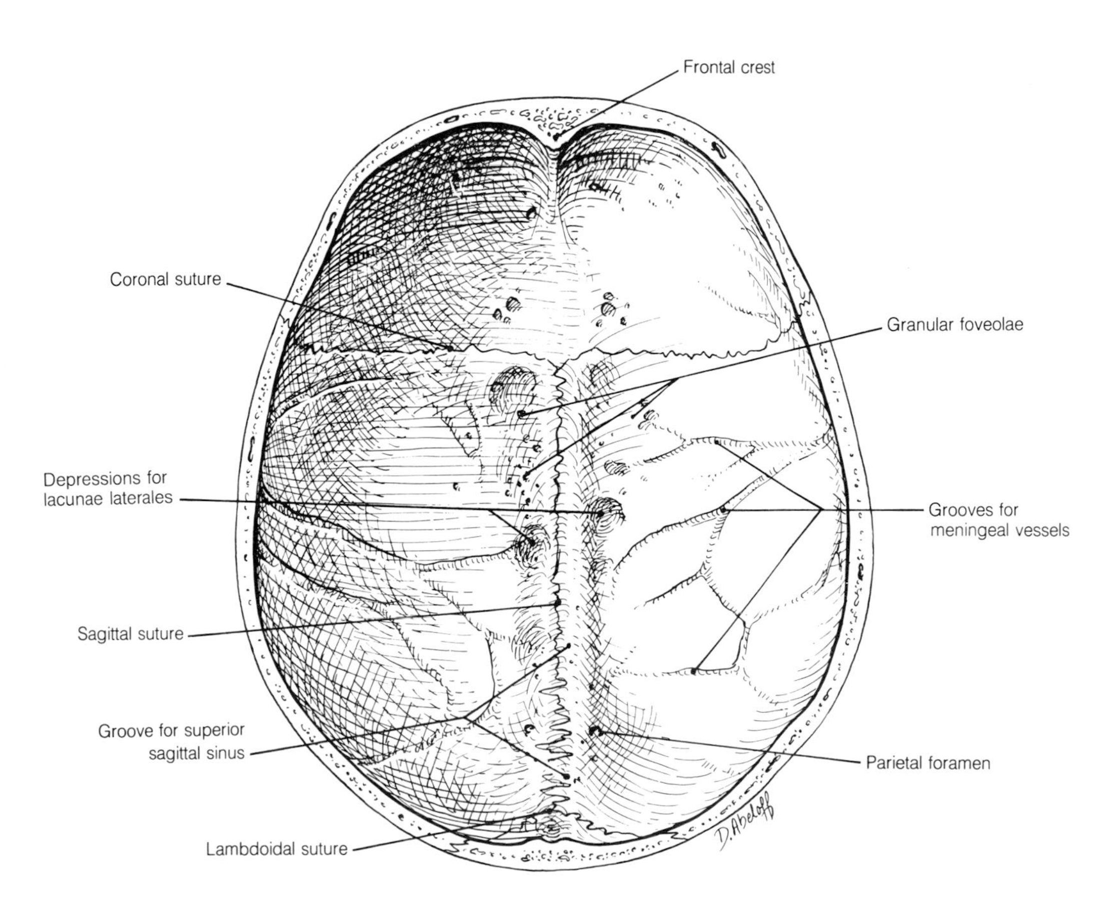

9-22 The interior of the skull cap.

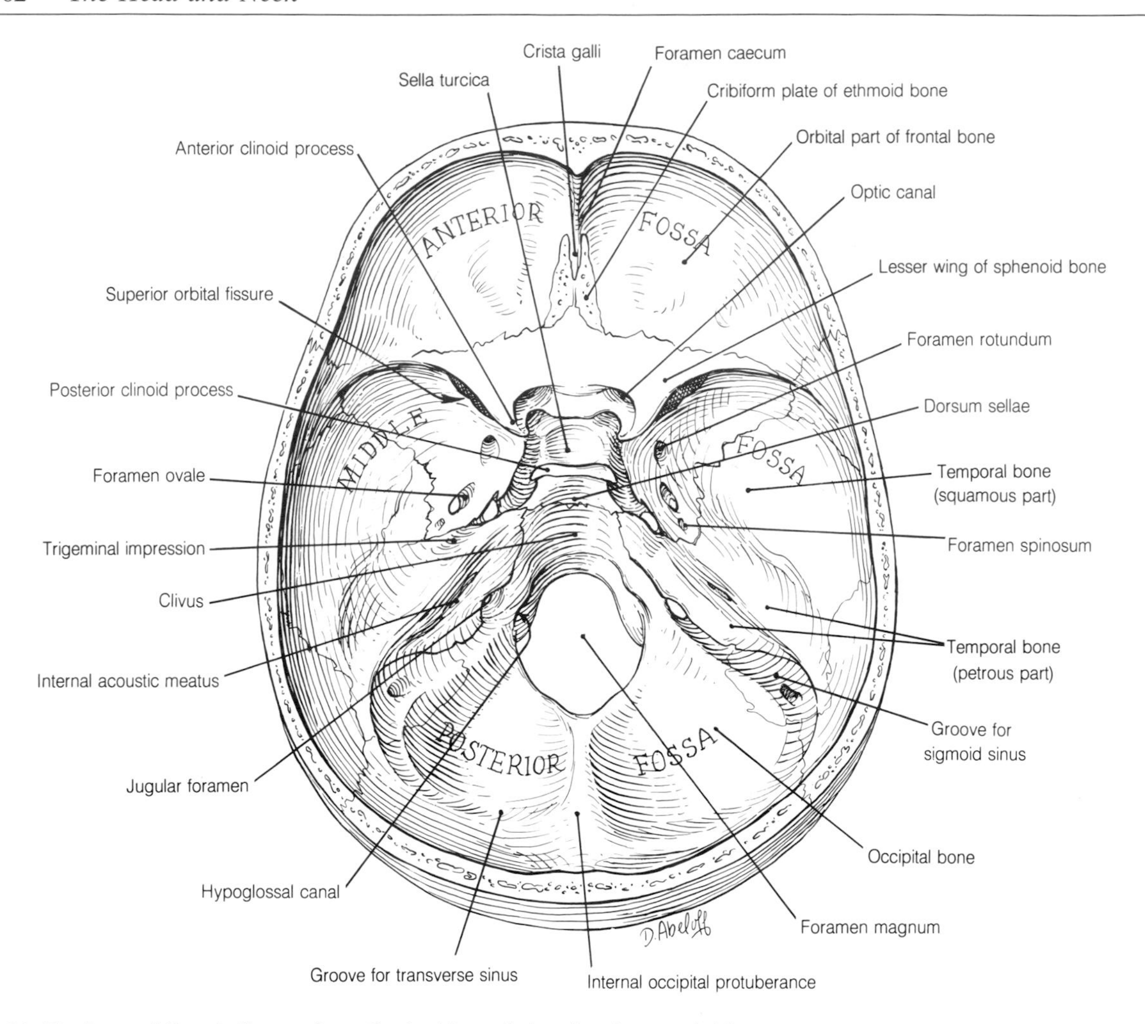

9-23 The base of the skull seen from the inside and showing the cranial fossae.

cribriform* **plate** of the **ethmoid*** **bone**. The perforations in the plate transmit the olfactory nerves and a small slit transmits the anterior ethmoidal nerve. The **crista galli*** projects upward from the plate. The posterior part of the floor is formed by the **lesser wings of the sphenoid bone** linked by **the jugum**.* The fossa houses the frontal lobes of the cerebral hemispheres and the olfactory bulbs which lie on the cribriform plate.

* Cribrum, *L.* = a sieve; Ethmos, *Gk.* = a sieve; Crista galli, *L.* = a cock's comb; Jugum, *L.* = a yoke, i.e., it joins the lesser wings of the sphenoid bone

The Middle Cranial Fossa

This is really a pair of fossae lying a step below the anterior fossa as is shown in Fig. 9-23. The two fossae are separated in the midline by the **body** of the sphenoid bone. Each fossa is bounded anteriorly by the **lesser wing of the sphenoid**, laterally by the **squamous part** of the temporal bone and posteriorly by a fairly well defined ridge on the superior margin of the **petrous part** of the temporal bone. The floor is formed by the **greater wing** of the sphenoid and by the squamous and petrous parts of the temporal bone.

The temporal lobes of the cerebral hemispheres nestle in the two fossae laterally, the floor of the forebrain, the optic chiasma, the termination of the internal carotid arteries, and the pituitary gland are related to the body of the sphenoid bone. The upper surface of the body is concavoconvex and is known as the **sella turcica***. A further midline depression in the bone is the **hypophyseal fossa**.

Additional features to note in Fig. 9-23 are:

1. The **anterior and posterior clinoid* processes** at the four corners of the sella turcica;
2. The **optic canal** surrounded by the two roots of the lesser wing of the sphenoid bone;
3. The **superior orbital fissure** between the lesser and greater wings of the sphenoid;
4. The foramina at the root of the greater wing of the sphenoid—these are the **foramina rotundum, ovale**, and **spinosum**;
5. A groove (or sometimes in part a canal) that extends from the foramen spinosum across the floor of the middle fossa and ascends its lateral wall where it divides into two—these grooves house the **middle meningeal vessels** and their frontal and parietal branches;
6. The **foramen lacerum*** between the tip of the petrous temporal bone and the body of the sphenoid. Opening into this foramen is the **carotid canal** which carries the internal carotid artery, however, the foramen itself transmits no major structure through its whole length and is partially obliterated by cartilage;
7. A shallow impression near the apex of the petrous temporal bone that houses the **trigeminal ganglion**.

The Posterior Cranial Fossa

The posterior fossa is bounded anteriorly by the petrous parts of the temporal bones and the posterior surface of the body of the sphenoid which is known as the **dorsum sellae.** Its walls are formed partly by the petrous parts of the temporal bones and partly by the **occipital bone** which also forms the floor. In life the fossa is almost completely roofed by a double layer of dura mater called the **tentorium cerebelli.** An anterior opening in this allows the passage of the midbrain. The fossa contains the pons, the medulla, and the cerebellum. In addition note in Fig. 9-23:

1. The large oval **foramen magnum** in the midline;
2. A short canal in the anterior margin of the foramen magnum—the **hypoglossal canal**;
3. An S-shaped sulcus at the adjacent margins of the petrous temporal and occipital bones which accommodates the **sigmoid* sinus** and extends downward to the **jugular foramen**;
4. A groove for the **transverse sinus** traveling posteriorly from the sigmoid sulcus to the **internal occipital protuberance**—the tentorium cerebelli is attached to its margins;
5. That the groove for the right transverse sinus is usually continuous superiorly with the groove for the superior sagittal sinus which was seen in Fig. 9-22;
6. That above and just anterior to the jugular foramen is the **internal acoustic meatus** in the petrous temporal bone;
7. The smooth bony surface sloping downward from the sella turcica to the foramen magnum which is formed by the **dorsum sellae** of the sphenoid bone and the **clivus** of the occipital bone.

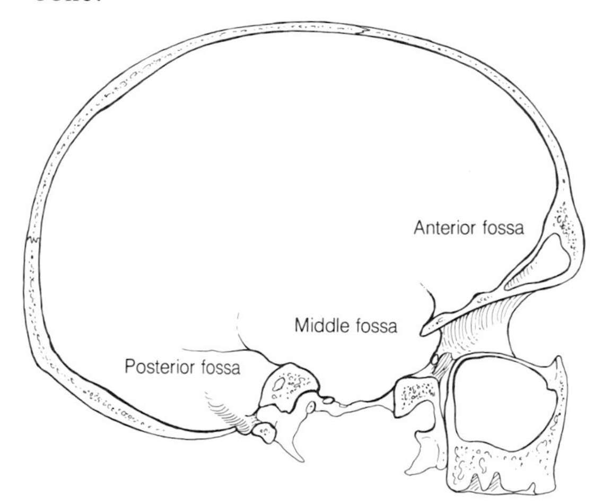

9-24 A paramedian section through the cranial fossae.

The Cranial Meninges

In the cranial cavity the brain is surrounded by the same meningeal layers that are found around the spinal cord and the corresponding layers are continuous with each other at the foramen magnum. However, some differences do exist and need explanation.

---

* Sella turcica, *L.* = Turkish saddle; Kline, *Gk.* = bed, the allusion is of course to a four-poster bed; Lacer, *L.* = rough or ragged, i. e., the edges of the foramen

* Sigma, *Gk.* = the letter *S* in the Greek alphabet

The Dura Mater

The fibrous dura mater is in almost all areas of the cranial cavity fused to the periosteum (endosteum) of the overlying cranial bones. There is, therefore, no epidural space as encountered in the vertebral canal. The combined dura and periosteum are often considered synonymous with dura mater and in fact so-called epidural hemorrhages occur between periosteum and bone. The true dura becomes continuous with that of the spinal cord at the foramen magnum. The periosteal or endosteal layer is reflected here and becomes continuous with the pericranium as it does at other foramina and fissures. In some regions double layers of true dura mater are pinched off from the periphery and extend into the cranial cavity as septa. The two major examples of this arrangement are the **falx* cerebri** and the **tentorium* cerebelli**.

* Falx, *L.* = a sickle; Tentorium, *L.* = a tent

The Falx Cerebri: This double layer of dura mater lies in the midline and separates the left from the right cerebral hemisphere. In Fig. 9-25 follow its attachment to the floor of the anterior cranial fossa at the crista galli, to the vault as far back as the internal occipital protuberance, and forward again along the center of the tentorium cerebelli. Between its anterior and posterior attachments it has a crescentic free margin.

The Tentorium Cerebelli: This fold of dura mater forms a roof over the posterior cranial fossa and the enclosed cerebellum. It is attached to the margins of the groove for the transverse sinus, to the superior border of the petrous temporal bone, and to the posterior clinoid process. The sharply concave free margin surrounds the midbrain between the middle and posterior cranial fossae and sweeps forward to the anterior clinoid process. It is at this site that rapid increases in intracranial pressure may cause a downward displacement of the brainstem and herniation of forebrain structures under the tentorium cerebelli. This may result in brain lacerations and hemorrhages.

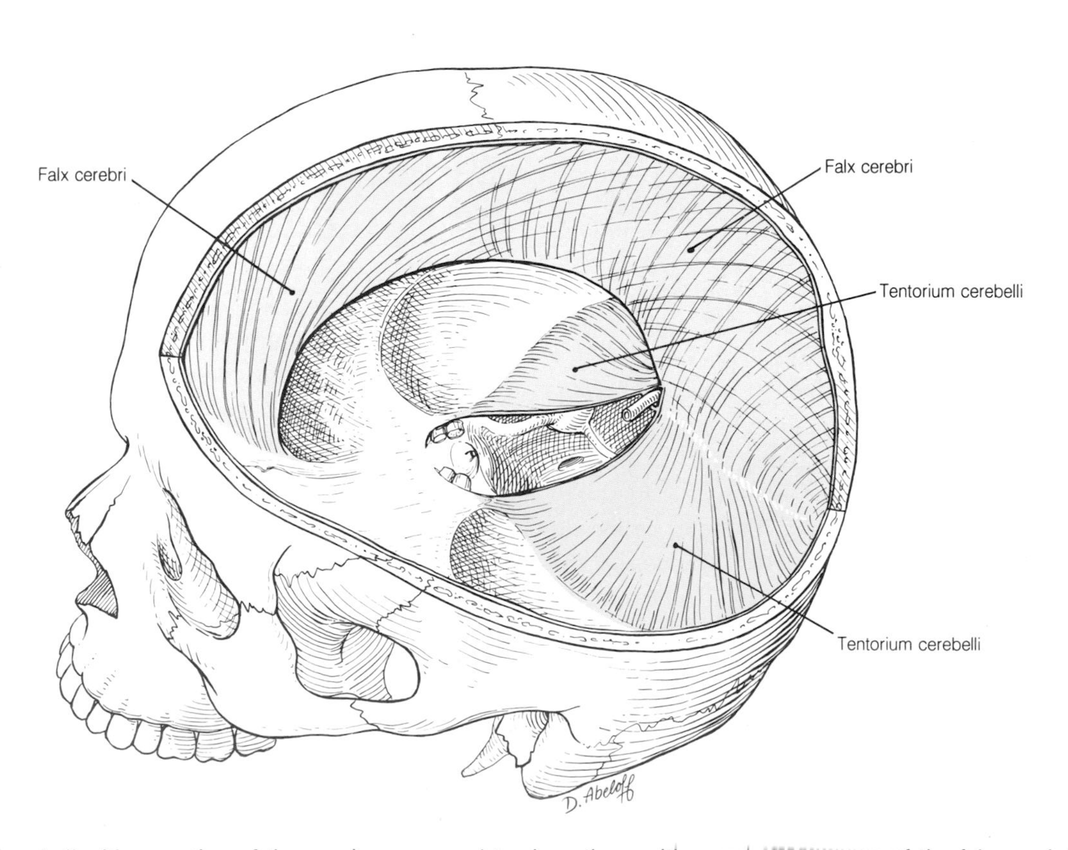

9-25 The skull with a portion of the cranium removed to show the position and attachments of the falx cerebri and tentorium cerebelli.

The Arachnoid Mater

This delicate membrane surrounds the brain and follows its general outline. It is separated from the dura mater by the **subdural space** and from the underlying pia mater by the **subarachnoid space** which contains cerebrospinal fluid. Where the arachnoid bridges major irregularities of the brain surface this space is expanded to form **subarachnoid cisterns.** These are the **cerebellomedullary, pontine, interpeduncular**, and **chiasmatic cisterns** and their positions should be identified in Fig. 9-26. The cerebellomedullary cistern can be penetrated by a needle introduced into the foramen magnum from the back of the neck. This approach is sometimes used to introduce radiopaque material above the site of a spinal cord lesion prior to radiographic examination.

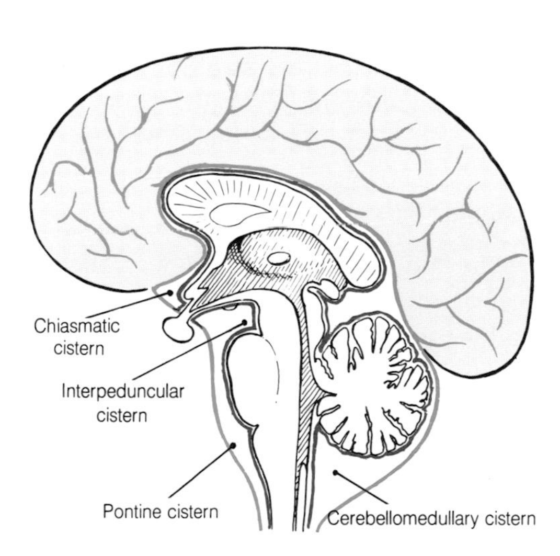

9-26 Median sagittal section through the brain to show the subarachnoid cisternae.

The Pia Mater

The vascular pia mater closely covers the surface of the brain. It dips down between the **gyri*** of the cerebral hemispheres and the **folia*** of the cerebellar hemispheres. It also continues as a sheath around the many small vessels that penetrate into the brain. In some regions the pia mater is invaginated into the ventricles of the brain to take part in the formation of the choroid plexuses. It is these structures that secrete cerebrospinal fluid into the ventricular system.

* Gyrus, *L.* = ring or circle; Folium *L.* = leaf

The Nerve and Blood Supply of the Meninges

The dura mater of the anterior and middle cranial fossae and the vault are supplied by the **trigeminal nerve** either directly from the trigeminal ganglion or through branches from the ophthalmic, maxillary, and mandibular nerves. The dura of the posterior fossa is supplied by branches of the **upper three cervical nerves** which enter the skull through the foramen magnum, the jugular foramen, and the hypoglossal canal.

The dense fibrous composition of the dura needs little blood supply and the many small meningeal branches received from the internal and external carotid arteries and the larger **middle meningeal artery** serve to supply the cranial bones more than the underlying dura.

The Cranial Venous Sinuses

The close association between dura and periosteum is lost at a number of places where the two layers are separated by venous sinuses. Venous sinuses also lie within the folds of dura mater already described. As well as receiving blood drained from the brain, these sinuses are also concerned with the reabsorption of cerebrospinal fluid from the subarachnoid space. Although some reabsorption probably occurs from the spinal subarachnoid space, the arachnoid mater in the cranial cavity is modified more obviously for this purpose at sites where it comes in close contact with the endothelial lining of a venous sinus. These regions are known as **arachnoid granulations**. Large accumulations of these are visible alongside the superior sagittal sinus where they project into venous **lacunae laterales** that communicate with the sinus. This arrangement is illus-

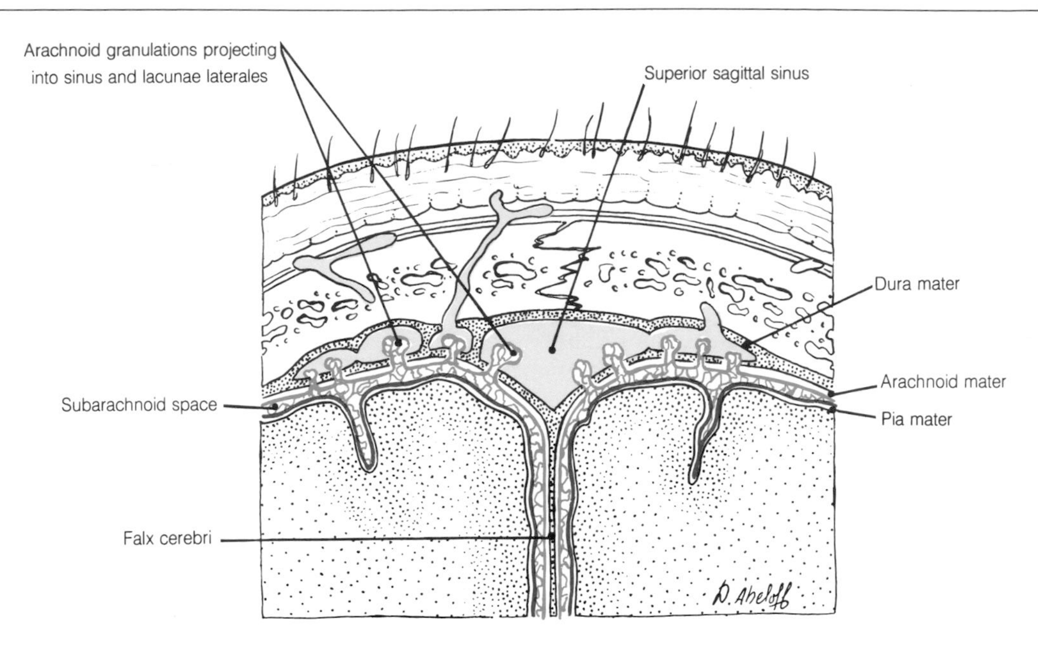

9-27 Coronal section through superior sagittal sinus to show arachnoid granulations.

trated in Fig. 9-27 where it can be seen that a sinus will also drain blood from the diploë and communicate with veins outside the skull through **emissary veins**.

The greater part of the blood reaching the sinuses is received from the **superficial cerebral veins** on the surface of the hemispheres, **cerebellar veins**, and the **great cerebral vein** which drains deeper parts of the brain.

The positions and communications of the main venous sinuses are seen in Fig. 9-28 and this illustration should be followed as the description of each sinus is read.

The Superior Sagittal Sinus

This sinus lies in the attached margin of the falx cerebri. It receives numerous superficial cerebral veins and shows several well marked accumulations of arachnoid granulations. Where the falx joins the tentorium cerebelli the sinus usually turns laterally and to the right in the attached margin of the tentorium. Here it becomes the **right transverse sinus.**

The Inferior Sagittal Sinus

This lies in the free margin of the falx cerebri. It runs posteriorly to join the **straight sinus** in the midline of the tentorium cerebelli.

The Straight Sinus

Lying in the midline of the tentorium cerebelli this sinus receives blood from both the inferior sagittal sinus and the **great cerebral vein** draining deeper parts of the brain. The straight sinus turns to the left to become the left transverse sinus. A communication between the straight and superior sagittal sinuses may be found at this point and is called the **confluence of sinuses.**

The Transverse Sinuses

These sinuses are found in the margins of the tentorium cerebelli where they are attached to the lateral walls of the cranial cavity. The left is commonly a continuation of the straight sinus and the right of the superior sagittal sinus. They run forward from the internal occipital protuberance to become the sigmoid sinuses at the posterior aspect of the petrous part of the temporal bone. The transverse sinuses are joined by cerebral and cerebellar veins.

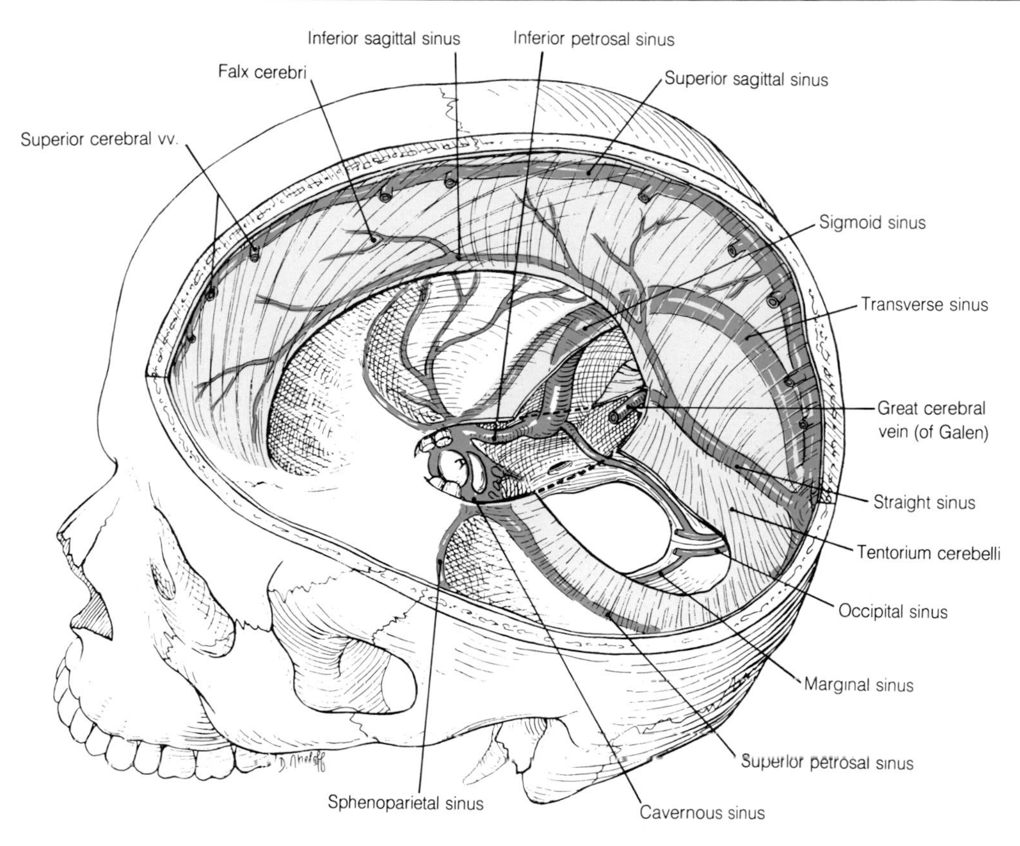

9-28 Showing the positions of the cranial venous sinuses.

The Sigmoid Sinuses

Each sigmoid sinus lies in an S-shaped groove in the petrous part of the temporal bone and in the occipital bone. The groove carries the sinus downward to the posterior part of the **jugular foramen.** At this point it becomes the **internal jugular vein.**

The Cavernous Sinuses

These two sinuses lie on either side of the body of the sphenoid bone between the dura of the middle cranial fossa and the periosteum covering the bone. This relationship is seen in Fig. 9-29. The two sinuses communicate with each other across the midline around the pituitary gland and the anterior part of each sinus abuts the superior orbital fissure where it is joined by the **ophthalmic veins** and the **central vein of the retina**. The **superficial middle cerebral vein** drains into the roof of the sinus as does the small **sphenoparietal sinus**. Through the foramen ovale the sinuses communicate with the **pterygoid plexus of veins** in the **infratemporal fossa** and they are drained posteriorly by the **superior** and **inferior petrosal sinuses** which in turn join the transverse sinuses and internal jugular veins, respectively. The posterior end of the sinus overlies the foramen lacerum and here the **internal carotid artery** comes to lie within the sinus. Just lateral to the posterior part of the sinus is the **trigeminal gan-**

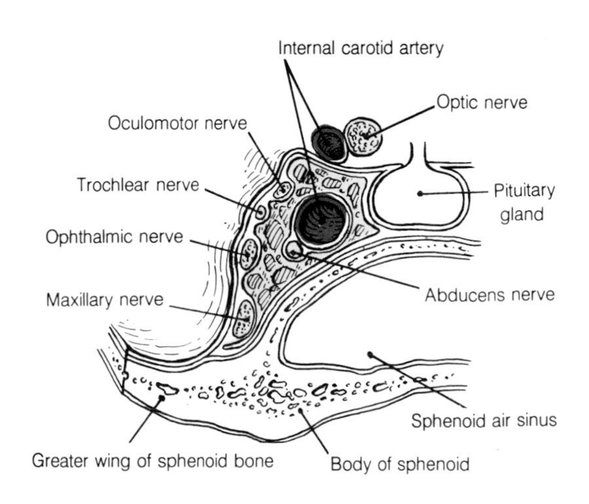

9-29 Coronal section through the cavernous sinus.

glion. The upper two divisions of the trigeminal nerve run forward in the lateral wall of the sinus. The third and fourth cranial nerves enter the roof of the sinus and the sixth enters its posterior wall along with the inferior petrosal sinus. Turn again to Fig. 9-29 and establish the positions that these structures take up. In the lateral wall are the **occulomotor, trochlear, ophthalmic**, and **maxillary nerves**. Within the sinus are the **internal carotid artery** and **the abducens nerve.** The **mandibular division** of the trigeminal nerve leaves the cranial cavity through the foramen ovale without becoming directly related to the sinus.

The communications of the cavernous sinus with the ophthalmic and facial veins and the pterygoid plexus and facial veins provide a possible pathway for infections on the face to spread to the sinus and thence to the superficial middle cerebral vein. Note that all structures passing through the sinus are surrounded by vascular endothelium.

The Basilar Venous Plexus

This plexus of veins lies on the clivus and is of importance in that it provides a communication between the internal vertebral venous plexus and the veins and venous sinuses within the cranial cavity.

Arteries of the Cranial Cavity

The brain is supplied with blood by the two **internal carotid arteries** and the two **vertebral arteries**. Of these the internal carotid arteries are the more important and the disabling or fatal result of their obstruction or rupture of their branches are well known as either a stroke or cerebrovascular accident. The occlusion of an internal carotid artery even for a few minutes leads to irreparable brain damage.

The Internal Carotid Artery in the Cranial Cavity

The internal carotid artery enters the skull through the carotid canal and appears in the middle cranial fossa through the foramen lacerum in the floor of the cavernous sinus. The diagram in Fig. 9-30 shows how the carotid canal opens into the "length" of the foramen lacerum and thus allows the artery to enter the base of the skull through one foramen and leave it through another. The artery then runs forward within the sinus and at the anterior end turns upward to pierce the roof of the sinus medial to the anterior clinoid process. Here it lies opposite the optic canal and gives off the **ophthalmic artery** before curving backward over the roof of the sinus. It then turns upward again lateral to the **optic chiasma**. At this point it terminates by dividing into the **anterior and middle cerebral arteries**, and the **posterior communicating artery**. The artery can be demonstrated radiographically and its rather complex course can be followed in the carotid arteriogram seen in Fig. 9-31. In this note the relationship of the artery to the carotid canal, the sphenoid air sinus, and thus the cavernous sinus. Its two main terminal branches can also be identified.

Throughout its course the internal carotid artery carries a plexus of postganglionic sympathetic fibers on its wall. This plexus is the main source of

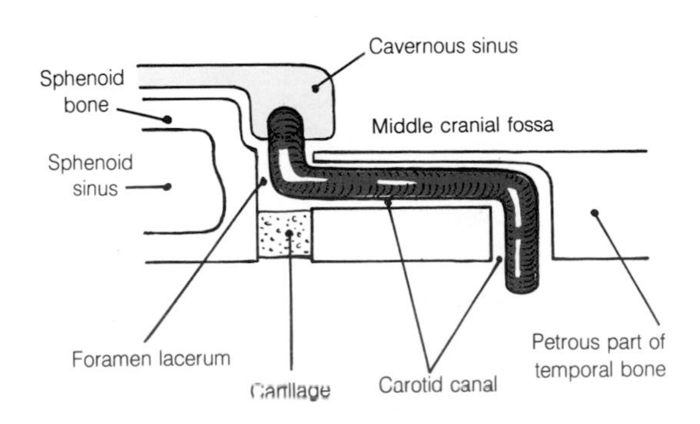

9-30 Diagram showing the course of the internal carotid artery through the carotid canal and the foramen lacerum.

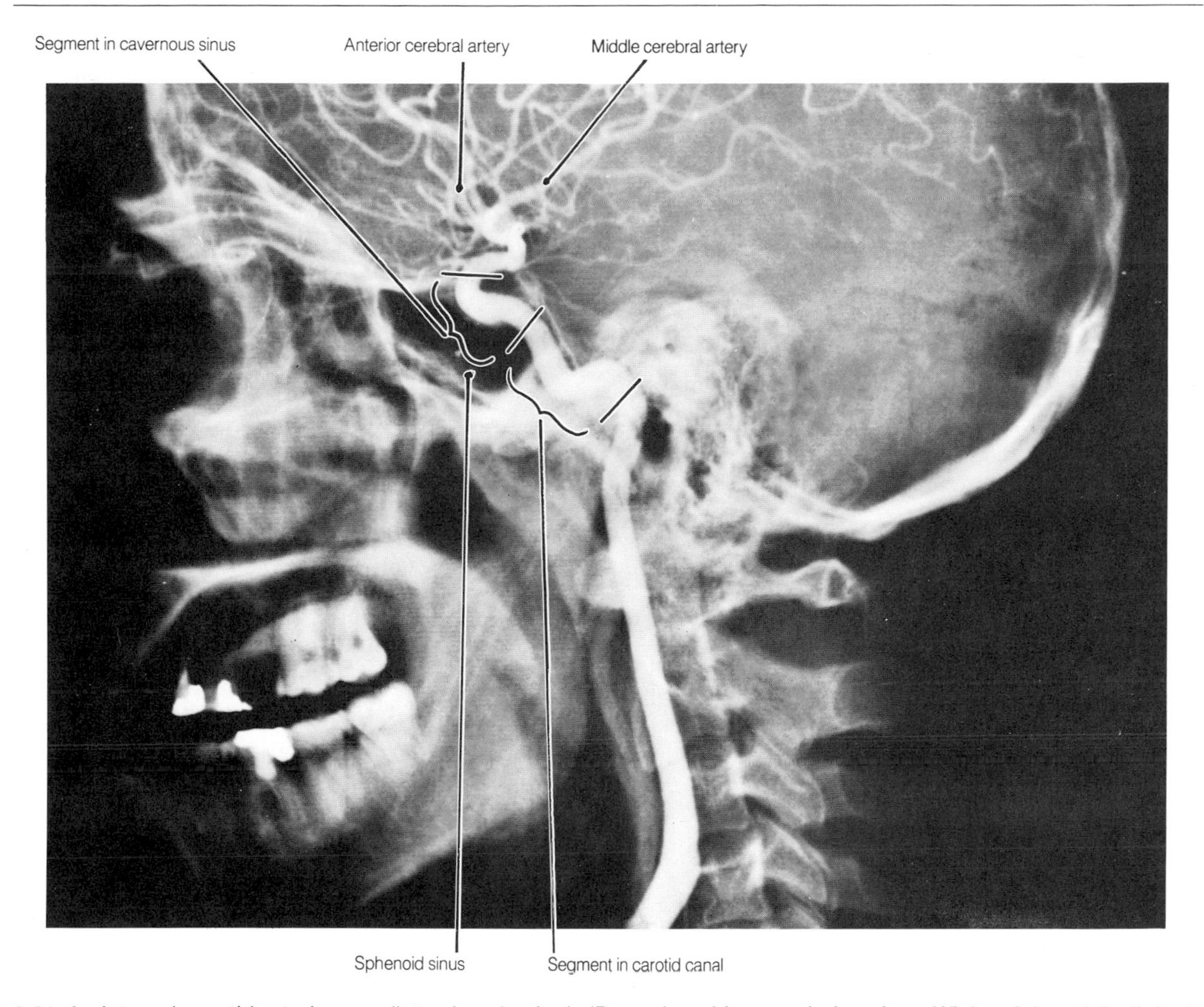

9-31 An internal carotid arteriogram (lateral projection). (Reproduced by permission, from Wicke: *Atlas of Radiologic Anatomy*, 4th Ed, Urban & Schwarzenberg, Baltimore-Munich, 1987).

sympathetic nerves for distribution to intracranial structures. While traveling through the cavernous sinus the artery gives off a small but important branch to the pituitary gland and over this part of its course is covered by the endothelium lining the sinus. The distribution of the terminal branches of the artery will be shown after the vertebral arteries have been described.

The Vertebral Artery in the Cranial Cavity

The vertebral artery enters the cranial cavity through the foramen magnum and gradually curls around to the anterior aspect of the brainstem and joins its fellow from the opposite side to form the single **basilar artery** at the lower border of the pons. The basilar artery ascends in the midline to the upper border of the pons where it divides into the **two posterior cerebral arteries**. The course of the vertebral artery in the neck and the posterior cranial fossa can be followed in the vertebral arteriogram shown in Fig. 9-32. Note its distinctive pattern as it passes through the vertebrarterial foramina of the atlas and axis. The cerebellar, basilar, and posterior cerebral arteries can all be identified.

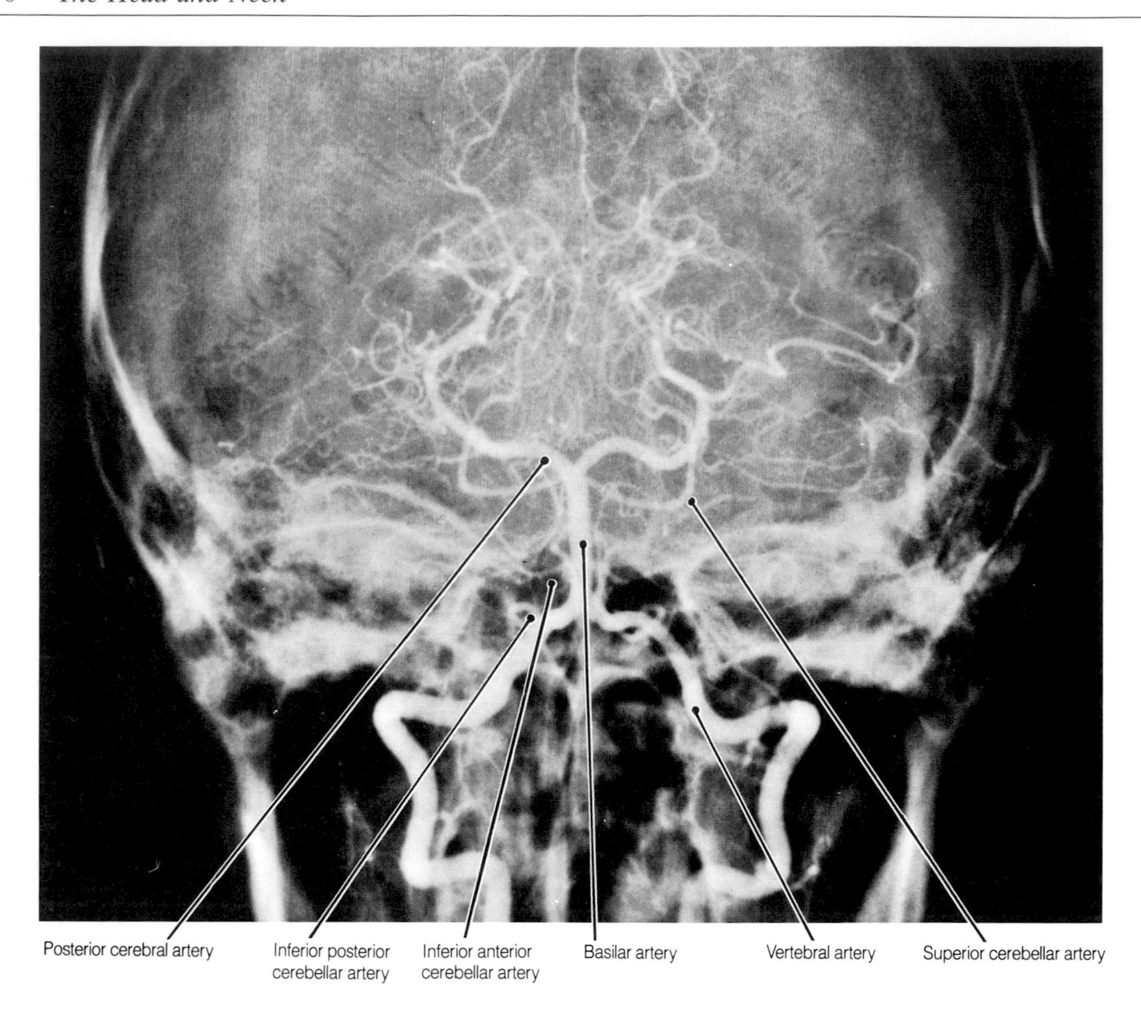

9-32 A vertebral arteriogram (a.p. projection). (Reproduced by permission, from Wicke: *Atlas of Radiologic Anatomy*, 4th Ed, Urban & Schwarzenberg, Baltimore-Munich, 1987).

The Circulus Arteriosus

By means of the **anterior communicating arteries** which unite the anterior cerebral arteries and the posterior communicating arteries which join the middle cerebral and posterior cerebral arteries, blood flowing from the vertebral and carotid arteries becomes continuous at a somewhat polygonal anastomosis termed the **circulus arteriosus**. In Fig. 9-33 the arrangement of this anastomosis is illustrated as are the important branches of the vertebral, basilar and carotid arteries. Note:

1. the **anterior communicating artery** uniting the two anterior cerebral arteries and completing the "circle";
2. the internal carotid arteries and their terminal branches, **the anterior and middle cerebral arteries**, and the **posterior communicating arteries**;
3. the **posterior cerebral arteries** which are the terminal branches of the basilar artery;
4. the **small central branches** of the three cerebral arteries;
5. the **superior cerebellar** and **anterior inferior cerebellar branches** of the basilar artery;
6. the **pontine and labyrinthine branches** of the basilar artery between its two cerebellar branches (the labyrinthine arteries supply the inner ear);
7. the **posterior inferior cerebellar** and **anterior spinal branches** of the vertebral arteries; and
8. the **oculomotor nerves** separating the posterior cerebral from the superior cerebellar arteries.

The circulus arteriosus lies in the interpuduncular cistern beneath the forebrain and embraces the **optic chiasma** and **infundibulum** of the pituitary gland. While there is some debate about the value

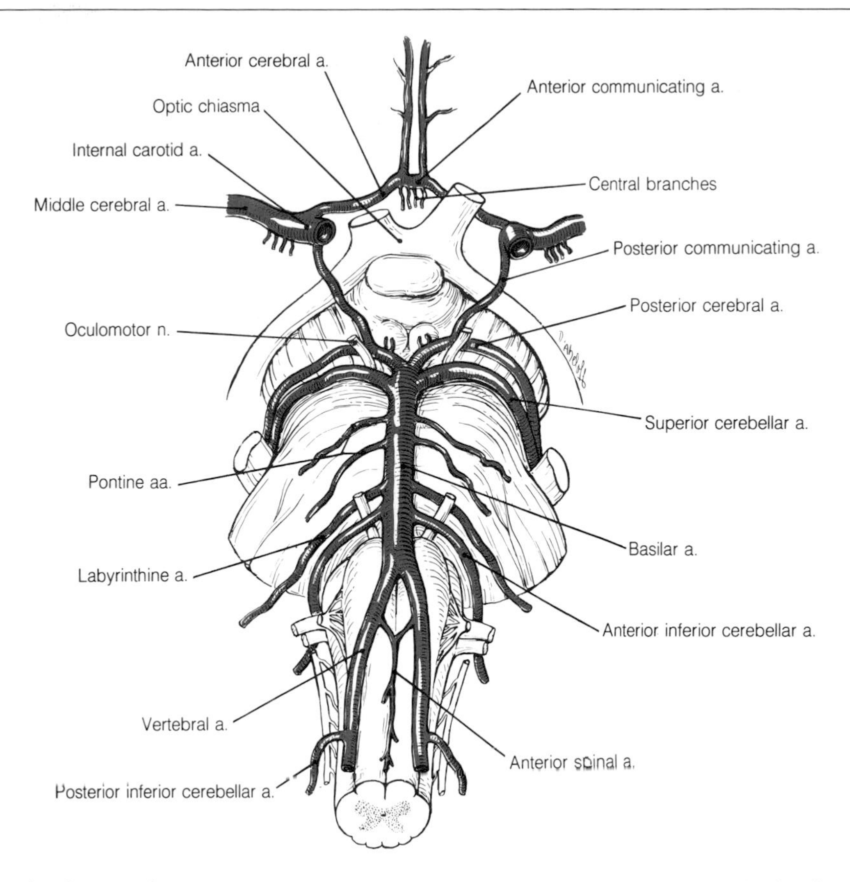

9-33 Showing the circulus arteriosus and the branches of distribution of the internal carotid, basilar, and vertebral arteries.

of this arrangement of vessels as a collateral circulation it may serve to equalize pressures in the branches of the carotid and vertebral arteries.

The Cerebral Arteries

The cerebral arteries supply the cerebral cortex from the surface and their cortical branches are the vessels, accompanied by cortical veins, that can be seen running beneath the arachnoid mater of the hemispheres. The deeper regions of the brain are supplied by the small but important central branches. Of these the most noteworthy is the "artery of cerebral hemorrhage," a branch of the middle cerebral artery that passes through the **internal capsule** where fibers of the motor pathway are closely congregated. A rupture or obstruction of this vessel leads to the classical picture of a stroke.

The Anterior Cerebral Artery

The anterior cerebral artery initially runs forward above the optic nerve before turning upward and backward over the corpus callosum in the longitudinal fissure. It is joined to its fellow by the

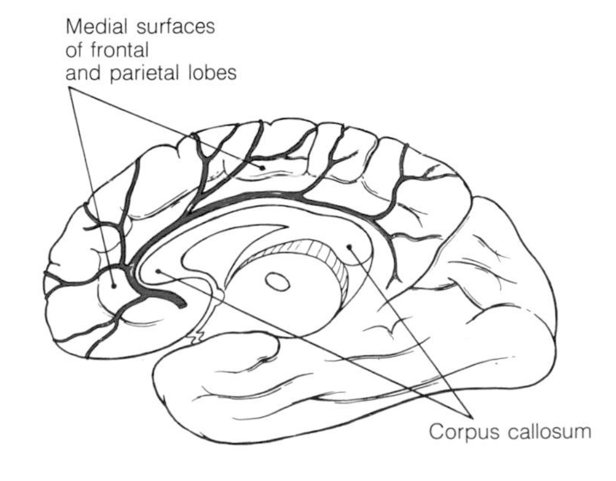

9-34 Diagram of the anterior cerebral artery and its distribution.

anterior communicating artery. The anterior cerebral artery supplies the orbital surface of the frontal lobe and medial surface of the cerebral hemisphere except over the occipital lobe (Fig. 9-34).

The Middle Cerebral Artery

The middle cerebral artery passes laterally in the **lateral cerebral sulcus** and then breaks up into branches which supply the lateral surface of the cerebral hemisphere (Fig. 9-35).

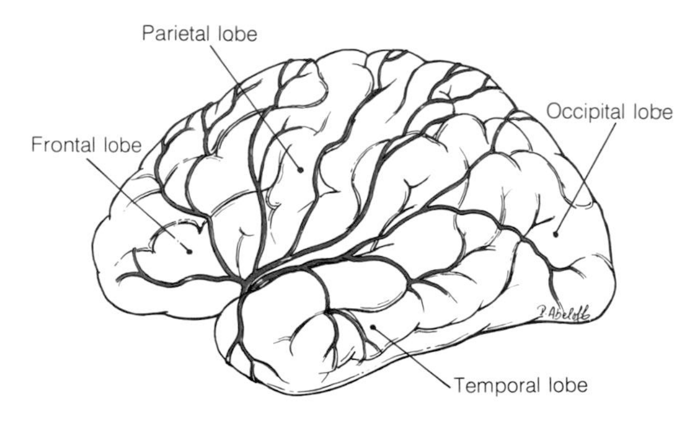

9-35 Diagram of the middle cerebral artery and its distribution.

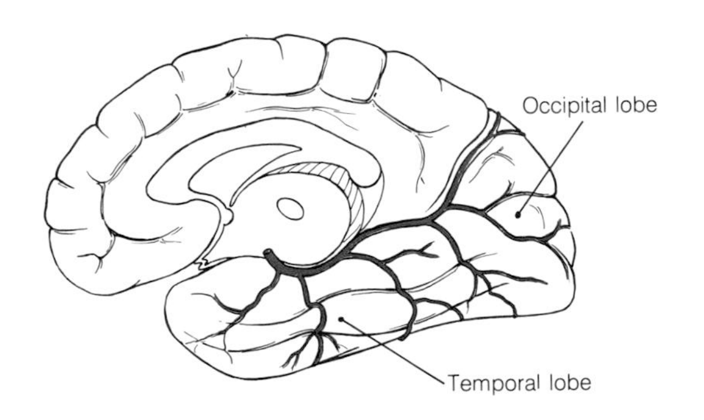

9-36 Diagram of the posterior cerebral artery and its distribution.

The Posterior Cerebral Artery

The posterior cerebral artery is joined by the posterior communicating branch of the internal carotid artery as it winds around the **cerebral peduncles.** On reaching the inferior surface of the occipital lobe, it provides branches to supply this and all but the lateral surface of the temporal lobe, and finally the medial surface of the occipital lobe (Fig. 9-36). It is worth noting here that this artery supplies the cortical areas around the calcarine sulcus which are concerned with visual perception. Because of this, interruption of its circulation leads to "cortical blindness", i. e. although the eyes are functioning normally the signals that they generate are not interpreted by the cortex.

The Cranial Nerves in the Cranial Cavity

A knowledge of the course of the cranial nerves through the cranial cavity is of great importance in neurological diagnosis. In the restricted space of this closed cavity relatively small lesions may be clinically localized by the effect that they have on the function of a nearby nerve. For this reason an opportunity is taken in this section to give brief reviews of the entire course and distribution of each cranial nerve so that they may be referred to as more detailed regional descriptions of these nerves are given.

Twelve pairs of cranial nerves arise from the brain and follow a longer or shorter course to their foramina of exit. While a number of these nerves subserve a single function, it must be realized that others carry a mixture of fibers–sensory or motor to visceral or somatic structures or to structures derived from the branchial arches. Each nerve may, therefore, have connections with a number of central nuclei (collections of cell bodies). However, a number of generalizations may still be made.

Motor fibers to skeletal muscles will have a cell body within the brainstem and an uninterrupted axon from this to the muscle they supply; i. e., there is no peripheral synapse or cell body.

Sensory fibers also run uninterrupted from the periphery to the point within the brainstem where they synapse with a second-order neuron. However, they bear a cell body outside the brainstem; i. e., although there is no peripheral synapse there is a peripheral cell body. These cell bodies are gathered together to form ganglia, e. g., the trigeminal or facial ganglion, and these ganglia correspond to the spinal ganglia of the spinal cord.

There is one exception to this rule. For some reason proprioceptive fibers of the fifth cranial nerve have their cell bodies within the brainstem.

There are two modifications to this rule. The retina and optic nerve and the olfactory bulb and olfactory tract are peripheral extensions of the brain. Thus fibers which they contain are not truly peripheral.

Parasympathetic fibers leaving the brainstem in cranial nerves are preganglionic fibers with a cell body within the brainstem. The preganglionic fiber synapses about the cell body of a postganglionic fiber in a peripherally located ganglion. The postganglionic fibers may return to a cranial nerve for further distribution. There are a number of these ganglia in the head, face, and neck. It will be found that some fibers that are not parasympathetic fibers pass through these ganglia, but they never synapse or have a cell body in the ganglion.

Sympathetic fibers found in the skull and face are always postganglionic. Their preganglionic fibers arise in the upper two or so segments of the thoracic spinal cord and ascend the sympathetic trunk to synapse around the cell bodies of postganglionic fibers in the cervical sympathetic ganglia. These are distributed on the branches of the internal and external carotid arteries.

A description of each cranial nerve follows. Central nuclei will be mentioned but their exact location and central connections are the province of neuroanatomy. The origin of each nerve from the brain can be seen in Fig. 9-37. The component fibers carried by the nerves will also be given and their nature should be understood. To aid this understanding a short list of definitions of terms to be used appears below.

Motor fibers are fibers leaving the brainstem to produce a peripheral effect in muscles or glands.

Sensory fibers are fibers carrying information from a wide variety of receptors to the brain.

Somatic fibers are motor or sensory fibers supplying structures derived from the embryonic somatopleure.

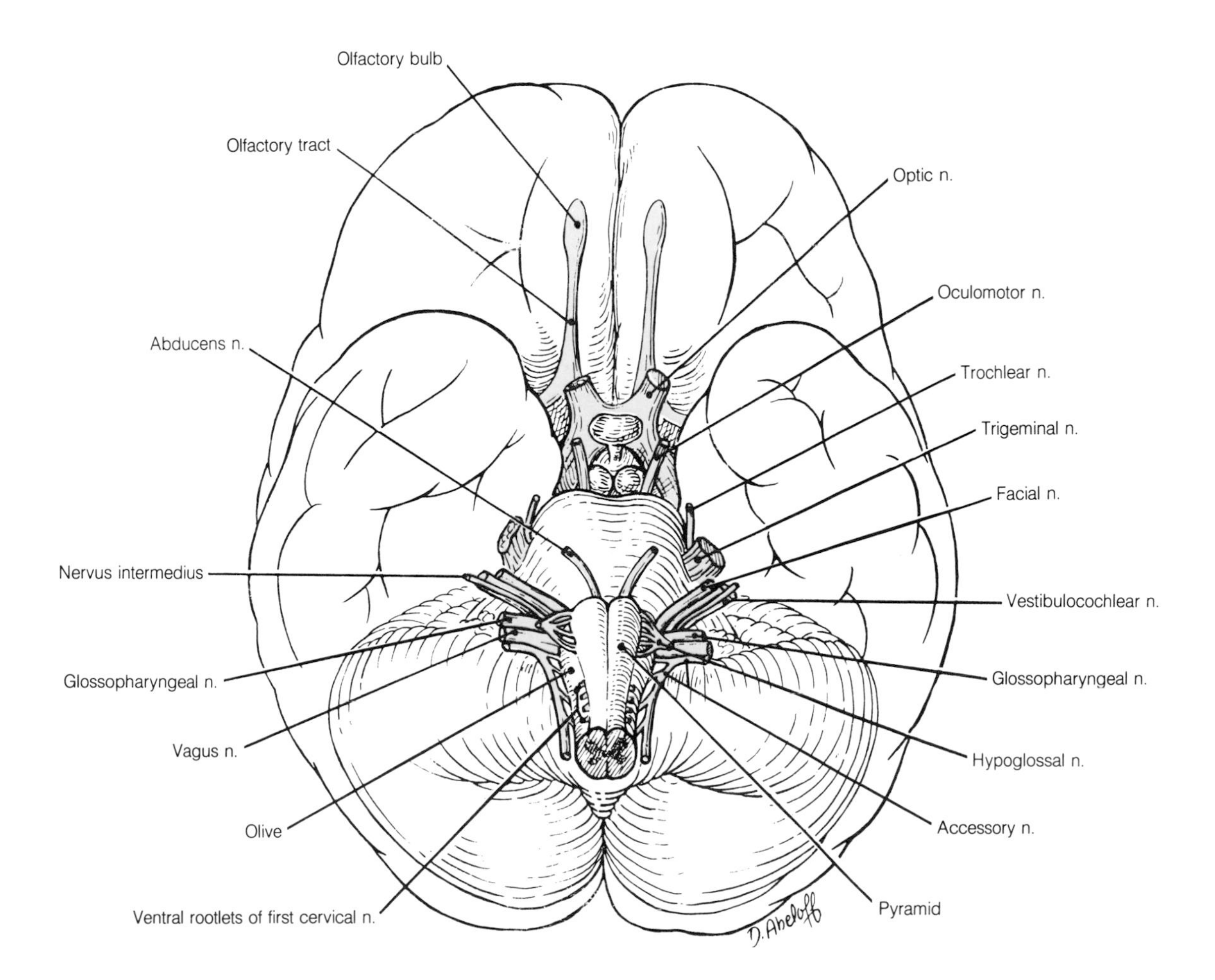

9-37 The inferior aspect of the brain showing the origins of the cranial nerves.

Visceral fibers are motor or sensory fibers supplying structures derived from the embryonic splanchnopleure.

Special sensory fibers are sensory fibers originating in special sense organs, e. g., taste (or gustatory) fibers. These may be visceral or somatic.

General sensory fibers are all those not special and may be visceral or somatic.

Branchial fibers are motor fibers to striated muscles derived from the branchial arches.

This nomenclature obviates some of the problems that arise from the use of the words afferent and efferent and some confusion that may follow from calling the branchial arches visceral arches. The terms defined can be combined, for example, to describe the autonomic fibers of the vagus nerve as visceral motor.

I. The Olfactory Nerve

The first neurons in this system are bipolar. Their peripheral processes and cell bodies lie in the olfactory mucous membrane in the roof of the nasal cavity. Their cranial processes pass up through the cribriform plate of the ethmoid bone to join the olfactory bulb in the anterior cranial fossa. Here they synapse with neurons of the olfactory tract which passes back to the region of the anterior perforated substance of the brain. The olfactory nerves are surrounded by prolongations of dura and arachnoid mater as they pass through the ethmoid bone.

II. The Optic Nerve

Like the olfactory bulb and tract, the retina and optic nerves are forward extensions of the brain. Rods and cones stimulate the bipolar cells which synapse with the cell bodies of second-order neurons in the retina whose axons go to form the optic nerve. Each optic nerve leaves the sclera of the eyeball just medial to its posterior pole and travels a slightly curved course to the optic canal at the back of the bony orbit. The nerve is surrounded by dura and arachnoid mater within the orbit. On entering the middle cranial fossa through the optic canal the two nerves converge and fuse with each other at the optic chiasma. Here about 50 percent of the fibers of each nerve cross over to the opposite side. From the optic chiasma the optic tracts pass backward to the two lateral geniculate bodies and after a further synapse fibers eventually reach the visual cortex.

Fibers of the first and second cranial nerves are classified as special somatic sensory fibers but in fact their developmental origins and history are very different.

III. The Oculomotor Nerve

The nerve leaves the midbrain at the medial aspect of the cerebral peduncle and passes forward to reach the roof of the cavernous sinus which it pierces. From here it runs forward in the lateral wall of the sinus to the superior orbital fissure. It enters the orbit through the fissure and divides into superior and inferior divisions. The superior division supplies the levator palpebrae superioris and the superior rectus muscles. The inferior division supplies the medial rectus, inferior rectus, and the inferior oblique muscles. In summary, therefore, the oculomotor nerve supplies all the extraocular muscles except the superior oblique and lateral rectus.

The extraocular muscles are derived from cranial myotomes and are classed as somatic muscles. The oculomotor fibers that supply them are, therefore, somatic motor fibers. In addition, the oculomotor nerve also carries preganglionic fibers of the parasympathetic autonomic system. These leave the inferior division of the nerve in the orbit and run to the ciliary ganglion. Here the preganglionic fibers synapse with the cell bodies of postganglionic fibers. These leave the ganglion in short ciliary nerves and on reaching the eyeball supply the ciliary muscle of the lens and the sphincter muscle of the pupil, both of which are smooth muscles. The component fibers of the oculomotor nerve are, therefore, somatic motor and visceral motor.

IV. The Trochlear Nerve

Although a somatic motor nerve the trochlear nerve is unusual in that it appears on the dorsal aspect of the brainstem. It arises from the midbrain and curls around this to lie alongside the oculomotor nerve and with it pierce the roof of the cavernous sinus. It then runs forward in the lateral wall of the sinus and enters the orbit through the superior orbital fissure. Almost immediately it supplies the superior oblique muscle. The nerve, therefore, carries somatic motor fibers.

V. The Trigeminal Nerve

Sensory and motor roots of this nerve arise separately from the pons but in close association with each other. The sensory root passes from the posterior to middle cranial fossa over the superior margin of the petrous part of the temporal bone. Just inside the middle fossa it expands as the trigeminal (sensory) ganglion. From the ganglion arise three divisions, the ophthalmic, the maxillary, and the mandibular divisions. The ophthalmic and maxillary divisions pass forward in the lateral wall of the cavernous sinus. The terminal branches of the ophthalmic division pass through the superior orbital fissure to enter the orbit. Here they supply sensory fibers to the lacrimal gland and nearby air sinuses. More peripheral branches leave the orbit anteriorly to supply the scalp, forehead, upper lid, and nose. The maxillary division leaves the skull through the foramen rotundum to supply parts of the nasal and oral cavities and the skin of the cheek and upper lip. The mandibular division passes out of the skull through the foramen ovale almost immediately after leaving the trigeminal ganglion. As it passes through the foramen it is joined by the small motor root. The combined nerves supply sensory fibers to the skin over the mandible, lower lip, and temporal region, and much of the oral cavity. Motor fibers supply the muscles of mastication, including the anterior belly of the digastric muscle, the mylohyoid muscle, and the tensor tympani and tensor veli palatini muscles. All these muscles are derivatives of the first branchial arch. Note that the mandibular division is the only division to carry branchial motor fibers.

The component fibers of the trigeminal nerve are, therefore, general somatic sensory and branchial motor. While this statement is strictly true, further reading will show that the branches of the trigeminal nerve are frequently used to distribute fibers derived from other cranial nerves.

VI. The Abducens Nerve

It seems unreasonable just for the sake of numerical orderliness to separate the abducens nerve from the oculomotor and trochlear nerves because all three are functionally related. The nerve arises from the ventral aspect of the brainstem at the junction of the pons and medulla and is the most medial of a number of nerves arising at this level. It has an unusual course in that it enters the inferior petrosal sinus in the posterior cranial fossa. This sinus leads it to the cavity of the cavernous sinus where it travels forward to the superior orbital fissure just lateral to the internal carotid artery which also lies in the sinus. On entering the orbit the nerve supplies the lateral rectus muscle. As with the trochlear nerve, the component fibers are somatic motor fibers. Note that each of the three motor nerves to the extraocular muscles will also contain somatic sensory fibers which carry proprioceptive information but these are transferred to branches of the trigeminal nerve in the orbit.

VII. The Facial Nerve

The facial nerve has a short course within the cranial cavity. It leaves the brainstem at the lower border of the pons just lateral to the root of the sixth nerve. Accompanied by the eighth nerve it enters the internal acoustic meatus and travels laterally through the petrous temporal bone until it meets the cavity of the middle ear. At this point it turns sharply backward forming a knee-shaped bend. Here lies its sensory ganglion, the geniculate* ganglion. On reaching the posterior wall of the middle ear it passes downward and leaves the skull through the stylomastoid foramen where it lies close to the parotid gland. Its subsequent course and branches in the face and scalp have already been described. This portion of the nerve supplies muscles derived from the second branchial arch, stapedius (by a small intracranial branch through the posterior wall of the middle ear), stylohyoid, the posterior belly of the digastric muscle, and the muscles of facial expression. Its fibers are, therefore, branchial motor fibers.

The Nervus Intermedius

This small nerve usually arises from the brainstem close to the facial nerve with which it becomes continuous in the petrous part of the temporal

* Genu, *L.* = knee, i. e., the ganglion lies at a knee-like bend in the facial nerve

bone. It contributes visceral motor (parasympathetic) fibers to the facial nerve which leave it again by two branches. These, the greater petrosal nerve and the chorda tympani nerve are given off within the petrous part of the temporal bone before the facial nerve leaves the skull.

The greater petrosal nerve leaves the facial nerve at the genu (its fibers do not synapse or have cell bodies here) and passes through a small slit into the middle cranial fossa. Running medially to the foramen lacerum it joins with the deep petrosal nerve which is formed from postganglionic sympathetic fibers of the plexus on the internal carotid artery. The combined nerve, now known as the nerve of the pterygoid canal, passes forward in this canal to the pterygopalatine ganglion. Here the preganglionic parasympathetic fibers synapse with the cell bodies of postganglionic fibers. These and the postganglionic sympathetic fibers are distributed to the glands of the nasopharynx, nasal cavity, hard and soft palate, and to the lacrimal gland.

The chorda tympani nerve leaves the facial nerve to enter the middle ear half a centimeter or so above the stylomastoid foramen. It runs across the inner aspect of the drum beneath the mucous membrane, enters the bone again and leaves the skull through the petrotympanic fissure. After a short course it joins the lingual branch of the mandibular division of the trigeminal nerve. Its fibers are transmitted in this nerve to the submandibular ganglion where they synapse. The postganglionic fibers pass to the submandibular gland or return to the lingual nerve and are distributed to the sublingual gland and to the many smaller collections of salivary glands in the mouth.

The chorda tympani also forms a pathway by which special visceral sensory fibers subserving taste pass from the anterior two-thirds of the tongue back to the facial nerve. These sensory fibers have their cell bodies in the geniculate ganglion and continue to the brainstem in the nervus intermedius.

The components of the facial nerve and nervus intermedius are, therefore, branchial motor (to muscles of the second arch), visceral motor (parasympathetic), and special visceral sensory (taste) fibers.

VIII. The Vestibulocochlear Nerve

The vestibulocochlear, or eighth nerve as it is often called, is a purely sensory nerve. The neurons it contains are bipolar. The cell bodies of the cochlear neurons lie in the spiral ganglion closely associated with the spiral organ (of Corti) in the inner ear. Their short peripheral processes are stimulated by receptors in the organ and their central processes end within the brainstem. Cell bodies and peripheral processes of the vestibular neurons also lie in the middle ear close to the vestibular apparatus and their central processes also end within the brainstem. Thus, although the cell bodies lie somewhat more peripherally, the pattern of a sensory nerve is retained. The central fibers are carried in the combined nerve along the internal acoustic meatus through the petrous part of the temporal bone. Leaving the internal acoustic meatus the nerve runs a short course in the posterior cranial fossa before joining the brainstem at the lower border of the pons.

The fibers in this nerve clearly serve organs of special sensation and because their origin has been thought to be in part from ectoderm, they are said to be special somatic sensory fibers.

IX. The Glossopharyngeal Nerve

The glossopharyngeal nerve arises as a series of rootlets from a sulcus on the medulla just posterior to the olive (the more lateral of the swellings on its ventral surface). The rootlets fuse and the nerve leaves the cranial cavity through the anterior part of the jugular foramen in the posterior cranial fossa. It is in the foramen that its two sensory ganglia are situated. Outside the skull the glossopharyngeal nerve runs between the internal and external carotid arteries before turning forward to pass between the superior and middle constrictor muscles of the pharynx. From here it distributes sensory fibers to the pharynx, tonsil, and posterior third of the tongue. The branches to this region of the tongue also carry taste fibers. Before entering the pharyngeal wall the glossopharyngeal nerve gives off a motor branch to the stylopharyngeus muscle, which is the sole muscular derivative of the third branchial arch.

By a rather complex course the glossopharyngeal nerve also distributes preganglionic parasympathetic (visceral motor) fibers. Just after leaving the skull a small branch called the tympanic nerve reenters and passing through the middle ear appears in the middle cranial fossa as the lesser petrosal nerve. Crossing the fossa it leaves with the mandibular division of the trigeminal nerve through the foramen ovale. Just below the base of the skull and between the pharynx and the mandibular nerve it reaches the otic ganglion where its fibers synapse. The postganglionic fibers join the auriculotemporal branch of the mandibular nerve for distribution to the parotid gland.

The fibers carried by the glossopharyngeal nerve are, therefore, general visceral sensory from pharynx, tongue, and tonsil, special visceral sensory (taste) from the tongue, visceral motor to the parotid gland, and branchial motor to the stylopharyngeus.

X. The Vagus Nerve

The rootlets that form the vagus nerve arise in series with those of the glossopharyngeal nerve just posterior to the medullary olive. The fused rootlets leave the jugular foramen. At and just below this foramen the vagus shows a superior and an inferior sensory ganglion. Shortly after leaving the skull the cranial root of the accessory nerve joins the vagus and contributes to it many of the branchial motor fibers which it distributes. In the neck the vagus descends between the internal and common carotid arteries and the internal jugular vein. Its subsequent course through the thorax and abdomen is described with these regions. While in the neck a pharyngeal branch contributes to the pharyngeal plexus and its fibers supply the muscles of the pharynx (except stylopharyngeus) and of the palate (except tensor veli palatini). By means of its superior laryngeal and recurrent laryngeal branches the vagus also supplies the mucous membrane and muscles of the larynx. All the muscles mentioned are branchial arch derivatives.

The vagus plays a major part in the distribution of the parasympathetic system. Visceral motor fibers are supplied to glands and smooth muscle of the pharynx, larynx, trachea, bronchi, lungs, and the alimentary canal as far as the left colic flexure. In addition, its cardiac branches modify the action of the heart. Visceral sensory fibers are carried to the brain from the thoracic and abdominal organs and from the larynx and lower pharynx. Some taste fibers run in the vagus from the few taste buds that lie between tongue and epiglottis.

In summary, therefore, the component fibers of the vagus include branchial motor fibers, visceral motor fibers (parasympathetic), general visceral sensory fibers, and special visceral sensory fibers. General somatic sensory fibers supplying a small area of the skin of the pinna are described but are of little importance. All sensory fibers have their cell bodies in one or the other of the two vagal sensory ganglia.

XI. The Accessory Nerve

The accessory nerve is formed in the cranial cavity by the union of a cranial root which arises from the brain stem and a spinal root which arises from the upper cervical segments of the cord. The rootlets that fuse to form the cranial root arise in series with those of the glossopharyngeal and vagus nerves. The nerve thus formed leaves the cranial cavity through the jugular foramen where it is united for a short distance with the spinal root. Its course soon terminates as it joins the vagus below the vagal ganglia. Its component fibers are branchial motor in function.

Although the spinal root comes into close anatomical association with the cranial root of the accessory nerve, it is not functionally related to it. It is formed by the union of rootlets arising from the upper four or five cervical segments. The nerve, thus formed passes up through the foramen magnum and joins the cranial root as the latter leaves through the jugular foramen. The two roots separate almost immediately after they leave the skull. The spinal root passes backward through the neck deep to the sternomastoid muscle and across the posterior triangle to disappear beneath the trapezius muscle. It supplies somatic motor fibers to the sternocleidomastoid and trapezius muscles.

XII. The Hypoglossal Nerve

This nerve is probably not a true cranial nerve but a composite of segmental nerve ventral roots supplying a number of "occipital" somites that have become incorporated into the base of the skull. The muscle derived from these somites has migrated to the floor of the mouth and developed into the musculature of the tongue. The nerve has followed the migrating muscle as is so often the case. Its component fibers are, therefore, classified as somatic motor fibers.

The hypoglossal nerve arises as a number of rootlets from the shallow sulcus between the pyramid and olive of the medulla and leaves the cranial cavity through the hypoglossal canal. The cranial opening of the canal is near the lip of the foramen magnum and the canal passes through the base of the occipital condyle. The nerve passes down the upper part of the neck between the internal carotid artery and the internal jugular vein. It then turns forward at the level of the occipital artery hooking around its sternomastoid branch. Passing forward over the loop of the lingual artery onto the surface of the hyoglossus muscle it breaks up to supply the tongue muscles. Apart from its lingual branches it also sends branches to supply the geniohyoid and thyrohyoid muscles and a descending branch, the superior root of the ansa cervicalis. The nerve fibers in these latter branches are not true hypoglossal nerve fibers. They are passed to the hypoglossal nerve from the ventral ramus of the first cervical nerve soon after the hypoglossal nerve leaves its canal.

Having completed the description of the cranial cavity, attention can be turned to the deeper regions of the face and the neck.

The Orbit

The two orbits are the bony cavities that contain the eyes and their associated structures. A good knowledge of the orbit and its contents is of importance even though the speciality of ophthalmology may not be followed. The large number of structures packed into this small and restricted space means that relatively trivial injuries and infections may rapidly deteriorate if not recognized early at the primary care level. For rather similar reasons more distant lesions often produce presenting signs and symptoms in this highly sensitive region.

In man, unlike many other animals, the orbit is complete except at its anterior opening. Look at the diagram in Fig. 9-38 and see that the cavities of the orbits are approximately pyramidal in shape and have a quadrangular base whose four borders form the margin of the orbital opening and an

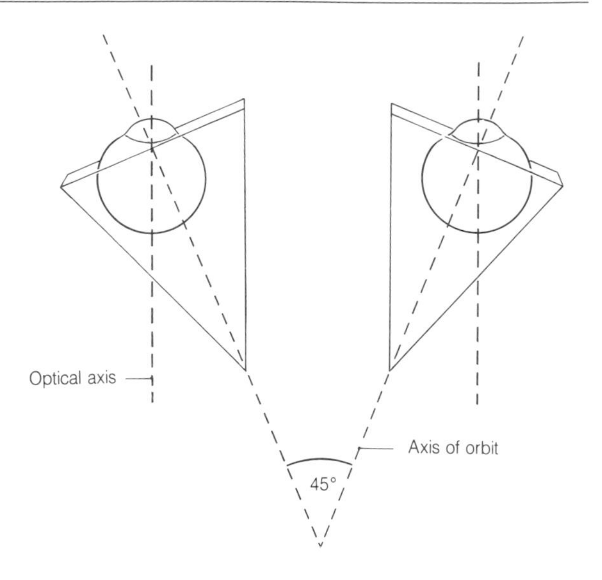

9-38 Diagram illustrating the orbital and optical axes.

apex which is directed posteriorly. The four sides that converge on the apex face superiorly, inferiorly, laterally and medially. The medial sides of each orbit are parallel to each other and as a result the axes of the bony orbits lie at about 45° to each other. However, in order to obtain distant binocular vision the optical axes of the eyes must

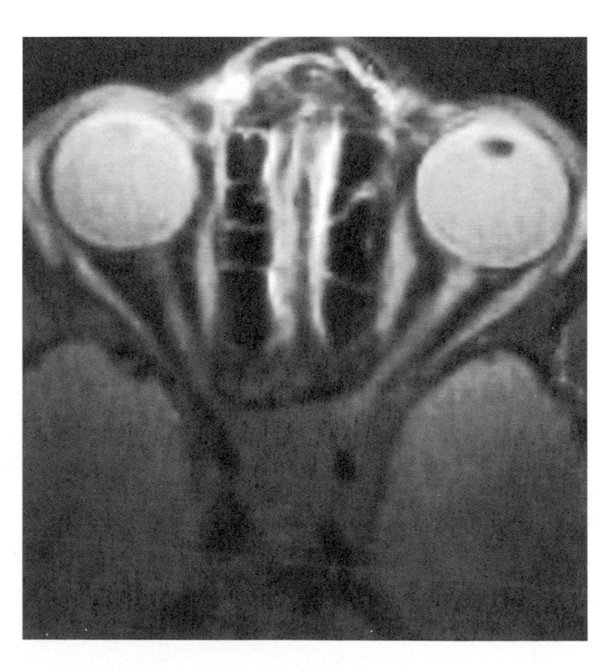

9-39 MR scan of the orbits of a four year old child. (Courtesy of M. A. Hall-Craggs, B.A., M.R.C.P., F.R.C.R., and The Hospital For Sick Children, Great Ormond Street, London).

be parallel to each other. This difference between the orbital and optical axes is important for an understanding of the actions of the muscles moving the eye. In Fig. 9-39 which is an MR scan of the orbits of a child the alignment of these, the eyeballs and the optic nerves on either side of the nasal cavities can be confirmed.

The Bony Orbit

The Orbital Margin

Using Fig. 9-40 follow the margin of the left orbit. The **superior margin** is formed by the frontal bone (blue). It shows either a notch or foramen for the passage of the supraorbital nerve onto the scalp. The **lateral margin** is formed in part by the frontal bone, but also by the frontal process of the zygomatic bone (red). The **inferior margin** is formed by the zygomatic bone together with the maxillary bone (yellow). Just below the midpoint of this margin is found the infraorbital foramen which transmits the infraorbital nerve and artery.

The **medial margin** is formed above by the frontal bone and below by the maxillary bone. Closer examination will show that the margin surrounds the small lacrimal bone. This presents a depression in which lies the lacrimal sac.

The Walls of the Orbit

The walls of the orbit can also be seen in Fig. 9-40 and this illustration should be referred to as the following description is read.

The **roof** is formed largely by the thin **orbital plate of the frontal bone** and is completed posteriorly by the **lesser wing of the sphenoid.** Between the latter bone and the **greater wing of the sphenoid** in the lateral wall lies the **superior orbital fissure**. The **optic canal** lies in the lesser wing of the sphenoid bone medial to the fissure. The roof separates the orbit from the **anterior cranial fossa** and may be excavated to a lesser or greater extent by the **frontal air sinus.**

The **lateral wall** is formed by the **zygomatic bone** and the **greater wing of the sphenoid**

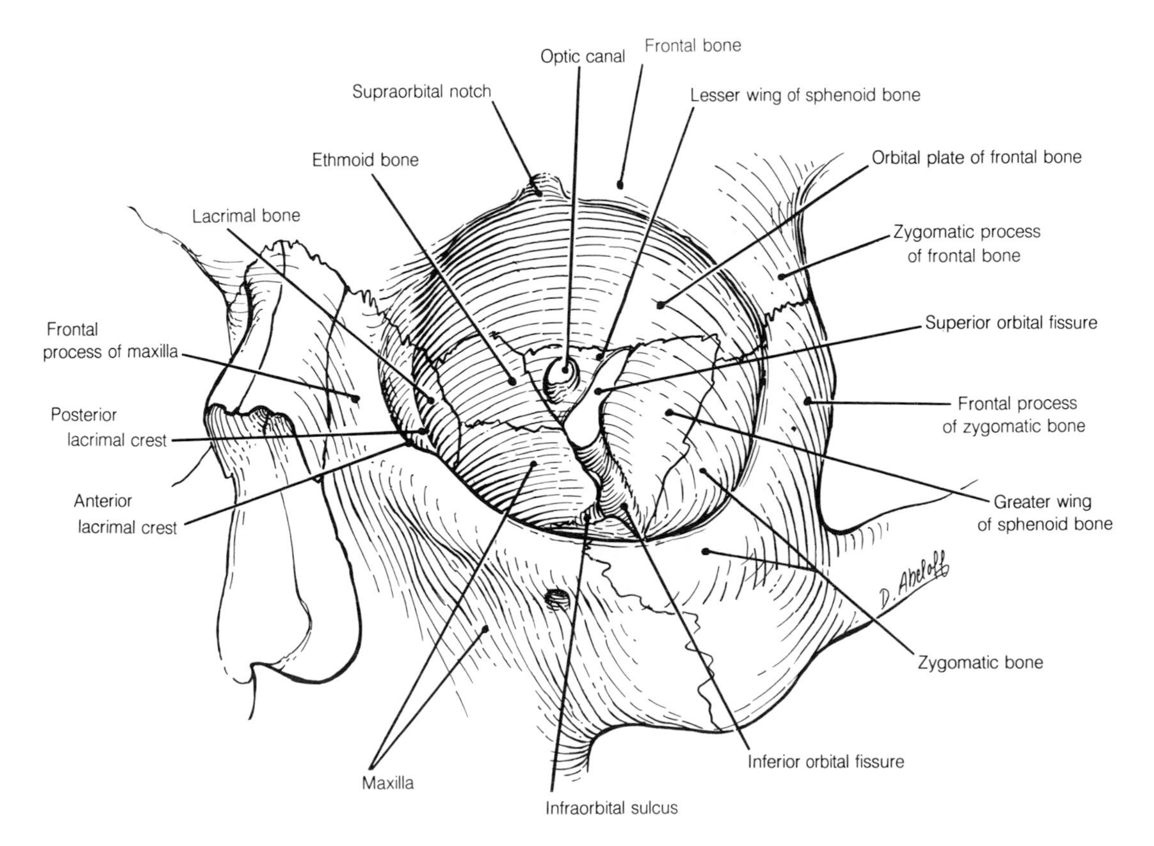

9-40 Showing the bones forming the margins and walls of the left orbit. Note that the skull has been turned slightly to the right to expose the medial wall.

bone. It is separated from the floor posteriorly by the **inferior orbital fissure** which communicates with the **infratemporal fossa**. The lateral wall itself separates the orbit from the **temporal region**.

The **floor** of the orbit is formed by the **zygomatic and maxillary bones**. Its main feature is an **infraorbital canal** or sulcus which communicates with the infraorbital foramen. The floor separates the orbit from the **maxillary air sinus**.

The **medial wall** of the orbit is formed by the **ethmoid and lacrimal bones**. The thin orbital plate of the ethmoid bone barely separates the orbit from the **ethmoid air cells**.

It is important to retain a picture of the major structures from which its walls separate the orbit. As can be seen in Fig. 9-41 through these walls tumors and infections may readily spread to or from the surrounding cavities.

The Contents of the Orbit

The orbit contains the eyeball and the optic nerve, the extrinsic muscles of the eyeball and the nerves supplying them, the branches of the ophthalmic division of the fifth cranial nerve, the ciliary ganglion, the ophthalmic artery and veins, and finally "packing" material of connective tissue and fat.

The Eyeball

The eyeball is embedded in orbital fat and supported in the orbit by a **suspensory ligament** which forms a hammock beneath it. The ligament is attached to the medial and lateral walls of the orbit. It is a thickened portion of a fine layer of fascia called the **fascia bulbi** which surrounds the

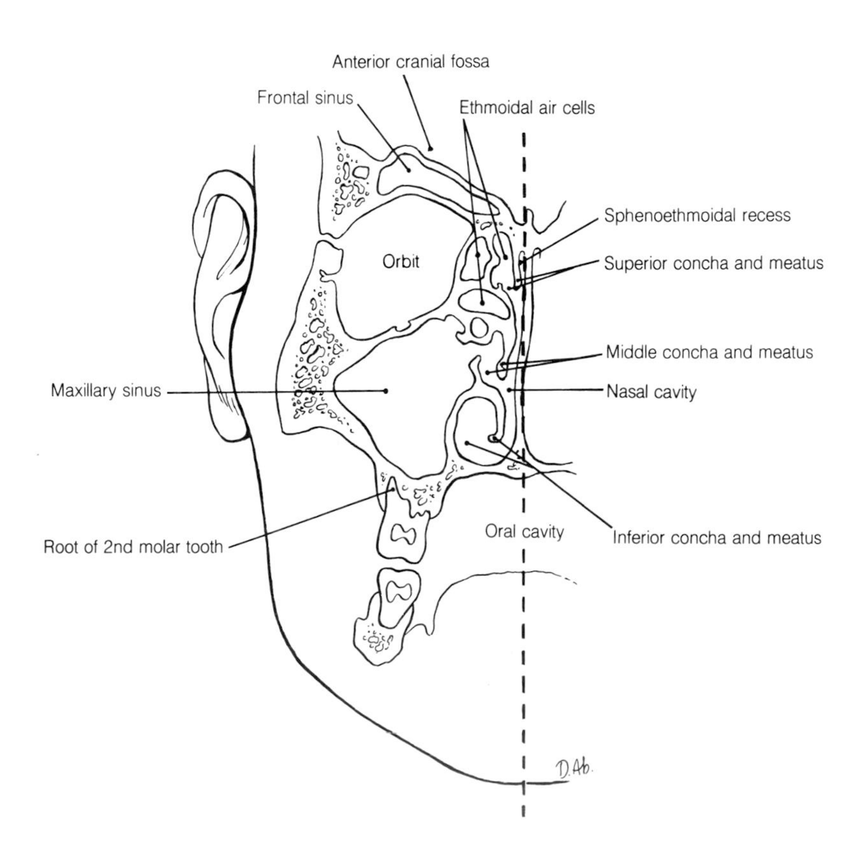

9-41 Coronal section through the facial skeleton.

eyeball, separates it from surrounding structures, and serves as a socket within which it moves.

The movements of the eyeball all take place about its center. Other movements are prevented by the attachments of the rectus muscles to the bony orbit, the orbital fat, and the forward pull of the two oblique muscles.

Fractures of the zygomatic bone are not uncommon, especially in contact sports. A common sign of displacement of the bone is double vision (diplopia) caused by depression of the lateral attachment of the suspensory ligament to the zygomatic bone where it forms part of the lateral wall of the orbit.

Muscles in the Orbit

The muscles in the orbit are known as the extraocular muscles or extrinsic muscles of the eye in contrast to the intrinsic muscles which control the lens and the size of the pupillary opening. Six of these muscles are attached to the eyeball and one to the upper lid.

Fig. 9-42 is a diagram of the right orbit which includes the optic foramen and the superior orbital fissure. Surrounding the optic foramen and the middle part of the superior orbital fissure is a fibrous ring or **common annular tendon**. Four of the muscles of the orbit arise from this ring, two more from just above it. The remaining muscle arises from just inside the lower medial angle of the orbital margin. Using this diagram and the illustrations in Fig. 9-43 identify the attachments of the extraocular muscles as they are described below.

The **levator palpebrae superioris** arises from the apex of the orbit superior to the annular tendon. It runs forward through the orbit just below the roof to be attached to skin of the upper lid and the tarsal plate. The deeper portion of the muscle is largely composed of smooth muscle fibers and is supplied by sympathetic nerves.

The **superior oblique muscle** also arises from above the ring just medial to the levator palpebrae. It passes forward to the upper medial angle of the orbital margin where it becomes tendinous. The tendon passes through a fibrous ring or pulley (trochlea) and then obliquely backward and laterally beneath the superior rectus to be attached behind the equator of the eyeball and to its upper lateral posterior quadrant. Note that the direction

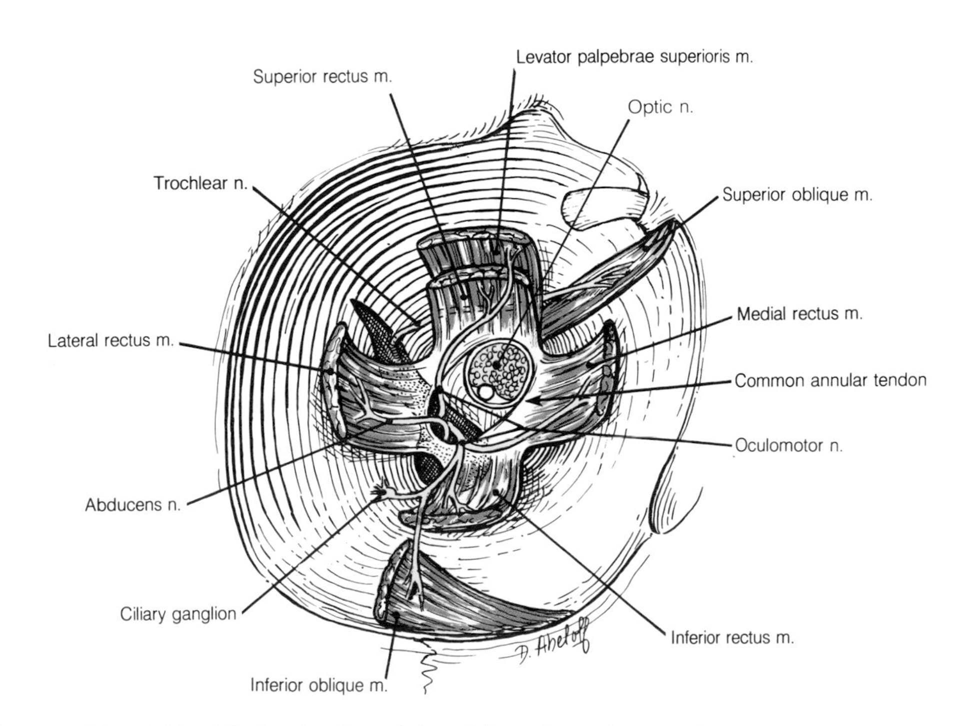

9-42 Diagram of the right orbit showing the origins of the extraocular muscles.

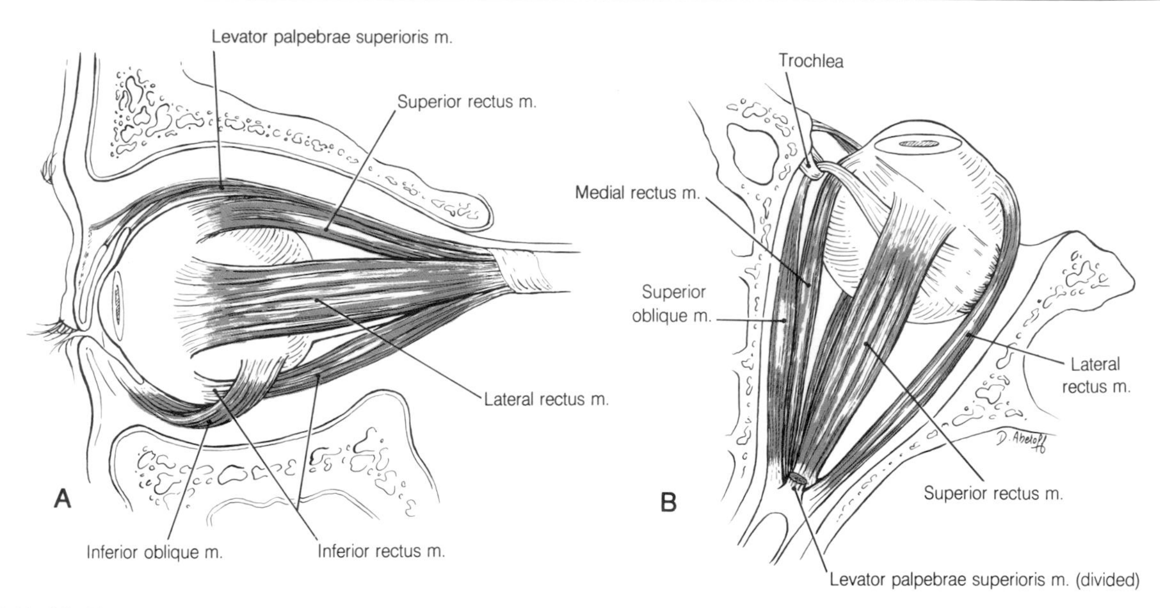

9-43 (*A*) The extraocular muscles of the left eye viewed from the left side. (*B*) The extraocular muscles of the right eye viewed from above.

in which the muscle acts on the eyeball is determined by the position of the pulley and not by its posterior attachment.

The **four rectus muscles, superior, inferior, medial, and lateral**, arise from the common annular tendon and run forward to embrace the eyeball and insert into it in front of the equator.

The **inferior oblique muscle** arises from the orbit just inside the inferior medial angle at a point vertically below the ring retaining the superior oblique tendon. The muscle runs backward and laterally below the inferior rectus to be attached behind the equator of the eyeball to its lower lateral posterior quadrant.

The nerve supply of the extraocular muscles can be quite simply stated and remembered. They are all supplied by the **oculomotor nerve** except the superior oblique muscle (passing through a pulley or **trochlea**) which is supplied by the **trochlear nerve** and the lateral rectus muscle (an **abductor** of the eye) which is supplied by the **abducens*** **nerve.**

Movements of the Eyeball

The most important movements of the eyeball occur about two axes, each passing through the center of the sphere. Movements about a horizontal axis lying on a coronal plane are **elevation** and **depression** and movements about a vertical axis are **adduction and abduction**. These two axes and the movements are illustrated in Fig. 9-44.

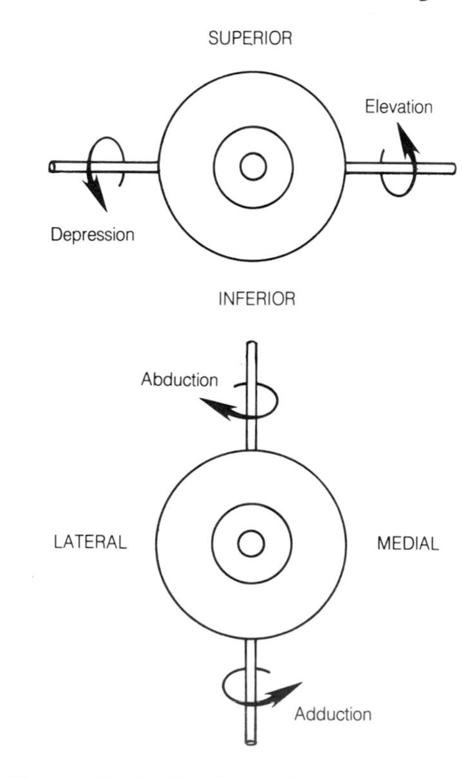

9-44 Diagram illustrating the vertical and horizontal axes about which the main movements of the eyeball take place.

* Abduco, *L.* = I lead away from

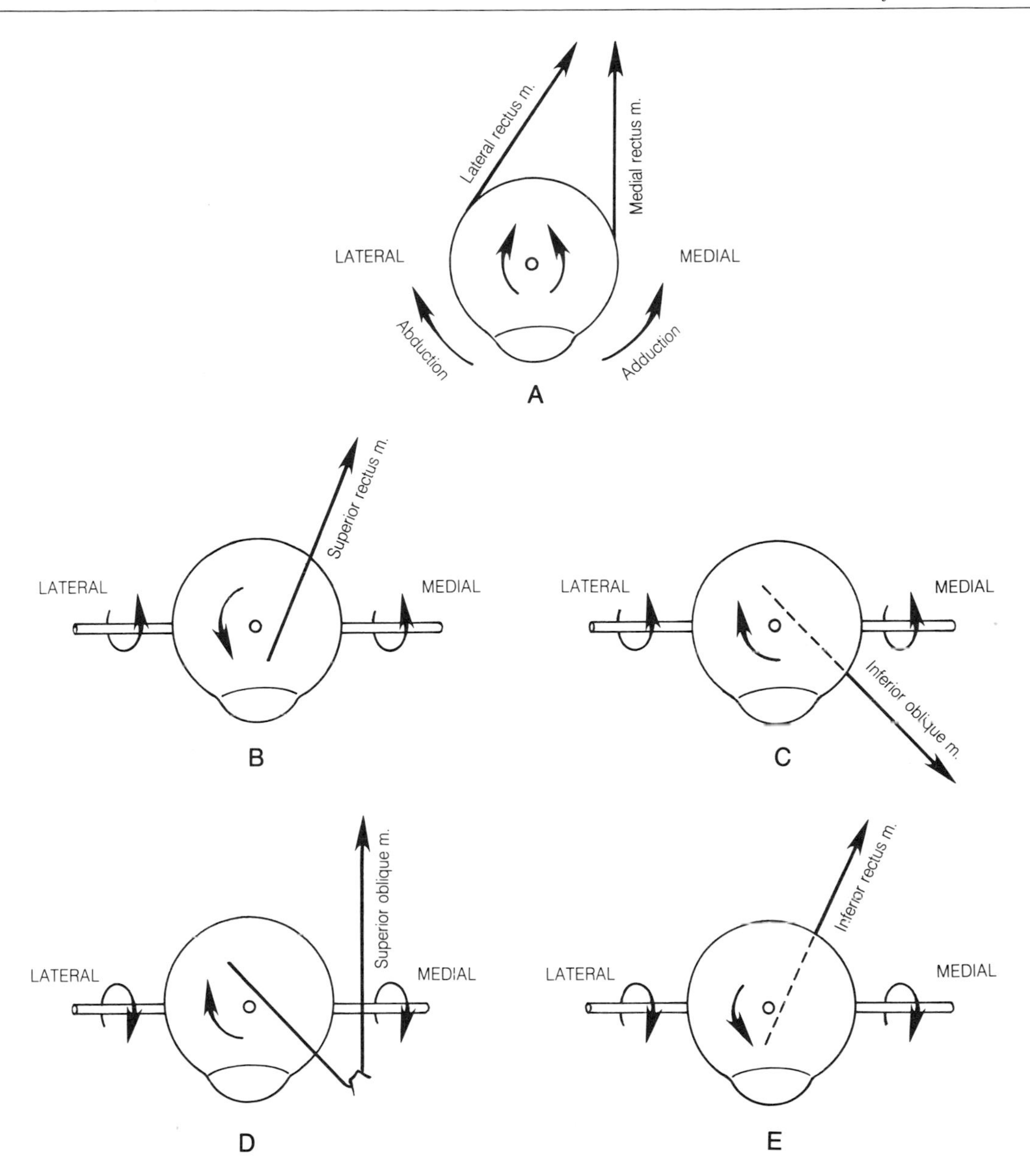

9-45 Diagrams illustrating the actions of the extraocular muscles of the right eye when viewed from above.

The medial and lateral recti rotate the eyeball around the **vertical axis** and are therefore able to produce adduction (medial rectus) and abduction (lateral rectus) when the gaze is directed forward. These actions are shown in Fig. 9-45 A.

Elevation will clearly be produced by the superior rectus as it rotates the eyeball around the **horizontal axis**, but because its origin lies medially in the orbit its pull lies **medial to the vertical axis** and it will also produce some adduction. However, the inferior oblique which arises from the front of the medial side of the orbit also produces elevation but, in addition, some abduction. As a result, balanced action of these two muscles produces pure elevation. The actions of the two muscles are illustrated in Fig. 9-45, B and C.

Depression is produced by the inferior rectus and superior oblique which act in a manner similar to that of the two elevators. Remember that the superior oblique will be pulling from the direction of its trochlea. How this is achieved can be worked out from Fig. 9-45, D and E.

Intermediate movements can be executed by a combination of the four cardinal movements. Any failure in the coordination of the muscles of the two eyes will lead to the perception of a double image, a condition known as diplopia.

The Nerves in the Orbit

The optic nerve enters the orbit through the optic canal. The lacrimal, frontal, and nasociliary branches of the ophthalmic nerve enter through the superior orbital fissure and together with these are found the oculomotor, trochlear, and abducens nerves. In addition to supplying extraocular muscles the oculomotor nerve also carries parasympathetic fibers to the ciliary ganglion which lies within the orbit close to the eyeball. Infraorbital and zygomatic branches of the maxillary nerve also enter the orbit, approaching it from the inferior orbital fissure. The manner in which these nerves gain entry to the orbit is summarized in Fig. 9-46. Note the relationship to the common annular tendon of the nerves passing through the superior orbital fissure and the fact that the oculomotor nerve has divided into a superior and inferior branch. Remember also that the anterior extremity of the cavernous sinus is directly applied to the cranial side of the superior orbital fissure and that it is in the wall and cavity of this sinus that the nerves reach the fissure. The sinus is in a position to receive the ophthalmic vein which also passes through the fissure.

The Optic Nerve

The **optic nerve** carries afferent fibers from the retina of the eyeball to the optic chiasma. It has, therefore, an orbital and intracranial course. It appears at the back of the eyeball slightly medial to the optical axis and runs medially and backward to the optic canal through which it travels in company with the **ophthalmic artery.** In Fig. 9-46 it can be seen that the nerve lies within the cone of extraocular muscles and that in the canal it is above and medial to the artery. In its orbital course the nerve also contains the **central artery and vein of the retina**. The artery is a branch of

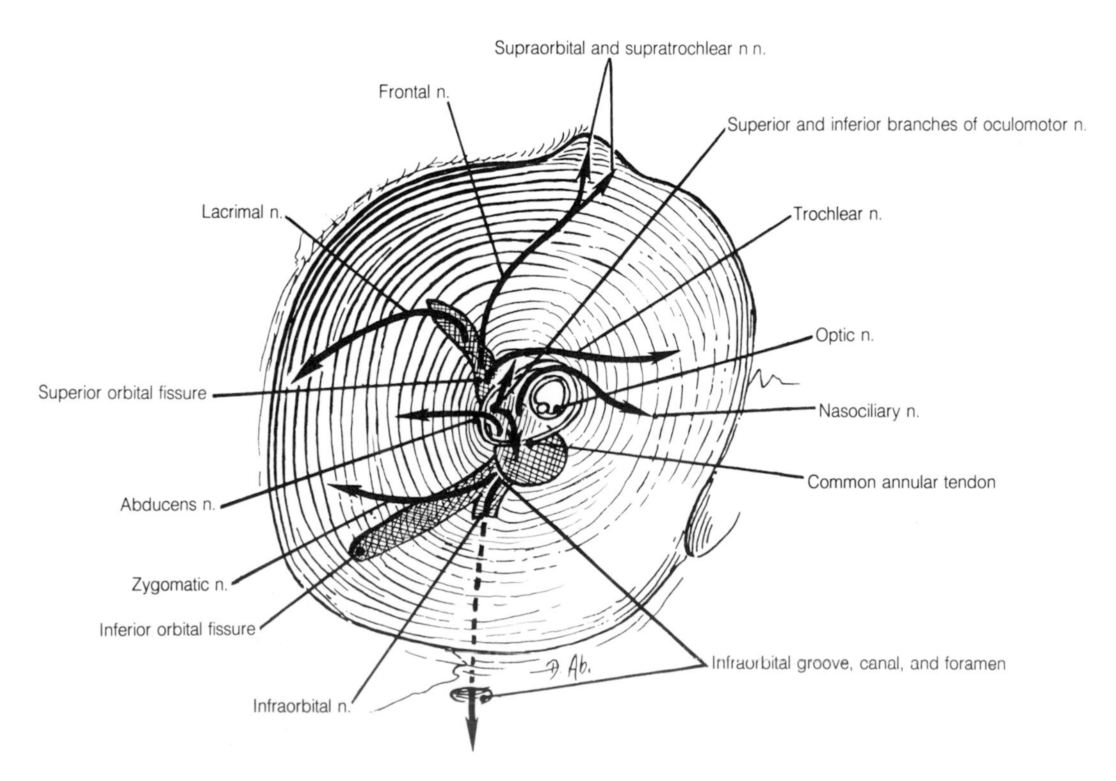

9-46 Diagram illustrating the mode of entry and distribution of nerves in the right orbit.

the ophthalmic artery and because it is virtually an end artery its obstruction causes immediate and complete blindness of the eye it supplies.

The optic nerve is not a true cranial nerve but a forward extension of the brain. As a result, it is surrounded by extensions of the meninges which fuse with the outer fibrous coat of the eyeball at the point of exit of the nerve. Any rise in intracranial pressure is transmitted to the subarachnoid space surrounding the optic nerve and may impede venous return along the retinal veins. This causes edema of the **optic disc** (papilledema) which can be seen when the retina is examined with the ophthalmoscope. Thus, some impairment of vision and edema of the disc may be an important and early sign of an increase in the pressure of the cerebrospinal fluid.

The Ophthalmic Nerve

The ophthalmic nerve approaches the superior orbital fissure in the lateral wall of the cavernous sinus. At this point it divides into the lacrimal, frontal, and nasociliary nerves. The course of the branches can be followed in Figs. 9-47 and 9-48.

The **lacrimal nerve** (Fig. 9-47) enters the orbit through the superior orbital fissure above the common annular tendon. Keeping to the lateral margin of the roof of the orbit it receives secretomotor fibers for the lacrimal gland from the zygomatic nerve and terminates by supplying the gland, conjunctiva, and the lateral part of the upper lid.

The **frontal nerve** (Fig. 9-47) enters the orbit through the superior orbital fissure above the common annular tendon. Running forward in the roof of the orbit it divides into the **supratrochlear** and **supraorbital nerves** which leave the orbit at its superior margin to supply the skin of the forehead and scalp. Branches are given off to the frontal air sinuses.

The **nasociliary nerve** (Fig. 9-48) enters the orbit through the superior orbital fissure within the

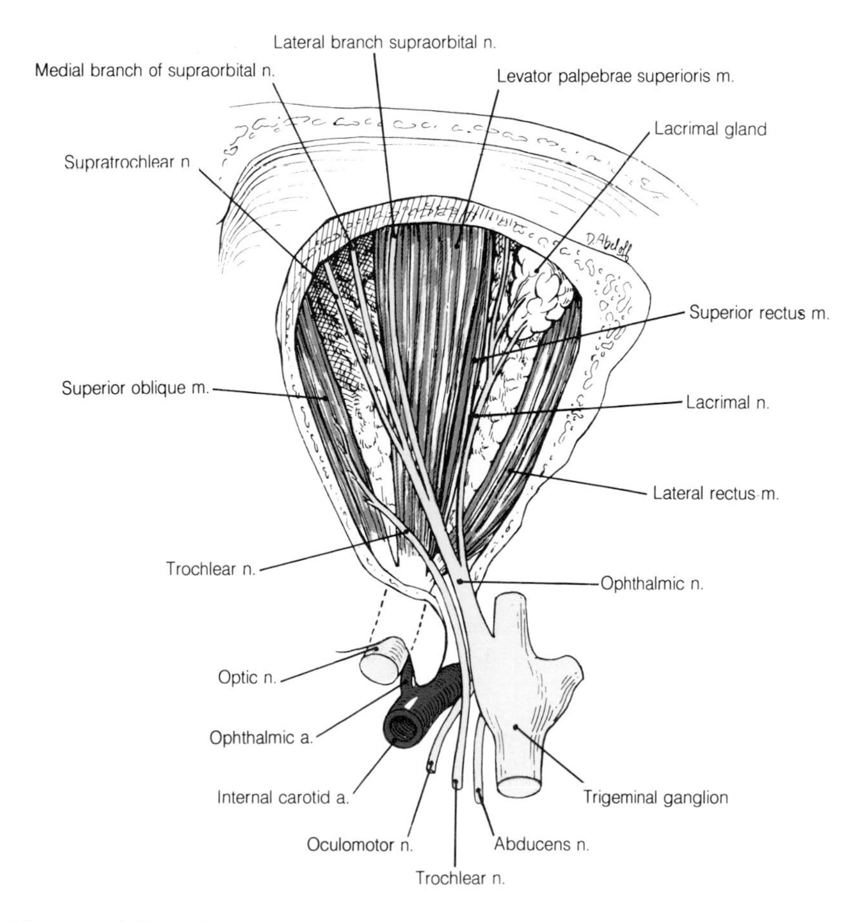

9-47 The right orbit exposed from above.

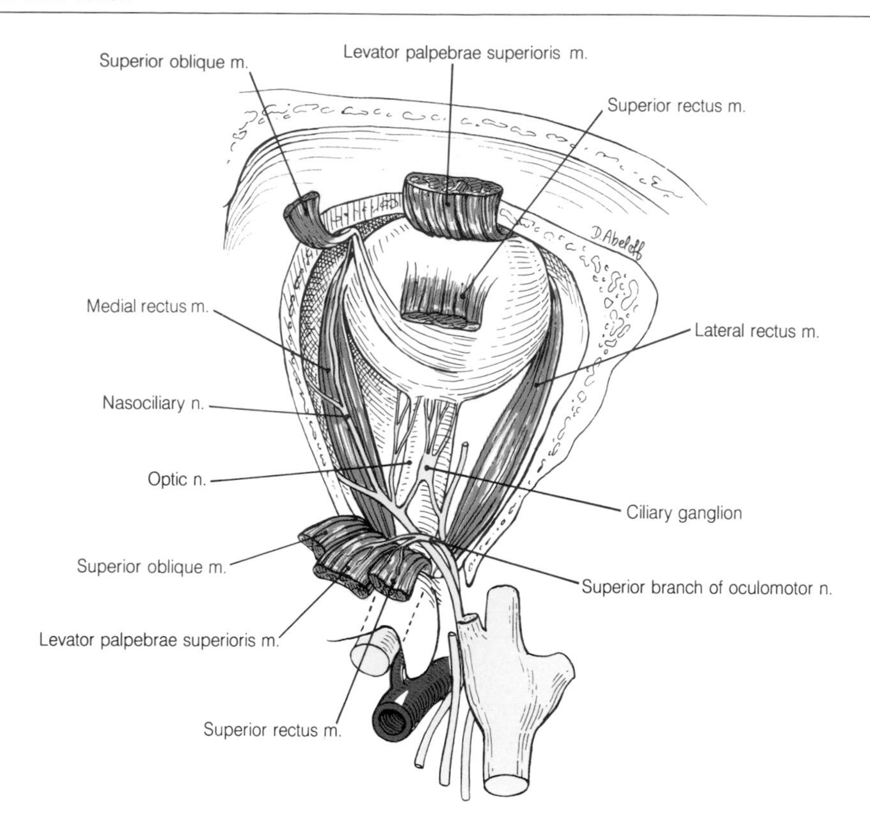

9-48 The right orbit exposed from above. The levator palpebrae superioris, superior rectus, and superior oblique muscles have been divided and reflected.

common annular tendon. Running medially across and superior to the optic nerve it divides into the anterior and posterior ethmoidal nerves and the infratrochlear nerve.

The **anterior ethmoidal nerve** supplies the ethmoidal air cells and appears in the anterior cranial fossa before passing through the nasal slit to enter the roof of the nose. Here it supplies nearby mucous membrane of the septum and lateral wall of the nose before becoming superficial and supplying the skin to one side of the midline of the nose by its **external nasal branch**.

The **posterior ethmoidal nerve**, which is not always present, supplies the sphenoid sinus and ethmoidal air cells.

The **infratrochlear nerve** supplies skin of the upper lid and adjacent conjunctiva and skin of the nose.

Additional important branches of the nasociliary nerve in the orbit are the **nasociliary root to the ciliary ganglion**, which is described later, and **long ciliary nerves** carrying sympathetic fibers to the dilator pupillae muscle and sensory fibers to the cornea.

The Nerves to the Extraocular Muscles

As with the ophthalmic nerve, the trochlear, oculomotor, and abducens nerves are led to the superior orbital fissure by the cavernous sinus. The trochlear nerve enters the orbit above the common annular tendon but the oculomotor and abducens nerves enter through it and remain within the cone of muscles.

The **trochlear nerve** (Fig. 9-47) enters the orbit through the superior orbital fissure above the common annular tendon and running medially, enters the upper border of the superior oblique muscle to supply it. Damage to this nerve leads to a paralysis of the superior oblique muscle and an inability to look downward and laterally with the affected eye.

The **oculomotor nerve** enters the orbit through

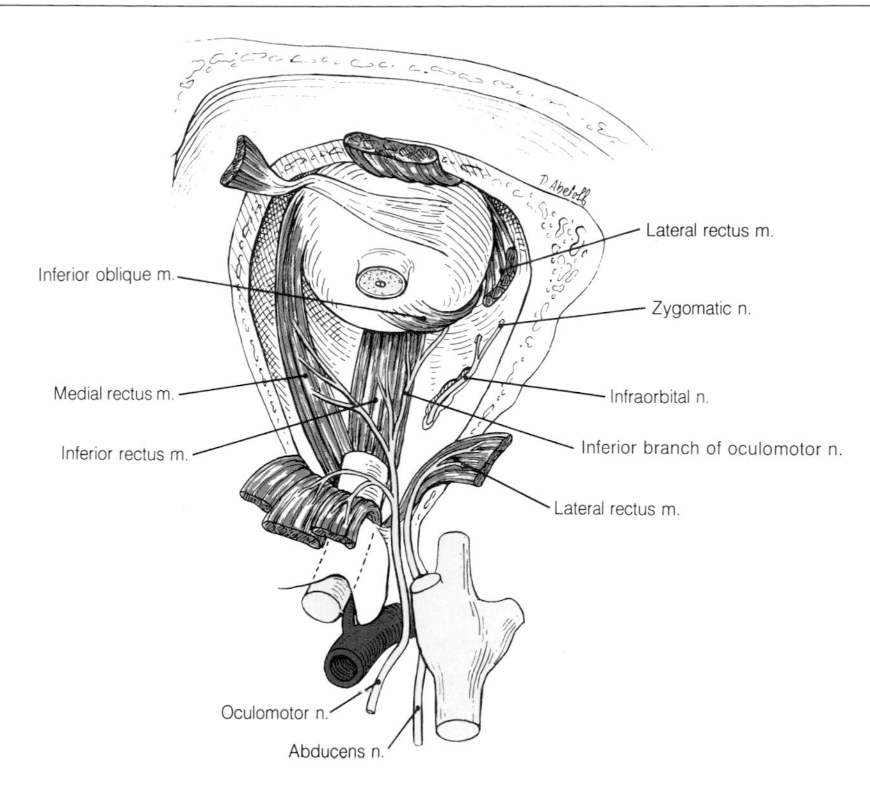

9-49 The right orbit exposed from above. The optic nerve has been divided and the eyeball is depressed.

the superior orbital fissure within the common annular tendon as a **superior and inferior branch**. The superior branch (Fig. 9-48) supplies the overlying superior rectus and levator palpebrae superioris muscles. The inferior branch (Fig. 9-49) supplies the medial rectus, inferior rectus, and inferior oblique muscles. The inferior branch also supplies an **oculomotor root** (parasympathetic) to the **ciliary ganglion**. Damage to the oculomotor nerve and paralysis of the muscles it supplies leads to drooping of the upper lid (ptosis) and an external squint or strabismus due to the unopposed action of the lateral rectus and superior oblique muscles. In addition, the loss of parasympathetic fibers leads to an inability to contract the ciliary muscles and the sphincter pupillae and, thus, a failure to accommodate the eye for close vision.

The **abducens nerve** (Fig. 9-49) enters the orbit through the superior orbital fissure within the common annular tendon. It supplies the lateral rectus muscle. Because of its long course to the orbit from its origin in the posterior cranial fossa, it is particularly susceptible to damage. This results in an internal strabismus or convergent squint.

The Maxillary Nerve

The maxillary nerve lies very close to the orbit in the **pterygopalatine fossa** and is continued into it as the **infraorbital nerve** (Fig. 9-49). In the fossa it gives off the **zygomatic branch** which together with the infraorbital nerve enters the orbit through the inferior orbital fissure.

The **zygomatic nerve** runs laterally in the orbit and branches leave through small foramina in the zygomatic bone to supply skin of the temporal region and face. A further **communicating branch** carries secretomotor (parasympathetic) fibers to the **lacrimal nerve** for transmission to the **lacrimal gland**.

The **infraorbital nerve** enters the infraorbital canal in the floor of the orbit and appears on the face just below the inferior orbital margin through the **infraorbital foramen**. The nerve supplies the upper canine and incisor teeth, the maxillary sinus, the skin and mucous membrane of the cheek, and the adjacent skin of the nose and lower lid.

The Ciliary Ganglion

The ciliary ganglion, about the size of a large pinhead, lies a few millimeters lateral to the optic nerve embedded within the orbital fat. It is a parasympathetic ganglion; i.e., the cell bodies and synapses it contains are part of the parasympathetic nervous system. However, like other parasympathetic ganglia it is traversed by sympathetic and sensory neurons which do not synapse nor have cell bodies. It is closely associated with both the **oculomotor** and **nasociliary nerves**. Also included in the figure above are the **long ciliary nerves** for these also enter the eyeball and are concerned with its function and sensory innervation. To understand the connections of the ganglion look at the diagram in Fig. 9-50 and note that:

1. Preganglionic parasympathetic fibers (*blue*) reach it by its oculomotor root. These fibers are carried from the Edinger-Westphal nucleus in the brainstem and reach the orbit in the oculomotor nerve. They synapse in the ganglion with the cell bodies of postganglionic fibers.
2. The postganglionic parasympathetic fibers (*yellow*) leave the ganglion by a number of fine short ciliary nerves to enter the back of the eyeball around the optic nerve.
3. Postganglionic sympathetic fibers (*red*) from the superior cervical ganglion reach the ganglion by way of a sympathetic root derived from the internal carotid plexus. These fibers also leave the ganglion in the short ciliary nerves.
4. Sensory fibers (*green*) reach the ganglion by way of the nasociliary root. This is a small communicating branch from the nearby nasociliary nerve. These fibers also leave the ganglion in the short ciliary nerves to supply the eyeball.
5. The long ciliary nerves which are branches of the nasociliary nerve carry both sensory and sympathetic fibers to the eyeball.

The extremely important functions of these fibers reaching the eyeball are tabulated below.

1. The parasympathetic fibers supply the sphincter pupillae muscle and the ciliary muscle. It is interruption of these fibers that causes the dilated pupil and failure to accommodate which follow damage to the oculomotor nerve.
2. The sympathetic fibers supply the vessels of the eyeball and the dilator pupillae muscle and for this reason the dilated pupil is associated with fear.
3. The sensory fibers supply the eyeball and the cornea. The integrity of the latter innervation is essential for the protection of the delicate corneal epithelium.

Accommodation

The process of accommodation is a reflex response to viewing objects only a short distance from the eyes as in reading. It involves the following three actions: 1. Convergence of the eyes to retain binocular vision. This is performed by the

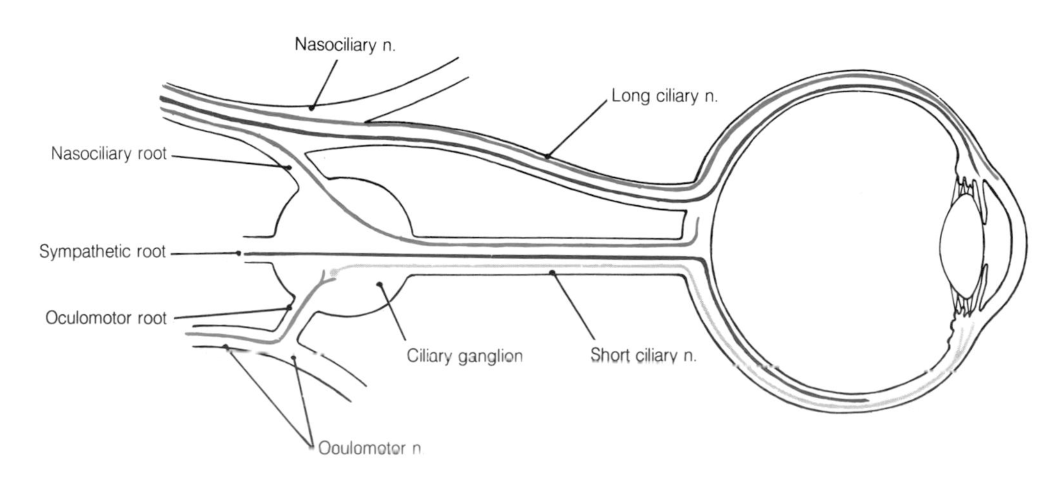

9-50 Diagram of the connections and distribution of the ciliary ganglion and long ciliary nerves.

medial recti. 2. Constriction of the pupil to increase the depth of focus. This action is comparable to reducing the size of the aperture of a camera and is carried out by the sphincter pupillae muscle. 3. Contraction of the ciliary muscle to reduce the peripheral tension on the lens and allow it to assume a more spherical shape and bring a nearby object into focus on the retina.

Note that each of the muscles responsible for these actions is supplied by fibers carried in the oculomotor nerve and hence the loss of accommodation that occurs when the nerve is damaged.

Vessels of the Orbit

The Ophthalmic Artery

The structures within the orbit, including the eyeball, are supplied by a single vessel called the **ophthalmic artery**. This artery is given off by the **internal carotid artery** as it emerges from the roof of the **cavernous sinus** close to the **optic canal**. The ophthalmic artery enters the orbit through the optic canal in company with and inferolateral to the optic nerve. Initially it runs laterally in the orbit but then turns medially and, crossing above the optic nerve in company with the **nasociliary nerve**, reaches the medial wall. From here it continues forward to terminate as the **supratrochlear** and **dorsal nasal arteries**.

The most important branch of the ophthalmic artery is the **central artery of the retina**. This small vessel enters the optic nerve and continues forward with the nerve to enter the eyeball and supply the retina. There is neither an effective anastomosis of this artery with other vessels reaching the eyeball nor are there anastomoses between the four terminal branches which supply the retina in quadrants. Any obstruction therefore of the main vessel or its branches will lead to total or partial blindness. The appearance and distribution of these branches in the retina can be seen through an ophthalmoscope and are illustrated in Fig. 9-51. The distribution of the remaining branches is illustrated in Fig. 9-52. In the figure note that the branches supply the eyeball through the **ciliary arteries**, the lacrimal gland and the lateral parts of both upper and lower lids through the **lacrimal artery**, the ethmoidal air sinuses and nasal cavities through the **anterior** and **posterior ethmoidal arteries**, the forehead and scalp through the **supratrochlear** and **supraorbital arteries**, and the re-

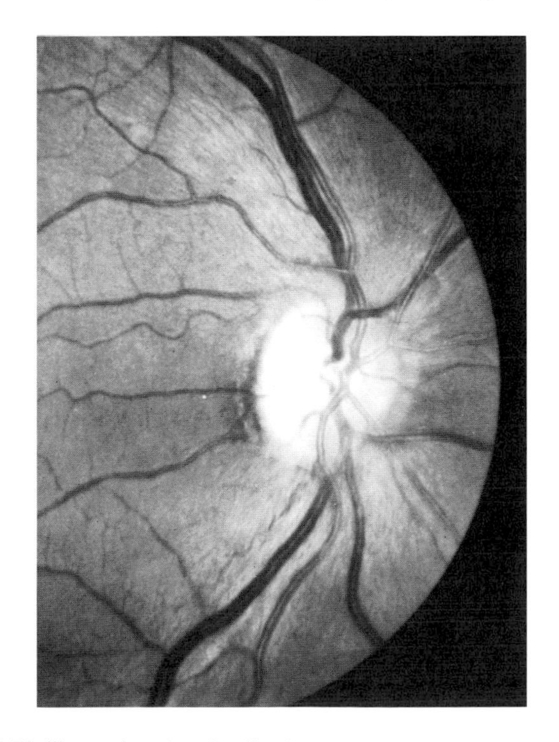

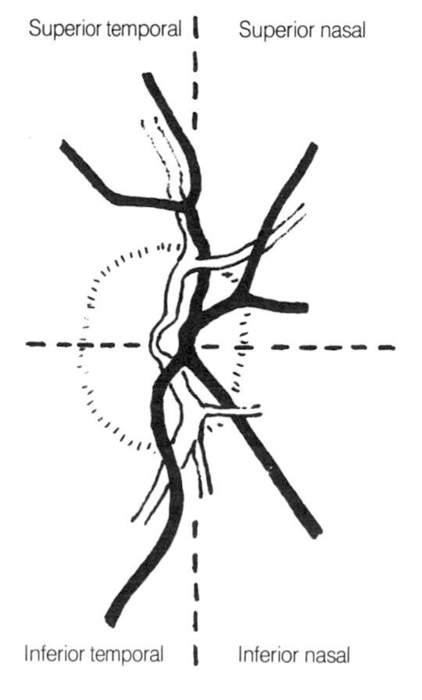

9-51 Illustrating the distribution of the central artery of the retina (*black* in diagram) to four quadrants of the retina. (Courtesy of Lois A. Young, M. D., Department of Ophthalmology, University of Maryland School of Medicine; Photographer, William Buie).

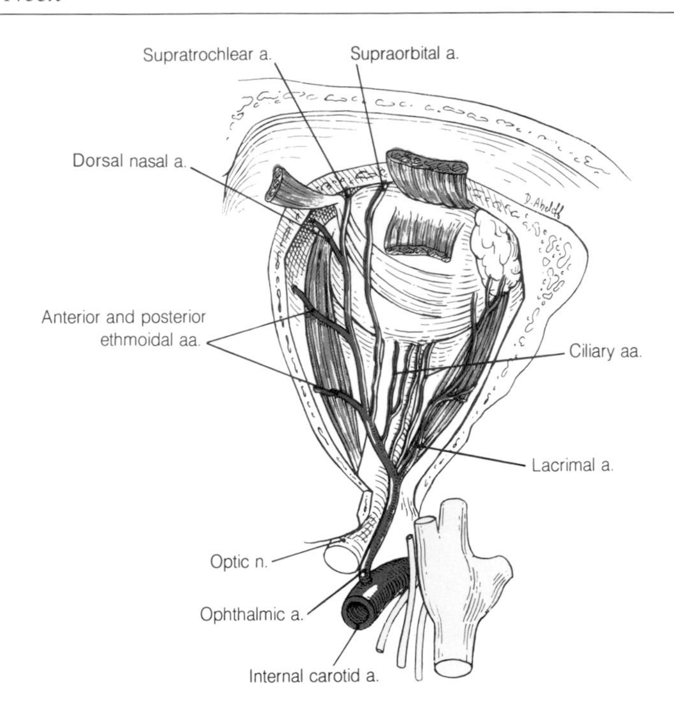

9-52 The right orbit exposed from above to show the course and distribution of the ophthalmic artery.

gion adjacent to the bridge of the nose through the **dorsal nasal artery**. Additional branches which are not illustrated supply the extraocular muscles.

The Ophthalmic Veins

The **superior ophthalmic vein** is formed by tributaries corresponding to the branches of the ophthalmic artery. It follows the artery back through the orbit but leaves trough the **superior orbital fissure** to open into the **cavernous sinus.**

The **inferior ophthalmic vein** is formed in the floor of the orbit and usually joins the superior vein.

The superior ophthalmic vein communicates with the **facial vein** and the inferior communicates with the **pterygoid plexus of veins** through the inferior orbital fissure. As neither ophthalmic vein possesses valves, blood draining from the face may reach the cavernous sinus and thus form a possible route for the spread of infection from the face into the cranial cavity.

The Eyeball

Within the orbit is suspended the eyeball. It is approximately spherical and has a diameter of about 2.5 cm. Projecting forward from this sphere is a segment of a sphere of lesser diameter. This is the translucent cornea.

The part of the eyeball that can be examined through the palpebral fissure is covered by conjunctiva which is reflected on to the upper and lower lids at the conjunctival fornices. The remainder of the eyeball is loosely surrounded by a fine fascial sheath. This sheath fuses with the underlying fibrous coat of the eyeball at the conjunctival fornices and at sites at which vessels, nerves, and tendons enter. The fascial sheath is also continuous with the fascial coverings of the extrinsic muscles. The whole is known as the **fascia bulbi**.

The wall of the eyeball is formed of three coats. The outer is the **fibrous coat** and is made up of the **sclera** which is continuous anteriorly with the **cornea** at the corneoscleral junction or **limbus**. An intermediate **vascular coat** comprises the **choroid, ciliary body**, and **iris**. The **internal coat** is the light-sensitive **retina**. The fibrous and vascular coats are penetrated posteriorly by the **optic nerve** whose fibers become continuous with the retina and whose dural covering fuses with the sclera.

The wall of the eyeball surrounds the chambers of the eye. The **anterior chamber** lies between the

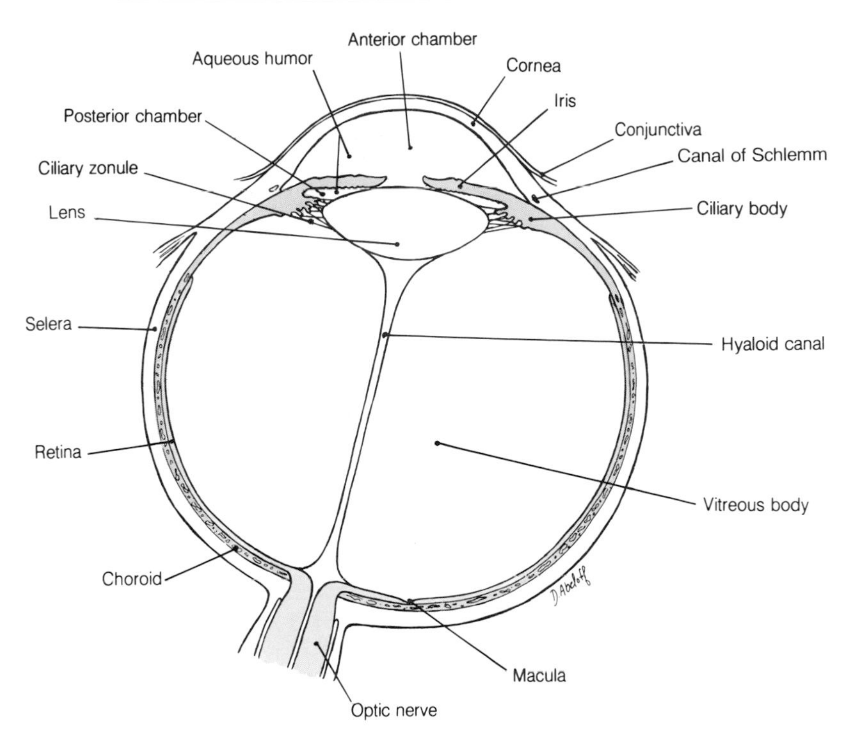

9-53 Semischematic illustration of a horizontal section through the right eyeball.

cornea and the iris and communicates through the pupil of the iris with the **posterior chamber** which lies between the iris and the **lens** with its surrounding **ciliary zonule**. Both these chambers contain **aqueous fluid**. The large **vitreous chamber** fills the remainder of the eyeball behind the lens and contains a gelatinous substance called the **vitreous*** **body**. Each of these structures should now be identified in Fig. 9-53 before they are described in more detail.

The Fibrous Coat

The **sclera** is a tough membrane of dense connective tissue surrounding about five-sixths of the eyeball. Its outer surface is white and forms "the white of the eye" that is seen through the conjunctiva covering the visible portion. Posteriorly the dural sheath of the optic nerve fuses with the sclera and where its nerve bundles penetrate the sclera a sieve-like region is formed. This is the **lamina cribrosa** which is the weakest area of the sclera. When intraocular pressure is raised, as in glaucoma, this region bulges outward and a "cupped" optic disc is seen through the ophthalmoscope. Anteriorly the sclera is continuous with the cornea at the **limbus.**

The **cornea** is also a tough membrane, but the flattened lamellae of fibrous tissue of which it is formed are quite transparent. It is not often realized that the cornea is a lens of fixed focal length and that it contributes substantially to refraction in the eye. The epithelium covering the outer surface of the cornea is supplied by the **nasociliary branch** of the **ophthalmic nerve** and the importance of this innervation and the secretion of tears in protecting the cornea has already been discussed.

The Vascular Coat or Uveal Tract

The **choroid** is an extremely vascular, thin, pigmented membrane that is loosely attached to the inner aspect of the sclera. Only at the optic disc is it firmly adherent. Anteriorly it blends with the ciliary body.

The **ciliary body** is concerned with the suspension of the lens and with increasing its power of refraction in the process of accommodation. It

* Vitreus, *L.* = made of glass

contains the **ciliary muscle** which is composed of smooth muscle fibers arranged in a circular and a radiating pattern. Extending from the ciliary body toward the lens are the **ciliary processes** which adhere to the fibers of the **ciliary zonule** (suspensory ligament of the lens).

The **iris** is a flat, highly vascular, and pigmented disc continuous with the anterior surface of the ciliary body. It is perforated centrally at the **pupil**. It contains the smooth muscle fibers of the **sphincter** and **dilator pupillae** which control the size of the pupil.

The Internal Coat

The **retina** is the light-sensitive membrane lining the vascular coat. It does not extend as far forward as the ciliary body but terminates at an irregular margin called the **ora serrata**. However, a thin membrane devoid of nervous tissue but including the pigmented layer is continued beyond the ora serrata to line the deep surface of the ciliary body and the iris. These are the **ciliary** and **iridial parts of the retina**. The retina is built up of a number of quite discrete layers, but outermost of which is a **layer of pigmented cells**. Immediately within this lie the light-sensitive cells called **rods** and **cones**. These cells are most numerous over the portion of the retina called the **macula lutea** which lies around the optical axis of the eyeball but become less so toward the periphery. Their absence from the region of the **optic disc** accounts for the "blind spot". At the **fovea centralis**, which lies at the center of the macula, only cones are found. Inside the layer of rods and cones there is a layer of bipolar nerve cells which synapse with the rods and cones peripherally and the cell bodies of the ganglion cells centrally. The central processes of the ganglion cells form the innermost layer of the retina and come together at the optic disc to form the optic nerve. Only a fine **internal limiting membrane** separates these nerve fibers from the **vitreous body**. The organization of the retina is illustrated in the simplified diagram shown in Fig. 9-54. Note that the receptor cells lie in the outer part of the retina.

When examined with an ophthalmoscope a number of important features of the retina can be identified. Look at Fig. 9-55 in which a typical view of the retina is shown. Identify the **optic disc**

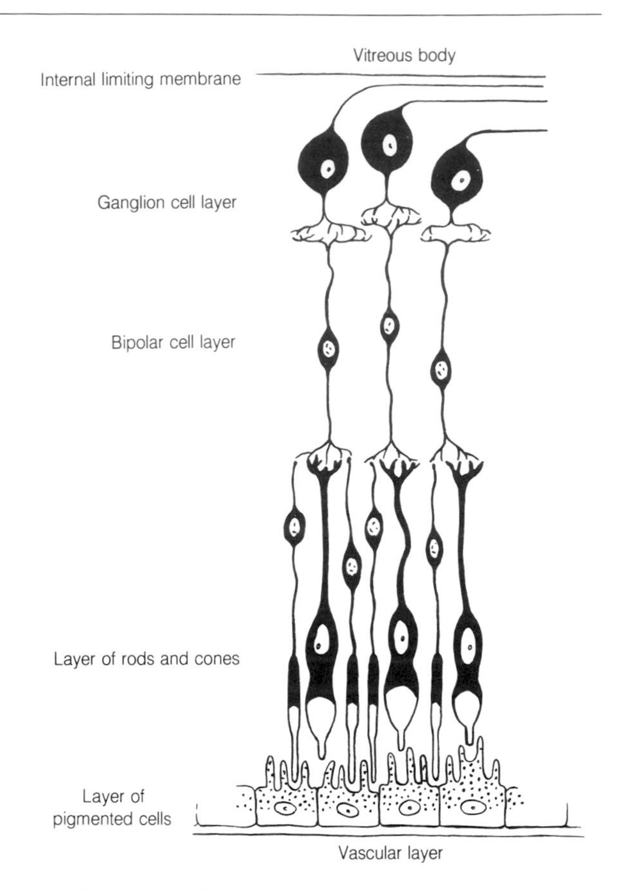

9-54 Diagram of the layers of the retina.

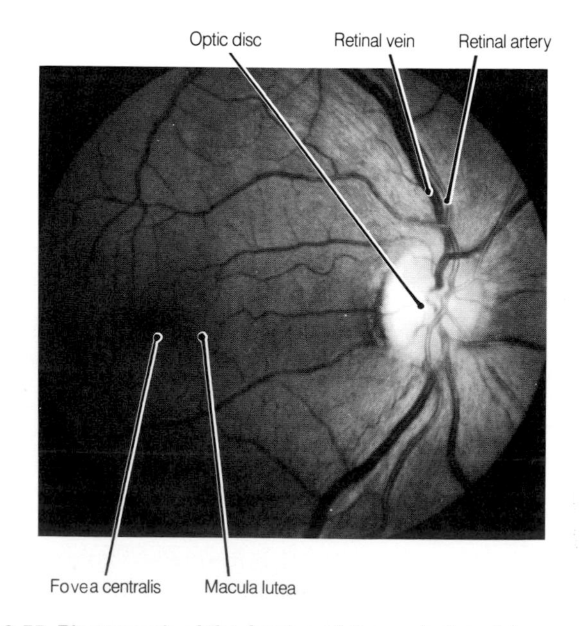

9-55 Photograph of the fundus of the eyeball as it is seen through the ophthalmoscope. (Courtesy of Lois A. Young, M.D., Department of Ophthalmology, University of Maryland School of Medicine; Photographer, William Buie).

which is much lighter than the rest of the retina. From the center of this radiate the terminal branches of the **central artery of the retina** each of which supply one quadrant. The retinal veins, which do not closely follow the arteries, converge on the disc. About three millimeters to the lateral side of the optic disc the yellowish **macula lutea** can be seen. Although free of major blood vessels, it has a profuse capillary blood supply. It is here that the rods and cones are most densely packed and vision is most acute. At the center of the macula lutea there is a shallow pit called the **fovea centralis**. This is the thinnest part of the retina and little but supporting glial cells overlie an area of closely packed cones.

The Aqueous Humor

The aqueous humor fills the anterior and posterior chambers of the eye. It is formed in the posterior chamber by active transport from the capillaries of the ciliary body, passes into the anterior chamber through the pupil and is absorbed at the angle between the cornea and iris into the **sinus venosus sclerae** (canal of Schlemm) (Fig. 9-53) and **anterior ciliary veins**. The aqueous humor carries nutrients to the avascular cornea and lens and is responsible for the maintenance of the intraocular pressure. Obstruction to its drainage leads to a rise in intraocular pressure as in the condition of glaucoma.

The Lens

The lens lies between the posterior chamber and the vitreous body. It is transparent, avascular, and biconvex in shape. It is formed of a series of concentrically arranged lamellae attached to anterior and posterior Y-shaped septa which pass into the substance of the lens. A nucleated columnar epithelium covers the anterior surface of the lens and the whole is surrounded by an elastic capsule.

The Vitreous Body

The vitreous body fills the remaining and greater part of the eyeball. It is a transparent, structureless gel. Between the lens and the optic disc it is traversed by the **hyaloid canal** which in fetal life carried the **hyaloid artery**. Anteriorly the vitreous body is condensed to form the **ciliary zonule** between the ciliary processes and the lens. Radially arranged fibers in this are more commonly known as the suspensory ligament of the lens.

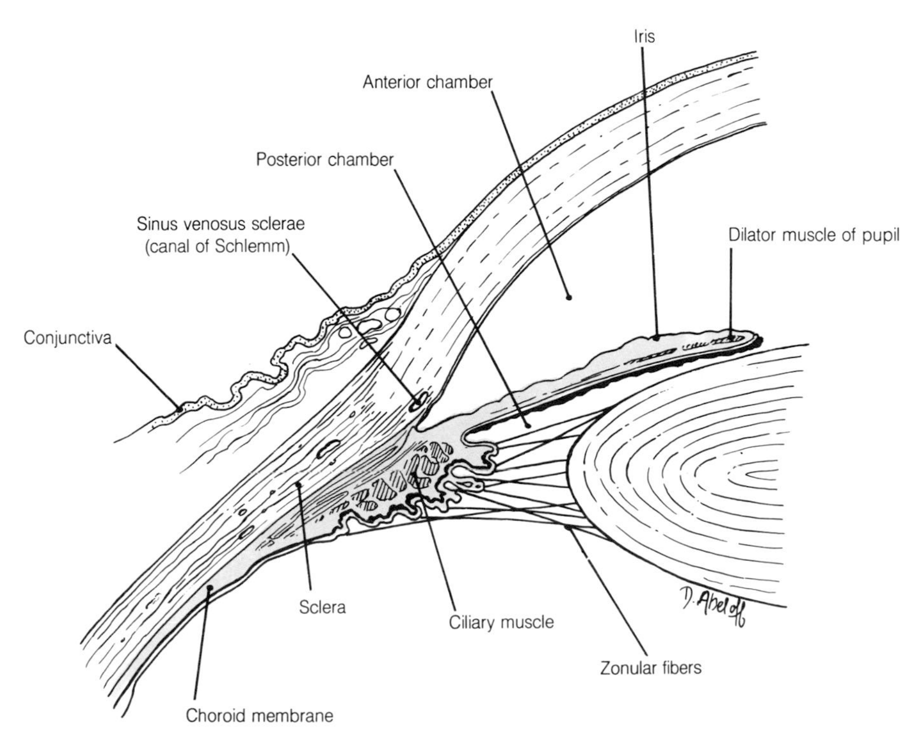

9-56 Semischematic illustration of the iridocorneal angle.

In Fig. 9-56 a more detailed illustration of the region around the ciliary body and the angle between the cornea and iris shows many of the structures that have been described.

The Blood Supply of the Eyeball

The eyeball is supplied by branches of the ophthalmic artery. The importance of the **central artery of the retina** in the supply of the internal nervous coat has already been discussed. The vascular coat is supplied by **long and short posterior ciliary arteries** and **anterior ciliary arteries** which reach the eyeball at the attachments of the extraocular muscles. An arterial circle is formed near the periphery of the iris and from this **iridial arteries** are derived. Much of the venous blood returns through a venous plexus to four **venae vorticosae** which leave the eyeball just behind its equator. The arrangement of these vessels is shown diagrammatically in Fig. 9-57.

The Parotid Region

The parotid region overlies in part both the temporal and infratemporal regions and has, therefore, an intermediate position between the superficial and deep structures of the face. The main structure within it is the **parotid gland** which has already been mentioned in reference to the distribution of the facial nerve. Also within the gland are the **retromandibular vein**, the **external carotid artery**, and the origins of its two terminal branches, the **maxillary** and **superficial temporal arteries**.

The Parotid Gland

The parotid gland is the largest of the paired salivary glands which open into the oral cavity. It lies wedged between the ramus of the mandible and the masseter anteriorly and the sternocleidomastoid posteriorly and overlaps both the muscles

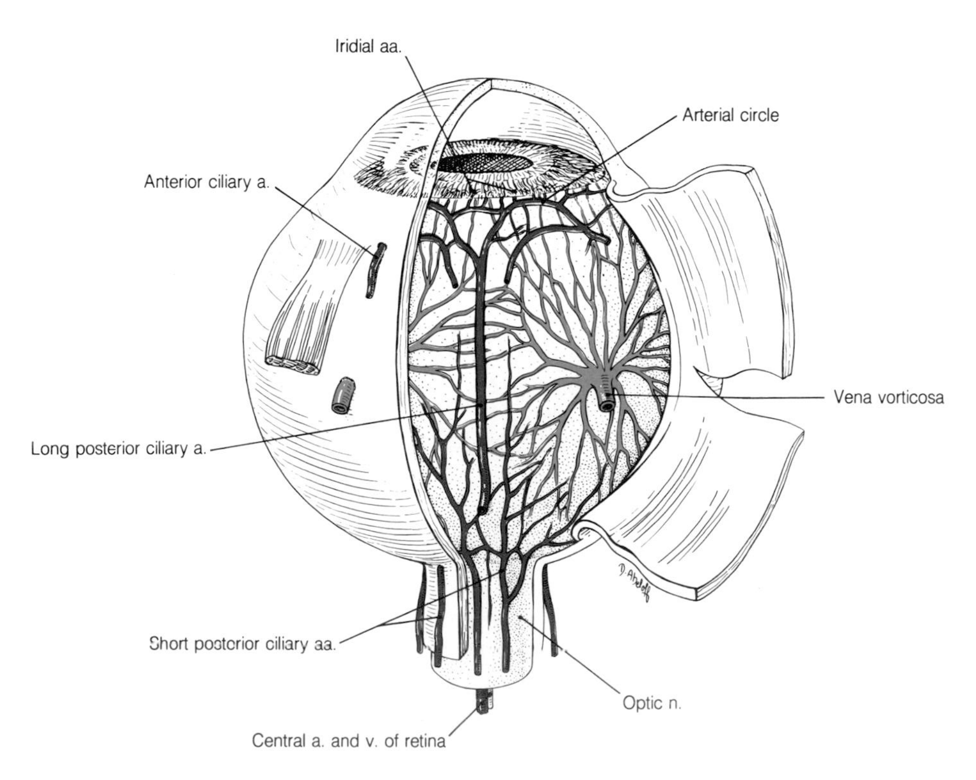

9-57 Semischematic illustration of the blood vessels of the eyeball.

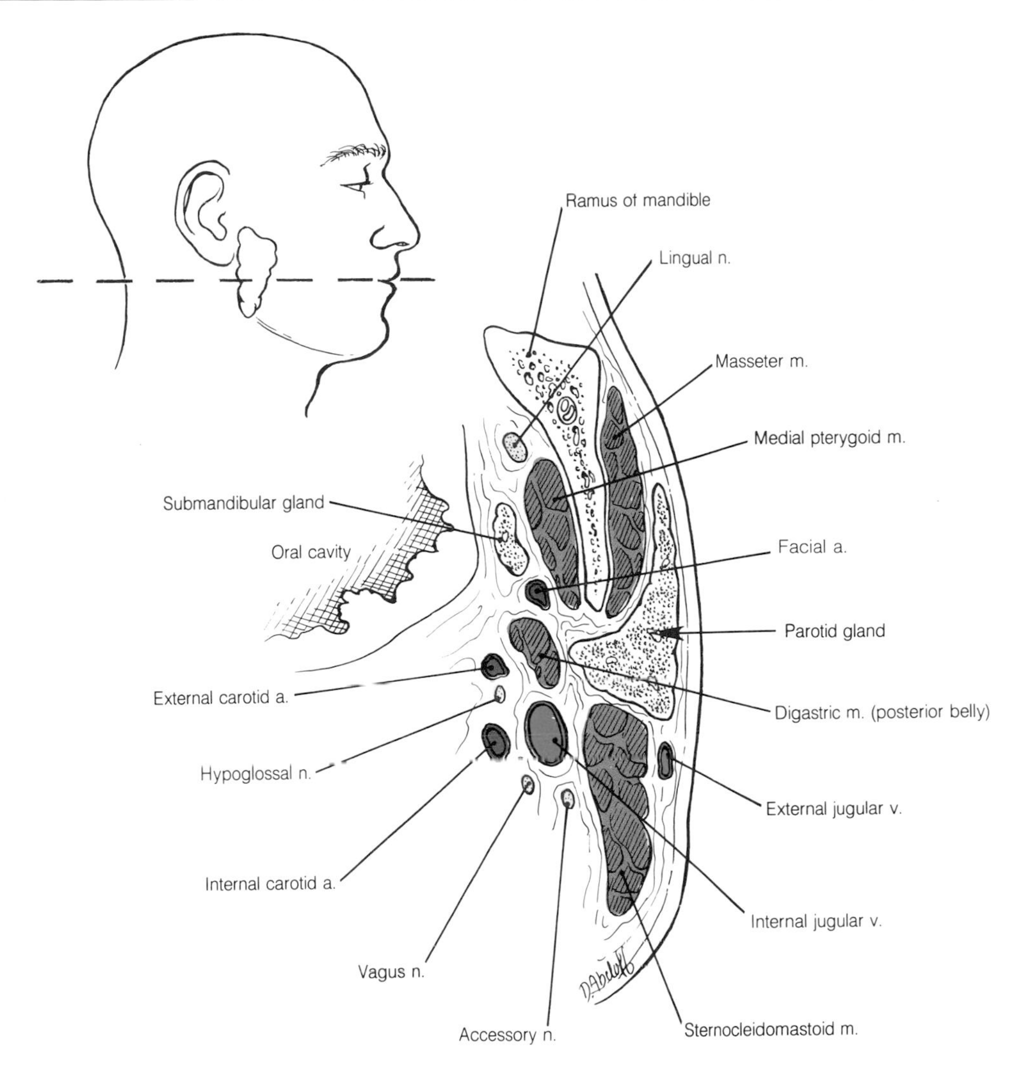

9-58 Horizontal section through the lower part of the ramus of the mandible to show the parotid gland and its nearby relations.

superficially. Its deepest part is related to the posterior belly of the digastric muscle. Confirm these relationships in Fig. 9-58 and also note how close its deep surface is to the internal jugular vein, internal carotid artery, and the nerves associated with them. The superior surface of the gland is related to the cartilaginous part of the external acoustic meatus and the temporomandibular joint and its more superficial part overlaps the zygomatic arch above these. Superficially the gland is covered by the skin and platysma muscle and branches of the great auricular nerve will cross it as seen in Fig. 9-59.

The superficial layer of the deep cervical fascia divides to enclose the whole gland in a fascial sheath which separates its lower part from the submandibular salivary gland. The pain of acute viral parotitis or mumps is probably due to the pressure within the sheath caused by the inflammation.

From the anterior border of the gland appears the **parotid duct**. This crosses the surface of the masseter and at its anterior border turns medially to pierce the buccal pad of fat and the buccinator muscle. It opens into the oral cavity opposite the crown of the second upper molar tooth. While crossing the masseter the parotid duct receives the

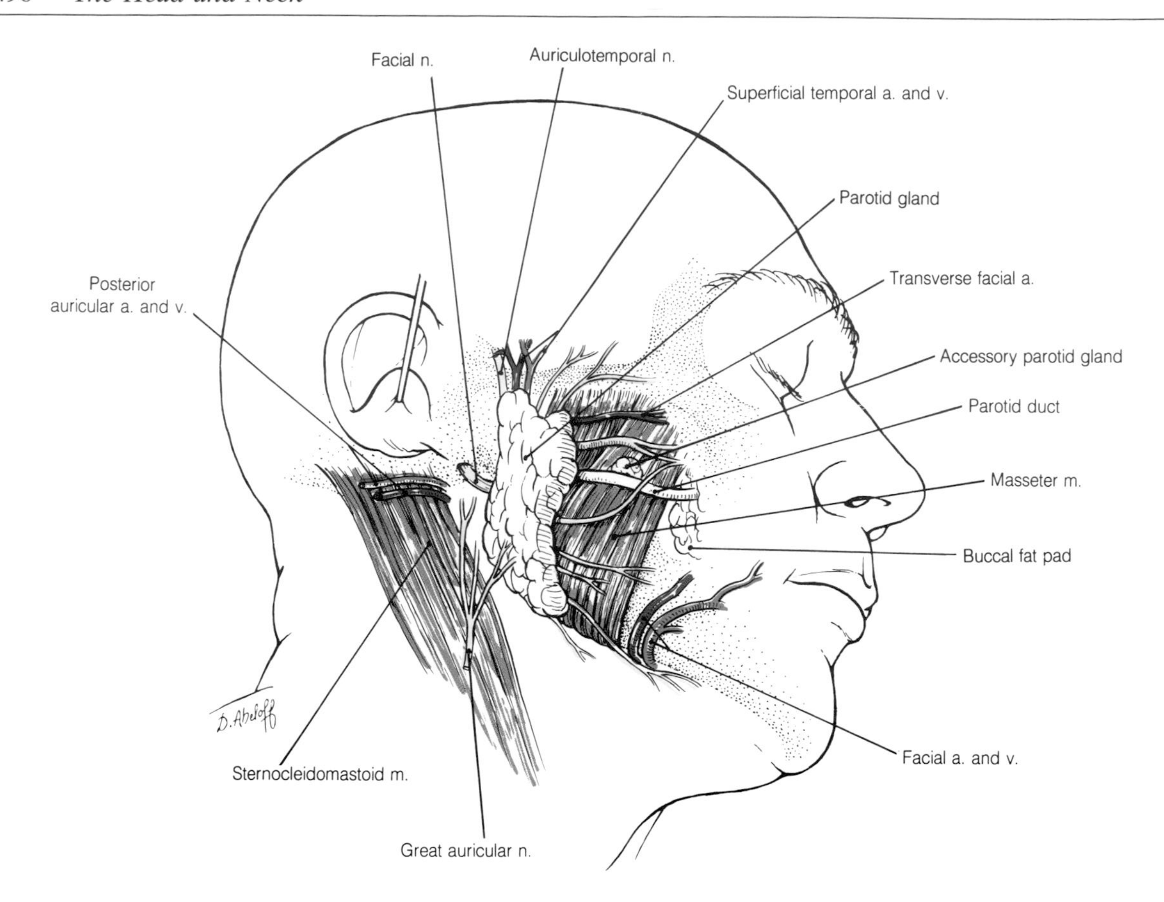

9-59 The parotid region exposed to show the parotid gland, the parotid duct, and related vessels and nerves.

duct of the **accessory parotid gland** which lies above the main duct. The duct and accessory gland can be seen in Fig. 9-59.

A number of important structures lie within the substance of the gland. Of these the most superficial and perhaps the most important is the **facial nerve**. In Fig. 9-59 this can be seen to enter the posterior border of the gland as a single trunk and appear at its anterior border as the branches of distribution that were described earlier with the face. The gland is also traversed vertically by the **retromandibular vein** and by the **external carotid artery** which enters its deep surface and divides into the **maxillary and superficial temporal arteries**. The maxillary artery passes forward on the medial side of the neck of the mandible to enter the infratemporal fossa. The superficial temporal artery gives off the **transverse facial artery** and then appears at the upper border of the gland.

The gland receives its blood supply from the external carotid artery and its nearby branches and veins drain to the retromandibular and external jugular veins. Lymph nodes lie both on the superficial surface and embedded in the gland and drain to the superficial and deep cervical nodes.

Postganglionic sympathetic fibers reach the gland from the plexuses on the external carotid and middle meningeal arteries. Postganglionic parasympathetic secretomotor fibers are carried to the gland from the **otic ganglion** by the **auriculotemporal nerve**. The otic ganglion and the source of the preganglionic fibers from the glossopharyngeal nerve are described with the infratemporal fossa.

Wounds of the cheek may damage the parotid duct and it is worth knowing that this lies on the middle third of a line drawn between the tragus of the ear and the midpoint between the nostril and angle of the mouth.

The Temporal and Infratemporal Regions

Before discussing these regions, the relevant osteology of the skull and mandible need more careful examination. In Fig. 9-60 find the **zygomatic process of the frontal bone** and follow its posterior border upward and backward. It will lead to the **inferior temporal line** which marks the upper limit of the temporal fossa and the attachment of the muscle **temporalis**. Sometimes a more faint line, the superior temporal line, can be seen above this; it gives attachment to the tough **temporal fascia**. If the line of the posterior border of the process is followed downward it will lead to the **frontal process of the zygomatic bone** from which it is then deflected posteriorly onto the **zygomatic arch**. This is formed by the zygomatic bone anteriorly and the temporal bone posteriorly. This arch bridges over the temporalis muscle and provides the superior attachment of the masseter muscle. The zygomatic arch blends with the squamous portion of the temporal bone and below its posterior end is the **external acoustic meatus**. The temporal lines, the frontal process of the zygomatic bone, and the zygomatic arch form the boundaries of the **temporal fossa**. The floor of the fossa can be seen to be formed by the parietal, temporal, frontal, and sphenoid bones. Where these bones approximate to each other is a landmark of physical anthropology, the pterion. This is also the surface marking of the anterior branch of the middle meningeal artery and the anterior end of the lateral sulcus of the brain.

If the lower border of the zygomatic arch is traced backward it will also lead to the **articular tubercle** and the **mandibular fossa** on the inferior surface of the temporal bone. As will be seen, these two structures form integral parts of the temporomandibular joint.

The temporal fossa leads downward into the **infratemporal fossa** which can also be seen in Fig. 9-60. This lies between the ramus of the mandible (omitted from the figure) and the lateral pterygoid plate of the sphenoid bone which projects downward from the base of the skull. It is roofed medially by the inferior surface of the

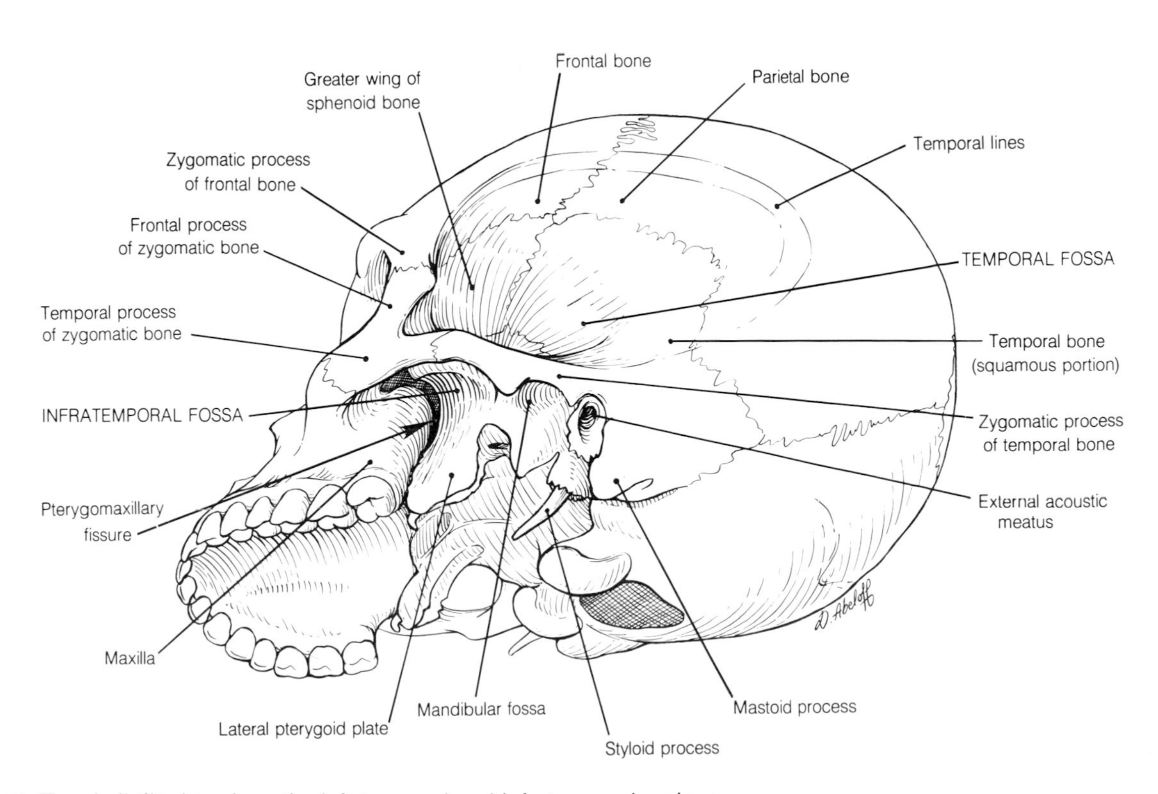

9-60 The skull tilted to show the left temporal and infratemporal regions.

greater wing of the sphenoid bone. Anteriorly lies the posterior surface of the maxilla. Posteriorly lie the temporomandibular joint, the carotid sheath, and the styloid process. Inferiorly, the space is continuous with the soft tissues of the neck. Note that between the lateral pterygoid plate and the maxilla there is an irregular gap called the **pterygomaxillary fissure**. It leads to the **pterygopalatine fossa**.

Some features of the **mandible** also need to be described and these are illustrated in Fig. 9-61. Each half possesses a curved horizontal **body** and these two half-arches are united anteriorly at the **symphysis menti**. In addition, a vertical **ramus** ascends from each body posteriorly. The ramus and the body are joined at the **angle of the mandible**. The ramus supports two processes. Anteriorly, there is a pointed **coronoid process**. Posteriorly lies the **condylar process**. The two processes are separated by the **mandibular notch**. The lateral surface of the ramus shows no distinctive feature other than markings for the attachment of the masseter muscle. The medial surface shows the **mandibular foramen** partly covered by a spine of bone called the **lingula**. The region of the angle shows a well-marked roughened area for the attachment of the **medial pterygoid muscle**.

The temporal and infratemporal fossae are occupied by the muscles of mastication, the maxillary artery and its branches, the pterygoid plexus of veins, branches of the mandibular division of the trigeminal nerve, the otic ganglion, and the chorda tympani nerve. The temporomandibular joint is an important and closely related structure.

The Muscles of Mastication

The muscles of mastication are all developed from mesoderm of the first branchial arch and are all supplied by the trigeminal nerve. In practice, the motor fibers of this nerve are contributed only to

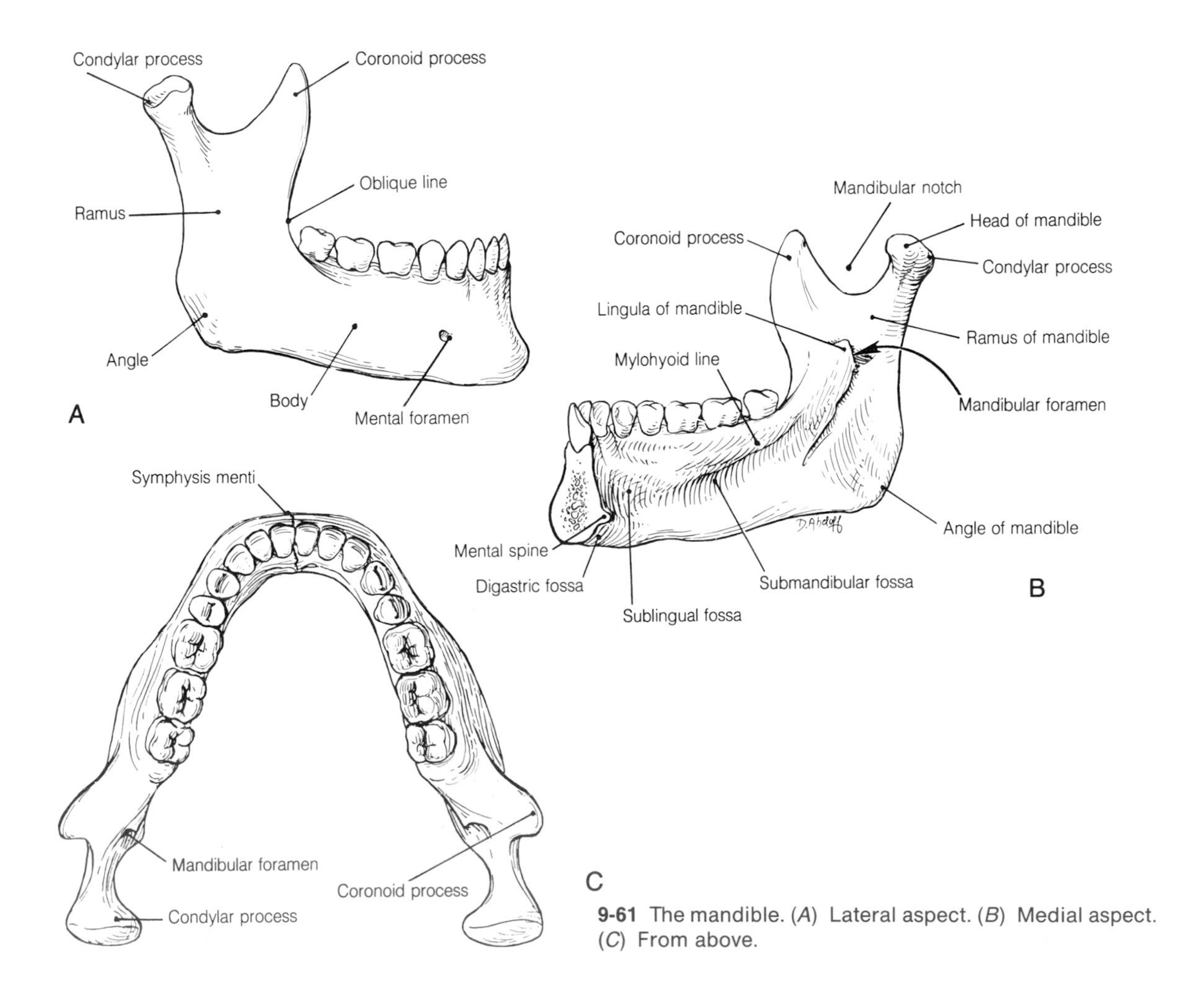

9-61 The mandible. (*A*) Lateral aspect. (*B*) Medial aspect. (*C*) From above.

the mandibular nerve and it can, therefore, be said that all the muscles are supplied by this nerve. Each of these muscles arises from the skull and is inserted into the mandible causing its movement at the temporomandibular joint. Their positions and attachments are illustrated in Fig. 9-62. As each muscle is described note the direction of its fibers as this determines its action.

Masseter

The masseter arises from the lower border and medial surface of the zygomatic arch and is attached to the lateral aspect of the angle and lower half of the ramus of the mandible. This powerful quadrilateral muscle elevates the mandible and occludes the teeth. It is supplied by the **masseteric nerve**, a branch of the mandibular nerve

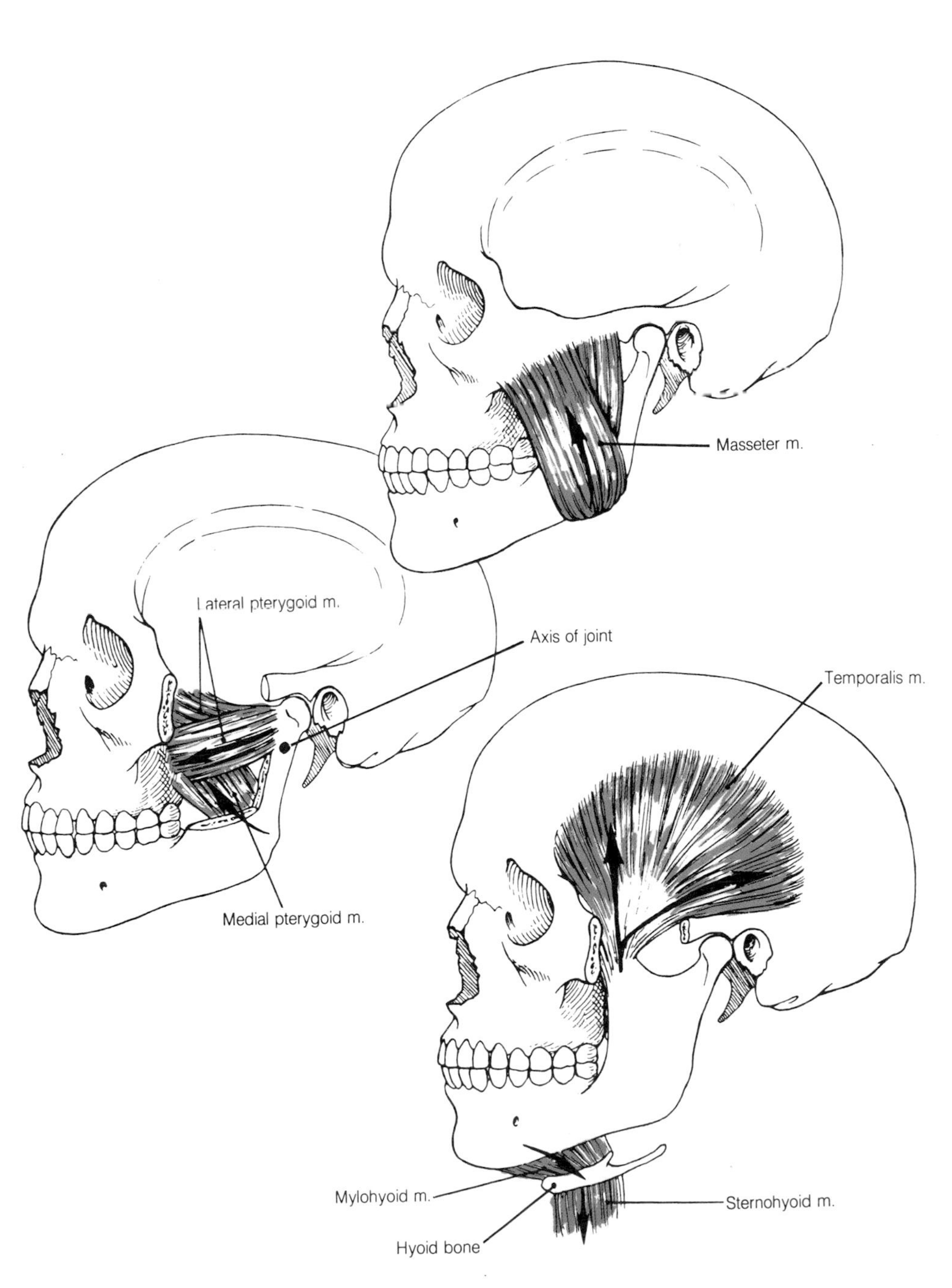

9-62 Illustrating the muscles of mastication.

which reaches the deep surface of the muscle through the mandibular notch.

Temporalis

The temporalis muscle arises from the temporal fossa and a tough layer of overlying temporal fascia. The fan-shaped belly narrows to a tendon which passes deep to the zygomatic arch to be attached to the borders of the coronoid process of the mandible and the anterior border of the ramus. The varied angulation of its fibers allows temporalis to both elevate the mandible and draw it backward or retract it. The muscle is supplied by **deep temporal nerves** which are branches of the mandibular nerve.

The Pterygoid Muscles

The two pterygoid muscles often lead to problems in the understanding of their attachments and actions. Both arise from the lateral pterygoid plate which lies anterior and medial to the temporomandibular joint and is deflected laterally guiding the muscles toward the mandible. Because their origins are anterior to the joint both may displace the mandible forward (protrusion). However, as the lateral pterygoid is attached to the mandible above the level of the axis of the movement of the joint, it will aid in depression while the medial pterygoid, whose fibers pass downward in front of this axis, will produce elevation. Asynchronous actions of the left and right pterygoid muscles produce the movement of chewing.

The **medial pterygoid muscle** arises largely from the medial surface of the lateral pterygoid plate. The plate guides the fibers downward, backward, and laterally to their attachment to the medial surface of the angle of the mandible below the level of the mandibular foramen. Its line of action lies in front of the axis of the temporomandibular joint and it, therefore, helps to elevate the mandible. The anterior position of the muscle also allows it to protrude the mandible. The muscle is supplied by the **medial pterygoid nerve**, a branch of the mandibular nerve.

The **lateral pterygoid muscle** arises by two heads, one from the lateral surface of the lateral pterygoid plate and one from the inferior aspect of the greater wing of the sphenoid that has been seen to roof the infratemporal fossa. The two heads are, of course, adjacent to each other. The muscle narrows toward its attachment to an anterior impression on the neck of the mandible just below the head and to the capsule and articular disc of the temporomandibular joint. It pulls both the head and the disc forward when the mouth is opened. The lateral pterygoid is supplied by the **lateral pterygoid nerve**, a branch of the mandibular nerve.

The Mandibular Nerve

In an earlier section, the mandibular nerve has been followed to its exit from the skull through the foramen ovale. This foramen leads the nerve into the infratemporal fossa where it is immediately joined by the motor root of the trigeminal nerve. It, therefore, contains, after this point, all the motor fibers distributed by the trigeminal nerve. The motor fibers innervate tensor veli palatini, tensor tympani, the four muscles of mastication, the anterior belly of the digastric muscle, and the mylohyoid muscle. On leaving the foramen ovale, the nerve is separated from the pharynx by the tensor veli palatini muscle. Superficially lies the upper head of the lateral pterygoid muscle. Before the nerve divides into an anterior and posterior division, the main trunk gives off two branches. These are a **meningeal branch**, which re-enters the cranial cavity through the foramen spinosum, and the **nerve to the medial pterygoid muscle**. A fine **communicating branch** also leaves this nerve to pass through the otic ganglion without synapse and supply the tensor tympani and tensor veli palatini muscles, which lie close by. After these branches have left the nerve, it divides into a small short anterior division and a longer and stouter posterior division.

Distribution of the Anterior Division of the Mandibular Nerve

All the branches of the anterior division contain motor fibers to muscles except the buccal nerve which is purely sensory. These branches are illustrated diagrammatically in Fig. 9-63. Note:

1. two **deep temporal nerves** which appear above the upper border of the lateral pterygoid to supply temporalis;

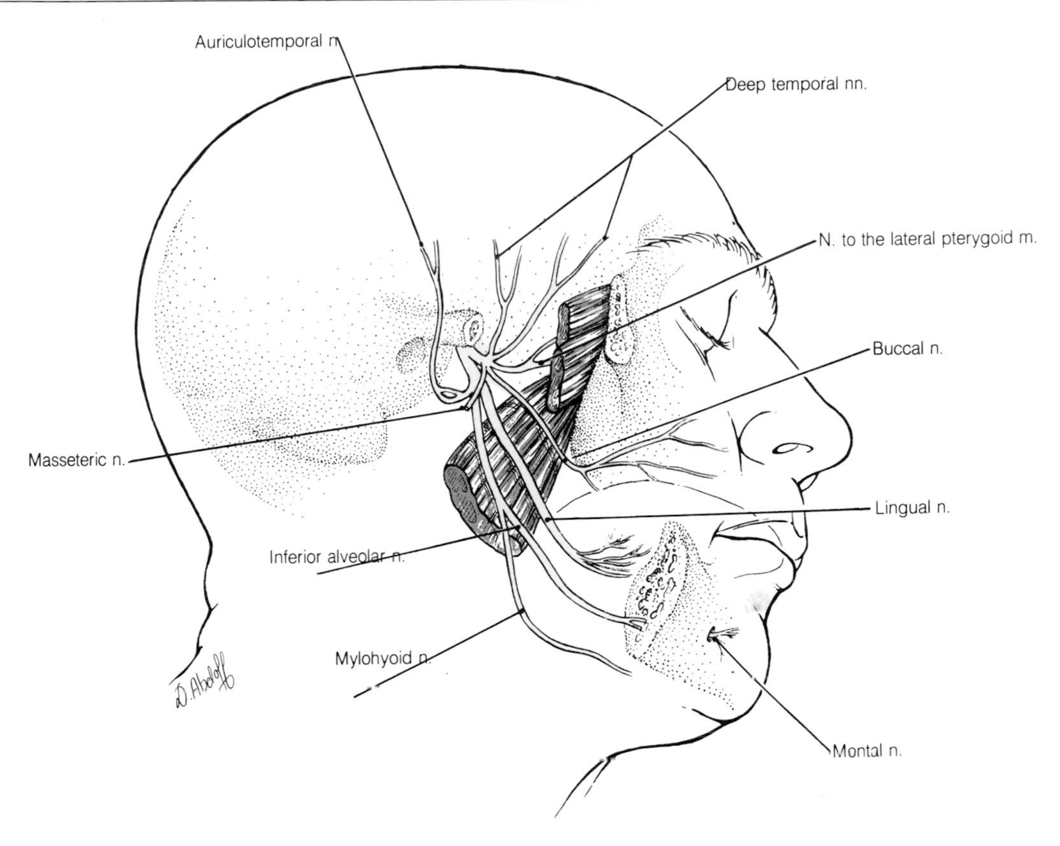

9-63 Showing the distribution of the mandibular nerve. The ramus of the mandible and the zygomatic arch have been removed.

2. the **masseteric nerve** which follows a similar course but turns laterally through the mandibular notch of the mandible to supply the masseter;
3. the **lateral pterygoid nerve** which enters the deep surface of that muscle; and
4. the **buccal nerve** which passes forward between the two heads of the lateral pterygoid to appear at the anterior border of the masseter. Here it supplies skin of the cheek, the underlying mucous membrane of the oral cavity and the buccal surface of the gums.

Distribution of the Posterior Division of the Mandibular Nerve

The posterior division is considerably larger than the anterior division. Its branches are all sensory and only one, the inferior alveolar nerve, carries motor fibers. Follow Fig. 9-63 closely as the branches are described.

The large **lingual nerve** appears at the lower border of the lateral pterygoid and runs over the superficial surface of the medial pterygoid to lie just deep to the mucous membrane lining the inner aspect of the mandible below the third lower molar tooth. Its subsequent course will be described with the tongue. While lying deep to the lateral pterygoid muscle the lingual nerve receives an important contribution from the facial nerve called the **chorda tympani**.

The **inferior alveolar nerve** runs parallel with the lingual nerve over the surface of the medial pterygoid but then leaves it to enter the **mandibular foramen** for distribution to the teeth of the lower jaw and, as the **mental nerve**, the skin over the chin. Just before entering the mandibular foramen the **mylohyoid nerve**, which supplies the mylohyoid muscle and the anterior belly of the digastric muscle, is given off.

The **auriculotemporal nerve** passes laterally behind the temporomandibular joint and ascends over the posterior root of the zygomatic arch to reach and supply the skin of the temporal region. Small branches of this nerve supply the pinna, the external acoustic meatus, the eardrum, and the temporomandibular joint. The auriculotemporal nerve also plays an important part in the distribution of postganglionic parasympathetic fibers from the otic ganglion to the parotid gland.

The Otic Ganglion

This is a small parasympathetic ganglion concerned with the secretomotor innervation of the parotid gland. It is associated functionally with the glossopharyngeal nerve. However, it lies deep in the infratemporal fossa just below the foramen ovale and thus topographically very close to the mandibular nerve which lies on its lateral side. In Fig. 9-64 follow the preganglionic parasympathetic fibers (*blue*) which leave the brainstem in the **glossopharyngeal nerve** and pass with it out of the skull but re-enter again in the **tympanic nerve** to reach the middle ear. From there the fibers run as the **lesser petrosal nerve** into the middle cranial fossa, and leaving the skull again through the foramen ovale, reach the otic ganglion as it lies between the mandibular nerve and tensor veli palatini. Despite this complex course, no synapse occurs. However, in the ganglion these preganglionic fibers synapse with postganglionic fibers (*yellow*) which are distributed to the parotid gland through the auriculotemporal nerve. Also passing through the ganglion but without synapse, are postganglionic sympathetic fibers for the parotid gland and motor fibers from the mandibular nerve which run to the tensor veli palatini and tensor tympani muscles. The sympathetic fibers reach the ganglion from the plexus on the surface of the nearby middle meningeal artery.

The Maxillary Artery

The maxillary artery was first encountered in the parotid gland where it arises a terminal branch of the external carotid artery. It has a wide distribution to the ear, dura, teeth, muscles of mastication, orbit, palate, and face. It terminates as the sphenopalatine artery, which supplies the nasal cavity. If its course is followed carefully in Fig. 9-65 and if possible with the aid of a skull as well, its branches can be understood more easily.

Lying initially within the substance of the parotid gland and anterior the the external acoustic meatus, it passes forward deep to the condylar process of the mandible. In this part of its course it gives off two small branches. These are a **deep auricular artery** which supplies the external acoustic meatus and outer surface of the eardrum and an **anterior tympanic artery** which reaches the middle ear to supply the inner surface of the drum. As the maxillary artery enters the infratemporal fossa, the **middle meningeal artery** leaves it to run upward deep to the lateral pterygoid muscle and enter the cranial cavity through the **foramen spinosum**. Here too the **inferior alveolar artery** leaves the maxillary artery and courses downward and forward over the medial pterygoid muscle in company with the inferior alveolar nerve. As does the nerve, it gives off a **mylohyoid branch** before entering the mandibular foramen to supply the teeth of the lower jaw.

The course of the maxillary artery through the infratemporal fossa is variable. It sometimes pas-

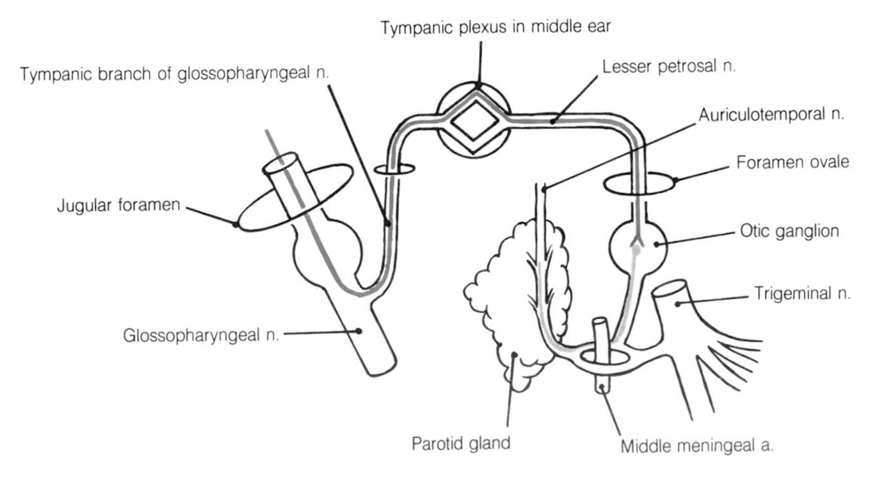

9-64 Diagram of the parasympathetic innervation of the parotid gland.

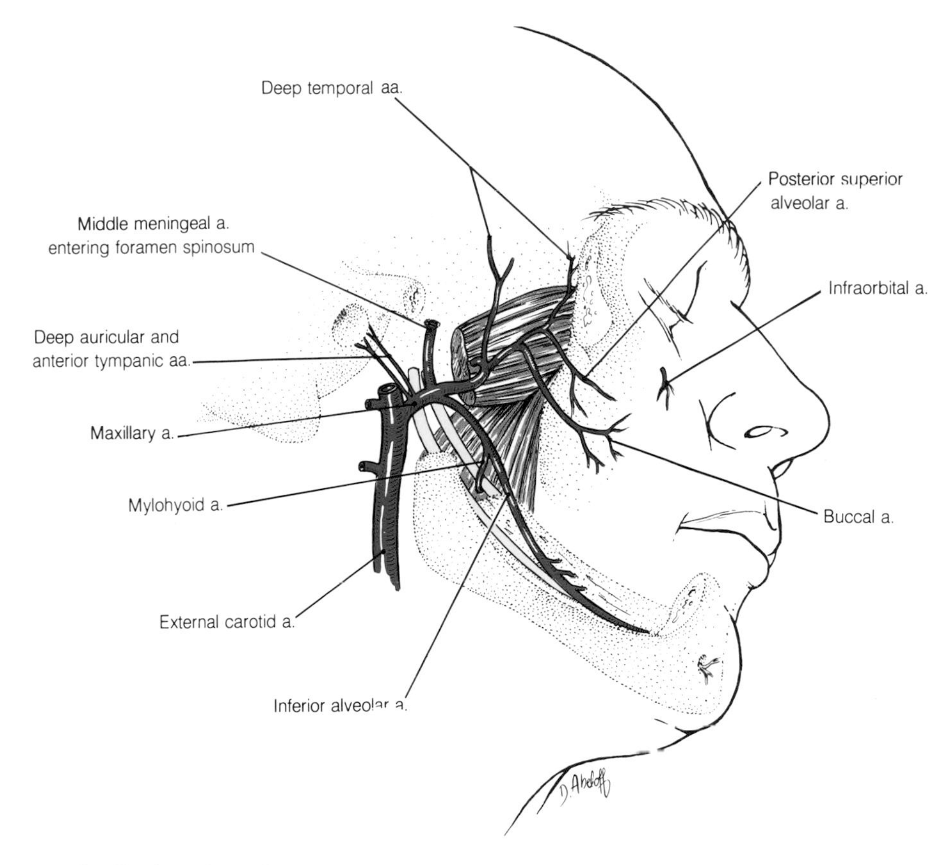

9-65 To show the distribution of the first and second parts of the maxillary artery.

ses superficial to and sometimes deep to the lateral pterygoid muscle. However, as it traverses the fossa it supplies **deep temporal, masseteric**, and **pterygoid branches** to the four muscles of mastication.

On reaching the anterior part of the fossa, the maxillary artery passes between the two heads of the lateral pterygoid muscle and enters the **pterygopalatine fossa** through the **pterygomaxillary fissure**. As it enters this fossa the artery lies close to the posterior surface of the maxilla and it is here that the **posterior superior alveolar artery** is given off. This vessel breaks up into several channels which enter small canals on the posterior aspect of the maxilla. These supply the molar and premolar teeth of the upper jaw.

Within the pterygopalatine fossa the maxillary artery gives off the **infraorbital artery** which has already been seen to appear on the face at the infraorbital foramen. It reaches this foramen by passing forward in the infraorbital canal in the floor of the orbit. While in the canal, **anterior superior alveolar branches** are given off to supply the upper incisor and canine teeth.

The remaining **palatine, pharyngeal**, and **sphenopalatine arteries** which arise from the maxillary artery in the pterygopalatine fossa will be described in more detail with the nasal cavity.

The branches of the maxillary artery that have been mentioned are now summarized below:

- Deep auricular artery
- Anterior tympanic artery
- Middle meningeal artery
- Inferior alveolar artery
  - Mylohyoid branch
- Deep temporal arteries
- Masseteric artery
- Pterygoid branches
- Posterior superior alveolar artery
- Infraorbital artery
  - Anterior superior alveolar branches.
- Palatine arteries
- Sphenopalatine artery
- Pharyngeal branches

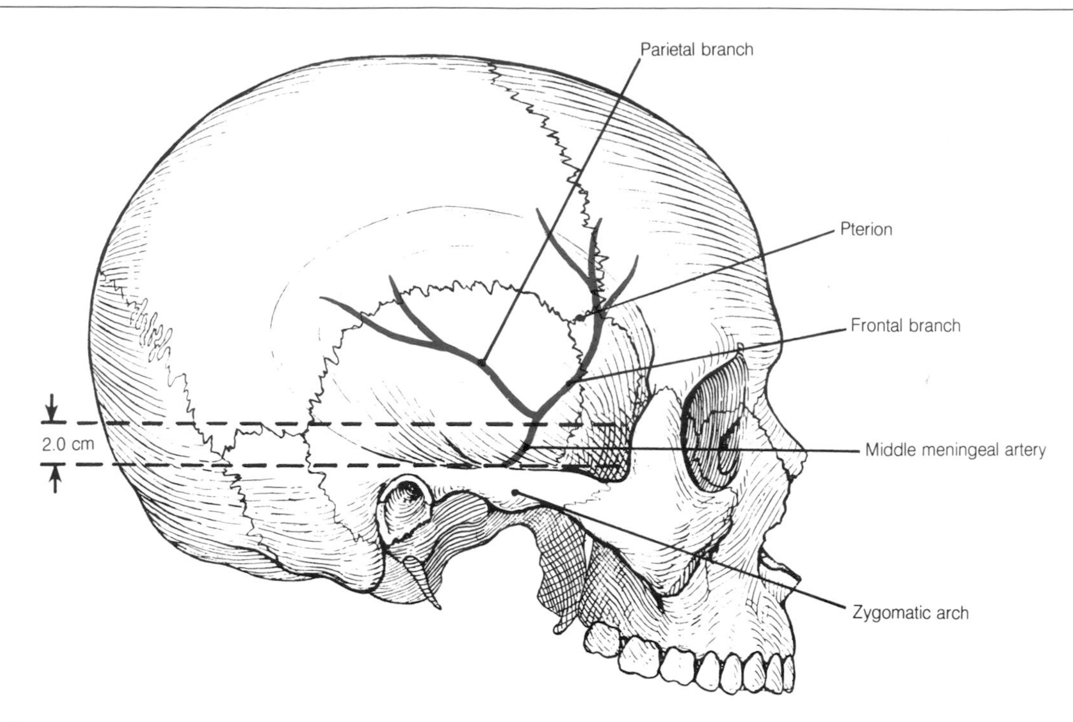

9-66 Showing the surface markings of the middle meningeal artery.

Surface Anatomy of the Middle Meningeal Artery

Fractures of theskull may pass across the course of the middle meningeal artery, and because the vessel is firmly adherent to the bone, it may tear and thus lead to an extradural hemorrhage. Accurate exposure of this artery may be urgently necessary to stop it bleeding. It enters the skull just above the level of the middle of the zygomatic arch and divides about 2.0 cm above this point. Its frontal branch continues upward to cross the pterion while its parietal branch sweeps immediately backward. These surface markings can be confirmed in Fig. 9-66.

The Pterygoid Plexus of Veins

This venous plexus lies among the muscles of mastication in the infratemporal fossa. Into it drain veins from the nasal and oral cavities and from the orbit. It communicates with the cavernous sinus and with the facial vein and, thus forms part of a pathway by which infection may spread from the face to a cranial venous sinus.

The Temporomandibular Joint

This is a synovial joint between the **head of the mandible** and the **mandibular fossa** and the adjacent **articular tubercle** on the inferior surface of the squamous part of the temporal bone. Because both bones taking part have been ossified in membrane rather than hyaline cartilage, their articular surfaces are covered by fibrocartilage. The joint is surrounded by a loose fibrous capsule which is attached near the articular margins and is lined by synovial membrane. The cavity of the joint is completely divided into two by a fibrous **articular disc** which is attached at its margin to the capsule. The disc separates the mandible from the temporal bone and increases the congruity of the two opposing articular surfaces. Anteriorly, both the disc and the capsule receive fibers from the lateral pterygoid muscle. The capsule is reinforced on its lateral aspect by the **lateral ligament.** More distant accessory or extrinsic ligaments are the **sphenomandibular ligament** and the **stylomandibular ligament**. The former is a thin band extending from the spine of the sphenoid bone to the lingula, the latter extends from the styloid process to the posterior border of the ramus of the mandible. The arrangement of these ligaments can be seen in Fig. 9-67, A and B.

The movements of the joint occur between the head of the mandible and the disc where the hingelike actions of **elevation** and **depression** of the mandible occur, and between the disc and the mandibular fossa and articular tubercle where forward gliding or **protrusion** and backward gliding or **retraction** of the mandible take place. When the mouth is opened the joints of both sides act synchronously and the movement is produced by a combination of depression and protrusion as the disc and condyle slide forward from the fossa down the sloping surface of the tubercle. Elevation and retraction lead to closure. These changes in relationship between the articular surfaces and the disc will be more easily understood after reference to Fig. 9-68, A and B. Asynchronous gliding movements at the two joints lead to the motion of chewing. Depression is produced by the lateral pterygoid muscles and other suprahyoid muscles,

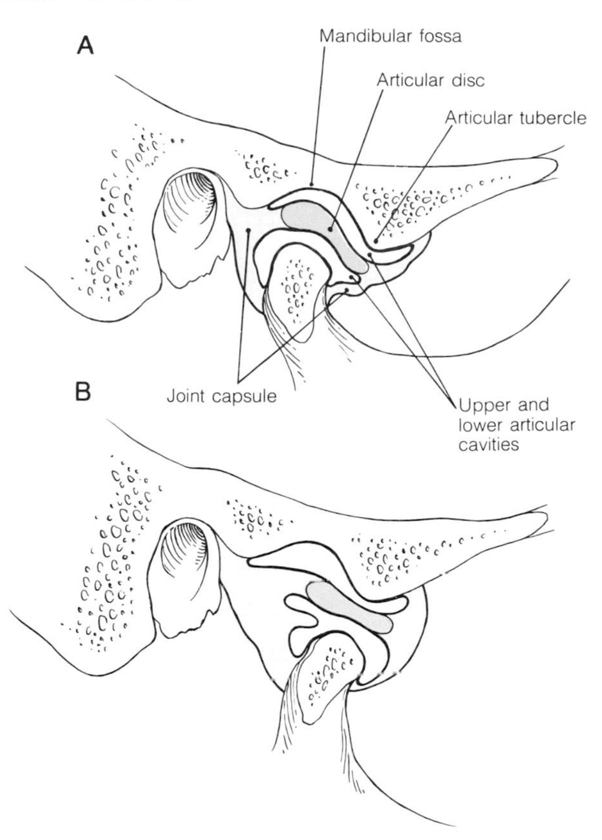

9-68 Showing the forward movement of the head of the mandible and articular disc that occurs between the closed (*A*) and open (*B*) position of the jaws.

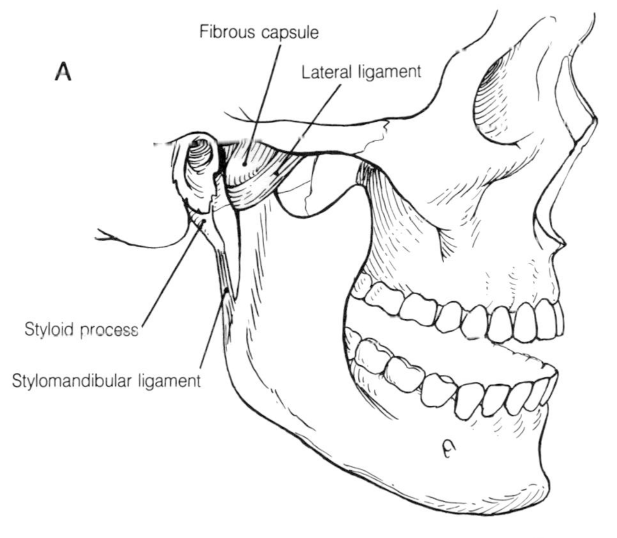

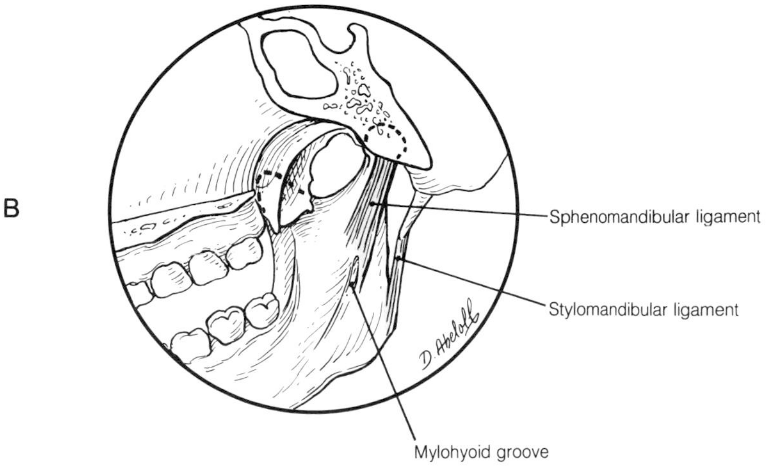

9-67 (*A*)The temporomandibular joint seen from the lateral side. (*B*) To show the sphenomandibular ligament on the medial side of the joint.

and elevation by the temporalis, masseter, and medial pterygoid muscles. Both pterygoid muscles protrude the mandible and in this movement the attachment of the lateral pterygoid to the disc and capsule of the joint ensures that these glide forward in company with the condyle. Retraction is carried out by the more horizontal posterior fibers of the temporalis muscle. The actions of these muscles of mastication can be reviewed again in Fig. 9-69.

The joint is supplied by the **auriculotemporal and masseteric nerves.** Anterior dislocation of the joint sometimes occurs and may even be precipitated by a yawn as the mandibular condyle slips forward beyond the summit of the articular tubercle.

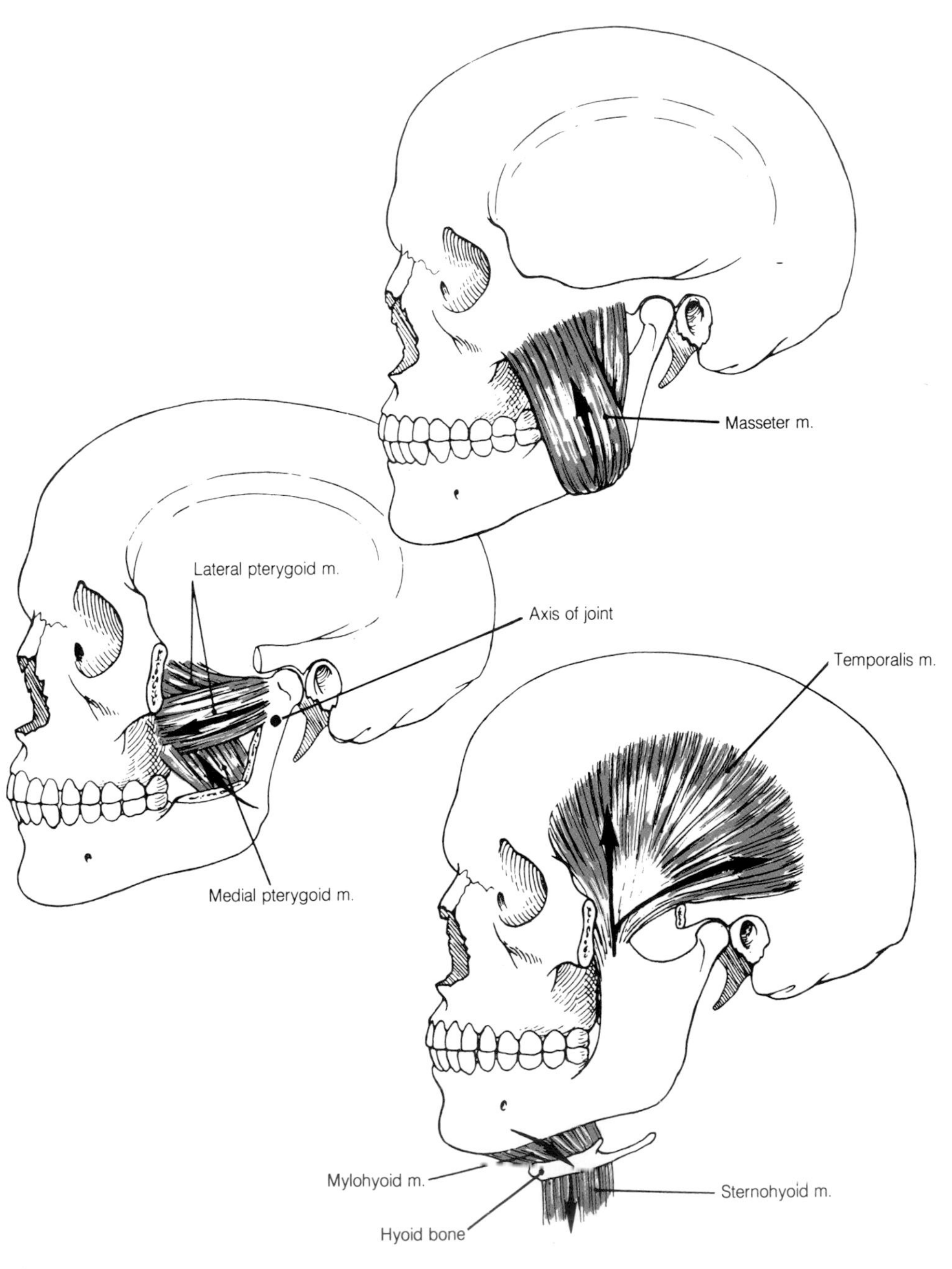

9-69 Illustrating the muscles of mastication.

Craniovertebral Joints

Before considering these articulations the relevant features of the bones involved will be reviewed.

The Occipital Bone

On the inferior surface of the occipital bone the two occipital condyles lie on either side of the foramen magnum. The articular surfaces are oval or bilobed and convex and their long axes converge slightly toward their anterior extremities. They are illustrated in Fig. 9-70.

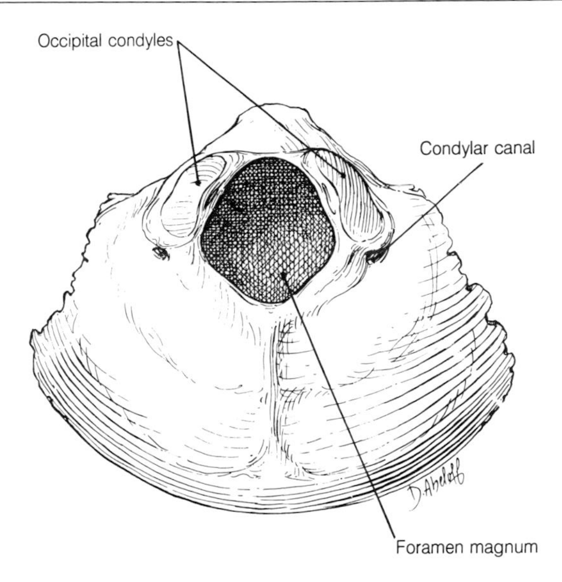

9-70 The occipital bone and occipital condyles viewed from below.

The Atlas

The first cervical vertebra consists of two lateral masses connected by an anterior and posterior arch. This vertebra has no body and the posterior arch bears a small posterior tubercle but no spine. In Fig. 9-71, note:

1. that on the upper surface of each lateral mass there is an oval facet (*A*) for articulation with an occipital condyle,
2. that on the inferior surface of each lateral mass there is a facet for articulation with a superior articular facet of the axis (*B*),
3. that at the center of the posterior aspect of the anterior arch there is a small facet for articulation with the dens of the axis (*C*), and
4. that on their medial aspects each lateral mass

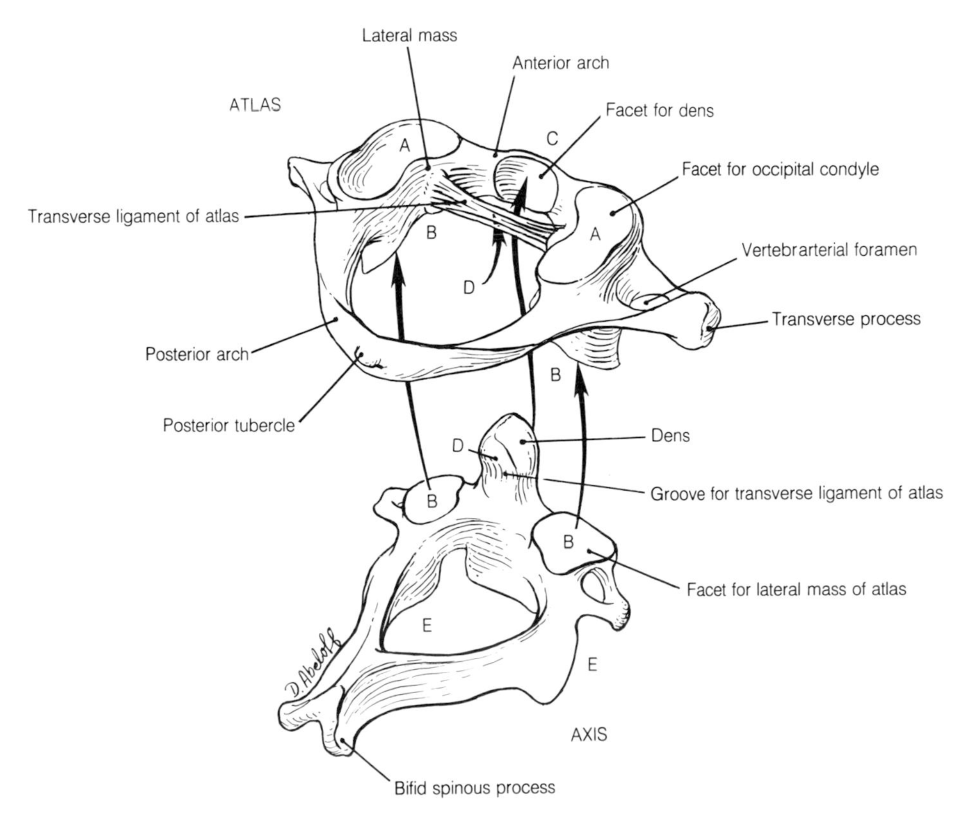

9-71 The atlas and axis separated to show their articulations.

bears a small tubercle for the attachment of the **transverse ligament of the atlas**. The ligament articulates with the dens at *D* where a synovial bursa is interposed.

The Axis

When viewed from below the axis is seen to have the characteristics of a typical cervical vertebra but its superior aspect is modified for articulation with the atypical atlas. In Fig. 9-71 it can be seen that its most distinguishing feature is the dens or odontoid process which is fused to the body of the axis and projects superiorly, posterior to the anterior arch of the atlas. On either side of the dens the body is obscured by a pair of lateral superior facets which articulate with the lateral masses of the atlas. Identify these in Fig. 9-71 and note that they lie anterior to the vertebrarterial foramina and do not correspond to typical superior articular processes. Typical inferior articular processes (*E*) are, however, present for articulation with the third cervical vertebra.

The Atlantooccipital Joints

Refer to Fig. 9-71 again. At *A* the **superior articular surfaces** on the lateral masses of the atlas articulate with the **occipital condyles** by a pair of synovial joints. These allow the movements of flexion, extension, and some lateral flexion but no rotation. Extending from the margins of the foramen magnum to the anterior and posterior arches of the atlas are the **anterior and posterior atlantooccipital membranes**. The anterior membrane is complete but the posterior is perforated by the vertebral artery. These membranes are illustrated in Fig. 9-72 A.

The Atlantoaxial Joints

The atlantoaxial articulation is comprised of three synovial joints. Two of these are between the **inferior articular surfaces** on the lateral masses of the atlas and the superiorly facing **anterior articular surfaces** on either side of the dens of the axis. These are indicated as *B-B* in Fig. 9-71. The third joint is between the anterior surface of the dens and the posterior surface of the anterior arch of the atlas *C-C* in Fig. 9-71. In addition, a synovial bursa lies between the posterior aspect of the dens and the **transverse ligament of the atlas**. This in practice forms a fourth articulation between axis and atlas and is indicated at *D-D* in Fig. 9-71. The transverse ligament lies between the lateral masses of the atlas and is crossed by a longitudinal band joining the body of the axis to the foramen magnum. Together the two ligaments are known as the **cruciform ligament**. An **apical ligament** and two laterally directed **alar ligaments** also link the top of the dens to the anterior margin of the foramen magnum. The positions of the ligaments joining the skull, atlas, and axis are shown in Fig. 9-72 A and B, and these should be used to note:

In Fig. A

1. The anterior atlantooccipital membrane
2. The anterior longitudinal ligament
3. The apical ligament between the tip of the dens and the margin of the foramen magnum
4. The horizontal element of the cruciform ligament called the transverse ligament of the atlas
5. The vertical element of the cruciform ligament
6. The membrana tectoria which is a superior continuation of the posterior longitudinal ligament and attached together with the cruciform ligament just inside the foramen magnum
7. The posterior atlantooccipital membrane which is perforated to allow the passage of the vertebral artery and the first cervical nerve
8. The ligamentum nuchae

In Fig. B

1. The cruciform ligament
2. The alar ligaments of the dens and their lateral attachments to the medial aspects of the occipital condyles. The dens and the medial attachments of the ligaments are hidden by the cruciform ligament

It is at the three atlantoaxial articulations that rotation of the head and atlas occurs on the axis.

The axis articulates with the third cervical vertebra through a pair of typical inferior articular surfaces and by a typical intervertebral joint between the adjacent bodies.

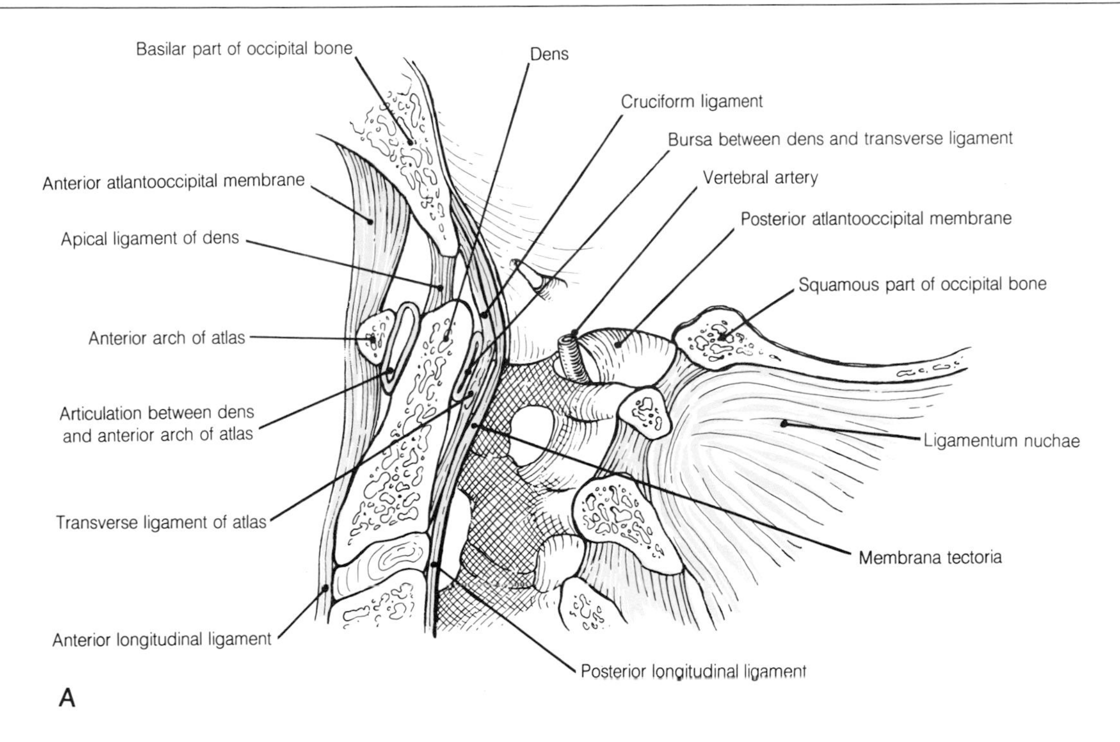

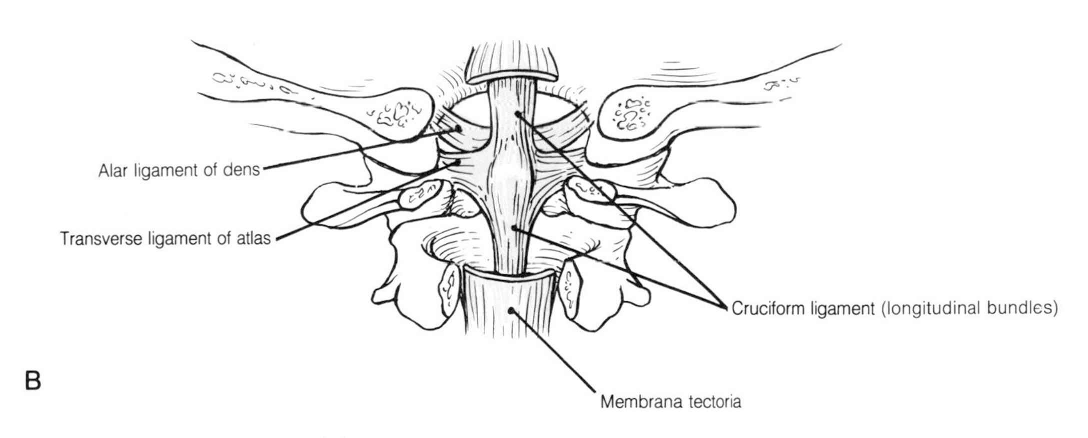

9-72 (*A*) Median section through the axis, atlas, and occipital bones showing the ligaments that unite them. (*B*) The cruciform ligament viewed from behind.

The Soft Tissues of the Neck

The soft tissues of the neck are mainly oriented in a vertical manner. Anteriorly and in the midline are found the respiratory and alimentary passages flanked by vascular and nervous structures. Centrally the neck is supported by the cervical vertebral column from which the spinal nerves emerge laterally. Posteriorly two muscle masses lie one on either side of the cervical spinous processes and the ligamentum nuchae and a more superficial layer formed by the trapezius and sternocleidomastoid muscles partially surrounds the whole. These soft tissues are enclosed in a sleeve of deep fascia and additional fascial sheaths partly or completely surround deeper structures. Examine Fig. 9-73 to gain a picture of the general disposition of the regions described above. In this particular section, which has been made through the seventh cervical vertebra, the thyroid gland par-

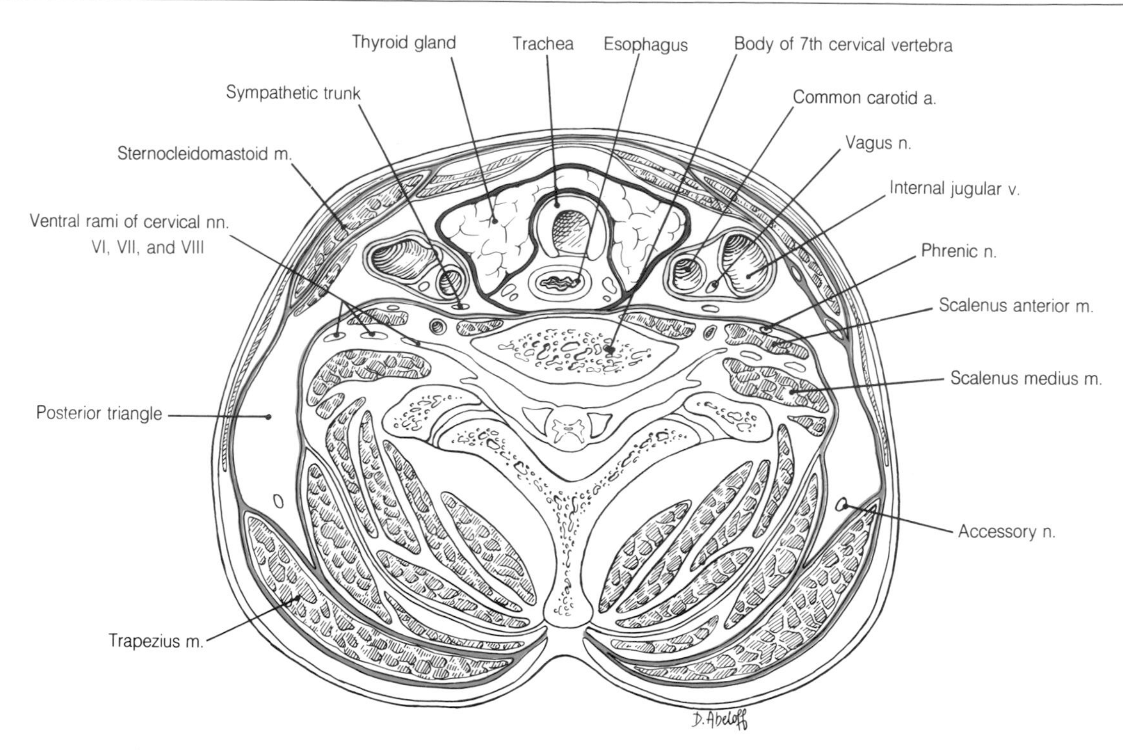

9-73 Horizontal section through the neck at the level of the body of the seventh cervical vertebra showing the superficial (*green*), prevertebral (*blue*), and pretracheal (*red*) layers of the deep cervical fascia and also the carotid sheath (*yellow*).

tially surrounds the trachea. Note also how far forward in the neck the body of the vertebra lies.

The Fasciae of the Neck

The **superficial fascia** of the neck, as elsewhere, consists of a layer of loose and adipose connective tissue lying between the skin and the deep fascia. Anteriorly it contains the platysma muscle.

Beneath the superficial fascia the neck is surrounded by a **superficial layer** of **deep fascia**. Turn again to Fig. 9-73 and follow this layer anteriorly from the ligamentum nuchae. Note how it splits to enclose trapezius and then forms the roof of the posterior triangle of the neck again dividing to enclose the sternocleidomastoid muscle. Between the anterior borders of the two sternocleidomastoid muscles it crosses the front of the neck as a single sheet covering the anterior triangle of the neck. Above, the superficial layer of deep fascia also splits to enclose the parotid and submandibular glands. A thickening of the fascia on the deep surface of the parotid gland is the **stylomandibular ligament**.

A **prevertebral layer** of fascia covers the prevertebral muscles and extending at first laterally and then posteriorly over these it forms the floor of the posterior triangle. This layer of fascia can be followed downward over the subclavian artery and trunks of the brachial plexus into the axillary sheath.

The **carotid sheath** is a loose condensation of deep fascia surrounding the common and internal cartoid arteries, the internal jugular vein, and vagus nerve.

The thin **pretracheal fascia** also surrounds the thyroid gland. The superior attachment of this layer to the cricoid cartilage of the larynx accounts for the movements that the thyroid and thyroid tumors show when the larynx is raised on swallowing.

The Posterior Triangle of the Neck

This is a descriptive triangle lying in the lateral aspect of the neck. Its apex lies on the occipital bone just behind the mastoid process where the attachments of the trapezius and sternocleidomastoid muscles meet. Its borders are formed by the opposing margins of the same muscles and its base consists of a short-length of clavicle left uncovered by the clavicular attachments of the muscles. It is in fact a very narrow triangle but by retracting its borders and exploring a little beyond its strict margins a very important junctional region between the neck, thorax, and upper limb is exposed. It is in the posterior triangle that the trunks of the brachial plexus that supply the skin and musculature of the upper limb appear. At the level of the first rib this nerve plexus is joined by the main artery and vein which supply and drain the limb and the combined neurovascular bundle proceeds into the arm through a space known as the axilla. It is the portion of the plexus lying in the posterior triangle that is rather liable to injury. Trauma is usually indirect and results from forcible abduction of the head and neck from the shoulder or forcible elevation of the arm on the trunk. These injuries may occur, for example, during a difficult obstetric delivery or in motorcycle accidents. Chronic compression syndromes can also occur in the region of the first rib and may be associated with the presence of an accessory cervical rib. Injury or obstruction may also involve the subclavian or axillary arteries and anastomoses between these vessels in the region of the neck and shoulder may allow an adequate collateral circulation to open up. Operations in this region range from a "simple" biopsy of a lymph node to a "block dissection" of nodes in the treatment of neoplastic disease. In both cases, the anatomical landmarks are essential and in the former procedure damage by the operator to any structure contained in the triangle, for example the accessory nerve, must be avoided at all costs. A brachial plexus block may also be produced in this region by the infiltration of a local anesthetic.

Examine now Fig. 9-74. To gain this exposure the skin, superficial fascia and the platysma have been removed together with the roof of the triangle which is formed by the superficial layer of the deep cervical fascia which spans the triangle between the bordering muscles. Note in this illustration:

1. The **external jugular vein** formed near the angle of the mandible and crossing the sternocleidomastoid muscle to drain into the **subclavian vein**.

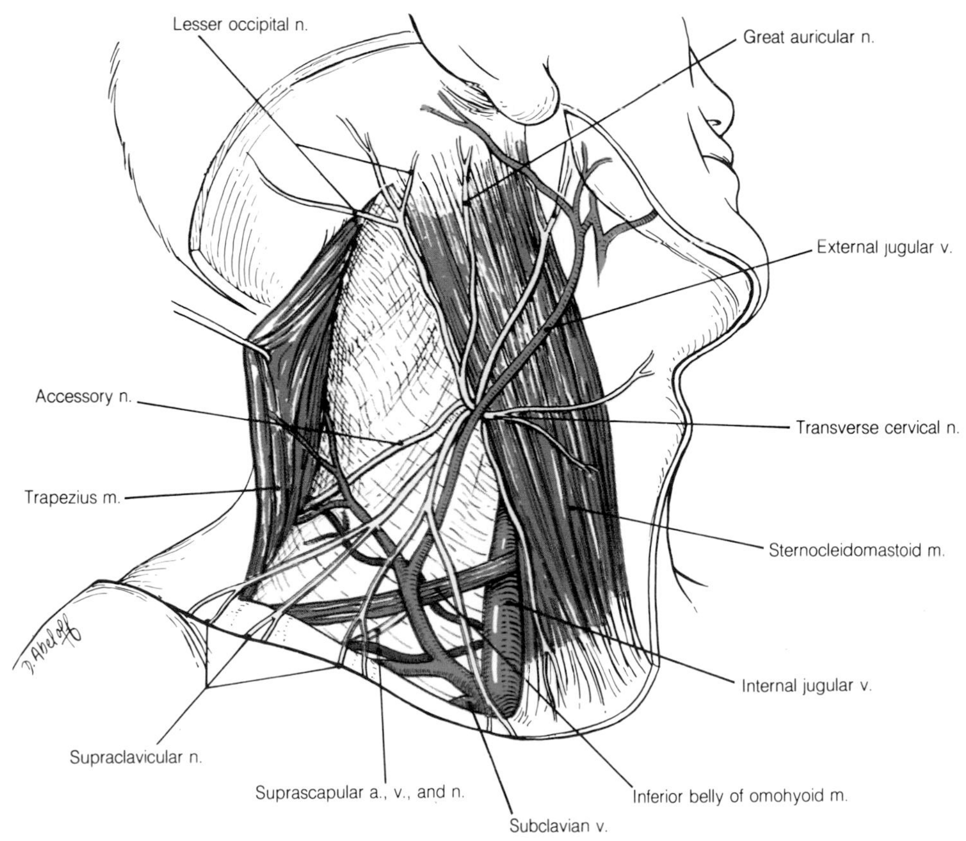

9-74 A superficial dissection of the right posterior triangle of the neck.

2. A point midway down the posterior border of sternocleidomastoid where a leash of cutaneous nerves appear. These are the **lesser occipital nerve** (C2), the **great auricular nerve** (C2 and C3), the **transverse cervical nerve** (C2 and C3), and the **supraclavicular nerves** (C3 and C4). All these cutaneous nerves, as their segmental levels suggest, are derived from the **ventral rami** of spinal nerves forming the **cervical plexus**.
3. The **internal jugular vein** beneath the posterior border of the sternocleidomastoideus and its junction with the subclavian vein at the anterior angle of the triangle.
4. The **transverse cervical** and **suprascapular arteries** appearing from beneath the internal jugular vein. Both these vessels are branches of the **thyrocervical trunk** of the **subclavian artery**.
5. The **suprascapular nerve** from the brachial plexus meeting the **suprascapular artery**.
6. On the floor of the triangle the **spinal accessory nerve** running from the sternocleidomastoid which it supplies to the trapezius which it also supplies.

If these more superficial structures are removed together with the floor of the triangle which is formed by the prevetrebral fascia, the prevertebral muscles and their relationship to the trunks of the brachial plexus and the subclavian artery can be seen as in Fig. 9-75. In this illustration note how the **trunks of the brachial plexus** and the **subclavian artery** appear between the **scalenus anterior** and **scalenus medius muscles**. Observe also the **phrenic nerve** crossing obliquely over the scalenus anterior muscle to which it is held by the prevertebral fascia.

Of the muscles associated with the posterior triangle, the trapezius, splenius capitis, and levator scapulae have been described elsewhere but the sternocleidomastoid and scalene muscles now need to be considered in more detail.

The Sternocleidomastoid Muscle

This muscle runs obliquely across the side of the neck and separates the posterior from the anterior triangle. It also covers the carotid sheath. Its proximal attachment to the skull is at the lateral surface of the mastoid process and from the immedi-

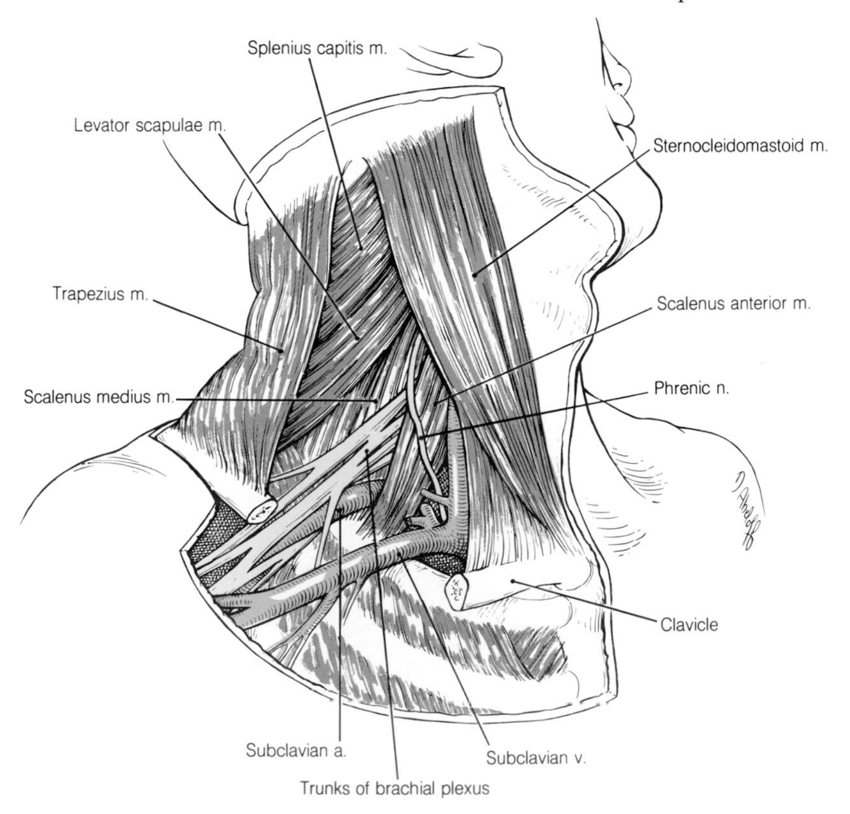

9-75 A deep dissection of the right posterior triangle of the neck.

ately adjacent portion of the superior nuchal line. Its name indicates its dual distal attachment. The more superficial sternal head is attached to the upper anterior surface of the manubrium sterni by a flattened tendon. Its deep clavicular head has a more fleshy attachment to the medial third of the upper surface of the clavicle. It is supplied by the **spinal accessory nerve** and by a branch of the **ventral ramus of the second cervical spinal nerve**. As with trapezius, the accessory fibers are motor and the cervical nerve fibers proprioceptive. The two muscles may act together or alone. The action of one muscle is to tilt the head toward the shoulder of the same side while rotating the face to the opposite side. Resisting this rotational component is a useful method of testing the accessory nerve. Together the muscles draw the head forward. When the head and cervical spine are held extended, the sternomastoid muscles can help to raise the thoracic cage in forced inspiration. This muscle is illustrated in Fig. 9-75.

The Scalene Muscles

Scalenus anterior arises from the anterior tubercles of the transverse processes of the third to sixth cervical vertebrae. Below it is attached to the **scalene tubercle** of the first rib. Its many important relations will be considered with the root of the neck.

The **scalenus medius** is rather larger than the anterior muscle. It arises from the posterior tubercles of the transverse processes of the lower five cervical vertebrae. The lower attachment of the muscle is to the first rib behind the groove for the subclavian artery. The **scalenus posterior** is not always clearly separated from the medius but its lower attachment is to the outer surface of the second rib.

The anterior and posterior tubercles of the cervical transverse processes from which the scalene muscles arise lie anterior and posterior to the groove for the spinal nerve and thus, as might be expected, the brachial plexus emerges between scalenus anterior and medius.

The scalene muscles all produce lateral flexion of the cervical spine. When the neck is stabilized they can be used as inspiratory muscles.

The scalene muscles are supplied by **branches of the ventral rami of the third to eighth cervical nerves**.

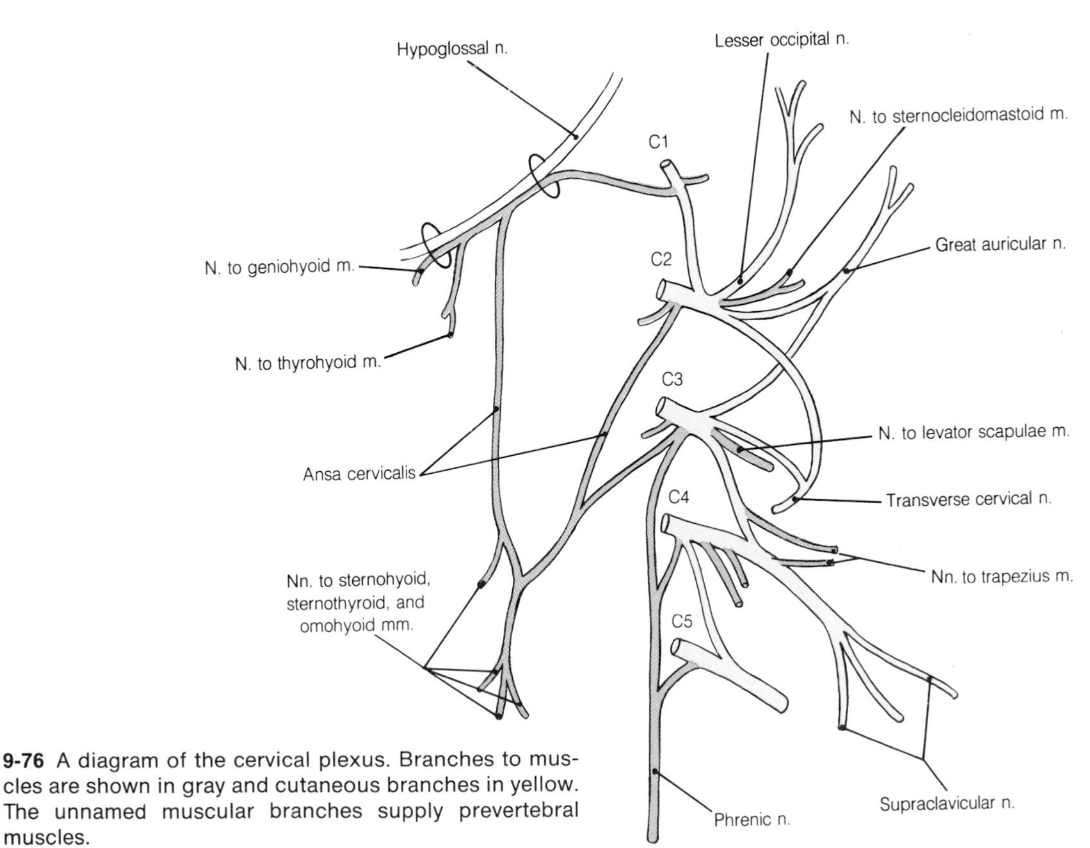

9-76 A diagram of the cervical plexus. Branches to muscles are shown in gray and cutaneous branches in yellow. The unnamed muscular branches supply prevertebral muscles.

The Cervical Plexus

A number of cutaneous nerves that supply the regions of the ear, neck, and the supraclavicular region together with motor nerves to muscles of the neck have been described as arising from the ventral rami of cervical nerves. While the motor nerves tend to arise directly from the rami the cutaneous nerves arise in a plexiform manner. In addition, a loop or ansa is formed for the supply of the infrahyoid muscles.

The arrangement of this plexus is illustrated diagrammatically in Fig. 9-76. Note particularly the fibers from the first nerve that temporarily join the **hypoglossal nerve**. These travel with the nerve to supply the geniohyoid and thyrohyoid muscles or leave it as the **superior root of the ansa cervicalis**. The loop is completed by contributions from the **second and third cervical nerves** which form the **inferior root**. Branches from the ansa supply the sternohyoid, sternothyroid, and omohyoid muscles. The contribution of the third, fourth, and fifth cervical nerves to the phrenic nerve can also be seen and of these, that from the fourth nerve is the most important. The first cervical ventral ramus like the first dorsal ramus has no cutaneous branches.

The Cutaneous Innervation of the Head and Neck

With the description of the cervical plexus and the appearance of its cutaneous branches in the posterior triangle completed, the innervation of the skin of the head and neck by cranial nerves and by dorsal and ventral rami of cervical nerves can be reviewed in Fig. 9-77. Note again that the greater part of the face and the scalp to the vertex are supplied by the three branches of the trigeminal nerve and that the anterior part of the neck is supplied by ventral rami of cervical nerves and the posterior part by dorsal rami of cervical nerves.

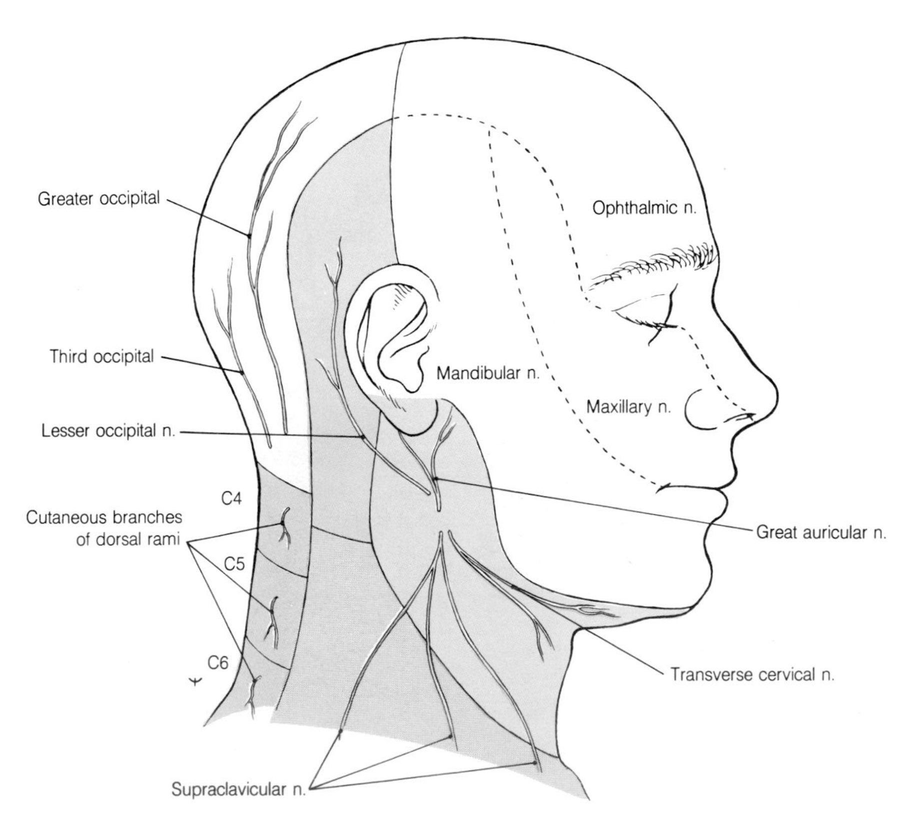

9-77 Diagram showing the cutaneous innvervation of the head and neck.

The Anterior Triangles

These triangles are bounded above by the inferior border of the mandible, laterally by the anterior borders of the sternocleidomastoid muscles and are separated from each other by an imaginary midline. Their apices join at the suprasternal notch. The roof of the triangle is formed by skin and superficial fascia containing the platysma muscle. Also in the roof are the transverse cervical nerve, the cervical branch of the facial nerve, and the anterior jugular vein. Removal of the roof exposes a series of flat strap-like muscles and the digastric muscle, and in the midline, the hyoid bone, the anterior aspect of the larynx, the trachea, and the isthmus of the thyroid gland. In Fig. 9-78, the skin and platysma have been removed but the arrangement of the remaining structures can be seen.

The Anterior Jugular Vein

The superficial veins in the anterior triangles seldom conform to a typical pattern, but an **anterior jugular vein** can usually be recognized which collects tributaries from the submandibular region and descends near the midline of the neck. On nearing the sternal attachment of sternocleidomastoid it turns laterally deep to this muscle to empty into the external jugular vein or the subclavian vein. An **anterior venous arch** frequently joins the left and right veins across the midline.

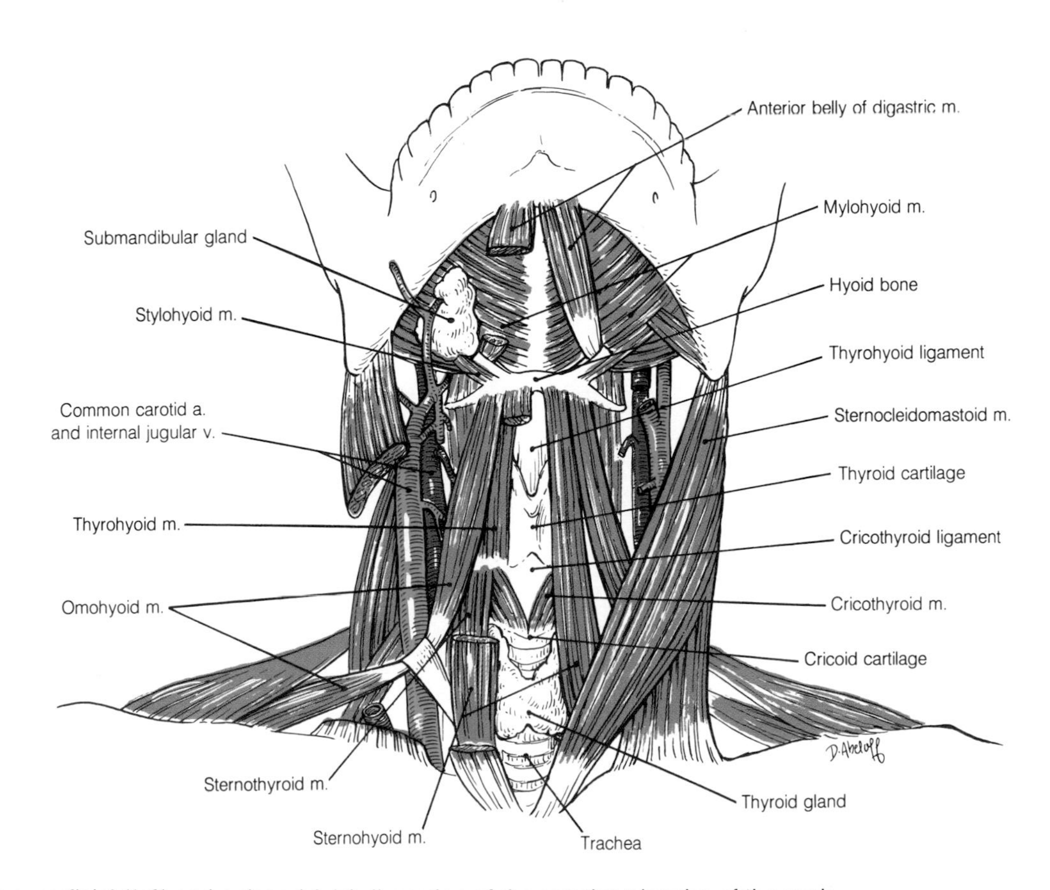

9-78 A superficial (*left*) and a deep (*right*) dissection of the anterior triangles of the neck.

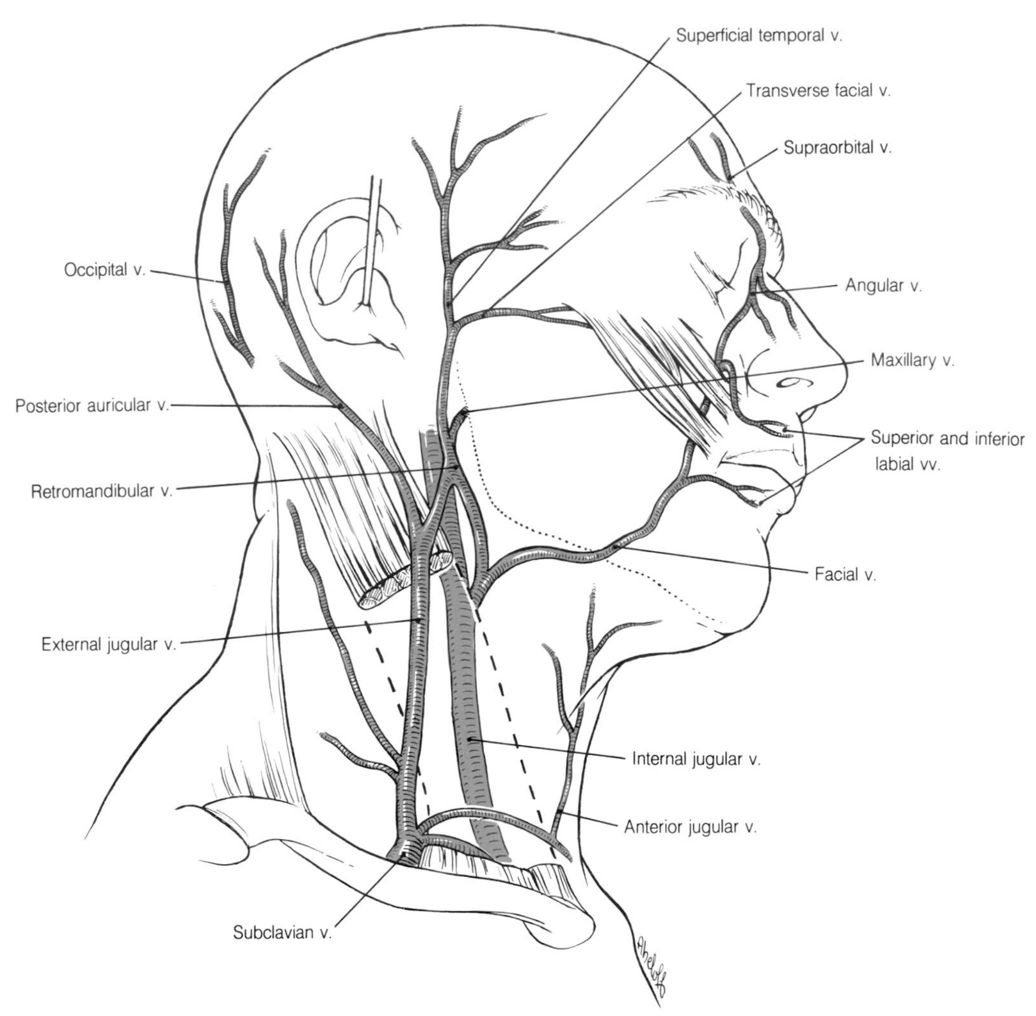

9-79 Showing the superficial veins of the head and neck.

In Fig. 9-79, the superficial venous drainage of the head and neck can be reviewed. The figure includes a number of veins already mentioned and others which accompany deeper arteries but drain into the superficial system. Note:

1. The **retromandibular vein** formed by the union of the **superficial temporal** and **maxillary veins**.
2. That the retromandibular vein divides into two branches. A posterior branch joins the **posterior auricular vein** to form the **external jugular vein**. An anterior branch joins the **facial vein** and the common trunk so formed drains into the **internal jugular vein**.
3. The **occipital vein** draining the posterior region of the sclap runs deeply to join the venous plexus in the suboccipital triangle.

Before considering the muscles encountered in the anterior triangle, the inner aspect of the body of the mandible and the hyoid bone must be examined.

The Mandible (Medial Aspect)

In Fig. 9-80, which shows the medial aspect of the right mandible, identify the **mylohyoid line**. Beginning just below the third molar tooth close to the attachment of the superior pharyngeal constrictor it descends obliquely and forward over the body of the mandible and becomes less distinct as it nears the **symphysis menti**. Just above this point is found a small tubercle called the **mental spine**. Just below this point on the inferior margin of the body there is a shallow oval depression called the **digastric fossa**.

The Hyoid Bone

The hyoid bone is a true bone and although

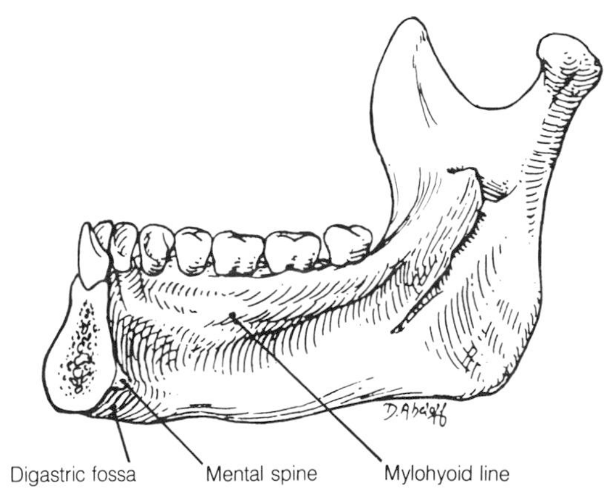

9-80 The medial aspect of the mandible.

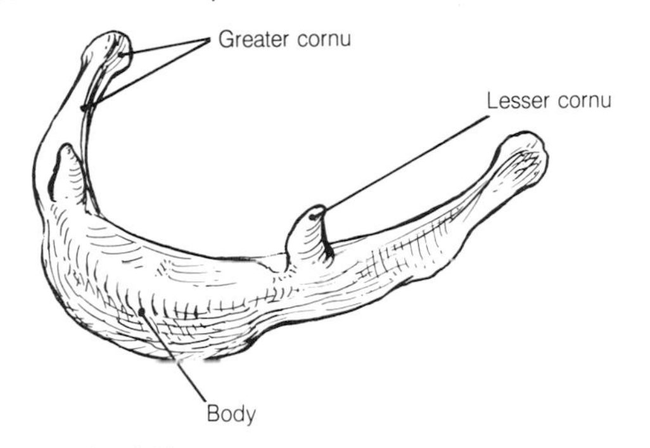

9-81 The hyoid bone.

closely associated with the larynx should not be confused with the laryngeal cartilages. When the head is relaxed it lies under cover of the mandible from which it is suspended by muscles. It is shaped like a U with its limbs pointing backward (see Fig. 9-81). Identify the **body**, the **greater cornua***, and the **lesser cornua.** They are initially separate units joined by cartilage but they fuse in middle or later life.

The Muscles of the Anterior Triangle

The muscles lying in the anterior triangle are conveniently described as a **suprahyoid** and an **infrahyoid** group. This grouping is also of functional significance as those muscles above the hyoid raise it and indirectly the larynx while those below depress it. These movements are of importance in the action of swallowing. The suprahyoid and infrahyoid muscles and their relationships in the neck are illustrated in Fig. 9-82.

* Cornu, *L.* = a horn

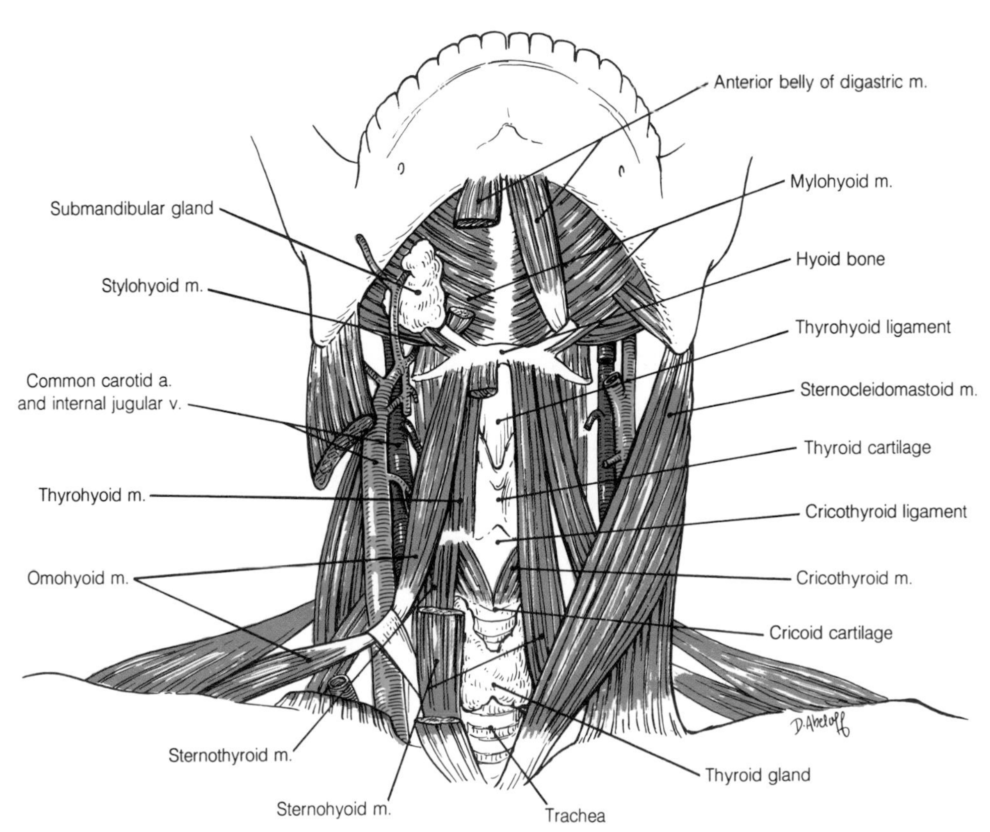

9-82 A superficial (*left*) and a deep (*right*) dissection of the anterior triangles of the neck.

Suprahyoid Muscles

As its name implies, the **digastric muscle** has two bellies. The anterior belly arises from a shallow **digastric fossa** on the lower border of the mandible close to the symphysis menti. It descends to the greater cornu of the hyoid bone where a tendon intermediate between it and the posterior belly is held by a fascial sling. The posterior belly ascends again to be attached to the **mastoid notch** on the medial surface of the mastoid process. When the hyoid is stabilized the muscle will depress the mandible, but when the mandible is fixed it will raise the hyoid. The anterior belly is supplied from the mandibular nerve by the **mylohyoid nerve**, a branch of the inferior alveolar nerve; the posterior belly is supplied by the **facial nerve**.

Lying deep to the digastric muscles and extending from hyoid to mandible the **mylohyoid muscle** forms an important part of the floor of the mouth and will be described in more detail under this heading. It is supplied by the **mylohyoid nerve**.

The small **geniohyoid muscle** lies deep to mylohyoid and extends from the **inferior part of the mental spine** to the body of the hyoid bone. It is supplied by the **first cervical nerve** by fibers carried to it by the hypoglossal nerve.

The **stylohyoid muscle** arises from the base of the styloid process and its slender belly passes downward and forward to be attached to the lateral extremity of the body of the hyoid bone. As a derivative of the second branchial arch, it is supplied by the **facial nerve**.

Each of the muscles is able to raise the hyoid bone and through means of the **thyrohyoid membrane** the larynx also.

The **stylohyoid ligament** is closely associated with the suprahyoid muscles. It extends from the tip of the **styloid process** of the temporal bone to the tip of the lesser cornu of the hyoid.

Infrahyoid Muscles

The strap-like **sternohyoid muscle** extends from the posterior aspect of the sternoclavicular joint and an adjacent area on the manubrium sterni to the body of the hyoid bone. It is supplied by the **ansa cervicalis**.

Lying lateral to sternohyoid, the **omohyoid** has two bellies. The **superior belly** descends from the body of the hyoid nearly parallel to the sternohyoid to reach a fascial sling above the medial end of the clavicle. An intermediate tendon passes through the sling and the **inferior belly** extends from this point backward across the base of the posterior triangle to become attached to the superior border of the scapula near the suprascapular notch. In its course it passes behind the sternocleidomastoid and over the carotid sheath. It is supplied by the **ansa cervicalis**.

The **thyrohyoid muscle** lies undercover of the upper ends of omohyoid and sternohyoid and extends from the lower border of the body of the hyoid to the oblique line on the thyroid cartilage. It is supplied by fibers of the **first cervical nerve** carried to it by the hypoglossal nerve.

The **sternothyroid** also lies beneath the sternohyoid and is in continuity with thyrohyoid at the oblique line on the thyroid cartilage from where it extends to the deep surface of the manubrium sterni. It is supplied by the **ansa cervicalis**.

With the exception of the thyrohyoid, the muscles of this group depress the larynx. The thyrohyoid although lying below the hyoid is attached to the larynx and if the hyoid is stabilized or raised its contraction will elevate the larynx. Because of their thin narrow bellies, the infrahyoid muscles are often referred to as the strap muscles.

The Carotid Sheath and Carotid Arteries

The removal of the parotid gland, and the sternocleidomastoid, digastric, and strap muscles of the neck exposes the **carotid sheath**. This extends from the thoracic inlet to the base of the skull and lies to the side of the midline respiratory and alimentary channels. It contains the **common and internal carotid arteries**, the **internal jugular vein, vagus nerve**, and the **ansa cervicalis**. The termination of the extracranial course of the internal carotid artery and the exit of the cranial nerves from the base of the skull in this region are still obscured by the styloid process and the origins of the muscles attached to it.

The Common Carotid Artery

On the left hand side the **common carotid artery** arises from the **arch of the aorta**. It enters the neck behind the left sternoclavicular joint and deep to the sternocleidomastoid muscle and ascends the side of the neck to the level of the upper border of the thyroid cartilage. Here it bifurcates to form the **internal and external carotid arteries**. On the right hand side the common carotid artery arises behind the right sternoclavicular joint as one of the two terminal branches of the **brachiocephalic trunk** (the other being the right subclavian artery). Its course in the neck is similar to the left artery. The terminal part of the common carotid artery is dilated at the point of origin of the internal carotid artery to form the **carotid sinus** whose wall contains sensory nerve endings responsive to changes in arterial pressure. Deep to the bifurcation the **carotid body** lies embedded in the tunica adventitia. This small collection of glomus cells functions as a chemoreceptor. Both the carotid sinus and carotid body are innervated by the **carotid sinus branch of the glossopharyngeal nerve**.

The External Carotid Artery

Beginning at the upper border of the thyroid lamina the **external carotid artery** ascends deep to the posterior belly of the digastric muscle and the stylohyoid muscle and on the surface of the pharynx to enter the parotid gland. It divides behind the neck of the mandible into the **superficial temporal artery** and the **maxillary artery** as has already been described. In addition to these two

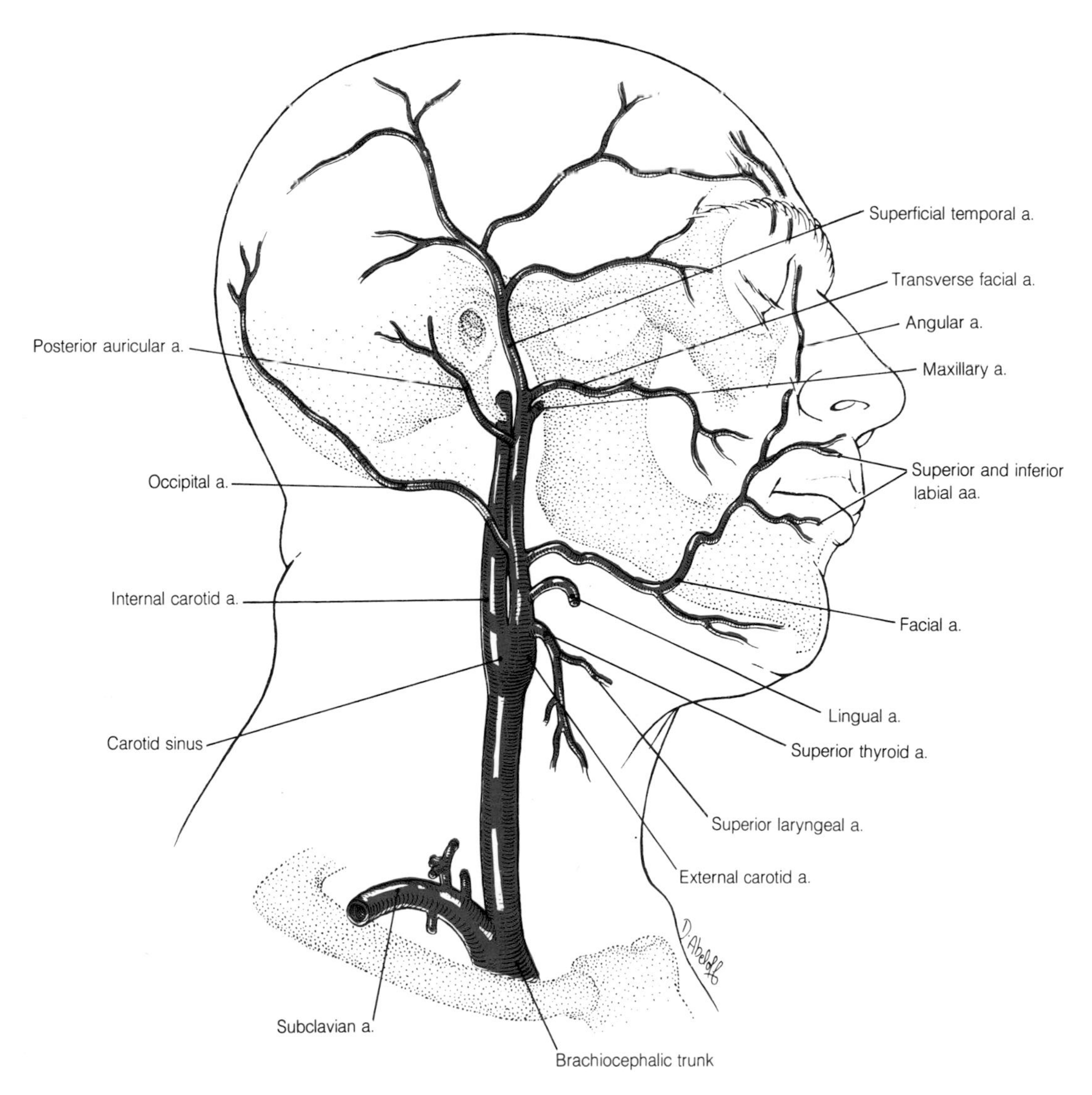

9-83 Showing the carotid arteries in the neck and the branches of the external carotid artery.

terminal branches, six further branches are given off. These are summarized below and their general course and distribution can be seen in Fig. 9-83. Each is described in more detail with the region it supplies.

The **superior thyroid artery** is the first anterior branch. It slopes steeply downard to the upper pole of the thyroid gland. It gives off the **superior laryngeal artery**.

The **lingual artery** arises anteriorly above the superior thyroid artery. It describes a typical loop before passing deep to the hyoglossus muscle and reaching the tongue.

The **facial artery**, arising above the lingual artery, passes deep to the posterior belly of the digastric muscle and the stylohyoid muscle. Now lying beneath the angle of the mandible it grooves the surface of the **submandibular gland** which it supplies and curls around the mandible onto the face at the anterior border of the masseter muscle. It gives off a **tonsillar branch** which pierces the superior constrictor of the pharynx.

The **occipital artery** arises at the same level as the facial artery but passes posteriorly deep to the digastric muscle to appear on the scalp above the suboccipital triangle.

The **posterior auricular artery** is small and passes deep to the parotid gland to supply the ear and the occipital region.

The **ascending pharyngeal artery** is also small. It arises close to the origin of the external carotid on its medial aspect. It ascends on the surface of the pharynx to the base of the skull.

The branches of the external carotid artery and the subsequent branches of these form a very free anastomosis across the midline and hemorrhage from the face and scalp is usually copious. Even temporary ligation of the external carotid artery prior to operations on the face and jaws does less than would be expected to reduce bleeding. On the other hand, the anastomosis between branches of the internal and external carotid arteries around the orbit is of little value following obstruction to the internal carotid artery.

The Internal Carotid Artery

Beginning at the bifurcation of the common carotid artery at the level of the upper border of the thyroid cartilage, the **internal carotid artery** ascends over the anterior surfaces of the transverse processes of the upper three cervical vertebrae on the surface of longus capitís to reach **the carotid canal** in the base of the skull. The pharynx lies medial to it and below and laterally it is under cover of the posterior belly of the digastric muscle and the sternocleidomastoid. Above this level the styloid process and the muscles attached to it lie laterally. There are no branches given off by this artery in the neck. Its importance lies in the fact that the two internal carotid arteries supply the greater part of the cerebral hemispheres, the eye, and the contents of the orbit. Obstruction will lead to ischemia of a hemisphere and a contralateral paralysis or hemiplegia. If time allows, the obstruction may be localized by arteriography and removed surgically.

Cranial Nerves and the Carotid Arteries

The internal and external carotid arteries are intimately related to the last four cranial nerves. To understand these relationships it is worth turning to the base of the skull and examining the foramina of entry and exit of these structures. With the aid of Fig. 9-84 identify the foramina and note as described below the structures which they convey.

1. Find the occipital condyle and, passing through its base, the hypoglossal canal that leads the hypoglossal nerve towards the structures leaving the jugular foramen.
2. The irregular jugular foramen transmits the glossopharyngeal, vagus, and accessory nerves, and posterior to these the internal jugular vein.
3. Immediately anterior to the jugular foramen lies the carotid canal carrying the internal carotid artery and on the surface of the artery the internal carotid nerve and plexus consisting of postganglionic sympathetic fibers derived from the superior cervical sympathetic ganglion.

It can be seen from this examination that the internal carotid artery, internal jugular vein, the last four cranial nerves and the sympathetic chain must lie very close to each other at the base of the skull. Note also the styloid process. This and the muscles attached to it separate the structures just described from the overlying parotid gland and the external carotid artery lying in it.

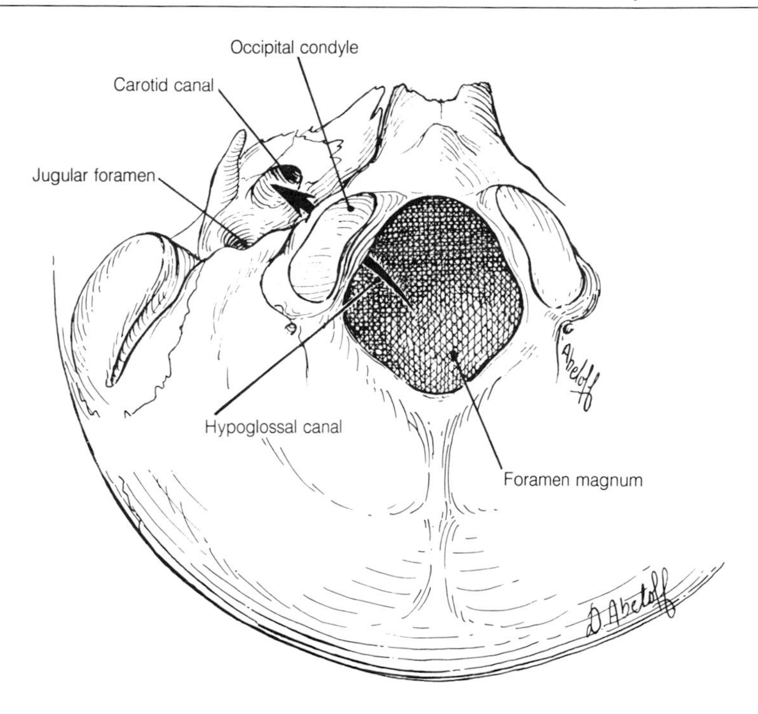

9-84 The base of the skull showing the proximity of the carotid canal, jugular foramen, and hypoglossal canal.

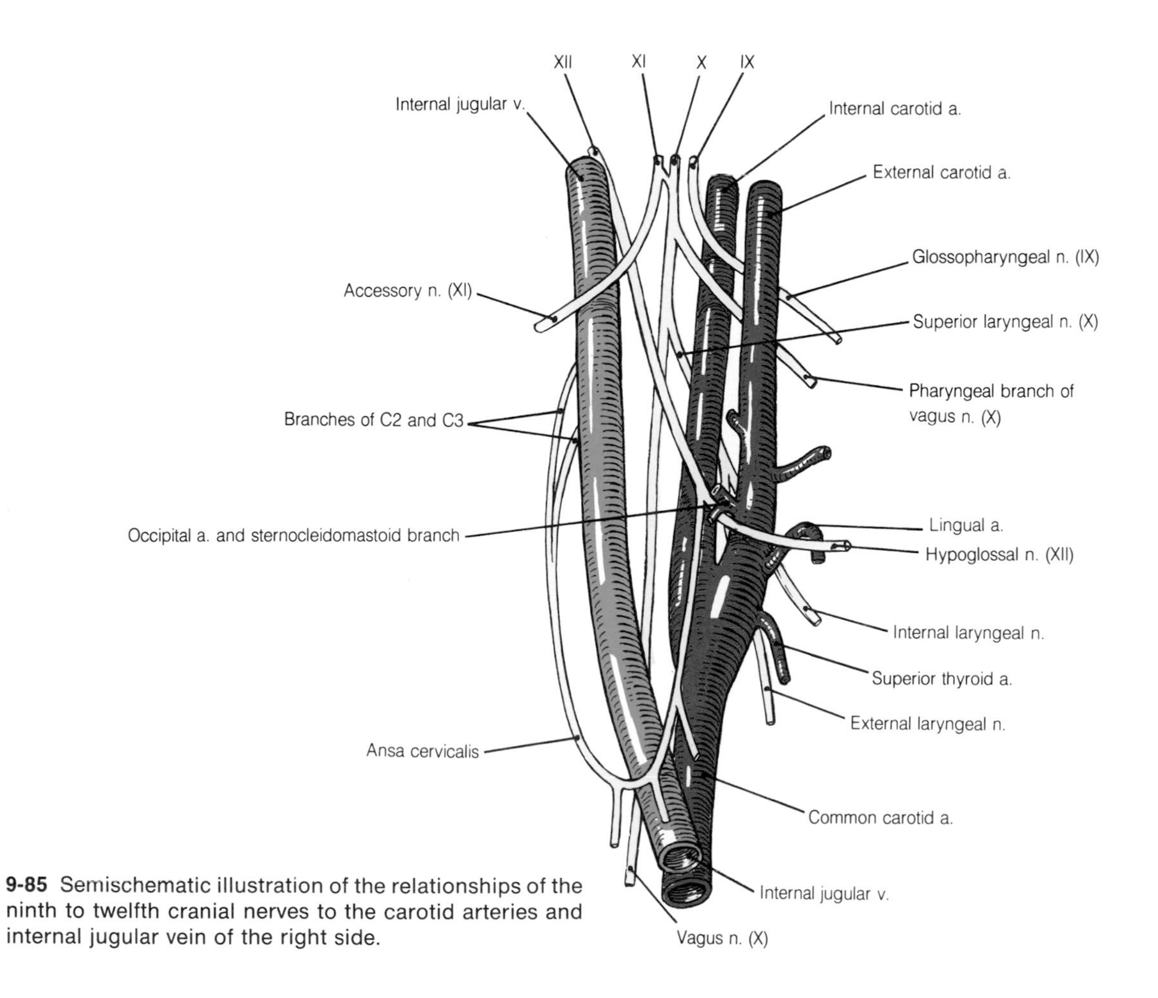

9-85 Semischematic illustration of the relationships of the ninth to twelfth cranial nerves to the carotid arteries and internal jugular vein of the right side.

The course of these structures in the neck can now be followed. Using Fig. 9-85, note that even though the structures have been artificially separated for clarity, they still have close relationships. Compare the top of this figure with Fig. 9-84. Above, the most posterior structure is the **internal jugular vein**. It is initially crossed by the **accessory nerve** and then gradually moves to the lateral side of the carotid vessels as it descends in the neck. The **hypoglossal nerve** begins deeply, medial to the internal jugular vein, descends behind the accessory nerve and gradually takes up a more superficial plane before crossing both internal and external carotid arteries on its way to the floor of the mouth and the tongue. Note how it passes over the loop of the lingual artery. This feature often simplifies its identification. The **accessory nerve** passes backward, crossing the internal jugular vein superficially. It supplies the overlying sternocleidomastoid muscle and, appearing at its posterior border, enters the posterior triangle. The **vagus nerve** maintains its position between the internal jugular vein and either the internal or common carotid artery throughout its course in the neck. Note how its **pharyngeal branch** to the **pharyngeal plexus** passes between the carotid arteries. The **superior laryngeal nerve**, a laryngeal branch of the vagus, passes deep to both arteries on the surface of the pharynx and terminates by dividing into the **internal and external laryngeal nerves. Cardiac branches** are also given off by the vagus as it passes down through the neck. Like the pharyngeal branch of the vagus, the **glossopharyngeal nerve** runs between the two carotid arteries. Note its small **motor branch** to the stylopharyngeus which is the only muscle supplied by the glossopharyngeal nerve. The glossopharyngeal nerve subsequently passes between the adjacent borders of the superior and middle pharyngeal constrictor muscles to reach the posterior part of the tongue. Medial to this close-packed collection of structures lies the pharynx and, although not illustrated, the internal jugular vein is surrounded by the deep cervical lymph nodes. Crossing the internal jugular vein in its lower part is the **ansa cervicalis**.

The Internal Jugular Vein

The internal jugular vein, a continuation of the sigmoid sinus, leaves the cranial cavity through the jugular foramen in company with the ninth, tenth, and eleventh cranial nerves and the **inferior petrosal sinus,** which joins the vein soon after its exit. At this point the vein is somewhat distended and called the **jugular bulb**. The vein descends in the neck at first posterior to the internal carotid artery and then lateral to this and the common carotid arteries. It terminates at its union with the subclavian vein to form the brachiocephalic vein behind the sternoclavicular joint.

The internal jugular vein drains blood from the cranial cavity and receives tributaries from the inferior petrosal sinus, the pharynx, the face, floor of the mouth, the tongue, and the thyroid gland.

Prevertebral Structures

Deep in the neck between the carotid sheath and the prevertebral fascia covering the longus colli and longus capitis muscles is the **cervical sympathetic trunk**.

The Sympathetic Trunk in the Neck

The sympathetic trunk ascends the neck vertically on the surface of the longus colli and longus capitis muscles and posterior to the carotid sheath. The trunk consists of a large **superior cervical ganglion** lying at the level of the second and third cervical vertebrae, a **middle cervical ganglion** lying close to the inferior thyroid artery, and the **cervicothoracic** or stellate ganglion. The latter may exist as a separate **inferior cervical ganglion** or be fused with the first thoracic ganglion. The ganglia are joined by connecting branches to complete the trunk.

The Distribution of the Sympathetic System in the Neck

The cell bodies of the preganglionic fibers that reach the cervical sympathetic ganglia lie in upper thoracic segments of the cord. These fibers ascend the trunk and synapse in the ganglia with the cell bodies of postganglionic fibers. As elsewhere in the sympathetic trunk, postganglionic fibers in the branches of the sympathetic ganglia are distributed to the peripheral nerves (including the cranial nerves in this region), on the surface of blood vessels, and by named and unnamed branches to the viscera. Remember that no white rami communicantes are present above the level of the first thoracic segment and that the distribution of postganglionic fibers to the cervical spinal nerves and the cranial nerves is by means of grey rami communicantes. The distribution of each ganglion is summarized below.

The superior cervical ganglion

Grey rami communicantes to the upper four cervical nerves and to the vagus and hypoglossal nerves
Branches to the carotid body and pharyngeal plexus
A cardiac branch
Branches running to the common, internal and external carotid arteries

The middle cervical ganglion

Grey rami communicantes to the fifth and sixth cervical nerves
A cardiac branch
Branches running to the thyroid and parathyroid glands along the inferior thyroid artery

The cervicothoracic ganglion

Grey rami communicantes to the seventh and eighth cervical and first thoracic nerves
A cardiac branch
Branches to the subclavian and vertebral arteries
A loop connecting the ganglion with the middle cervical ganglion that runs anterior to the subclavian artery and is called the **ansa subclavia**

The distribution of the cardiac branches of the ganglia is described with the cardiac plexuses in the thorax (Chapter 4).

The sympathetic trunk enters the cranial cavity on the surface of the internal carotid artery as the **internal carotid nerve** which continues upward from the superior cervical ganglion.

The Anterior Vertebral Muscles

The sympathetic trunk has been described as ascending in the neck on the surface of the **longus colli** and **longus capitis** muscles. These form a flat sheet of muscle anterior to the cervical and upper three thoracic vertebrae. The longus capitis is attached superiorly to the basilar part of the occipital bone. Two further short muscles extend from the lateral mass and transverse process of the atlas to the occipital bone. These are the **rectus capitis anterior and lateralis**.

Each of these anterior vertebral muscles is supplied by **ventral rami of cervical nerves**.

The Midline Structures of the Face and Neck

In this section structures forming part of the alimentary and respiratory passages in the neck and others closely associated with them will be described.

Although initially separated from each other, the respiratory and alimentary tracts become common in the middle part of the pharynx only to separate again below this point. The alimentary tract continues as the laryngopharynx and esophagus and the respiratory tract as the larynx and cervical part of the trachea. Closely applied to the trachea is the thyroid gland. The general arrangement of these strucutres can be seen in Fig. 9-86.

The Pharynx

The pharynx lies in front of the cervical vertebral column and prevertebral fascia and behind the nasal cavity, oral cavity, and larynx. Hence, although a continuous passage, it is described in three parts: The **nasopharynx, the oropharynx**, and the **laryngopharynx**. The pharynx can be thought of as a vertical fibromuscular tube with a number of deficiencies in its walls. The superior end of the tube is applied to the base of the skull and is thus occluded. The inferior opening, which lies at the level of the lower border of the cricoid cartilage of the larynx, is continuous with the esophagus. The anterior wall is defective where it communicates with the nasal cavities at the

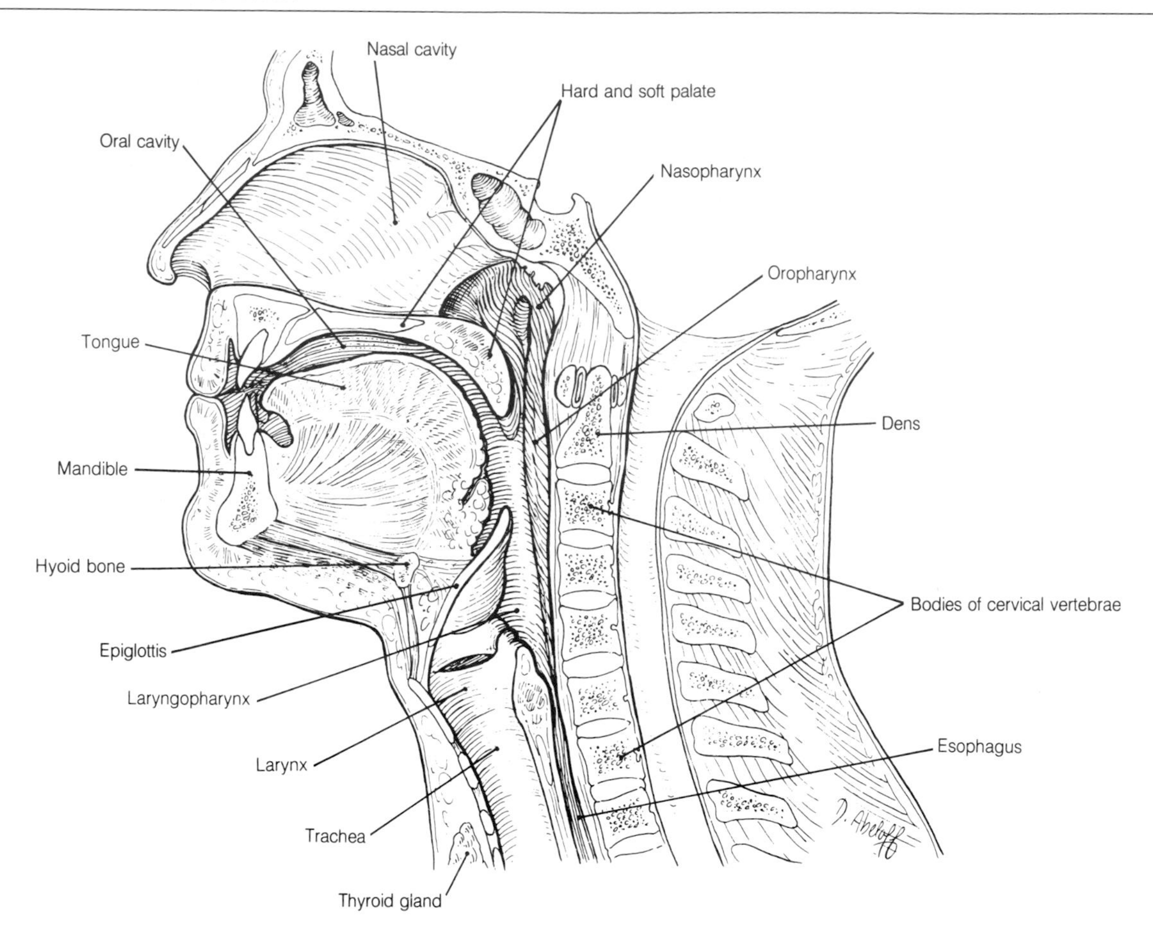

9-86 A median section through the face and neck showing the arrangement of the main parts of the respiratory and alimentary tracts.

choanae. Below this lies an anterior communication with the oral cavity called the **oropharyngeal isthmus**. The lowest anterior opening is the **laryngeal inlet**. The lateral walls are complete except those of the nasopharynx which are perforated by the two auditory tubes which connect the nasopharynx with the middle ear cavities. The posterior wall shows no defects.

It can be deduced from this description that the nasopharynx is a purely respiratory tube, that the oropharynx is shared by both respiratory and alimentary systems, and that the laryngopharynx is alimentary. In keeping with this, the epithelium of the nasopharynx is respiratory in type, whereas in the remaining parts, which are subjected to aliment, it is stratified squamous in nature. Outside the mucous membrane lies the **pharyngobasilar fascia** which separates the mucous membrane from an incomplete muscular layer formed by the three pharyngeal constrictor muscles. These are in turn surrounded by the thin **buccopharyngeal fascia** which allows the pharynx expansion and mobility. The three constrictor muscles form a continuous overlapping layer posteriorly but diverge from each other as they extend toward their anterior attachments. At each overlap the lower muscle lies outside the upper. In Fig. 9-87 follow the attachments of each muscle as it is described.

The Superior Constrictor

The superior constrictor has a continuous anterior attachment to the lower part of the free border of the medial pterygoid plate, to the pterygoid hamulus, to the pterygomandibular raphe, and to the posterior end of the mylohyoid line. It should

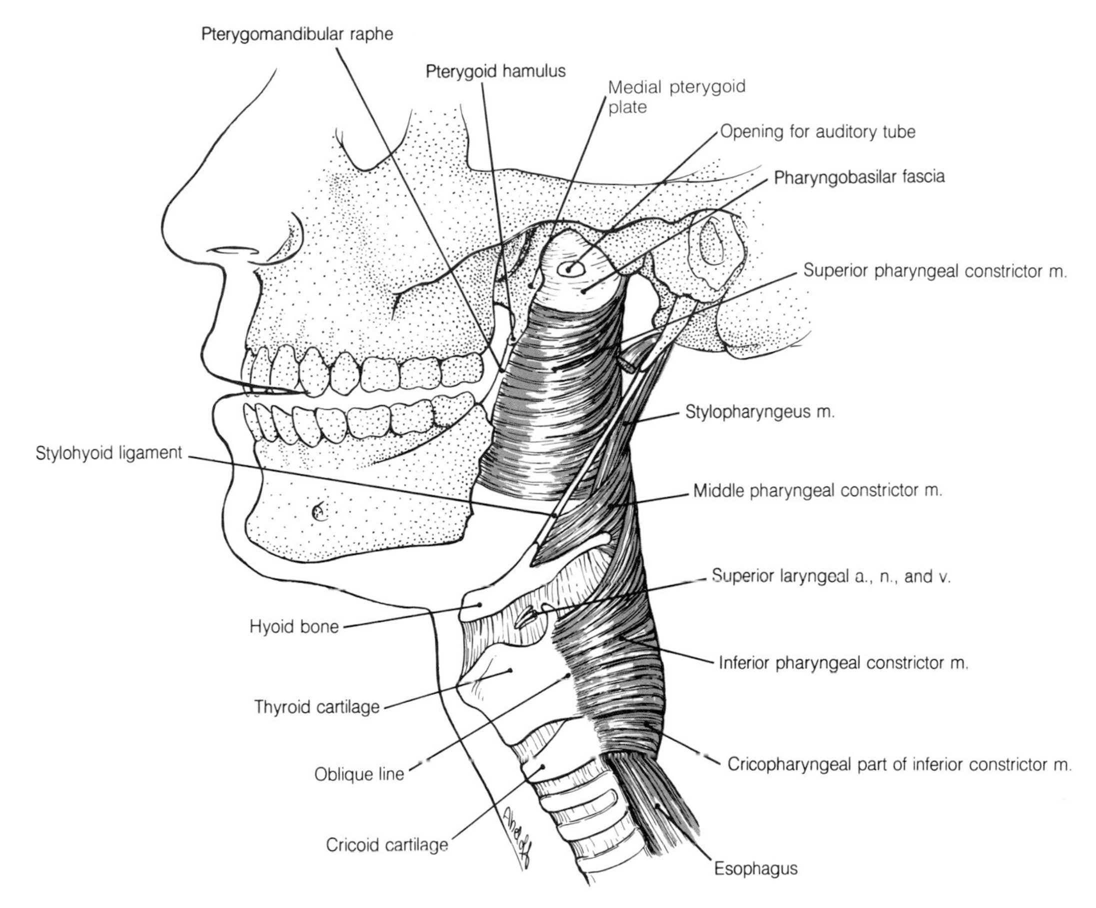

9-87 Showing the pharyngeal constrictor muscles and their anterior attachments. The ramus of the mandible and the lateral pterygoid plate have been removed.

be remembered that the buccinator is attached anteriorly to the raphe and there is thus a continuous muscular sheet extending from the lips, through the cheeks, and into the oropharynx.

From its anterior attachments the muscle sweeps around posteriorly to be attached in the midline to the **pharyngeal tubercle of the occipital bone** on the base of the skull, and below this, to a vertical fibrous band called the **pharyngeal raphe**. The space between the upper border of the muscle, the base of the skull, and the upper free edge of the medial pterygoid plate is filled in by the firm **pharyngobasilar fascia** which is perforated by the **auditory tube**.

The Middle Constrictor

The middle constrictor arises anteriorly from the stylohyoid ligament and the lesser and greater cornua of the hyoid bone. From this V-shaped origin the muscle fans out to enclose the superior constrictor posteriorly and reach the pharyngeal raphe. The lower fibers of the muscle arch downward to the level of the larynx. Through the gap between the superior and middle constrictors pass the **stylopharyngeus muscle** and the **glossopharyngeal nerve**.

The Inferior Constrictor

The inferior constrictor arises from the oblique line on the lamina of the thyroid cartilage, the cricoid cartilage, and a small tendinous arch between the two. Again the upper fibers arch upward and overlap the middle constrictor. The lower fibers, however, which arise from the cricoid cartilage, run almost horizontally, and are sometimes named the **cricopharyngeus muscle**.

Unlike the rest of the muscle, these fibers are supplied by the **external laryngeal nerve**. It is above these fibers that the pharyngeal mucous membrane sometimes pouches out to form a pharyngeal diverticulum.

With the exception of cricopharyngeus, the constrictor muscles of the pharynx are supplied by branches of the pharyngeal plexus.

The Stylopharyngeus and Salpingopharyngeus Muscles

These two small muscles are closely associated with the pharynx although both are largely attached to the larynx.

The **stylopharyngeus** arises from the medial aspect of the styloid process, descends close to the superior pharyngeal constrictor, and passes with the glossopharyngeal nerve between the superior and middle constrictor muscles. Its lower attachment is to the posterior border of the thyroid cartilage. It is supplied by the **glossopharyngeal nerve**.

The **salpingopharyngeus** arises from the expanded pharyngeal end of the cartilage of the auditory tube. It descends inside the superior constrictor muscle to the attached like stylopharyngeus to the thyroid cartilage. It is innervated from the **pharyngeal plexus**.

Internal Features of the Pharynx

The internal aspect of the pharynx can best be illustrated by exposing it through its posterior wall as in Fig. 9-88. In practice much of it can be examined directly or by the use of a mirror. Especially in children, the mucous membrane of the posterior wall of the nasopharynx may be elevated

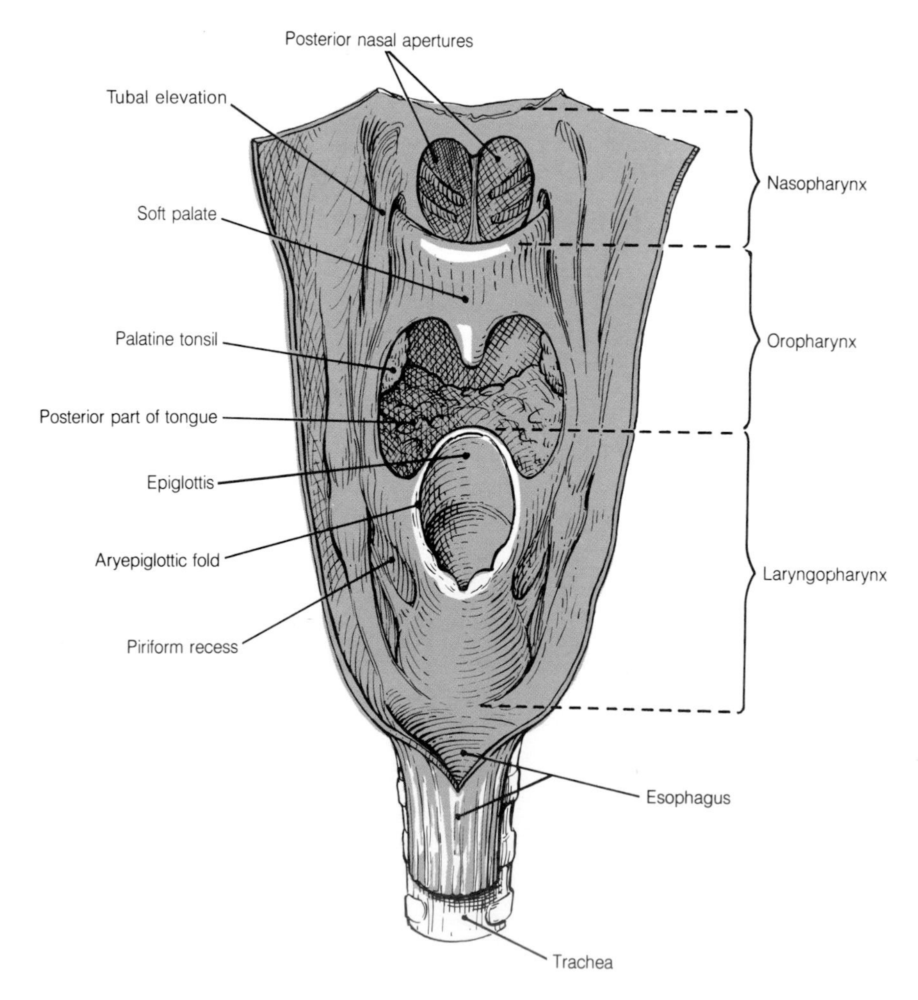

9-88 The pharynx opened from behind to show the anterior wall.

by the underlying **pharyngeal tonsil** or "adenoids." This may lead to obstruction of the airway and result in mouthbreathing and deformities of the hard palate and upper dental arch. Laterally, the openings of the auditory tubes are marked by the **tubal elevations** (tori tubarii) caused by the superior and posterior projection of their cartilaginous walls. A fold of mucous membrane overlying the salpingopharyngeus muscle descends from these elevations. The levator veli palatini muscles, which lie within the muscular wall of the pharynx, form additional folds behind these.

Within the oropharynx lie the **palatine tonsils.** These are described in more detail elsewhere. The posterior third of the tongue forms the anterior wall of the oropharynx and when depressed by a laryngoscope allows the **valleculae** to be seen. These are shallow fossae lying on either side of a midline fold of mucous membrane extending from the tongue to the epiglottis. The **glossoepiglottic fold**.

Laryngoscopy also allows inspection of the epiglottis and laryngeal inlet. On either side of the laryngeal inlet are the **piriform recesses**. These recesses lie between the mucous membrane covering the aryepiglottic folds and that covering the thyroid cartilage and thyrohyoid membrane. Beneath the mucous membrane lies the **internal laryngeal nerve**. The importance of these fossae lies in the facts that particles of food may become lodged here and that malignant changes may occur in the lining epithelium with few initial symptoms.

The Nerve Supply of the Pharynx

The pharynx is innervated through the **pharyngeal plexus**. This plexus, lying over the middle constrictor, is contributed to by the **vagus**, the **glossopharyngeal nerve**, and **sympathetic fibers** from the superior cervical ganglion. In addition, sensory fibers from the **maxillary nerve** via the **pterygopalatine ganglion** supply the nasopharynx with sensation. Of the nerves contributing to the pharyngeal plexus, the vagus is motor to the pharyngeal constrictor muscles. (These are all striated muscles derived from branchial arches.) The glossopharyngeal nerve supplies sensory fibers to the mucous membrane below the nasopharynx and the sympathetic fibers supply the blood vessels.

The Vessels of the Pharynx

Many vessels contribute to the blood supply of the pharynx. Most are nearby branches of the external carotid artery and include the **ascending pharyngeal artery**, the **facial and lingual arteries**, and the **laryngeal arteries**. The **inferior thyroid arteries** also supply its lower part. Blood drains to venous plexuses on the external surface of the pharynx which empty into the internal jugular vein. Lymphatics run directly to deep cervical nodes.

The Auditory Tube

The auditory tube connects the nasopharynx with the **tympanic cavity**. Its anterior two-thirds forms a narrow slit supported by the **cartilage of the auditory tube** which partly surrounds it. Its posterior third runs in a bony canal in the petrous part of the temporal bone. The tube is directed downward and medially from the middle ear and enters the nasopharynx by piercing the pharyngobasilar fascia above the superior constrictor muscle. Here the cartilage of the tube raises the mucous membrane of the pharynx to form the **tubal elevation**.

The mucous membrane lining the tube is continuous with that of the tympanic cavity and nasopharynx. The cartilaginous part is normally closed and as a result, changes in atmospheric pressure are not immediately transmitted to the tympanic cavity. The pressures are equalized by swallowing or yawning. The tube may also become obstructed by the swelling of the mucous membrane which accompanies an upper respiratory infection. Because of this, middle ear infections may become a secondary complication. Note in Fig. 9-89 the manner in which the cartilage is

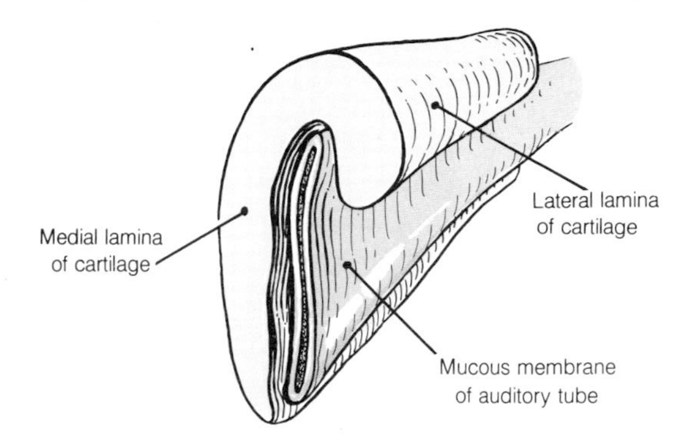

9-89 Semischematic drawing to show the shape of the cartilage of the left auditory tube.

folded over the tube to form a shallow **lateral lamina** and a deep **medial lamina**. The contraction of the tensor veli palatini muscle which is attached to it may be the reason why the tube opens on swallowing.

The Nose

The nose includes that portion projecting from the face, the **external nose**, and the **nasal cavities** which lie within a bony and cartilaginous skeleton and are separated from each other by a midline **septum**. The nasal cavities communicate externally by means of the nostrils or **nares** and posteriorly with the nasopharynx by means of the **choanae**. In addition, it will be seen that the nasolacrimal duct and a number of air sinuses open into the cavities. These cavities function as a respiratory pathway where the air is warmed, moistened, and cleaned of particulate matter. They also serve the important functions of olfaction and much of what is commonly accepted as taste.

The Skeleton of the Nose

The skeleton of the external portion of the nose is largely cartilaginous. The bony portion can be seen in Fig. 9-90 to consist of a pair of nasal bones superiorly, while the remainder of the bony anterior nasal aperture is completed by the frontal processes and notched medial borders of the two maxillary bones. The shape of the remainder of the external nose is maintained by the midline septal cartilage and lateral nasal and alar cartilages.

The Nasal Cavities

Each nasal cavity can be considered to have a roof, a floor, a lateral wall, and a medial wall which is the nasal septum. These details of osteology should aid in relating the nasal cavities (sources of infection) to important surrounding cavities such as the anterior cranial fossa, the orbit, and the paranasal air sinuses. In Fig. 9-91 the bones surrounding the nasal cavity are shown in a semischematic manner and this illustration should be used with the following description of the bony cavity.

The **floor** is formed anteriorly by the **palatine process of the maxilla** and posteriorly by the **horizontal plate of the palatine bone**. It separates the cavity from the mouth.

The **roof** is formed by the **nasal, frontal, ethmoid**, and **sphenoid bones**. Above lies the an-

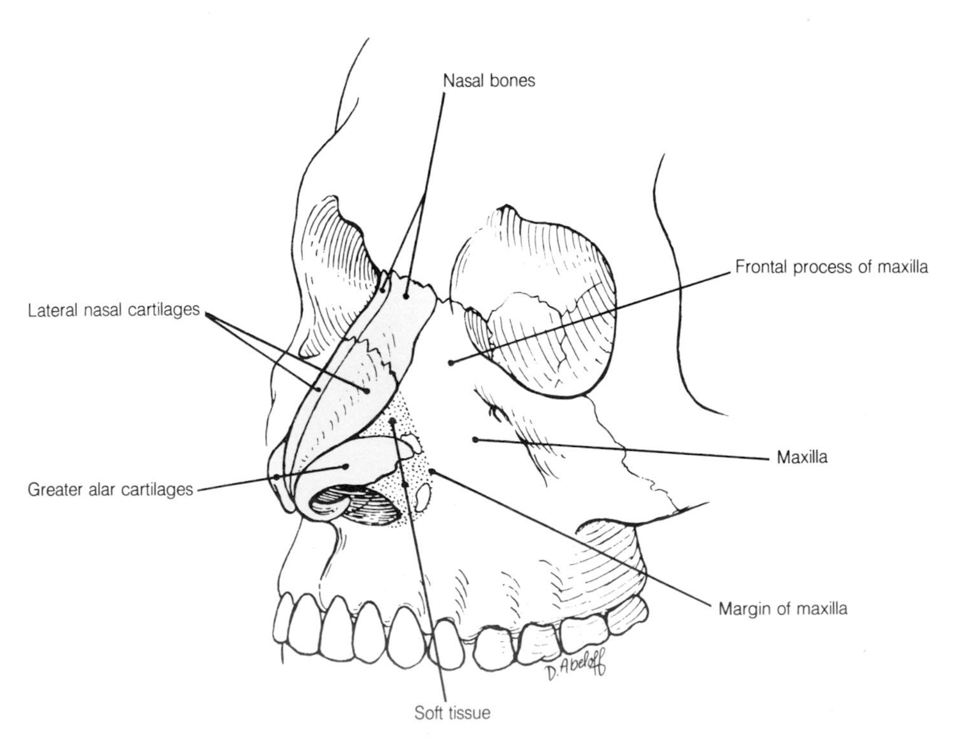

9-90 Illustrating the bony and cartilaginous components of the external nose.

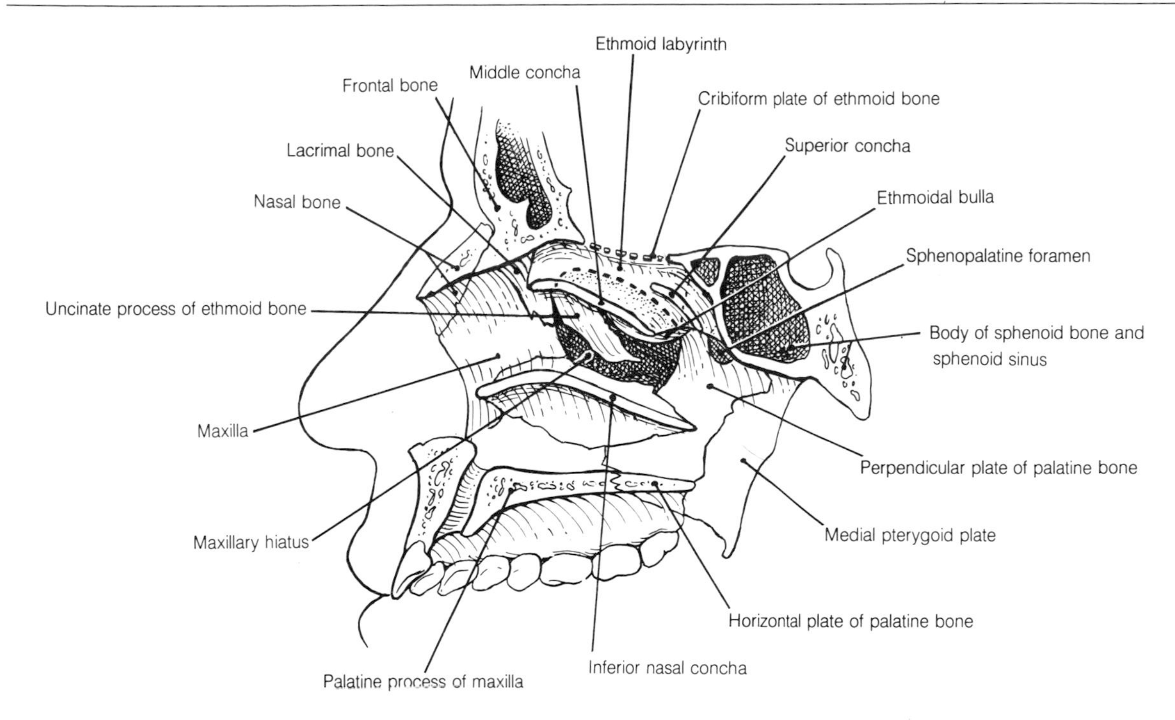

9-91 Showing the bony skeleton of the roof, lateral wall, and floor of the nasal cavity (only the roots of the conchae are shown).

terior cranial fossa and within the sphenoid bone is the sphenoid sinus.

The **lateral wall** is complicated by the overlapping and superimposition of bones. Look at Fig. 9-91. In this the lateral wall is seen to be formed by the **maxillary, palatine**, and **sphenoid bones**. Note the large **maxillary hiatus** and the small **sphenopalatine foramen**. The lateral wall is completed by the superimposition of the **lacrimal** and **ethmoid bones**, and an additional bone called the **inferior nasal concha**. Note that the maxillary

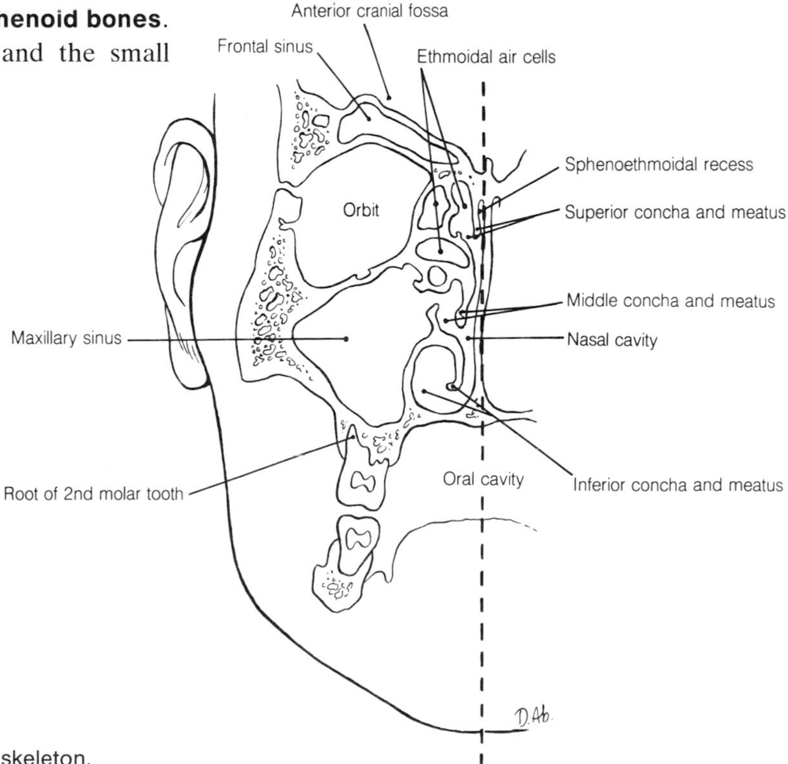

9-92 Coronal section through the facial skeleton.

hiatus is now reduced to a slit. One of the features of the lateral wall is the presence of three **conchae**. These are horizontal projections which curve downward at their medial edges. The upper two are part of the ethmoid bone and the lowest is a separate bone. They divide the nasal cavity into four passages: the **inferior, middle and superior meatuses** and, above the superior concha, the small **sphenoethmoidal recess**. In Fig. 9-91 the conchae have been removed and only their bases are shown, but in Fig. 9-92, which illustrates a coronal section through the facial skeleton and nasal cavity, the conchae and the meatuses can be seen. Note also the surrounding cavities; i.e. the anterior cranial fossa and frontal sinus above, the orbit and maxillary sinus to the side, and the oral cavity below.

The exact details of the bones forming the lateral wall of the nose are not of great importance but careful note should be made of the ducts of a number of nearby cavities that open into the meatuses. In Fig. 9-93 the bony lateral wall of the nasal cavity is shown covered with mucous membrane. The conchae have again been partially removed. In this illustration note that:

1. Below the inferior concha is the opening of the **nasolacrimal duct** which drains the conjunctival sac;
2. Below and hidden by the middle concha is the **hiatus semilunaris**. It lies between the **bulla** and **uncinate process** of the ethmoid bone and into it open the **frontonasal duct** anteriorly followed by the **anterior ethmoidal air sinuses** and the **maxillary sinus**;
3. The **bulla** is formed by the underlying **middle ethmoidal air sinuses** and the open on its surface;
4. The superior concha is short and so is the superior meatus. The **posterior ethmoidal air sinuses** open into the superior meatus;
5. Between the roof of the cavity and the superior concha lies the **sphenoethmoidal recess.** Into this region the posteriorly situated **sphenoidal sinus** opens.

The **medial wall** of each nasal cavity is the septum. It is partly bony and partly cartilaginous. As can be seen in Fig. 9-94, its bony portion is formed by the **perpendicular plate of the ethmoid bone** and by a thin sheet of bone called the **vomer**.* The cartilaginous portion is formed by the **septal cartilage**. The septum not infrequently deviates from the midline, sometimes to the extent that it interferes with function and requires surgical realignment.

The Mucous Membrane of the Nasal Cavity

The nasal cavity is lined by a variety of epithelia. The area just inside the anterior nares is called the **vestibule** and lined by hairy skin. The remainder, except for a small area of the roof, septum, and

* Vomer, *L.* = a ploughshare

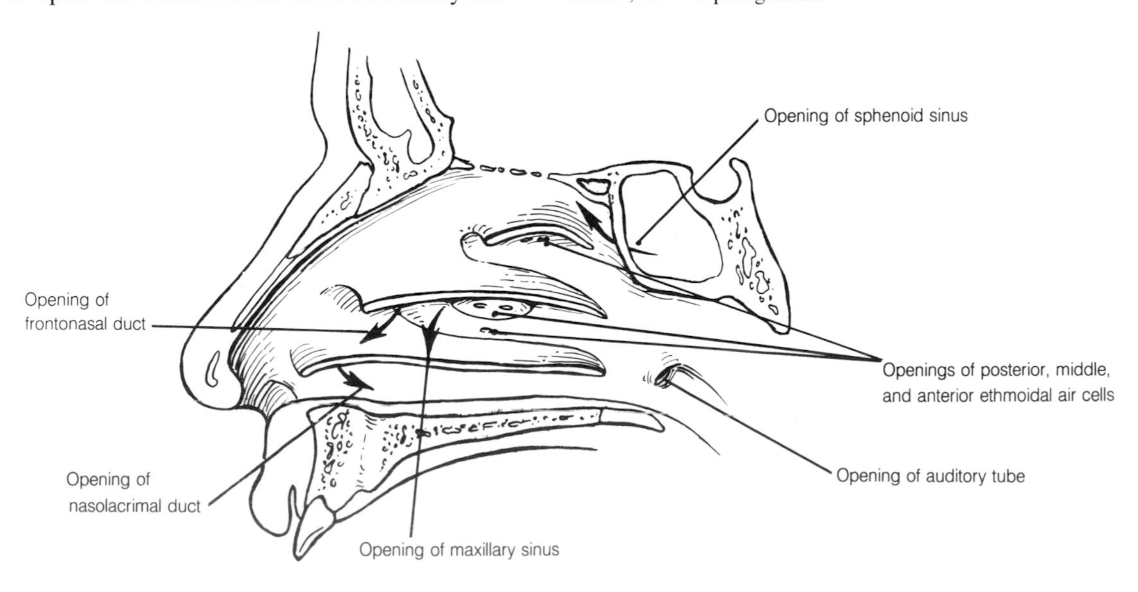

9-93 The lateral wall of the nose. Portions of the superior, middle, and inferior conchae have been removed.

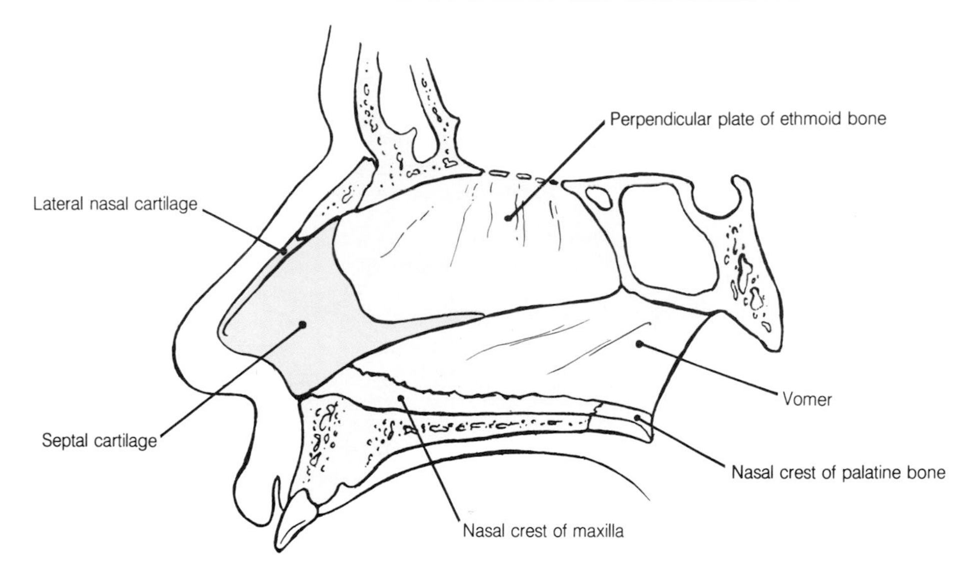

9-94 Showing the bones and cartilage of the nasal septum.

lateral wall which is specialized for olfaction, is lined by a ciliated columnar respiratory epithelium which contains many goblet cells. Beneath the epithelium lie lymphoid tissue, further mucous and serous glands, and a rich vascular plexus, hence, the filtering, moistening, and warning function of the nose. At the openings of their ducts into the nasal cavity the mucous membrane is continuous with the mucous membrane of the surrounding air sinuses and the conjunctival sacs. A small area of mucous membrane over the upper septum, roof, and lateral wall of the cavity is modified for olfaction. Here the epithelium is formed from supporting cells between which lie the bipolar nerve cells called olfactory cells. The distal processes of the cells terminate as the microscopic "hairs" which project from the surface of the mucous membrane. The sensations generated by these cells play a large part in the sensation of taste (gustation) as well as that of smell (olfaction).

The Blood Supply of the Nasal Cavity

The blood supply of both the medial and lateral walls of the cavity is provided by branches of the

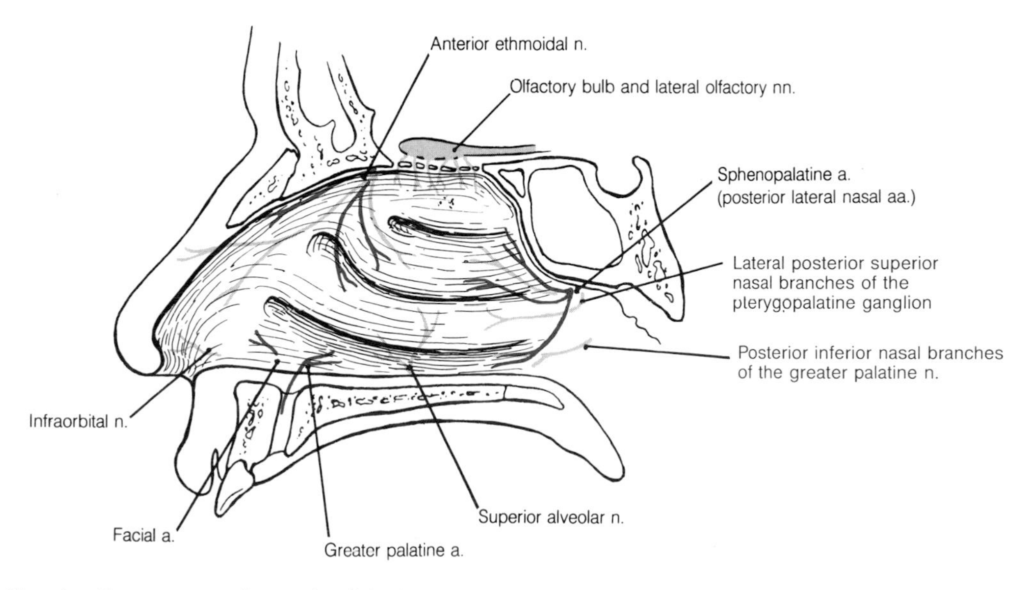

9-95 Showing the nerves and vessels of the lateral wall of the nose.

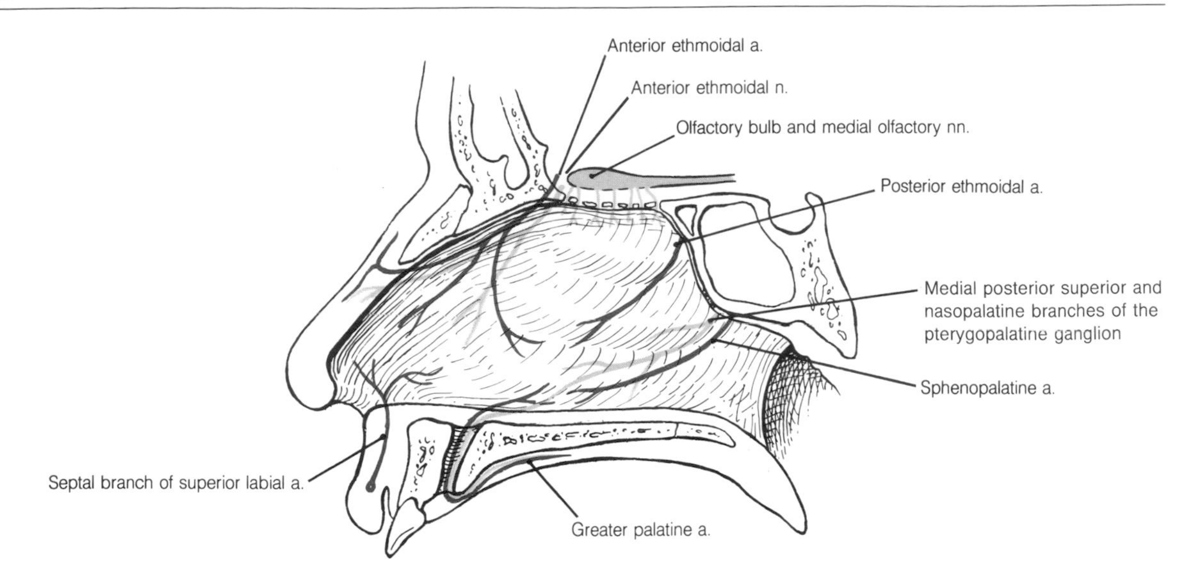

9-96 Showing the nerves and vessels of the nasal septum.

sphenopalatine artery, the **anterior and posterior ethmoidal arteries**, the **greater palatine artery**, and the **superior labial branch of the facial artery**. It is rupture of branches of the last vessel that commonly gives rise to nosebleeds (epistaxis). The regions supplied by these vessels are indicated in Fig. 9-95 and 9-96.

A plexus of veins beneath the mucous membrane drains into the sphenopalatine, facial, and ophthalmic veins.

The Nerve Supply of the Nasal Cavity

The specialized olfactory epithelium contains the bipolar olfactory cells. The central processes of these neurons pass through the cribriform plate of the ethmoid bone as the **olfactory nerves**. These enter the **olfactory bulb** where they synapse with neurons of the **olfactory tract**. Fractures of the base of the skull which pass through the cribriform plate may tear the meningeal sleeves that surround the olfactory nerves and so give rise to leakage of cerebrospinal fluid into the nose.

General sensation for the mucous membrane of the upper and anterior extents of the lateral wall and septum is supplied by branches of the **anterior ethmoidal nerve**, itself a branch of the **ophthalmic nerve**. The lower and posterior extents are supplied by the **maxillary nerve** through branches from the pterygopalatine ganglion. In Figs. 9-95 and 9-96 these can be seen on the lateral wall to be the **lateral posterior superior branches** and **posterior inferior branches** of the **greater palatine nerve** and on the septum to be **medial posterior superior branches.** Of these the longest is called the **nasopalatine nerve** as it passes through the incisive canal to provide some supply to the palate. The skin of the vestibule is supplied by the **infraorbital nerve**. Autonomic innervation of the mucosa is channelled into the nose through all the branches of the **pterygopalatine ganglion**.

A more detailed description of the maxillary artery, maxillary nerve, and the pterygopalatine ganglion will follow.

The Pterygopalatine Fossa

The key to the blood supply and nerve supply of a large part of the nasal cavity lies in this small bony space. Its significance can only be fully realized by the examination of the skull with the aid of a soft probe such as a pipe cleaner. The fossa can be approached through the **pterygomaxillary fissure**, i.e., the fissure between the lateral pterygoid plate and the maxilla, and this can be seen in Fig. 9-97. The fossa itself lies anterior to the **pterygoid plates of the sphenoid** and posterior to the **maxilla** and the **perpendicular plate of the palatine bone** forms its medial wall and separates it from the more posterior part of the nasal cavity.

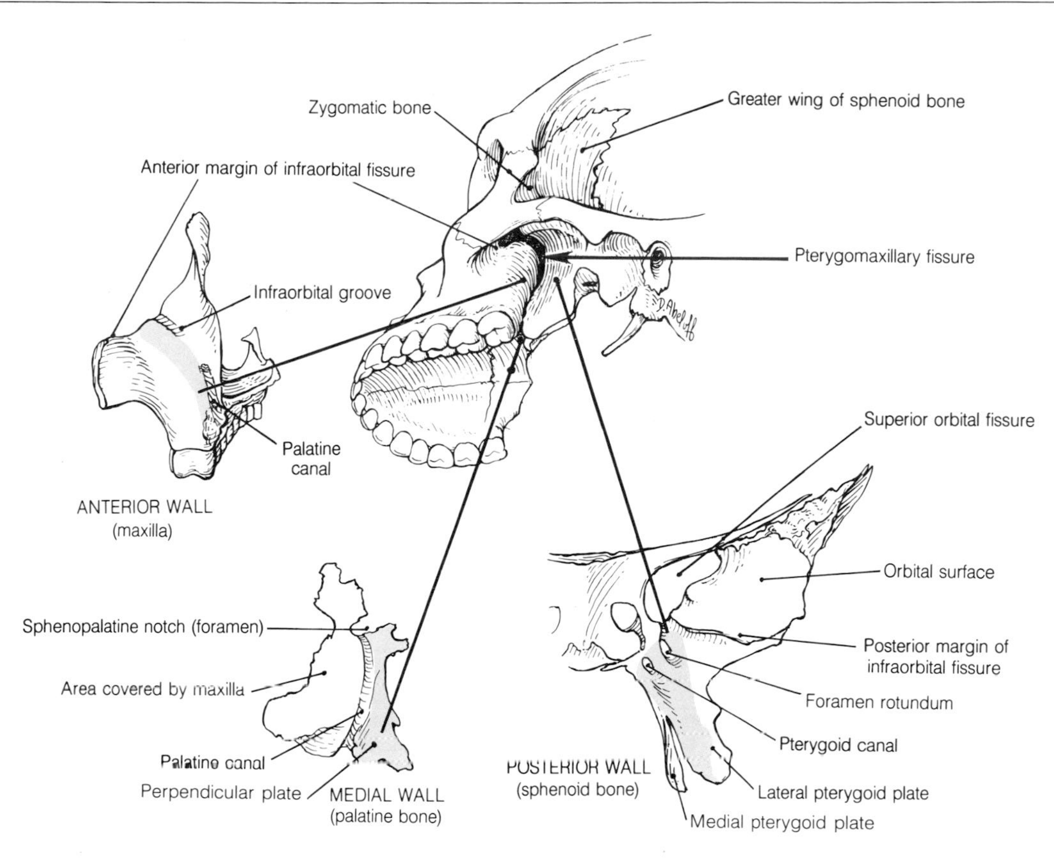

9-97 The anterior, medial, and posterior bony walls of the left pterygopalatine fossa separated to show its communications.

A number of foramina, fissures, and canals carry nervous and vascular structures to the fossa for distribution to the nose, nasopharynx, palate, orbit, and upper dentition.

In Fig. 9-97 the surfaces of the bones forming the posterior, medial, and anterior walls of the fossa are illustrated separately so that the communications of the fossa can now be visualized as this section is read.

The **maxillary artery** has already been seen to enter the fossa through the pterygomaxillary fissure. On the posterior wall the **foramen rotundum** carries the **maxillary nerve** into the fossa and the **pterygoid canal** carries the **nerve of the pterygoid canal**. This was formed near the foramen lacerum by fusion of the deep petrosal nerve and the greater petrosal nerve. As will be seen, branches of these nerves and vessels leave the fossa for further distribution through the **sphenopalatine foramen** for the nasal cavity, the **palatine canal** for the palate, the **infraorbital canal** for the orbit and face, and the **inferior orbital fissure** for the upper teeth.

The Maxillary Artery

The maxillary artery enters the pterygomaxillary fissure to reach the pterygopalatine fossa. It gives off a **posterior superior alveolar artery** and the **infraorbital artery** just as it enters he fossa. The former artery branches to enter small **alveolar foramina** on the posterior surface of the maxilla. The latter passes across the inferior orbital fissure to reach the floor of the orbit. Here it enters the **infraorbital canal** or **groove** which caries it and the infraorbital nerve to the **infraorbital foramen** and, thence, to the face. Within the fossa the maxillary artery gives off the **descending palatine artery** which runs with the greater palatine nerve down the greater palatine canal to reach the hard palate. While in the canal one or two **lesser palatine arteries** are given off for the supply of the soft palate and tonsil and it continues as the

greater palatine artery. The greater palatine artery emerges from its canal at the **greater palatine foramen** and runs forward across the hard palate to the **incisive foramen**. Here it enters the nasal cavity to anastomose with the **sphenopalatine artery** on the septum. The maxillary artery terminates as the **sphenopalatine artery** which passes through the **sphenopalatine foramen** to contribute to the blood supply of the lateral wall of the nose and the septum. This it does by **posterior lateral nasal arteries** and, after crossing over the roof of the cavity, by **septal branches.** The distribution of the last part of the maxillary artery is summarized below and illustrated in Fig. 9-98.

- Maxillary Artery
  - Posterior superior alveolar arteries
  - Infraorbital artery
    - Anterior superior alveolar arteries
  - Descending palatine artery
    - Greater palatine artery
    - Lesser palatine arteries
  - Sphenopalatine artery
    - Posterior lateral nasal arteries
    - Septal branches

The Maxillary Nerve

The maxillary nerve leaves the middle cranial fossa through the foramen rotundum and enters the pterygopalatine fossa. Here it gives off the **zygomatic nerve**, the **posterior superior alveolar nerves**, and the **infraorbital nerve**.

The zygomatic nerve reaches the lateral wall of the orbit through the inferior orbital fissure. It terminates as the **zygomaticotemporal** and **zygomaticofacial branches** which pierce the zygomatic bone to supply skin of the face. In the orbit a communication of this nerve with the **lacrimal nerve** carries secretomotor fibers from the pterygopalatine ganglion to the lacrimal gland.

The **posterior superior alveolar nerves** enter the posterior surface of the maxilla to supply the upper molar teeth and gums and the adjacent cheek.

The **infraorbital nerve** reaches the floor of the orbit through the **inferior orbital fissure** and runs forward in the **infraorbital canal**.

Within the canal the **middle and anterior superior alveolar nerves** are given off for the supply of the remaining upper teeth. The infraorbital nerve

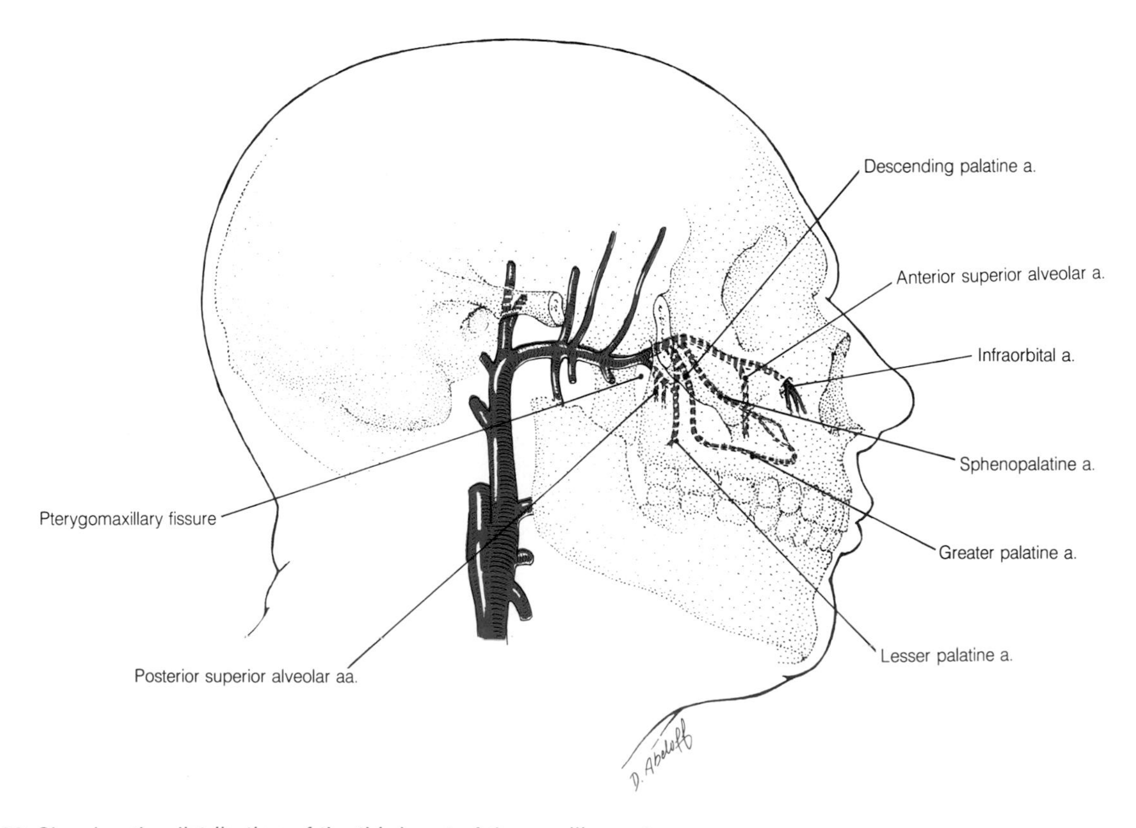

9-98 Showing the distribution of the third part of the maxillary artery.

passes to the face through the **infraorbital foramen** and terminates as branches supplying the lower eyelid, the side of the nose, the cheek, and the upper lip.

All three superior alveolar nerves reach the alveolar portion of the maxilla by passing in close relationship to the wall of the maxillary sinus. As a result irritation of these nerves by inflammation in the sinus may simulate toothache.

In addition to these branches, the maxillary nerve gives off one or two substantial **ganglionic branches** to the **pterygopalatine ganglion** which lies just below the nerve as it enters the pterygopalatine fossa. It is through branches from this ganglion that sensory fibers of the maxillary nerve reach the nose, palate, and nasopharynx. The branches of the maxillary nerve are summarized below and illustrated in Fig. 9-99.

Maxillary nerve
- Ganglionic branches to the pterygopalatine ganglion
- Zygomatic nerve
  - Zygomaticotemporal branch
  - Zygomaticofacial branch
  - Communicating branch to lacrimal nerve
- Posterior superior alveolar nerves
- Infraorbital nerve
  - Middle superior alveolar nerve
  - Anterior superior alveolar nerve
  - Palpebral, nasal, buccal, and labial branches

The Pterygopalatine Ganglion

This ganglion, like the ciliary, otic and submandibular ganglia, is a parasympathetic autonomic ganglion. That is, the cell bodies that is contains are those of postganglionic parasympathetic fibers. However, like other parasympathetic ganglia, fibers of other functions pass through it without synapse or cell bodies. The ganglion lies in the pterygopalatine fossa just lateral to the

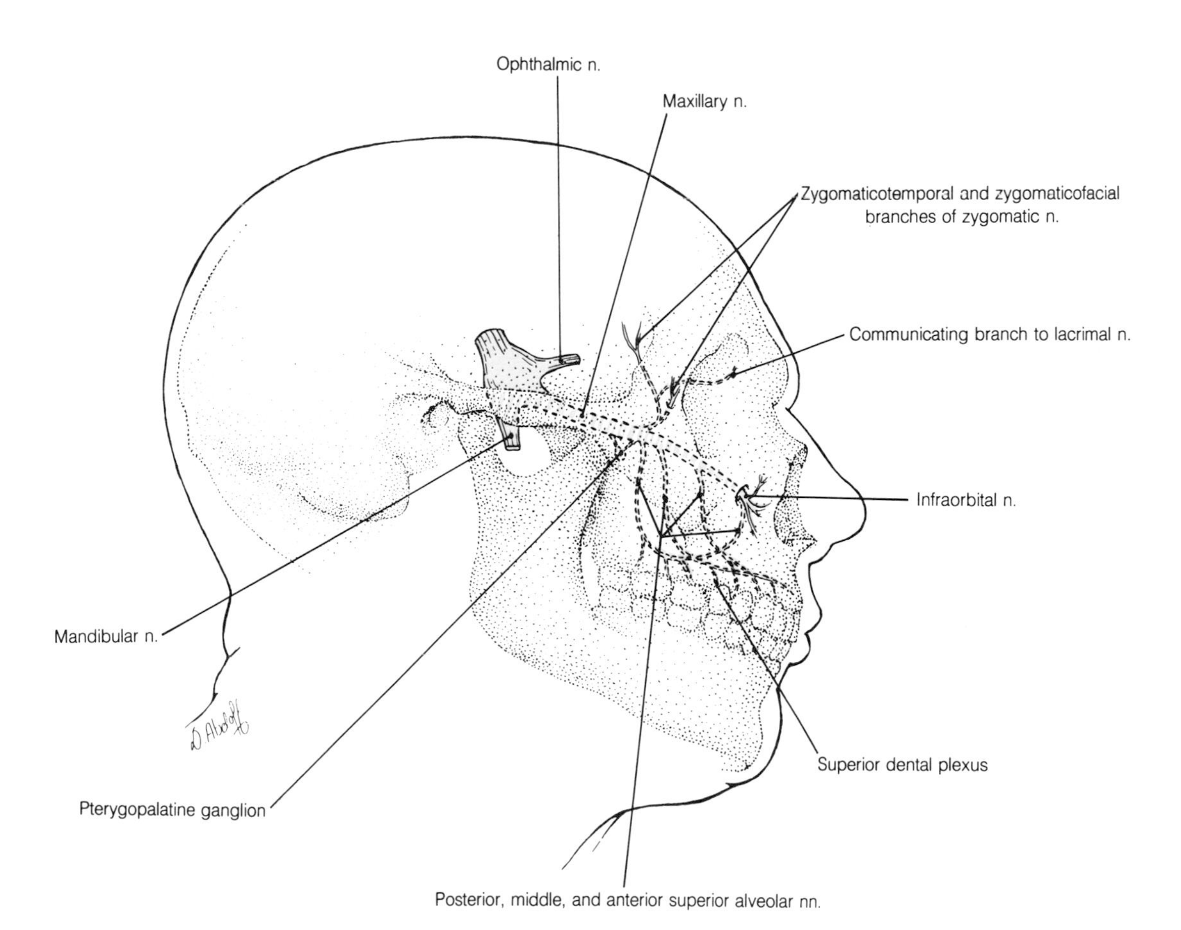

9-99 To show the distribution of the maxillary nerve.

sphenopalatine foramen, poised, as it were, to pass its branches into the nose, nasopharynx and palatine canal. The connections of the ganglion are illustrated diagrammatically in Fig. 9-100, A and B. Preganglionic parasympathetic fibers (*blue*) are derived from the **facial nerve** through its branch the **greater petrosal nerve**. This nerve leaves the petrous part of the temporal bone in the middle cranial fossa and at the foramen lacerum joins the **deep petrosal nerve** to form the **nerve of the pterygoid canal**. It is the parasympathetic fibers in this nerve that synapse with the cell bodies of the **postganglionic fibers** (*yellow*). **Sympathetic fibers** (*red*) from the **carotid plexus** run in the **deep petrosal nerve** and also reach the ganglion along the nerve of the pterygoid canal. These sympathetic fibers are, of course, postganglionic and have their cell bodies in the superior cervical ganglion. **Fibers of common sensation** (*green*) are passed to the ganglion through the **ganglionic branches** of the **maxillary nerve**. Each of these varieties of fiber are distributed together by branches of the ganglion which are described below. The parasympathetic fibers are secretomotor to the many nasal and palatine glands and, as already described, some pass to the maxillary nerve for distribution to the lacrimal gland through its zygomatic branch. The sympathetic fibers are vasomotor and the sensory fibers supply the palate, tonsil, nasal cavities, and the roof of the nasopharynx.

The **greater palatine nerve** descends through the **greater palatine canal** to appear on the hard palate at the **greater palatine foramen**. It supplies the mucous membrane of the palate and adjacent gums. While in the greater palatine canal it

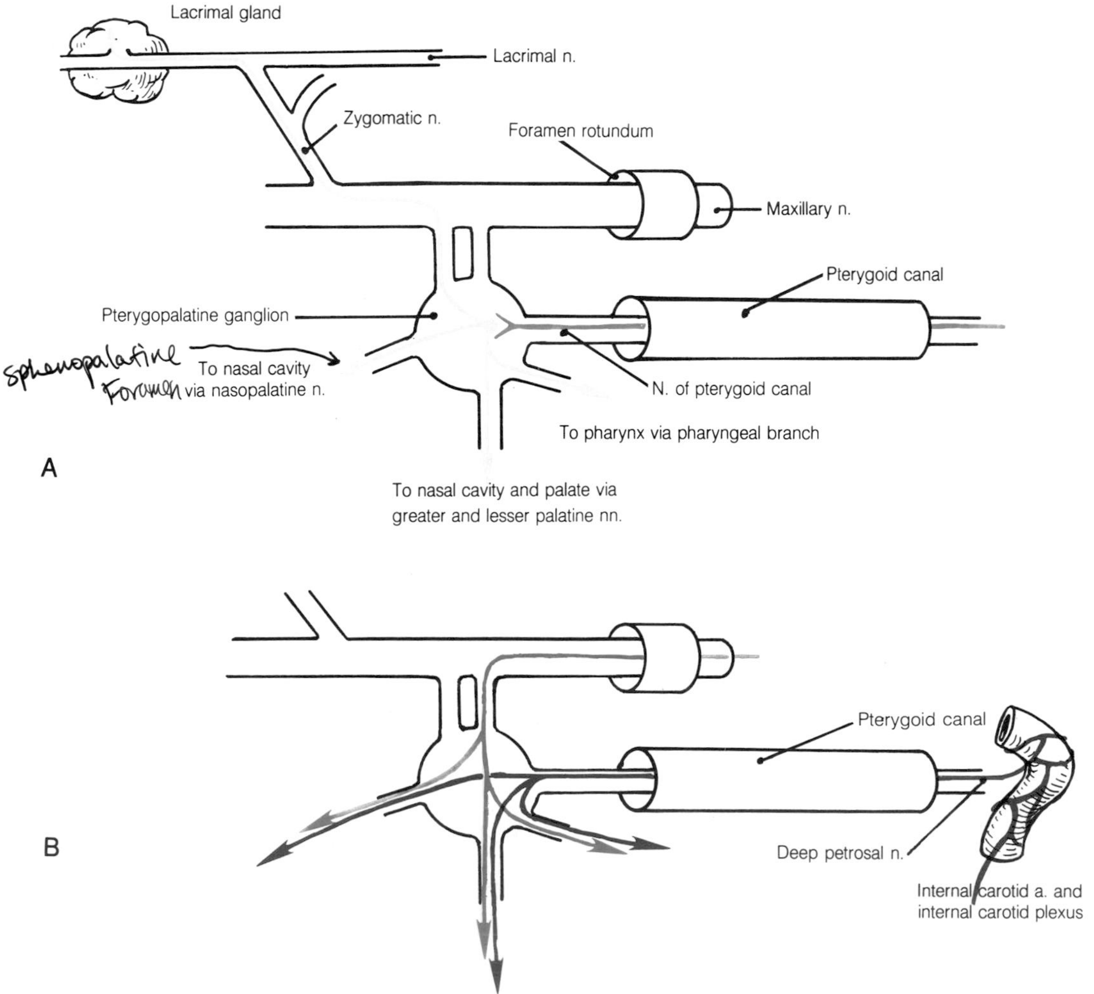

9-100 Diagrams illustrating the connections and distribution of parasympathetic fibers (*A*) and sympathetic and sensory fibers (*B*) of the pterygopalatine ganglion.

also gives off **posterior inferior nasal branches** which pierce the palatine bone to help supply the lateral wall of the nose.

The **lesser palatine nerves** begin their course in the greater palatine canal but emerge through the **lesser palatine foramina** to pass backward and supply the soft palate and tonsilar region.

Posterior superior nasal nerves pass through the sphenopalatine foramen to supply the lateral wall of the nasal cavity or run across the roof to supply the septum. Of the latter the **nasopalatine nerve** runs downward and forward to the **incisive foramen** through which is passes to supply the anterior part of the palate.

A small **pharyngeal branch** runs posteriorly from the ganglion to supply the nasopharynx.

The Paranasal Air Sinuses

Centered around the nasal cavity a number of the bones of the face and skull are themselves hollowed out to form cavities which communicate with the nasal cavity. These air sinuses are lined with respiratory-type mucous membrane. The function of the cavities is unknown although they would appear to lighten the facial bones and give additional resonance to the voice. Their impor-

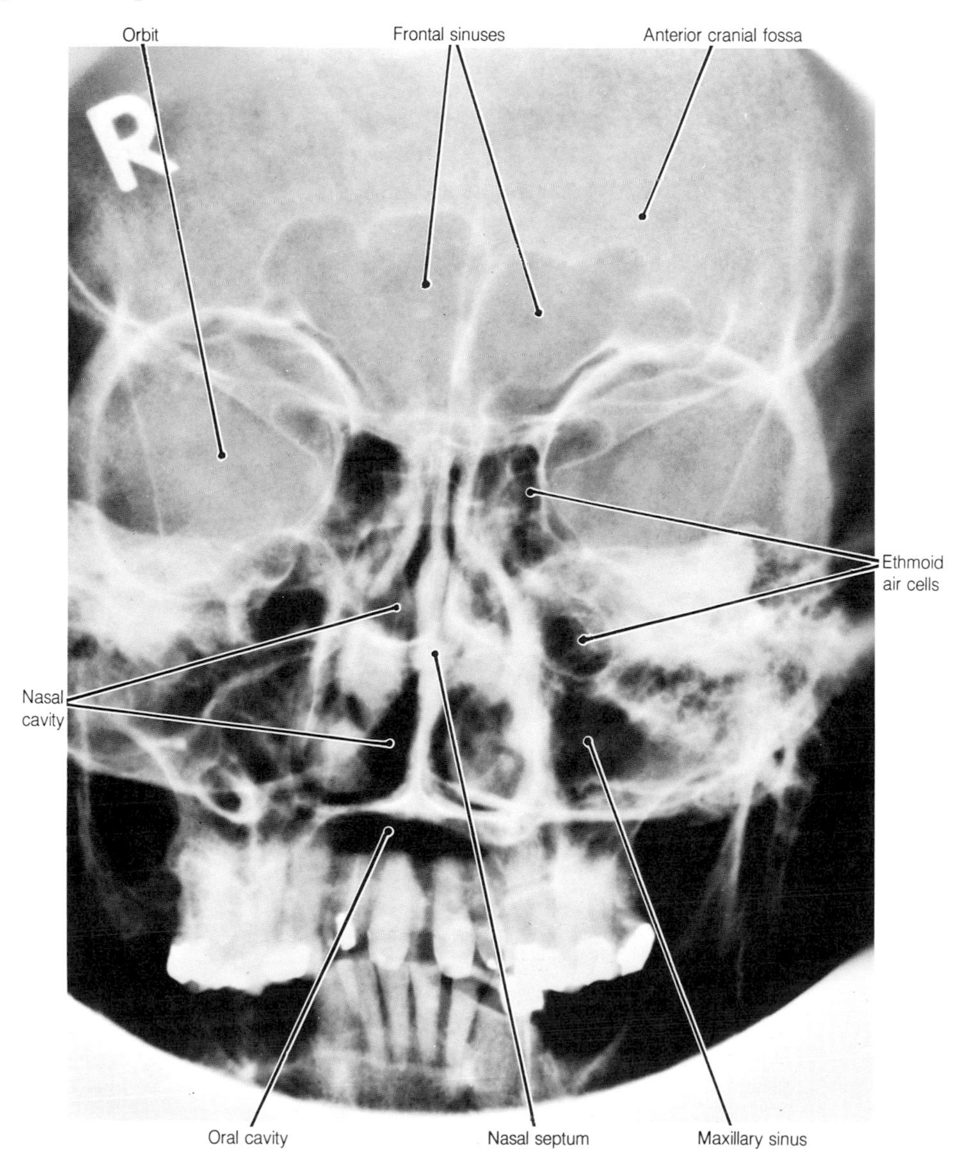

9-101 Radiograph of the paranasal sinuses (p.a. projection) (Reproduced by permission, from Wicke: *Atlas of Radiologic Anatomy,* 4th Ed, Urban & Schwarzenberg, Baltimore-Munich, 1987).

tance lies more in their involvement in upper respiratory infections which is exacerbated by their narrow drainage passages and their relationships to surrounding structures. Paranasal sinuses are found in the frontal, ethmoid, maxillary, and sphenoid bones. They are absent or rudimentary at birth and only begin to enlarge with the eruption of the second dentition.

The **frontal sinuses** lie in the frontal bone above the orbit. Their size is very variable. They drain into the **middle meatus** of the nose through the **frontonasal duct**. They are supplied by the supraorbital nerve and inflammation of a sinus may lead to acute frontal headache.

The **ethmoidal sinuses** are cavernous, forming an anterior, middle, and posterior group of cells. These open into the **middle and superior meatuses**. Their relationship to the medial wall of the orbit is important, as the thin plate of bone that separates them from this cavity may be fractured and allow infection to enter.

The **sphenoidal sinuses** lie within the body of the sphenoid and open by two orifices into the **sphenoethmoidal recess**.

The **maxillary sinuses** are the largest. They lie in the maxilla, below the orbit, lateral to the nasal cavity and above the roots of the first two molar teeth. Pain originating in the sinus and pain from an infected tooth may be confused with each other. The sinuses open into the **hiatus semilunaris** by an opening that lies above the level of the floor of the sinus. Recurrent infection may impair the action of the cilia of the lining cells and, thus, further diminish effective drainage.

The positions of the sinuses and their relationships to surrounding cavities can be seen in the radiograph reproduced in Fig. 9-101.

The Oral Cavity

The oral cavity is divided into a peripheral portion called the **vestibule** and an inner **oral cavity**

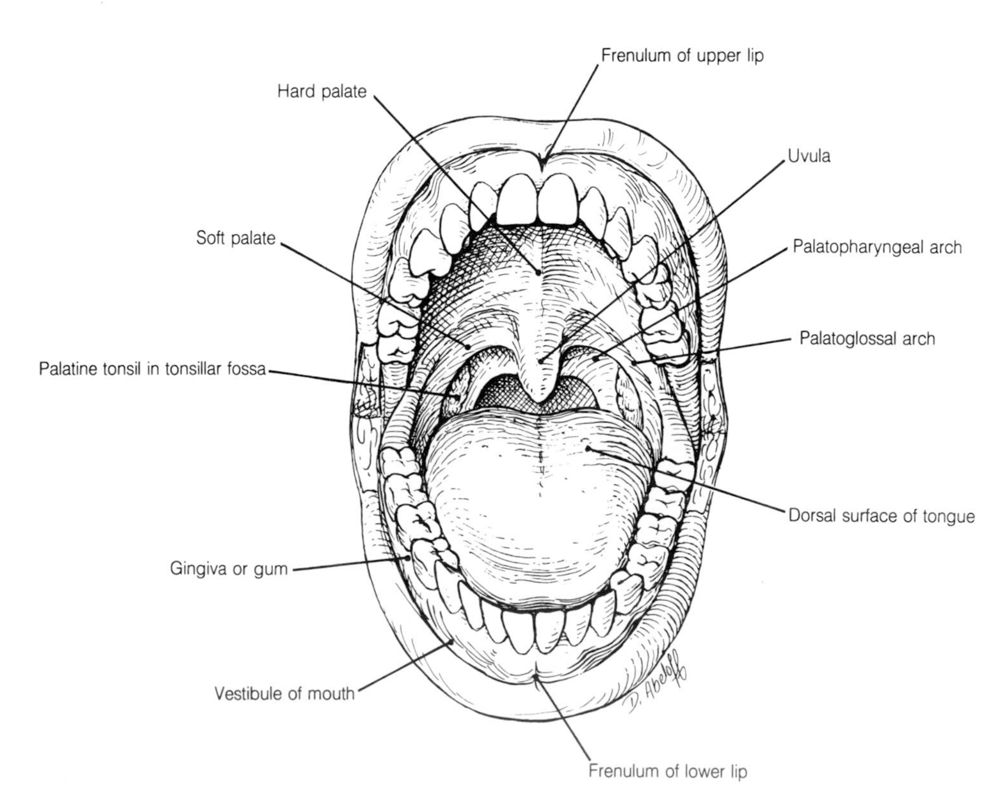

9-102 To show the boundaries of the oral cavity.

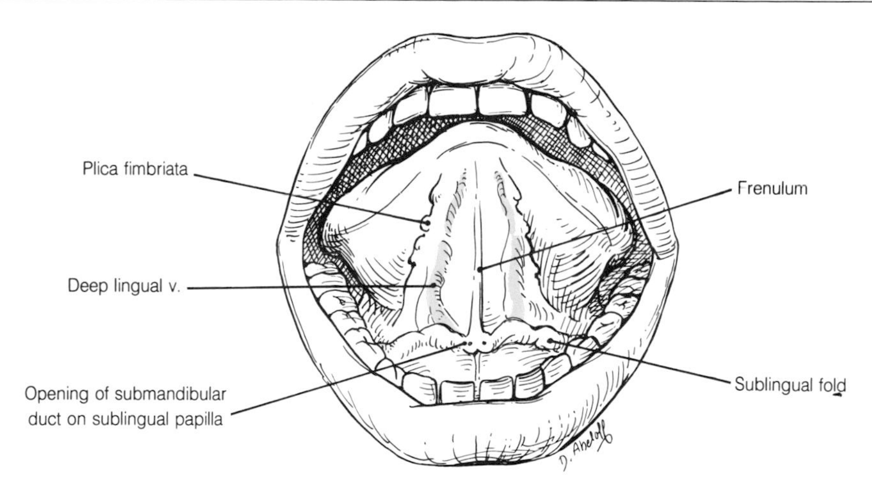

9-103 Showing details of the ventral surface of the tongue and of the floor of the mouth.

proper. The vestibule is the narrow slit-like space lying between the lips and cheeks externally and the gums and teeth internally. This space communicates with the exterior through the **oral fissure** between the lips and, when the teeth are occluded, with the oral cavity proper through small spaces behind the third molar teeth. The parotid duct opens into the vestibule on the surface of the cheek opposite the crown of the second upper molar tooth.

Look at Fig. 9-102 and see that the oral cavity proper is bounded anteriorly and laterally by the teeth and gums, its roof is formed by the hard and soft palate and the anterior two-thirds of the tongue occupies most of its floor. Posteriorly the oral cavity communicates with the pharynx by means of the **oropharyngeal isthmus** which passes between the palatoglossal arch and the tongue. If the tongue is raised as in Fig. 9-103, a midline fold of mucous membrane, the **frenulum of the tongue,** can be seen connecting its anterior part to the floor of the mouth. On either side of the frenulum the **submandibular ducts** open onto the floor of the mouth at the **sublingual papillae** and the ducts themselves form the **sublingual folds** running laterally and backward from the papillae. The **sublingual glands** lie deep to these folds and their many small ducts open onto its surface.

In addition to the three named pairs of large salivary glands there are many small glands scattered through the mucous membrane of the oral cavity. This mucous membrane is closely adherent to the underlying muscles of the cheek, to the lips and tongue, and, over the hard palate it is almost inseparable from the periosteum. The mucous membrane of the vestibule is supplied by the **infraorbital branch of the maxillary nerve** above and below by the **buccal and mental branches of the mandibular nerve**.

The Lips

The upper and lower lips form a muscular valve which surrounds the oral fissure and retains saliva and food within the oral cavity. Their efficient closure is also essential for the production of the consonants "P" and "B." They are covered by skin externally and mucous membrane internally and contain the orbicularis oris muscle, the superior and inferior labial vessels and nerves, and many small salivary glands. The lips are frequently lacerated and good cosmetic and functional results require careful repair.

The Teeth

While the care of teeth is ascribed to the dental surgeon their condition may reflect more generalized disease and their eruption and loss are mileposts in the maturation of a child and of considerable concern to an anxious mother.

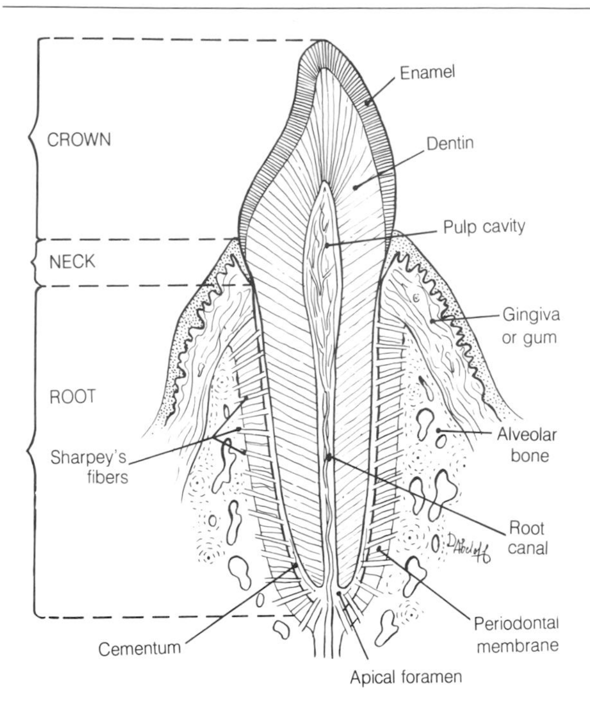

9-104 Diagrammatic section of an incisor tooth and its supporting tissues.

Dental Morphology

The main features of a tooth and its supporting alveolar bone are seen in Fig. 9-104, and the following should be noted.

1. At approximately the gum margin a tooth can be divided into a **crown** above this level and a **root** below.
2. The greater part of the tooth is formed by hard, avascular **dentin**.
3. The dentin of the crown is covered by a thin layer of extremely hard **enamel** and that of the root by an equally thin layer of bone-like **cementum**.
4. Within the dentin lies the **pulp cavity**. The **pulp** is a loose connective tissue which carries vessels and nerves to the inner surface of the **dentin**.
5. At the **apical foramen**, the pulp is continuous with the **periodontal ligament**. It is by this ligament and the strong bundles of collagen fibers it contains (Sharpey's fibers) that the tooth is anchored to the surrounding **alveolar bone**.

In the adult, each half jaw has two **incisors**, a **canine**, two **premolar**, and three **molar teeth**. Each variety of tooth has a characteristic shape displayed by both its crown and root and this is adapted to its function.

The **incisors** have a linear cutting edge, a chisel-shaped crown, and a single root.

The **canine teeth** have a conical crown and a large single root which is out of proportion to the size of the crown and probably an evolutionary vestige of a much larger tooth.

Premolar teeth, which are often known as bicuspids, have two cusps and may have one or two roots.

The **molar teeth** decrease in size from first to last as do the number of cusps on their crowns. They are used for chewing and grinding. Upper molars usually have three roots and lower molars two.

The Nerve and Blood Supply of the Teeth

The pulp and periodontal membrane of the teeth are supplied by plexuses derived from branches of the **superior alveolar nerves** in the upper jaw and of the **inferior alveolar nerve** in the lower jaw. The surrounding gums are supplied by the nerves to the mucous membrane of the nearby cheek and oral cavity. **Superior and inferior alveolar branches of the maxillary artery** accompany the nerves.

Occlusion

The teeth are arranged in an upper and lower dental arch. When the two arches are approximated to give maximum contact, the teeth are described as being in centric occlusion. In this position the teeth of the upper arch lie slightly outside those of the lower.

The Deciduous Teeth

At birth no teeth have erupted and the jaws contain only the crowns of the developing deciduous dentition. This dentition consists of two **incisors**, one **canine tooth, and two molar teeth** in each half jaw. The lower central incisors are, at six months, the first to erupt and are followed by the remaining incisors over the next three months. The first molars appear at about one year and the canine teeth at eighteen months. The dentition is completed by the eruption of the second molar teeth at two years.

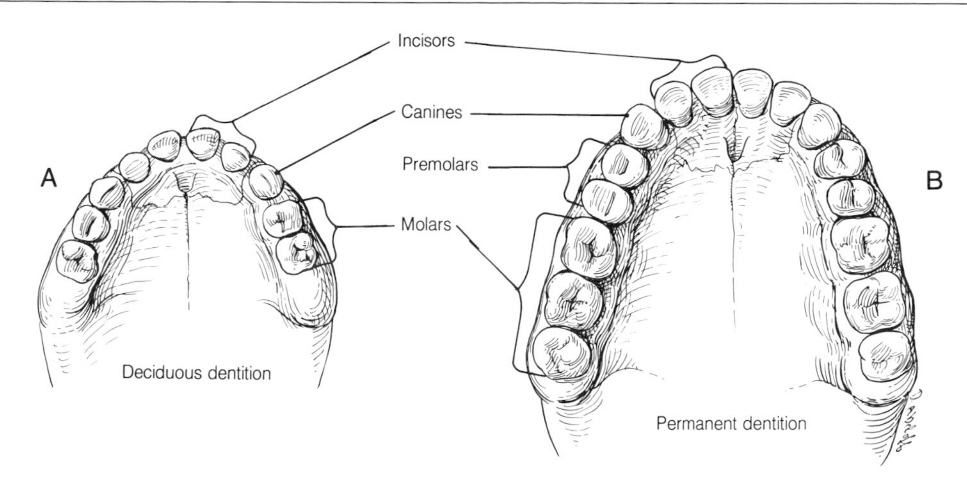

9-105 A comparison of the upper dental arcades of a child (*A*) and an adult (*B*).

The Permanent Dentition

In the permanent dentition each half jaw contains **two incisors, one canine tooth, two premolars, and three molars**. At five years the crowns of all but the third molar tooth have formed below the level of the deciduous teeth. At six years the first molar tooth erupts behind the second deciduous molar. This is followed by the incisors, canines, and by the twelfth year the premolars, which replace the deciduous molars. Thus, all the deciduous teeth are shed by the age of twelve. The second molars appear soon after this, but the third molars seldom appear before the eighteenth year and sometimes remain permanently unerupted. As permanent teeth are lost the surrounding alveolar bone is resorbed reducing the depth of the jaw.

Typical times of eruption of the deciduous and permanent teeth are given below and the superior dental arch of a child and an adult are illustrated in Fig. 9-105, A and B.

Deciduous Teeth	
Central incisors	6– 8 mos
Lateral incisors	8–10 mos
First molars	12–16 mos
Canines	16–20 mos
Second molars	20–30 mos

Permanent Teeth	
First molar	6– 7 yrs
Central incisors	6– 8 yrs
Lateral incisors	7– 9 yrs
Canines	9–12 yrs
First and second premolars	10–12 yrs
Second molars	11–13 yrs
Third molars	17–21 yrs

To summarize the maturation of the deciduous and permanent dentition:

1. The first teeth to appear are the deciduous incisors.
2. The last deciduous tooth to appear is the second molar at about two years.
3. There are no premolar teeth in the deciduous dentition.
4. The first permanent molar tooth appears behind the second deciduous molar during the sixth year.
5. The deciduous molars are replaced by the permanent premolar teeth.
6. All deciduous teeth are replaced by permanent teeth by the twelfth year.
7. The last permanent tooth to appear is the third molar and this may be delayed indefinitely.

The Tongue

The tongue is a mobile muscular organ covered by epithelium. Its anterior two-thirds lie in the oral cavity and its posterior third lies in the oropharynx. While its tip is free and may be protruded beyond the lips, its ventral surface or root is anchored by musculature to the hyoid bone and mandible. It is used in tasting, in the preparation and swallowing of food, and in speech.

The Epithelium of the Tongue

The anterior two-thirds of the dorsal surface of the tongue are covered by a rough stratified squamous epithelium. The roughness is due to the presence of fine **filiform** and **fungiform papillae**. The latter are concentrated near the periphery and bear taste buds. The ventral surface, like the floor of the mouth, is covered by smooth shiny epithelium. Anteriorly this forms a midline fold, **the frenulum**. This can be clearly seen when the tip of the tongue is elevated. In Fig. 9-106 follow the frenulum to the floor of the mouth and identify the **sublingual papillae** on which lie the orifices of the **submandibular ducts**. Directed laterally from the papillae are the **sublingual folds**. Parallel to and on either side of the frenulum are the **deep lingual veins** and irregular folds of mucous membrane called the **plicae fimbriatae**.

As can be seen in Fig. 9-107, a shallow inverted V-shaped groove, the **sulcus terminalis**, separates the anterior two-thirds of the dorsal surface

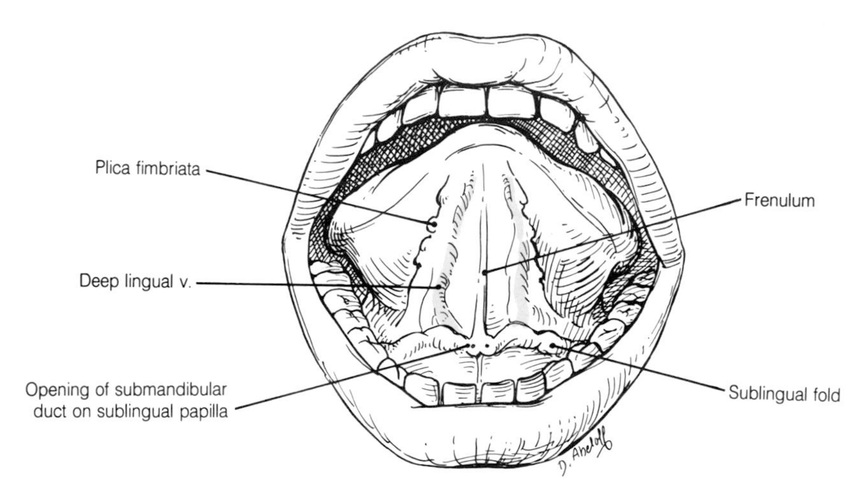

9-106 Showing details of the ventral surface of the tongue and of the floor of the mouth.

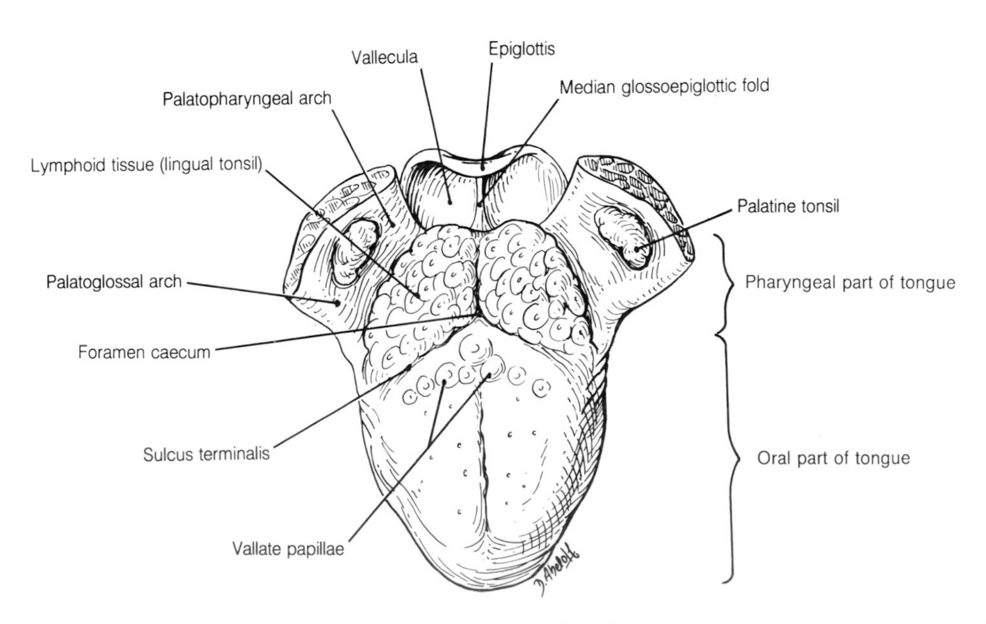

9-107 Showing the dorsum of the tongue. Note that the soft palate has been removed and the palatoglossal and palatopharyngeal arches separated.

of the tongue from that of the posterior third. Where the limbs of the V meet there is a small pit called the **foramen cecum** which indicates the site of the proximal end of the embryonic **thyroglossal duct**. In front of the sulcus there is a row of large papillae each surrounded by a circular groove. These are known as the **circumvallate*** **papillae** and are plentifully supplied with taste buds which lie in the walls of the fossa which surrounds each of them.

Behind the sulcus terminalis the epithelium is smooth but irregular, owing to the presence of underlying glands and lymphatic tissue. The latter is sometimes named the **lingual tonsil**.

Posterior to the tongue the epithelium is reflected onto the anterior surface of the epiglottis forming the two **valleculae** separated by a **median glossoepiglottic fold**.

The Nerve Supply of the Mucous Membrane of the Tongue

The innervation of the mucous membrane of the tongue is related to the first and third branchial arches from which it is developed. Nerves supplying structures derived from these arches are the **trigeminal nerve** (first arch) and the **glossopharyngeal nerve** (third arch). The second arch takes no definitive part in the formation of the mucous membrane, but a branch of the facial nerve, called the **chorda tympani**, contributes to the innervation of its anterior two-thirds. Although the sulcus terminalis represents the line of fusion of the first and third arch components, the glossopharyngeal nerve extends its field to include the circumvallate papillae which lie just anterior to it. It must be stressed that the musculature of the tongue is developed from myotomes of occipital somites whose segmental nerves have become fused to form the **hypoglossal nerve**.

The mucous membrane of the tongue is supplied by nerve fibers of common sensation (pain, touch, and temperature), fibers for the special sense of taste, and secretomotor fibers for its many small salivary glands. For the posterior third, including the circumvallate papillae, all these fibers are carried in the **glossopharyngeal nerve**.

* Vallum, *L.* = a palisade or fortification

For the anterior two-thirds, fibers of common sensation are carried back to the brainstem by the **lingual nerve** and their cell bodies lie in the **trigeminal ganglion**. Taste fibers start in the lingual nerve but are transferred in the **chorda tympani** to the **facial nerve**. Their cell bodies lie in the **geniculate ganglion** the sensory ganglion of the facial nerve. Parasympathetic preganglionic secretomotor fibers leave the brain with the facial nerve (nervus intermedius), are transferred in the chorda tympani to **the lingual nerve**, and leave the lingual nerve to synapse in the **submandibular ganglion**. Postganglionic fibers return to the lingual nerve for distribution to the glands of the tongue.

Sympathetic vasomotor fibers reach all parts of the tongue from the superior cervical ganglion via the lingual artery and its branches.

The Muscles of the Tongue

The muscles of the tongue are divided into intrinsic and extrinsic muscles. The intrinsic muscles lie wholly within the tongue and have no bony attachment. They are arranged as longitudinal, transverse, and vertical intersecting sheets on each side of a vertical midline **fibrous septum** and are concerned with changing the shape of the tongue. The extrinsic muscles arise from bony attachments to the palate, styloid process, hyoid bone, and mandible. They are concerned with changes in position and each is paired. The positions of the muscles are shown in Fig. 9-108 and this illustration should be studied as a description of each muscle is given.

The **genioglossus** arises from the superior part of the mental spine on the deep surface of the symphysis menti. It forms the greater part of the mass of the tongue and its fibers fan out to be attached to the mucous membrane from the tip to the base. The main function of the genioglossus is to pull the tongue forward and protrude its tip.

The **hyoglossus** arises from the body and greater horn of the hyoid bone and extends upward to the side of the tongue. Its action is to depress the tongue. Its superficial and deep surfaces are related to a number of important structures which will be considered in detail with the floor of the mouth.

The **palatoglossus** extends from the soft palate

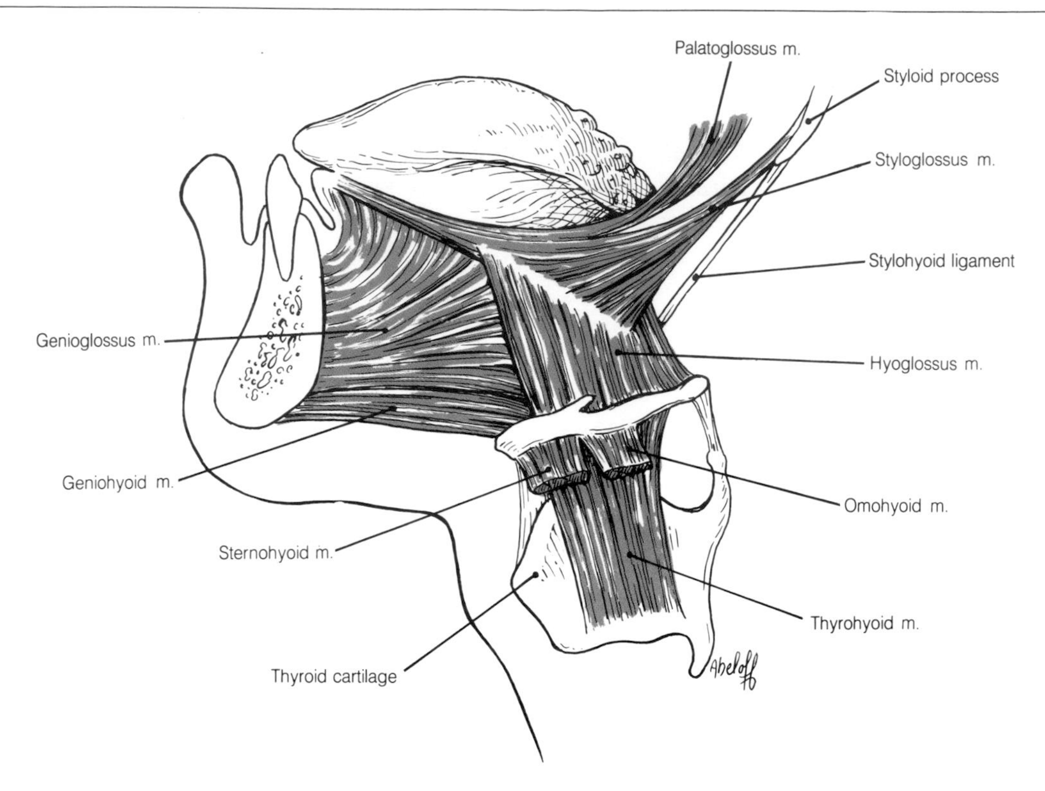

9-108 Illustrating the extrinsic muscles of the tongue.

downward to the side of the tongue forming the palatoglossal fold of the oropharyngeal isthmus. Its contraction elevates the tongue toward the palate and approximates the palatoglossal folds. In this manner it closes off the oral cavity from the oropharynx.

The **styloglossus** arises from the styloid process and passing downward and forward enters the side of the tongue by intermingling with the hyoglossus muscle. Its action is to pull the tongue upward and backward.

The Nerve Supply of the Muscles of the Tongue

With the exception of the **palatoglossus muscle**, which is supplied from the **pharyngeal plexus** by vagal fibers, all the intrinsic and extrinsic muscles are supplied by the **hypoglossal nerve**. The course of this nerve is considered in more detail with the floor of the mouth. It should be noted, however, that when this nerve is damaged efforts to protrude the tongue lead to deviation of its tip to the side of the damaged nerve. This is a useful clinical test of the hypoglossal nerve.

The Blood Vessels of the Tongue

The tongue is supplied by the **lingual artery**. This vessel is an anterior branch of the external carotid artery arising between the superior thyroid and facial arteries at a point just behind the posterior margin of the hyoglossus muscle. After forming its characteristic loop which is crossed by the hypoglossal nerve it passes deep to the hyoglossus muscle and, after ascending to reach the dorsal aspect of the tongue, runs forward to the tip. **Dorsal lingual branches** supply the posterior part of the dorsum and a further small branch supplies the sublingual gland. Lingual veins drain alongside the artery. Deep veins, visible on the ventral surface of the tongue pass superficially over the hyoglossus muscle. All terminate directly or indirectly in the internal jugular vein. The course of these vessels will be better appreciated when the floor of the mouth and submandibular regions are described.

The Lymphatic Drainage of the Tongue

The lymphatic drainage of the tongue is of importance in that it determines which lymph nodes become involved in the spread of a carcinoma. The most important feature is that thc tip, the central region, and the posterior third are drained bilaterally and thus metastases *may* be found in nodes on the side opposite to the lesion. The course of the vessels and the site of the nodes are illustrated diagrammatically in Fig. 9-109. All the intermediate nodes ultimately drain to the deep cervical nodes.

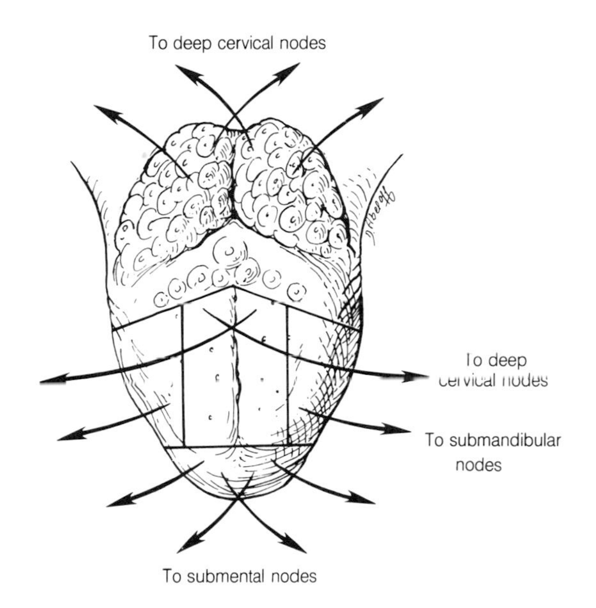

9-109 A diagrammatic illustration of the lymphatic drainage of the tongue.

The Floor of the Mouth and Submandibular Region

The tongue is suspended between the left and right halves of the mandible by a sling formed by the mylohyoid and geniohyoid muscles and the hyoid bone. This sling which forms a boundary between the mouth and the neck has considerable mobility as the hyoid is free to move upward and downward and backward and forward. In Fig. 9-110 this sling is viewed from above.

The Mylohyoid Muscle

Arising from the **mylohyoid line** on the internal surface of the **body of the mandible** the fibers of mylohyoid slope downward, forward, and medially. The anterior fibers are inserted into a **midline raphe** extending from the **symphysis menti** to the **hyoid bone**. The more posterior fibers are attached to the **body of the hyoid bone**. Its superficial surface is related to the **anterior belly of the digastric muscle, the superficial part of the submandibular gland**, and the **facial vessels**; these

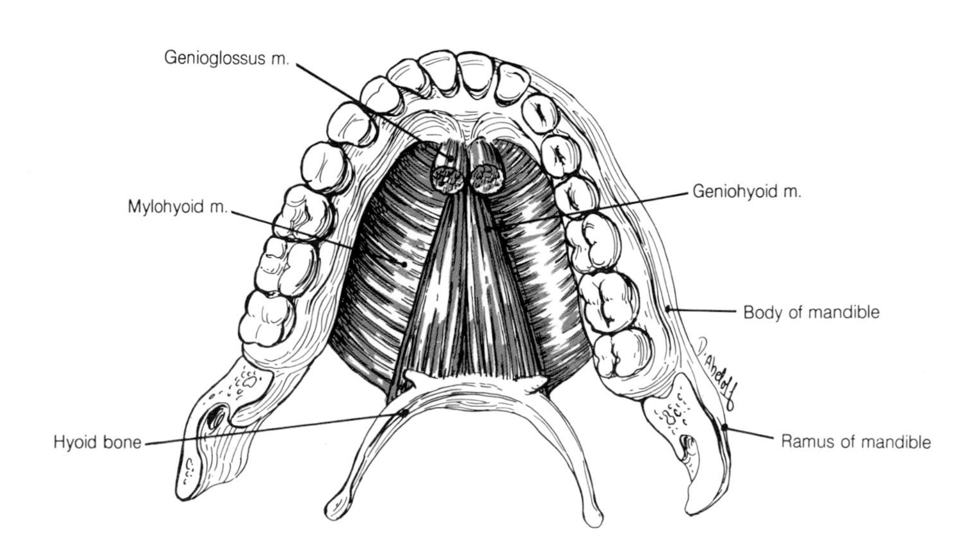

9-110 To show the muscles supporting the floor of the mouth.

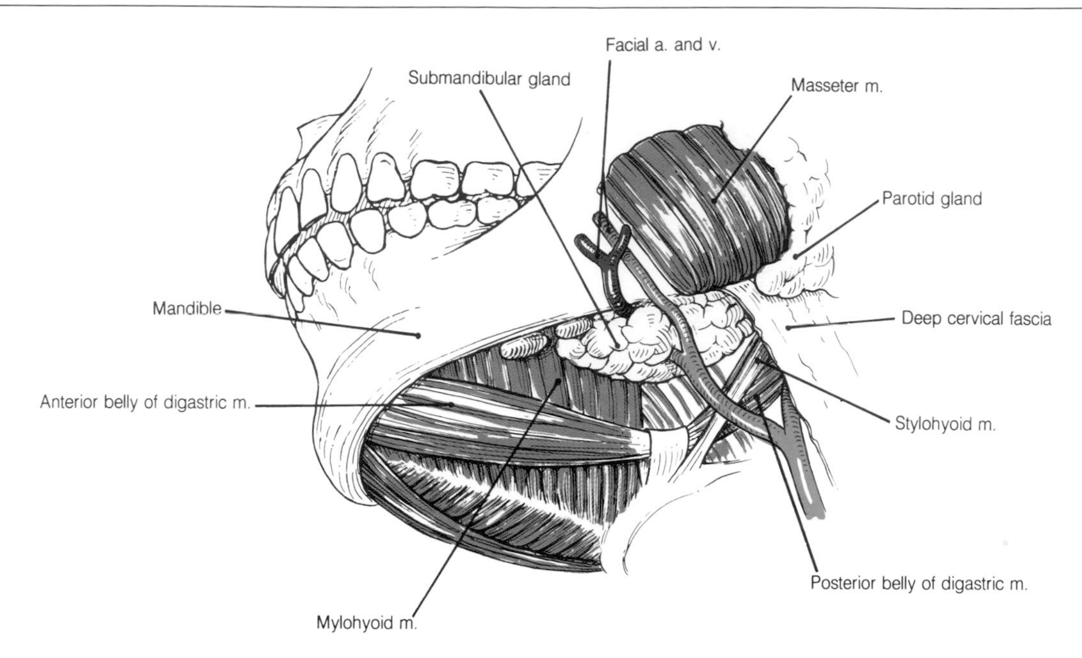

9-111 The submandibular region (superficial structures).

relationships can be seen in Fig. 9-111. The mylohyoid has already been grouped with the suprahyoid muscles and its action is to raise the hyoid bone and in so doing raise the floor of the mouth and press the tongue against the hard palate.

The muscle is supplied by the **mylohyoid branch of the inferior alveolar nerve**. This small branch is given off just before the inferior alveolar nerve enters the **mandibular foramen** and reaches the muscle in company with the **mylohyoid branch of the inferior alveolar artery**.

The mylohyoid muscle serves as a partition between the floor of the mouth and the neck. Look carefully at Fig. 9-112 which shows the origin of the muscle from the mylohyoid line. Note in this the fossa for the superficial part of the submandibular gland **below** this attachment and the fossa for the sublingual gland **above** it. Note also the small gap between the attachment of the pterygomandibular raphe and the mylohyoid muscle. It is here that the lingual nerve gains access to the floor of the mouth above mylohyoid. At this point, also, the nerve lies just beneath the mucous membrane of the oral cavity.

The Superficial Part of the Submandibular Gland

The submandibular salivary gland consists of a larger **superficial part** which is continuous around

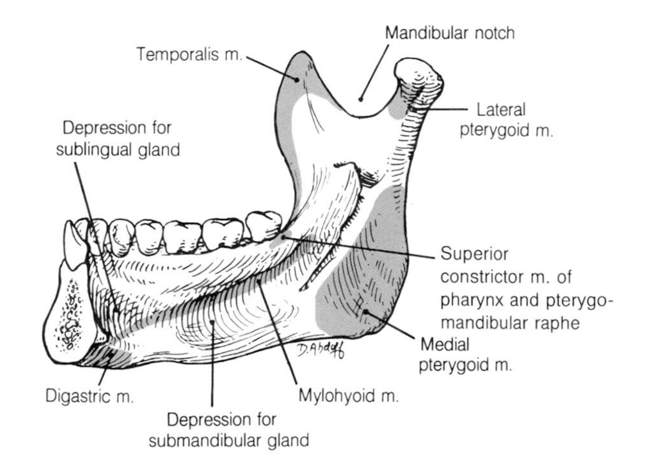

9-112 The medial aspect of the right half of the mandible showing muscle attachments.

the posterior border of the mylohyoid with a smaller **deep part**. The superficial part lies wedged between the **body of the mandible** and the underlying **suprahyoid muscles**. Laterally it is related to the submandibular fossa, a shallow depression on the inner surface of the mandible below the posterior end of the mylohyoid line. Medially it is related to the mylohyoid anteriorly and to the styloglossus and stylohyoid ligament posteriorly. It is separated from the parotid gland by the stylomandibular ligament, a thickening in the deep fascia of the neck which encloses that gland. Running over the upper border of the gland from its deep medial to superficial lateral surface is the **facial artery.** In this manner the artery reaches the lower border of the mandible close to the anterior margin of the masseter muscle (Fig. 9-111). A superficial inferior surface is covered by skin and platysma.

If the mylohyoid muscle is removed the underlying hyoglossus muscle is exposed. On the superficial surface of this muscle are found the deep portion of the submandibular gland, the submandibular duct, the lingual nerve, the submandibular ganglion, the hypoglossal nerve, and the veins from the tongue that lie alongside it. The position and course of these structures should be studied in Fig. 9-113 as each is described.

The Hyoglossus Muscle

The hyoglossus muscle is rectangular in shape and arises from the **greater horn of the hyoid bone and an adjacent portion of the body**. Its fibers ascend with a forward inclination to join the side of the tongue where they blend with the **styloglossus**. When the hyoid bone is fixed its action is to depress the tongue. It is supplied by the **hypoglossal nerve**.

The Deep Part of the Submandibular Gland

The deep portion of the submandibular gland is continuous with the superficial portion around the posterior border of the mylohyoid muscle and lies in the space between that muscle and the underlying styloglossus and hyoglossus. The lingual nerve lies above it and the hypoglossal nerve below. The **submandibular duct** emerges from beneath the anterior border of the gland and runs forward on the surface of hyoglossus; it then passes between the sublingual gland and the genioglossus muscle to enter the floor of the mouth at the **sublingual papilla**. Both parts of the gland receive their blood supply from the facial and lingual arteries. **Postganglionic parasympathetic secretomotor fibers** reach it from the **submandibular ganglion**.

The Lingual Nerve

The lingual nerve arises from the posterior trunk of the mandibular nerve in the intratemporal fossa. It is soon joined by the **chorda tympani branch of the facial nerve.** It has already been described in the infratemporal fossa as it appears below the lateral pterygoid muscle and runs forward with the inferior alveolar nerve over the surface of the medial pterygoid. Slipping over the anterior border of the medial pterygoid and below the superior constrictor it lies for a short distance between the mandible and the mucous membrane of the mouth just blow the third molar tooth (see Fig. 9-112). From here it passes across styloglossus to reach the lateral surface of hyoglossus. Initially above the deep portion of the submandibular

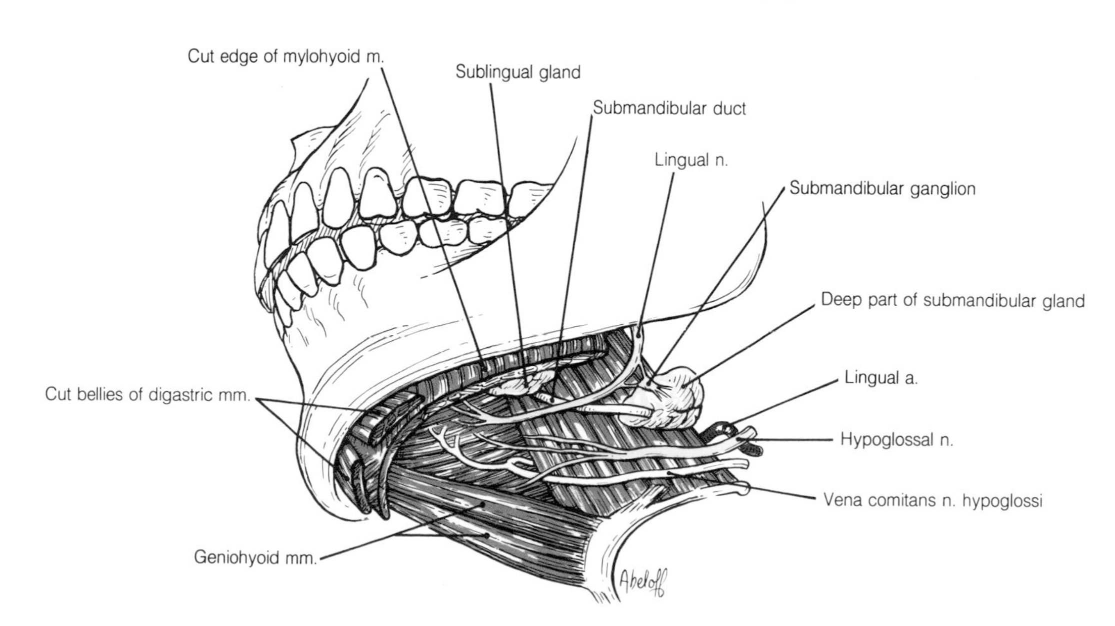

9-113 The submandibular region (deep structures).

gland the nerve sweeps downward and crosses the duct superficially. It then ascends on the medial side of the duct and breaks up into its terminal branches to the mucous membrane of the tongue.

The Submandibular Ganglion

The submandibular ganglion is suspended by fine nerve filaments from the lingual nerve as it lies on the hyoglossus above the deep part of the submandibular gland. It is a **parasympathetic ganglion** and the origin and distribution of the parasympathetic fibers is illustrated diagrammatically in Fig. 9-114. It receives preganglionic fibers (*blue*) from the facial nerve via the chorda tympani and lingual nerves and the postganglionic fibers (*yellow*) are distributed by small branches to the submandibular gland. Some fibers pass back to the lingual nerve and are carried by this to the sublingual gland and the many small anterior lingual and oral glands. Postganglionic sympathetic fibers also reach the ganglion from the surface of the facial artery. These have arisen in the superior cervical ganglion and are vasomotor to the submandibular and sublingual glands.

The Hypoglossal Nerve

The hypoglossal nerve has been seen to descend from the hypoglossal canal between the internal jugular vein and the internal carotid artery. It emerges from between these vessels and turns forward to cross the internal and external carotid arteries and the loop of the lingual artery. This course leads it onto the superficial surface of hyoglossus. Here it runs forward below the deep part of the submandibular gland, the submandibular duct, and the lingual nerve. Passing off hyoglossus onto genioglossus it breaks up into its terminal muscular branches.

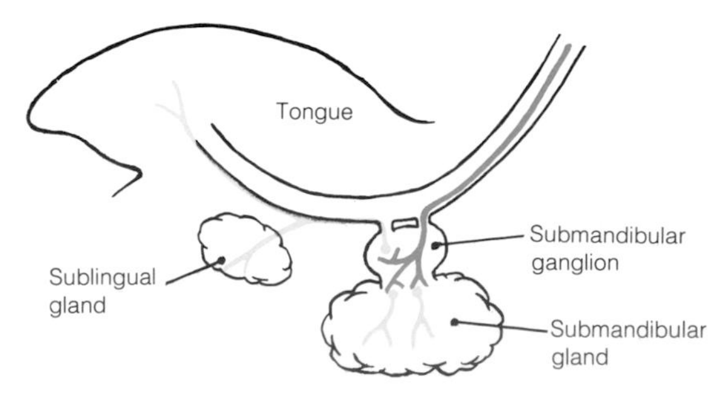

9-114 Diagram illustrating the parasympathetic innervation of the submandibular and sublingual glands.

It will be remembered that the hypoglossal nerve supplies all the muscles of the tongue with the exception of palatoglossus. In addition, it carries fibers from the first cervical spinal nerve which are distributed via its contribution to the ansa cervicalis and its branches to the thyrohyoid and geniohyoid muscles. The deep lingual vein and a vein from the sublingual gland join and run posteriorly along the course of the hypoglossal nerve on the hyoglossus. Because of this association the vein is called the **vena comitans nervi hypoglossi**; it drains into the facial vein or directly into the internal jugular vein.

Structures Deep to the Hyoglossus Muscle

The hyoglossus muscle lies superficial to the intrinsic muscles of the tongue, the genioglossus and, in its lower part, the middle constrictor muscle of the phyranx. Passing deep to its posterior border are found the **glossopharyngeal nerve**, the **stylohyoid ligament**, and the **lingual artery**. The distribution of the nerve and artery to the tongue has been described. The stylohyoid ligament is a fibrous cord extending from the tip of the styloid process to the lesser cornu of the hyoid bone and thus links two bony derivatives of the second branchial arch. It occasionally becomes ossified.

The Sublingual Gland

The most superficial structure in the floor of the mouth is the sublingual gland. It lies anteriorly beneath the mucous membrane and above the anterior part of mylohyoid muscle. Laterally it is adjacent to the body of the mandible in which the shallow **sublingual** fossa is found. Medially it is separated from the genioglossus by the lingual nerve and submandibular duct. Its many small ducts open into the floor of the mouth over the **sublingual fold** (see Fig. 9-106). Its blood supply is provided by the lingual artery through the small **sublingual artery**. Postganglionic parasympathetic secretomotor fibers reach it from the submandibular ganglion. The main relations of the sublingual gland can be established in the coronal section through the region shown in Fig. 9-115 and this illustration can also be used to review the structures lying deep and superficial to the mylohyoid muscle.

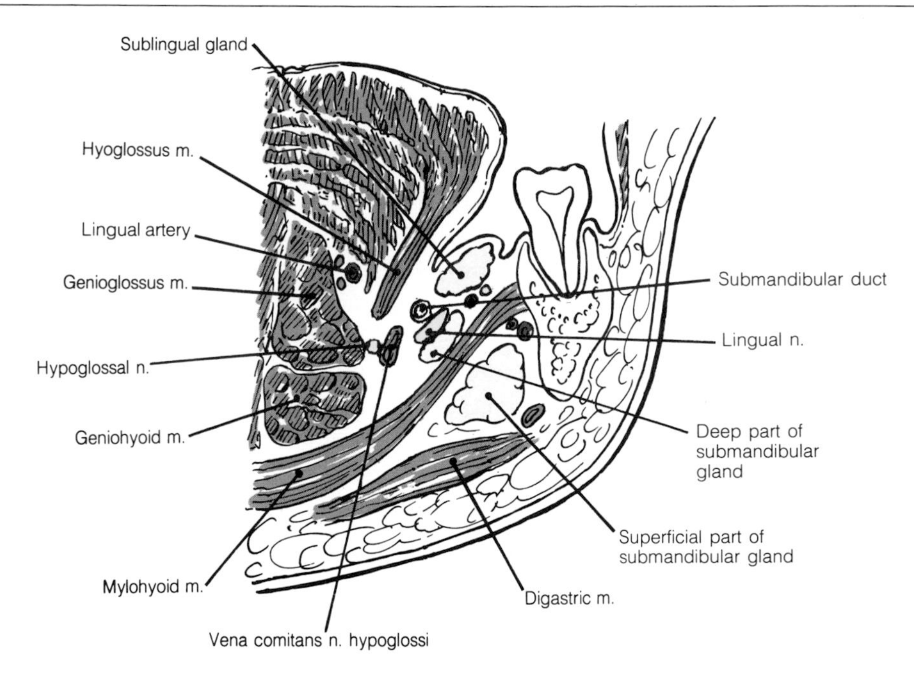

9-115 Coronal section through the submandibular and sublingual regions.

The Oropharyngeal Isthmus, Soft Palate, and Tonsils

The Oropharyngeal Isthmus

The oral cavity communicates with the oropharynx through the oropharyngeal isthmus. Look at Fig. 9-116 and see that the isthmus is bounded above by the **soft palate**, laterally by the palatoglossal arches, and below by the dorsum of the tongue. The palatine tonsils lie just beyond the palatoglossal arches.

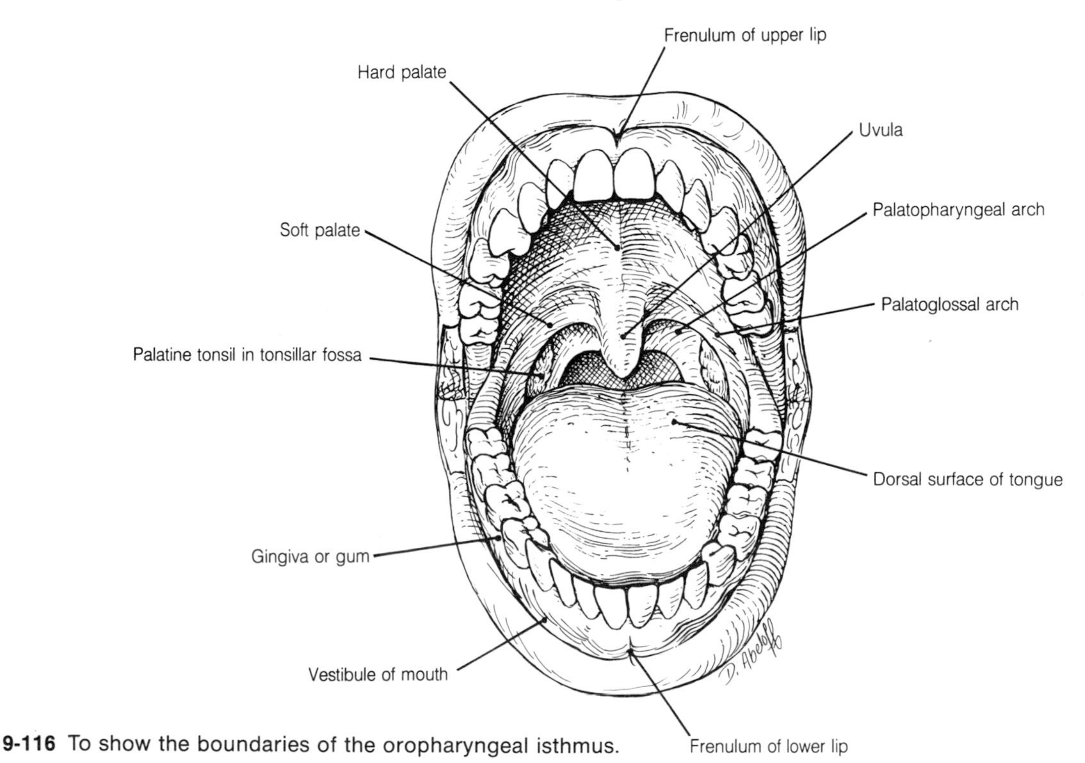

9-116 To show the boundaries of the oropharyngeal isthmus.

The Soft Palate

The soft palate is attached to the posterior border of the **hard palate**. It is a mobile muscular aponeurosis whose free border projects between the **nasopharynx** and **oropharynx**. It is covered by a mucous membrane which is continuous with that of the hard palate on its oral surface and with that of the nasal cavities on its superior surface. Its oral surface is thickened by the presence of many mucous glands. The posterior border of the soft palate shows a conical midline projection called the **uvula** and two lateral folds arching downward to the base of the tongue. The anterior fold is the **palatoglossal arch** and covers the palatoglossus muscle. The posterior arch is the **palatopharyngeal arch** and covers **palatopharyngeus**. Between the arches lies the **tonsilar fossa** containing the **palatine tonsil**.

To each side of the palatine aponeurosis are attached four muscles. Their orderly function is essential to the act of swallowing. In addition, contraction of the levator and tensor muscles which arise in part from the cartilaginous portion of the auditory tube probably opens the auditory tube and allows the pressure of air trapped in the middle ear cavity to equalize with atmospheric pressure. The muscles are illustrated in Fig. 9-117 and described below.

The **levator veli palatini** muscle arises from the tip of the petrous temporal bone and the lateral side of the auditory tube and descends beneath the mucous membrane of the nasopharynx to reach the soft palate. It and its fellow form a sling which raises the soft palate.

The **tensor veli palatini** arises from the scaphoid fossa on the pterygoid part of the sphenoid bone and from the cartilage of the auditory tube. It descends outside the superior constrictor of the pharynx medial to the otic ganglion and mandibular nerve. Its tendon passes around the pterygoid hamulus making a 90° bend and spreads out to form the greater part of palatine aponeurosis. The muscle tightens and flattens the soft palate.

The **palatoglossus** is a small muscle extending from the palate to the tongue and its belly lies in the palatoglossal arch. Its action is to raise the tongue and approximate the palatoglossal arch to its fellow. In so doing it helps to close the oropharyngeal isthmus.

The **palatopharyngeus** arises from the aponeurosis, descends in the palatopharyngeal arch, and is attached to both the pharyngeal wall and the posterior border of the thyroid cartilage. Contraction of the muscles narrows the palatopharyngeal arch and raises both the larynx and

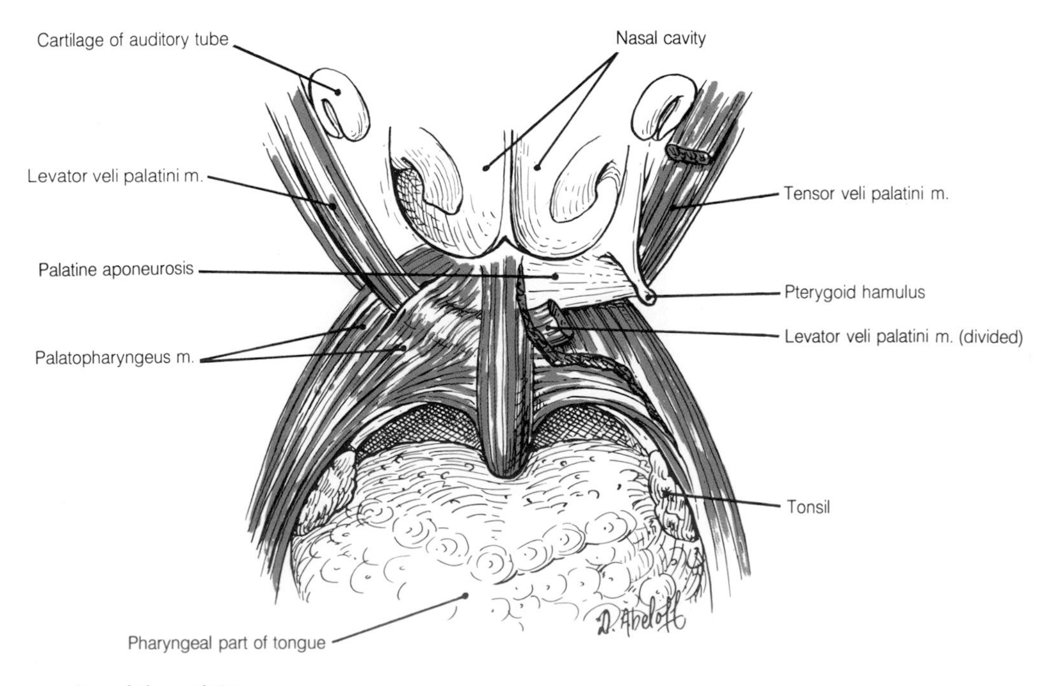

9-117 The muscles of the palate.

pharynx toward the oropharyngeal portion of the tongue.

With the exception of tensor veli palatini which is supplied by the **mandibular nerve** these muscles are supplied by branchial motor fibers carried to the pharyngeal plexus by the **vagus**.

The Palatine Tonsils

The palatine tonsils consist of a mass of lymphoid tissue lying in the **tonsillar fossae** between the palatoglossal and palatopharyngeal arches. They are covered by stratified squamous mucous membrane which has a pitted appearance. The pits lead to the **tonsillar crypts** which are blindly ending recesses. A deep **supratonsillar cleft** separates the uppermost portion from the main part of the tonsil. The lateral surface is covered by a fibrous capsule which is adjacent to the superior constrictor muscle of the pharynx. It is supplied by tonsillar branches of the facial artery and drained by the paratonsillar vein. Lymphatics pass to the **jugulodigastric deep cervical nodes** and these nodes are typically enlarged in tonsillitis. The tonsillar region receives its nerve supply from the **lesser palatine** and **glossopharyngeal nerves**. Tonsillectomy was at one time commonly performed on children of school age. Severe hemorrhage from tears of the paratonsillar vein were a constant hazard. In fact, most enlarged tonsils regress at puberty and the operation is now only performed with strong indications for its necessity.

At this point attention should be drawn to the ring of lymphoid tissue which lies around the oropharynx and nasopharynx. Contributing to the ring are the lingual tonsils. i.e., the lymphoid tissue beneath the mucous membrane of the pharyngeal part of the tongue, the palatine tonsils, and the pharyngeal tonsils or adenoids. Smaller intermediate collections of lymphoid tissue such as the tubal tonsil complete the ring. The significance of this arrangement is obscure.

The description of the tongue, the floor of the mouth, the oropharyngeal isthmus, the soft palate, and pharynx now allows the act of swallowing to be discussed.

Deglutition

Deglutition or swallowing can be considered to occur in a voluntary and an involuntary phase. In the mouth the food is prepared into a bolus by chewing and the addition of saliva. By means of the intrinsic muscles of the tongue the bolus is pressed against the hard palate and forced backward toward the soft palate which is at this time depressed. Elevation of the posterior third of the tongue and approximation of the palatoglossal arches ejects the bolus through the oropharyngeal isthmus into the oropharynx. At this point the involuntary stage begins. The soft palate is elevated and tensed in order to seal off the nasopharynx. To separate the respiratory from the alimentary tracts the larynx is raised behind the hyoid bone and the root of the tongue while the aryepiglottic folds around the laryngeal inlet are approximated and the epiglottis is depressed. The pharynx is also raised toward the bolus which by successive contraction of the pharyngeal constrictors is passed to the lowest part of the pharynx and thence to the esophagus.

Clearly the orderly sequence in which the involuntary movements of swallowing occur require that the cranial nerves concerned in the reflexes be intact. Dysphagia or difficulty in swallowing may follow lesions involving the ninth and tenth cranial nerves.

It is of interest that although the later stages of swallowing are involuntary, some control can be gained by patients who have had a total laryngectomy and have been taught to swallow air and eject it under pressure in short bursts. This allows a remarkable approximation to normal speech.

The Larynx

The larynx lies below the oropharynx and the pharyngeal portion of the tongue. It opens into the laryngopharynx above and is continuous with the trachea below. It is primarily a sphincter which is able to separate the respiratory from the alimentary tract and its adaptation for phonation is a secondary feature. It consists of a skeleton of cartilages and membranes built around the first part of the respiratory tract. The shape and position of the skeletal elements are maintained or altered by a group of intrinsic muscles.

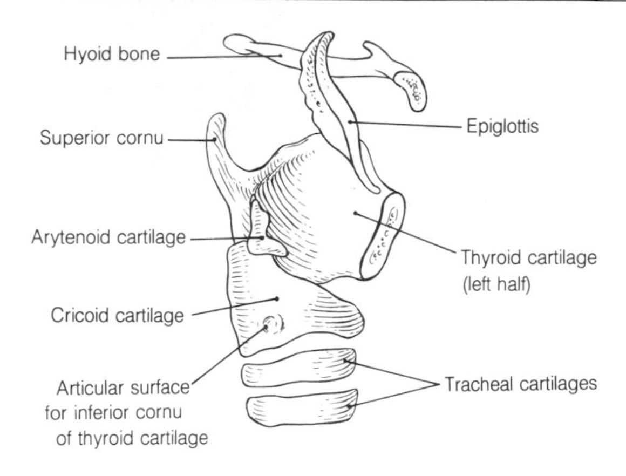

9-118 The laryngeal cartilages as they lie in the larynx. The hyoid bone is also shown.

The Laryngeal Cartilages

In Figs. 9-118 and 9-119 the laryngeal cartilages are shown in articulated and exploded views. These illustrations should be followed as each cartilage is described.

The **cricoid cartilage** forms the foundation for the laryngeal skeleton. It is a complete and inexpansible ring, shallow anteriorly and deep posteriorly, hence, it is often likened to a signet ring. Below it lies the trachea. The two **arytenoid cartilages** articulate with its deep posterior part or **lamina**. Above lies **the thyroid cartilage** with whose **inferior cornua** it articulates by two small **synovial joints**.

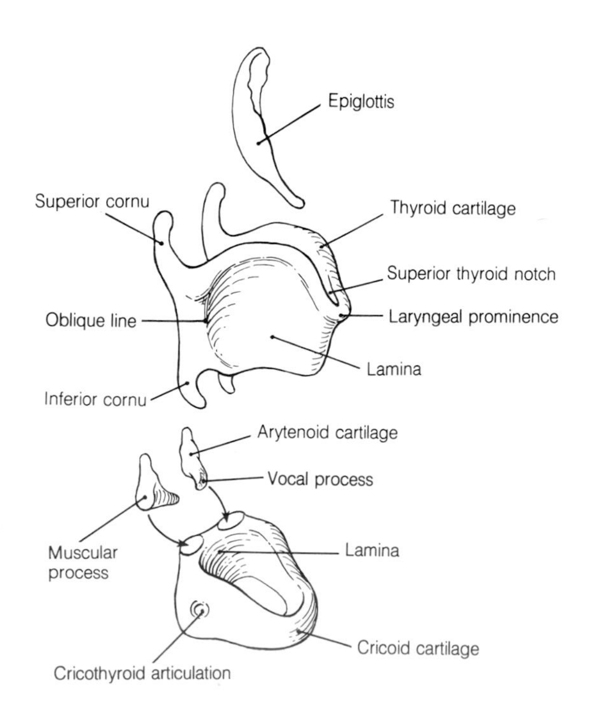

9-119 The laryngeal cartilages separated from each other.

The **arytenoid cartilages** are somewhat pyramidal in shape. A forwardly projecting **vocal process** gives attachment to the **vocal fold** and a lateral **muscular process** serves for the attachment of muscles. At its synovial articulation with the cricoid it is able to rotate about a vertical axis and to slide up and down the cricoid lamina.

The **thyroid cartilage** consists of two **laminae** which are united anteriorly at the **laryngeal prominence** or "Adam's apple" and diverge posteriorly. The midline **superior thyroid notch** lies between the upper borders of the laminae. Projecting upward and downward from its posterior free borders are the **superior and inferior cornua**. The inferior cornua articulate with the cricoid cartilage and the superior cornua are connected to the tips of the **greater cornua of the hyoid bone** by the **lateral thyrohyoid ligaments**. On each side the laminae display an **oblique line** to which the muscles sternothyroid and thyrohyoid and the inferior constrictor of the pharynx are attached.

The **epiglottis**, a gently curved leaf-like structure, is attached inferiorly to the internal aspect of the upper border of the thyroid cartilage. Its free upper portion projects superiorly above the level of the hyoid bone.

The laryngeal cartilages are formed of **hyaline cartilage** with the exception of the epiglottis which is formed of **elastic cartilage**.

The Laryngeal Membranes

Beneath the lining mucous membrane, the cartilages of the larynx are linked by membranes to each other and to the hyoid bone above and the first tracheal ring below. Thickenings in these membranes are described as ligaments. In Fig. 9-120 note the following features:

1. The upper border of the thyroid cartilage is joined to the deep surface of the body of the hyoid bone and to the greater horn by the **thyrohyoid membrane**.
2. This membrane is thickened in the midline anteriorly to form the **median thyrohyoid liga-**

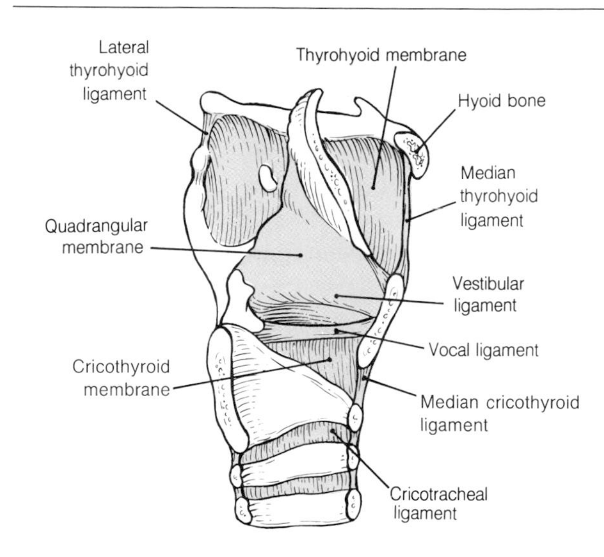

9-120 Median section of the larynx, hyoid bone, and trachea.

ment and laterally to form the **lateral thyrohyoid ligaments**.

3. The lower border of the cricoid cartilage is joined to the first ring of the trachea by the **cricotracheal ligament**.
4. The cricoid, thyroid, and arytenoid cartilages are joined by the **cricothyroid membrane** (cricovocal membrane). Its upper free border extends from the vocal process of the arytenoid cartilage to the deep surface of the thyroid where it converges on its fellow in the midline. This free upper border is the **vocal ligament**.
5. A midline thickening in the cricothyroid membrane connects the adjacent borders of the cricoid and thyroid cartilages and is called the **median cricothyroid ligament**.
6. A less distinct membrane joins the arytenoid cartilage to the epiglottis. This is the **quadrangular membrane** and its lower free border forms the **vestibular ligament**.

The Muscles of the Larynx

Like the tongue the larynx can be said to have extrinsic and intrinsic muscles. The former stabilize or move the larynx as a whole while the latter may alter the position of one part relative to the others. However, many of the muscles that move the larynx are not, in fact, attached to laryngeal cartilages but to the hyoid bone and thus indirectly to the thyroid cartilage via the thyrohyoid membrane. Among these should be considered the **suprahyoid and infrahyoid** muscles. More intimately connected to the larynx are the **stylopharyngeus, salpingopharyngeus**, and **palatopharyngeus**, which all find attachment to the posterior border of the thyroid cartilage. Finally, the **sternothyroid** and **thyrohyoid muscles**, both of which are attached to the oblique line of the thyroid cartilage, can be considered as true extrinsic muscles. With the help of Figs. 9-121, 9-122, and 9-123, the actions of the intrinsic muscles will be more easily appreciated in the subsequent discussion. Note that the naming of the muscles defines their attachments.

The **cricothyroid muscle** (Fig. 9-121) lies superficially on the anterior aspect of the larynx. Arising anteriorly from the arch of the cricoid it fans out to be attached to the lower border of the thyroid cartilage.

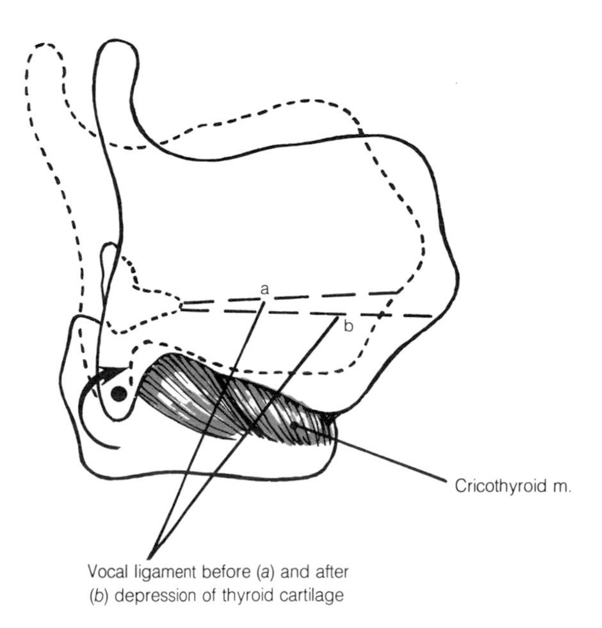

9-121 Illustrating the position and action of the cricothyroid muscle.

The **posterior cricoarytenoid muscle** (Fig. 9-122) arises from the posterior aspect of the lamina of the cricoid and extends laterally and upward to be inserted into the muscular process of the arytenoid cartilage.

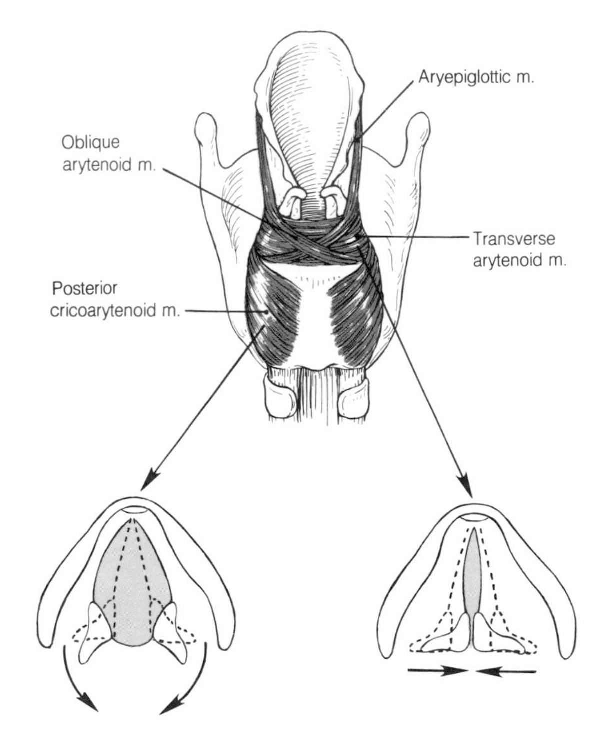

9-122 Illustrating the positions and actions of the posterior cricoarytenoid and transverse arytenoid muscles.

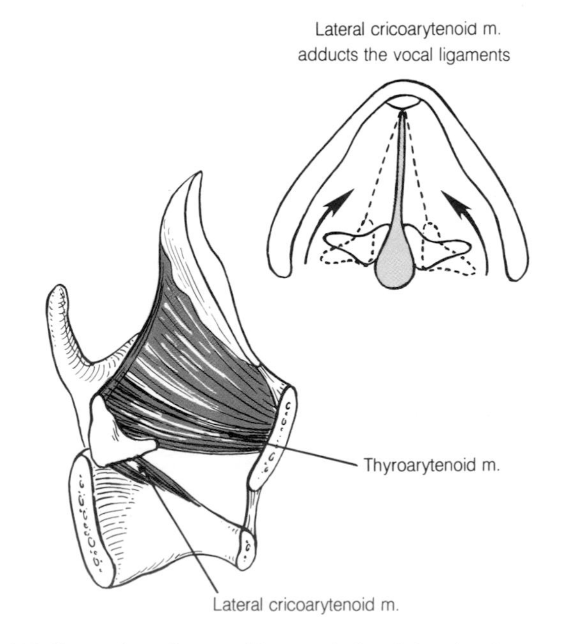

9-123 Ilustrating the positions of the lateral cricoarytenoid and thyroarytenoid muscles. The action of the lateral cricoarytenoid muscle is also shown.

The **lateral cricoarytenoid muscle** (Fig. 9-123) arises from the upper border of the cricoid arch. Its fibers pass backward beneath the lower border of the thyroid to reach the muscular process of the arytenoid cartilage.

The **transverse arytenoid muscle** (Fig. 9-122) spans the gap between the two arytenoid cartilages.

The **thyroarytenoid muscle** (Fig. 9-123) lies within the larynx and lateral to the vocal fold. It is attached anteriorly to the posterior aspect of the thyroid close to the midline and posteriorly to the anterior aspect of the arytenoid cartilage. A discrete bundle of its lower fibers is attached to the vocal process and is named the **vocalis muscle**. Upper fibers of this muscle also extend into the aryepiglottic fold and are called the **thyroepiglotticus**.

The **oblique arytenoid muscles** (Fig. 9-122) are a pair of crossed bundles lying on the superficial surface of the transverse arytenoid muscle. Like this muscle they also link the two arytenoid cartilages together. These fibers are continued into the aryepiglottic folds as the muscle **aryepiglotticus**.

The Actions of the Laryngeal Muscles

The intrinsic muscles serve two purposes. The first is to act as sphincters for the protection of the respiratory tract or the raising of intrapulmonary pressure, and the second is to adjust the vocal ligaments in respiration and phonation. The actions of the individual muscles can be defined more exactly and may contribute to both of these purposes.

The arytenoid cartilages are able to rotate about a vertical axis and thus swing the vocal processes medially or laterally. They may also glide up and down the sloping shoulders of the cricoid lamina, thus, narrowing or widening the gap between them. Rotation of the vocal process medially approximates the vocal ligaments and this movement is carried out by the lateral cricoarytenoid muscles which pull forward on the muscular processes (Fig. 9-123). The posterior cricoarytenoid muscles pull backward on the muscular processes, swing the vocal processes laterally and separate the vocal ligaments (Fig. 9-122). The transverse and oblique arytenoid muscles draw the two arytenoid cartilages together (Fig. 9-122). The diagrams in Figs. 9-122 and 9-123 show the effects that these movements will have on the space between the vocal folds.

The thyroid cartilage articulates by its inferior cornu with the posterior part of the cricoid cartilage. Movement at this joint allows the anterior borders of the two cartilages to come together or separate. In Fig. 9-121 it can be seen how the anteriorly placed cricothyroid muscles can pull the thyroid cartilage down toward the cricoid cartilage and thus lengthen and increase the tension of the vocal ligaments. The opposite effect is achieved by the thyroarytenoid muscles as their contraction shortens the vocal ligaments.

It is worth drawing attention to the triangular area on the lamina of the cricoid cartilage between the two posterior cricoarytenoid muscles. This is not bare but serves for the attachment of the longitudinal muscle fibers of the esophagus.

The extrinsic muscles are concerned with the movement of the larynx as a whole and are particularly involved in elevation of the larynx during swallowing and the subsequent restoration of its position. Through the thyrohyoid membrane the digastric, mylohyoid, stylohyoid, and geniohyoid muscles are able to raise the larynx. The muscles stylopharyngeus and palatopharyngeus also raise the larynx together with the pharynx. The larynx is depressed by the sternohyoid, sternothyroid, and omohyoid muscles.

With the exception of the cricothyroid muscles, which are supplied by the **external laryngeal nerve**, all the intrinsic muscles of the larynx are supplied by the **recurrent laryngeal nerve**.

Accidental damage to a recurrent laryngeal nerve during a thyroidectomy will lead to a "closure" of the ipsilateral vocal fold and a hoarse voice. Damage to both nerves reduces the voice to a whisper and may produce respiratory distress.

The Mucous Membrane of the Larynx

The mucous membrane of the larynx is, as it were, draped over the cartilages and membranes of the larynx and gives rise to a number of features which are illustrated in Figs. 9-124 and 9-125. It falls in front of the epiglottis to form the **valleculae** and lateral to the aryepiglottic folds to form the **piriform recesses** which lie between the folds and the laminae of the thyroid cartilage. Within the larynx the mucous membrane is tucked under the vestibular ligament to form the **laryngeal ventricle** between the **vestibular and vocal folds**. The mucous membrane is firmly adherent to the posterior aspect of the epiglottis and over the vocal

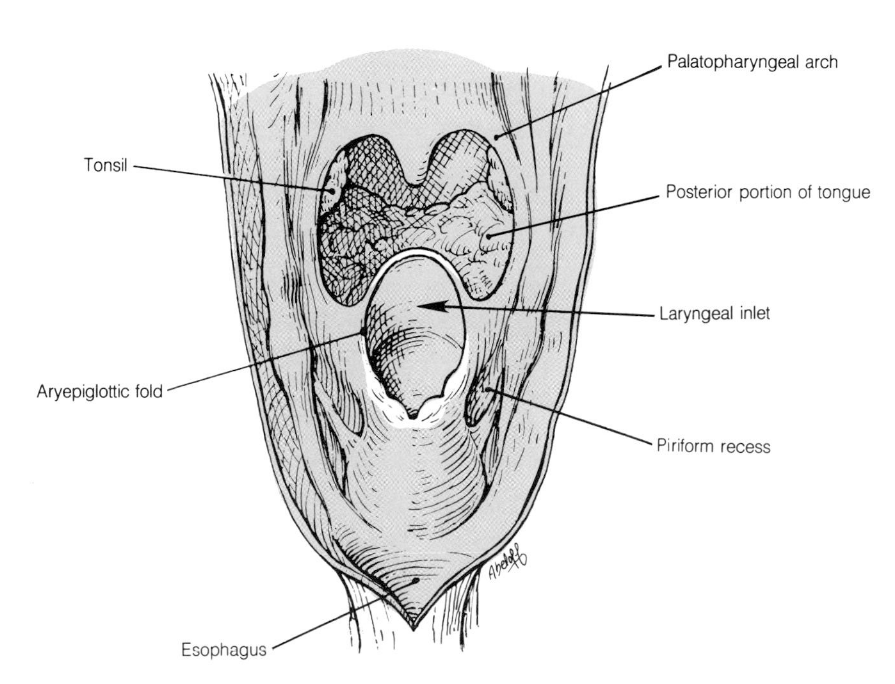

9-124 The larynx viewed from behind and showing the laryngeal inlet and the piriform recesses.

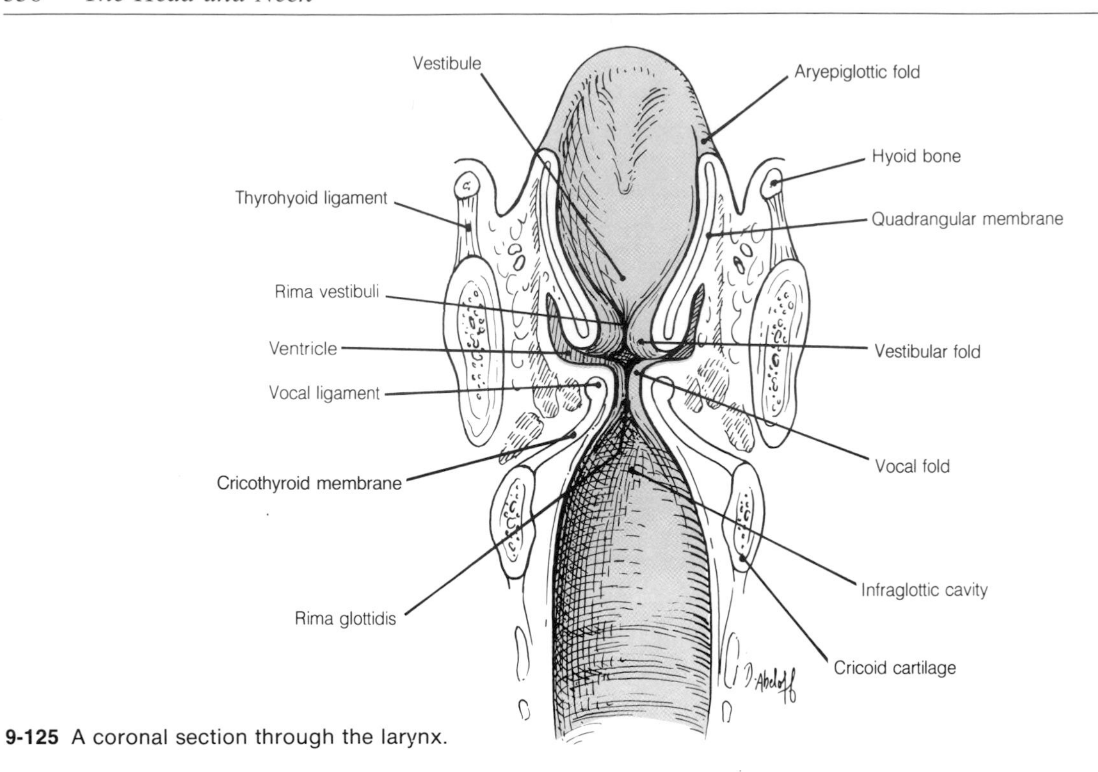

9-125 A coronal section through the larynx.

folds. Elsewhere it is loose and subject to acute swelling which produces edema of the glottis and respiratory obstruction. The epithelium is stratified squamous over the vestibular folds and above this level. Below, except over the vocal folds where it is again stratified squamous, the epithelium is of the respiratory type and contains ciliated columnar cells and many mucous glands.

The mucous membrane of the larynx is supplied by the **internal and recurrent laryngeal nerve**. The internal laryngeal nerve pierces the thyrohyoid membrane with the **superior laryngeal artery**, a branch of the **superior thyroid artery**, and supplies the mucous membrane above the vocal folds. The recurrent laryngeal nerve passes beneath the inferior constrictor of the pharynx with the **inferior laryngeal artery**, a branch of the **inferior thyroid artery**, and supplies the mucous membrane below the vocal folds. Veins accompanying the arteries pass back to the **superior, middle, and inferior thyroid veins**. Lymphatics also accompany the two arteries to reach deep cervical nodes alongside the internal jugular vein.

To summarize the nerves and vessels of the larynx:

1. All the intrinsic muscles except the cricothyroid muscle are supplied by the recurrent laryngeal nerve.
2. The cricothyroid is supplied by the external laryngeal nerve.
3. The mucous membrane above the vocal folds is innervated by the internal laryngeal nerve and supplied with blood by the superior laryngeal artery.
4. The mucous membrane below the vocal folds is innervated by the recurrent laryngeal nerve and supplied with blood by the inferior laryngeal artery.
5. Veins and lymphatics from each of these regions pass back along the course of the arteries. The veins drain into the superior, middle, and inferior thyroid veins and the lymphatics reach deep cervical lymph nodes.

The Cavity of the Larynx

The **inlet of the larynx** allows communication between the pharynx and the larynx. It is bounded anteriorly by the **epiglottis**, laterally by the **aryepiglottic folds**, and posteriorly by an **interarytenoid fold**. The shape of the cavity can be best appreciated by an examination of the coronal section seen in Fig. 9-125. In this illustration note:

1. The inlet leads to the **vestibule** which extends down to the **vestibular folds**.
2. The cavity narrows at the **vestibular folds** but

widens again between the vestibular and **vocal folds** at the **laryngeal ventricle**.

3. An anterior and upward extension of this recess is the **saccule**.
4. The cavity narrows again at the **vocal folds** which contain the vocal ligaments and the space between these is the **rima*** **glottidis**.
5. Below the vocal folds the larynx widens and assumes a more circular shape at the **infraglottic cavity** before it becomes continuous with that of the trachea.

Foreign bodies entering the larynx may be arrested at the laryngeal inlet or, if small enough to pass further, may reach the narrow rima glottidis. Acute respiratory obstruction is most rapidly relieved by a laryngotomy performed through the cricothyroid membrane, but a more satisfactory procedure is a tracheotomy in which an opening is formed in the trachea below the cricoid cartilage and through which a curved tube may be passed.

The Trachea in the Neck

The trachea is a membraneous tube reinforced by incomplete C-shaped rings of hyaline cartilage and it is lined by respiratory epithelium. The rings are completed posteriorly by the **trachealis muscle** (smooth muscle). The trachea extends from the cricoid cartilage at the **level of the sixth cervical vertebra** to its bifurcation in the thorax at the **level of the fourth or fifth thoracic vertebra** and is about 10 cm in length. However, laryngeal and respiratory movements can produce rapid changes in these levels.

The trachea has a very superficial position in the neck, being covered by skin and fascia above and overlapped inferiorly by the **sternohyoid and sternothyroid muscles**. Its second and third rings are crossed by the **isthmus of the thyroid gland**. In the infant the **left brachiocephalic vein** may cross it above the level of the manubrium, a point to be remembered when performing a tracheostomy on a child. Laterally the trachea is embraced by the **lobes of the thyroid gland** which when enlarged may distort its lumen. Posteriorly it is related to the underlying **esophagus** and the **recurrent laryngeal nerves** which ascend in the grooves between the trachea and esophagus. Look now at Fig. 9-126 which shows many of these relationships, and also serves as a good review of the soft tissues of the neck seen at the level of the seventh cervical vertebra.

The trachea is supplied in the neck by the nearby **inferior thyroid arteries** and veins return to the **inferior thyroid veins**. Lymphatics pass to pretracheal and paratracheal lymph nodes. The trachea, like the bronchial tree, has both a sympathetic and parasympathetic nerve supply. It should be remembered that stimulation of the sympathetic supply causes relaxation of the tracheal smooth muscle, while parasympathetic activity causes contraction.

* Rima, *L.* = crack or fissure

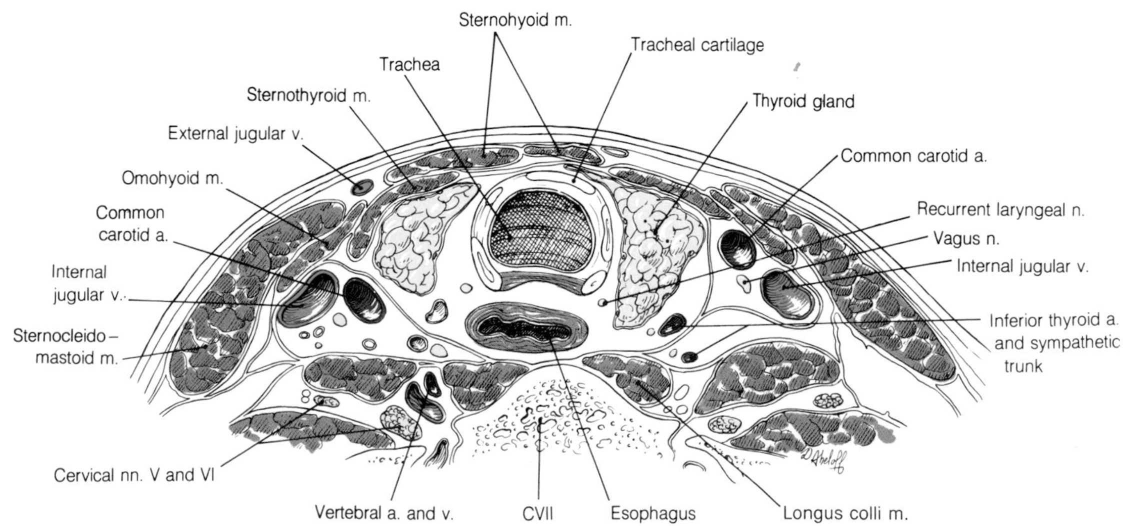

9-126 Horizontal section through the neck.

The Thyroid Gland

The thyroid gland is a large endocrine gland which through the hormones thyroxine and triiodothyronine regulates the metabolic rate. It lies anteriorly in the neck and partly encircles the larynx and trachea to which it is attached by the pretracheal fascia. It consists of two **lobes** which are joined across the midline by the **isthmus**. The lobes and isthmus are surrounded by a thin **fibrous capsule** beneath which a profuse network of vessels lies. The medial surfaces of the pear-shaped lobes lie on the side of the larynx and trachea and extend from the lamina of the thyroid cartilage to the level of the sixth tracheal ring. When enlarged, they may pass into the superior mediastinum. A posterolateral surface is related to the carotid sheath and the lobes are covered anteriorly by the infrahyoid muscles. The isthmus lies over the second and third tracheal rings and must, therefore, be avoided or divided if a tracheostomy is to be performed. A small **pyramidal lobe** is often found extending upward from the isthmus. This represents the lower end of the **thyroglossal duct**. The main relations of the thyroid can be confirmed by looking at Fig. 9-126. Its appearance and position in the neck can be seen in Fig. 9-127.

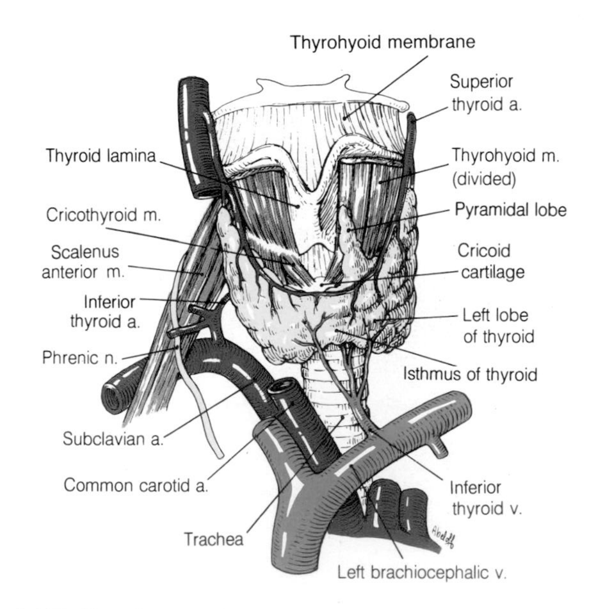

9-127 Showing the thyroid gland, its main relations, and its blood supply.

As an endocrine gland the thyroid has a very rich blood supply. This is provided by the two superior and two inferior thyroid arteries. The **superior thyroid artery** is the first anterior branch of the **external carotid artery**. It descends to reach the apex of the lobe where it divides into an **anterior and posterior branch**. The anterior branch anastomoses with that of the opposite side at the upper border of the isthmus. The **inferior thyroid artery** is a branch of the **thyrocervical trunk** which arises from the **subclavian artery** near the medial border of the scalenus anterior muscle. At first ascending this border it then passes medially behind the carotid sheath to reach the lower part of the lobe. Near to the gland the inferior thyroid artery becomes closely related to the **recurrent laryngeal nerve**. As ligature of the thyroid arteries precedes partial or total excision of the gland, it stands in danger of being accidentally severed or included in a ligature. Paralysis of all the muscles of that side of the larynx except for the cricothyroid will follow. Occasionally a single **thyroidea ima artery** arising from the brachiocephalic trunk also supplies the gland.

The thyroid is drained by **superior and middle thyroid veins** which end in the **internal jugular vein**. Several **inferior thyroid veins** also descend from the thyroid over the trachea and open into the **brachiocephalic veins**. The two arteries and the inferior thyroid veins are shown in Fig. 9-127. Lymphatics from the thyroid pass to the tracheal, brachiocephalic, or deep cervical nodes.

Sympathetic vasomotor fibers reach the thyroid from the cervical ganglia. While there is evidence for some neural control of thyroid secretion by the hypothalamus, the mechanism is not clear.

The Parathyroid Glands

The parathyroid glands are four small endocrine glands lying on or embedded in the posterior surface of the lobes of the thyroid. Their hormone is concerned with the regulation of ionized calcium in the body fluids. They are supplied by both the superior and inferior thyroid arteries.

Accidental removal of these small glands during thyroid surgery is followed by an increase in neuromuscular excitability and muscle spasms.

The Root of the Neck

A description of the root of the neck allows the inclusion of a number of structures not described elsewhere and a review of an important junctional region uniting the neck, thorax and upper limb. Figure 9-128 should be referred to repeatedly as the structures in this region are described.

The region is encircled by the thoracic inlet. To this are attached the scalene muscles and of these the scalenus anterior forms a useful landmark. On the left hand side of the illustration note the subclavian artery and the trunks of the brachial plexus appearing between its posterior border and the anterior border of scalenus medius. The subclavian vein passes into the thorax anterior to the muscle. On either side of the midline the right vagus and the origin of the right recurrent laryngeal nerve are shown on the right side of the figure and the termination of the thoracic duct on the left. The two vertebral arteries can also be seen passing posteriorly to reach the vertebrarterial foramen of the sixth cervical vertebra. In the midline lie the trachea, esophagus, and the recurrent laryngeal nerves with the common carotid arteries passing upward on each side. Note particularly the apices of the lungs as they extend superiorly above the level of the first rib.

The Subclavian Artery

Branches of the subclavian artery have been encountered in the neck, upper limb, and thorax. It is now possible to link them to their parent vessel.

The **left subclavian artery** arises directly from the arch of the aorta. The right subclavian artery arises with the right common carotid artery from the brachiocephalic trunk. The artery is described in three parts. The **first part** is that before it passes behind the scalenus anterior muscle, the **second part** is that posterior to the muscle, and the **third part** is that lying on the superior surface of the first rib. As the vessel approaches the first rib, it arches over the cupula of the pleura and the apex of the lung.

The branches of the subclavian artery are summarized below:
- the vertebral artery,
- the internal thoracic artery,
- the thyrocervical trunk,
- the costocervical trunk, and
- the dorsal scapular artery.

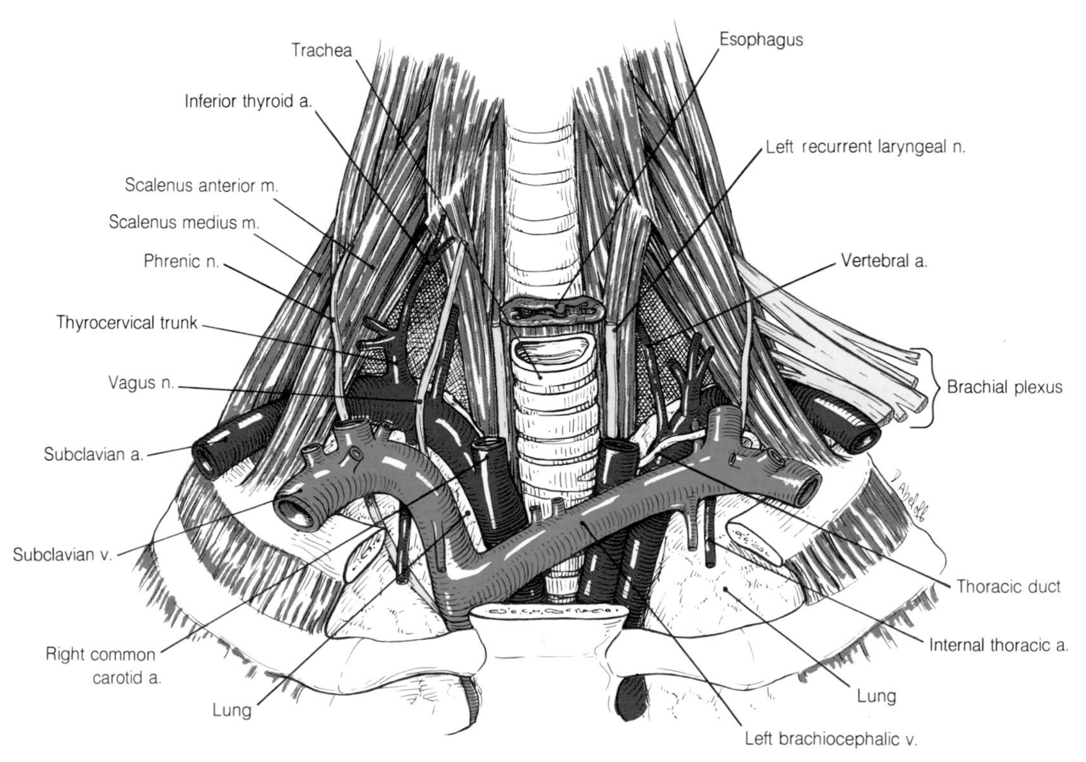

9-128 Illustration of the structures at the root of the neck.

The Vertebral Artery

The vertebral artery arises posteriorly from the first part of the subclavian artery. It passes superiorly to reach the vertebrarterial foramen of the sixth cervical vertebra and ascends through the remaining foramina. Emerging from the first vertebrarterial foramen, it passes behind the lateral mass of the atlas, pierces the posterior altantooccipital membrane and enters the cranial cavity through the foramen magnum. Its branches in the neck are muscular and spinal.

The Internal Thoracic Artery

This vessel arises from the first part of the subclavian artery and descends into the thorax. Its course and branches are described in Chap. 4.

The Thyrocervical Trunk

This is a short trunk arising from the subclavian artery at the medial border of scalenus anterior. It divides into the **inferior thyroid artery** which ascends the border of the muscle before turning medially behind the carotid sheath to reach the thyroid gland and the **suprascapular and transverse cervical arteries** which run laterally across scalenus anterior to reach the posterior triangle.

The Costocervical Trunk

This trunk arises from the second part of the subclavian artery and arches backward over the pleura to reach the neck of the first rib. Here it divides into the **highest intercostal artery** which supplies the first two intercostal spaces and the **deep cervical artery** which ascends in the deep muscles of the neck.

The Dorsal Scapular Artery

The dorsal scapular artery arises from the third part of the subclavian artery. It passes laterally across the posterior triangle to join the **dorsal scapular nerve** with which it descends the medial border of the scapula deep to the rhomboid muscles.

The Subclavian Vein

This vein begins at the lateral border of the first rib as a continuation of the axillary vein. It passes anterior to the scalenus anterior muscle and joins the **internal jugular vein** to form the **brachiocephalic vein**. Its main tributary is the **external jugular vein**. Veins accompanying the branches of the subclavian artery more often drain into the brachiocephalic veins.

The Thoracic Duct

The **thoracic duct** enters the root of the neck behind the left subclavian artery. It then arches laterally over the pleura and crosses anterior to the vertebral artery and vein and the thyrocervical trunk before opening into the junction of the left subclavian and internal jugular veins. A part of this course can be seen in Fig. 9-128. The **right lymphatic duct** ascends from the thorax to join the venous system in a similar manner on the right side.

The Right Recurrent Laryngeal Nerve

The right recurrent laryngeal nerve leaves the vagus in the root of the neck and hooks beneath the subclavian artery on its way to the side of the trachea. From here, like the left nerve, it ascends between the trachea and esophagus to enter the larynx beneath the inferior constrictor of the pharynx.

Cervical Compression Syndromes

The subclavian artery and the roots of the brachial plexus pass through the very narrow slit formed by

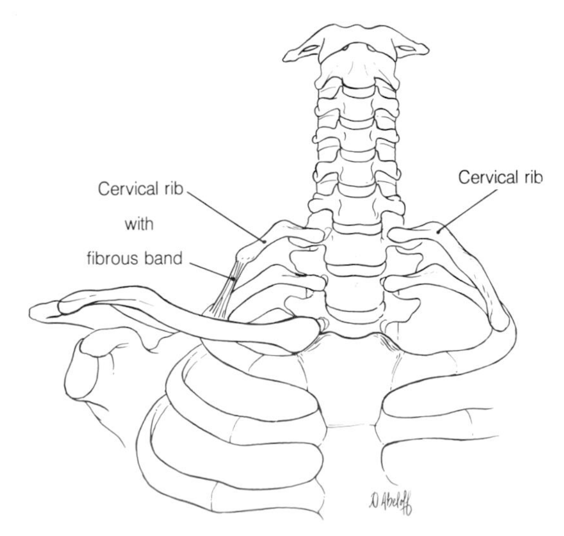

9-129 Showing examples of cervical ribs articulating with the seventh cervical vertebra.

the scalenus anterior and medius muscles and the first rib. At this site both structures are subject to compression which leads to the neurological symptoms of pain, numbness, and tingling in the hand and vascular disturbances, particularly when the limb is held to the side. These conditions are often associated with a cervical rib above which both the subclavian artery and the plexus may pass. However, a cervical rib does not always give rise to these symptoms and the symptoms may be found in its absence. Examples of cervical ribs are illustrated in Fig. 9-129.

The Ear and Vestibular Apparatus

The organ of hearing and the vestibular apparatus which receives information about the position and movement of the head are closely associated anatomically and are supplied by the same cranial nerve, the vestibulocochlear nerve. Both organs lie buried in the temporal bone and their relationships to the bone are illustrated in Fig. 9-130. Note in the figure the region called the suprameatal triangle which becomes acutely tender when there is an underlying infection of the middle ear. To aid in the description of this small but complex region it is divided into the external ear, the middle ear and the internal ear. The external ear is separated from the middle ear by the tympanic membrane or eardrum and the inner ear is separated from the middle ear by a bony partition.

The External Ear

Included in the description of the external ear are the **auricle** and the **external acoustic meatus**.

The Auricle

The auricle is formed from a double layer of skin reinforced by an irregular cartilaginous plate. It is this plate that produces the characteristic features of the auricle which are shown in Fig. 9-131. The cartilage is absent from the lobule. The auricle is attached to the skull by both ligaments and small muscles and leads to the external acoustic meatus.

The External Acoustic Meatus

The external acoustic meatus extends from the auricle to the tympanic membrane and is about

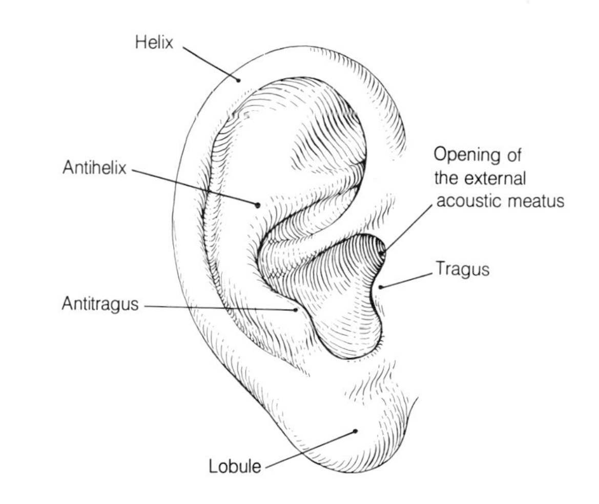

9-131 The auricle.

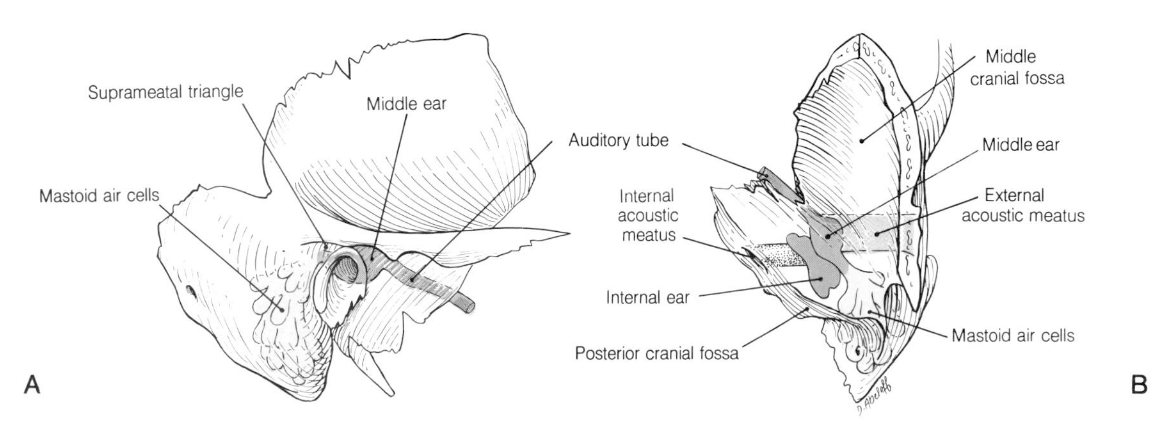

9-130 Showing the relationships of the outer, middle, and inner ear to the temporal bone seen from the lateral aspect (*A*) and superior aspect (*B*).

2.5 cm in length. It is lined by a layer of thin skin which also covers the outer surface of the tympanic membrane. This skin which is rather sensitive is supplied by branches of the **auriculotemporal nerve**. Beneath the skin lie the wax secreting **ceruminous glands**. The outer third of the meatus is almost surrounded by cartilage which is continuous with the auricular cartilage. This is joined by fibrous tissue to the bony inner two-thirds of the meatus. The acoustic meatus is not straight and it is narrowed at the junction of its cartilaginous and osseous portions.

The meatus and the tympanic membrane can be examined by the introduction of an otoscope. The best view of the tympanic membrane is obtained if the auricle is gently drawn upward and backward. The examiner should remember that the external meatus of a child is very short.

The Middle Ear

The cavity of the middle ear lies in the temporal bone. It is lined by a mucous membrane which is continuous anteriorly with that of the **auditory tube** and **nasopharynx** and posteriorly with that of the **mastoid air cells**. The cavity formed by this mucous membrane contains only air and all other structures to be described in the middle ear lie outside the mucous membrane.

In shape the bony cavity is irregular but could be likened to a flat box whose lid represents the lateral wall of the cavity and the bottom of the box the medial wall. The sides of the box represent the roof, the floor, and the anterior and posterior walls, but each of these is very narrow. The tympanic membrane lies in the lateral wall and the inner ear lies deep to the medial wall. Between the lateral and medial wall a chain of three **ossicles** articulate. The general shape of the cavity and the position of the ossicles can be seen in Fig. 9-132. Note in this that the cavity extends well above the level of the **tympanic membrane** into the **epitympanic recess** and that this space houses the **head of the malleus** and the **incus**.

Each of the walls, the roof and the floor have particular features and relationships which must now be described.

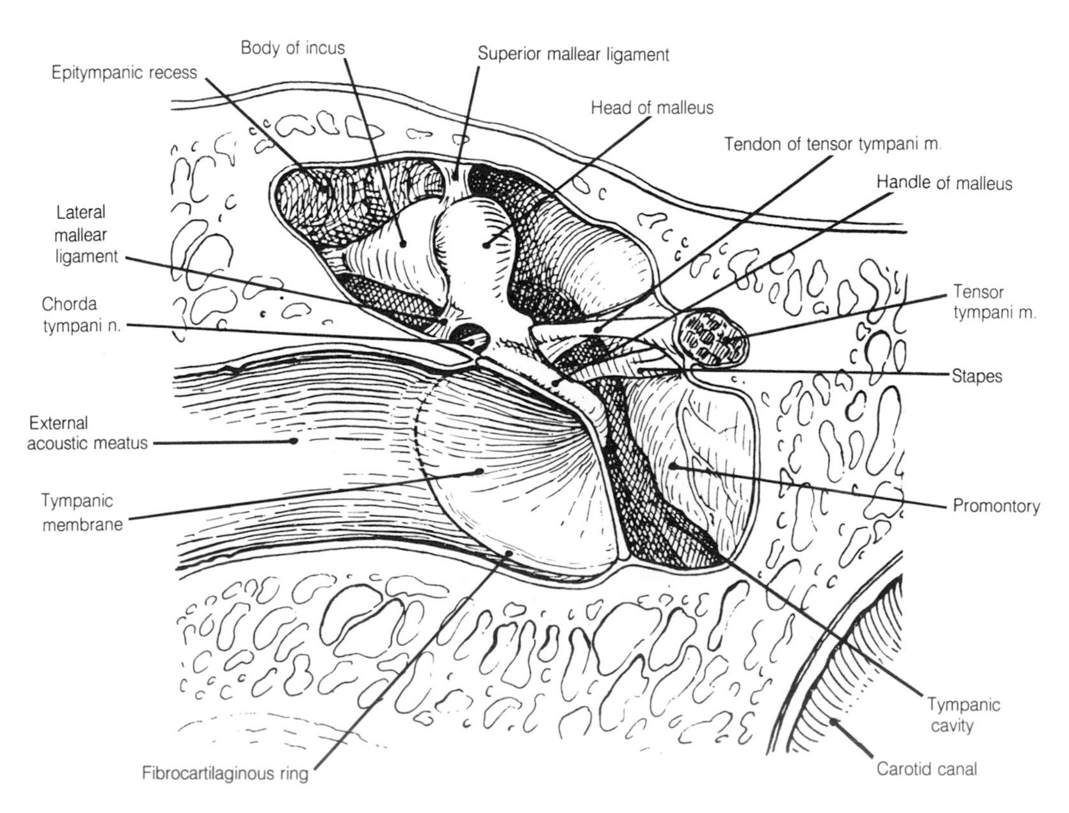

9-132 A coronal section of the middle ear and tympanic membrane.

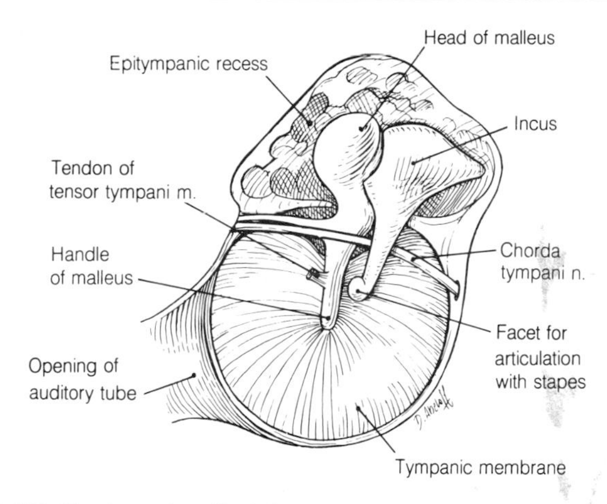

9-133 The lateral wall of the middle ear.

The Lateral Wall

The lateral wall is bony above but below it is completed by the oval **tympanic membrane**. To the inner aspect of this is applied the **handle of the malleus**. The tympanic membrane and the malleus are crossed by the **chorda tympani** branch of the **facial nerve**. These structures can all be seen in the diagram in Fig. 9-133.

The Posterior Wall

The posterior wall is absent superiorly where the **aditus to the mastoid antrum** leaves the epitympanic recess. Below the aditus there is a small conical projection called the **pyramid**. From this issues the **stapedius muscle** which is attached to the neck of the **stapes**. Lateral to the pyramid is the opening of the canal that carries the chorda tympani and medially a vertical prominence beneath which is the descending portion of the **facial canal**. Most of these features can be seen in the diagram of the medial wall shown in Fig. 9-134.

The Medial Wall

The medial wall separates the middle ear from the cochlea and the vestibule of the inner ear and each of these structures forms elevations on it. A diagram of the medial wall is shown in Fig. 9-134 and this should be used to identify the following features. The most obvious feature is the **promontory** which bulges from its center. This is formed by the first turn of the **cochlea**. Above and posterior to the promontory is a small opening called

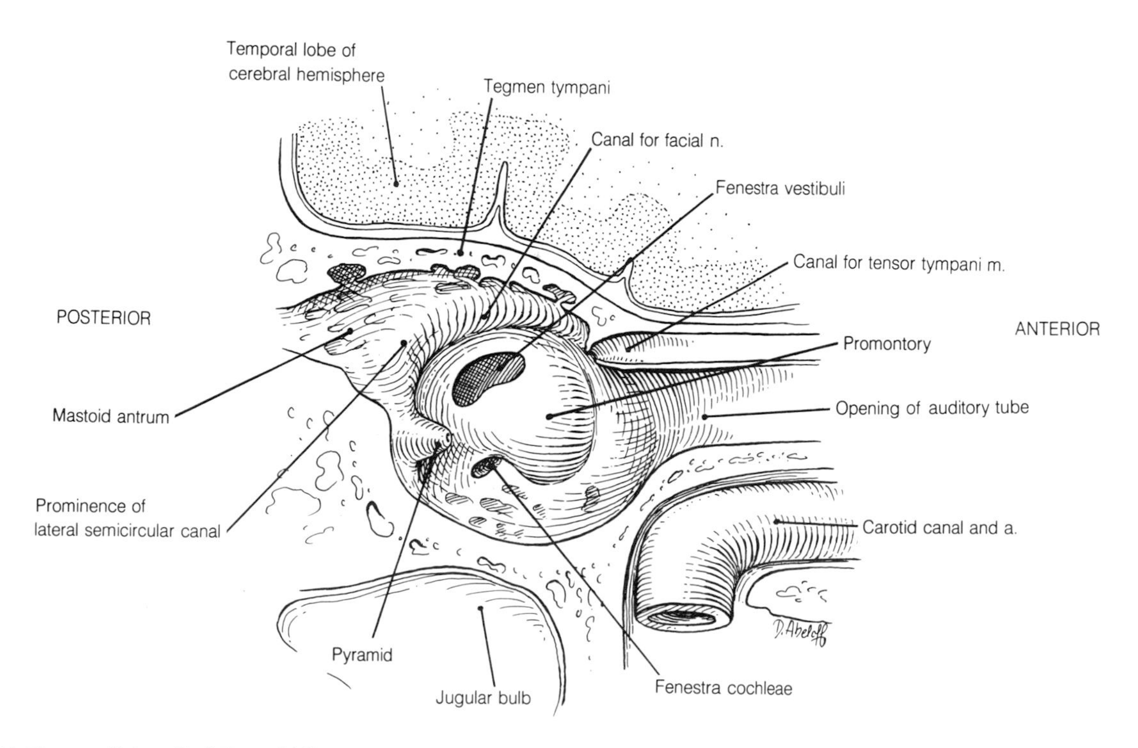

9-134 The medial wall of the middle ear.

the **fenestra*** **vestibuli** which is occluded by the **base of the stapes**. Below and posterior to the promontory there is another small opening called the **fenestra cochleae**. This is closed by the **secondary tympanic membrane**. Near the roof is a horizontal prominence that covers the facial canal and above this a further prominence is produced by the **lateral semicircular canal**. On the surface of the promontory lie the **tympanic nerve** and **tympanic plexus**.

The Anterior Wall

Two canals open onto the superior part of the anterior wall. The upper transmits the **tensor tympani muscle** whose tendon is attached to the handle of the malleus, the lower and larger is the **auditory tube**. Beneath this a thin plate of bone separates the cavity from the **carotid canal** and **internal carotid artery** (Fig. 9-134). This region is pierced by caroticotympanic nerves from the carotid plexus and tympanic branches of the carotid artery.

The Roof

The main feature to note about the roof of the middle ear is the thin plate of bone called the **tegmen tympani** which separates it from the middle cranial fossa and the overlying temporal pole of the cerebrum (Fig. 9-134). It is through this thin roof that infection of the middle ear (otitis media) may spread and form an intracranial abscess. Fractures of the base of the skull often pass through the roof and rupture the tympanic membrane. If the meninges are also torn, blood and cerebrospinal fluid will flow from the external acoustic meatus.

The Floor

Like the roof, the floor is also thin and separates the cavity from the **jugular bulb** (Fig. 9-134). The floor is pierced by the tympanic nerve which runs to the tympanic plexus on the promontory of the medial wall.

In summary, the important relations of the walls of the middle ear are:

The lateral wall	the tympanic membrane the external auditory meatus
The posterior wall	the mastoid antrum the facial nerve
The medial wall	the internal ear the facial nerve
The anterior wall	the auditory canal the carotid canal
The roof	the middle cranial fossa, meninges, and temporal lobe
The floor	the jugular bulb

The Mastoid Antrum and Air Cells

The **aditus to the antrum** leaves the posterior wall of the middle ear and leads to the **mastoid antrum**, an air sinus in the petrous part of the temporal bone. From this extend further air cells which to a lesser or greater extent excavate the mastoid process. These are the **mastoid air cells**. Unlike other air sinuses, the mastoid antrum is well developed at birth; the mastoid process and air cells are not.

The cavities of the mastoid air cells, the antrum, and the middle ear are in communication with each other and with the nasopharynx through the auditory tube. Upper respiratory infections may, therefore, spread to each of these cavities and, with obstruction of the narrow auditory tube by swelling, drainage of the products of the infection may occur through rupture of the tympanic membrane.

The Tympanic Membrane

The tympanic membrane has already been described as oval in shape (10 mm × 8 mm) and lying in the lateral wall of the middle ear.

Note that it is placed obliquely so that its lateral surface faces both forward and downward. The lateral surface is also concave. The convexity of the medial surface brings the center of the membrane to within 2 mm of the promontory.

Its structure is fibrous and it is covered on its

* Fenestra, *L.* = a window; compare with fenêtre (*Fr.*) and fenster (*Germ.*)

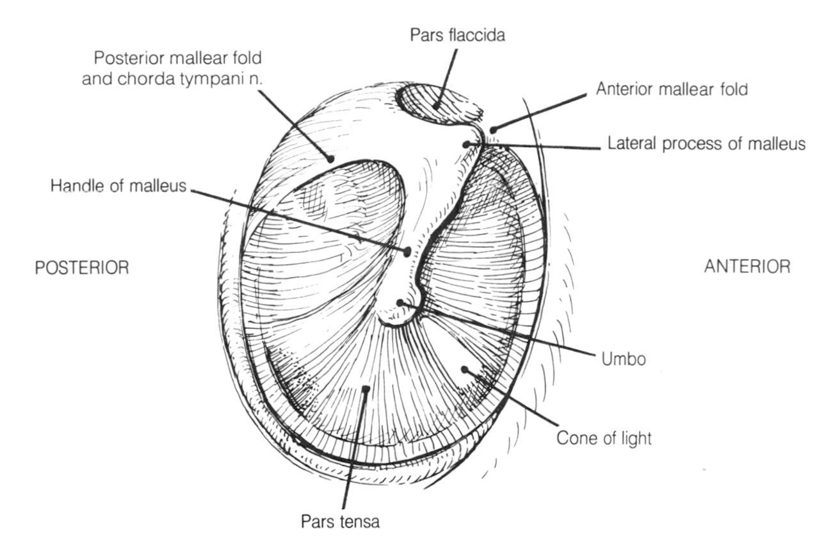

9-135 The lateral surface of the tympanic membrane.

lateral aspect by thin skin and on its medial aspect by the mucous membrane lining the middle ear cavity. The fibrous layer is thickened at its periphery to form a ring which rests in a groove in the temporal bone. The groove, however, is deficient above and adjacent to this deficiency the tympanic membrane is lax and is known as the **pars flaccida** while the remainder is called the **pars tensa**. At the margins of the pars flaccida two **mallear folds** extend from the ends of the sulcus to the malleus whose handle is firmly attached to the medial surface of the membrane above the central boss or **umbo**. The diagram of the lateral surface of the tympanic membrane shown in Fig. 9-135 can be used to identify these structures and compared with the photograph of the membrane shown in Fig. 9-136. Remember that the appearance of the malleus is seen through the membrane and that the membrane is concave when viewed from this aspect. This concavity produces a bright cone of reflected light at the anterior inferior quadrant. This is lost if a rise in

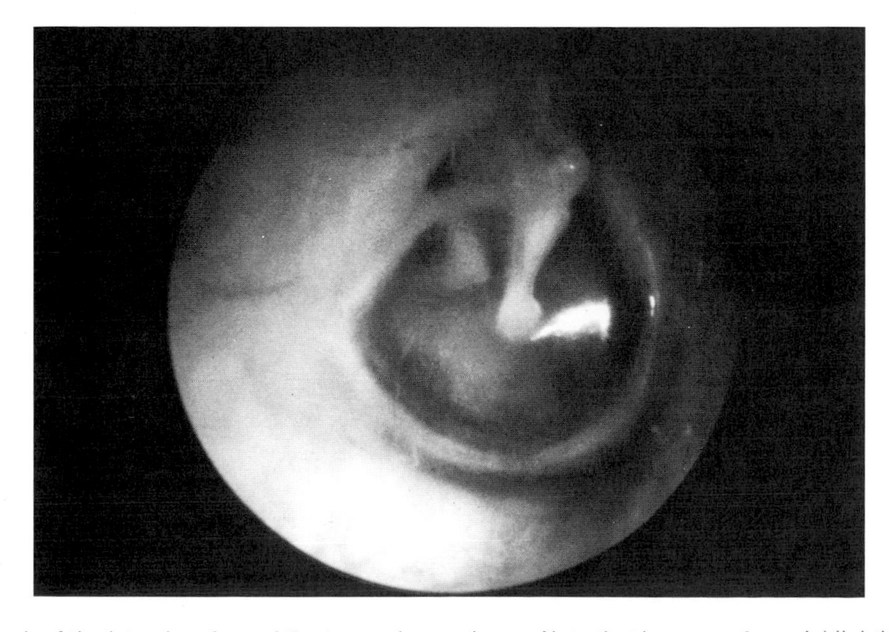

9-136 Photograph of the lateral surface of the tympanic membrane. Note that because of special lighting the "cone of light" lies more anteriorly than when seen with the clinical otoscope. (Courtesy of William C. Gray, M.D., Division of Otolaryngology, University of Maryland School of Medicine.)

the pressure in the middle ear causes the drum to bulge toward the examiner.

The Ossicles

The middle ear cavity contains a chain of three small bones, the **malleus**,* the **incus**,* and the **stapes**.* The details of these bones can be seen in Fig. 9-137. The **handle of the malleus** is attached to the medial surface of the tympanic membrane and the **base of the stapes** to the margins of the fenestra vestibuli. The intermediate incus articulates with the **head of the malleus** and the **body of the stapes** by small synovial joints. The tendon of tensor tympani is attached to the handle of the malleus and that of stapedius to the body of the stapes. Remember that the ossicles lie outside the mucous membrane of the middle ear. The ossicles transmit movements of the tympanic membrane produced by sound waves to the fluid which fills the scala vestibuli of the cochlea. The movements of this column of fluid are in turn accommodated by reciprocal movement of **the secondary tympanic membrane** that covers the fenestra cochleae. The two small skeletal muscles attached to the ossicles damp their movements. The **tensor tympani** arises from the cartilaginous part of the auditory tube and enters the middle ear in a bony canal above the bony part of the tube. It is attached to the handle of the **malleus** and supplied by the **mandibular nerve** through its branch to the medial pterygoid muscle. The **stapedius muscle** arises within the pyramid and is attached to the body of the stapes. It is supplied by the **facial nerve**.

* Malleus, *L.* = a hammer; Incus, *L.* = an anvil; Stapes, *L.* = a stirrup

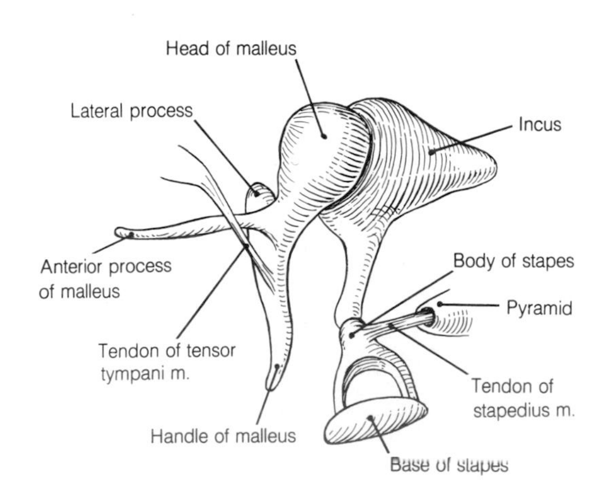

9-137 The auditory ossicles.

The Facial Nerve in the Temporal Bone

The facial nerve has been seen to enter the **internal acoustic meatus** and leave the **stylomastoid foramen**. It is only now that its course and branches in the temporal bone between these two points can be adequately described and followed in Fig. 9-138.

The **facial nerve** and the **nervus intermedius** enter and pass laterally in the internal auditory meatus together with the **vestibulocochlear nerve** and the **labyrinthine artery**. At the bottom of the canal the now fused motor and sensory roots enter the **facial canal**. The nerve continues laterally in the canal above the internal ear until the medial wall of the middle ear is reached. Here it turns sharply backward above the promontory. It then passes downward in the medial wall of the **aditus ad antrum** and leaves the temporal bone at the **stylomastoid foramen**. The **geniculate ganglion**, the sensory ganglion of the facial nerve, lies at the sharp bend or genu that the nerve makes on reaching the middle ear. During its course through the temporal bone the facial nerve gives off the following three branches:

the greater petrosal nerve,
the nerve to stapedius, and
the chorda tympani.

The **greater petrosal nerve** leaves the facial nerve at the geniculate ganglion and passes medially and forward in the temporal bone to leave it at a hiatus in the middle cranial fossa. It continues in the fossa toward the foramen lacerum where it is joined by the **deep petrosal nerve** to become the **nerve of the pterygoid canal**.

The **nerve to stapedius** runs in a small canal to reach the muscle in the pyramid.

The **chorda tympani** is given off only a short distance above the stylomastoid foramen but runs upward in a bony canal to the posterior wall of the middle ear. Here it crosses the deep surface of the tympanic membrane and handle of the malleus to enter a further canal that leads to the **petrotympanic fissure**. Just outside the skull it joins the lingual nerve in the infratemporal fossa.

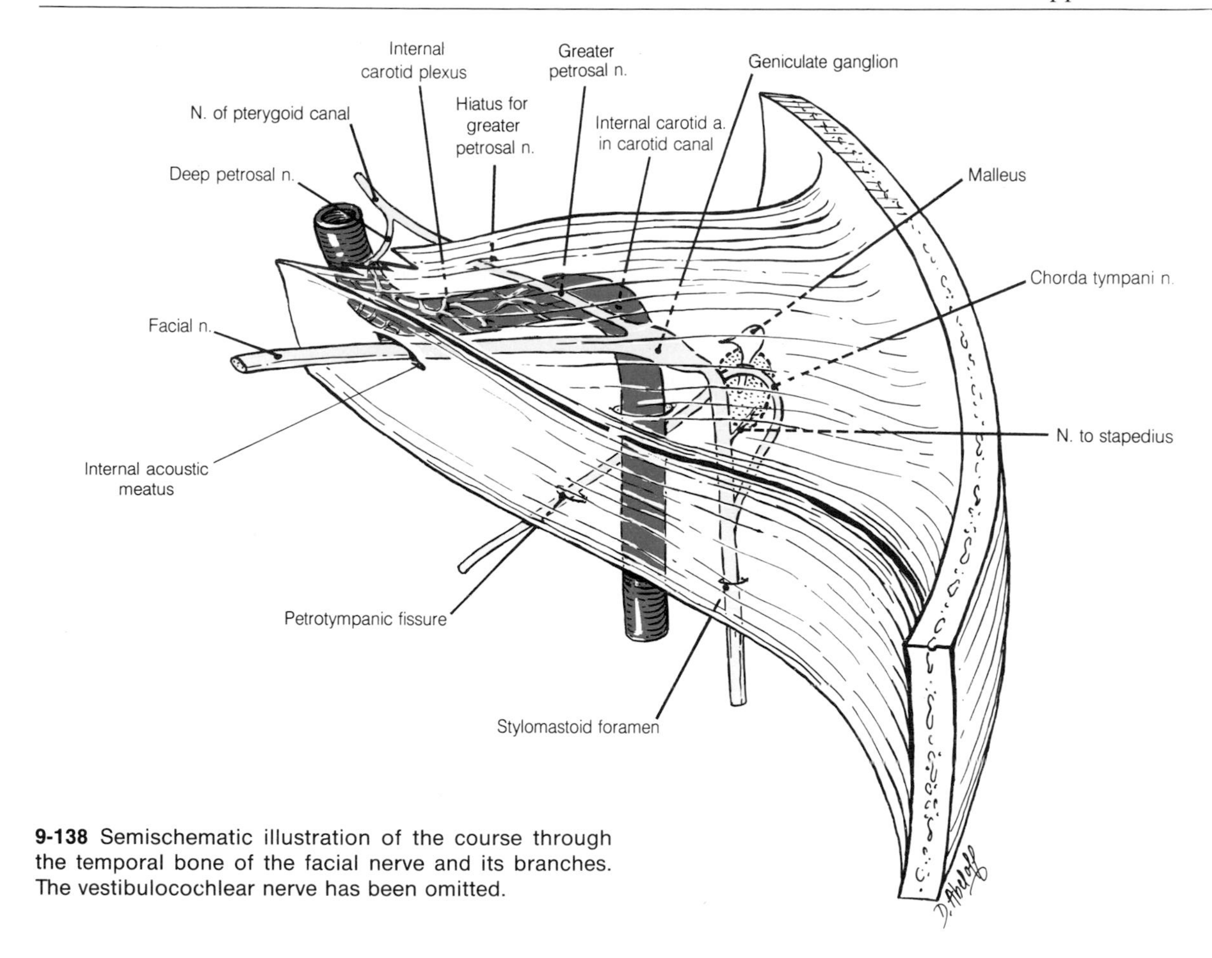

9-138 Semischematic illustration of the course through the temporal bone of the facial nerve and its branches. The vestibulocochlear nerve has been omitted.

The Tympanic Nerve and Tympanic Plexus

The **tympanic branch of the glossopharyngeal nerve** has been described as contributing preganglionic parasympathetic fibers to the **otic ganglion**. The tympanic nerve leaves the glossopharyngeal nerve just outside the cranium but reenters it through a small canal between the jugular foramen and the carotid canal. This leads it to the floor of the middle ear which is penetrates to reach the promontory. Here together with caroticotympanic branches of the internal carotid plexus it forms the **tympanic plexus**. From the tympanic plexus arise branches supplying the mucous membrane of the middle ear, the auditory tube, and the mastoid air cells as well as the **lesser petrosal nerve**. This nerve leaves the middle ear and appears on the anterior surface of the temporal bone through a hiatus below that for the greater petrosal nerve. It runs across the floor of the middle cranial fossa to leave through the foramen ovale and join the **otic ganglion**.

The Internal Ear

The internal ear lies within the hard petrous part of the temporal bone medial to the middle ear. It consists of the **bony labyrinth** and the **membra-**

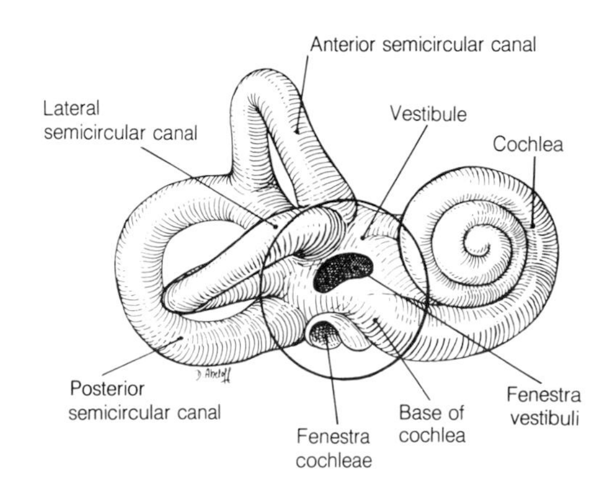

9-139 The bony labyrinth illustrated as a cast of its cavities. Those regions lying adjacent to the medial wall of the middle ear are surrounded by a circle.

nous labyrinth which it contains. The two labyrinths are separated by a space containing a fluid called **perilymph**.

The Bony Labyrinth

The bony labyrinth is a series of intercommunicating cavities in the bone which are lined by periosteum. Look at Fig. 9-139 in which these cavities are represented in the form of a cast. Identify the **cochlea**, the central **vestibule** with two openings, the **fenestra vestibuli**, and the **fenestra cochleae**. From the vestibule arise the **three semicircular canals**. Note that inside the circle are the structures that have already been recognized on the medial wall of the middle ear. These are the **two fenestrae**, the **base of the cochlea** forming the **promontory**, and one limb of the **lateral semicircular canal**. The cochlea contains a spiraling shelf of bone projecting from a central support in rather a "helter skelter" manner. This is called the **modiolus**.

The **perilymph** fills the space between the bony and membranous labyrinths. Its compositions is very similar to that of cerebrospinal fluid but the manner in which it is formed and absorbed is not precisely known.

The Membranous Labyrinth

The membranous labyrinth is a series of intercommunicating sacs and ducts filled with a fluid called **endolymph**. In Fig. 9-140 these are superimposed on an outline of the bony labyrinth. Identify in this

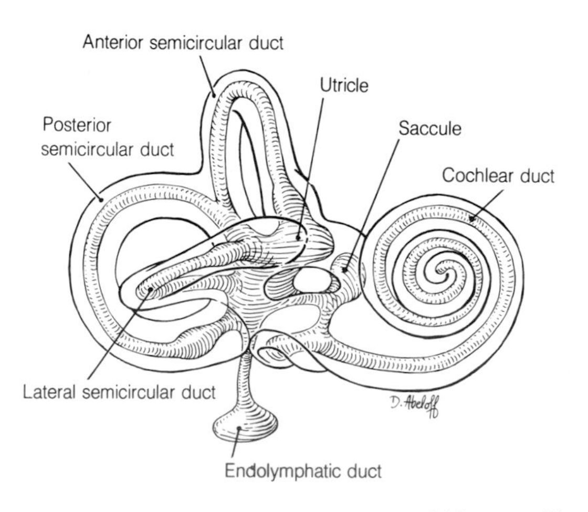

9-140 The membranous labyrinth shown within an outline of the bony labyrinth.

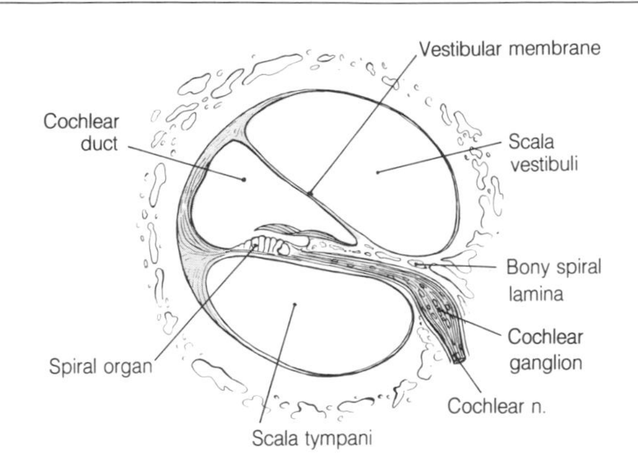

9-141 A transverse section through the spiral canal of the cochlea.

the **cochlear duct** leading to the fenestra cochleae, the **saccule**, the **endolymphatic duct and sac**, and the **utricle** lying in the vestibule and the **semicircular ducts** in the semicircular canals. Note that the membranous labyrinth by no means fills the cavity and that the semicircular ducts have only about one-quarter of the diameter of the semicircular canals.

The **cochlear duct** contains the spiral organ (of Corti) in which the receptors of the auditory apparatus lie. In Fig. 9-141 the spiral organ can be seen to lie between the **scala*** **vestibuli** and the **scala tympani**. Both the latter spaces are filled with perilymph and communicate with each other at the tip of the cochlea.

The **saccule and utricle** contain receptor organs which respond to both linear acceleration and the static pull of gravity.

The **receptors of the semicircular ducts** respond to rotational acceleration in the three different planes in which the ducts lie. These planes and their relationship to the temporal bone are illustrated in Fig. 9-142.

The **endolymphatic duct** leaves the bony labyrinth to open into the blindly ending **endolymphatic sac** which lies under the dura mater covering the posterior surface of the petrous temporal bone.

The **endolymph** which fills the membranous labyrinth has a composition similar to that of intracellular fluid. It is not exactly known which cell type is responsible for its production.

---

* Scalae (pl.), *L.* = a flight of stairs; in this case a spiral staircase

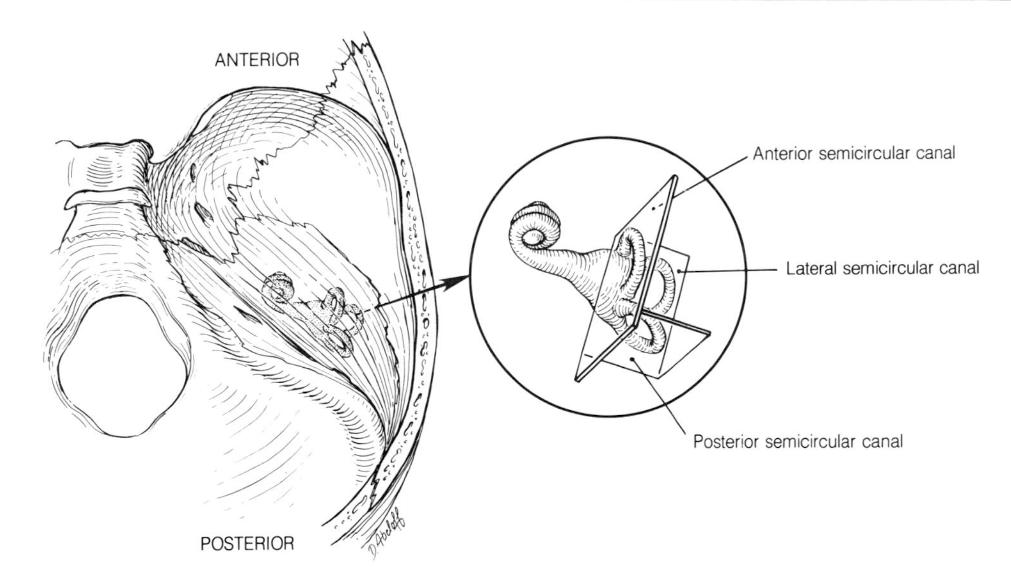

9-142 The shape of the internal ear superimposed on the temporal bone and (*insert*) the planes on which the semicircular canals lie.

The Vestibulocochlear Nerve

Near the lateral end of the internal auditory meatus the vestibulocochlear nerve divides into its anterior **cochlear root** and posterior **vestibular root**. The vestibular root enlarges to form the **vestibular ganglion**; its peripheral extension enters the bone to supply the receptors in the semicircular ducts, the saccule, and the utricle. The cochlear root also enters the bone to reach the **spiral ganglion** which lies at the base of the shelf of bone projecting from the modiolus (Fig. 9-141). Its peripheral extensions supply the spiral organ.

Both the vestibular and spiral ganglia contain bipolar nerve cells whose peripheral processes extend to receptor organs and whose proximal processes run back to enter the brainstem.

The Labyrinthine Artery

The labyrinthine artery is a branch of the basilar artery which enters the internal acoustic meatus. At the bottom of this it divides into two branches for the supply of the vestibule and cochlea.

Index

Illustrations are indicated by an f after the number.

D

G